Major Geological, Climatic, and Biological Events

ximate time since ning of each interval in millions rs before the present

S0-ACZ-540

Approximate time	Events
0.01 1.8	There were extensive and repeated periods of glaciation in the Northern Hemisphere. The Neogene midlatitude savanna faunas became extinct, and hominins expanded throughout the Old World. Near the end of the interval, hominins reached the New World. There was an extinction of many large mammals, especially in the New World and in Australia. The remaining carnivorous flightless birds also became extinct.
5.3 23 34	Cooler and more arid climates persisted, resulting from mountain uplift and the formation of the Isthmus of Panama near the end of the interval. The Arctic ice cap formed by the end of the interval. The first grasslands spread in the middle latitudes. Modern families of mammals and birds radiated, and marine mammals and birds diversified in the oceans. The first hominins were seen near the end of the interval.
55 65.5	Global climate was warm in the early part of the interval, with forests above the Arctic Circle, but later in the interval temperatures fell in the higher latitudes, with the formation of the Antarctic ice cap. Mammals diversified into larger body sizes and a greater variety of adaptive types, including predators and herbivores. Mammal radiations included archaic forms, now extinct, and the earliest members of living orders. Giant carnivorous flightless birds were common as predators.
146	Further separation of the continents occurred, including the breakup of the southern continent, Gondwana. Teleost fishes radiated, but marine reptiles flourished. Angiosperms first appeared, and rapidly diversified to become the dominant land plants by the end of the period. Dinosaurs remained the dominant tetrapods, but small mammals, including the first therians, diversified. Air space and shorelines were shared by birds and pterosaurs, and the first snakes appeared. A major mass extinction at the end of the period, defining the end of the Mesozoic, claimed dinosaurs, pterosaurs, and marine reptiles, as well as many marine invertebrates.
200	The world continent began to break up, with the formation of the Atlantic Ocean. Marine invertebrates began to take on a modern aspect with the diversification of predators, modern sharks and rays appeared, and marine reptiles diversified. Conifers and other gymnosperms were the dominant terrestrial vegetation, and insects diversified. Dinosaurs diversified while mammals remained small and relatively inconspicuous. The first birds, lizards, and salamanders were seen at the end of the period.
251	The world continent was relatively high, with few shallow seas. No evidence of glaciation existed, and the interior of the continent was arid. Seed fern terrestrial vegetation was replaced by conifers in the later part of the period. Mammal-like reptiles declined, while archosaurian reptiles (including dinosaur ancestors) diversified. Remaining large nonamniote tetrapods now all specialized aquatic forms. First appearances by the end of the period included true mammals, dinosaurs, pterosaurs, marine reptiles, crocodiles, lepidosaurs, froglike amphibians, and teleost fishes.
299	A single world continent, Pangea, was formed at the end of the period. Glaciation ceased early in the period. The large terrestrial nonamniote tetrapods declined and the amniotes radiated. Amniote diversification included the ancestors of modern reptiles and the ancestors of mammals, the mammal-like reptiles, which were the dominant large terrestrial tetrapods. The first herbivorous tetrapods were known. The largest known mass extinction event occurred on both land and sea at the end of the period. It coincided with low levels of atmospheric O_2 and marked the end of the Paleozoic.
359	There was a major glaciation in the second half of the period, with low atmospheric levels of CO_2 and high levels of O_2. Coal swamps were prevalent in the then-tropical areas of North America and Europe. Major radiation of insects, including flying forms. Diversification of jawed fishes, including sharklike forms and primitive bony fishes, and first appearance of modern types of jawless fishes. Extensive radiation of non-amniote tetrapods, with the appearance of the first amniotes (including the earliest mammal-like reptiles) by the late part of the period.
416	There was major mountain building in North America and Europe. Major freshwater basins preserved, containing the first tetrapods at the end of the period in equatorial regions. About the same time there were the first forests with tall trees on land, and terrestrial arthropods diversified. Both jawed and jawless fishes diversified, but both experienced major extinctions toward the end of the period, with the disappearance of the ostracoderms, the armored jawless fishes.
444	The extensive shallow seas continued, but on dry land there was the first evidence of vascular plants and arthropods. Jawless fishes radiated, and jawed fishes (sharklike forms) were now definitely known.
488	There were widespread shallow seas over the continents, and the global climate was equable until a sharp glaciation at the end of the period. First evidence of complex plants on land. Major radiation of marine animals, including the first well-known jawless fishes and fragmentary evidence of jawed fishes.
542	Continental masses of the late Proterozoic now broken up in to smaller blocks, covered by shallow seas. Explosive radiation of animals at the beginning of the period, with first appearance of forms with shells or other hard coverings. First appearance of chordates and great diversification of arthropods, including trilobites. First vertebrates appeared early in the period.
2,500	Formation of large continental masses. Oxygen first appears in the atmosphere. First eukaryotic organisms appeared around 2 billion years ago. Major diversification of life at 1 billion years ago, with multicellular organisms, including algae. First animals appeared around 600 million years ago, just after a major glaciation.
4,527	Formation of the Earth. Major bombardment of the Earth by extraterrestrial bodies, precluding formation of life until 4 billion years ago (first fossils known at 3.8 billion years ago). Small continents. Hydrosphere definite at 3.8 billion years, atmosphere without free oxygen.

Vertebrate Life

EIGHTH EDITION

F. Harvey Pough
Rochester Institute of Technology

Christine M. Janis
Brown University

John B. Heiser
Cornell University

Benjamin Cummings

San Francisco   Boston   New York
Cape Town   Hong Kong   London   Madrid   Mexico City
Montreal   Munich   Paris   Singapore   Sydney   Tokyo   Toronto

Editor-in-Chief: Beth Wilbur
Senior Acquisitions Editor: Star MacKenzie
Exec. Dir. of Development: Deborah Gale
Project Editor: Dusty Freeman, TBC Project Management
Editorial Assistant: Erin Mann
Executive Managing Editor: Erin Gregg
Managing Editor: Michael Early
Senior Production Supervisor: Shannon Tozier
Production Supervisor: Camille Herrera
Production and Composition Service: Katie Ostler and Sandy Reinhard, Black Dot Group
Interior Designer: Elm Street Publishing Services

Cover Design and Production: Marilyn Perry
Illustrators: Dartmouth Publishing
Photo Researcher: Maureen Spuhler
Image Permissions Coordinator: Fran Toepfer
Manufacturing Buyer: Michael Penne
Director of Marketing: Christy Lawrence
Exec. Marketing Manager: Lauren Harp
Text Printer: Courier Westford
Cover Printer: Phoenix Color
Cover Photo Credit: Tom Brakefield/Digital Vision/Getty Images

About the Authors

F. Harvey Pough is a Professor of Biology at the Rochester Institute of Technology. He began his biological career at the age of fourteen when he and his sister studied the growth and movements of a population of eastern painted turtles in Rhode Island. His research now focuses on organismal biology, blending physiology, morphology, behavior, and ecology in an evolutionary perspective. Undergraduate students regularly participate in his research and are coauthors of many of his publications. He especially enjoys teaching undergraduates and has taught courses in vertebrate zoology, functional ecology, herpetology, environmental physiology, and the organismal biology of humans. Currently he is teaching a year-long introductory biology course. He has published more than a hundred papers reporting the results of field and laboratory studies of turtles, snakes, lizards, frogs, and tuatara that have taken him to Australia, New Zealand, Fiji, Mexico, Costa Rica, Panama, and the Caribbean, as well as most parts of the United States. He is a Fellow of the American Association for the Advancement of Science and a Past-President of the American Society of Ichthyologists and Herpetologists.

Christine M. Janis is a Professor of Biology at Brown University, where she teaches comparative anatomy and vertebrate evolution. A British citizen, she obtained her bachelor's degree at Cambridge University and then crossed the pond to get her Ph.D. at Harvard University. She is a vertebrate paleontologist with a particular interest in mammalian evolution and faunal responses to climatic change. She first became interested in vertebrate evolution after seeing the movie *Fantasia* at the impressionable age of seven. That critical year was also the year that she began riding lessons, and she has owned at least one horse since the age of 12. Many years later, she is now an expert on ungulate (hoofed mammal) evolution, and is currently the president of the Society for the Study of Mammalian Evolution. She attributes her life history to the fact that she has failed to outgrow either the dinosaur phase or the horse phase.

John B. Heiser was born and raised in Indiana and completed his undergraduate degree in biology at Purdue University. He earned his Ph.D. in ichthyology from Cornell University for studies of the behavior, evolution, and ecology of coral reef fishes, research which he continues today with colleagues specializing in molecular biology. For fifteen years, he was Director of the Shoals Marine Laboratory operated by Cornell University and the University of New Hampshire on the Isles of Shoals in the Gulf of Maine. While at the Isles of Shoals, his research interests focused on opposite ends of the vertebrate spectrum—hagfish and baleen whales. J.B. enjoys teaching vertebrate morphology, evolution, and ecology, both in the campus classroom and in the field, and is recipient of the Clark Distinguished Teaching Award from Cornell University. His hobbies are natural history, travel and nature photography, and videography, especially underwater using scuba. He has pursued his natural history interests on every continent and all the world's major ocean regions. Because of his experience, he is a popular ecotourism leader, having led Cornell Adult University groups to the Caribbean, Sea of Cortez, French Polynesia, Central America, the Amazon, Borneo, Antarctica, and Spitsbergen in the High Arctic.

Brief Contents

Contents

Preface

The theme of *Vertebrate Life* is organismal biology—that is, how the anatomy, physiology, ecology, and behavior of animals interact to produce organisms that function effectively in their environments and how lineages of organisms change through evolutionary time. We made several organizational changes in the seventh edition to emphasize that functional approach. Comments from colleagues indicate that those changes were effective, and we have retained that organization in this edition.

The eighth edition emphasizes advances in our understanding of vertebrates since the seventh edition was published. Several themes have been greatly expanded:

- **Molecular biology.** Molecular studies have produced new information about phylogenetic relationships that illuminates events as distant as the origin of gnathostome lineages and as recent as the separation of modern orders of mammals.
- **Fossil evidence.** Newly described fossils have expanded our understanding of the evolutionary diversity of vertebrate lineages and, particularly, our own human lineage.
- **Climate change.** The ever-increasing evidence of global warming has important implications for the biology and conservation of vertebrates, and new information about atmospheric conditions during the Paleozoic and Mesozoic eras sheds light on vertebrate diversification.
- **Evo-Devo.** The field of evolutionary development is helping us to understand the mechanisms responsible for the appearance of new structures and functions, including feathers, the middle ear, and lactation.
- **Conservation.** As the pace of extinction quickens, specific situations raise acute concerns: the global decline in amphibian populations, part of which can be traced to the worldwide spread of a fungal infection; the threats posed to fisheries by fish-farming and transgenic fishes; the difficulty of preserving large animals that require huge home ranges, and especially the difficulties associated with large predators, such as tigers, that sometimes eat people.

- **Access to Information.** The expansion of electronic databases and the accessibility of online resources give students increased access to the primary literature and authoritative secondary sources, and we have increased the citations of printed and online journals and of websites to encourage students to explore these sources.

Amid all of these changes, the element of *Vertebrate Life* that has always been most important to the authors has remained constant: We are biologists because we care enormously about what we do and the animals we work with. We are deeply committed to passing on the fascination and sheer joy that we have experienced to new generations of biologists and to providing information and perspectives that will help them with the increasingly difficult task of ensuring that the enormous vigor and diversity of vertebrate life does not vanish.

NEW! to the eighth edition we will provide images from the text in jpeg and PowerPoint. They can be downloaded from the Instructor Resource Center at www.pearsonhighered.com.

Acknowledgments

We are very grateful to the excellent production team assembled by Benjamin Cummings for this edition: Editor, Star MacKenzie; Project Editor, Dusty Friedman; Photo Researcher, Maureen Spuhler; and, with Black Dot Group, Production Editor, Katrina Ostler and Copy Editor, Carla Breidenbach. Their mastery of every step on the enormously complex path from a manuscript to a bound copy of a book has been enormously comforting to the authors.

We are especially pleased by the return of Jennifer Kane as the artist for this edition. Jennifer first met *Vertebrate Life* when she was a student, and she brings that perspective to her work. Jennifer combines the ability to render anatomical information accurately with an empathy for vertebrates that allows her to produce drawings so lifelike that they appear ready to walk off the page.

Writing a book with a scope as broad as this one requires the assistance of many people. We list below the colleagues who generously provided comments, suggestions, and photographs and who responded to our requests for information.

Jon Baskin, *Texas A & M University*
Larry Buckley, *Rochester Institute of Technology*
Brooks Burr, *Southern Illinois University*
John Cadle, *California Academy of Sciences*
Allison Cree, *University of Otago*
Kevin De Queiroz, *National Museum of Natural History*
Mark Dimmitt, *Arizona-Sonora Desert Museum*
Colleen Farmer, *University of Utah*
Nick Geist, *Sonora State University*
Virginia Hayssen, *Smith College*
Gene Helfman, *Institute of Ecology, University of Georgia*

Kerry Kilburn, *Old Dominion University*
Karen Lips, *Southern Illinois University*
Kevin McGraw, *Arizona State University*
Alan Savitzky, *Old Dominion University*
Marcus Simons, *New Zealand Department of Conservation*
Douglas Woodhams, *James Madison University*
Sam Young, *James Cook University*

We also especially appreciate the work of reviewers of this text: Brooks M. Burr, *Southern Illinois University;* Virginia Hayssen, *Smith College;* Kerry S. Kilburn, *Old Dominion University;* Kevin McGraw, *Arizona State University.*

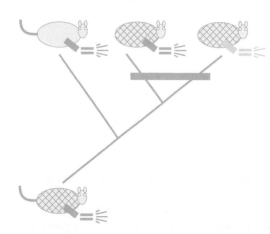

PART I

Vertebrate Diversity, Function, and Evolution

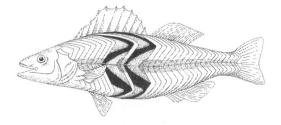

The more than 57,000 living species of vertebrates inhabit nearly every part of the Earth, and other kinds of vertebrates that are now extinct lived in habitats that no longer exist. Increasing knowledge of the diversity of vertebrates was a product of the European exploration and expansion that began in the fifteenth and sixteenth centuries. In the middle of the eighteenth century, Swedish naturalist Carolus Linnaeus developed a binominal classification to catalog the varieties of animals and plants. Despite some problems in reflecting evolutionary relationships, the Linnaean system remains the basis for naming living organisms today.

A century later, Charles Darwin and Alfred Russell Wallace explained the diversity of plants and animals as the product of natural selection and evolution. In the early twentieth century, their work was coupled with the burgeoning information about mechanisms of genetic inheritance. This combination of genetics and evolutionary biology, known as the New Synthesis, or neo-Darwinism, continues to be the basis for understanding the mechanics of evolution. Methods of classifying animals have also changed during the twentieth century; and classification, which began as a way of trying to organize the diversity of organisms, has become a powerful tool for generating testable hypotheses about evolution.

Vertebrate biology and the fossil record of vertebrates have been at the center of these changes in our view of life. Comparative studies of the anatomy, embryology, and physiology of living vertebrates have often supplemented the fossil record. These studies reveal that evolution acts by changing existing traits. All vertebrates share basic characteristics that are the products of their common ancestry, and the process of evolution can be analyzed by tracing the modifications of these characters. Thus, an understanding of vertebrate form and function is basic to understanding the evolution of vertebrates and the ecology and behavior of living species.

CHAPTER 1

The Diversity, Classification, and Evolution of Vertebrates

Evolution is central to vertebrate biology because it provides a principle that organizes the diversity we see among living vertebrates and helps to fit extinct forms into the context of living species. Classification, initially a process of attaching names to organisms, has become a method of understanding evolution and planning strategies for conservation. Current views of evolution stress natural selection operating at the level of individuals as the predominant mechanism that produces changes in a population over time. The processes and events of evolution are intimately linked to the changes that have occurred on Earth during the history of vertebrates. These changes have resulted from the movements of continents and the effects of those movements on climates and geography. In this chapter, we present an overview of the scene, the participants, and the events that have shaped the biology of vertebrates.

1.1 The Vertebrate Story

Mention "animal" and most people will think of a vertebrate. Vertebrates are often abundant and conspicuous parts of people's experience of the natural world. Vertebrates are also very diverse: The more than 57,000 **extant** (currently living) species of vertebrates range in size from fishes weighing as little as 0.1 gram when fully mature to whales weighing over 100,000 kilograms. Vertebrates live in virtually all the habitats on Earth. Bizarre fishes, some with mouths so large they can swallow prey larger than their own bodies, cruise through the depths of the sea, sometimes luring prey to them with glowing lights. Fifteen kilometers above the fishes, migrating birds fly over the crest of the Himalayas, the highest mountains on Earth.

The behaviors of vertebrates are as diverse and complex as their body forms. Vertebrate life is energetically expensive, and vertebrates get the energy they need from food they eat. Carnivores eat the flesh of other animals and show a wide range of methods of capturing prey. Some predators search the environment to find prey, whereas others wait in one place for prey to come to them. Some carnivores pursue their prey at high speeds, and others pull prey into their mouths by suction. Many vertebrates swallow their prey intact, sometimes while it is alive and struggling, but other vertebrates have very specific methods of dispatching prey. Venomous snakes inject complex mixtures of toxins, and cats (of all sizes, from house cats to tigers) kill their prey with a distinctive bite on the neck.

Herbivores eat plants. Plants cannot run away when an animal approaches, so they are easy to catch, but they are hard to chew and digest and frequently contain toxic compounds. Herbivorous vertebrates show an array of specializations to deal with the difficulties of eating plants. Elaborately sculptured teeth tear apart tough leaves and expose the surfaces of cells, but the cell walls of plants contain cellulose, which no vertebrate can digest. Herbivorous vertebrates rely on microorganisms living in their digestive tracts to digest cellulose. In addition, these endosymbionts (organisms that live inside another organism) detoxify the chemical substances that plants use to protect themselves.

Reproduction is a critical factor in the evolutionary success of an organism, and vertebrates show an astonishing range of behaviors associated with mating and reproduction. In general, males court females and females care for the young, but these roles are reversed in many species of vertebrates. At the time of birth or hatching, some vertebrates are entirely self-sufficient and never see their parents, whereas other vertebrates (including humans) have extended periods of obligatory parental care. Extensive parental care is found in seemingly unlikely groups of vertebrates—fishes that incubate eggs in their mouths, frogs that carry their tadpoles to water and then return to feed them, and birds that feed their nestlings a fluid called crop milk that is very similar in composition to mammalian milk.

The diversity of living vertebrates is enormous, but the species now living are only a small proportion of the species of vertebrates that have existed. For each living species, there may be more than a hundred extinct species, and some of these have no counterparts among living forms. For example, the dinosaurs that dominated the Earth for 180 million years are so entirely different from living animals that it is hard to reconstruct the lives they led. Even mammals were once more diverse than they are now. The Pleistocene epoch saw giants of many kinds—ground sloths as big as modern rhinoceroses and raccoons as large as bears. The number of species of terrestrial vertebrates probably reached its maximum in the middle Miocene, 12 to 14 million years ago, and has been declining since then.

The story of vertebrates is fascinating. Where they originated, how they evolved, what they do, and how they work provide endless intriguing details. In preparing to tell this story, we must introduce some basic information, including what the different kinds of vertebrates are called, how they are classified, and what the world was like as the story of vertebrates unfolded.

■ Major Groups of Vertebrates

Two major groups of vertebrates are distinguished on the basis of an innovation in embryonic development, the appearance of three membranes formed by tissues that come from the embryo itself (see Chapter 9 Section, 9.3). One of these membranes, the amnion, surrounds the embryo, and animals with this structure are called **amniotes**. The division between non-amniotes and amniotes corresponds roughly to

aquatic and terrestrial vertebrates, although many amphibians and a few fishes lay non-amniotic eggs in nests on land.

Among the amniotes, we can distinguish two major evolutionary lineages—the sauropsids (reptiles, including birds) and the synapsids (mammals). These lineages separated from each other in the Late Devonian period, before vertebrates had developed many of the characters we see in extant species. As a result, synapsids and sauropsids represent parallel but independent origins of basic characters such as lung ventilation, kidney function, insulation, and temperature regulation.

Figure 1–1 shows the major kinds of vertebrates and the approximate numbers of living species. In the following sections, we briefly describe the different kinds of vertebrates.

■ Non-Amniotes

The embryos of non-amniotes are enclosed and protected by membranes that are produced by the reproductive tract of the female. This is the condition seen among the invertebrate relatives of vertebrates, and it is retained in the primarily aquatic living vertebrates, the fishes and amphibians.

Hagfishes and Lampreys—Myxinoidea and Petromyzontoidea Lampreys and hagfishes are elongate, limbless, scaleless, and slimy and have no internal bony tissues. They are scavengers and parasites and are specialized for those roles. Hagfishes (about 70 species) are marine, living on the seabed at depths of 100 meters or more. In contrast, many of the 38 species of lampreys are migratory forms that live in oceans and spawn in rivers.

Hagfishes and lampreys are unique among living vertebrates because they lack jaws; this feature makes them important in the study of vertebrate evolution. They have traditionally been grouped as agnathans (Greek *a* = without and *gnath* = jaw) or cyclostomes (Greek *cyclo* = round and *stoma* = mouth), but they are probably not closely related to each other and instead represent two independent (i.e., separate) evolutionary lineages. Hagfishes lack many of the features characterizing most vertebrates; for example, they have no trace of vertebrae and sometimes are classified in the Craniata (animals with a skull) but not in the Vertebrata (animals with a vertebral column). In contrast, lampreys have rudimentary vertebrae as well as many other characters they share with jawed vertebrates. The jawless condition of both lampreys and hagfishes, however, is ancestral.

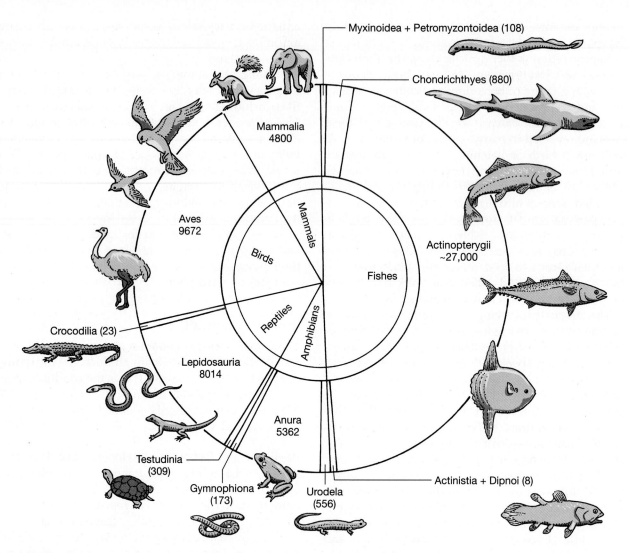

▲ **FIGURE 1–1** Diversity of vertebrates. Areas in the diagram correspond to approximate numbers of living species in each group. (These are estimates, and the numbers change frequently as new species are described.) Common names are in the center circle, and formal names for the groups are on the outer circle.

Sharks, Rays, and Ratfishes–Chondrichthyes The name Chondrichthyes (Greek *chondro* = cartilage and *ichthyes* = fish) refers to the cartilaginous skeletons of these fishes. Extant sharks and rays form a group called the Neoselachii (Greek *neo* = new and *selach* = shark), but the two kinds of fishes differ in body form and habits. Sharks have a reputation for ferocity that most of the 403 species would have difficulty living up to. Many sharks are small (15 centimeters or less); and the largest species, the whale shark (which grows to 10 meters), is a filter feeder that subsists on plankton it strains from the water. The approximately 534 species of rays are dorsoventrally flattened and frequently bottom dwellers that swim

with undulations of their extremely broad pectoral fins.

There are only about 33 species in the second group of chondrichthyans, the ratfishes or chimaeras. The name of the group, Holocephalii (Greek *holo* = whole and *cephal* = the head), refers to the single gill cover that extends over all four gill openings. These are bizarre marine animals with long, slender tails and bucktoothed faces that look rather like rabbits. They live on the seafloor and feed on hardshelled prey, such as crustaceans and mollusks.

Bony Fishes–Osteichthyes Bony fishes, the Osteichthyes (Greek *osteo* = bone and *ichthyes* = fish), are so diverse that any attempt to characterize them briefly is doomed to failure. Two broad categories can be recognized: the ray-finned fishes (actinopterygians; Greek *actino* = ray and *ptero* = wing or fin) and the lobe-finned or fleshy-finned fishes (sarcopterygians; Greek *sarco* = flesh).

The ray-finned fishes have radiated extensively in fresh and salt water. More than 27,000 species of ray-finned fishes have been named, and several thousand additional species may await discovery. A single project, the Census of Marine Life, is describing 150 to 200 previously unknown species of ray-finned fishes annually. Two major groups can be distinguished among actinopterygians. The Chondrostei (bichirs, sturgeons, and paddlefishes) includes about 43 species that are survivors of an early radiation of bony fishes. Bichirs (about 16 species) are swamp- and river-dwellers from Africa; they are known as African reed fish in the aquarium trade. Sturgeons (27 species) are large fishes with protrusible, toothless mouths that are used to suck food items from the bottom. Sturgeons are the source of caviar—eggs are taken from the female before they are laid. Of course, this kills the female sturgeon, and many species have been driven close to extinction by overfishing. Paddlefish (two species, one in the Mississippi drainage of North America and the other in the Yangtze River of China) have a paddlelike snout with organs that locate prey by sensing electrical fields.

The Neopterygii, the modern radiation of ray-finned fishes, can be divided into three lineages. Two of these—the gars (Lepisosteiformes, seven species) and the bowfin (one species, *Amia calva*)—are relicts of earlier radiations. These fishes have cylindrical bodies, thick scales, and jaws armed with sharp teeth. They seize prey in their mouths with a sudden rush or gulp, and they lack the specializations of the jaw apparatus that allow later bony fishes to use more complex feeding modes.

The third lineage of neopterygians, the Teleostei, includes more than 27,000 species of fishes covering every imaginable combination of body size, habitat, and habits. Most familiar fishes are teleosts—the trout, bass, and panfish that anglers seek; the sole (a kind of flounder) and swordfish featured by seafood restaurants; and the salmon and tuna whose by-products find their way into canned catfood. Modifications of the body form and jaw apparatus have allowed many teleosts to be highly specialized in their swimming and feeding habits.

In one sense, only eight species of lobe-finned fishes survive, the six species of lungfishes (Dipnoi) found in South America, Africa, and Australia and the two species of coelacanth (Actinistia), one from deep waters off the east coast of Africa and a second species recently discovered near Indonesia. These are the living fishes most closely related to terrestrial vertebrates, and a more accurate view of sarcopterygian diversity includes their terrestrial descendants—amphibians, mammals, turtles, lepidosaurs (the tuatara, lizards, and snakes), crocodilians, and birds. From this perspective, bony fishes include two major evolutionary radiations, each containing more than 25,000 living species.

Salamanders, Frogs, and Caecilians—Urodela, Anura, and Gymnophiona These three groups of vertebrates are popularly known as amphibians (Greek *amphi* = double and *bios* = life) in recognition of their complex life histories, which often include an aquatic larval form (the larva of a salamander or caecilian and the tadpole of a frog) and a terrestrial adult. All amphibians have bare skins (i.e., lacking scales, hair, or feathers) that are important in the exchange of water, ions, and gases with their environment. Salamanders (556 species) are elongate animals, mostly terrestrial, and usually with four legs; anurans (frogs, toads, treefrogs—5362 species in all) are short-bodied, with large heads and large hind legs used for walking, jumping, and climbing; and caecilians (about 173 species) are legless aquatic or burrowing animals.

■ Amniotes

An additional set of membranes associated with the embryo appeared during the evolution of vertebrates. They are called fetal membranes because they are derived from the embryo itself rather than from the reproductive tract of the mother. The amnion is one of these membranes, and vertebrates with an amnion are called amniotes. In general, amniotes are more terrestrial than non-amniotes; but, there are also secondarily aquatic species of amniotes (such as sea turtles and whales), as well as many species of salamanders and frogs that spend their entire lives on land despite being non-amniotes. However, many features distinguish non-amniotes (fishes and amphibians) from amniotes (mammals and reptiles, including birds), and we will use the terms to identify which of the two groups are being discussed.

By the Late Devonian, amniotes were well established on land. They ranged in size from lizardlike animals a few centimeters long through cat- and dog-size species to the cow-size parieasaurs. Some were herbivores, others were carnivores. In terms of their physiology, however, we can infer that they retained ancestral characters. They had scale-covered skins without an insulating layer of hair or feathers, a simple kidney that could not produce highly concentrated urine, simple lungs, and a heart in which the ventricle was not divided by a septum.

Terrestrial life requires efficient systems to extract oxygen from air and transport it via the circulatory system to the tissues, to eliminate waste products while retaining water, and to keep body temperature stable as the external temperature changes. These systems evolved in both lineages; but, because they evolved independently, the lungs, hearts, kidneys, and body coverings of sauropsids and synapsids are different.

Sauropsid Amniotes The sauropsid lineage contains the extant species we call reptiles: turtles, the scaly reptiles (tuatara, lizards, and snakes), crocodilians, and birds. Extinct sauropsids include the forms that dominated the world during the Mesozoic era—dinosaurs and pterosaurs (flying reptiles) on land and a variety of marine forms, including ichthyosaurs and plesiosaurs, in the oceans.

Turtles—Testudinia The approximately 309 species of turtles (Latin *testudo* = a turtle) are probably the most immediately recognizable of all vertebrates. The shell that encloses a turtle has no exact duplicate among other vertebrates, and the morphological modifications associated with the shell make turtles extremely peculiar animals. They are, for example, the only vertebrates with the shoulders (pectoral girdle) and hips (pelvic girdle) inside the ribs.

Tuatara, Lizards, and Snakes—Lepidosauria These three kinds of vertebrates can be recognized by their scale-covered skin (Greek *lepido* = scale and *saur* = lizard), as well as by characteristics of the skull. The two species of tuatara, stocky-bodied animals found only on some islands near New Zealand, are the sole living remnants of an evolutionary lineage of animals called Sphenodontida, which was more diverse in the Mesozoic. In contrast, lizards (about 5000 species) and snakes (about 3015 species of highly specialized lizards) are now at the peak of their diversity.

Alligators and Crocodiles—Crocodilia These impressive animals, which draw their name from *crocodilus*, the Greek word for crocodile, are in the same evolutionary lineage (the Archosauria) as dinosaurs and birds.

The 23 species of crocodilians, as they are known collectively, are semiaquatic predators, with long snouts armed with numerous teeth. They range in size from the saltwater crocodile, which has the potential to grow to a length of 7 meters, to dwarf crocodiles and caimans that are less than a meter long. Their skin contains many bones (osteoderms; Greek *osteo* = bone and *derm* = skin) that lie beneath the scales and provide a kind of armor plating. Crocodilians are noted for the parental care they provide for their eggs and young.

Birds—Aves The birds (Latin *avis* = a bird) are a lineage of dinosaurs that evolved flight in the Mesozoic. They have diversified into more than about 9672 species. Feathers are characteristic of extant birds, and feathered wings are the structures that power a bird's flight. Recent discoveries of dinosaur fossils with traces of feathers show that feathers evolved before flight. This offset in the time of appearance of feathers and flight illustrates an important principle: the function of a trait in an extant species is not necessarily the same as its function when it first appeared. In other words, current utility is not the same as evolutionary origin. The original feathers were almost certainly structures that were used in courtship displays, and their modification as airfoils and for streamlining in birds is a secondary event.

Synapsid Amniotes The synapsid lineage contains the three kinds of extant mammals: the monotremes (prototheria; the platypus and echidna), marsupials (metatherians), and placentals (eutherians). Extinct synapsids include forms that diversified in the Paleozoic era—pelycosaurs and therapsids—and the rodentlike multituberculates of the Late Mesozoic.

Mammals—Mammalia The living mammals (Latin *mamma* = a teat) can be traced to an origin in the late Paleozoic, from some of the earliest fully terrestrial vertebrates. Extant mammals include about 4800 species, most of which are placental mammals. Both placentals and marsupials possess a placenta, a structure that transfers nutrients from the mother to the embryo and removes the waste products of the embryo's metabolism. Placentals have a more extensive system of placentation and a long gestation period, while marsupials have a short gestation period and give birth to very immature young that continue their development attached to a nipple, often in an external pouch on the mother's abdomen. Marsupials dominate the mammalian fauna only in Australia. Kangaroos, koalas, and wombats are familiar Australian marsupials. The strange monotremes, the platypuses and the echidnas—also from Australia—are mammals whose young are hatched from eggs. All mammals, including monotremes, feed their young with milk.

■ New Species

New species of vertebrates are described weekly—this is why we use the words "approximately" and "about" when we cite the numbers of species. About 200 new species of fish are described annually, and the Census of Marine Life program alone is describing an average of two new species of fish per week. In 2002, a molecular analysis of the relationships of rhacophorid frogs on Madagascar increased the number of species from 18 to more than 100.

Many of the newly described species of vertebrates are small and occur in remote areas, but new species of large and conspicuous vertebrates are still being described (**Figure 1–2**).

- In 2005, a new species of mangabey monkey was discovered in Tanzania, and subsequent study showed that it was so different from related species that a new genus, *Rungwecebus*, was created for it (Jones et al. 2005, Davenport et al. 2006). This monkey, which is as big as a medium-size dog, has a loud vocalization and occurs in forest adjacent to cultivated areas. In fact, the type specimen was captured in a trap set by a farmer in a field of maize (Figure 1–2a).
- In 2007, the clouded leopard that occurs on the island of Borneo (Figure 1–2b) was identified as a new species, *Neofelis diardi* (Wilting et al. 2007).
- Since 2000, three new species of whales have been described—two rorquals (*Balaenoptera omurai* from the Indo-Pacific region and *B. edeni*, which has a worldwide distribution) and a new right whale (*Eubalaena japonica*) from the North Pacific (Rosenbaum et al. 2000, Wado et al. 2003). These animals are from 11 to 17 meters long,

(a)

(b)

▲ **FIGURE 1–2** Two new species of vertebrates that have been described recently. (a) An adult male highland mangabey, *Rungwecebus kipunji*. (b) The Bornean clouded leopard, *Neofelis diardi*.

1.2 Classification of Vertebrates

The diversity of vertebrates (more than 57,000 living species and perhaps 100 times that number of species now extinct) makes the classification of vertebrates an extraordinarily difficult task. Yet classification has long been at the heart of evolutionary biology. Initially, classification of species was seen as a way of managing the diversity of organisms, much as an office filing system manages the paperwork of the office. Each species could be placed in a pigeonhole marked with its name; when all species were in their pigeonholes, the diversity of vertebrates would have been encompassed. This approach to classification was satisfactory as long as species were regarded as static and immutable: once a species was placed in the filing system, it was there to stay.

Acceptance of the fact that species evolve has made that kind of classification inadequate. Now biologists must express evolutionary relationships among species by incorporating evolutionary information in the system of classification. Ideally, a classification system should not only attach a label to each species, it should also encode the evolutionary relationship between that species and other species. Modern techniques of systematics (the evolutionary classification of organisms) have become methods for generating testable hypotheses about evolution.

Classification and Names

Our system of naming species is pre-Darwinian. It traces back to methods established by the naturalists of the seventeenth and eighteenth centuries, especially those of Carl von Linné, a Swedish naturalist, better known by his Latin pen name, Carolus Linnaeus. The Linnaean system employs binominal nomenclature to designate species and arranges species into hierarchical categories (**taxa**, singular *taxon*) for classification. This system is incompatible in some respects with evolutionary biology (de Queiroz and Gauthier 1992, Pennisi 2001), but it is still widely used.

Binominal Nomenclature

The scientific naming of species became standardized when Linnaeus's monumental work, *Systema Naturae* (*The System of Nature*), was published in sections between 1735 and 1758. Linnaeus attempted to give an identifying name to every known species of plant and animal. His method assigns a binominal (two-word) name to each species. Familiar examples include *Homo sapiens* for human beings (Latin *hom* = human and *sapien* = wise), *Passer domesticus* for the house sparrow (Latin *passer* = sparrow and *domesticus* = belonging to the house), and *Canis familiaris* for the domestic dog (Latin *canis* = dog and *familiaris* = of the family).

Why use Latin words? Latin was the early universal language of European scholars and scientists. It has provided a uniform usage that scientists, regardless of their native language, continue to recognize worldwide. The same species may have different colloquial names, even in the same language. For example, *Felis concolor* (Latin for "the uniformly colored cat") is known in various parts of North America as cougar, puma, mountain lion, American panther, painter, and catamount. In Central and South America it is called león colorado, onça-vermelha, poema, guasura, or yaguá-pitá. But biologists of all nationalities recognize the name *Felis concolor* as referring to a specific kind of cat.

Hierarchical Groups: The Higher Taxa

Linnaeus and other naturalists of his time developed what they called a natural system of classification.

The **species** is the basic level of biological classification, but the definition of a species has been contentious, partly because criteria that have been used to identify extant species (e.g., reproductive isolation from other species) don't work for fossil species and don't always correspond to genetic differences. Similar species are grouped together in a **genus** (plural *genera*), based on characters that define the genus. The most commonly used characters were anatomical because they can be most easily preserved in museum specimens. Thus all doglike species—various wolves, coyotes, and jackals—were grouped together in the genus *Canis* because they all share certain anatomical features, such as an erectile mane on the neck and a skull with a long, prominent sagittal crest on the top of the skull from which massive temporal (jaw-closing) muscles originate. Linnaeus's method of grouping species was functional because it was based on anatomical (and to some extent on physiological and behavioral) similarities and differences. Linnaeus lived before there was any knowledge of genetics and the mechanisms of inheritance, but he used characters that we understand today are genetically determined biological traits that generally express the degree of genetic similarity or difference among groups of organisms. Genera are placed in families, families in orders, orders in classes, and animal classes in phyla (singular *phylum*).

1.3 Phylogenetic Systematics

All methods of classifying organisms, even pre-Linnaean systems, are based on similarities among the included species, but some similarities are more significant than others. For example, nearly all vertebrates have paired limbs, but only a few kinds of vertebrates have mammary glands. Consequently, knowing that the species in question have mammary glands tells you more about the closeness of their relationship than knowing that they have paired limbs. You would thus give more weight to the presence of mammary glands than to paired limbs.

A way to assess the relative importance of different characteristics was developed in the mid-twentieth century by Willi Hennig, who introduced a method of determining evolutionary relationships called **phylogenetic systematics** (Greek *phyla* = tribe and *genesis* = origin). An evolutionary lineage is a **clade** (from *cladus*, the Greek word for a branch), and phylogenetic systematics is also called **cladistics**. Cladistics recognizes only groups of organisms that are related by common descent. The application of cladistic methods has made the study of evolution rigorous. The groups of organisms recognized by cladistics are called natural groups, and they are linked in a nested series of ancestor-descendant relationships that trace the evolutionary history of the group. Hennig's contribution was to insist that these groups can be identified only on the basis of **derived characters**.

Derived means "different from the ancestral condition." A derived character is called an **apomorphy** (Greek *apo* = away from [i.e., derived from] and *morph* = form). For example, the feet of terrestrial vertebrates have distinctive bones—the carpals, tarsals, and digits. This arrangement of foot bones is different from the ancestral pattern seen in lobe-finned fishes, and all lineages of terrestrial vertebrates had that derived pattern of foot bones at some stage in their evolution. (Many groups of terrestrial vertebrates—horses, for example—have subsequently modified the foot bones, and some, such as snakes, have lost the limbs entirely. The significant point is that those evolutionary lineages include species that had the derived terrestrial pattern.) Thus, the terrestrial pattern of foot bones is a **shared derived character** of terrestrial vertebrates. In cladistic terminology, shared derived characters are called **synapomorphies** (Greek *syn* = together).

Of course, organisms also share ancestral characters, that is, characters that they have inherited unchanged from their ancestors. These are called **plesiomorphies** (Greek *plesi* = near in the sense of "similar to the ancestor"). Terrestrial vertebrates have a vertebral column, for example, that was inherited from lobe-finned fishes. Hennig called shared ancestral characters **symplesiomorphies** (*sym*, like *syn*, is a Greek root that means "together"). Symplesiomorphies tell us nothing about degrees of relatedness. The principle that only *shared derived* characters can be used to determine evolutionary relationships is the core of cladistics.

The conceptual basis of cladistics is straightforward, although applying cladistic criteria to real organisms can become very complicated. To illustrate cladistic classification, consider the examples presented in **Figure 1–3**. Each of the three **cladograms** (diagrams showing hypothetical sequences of branching during evolution) illustrates a possible evolutionary relationship for the three taxa (plural of taxon, which means a species or group of species), identified as 1, 2, and 3. To make the example a bit

more concrete, we can consider three characters: the number of toes on the front foot, the skin covering, and the tail. For this example, let's say that in the ancestral character state there are five toes on the front foot, and in the derived state there are four toes. We'll say that the ancestral state is a scaly skin, and the derived state is a lack of scales. As for the tail, it is present in the ancestral state and absent in the derived state.

Figure 1–3 shows the distribution of those three character states in the three taxa. The animals in taxon 1 have five toes on the front feet, lack scales, and have a tail. Animals in taxon 2 have five toes, scaly skins, and no tails. Animals in taxon 3 have four toes, scaly skins, and no tails.

How can we use this information to decipher the evolutionary relationships of the three groups of animals? Notice that the derived number of toes occurs only in taxon 3, and the derived tail condition (absent) is found in taxa 2 and 3. The most **parsimonious** phylogeny (i.e., the evolutionary relationship requiring the fewest number of changes) is represented by the leftmost diagram in Figure 1–3. Only three changes are needed to produce the derived character states:

1. In the evolution of taxon 1, scales are lost.
2. In the evolution of the lineage including taxon 2 + taxon 3, the tail is lost.
3. In the evolution of taxon 3, a toe is lost from the front foot.

The other two phylogenies shown in Figure 1–3 are possible, but they would require tail loss to occur independently in taxon 2 and in taxon 3. Any change in a structure is an unlikely event, so the most plausible phylogeny is the one requiring the fewest changes. The second and third phylogenies each require four evolutionary changes, so they are less parsimonious than the first phylogeny we considered.

A phylogeny is a hypothesis about the evolutionary relationships of the groups included. Like any scientific hypothesis, it can be tested when new data become available. If it fails that test, it is falsified; that is, it is rejected, and a different hypothesis (a different cladogram) takes its place. The process of testing hypotheses and replacing those that are falsified is a continuous one, and changes in the cladograms in successive editions of this book show where new information has generated new hypotheses. The most important contribution of phylogenetic systematics is that it enables us to frame testable hypotheses about the sequence of events during evolution.

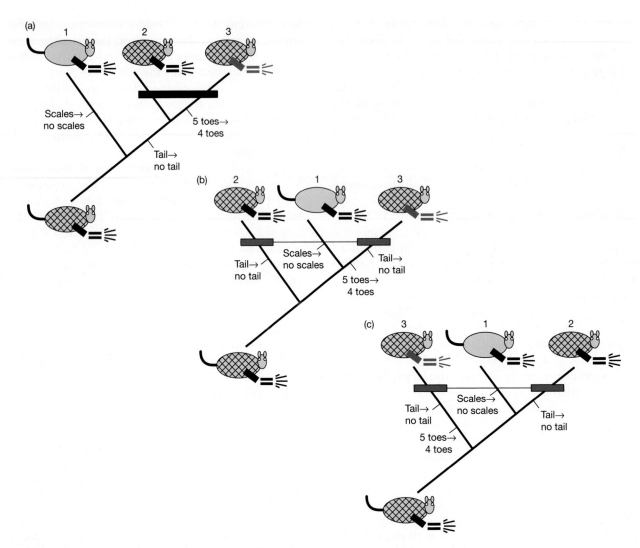

▲ **FIGURE 1–3** Three cladograms showing the possible evolutionary relationships of three taxa. Bars connect derived characters (apomorphies). The black bar shows a shared derived character (a synapomorphy) of the lineage that includes taxa 2 and 3. Colored bars represent two independent origins of the same derived character state that must be assumed to have occurred if there was no apomorphy in the most recent common ancestor of taxa 2 and 3. The labels identify changes from the ancestral character state to the derived condition. Cladogram (a) requires a total of three changes from the ancestral condition to explain the distribution of characters in the extant taxa, whereas cladograms (b) and (c) require four changes. Because cladogram (a) is more parsimonious (i.e., requires the smallest number of changes) it is considered to be the most likely sequence of changes.

So far we have avoided a central issue of phylogenetic systematics: How do scientists know which character state is ancestral (plesiomorphic) and which is derived (apomorphic)? That is, how can we determine the direction (**polarity**) of evolutionary transformation of the characters? For that, we need additional information. Increasing the number of characters we are considering can help, but comparing the characters we are using with an **outgroup** that consists of the closest relatives of the **ingroup** (i.e., the organisms we are studying) is the preferred method. A well-chosen outgroup will possess ancestral character states compared to the ingroup. For example, lobe-finned fishes are an appropriate outgroup for terrestrial vertebrates.

1.4 The Problem with Fossils: Crown and Stem Groups

Evolutionary lineages must have a single evolutionary origin; that is, they must be monophyletic (Greek *mono* = one, single) and include all the descendants of that ancestor. The cladogram depicted in **Figure 1–4** is a hypothesis of the evolutionary relationships of the major living groups of vertebrates. A series of dichotomous branches extends from the origin of vertebrates to the groups of extant vertebrates. Cladistic terminology assigns names to the lineages originating at each branch point. This process produces a nested series of groups, starting with the most inclusive. For example, the Gnathostomata includes all vertebrate animals that have jaws; that is, every taxon above number 3 in Figure 1–4 is included in the Gnathostomata; every taxon above number 4 is included in the Osteichthyes (bony fish), and so on. Because the lineages are nested, it is correct to say that humans are both gnathostomes and osteichthyans. After number 7, the cladogram divides into Lissamphibia and Amniota, and humans are in the Amniota lineage. The cladogram divides again above number 10 into two lineages, the Sauropsida and Synapsida lineages. Humans are in the Eutheria, which is in the synapsid lineage.

This method of tracing ancestor-descendant relationships allows us to decipher evolutionary pathways that extend from fossils to living groups, but a difficulty arises when we try to find names for groups that include fossils (Kemp 1999, Budd 2001). The derived characters that define the extant groups of vertebrates did not necessarily appear all at the same time. On the contrary, evolution usually acts by gradual and random processes, and derived characters appear in a stepwise fashion. The extant members of a group have all of the derived characters of that group because that is how we define the group today; but, as you move backward through time to fossils that are ancestral to the extant species, you encounter forms that have a mosaic of primitive and derived characters. The further back in time you go, the fewer derived characters the fossils have.

What can we call the parts of lineages that contain these fossils? They are not included in the extant groups because they lack some of the derived characters of those groups, but the fossils in the lineage are more closely related to the extant group than they are to animals in other lineages.

The solution to this problem lies in naming two types of groups: crown groups and stem groups. The crown groups are defined by the extant species, the ones that have all the derived characters. The stem groups are the extinct forms that preceded the point at which the first member of the crown group branched off. Basically, stem groups contain fossils with some derived characters, and crown groups contain extant species plus those fossils that have all of the derived characters of the extant species. Stem groups are **paraphyletic** (Greek *para* = beside, beyond); that is, they do not contain all of the descendants of the ancestor of the stem group plus the crown group because the crown group is excluded by definition.

1.5 Evolutionary Hypotheses

Phylogenetic systematics is based on the assumption that organisms grouped together share a common heritage, which accounts for their similarities. Because of that common heritage, we can use cladograms to ask questions about evolution. By examining the origin and significance of characters of living animals, we can make inferences about the biology of extinct species (see Swiderski 2001). For example, the phylogenetic relationship of crocodilians, dinosaurs, and birds is shown in **Figure 1–5**. We know that both crocodilians and birds display extensive parental care of their eggs and young. Some fossilized dinosaur nests contain remains of baby dinosaurs, suggesting that at least some dinosaurs may also have cared for their young. Is that a plausible inference?

Obviously there is no direct way to determine what sort of parental care dinosaurs had. The intermediate lineages in the cladogram (pterosaurs and dinosaurs) are extinct, so we cannot observe their reproductive behavior. However, the phylogenetic diagram in Figure 1–5 provides an indirect way to approach the question. Both crocodilians and birds, the closest living relatives of the dinosaurs, do have parental care. Looking at living representatives of more distantly related lineages (outgroups), we see that parental care is not universal among fishes, amphibians, or sauropsids other than crocodilians. With that information, the most parsimonious explanation of the occurrence of parental care in both crocodilians and birds is that it had evolved in that lineage *before* the crocodilians separated from dinosaurs + birds. (We cannot prove that parental care did not evolve separately in crocodilians and in birds, but one change to parental care is more likely than two changes.) Thus, the most parsimonious hypothesis is that parental care is a derived character of the evolutionary lineage containing crocodilians + dinosaurs + birds (the Archosauria). That

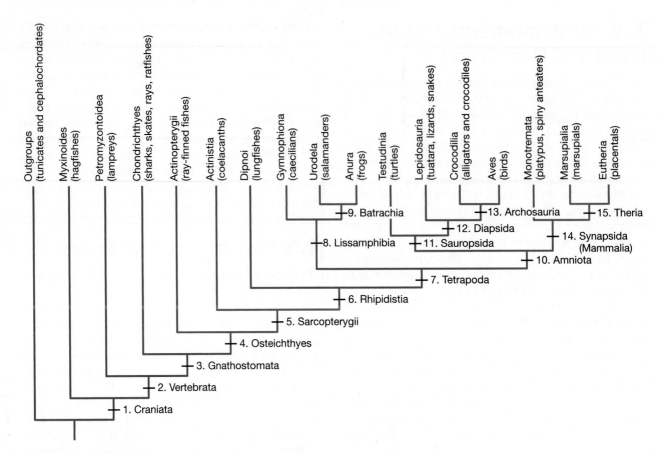

▲ **FIGURE 1–4** Phylogenetic relationships of extant vertebrates. This diagram shows the probable relationships among the major groups of extant vertebrates. Note that the cladistic groupings are nested progressively; that is, all placental mammals are therians, all therians are synapsids, all synapsids are amniotes, all amniotes are tetrapods, and so on.

Legend: Selected derived characters of the groups identified by numbers. More details are provided in subsequent chapters. **1. Craniata**—All living animals with a cranium: distinct head region skeleton incorporating anterior end of notochord, brain consisting of three regions, paired kidneys, gill bars made of cartilage, neural crest tissue. **2. Vertebrata**—All living craniates except hagfishes: vertebrae or vertebral elements (arcualia). **3. Gnathostomata**—All living vertebrates except lampreys: jaws formed from mandibular arch, teeth containing dentine. **4. Osteichthyes**—All living gnathostomes except the Chondrichthyes: presence of lung or swimbladder derived from the gut, unique pattern of dermal bones of the head and shoulder region. **5. Sarcopterygii**—All living osteichthyans except ray-finned fishes: unique supporting skeleton in fins. **6. Rhipidistia**—All living sarcopterygians except coelacanths: teeth with a distinctive pattern of folded enamel, a distinctive pattern of dermal skull bones. **7. Tetrapoda**—All living rhipidistians except lung-fishes: limbs with carpals, tarsals, and digits. **8. Lissamphibia**—All living amphibians: structure of the skin and elements of the inner ear. **9. Batrachia**—Salmanders and frogs: characteristics of the ears and loss of dermal scales.

10. Amniota—All living tetrapods except amphibians: shelled egg with a distinctive arrangement of extraembryonic membranes (the amnion, chorion, and allantois). **11. Sauropsida**—All living amniotes except synapsids (mammals and their extinct relatives): tabular and supratemporal bones small or absent, beta keratin present. **12. Diapsida**—All living sauropsids except turtles: skull with both dorsal and ventral temporal openings (fenestrae). **13. Archosauria**—All living diapsids except lepidosaurs: presence of a fenestra anterior to the orbit of the eye. **14. Synapsida (Mammalia)**—All living mammals and their extinct relatives: only a lower temporal fenestra present. **15. Theria**—All living mammals except monotremes: tribosphenic molar, a mobile scapula with loss of the coracoid bone.

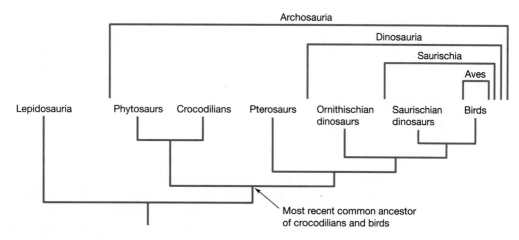

▲ **FIGURE 1–5** Using a cladogram to make inferences about behavior. The cladogram shows the relationships of the Archosauria, the evolutionary lineage that includes living crocodilians and birds. (Phytosaurs were crocodile-like animals that disappeared at the end of the Triassic, and pterosaurs were the flying reptiles of the Jurassic and Cretaceous.) Both extant groups—crocodilians and birds—display extensive parental care of eggs and young. The most parsimonious explanation of this situation assumes that parental care is an ancestral character of the archosaur lineage.

means we are probably correct when we interpret the fossil evidence as showing that dinosaurs did have parental care.

Figure 1–5 also shows how cladistics has made talking about restricted groups of animals more complicated than it used to be. Suppose you wanted to refer to just the two lineages of animals that are popularly known as dinosaurs—ornithischians and saurischians. What could you call them? Well, if you call them dinosaurs, you're not being PC (phylogenetically correct), because the Dinosauria lineage includes birds. So if you say dinosaurs, you are including ornithischians + saurischians + birds, even though any seven-year-old would understand that you are trying to restrict the conversation to extinct Mesozoic animals.

In fact, in cladistic terminology there is no correct taxonomic name for just the animals popularly known as dinosaurs. That's because cladistics recognizes only monophyletic lineages, and a monophyletic lineage includes an ancestral form *and* all its descendants. The most recent common ancestor of ornithischians, saurischians, and birds in Figure 1–5 lies at the intersection of the lineage of ornithischians with saurischians + birds, so Dinosauria is a monophyletic lineage. If birds are omitted, however, all the descendants of the common ancestor are no longer included; and ornithischians + saurischians minus birds does not fit the definition of a monophyletic lineage. It would be called a paraphyletic group. The stem groups discussed in the previous section are paraphyletic because they do not include all of the descendants of the fossil forms.

Biologists who are interested in how organisms live often want to talk about paraphyletic groups. After all, the dinosaurs (in the popular sense of the word) differed from birds in many ways. The only correct way of referring to the animals popularly known as dinosaurs is to call them nonavian dinosaurs, and you will find that and other examples of paraphyletic groups later in the book. Sometimes even this construction does not work because there is no appropriate name for the part of the lineage you want to distinguish. In this situation we will use quotation marks (e.g., "ostracoderms") to indicate that the group is paraphyletic.

Another important bit of terminology is **sister group**. The sister group is the monophyletic lineage most closely related to the monophyletic lineage being discussed. In Figure 1–5, for example, the lineage that includes crocodilians and phytosaurs is the sister group of the lineage that includes pterosaurs + ornithischians + saurischians + birds. Similarly, pterosaurs are the sister group of ornithischians + saurischians + birds, ornithischians are the sister group of saurischians + birds, and saurischians are the sister group of birds.

■ Determining Phylogenetic Relationships

We've established that the derived characters systematists used to group species into higher taxa must be inherited through common ancestry. That is, they are **homologous** (Greek *homo* = same) similarities. In principle, that notion is straightforward; but in practice, the determination of common ancestry can be complex. For example, birds and bats have wings that are modified forelimbs, but the wings were not inherited from a common ancestor with wings. The evolutionary lineages of birds (Sauropsida) and bats (Synapsida) diverged long ago, and wings evolved independently in the two groups. This process is called **convergent evolution. Parallel evolution** describes the situation in which species that have diverged relatively recently develop similar specializations. The long hind legs that allow the North American kangaroo rats and the African jerboa to jump are an example of parallel evolution in these two lineages of rodents. A third mechanism, **reversal**, can produce similar structures in distantly related organisms. Sharks and cetaceans (porpoises and whales) have very similar body forms, but they arrived at that similar form from different directions. Sharks retained an ancestral aquatic body form, whereas cetaceans arose from a lineage of terrestrial mammals with well-developed limbs that returned to an aquatic environment and reverted to the aquatic body form.

Convergence, parallelism, and reversal are forms of **homoplasy** (Greek *homo* = same and *plas* = form, shape). Homoplastic similarities do not indicate common ancestry. Indeed, they complicate the process of deciphering evolutionary relationships. Convergence and parallelism give an appearance of similarity (as in the wings of birds and bats) that is not the result of common evolutionary origin. Reversal, in contrast, conceals similarity (e.g., between cetaceans and their four-legged terrestrial ancestors) that is the result of common evolutionary origin.

■ Phylogeny and Conservation

Combining genetic analysis with cladistic analyses can provide an important tool for biologists concerned with conservation. For example, the new species of mammals described in Section 1.1 were identified by comparing their DNA with the DNA of related species. When a genetic difference is large, it means that the two forms have been reproductively isolated from each other and have followed different evolutionary trajectories. From a conservationist's perspective, lineages that have evolved substantial genetic differences are Evolutionarily Significant Units (ESUs), and management plans should protect the genetic diversity of ESUs.

For example, a genetic study published in 2007 revealed that the clouded leopards on the islands of Borneo and Sumatra (*Neofelis diardi*) and those on the Asian mainland (*Neofelis nebulosa*) separated between 1.4 and 2.8 million years ago, and the three forms have been following independent evolutionary trajectories since then (Wilting et al. 2007). The genetic distance between the mainland form and the island forms is as great as the difference between lions and tigers (**Figure 1–6a,b**). Furthermore, the island populations are reproductively isolated from each other, and the clouded leopards on Borneo and Sumatra are genetically distinct. Thus, the three forms represent three ESUs, and conservation plans should treat the mainland species and the two island species separately. Before this study the three forms were grouped together, and a substantial portion of the genetic diversity of clouded leopards was unrecognized.

Genetic analyses do not always provide a clear answer to questions about conservation, however. A comparison of polar bears (*Ursus maritimus*) with brown bears (*Ursus arctos*) reveals a more complicated situation (**Figure 1–6c–e**). Three lineages of brown bears that differ in geographic distribution can be identified (Talbot and Shields 1996). One lineage includes bears found in Eurasia and western Alaska, a second includes brown bears from eastern Alaska, and the third is composed of brown bears from the islands of southeastern Alaska. Polar bears group with this last lineage. In other words, despite the great differences in appearance (Figure 1–6d,e), ecology, and behavior that distinguish polar bears from brown bears, polar bears are genetically contained within a lineage of brown bears. Thus, decisions about conservation cannot be based solely on genetic distance, because no one would argue that polar bears do not constitute an ESU that should have its own conservation plan.

1.6 Earth History and Vertebrate Evolution

Since their origin in the early Paleozoic, vertebrates have been evolving in a world that has changed enormously and repeatedly. These changes have affected vertebrate evolution directly and indirectly. Understanding the sequence of changes in the positions of continents, and the significance of those positions regarding climates and interchange of faunas, is central to understanding the vertebrate

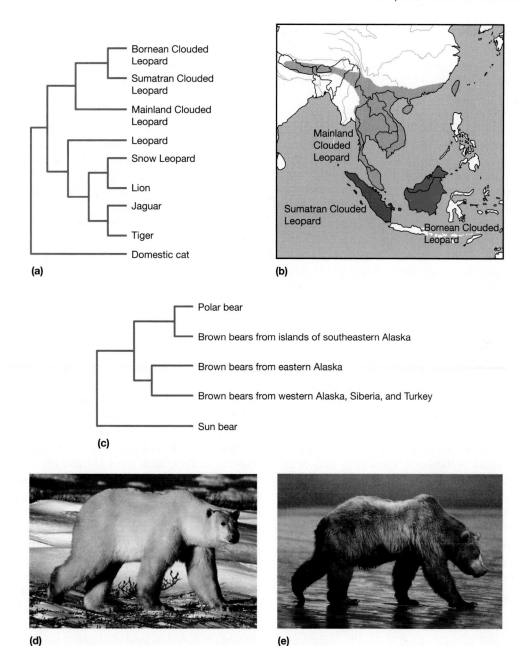

(a)

(b)

(c)

(d)

(e)

▲ **FIGURE 1–6** Examples of genetic analyses for conservation. (a and b) The two island forms of *Neofelis diardi* are more closely related to each other than either is to the mainland form, *Neofelis nebulosa*. (That is, the two island forms have a recent common ancestor.) Nonetheless, genetic differences distinguish the leopards on Sumatra from those on Borneo. (c-e) Despite their great differences in appearance, polar bears (*Ursus maritimus*) are genetically like the brown bears (*Ursus arctos*) found in the islands of southeastern Alaska.

story. These events are summarized inside the front cover of the book, and Chapters 7, 15, and 19 give details.

The history of the Earth has occupied three geological **eons**: the Archean, Proterozoic, and Phanerozoic. Only the Phanerozoic, which began 540 million years ago, contains vertebrate life, and it is divided into three geological **eras**: the Paleozoic (Greek *paleo* = ancient and *zoo* = animal), Mesozoic (Greek *meso* = middle), and Cenozoic (Greek *cen* = recent). These eras are divided into **periods**, which can be further subdivided in a variety of ways, such as the subdivisions called **epochs** within the Cenozoic period from the Paleocene to the Recent.

Movement of landmasses, called continental drift, has been a feature of Earth's history at least since the Proterozoic, and the course of vertebrate evolution has been shaped extensively by continental movements. By the early Paleozoic, roughly 540 million years ago, a recognizable scene had appeared. Seas covered most of the Earth as they do today, large continents floated on the Earth's mantle, life had become complex, and an atmosphere of oxygen had formed, signifying that the photosynthetic production of food resources had become a central phenomenon of life.

The continents still drift today—North America is moving westward and Australia northward at approximately 4 centimeters per year (about the rate at which fingernails grow). Because the movements are so complex, their sequence, their varied directions, and the precise timing of the changes are difficult to summarize. When the movements are viewed broadly, however, a simple pattern unfolds during vertebrate history: fragmentation, coalescence, fragmentation.

Continents existed as separate entities over 2 billion years ago. Some 300 million years ago, all of these separate continents combined to form a single landmass known as Pangaea, which was the birthplace of terrestrial vertebrates. Persisting and drifting northward as an entity, this huge continent began to break apart about 150 million years ago. Its separation occurred in two stages: first into Laurasia in the north and Gondwana in the south, and then into a series of units that have drifted and become the continents we know today.

The complex movements of the continents through time have had major effects on evolution of vertebrates. Most obvious is the relationship between the location of landmasses and their climates. At the end of the Paleozoic, much of Pangaea was located on the equator, and this situation persisted through the middle of the Mesozoic. Solar radiation is most intense at the equator, and climates at the equator are correspondingly warm. During the late Paleozoic and much of the Mesozoic, large areas of land enjoyed tropical conditions. Terrestrial vertebrates evolved and spread in these tropical regions. By the end of the Mesozoic, much of Earth's landmass had moved out of equatorial regions; and, by the mid-Cenozoic, most terrestrial climates in the higher latitudes of the Northern and Southern Hemispheres were temperate instead of tropical.

A less obvious effect of the position of continents on terrestrial climates comes from changes in patterns of oceanic circulation. For example, the Arctic Ocean is now largely isolated from the other oceans, and it does not receive warm water via currents flowing from more equatorial regions. High latitudes are cold because they receive less solar radiation than do areas closer to the equator, and the Arctic Basin does not receive enough warm water to offset the lack of solar radiation. As a result, the Arctic Ocean is permanently frozen, and cold climates extend well southward across the continents. The cooling of climates in the Northern Hemisphere at the end of the Eocene epoch, around 45 million years ago, may have been a factor leading to the extinction of archaic mammals, and it is partly the result of changes in oceanic circulation at that time.

Another factor that influences climates is the relative level of the continents and the seas. At some periods in Earth's history, most recently in the Late Mesozoic and again in the first part of the Cenozoic, shallow seas have flooded large parts of the continents. These **epicontinental seas** extended across the middle of North America and the middle of Eurasia during the Cretaceous period and early Cenozoic era. Water absorbs heat as air temperature rises, and then releases that heat as air temperature falls. Thus, areas of land near large bodies of water have maritime climates—they do not get very hot in summer or very cold in winter, and they are usually moist because water that evaporates from the sea falls as rain on the land. Continental climates, which characterize areas far from the sea, are usually dry with cold winters and hot summers. The draining of the epicontinental seas at the end of the Cretaceous probably contributed to the demise of the dinosaurs by making climates in the Northern Hemisphere more continental.

In addition to changing climates, continental drift has formed and broken land connections between the continents. Isolation of different lineages of vertebrates on different landmasses has produced dramatic examples of the independent evolution of similar types of organisms. These are well shown by mammals in the mid-Cenozoic, a period when the Earth's continents reached their greatest separation during the history of vertebrates.

Much of evolutionary history appears to depend on whether a particular lineage of animals was in the right place at the right time. This random element of evolution is assuming increasing prominence as more detailed information about the times of extinction of old groups and radiation of new groups suggests that competitive replacement of one group by another is not the usual mechanism of large-scale evolutionary change. The movements of continents and their effects on climates and the isolation or dispersal of animals are taking an increasingly central role in our understanding of vertebrate evolution. On a continental scale, the advance and retreat of glaciers in the Pleistocene caused homogeneous habitats to split and merge repeatedly, isolating populations of widespread species and leading to the evolution of new species.

Summary

The more than 57,000 species of living vertebrates span a size range from less than a gram to more than 100,000 kilograms. They live in habitats from the bottom of the sea to the tops of mountains. This extraordinary diversity is the product of more than 500 million years of evolution, and the vast majority of species fall into one of the two major divisions of bony fishes (Osteichthyes)—the aquatic ray-finned fishes (Actinopterygii) and the primarily terrestrial lobe-finned fishes and tetrapods (Sarcopterygii), each of which contains more than 25,000 extant species.

Phylogenetic systematics, usually called cladistics, classifies animals on the basis of shared derived character states. Natural evolutionary groups can be defined only by these derived characters; retention of ancestral characters does not provide information about evolutionary lineages. Application of this principle produces groupings of animals that reflect evolutionary history as accurately as we can discern it and forms a basis for making hypotheses about evolution and for designing management plans that conserve the genetic diversity of evolutionary lineages.

The Earth has changed dramatically during the half-billion years of vertebrate history. Continents were fragmented when vertebrates first appeared; coalesced into one enormous continent, Pangaea, about 300 million years ago; and began to fragment again about 150 million years ago. This pattern of fragmentation, coalescence, and fragmentation has resulted in isolation and renewed contact of major groups of vertebrates on a worldwide scale.

Additional Readings

Budd, G. 2001. Climbing life's tree. *Nature* 412:48.

Census of Marine Life. (The Census of Marine Life is a growing global network of researchers in more than 80 nations engaged in a ten-year initiative to assess and explain the diversity, distribution, and abundance of marine life in the oceans—past, present, and future.) <www.coml.org>

Davenport, T. R. B. et al. 2006. A new genus of African monkey, *Rungwecebus*: Morphology, ecology, and molecular phylogenetics. *Science* 312:1378–1381.

de Queiroz, K., and J. Gauthier. 1992. Phylogenetic taxonomy. *Annual Review of Ecology and Systematics* 23:449–480.

Hennig, W. 1966. *Phylogenetic Systematics*. Urbana, IL: University of Illinois Press.

Jones, T. et al. 2005. The highland mangabey *Lopjocebus kipunji*: A new species of African monkey. *Science* 308:1161–1164.

Kemp. T. S. 1999. *Fossils and Evolution*. Oxford, UK: Oxford University Press.

Pennisi, E. 2001. Linnaeus's last stand? *Science* 291:2304–2307.

Rosenbaum, H. C. et al. 2000. World-wide genetic differentiation of *Eubalaena*: Questioning the number of right whale species. *Molecular Ecology* 9:1793–1802.

Swiderski, D. L. (ed.) 2001. Beyond reconstruction: Using phylogenies to test hypotheses about vertebrate evolution. *American Zoologist* 41:485–607.

Talbot, S. L. and G. F. Shields. 1996. Phylogeography of brown bears (*Ursus arctos*) of Alaska and paraphyly within the Ursidae. *Molecular Phylogenetics and Evolution* 5:477–494.

Wado, S. et al. 2003. A newly discovered species of living baleen whale. *Nature* 426:278–281.

Wilting, A. et al. 2007. Clouded leopard phylogeny revisited: Support for species recognition and population division between Borneo and Sumatra. *Frontiers in Zoology* 4:15. <http://www.frontiersinzoology.com/content/4/1/15>

CHAPTER 2

Vertebrate Relationships and Basic Structure

In this chapter, we explain the structures that are characteristic of vertebrates, discuss the relationship of vertebrates to other members of the animal kingdom, and describe the systems that make vertebrates functional animals. We need an understanding of the fundamentals of vertebrate design to appreciate the changes that have occurred during their evolution and to trace homologies between primitive vertebrates and derived ones.

2.1 Vertebrates in Relation to Other Animals

Vertebrates are a diverse and fascinating group of animals. Because we are vertebrates ourselves, that statement may seem chauvinistic, but vertebrates are remarkable in comparison with most other animal groups. Vertebrates are in the subphylum Vertebrata of the phylum Chordata. About 30 other animal phyla have been named, but only the phylum Arthropoda (insects, crustaceans, spiders, etc.) rivals the vertebrates in diversity of forms and habitat. And it is only in the phylum Mollusca (snails, clams, and squid) that we find animals (such as octopus and squid) that approach the very large size of some vertebrates and have a capacity for complex learning.

The tunicates (subphylum Urochordata) and cephalochordates (subphylum Cephalochordata) (these animals are described later in the chapter) are placed with vertebrates in the phylum Chordata. Within the chordates, cephalochordates and verte-brates are probably more closely related to each other than either is to tunicates. Chordates are united by several shared derived features, which are seen in all members of the phylum at some point in their lives, include a notochord (a dorsal stiffening rod that gives the phylum Chordata its name); a dorsal hollow nerve cord; a segmented, muscular postanal tail (i.e., extending beyond the gut region); and an endostyle. The endostyle is a ciliated, glandular groove on the floor of the pharynx that secretes mucus for trapping food particles during filter feeding. It is generally homologous with the thyroid gland of vertebrates, an endocrine gland involved with regulating metabolism. Chordates are also characterized by a pharynx (throat region) containing gill slits, which are used for filter feeding in nonvertebrate chordates and respiration in primarily aquatic vertebrates (i.e., fishes), but these structures are also seen in some other deuterostomes (the larger grouping to which chordates belong) and may be a primitive feature for the group.

Although chordates are all basically bilaterally symmetrical animals (i.e., one side is the mirror image of the other), they share an additional type of left-to-right asymmetry within the body that is determined by the same genetic mechanism (Boorman and Shimeld 2002). This type of left-to-right asymmetry is apparent in ourselves: for example, in the positioning of the heart on the left-hand side and most of the liver on the right-hand side. Rare individuals have a condition termed "situs inversus," in which the major body organs are reversed.

The relationship of chordates to other kinds of animals is revealed by anatomical, biochemical, and embryological characters as well as by the fossil record. **Figure 2–1** shows the relationships of animal phyla. Vertebrates superficially resemble other active animals, such as insects, in having a distinct head end, jointed legs, and bilateral symmetry. However, and perhaps surprisingly, the phylum Chordata is closely related to the phylum Echinodermata (starfishes, sea urchins, and the like), which are marine forms without distinct heads and with pentaradial (fivefold and circular) symmetry as adults.

The chordates, echinoderms, and a couple of other phyla (hemichordates and xenoturbellids) are linked as **deuterostomes** (Greek *deutero* = second and *stoma* = mouth) by several unique embryonic features, such as the way in which their eggs divide after fertilization (egg cleavage), their larval form, and some other features discussed later. Hemichordates are a small, rather obscure phylum of marine animals containing

the earthwormlike acorn worms and the fernlike pterobranchs. Xenoturbellids are small marine wormlike forms (only two species are known), which have recently been identified as deuterostomes by molecular data. Hemichordates were long considered the sister group of chordates because both groups have pharyngeal slits, and hemichordates also have features of the pharynx that can be interpreted as the precursor to an endostyle. However, we now consider these to be primitive deuterostome features. Although modern echinoderms lack pharyngeal slits, some extinct echinoderms appear to have had them (note that the diversity of extinct echinoderms was much greater than that of the living forms). Furthermore, early echinoderms were bilaterally symmetrical, meaning that the fivefold symmetry of modern echinoderms is probably a derived character of that lineage (see Gee 1996). Molecular

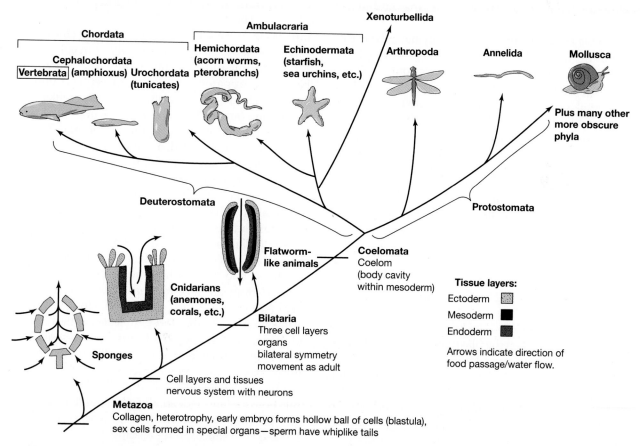

▲ **FIGURE 2–1** A simplified phylogeny of the animal kingdom (metazoans). There are a total of about 30 phyla today (Chordata, Echinodermata, Annelida, etc. represent phyla). Approximately 15 additional phyla are known from the early Paleozoic era, and became extinct at the end of the Cambrian period.

data currently unite echinoderms and hemichordates as the Ambulacraria, with xenoturbellids more closely related to these phyla than to the chordates (Figure 2–1) (see Bourlat et al. 2006).

To consider how deuterostomes are related to other animals, we will start at the bottom of the tree and work upward. All animals (metazoans) are multicellular and share common embryonic and reproductive features: the embryo initially forms a hollow ball of cells (the blastula), they have sex cells formed in special organs, and they have motile sperm with whiplike tails.

Animals more derived than sponges have a nervous system, and their bodies are made of distinct layers of cells, or germ layers, that are laid down early in development at a stage called gastrulation. Gastrulation occurs when the hollow ball of cells forming the blastula folds in upon itself, producing two distinct layers of cells and an inner gut with an opening to the outside at one end. The outer layer of cells is the **ectoderm** (Greek *ecto* = outside and *derm* = skin), and the inner layer forms the **endoderm** (Greek *endo* = within).

Jellyfishes and related animals only have these two layers of body tissue, making them diploblastic (Greek *diplo* = two and *last* = a bud or sprout). Animals more derived than jellyfishes and their kin add an additional, middle cell layer of **mesoderm** (Greek *mesos* = middle), making them triploblastic (Greek *triplo* = three). Triploblasts also have a gut that opens at both ends (i.e., with a mouth and an anus) and are bilaterally symmetrical with a distinct anterior (head) end at some point in their life. The mesoderm forms the body's muscles, and only animals with a mesoderm are able to be motile as adults; larval forms do not need muscles because they are small enough to be powered by hairlike cilia on the outer surface.

The **coelom**, an inner body cavity that forms as a split within the mesoderm, is another derived character of most, but not all, triploblastic animals. Coelomate animals (i.e., animals with a coelom) are split into two groups on the basis of how the mouth and anus form. When the blastula folds in on itself to form a gastrula, it leaves an opening to the outside called the blastopore (Latin *porus* = a small opening). In jellyfish, the blastopore is the only opening into the interior of the body, and it serves as both mouth and anus. During the embryonic development of coelomates, a second opening develops. In the lineage called protostomes (Greek *proto* = first and *stome* = mouth), the blastopore (which was the first opening in the embryo) becomes the mouth, whereas in deuterostomes the second opening becomes the

mouth and the blastopore becomes the anus. The way that the coelom forms during development also differs between protostomes and deuterostomes, and molecular data now confirms the separate identity of these two groups. Mollusks (snails, clams, and squid), arthropods (insects, crabs, and spiders), annelids (earthworms), and many other phyla are protostomes; chordates, hemichordates, and echinoderms are deuterostomes (see Figure 2–1).

■ Nonvertebrate Chordates

The two groups of extant nonvertebrate chordates are small marine animals. More types of nonvertebrate chordates may have existed in the past, but these soft-bodied animals are rarely preserved as fossils. Some possible Early Cambrian primitive chordates are described at the end of this section.

Urochordates Present-day **tunicates** (subphylum Urochordata) are marine animals that filter particles of food from the water with a basketlike perforated pharynx. There are about 2000 living species, and all but 100 or so are sedentary as adults, attaching themselves to the substrate either singly or in colonies.

Most adult tunicates (also known as sea squirts) bear little apparent similarity to cephalochordates and vertebrates). However, their tadpolelike, free-swimming larvae (**Figure 2–2a**) look more like forms that belong within the phylum Chordata. Tunicate larvae have a notochord, a dorsal hollow nerve cord, and a muscular postanal tail that moves in a fishlike swimming pattern. Most species have a brief free-swimming larval period (a few minutes to a few days) after which the larvae metamorphose into sedentary adults attached to the substrate.

A popular and long-held theory was that the earliest chordates would have been like tunicates (sessile as adults), and then cephalochordates and vertebrates evolved from an ancestor that resembled a tunicate larva (see Lacalli 2004 for a review). However, it now seems more likely that those tunicates with the sessile adult stage are the derived forms, having secondarily lost many chordate features as adults, and that it is the living species that remain free-swimming as adults that most resemble the ancestral chordate. The ancestral chordate (and indeed, the ancestral deuterostome) was probably a free-swimming wormlike creature that used gill slits for filter feeding.

Cephalochordates The subphylum Cephalochordata contains some 22 species, all of which are small, superficially fishlike marine animals usually less

(a) Free-swimming larval tunicate (Urochordata)

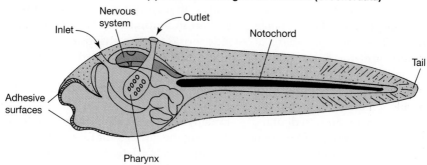

(b) Sessile adult tunicate (Urochordata)

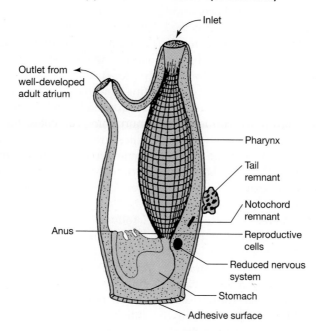

**(c) The lancelet, amphioxus (Cephalochordata)
(posterior myomeres removed)**

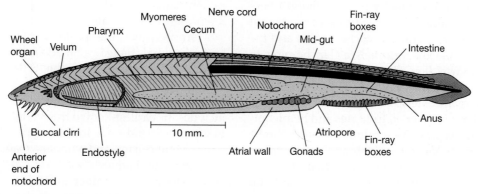

▲ **FIGURE 2–2** Nonvertebrate chordates. Tunicates have a free-swimming larva (a) that metamorphoses into a sessile adult (b), whereas amphioxus (c) is free-swimming throughout its life.

than 5 centimeters long. The best-known cephalo-chordate is the lancelet (*Branchiostoma lanceolatum*), more commonly known as **amphioxus**. (Greek *amphi* = both and *oxy* = sharp; *amphioxus* means "sharp at both ends," an appropriate term for an animal in which the front and rear ends are nearly the same shape because it lacks a distinct head.) Lancelets are widely distributed in marine waters of the continental shelves and are usually burrowing, sedentary animals as adults. In a few species, the adults retain the active, free-swimming behavior of the larvae.

A notable characteristic of amphioxus is its fish-like locomotion. This results from a feature shared with vertebrates: **myomeres**—blocks of striated muscle fibers arranged along both sides of the body and separated by sheets of connective tissue. (Tunicate larvae have banded muscles in their tails but don't have distinct myomeres or any muscle in the body region.) Sequential contraction of myomeres bends the body from side to side, resulting in forward or backward propulsion. The notochord acts as an incompressible elastic rod, extending the full length of the body and preventing the body from shortening when the myomeres contract. While the notochord of vertebrates ends midway through the head region, the notochord of amphioxus extends from the tip of the snout to the end of the tail, projecting well beyond the region of the myomeres at both ends. This anterior elongation of the notochord apparently is a specialization that aids in burrowing.

Figure 2–2c shows some details of the internal structure of amphioxus. Amphioxus and vertebrates differ in the use of the pharyngeal slits. Amphioxus has no gill tissue associated with these slits; its body is small enough that oxygen uptake and carbon dioxide loss occur by diffusion over the body surface. Instead, the gill slits are used for filter feeding. Water is moved over the gill slits by cilia on the gill bars between the slits, aided by the features of the buccal (mouth region, from Latin *bucc* = cheek) cirri and the wheel organ, while the velum is a flap helping to control the one-way flow of water.

In addition to the internal body cavity, or coelom, amphioxus has an external body cavity called the atrium, which is also seen in tunicates and hemichordates—and thus is probably a primitive deuterostome feature—but is absent from vertebrates. (This atrium is not the same as the atrium of the vertebrate heart; the word *atrium* [plural *atria*] comes from the Latin term for an open space). The atrium of amphioxus is formed by outgrowths of the body wall (metapleural folds), which enclose the body ventrally. Imagine yourself wearing a cape and then extending your arms until there is a space between the cape and your body—that space would represent the position of the atrium in amphioxus. The atrium opens to the outside world via the atriopore, opening in front of the anus. The atrium appears to work in combination with the beating of the cilia on the gill bars and the wheel organ in the head to control passage of substances through the pharynx and is probably functionally associated with the primitive chordate feature of using the gill slits for filter feeding.

Cephalochordates have several derived characters that are shared with vertebrates but absent from tunicates. In addition to the myomeres, amphioxus has a circulatory system similar to that of vertebrates, with a dorsal aorta and a ventral heartlike structure that forces blood through the gills. Additionally, although amphioxus lacks a distinct kidney, both it and vertebrates share specialized excretory cells called podocytes, and amphioxus has a vertebrate-like tail fin. Amphioxus and vertebrates also share some unique embryonic features (see Gans 1989). These morphological characters indicate that cephalocordates are the sister group of vertebrates. However, some biologists argue that tunicates are the sister group of vertebrates, primarily on molecular grounds (see Schubert et al. 2006).

Cambrian Chordates   The best-known early chordate-like animal is *Pikaia*. About 100 specimens of this animal are known from the Middle Cambrian Burgess Shale in British Columbia (**Figure 2–3a**). The most obvious chordate features of *Pikaia* are myomeres and a notochord running along the posterior two-thirds of the body. *Pikaia* is often considered to be a cephalochordate but may represent some other type of nonvertebrate chordate. Note that, unlike the condition in amphioxus, the myomeres are straight rather than V-shaped, and there is scant evidence for the presence of gill slits.

More recently some spectacular fossil finds of soft-bodied animals have been made from the Early Cambrian of southern China, the 520-million-year-old Chengiang Fauna; this is a good 10 million years younger than the Burgess Shale. This fossil deposit includes the earliest-known true vertebrates (described in Chapter 3) and some intriguing possible early chordates, vetulicolians and yunnanozoans. (Note, however, that the flattened "road kill–like" nature of the specimens makes it rather difficult to interpret their structure.) Vetulicolians have an apparent endostyle, an expanded pharngeal region with gill slits, and a possibly segmented body region, but they appear to lack a notochord (**Figure 2–3b**), They have been

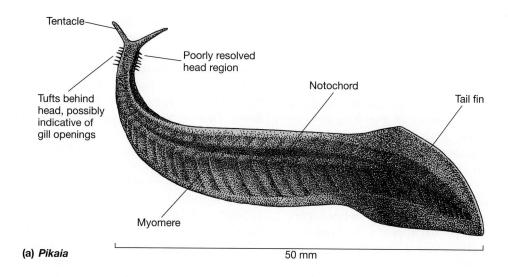

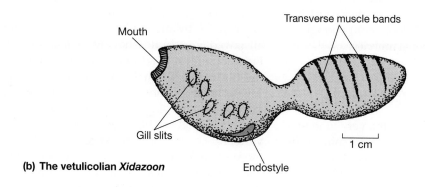

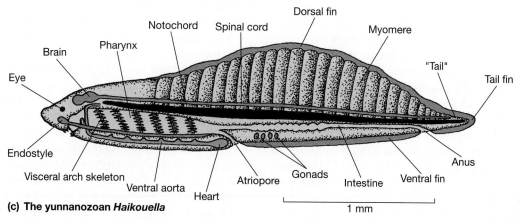

▲ **FIGURE 2–3** Some early chordates or chordate relatives. (a) *Pikaia*, from the Middle Cambrian Burgess Shale of British Columbia. (b) The vetulicolian *Xidazoon* and (c) the yunnanozoan *Haikouella*, both from the Early Cambrian Chengiang Fauna of southern China.

proposed as stem deuterostomes, stem chordates, or even stem tunicates (because most tunicates lack the notochord in the adult form) (Lacalli 2002).

Yunnanozoans are more derived forms with the chordate features of myomeres, a notochord, and a pharynx apparently enclosed in an atrium. Study of over 300 individual specimens of the yunnanozoan *Haikouella* (see **Figure 2–3c**) by Jon Mallatt and Jun-yuan Chen (2003) has revealed that this animal has a number of derived features that place it as the sister group to vertebrates (although, inevitably, some other researchers disagree with this interpretation). These features include a large brain, clearly defined eyes, thickened branchial bars (that appear to be made of a type of cartilage similar to that of lamprey larvae), and an upper lip like that of larval lampreys. The endostyle and tentacles surrounding the mouth suggest that this animal was a suspension feeder, like amphioxus. However, *Haikouella* already appears to have developed a vertebrate-like muscular pharynx, as evidenced by the thickened branchial bars which look stout enough to support gill tissue and pharyngeal muscles.

2.2 Definition of a Vertebrate

The term *vertebrate* is derived from the vertebrae that are serially arranged to make up the spinal column, or backbone. In ourselves, as in other land vertebrates, the vertebrae form around the notochord during development and also encircle the nerve cord. The bony vertebral column replaces the original notochord after the embryonic period. In many fishes the vertebrae are made of cartilage rather than bone.

All vertebrates have the uniquely derived feature of a **cranium**, or skull, which is a bony, cartilaginous, or fibrous structure surrounding the brain. They also have a prominent head, containing complex sense organs. Although many of the genes that determine head development in vertebrates can also be found in amphioxus, the anterior portion (the first three segments) of the vertebrate head does seem to be a new feature of vertebrates (Northcutt 2005).

However, not all animals included within the traditional subphylum Vertebrata have vertebrae. Among living **agnathans** (jawless vertebrates), hagfishes lack vertebral elements entirely, and lampreys have only cartilaginous rudiments (arcualia) flanking the nerve cord. Fully formed vertebrae, with a **centrum** (plural = *centra*) surrounding the notochord, are found only in **gnathostomes** (jawed vertebrates) (see Chapter 3), and many jawed fishes retain a functional notochord as adults. Because of the lack of vertebrae in hagfishes, some people prefer the term **Craniata** to Vertebrata for the subphylum. However, we continue to use the familiar term *vertebrate* in this book.

Two embryonic features may account for many of the differences between vertebrates and other chordates. One is the ***Hox* gene complex** (homeobox genes) that characterizes animals. *Hox* genes regulate the expression of a hierarchy of other genes that control the process of development along the long axis of the body from front to back. Jellyfishes (and possibly also sponges) have one or two *Hox* genes, the common ancestor of protostomes and deuterostomes probably had seven, and more derived metazoans have up to thirteen. However, vertebrates are unique in having undergone the duplication of the entire *Hox* complex.

There appears to have been one duplication event at the start of vertebrate evolution: amphioxus has a single *Hox* cluster, while the living jawless vertebrates have two. A second duplication event had taken place by the evolution of gnathostomes, with all jawed vertebrates having at least four clusters. Finally there were additional duplications within the ray-finned bony fishes, and teleosts may have up to seven clusters. More complex animals usually have a greater amount of genetic material, and it is thought that the doubling of this gene sequence at the start of vertebrate evolution made possible the evolution of a more complex type of animal. However, Donoghue and Purnell (2006) argue that this hypothesis may be overly simplistic, as discussed further in Chapter 3 with the issue of the origin of gnathostomes. A possibly more genetic innovation in vartebrates, that may account for their complexity, is the acquisition of an additional 41 MicroRNAs (while other chordates only have 2 or 3) (Heimberg et al 2008).

The second embryonic feature of vertebrates is the development of a type of tissue called **neural crest** that forms many new structures in vertebrates, especially in the head region (Northcutt and Gans 1983) (a more detailed description is given in Section 2.3). Neural crest can be considered as the most important item in the origin of the vertebrate body plan, representing a (fourth) germ layer that is unique to vertebrates and is on a par with ectoderm, endoderm, and mesoderm (Hall 2000, Holland and Chen 2001).

Neural crest cells originate at the lateral boundary of the neural plate, the embryonic structure that makes the nerve cord, and they later migrate throughout the body to form a variety of structures. A similar population of cells, with a similar genetic expression, can also be found in amphioxus, but here the cells do not migrate and do not change into different cell types.

Recently cells resembling migratory neural crest cells have been identified in the larval stage of one tunicate species, where they differentiate into pigment cells (Jeffrey et al. 2004); note that pigment cells in vertebrates are derived from neural crest. These cells in nonvertebrate chordates may represent the precursor to the vertebrate condition of neural crest (Holland and Chen 2001). Note that if the Cambrian chordate *Haikouella* has been correctly interpreted as having eyes and a muscular pharynx, these features would imply the presence of neural crest in this animal.

Embryonic tissue that may be related to neural crest forms the epidermal placodes (i.e., thickenings), which give rise to the complex sensory organs of vertebrates, including the nose, eyes, and inner ear. Some placode cells migrate caudally to contribute, along with the neural crest cells, to the lateral line system and to the cranial nerves that innervate it.

The brains of vertebrates are larger than the brains of primitive chordates and have three parts—the forebrain, midbrain, and hindbrain. The brain of amphioxus is not obviously divided, but genetic studies show that it may be homologous to the vertebrate brain with the exception of the front part of the vertebrate forebrain (the **telencephalon**) (Zimmer 2000). The telencephalon is the portion of the brain that contains the cerebral cortex, the area of higher processing in vertebrates.

2.3 Basic Vertebrate Structure

This section serves as an introduction to vertebrate anatomical structure and function. The heart of this section is in **Table 2.1** and **Figure 2–4**, which contrast the basic vertebrate condition with that of a nonvertebrate

TABLE 2.1 Comparison of features in nonvertebrate chordates and primitive vertebrates

Generalized Nonvertebrate Chordate	Primitive Vertebrate
(Based on features of the living cephalochordate amphioxus)	(Based on features of the living jawless vertebrates—hagfishes and lampreys)
A. Brain and Head End	
Notochord extends to tip of head (may be derived condition). No cranium (skull).	Head extends beyond tip of notochord. Cranium—skeletal supports around brain, consisting of capsules surrounding the main parts of the brain and their sensory components plus underlying supports.
Simple brain (= cerebral vesicle), no specialized sense organs (except photoreceptive frontal organ, probably homologous with the vertebrate eye).	Tripartite brain and multicellular sense organs (eye, nose, inner ear).
Poor distance sensation (although the skin is sensitive).	Improved distance sensation: in addition to the eyes and nose, also have a lateral line system along the head and body that can detect water movements (poorly developed lateral line system on the head is found only in hagfishes).
No electroreception.	Electroreception may be a primitive vertebrate feature (but absent in hagfishes, possibly lost).
B. Pharynx and Respiration	
Gill arches used for filter feeding (respiration is by diffusion over the body surface).	Gill arches (= pharyngeal arches) support gills that are used primarily for respiration.
Numerous gill slits (up to 100 on each side).	Fewer gill slits (6–10 on each side), individual gills with highly complex internal structure (gill filaments).
Pharynx not muscularized (except in wall of atrium, or external body cavity).	Pharynx with specialized (branchiomeric) musculature.
Water moved through pharynx and over gills by ciliary action.	Water moved through pharynx and over gills by active muscular pumping.
Gill arches made of collagen-like material (musculoscleroproteins).	Gill arches made of cartilage (allows for elastic recoil—aids in pumping).

(continued)

TABLE 2.1 *(Continued)*

Generalized Nonvertebrate Chordate	Primitive Vertebrate
C. Feeding and Digestion	
Gut not muscularized: food passage by means of ciliary action.	Gut muscularized: food passage by means of muscular peristalsis.
Digestion of food is intracellular: individual food particles taken into cells lining gut.	Digestion of food is extracellular: enzymes poured onto food in gut lumen, then breakdown products absorbed by cells lining gut.
No discrete liver and pancreas: structure called the midgut cecum or diverticulum is probably homologous to both.	Discrete liver and pancreatic tissue.
D. Heart and Circulation	
Ventral pumping structure (no true heart, just contracting regions of vessels; = sinus venosus of vertebrates). Also accessory pumping regions elsewhere in the system.	Ventral pumping heart only (but accessory pumping regions retained in hagfishes). Three-chambered heart (listed in order of blood flow): sinus venosus, atrium, and ventricle.
No neural control of the heart to regulate pumping.	Neural control of the heart (except in hagfishes).
Circulatory system open: large blood sinuses; capillary system not extensive.	Circulatory system closed: without blood sinuses (some remain in hagfishes and lampreys) and with an extensive capillary system.
Blood not specifically involved in the transport of respiratory gases (O_2 and CO_2 mainly transported via diffusion). No red blood cells or respiratory pigment.	Blood specifically involved in the transport of respiratory gases. Have red blood cells containing the respiratory pigment hemoglobin (binds with O_2 and CO_2 and aids in their transport).
E. Excretion and Osmoregulation	
No specialized kidney. Coelom coelom filtered by solenocytes (flame cells) that work by creating negative pressure within cell. Cells empty into the atrium (false body cavity) and then to the outside world via the atriopore.	Specialized glomerular kidneys; segmental structures along dorsal body wall; works by ultrafiltration of blood. Empty to the outside via the archinephric ducts leading to the cloaca.
Body fluids same concentration and ionic composition as seawater. No need for volume control or ionic regulation.	Body fluids more dilute than seawater (except for hagfishes). Kidney important in volume regulation, especially in freshwater environment. Monovalent ions regulated by the gills (also the site of nitrogen excretion), divalent ions regulated by the kidney.
F. Support and Locomotion	
Notochord provides main support for body muscles.	Notochord provides main support for body muscles, vertebral elements around nerve cord at least in all vertebrates except hagfishes.
Myomeres with simple V-shape.	Myomeres with more complex W-shape.
No lateral fins; no median fins besides tail fin.	Primitively, no lateral fins. Caudal (tail) fin has dermal fin rays. Dorsal fins present in all except hagfishes.

chordate such as amphioxus. These same systems will be further discussed for more derived vertebrates in later chapters: our aim is to provide a general introduction to the basics of vertebrate design. More comprehensive detail can be found in books such as Hildebrand and Goslow (2001), Kardong (2001), and Liem et al. (2001).

At the whole-animal level, an increase in body size and increased activity distinguish vertebrates from nonvertebrate chordates. Early vertebrates generally had body lengths of 10 centimeters or more, which is about an order of magnitude larger than nonvertebrate chordates. Because of their relatively large size, vertebrates need specialized systems to carry out processes that are accomplished by diffusion or ciliary action in smaller animals. Vertebrates are also more active animals than other chordates, so they need organ systems that can carry out physiological processes at a greater rate (see Gans 1989). The transition from nonvertebrate chordate to vertebrate was probably related to the adoption of a more actively predaceous mode of life, as evidenced by the features of the vertebrate head (largely derived from neural crest tissue) that would enable suction feeding with a muscular pharynx, and a bigger brain and more complex sensory organs for perceiving the

▶ **FIGURE 2–4** Design of a generalized amphioxus-like nonvertebrate chordate. (a) compared with that of a hypothetical primitive vertebrate (b). (Note that the myomeres actually extend all the way down the body, see Figure 2–10).

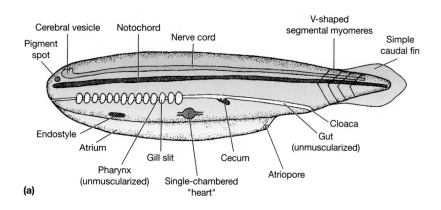

Amphioxus-like nonvertebrate chordate

(a)

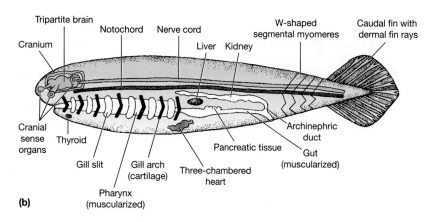

Hypothetical primitive vertebrate

(b)

environment (Northcutt 2005). If the Cambrian animal *Haikouella* (see Figure 2–3c) is correctly interpreted as the sister taxon to vertebrates, then it seems likely that the initial shift on the path to full vertebrate status was a change in the mode of respiration: *Haikouella* has a muscular pharynx, suggesting active use of the gills for respiration, but it also retains the amphioxus-like apparatus for filter feeding (Butler 2006).

Vertebrates are characterized by mobility, and the ability to move requires muscles and a skeleton. Mobility brings vertebrates into contact with a wide range of environments and objects in those environments, and a vertebrate's external protective covering must be tough but flexible. Bone and other mineralized tissues that we consider characteristic of vertebrates had their origins in this protective integument.

Embryology

Studying embryos can show how systems develop and how the form of the adult is related to functional and historical constraints during development. Modern scientists no longer adhere rigidly to the biogenetic law that "ontogeny recapitulates phylogeny" (i.e., the idea that the embryo faithfully passes through its ancestral evolutionary stages in the course of its development) proposed by nineteenth-century embryologists such as von Baer and Haeckel. Nevertheless, embryology can provide clues about the ancestral condition and about homologies between structures in different animals (Northcutt 1990).

The development of vertebrates from a single fertilized cell (the zygote) to the adult condition will be summarized only briefly. This is important background information for many studies, but a detailed treatment is beyond the scope of this book. Note, however, that there is an important distinction in development between vertebrates and invertebrates: invertebrates develop from cell lineages whose fate is predetermined, but vertebrates are much more flexible in their development and use inductive interactions between developing structures to determine the formation of different cell types and tissues (Hall 2005).

We saw earlier that all animals with the exception of sponges are formed of distinct tissue layers, or germ layers. The fates of germ layers have been very conservative throughout vertebrate evolution. The outermost germ layer, the ectoderm, forms the adult superficial layers of skin (the epidermis); the linings of the most anterior and most posterior parts of the digestive tract; and the nervous system, including most of the sense organs (such as the eye and the ear). The innermost layer, the endoderm, forms the rest of the digestive tract's lining, as well as the lining of glands associated with the gut—including the liver and the pancreas—and most respiratory surfaces of vertebrate gills and lungs. Endoderm also forms the taste buds and the thyroid, parathyroid, and thymus glands (Graham 2001).

The middle layer, the mesoderm, is the last of the three layers to appear in development, perhaps reflecting the fact that it is the last layer to appear in animal evolution. It forms everything else: muscles, skeleton (including the notochord), connective tissues, and circulatory and urogenital systems. A little later in development, there is a split within the originally solid mesoderm layer, forming a coelom or body cavity. The coelom is the cavity containing the internal organs, and it is divided into the **pleuroperitoneal cavity** (around the viscera) and the **pericardial cavity** (around the heart). These cavities are lined by thin sheets of mesoderm—the **peritoneum** (= the **pericardium** around the heart). The gut is suspended in the peritoneal cavity by sheets of peritoneum called **mesenteries**.

Neural crest forms many of the structures in the anterior head region, including some bones and muscles that were previously thought to be formed by mesoderm. It also forms almost all of the peripheral nervous system (i.e., that part of the nervous system outside of the brain and the spinal cord) and contributes to portions of the brain. Some structures in the body that are new features of vertebrates are also formed from neural crest. These include the adrenal glands, pigment cells in the skin, secretory cells of the gut, and smooth muscle tissue lining the aorta.

Figure 2–5 shows a stage in early embryonic development in which the ancestral chordate feature of pharyngeal pouches in the head region makes at least a fleeting appearance in all vertebrate embryos. In fish, the grooves between the pouches (the pharyngeal clefts) perforate to become the gill slits, whereas in land vertebrates these clefts disappear in later development. The linings of the pharyngeal pouches give rise to half a dozen or more glandular structures often associated with the lymphatic system, including the thymus gland, parathyroid glands, carotid bodies, and tonsils.

The dorsal hollow nerve cord typical of vertebrates and other chordates is formed by the infolding and subsequent pinching off and isolation of a long ridge of ectoderm running dorsal to the developing notochord. The notochord itself appears to contain the developmental instructions for this critical embryonic event, which is probably why the notochord is retained in the embryos of vertebrates (such as us) that no longer have the complete structure in the adult. The cells that will form the neural crest arise next to the developing nerve cord (the neural tube) at this stage. Slightly later in development, these neural crest cells disperse laterally and ventrally, ultimately settling and differentiating throughout the embryo.

Embryonic mesoderm becomes divided into three distinct portions, as shown in Figure 2–5, with the result that adult vertebrates are a strange mixture of segmented and unsegmented components. The dorsal (upper) part of the mesoderm, lying above the gut and next to the nerve cord, forms an epimere, a series of thick-walled segmental buds, (**somites**) that extends from the head end to the tail end. The ventral (lower) part of the mesoderm, surrounding the gut and containing the coelom, is thin-walled and unsegmented and is called the **lateral plate** (the hypomere). Small segmental buds linking the somites and the lateral plate are called nephrotomes (the mesomere or the intermediate mesoderm). The nervous system also follows this segmented versus unsegmented design, as will be discussed later.

The segmental somites will eventually form the dermis of the skin, the striated muscles of the body that are used in locomotion, and portions of the skeleton (the vertebral column, ribs, and portions of the back of the skull). Some of these segmental muscles later migrate ventrally from their originally dorsal (**epaxial**) position to form the layer of striated muscles on the underside of the body (the **hypaxial** muscles), and from there they form the muscles of the limbs in **tetrapods** (four-footed land

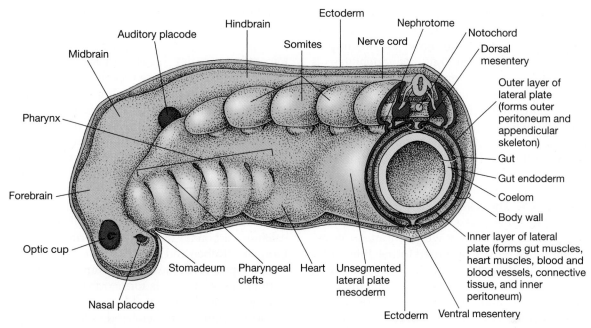

▲ **FIGURE 2–5** Three-dimensional view of a portion of a generalized vertebrate embryo at the developmental stage (called the pharyngula) when the developing gill pouches appear. The ectoderm is stripped off the left side, showing segmentation of the mesoderm in the trunk region and pharyngeal development. The stomadeum is the developing mouth.

vertebrates). The lateral plate forms all the internal, nonsegmented portions of the body, such as the connective tissue, the blood vascular system, the mesenteries, the peritoneal and pericardial linings of the coelomic cavities, and the reproductive system. It also forms the smooth muscle of the gut and the **cardiac** (heart) muscle. The nephrotomes form the kidneys (which are elongated segmental structures in the primitive vertebrate condition), the kidney drainage ducts (the **archinephric ducts**), and the gonads.

Some exceptions exist to this segmented versus nonsegmented division of the vertebrate body. The locomotory muscles, both **axial** (within the trunk region) and **appendicular** (within the limbs), and the axial skeleton are derived from the somites. Curiously, however, the limb bones (with the exception of the scapula in the shoulder girdle, which is formed from the somite) are derived from the lateral plate, as are the tendons and ligaments of the appendicular muscles, even though they essentially form part of the segmented portion of the animal. The explanation for this apparent anomaly may lie in the fact that limbs are add-ons to the basic limbless vertebrate design, as seen in the living jawless vertebrates.

Other peculiarities are found in the expanded front end of the head of vertebrates, which has a complex pattern of development and does not follow the simple segmentation of the body (Northcutt 1990). The head mesoderm contains only somites (no lateral plate), which give rise to the striated eye muscles and **branchiomeric** muscles powering the pharyngeal arches (gills and jaws). Within the brain, the anteriormost part of the forebrain (the front of the telencephalon) and the midbrain are not segmented, but the rest of the forebrain and the hindbrain show segmental divisions during development (Redies and Puelles 2001).

■ Adult Tissue Types

There are five kinds of tissue in vertebrates: epithelial, connective, vascular (i.e., blood), muscular, and nervous. These tissues are combined to form larger units called organs, which often contain most or all of the five basic tissues.

A fundamental component of most animal tissues is the fibrous protein **collagen**. Collagen is primarily a mesodermal tissue: in addition to the softer tissues of organs, it also forms the organic matrix of bone and the tough tissue of tendons and ligaments. Vertebrates have a unique type of fibrillar collagen that may be responsible for their ability to form an internal skeleton (Boot-Handford and Tuckwell 2003).

Collagen is stiff and does not stretch easily. In some tissues, collagen is combined with the protein **elastin,** which can stretch and recoil. Another important fibrous protein, seen only in vertebrates, is **keratin**. While collagen forms structures within the mesoderm, keratin is primarily an ectodermal tissue. Keratin is mainly found in the epidermis (outer skin) of tetrapods, making structures such as hair, scales, feathers, claws, horns, beaks, etc., but it also forms the horny toothlike structures of the living jawless vertebrates.

The Integument The external covering of vertebrates, the integument, is a single organ, making up 15 to 20 percent of the body weight of many vertebrates and much more in armored forms. It includes the skin and its derivatives, such as glands, scales, dermal armor, and hair. The skin protects the body and receives information from the outside world. The major divisions of the vertebrate skin are the **epidermis** (the superficial cell layer derived from embryonic ectoderm) and the unique vertebrate **dermis** (the deeper cell layer of mesodermal and neural crest origin). The dermis extends deeper into a subcutaneous tissue (hypodermis) that is derived from mesoderm and overlies the muscles and bones.

The epidermis forms the boundary between the vertebrate and its environment, and is of paramount importance in protection, exchange, and sensation. It often contains secretory glands and may play a significant role in osmotic and volume regulation. The dermis, the main structural layer of the skin, includes many collagen fibers that help to maintain its strength and shape. The dermis contains blood vessels, and blood flow within these vessels is under neural and hormonal control (e.g., as in human blushing, when the vessels are dilated and blood rushes to the skin).

The dermis also houses **melanocytes** (pigment cells containing melanin and that are derived from the neural crest) and smooth muscle fibers, such as the ones in mammals that produce skin wrinkling around the nipples. In tetrapods, the dermis houses most of the sensory structures and nerves associated with sensations of temperature, pressure, and pain.

The hypodermis, or subcutaneous tissue layer, lies between the dermis and the fascia overlying the muscles. This region contains collagenous and elastic fibers and is the area in which subcutaneous fat is stored by birds and mammals. The subcutaneous striated muscles of mammals, such as those that enable them to flick the skin to get rid of a fly, are found in this area.

Mineralized Tissues Vertebrates have a unique type of mineral called hydroxyapatite, a complex compound of calcium and phosphorus. Hydroxyapatite is more resistant to acid than is calcite (calcium carbonate), which forms the shells of mollusks. The evolution of this unique calcium compound in vertebrates may be related to the fact that vertebrates rely on anaerobic metabolism during activity, producing lactic acid that lowers blood pH. A skeleton made of hydroxyapatite may be more resistant to acidity of the blood during anaerobic metabolism than is the calcium carbonate (calcite) that forms the shells of mollusks (Ruben and Bennett 1987).

Six types of tissues can become mineralized in vertebrates, and each is formed from a different cell lineage in development. Most of these tissues are found only in the mineralized condition after they have been laid down in development. An exception to this is **cartilage,** which is an important structural tissue in vertebrates and many invertebrates but is not usually mineralized. Cartilage has been induced to mineralize in vitro in lampreys, squid, whelks, and horseshoe crabs (Hall 2005), but mineralized cartilage has only been found in vivo in jawed vertebrates, where it forms the main mineralized internal skeletal tissue of sharks. (Sharks and other cartilaginous fishes appear to have secondarily lost true bone.) Some fossil jawless vertebrates also apparently had internal calcified cartilage, probably evolved independently from the condition in sharks (Janvier and Arsenault 2002).

Note that the cartilage and bone of jawed vertebrates is formed from an initial proteinaceous matrix made of collagen, but the cartilage of the living jawless fishes is noncollagen-based, formed from a matrix protein similar to that of the cartilage of amphioxus and other invertebrates (Donoghue and Sansom 2002).

Bone is the mineralized tissue of the internal skeleton of bony fishes and tetrapods. Bone may replace cartilage in development, as it does in our own skeletons, but bone is not simply cartilage to which minerals have been added. Rather, it is composed of different types of cells—osteocytes (Greek *osteo* = bone and cyte = cell), which are called osteoblasts (Greek *blasto* = a bud) while they are actually making the bone; in contrast, chondrocytes form cartilage. The cells that form bone and cartilage cells are derived from the mesoderm, except in the region in the front of the head, where they are derived from neural crest tissue. Bone and mineralized cartilage are both around 70 percent mineralized.

The other types of mineralized tissues are found in association with the teeth and the mineralized exoskeleton of primitive vertebrates. The enamel and dentine that form our teeth are the most mineralized of the tissues—about 99 percent and 90 percent mineral, respectively. This high degree of mineralization explains why teeth are more likely to be found as fossils than are bones. The cells that form dentine (odontoblasts) are derived from neural crest tissue, and those that form enamel (amyloblasts) are derived from the ectoderm.

A fifth type of vertebrate hard tissue, enameloid, is the enamel-like tissue that is the primitive vertebrate condition and is seen today in most fishes. True enamel, seen primarily in tetrapods and their lobe-finned fish ancestors, is formed from ectodermal cells, whereas enameloid resembles enamel in its degree of hardness and its position on the outer layer of teeth or dermal scales, but it is apparently mesodermally derived. Both enamel and enameloid may have evolved independently on a number of occasions (Donoghue et al. 2006). The final type of hard tissue is **cementum,** a bonelike substance that fastens the teeth in their sockets in some vertebrates, including mammals, and that may grow to become part of the tooth structure itself.

Vertebrate mineralized tissues are composed of a complex matrix of collagenous fibers, cells that secrete a proteinaceous tissue matrix, and crystals of hydroxyapatite. The hydroxyapatite crystals are aligned on the matrix of collagenous fibers in layers with alternating directions, much like the structure of plywood. This combination of cells, fibers, and minerals gives bone its complex latticework appearance that combines strength with relative lightness and helps to prevent cracks from spreading.

Bone remains highly vascularized even when it is mineralized (ossified). The vascular nature of bone allows bone to remodel itself. Old bone is eaten away by specialized blood cells (osteoclasts, from the Greek *clast*=broken), which are derived from the same cell lines as the macrophage white blood cells that engulf foreign bacteria in the body. Osteoblasts enter behind the osteoclasts and deposit new bone. In this way, a broken bone can mend itself and bones can change their shape to suit the mechanical stresses imposed on the animal. This is why exercise builds up bone and why astronauts lose bone in the zero gravity of space. Calcified cartilage, as seen in the skeleton of sharks, is unable to remodel itself because it does not contain blood vessels.

There are two main types of bone in vertebrates: **dermal bone,** which, as its name suggests, is formed in the skin without a cartilaginous precursor; and **endochondral bone,** which is formed inside cartilage. Dermal bone (**Figure 2–6a**) is the primitive type of vertebrate bone first seen in the fossil jawless vertebrates called **ostracoderms,** which are described in Chapter 3. Only in the bony fishes and tetrapods is the endoskeleton composed primarily of bone. In these vertebrates, the endoskeleton is initially laid down in cartilage and is replaced by bone later in development.

Dermal bone originally was formed around the outside of the body, like a suit of armor (*ostracoderm* means "shell-skinned"), forming a type of exoskeleton. We think of vertebrates as possessing only an endoskeleton, but most of our skull bones are dermal bones, and they form a shell around our brains. The

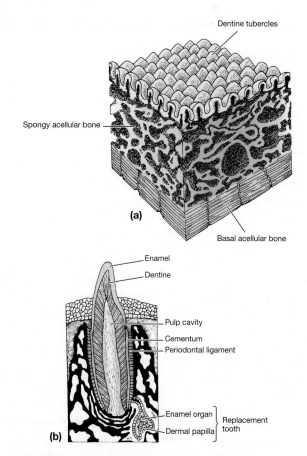

▲ **FIGURE 2–6** Organization of vertebrate mineralized tissues. (a) Three-dimensional block diagram of dermal bone from an extinct jawless vertebrate (heterostracan ostracoderm). (b) Section through a developing tooth (shark scales are similar).

▶ **FIGURE 2–7** Vertebrate skeletons.
(a) the originally dermal bone exoskeleton
with (b) the originally cartilaginous bone
endoskeleton. (The animal depicted is a
primitive extinct bony fish.)

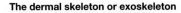

The dermal skeleton or exoskeleton

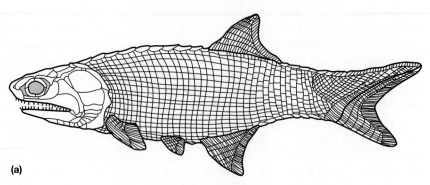

(a)

The endodermal skeleton or endoskeleton

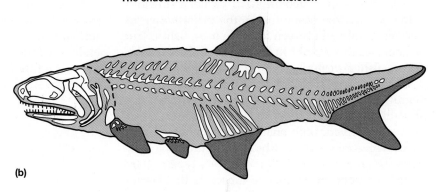

(b)

endoskeletal structure of vertebrates initially consisted only of the braincase and was originally formed from cartilage. Thus, the condition in many early vertebrates was a bony exoskeleton and a cartilaginous endoskeleton (**Figure 2–7**).

Teeth form from a type of structure called a dermal papilla so form only in the skin, usually over dermal bones. When the tooth is fully formed, it erupts through the gum line. Replacement teeth may start to develop to one side of the main tooth even before its eruption. The basic structure of the teeth of jawed vertebrates is like the structure of **odontodes**, which were the original toothlike components of primitive vertebrate dermal armor. Teeth are composed of an inner layer of dentine and an outer layer of enamel or enameloid around a central pulp cavity (**Figure 2–6b**). Shark scales (dermal denticles) have a similar structure. Although teeth and dermal elements like odontodes are similar in terms of their tissue composition, they are probably not homologous, as was once thought: although teeth are usually considered to be part of the dermal exoskeleton, it turns out that endoderm is an essential prerequisite for their development, which is not the case for dermoskeletal elements.

■ The Skeletomuscular System

The basic endoskeletal structural features of chordates are the notochord, acting as a dorsal stiffening rod running along the length of the body, and some sort of gill skeleton that keeps the gill slits patent (open). Vertebrates initially added the cranium surrounding the brain. Later vertebrates added the dermal skeleton of external plates and the axial skeleton (vertebrae, ribs, and median fin supports), and still later vertebrates added the appendicular skeleton (bones of the limb skeleton and limb girdles).

The Cranial Skeleton and Musculature The skull, or cranium, is formed by three basic components: the **chondrocranium** (Greek *chondr* = cartilage [literally "gristle"] and *cran* = skull) surrounding the brain; the **splanchnocranium** (Greek *splanchn* = viscera) forming the gill supports; and the **dermatocranium** (Greek *derm* = skin) forming in the skin as an outer cover and not seen in the earliest vertebrates.

The splanchnocranial components of the vertebrate skeleton are rather confusingly known by various different names. In general terms they are known as "gill arches" because they support the gill tissue and muscles. In all extant vertebrates, the anterior elements of the splanchnocranium are specialized into nongill-bearing structures, such as the jaws of gnathostomes (jawed vertebrates). More technical terms for these structures are *pharyngeal arches* (as they form in the pharynx region) and *branchial arches* (just a fancier way of saying "gill arches"; Greek *branchi* = gill). Yet another term for these structures is *visceral arches*, as the splanchnocranium is also known as the visceral skeleton. A common usage is to call these structures "pharyngeal arches" in development, and then "branchial arches" (or, more colloquially, "gill arches") in the adult, especially for those arches that actually do bear gill tissue (i.e., arches 3–7). We will adhere to that terminology here. A further term associated with the pharyngeal region, *aortic arches,* refers not to the gill skeleton but to the segmental arteries that supply the gill arches.

The vertebrate chondrocranium and splanchnocranium are formed primarily from neural crest tissue. However, note that a splanchnocranium of sorts (a gill skeleton) is present in cephalochordates and hemichordates; this skeleton is probably secondarily lost in tunicates and was perhaps also present in early echinoderms. Thus the splanchocranium precedes the origin of vertebrates and neural crest. (In vertebrate development, pharyngeal arches will form in the absence of neural crest tissue, although they do not chondrify [Graham 2001], and the gill supports of other deuterostomes are derived from the endoderm [Rychel et al. 2006].) In the primitive vertebrate condition, the chondrocranium and splanchnocranium are formed from cartilage; but, in the adults of some bony fishes and tetrapods, they are made of endochondral bone. The dermatocranium is made from dermal bone, which is formed in a membrane rather than in a cartilaginous precursor (and thus also is known as membrane bone), and is cartilaginous only as a rare secondary condition in some fishes, such as sturgeons. **Figure 2–8** shows a diagrammatic representation of the structure and early evolution of the vertebrate cranium, and **Figure 2–9** illustrates some vertebrate crania in more detail.

There are two main types of striated muscles in the head of vertebrates: the extrinsic eye muscles and the branchiomeric muscles. Six muscles (seven in lampreys) in each eye rotate the eyeball in all vertebrates except hagfishes, in which their absence may represent secondary loss. Like the striated muscles of the body, these muscles are innervated by somatic motor nerves.

The branchiomeric muscles are associated with the splanchnocranium and are used to suck water into the mouth during feeding and respiration. Branchiomeric muscles have a distinctive type of innervation. Striated muscles are usually innervated by motor nerves exiting from the ventral part of the spinal cord or brain, but the branchiomeric muscles are innervated by nerves from the brain that exit dorsally. The reason for this different pattern is not clear, but it emphasizes the way in which the vertebrate head is unusual in its structure and development compared to the rest of the body.

The Axial Skeleton and Musculature The notochord is the original "backbone" of all chordates, although it is never actually made of bone. Vertebrae made of cartilage or bone that originally surround and later replace the notochord (in both ontogeny and phylogeny) are not found in the most primitive vertebrates, and the structure of vertebrae is covered in Chapter 3.

The notochord is made up of a core of large, closely spaced cells packed with incompressible fluid-filled vacuoles that make the notochord rigid. The notochord is wrapped in a complex fibrous sheath that is the site of attachment for segmental muscles and connective tissues. In all vertebrates, the notochord ends anteriorly just posterior to the pituitary gland and continues posteriorly to the tip of the fleshy portion of the tail. The original form of the notochord is lost in adult tetrapods, but portions remain as components of the intervertebral discs between the vertebrae.

The axial muscles are comprised of myomeres that are complexly folded in three dimensions so that each one extends anteriorly and posteriorly over several body segments (**Figure 2–10**). Sequential muscle blocks overlap and produce undulation of the body when they contract. In amphioxus, myomeres have a simple V shape, whereas in vertebrates they have a W shape. The myomeres of jawed vertebrates are divided into epaxial (dorsal) and hypaxial (ventral) portions by a sheet of fibrous tissue called the horizontal septum.

The segmental pattern of the axial muscles is clearly visible in fishes. It is easily seen in a piece of raw or cooked fish where the flesh flakes apart in zigzag blocks, each block representing a myomere. This pattern is similar to the fabric pattern of interlocking V shapes known as herringbone (perhaps it would be better termed "herring-muscle"). In tetrapods, the pattern is less obvious, but the segmental pattern can be observed on the washboard (or six-pack) stomach of body builders, where each ridge of the washboard represents a segment of the rectus abdominus muscle (a hypaxial muscle of tetrapods).

Locomotion   Many small aquatic animals, especially larval forms, move by using cilia (small projections from the surface of a cell) to beat against the water. However, ciliary propulsion works only at very small body sizes. Larger chordates use the serial contraction of segmental muscle bands in the trunk and tail for locomotion, a feature that possibly first appeared as a startle response in larvae. The notochord stiffens the body so it bends from side to side as the muscles contract. Without the notochord, contraction of these muscles would result only in shortening of the body, not in propulsion.

Most fishes still use this basic type of locomotion today. The paired fins of jawed fishes are generally used for steering, braking, and providing lift, but not for propulsion except in some specialized fishes such as skates and rays that have winglike pectoral fins and in some derived bony fishes (teleosts) such as seahorses and coral reef fish.

■ Energy Acquisition and Support of Metabolism

Food energy gleaned from the environment must be processed by the digestive system to release energy and nutrients that must then be carried to the tissues. Oxygen is required for the process of energy release, and the functions of gas-exchange surfaces and the circulatory system are closely intertwined with those of the digestive system.

Feeding and Digestion   Feeding includes getting food into the oral chamber (i.e., the mouth), oral or pharyngeal processing ("chewing" in the broad sense— although only mammals truly chew their food), and swallowing. Digestion includes the breakdown of complex compounds into small molecules that are absorbed across the wall of the gut. Both feeding and digestion are two-part processes; each has a physical component and a chemical component, although physical components dominate in feeding and chemical components dominate in digestion.

Vertebrate ancestors probably filtered small particles of food from the water, as amphioxus and larval lampreys still do. Most vertebrates are particulate feeders; they take in their food as bite-size pieces rather than as tiny particles. Vertebrates have a larger gut volume than amphioxus and gut muscles that move the food by rhythmical muscular contractions (peristalsis). Vertebrates digest their food by secreting digestive enzymes produced by the liver and the pancreas into the gut, while amphioxuses digest food within the gut cells themselves. The pancreas also secretes the hormone insulin, which is involved in the regulation of glucose metabolism and blood sugar levels.

In the primitive vertebrate condition, there is no stomach, no division of the intestine into small and large portions, and no distinct rectum. The intestine opens to the **cloaca,** a common opening in most vertebrates for the urinary, reproductive, and digestive systems. (Note that the traditional "postanal tail" of chordates should really be called the "postcloacal tail": a distinct anus, separate from the urogenital systems, is only present in mammals.)

Respiration and Ventilation   Ancestral chordates probably relied on oxygen absorption and carbon dioxide loss by diffusion across a thin skin (cutaneous respiration). This is the mode of respiration of the small, sluggish amphioxus.

Cutaneous respiration is important for many vertebrates (e.g., modern amphibians), but the combination of large body size and high levels of activity make specialized gas-exchange structures essential for most vertebrates. Gills are effective in water, whereas lungs work better in air. Both gills and lungs have large internal

▶ **FIGURE 2–8** Diagrammatic view of the form and early evolution of the cranium of vertebrates. The primitive condition (a) was to have a chondrocranium formed from the paired sensory capsules, one pair for each part of the tripartite brain, with the underlying support provided by paired anterior trabeculae (at least in jawed vertebrates) and parachordals flanking the notochord posteriorly. The splanchnocranium was probably primitively made up of seven pairs of pharyngeal arches supporting six gill openings, without any anterior specializations. In the lamprey (b), the mandibular (second segment arch, but termed *arch 1* because there is no arch in the first segment) pharyngeal arch becomes the velum and other supporting structures in the head, and the remainder of the splanchnocranium forms a complex branchial basket on the outside of the gills (possibly in association with the unique mode of tidal gill ventilation). Above the level of the lamprey, the chondrocranium and splanchnocranium are surrounded with a dermatocranium of dermal bone, as first seen in ostracoderms (c). In gnathostomes (d,e) the pharyngeal arches of the mandibular (second) and hyoid (third) head segments become modified to form the jaws and jaw supports. The dermatocranium is lost in chondrichthyes (d), and in osteichthyes (e) it forms in a characteristic pattern still visible in us, including a bony operculum covering the gills and aiding in ventilation in bony fishes.

Lateral Views

Cross-sectional Views
(at level of dotted line in lateral views)

(e) Osteichthyan

Dermal operculum covering gills (connects to shoulder girdle)

Jaws encased in dermal bone

Chondrocranium
Original upper jaw
Dermal bone upper jaw
Dermal bone also forms palate
Original lower jaw
Dermal bone lower jaw
Hyoid elements
Gular bones

(d) Chondrichthyan

Chondrocranium

Gill arch 1 (= mandibular arch) forms jaws and holds teeth

Gill arch 2 (= hyoid arch) forms jaw and tongue support

Gill slit that was here was squeezed out and now forms the spiracle

Gill arches are now more complexly hinged

Chondrocranium and splanchnocranium may now be made of bone

Dermatocranium lost

Chondrocranium
Upper jaw
Teeth
Lower jaw
Hyoid elements

(c) Ostracoderm

Dermal shield surrounds rest of cranium

Splanchnocranium elements may form oral cartilages in this region

Evolution of jaws

Chondrocranium (cartilaginous)
Dermal head shield (bony)
Splanchnocranium (cartilaginous)

(b) Lamprey

Chondrocranium distorted up and back

Anterior gill arches (1 and 2) form supports for oral hood and tongue

Posterior gill arches (3-7 plus 3 additional) form complex branchial basket

Evolution of dermal bone

Otic capsule
Parachordal
Lingual cartilage
Notochord
Branchial basket

(a) Basic primitive vertebrate design

Optic capsule (surrounds eyes and midbrain)

Otic capsule (surrounds inner ear and hindbrain)

Nasal capsule (surrounds nose and forebrain)

Parachordals (posterior underlying support)

Anterior trabecula (anterior underlying support)

1 2 3 4 5 6 7
Gill openings in here

Zig-zag shape of gill arches allows muscles to change the volume of the pharynx

Otic capsule
Parachordal
Notochord (extends to front of otic capsule)
Gill arch 2

Ventral view of basic primitive vertebrate

Optic capsule
Otic capsule
Notochord
Nasal capsule
Parachordal
Anterior trabecula

Chondrocranium

Splanchnocranium

Dermatocranium

▶ **FIGURE 2–9** The crania of three primitive vertebrates. (a) the chondrocranium and splanchnocranium of a lamprey compared to (b) the chondrocranium and splanchnocranium of a living cartilaginous vertebrate (a shark), and (c) the dermatocranium of a primitive bony fish.

Chondrocranium and splanchnocranium of a lamprey

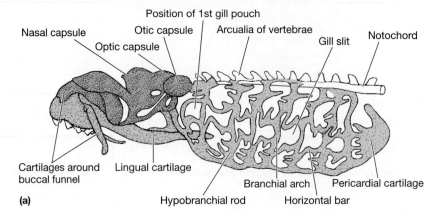

(a)

Chondrocranium and splanchnocranium of a shark

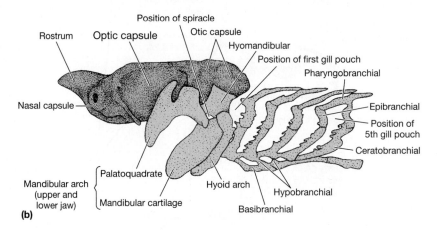

(b)

Dermatocranium of a primitive bony fish

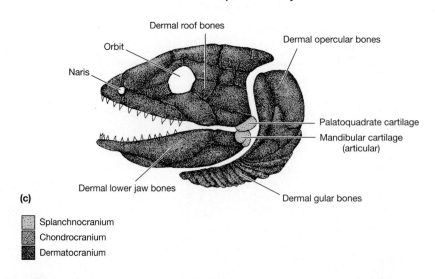

(c)

Splanchnocranium
Chondrocranium
Dermatocranium

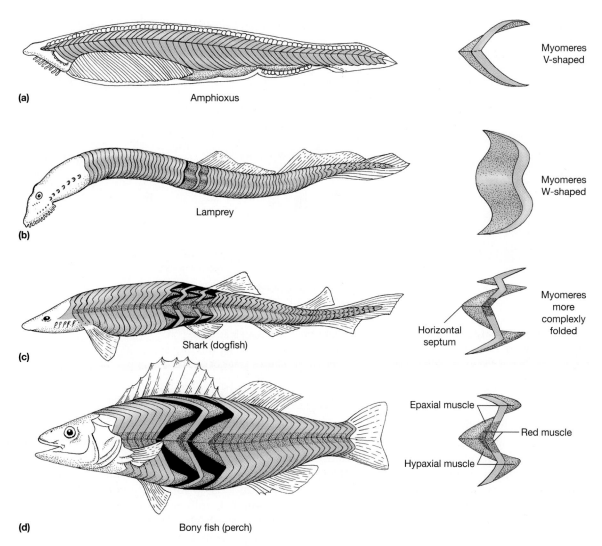

▲ FIGURE 2–10 Chordate body muscles (myomeres). (a) amphioxus (nonvertebrate chordate), (b) lamprey (jawless vertebrate), and jawed vertebrates, (c) shark and (d) bony fish.

surface areas that allow oxygen to diffuse from the surrounding medium (water or air) into the blood.

Cardiovascular System Blood carries oxygen and nutrients through the vessels to the cells of the body, removes carbon dioxide and other metabolic waste products, and stabilizes the internal environment. Blood also carries hormones from their sites of release to their target tissues.

Blood is a fluid tissue composed of liquid plasma, red blood cells (erythrocytes) that contain the iron-rich protein hemoglobin, and several different types of white blood cells (leukocytes) that are part of the immune system. Cells specialized to promote clotting of blood (thrombocytes) are present in all verte-

brates except mammals, in which they are replaced by noncellular platelets.

The blood of vertebrates is in a closed circulatory system; that is, one in which the arteries and veins are connected by capillaries. Arteries carry blood away from the heart, and veins return blood to the heart (**Figure 2–11**). Blood pressure is higher in the arterial system than in the venous system, and arteries have thicker walls than veins, with a layer of smooth muscle.

Interposed between the smallest arteries (arterioles) and the smallest veins (venules) are the capillaries, which are the sites of exchange between blood and tissues. Their walls are only one cell layer thick so diffusion is rapid, and capillaries pass close to

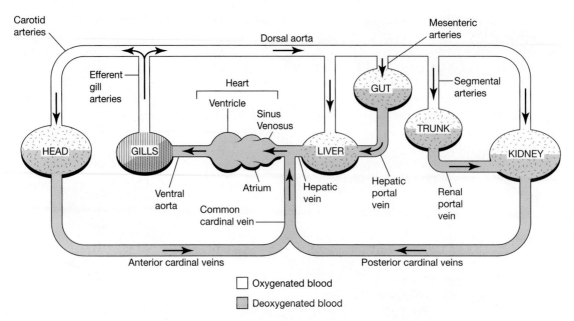

▲ **FIGURE 2–11** Diagrammatic plan of vertebrate cardiovascular circuit. All vessels are paired on the left and right sides of the body except for the midline ventral aorta and dorsal aorta. Note that the cardinal veins actually run dorsally in the real animal, flanking the carotid arteries (anterior cardinals) or the dorsal aorta (posterior cardinals).

every cell. Collectively the capillaries provide an enormous surface area for the exchange of gases, nutrients, and waste products. Capillaries form dense beds in metabolically active tissues and are sparsely distributed in tissues with low metabolic activity.

Blood flow through capillary beds is regulated by precapillary sphincter muscles. Arteriovenous **anastomoses** connect some arterioles directly to venules, allowing blood to bypass a capillary bed. Normally only a fraction of the capillaries in a tissue have blood flowing through them. When the metabolic activity of a tissue increases—when a muscle becomes active, for example—waste products of metabolism stimulate precapillary sphincters to dilate, increasing blood flow to that tissue.

Blood vessels that lie between two capillary beds are called portal vessels. The hepatic portal vein, seen in all vertebrates, lies between the capillary beds of the gut and the liver (see Figure 2–11). Substances absorbed from the gut are transported directly to the liver, where toxins are rendered harmless and some nutrients are processed or removed for storage. Most vertebrates also have a renal portal vein between the veins returning from the tail and posterior trunk and the kidneys (see Figure 2–11). The renal portal system is not well developed in jawless vertebrates and has been lost in mammals. The renal portal system presumably brings the waste

metabolites returning from the axial muscles that are used in locomotion directly to the kidney, where they are excreted.

The vertebrate heart is a muscular tube folded on itself and is primitively constricted into three sequential chambers: the **sinus venosus**, the **atrium**, and the **ventricle**. Our so-called four-chambered heart represents the combination of an atrium and a ventricle, both divided into two halves (left and right).

The sinus venosus is a thin-walled sac with few cardiac muscle fibers. It is filled by pressure in the veins, aided by pulsatile reductions in pressure in the pericardial cavity surrounding the heart as the heart beats. Suction produced by muscular contraction draws blood anteriorly into the atrium, which has valves at each end that prevent backflow. The ventricle is thick-walled, and the muscular walls have an intrinsic pulsatile rhythm (in all vertebrates except hagfishes), which can be speeded up or slowed down by the nervous system. Contraction of the ventricle forces the blood into the ventral aorta. Mammals no longer have a distinct structure identifiable as the sinus venosus; rather, it is incorporated into the wall of the right atrium as the sinoatrial node, which controls the basic pulse of the heartbeat.

The basic vertebrate circulatory plan consists of a heart that pumps blood into the single midline ventral aorta. Paired sets of aortic arches (originally six in number, one of each pair supplying each side of the head) branch from the ventral aorta (**Figure 2–12**). In the original vertebrate circulatory pattern, retained in fishes, the aortic arches lead to the gills, where the blood is oxygenated and returns to the dorsal aorta. The dorsal aorta is paired above the gills, and the anterior portions run forward to the head as the carotid arteries. Behind the gill region, the two vessels unite into a single dorsal aorta that carries blood posteriorly.

The dorsal aorta is flanked by paired cardinal veins that return blood to the heart (see Figure 2–11). Anterior cardinal veins (the jugular veins) draining the head and posterior cardinal veins draining the body unite on each side in a common cardinal vein that enters the atrium of the heart. In lungfish and tetrapods, the posterior cardinal veins are essentially replaced by a single midline vessel, the posterior vena cava. Blood is also returned separately to the heart from the gut and liver via the hepatic portal system.

Excretory and Reproductive Systems The excretory and reproductive systems are formed from the nephrotome or intermediate mesoderm, which forms the embryonic nephric ridge (**Figure 2–13**). The kidneys are segmental in structure, whereas the gonads (the organs that produce **gametes** [sex cells]—**ovaries** in females and **testes** in males) are unsegmented. While the gonads are derived from the mesoderm, the gametes themselves are formed in the endoderm, and then migrate up through the dorsal mesentery (see Figure 2–5) to enter the gonads. The archinephric duct drains urine from the kidney to the cloaca and from there to the outside world. In jawed vertebrates, this duct is also used for the release of sperm by the testes.

The kidneys dispose of waste products, primarily nitrogenous waste from protein metabolism, and regulate the body's water and minerals—especially sodium, chloride, calcium, magnesium, potassium, bicarbonate, and phosphate. In tetrapods the kidneys are responsible for almost all these functions, but in fishes and amphibians the gills and skin also play important roles (see Chapter 4). The original role of the kidney in vertebrates may have been regulation of divalent ions such as calcium and phosphate.

The kidney of fishes is a long, segmental structure extending the entire length of the dorsal body wall. In all vertebrate embryos, the kidney is composed of three portions: pronephros, mesonephros, and metanephros (see Figure 2–13). The pronephros is functional only in the embryos of living vertebrates and possibly in adult hagfishes. The kidney of adult fishes and amphibians includes the mesonephric and metanephric portions and is known as an **opisthonephric kidney**. The compact bean-shaped kidney seen in adult amniotes (the **metanephric kidney**) includes only the metanephros, drained by a new tube, the **ureter**, derived from the basal portion of the archinephric duct.

The basic units of the kidney are microscopic structures called **nephrons**. Vertebrate kidneys work by ultrafiltration: high blood pressure forces water, ions, and small molecules through tiny gaps in the capillary walls. Nonvertebrate chordates lack true kidneys. Amphioxus has excretory cells called solenocytes associated with the pharyngeal blood vessels that empty individually into the false body cavity (the atrium). The effluent is discharged to the outside via the atriopore. The solenocytes of amphioxus are thought to be homologous with the podocytes of the vertebrate nephron, which are the cells that form the wall of the renal capsule.

Reproduction is the means by which gametes (eggs and sperm) are produced, released, and combined with gametes from a member of the opposite sex to produce a fertilized zygote. Vertebrates usually have two sexes, and sexual reproduction is the norm—although unisexual species occur among fishes, amphibians, and lizards.

The gonads are paired and usually lie on the posterior body wall behind the peritoneum (the lining of the body cavity); it is only among mammals that the testes are found outside the body in a scrotum. Ovaries contain large primary sex cells called follicles. As they mature, the follicular cell layer becomes much larger, nurturing the developing egg (ovum; plural, *ova*), stimulating the development of yolk in the egg, and producing the hormone estrogen. When the eggs mature, the follicle ruptures, releasing the completed egg (ovulation). The testes are composed of interconnecting seminiferous tubules where sperm develop. In these tubules, sperm are supported, nourished, and conditioned by cells—the supporting, or Sertoli, cells—that remain permanently attached to the tubule walls. The testes also produce the hormone testosterone.

In the primitive vertebrate condition, which is retained in living jawless vertebrates, there is no special tube or duct for the passage of the gametes. Rather, the sperm or eggs erupt from the gonad and move through the coelom to pores that open to the base of the archinephric ducts. In jawed vertebrates, however, the gametes are always transported to the cloaca via specialized paired ducts (one for each gonad). In males, sperm are released directly into the archinephric ducts that drain the kidneys in nonamniotes and embryonic

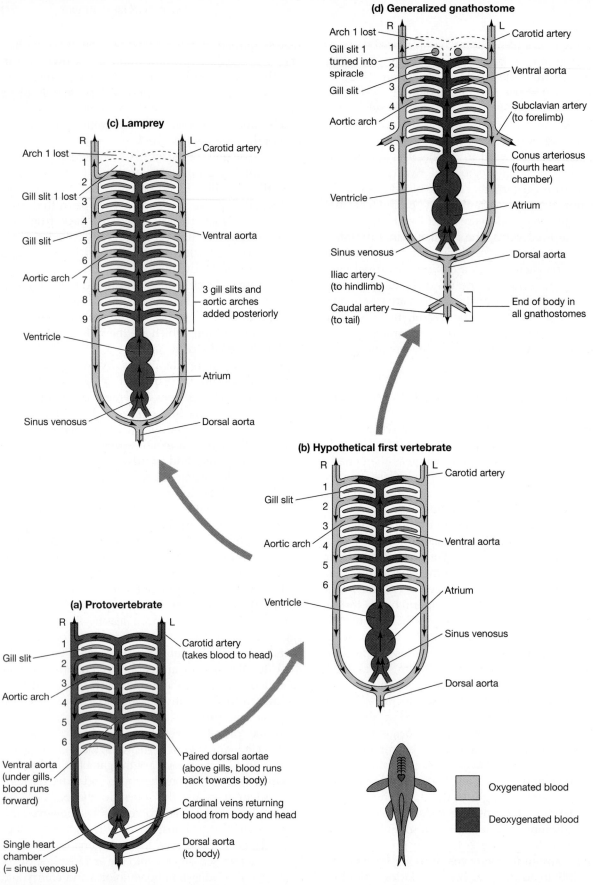

(d) Generalized gnathostome

Arch 1 lost
Gill slit 1 turned into spiracle
Gill slit
Aortic arch
Carotid artery
Ventral aorta
Subclavian artery (to forelimb)
Conus arteriosus (fourth heart chamber)
Ventricle
Atrium
Sinus venosus
Dorsal aorta
Iliac artery (to hindlimb)
Caudal artery (to tail)
End of body in all gnathostomes

(c) Lamprey

Arch 1 lost
Carotid artery
Gill slit 1 lost
Gill slit
Aortic arch
Ventral aorta
3 gill slits and aortic arches added posteriorly
Ventricle
Atrium
Sinus venosus
Dorsal aorta

(b) Hypothetical first vertebrate

Gill slit
Aortic arch
Carotid artery
Ventral aorta
Ventricle
Atrium
Sinus venosus
Dorsal aorta

(a) Protovertebrate

Gill slit
Aortic arch
Carotid artery (takes blood to head)
Ventral aorta (under gills, blood runs forward)
Paired dorsal aortae (above gills, blood runs back towards body)
Cardinal veins returning blood from body and head
Single heart chamber (= sinus venosus)
Dorsal aorta (to body)

Oxygenated blood
Deoxygenated blood

40

◀ **FIGURE 2–12** Diagrammatic view of the form and early evolution of the heart and aortic arches of vertebrates. In the protovertebrate (a) and the earliest true vertebrate condition (b), there were probably six pairs of aortic arches, just as are seen in the embryos of all living vertebrates, although arch 1 is never seen in the adults. These arches were used in association with feeding, not for respiration. In lampreys (c), additional aortic arches are added posteriorly to accommodate more gill openings. In gnathostomes (d), the subclavian and iliac arteries are added to the main circulatory system, supplying the forelimbs and hindlimbs, respectively. A fourth chamber is also added to the heart, the conus arteriosus, which damps out the pulsatile component of the blood flow.

amniotes. In females, the egg is still released into the coelom but is then transported via a new structure, the **oviduct.** The oviducts produce substances associated with the egg, such as the yolk or the shell. The oviducts can become enlarged and fused in various ways to form a uterus or paired uteri in which eggs are stored or young develop.

Vertebrates may deposit eggs that develop outside the body or retain the eggs within the mother's body until embryonic development is complete. Shelled eggs must be fertilized in the oviduct before the shell and albumen are deposited. Vertebrates that lay shelled eggs, and vertebrates that are viviparous, must have some sort of intromittent organ—such as the pelvic claspers of sharks and the amniote penis—by which sperm are inserted into the female's reproductive tract.

■ Coordination and Integration

The nervous and endocrine systems respond to conditions inside and outside an animal, and together they control the actions of organs and muscles, coordinating them so they work in concert.

General Features of the Nervous System Individual cells called neurons are the basic units of the nervous system. In jawed vertebrates, the axons of neurons are encased in a fatty insulating coat, the myelin sheath, which increases the conduction velocity of the nerve impulse. Generally, the extensions of the neurons, called axons, are collected together like wires in a cable (**Figure 2–14**). Such collections of axons in the peripheral nervous system (PNS; i.e., that which is not within the brain or spinal cord) are called nerves; within the central nervous system (CNS; i.e., within the brain and nerve cord) they are called tracts. A group of cell bodies (the portion of a nerve cell that contains the nucleus) is known as a ganglion in the PNS and a nucleus in the CNS. Neurons communicate with other neurons via short processes called dendrites.

The nerve cord (= spinal cord) is composed of a hollow tube with the cell bodies (gray matter) on the inside and the myelin-covered (in jawed vertebrates) axons (white matter) on the outside. The nerves of the PNS are segmentally arranged, exiting from the spinal cord between the vertebrae. The spinal cord receives sensory inputs, integrates them with other

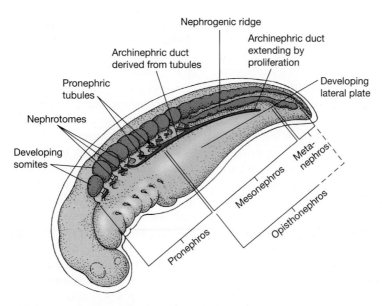

▲ **FIGURE 2–13** Kidney development in a generalized vertebrate embryo, showing the nephrotome regions.

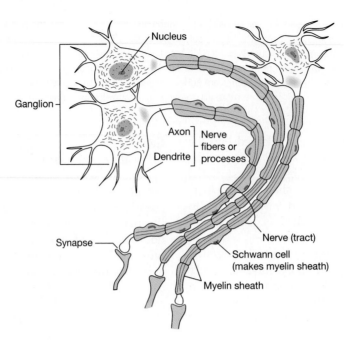

▲ **FIGURE 2–14** Generalized vertebrate neurons.

portions of the CNS, and sends impulses that cause muscles to contract. The spinal cord has considerable autonomy in many vertebrates. Even such complex movements as swimming are controlled by the spinal cord rather than the brain, and fishes continue coordinated swimming movements when the brain is severed from the spinal cord. The familiar knee-jerk reaction is produced by the spinal cord as a reflex arc. The trend in vertebrate evolution has been to develop more complex connections within the spinal cord and between the spinal cord and the brain.

Vertebrates are unique among animals in having a dual type of nervous system: the **somatic nervous system** (the so-called voluntary nervous system) and the **visceral nervous system** (the so-called involuntary nervous system). This dual nervous system mimics the dual type of development of the mesoderm. The somatic nervous system innervates the structures derived from the segmented portion (the somites), including the striated muscles that we can move consciously (e.g., the limb muscles), and relays information from sensation that we are usually aware of (e.g., from temperature and pain receptors in the skin). The visceral nervous system innervates the smooth and cardiac muscles that we usually cannot move consciously (e.g., the gut and heart muscles) and relays information from sensations that we are not usually aware of, such as the receptors monitoring the levels of carbon dioxide in the blood.

Each spinal nerve complex is thus made up of four types of fibers: somatic motor fibers to the body; somatic sensory fibers from the body wall; visceral motor fibers to the muscles and glands of the gut and to the blood vessels of both the gut and peripheral structures like the skin; and visceral sensory fibers from the gut wall and blood vessels. The motor portion of the visceral nervous system is known as the **autonomic nervous system.** In more derived vertebrates, such as mammals, this system becomes divided into two portions: the **sympathetic nervous system** (usually acting to speed things up) and the **parasympathetic nervous system** (usually acting to slow things down).

Vertebrates also have nerves that emerge directly from the; these **cranial nerves** (10 pairs in the primitive vertebrate condition, 12 in amniotes) are identified by Roman numerals. Some of these nerves, such as the ones supplying the nose (the olfactory nerve, I) or the eyes (the optic nerve, II), are not true nerves at all, but outgrowths of the brain. Somatic motor fibers in cranial nerves innervate the muscles that move the eyeballs and the branchiomeric muscles that power the jaws and gill arches. There is also a system of special visceral sensory fibers that convey information to the brain from the taste buds. We are unaware of most visceral sensory sensations, but taste is obviously a conscious phenomenon. The special sensory nerves that supply the lateral line in fishes are also from the cranial nerves.

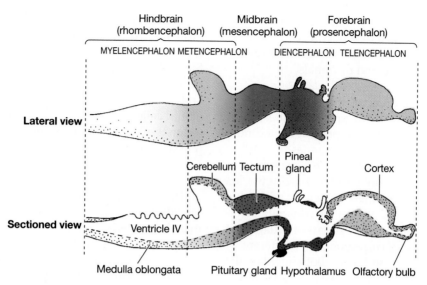

Hindbrain (rhombencephalon) | Midbrain (mesencephalon) | Forebrain (prosencephalon)

MYELENCEPHALON METENCEPHALON | DIENCEPHALON TELENCEPHALON

Lateral view

Cerebellum Tectum | Pineal gland | Cortex

Sectioned view

Ventricle IV

Medulla oblongata | Pituitary gland Hypothalamus Olfactory bulb

▲ **FIGURE 2–15** The generalized vertebrate brain.

The vagus nerve (cranial nerve X) ramifies through all but the most posterior part of the trunk, carrying the visceral motor nerve supply to various organs. People who break their necks may be paralyzed from the neck down (i.e., lose the function of their skeletal muscles) but still may retain their visceral functions (workings of the gut, heart, etc.) because the vagus nerve is independent of the spinal cord and exits above the break.

Brain Anatomy and Evolution All chordates have some form of a brain, as a thickening of the front end of the notochord. The brain of all vertebrates is a tripartite (three-part) structure (**Figure 2–15**), and the telencephalon (front part of the forebrain) and the olfactory receptors are probably true new features in vertebrates (Northcutt 2005). In the most primitive condition, the forebrain is associated with the sense of smell, the midbrain with vision, and the hindbrain with balance and detection of vibrations (hearing, in the broad sense). These portions of the brain are associated with the nasal, optic, and otic capsules of the chondrocranium, respectively (see Figure 2–8).

The hindbrain has two portions. The most posterior, the *myelencephalon,* or medulla oblongata, controls functions such as respiration and acts as a relay station for receptor cells from the inner ear. The anterior portion, the *metencephalon,* develops an important dorsal outgrowth, the *cerebellum*—present as a distinct structure only in jawed vertebrates among living forms. The cerebellum coordinates and regu-

lates motor activities whether they are reflexive (such as maintenance of posture) or directed (such as escape movements). The midbrain develops in conjunction with the eyes and receives input from the optic nerve, although in mammals the forebrain has taken over much of the task of vision.

The forebrain also has two parts. The posterior region is the **diencephalon,** which acts as a major relay station between sensory areas and the higher brain centers. The **pituitary gland,** an important endocrine organ, is a ventral outgrowth of the diencephalon. The floor of the diencephalon (the **hypothalamus**) and the pituitary gland form the primary center for neural-hormonal coordination and integration. Another endocrine gland, the **pineal organ,** is a dorsal outgrowth of the diencephalon. Its original function is thought to have been as a median photoreceptor. Many early tetrapods had a hole in the skull over the pineal gland to admit light, and this condition is still seen today in some reptiles (e.g., the tuatara).

The most anterior region of the adult forebrain, the **telencephalon,** develops in association with the olfactory capsules and coordinates inputs from other sensory modalities. In various ways in different vertebrate groups, the telencephalon becomes enlarged—in which condition it is also known as the **cerebrum** or cerebral hemispheres. Tetrapods develop an area in the cerebrum called the **neocortex** or neopallium, which becomes the primary seat of sensory integration and nervous control. Bony fishes also evolved a larger, more complex telencephalon—but by a completely different mechanism. Sharks and, perhaps surprisingly, hagfishes have also independently evolved relatively large forebrains, although a large cerebrum is primarily a feature of birds and mammals.

The Sense Organs   We think of vertebrates as having five senses—taste, touch, sight, smell, and hearing—but this list does not reflect the primitive condition, nor all the senses of living vertebrates. Complex, multicellular sense organs that are formed from epidermal placodes and tuned to the sensory worlds of the species that possess them are a derived feature of vertebrates. Water and air have very different physical properties, and the sensory systems of aquatic and terrestrial vertebrates are correspondingly different.

The senses of smell and taste both involve the detection of dissolved molecules by specialized receptors. We think of these two senses as being closely interlinked; for example, our sense of taste is poorer if our sense of smell is blocked by having a cold. However, the two senses are actually very different in their innervation. Smell is a somatic sensory system—sensing items at a distance, with the sensations being received in the forebrain. Taste is a visceral sensory system—sensing items on direct contact, with the sensations being received initially in the hindbrain.

The receptor field of the vertebrate eye is arrayed in a hemispherical sheet, the **retina**, which originates as an outgrowth of the brain. The retina contains two types of light-sensitive cells, **cones** and **rods**, which are distinguished from each other by morphology, photochemistry, and neural connections.

The capacity to perceive electrical impulses generated by the muscles of other organisms is also a form of distance reception, but one that works only in water. Electroreception was probably an important feature of early vertebrates and is seen today primarily in fishes. Related to electroreception is the ability seen in many fishes today to produce electric discharge for communication with other individuals or for protection from predators.

Originally the inner ear detected an animal's position in space, and it retains that function today in both aquatic and terrestrial vertebrates. The inner ear is also used for hearing (reception of sound waves) in tetrapods and in a few derived fishes. The basic sensory cell in the inner ear is the hair cell, which detects the movement of fluid resulting from a change of position or the impact of sound waves. In the lateral line system of fishes and aquatic amphibians, hair cells are aggregated into **neuromast organs,** which detect the movement of water around the body (see Chapter 4).

The inner ear contains the vestibular apparatus (also known as the membranous labyrinth), which includes the organs of balance and, in tetrapods only, the **cochlea** (organ of hearing). The vestibular apparatus is enclosed within the otic capsule of the skull and consists of a series of sacs and tubules containing a fluid called endolymph. Sound waves are transmitted to the inner ear, where they create waves of compression that pass through the endolymph. These waves stimulate the vestibular sensory cells, which are variants of the basic hair cell.

The lower parts of the vestibular apparatus, the *sacculus* and *utriculus*, house sensory organs called maculae, which contain tiny crystals of calcium carbonate resting on hair cells. Sensations from the maculae tell the animal which way is up and detect linear acceleration. The upper part of the vestibular apparatus contains the *semicircular canals* (SSC). Sensory areas in swellings at the end of each canal (*ampullae*) detect angular acceleration through cristae, hair cells embedded in a jellylike substance, by monitoring the displacement of endolymph during motion. Jawed vertebrates have three semicircular canals on each side of the head, hagfishes have one, and lampreys and fossil jawless vertebrates have two (**Figure 2–16**).

We often fail to realize the importance of the vestibular senses in ourselves because we usually depend on vision to determine position. However, we

Lamprey (2 semicircular canals)

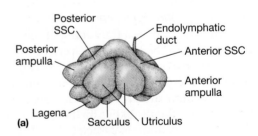

(a)

**Generalized gnathostome (shark)
(3 semicircular canals)**

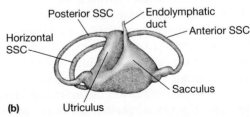

(b)

▲ **FIGURE 2–16** Design of the vestibular apparatus in fishes. The lamprey (a) has two semicircular canals, whereas gnathostomes (represented by a shark, b) have three semicircular canals.

can sometimes be fooled, as when sitting in a stationary train or car and thinking that we are moving, only to realize from the lack of input from our vestibular system that it is the vehicle *next* to us that is moving.

The Endocrine System The endocrine system transfers information from one part of the body to another via the release of a chemical messenger (**hormone**) that produces a response in the target cells. The time required for an endocrine response ranges from seconds to hours. Hormones are produced in discrete endocrine glands, whose primary function is hormone production and excretion (e.g., the pituitary, thyroid, thymus, and adrenals), and by organs with other major bodily functions—such as the gonads, kidneys, and gastrointestinal tract.

An interesting story has recently emerged concerning the parathyroid gland. This gland is found on either side of the thyroid gland in the throat of tetrapods and is important for regulation of calcium in the terrestrial environment. Recent developmental and genetic studies have shown that this gland is homologous with the internal gill buds of fishes that are also involved in calcium regulation (Okabe and Graham 2004).

The trend in the evolution of vertebrate endocrine glands has been consolidation from scattered clusters of cells or small organs in fishes to larger, better-defined organs in amniotes. Endocrine secretions are predominantly involved in controlling and regulating energy use, storage, and release, as well as in allocating energy to special functions at critical times.

■ The Immune System

Vertebrates have adaptive immunity, a type of immune system different from invertebrates (Pancer and Cooper 2006). While all animals have innate, specifically genetically encoded responses to pathogens, vertebrates additionally have evolved lymphocytes (a type of white blood cell), which provide a system of adjustable antigen recognition. Interestingly, the systems are somewhat different in jawless and jawed vertebrates. While gnathostomes generate lymphocyte receptors via immunoglobulin gene segments, lamprey and hagfish employ leucine-rich repeat molecules, and they lack a thymus gland, which produces lymphocytes in gnathostomes. The agnathan condition is probably the primitive condition for vertebrates, as it more closely resembles the mode of antipathogen responses in invertebrates.

Summary

Vertebrates are members of the phylum Chordata, a group of animals whose other members, tunicates and cephalochordates (amphioxus), are quite different from most modern vertebrates, being small, marine, and sluggish or entirely sessile as adults. Chordates share with many derived animal phyla the features of being bilaterally symmetrical, with a distinct head and tail end. Both embryological and molecular evidence show that chordates are related to other sessile marine animals, such as echinoderms (starfishes, etc.).

Chordates are generally distinguished from other animals by the presence of a notochord, a dorsal hollow nerve chord, a muscular postanal tail, and an endostyle (homologous to the vertebrate thyroid gland). Vertebrates appear to be most closely related to cephalochordates among nonvertebrate chordates, and most of the differences in structure and physiology reflect an evolutionary change to larger body size,

a greater level of activity, and a switch from filter feeding to predation.

Vertebrates have the unique features of an expanded head with multicellular sense organs and a cranium housing an enlarged, tripartite brain. The features that distinguish vertebrates from other chordates appear to be related to two critical embryonic innovations: a doubling of the *Hox* gene complex and the development of neural crest tissue. The complex activities of vertebrates are supported by a complex morphology. Study of embryology can throw light on how complex structures are formed and develop. The unique neural crest cells form many of the derived characters of vertebrates, especially those of the new anterior portion of the head.

An adult vertebrate can be viewed as a functional group of interacting systems involved in protection, support and movement, acquisition of energy, excretion, reproduction, coordination, and integration. These systems underwent profound functional and structural changes at several key points in vertebrate evolution. The most important transition was from the prevertebrate condition—as represented today by the cephalochordate amphioxus—to the vertebrate condition, shown by the living jawless vertebrates. Other important transitions, to be considered in later chapters, include the shift from jawless to jawed vertebrates and from fish to tetrapod.

Additional Readings

Boorman, C. J., and S. M. Shimeld. 2002. The evolution of left-right asymmetry in chordates. *BioEssays* 24:1004–1011.

Boot-Handford, R. P., and D. S. Tuckwell. 2003. Fibrillar collagen: the key to vertebrate evolution? A tale of molecular incest. *BioEssays* 25:142–151.

Bourlat, S. J. et al. 2006. Deuterostome phylogeny reveals monophyletic chordates and the new phylum Xenoturbellida. *Nature* 444:85–88.

Butler, A. B. 2006. The serial transformation hypothesis of vertebrate origins: Comment on "the New Head hypothesis revisited." *Journal of Experimental Zoology (Mol Dev Evol)* 306B:419–424.

Donoghue, P. C. J., and M. A. Purnell. 2006. Genome duplication, extinction, and vertebrate evolution. *Trends in Ecology and Evolutionary Biology* 20:312–319.

Donoghue, P. C. J., and I. J. Sansom. 2002. Origin and early evolution of vertebrate skeletonization. *Microscopy Research and Technique* 59:352–372.

Donoghue, P. C. J. et al. 2006. Early evolution of vertebrate skeletal tissues and cellular interactions, and the canalization of skeletal development. *Journal of Experimental Zoology (Mol Dev Evol)* 306B:278–294.

Gans, C. 1989. Stages in the origin of vertebrates: Analysis by means of scenarios. *Biological Reviews* 64:221–268.

Gee, H. 1996. *Before the Backbone*. London: Chapman & Hall.

Graham, A. 2001. The development and evolution of the pharyngeal arches. *Journal of Anatomy* 199:133–141.

Hall, B. K. 2000. The neural crest as a fourth germ layer and vertebrates as quadroblastic and triploblastic. *Evolution and Development* 2:3–5.

Hall, B. K. 2005. Consideration of the neural crest and its skeletal derivatives in the context of novelty/innovation. *Journal of Experimental Zoology (Mol Dev Evol)* 304B:548–557.

Hildebrand, M., and G. E. Goslow. 2001. *Analysis of Vertebrate Structure*, 5th ed., New York: Wiley.

Holland, N. D., and J. Chen. 2001. Origin and early evolution of vertebrates: New insights from advances in molecular biology, anatomy, and palaeontology. *BioEssays* 23:142–151.

Janvier, J., and M. Arsenault. 2002. Calcification of early vertebrate cartilage. *Nature* 417:609.

Jeffrey, W. R. et al. 2004. Migratory neural-crestlike cells form body pigmentation in a urochordate embryo. *Nature* 431:696–699.

Kardong, K. V. 2001. *Vertebrates—Comparative Anatomy, Function, Evolution*, 3rd ed. Dubuque, IA: Wm. C. Brown.

Lacalli, T. C. 2002. Vetulicolians—are they deuterosomes? Chordates? *BioEssays* 24:208–211.

Lacalli, T. C. 2004. Protochordate body plan and the evolutionary role of larvae: Old controversies resolved? *Canadian Journal of Zoology* 83:216–224.

Liem, K. F. et al. 2001. *Functional Anatomy of the Vertebrates—an Evolutionary Perspective*, 3rd ed. Philadelphia: Saunders College Publishing.

Mallatt, J., and J.-Y. Chen. 2003. Fossil sister group of craniates: Predicted and found. *Journal of Morphology* 248:1–31.

Northcutt, R. G. 1990. Ontogeny and phylogeny: A re-evaluation of conceptual relationships and some applications. *Brain, Behavior and Evolution* 36:116–140.

Northcutt, R. G. 2005. The New Head hypothesis revisited. *Journal of Experimental Zoology (Molecular, Developmental, Evolutionary)* 304B:274–287.

Northcutt, R. G., and C. Gans. 1983. The genesis of neural crest and epidermal placodes: A reinterpretation of vertebrate origins. *Quarterly Review of Biology* 58:1–28.

Okabe, M., and A. Graham. 2004. The origin of the parathyroid gland. *Proceedings of the National Academy of Sciences* 101:17716–17719.

Pancer, Z., and M. D. Cooper. 2006. The evolution of adaptive immunity. *Annual Reviews of Immunology* 24:497–518.

Redies, C., and L. Puelles. 2001. Modularity in vertebrate brain development and evolution. *BioEssays* 23:1100–1111.

Ruben, J. A., and A. F. Bennett. 1987. The evolution of bone. *Evolution* 41:1187–1197.

Rychel, A. L. et al. 2006. Evolution and development of the chordates: Collagen and pharyngeal cartilage. *Molecular Biology and Evolution* 23:541–549.

Schubert, M. et al. 2006. Amphioxus and tunicates as evolutionary model systems. *Trends in Ecology and Evolutionary Biology* 21:269–277.

Zimmer, C. 2000. In search of vertebrate origins: Beyond brain and bone. *Science* 287:1576–1579.

CHAPTER 3

Early Vertebrates: Jawless Vertebrates and the Origin of Jawed Vertebrates

The earliest vertebrates represented an important advance over the nonvertebrate chordate filter feeders. Their most conspicuous new feature was a distinct head end, containing a tripartite brain enclosed by a cartilaginous cranium (skull) and complex sense organs. They used the newly acquired pharyngeal musculature that powered the gill skeleton to draw water into the mouth and over the gills, which were now used for respiration rather than for filter feeding. Early vertebrates were active predators rather than sessile filter feeders. Many of them also had external armor made of bone and other mineralized tissues. We know a remarkable amount about the anatomy of some of these early vertebrates because the internal structure of their bony armor reveals much of their soft anatomy.

Gnathostomes represented an advance in the vertebrate design for high levels of activity and predation. Jaws themselves are homologous with the structures that form the gill arches and probably first evolved as devices to improve the strength and effectiveness of gill ventilation. Later they were modified for seizing and holding prey. In this chapter we trace the earliest steps in the radiation of vertebrates, beginning more than 500 million years ago, and discuss the biology of both the Paleozoic agnathans (the ostracoderms) and the extant forms (hagfishes and lampreys). We also consider the transition from the jawless condition to the jawed one and the biology of some of the early types of jawed fishes that did not survive the Paleozoic era (placoderms and acanthodians).

3.1 Reconstructing the Biology of the Earliest Vertebrates

■ The Earliest Evidence of Vertebrates

Until very recently our oldest evidence of vertebrates consisted of fragments of the dermal armor of the jawless vertebrates known as **ostracoderms**. These animals were very different from any vertebrate alive today. They were essentially fishes encased in bony armor (Greek *ostrac* = shell and *derm* = skin), quite unlike the living jawless vertebrates that lack bone completely. Bone fragments that can be assigned definitively to vertebrates are known from the Ordovician period, some 480 million years ago, although there are some pieces of mineralized material that are tentatively believed to represent vertebrates from the Late Cambrian period, around 500 million years ago. This was about 80 million years before whole-body vertebrate fossils became abundant, in the Late Silurian period.

More recent finds of early soft-bodied vertebrates from the Early Cambrian of China (Shu et al. 1999) extend the fossil record back by another 40 million years or so. Two different types of Early Cambrian vertebrates, *Myllokumingia* and *Haikouichthys*, are found in the same fossil deposit, the Chengiang Fauna, which also yielded the possible vertebrate relative *Haikouella* (see Chapter 2). Both of these vertebrates were small, fish-shaped, and about 3 centimeters long (**Figure 3–1a**). Evidence of a cranium and W-shaped myomeres mark these animals as true

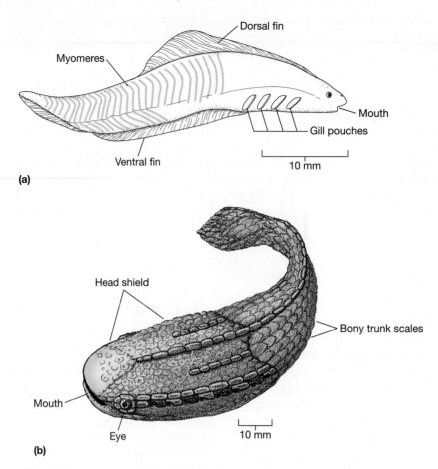

▲ **FIGURE 3–1** Some of the earliest vertebrates. (a) The Early Cambrian *Myllokumingia* from China. (b) The Ordovician pteraspid ostracoderm *Astraspis* from North America.

vertebrates. However, unlike ostracoderms, they lack any evidence of bone or mineralized scales. Both animals had a dorsal fin and ribbonlike ventral fins. Despite the great age of these early vertebrates, some researchers consider them to be more derived than present-day hagfishes. They have a dorsal fin, which is absent from hagfishes and amphioxus, as well as evidence of paired sensory structures in the head and segmental structures flanking the notochord that may have been lampreylike arcualia. *Haikouichthys* appears to have lampreylike cartilaginous gill supports (a branchial basket), and thus may be the sister taxon to the lampreys.

The next good evidence of early vertebrates is from the Early Ordovician; several sites scattered around the world have yielded vertebrate fossils in the form of bone fragments, suggesting that by this time ostracoderms had diversified and had a worldwide distribution. The earliest vertebrates represented by complete articulated fossils are from the Late Ordovician of Bolivia, Australia, and North America. These were armored, torpedo-shaped jawless fishes, ranging from 12 to 35 centimeters in length. These early fishes were externally armored with many small, close-fitting, polygonal bony plates 3 to 5 millimeters long. These plates abutted each another in the head and gill region, forming a head shield. Posteriorly they overlapped, as do scales in extant fishes. These bony plates show the presence of sensory canals, special protection around the eye, and—in the reconstruction of the North American *Astraspis* (**Figure 3–1b**)—as many as eight gill openings on each side of the head. Similar diverse assemblages of early vertebrates, are found in Late Silurian formations (about 400 million years old) in the United Kingdom, Norway, and North America.

The Ordovician was also a time for great radiation and diversification among marine invertebrates, following the extinctions at the end of the Cambrian. The early radiation of vertebrates involved both jawed and jawless groups, although the first evidence of jawed fish is in the Middle Ordovician, a little later than the first ostracoderms.

■ The Origin of Bone and Other Mineralized Tissues

Mineralized tissues composed of hydroxyapatite are a major new feature of vertebrates. Enamel (or enameloid) and dentine, which occur primarily in the teeth among living vertebrates, are at least as old as bone and were originally found in intimate association with bone in the dermal armor of ostracoderms. However, mineralized tissues did not appear at the start of vertebrate evolution and are lacking in the extant jawless vertebrates.

The origin of vertebrate mineralized tissues remains a puzzle: The earliest-known types were no less complex in structure than the mineralized tissues of living vertebrates. The basic units of mineralized tissue appear to be odontodes, little toothlike elements formed in the skin. They consist of projections of dentine, covered in some cases with an outer layer of enameloid, with a base of bone. Our own teeth are very similar to these structures and are probably homologous with them (see Figure 2–6 in Chapter 2).

Odontodes occur in almost unmodified form as the sharp denticles in the skin of sharks—and the larger scales, plates, and shields on the heads of many ostracoderms and early bony fishes are interpreted as aggregations of these units. Note that these bony elements would not have been external to the skin like a snail's shell. Rather, they were formed within the dermis of the skin and overlain by a layer of epidermis, as with our own skull bones. The primitive condition for vertebrate bone is to lack cells in the adult form. This type of acellular bone is also known as aspidin. Cellular bone is found only in gnathostomes and in some derived ostracoderms (osteostracans).

What could have been the original selective advantage of mineralized tissues in vertebrates? The head shields of ostracoderms were originally thought to be defensive structures against attack from predators—particularly the large, scorpion-like creatures known as eurypterids. However, the detailed structure of the bony tissues suggests that they had a more complex function than mere protection. The bony head shield may have originated as a

protective and insulating coating around the electroreceptors that enhanced detection of their prey. Subsequently, regulation of phosphorus, which is a relatively rare element in natural environments, may have been one of the early selective forces involved in the evolution of bone. Although we think of a skeleton as primarily supportive or protective, bone also serves as a store of calcium and phosphorus. Mineral regulation involves deposition and mobilization of calcium and phosphorus ions. The explanations proposed for the origin of bone and other mineralized tissues—protection, electroreception, and mineral storage—are not mutually exclusive. They all may have been involved in the evolutionary origin of these complex tissues. Donoghue and Sansom (2002) reviewed these different hypotheses and came to the conclusion that the original notion of protection may be the most viable one.

■ The Problem of Conodonts

Curious microfossils known as **conodont** elements are widespread and abundant in marine deposits from the Late Cambrian to the Late Triassic periods. Conodont elements are small (generally less than 1 millimeter long) spinelike or comblike structures composed of apatite (the particular mineralized calcium compound that is characteristic of vertebrate hard tissues). They were described as skeletal parts of marine algae by some paleontologists and as marine invertebrates by others. However, recent studies of conodont mineralized tissues have shown that they are similar in their microstructure to dentine, which is a uniquely vertebrate tissue. Thus conodont elements are considered to be the toothlike elements of true vertebrates. This interpretation has been confirmed by the discovery of impressions of complete conodont animals with vertebrate features (such as a notochord, a cranium, myomeres, and large eyes) and with conodont elements arranged within the pharynx in a complex apparatus (**Figure 3–2**).

We do not as yet understand quite how conodont teeth fit into the evolutionary picture. They may represent an early experiment in tissue mineralization prior to the evolution of the dermal skeleton in other vertebrates. The ability to form mineralized tissues in the skin may be a primitive vertebrate feature, but the production of teeth, or toothlike structures, may have evolved independently several different times among vertebrates, as will be further discussed later in the chapter.

▶ **FIGURE 3–2** Conodonts. (a) *Clydagnathus*. (b) Close-up of the feeding apparatus (conodont elements) inside the head of *Idiognathus*. The anterior S elements appear to be designed for grasping, and the posterior P elements appear to be designed for crushing.

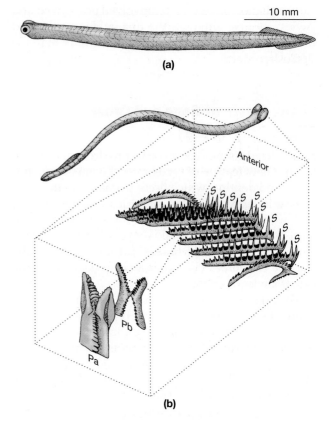

10 mm

(a)

(b)

Accepting conodonts as vertebrates has changed our ideas about early vertebrate interrelationships, particularly the importance—in the evolutionary hierarchy—of having mineralized tissues. The tooth-like structures of conodonts and the bony dermal skeletons of ostracoderms are now thought to make these animals more derived types of vertebrates than the soft-bodied, jawless fishes living today (see **Figure 3–3**).

■ The Environment of Early Vertebrate Evolution

By the Late Silurian, ostracoderms and early jawed fishes were abundant in both freshwater and marine environments. Under what conditions did the first vertebrates evolve? Researchers originally supposed that vertebrates evolved in fresh water. This was because the vertebrate kidney is clearly advantageous

▶ **FIGURE 3–3** Phylogenetic relationships of vertebrates within the Chordata. This diagram depicts probable relationships among primitive vertebrates, including living and extinct jawless vertebrates and the earliest jawed vertebrates. Quotation marks indicate paraphyletic groups. Black lines show relationships only; they do not indicate times of divergence nor the unrecorded presence of taxa in the fossil record. Dark bars show times from which fossils have been found, whereas lightly shaded bars indicate ranges of time when the taxon is known to be present but is unrecorded (or poorly recorded) in the fossil record. The numbers indicate derived characters that distinguish the lineages (as described in the legend).

Legend: 1. Chordata—notochord; hollow dorsal nerve cord; postanal tail; segmental muscle bands in at least the tail region; endostyle; brain of some sort at rostral end of nerve cord, including an adenohypophysis. **2.** Somites in embryo; myomeres in trunk and tail region; excretory system with podocyte cells; lateral plate mesoderm; embryological induction of neural tissues by the notochord; caudal fin fold, ventral to dorsal pattern of blood circulation through the gills. **3.** Craniata (Vertebrata in the traditional usage) cranium incorporating anterior end of the notochord and enclosing brain and paired sensory organs; distinct head region with these characters—tripartite division of brain with a telencephalon, olfactory receptors, and cranial nerves differentiated from neural tube, paired optic, auditory, and probably olfactory organs, one or more semicircular canals; neural crest cells and structures derived from these tissues (including cartilaginous gill skeleton); atrium lost; muscularized hypomere (lateral plate mesoderm); W-shaped

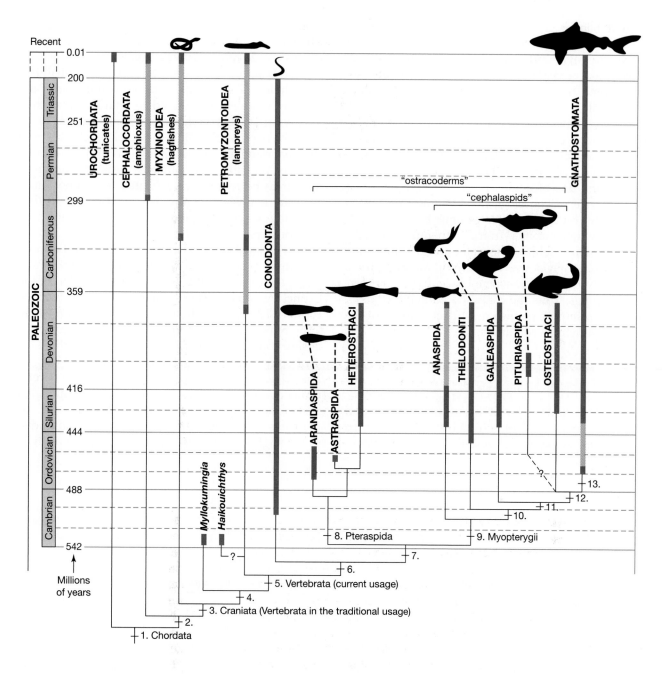

myomeres, innervated by ventral root spinal nerves; unique type of fibrillar collagen; distinctive endocrine glands; lateral line system; well-developed 3-chambered heart; capillaries, and blood vessels lined with epithelium; paired kidneys with an archinephric duct. **4.** Dorsal fin. **5.** Vertebrata (current usage)—presence of arcualia (vertebral rudiments surrounding the nerve cord); eyes well developed with extrinsic eye musculature; pineal eye; hypoglossal nerve; Mauthner neurons in the brain stem; sensory-line neuromasts; sensory lines on head and body; capacity for electroreception;

two semicircular canals; autonomic innervation of the heart; renal collecting ducts; spleen or splenic tissue; three (versus one) types of granular white blood cells; dilute body fluids; blood comprises more than 10% of body volume; ion transport in gills; higher metabolic rate; pituitary control of pigment cells and gametogenesis. **6.** Physiological capacity to form mineralized tissues in the dermis; horny teeth lost. **7.** Dermal-bone head shield; olfactory tract; cerebellum. **8.** Pteraspida—paired nasal openings; dermal skeleton with characteristic three layers, including spongy bone. **9.** Myopterygii—

paired lateral fin folds, dorsal and anal fin; endolymphatic duct in inner ear opens to surface (lost in gnathostomes above level of Chondrichthyes). **10.** Sclerotic ossicles. **11.** Perichondral bone (at least in the head). **12.** Presence of calcified cartilage in endoskeleton; cellular dermal bone; three-layered exoskeleton; slitlike (versus pouched) gills; pectoral fins with a narrow, concentrated base; large orbits; large head vein (dorsal jugular).
13. Gnathostomata—jaws and many other derived characters (see text), including paired nasal openings.

in that habitat, acting to rapidly eliminate excess water entering the body by osmosis. However, a marine origin of vertebrates is now widely accepted; it is considered that the kidney was fortuitously preadapted for this role in fresh water rather than specifically evolved for that purpose. Some researchers have proposed an estuarine origin of vertebrates, in a position of half-way salinity between seawater and fresh water (Ditrich 2007), but such scenarios do not account for the evidence listed below.

The first line of evidence for a marine origin of vertebrates is paleontological—the earliest vertebrate fossils are found in marine sediments. The second line of evidence comes from comparative physiology—all nonvertebrate chordates and deuterostome invertebrate phyla are exclusively marine forms, with body fluids in the same concentration as their surroundings. Thus the concentrated body fluids of the very primitive hagfish most likely represent the original vertebrate condition.

3.2 Extant Jawless Fishes

The extant jawless vertebrates—hagfishes and lampreys—once were placed with the ostracoderms in the class "Agnatha" because they lack the gnathostome features of jaws and two sets of paired fins. They also have other primitive features, such as lack of specialized reproductive ducts.

However, it is now clear that the "Agnatha" is a paraphyletic assemblage, and that ostracoderms are actually more closely related to gnathostomes than are living jawless vertebrates. Hagfishes and lampreys have often been linked as cyclostomes because they have round, jawless mouths, but this grouping is also probably paraphyletic because lampreys appear to be more closely related to gnathostomes than are hagfishes (Figure 3–3 and **Figure 3–4**) (but see later discussion about controversies in their relationships).

Both hagfishes and lampreys appear to be more primitive than the armored ostracoderms of the Paleozoic, so we will look at them before considering the extinct agnathans. The fossil record of the modern types of jawless vertebrates is sparse. Lampreys are known from the Late Devonian period—*Priscomyzon*, a short-bodied form, from South Africa; the Late Carboniferous period—*Hardistiella* from Montana and *Mayomyzon* from Illinois; and the Early Cretaceous period—*Mesomyzon* from Southern China. All these fossil lampreys appear to have been specialized parasites like the living forms. *Myxinikela* (an undisputed hagfish) and a second possible hagfish relative,

▶ **FIGURE 3–4** Simplified cladogram of vertebrates within the Chordata. Only living taxa and major extinct groups are shown. Quotation marks indicate paraphyletic groups. A dagger indicates extinct groups.

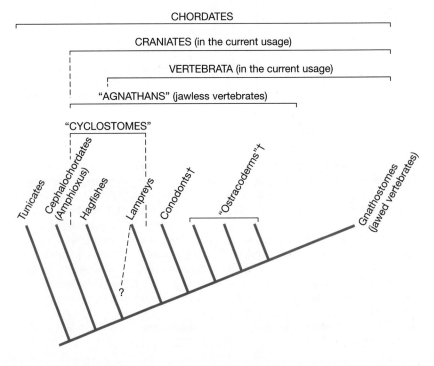

Gilpichthys, have been found in the same Carboniferous deposits as *Mayomyzon*.

■ Hagfishes—Myxinoidea

There are around 40 recognized species of hagfishes in two major genera (*Eptatretus* and *Myxine*). Adult hagfishes (**Figure 3–5**) are generally around half a meter in length, and they are elongated, scaleless, and pinkish to purple in color. Hagfishes are entirely marine, with a nearly worldwide distribution except for the polar regions. They are primarily deep-sea, cold-water inhabitants. They are the major scavengers of the deep-sea floor, drawn by their sense of smell to carcasses in large numbers.

A unique feature of hagfishes is the large mucus glands that open through the body wall to the outside. These so-called slime glands secrete enormous quantities of mucus and tightly coiled proteinaceous threads. The threads straighten on contact with seawater to entrap the slimy mucus close to the hagfish's body. An adult hagfish can produce enough slime within a few minutes to turn a bucket of water into a gelatinous mess (Martini 1998). This obnoxious behavior is apparently a deterrent to predators. When danger has passed, the hagfish makes a knot in its body and scrapes off the mass of mucus, then sneezes sharply to blow its nasal passage clear.

Hagfishes lack any trace of vertebrae, which is one reason they are placed as the sister group of all other vertebrates. Their internal anatomy shows many additional primitive features. For example, the kidneys are simple, and there is only one semicircular canal on each side of the head. Hagfishes have a single terminal nasal opening that connects with the pharynx via a broad tube. The eyes are degenerate or rudimentary and covered with a thick skin. The mouth is surrounded by six tentacles that can be spread and swept to and fro by movements of the head when the hagfish is searching for food. Two horny plates in the mouth bear sharp toothlike structures. These tooth plates lie to each side of a protrusible tongue and spread apart when the tongue is protruded. When the tongue is retracted, the plates fold together, and the teeth interdigitate in a pincerlike action. The feeding apparatus of hagfishes has been described as "extremely efficient at reeling in long worms, because the keratin plates alternately flick in and out of the oral cavity" (Mallatt 1985).

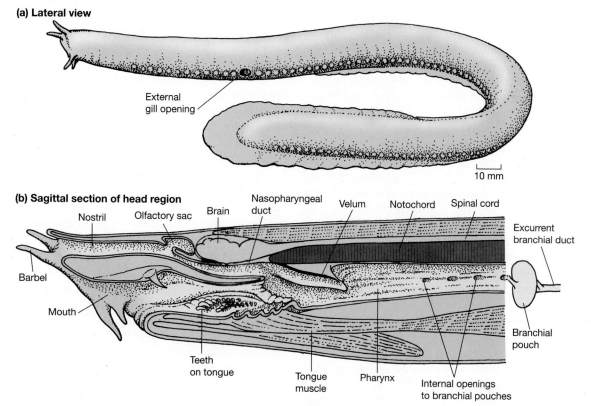

(a) Lateral view

External gill opening

10 mm

(b) Sagittal section of head region

Nostril · Olfactory sac · Brain · Nasopharyngeal duct · Velum · Notochord · Spinal cord · Excurrent branchial duct · Barbel · Mouth · Teeth on tongue · Tongue muscle · Pharynx · Internal openings to branchial pouches · Branchial pouch

▲ **FIGURE 3–5** Hagfishes.

Hagfishes attack dead or dying vertebrate prey. Once attached to the flesh, they can tie a knot in their tail and pass it forward along their body until they are braced against their prey and can tear off the flesh in their pinching grasp. They often begin by eating only enough outer flesh to enter the prey's coelomic cavity, where they dine on soft parts. Once a food parcel reaches the hagfish's gut, it is enfolded in a mucoid bag secreted by the gut wall. This membrane is permeable to digestive enzymes, and the products of digestion diffuse out to be absorbed by the gut. The indigestible parts of the prey are excreted still enclosed in the mucoid bag. The functional significance of this curious feature is unknown.

Different genera and species of hagfishes have different numbers of external gill openings. From 1 to 15 openings occur on each side. The external openings occur as far back as the midbody, although the pouchlike gill chambers are just posterior to the head. In some genera, the long tubes leading from the gills fuse, thus reducing the number of external openings. The posterior position of the gill openings may be related to burrowing.

Hagfishes have large blood sinuses and very low blood pressure. In contrast to all other vertebrates, hagfishes have accessory hearts in the liver and tail regions in addition to the true heart near the gills. These hearts are aneural, meaning that their pumping rhythm is intrinsic to the hearts themselves rather than coordinated via the central nervous system. In all these features, hagfishes resemble the primitive condition of amphioxus—although, like other vertebrates, their blood does have red blood cells containing hemoglobin, and the true heart has three chambers.

In most species, female hagfishes outnumber males by a hundred to one; the reason for this strange sex ratio is unknown. Examination of the gonads suggests that at least some species are hermaphroditic, but nothing is known of mating. The yolky eggs, which are oval and over a centimeter long, are encased in a tough, clear covering that is secured to the sea bottom by hooks. The eggs are believed to hatch into small, completely formed hagfishes, bypassing a larval stage. Unfortunately, almost nothing is known of the embryology or early life history of any hagfish because few eggs have been available for study. However, some hagfish embryos have recently been examined, and it has been determined that hagfishes possess neural crest like other vertebrates (Ota et al. 2007).

Hagfishes are never caught much above the bottom of the ocean, and they are often found in deep regions of the continental shelf. Some live in colonies, each individual in a mud burrow marked in some species by a volcano-like mound at the entrance. Hagfishes probably live a molelike existence, finding their prey beneath the ooze and at its surface. They must be active when out of their burrows, for they are quickly attracted to bait and fishes caught in gill nets. The small amount of morphological differences between populations indicates that hagfishes are not wide ranging, but rather tend to live and breed locally.

An increased economic interaction between hagfishes and humans over the past two decades is but one example of how vertebrate species are threatened by a burgeoning and highly consumptive human society. Fishermen using stationary gear such as gill nets have long been pulling in fishes damaged beyond sale by scavenging hagfishes. It is not surprising that the fishermen responded quickly when the use of hagfish skins for leather made hagfishes as valuable as the fish they had previously been seeking. Almost all so-called eel-skin leather products are made from hagfish skin. Worldwide demand for this leather has eradicated economically harvestable hagfish populations in Asian waters and in some sites along the West Coast of North America. Current fishing efforts are focusing on South America and the North Atlantic, where hagfishes may still be abundant. By one estimate, the Gulf of Maine contains a population density of hagfishes of 500,000 per square kilometer (Martini 1998).

Human exploitation of natural resources, such as fisheries, typically depletes stocks because no attention is given to the biology of the resource and its renewable, sustainable characteristics. For example, we do not know how long hagfishes live; how old they are when they first begin to reproduce; exactly how, when, or where they breed; where the youngest juveniles live; what are the diets and energy requirements of free-living hagfishes; or virtually any of the other information needed for good management. As a result, eel-skin wallets will probably become as rare as items made of whalebone (baleen), tortoiseshell, and ivory.

■ Lampreys—Petromyzontoidea

There are around 40 species of lampreys in two major genera (*Petromyzon* and *Lampetra*). Although lampreys are similar to hagfishes in size and shape (**Figure 3–6**), they are radically different in certain other respects. They have many features lacking in hagfishes but shared with gnathostomes. Perhaps

(a) Lateral view of an adult

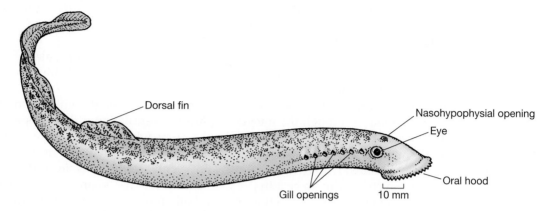

Dorsal fin

Nasohypophysial opening

Eye

Oral hood

Gill openings

10 mm

(b) Sagittal section of head region

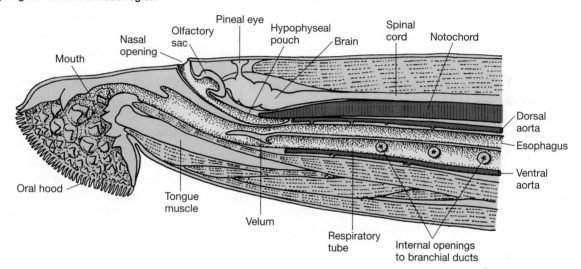

Pineal eye

Olfactory sac

Hypophyseal pouch

Nasal opening

Brain

Spinal cord

Notochord

Mouth

Dorsal aorta

Esophagus

Ventral aorta

Oral hood

Tongue muscle

Velum

Respiratory tube

Internal openings to branchial ducts

(c) Larval lamprey (ammocoete)

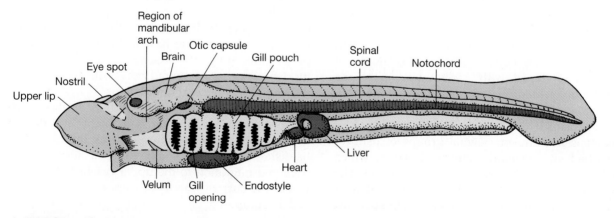

Region of mandibular arch

Otic capsule

Brain

Gill pouch

Spinal cord

Notochord

Eye spot

Nostril

Upper lip

Velum

Gill opening

Endostyle

Heart

Liver

▲ **FIGURE 3–6** Lampreys.

most important, they have vertebral structures (arcualia)—although these cartilaginous skeletal elements are minute and homologous only with the neural arches of gnathostome vertebrae.

Most lampreys are parasitic on other fishes, although some small, freshwater species have non-feeding adults. They attach to the body of another vertebrate (usually a larger bony fish) by suction and rasp a shallow, seeping wound through the integument of the host. The round mouth is located at the bottom of a large fleshy funnel (the oral hood), the inner surface of which is studded with horny conical spines. The oral hood, which appears to be a hypertrophied upper lip, is a unique derived structure in lampreys. The protrusible tonguelike structure is covered with similar spines, and together these structures allow tight attachment and rapid abrasion of the host's integument. This tongue is not homologous with the tongue of gnathostomes because it is innervated by a different cranial nerve (the trigeminal nerve, V, rather than the hypoglossal nerve, XII). An oral gland secretes an anticoagulant that prevents the victim's blood from clotting. Feeding is probably continuous when a lamprey is attached to its host. The bulk of an adult lamprey's diet consists of body fluids of fishes. The digestive tract is straight and simple, as one would expect for an animal that feeds on such a rich and easily digested diet as blood and tissue fluids. Lampreys generally do not kill their hosts but do leave a weakened animal with an open wound. At sea, lampreys feed on several species of whales and porpoises in addition to fishes. Swimmers in the Great Lakes, after having been in the water long enough for their skin temperature to drop, have reported initial attempts by lampreys to attach to their bodies.

Lampreys are unique among living vertebrates in having a single nasal opening situated on the top of the head, combined with a duct leading to the hypophysis (pituitary) and known as a nasohypophysial opening. Development of this structure involves distortion of the front of the head, and its function is not known. Several groups of ostracoderms had an apparently similar structure, which may have evolved convergently in those groups. The eyes of lampreys are large and well developed, as is the pineal body, which lies under a pale spot just posterior to the nasal opening. In contrast to hagfishes, lampreys have two semicircular canals on each side of the head—a condition shared with the extinct ostracoderms. In addition, the heart is innervated by the parasympathetic nervous system (the vagus nerve, X). Chloride cells in the gills and well-developed kidneys regulate ions, water, and nitrogenous wastes, as well as overall concentration of body fluids, allowing the lamprey to exist in a variety of salinities.

Lampreys have seven pairs of gill pouches that open to the outside just behind the head. In all other fishes, water is drawn into the mouth and then pumped out over the gills in continuous or **flow-through ventilation**. Adult lampreys spend much of their time with their suckerlike mouths affixed to the bodies of other fishes, and during this time they cannot ventilate the gills in a flow-through fashion. Instead they use in-and-out **tidal ventilation**: water is both drawn in and expelled through the gill slits. A flap called the velum prevents water from flowing out of the respiratory tube into the mouth. The lamprey's mode of ventilation is not very efficient at oxygen extraction; however, it is a necessary compromise given their specialized parasitic mode of feeding.

Lampreys are primarily found in northern latitude temperate regions, although a few species are known from southern temperate latitudes. Nearly all lampreys are **anadromous**; that is, they live as adults in oceans or big lakes and ascend rivers and streams to breed. Some of the most specialized species live only in freshwater, and do all of their feeding as larvae. The adults act solely as a reproductive stage in the life history of the species. Anadromous species that spend some of their life in the sea attain the greatest size, although 1 meter is about the upper limit. The smallest species are less than one-fourth that size. Little is known of the habits of adult lampreys because they are generally observed only during reproductive activities or when captured with their host. Despite their anatomically well-developed senses, no clear picture has emerged of how a lamprey locates or initially attaches to its prey. In captivity, lampreys swim sporadically with exaggerated, rather awkward lateral undulations.

Female lampreys produce hundreds to thousands of eggs, about a millimeter in diameter and devoid of any specialized covering such as that found in hagfishes. Lampreys spawn after a temperature-triggered migration to the upper reaches of streams where the current flow is moderate and the stream bed is composed of cobbles and gravel. Male and female lampreys construct a nest by attaching themselves by their mouths to large rocks and thrashing about violently. Smaller rocks are dislodged and carried away by the current. The nest is complete when a pit is rimmed upstream by large stones, downstream by a mound of smaller stones and sand that produces eddies. Water in the nest is oxygenated by this turbulence but does not flow strongly in a single direction. The female attaches to one of the upstream rocks, laying eggs, and the male wraps around her, fertilizing them as they are extruded, a process that may take two days. Adult lampreys die after breeding once.

The larvae hatch in about two weeks. The larvae are radically different from their parents and were originally described as a distinct genus, *Ammocoetes* (see Figure 3–6c). This name has been retained as a vernacular name for the larval form. A week to ten days after hatching, the tiny 6- to 10-millimeter-long ammocoetes leave the nest. They are wormlike organisms with a large, fleshy oral hood and non-functional eyes hidden deep beneath the skin. Currents carry the ammocoetes downstream to backwaters and quiet banks, where they burrow into the soft mud or sand and spend three to seven years as sedentary filter feeders. The protruding oral hood funnels water through the muscular pharynx, where food particles are trapped in mucus and carried to the esophagus. An ammocoete may spend its entire larval life in the same bed of sediment, with no major morphological or behavioral change until it is 10 centimeters or more in length and several years old. Metamorphosis begins in midsummer and produces a silver-gray juvenile ready to begin its life as a parasite. Downstream migration to a lake or the sea may not occur until the spring following metamorphosis. Adult life is usually no more than two years, and many lampreys return to spawn after one year. During the past 100 years, humans and lampreys have increasingly been at odds. Although the sea lamprey, *Petromyzon marinus*, seems to have been indigenous to Lake Ontario, it was unknown from the other Great Lakes of North America before 1921. The St. Lawrence River—flowing from Lake Ontario to the Atlantic Ocean—was no barrier to colonization by sea lampreys, and the rivers and streams that fed into Lake Ontario held landlocked populations. During their spawning migrations, lampreys negotiate waterfalls by slowly creeping upward using their sucking mouth, but the 50-meter height of Niagara Falls (between Lake Ontario and Lake Erie) was too much for even the most amorous lampreys. Even after the Welland Canal connected Lakes Erie and Ontario in 1829, lampreys did not immediately invade Lake Erie; it took a century for lampreys to establish themselves in Lake Erie's drainage basin.

From the 1920s to the 1950s, lampreys expanded rapidly across the entire Great Lakes basin. Once they reached the upper end of Lake Erie, lampreys quickly gained access to the other lakes. There they found suitable conditions, and by 1946 they inhabited all the Great Lakes. Lampreys were able to expand unchecked until sporting and commercial interests became alarmed at the reduction of economically important fish species, such as lake trout, turbot, and lake whitefish. Chemical lampricides as well as electrical barriers and mechanical weirs at the mouths of spawning streams have been employed to bring the Great Lakes lamprey populations down to their present level. Although the populations of jawed fish species, including those of commercial value, are recovering, it may never be possible to discontinue these costly antilamprey measures. Human mismanagement (or initial lack of management) of lampreys has been to our own disadvantage. The story of the demise of the Great Lakes fishery is but one of hundreds in the recent history of vertebrate life in which human failure to understand and appreciate the interlocking nature of the biology of our nearest relatives has led to gross changes in our environment. Introduction of exotic (not indigenous) species is a primary cause for the decline of many vertebrate species worldwide, especially in aquatic habitats.

■ The Importance of Extant Primitive Vertebrates in Understanding Ancient Ones

The fossil record of the first vertebrates reveals little about their pre-Silurian evolution, and it yields no undisputed clues about the evolution of vertebrate structure from the condition in nonvertebrate chordates. However, hagfishes and lampreys provide examples of surviving primitive vertebrates, representatives of the early agnathous radiation.

Although the biology and development of hagfishes is less well known than that of lampreys, hagfishes are especially important to our understanding of the earliest vertebrate condition. Hagfishes retain more primitive features than any other known vertebrate, living or fossil, and they have sometimes been excluded from the vertebrates. Recent work on hagfish embryos has confirmed the presence of true vertebrate neural crest tissue (Ota et al. 2007), confirming their status as vertebrates.

The majority of molecular studies link hagfish and lampreys as sister taxa, but a confounding problem here is that we cannot get molecules from primitive nonvertebrate chordates (such as the Cambrian *Haikouella*) or from early vertebrates (such as the Cambrian vertebrate *Haikouichthys* and the huge diversity of ostracoderms). Such missing data can bias the computer programs that create the phylogenies from the various characters (e.g., gene sequences), grouping taxa by a statistical artifact known as "long branch attraction" that does not represent the true phylogenetic relationship. We do not claim here that the molecular findings in the case of lampreys and hagfishes are artifacts, but the difference in the results obtained from morphological data versus molecular data is troubling.

Table 3.1 describes some morphological features of hagfishes that are more primitive than those of other living vertebrates, including lampreys. *Myxine* and its relatives appear to be considerably more derived than other hagfishes, and comparative studies offer the most promising path to understanding what is primitive about hagfishes and what is not. Many apparently primitive features of hagfishes, such as their virtual lack of eyes, are probably secondary features associated with mud-burrowing habits, but two features of the physiology of hagfishes suggest that they occupy a very primitive phylogenetic position. First, hagfish body fluids have the same concentration as seawater, whereas all other living craniates have dilute body fluids. Body fluids as concentrated as seawater would preclude survival in freshwater

habitats, but all major groups of ostracoderms had freshwater representatives—implying that they had dilute body fluids.

A second primitive feature is the taste sensory system. In lampreys and gnathostomes, the taste buds are in the oropharyngeal epithelium and innervated by cranial nerves VII, IX, and X. In hagfishes, however, specialized sensory buds are found in the epidermis and are innervated mainly by cranial nerve V and by spinal nerves, and the brain area that receives input from the taste sensors is also different from other chordates (Braun 1996). This implies that the hagfishes' system evolved independently from the sense of taste in lampreys and gnathostomes, suggesting an early divergence for the hagfish lineage.

While the majority of morphologists would disagree with molecular biologists and support the notion of the primitive nature of the hagfishes, some morphologists do argue for cyclostome monophyly. Jon Mallatt (1996, Mallatt and Sullivan 1998) notes in both hagfishes and lampreys the similarity of the elongated pharyngeal cavity, housing a powerful muscular tongue supported by a cartilaginous lingual apparatus, and a respiratory flap called the velum. An alternative interpretation is that these oral features perhaps represent retained primitive features

---

TABLE 3.1 Features of hagfishes that are more primitive than those of lampreys and gnathostomes

1. Have very small, paired eyes, and lack a pineal eye; extrinsic eye muscles and associated cranial nerves to move the eyes are lacking (but could represent secondary loss due to burrowing lifestyle).

2. Cranium made of a sheath of fibrous tissue rather than of cartilage.

3. Lack electroreception and apparently lack lateral line sensory system, but canals without neuromasts are present in the heads of some species.

4. A single semicircular canal only (organs of balance, one on each side of the head) in the inner ear (versus two in lampreys and ostracoderms, and three in gnathostomes)

5. Lack a hypoglossal nerve (carries motor neurons to tongue), Mauthner neurons in brain stem (responsible for the rapid "startle response"), and autonomic innervation of the heart.

6. System of taste reception is different from other vertebrates.

7. Circulatory system retains accessory "hearts," and only one type of granular white blood cell (three types in lampreys and gnathostomes).

8. Lack any type of spleen or splenic precursor.

9. Lack cartilaginous vertebral elements and muscles in the midline fins.

10. Lack renal collecting ducts in the kidney.

11. Have an open connection between the pericardial cavity and the coelom (closed in lampreys and gnathostomes).

12. Body fluids have the same concentration as seawater.

inherited from the earliest vertebrates. This appears to be the case for the pouched gills, which are shared with the more primitive ostracoderms. Alternatively, the features noted by Mallatt may have evolved convergently in the two different lineages, in association with the acquisition of an eel-like body shape. This elongated shape in turn may be related to the locomotor demands of active swimming at a relatively large size (i.e., considerably larger than amphioxus) without paired fins or a dermal shield to stabilize the head end of the animal.

3.3 The Radiation of Paleozoic Jawless Vertebrates— "Ostracoderms"

Ostracoderms encompass several distinct lineages, but our understanding of exactly how these different lineages are related to one another and to living vertebrates has changed considerably over the past decade (Janvier 1996, Maisey 1996, Donoghue et al. 2000). Most ostracoderms are characterized by the presence of a covering of dermal bone, usually in the form of an extensive armored shell or **carapace**, but sometimes in the form of smaller plates or scales (e.g., anaspids), and some are relatively naked (e.g., thelodonts).

The ostracoderms represent a paraphyletic assemblage because some more derived types are clearly more closely related to the gnathostomes (jawed vertebrates) than others (see Figure 3–3). Ostracoderms are more derived than extant agnathans: they had dermal bone, and impressions on the underside of the dorsal head shield suggest that they had the derived (i.e., more gnathostome-like) features of a cerebellum in the hindbrain and an olfactory tract connecting the olfactory bulb with the forebrain. Living agnathans lack a cerebellum, and their olfactory bulbs are incorporated within the rest of the forebrain rather than placed more anteriorly and linked to the head via the olfactory tract (= cranial nerve I).

Ostracoderms ranged in length from about 10 centimeters to more than 50 centimeters. Although they lacked jaws, some apparently had various types of movable mouth plates that lack analogs in any living vertebrates. These plates were arranged around a small, circular mouth that appears to have been located farther forward in the head than the larger, more gaping mouth of jawed vertebrates. They were once thought to be filter feeders, but most species

probably ate small, soft-bodied prey. Although most ostracoderms had some sort of midline dorsal fin, and while many heterostracans and anaspids had some sort of anterior, paired, finlike projections, only the more derived osteostracans had true pectoral fins, with an accompanying pectoral girdle and endoskeletal fin supports (Coates 2003). As in living jawless vertebrates, the notochord must have been the main axial support throughout adult life. **Figure 3–7** depicts some typical ostracoderms.

During the Late Silurian and the Devonian, most major known groups of extinct agnathans coexisted with early gnathostomes. Approximately 50 million years of coexistence makes it highly unlikely that ostracoderms were pushed into extinction by the radiation of gnathostomes. Agnathous and gnathostomatous vertebrates appear to represent two different basic types of animals, possibly exploiting different types of resources. The initial reduction of ostracoderm diversity at the end of the Early Devonian may be related to a lowering of global sea levels, with the resulting loss of coastal marine habitats. The extinction of the ostracoderms in the Late Devonian occurred at the same time as mass extinctions among many marine invertebrates and was probably a chance event. Gnathostomes also suffered in the Late Devonian mass extinctions, and an entire lineage (the placoderms) became extinct at the end of the Devonian.

3.4 The Basic Gnathostome Design

Gnathostomes were once viewed as an entirely separate radiation from the agnathans, but we now consider that they originated from within the agnathan radiation. Gnathostomes are considerably more derived than agnathans, not only in their possession of jaws but also in many other ways.

Jaws allow a variety of new feeding behaviors, including the ability to grasp objects firmly, and along with teeth enable the animal to cut food to pieces small enough to swallow or to grind hard foods. New food resources became available when vertebrates evolved jaws: Herbivory was now possible, and many gnathostomes became larger than contemporary jawless vertebrates. A grasping, movable jaw also permits manipulation of objects: jaws are used to dig holes, to carry pebbles and vegetation to build nests, and to grasp mates during courtship and juveniles during parental care. Gnathostomes are first known with certainty from the Early Silurian, but isolated sharklike scales suggest that they date

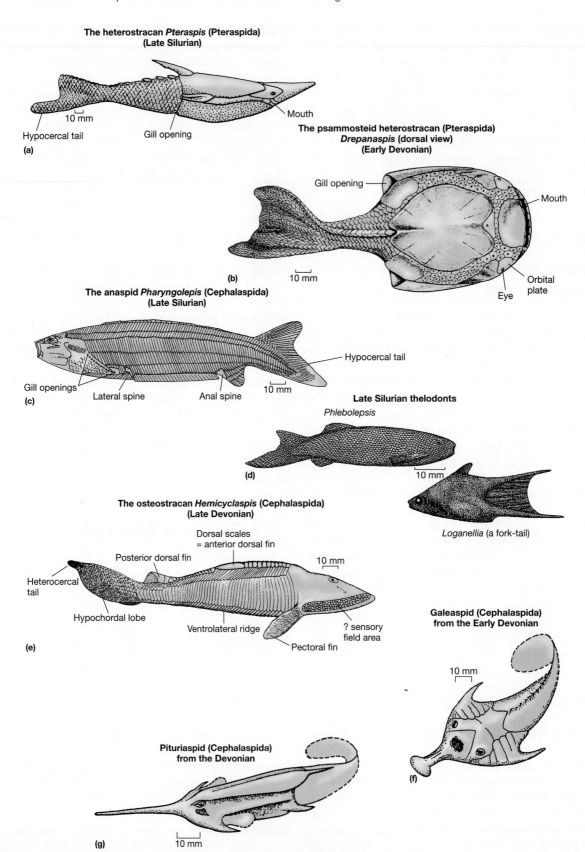

The heterostracan *Pteraspis* (Pteraspida)
(Late Silurian)

10 mm

Hypocercal tail
Gill opening
Mouth
(a)

The psammosteid heterostracan (Pteraspida)
Drepanaspis (dorsal view)
(Early Devonian)

Gill opening
Mouth
Orbital plate
Eye
(b)
10 mm

The anaspid *Pharyngolepis* (Cephalaspida)
(Late Silurian)

Hypocercal tail
Gill openings
Lateral spine
Anal spine
10 mm
(c)

Late Silurian thelodonts

Phlebolepis

10 mm
(d)

Loganellia (a fork-tail)

The osteostracan *Hemicyclaspis* (Cephalaspida)
(Late Devonian)

Dorsal scales
= anterior dorsal fin
Posterior dorsal fin
10 mm
Heterocercal tail
Hypochordal lobe
Ventrolateral ridge
? sensory field area
Pectoral fin
(e)

Galeaspid (Cephalaspida)
from the Early Devonian

10 mm
(f)

Pituriaspid (Cephalaspida)
from the Devonian

(g)
10 mm

▲ **FIGURE 3–7** Ostracoderm diversity. Pteraspida and common Cephalaspida.

back to the Middle Ordovician. The difference between gnathostomes and agnathans is traditionally described as the possession of jaws that bear teeth in most forms and two sets of paired fins or limbs (pectoral and pelvic). The gnathostome body plan (**Figure 3–8**) reveals that they are characterized by many other features, which imply that gnathostomes represent a basic step up in level of activity and complexity from the jawless vertebrates. These features include improvements in locomotor and predatory abilities and in the sensory and circulatory systems. Just as vertebrates show a duplication of the *Hox* gene complex in comparison to nonvertebrate chordates, living jawed vertebrates show evidence of a second duplication event. Gene duplication would have resulted in a greater amount of genetic material, perhaps necessary for building a more complex type of animal. However, Donoghue and Purnell (2006) advise caution about making such assumptions about the relationship between genetic complexity and morphological complexity based on living animals alone. As can be seen in Figure 3–3, a number of extinct ostracoderm taxa lie between living

agnathans and gnathostomes, and many features seen today only in gnathostomes may have been acquired in a more steplike fashion throughout ostracoderm evolution.

Extant gnathostomes have teeth on their jaws, but teeth must have evolved after the jaws were in place because early members of the most primitive jawed fish—the placoderms, which we will encounter later in this chapter—do not have teeth (see Section 3.5). Curiously, however, more derived placoderms seem to have evolved teeth independently of the gnathostome condition (Smith and Johanson 2003). Another reason to dissociate the evolution of jaws from the evolution of teeth is the presence of pharyngeal toothlike structures in various jawless vertebrates. These are most notable in the conodonts previously discussed (see Figure 3–2) but also have been reported in thelodonts, poorly known ostracoderms that lack a well-mineralized skeleton. Developmental studies support the notion that the first teeth were not oral structures but were embedded in the pharynx (McCollum and Sharpe 2001).

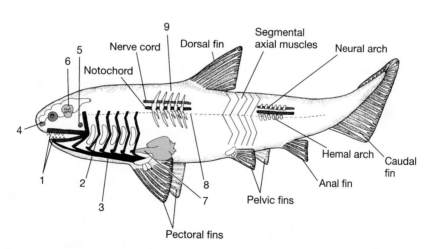

▲ **FIGURE 3–8** Generalized jawed vertebrate (gnathostome) showing derived features compared to the jawless vertebrate (agnathan) condition.

Legend: 1. Jaws (containing teeth) formed from the mandibular gill arch. **2.** Gill skeleton consists of jointed branchial arches and contains internal gill rakers that stop particulate food from entering the gills. Gill musculature is also more robust. **3.** Hypobranchial musculature allows strong suction in inhalation and suction feeding. **4.** Two distinct olfactory tracts, leading to two distinct nostrils. **5.** Original first gill slit squeezed to form the spiracle, situated between mandibular and hyoid arches. **6.** Three semicircular canals in the inner ear (addition of horizontal canal). **7.** Addition of a conus arteriosus to the heart, between the ventricle and the ventral aorta. (Note that the position of the heart is actually more anterior than shown here, right behind the most posterior gill arch.) **8.** Horizontal septum divides trunk muscle into epaxial (dorsal) and hypaxial (ventral) portions. It also marks the position of the lateral line canal, containing the neuromast sensory organs. **9.** Vertebrae now have centra (elements surrounding the notochord) and ribs, but note that the earliest gnathostomes have only neural and hemal arches, as shown in the posterior trunk.

Bony fishes and tetrapods have teeth embedded directly into the jawbones (**Figure 3–9**). However, because teeth form from a dermal papilla, they can be embedded only in dermal bones. Cartilaginous fishes such as sharks and rays lack dermal bones, and their teeth form within the skin, resulting in a tooth whorl that rests on the jawbone but is not actually embedded in it. This condition is probably the primitive one for all gnathostomes more derived than placoderms because it is also seen in the extinct acanthodians, which are the sister group of bony fishes (see Section 3.5).

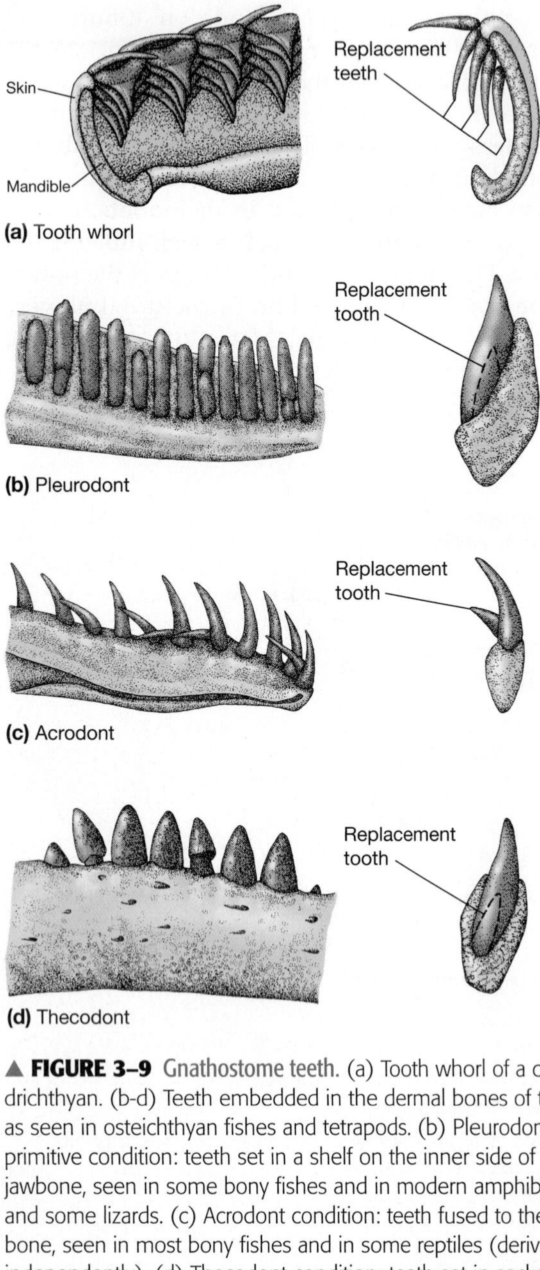

Adding jaws and hypobranchial muscles (which are innervated by spinal nerves) to the existing branchiomeric muscles (innervated by cranial nerves) allowed vertebrates to add powerful suction to their feeding mechanisms (**Figure 3–10a**). Another gnathostome feature is the presence of two distinct olfactory tracts leading to two distinct nostrils. Agnathans usually have only a single nasal opening, either at the front of the head (as in hagfishes) or on the top of the head (as in lampreys and Cephalapsida). (Heterostracans also had two nostrils, but this condition was not identical to the gnathostome one and was independently derived from it.) The cranium of gnathostomes has also been elongated both anteriorly and posteriorly over the agnathan condition.

Progressively more complex vertebrae are also a gnathostome feature. Vertebrae initially consisted of arches flanking the nerve cord (= neural arches, homologous with the arcualia of lampreys) with matching arches below the notochord (= hemal arches, which may be present in the tail only). (See the posterior portion of the trunk in Figure 3–8.) More derived gnathostomes had a vertebral centrum or central elements with attached ribs (**Figure 3–10b**). Still more complete vertebrae support the notochord and eventually replace it as a supporting rod for the axial muscles used for locomotion (mainly in tetrapods). Note that well-developed centra were not a feature of the earliest jawed fishes and were never seen in the two extinct groups of fishes—placoderms and acanthodians.

Ribs are another new feature in gnathostomes. They lie in the connective tissue between successive segmental muscles, providing increased anchorage for axial muscles. There is now a clear distinction between the epaxial and hypaxial blocks of the axial muscles, divided by a horizontal septum made of thin fibrous tissue that runs the length of the animal. The lateral line canal—containing the organs that sense vibrations in the surrounding water—lies in the plane of this septum, perhaps reflecting improved integration between locomotion and sensory feedback. In the inner ear there is a third (horizontal) semicircular canal, which may reflect an improved ability to navigate and orient in three dimensions.

Features such as jaws and ribs can be observed in fossils, so we know that they are unique to gnathostomes. However, we cannot know for sure whether

▲ **FIGURE 3–9** Gnathostome teeth. (a) Tooth whorl of a chondrichthyan. (b-d) Teeth embedded in the dermal bones of the jaw as seen in osteichthyan fishes and tetrapods. (b) Pleurodont, the primitive condition: teeth set in a shelf on the inner side of the jawbone, seen in some bony fishes and in modern amphibians and some lizards. (c) Acrodont condition: teeth fused to the jawbone, seen in most bony fishes and in some reptiles (derived independently). (d) Thecodont condition: teeth set in sockets and held in place by peridontal ligaments, seen in archosaurian reptiles and mammals (derived independently).

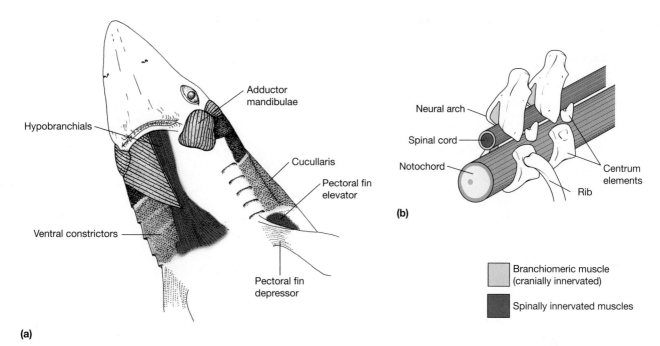

(a)

(b)

▲ **FIGURE 3–10** Some gnathostome specializations. (a) Ventral view of a dogfish, showing both branchiomeric muscles and the new, spinally innervated, hypobranchial muscles. (b) View of generalized gnathostome vertebral form, showing central elements and ribs.

other new features within the soft anatomy characterize gnathostomes alone or whether they were adopted somewhere within the ostracoderm lineage. We can surmise that some features of the nervous system that are seen only in gnathostomes among living vertebrates were acquired by the earliest ostracoderms. For example, impressions on the inner surface of the dermal head shield reveal the presence of a cerebellum in the brain and an olfactory tract, although only gnathostomes have two distinct olfactory tracts leading to widely separated olfactory bulbs. We can also see that no ostracoderm possessed the third semicircular canal.

Other new features of the nervous system in extant gnathostomes include insulating sheaths of myelin on the nerve fibers that increase the speed of nerve impulses (although, as discussed later, myelin sheaths may be absent in placoderms). The heart of gnathostomes has an additional small chamber in front of the pumping ventricle, the **conus arteriosus**, which acts as an elastic reservoir. Its presence is probably due to the strong ventricular pumping and high blood pressures of gnathostomes and the need to smooth out the pulsatile nature of the flow of blood. However, it has been determined that some ostracoderms (thelodonts) possessed a stomach, seen only in jawed vertebrates among living forms. **Table 3.2** summarizes derived features of living gnathostomes, most of which are undeterminable in fossils.

The Origin of Fins

Guidance of a body in three-dimensional space is complicated. Fins act as hydrofoils, applying pressure to the surrounding water. Because water is practically incompressible, force applied by a fin in one direction against the water produces a thrust in the opposite direction. A tail fin increases the area of the tail, giving more thrust during propulsion. Rapid adjustments of the body position in the water may be especially important for active, predatory fishes like the early gnathostomes, and the unpaired fins in the midline of the body (the dorsal and anal fins) control the tendency of a fish to roll (rotate around the body axis) or yaw (swing to the right or left) (**Figure 3–11**). The paired fins (pectoral and pelvic fins) can control the pitch (tilt the fish up or down) and act as brakes, and they are occasionally specialized to provide thrust during swimming (as with the enlarged pectoral fins of skates and rays).

Fins have nonlocomotor functions as well. Spiny fins are used in defense, and they may become systems

TABLE 3.2 Derived features of gnathostomes (in addition to those depicted in figure 3–8)

Cranial Characters

1. Cranium enlarged anteriorly to the end in a precerebral fontanelle.

2. Cranium elongated posteriorly, so that one or more occipital neural arches are incorporated in the rear of the skull.

3. Development of a postorbital process on the cranium, separating the functions of supporting the jaws and enclosing the eyes.

4. Intrinsic musculature in the eye for focusing the lens.

Internal Anatomical Characters

5. Atrium lies posterodorsally (versus laterally) to ventricle.

6. Renal portal vein present.

7. Spiral valve primitively formed within intestine (increases available area for food absorption).

8. Pancreas with both endocrine and exocrine functions.

9. Stomach and a distinct spleen.

10. Male gonads linked by ducts to excretory (archinephric) duct; female gonads with distinct oviducts.

11. Two (versus one) contractile actin proteins (one specific to striated muscle and one specific to smooth muscle).

12. Cartilage based in a proteinaceous matrix of collagen.

Sensory Characters

13. Nerves enclosed in myelinated sheaths.

14. Large, distinct cerebellum in the hindbrain.

15. Thicker spinal cord with "horns" of gray matter in cross-section.

16. Dorsal and ventral spinal nerve roots linked to form compound spinal nerves.

17. Unique and evolutionarily conservative pattern of head lateral line canals (lost in some adult amphibians and amniotes).

18. Lateral line on trunk region flanked or enclosed by specialized scales (lateral line lost in some adult amphibians and amniotes).

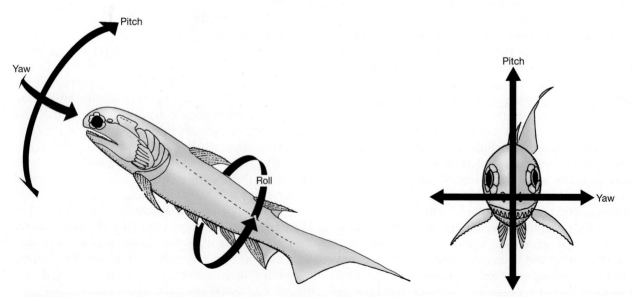

▲ **FIGURE 3–11** A primitive jawed fish (the acanthodian, *Climatius*). Views from the side and front illustrate pitch, yaw, and roll and the fins that counteract these movements.

to inject poison when combined with glandular secretions. Colorful fins are used to send visual signals to potential mates, rivals, and predators.

Even before the gnathostomes appeared, fish had structures that served the same purpose as fins. Many ostracoderms had spines or enlarged scales derived from dermal armor that acted like immobile fins. Some anaspids had long finlike sheets of tissue running along the flanks, and osteostracans had pectoral fins.

Most gnathostome fishes have a well-developed heterocercal caudal fin. A notochord that turns abruptly up or down increases the depth of the caudal fin and allows it to exert the force needed for rapid acceleration. All fishes with a fin-strengthening upturned or downturned axial skeleton tip have a noncollapsible caudal fin that is effective for burst swimming. Burst swimming is important in predator avoidance and can save energy when bursts of acceleration are alternated with glides.

3.5 The Transition from Jawless to Jawed Vertebrates

In Chapter 2 we saw that the branchial arches were a fundamental feature of the vertebrate cranium, providing support for the gills. It has long been known that vertebrate jaws are made of the same material as the skeletal elements that support the gills (cartilage derived from the neural crest), and they clearly develop from the first (mandibular) arch of this series in vertebrates. (Note that while the word "mandible" commonly refers to the lower jaw only, the mandibular arch encompasses both upper and lower jaws.)

It is helpful at this stage to envisage the vertebrate head as a segmented structure, with each branchial arch corresponding to a segment (**Figure 3–12**). The mandibular arch that forms the gnathostome jaw is actually formed within the second head segment; jaw supports are formed from the hyoid arch of the third head segment; and the more posterior branchial

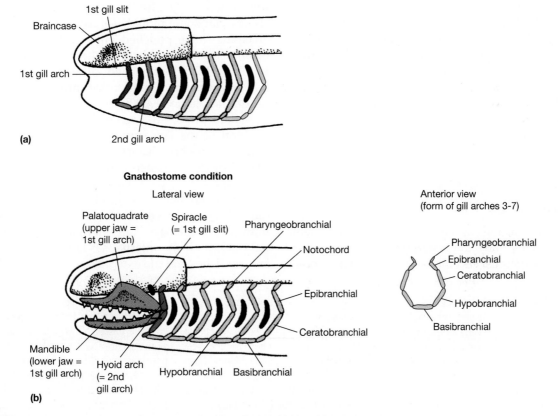

▲ **FIGURE 3–12** Evolution of the vertebrate jaw from anterior branchial arches. Blue shading indicates splanchnocranium elements (branchial arches and their derivatives).

arches that form the gill supports (arches 3 through 7) are formed in head segments 4 through 8. The branchial arch numbering does not match the numbering of the head segments because no evidence exists for a branchial arch functioning as a gill support structure in the first (premandibular) segment at any point in vertebrate history. However, some form of homolog to the pharyngeal arch equivalent for the first head segment may be found in the anterior trabeculae that underlie the front part of the braincase in gnathostomes (see Figure 2–8).

No living vertebrate has a pair of gill-supporting branchial arches in such an anterior position as the jaws. However, all vertebrates have some structure in this position (i.e., in the second head segment) that appears to represent the modification of an anterior pair of pharyngeal arches; these are the jaws in gnathostomes and velar (= supporting the velum) cartilages in lampreys and hagfishes. Thus it has been proposed that the common ancestor of living vertebrates had an unmodified pair of branchial arches in this position, with a fully functional gill slit lying between the first and second arches, and that living jawless and jawed vertebrates are both divergently specialized from this condition. There is no trace of a gill slit between arches 1 and 2 in living jawless vertebrates; but, in many living cartilaginous fishes, there is a small hole called the spiracle in this position, which is now used for water intake. Figure 3–12 summarizes the differences in the gill arches between jawed vertebrates and their presumed jawless ancestor and illustrates the major components of the hinged gnathostome gill arches (as seen in arches 3 through 7).

A Problem Posed by the Gills of Early Vertebrates

Because ostracoderms are now viewed as "stem gnathostomes," any understanding of the origin of gnathostomes must encompass the view that at some point a jawless vertebrate was transformed into a jawed one. However, for much of the last century, researchers considered jawless and jawed vertebrates as two separate evolutionary radiations. This was because of the apparent nonhomology of their branchial arches: at least in living jawless vertebrates (the situation is less clear for the fossil ones), the gill arches lie lateral to the gill structures, while in jawed vertebrates they lie medially (i.e., internal to the gills). Additionally extant jawless vertebrates have pouched gills with small, circular openings that are different from the flatter, more lens-shaped openings between the gills of gnathostomes. These differences in structure were once interpreted as evidence that jawless and jawed vertebrates represent two separate radiations from an originally gill-less ancestor.

This issue was addressed by Jon Mallatt (1985, 1996, 1998). He noted that the detailed structure of lamprey gills was actually more similar to the condition in sharks than the condition in hagfishes (**Figure 3–13**), implying a commonality of gill anatomy between gnathostomes and at least lampreys among jawless vertebrates. (The different gill anatomy in hagfishes is puzzling and may represent a highly specialized condition rather than the primitive one.) Note also that fossil record evidence shows that some more basal ostracoderms (such as anaspids) have pouched gills while more derived ones (osteostracans) had gnathostome-like slit gills, implying that the pouched gill condition is primitive and was later transformed into the slit gill condition (Janvier et al. 2006).

Mallatt proposed a reconciliation of the different position of the jaw supports as follows: hagfishes and lampreys both have some evidence of internally placed brachial elements in the form of velar cartilages, and some living sharks have evidence of small external branchial cartilages (see Figure 3–13). Perhaps the original vertebrate condition was to have both sets of cartilages (an internal set and an external one), with the internal set largely lost in living jawless fishes and the external set lost in jawed ones.

▶ **FIGURE 3–13** Jon Mallatt's scenario of the evolution of vertebrate gills and jaws. Animals are portrayed in ventral view (a = afferent artery carrying deoxygenated blood to the gills; e = efferent artery carrying oxygenated blood away from the gills). Arrows show the direction of water flow. (a) Probable ancestral vertebrate condition, with seven unmodified branchial arches (internal and external) and six pouched gills. (b) Hagfishes. Branchial arches 1–4 form the velum; gill openings 1–3 lost; additional gill pouches and (external) branchial arches added posteriorly. Note the difference from all other vertebrates in the position of the afferent artery and the respiratory filaments, and the fact that water passes within the gill pouch rather than between pouches. (c) Lamprey. First branchial arch (internal) modified to form velum; first gill opening lost; additional gill pouches and arches added posteriorly; external branchial arches form a complex branchial basket. (d) Hypothetical gnathostome ancestor. Gills now slitlike rather than pouched; internal branchial arches enlarged. (e) Early true gnathostome. Internal branchial arches 1 (mandibular) and 2 (hyoid) form the jaws and the jaw supports; first gill opening becomes the spiracle; external branchial arches 1–2 lost, 3–7 reduced.

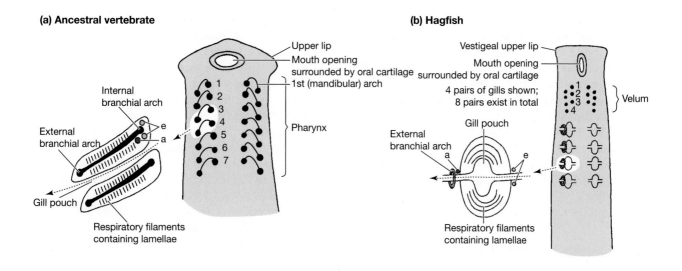

(a) Ancestral vertebrate

Upper lip
Mouth opening
surrounded by oral cartilage
1st (mandibular) arch

Internal
branchial arch

External
branchial arch

e
a

Gill pouch

Respiratory filaments
containing lamellae

Pharynx

1
2
3
4
5
6
7

(b) Hagfish

Vestigeal upper lip
Mouth opening
surrounded by oral cartilage

4 pairs of gills shown;
8 pairs exist in total

Gill pouch

External
branchial arch

a

e

Velum

1
2
3
4

Respiratory filaments
containing lamellae

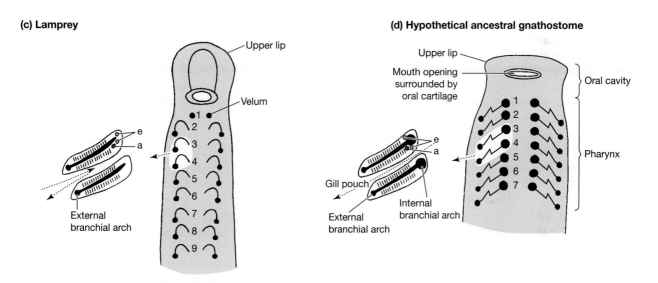

(c) Lamprey

Upper lip

Velum

e
a

External
branchial arch

1
2
3
4
5
6
7
8
9

(d) Hypothetical ancestral gnathostome

Upper lip
Mouth opening
surrounded by
oral cartilage

Oral cavity

e
a

Gill pouch

External
branchial arch

Internal
branchial arch

Pharynx

1
2
3
4
5
6
7

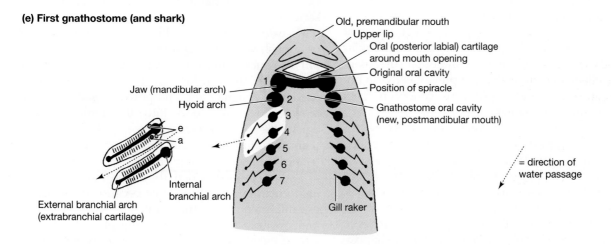

(e) First gnathostome (and shark)

Old, premandibular mouth
Upper lip
Oral (posterior labial) cartilage
around mouth opening
Original oral cavity
Position of spiracle
Gnathostome oral cavity
(new, postmandibular mouth)

Jaw (mandibular arch)
Hyoid arch

e
a

External branchial arch
(extrabranchial cartilage)

Internal
branchial arch

Gill raker

1
2
3
4
5
6
7

= direction of
water passage

The strengthened (and jointed) internal branchial arches in gnathostomes may be related to the more powerful mode of gill ventilation in these vertebrates, as will be discussed shortly. Note, however, that some researchers have concluded that as the lamprey velar cartilages are in the homologous position to gnathostome jaws (i.e., internal branchial arches of the mandibular segment), that jaws must be derived from velar cartilages, and that the gnathostome mandibular arch was never originally a branchial arch involved in gill ventilation (e.g., Janvier 1996).

Stages in the Origin of Jaws

In recent years, molecular developmental biology has provided fresh insights into the issue of the origin of jaws. It is worth noting that some of the controversies described above relate to a couple of misconceptions about the probable processes of evolution. First, we should consider that evolutionary transformation does not involve changing one adult structure into another adult structure: instead, morphology is altered via developmental changes and shifts in developmental timing. Secondly, just because the lamprey lies below the position of gnathostomes on the cladogram, this does not mean that it represents some sort of primitive vertebrate condition. On the contrary, lampreys, with their bizarrely hypertrophied upper lip, are highly derived in their own right.

Molecular developmental biology shows that the same genes are expressed in the mandibular segment in both lampreys and gnathostomes, thus elements in this position are likely to be homologous. The lamprey velum and velar cartilages, comprised of mandibular segment tissue, are highly specialized structures, and the lamprey upper lip is comprised of a strange mixture of material from the mandibular (second) segment and the premandibular (first) segment (Shigetani et al. 2005). Lampreys clearly do not represent some sort of primitive condition from which gnathostomes might be derived.

While the complex evolutionary scenarios concerning internal and external branchial arches described above may still have some validity, the "outside versus inside" position of the branchial arches may not be such a mystery if one considers developmental issues. A relatively simple change in the direction of the streams of neural crest tissue that form the arches in development might account for this difference in positioning in the adult (Kimmel et al. 2001). Exactly *how*

this might happen in ontogeny is an issue for the developmental biologists. What is perhaps of more interest to evolutionary morphologists is the reason *why* such a shift in branchial arch position should have occurred and why jaws evolved at all.

The notion that the more derived, predatory vertebrates should convert gill arches into toothed jaws has been more or less unquestioned. It is a common assumption that jaws are superior devices for feeding, and thus more derived vertebrates were somehow bound to obtain them. However, this simplistic approach does not address the issue of how the evolutionary event might actually have taken place: What use would a protojaw be prior to its full transformation? And even if early vertebrates had needed some sort of superior mouth design, why modify a branchial gill-supporting arch, which initially was located some distance behind the mouth opening? Why not just modify the existing cartilages and plates surrounding the mouth? Living agnathans have specialized oral cartilages, and various ostracoderms apparently had oral plates. Additionally, as we previously noted, gnathostome teeth must have evolved after jaws—so what use could a toothless jaw be? (A movie entitled "Gums" wouldn't be very frightening.)

Mallatt (1996, 1998) proposed a novel explanation of the origin of jaws based on the hypothesis that jaws were initially important for gill ventilation rather than predation (summarized in Figure 3–13). Numerous features of gnathostomes suggest that they are more active than jawless vertebrates and have greater metabolic demands. One derived gnathostome feature associated with such high activity is the powerful mechanism for pumping water over the gills. Gnathostomes have a characteristic series of internal branchial muscles as well as the new, external hypobranchials. These muscles not only push water through the pharynx in exhalation but also suck water into the pharynx during inhalation. Gnathostome fishes can generate much stronger suction than agnathans, and powerful suction is also a way to draw food into the mouth. Living agnathans derive a certain amount of suction from their pumping velum; but, this pump mostly pushes (rather than sucks) water, and its action is weak.

Mallatt proposed that the mandibular branchial arch enlarged into protojaws because it played an essential role in forceful ventilation—rapidly closing and opening the entrance to the pharynx. During strong exhalation, as the pharynx squeezed water back across the gills, water was kept in the mouth by bending the mandibular arch sharply shut. Next, during forceful inspiration, the mandibular arch was rapidly straightened to reopen the pharynx and allow water to enter. To accommodate the forces of the powerful muscles that bent the arch (the adductor mandibulae) and straightened it (the hypobranchial muscles), the mandibular arch enlarged and became more robust. The advantage of using the mandibular arch for this process would be that the muscles involved were of the same functional series as the other ventilatory muscles, and their common origin would ensure that all of the muscles were controlled by the same nerve circuits. Perhaps that is why the mandibular arch, rather than the more anterior oral cartilages that were not part of the branchial series derived from neural crest, became the jaws of gnathostomes.

3.6 Extinct Paleozoic Jawed Fishes

With jaws that can grasp prey, and muscles that produce powerful suction, the stage was set for gnathostomes to enter a new realm of feeding—attacking large, actively swimming prey. By the Devonian, gnathostomes were well known as entire body fossils, by which time they can be divided into four distinctive clades: two extinct groups—placoderms and acanthodians—and two groups that survive today—chondrichthyans (cartilaginous fishes) and osteichthyans (bony vertebrates).

Placoderms were highly specialized, armored fishes that appear to be basal to other gnathostomes. The cartilaginous fishes, which include sharks, rays, and ratfishes, evolved distinctive specializations of dermal armor, internal calcification, jaw and fin mobility, and reproduction. The acanthodians and the bony fishes may be closely related and are sometimes grouped together as the teleostomes. Bony fishes evolved endochondral bone in their internal skeleton, a distinctive dermal head skeleton that included an operculum covering the gills, and an internal air sac forming a lung or a swim bladder.

The bony fishes include the ray-finned fishes (actinopterygians), which comprise the majority of living fishes, and the lobe-finned fishes (sarcopterygians). Only a few lobe-finned fishes survive today (lungfishes and the coelacanths), but they were more diverse in the Paleozoic and are the group that gave rise to the tetrapods. Note that, in the proper cladistic sense, Osteichthyes include tetrapods. Bony fishes by themselves constitute a paraphyletic group because

their common ancestor is also the ancestor of tetrapods, and the same is true of the Sarcopterygii.

Before studying the extant groups of jawed fishes, we turn to the placoderms and acanthodians to examine the variety of early gnathostomes. **Figure 3–14** and **Figure 3–15** show the interrelationships of gnathostome fishes. Living and extinct groups of chondrichthyans are discussed in Chapter 5, and osteichthyans are discussed in Chapter 6.

▓ Placoderms—The Armored Fishes

As the name placoderm (Greek *placo* = plate and *derm* = skin) implies, placoderms were covered with a thick, often ornamented bony shield over the anterior one-half to one-third of their bodies. Unlike the ostracoderm condition, the placoderm bony shield was divided into separate head and trunk portions, linked by a mobile joint that allowed the head to be lifted up during feeding (**Figure 3–16a** on page 73). The endoskeleton was mineralized by perichondral bone, but endochondral bone appears to be limited to osteichthyans (Donoghue et al. 2006). Not all types of placoderms had pelvic fins, and it is speculated that the origin of pelvic fins in placoderms may have been separate from their origin in other gnathostomes (Coates 2003).

Placoderms are known from the Early Silurian to the end of the Devonian and were the most diverse vertebrates of their time in terms of both number of species and types of morphological specializations. Like the ostracoderms, they suffered massive losses in the Late Devonian extinctions; but, unlike any ostracoderm group, a few placoderm lineages continued for a further 5 million years, until the very end of the period. Ancestral placoderms were primarily marine, but a great many lineages became adapted to freshwater and estuarine habitats.

A recent discovery of a placoderm from the Late Devonian of Australia preserves some soft tissue structure that lends more credence to the hypothesis that placoderms are more primitive than any other gnathostomes (Trinajstic et al. 2007). The segmental muscles (myomeres) resemble those of lampreys in being only weakly W-shaped and not distinctly separated into epaxial and hypaxial portions (see Figure 3–8, feature number 8). Additionally, placoderms have the lampreylike feature of seven muscles moving each eye, rather than the six seen in extant gnathostomes, and their nerves appear to lack myelin sheaths.

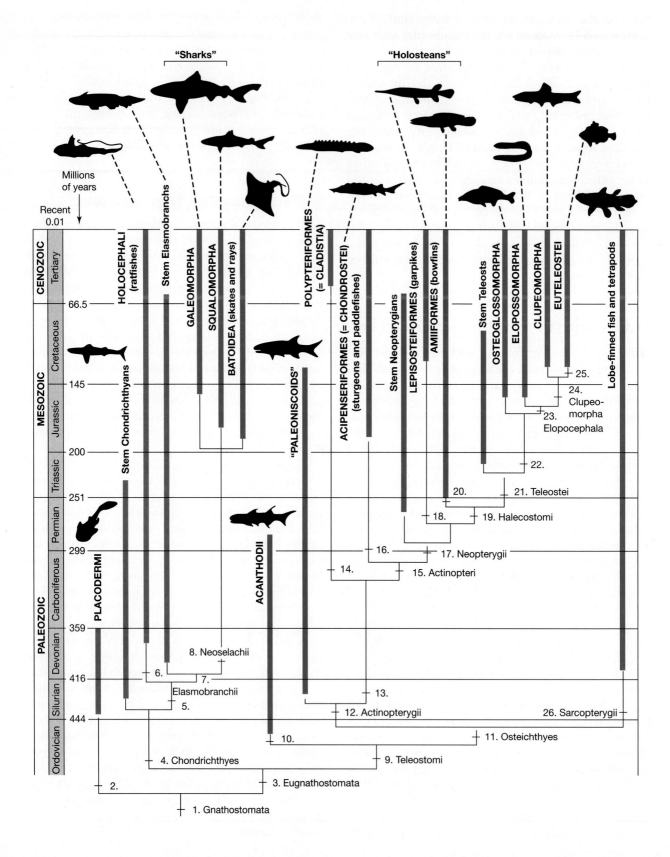

◄ **FIGURE 3–14** Phylogenetic relationships of jawed fishes. This diagram depicts the probable interrelationships among the major groups of basal gnathostomes. The black lines show interrelationships only; they do not indicate times of divergence nor the unrecorded presence of taxa in the fossil record. The numbers at the branch points indicate derived characters that distinguish the lineages. Only the best-corroborated relationships are shown.

Legend: 1. Gnathostomata–jaws formed of bilateral palatoquadrate (upper) and mandibular (lower) cartilages of the mandibular pharyngeal arch, modified hyoid gill arch; branchial arches internal to gill tissue, containing four elements on each side plus one unpaired ventral median element; three semicircular canals; internal supporting girdles associated with pectoral and pelvic fins; and many features of the soft anatomy. **2.** Placodermi–tentatively placed as the sister group of all other gnathostomes, but see Gardiner (1984) for a different view. Specialized joint between head and trunk dermal shield, a unique arrangement of dermal skeletal plates of the head and shoulder girdle, a unique pattern of lateral line canals on the head, and semidentine in the dermal bones. **3.** Chondrichthyes plus Teleostomi (Eugnathostomata)–epihyal element of second visceral arch modified as the hyomandibula, which is a supporting element for the jaw; true teeth rooted to the jaw (Osteichthyes) or in a tooth whorl; fusion of the nasal capsules to the rest of the chondrocranium; axial musculature divided into distinct epaxial and hypaxial components; myelinated nerves; six (reduced from seven) external eye muscles. **4.** Chondrichthyes–unique perichondral and endochondral mineralization (prismatic plates of apatite), placoid scales, unique teeth and tooth replacement mechanisms, distinctive characters of the basal and radial elements of the fins, inner ear labyrinth opens externally via the endolymphatic duct, distinctive features of the endocrine system. **5.** Neoselachii plus Holocephali–claspers on male pelvic fins and at least four additional fin-support characters. **6.** Holocephali–hyostylic jaw suspension, gill arches beneath the braincase, pectoral fin with two main basal elements, dorsal fin articulates with anterior elements of the axial skeleton. **7.** Elasmobranchii–pectoral fin with three main basal elements, shoulder joint narrowed, basibranchial separated by gap from basihyal (basal element of hyoid arch). **8.** Neoselachii–pectoral fin with three basal elements, the anteriormost of which is supported by the shoulder girdle; plus characteristics of the nervous system, cranium, and gill arches. **9.** Teleostomi–gill structures not attached to interbranchial septum, bony opercular covers, branchiostegal rays. **10.** Acanthodii–fin spines on anal and paired fins as well as on dorsal fins; paired intermediate fin spines between pectoral and pelvic fins.

11. Osteichthyes–a unique pattern of dermal head bones, including dermal marginal mouth bones with rooted teeth, a unique pattern of ossification of the dermal bones of the shoulder girdle, presence of lepidotrichia (fin rays), differentiation of the muscles of the branchial region, presence of a lung or swim bladder derived from the gut, medial insertion of the mandibular muscle on the lower jaw. **12.** Actinopterygii–basal elements of pectoral fin enlarged, median fin rays attached to skeletal elements that do not extend into fin, single dorsal fin, scales with unique arrangement, shape, interlocking mechanism, and histology (outer layer of complexly layered enameloid called ganoine). **13.** Polypteriformes plus Actinopteri (plus fossils such as *Moythomasia* and *Mimia*)–acrodon (a specialized dentine) forms a cap on the teeth, details of posterior braincase structure, specific basal elements of the pelvic fin are fused, and numerous features of the soft anatomy of extant forms that cannot be verified for fossils. **14.** Polypteriformes (Cladistia)–unique dorsal-fin spines, facial bone fusion, and pectoral-fin skeleton and musculature. **15.** Actinopteri–derived characters of the dermal elements of the skull and pectoral girdle and fins, a spiracular canal formed by a diverticulum of the spiracle penetrating the postorbital process of the skull, other details of skull structure, three cartilages or ossifications in the hyoid. Swim bladder connects dorsally to the foregut, fins edged by specialized scales (fulcra). **16.** Acipenseriformes (Chondrostei)–fusion of premaxillae, maxillae, and dermopalatine bones of snout; unique anterior palatoquadrate symphysis. **17.** Neopterygii–rays of dorsal and anal fins reduced to equal the number of endoskeletal supports, upper lobe of caudal fin containing axial skeleton reduced in size to produce a nearly symmetrical caudal fin, upper pharyngeal teeth consolidated into tooth-bearing plates, characters of pectoral girdle and skull bones. **18.** Lepisosteiformes (Ginglymodi)–vertebrae with convex anterior faces and concave posterior faces (opisthocoelous), toothed infraorbital bones contribute to elongate jaws. See character state 19. **19.** Halecostomi–modifications of the cheek, jaw articulation, and opercular bones, including a mobile maxilla. Relationships of the Lepisosteiformes, Amiiformes, their fossil relatives, and the Teleostei are currently subject to many differing opinions with no clear resolution based on unique shared derived characters. More conservative phylogenies than those here would represent their relationships as unresolved. Others would unite the lepisosteiformes and the amiiformes as the "Holostei." **20.** Amiidae (recent Amiiformes)–jaw articulation formed by means of a bone called the symplectic in addition to the quadrates. **21.** Teleostei–elongate posterior neural arches (uroneurals) contributing to the stiffening of the upper lobe of the internally asymmetrical caudal fin (the caudal is externally symmetrical, [= homocercal], at least primitively in teleosts), unpaired ventral pharyngeal tooth plates on basibranchial elements, premaxillae mobile, details of skull foramina, jaw muscles, and axial and pectoral skeleton. **22.** Recent Teleosts–presence of an endoskeletal basihyal, four pharyngobranchials and three hypobranchials, median tooth plates overlying basibranchials and basihyals. **23.** Elopocephala–two uroneural bones extend anteriorly to the second ural (tail) vertebral centrum; abdominal and anterior caudal epipleural intermuscular bones present. **24.** Clupeomorpha–pharyngeal tooth plates fused with endoskeletal gill-arch elements, neural arch of first caudal centrum reduced or absent, distinctive patterns of ossification and articulation of the jaw joint. **25.** Euteleostei–this numerically dominant group of vertebrates is poorly characterized, with no known [shared derived character] unique to this group that is present in all or perhaps even in most forms. However, the following have been used in establishing monophyly: presence of an adipose fin posteriorly on the mid-dorsal line, presence of nuptial tubercles on the head and body, and paired anterior membranous outgrowths of the first uroneural bones of the caudal fin. (These characters are usually lost in the most derived euteleosts.) The nature of these characters leads to a lack of consensus on the interrelationships of the basal clupeocephalids, although the group's monophyly is still generally accepted. **26.** Sarcopterygii–fleshy pectoral and pelvic fins have a single basal skeletal element, muscular lobes at the bases of those fins, enamel (versus enameloid) on surfaces of teeth, cosmine (unique type of dentine) in body scales, unique characters of jaws, articulation of jaw supports, gill arches, and shoulder girdles.

▶ **FIGURE 3–15** Simplified cladogram of gnathostomes, showing living taxa and major extinct groups only. Extinct taxonomic groups are indicated by a dagger (†).

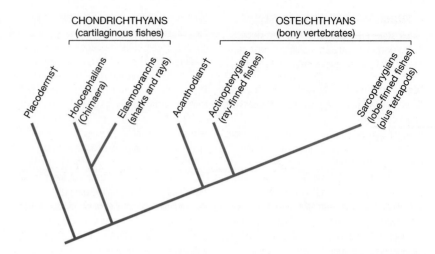

Placoderms have no modern analogues, and their massive external armor makes interpreting their lifestyle difficult, although they were apparently primarily benthic (i.e., deep-living) fishes. Most placoderms lacked true teeth; their toothlike structures (tooth plates, see Figure 3–16a) were actually projections of the dermal jawbones that were subject to wear and breakage without replacement. However, some later placoderms did possess true teeth that apparently evolved convergently with the condition in other gnathostomes. Placoderms also had some of their own unique specializations, apart from the nature of their head shield: for example, they had a unique type of cellular hard tissue (semidentine) in their dermal bones. In some species, pelvic appendages were found on some individuals but not in others, suggesting that these appendages were different in males and females. A similar situation occurs among living chondrichthyans—only males have pelvic appendages, which they use for internal fertilization. We can infer from this that the placoderms, like living chondrichthyans, had internal fertilization and probably complex courtship behaviors.

More than half of the known placoderms, nearly 200 genera, belonged to the predatory arthrodires (Greek *arthros* = a joint and *dira* = the neck). As their name suggests, they had specializations of the joint between the head shield and the trunk shield, allowing an enormous head-up gape, probably increasing both predatory and respiratory efficiency. *Dunkleosteus* was a voracious, 10-meter-long predatory arthrodire of the Devonian (**Figure 3–16c**).

There were several other types of placoderms, most of them flattened, bottom-dwelling forms. The antiarchs, such as *Bothriolepis*, looked rather like armored catfishes (**Figure 3–16b**). Their pectoral fins

were also encased in the bony shield, so that their front fins looked more like those of a crab. One specimen of this fish was preserved with hints of internal paired structures that were interpreted as possible lungs. The notion that placoderms had lungs seems to have caught popular attention, and you can see this asserted as fact in many books and websites. Some people have gone so far as to infer that, if placoderms had lungs, lungs must have been a primitive gnathostome feature lost in chondrichthyans. However, experts on placoderms do not believe that the paired structures were lungs, and this interpretation means that lungs are a derived character of bony fishes.

The range of body forms seen among placoderms includes two groups that resemble extant kinds of chondrichthyans—the ptyctodontids (**Figure 3–16d**), which had a body form similar to the chimaeras or ratfish, and the rhenanids (**Figure 13–16e**), which resemble extant skates.

■ Acanthodians

Acanthodians are so named because of the stout spines (Greek *acantha* = spine) anterior to their well-developed dorsal, anal, and often numerous paired fins. They are distinguished from other vertebrates by the presence of up to six pairs of ventrolateral fins in addition to the pectoral and pelvic fins of gnathostomes (**Figure 3–17**). Had land vertebrates evolved from these fishes, they might have been hexapods (six-legged) or octopods (eight-legged) rather than tetrapods (four-legged)! Acanthodians ranged from the Late Ordovician through the Early Permian periods, with their major diversity in the Early Devonian. The earliest forms were marine, but by the Devonian they were predominantly a freshwater group.

▼ FIGURE 3–16 Placoderms.

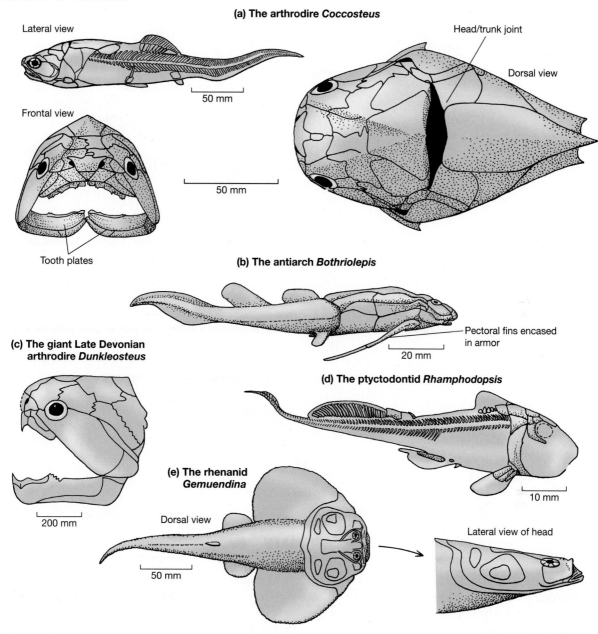

(a) The arthrodire *Coccosteus*

Lateral view

50 mm

Head/trunk joint

Dorsal view

Frontal view

50 mm

Tooth plates

(b) The antiarch *Bothriolepis*

Pectoral fins encased in armor

20 mm

(c) The giant Late Devonian arthrodire *Dunkleosteus*

200 mm

(d) The ptyctodontid *Rhamphodopsis*

10 mm

(e) The rhenanid *Gemuendina*

Dorsal view

50 mm

Lateral view of head

Acanthodians had slender bodies with a heterocercal tail fin, suggesting a preference for midwater conditions, in contrast to the primarily bottom-dwelling placoderms, They were usually not more than 20 centimeters long, although some species are known that were 2 meters long. They had large heads with wide-gaping mouths, and most had a basic fusiform fish shape. The acanthodids, the only group to survive into the Permian, became more eel-like in body form, lost their teeth, evolved long gill rakers, and were probably plankton-eating filter feeders.

Most workers now accept acanthodians as the sister group of the Osteichthyes. The acanthodians + osteichthyans (sometimes grouped as the Teleostomei) are diagnosed by a number of derived features (see Figure 3–14). However, acanthodians also bear a resemblance to chondrichthyans in having spines associated with their fins, and by having teeth (if present) forming a sharklike tooth whorl rather than being embedded in the jaws. Both of these characters were probably primitive gnathostome features.

▶ **FIGURE 3–17** Acanthodians.

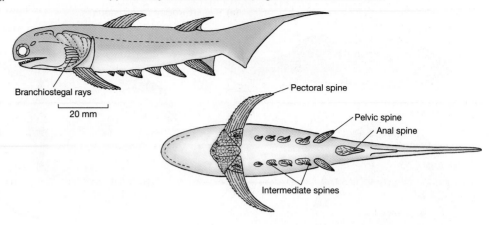

(a) The Early Devonian *Climatius*, a generalized ancestral form

Branchiostegal rays

20 mm

Pectoral spine

Pelvic spine

Anal spine

Intermediate spines

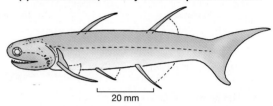

(b) *Ischnacanthus*, an Early Devonian predaceous form

20 mm

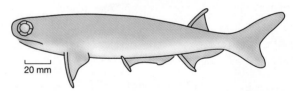

(c) *Acanthodes*, a Permian filter-feeding form

20 mm

Summary

Fossil evidence indicates that vertebrates evolved in a marine environment. Jawless vertebrates are first known from the Early Cambrian, and there is evidence that the first jawed vertebrates (gnathostomes) evolved as long ago as the Early Ordovician. The first vertebrates would have been more active than their ancestors, with a switch from filter feeding to more active predation and with a muscular pharyngeal pump for gill ventilation. Bone is a feature of many early vertebrates, although it was absent from the Early Cambrian forms and is also absent in the living jawless vertebrates—the hagfishes and lampreys. The first mineralized tissues were seen in the teeth of conodonts, enigmatic animals that have only recently been considered true vertebrates. Bone was first found with accompanying external layers of dentine and enamel-like tissue in the dermal armor of early jawless fishes called the ostracoderms. Current explanations for the original evolutionary use of bone include protection, a store for calcium and phosphorus, and housing for electroreceptive sense organs.

Among the living jawless forms, hagfishes are more primitive in their anatomy than the lampreys and all other known vertebrates, living or fossil. Lampreys, in turn, are more primitive than the extinct armored ostracoderms. Ostracoderms are widely known from the Silurian and Early Devonian, and none survived past the end of the Devonian. Ostracoderms were not a unified evolutionary group: some forms (Cephalapsida) were more closely related to gnathostomes than were other forms, sharing with gnathostomes the feature of a pectoral fin. Ostracoderms and gnathostomes flourished together for 50 million years; thus there is little evidence to support the idea that jawed vertebrates outcompeted and replaced jawless ones.

Just as the evolution of vertebrates from nonvertebrate chordates represented a step up in anatomical and physiological design, so did the evolution of jawed vertebrates from jawless ones. Jaws may have evolved initially to improve gill ventilation rather than to bite prey. In addition to jaws, gnathostomes have a number of derived anatomical features (such as true vertebrae, ribs, and a complete lateral-line sensory system), suggesting a sophisticated and powerful mode of locomotion and sensory feedback.

The early radiation of jawed fishes, first known in detail from the fossil record in the Late Silurian, included four major groups. Two groups, the chondrichthyans (cartilaginous fishes) and osteichthyans (bony fishes), survive today. Osteichthyans were the forms that gave rise to tetrapods in the Late Devonian. The other two groups, placoderms and acanthodians, are now extinct. Placoderms did not survive past the Devonian period, while acanthodians survived almost until the end of the Paleozoic. Placoderms were armored fishes, superficially like the ostracoderms in their appearance, and were the most diverse fishes of the Devonian period. They are considered to be the sister group to all other gnathostomes. Acanthodians probably form the sister taxon to the osteichthyans and have the unique feature of multiple pairs of ventral fins rather than the usual gnathostome complement of just two pairs, the pectoral and pelvic fins.

Additional Readings

Braun, C. B. 1996. The sensory biology of the living jawless fishes: A phylogenetic assessment. *Brain, Behavior and Evolution* 48:262–276.

Coates, M. I. 2003. The evolution of paired fins. *Theory Biosci.* 122:266–287.

Ditrich, J. 2007. The origin of vertebrates: A hypothesis based on kidney development. *Zoological Journal of the Linnean Society* 150:435–441.

Donoghue, P. C. J. et al. 2000. Conodont affinity and chordate phylogeny. *Biological Reviews* 75:191–251.

Donoghue, P. C. J. et al. 2006. Early evolution of vertebrate skeletal tissues and cellular interactions, and the canalization of skeletal development. *Journal of Experimental Zoology (Mol Dev Evol)* 306B:278–294.

Donoghue, P. C. J., and Purnell, M. A. 2006. Genome duplication, extinction, and vertebrate evolution. *Trends in Ecology and Evolutionary Biology* 20:312–319.

Donoghue, P. C. J., and Sansom, I. J. 2002. Origin and early evolution of vertebrate skeletonization. *Microscopy Research and Technique* 59:352–372.

Forey, P., and P. Janvier. 1994. Evolution of the early vertebrates. *American Scientist* 82:554–565.

Gardiner, G. G. 1984. The relationship of placoderms. *Journal of Vertebrate Paleontology* 4:379–395.

Janvier, P. 1996. *Early Vertebrates*. Oxford Monographs on Geology and Geophysics—33. Oxford, UK: Clarendon Press.

Janvier, P. et al. 2006. Lampreylike gills in a gnathostome-related Devonian jawless vertebrate. *Nature* 440:1183–1185.

Kimmel, C. B. et al. 2001. Neural crest patterning and the evolution of the jaw. *Journal of Anatomy* 199:105–119.

Long, J. A. 1995. *The Rise of Fishes*. Baltimore, MD: Johns Hopkins University Press.

Maisey, J. G. 1996. *Discovering Fossil Fishes*. New York: Henry Holt and Company.

Mallatt, J. 1985. Reconstructing the life cycle and the feeding of ancestral vertebrates. In R. E. Foreman et al. (eds.), *Evolutionary Biology of Primitive Fishes*. New York: Plenum, 59–68.

Mallatt, J. 1996. Ventilation and the origin of jawed vertebrates: A new mouth. *Zoological Journal of the Linnean Society* 117:329–404.

Mallatt, J. 1998. Crossing a major morphological boundary: The origin of jaws in vertebrates. *Zoology—Analysis of Complex Systems* 100:128–140.

Mallatt, J., and J. Sullivan. 1998. 28S and 18S rDNA sequences support the monophyly of lampreys and hagfishes. *Molecular Biology and Evolution* 15:1706–1718.

Martini, F. H. 1998. Secrets of the slime hag. *Scientific American* 279(4):70–75.

McCollum, M., and P. T. Sharpe. 2001. Evolution and development of teeth. *Journal of Anatomy* 199:153–159.

Ota, K. G. et al. 2007. Hagfish embryology with reference to the evolution of neural crest. *Nature* 446:672–675.

Purnell, M. A. 2001. Scenarios, selection and the ecology of early vertebrates. In P. E. Ahlberg (ed.), *Major Events in Vertebrate Evolution: Palaeontology, Phylogeny, Genetics, and Development. Systematics Association Special Volume Series*, 61:187–208. London, UK: Taylor and Francis.

Shigetani, Y. et al. 2005. A new evolutionary scenario for the vertebrate jaw. *BioEssays* 27:331–338.

Shu, D.-G. et al. 1999. Lower Cambrian vertebrates from south China. *Nature* 402:42–46.

Smith, M. M., and M. I. Coates. 2000. Evolutionary origins of teeth and jaws. In M. F. Teaford, M. M. Smith and M. W. J. Ferguson (eds.), *Development, Function and Evolution of Teeth*. Cambridge, UK: Cambridge University Press, 133–151.

Smith, M. M., and Z. Johanson. 2003. Separate evolutionary origins of teeth from evidence in fossil jawed vertebrates. *Science* 299:1235–1236.

Trinajstic, K. et al. 2007. Exceptional preservation of nerve and muscle tissues in Late Devonian placoderm fish and their evolutionary implications. *Biology Letters* 3:197–200.

PART II

Non-Amniotic Vertebrates: Fishes and Amphibians

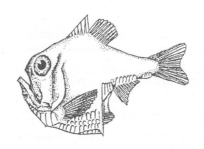

Vertebrates originated in the sea, and more than half of the species of living vertebrates are the products of evolutionary lineages that have never left an aquatic environment. Water now covers 73 percent of the Earth's surface (the percentage has been higher in the past) and provides habitats extending from deep oceans, lakes, and mighty rivers to fast-flowing streams and tiny pools in deserts. Fishes have adapted to all these habitats, and there are more than 27,000 species of extant cartilaginous and bony fishes.

Life in water poses challenges for vertebrates but offers many opportunities. Aquatic habitats are some of the most productive on Earth, and energy is plentifully available in many of them. Some aquatic habitats (coral reefs are an example) have enormous structural complexity, whereas others (like the open ocean) have virtually none. The diversity of fishes reflects specializations for this variety of habitats.

The diversity of fishes and the habitats in which they live have offered unparalleled scope for variations in life history. Some fishes produce millions of eggs that are released into the water to drift and develop on their own; other species of fishes produce a few eggs and guard both the eggs and the young, and numerous fishes give birth to young that require no parental care. Males of some species of fishes are larger than females; in others the reverse is true. Some species have no males at all, and a few species of fishes change sex partway through life. Feeding mechanisms have been a central element in the evolution of fishes, and the specializations of modern fishes range from species that swallow prey longer than their own bodies to species that rapidly extend their jaws like a tube to suck up minute invertebrates from tiny crevices. In the Devonian period, vertebrates entered a new world as fishlike forms emerged onto the land and occupied terrestrial environments. In this part of the book, we consider the evolution of this extraordinary array of vertebrates.

CHAPTER 4

Living in Water

Although life evolved in water and the earliest vertebrates were aquatic, the physical properties of water create some difficulties for aquatic animals. To live successfully in open water, a vertebrate must adjust its buoyancy to remain at a selected depth and force its way through a dense medium to pursue prey or to escape its own predators. Heat flows rapidly between an animal and the water around it, and it is difficult for an aquatic vertebrate to maintain a body temperature that is different from water temperature. (That phenomenon was dramatically illustrated when the *Titanic* sank—in the cold water of the North Atlantic, most of the victims died from hypothermia rather than by drowning.) Ions and water molecules move readily between the external environment and an animal's internal body fluids, so maintaining a stable internal environment can be difficult. On the plus side, ammonia is extremely soluble in water so disposal of nitrogenous waste products is easier in aquatic environments than on land. The concentration of oxygen in water is lower than it is in air, however, and the density of water imposes limits on the kinds of gas exchange structures that can be effective. Despite these challenges, many vertebrates are entirely aquatic. Fishes in particular, especially the bony fishes, have diversified into an enormous array of sizes and ways of life. In this chapter we will examine some of the challenges of living in water and the ways aquatic vertebrates (especially fishes) have responded to them.

4.1 The Aquatic Environment

Seventy-three percent of the surface of the Earth is covered by freshwater or salt water. Most of this water is held in the ocean basins, which are populated everywhere by vertebrates. Freshwater lakes and rivers hold a negligible amount of the water on Earth—about 0.01 percent. This is much less than the water tied up in the atmosphere, ice, and groundwater, but freshwater habitats are exceedingly rich biologically, and nearly 40 percent of all bony fishes live in freshwater.

Water and air are both liquids, but they have different physical properties that make them drastically different environments for vertebrates to live in (**Box 4–1**). In air, for example, gravity is an important force acting on an animal; but fluid resistance to movement (air resistance) is trivial for all but the fastest-flying birds. In water, the opposite relationship holds—gravity is negligible, but fluid resistance to movement is a major factor with which vertebrates must contend. Although each major clade of aquatic vertebrates solved environmental challenges in somewhat different ways, the basic specializations needed by all aquatic vertebrates are the same.

▧ Obtaining Oxygen in Water—Gills

Most aquatic vertebrates have gills, which are specialized structures where oxygen and carbon dioxide are exchanged. Teleosts are derived ray-finned fishes, and this group includes the majority of species of extant freshwater and marine fishes. The gills of

| BOX 4–1 | Water—A Nice Place to Visit, But Would You Want to Live There? |

Water and air are both fluids in which animals live, but the different physical properties of the two fluids make aquatic and terrestrial environments quite different places. Compared to air, water has high density, high viscosity, low oxygen content, high heat capacity and heat conductivity, and high electrical conductivity. These physical characteristics are reflected in the sizes and shapes of aquatic and terrestrial animals and in their physiology and behavior.

Density Water is more than 800 times as dense as air. A liter of water weighs 1 kilogram, whereas a liter of air weighs about 1.25 grams. Because of its density, water supports an animal's body. Aquatic animals do not need weight-bearing skeletons because they are close to neutral buoyancy in water. Aquatic vertebrates can grow larger than terrestrial forms because gravity has little effect on their body structure.

Viscosity Water is approximately 18 times as viscous as air. Viscosity is a measure of how readily a fluid flows across a surface—the higher the viscosity, the slower the fluid flows. (To visualize the effect of viscosity, think of how rapidly water flows compared to a more viscous fluid such as syrup.)

The effect of the high density and high viscosity of water can be seen in the streamlined shape of most aquatic animals compared to the lack of streamlining of terrestrial animals. Only the fastest birds need to worry about air resistance, but even slow-moving fish must be streamlined.

The path followed by fluid during breathing also reflects the differences in density and viscosity between water and air. Because air is light and flows easily, it can be pumped in and out of closed sacs (the lungs) during respiration. (This is called tidal ventilation because the fluid moves in and out.) The density and viscosity of water are too high for tidal ventilation to be feasible, and the gills of sharks and teleosts employ a one-way flow of water.

Oxygen Content Oxygen makes up approximately 20.9 percent of the volume of air. In other words, there are 209 milliliters (ml) of oxygen in a liter of air. The oxygen content of water varies, but is never more than 50 ml of oxygen per liter of water and is often 10 ml or less. The low oxygen content of water compared to air is an additional reason why fishes do not use tidal ventilation.

Heat Capacity and Heat Conductivity The specific heat of water (defined as the energy needed to produce a 1-degree centigrade change in temperature in 1 gram of the fluid) is nearly 3400 times that of air, and water conducts heat almost 24 times as fast as air. Those differences are reflected in the thermal (temperature) characteristics of aquatic compared to terrestrial environments. The high specific heat of water means that the water temperature in a pond changes far less from day to night than does the air temperature on the shore of the pond. Thus, an animal living in the pond has a more stable thermal regime than an animal living on the shore. However, because the heat conductivity of water is so high, the temperature at any given depth scarcely varies from place to place. If the water gets too hot, the only route of escape for an aquatic animal is to go to deeper water. That could require a dangerous long-distance movement for an animal living near the shore, even assuming that deep water is available. In contrast, temperature varies substantially over short distances on land because the low heat conductivity of air allows temperature differences to develop between sunny and shady spots. Thus terrestrial animals have a mosaic of temperatures to choose from, whereas aquatic animals have much less temperature variation in their habitats.

Electrical Conductivity Water is an electrical conductor, and aquatic animals can use electricity to detect the presence of other animals and as an offensive or defensive weapon to stun their prey or predators. Because air does not conduct electricity (at least not within the range of voltages that animals can generate), terrestrial animals cannot use electricity in these ways.

teleosts are enclosed in pharyngeal pockets called the opercular cavities (**Figure 4–1**). The flow of water is usually unidirectional—in through the mouth and out through the gills. Flaps just inside the mouth and flaps at the margins of the **operculae** (singular *operculum*; gill covers) of bony fish act as valves to prevent backflow. The respiratory surfaces of the gills are delicate projections from the lateral side of each gill arch. Two columns of gill filaments extend from each gill arch. The tips of the filaments from adjacent arches meet when the filaments are extended. As water leaves the buccal cavity, it passes over the filaments. Gas exchange takes place at the numerous microscopic projections from the filaments called **secondary lamellae**.

The pumping action of the mouth and opercular cavities (**buccal pumping**) creates a positive pressure across the gills so that the respiratory current is only

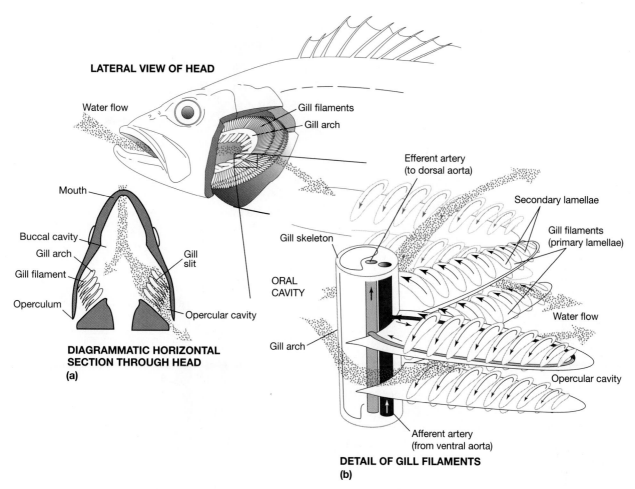

LATERAL VIEW OF HEAD

Water flow

Gill filaments

Gill arch

Mouth

Buccal cavity

Gill arch

Gill filament

Operculum

Gill slit

Opercular cavity

DIAGRAMMATIC HORIZONTAL SECTION THROUGH HEAD
(a)

Efferent artery (to dorsal aorta)

Secondary lamellae

Gill filaments (primary lamellae)

Gill skeleton

ORAL CAVITY

Gill arch

Water flow

Opercular cavity

Afferent artery (from ventral aorta)

DETAIL OF GILL FILAMENTS
(b)

▲ **FIGURE 4–1** Anatomy of bony fish gills. (a) Position of gills in head and general flow of water; (b) countercurrent flow of water (green arrows) and blood (black arrows) through the gills.

slightly interrupted during each pumping cycle. Some filter-feeding fishes and many open-ocean fishes—such as mackerel, certain sharks, tunas, and swordfishes—have reduced or even lost the ability to pump water across the gills. These fish create a respiratory current by swimming with their mouths open, a method known as **ram ventilation**, and these fishes must swim perpetually. Many other fishes rely on buccal pumping when they are at rest and switch to ram ventilation when they are swimming.

The arrangement of blood vessels in the gills maximizes oxygen exchange. Each gill filament has two arteries, an afferent vessel running from the gill arch to the filament tip and an efferent vessel returning blood to the arch. Each secondary lamella is a blood space connecting the afferent and efferent vessels (**Figure 4–2**). The direction of blood flow through the

lamellae is opposite to the direction of water flow across the gill. This arrangement, known as a **countercurrent exchanger**, assures that as much oxygen as possible diffuses into the blood. Open water fishes such as tunas, which sustain high levels of activity for long periods, have skeletal tissue reinforcing the gill filaments, large gill exchange areas, and a high oxygen-carrying capacity per milliliter of blood compared with sluggish bottom-dwelling fishes, such as toadfishes and flat fishes (**Table 4.1**).

Obtaining Oxygen from Air—Lungs and Other Respiratory Structures

Although the vast majority of fishes depend on gills to extract dissolved oxygen from water, fishes that live in water with low oxygen levels cannot obtain

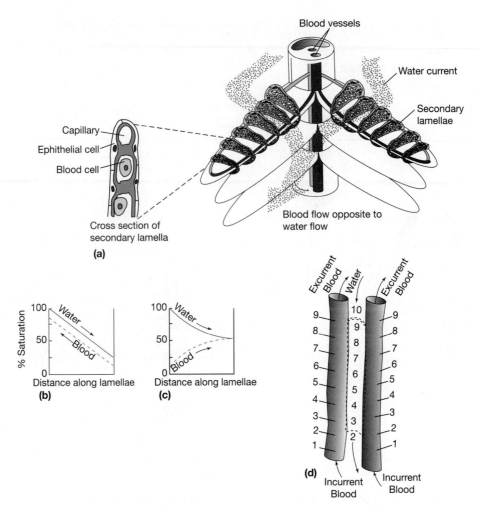

▲ **FIGURE 4–2** Countercurrent exchange in the gills of bony fishes. (a) The direction of water flow across the gill opposes the flow of blood through the secondary lamellae. Blood cells are separated from oxygen-rich water only by the thin epithelial cells and capillary wall, as shown in the cross section of a secondary lamella. (b) Countercurrent flow results in a high oxygen concentration in the blood leaving the gills. (c) If water and blood flowed in the same direction, the blood leaving the gills would have a lower oxygen concentration. (d) Relative oxygen content of blood in secondary lamellae and the water passing over them.

TABLE 4.1 The relation among general level of activity, rate of oxygen consumption, respiratory structures, and blood characteristics in fishes of three activity levels

Activity Level	Species of Fish	Oxygen Consumption (ml O_2/g · h)	No. Secondary Gill Lamellae (mm^{-1} of primary gill lamella)	Gill Area (mm^2/gm body mass)	Oxygen Capacity (ml O_2/100 ml blood)
High	Mackerel* (*Scomber*)	0.73	31	1160	14.8
Intermediate	Porgy (*Stenotomus*)	0.17	26	506	7.3
Sluggish	Toadfish[†] (*Opsanus*)	0.11	11	197	6.2

*Modified carangiform swimmers; swim continuously.
[†]Benthic fish.

enough oxygen via gills alone. These fish supplement the oxygen they get from their gills with additional oxygen obtained from the air via lungs or accessory air respiratory structures.

The accessory surfaces used to take up oxygen from air include enlarged lips that are extended just above the water surface and a variety of internal structures into which air is gulped. The anabantid fishes of tropical Asia (including the bettas and gouramies seen in pet stores) have vascularized chambers in the rear of the head, called labyrinths. Air is sucked into the mouth and transferred to the labyrinth, where gas exchange takes place. Many of these fishes are facultative air breathers; that is, they switch oxygen uptake from their gills to accessory respiratory structures when oxygen in the water becomes low. Others, like the electric eel and the anabantids, are obligatory air breathers. The gills alone cannot meet the respiratory needs of these fishes, even if the water is saturated with oxygen, and they drown if they cannot reach the surface to breathe air.

We think of lungs as being the respiratory structures used by terrestrial vertebrates, as indeed they are, but lungs first appeared in fishes and preceded the evolution of tetrapods by millions of years. Lungs may have evolved in freshwater placoderms during the Silurian and Devonian periods; and, if this interpretation is correct, lungs are an ancestral character for both bony fishes and their tetrapod descendants. Lungs develop embryonically as outpocketings (evaginations) of the pharyngeal region of the digestive tract, originating from its ventral or dorsal surface. The lungs of bichirs (a group of air-breathing fishes from Africa), lungfish, and tetrapods originate from the ventral surface of the gut, whereas the lungs of gars (a group of primitive bony fish) and the lungs of the derived bony fish known as teleosts originate embryonically from its dorsal surface.

Lungs used for gas exchange need a large surface area, which is provided by ridges or pockets in the wall. This design is known as an alveolar lung, and it is found in gars, lungfishes, and tetrapods. Increasing the volume of the lung by adding a second lobe is another way to increase surface area, and the lungs of lungfishes and tetrapods consist of two symmetrical lobes. (Bichirs have non-alveolar lungs with two lobes, but one lobe is much smaller than the other; gars have single-lobed alveolar lungs.)

◼ Adjusting Buoyancy

Holding a bubble of air inside the body changes the buoyancy of an aquatic vertebrate, and bichirs and teleost fishes use the lungs primarily as swim bladders that regulate a fish's position in the water. Swim bladders have smooth walls because there is no need for an expanded surface area.

Many bony fishes are neutrally buoyant (i.e., have the same density as water). These fish do not have to swim to maintain their vertical position in the water column. The only movement they make when at rest is backpedaling of the pectoral fins to counteract the forward thrust produced by water as it is ejected from the gills and a gentle undulation of the tail fin to keep them level in the water. Fishes capable of hovering in the water like this usually have well-developed swim bladders.

The swim bladder is located between the peritoneal cavity and the vertebral column (**Figure 4–3**). The bladder occupies about 5 percent of the body volume of marine teleosts and 7 percent of the volume of freshwater teleosts. The difference in volume corresponds to the difference in density of salt water and freshwater—salt water is denser, so a smaller swim bladder is needed. The swim bladder wall, which is composed of interwoven collagen fibers, is virtually impermeable to gas.

Neutral buoyancy produced by a swim bladder works as long as a fish remains at one depth, but if a fish swims vertically up or down, the hydrostatic pressure that the surrounding water exerts on the bladder changes, which in turn changes the volume of the bladder. For example, when a fish swims deeper, the additional pressure of the water column above it compresses the gas in its swim bladder, making the bladder smaller and reducing the buoyancy of the fish. When the fish swims toward the surface, water pressure decreases, the swim bladder expands, and the fish becomes more buoyant. To maintain neutral buoyancy, a fish must adjust the volume of gas in the swim bladder as it changes depth.

The fish regulates the volume of its swim bladder by secreting gas into the bladder when it swims down and removing gas when it swims up. Primitive teleosts—such as bony tongues, eels, herrings, anchovies, salmons, and minnows—retain a connection, the pneumatic duct, between the gut and swim bladder (see Figure 4–3a). These fishes are called **physostomous** (Greek *phys* = bladder and *stom* = mouth), and goldfish are a familiar example of this group. Because they have a connection between the gut and the swim bladder, they can gulp air at the surface to fill the bladder and can burp gas out to reduce its volume.

The pneumatic duct is absent in adult teleosts from more derived clades, a condition termed

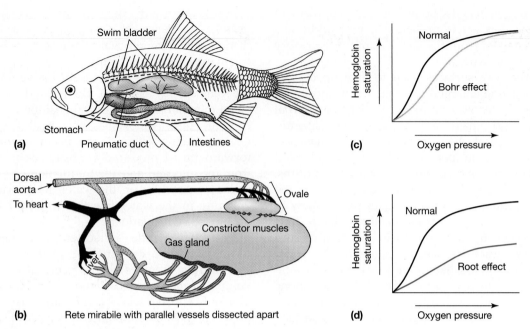

▲ **FIGURE 4–3** Swim bladder of bony fishes. (a) The swim bladder is in the coelomic cavity just beneath the vertebral column. This is a physostomous fish, in which the swim bladder retains its ancestral connection to the gut via the pneumatic duct. (b) The vascular connections of a physoclistous swim bladder, which has lost its connection to the gut. (c) The Bohr effect is a reduction in the affinity of hemoglobin for oxygen in the presence of acid. By creating a Bohr effect, the gas gland causes hemoglobin to release oxygen (i.e., to bind less oxygen). (d) The Root effect is a reduction in the maximum amount of oxygen that hemoglobin can bind. By creating a Root effect, the gas gland prevents oxygen in the gland from binding to hemoglobin in the blood. As a result, oxygen remains in the gland.

physoclistic (Greek *clist* = closed). Physoclists regulate the volume of the swim bladder by secreting gas from the blood into the bladder. Both physostomes and physoclists have a gas gland, which is located in the anterior ventral floor of the swim bladder (see Figure 4–3b). Underlying the gas gland is an area with many capillaries arranged to give countercurrent flow of blood entering and leaving the area. This structure, which is known as a **rete mirabile** ("wonderful net"; plural *retia mirabilia*), moves gas (especially oxygen) from the blood to the gas bladder. It is remarkably effective at extracting oxygen from the blood and releasing it into the swim bladder, even when the pressure of oxygen in the bladder is many times higher than its pressure in blood. Gas secretion occurs in many deep-sea fishes despite the hundreds of atmospheres of gas pressure within the bladder.

The gas gland secretes oxygen by releasing lactic acid and carbon dioxide, which acidify the blood in the rete mirabile. Acidification causes hemoglobin to release oxygen into solution (the Bohr and Root effects). Because of the anatomical relations of the rete mirabile, which folds back upon itself in a countercurrent multiplier arrangement, oxygen released from the hemoglobin accumulates and is retained within the rete until its pressure exceeds the oxygen pressure in the swim bladder. At this point oxygen diffuses into the bladder, increasing its volume. The maximum multiplication of gas pressure that can be achieved is proportional to the length of the capillaries of the rete mirabile, and deep-sea fishes have very long retia. A large Root effect is characteristic only of the blood of ray-finned fishes, and it is essential for the function of the gas gland.

Physoclists have no connection between the swim bladder and the gut, so they cannot burp to release excess gas from the bladder. Instead, physoclists open a muscular valve, called the ovale, located in the posterior dorsal region of the bladder adjacent to a capillary bed. The high internal pressure of oxygen in the bladder causes it to diffuse into the blood of this capillary bed when the ovale sphincter is opened.

Cartilaginous fishes (sharks, rays, and ratfish) do not have swim bladders. Instead, these fish use the liver to create neutral buoyancy. The average tissue densities of sharks with their livers removed are heavier than water—1.06 to 1.09 grams per milliliter. The liver of a shark, however, is well known for its high oil content (shark-liver oil). Shark liver tissue has a density of only 0.95 grams per milliliter, which is lighter than water, and the liver may contribute as much as 25 percent of the body mass. A 4-meter tiger shark (*Galeocerdo cuvieri*) weighing 460 kilograms on land may weigh as little as 3.5 kilograms in the sea. Not surprisingly, bottom-dwelling sharks, such as nurse sharks, have livers with fewer and smaller oil vacuoles in their cells, and these sharks are negatively buoyant.

Nitrogen-containing compounds in the blood of cartilaginous fishes also contribute to their buoyancy. Urea and trimethylamine oxide in the blood and muscle tissue provide positive buoyancy because they are less dense than an equal volume of water. Chloride ions, too, are lighter than water and provide positive buoyancy, whereas sodium ions and protein molecules are denser than water and are negatively buoyant. The net effect of these solutes is a significant positive buoyancy.

Many deep-sea fishes have deposits of light oil or fat in the gas bladder, and others have reduced or lost the gas bladder entirely and have lipids distributed throughout the body. These lipids provide static lift, just like the oil in shark livers. Because a smaller volume of the bladder contains gas, the amount of secretion required for a given vertical descent is less. Nevertheless, a long rete mirabile is needed to secrete oxygen at high pressures, and the gas gland in deep-sea fishes is very large. Fishes that migrate over large vertical distances depend more on lipids such as wax esters than on gas for buoyancy, whereas their close relatives that do not undertake such extensive vertical movements depend more on gas for buoyancy.

Air in the lungs of air-breathing aquatic vertebrates reduces their density. Unlike most fishes, air-breathing vertebrates must return to the surface at intervals, so they do not hover at one depth in the water column. Deep-diving animals, such as elephant seals and some whales and porpoises, face a different problem, however. These animals dive to depths of 1000 meters or more and are subjected to pressures more than 100 times higher than at the surface. Under those conditions, nitrogen would be forced from the air in the lungs into solution in the blood and carried to the tissues at high pressure.

When the animal rose toward the surface, the nitrogen would be released from solution. If the animal moved upward too fast, the nitrogen would form bubbles in the tissues—this is what happens when human deep-sea divers get "the bends" (decompression sickness). Specialized diving mammals avoid the problem by allowing the thoracic cavity to collapse as external pressure rises. Air is forced out of the lungs as they collapse, reducing the amount of nitrogen that diffuses into the blood. Even these specialized divers would have problems if they made repeated deep dives, however; a deep dive is normally followed by a period during which the animal remains near the surface and makes only shallow dives until the nitrogen level in its blood has equilibrated with the atmosphere.

4.2 Water and the Sensory World of Fishes

Water has properties that influence the behaviors of fishes and other aquatic vertebrates. Light is absorbed by water molecules and scattered by suspended particles. Objects become invisible at a distance of a few hundred meters even in the very clearest water, whereas distance vision is virtually unlimited in clear air. Fishes supplement vision with other senses, some of which can operate only in water. The most important of these aquatic senses is mechanical and consists of detecting water movement via the lateral line system. Small currents of water can stimulate the sensory organs of the lateral line because water is dense and viscous. Electrical sensitivity is another sensory mode that depends on the properties of water and does not operate in air. In this case it is the electrical conductivity of water that is the key. Even vision is different in water and air because of the different refractive properties of the two media.

■ Vision

Vertebrates generally have well-developed eyes, but the way an image is focused on the retina is different in terrestrial and aquatic animals. Air has an index of refraction of 1.00, and light rays bend as they pass through a boundary between air and a medium with a different refractive index. The amount of bending is proportional to the difference in indices of refraction. Water has a refractive index of 1.33, and the bending of light as it passes between air and water causes underwater objects to appear closer to an observer in air than they really are. The corneas of the eyes of terrestrial and aquatic vertebrates have an index of refraction of about 1.37, so light is bent as it passes through the air-cornea interface. As a result, the cornea of a terrestrial vertebrate plays a substantial role in focusing an image on the retina. This relationship does not hold in water, however, because the refractive index of the cornea is too close to that of water for the cornea to have much effect in bending light. The lens plays the major role in focusing light on the retina of an aquatic vertebrate, and fish have spherical lenses with high refractive indexes. The entire lens is moved toward or away from the retina to focus images of objects at different distances from the fish. Terrestrial vertebrates have flatter lenses, and muscles in the eye change the shape of the lens to focus images. Aquatic mammals such as whales and porpoises have spherical lenses like those of fishes.

■ Taste

Fishes have taste-bud organs in the mouth and around the head and anterior fins. In addition, receptors of general chemical sense detect substances that are only slightly soluble in water, and olfactory organs on the snout detect soluble substances. Sharks and salmon can detect odors at concentrations of less than 1 part per billion. Homeward migrating salmon are directed to their stream of origin from astonishing distances by a chemical signature from the home stream that was permanently imprinted when they were juveniles. Plugging the nasal olfactory organs of salmon destroys their ability to home.

■ Touch

Mechanical receptors provide the basis for detection of displacement—touch, sound, pressure, and motion. Like all vertebrates, fishes have an internal ear (the labyrinth organ, not to be confused with the organ of the same name that assists in respiration in anabantid fishes) that detects changes in speed and direction of motion. Fishes also have gravity detectors at the base of the semicircular canals that allow them to distinguish up from down. Most terrestrial vertebrates also have an auditory region of the inner ear that is sensitive to sound-pressure waves. These diverse functions of the labyrinth depend on basically similar types of sense cells, the hair cells (**Figure 4–4**). In fishes and aquatic amphibians, clusters of hair cells and associated support cells form **neuromast organs** that are dispersed over the surface of the head and body. In jawed fishes, neuromast organs are often located in a series of canals on the head, and one or more canals pass along the sides of the body onto the tail. This surface receptor system of fishes and aquatic amphibians is referred to as the **lateral line system**. Lateral line systems are found only in aquatic vertebrates because air is not dense enough to stimulate the neuromast organs. Amphibian larvae have lateral line systems, and lateral lines are retained in adults of aquatic amphibians such as African clawed frogs and mudpuppies but are lost in the adults of terrestrial species. Terrestrial vertebrates that have secondarily returned to the water, such as whales and porpoises, do not have lateral line systems.

■ Detecting Water Displacement

Neuromasts of the lateral line system are distributed in two configurations—within tubular canals or exposed in epidermal depressions. Many kinds of fishes have both arrangements. Hair cells have a **kinocilium** placed asymmetrically in a cluster of **microvilli**. Hair cells are arranged in pairs with the kinocilia positioned on opposite sides of adjacent cells. A neuromast contains many such hair cell pairs. Each neuromast has two afferent nerves: one transmits impulses from hair cells with kinocilia in one orientation, and the other carries impulses from cells with kinocilia positions reversed by 180 degrees. This arrangement allows a fish to determine the direction of displacement of the kinocilia.

All kinocilia and microvilli are embedded in a gelatinous structure, the **cupula**. Displacement of the cupula causes the kinocilia to bend. The resultant deformation either excites or inhibits the neuromast's nerve discharge. Each hair cell pair, therefore, signals the direction of cupula displacement. The excitatory output of each pair has a maximum sensitivity to displacement along the line joining the kinocilia, and falling off in other directions. The net effect of cupula displacement is to increase the firing rate in one afferent nerve and to decrease it in the other nerve. These changes in lateral line nerve firing rates thus inform a fish of the direction of water currents on different surfaces of its body.

Water currents of only 0.025 millimeters per second are detected by the exposed neuromasts of the African clawed frog, *Xenopus laevis*, with maximum response to currents of 2 or 3 millimeters per second. Similar responses occur in fishes. The lateral line organs also respond to low-frequency sound, but controversy exists as to whether sound is a natural lateral line stimulus. Sound induces traveling pressure

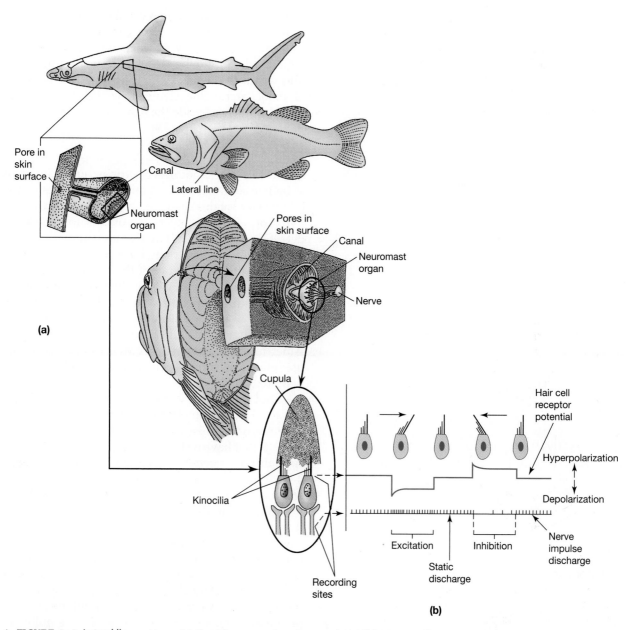

(a)

(b)

▲ **FIGURE 4–4** Lateral line systems. (a) Semidiagrammatic representations of the two configurations of lateral line organs in fishes. (b) Hair cell deformations and their effect on hair cell transmembrane potential (receptor potential) and afferent nerve-cell discharge rates. Deflection of the kinocilium (dark line) in one direction reduces discharge rate, and deflection in the opposite direction increases discharge rate.

waves in the water and also causes local water displacement as the pressure wave passes. It is difficult to be sure whether neuromast output results from the water motions on the body surface or the sound's compression wave.

Several surface-feeding fishes and African clawed frogs provide vivid examples of how the lateral line organs act under natural conditions. These animals find insects on the water surface by detecting surface waves created by the prey's movements. Each neuromast group on the head of the killifish, *Aplocheilus lineatus*, provides information about surface waves coming from a different direction (**Figure 4–5**). All groups, however, have overlapping stimulus fields. Removing a neuromast group from one side of the head disturbs the directional response to stimuli,

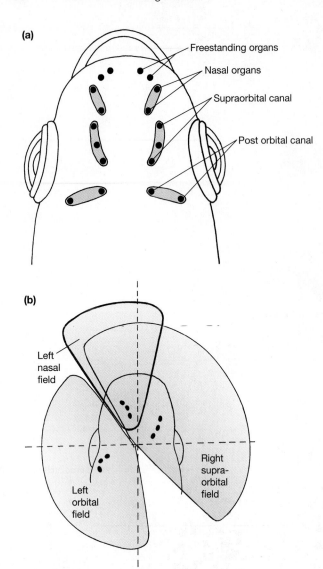

(a)

Freestanding organs

Nasal organs

Supraorbital canal

Post orbital canal

(b)

Left nasal field

Left orbital field

Right supra-orbital field

▲ **FIGURE 4–5** Distribution of the lateral line canal organs. (a) on the dorsal surface of the head of the killifish *Fundulus notatus*. (b) The perceptual fields of the head canal organs in another killifish, *Aplocheilus lineatus*. The wedge-shaped areas indicate the fields of view for each group of canal organs. Note that fields overlap on opposite sides as well as on the same side of the body, allowing the lateral line system to localize the source of a water movement.

showing that a fish combines information from groups on both sides of the head to interpret water movements.

The large numbers of neuromasts on the heads of some fishes might be important for sensing vortex trails in the wakes of adjacent fishes in a school. Many of the fishes that form extremely dense schools (herrings, atherinids, mullets) lack lateral line organs along the flanks and retain canal organs only on the head. These well-developed cephalic canal organs concentrate sensitivity to water motion in the head region, where it is needed to sense the turbulence into which the fish is swimming, and the reduction of flank lateral line elements would reduce noise from turbulence beside the fish.

■ Electrical Discharge

Unlike air, water conducts electricity, and seawater is a better conductor of electricity than freshwater because it contains dissolved salts. The high conductivity of seawater makes it possible for sharks to detect the electrical activity that accompanies muscle contractions of their prey. Electricity can also be a weapon—the torpedo ray of the Mediterranean, the electric catfish from the Nile River, and the electric eel of South American rivers can discharge enough electricity to stun prey animals and deter predators. The weakly electric knifefish (Gymnotidae) of South America and elephant fish (Mormyridae) of Africa use electrical signals for courtship and territorial defense.

All of these electric fish use modified muscle tissue to produce the electrical discharge. The cells of such modified muscles, called **electrocytes**, are muscle cells that have lost the capacity to contract and are specialized for generating an ion current flow (**Figure 4–6**). When at rest, the membranes of muscle cells and nerve cells are electrically charged, with the intracellular fluids about 84 millivolts more negative than the extracellular fluids. The imbalance is primarily due to sodium ion exclusion. When the cell is stimulated, sodium ions flow rapidly across the smooth surface, sending its potential to a positive 67 millivolts. Only the smooth surface depolarizes; the rough surface remains at −84 millivolts, so the potential difference across the cell is 151 millivolts (from −84 to +67 millivolts). Because electrocytes are arranged in stacks like the batteries in a flashlight, the potentials of many layers of cells combine to produce high voltages. The South American electric eel has up to 10,000 layers of cells and can generate potentials in excess of 600 volts.

Most electric fish are found in tropical freshwaters of Africa and South America. Few marine forms can generate specialized electric discharges—among marine cartilaginous fishes only the torpedo ray (*Torpedo*), the ray genus *Narcine*, and some skates are electric; and among marine teleosts, only the stargazers (family Uranoscopidae) produce specialized discharges.

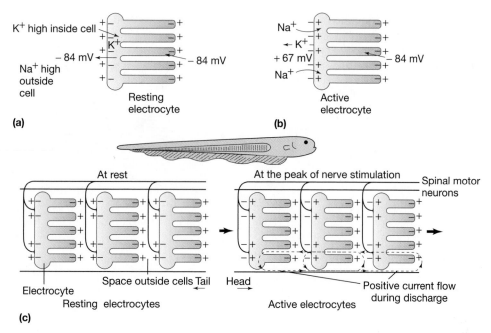

▲ **FIGURE 4–6** Electrical fishes. Some fishes use transmembrane potentials of modified muscle cells by electric fishes to produce a discharge. In this diagram the smooth surface is on the left and the rough surface on the right. Only the smooth surface is innervated. (a) At rest, K+ (potassium ion) is maintained at a high internal concentration and Na+ (sodium ion) at a low internal concentration by the action of a Na+/K+ cell-membrane pump. Permeability of the membrane to K+ exceeds the permeability to Na+. As a result, K+ diffuses outward faster than Na+ diffuses inward (arrow) and sets up the −84 mV resting potential. (b) When the smooth surface of the cell is stimulated by the discharge of the nerve, sodium diffuses into the cell and potassium diffuses out of the smooth surface, changing the potential to +67 mV. The rough surface does not depolarize and retains a −84 mV potential, creating a potential difference of 151 mV across the cell. (c) By arranging electrocytes in series to sum the potentials of individual cells, some electric fishes can generate very high voltages. Electric eels, for example, have 10,000 electrocytes in series and produce potentials in excess of 600 volts.

▪ Electroreception by Sharks and Rays

Many fishes, especially sharks, are able to detect electric fields. Sharks have structures known as the **ampullae of Lorenzini** on their heads, and rays have them on the pectoral fins as well. The ampullae are sensitive electroreceptors (**Figure 4–7**). The canal connecting the receptor to the surface pore is filled with an electrically conductive gel, and the wall of the canal is nonconductive. Because the canal runs for some distance beneath the epidermis, the sensory cell can detect a difference in electrical potential between the tissue in which it lies (which reflects the adjacent epidermis and environment) and the distant pore opening. Thus, it can detect electrical fields, which are changes in electrical potential in space. The ancestral electroreceptor cell was a modification of the hair cells of the lateral line.

Electroreceptors of sharks respond to minute changes in the electrical field surrounding an animal.

They act like voltmeters, measuring a difference in electric potentials at discrete locations across the body surface. Voltage sensitivities are remarkable: ampullary organs have thresholds lower than 0.01 microvolts per centimeter, a level of detection achieved only by the best voltmeters. Sharks use their electrical sensitivity to detect prey. All muscle activity generates electrical potential: motor nerve cells produce extremely brief changes in electrical potential, and muscular contraction generates changes of longer duration. In addition, a steady potential issues from an aquatic organism as a result of the chemical imbalance between the organism and its surroundings. A shark can locate and attack a hidden fish by relying only on this electrical activity (**Figure 4–8**).

The glycoprotein gel that fills the canals can respond to temperature changes of less than 0.001°C. Many sharks track temperature gradients at sea, and they may use this ability to locate boundaries between

▶**FIGURE 4–7** Ampullae of Lorenzini. (a) Distribution of the ampullae on the head of a spiny dogfish, *Squalus acanthius*. Open circles represent the surface pores; the black dots are positions of the sensory cells. (b) A single ampullary organ consists of a sensory cell connected to the surface by a pore filled with a substance that conducts electricity.

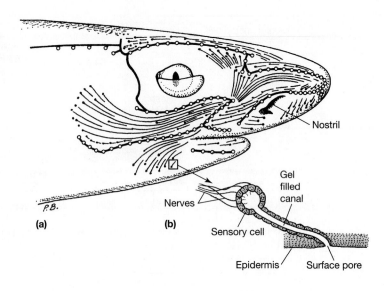

▶**FIGURE 4–8** Electrolocation capacity of sharks. (a) A shark can locate a live fish concealed from sight beneath the sand. (b) The shark can still detect the fish when it is covered by an agar shield that blocks olfactory cues but allows the electrical signal to pass. (c) The shark follows the olfactory cues (displaced by the agar shield) when the live fish is replaced by chopped bait that produces no electrical signal. (d) The shark is unable to detect a live fish when it is covered by a shield that blocks both olfactory cues and the electrical signal. (e) The shark attacks electrodes that give off an electrical signal duplicating a live fish without producing olfactory cues. These experiments indicate that when the shark was able to detect an electrical signal, it used that to locate the fish—and it was also capable of homing on a chemical signal when no electrical signal was present. This dual system allows sharks to find both living and dead food items.

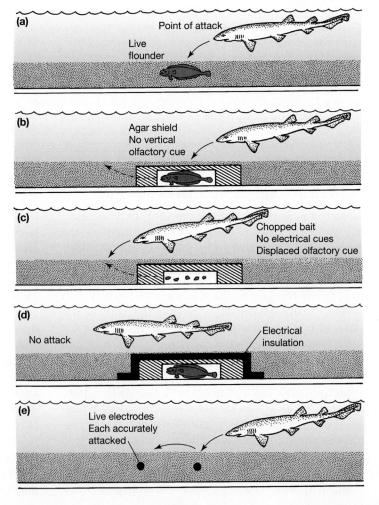

water masses of different temperatures where concentrations of prey fishes are likely to be found.

Sharks may use electroreception for navigation as well. The electromagnetic field at the Earth's surface produces tiny voltage gradients, and a swimming shark could encounter gradients as large as 0.4 millivolts per centimeter—well above the level that can be detected by ampullary organs. In addition, ocean currents generate electrical gradients as large as 0.5 millivolts per centimeter as they carry ions through the Earth's magnetic field.

■ Electrolocation by Teleosts

Unusual arrangements of electrocytes are present in several species of fishes that do not produce electric shocks. In these fishes—which include the knifefishes (Gymnotidae) of South America and the elephant fish (Mormyridae) of Africa—the discharge voltages are too small to be of direct defensive or offensive value. Instead, weakly electric teleost fishes use their discharges for electrolocation and social communications. When a fish discharges its electric organ, it creates an electric field in its immediate vicinity (**Figure 4–9**). Because of the high energy costs of maintaining a continuous discharge, electric fishes produce a pulsating discharge. Most weakly electric teleost fishes pulse at rates between 50 and 300 cycles per second, but the knifefishes of South America reach 1700 cycles per second, which is the most rapid continuous firing rate known for any vertebrate muscle or nerve. African and South American electric fishes are mostly nocturnal and usually live in turbid waters where vision is limited to short distances even in daylight.

The electric field from even weak discharges may extend outward for a considerable distance in freshwater because electric conductivity is relatively low. The electric field the fish creates will be distorted by the presence of conductive and resistant objects. Rocks are highly resistive, whereas other fishes, invertebrates, and plants are conductive. Distortions of the field cause a change in the distribution of electric potential across the fish's body surface. An electric fish detects the presence, position, and movement of objects by sensing where on its body maximum distortion of its electric field occurs.

The skin of weakly electric teleosts contains special sensory receptors: ampullary organs and tuberous organs. These organs detect tonic (steady) and phase (rapidly changing) discharges, respectively. Electroreceptors of teleosts are modified lateral line neuromast receptors. Like lateral line receptors, they have double innervation—an afferent channel that sends impulses

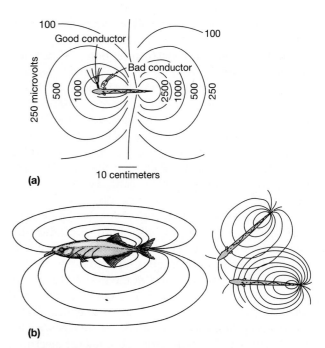

(a)

(b)

▲ **FIGURE 4–9** Electrolocation by fishes. (a) Distortion of the electrical field (lines) surrounding a weakly electric fish by conductive and nonconductive objects. (Reference electrode 150 cm lateral to fish.) Conductive objects concentrate the field on the skin of the fish, where the increase in electrical potential is detected by the electroreceptors. Nonconductive objects spread the field and diffuse potential differences along the body surface. (b) The electrical field surrounding a weakly electric fish can be modulated by variations in the electric organ discharge for communication (left). When two electric fish swim close enough to one another, they offset the frequencies of their discharges to avoid interference (right).

to the brain and an efferent channel that causes inhibition of the receptors. During each electric organ discharge, an inhibitory command is sent to the electroreceptors, and the fish is rendered insensitive to its own discharge. Between pulses, electroreceptors report distortion in the electric field or the presence of a foreign electric field to the brain. Electric organ discharges vary with habits and habitat. Species that form groups or live in shallow, narrow streams generally have discharges with high frequency and short duration. These characteristics reduce the chances of interference from the discharges of neighbors. Territorial species, in contrast, have long electric organ discharges. Electric organ discharges vary from species to species. In fact, some species of electric fishes were first identified by their electric organ discharges, which were recorded by placing electrodes in water that was too murky for any fish to be visible. During the breeding season, electric organ discharges distinguish immature individuals, females with eggs, and sexually active males of some species.

Electrogenesis and electroreception are not restricted to a single group of aquatic vertebrates, and monotremes (the platypus and the echidna, early offshoots off the main mammalian lineages that still lay eggs) use electroreception to detect prey. Electrosensitivity was probably an early feature of vertebrate evolution. The brain of the lamprey responds to electric fields, and it seems likely that the earliest vertebrates had electroreceptive capacity. All fishlike vertebrates of lineages that evolved before the neopterygians have electroreceptor cells. These cells, which have a prominent kinocilium, fire when the environment around the kinocilium is negative relative to the cell. Their impulses pass to the midline region of the posterior third of the brain. Electrosensitivity was apparently lost in neopterygians, and teleosts have at least two separate new evolutions of electroreceptors. Electrosensitivity in teleosts is distinct from that of other vertebrates: teleost electroreceptors *lack* a kinocilium and fire when the environment is *positive* relative to the cell, and nerve impulses are sent to the lateral portions of the brain rather than to the midline.

4.3 The Internal Environment of Vertebrates

Seventy to eighty percent of the body mass of most vertebrates is water, and the chemical reactions that release energy or synthesize new chemical compounds take place in an aqueous environment. The body fluids of vertebrates contain a complex mixture of ions and other solutes. Some ions are cofactors that control the rates of metabolic processes; others are involved in the regulation of pH, the stability of cell membranes, or the electrical activity of nerves. Metabolic substrates and products must diffuse from sites of synthesis to the sites of utilization. Almost everything that happens in the body tissues of vertebrates involves water, and maintaining the concentrations of water and solutes within narrow limits is a vital activity. Water sounds like an ideal place to live for an animal that is itself mostly made of water, but in some ways an aquatic environment can be too much of a good thing. Freshwater vertebrates—especially fishes and amphibians—face the threat of being flooded with water that flows into them from their environment, and saltwater vertebrates must prevent the water in their bodies from being sucked out into the sea.

Temperature, too, is a critical factor for living organisms because chemical reactions are temperature sensitive. In general, the rates of chemical reactions increase as temperature increases, but not all reactions have the same sensitivity to temperature. Furthermore, the permeability of cell membranes and other features of the cellular environment are sensitive to temperature. A metabolic pathway is a series of chemical reactions in which the product of one reaction is the substrate for the next, yet each of these reactions may have a different sensitivity to temperature, so a change in temperature can mean that too much or too little substrate is produced to sustain the next reaction in the series. To complicate the process of regulation of substrates and products even more, the chemical reactions take place in a cellular milieu that itself is changed by temperature. Clearly, the smooth functioning of metabolic pathways is greatly simplified if an organism can limit the range of temperatures its tissues experience.

Water temperature is more stable than air temperature because water has a much higher heat capacity than air. The stability of water temperature simplifies the task of maintaining a constant body temperature, as long as the body temperature the animal needs to maintain is the same as the temperature of the water around it. An aquatic animal has a hard time maintaining a body temperature different from water temperature, however, because water conducts heat so well. Heat flows out of the body if an animal is warmer than the surrounding water and into the body if the animal is cooler than the water.

In the following sections, we discuss in more detail how and why vertebrates regulate their internal environments and the special problems faced by aquatic animals.

4.4 Exchange of Water and Ions

An organism can be described as a leaky bag of dirty water. That is not an elegant description, but it accurately identifies the two important characteristics of a living animal—it contains organic and inorganic substances dissolved in water, and this fluid is enclosed by a permeable body surface. Exchange of matter and energy with the environment is essential to the survival of the organism, and much of that exchange is regulated by the body surface. Water molecules and ions pass through the skin quite freely, whereas larger molecules move less readily. The significance of this differential permeability is particularly conspicuous in the case of aquatic vertebrates, but it

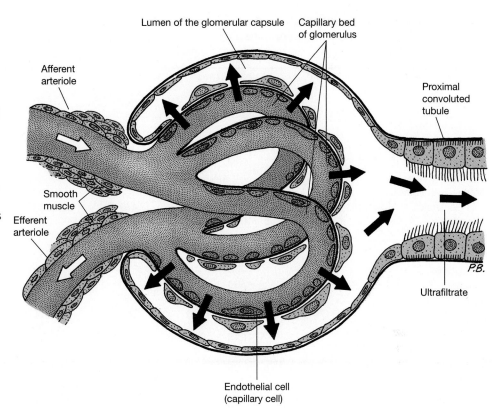

▶ **FIGURE 4–10** Detail of a typical mammalian glomerulus. Blood pressure forces an ultrafiltrate of the blood through the walls of the capillary into the lumen of the glomerular capsule. The blood flow to each glomerulus is regulated by smooth muscles that can close off the afferent and efferent arterioles to adjust the glomerular filtration rate of the kidney as a whole. The ultrafiltrate, which consists of water, ions, and small molecules, passes from the glomerular capsule into the proximal convoluted tubule, where the process of adding and removing specific substances begins.

Labels in figure: Lumen of the glomerular capsule; Capillary bed of glomerulus; Afferent arteriole; Proximal convoluted tubule; Smooth muscle; Efferent arteriole; Endothelial cell (capillary cell); Ultrafiltrate; P.B.

applies to terrestrial vertebrates as well. Vertebrates use both active and passive exchange to regulate their internal concentrations in the face of varying external conditions.

The Vertebrate Kidney

An organism can tolerate only a narrow range of concentrations of the body fluids and must eliminate waste products before they reach harmful levels. The molecules of ammonia that result from the breakdown of protein are especially important because they are toxic. Vertebrates have evolved superb capacities for controlling water balance and excreting wastes, and the kidney plays a crucial role in these processes.

The adult vertebrate kidney consists of hundreds to millions of tubular **nephrons**, each of which produces urine. The primary function of a nephron is removing excess water, salts, waste metabolites, and foreign substances from the blood. In this process, the blood is first filtered through the **glomerulus**, a structure unique to vertebrates (**Figure 4–10**). Each glomerulus is composed of a leaky arterial capillary tuft encapsulated within a sievelike filter. Arterial blood pressure forces fluid into the nephron to form an **ultrafiltrate**,

composed of blood minus blood cells and larger molecules. The ultrafiltrate is then processed to return essential metabolites (glucose, amino acids, and so on) and water to the general circulation. The fluid that is left after this processing is urine.

Regulation of Ions and Body Fluids

The salt concentrations in the body fluids of many marine invertebrates are similar to those in seawater, as are those of hagfishes (**Table 4.2**). It is likely that the first vertebrates also had ion levels similar to those in seawater. In contrast, salt levels are greatly reduced in the blood of all other vertebrates, a characteristic shared only with invertebrates that have penetrated estuaries, fresh waters, or the terrestrial environment.

In the context of body fluids, **solute** means a small molecule that is dissolved in water or blood plasma. Salt ions, urea, and some small carbohydrate molecules are the solutes primarily involved in the regulation of body fluid concentrations. The presence of solutes lowers the kinetic activity of water. Therefore, water flows from a dilute solution (high kinetic activity of water) to a more concentrated solution (low kinetic activity)—a phenomenon called **osmosis**. Seawater has a concentration of approximately 1000 millimoles per

TABLE 4.2 Representative concentrations of sodium and chloride and osmolality of the blood in vertebrates and marine invertebrates. Concentrations are expressed in millimoles per kilogram of water; all values are reported to the nearest 5 units.

Type of Animal	mmol	Na$^+$	Cl$^-$	Other Major Osmotic Factor
Seawater	~1000	475	550	
Freshwater	<10	~5	~5	
Marine invertebrates				
Coelenterates, mollusks, etc.	~1000	470	545	
Crustacea	~1000	460	500	
Marine vertebrates				
Hagfishes	~1000	535	540	
Lamprey	~300	120	95	
Teleosts	<350	180	150	
Coelacanth	<1000 to 1180	180	200	Urea 375
Elasmobranch (bull shark)	1050	290	290	Urea 360
Holocephalian	~1000	340	345	Urea 280
Freshwater vertebrates				
Polypterids	200	100	90	
Acipenserids	250	130	105	
Primitive neopterygians	280	150	130	
Dipnoans	240	110	90	
Teleosts	<300	140	120	
Elasmobranch (bull shark)	680	245	220	Urea 170
Elasmobranch (freshwater rays)	310	150	150	
Amphibians*	~250	~100	~80	
Terrestrial vertebrates				
Reptiles	350	160	130	
Birds	320	150	120	
Mammals	300	145	105	

*Ion levels and osmolality are variable but tend toward 200 mmol · kg^{-1} in freshwater.

kilogram of water (mmol·kg^{-1}). Most marine invertebrates and hagfishes have body fluids that are in osmotic equilibrium with seawater; that is, they are **isosmolal** to seawater. Body fluid concentrations in marine teleosts and lampreys are between 300 and 350 mmol·kg^{-1}. Therefore, water flows outward from their blood to the sea (i.e., from a region of high kinetic activity of water to a region of lower kinetic activity). Cartilaginous fishes retain urea and other nitrogen-containing compounds, raising the osmolality of their blood slightly above that of seawater so water flows from the sea into their bodies. These osmolal differences are specified by the terms **hyposmolal** (lower solute concentrations than the surrounding water, as seen in marine teleosts and lampreys) and **hyperosmolal** (higher solute concentrations than the surrounding water, as seen in coelacanths and cartilaginous fishes).

Salt ions, such as sodium and chloride, can also diffuse through the surface membranes of an animal, so the water and salt balance of an aquatic vertebrate in seawater is constantly threatened by outflow of water and inflow of salt and in freshwater by inflow of water and outflow of salts.

Most fishes are **stenohaline** (Greek *steno* = narrow and *haline* = salt); that means that they inhabit either freshwater or seawater and tolerate only modest changes in salinity. Because they remain in one environment, the magnitude and direction of the osmotic gradient to which they are exposed is stable. Some fishes, however, move between freshwater and seawater and tolerate large changes in salinity. These fishes are called **euryhaline** (Greek *eury* = wide), and water and salt gradients are reversed in euryhaline species as they move from one medium to the other.

Freshwater Organisms—Teleosts and Amphibians

Several mechanisms are involved in the salt and water regulation of vertebrates that live in freshwater. The body surface of fishes has low permeability to water and to ions. However, fishes cannot entirely prevent osmotic exchange. Gills, which must be permeable to oxygen and carbon dioxide, are also permeable to water. As a result, most water and ion movements take place across the gill surfaces. Water is gained by osmosis, and ions are lost by diffusion. A freshwater teleost does not drink water because osmotic water movement is already providing more water than it needs—drinking would only increase the amount of water it had to excrete via the kidneys. To compensate for this influx of water, the kidney of a freshwater fish or amphibian produces a large volume of urine. Salts are actively reabsorbed to reduce salt loss. Indeed, urine processing in a freshwater teleost provides a simple model of vertebrate kidney function.

The large glomeruli of freshwater teleosts produce a copious flow of urine, but the glomerular ultrafiltrate is isosmolal to the blood and contains essential blood salts (**Figure 4–11**). To conserve salt, ions are reabsorbed across the **proximal** and **distal convoluted tubules**. (Reabsorption is an active process that consumes metabolic energy.) Because the distal convoluted tubule is impermeable to water, water remains in the tubule and the urine becomes less concentrated as ions are removed from it. Ultimately, the urine becomes hyposmolal to the blood. In this way the water that was absorbed across the gills is removed and ions are conserved. Nonetheless, some ions are lost in the urine in addition to those lost by diffusion across the gills. Salts from food compensate for some of this loss, and teleosts have cells in the gills (**chloride cells**) that actively transport chloride ions. In freshwater, these cells take up chloride ions from the water. The chloride ions are moved by active transport against a concentration gradient, and this process requires energy. Sodium ions also enter the gills, passively following the chloride ions.

Freshwater amphibians face the same osmotic problems as freshwater fishes. The entire body surface of amphibians is involved in the active uptake of ions from the water. Like freshwater fishes, aquatic amphibians do not drink because the osmotic influx of water more than meets their needs. Also like fishes, aquatic amphibians lose ions by diffusion. The skin of these amphibians contains cells that actively take up ions from the surrounding water. Acidity inhibits this active transport of ions in both amphibians and fishes, and inability to maintain internal ion concentrations is one of the causes of death of these animals in habitats acidified by acid rain and snow.

Marine Organisms—Teleosts and Other Fishes

The osmotic and ionic gradients of vertebrates in seawater are basically the reverse of those experienced by freshwater vertebrates. Seawater is more concentrated than the body fluids of vertebrates, so there is a net outflow of water by osmosis and a net inward diffusion of ions.

Teleosts The integument of marine fishes, like that of freshwater teleosts, is highly impermeable so that most osmotic and ion movements occur across the gills (**Figure 4–12**). The kidney glomeruli are small, and the glomerular filtration rate is low. Little urine is formed, and the water lost in urine is reduced. Marine teleosts lack a water-impermeable distal convoluted tubule. As a result, urine leaving the nephron is less copious but more concentrated than that of freshwater teleosts, although it is always hyposmolal to blood. To compensate for osmotic dehydration, marine teleosts do something unusual—they drink seawater. Sodium and chloride ions are actively absorbed across the lining of the gut, and water flows by osmosis into the blood. Estimates of seawater consumption vary, but many species drink in excess of 25 percent of their body weight per day and absorb 80 percent of this ingested water. Of course, drinking seawater to compensate for osmotic water loss increases the influx of sodium and chloride ions. To compensate for this salt load, chloride cells in the gills actively pump chloride ions outward against a large concentration gradient.

Hagfishes and Cartilaginous Fishes Hagfishes have few problems with ion balance because they regulate only divalent ions and reduce osmotic water movement by being nearly isosmolal to seawater. Cartilaginous fishes and coelacanths also minimize osmotic flow by maintaining the internal concentration of the body fluid close to that of seawater. These animals retain nitrogen-containing compounds (primarily urea and trimethylamine oxide) to produce osmolalities that are usually slightly hyperosmolal to

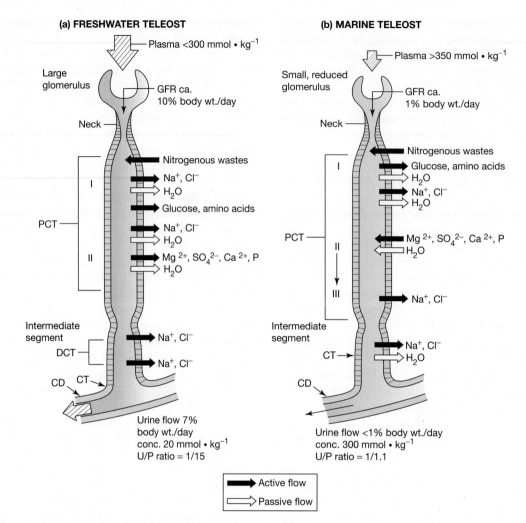

(a) FRESHWATER TELEOST

Plasma <300 mmol • kg^{-1}

Large glomerulus

GFR ca. 10% body wt./day

Neck

Nitrogenous wastes
Na$^+$, Cl$^-$
H$_2$O

I

Glucose, amino acids

PCT

Na$^+$, Cl$^-$
H$_2$O

II

Mg^{2+}, SO$_4^{2-}$, Ca^{2+}, P
H$_2$O

Intermediate segment

DCT

Na$^+$, Cl$^-$

Na$^+$, Cl$^-$

CD   CT

Urine flow 7% body wt./day
conc. 20 mmol • kg^{-1}
U/P ratio = 1/15

(b) MARINE TELEOST

Plasma >350 mmol • kg^{-1}

Small, reduced glomerulus

GFR ca. 1% body wt./day

Neck

I

Nitrogenous wastes
Glucose, amino acids
H$_2$O
Na$^+$, Cl$^-$
H$_2$O

PCT

II

Mg^{2+}, SO$_4^{2-}$, Ca^{2+}, P
H$_2$O

III

Na$^+$, Cl$^-$

Intermediate segment

CT

Na$^+$, Cl$^-$
H$_2$O

CD

Urine flow <1% body wt./day
conc. 300 mmol • kg^{-1}
U/P ratio = 1/1.1

Active flow
Passive flow

▲ **FIGURE 4–11** Kidney structure and function of marine and freshwater teleosts. Arrows pointing into the lumen of the kidney tubules show movement of substances into the forming urine, and arrows pointing outward show movement from the urine back into the body fluids. Dark arrows show active movements (i.e., those that involve a transport system) and light arrows show passive flow down an activity gradient. GFR is the glomerular filtration rate; that is, the rate at which the ultrafiltrate is formed. It is expressed as percentage of body weight per day. Freshwater teleosts are flooded by water that they must excrete; consequently they have high GFRs. Marine teleosts have the opposite problem; they lose water by osmosis to their surroundings and must conserve water in the kidney—consequently they have low GFRs. PCT is the proximal convoluted tubule. (Two segments [I and II] of the PCT are recognized in both freshwater and marine teleosts. Segment III of the PCT of marine fishes is sometimes equated with the DCT [distal convoluted tubule] of freshwater fishes.) Substances a fish needs to conserve (primarily glucose and amino acids) are actively removed from the ultrafiltrate in the PCT, and nitrogenous waste products (ammonia and urea) are actively added to the forming urine. Freshwater fishes actively remove divalent ions (magnesium, sulfate, calcium, and phosphorus) from the forming urine, whereas marine fishes actively excrete those ions into the urine. Sodium and chloride are also removed from the forming urine. That makes sense for freshwater fish because they are trying to conserve those ions, but it is surprising for marine fishes because they are battling an influx of excess sodium and chloride from seawater. The explanation of that paradox for marine fishes is that movement of sodium and chloride ions is needed to produce a passive flow of water back into the body. Freshwater fish continue to reabsorb sodium and chloride in the CT (collecting tubule), but the walls of their CTs are not permeable to water so there is no passive uptake at this stage. In contrast, the CTs of marine fishes are permeable to water, allowing further recovery of water at this point. The net effect of the differences in GFR, inward and outward movements, and permeabilities allows freshwater fishes to produce copious amounts of dilute urine that rids them of excess water and marine fishes to produce scanty amounts of more concentrated urine that conserves water.

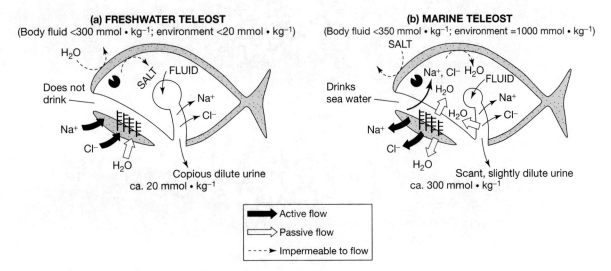

▲ **FIGURE 4–12** Water and salt regulation by freshwater and marine teleosts. The body fluids of freshwater fishes are more concentrated than the water surrounding them; consequently they gain water by osmosis and lose sodium and chloride by diffusion. They do not drink water; they actively absorb sodium and chloride through the gills, and they have kidneys with large glomeruli that produce a large volume of dilute urine. Marine fishes are less concentrated than the water they live in; consequently they lose water by osmosis and gain sodium and chloride by diffusion. Marine fish drink water and actively excrete the sodium and chloride through the gills. They have kidneys with small glomeruli that produce small volumes of more concentrated urine.

seawater (see Table 4.2). As a result, cartilaginous fishes gain water by osmotic diffusion across the gills and do not need to drink seawater. This net influx of water permits large kidney glomeruli to produce high filtration rates and therefore rapid elimination of metabolic waste products from the blood. Urea is very soluble and diffuses through most biological membranes, but the gills of cartilaginous fishes are nearly impermeable to urea and the kidney tubules actively reabsorb it. With internal ion concentrations that are low relative to seawater, cartilaginous fishes experience ion influxes across the gills as do marine teleosts. Unlike the gills of marine teleosts, those of cartilaginous fishes have low ion permeabilities (less than 1 percent those of teleosts). Cartilaginous fishes generally do not have highly developed salt-excreting cells in the gills. Rather, they achieve ion balance by secreting from the rectal gland a fluid that is approximately isosmolal to body fluids and seawater but contains higher concentrations of sodium and chloride ions than do the body fluids.

Freshwater Sharks and Rays and Marine Amphibians
Some cartilaginous fishes are euryhaline—sawfishes, some stingrays, and bull sharks are examples. In seawater, bull sharks retain high levels of urea; but in freshwater their blood urea levels decline. Stingrays

in the family Potamotrygonidae spend their entire lives in freshwater and have very low blood urea concentrations. Their blood sodium and chloride ion concentrations are 35 to 40 percent below those in sharks that enter freshwater and are only slightly above levels typical of freshwater teleosts (see Table 4.2). The potamotrygonids may have lived in the Amazon basin for tens of millions of years, and their reduced salt and water gradients may reflect long adaptation to freshwater. When exposed to increased salinity, potamotrygonids do not increase the concentration of urea in the blood as euryhaline species do, even though the enzymes required to produce urea are present. Apparently their long evolution in freshwater has led to an increase in the permeability of their gills to urea and reduced the ability of their kidney tubules to reabsorb it.

Most amphibians are found in freshwater or terrestrial habitats. One of the few species that occurs in salt water is the crab-eating frog, *Fejervarya* (formerly *Rana*) *cancrivora*. This frog inhabits intertidal mudflats in Southeast Asia and is exposed to 80 percent seawater at each high tide. During seawater exposure, the frog allows its blood ion concentrations to rise, and thus reduces the ionic gradient. In addition, ammonia is removed from proteins and rapidly converted to urea, which is released into the blood.

Blood urea rises from 20 to 30 mmol · kg^{-1}, and the frogs become hyperosmolal to the surrounding water. In this sense, the crab-eating frog acts like a shark and absorbs water osmotically. Frog skin, unlike that of sharks, is permeable to urea; and urea is thus rapidly lost. To compensate for this loss, the activity of the urea-synthesizing enzymes is very high. Tadpoles of the crab-eating frog, like most tadpoles, lack urea-synthesizing enzymes until late in their development. Thus, tadpoles of crab-eating frogs must use a method of osmoregulation different from that of adults. The tadpoles have salt-excreting cells in the gills; and, by pumping ions outward as they diffuse inward, the tadpoles maintain their blood hyposmolal to seawater in the same manner as do marine teleosts.

■ Nitrogen Excretion by Aquatic Vertebrates

Carbohydrates and fats are composed of carbon, hydrogen, and oxygen, and the waste products from their metabolism are carbon dioxide and water molecules that are easily voided. Proteins and nucleic acids are another matter, for they contain nitrogen. When protein is metabolized, the nitrogen is enzymatically reduced to ammonia through a process called deamination. Ammonia is very soluble in water and diffuses readily, but it is also extremely toxic. Rapid excretion of ammonia is therefore crucial. Differences in how ammonia is excreted are partly a matter of the availability of water and partly the result of differences among phylogenetic lineages. Nitrogen is eliminated by most vertebrates as ammonia, as urea, or as uric acid. Many vertebrates excrete all three of these substances, but the proportions of the three compounds differ among the groups of vertebrates.

Many aquatic invertebrates excrete ammonia directly, as do vertebrates with gills, permeable skins, or other permeable membranes that contact water. Excretion of nitrogenous wastes as ammonia is called **ammonotelism**, excretion as urea is **ureotelism**, and excretion as uric acid is **uricotelism**. Urea is synthesized from ammonia in a cellular enzymatic process called the **urea cycle**. Urea synthesis requires more energy than does ammonia production, but urea is less toxic than ammonia.

Urea has two advantages. First, it is retained by some marine vertebrates to counter osmotic dehydration. A second function of urea synthesis is the detoxification of ammonia when there is not enough

water available to allow it to be excreted as fast as it is produced. Because urea is not very toxic, it can be concentrated in urine, thus conserving water.

Ureotelism probably evolved independently several times. Perhaps ureotelism developed in early freshwater fishes to avoid osmotic dehydration as they reinvaded the sea. Cartilaginous fishes and coelacanths illustrate the use of urea and other nitrogen-containing compounds to regulate internal osmotic concentrations at high levels. Ureotelism was an important factor for the evolution of terrestrial vertebrates, because it allowed nitrogen to be retained in a detoxified state until water was available to excrete it.

4.5 Responses to Temperature

Vertebrates occupy habitats from cold polar latitudes to hot deserts. To appreciate this adaptability, we must consider how temperature affects a vertebrate such as a fish or amphibian that has little capacity to maintain a difference between its body temperature and the temperature of the water around it (a poikilotherm). Organisms have been called bags of chemicals catalyzed by enzymes. This description emphasizes that living systems are subject to the laws of physics and chemistry just as nonliving systems are. Because temperature influences the rates at which chemical reactions proceed, temperature vitally affects the life processes of organisms. The rates of most chemical reactions increase or decrease when the temperature changes. The ratio of the rate at one temperature compared to the rate at a temperature 10°C higher or lower is called Q_{10}. Because Q_{10} is the *ratio* of the rates, a Q_{10} of 1.0 means that the rate stays the same, a Q_{10} greater than 1 means the rate increases, and a Q_{10} less than 1.0 indicates the rate decreases (**Figure 4–13**). A Q_{10} can be used to describe the effect of temperature on biological processes at all levels of biological organization from whole animals down to molecules.

The **standard metabolic rate** (SMR) of an organism is the minimum rate of oxygen consumption needed to sustain life. That is, the SMR includes the costs of ventilating the lungs or gills, of pumping blood through the circulatory system, of transporting ions across membranes, and of all the other activities necessary to maintain the integrity of an organism. The SMR does not include the costs of activities like locomotion or the cost of growth. The SMR is temperature sensitive, and that means the energy cost of living is affected by changes in body temperature. If the SMR of a fish is 2 joules per

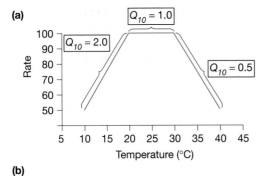

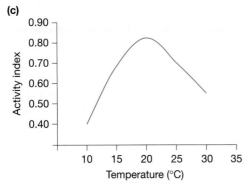

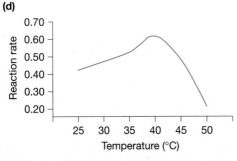

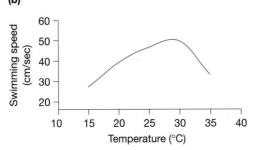

▲ **FIGURE 4–13** The effect of temperature on organisms.
(a) A hypothetical reaction for which the rate initially increases
and then falls as temperature rises. The Q_{10} of a reaction can
be calculated between any two temperatures that are far
enough apart to be biologically meaningful by using the
formula $Q_{10} = (R_2/R_1)^{10/(T_2-T_1)}$, where R_1 and R_2 are the rates
at temperatures T_1 and T_2, respectively. (b) The maximum
swimming speed of a goldfish. (c) Spontaneous activity by
goldfish. (d) The activity of the enzyme lactic dehydrogenase
from a lungfish.

minute at 10°C and the Q_{10} is 2, the fish will use
4 joules per minute at 20°C and 8 joules per minute
at 30°C. That increase in energy use translates to a
corresponding increase in the amount of food the
fish must eat.

■ Controlling Body Temperature— Ectothermy and Endothermy

Because the rates of many biological processes are
affected by temperature, it would be advantageous
for any animal to be able to control its body temperature. However, the high heat capacity and heat conductivity of water make it difficult for most fishes or
aquatic amphibians to maintain a temperature difference between their bodies and their surroundings.
Air has both a lower heat capacity and a lower conductivity than water, and the body temperatures of
most terrestrial vertebrates are at least partly independent of the air temperature. Some aquatic vertebrates also have body temperatures substantially
above the temperature of the water around them.
Maintaining these temperature differences requires
thermoregulatory mechanisms, and these are well
developed among vertebrates.

The classification of vertebrates as poikilotherms
(Greek *poikilo* = variable and *therm* = heat) and
homeotherms (Greek *homeo* = the same) was widely
used through the middle of the twentieth century,
but this terminology has become less appropriate
as our knowledge of the temperature-regulating
capacities of a wide variety of animals has become
more sophisticated. Poikilothermy and homeothermy
describe the variability of body temperature, and
these terms cannot readily be applied to groups of
animals. For example, mammals have been called
homeotherms and fishes poikilotherms, but some
mammals allow their body temperatures to drop
20°C or more from their normal levels at night and in
the winter, whereas many fishes live in water that
changes temperature less than 2°C in an entire year.
That example presents the contradictory situation of a
homeotherm that experiences 10 times as much variation in body temperature as a poikilotherm.

Complications like these make it very hard to use
the words *homeotherm* and *poikilotherm* rigorously.
Most biologists concerned with temperature regulation prefer the terms *ectotherm* and *endotherm*. These
terms are *not* synonymous with poikilotherm and
homeotherm because, instead of referring to the variability of body temperature, they refer to the sources
of energy used in thermoregulation. **Ectotherms**

(Greek *ecto* = outside) gain their heat largely from external sources—by basking in the sun, for example, or by resting on a warm rock. **Endotherms** (Greek *endo* = inside) largely depend on metabolic production of heat to raise their body temperatures. The source of heat used to maintain body temperature is the major difference between ectotherms and endotherms. Terrestrial ectotherms (like lizards and turtles) and endotherms (like birds and mammals) all have activity temperatures ranging from 30°C to 40°C.

Endothermy and ectothermy are not mutually exclusive mechanisms of temperature regulation, and many animals use them in combination. In general, birds and mammals are endothermal, but some species make extensive use of external sources of heat. For example, roadrunners are predatory birds living in the deserts of the southwestern United States and adjacent Mexico. On cold nights, roadrunners become hypothermic, allowing their body temperatures to fall from the normal level of 38 or 39°C down to 35°C or less. In the mornings they bask in the sun, raising the feathers on their backs to expose an area of black skin. Calculations indicate that a roadrunner can save 132 joules per hour by using solar energy instead of metabolism to raise its body temperature.

Deviations from general patterns of temperature regulation go the other way as well. Snakes are normally ectothermal, but the females of several species of python coil around their eggs and produce heat by rhythmic contraction of their trunk muscles. The rate of contraction increases as air temperature falls, and a female Indian python is able to maintain her eggs close to 30°C at air temperatures as low as 23°C. This heat production entails a substantial increase in the python's metabolic rate—at 23°C, a female python uses about 20 times as much energy when she is brooding as she does normally. Thus, generalizations about the body temperatures and thermoregulatory capacities of vertebrates must be made cautiously, and the actual mechanisms used to regulate body temperature must be studied carefully.

■ Regional Heterothermy–Warm Fishes

Regulation of body temperature is not an all-or-nothing phenomenon for vertebrates. **Regional heterothermy** is a general term used to refer to different temperatures in different parts of an animal's body. Dramatic examples of regional heterothermy are found in several fishes that maintain some parts of their bodies at temperatures 15°C warmer than the water in which they are swimming. That's a remarkable accomplishment for a fish because each time the

blood passes through the gills it comes into temperature equilibrium with the water. Thus, to raise its body temperature by using endothermal heat production, a fish must prevent the loss of heat to the water via the gills.

The mechanism used to retain heat is a countercurrent system of blood flow in retia mirabilia. As cold arterial blood from the gills enters the warm part of the body, it flows through a rete and is warmed by heat from the warm venous blood that is leaving the tissue. This arrangement is found in some sharks, especially species in the family Lamnidae (including the mako, great white shark, and porbeagle), which have retia mirabilia in the trunk. These retia retain the heat produced by activity of the swimming muscles, with the result that those muscles are kept 5°C to 10°C warmer than water temperature.

Scombroid fishes, a group of teleosts that includes the mackerels, tunas, and billfishes (swordfish, sailfish, spearfish, and marlin), have also evolved endothermal heat production. Tuna have an arrangement of retia that retains the heat produced by myoglobin-rich swimming muscles located close to the vertebral column (**Figure 4–14**). The temperature of these muscles is held near 30°C at water temperatures from 7°C to 23°C. Additional heat exchangers are found in the brains and eyes of tunas and sharks, and these organs are warmer than water temperature but somewhat cooler than the swimming muscles.

The billfishes have a somewhat different arrangement in which only the brain and eyes are warmed, and the source of heat is a muscle that has changed its function from contraction to heat production. The superior rectus eye muscle of these billfishes has been extensively modified. Mitochondria occupy more than 60 percent of the cell volume, and changes in cell structure and biochemistry result in the release of heat by the calcium-cycling mechanism that is usually associated with contraction of muscles. A related scombroid, the butterfly mackerel, has a thermogenic organ with the same structural and biochemical characteristics found in billfishes, but in the mackerel it is the lateral rectus eye muscle that has been modified.

An analysis of the phylogenetic relationships of scombroid fishes by Barbara Block and her colleagues (Block et al. 1993) suggests that endothermal heat production has arisen independently three times in the lineage—once in the common ancestor of the living billfishes (by modification of the superior rectus eye muscle), once in the butterfly mackerel lineage (modification of the lateral rectus eye muscle), and a third time in the common ancestor of tunas and bonitos (involving the development of countercurrent

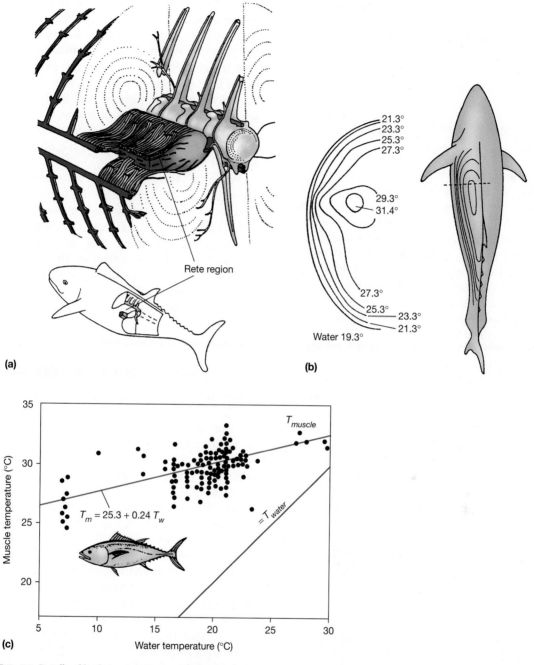

(a)

(b)

(c)

T_{muscle}

$T_m = 25.3 + 0.24\, T_w$

$= T_{water}$

Muscle temperature (°C)

Water temperature (°C)

Rete region

Water 19.3°

▲ **FIGURE 4–14** Details of body temperature regulation by the blue fin tuna. (a) The red muscle and retia are located adjacent to the vertebral column. (b) Cross-sectional views showing the temperature gradient between the core (at 31.4°C) and water temperature (19.3°C). (c) Core muscle temperatures of blue fins compared to water temperature.

heat exchangers in muscle, viscera, and brain, and development of red muscle along the horizontal septum of the body).

The ability of these fishes to keep parts of the body warm may allow them to venture into cold water that would otherwise interfere with body

functions. Block (1993) has pointed out that modification of the eye muscles and the capacity for heat production among scombroids is related to the temperature of the water in which they swim and capture prey. The metabolic capacity of the heater cells of the butterfly mackerel, which is the species that occurs in the coldest water, is the highest of all vertebrates. Swordfishes, which dive to great depths and spend several hours in water temperatures of 10°C or less, have better-developed heater organs than do marlins, sailfishes, and spearfishes, which spend less time in cold water.

■ Warm Bodies, Cold Seas—Marine Mammals and Sea Turtles

The temperature equilibration of blood and water that occurs in the gills is the primary obstacle to whole-body endothermy for fish. Countercurrent systems allow them to keep critical parts of their bodies warm, but other parts are at water temperature. Air-breathing aquatic tetrapods avoid that problem because they have lungs instead of gills. In addition, fully aquatic mammals have a layer of insulation that helps to retain metabolic heat in the body. As a result, marine mammals can maintain their entire bodies at normal mammalian temperatures even though some of them spend their lives in water that is 30 degrees or more below body temperature.

Mammals are endotherms, and their high metabolic rates combined with the muscular metabolism that accompanies activity release heat that warms the body. A furry body covering (the **pelage**) traps the metabolic heat in the dead air spaces between hairs and reduces the movement of heat out of the body, just as the air trapped between strands of fiberglass insulations reduces the movement of heat through the wall of a building. A fur coat works very well in air but not as well in water. If water displaces the air between the hairs, the coat loses its insulative properties. Many semiaquatic mammals, such as otters and beavers, have water-repellent pelts that retain air very well, but a fur-covered aquatic mammal must groom its coat on land frequently to renew the layer of trapped air.

The most specialized marine mammals, cetaceans (whales and porpoises) and pinnipeds (seals, sea lions, and walruses) use **blubber** (a layer of fat beneath the skin) rather than fur for insulation. Blubber is an extremely effective insulator, so good that some seals risk death by overheating if they undertake prolonged strenuous activity on land. Even in water, strenuous activity can lead to overheating. Aquatic mammals have countercurrent exchange systems in their flippers that allow them to retain heat in the body or release it to the ocean. The venous blood returning from the flipper is cold because it has passed through a flat structure with a large surface

area that is in contact with the frigid ocean water. Those are ideal conditions for heat exchange, and the blood that leaves the flippers is very close to water temperature. When a marine mammal needs to retain heat in the body, blood returning from the flippers flows through veins that are closely associated with the arteries that carry blood from the body to the flippers. Cold venous blood is heated by warm arterial blood flowing out from the core of the body. By the time the venous blood reaches the body, it is nearly back to body temperature.

When a bout of rapid swimming produces enough heat to increase body temperature above normal, the animal changes the route that venous blood takes as it returns from the flipper. Instead of flowing through veins that are pressed closely against the walls of the arteries, the returning blood is shunted through vessels distant from the arteries. The arterial blood is still hot when it reaches the flipper, and that heat is dissipated into the water, cooling the animal.

■ Body Size and Surface-to-Volume Ratio

Body size is an extremely important element in the exchange occurring between an organism and its environment. For objects of the same shape, volume increases as the cube of linear dimensions, whereas surface area increases only as the square of linear dimensions. Consider a cube that is 1 centimeter (cm) on each side. Each face of the cube is 1 cm² (that is, 1 cm · 1 cm) and a cube has six faces (numbered 1 through 6 if the cube comes from a set of dice), so the total surface area is 6 cm². The volume of that cube is one cubic centimeter (1 cm · 1 cm · 1 cm), and the ratio of surface to volume is 6 cm²/1 cm³.

If we double the linear dimensions of the cube, each side becomes 2 centimeters long. Each face has a surface area of 4 cm² (that is, 2 cm · 2 cm), and the total surface area is 24 cm² (6 · 4 cm²). The volume of this larger cube is 8 cm³ (that is, 2 cm · 2 cm · 2 cm), and the ratio of surface to volume is 24 cm²/8 cm³, which reduces to 3 cm²/1 cm³. Thus the larger cube has only half as much surface area per unit volume as the smaller cube. A cube 3 centimeters on a side has a surface-to-volume ratio of 54 cm²/27 cm³ (or, 2 cm²/1 cm³); a cube 4 cm on a side has a ratio of 96 cm²/64 cm³ (or, 1.5 cm²/1 cm³).

The pattern emerging from these calculations is shown in **Figure 4–15**: as an object gets larger, it has progressively less surface area in relation to its volume. Exchange between an animal and its environment

▶**FIGURE 4–15** Linear dimensions, surface area, and volume. A cube illustrates how the surface-to-volume ratio changes with size. As the length of the side of the cube is increased from 1 cm to 10 cm, total surface area of the cube increases as the square of that length, whereas the volume of the cube increases as the cube of the length of a side. Because volume increases more rapidly than surface area, the surface-to-volume ratio of the cube decreases as the size of the cube increases. Functionally this means that as an object becomes larger, it has less surface area relative to its volume. Thus the rate of exchange with the environment decreases. For example, if you take two cubes—one that is 1 cm on a side and the other 10 cm on a side—then heat them to the same temperature and put them side by side on a table, the small cube will cool to room temperature faster than the large cube.

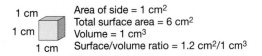

1 cm
1 cm
1 cm

Area of side = 1 cm^2
Total surface area = 6 cm^2
Volume = 1 cm^3
Surface/volume ratio = 1.2 cm^2/1 cm^3

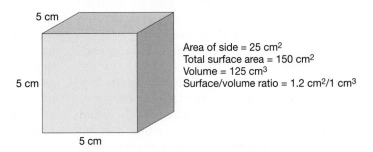

5 cm
5 cm
5 cm

Area of side = 25 cm^2
Total surface area = 150 cm^2
Volume = 125 cm^3
Surface/volume ratio = 1.2 cm^2/1 cm^3

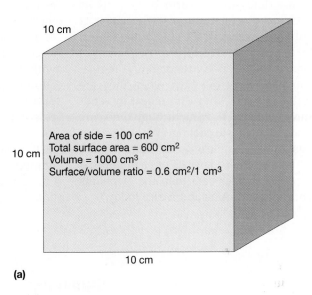

10 cm
10 cm
10 cm

Area of side = 100 cm^2
Total surface area = 600 cm^2
Volume = 1000 cm^3
Surface/volume ratio = 0.6 cm^2/1 cm^3

(a)

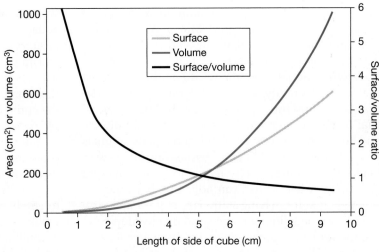

(b)

occurs through its body surface, and larger species have proportionally less area for exchange in relation to the volume (or mass) of their bodies. The biological significance of that relationship lies in the conclusion that bigger species exchange energy with the environment less rapidly than smaller species merely because of the difference in surface-to-volume ratio.

Simply being big gives an animal some independence of external temperature because heat cannot flow rapidly in or out of a large body through its relatively small surface. The enormous dinosaurs that lived in the Jurassic and Cretaceous periods would have had very stable body temperatures just by virtue of their size, and it would take many days for a brontosaurus-like dinosaur to warm or cool as its environment changed temperature. Even elephants (which are only one-twentieth the size of the largest dinosaurs) are big enough to feel the consequences of surface-to-volume ratio in body temperature regulation. Elephants can easily overheat when they are active. When that happens, they dump heat by sending large volumes of blood flowing through their ears and waving them to promote cooling. (We know from cave paintings made by Pleistocene humans that mammoths, the far northern species of elephant that is now extinct, had smaller ears than the elephants living in warm climates in Asia and Africa today.)

Being big makes temperature regulation in water easier, as leatherback sea turtles dramatically illustrate. Leatherbacks (*Dermochelys coriacea*) are the largest living turtles, reaching adult body masses of 850 kilograms or more. They are also the most specialized sea turtles, having lost the bony shell that covers most turtles and replaced it with a thick, leatherlike external body covering. Leatherback turtles are **pelagic** (live in the open ocean); their geographic range extends north to Alaska and halfway up the coast of Norway and south past the southern tip of Africa almost to the tip of South America. Water temperatures in these areas are frigid, and the body temperatures of the turtles are as much as 18°C higher than the water. Turtles have metabolic rates much lower than those of mammals. Nonetheless, the combination of large body size and a correspondingly small surface-to-volume ratio with countercurrent heat exchangers in the flippers allows leatherback sea turtles to retain the heat produced by muscular activity. Large body size is an essential part of the turtle's temperature-regulating mechanism. Other sea turtles are half the size of leatherbacks or less, and their geographic ranges are limited to warm water because they are not big enough to maintain a large difference between their body temperatures and water temperature.

Summary

The properties of water offer both advantages and disadvantages for aquatic vertebrates. Water is some 800 times more dense than air, and vertebrates in water are close to neutral buoyancy. That means the skeletons of aquatic vertebrates do not have to resist the force of gravity, and the largest aquatic vertebrates are substantially larger than the largest terrestrial forms. Animal body tissues (especially muscle and bone) are denser than water, and aquatic animals offset that weight with lighter tissues (air-filled swim bladders and lungs or oily livers) to achieve neutral buoyancy. A teleost fish can adjust its buoyancy so precisely that it can hang stationary in the water with only a backpedaling of its pectoral fins to counteract the forward propulsion generated by water leaving the gills and a gentle undulation of the tail fin to stay level in the water.

The density and viscosity of water create problems for animals trying to move through water or to move water across gas-exchange surfaces. Even slow-moving aquatic vertebrates must be streamlined, and a tidal respiratory system (moving the respiratory fluid in and out of a lung) takes too much energy to be practical for vertebrates that breathe water. These animals either have flow-through ventilation (the gills of most fishes) or use the entire body surface for aquatic gas exchange (amphibians).

Water is less transparent than air and vision is often limited to short distances, but the density and electrical conductivity of water make mechanical and electrical senses effective. Sensors that are exquisitely sensitive to the movement of water are distributed over the bodies of fishes and aquatic amphibians. In addition, cartilaginous fishes can find hidden prey by detecting the electrical discharges from contracting muscles as the prey breathes. Other fishes create electrical fields in the water to detect the presence of prey or predators, and some use powerful electric organ discharges to stun prey or deter predators.

Aquatic animals are continuously gaining or losing water and ions to their surroundings. Water flows from areas of high kinetic activity (dilute solutions) to low kinetic activity (concentrated solutions), and ions move down their own activity gradients. The body fluids of freshwater fishes are more concentrated than the surrounding water, so they are flooded by an inward flow of water and further diluted by an outward diffusion of ions. Marine fishes are less concentrated than seawater, so they must contend with an outward flow of water and an inward diffusion of ions. Some marine fishes (especially cartilaginous fishes and coelacanths) accumulate urea in the body to raise their internal osmotic concentration close to that of seawater.

Ammonia is a waste product from deamination of proteins. It is toxic but very soluble in water, so it is easy for aquatic vertebrates to excrete. Urea and uric acid are less toxic compounds vertebrates use to dispose of waste nitrogen. Some vertebrates produce mixtures of all three compounds, changing the proportions as the availability of water changes.

Water has a high heat capacity and conducts heat readily. Because of these properties, water temperature is more stable than air temperature, and it is hard for an aquatic animal to maintain a difference between its own body temperature and the temperature of the water surrounding it. Fish have a particularly difficult time maintaining a body temperature different from water because the temperature of their blood comes into equilibrium with water temperature as the blood passes through the gills. Nonetheless, some fish use countercurrent heat exchange systems to keep parts of their bodies at temperatures well above water temperature. The largest extant sea turtle, the leatherback, is able to maintain a body temperature 18°C above water temperature. Part of the temperature difference can be traced to countercurrent heat exchangers, which minimize loss of heat from the flippers; the enormous body size of these turtles is also a critical factor. Body surface area increases as the square of linear dimensions, whereas body volume increases as the cube of linear dimensions. Consequently, large animals exchange heat energy with the environment more slowly than do smaller animals. Merely being big confers a degree of stability to the internal environment of an animal.

Additional Readings

Berenbrink, M. et al. 2005. Evolution of oxygen secretion in fishes and the emergence of a complex physiological system. *Science* 307:1752–1757.

Block, B. A. et al. 1993. Evolution of endothermy in fish: Mapping physiological traits on a molecular phylogeny. *Science* 260:210–214.

Stoddard, P. K. 1999. Predation enhances complexity in the evolution of electric fish signals. *Nature* 400:254–256.

von der Emde, G. 1998. Electroreception. In D. H. Evans (ed.) *The Physiology of Fishes*. Boca Raton, FL: CRC Press: 313–314.

Westby, G. W. M. 1988. The ecology, discharge diversity, and predatory behavior of gymnotiform electric fish in the coastal streams of French Guiana. *Behavioral Ecology and Sociobiology* 22:341–354.

Withers, P. C. et al. 1994. Buoyancy role of urea and TMAO in an elasmobranch fish, the Port Jackson shark, *Heterodontus portjacksoni*. *Physiological Zoology* 67:693–705.

CHAPTER 5

Radiation of the Chondrichthyes

The appearance of jaws and internally supported, paired appendages described in Chapter 3 was a major event in vertebrate history. The diversity of predatory specializations available to a vertebrate with jaws and precise steering is great, and the appearance of these characters signaled a new radiation of vertebrates. The cartilaginous fishes (sharks, rays, and ratfishes) are the descendants of one clade of this radiation, and they combine derived characters such as a cartilaginous skeleton with a generally primitive anatomy. Sharks have undergone three major radiations, which can be broadly associated with increasingly specialized feeding mechanisms, and extant sharks and rays are a diverse and successful group of fishes. In this chapter, we consider the origins and success of extant cartilaginous fishes.

5.1 Chondrichthyes—The Cartilaginous Fishes

The sharks and their relatives make a definite first appearance in the fossil record in the Early Devonian period (although isolated scales appear earlier) with a unique combination of derived and ancestral characters. Bone, found in many agnathans (including those possibly ancestral to gnathostomes), is absent from the endoskeleton of Chondrichthyes. Mineralization of the axial and appendicular skeleton takes place in the superficial layers of cartilage matrix as globular or stellate deposits of crystalline calcium. These deposits are visible as areas of calcification in the surface of a cartilaginous skeletal structure. This condition is known as tesserate or prismatic endoskeletal calcification. This unique mode of endoskeletal mineralization is the defining synapomorphy for Chondrichthyes. (The presence of male intromittent structures—pelvic appendage claspers—is a possible but unconfirmed second synapomorphy. [Grogan and Lund 2004].)

Chondrichthyes present a mosaic of ancestral and derived vertebrate characters: The loss of bone is a derived character that is probably associated with lightening the body and making it more maneuverable, and some of the organ and physiological systems of chondrichthyans have evolved to levels surpassed by few other extant vertebrates. At the same time, these fishes retain many primitive anatomical characters; sharks have long been used to exemplify an ancestral vertebrate body form. Extant forms can be divided into two groups: those with a single gill opening on each side of the head and those with multiple gill openings on each side (Table 5.1).

The clade with one gill opening is the Holocephali (Greek *holo* = whole and *cephalo* = head), so named for the undivided appearance of the head that results from having a single gill opening. The common names of this group—ratfish, rabbitfish, and chimera— come from their bizarre form: a long flexible tail, a fishlike body, and a head with big eyes and buckteeth that resembles a caricature of a rabbit. The Neoselachii (Greek *neo* = new and *selach* = shark) have multiple gill openings on each side of the head and include the sharks (most often cylindrical forms with

TABLE 5.1 Classification of chondrichthyes, the cartilaginous fishes

Neoselachii (sharks, skates, and rays)
Galeomorpha (Galea): About 279 species of sharklike fishes with an anal fin, encompassing sand tigers, mackerel sharks, megamouth shark, thresher sharks, basking shark, requiem sharks, and possibly hornsharks, wobbegons, nurse sharks, and whale sharks, from less than 1 m to definitely 12 m and possibly 18 m.
"Squalomorpha": About 124 species of sharklike fishes without an anal fin (and, according to some systematists, all skates and rays, which also lack an anal fin). The sharklike forms encompass most deep-sea sharks, dogfish sharks, angel sharks, and saw sharks, 15 cm to more than 7 m. (The "Squalomorpha" lineage is not monophyletic, but the relationships of the four to six lineages it contains are not yet understood.)
Batoidea (skates and rays): At least 534 species of electric rays, stingrays, manta rays, a large number of skates, and a few entirely freshwater species, from under 1 m to over 6 m in length and over 6 m wide.

Holocephali (ratfishes)
Chimaeriformes: Three families with about 33 species of mostly deepwater fishes: the plownose chimaeras (Callorhynichidae), shortnose chimaeras (Chimaeridae), and longnose chimaeras (Rhinochimeridae), 60 cm to nearly 1.5 m.

five to seven gill openings on each side of the head), as well as skates and rays (flattened forms with five pairs of gill openings on the ventral surface of the head).

5.2 Evolutionary Specializations of Chondrichthyes

Despite a rather good fossil record, the phylogeny of cartilaginous fishes remains unclear. Early Chondrichthyes, like extant species, were diverse in form and habitats. In the Late Devonian, sharks were found primarily in freshwater habitats in contrast to their primarily marine distribution today. Their initial radiation from a common ancestor emphasized changes in teeth, jaws, and fins that evolved at different rates in different lineages. A derived dentition is found with an ancestral fin structure in some lineages, whereas the opposite combination is seen in other lineages. Thus, fossil chondrichthyans display mosaics of ancestral and derived characters.

Through time, different lineages of Chondrichthyes developed similar but not identical modifications in feeding and locomotor structures. This pattern of similar adaptations appearing independently and repeatedly in related lineages is an example of parallel evolution. In the following sections, we will trace three radiations of chondrichthyans, focusing on the fusiform (torpedo-shaped) predators popularly known as sharks, and describing progressive changes in tooth and jaw structure and in the form of the fins and tail. This analysis focuses on the parallel evolution of

characters in multiple lineages during the Paleozoic and Mesozoic eras. Next we will examine skates and rays, and finally we will describe the bizarre and poorly known forms called ratfishes.

5.3 The Paleozoic Chondrichthyan Radiation

The stem chondrichthyans are identified by the form of the teeth common to most members—basically three-cusped with little root development (**Figure 5-1**). Although there is evidence of bone around their bases, the teeth are primarily composed of dentine capped with an enameloid coat. The central cusp is the largest in *Cladoselache*, the best-known genus, and smallest in *Xenacanthus*, a more specialized form.

Cladoselache was sharklike in appearance, about 2 meters long when fully grown, with large fins and mouth and five separate external gill openings on each side of the head. The mouth opened terminally, and ligaments attached the palatoquadrate tightly to the chondrocranium. The jaw also obtained some support from the second branchial arch, the hyoid arch. The term **amphistylic** (Greek *amphi* = both and *styl* = pillar or support) is applied to this mode of multiple sites of upper jaw suspension, which may have been the ancestral condition for gnathostomes. The gape was large, the jaws extending well behind the rest of the skull. The three-pronged teeth were probably especially efficient for feeding on soft-bodied prey, such as fishes or cephalopods that could

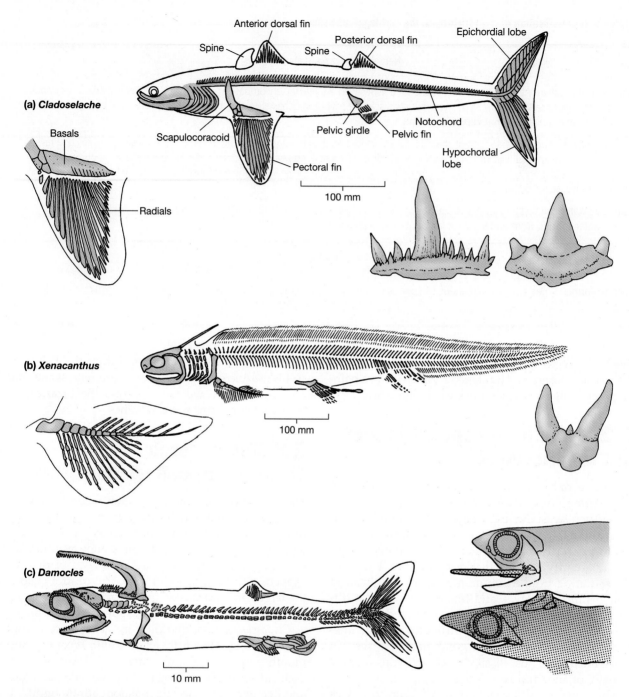

▲ **FIGURE 5–1** Early chondrichthyans. (a) *Cladoselache*. (b) *Xenacanthus*, a freshwater form with details of its primitive (archipterygial) pectoral fin structure and peculiar teeth. (c) *Left*—male *Damocles serratus*, a 15-centimeter shark from the Late Carboniferous period showing sexually dimorphic nuchal spine and pelvic claspers; *right*—male (below) and female (above) as fossilized, possibly in courtship position.

(a) **(b)**

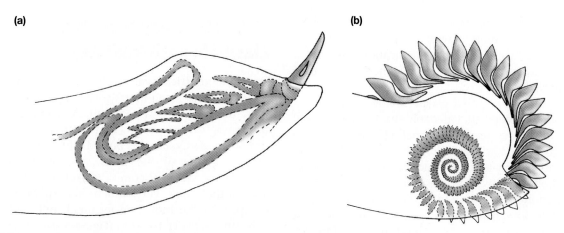

▲ **FIGURE 5–2** Tooth replacement by chondrichthyans. (a) Cross section of the jaw of an extant shark, showing a single functional tooth backed by a band of replacement teeth in various stages of development. (b) Lateral view of the symphysial (middle of the lower jaw) tooth whorl of the early chondrichthyan edestoid *Helicoprion*, showing the chamber into which the lifelong production of teeth spiraled.

be swallowed whole or severed by the knife-edged cusps.

As teeth are used, they become worn; cusps break off, and cutting edges grow dull. In sharks, each tooth on the functional edge of the jaw is but one member of a tooth whorl, attached to a ligamentous band that courses down the inside of the jaw cartilage deep below the fleshy lining of the mouth (**Figure 5–2**).

Aligned in each whorl in a file directly behind the functional tooth are a series of developing teeth. Essentially the same dental apparatus is present in all sharks, living and extinct. Tooth replacement is rapid: young sharks replace each lower-jaw tooth as often as every 8.2 days and each upper-jaw tooth every 7.8 days.

The body of *Cladoselache* was supported only by a notochord, but cartilaginous neural arches gave added protection to the spinal cord. The fins of *Cladoselache* consisted of two dorsal fins, paired pectoral and pelvic fins, and a well-developed forked tail. The first and sometimes second dorsal fins were preceded by stout spines, triangular in cross section and thought to have been covered by soft tissue during the life of the shark. The dorsal fins were broad triangles with an internal structure consisting of a triangular basal cartilage and a parallel series of long radial cartilages extending to the margin of the fin. The pectoral fins were larger but similar in construction.

Among the early radiations of sharks, almost every type seems to have had a different sort of internal pectoral fin arrangement. However, all possessed basal elements that anchored the pectoral fins in place. The pectoral fins appear to have had little capacity for altering their angle of contact with the water. The pelvic fins were smaller, but otherwise much like the pectorals. Some species had pelvic fins with claspers (male copulatory organs). No anal fin is known, and it is also lacking in many extant sharks.

The caudal fin of *Cladoselache* is distinctive (see Figure 5–1a). Externally symmetrical, its internal structure was asymmetrical and contained elements resembling the hemal arches that protect the caudal blood vessels in extant sharks. Long, unsegmented radial cartilages extended into the hypochordal (lower) lobe of the fin. At the base of the caudal fin were paired lateral keels that are identifying characteristics of extant rapid pelagic (open-water) swimmers.

The skin had only a few scales, which were limited to the fins, around the eye, and within the mouth behind the teeth. These scales resembled the teeth—cusps of dentine covered with an enamel-like substance, all of which enveloped a cellular core or pulp cavity. Scales of this sort are called **placoid scales**. Each scale of these most ancient sharks had several pulp cavities corresponding to the several cusps.

We can piece together the lives of many of the early chondrichthyans from their morphology and fossil localities. Most early sharks, and *Cladoselache* in particular, were probably pelagic predators that swam after their prey in a sinuous manner, engulfing it whole or slashing it with daggerlike teeth. The lack of body denticles and calcification suggests a tendency to reduce weight and thereby increase buoyancy.

The presence of pelvic claspers shows that at least some, and perhaps all, Paleozoic sharks had internal fertilization. Two species of small (15 cm) sharklike forms from the Early Carboniferous of Montana may show how complex reproductive behavior was in early chondrichthyans. Males of these species can be identified by pelvic claspers, a sharp rostrum, and an enormous forwardly curved middorsal spine just behind the head (see Figure 5–1c), whereas females had neither claspers nor spines. One of the fossils may be a pair that died in a precopulatory courtship position, with the female grasping the male's dorsal spine in her jaws.

Another early group of chondrichthyan fishes, the edestoids, had large pectoral fins and stiff, symmetrical, deeply forked tails. Among extant sharks, these features are characteristic of fast-swimming oceanic species. The edestoids had a peculiar dentition (see Figure 5–2b). Most of the tooth whorls were greatly reduced, but the central (symphysial) tooth row of the mandible was tremendously enlarged, and each tooth interlocked with adjacent teeth at its base. Several members of this tooth whorl were functional at the same time. Some forms had blunt teeth for crushing shelled prey, and others had teeth with knife-blade cusps. The mandibular tooth row bit against small, flat teeth in the palatoquadrate of the upper jaw. Most edestoids replaced their teeth rapidly, the oldest worn teeth being shed from the tip of the mandible. In contrast, *Helicoprion* retained all its teeth in a specialized chamber into which the lifelong production of teeth spiraled. Perhaps the teeth in this chamber provided a solid foundation for the functional teeth.

One of the score or so of genera produced in the early radiation of chondrichthyans was *Xenacanthus*, which had a braincase, jaws, and jaw suspension very similar to those of *Cladoselache*. But there the resemblance ends. The xenacanths were freshwater bottom-dwellers with very robust fins and heavily calcified, cartilaginous skeletons that would have decreased their buoyancy. The xenacanths appeared in the Devonian and survived until the Triassic period, when they died out without leaving direct descendants. Details of their gills and fin skeletons indicate that xenacanths are at the base of the elasmobranch lineage.

5.4 The Early Mesozoic Elasmobranch Radiation

Further chondrichthyan evolution involved changes in feeding and locomotor systems. Species exhibiting these modifications appear in the Carboniferous, and this radiation of stem chondricthyans flourished until the Late Cretaceous period. *Hybodus* was a well-known genus of the Late Triassic through the Cretaceous. We have complete skeletons 2 meters in length that look very much like modern sharks except that the mouth is terminal, not underslung beneath a sensory rostrum (**Figure 5–3a**).

The heterodont dentition (that is, different-shaped teeth along the jaw) of hybodont sharks (*Hybodus* and its relatives) seems pivotal to their success. The anterior teeth had sharp cusps and appear to have been used for piercing, holding, and slashing softer foods. The posterior teeth were stout, blunt versions of the anterior teeth in batteries consisting of several teeth from each individual tooth whorl. The living hornsharks of the genus *Heterodontus* have similar dentition (**Figure 5–3b**). Hornsharks feed on small fishes, crabs, shrimp, sea urchins, clams, mussels, and oysters. The sharp anterior teeth seize and kill soft-bodied food, and the pavementlike posterior teeth crush the shells of crustaceans and mollusks.

The hybodont sharks also showed advances in the structure of the pectoral and pelvic fins that made them more mobile than the broad-based fins of Paleozoic sharks. Both pairs of fins were supported on narrow stalks formed by three narrow, platelike basal cartilages that replaced the long series of basals seen in earlier sharks. The narrow base allowed the fin to be rotated to different angles as the shark swam up or down. The blade of the fin also changed: The cartilaginous radials were segmented and did not extend to the fin margin. Proteinaceous, flexible fin rays called **ceratotrichia** extended from the outer radials to the margin of the fin. Muscles within the fin could curve it from front to back and from base to tip. The mobility and flexibility of these fins allowed them to be used for steering in ways that seem impossible for the more rigid fin construction characteristic of *Cladoselache*. By assuming different shapes, the pectoral fins could produce lift anteriorly, aid in turning, or stabilize straight-line movement.

Along with changes in the paired fins, the caudal fin assumed new functions, and an anal fin appeared. Caudal fin shape was altered by reduction of the hypochordal lobe, division of its radials, and addition

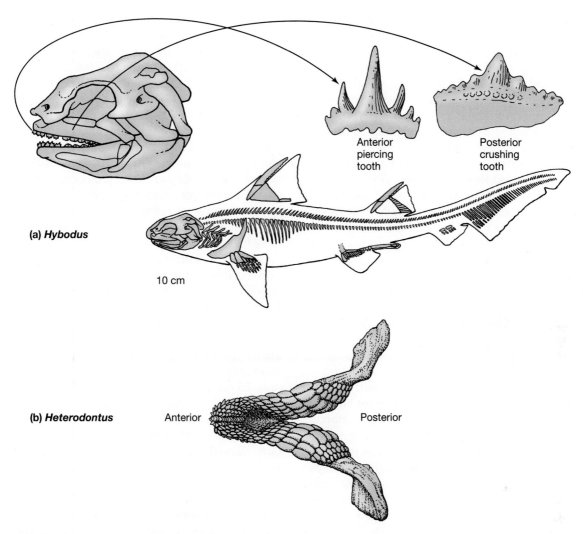

▲ **FIGURE 5–3** Fossil eleasmobranchs. (a) The fossil elasmobranch *Hybodus*. (b) Palatal view of upper jaw (palatoquadrate) of the extant hornshark, *Heterodontus*, with a dentition similar to that of many sharks during the late Paleozoic and early Mesozoic.

of flexible ceratotrichia. This tail-fin arrangement is generally known as **heterocercal** (Greek *hetero* = different and *kerkos* = tail), although **epicercal** (Greek *epi* = above) is a more precise name for this tail shape—known from as far back as some Paleozoic jawless fishes (Chapter 3). The value of the elasmobranch heterocercal tail lies in its flexibility (because of the more numerous radial skeletal elements) and the control of shape made possible by the intrinsic musculature. When it was undulated from side to side, the fin twisted so that the flexible lower lobe trailed behind the stiff upper one. This distribution of force produced forward and upward thrust that could lift a shark from a resting position or counteract its tendency to sink as it swam horizontally.

Other morphological changes in sharks of the second major radiation include the appearance of a complete set of hemal arches that protected the arteries and veins running below the notochord; well-developed ribs; and narrow, more pointed dorsal-fin spines closely associated with the leading edges of the dorsal fins. These spines were ornamented with ridges and grooves and studded with barbs on the posterior surface, suggesting that they were used in defense. Claspers are found in all species, leaving little doubt that they had elaborate courtship and internal fertilization.

Hybodus and its relatives resembled their presumed *Cladoselache*-like ancestors in having terminal mouths, an amphistylic jaw suspension, unconstricted notochords, and multicusped teeth, but a direct line cannot be drawn between the two in time or in morphology. Some forms considered related to *Hybodus* had *Cladoselache*-like dentition combined with pectoral fins that had three basal supporting elements and were nearly immobile; others developed a very tetrapod-like support for highly mobile pectoral fins. Their caudal fin was reduced, and they probably moved around on the seafloor using their limblike pectoral fins. Another form, known only from a 5-centimeter juvenile, had a paddle-shaped rostrum one-third its body length. Other types were 2.5-meter giants with blunt snouts and enormous jaws. Despite their variety and success during the Mesozoic, this second radiation of chondrichthyans became increasingly rare and disappeared at the end of the Mesozoic or in the early Cenozoic era.

5.5 The Extant Radiation—Sharks, Skates, and Rays

The first representatives of the extant radiation of chondrichthyans appeared at least as early as the Triassic. By the Jurassic period, sharks of modern appearance had evolved, and a surprising number of Jurassic and Cretaceous genera are still extant. The most conspicuous difference between most members of the earlier radiations and extant sharks is the rostrum or snout that overhangs the ventrally positioned mouth in most extant forms. Less obvious, but of major importance, was the development of solid, calcified vertebrae which constricted—and in some species even replaced—much of the notochord. A third innovation of the extant chondrichthyans is a thicker and more structurally complex enamel-like material on the teeth than seen in earlier groups.

■ Sharks

The technical characters distinguishing the clades of extant sharks are subtle. There are about 403 species of pleurotreme (Greek *pleur* = the side and *trem* = a hole) neoselachians—the sharks with gill openings on the sides of the head (**Figure 5–4**). Although similar in overall appearance, the extant pleurotreme sharks actually come from at least two lineages (see Figure 3–14), and recent molecular studies suggest that four or more related lineages may be included. More ancestral in their general anatomy—especially the smaller size of their brains—are the approximately 124 species of squaloid sharks and their numerous relatives. (Some authorities group these

animals as the Squalea, but other specialists do not believe that they form a monophyletic lineage.) Squaloids include the spiny and green dogfish, the cookie-cutter shark, and the basking and megamouth sharks. These species usually live in cold, deep water. The other 280 or so species of sharks are almost all members of the galeoid (Galea) lineage that includes the whale shark, the mackerel sharks (including the great white shark), the carcharhinid or requiem sharks, the hammerhead sharks, and perhaps the hornshark, nurse, and carpet sharks as well. Galeoid sharks are the dominant carnivores of shallow, warm, species-rich regions of the oceans.

Throughout their evolutionary history, sharks have been consummate carnivores. In the third radiation of the chondrichthyans in the mid-Mesozoic, derived locomotor, trophic, sensory, and behavioral characteristics produced forms that still dominate the top levels of marine food webs today. Sharks show enormous diversity in size. A typical shark is about two meters long, the largest extant forms attain at least 12 meters and perhaps grow to 18 meters, and a few interesting miniature forms are only 25 centimeters long and inhabit mostly deeper seas off the continental shelves.

Despite their range in size, all extant sharks have common skeletal characteristics that earlier radiations lacked. The cartilaginous vertebral centra of extant sharks are distinctive. Between centra, spherical remnants of the notochord fit into depressions on the opposing faces of adjacent vertebrae. Thus, the axial skeleton can flex from side to side with rigid central elements of calcified cartilage swiveling on ball-bearing joints of notochordal remnants. In addition to the neural and hemal arches, extra elements that are not found in the axial skeleton of other vertebrates (the intercalary plates) protect the spinal cord above and the major arteries and veins below the centra.

Shark placoid scales also changed in the third radiation. Although scales of the same general type are known from earlier chondrichthyans, they are often in clusters or fused into larger plates. The shagreen (sharkskin) body covering of modern sharks is a unique armor that is flexible yet very protective. The placoid scales of extant sharks have a single cusp and a single pulp cavity. The size, shape, and arrangement of these placoid scales reduce turbulence in the flow of water next to the body surface and increase the efficiency of swimming. (Swimsuits duplicating the surface properties of sharkskin appear to cut a few hundredths of a second from a swimmer's time—enough to be the margin of victory

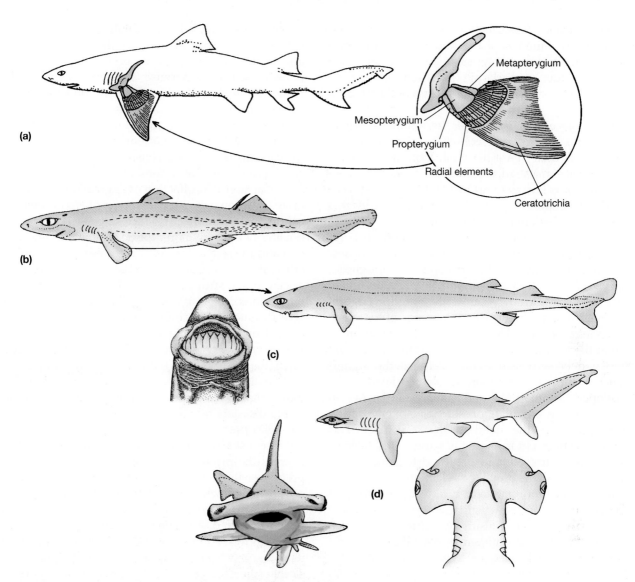

▲ **FIGURE 5–4** Representative extant sharks. (a) *Negaprion brevirostris*, the lemon shark, grows to a length of 2 meters. The internal anatomy of the pectoral girdle and fin are shown superimposed in their correct relative positions. (b) *Etmopterus vierens*, the green lantern shark, is a bioluminescent miniature shark only 25 centimeters long, yet it feeds on much larger prey items. (c) *Isistius brasiliensis*, the cookie-cutter shark, is another miniature species whose curious mouth (left) is able to take chunks from fish and cetaceans much larger than itself. (d) Hammerhead shark (*Sphyrna*) in lateral, ventral, and frontal views. Some species of hammerheads grow to lengths of 3 meters or more.

in world-level competition.) Individual scales in ancestral sharks often fused to form larger scales as the shark grew larger, whereas extant sharks add more and larger scales to their skin as they grow.

Sensory Systems and Prey Detection The sensory systems of extant sharks, skates, and rays are refined and diverse. In Chapter 4 we described the extraordi-

nary sensitivity of the neuromast organs and ampullae of Lorenzini to electrical potentials and minute temperature differences. Sharks use their thermal and electrosensitivities to locate boundaries between water masses of different temperatures where concentrations of prey species are likely to be found, to detect hidden prey, and to navigate through the open sea. Ampullae of Lorenzini, composed of soft tissues, do not normally fossilize, but their concentration on the rostrum of modern sharks and the near universal appearance of the rostrum as a new character of the extant radiation of sharks makes a strong case for the role of electrosensitivity in the success of the modern forms.

Chemoreception is another important sense. In fact, sharks have been described as swimming noses, so acute is their sense of smell. Experiments have shown that some sharks respond to chemicals in concentrations as low as 1 part in 10 billion! The strange heads of hammerhead sharks of the genus *Sphyrna* (see Figure 5–4) may enhance the directionality of their olfactory apparatus by placing the nostrils far apart on the odd lateral expansions of their heads, but the anatomy of the scalloped hammerhead, *Sphyrna lewini*, does not support this interpretation. In this species, grooves across the anterior edge of the head lead to the incurrent openings of the olfactory sacs so water is channeled into the sacs from all along the leading edge of the head rather than from two widely separated points. Perhaps the significance of the wide, flat hammer-shaped head of the scalloped hammerhead and similar sharks lies in spreading the ampullae of Lorenzini over a larger area than does the standard shark rostrum. This arrangement might increase the sensitivity to electrical impulses from buried prey and to minute geomagnetic gradients used for navigation.

Vision is important to the feeding behavior of sharks, and vision at low light intensities is especially well developed. This sensitivity is due to a rod-rich retina and cells with numerous platelike crystals of guanine that are located just behind the retina in the eye's choroid layer. Collectively called the tapetum lucidum, these cells contain shiny crystals of guanine that act like mirrors to reflect light back through the retina and increase the chance that photons will be absorbed. (A tapetum lucidum is found in many nocturnal animals and accounts for the eyeshine of animals seen in the headlights of a car.) The tapetum lucidum of sharks is beneficial at night and in the depths of the sea, but it has obvious disadvantages in the bright light of the sea surface at midday. In this situation, cells containing the dark pigment melanin expand over the reflective surface to absorb light passing through the retina. With so many sophisticated sensory systems, it is not surprising that the brains of many species of sharks are proportionately heavier than the brains of other fishes and approach the brain-to-body-mass ratios of some tetrapods.

Anecdotal and circumstantial evidence suggests that sharks use their sensory modalities in an ordered sequence to locate, identify, and attack prey. Olfaction is often the first of the senses to alert a shark to potential prey, especially when the prey is wounded or otherwise releasing body fluids. Although little is known about the olfactory landscape of the oceans, a shark presumably employs its sensitive sense of smell to swim upcurrent through an increasing odor gradient toward its source. Because of its exquisite sensitivity, a shark can use smell as a long-distance sense.

Not as useful as olfaction over great distances, but more directional over a wide range of environmental conditions, is another distance sense—the vibration sensitivity of mechanoreception. The lateral line system and the sensory areas of the inner ear are highly efficient in detecting vibrations such as those produced by a struggling fish. The effectiveness of mechanoreception in drawing sharks from considerable distances to a sound source has been demonstrated by using underwater speakers to broadcast vibrations like those produced by a struggling fish. The same phenomenon has been demonstrated unintentionally in macabre sea-rescue operations in which sharks were apparently attracted to survivors of sea disasters by vibrations from the rotors of rescue helicopters.

Once a shark is close to the stimulus, vision takes over as the primary mode of prey detection. If the prey is easily recognized, a shark may proceed directly to an attack. Unfamiliar prey is treated differently, as studies aimed at developing shark deterrents have discovered. A circling shark may suddenly turn and rush toward unknown prey. Instead of opening its jaws to attack, however, the shark bumps or slashes the surface of the object with its rostrum. Opinions differ on whether this is an attempt to determine texture through mechanoreception, make a quick electrosensory appraisal, or use the rough placoid scales to abrade the surface and release fresh olfactory cues.

Following further circling and apparent evaluation of all sensory cues from the potential prey, the shark may either wander off or attack. In the latter case, the rostrum is raised and the jaws protrude. In the last moments before contact, some sharks draw an opaque eyelid (the nictitating membrane) across each eye to protect it. At this point, it appears that sharks shift entirely to electroreception to track prey. This hypothesis was developed while studying the attacks by large sharks on bait suspended from boats. Divers in submerged protective cages watched the attacks. After occluding their eyes with the nictitating membranes or approaching so close that the

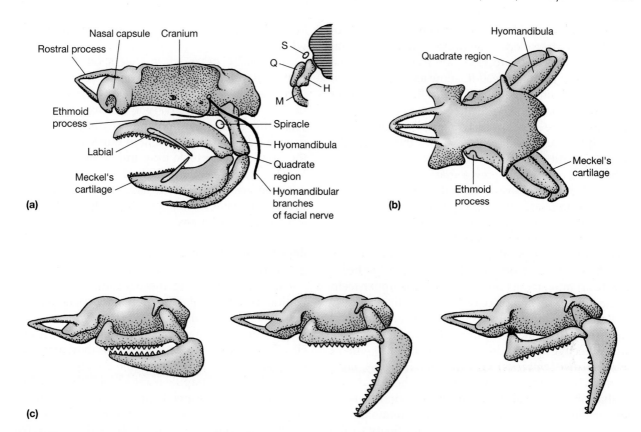

▲ FIGURE 5–5 Anatomical relationships of the jaws and chondrocranium of extant hyostylic sharks. (a) Lateral and cross-sectional views of the head skeleton of *Scyllium* with the jaws closed. (S = spiracle; Q = quadrate region of the palatoquadrate; H = hyomandibula; M = mandible) (b) Dorsal view of *Carcharhinus*. (c) During jaw opening and upper jaw protrusion, the hyomandibula rotates from a position at a 45° angle to the long axis of the cranium to a 90° angle to that axis.

rostrum blocks visual contact with the bait, attacking sharks frequently veered away from the bait and bit some inanimate object near the bait (including the observer's cage, much to the dismay of the divers). Apparently, the unnatural environment of the bait stations included strong influences on local electric fields from the cages, and the sharks mistakenly attacked these artificial sources of electrical activity.

Jaws and Feeding Mobility within the head skeleton, known as **cranial kinesis**, allows consumption of large food items. Cranial kinesis permits inclusion of large items in the diet without excluding smaller foodstuff. Extant sharks have a derived type of **hyostylic** jaw suspension. In this condition, an enlarged hyomandibular cartilage (hyomandibula), which braces the posterior portion of the palato-quadrate, attaches firmly but movably to the side of the cranium (**Figure 5–5**). A second connection to the chondrocranium is via paired palatoquadrate projections (the orbital or ethmoid processes) extending dorsally between the eyes and attached to the braincase by elastic ligaments. Hyostyly permits multiple jaw positions, including protrusion of the upper jaw, each appropriate to different feeding opportunities.

The right and left halves of the pectoral girdle are fused together ventrally into a single U-shaped scapu-locoracoid cartilage. Hypobranchial muscles running from the ventral coracoid portion to the front of the lower jaw open the mouth. The advantages of the jaws of extant sharks are displayed when the upper jaw is protruded. Muscles swing the hyomandibula laterally and anteriorly to increase the distance between the right and left jaw articulations and thereby increase the volume of the orobranchial chamber. This expansion, which sucks water and food forcefully into the mouth, is not possible with the more primitive amphistylic jaw suspension because the palato-quadrate is tightly attached to the chondrocranium.

With hyomandibular extension, the palato-quadrate is protruded to the limits of the elastic ligaments on its orbital processes. Protrusion drops the mouth away from the head to allow a shark to bite an organism much larger than itself. The dentition of the palatoquadrate is specialized for attacking prey too large to be swallowed whole—the teeth on the palatoquadrate are stouter than those in the mandible and often recurved and strongly serrated. When feeding on large prey, a shark opens its mouth, sinks its lower and upper teeth deeply into the prey, and then protrudes its upper jaw ever more deeply into the slash initiated by the teeth. As the jaws reach their maximum initial penetration, the shark throws its body into exaggerated lateral undulations, which results in a violent side-to-side shaking of the head. The head movements bring the serrated upper teeth into action to sever a piece of flesh from the victim.

Rare observations of sharks feeding under natural conditions indicate that these fishes are versatile and effective predators. The great white shark (*Carcharodon carcharias*) kills mammalian prey, such as seals, by exsanguination—bleeding them to death. A shark holds a seal tightly in its jaws until it is no longer bleeding and then bites down, removing an enormous chunk of flesh. The carcass floats to the surface, and the shark returns to it for another bite. A white shark may feed rather leisurely on an acceptable carcass and defend it from other white sharks with typical fishlike side-by-side tail slaps, as well as tail lobbing (slapping the tail down on the surface of the water) and breaching (leaping out of the water and landing with a splash on the side of the body)—behaviors more familiar among dolphins and whales.

Attacks by great white sharks on sea lions are quite different from those on seals because sea lions—unlike seals—have powerful front flippers that can be used effectively in defense. Sharks often seize and release sea lions repeatedly until they die from blood loss. White sharks quickly release prey they find unacceptable after initial mouthing. Klimley (1994) suggested that lack of blubber is a primary rejection criterion. This behavior may explain why great white sharks seize but then release and ignore sea otters and humans.

Reproduction Much of the success of the extant grade of neoselachians may be attributed to their sophisticated breeding mechanisms. Internal fertilization is universal. The pelvic claspers of males have a solid skeletal structure that may increase their effectiveness. During copulation (**Figure 5–6a**), a single clasper is bent at 90 degrees to the long axis of the body, so the dorsal groove on the clasper lies directly under the cloacal papilla from which sperm are emitted. The flexed clasper is inserted into the female's cloaca and locked there by an assortment of barbs, hooks, and spines near the clasper's tip. Sperm from the genital tract are ejaculated into the clasper groove. Simultaneously, a muscular subcutaneous sac extending anteriorly beneath the skin of the male's pelvic fins contracts. This siphon sac has a secretory lining and is filled with seawater by the pumping of the male's pelvic fins before copulation. Seminal fluid from the siphon sac washes sperm

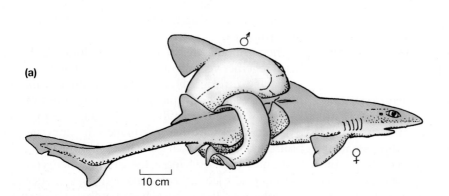

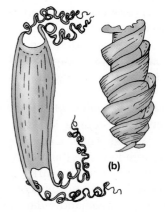

▲ **FIGURE 5–6 Reproduction by sharks.** (a) Copulation in the European spotted catshark *Scyliorhinus*. Only a few other species of sharks and rays have been observed *in copulo*, but all assume postures so that one of the male's claspers can be inserted into the female's cloaca. (b) (Not to same scale.) The egg cases of two oviparous sharks, *Scyliorhinus* (left) and *Heterodontus* (right).

down the groove into the female's cloaca, from which point the sperm swim up the female's reproductive tract.

Male sharks of small species secure themselves *in copulo* by wrapping around the female's body. Large sharks swim side by side, their bodies touching, or enter copulation in a sedentary position with their heads on the substrate and their bodies angled upward. Many male sharks and skates bite the female's flanks or hold on to one of her pectoral fins with their jaws, and the skin on the back and flanks of a female of these species may be twice as thick as the skin of a male the same size.

Reproductive strategies among vertebrates illustrate a trade-off between quantity and quality of offspring. At one extreme are species like the cod, which produce millions of small eggs. A mother cod invests little time or energy in an individual egg, but she produces so many offspring that some survive to become adults. At the opposite end of the scale are animals (humans, for example) that produce very few young and invest a great deal of time and energy to ensure the survival of each one.

With the evolution of internal fertilization, sharks adopted a reproductive strategy favoring the production of a small number of offspring that are retained, protected, and nourished for varying periods within the female's body. This mode of reproduction requires a significant investment of energy by the female, and it succeeds when adults have long life expectancies.

The energy a female invests in nourishing the embryo can be in the form of an egg yolk (as in birds), or it can be delivered to the embryo by the reproductive tract of the female (as in placental mammals). Two sets of terminology are used to describe how embryos are nourished and how young are produced. **Lecithotrophy** (Greek *lecith* = egg and *troph* = nourishment) refers to the situation in which yolk supplies most of the nourishment for the embryo, whereas **matrotrophy** (Greek *matro* = mother) means that the reproductive tract of the female supplies most of the energy. (Lecithotrophy and matrotrophy are at the ends of a continuum, and many vertebrates use a combination of lecithotrophy and matrotrophy in which the embryo receives some nourishment from a yolk and some from the reproductive tract of the mother.) An older but still useful pair of terms describes the way in which a baby is produced: oviparity means the baby hatches from an egg that is deposited outside the body of the mother (as in birds, for example) and viviparity means that a fully developed baby is born (as in placental

mammals). The embryo of an oviparous species is lecithotropic, but viviparous species include a range of modes of fetal nourishment extending from lecithotrophy to matrotropy. There also are species of vertebrates in which the young hatch and begin feeding while they are still within the oviduct, consuming cells from the walls of the oviduct or eating unhatched eggs containing what would have been their siblings.

Sharks display both oviparity and viviparity. Most oviparous sharks produce large eggs with large yolks (the size of a chicken yolk or larger). A specialized structure at the anterior end of the oviduct, the nidimental gland, secretes a proteinaceous case around the fertilized egg. Protuberances on the cases become tangled with vegetation or wedged into protected sites on the substrate (see **Figure 5–6b**). The embryo obtains nutrition exclusively from the yolk during the six- to ten-month developmental period. Movements of the embryo produce a flow of water through openings in the shell that flushes out organic wastes and brings in dissolved oxygen. The young are generally miniature replicas of the adults when they hatch and seem to live much as they do when mature.

A significant step in the evolution of shark reproduction was prolonged retention of the fertilized eggs in the reproductive tract. The eggs often hatch within the oviducts, and the young may spend as long in their mother after hatching as they did within the shell, eventually emerging as miniatures of the adults. Most sharks with this reproductive mode have about a dozen young at a time.

Sharks have independently evolved matrotrophy several times in very different ways. In sand tiger sharks (genus *Odontaspis*) and the white shark (*Carcharodon carcharias*), embryos feed on their siblings and eggs that may continue to be ovulated until only a few uterine rivals are born. Some sharks develop long spaghetti-like extensions of the oviduct walls that penetrate the mouth and gill openings of the internally hatched young and secrete a milky nutritive substance. The most common and most complex form of viviparity is found among sharks that develop a yolk sac placenta that allows an embryo to obtain nourishment from the maternal uterine bloodstream via its highly vascular yolk sac. This mode of reproduction is called **placentotrophic matrotrophy** (Latin *placenta* = a round flat cake). No matter which form of nourishment has brought young sharks to their free-living size, there is no evidence of parental care once the eggs are laid or the young are born.

Social Behavior Sharks have long been considered solitary and asocial, but this view is changing. Accumulating field observations, often from aerial surveys or by scuba divers in remote areas, indicate that sharks of many species aggregate in great numbers periodically, perhaps annually. More than 60 giant basking sharks have been observed milling together and occasionally circling in head-to-tail formations in an area off Cape Cod in summer, and an additional 40 individuals were nearby. Over 200 hammerhead sharks have been seen near the surface off the eastern shore of Virginia in successive summers. Divers on seamounts that reach to within 30 meters of the surface in the Gulf of California have observed enormous aggregations of hammerheads schooling in an organized manner around the seamount tip. Some of these hammerhead observations include behavior thought to be related to courtship. The schools are initially formed by large numbers of females, the largest females aggressively securing a central position within the schools. Eventually males arrive, dash into the center of the schools, and mate with the largest females. More than 1000 individuals of the blue shark have been observed near the surface over canyons on the edge of the continental shelf off Ocean City, Maryland. Fishermen are all too familiar with the large schools of spiny dogfishes that seasonally move through shelf regions, ruining fishing by destroying gear, consuming bottom fishes and invertebrates, and displacing commercially more valuable species. These dogfish schools are usually made up of individuals that are all the same size and the same sex. The distribution of schools is also peculiar: female schools may be inshore and males offshore, or male schools may all be north of some point and the females south. Our understanding of these phenomena is slim, but it is clear that not all sharks are solitary all the time and aggregations are often related to reproduction.

Life History and Conservation of Sharks Whatever their social behavior and reproductive mode, sharks produce relatively few young during an individual female's lifetime. This is a life-history pattern that depends on high rates of survival for young animals and long life expectancies for adults. Internal fertilization and the life-history characteristics that accompany it evolved in sharks fully 350, perhaps 400 million years ago and have been a successful strategy throughout the world's oceans ever since. Now alterations of the habitat and heavy predation by humans threaten the survival of many species.

Although young sharks are relatively large compared with other fishes, they are subject to predation, especially by other sharks. Many species depend on protected nursery grounds—usually shallow inshore waters, which are the areas most subject to human disturbance and alteration. In addition, adult sharks are increasingly falling prey to humans. A rapid expansion in recreational and commercial shark fishing worldwide threatens numerous species of these long-lived, slowly reproducing top predators. (See Helfman 2007 for a comprehensive treatment of this phenomenon.)

People like to eat sharks, whether they know what they are eating or not. Spiny dogfish (*Squalus acanthias*) has long been served up as "fish and chips" in Europe and began to make its unheralded way into the American prepared food market in the 1980s. "Mako" shark steaks (sliced from sharks of two genera, *Isurus* and *Lamna*) have become an alternative to swordfish in the fresh seafood cases of nearly every American supermarket. Europe has already seen drastic drops in the populations of these sharks from commercial fisheries. For example, Norway targeted *Lamna* for intensive fishing and initially harvested as much as 8060 tons in a single year from the northeastern Atlantic. However, within seven years, the catch had fallen to 207 tons, and since the 1970s the Norwegians have been unable to catch 100 tons per year. Despite such histories, the United States did not begin regulation of any shark fishery until 1993 and did nothing to control the spiny dogfish catch until years later. By the year 2000, emergency seasonal bans on all dogfish catches were being implemented as annual harvest quotas were filled months before the quota year ended. Standardized annual fisheries surveys have shown that, in spite of the claims of large numbers of dogfish made by the fishing industry, the biomass of female dogfish has dropped significantly. These same assessments found that, from 1997 to 2003, dogfish "pups" had become rare or virtually absent.

At about the same time as domestic consumption of shark increased, an even more powerful economic force exploded on the scene—export of shark fins to Asian markets. Shark-fin soup, which is reputed to have medicinal properties, may fetch $90 or more a bowl in restaurants in Asia. Dried shark fins sell at wholesale in Hong Kong for more than $500 a kilogram; in the United States, fresh wet fins sell for $200 or more per kilogram. The success of Asian economies during the 1980s and 1990s created an almost unlimited market for fins. Because the rest of the carcass was worthless by comparison, shark

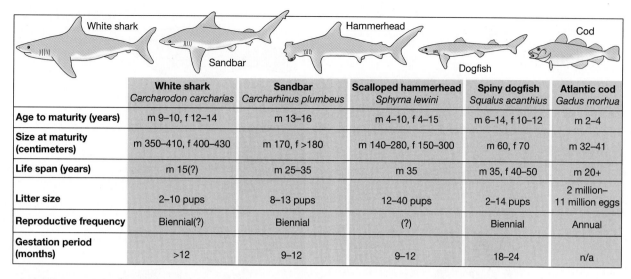

	White shark *Carcharodon carcharias*	Sandbar *Carcharhinus plumbeus*	Scalloped hammerhead *Sphyrna lewini*	Spiny dogfish *Squalus acanthius*	Atlantic cod *Gadus morhua*
Age to maturity (years)	m 9–10, f 12–14	m 13–16	m 4–10, f 4–15	m 6–14, f 10–12	m 2–4
Size at maturity (centimeters)	m 350–410, f 400–430	m 170, f >180	m 140–280, f 150–300	m 60, f 70	m 32–41
Life span (years)	m 15(?)	m 25–35	m 35	m 35, f 40–50	m 20+
Litter size	2–10 pups	8–13 pups	12–40 pups	2–14 pups	2 million– 11 million eggs
Reproductive frequency	Biennial(?)	Biennial	(?)	Biennial	Annual
Gestation period (months)	>12	9–12	9–12	18–24	n/a

▲ **FIGURE 5–7** Life history parameters of sharks. Some shark life-history parameters compared to that of another important object of intensive fisheries, the Atlantic cod. Their life-history characteristics make sharks vulnerable to over exploitation. (m = male; f = female)

finning—the practice of catching sharks, cutting off the fins, and throwing the rest of the animal, dead or alive, back into the sea—became a worldwide phenomenon. This wasteful and cruel business moved into American waters in a very big way. Long lining is a method of fishing that relies on laying out lines thousands of feet long with baited hooks hanging from them. By the mid-1990s, long liners were setting millions of hooks along the Eastern and Gulf coasts of the United States annually (8.9 million in 1995 alone). In 1997, long liners in Hawaii caught over 100,000 sharks and discarded 98.6 percent of the mass of the catch. Shark, skate, and ray catches worldwide more than quadrupled to over 800,000 metric tons each year (more than 70 million individuals). It is little wonder that in the United States alone, various species of coastal sharks have experienced population reductions of 50 to 85 percent over the past 25 years. In the Northwest Atlantic, all of eleven large coastal species and seven oceanic species of sharks for which there are long line catch data have declined by more than 50 percent (some by as much as 89 percent) in eight to fifteen years (Baum et al. 2003).

Sharks receive very little attention from bodies that regulate international fisheries, and most nations have no effective fisheries management at all. Many commercially important sharks are pelagic, migrating across multiple political boundaries and spending significant portions of their lives in international waters where there is no political jurisdiction. As a result, effective preservation measures are hard to implement. In 2000, the U.S. House of Representatives approved a nationwide ban on shark finning; but, even before the bill was introduced in the Senate, long liners were simply moving their base of operations from Hawaii to islands out of American jurisdiction. In late 2003, the United Nations General Assembly passed a resolution asking nations to ban shark finning. Unfortunately, the resolution does not include any provisions for enforcing a ban.

Because of their biological characteristics, all sharks are particularly susceptible to near extirpation by fishing. They grow slowly, mature late in their lives, have few young at a time, and, because of the great amount of energy invested in those young, females do not reproduce every year (**Figure 5–7**).

Relatively few individuals occur in any area, except perhaps at times of breeding or other social aggregations. For example, only nine to fourteen individual great white sharks were observed over a period of five years in the South Farallon Islands near San Francisco. The same individuals returned each fall when prey numbers were high. When only four great white sharks were killed near the sea lion colony that attracted them, the number of attacks on seals and sea lions in the area dropped by half for the next two years. Fisheries management specialists say that there is little chance that populations of many species of overfished sharks will show significant recovery in less than half a century, even with strict fishing limitations.

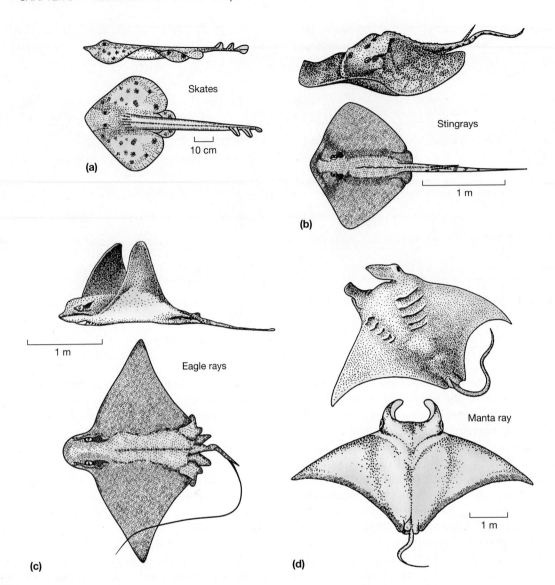

▲ **FIGURE 5–8** Representative extant skates and rays. (a, b) benthic forms; (c, d) pelagic forms. (a) *Raja*, a typical skate. (b) *Dasyatis*, a typical ray. The following pelagic batoids are closely related to the rays: (c) *Aetobatus* is representative of the eagle rays, wide-ranging shelled invertebrate predators. (d) *Manta* is representative of its family of gigantic fishes that feed exclusively on zooplankton. Extensions of the pectoral fins anterior to the eyes (the horns of these devil rays) help funnel water into the mouth during filter feeding.

▓ Skates and Rays

The hypotremate (Greek *hypo* = below and *trema* = opening) neoselachians—the skates and rays—are more diverse than sharks. Approximately 534 extant species of skates and rays are currently recognized (**Figure 5–8**). The suite of specializations characteristic of skates and rays relates to their early assumption of a bottom-dwelling, durophagous (Latin *duro* = hard and Greek *phagus* = to eat) habit. The teeth are almost all flat-crowned plates that form a pavementlike dentition. The mouth is often highly and rapidly protrusible to provide powerful suction used to dislodge shelled invertebrates from the substrate. The dorsal-ventral flattening of their bodies and lateral extension of their pectoral fins provides a large surface over which ampullae of Lorenzini are distributed. Expansion of the anterior part of the body is also found in several groups of sharks, including hammerheads, and may increase the ability to detect electrical and magnetic fields.

Skates and rays have a long history of phylogenetic isolation from the lineages of extant sharks but appear to be derived from the squaloids. A handful of specialized shark species probably belonging to the squatinoid lineage of the Squalea are bottom-living specialists and illustrate what an intermediate stage in the transition to skates and rays may have been like. Recent molecular studies, some not yet published, indicate that skates and rays are not closely related to any living sharks but form a distinct clade of neoselachian evolution that split from the ancestors of all modern sharks. Not all systematists accept this molecular evidence, and the relationship between rays and sharks remains unresolved (McEachran and Aschliman 2004).

Skates are distinguished from rays by their tails and modes of reproduction. Skates have an elongate but thick tail stalk supporting two dorsal fins and a terminal caudal fin (see Figure 5–8a). A typical ray (see Figure 5–8b) has a whiplike tail stalk with fins replaced by one or more enlarged, serrated, and venomous dorsal barbs. Derived forms, such as stingrays (family Dasyatidae), have a few greatly elongated and venomous spines derived from modified placoid scales at the base of the tail. The electric skates (family Rajidae) have specialized tissues in their long tails that are capable of emitting a weak electric discharge. Each species appears to have a unique pattern of discharge, and the discharges may identify conspecifics (members of the same species) in the gloom of the seafloor. The electric rays and torpedo rays (family Torpedinidae) have modified gill muscles, producing electrical discharges of up to 200 volts that are used to stun prey. Skates are oviparous, laying eggs enclosed in horny shells popularly called "mermaid's purses," whereas rays are viviparous.

Skates and rays are similar in being dorsoventrally flattened, with radial cartilages extending to the tips of the greatly enlarged pectoral fins. The anteriormost basal elements fuse with the chondrocranium in front of the eye and with one another in front of the rest of the head. Skates and rays swim by undulating these massively enlarged pectoral fins. The placoid scales so characteristic of the integument of a shark are absent from large areas of the bodies and pectoral fins of skates and rays. The few remaining denticles are often greatly enlarged to form scales called bucklers along the dorsal midline.

Skates and rays are primarily benthic invertebrate feeders that occasionally capture small fishes. Many skates and rays rest on the seafloor and cover themselves with a thin layer of sand. They spend hours partially buried and nearly invisible except for their prominent eyes, surveying their surroundings. The largest rays, like the largest sharks, are plankton strainers. Devilfishes or manta rays of the family Mobulidae are up to 6 meters in width (see Figure 5–8d). These highly specialized rays swim through the open sea with winglike motions of the pectoral fins, filtering plankton from the water as they go.

The dentition of many benthic rays is sexually dimorphic. Different dentitions coupled with the generally larger size of females might reduce competition for food resources between the sexes, but no difference in stomach contents has been found. Since the male uses its teeth to hold or stimulate a female before and during copulation, sexual selection may be at work. Males of the stingray *Dasyatis sabina* have blunt teeth like those of females for most of the year; but, during the breeding season, males grow sharp-cusped teeth that are used for courtship.

5.6 Holocephali—the Bizarre Chondrichthyans

The approximately 33 extant forms of ratfishes, none much over a meter in length, have a soft anatomy more similar to sharks and rays than to any other extant fishes (**Figure 5–9**). They have long been grouped with neoselachians as Chondrichthyes because of these shared specializations, but they have a bizarre suite of unique features as well (Didier 2004). Generally found in water deeper than 80 meters and thus not well known in the wild, the Holocephali move into shallow water to deposit their 10-centimeter horny-shelled eggs. Several holocephalians have elaborate rostral extensions of unknown function. The little we know of their natural history indicates that most species feed on shrimp, gastropod mollusks, and sea urchins. Their locomotion is produced by lateral undulations of the body that throw the long tail into sinusoidal waves and by fluttering movements of the large, mobile pectoral fins. The solidly fused nipping and crushing tooth plates grow throughout life, adjusting their height to the wear they suffer. Of special interest are the armaments—a poison gland associated with the stout dorsal spine in some species and a macelike cephalic clasper of males. Males have been observed using the spine-encrusted cephalic clasper to pin the female's pectoral fin against his forehead while attempting copulation.

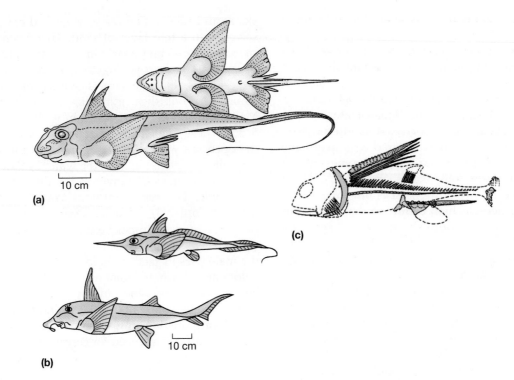

▲ **FIGURE 5–9** Holocephalians.   (a) A chimaera, the spotted ratfish (*Hydrolagus colliei*), is an extant holocephalan. (b) Representatives of two other extant groups: plownose chimaera (lower) and longnose chimaera (upper). (c) *Phomeryele*, a Carboniferous iniopterygian shark favored by many paleontologists as a relative of the extant holocephalians.

Since the 1950s, Rainer Zangerl and coworkers have described a group of bizarre and obviously specialized forms of Paleozoic chondrichthyians (see Figure 5–9c). These Iniopterygia (Greek *inio* = back of the neck and *ptero* = wing) have characteristics that Zangerl considered to be evidence for a link between the earliest sharks and holocephalians. As in extant holocephalians, the palatoquadrate is fused to the cranium (known as **autostylic** jaw sus-pension), but the teeth, unlike those of extant rat-fishes, are in replacement families like those of neoselachians. The earliest close relative of the mod-ern holocephalians is now thought to be of Late Car-boniferous age, too old to be descended from the contemporaneous iniopterygians. Current thinking is that both iniopterygians and holocephalians arose from ancestral sharks following a somewhat parallel line in evolution.

Summary

The chondrichthyans first appear in the fossil record in the Late Silurian period and are distinguished by an entirely cartilaginous endoskeleton with a unique form of prismatic mineralization. Three radiations of sharklike chondrichthyans can be traced through increasingly sophisticated characters of the jaws and fins. Paleozoic forms had broad-based fins that were probably immovable and an upper jaw that was firmly fixed to the skull, limiting the gape of the mouth. In the early Mesozoic, a second radiation pro-duced species in which the bases of the pectoral and pelvic fins were narrow, allowing them to swivel, and the fins had intrinsic muscles that could change their curvature. The third radiation of sharks appeared in the middle of the Mesozoic, and descen-dants of these forms remain the dominant predators in shallow seas. Further modifications of the skull allow the jaws to be protruded, and a shark can bite chunks of flesh from prey too large to be swallowed whole.

Skates and rays may or may not be derived from the extant shark radiation but are adapted for life on the sea bottom. They are dorsoventrally flattened, with eyes and spiracles (remnants of gill slits) on tops of their heads. Many skates and rays lie buried in sand and ambush their prey, but the largest species—the manta rays—are plankton feeders.

The holocephalians (ratfishes or chimeras) are a small group of bizarre fishes that probably branched off from ancestral chondrichthyans in the Paleozoic. They generally occur in deep water, and little is known of their natural history and behavior.

Neoselachians probably evolved internal fertilization in the Paleozoic. The life histories of most species are based on producing a few relatively large young at a time. This reproductive strategy depends on high survival of young and long life expectancies for adults. It worked well for 350, perhaps 400 million years, but within the past 50 years, loss of coastal habitat and outrageous overfishing have brought many species of sharks to the edge of extinction.

Additional Readings

Barton, M. 2007. *Bond's Biology of Fishes*, 3rd ed. Belmont, CA: Thomson Brooks/Cole.

Baum, J. K. et al. 2003. Collapse and conservation of shark populations in the Northwest Atlantic. *Science* 299:389–392.

Carroll, R. L. 1987. *Vertebrate Paleontology and Evolution*. New York, NY: Freeman.

Didier, D. A. 2004. Phylogeny and classification of extant Holocephali. In Carrier, J. C. et al. (eds.). 2004. *Biology of Sharks and Their Relatives*. Boca Raton, FL: CRC Press: 115–118.

Douady, C. J. et al. 2002. Molecular phylogenetic evidence refuting the hypothesis of Batoid (rays and skates) as derived sharks. *Molecular Phylogenetics and Evolution* 26: 215–221.

Grogan, E. D. and R. Lund. 2004. The origin and relationships of early Chondrichthyes. In Carrier, J. C. et al. (eds.). 2004. *Biology of Sharks and Their Relatives*. Boca Raton, FL: CRC Press: 3–32

Hamlett, W. C. (ed.). 1999. *Sharks, Skates and Rays: The Biology of Elasmobranch Fishes*. Baltimore, MD: The Johns Hopkins University Press.

Helfman G. S. 2007. *Fish Conservation*. Washington, DCS: Island Press.

Helfman, G. S. et al. 2007. *The Diversity of Fishes*, 2nd ed. Malden, MA: Blackwell.

Janvier, P. 1996. *Early Vertebrates*. Oxford, UK: Clarendon Press.

Kenney, R. D. et al. 1985. Shark distributions off the Northeast United States from marine mammal surveys. *Copeia* 1985:220–223.

Klimley, A. P. 1994. The predatory behavior of the white shark. *American Scientist* 82:122–133.

Klimley, A. P. 1999. Sharks beware. *American Scientist* 87:488–491.

Klimley, A. P. 2003. *The Secret Life of Sharks*. New York, NY: Simon and Schuster.

Klimley, A. P., and D. G. Ainley. 1996. *Great White Sharks: The Biology of Caraharodon carcharias*. San Diego, CA: Academic Press.

Maisey, J. G. 1996. *Discovering Fossil Fishes*. New York, NY: Henry Holt.

McEachran, J. D. and N. Aschliman. 2004. Phylogeny of Batoidea. in Carrier, J. C. et al. (eds.). 2004. *Biology of Sharks and Their Relatives*. Boca Raton, FL: CRC Press: 79–114.

Miller, R. F. et al. 2003. The oldest articulated chondrichthyan from the Early Devonian period. *Nature* 425:501–504.

Moyle, P. B., and J. J. Cech, Jr. 2000. *Fishes: An Introduction to Ichthyology*, 4th ed. Upper Saddle River, NJ: Prentice Hall.

Nelson, J. S. 2006. *Fishes of the World*, 4th ed. New York, NY: Wiley.

CHAPTER 6

Dominating Life in Water: The Major Radiation of Fishes

By the end of the Silurian period, the agnathous fishes had diversified and the cartilaginous gnathostomes were in the midst of their first radiation. The stage was set for the appearance of the largest extant group of vertebrates, the bony fishes. The first fossils of bony fishes (Osteichthyes) occur in the Late Silurian. Osteichthyans are well represented from the Early Devonian period. Their radiation was in full bloom by the middle of the Devonian, with two major groups diverging: the ray-finned fishes (Actinopterygii) and lobe-finned fishes (Sarcopterygii). Specialization of feeding mechanisms is a key feature of the evolution of these major groups of vertebrates. Increasing mobility among the bones of the skull and jaws allowed ray-finned fishes in particular to exploit a wide range of prey types and predatory modes.

Specializations in locomotion, habitat, behavior, and life histories accompanied the specializations of feeding mechanisms.

6.1 The Appearance of Bony Fishes

The Devonian is known as the Age of Fishes because all major lineages of fishes, extant and extinct, coexisted in the fresh and marine waters of the planet for its 48-million-year duration (Figure 6–1). Most groups of gnathostome fishes either made their first appearance or significantly diversified during this period. Among them was the most species-rich and morphologically diverse lineage of vertebrates, the Osteichthyes or bony fishes (see Figure 3–14).

▶ **FIGURE 6–1** Three major radiations of fishes. Cartilaginous and bony fishes have undergone three periods of radiation and diversification. Several extant forms survive from the first of these radiations, which occurred mainly in the Late Paleozoic era. Among cartilaginous fishes the ratfishes (Holocephali) are remnants of this radiation, and the surviving bony fishes include both extant types of aquatic sarcopterygians (the lungfishes and coelacanths) and the bichirs and reedfishes (Polypteriformes), which are primitive ray-finned fishes. The Mesozoic radiations produced the primitive sharks described in Chapter 5, but there are no extant cartilaginous fish lineages from that radiation. Gars (Lepisosteiformes) and the bowfin (Amiiformes) are survivors of Mesozoic lineages of bony fishes. The Cenozoic radiation saw the origin of extant lineages of sharks and rays, as well as the modern lineages of sturgeons and paddlefishes (Acipenseriformes) and the teleosts (Teleostei), which—with more than 24,000 extant species—form the vast majority of freshwater and marine fishes. (Name of cartilaginous fishes are shown in boldface.)

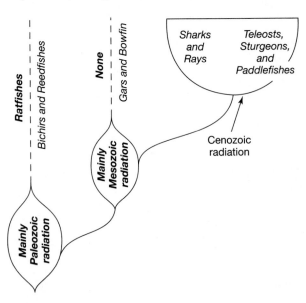

Earliest Osteichthyes and the Major Groups of Bony Fishes

Fragmentary remains of bony fishes are known from the Late Silurian, but not until the Devonian are more complete remains found. These animals resemble acanthodians in their cranial structure, and these similarities suggest a common ancestor for acanthodians and osteichthyans in the Early Silurian. Remains of the bony fishes representing a radiation of forms already in full bloom appear in the Early to Middle Devonian. Two major and distinctive types of osteichthyans had unique locomotor and feeding characters and were the dominant fishes during the Devonian. In the sarcopterygian lineage, the rays of the paired fins extend from a central shaft of bones in a featherlike or leaflike manner to support the fin web. In the actinopterygian lineage, in contrast, these rays spread outward like a fan from bones at the base of the fins. The derived modern bony fishes are the teleosts, the largest group of extant vertebrates that arose among the actinopterygians (Table 6.1). Fossils of the two basic types of osteichthyans are abundant from the Middle Devonian onward. The Sarcopterygii (Figure 6–2a–d) and the Actinopterygii (Figure 6–2e) are sister groups.

Shared derived characters of these bony fishes include patterns of lateral line canals, similar opercular and pectoral girdle dermal bone elements, and fin webs supported by bony dermal rays. A fissure allowed movement between the anterior and posterior halves of the chondrocranium in many forms. The presence of bone is not a unifying character of

TABLE 6.1 Classification and geographic distribution of osteichthyes, the bony fishes. Major groups are shown in the shaded bars. Only the evolutionarily or numerically most important groups are listed. The subdivision of the Neopterygii varies greatly from author to author. Groups in quotation marks are not monophyletic, but relationships are not yet understood. Names in square brackets are alternative names for the groups.

Sarcopterygii (fleshy-finned fishes and tetrapods)
Actinistia [Coelacanthiformes]: 2 species of coelacanths from the western Indian Ocean and central Indonesia, deep-water marine, 1 m to 1.5 m.
Dipnoi: 6 species of lungfishes from the Southern Hemisphere, freshwater, less than 1 m to 1.8 m.
Tetrapoda: More than 40,000 species of terrestrial and secondarily aquatic vertebrates.
Actinopterygii (ray-finned fishes)
Polypteriformes [Cladistia]: At least 16 species of bichirs and the reedfish, Africa, freshwater, less than 30 cm to 90 cm.
Acipenseriformes [Chondrostea]: 27 species of sturgeons and paddlefishes, Northern Hemisphere, coastal and freshwater, about 2 m to at least 4.2 m.
Neopterygii
Lepisosteiformes [Ginglymodi]: 7 species of gars, North and Central America, fresh and brackish water, less than 1 m to about 3 m.
Amiiformes: 1 species, the bowfin, North America, freshwater, up to 90 cm.
Teleostei
Osteoglossomorpha: At least 219 species of bony tongues, worldwide, mostly tropical freshwater, 10 cm to at least 2.5 m.
Elopomorpha: At least 856 species of tarpons and eels, worldwide, mostly marine, 1 m to 4 m.
Clupeomorpha: About 364 species herrings and anchovies, worldwide, especially marine, 6 cm to 1 m.
Euteleostei
Ostariophysi: More than 7,931 species of catfish and minnows, worldwide, mostly freshwater, 1 cm to 5 m.
"Protacanthopterygii:" About 366 species of trouts, salmons, and their relatives, temperate Northern and Southern Hemisphere, freshwater from about 7 cm to at least 1.4 m.
"Stem Neoteleosts:" About 916 species of lanternfishes and their relatives, worldwide, mostly mesopelagic or bathypelagic marine, 7 cm to 1.8 m.
Paracanthopterygii: About 1,340 species of cods and anglerfishes, Northern Hemisphere, marine and freshwater, 6 cm to 2 m.
Acanthopterygii
Atherinomorpha: About 1624 species of silversides, killifishes, and their relatives, worldwide, surface-dwelling, freshwater and marine, under 4 cm to about 2 m.
Perciformes: More than 13,173 species of perches and their relatives, worldwide, primarily marine, 8 to 10 mm to 3 m.

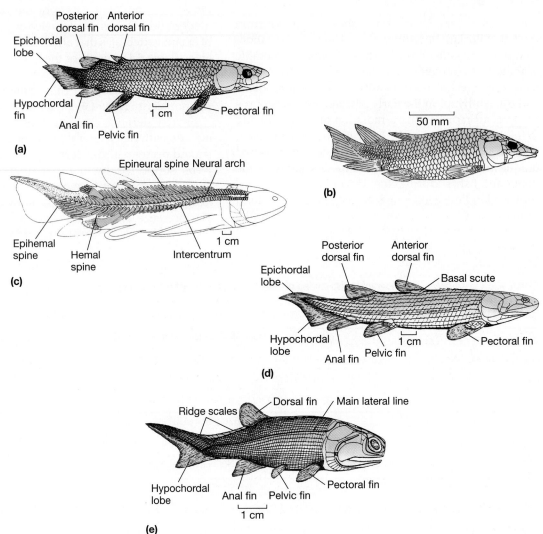

▶ **FIGURE 6–2** Primitive osteichthyans. Dipnoans: (a) relatively unspecialized dipnoan *Dipterus*, Middle Devonian; (b) long-snouted dipnoan *Griphognathus*, Late Devonian. Other Sarcopterygian fishes: (c) porolepiform *Holoptychius*, which was laterally compressed in life, Late Devonian to Early Carboniferous; (d) cylindrical osteolepiform *Osteolepis*, Middle Devonian; (e) typical early actinopterygian *Moythomasia*, Late Devonian.

Osteichthyes because agnathans, placoderms, and acanthodians also have bone, and the absence of bone in chondrichthyans is derived. What *is* unique to the bony fishes is endochondral bone (bone that replaces cartilage ontogenetically). In addition to endochondral bone ossification, osteichthyes retained the more primitive dermal and perichondral bone-forming mechanisms as well.

Many of the names that identify groups of fishes refer to bone; that is, they contain *ost* or *ostei* as part of the name (Greek *osteo* = bone). These names were coined before the distribution of bone in primitive vertebrates was recognized. Likewise, the various names long in use for the extant actinopterygian subgroups imply an increase in the ossification of the skeleton as an evolutionary trend. For example, chondrosteans ("cartilaginous bony fishes," the sturgeons and paddlefishes) are the sister group of a radiation often called the holosteans ("entirely bony fishes," the gars and *Amia*), and this radiation culmi-

nated in teleosteans ("final bony fishes"). The fossil record does not reveal a regular sequence of increasing ossification, however. On the contrary, a tendency to reduce ossification, especially in the skull and scales, is apparent when the full array of early osteichthyans is compared with their derived descendants.

Although it is not usually fossilized, a gas-filled bladder (an outgrowth of the anterior portion of the alimentary tract) may have been a universal character of Osteichthyes. Even some fossil jawless fishes appear to have such an outgrowth, so the character probably is ancestral for Osteichthyes. The gas bladder has been modified to become the lungs of sarcopterygians and a buoyancy adjustment organ in actinopterygians.

Although the monophyletic relationship of the ray-finned fishes as a whole is supported by several shared derived characters, the relationships among early osteichthyans are disputed, and no phylogenetic hypothesis is yet widely accepted.

■ Evolution of the Sarcopterygii

Primitive sarcopterygians were 20 to 70 centimeters long and cylindrical in shape. They had two dorsal fins, an epichordal lobe on the heterocercal caudal fin (a fin area supported by the dorsal side of the vertebral column), and paired fins that were fleshy, scaled, and had a bony central axis.

The jaw muscles of sarcopterygian fishes were massive by comparison with those of actinopterygians, and the size of these muscles produced skull characters that set sarcopterygians apart from the actinopterygians. Finally, the early sarcopterygians were entirely coated with a peculiar layer of dentine-like material, **cosmine**, that spread across the sutures between dermal bones and appears to have been reabsorbed and then rebuilt periodically.

The Dipnoi (lungfishes) have unique derived features clearly indicating that this lineage is monophyletic. The other sarcopterygian fishes have been variously combined as a single taxon, the Crossopterygii (now considered by most workers to be paraphyletic) or two separate lineages, rhipidistians (entirely extinct forms) and actinistians (for the sole surviving genus *Latimeria*, the extant coelacanths and their undisputed fossil relatives). Even this arrangement has been brought into question by the splitting of the rhipidistians into several different lineages, including one (the Porolepiformes) that is related to lungfishes and another (the Osteolepiformes) that is more closely related to tetrapods.

6.2 Extant Sarcopterygii—Lobe-Finned Fishes

Although they were abundant in the Devonian, the number of primarily aquatic sarcopterygians dwindled in the late Paleozoic and Mesozoic eras. (All terrestrial vertebrates are sarcopterygians, of course, so the total number of extant sarcopterygians is enormous.) The early evolution of sarcopterygians resulted in a significant radiation in fresh and marine waters. Today only four nontetrapod genera remain: the dipnoans or lungfishes (*Neoceratodus* in Australia, *Lepidosiren* in South America, and *Protopterus* in Africa; **Figure 6–3**), and the actinistian *Latimeria* (the coelacanths) in waters 100 to 300 meters deep off East Africa and central Indonesia (**Figure 6–4**). We will discuss fossil sarcopterygian fishes in more detail with their sister group, the tetrapods, in Chapter 9.

■ Dipnoans

Extant Dipnoi are distinguishable by the lack of articulated tooth-bearing premaxillary and maxillary bones and the fusion of the palatoquadrate to the undivided cranium (**autostylic** jaw suspension), all of which are derived conditions within the Osteichthyes. Teeth are scattered over the palate and fused into tooth ridges along the lateral palatal margins. Powerful adductor muscles of the lower jaw spread upward over the chondrocranium. Throughout their evolution, this **durophagous** (feeding on hard foods) crushing

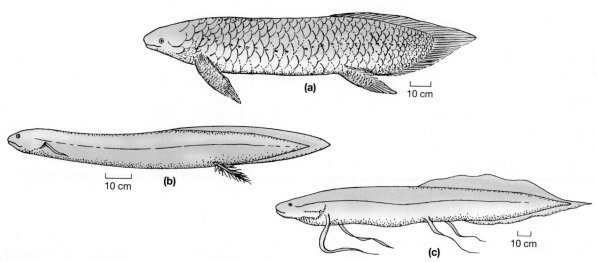

▲ **FIGURE 6–3** Extant dipnoans. (a) Australian lungfish, *Neoceratodus forsteri*. (b) South American lungfish, *Lepidosiren paradoxa*, male. (c) African lungfish, *Protopterus*.

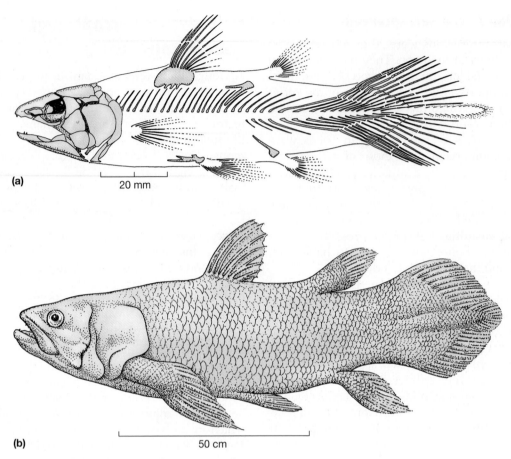

▲ **FIGURE 6–4** Representative Actinistia (coelacanths). (a) *Rhabdoderma*, a Carboniferous actinistian. (b) *Latimeria chalumnae*, the best known of the extant coelacanths.

apparatus has persisted. The earliest dipnoans were marine. During the Devonian, lungfishes evolved a body form quite distinct from the other Osteichthyes. The dorsal, caudal, and anal fins fused into one continuous fin that extends around the entire posterior third of the body. In addition, the caudal fin changed from heterocercal to symmetrical, and the mosaic of small dermal bones of the earliest dipnoan skulls (often covered by a continuous sheet of cosmine) changed to a pattern of fewer large elements without the cosmine cover. Most of this transformation can be explained as a result of **paedomorphosis** (Greek *paedo* = child and *morph* = form; the appearance of juvenile characters in an adult). For a time after their discovery 150 years ago, lungfishes were considered to be specialized salamanders—which they do superficially resemble, especially newly hatched lungfishes that have external gills like some aquatic salamanders.

The Australian lungfish, *Neoceratodus forsteri*, is morphologically most similar to Paleozoic and Mesozoic dipnoans. Like all other extant lungfish,

Neoceratodus is restricted to fresh waters; naturally occurring populations are limited to southeastern Queensland. The Australian lungfish may attain a length of 1.5 meters and a reported weight of 45 kilograms. It swims by body undulations or slowly walks across the bottom of a pond on its pectoral and pelvic appendages. Chemical senses seem important to lungfishes, and their mouths contain numerous taste buds. The nasal passages are located near the upper lip, with the incurrent openings on the rostrum just outside the mouth and the excurrent openings within the oral cavity. Thus, gill ventilation draws water across the nasal epithelium. *Neoceratodus* respires almost exclusively via its gills and uses its single lung only when stressed. Little is known of its behavior.

Australian lungfishes have a complex courtship that may include male territoriality, and they are selective about the vegetation upon which they lay their adhesive eggs; however, no parental care has been observed after spawning. The jelly-coated eggs, 3 millimeters in diameter, hatch in three to four weeks, but the young are elusive and nothing is known of their juvenile life.

Surprisingly little is known about the single South American lungfish, *Lepidosiren paradoxa*, but the closely related African lungfishes, *Protopterus*, with four recognized species, are better known. All are thin-scaled, heavy-bodied, elongate fishes and have unique filamentous and highly mobile paired appendages. These two genera are distinguished by different numbers of weakly developed gills. Because their gills are very small, these lungfishes rely on their lungs to obtain oxygen and drown if they are prevented from breathing air. The gills are important in eliminating carbon dioxide, which is very soluble in water. *Lepidosiren* males develop vascularized extensions on their pelvic fins during the breeding season, probably used to supply oxygen from the male's blood to the young in the nest cavity.

Although the skeletons of lungfishes are mostly cartilaginous, their tooth plates are heavily mineralized and fossilize readily, and a peculiar feature of some species of African lungfishes, estivation, considerably increases their chance of fossilization. Similar in some ways to hibernation, estivation is induced by heat or drying of the habitat rather than by cold. African lungfishes frequent areas that flood during the wet season and bake during the dry season—habitats not available to actinopterygians except by immigration during floods. The lungfishes enjoy the flood periods, feeding heavily and growing rapidly. When flood waters recede, the lungfish digs a vertical burrow in the mud. The burrow ends in an enlarged chamber that varies in depth in proportion to the size of the animal—the deepest chambers are less than a meter. As drying proceeds, the lungfish becomes more lethargic and breathes air from the burrow opening. Eventually the water in the burrow dries up, and the lungfish enters the final stages of estivation—folded into a U-shape with its tail over its eyes. Heavy secretions of mucus that the fish has produced since entering the burrow dry and form a protective envelope around its body. Only an opening at its mouth remains to permit breathing.

Although the rate of energy consumption during estivation is very low, metabolism continues, using muscle proteins as an energy source. When the rains return, the withered and shrunken lungfish becomes active and feeds voraciously. In less than a month, it regains its previous size. Lungfishes normally spend less than six months in estivation, but they have been revived after four years of enforced estivation.

Estivation is an ancient trait of dipnoans. Fossil burrows containing lungfish tooth plates have been found in Carboniferous- and Permian-era deposits of North America and Europe. Without the unwitting assistance of the lungfishes, which initiated fossilization by burying themselves, such fossils might not exist.

▓ Actinistians

Actinistians are unknown before the Middle Devonian. Their hallmarks are fin webs that originate from elongate muscular bases (lobes)—except for the first dorsal fin, which lacks a lobe—and a unique symmetrical three-lobed tail with a central fleshy lobe that ends in a fringe of rays (see Figure 6–4). Actinistians also differ from all other sarcopterygians in their head bones (among other elements, they lack a maxilla, a condition that has been derived in parallel to lungfishes), in details of the fin structure, and in the presence of a curious rostral organ. While other bony fishes radiated into a wide variety of body forms, the actinistians retained their peculiar shape. Following rapid evolution during the Devonian, the actinistians show a history of stability. Late Devonian actinistians differ only slightly from Cretaceous fossils and the extant species, mostly in the degree of skull ossification. Some early actinistians lived in shallow fresh waters, but the fossil remains of actinistians during the Mesozoic are largely marine.

Fossil actinistians are not known after the Cretaceous period; until about 70 years ago they were thought to be extinct, but in 1938 an African fisherman bent over an unfamiliar catch from the Indian Ocean and nearly lost his hand to its ferocious snap. Imagine the astonishment of the scientific community when J. L. B. Smith of Rhodes University announced that the catch was an actinistian. This large fish was so similar to Mesozoic fossil coelacanths that its systematic position was unquestionable. Smith named this living fossil *Latimeria chalumnae* in honor of his former student Marjorie Courtenay Latimer, who saw the strange catch, recognized it as unusual, and brought the specimen to his attention.

Despite public appeals, no further specimens of *Latimeria* were captured until 1952. Since then more than 150 specimens, ranging in size from 75 centimeters to slightly over 2 meters and weighing from 13 to 80 kilograms, have been caught in the Comoro Archipelago or in nearby Madagascar or Mozambique. Coelacanths are hooked near the bottom, usually in 260 to 300 meters of water about 1.5 kilometers offshore. Strong and aggressive, *Latimeria chalumnae* is steely blue-gray with irregular white spots and reflective golden eyes. The reflection comes from a tapetum lucidum that enhances visual ability in dim light. The swim bladder is filled with fat and has ossified walls. The rostral organ, a large cavity in the midline of the snout, communicates with the exterior by three pairs of rostral tubes enclosed by canals in the wall of the chondrocranium. These tubes are filled with gelatinous material and open to the surface through a series of six pores. The rostral organ is almost certainly an electroreceptor. *Latimeria chalumnae* is a predator—stomachs have contained fishes and squid.

A fascinating glimpse of the life of the coelacanth was reported by Hans Fricke and his colleagues, who used a small submarine to observe the fishes. They saw six coelacanths at depths between 117 and 198 meters off a short stretch of the shoreline of one of the Comoro Islands. Coelacanths were seen only in the middle of the night and only on or near the bottom. Unlike extant lungfishes, the coelacanths did not use their paired fins as props or to walk across the bottom. However, when they swam the pectoral and pelvic appendages were moved in the same sequence as tetrapods move their limbs.

In 1927, D. M. S. Watson described two small skeletons from inside the body cavity of *Undina*, a Jurassic coelacanth, and suggested that coelacanths were viviparous. In 1975, dissection of a 1.6-meter *Latimeria* confirmed this prediction, revealing five young, each 30 centimeters long and at an advanced stage of embryonic development. Several other females with young inside in advanced stages of development have now been collected. Internal fertilization must occur, but how copulation is achieved is unknown, since males show no specialized copulatory organs.

In 1998 an announcement was made that astounded ichthyologists: Another coelacanth had been discovered 10,000 kilometers to the east of the Comoro Islands. Two specimens were caught in shark-fishing nets set 100 to 150 meters deep off the northeast tip of Sulawesi, a large central Indonesian island near Borneo. Subsequently named *Latimeria menadoensis* after the nearest major city, the new coelacanth appears from DNA data to have separated from the East African species 1.8 to 11.0 million years ago. The Comoro Islands are volcanic and much younger than the date of divergence of the two species of *Latimeria*, and it is clear that there was an ancestral population of coelacanths elsewhere. That population may still exist, awaiting discovery. If that discovery occurs, it will probably begin once again with local fishers and an ichthyologist's visit to their fish market.

Despite some excellent fossils and numerous specimens of the extant coelacanths, there has never been stable agreement about the evolutionary relationships of the Actinistia with other gnathostomes. Workers have disagreed about coelacanths more than about most other vertebrate taxa because *Latimeria* has a puzzling combination of derived morphological and physiological characters. Some of its characters are similar to Chondrichthyes (a high internal urea concentration, which is probably an ancestral gnathostome feature rather than a derived character of the extant lineages), others to Dipnoi (absence of a maxilla, sequence of movements of the paired appendages during locomotion), and others to Actinopterygii (a one-to-one ratio of basal elements to fin rays). In addition, *Latimeria* have a curious collection of unique features, including an intracranial joint between the anterior and posterior halves of the braincase and the electrosensory rostral organ. Currently most workers agree that coelacanths are the sister group of the lineage that produced the lungfishes and the tetrapods.

6.3 Evolution of the Actinopterygii

Stem actinopterygians include a variety of diverse taxa, formerly placed in a group of extinct fishes, the "paleoniscoids," which is no longer considered to be monophyletic. Although fragmentary fossils of Late Silurian actinopterygians exist, complete fossil skeletons are not found earlier than the Middle to Late Devonian. Early actinopterygians were small fishes (usually 5 to 25 centimeters long, although some were over a meter) with a single dorsal fin and a strongly heterocercal, forked caudal fin with little fin web. Paired fins with long bases were common, but several taxa had pectoral fins with expanded muscular bases (see Figure 6–2e). The interlocking scales were thick like those of sarcopterygians but differed in structure and in growth pattern. In sarcopterygians the hard outer coating of the scales is cosmine (which is derived from dentine), whereas in actinopterygians it is ganoine (derived from enamel). Parallel arrays of closely packed radial bones supported the bases of the fins. The number of bony rays supporting the fin membrane was greater than the number of supporting radials, and these rays were clearly derived from elongated scales aligned end to end.

▣ Biology of Early Actinopterygians

Ray-finned fishes represent the largest radiation of vertebrates with some 27,000 extant species, and they have a variety of derived characters (**Box 6–1**).

BOX 6–1   Brainy Fish

Because tetrapods evolved from fish, we often think of fishes as being more primitive than animals like ourselves, but, of course, different lineages of fishes have been evolving from our common ancestor as long as we have. Ray-finned fishes have many unique, highly derived characteristics, such as the additional doubling of the *Hox* genes described in Chapter 2.

Actinopterygians are derived with respect to all other vertebrates in the way they enlarge the forebrain. The normal condition among vertebrates with enlarged brains is to make space by growing the larger cerebral hemispheres in upon themselves (**Figure 6–5**). This condition is seen among tetrapods and also occurs independently in some other vertebrates, such as sharks. Actinopterygians, on the other hand, make their brains large in the opposite way. Rather than *inverting* the growing cerebral hemispheres, they *evert* them, so that their forebrains are effectively inside out with respect to those of other vertebrates.

An ingenious explanation for this condition has been proposed by Glenn Northcutt (Northcutt 2002). He points out that actinopterygians differ from other fishes in having many species of small body size (most Devonian species were less than 15 centimeters in length). Living actinopterygians also have tiny larvae, with eyes that are large in proportion to the size of their heads. This combination of small head size and large eyes may create space-packing problems within the developing head, resulting in this peculiar method of brain development.

Other strange brains are found among the elephantnose fishes (osteoglossomorphs in the family Mormyridae). These fishes have a brain-to-body-size ratio that rivals the condition in humans; but, instead of the forebrain being the part that has been enlarged, they possess an enormously enlarged cerebellum in the hindbrain (Nilsson 1996). This large cerebellum is probably related to their powers of processing the electrical signals they use for navigation and predation in murky waters.

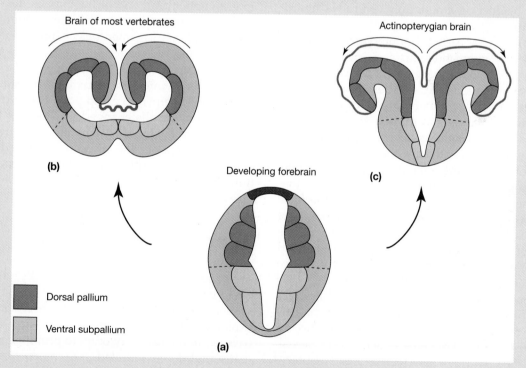

▲ Figure 6–5 **Forebrains of fishes.** Schematic drawings of transverse sections through the vesicle of the developing forebrain. (a) The primitive condition found in craniates with small brains. (b) The invaginating cerebral hemispheres of an amphibian compared to (c) the evaginating hemispheres of an actinopterygian.

Specializations of the early ray-finned fishes for locomotion and feeding deserve special attention. Primitive actinopterygians were diverse and successful in freshwater and marine habitats from 380 to 280 million years ago. The different types share numerous characters, but these are all primitive relative to later members of the Actinopterygii and are thus of no use in defining an evolutionary lineage. Near the end of the Paleozoic, the ray-finned fishes showed signs of change. The upper and lower lobes of the caudal fin were often nearly symmetrical, and all fin membranes were supported by fewer bony rays—the dorsal and anal fins had about one bony ray for each internal supporting radial. This morphological reorganization probably increased the flexibility of the fins.

The dermal armor of late Paleozoic ray-finned fishes was also reduced compared to that of their ancestors. The changes in fins and armor may have been complementary—more mobile fins mean more versatile locomotion, and increased ability to avoid predators may have permitted a reduction in heavy armor. This reduction of weight could have further stimulated the evolution of increased locomotor ability, which was probably enhanced by perfection of the swim bladder as a delicately controlled hydrostatic device.

The lower jaw of early actinopterygians was supported by the hyomandibula, and in most forms the jaw was snapped closed in a scissors action by the adductor mandibulae muscles. The adductor mandibulae muscle originated in a narrow enclosed cavity between the dermal maxilla and the palatoquadrate (which is tight against the cranium). The muscle's insertion was toward the rear of the lower jaw, near the jaw's articulation with the quadrate. This arrangement produced a lever system that moved the jaw rapidly (a quick snap), but the force created was low. The close-knit dermal bones of the cheeks permitted little expansion of the **orobranchial chamber** (the mouth and gills) beyond that required for respiration, and there was no space for a larger adductor mandibulae muscle.

The food-gathering apparatus of several lineages of bony fish underwent radical changes in the Permian. In the Late Permian, one lineage produced a new clade of actinopterygians—the neopterygians—distinguished by a new jaw mechanism. Neopterygians have jaws with a short maxilla, with its posterior end freed from the other bones of the cheek. Because the cheek was no longer solid, the nearly vertically oriented hyomandibula could swing out laterally when the mouth opened, rapidly increasing the volume of the orobranchial chamber and producing a powerful suction that drew prey into the mouth. The crushing power of the sharply toothed jaw could be increased because the adductor muscle, no longer limited in size by a solid bony cheek, expanded dorsally through the space opened by the freeing of the maxilla. In addition, an extra lever arm developed at the site of insertion of the adductor muscle on the lower jaw. This lever arm, called a **coronoid process**, appears on different jaw bones in different lineages of bony fishes. In most modern fishes, the coronoid process is on either the dentary or the angular bone. A coronoid process adds torque, and thus power, to the jaws of many forceful biters.

The neopterygians first appeared in the Late Permian (**Figure 6–6**) and became the dominant fishes of the Mesozoic. During the Late Triassic period, stem neopterygians gave rise to fishes with further feeding and locomotor specializations. In derived neopterygians, most notably in the teleosts, the bones of the gill cover (operculum) became connected to the mandible so that expansion of the orobranchial chamber aided in opening the mouth. The anterior articulated end of the maxilla developed a ball-and-socket joint with the neurocranium. Because of its ligamentous connection to the mandible, the free posterior end of the maxilla was rotated forward as the mouth opened (see Figure 6–6b). This pointed the maxilla's marginal teeth forward and helped to grasp prey. The folds of skin covering the maxilla changed the shape of the gape from a semicircle to a circular opening. These changes increased the suction produced during opening of the mouth. The result was greater directionality of suction and elimination of a possible side-door escape route for small prey.

■ Specializations of the Teleosts

Although teleosts probably evolved in the sea, they soon radiated into freshwater. By the Late Cretaceous, most of the more than 400 families of modern teleosts had evolved. Teleosts continued and expanded the changes in fins and jaws that contributed to the diversity of more primitive neopterygians and have radiated into more than 27,000 species occupying almost every ecological niche imaginable for an aquatic vertebrate—including some kinds of fish that spend substantial amounts of time on land.

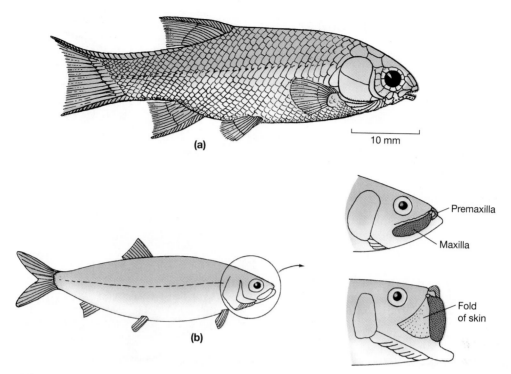

▲ **FIGURE 6–6** Teleosts. (a) *Acentrophorus* of the Permian illustrates an early member of the late Paleozoic neopterygian radiation. (b) *Leptolepis,* an Early Jurassic teleost with enlarged mobile maxillae that form a nearly circular mouth when the jaws are fully opened. Membranes of skin close the gaps behind the protruded bony elements. Modern herrings have a similar jaw structure.

Specializations of the Jaws Most of the main themes in teleost evolution involve changes in the jaws, from simple prey-grabbers to highly sophisticated suction devices. Suction is important in prey capture in water because a rapid approach by a predator pushes a wave of water in front of its head toward the prey. This wave flows around and away from the mouth and could push prey away from the grasp of the predator's jaws. Neopterygians had solved this problem by rapidly increasing the volume of the orobranchial chamber, creating a flow of water that carries prey into the mouth. Feeding specializations in the earliest teleosts involved only a slight loosening of the premaxillary bones so that they moved during jaw opening to accentuate the round mouth shape. One early clade of teleosts showed an enlargement of the free-swinging posterior end of the maxilla to form a nearly circular mouth when the jaws were fully opened. Later in the radiation of teleosts, distinctive changes in the jaw apparatus permitted a wide variety of feeding modes based on the speed of opening the jaws and the powerful suction produced by the highly integrated movements of jaw, gill cover, and cranium.

In addition to rapid and forceful suction, many teleosts have great mobility in the skeletal elements that rim the mouth. This mobility allows the grasping margins of the jaws to be extended forward from the head, often at remarkable speed. The functional result, called **protrusible** jaws, has evolved three or four times in different teleost clades.

Although jaw protrusion is generally associated with perches and their many relatives (perciform fishes, **Figure 6–7**), it also occurs in the related silversides and their kin (atherinomorphs), in the cods and anglerfishes (paracanthopterygians), and in the minnows (cypriniform ostariophysans). Jaw morphology differs significantly among fishes with protrusible jaws, which clearly shows that these groups have evolved protrusion independently.

All jaw protrusion mechanisms involve complex ligamentous attachments that allow the ascending processes of the premaxilla to slide forward on top of the cranium without dislocation. In addition, since no muscles are in position to pull the premaxillae forward, they must be pushed by leverage from behind. Two sources provide the necessary leverage. First, opening the lower jaw may protrude the

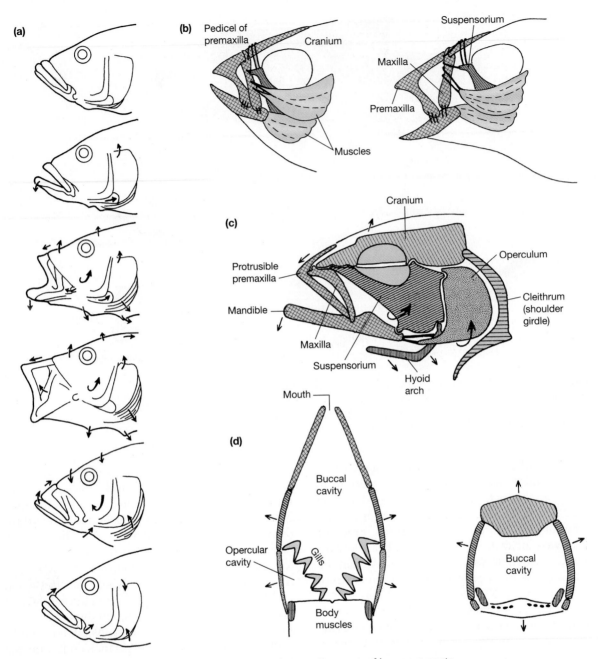

▲ **FIGURE 6–7** Jaw protrusion in suction feeding. (a) *Top to bottom:* Sequence of jaw movements in an African cichlid fish, *Serranochromi*. (b) Muscles, ligaments, and bones involved in movements during premaxillary protrusion. (c) Skeletal movements and ligament actions during jaw protrusion. (d) Frontal section (*left*) and cross section (*right*) of buccal expansion during suction feeding.

premaxillae through ligamentous ties between the mandible and the posterior tip of the premaxillae (see Figure 6–7b). Second, leverage can be provided by complex movements of the maxillae, which become isolated from the rim of the mouth by long, posterior projections of the premaxillae that often bear teeth.

The independent movement of the protrusible upper jaw also permits the mouth to be closed by extension of the premaxillary bones while the orobranchial cavity is still expanded, trapping prey in the mouth.

With so many groups converging on a function as complex as jaw protrusion, one would expect that it would have great functional significance. Surprisingly, no single hypothesis of the advantage of jaw protrusion has much experimental support. Protrusion may have enhanced the hydrodynamic efficiency of the circular mouth opening of primitive teleosts, but this hypothesis seems insufficient to produce the complex anatomical changes needed. Protrusion may aid in gripping prey. Perhaps jaw protrusion allows the jaws to be fitted to the substrate during feeding while the body remains in the horizontal position required for rapid escape from the fish's own predators.

Further advantages of protrusible jaws may lie in the functional independence of the upper jaws relative to other parts of the feeding apparatus. Some fishes, such as silversides and killifishes, can greatly protrude, moderately protrude, or not protrude the upper jaw while opening the mouth and creating suction. These modulations direct the mouth opening and direction of suction ventrally, straight ahead, or dorsally, allowing the fish to feed from substrate, water column, or surface with equal ease. Perhaps the most broadly applicable hypothesis for jaw protrusion is that shooting out the jaws in front of the head increases the predator's approach velocity by 39 percent to 89 percent in the crucial last instant of its attack.

Pharyngeal Teeth Powerful mobile pharyngeal jaws have evolved several times among actinopterygians. Ancestrally, ray-finned fishes had numerous dermal tooth plates in the pharynx. These plates were aligned with (but not fused to) both dorsal and ventral skeletal elements of the gill arches. A general trend of fusion of these tooth plates to one another and to a few gill arch elements above and below the esophagus can be traced in the Neopterygii. These consolidated pharyngeal jaws were not very mobile ancestrally; they were used primarily to hold and manipulate prey in preparation for swallowing it whole. In the minnows and their relatives the suckers (both groups are ostariophysan fishes), the primary jaws are toothless but protrusible. The pharyngeal jaws are greatly enlarged and close against a horny pad on the base of the skull. These specializations allow extraction of nutrients from thick-walled plant cells, and the minnows and suckers represent one of the largest radiations of herbivores among vertebrates.

In the most derived teleosts, the muscles associated with the branchial skeletal elements supporting the pharyngeal jaws have undergone radical evolution, resulting in a variety of powerful movements of the pharyngeal jaw tooth plates. Not only are the movements of these second jaws completely unrelated to the movements and functions of the primary jaws, but in a variety of derived teleosts the upper and lower tooth plates of the pharyngeal jaws move quite independently of each other. At least some moray eels can extend their pharyngeal jaws out of their throats and into their oral cavity to grasp struggling prey and pull it back into the throat and esophagus (Mehta and Wainwright 2007). With so many separate systems to work with, it is little wonder that some of the most extensive adaptive radiations among teleosts have been in fishes endowed with protrusible primary jaws and specialized mobile pharyngeal jaws.

Specializations of the Fins The caudal fin of adult actinopterygians is supported by a few enlarged and modified hemal spines, called hypural bones, that articulate with the tip of the abruptly upturned vertebral column. In general, the number of hypural bones decreases during the transition from the earliest actinopterygians to the more derived teleosts. Modified posterior neural arches—the uroneurals—add further support to the dorsal side of the tail. These uroneurals are a derived character of teleosts. Thus supported, the caudal fin of teleosts is symmetrical and flexible. This type of caudal structure is known as homocercal. Along with a swim bladder that adjusts buoyancy, a homocercal tail allows a teleost to swim horizontally without using its paired fins for control, as sharks must. Some studies suggest that the action of the symmetrical tail may be more complex than had been realized previously (Lauder 1994). In burst and sprint swimming, the tail produces a symmetrical force; but, during steady-speed swimming, intrinsic muscles in the tail may produce an asymmetric action that increases maneuverability without requiring use of the lateral fins. Relieved of responsibility for controlling lift, the paired fins of teleosts are flexible, mobile, and diverse in shape, size, and position. They have become specialized for activities that include food gathering, courtship, sound production, walking, and flying.

As we saw among earlier groups, improvements in locomotion were accompanied by reduction of armor. Modern teleosts are thin-scaled by Paleozoic and Mesozoic standards, and many lack scales entirely. The few heavily armored exceptions generally show a secondary reduction in locomotor abilities.

6.4 Extant Actinopterygii– Ray-Finned Fishes

With an estimated 27,000 extant species so far described—and new species being discovered regularly—the extant actinopterygians present a fascinating, even bewildering, diversity of forms of vertebrate life. Because of their numbers, we are forced to survey them briefly, focusing on the primary characteristics of selected groups and their evolution.

The study of the phylogenetic relationships of actinopterygians entered its current active state in 1966 with a major revised scheme of teleostean phylogeny proposing several new relationships. In the following decades our understanding of the interrelationships of ray-finned fishes has grown phenomenally. Nevertheless, some of the relationships are uncertain and should be considered hypotheses (Lauder and Liem 1983, Nelson 2006, Stiassny et al. 1996).

Why are there so many actinopterygians? For one thing, aquatic habitats cover 73 percent of the planet.

But why then are the chondrichthyans and the sarcopterygian fishes relatively species poor? Recent studies by A. R. McCune and her colleagues (1996–1998) indicate that both fossil and extant actinopterygians may form new species more rapidly than birds or even some insects. Fish species evolving in the isolation of lakes appear to form new species especially rapidly.

▨ Polypteriformes and Acipenseriformes

Although primitive actinopterygians were replaced during the early Mesozoic by neopterygians, a few specialized forms of primitive ray-finned fishes have survived. The most primitive surviving lineage of actinopterygian fishes is the Polypteriformes, the bichirs and reedfish (**Figure 6–8**). The extant Polypteriformes are 11 species of elongate, heavily armored fishes from Africa. The name Polypteriformes (Greek *poly* = many, *ptery* = fin, and *form* = shape) refers to the peculiar flaglike dorsal

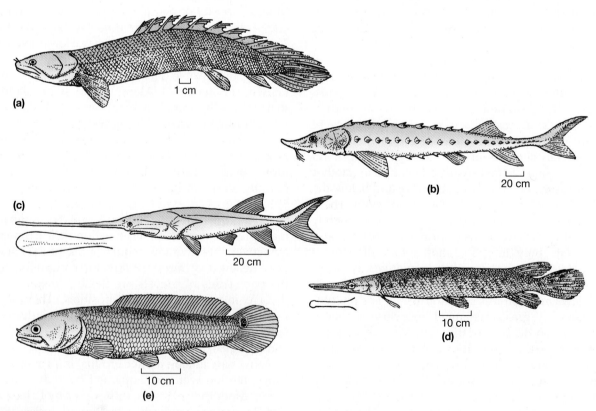

▲ **FIGURE 6–8** Extant nonteleostean actinopterygian fishes and primitive neopterygians. Actinopterygians: (a) *Polypterus,* a bichir; (b) *Acipenser,* a sturgeon; and (c) *Polyodon spathula,* one of two extant species of paddlefishes. Primitive neopterygians: (d) *Lepisosteus,* a gar; and (e) *Amia calva,* the bowfin. The drawings are not to scale.

finlets and the fleshy bases of their pectoral fins that are among their unique specializations, but little is known about their natural history and the significance of these features. They are less than a meter in length, slow-moving fishes with modified heterocercal tails. Polypteriformes differ from other extant stem actinopterygians in having well-ossified skeletons. In addition to a full complement of dermal and endochondral bones, polypteriformes are covered by thick, interlocking, multilayered scales. These scales are covered with a coat of ganoine, an enamel-like tissue characteristic of primitive actinopterygians, and are called ganoid scales. Larval bichirs (*Polypterus*) have external gills, possibly an ancestral condition for Osteichthyes. *Erpetoichthys*, the reedfish, is eel-like although armored with a full complement of ganoid scales. All polypteriformes are predatory. They also have ventrally placed lungs, which is a primitive condition.

Acipenseriformes (sturgeons and paddlefishes) includes two extant and two fossil families of specialized actinopterygians (see Figure 6–8b,c). The 24 species of sturgeons, family Acipenseridae (Latin *acipenser* = sturgeon), are large (1 to 6 meters), active, benthic fishes. They lack endochondral bone and have lost much of the dermal skeleton of more primitive actinopterygians. Sturgeons have a strongly heterocercal tail armored with a specialized series of scales extending from the dorsal margin of the caudal peduncle (the elongated base of the tail where it connects to the trunk) and continuing along the upper edge of the caudal fin. These scales are an ancestral character of the earliest Paleozoic actinopterygians. Most sturgeons have five rows of enlarged armorlike scales along the body. The protrusible jaws of sturgeons make them effective suction feeders. The mode of jaw protrusion is unique and derived independently from that of teleosts. Protrusible jaws plus adaptations for benthic life are the basis for considering sturgeons specialized compared to early fossil actinopterygians.

Sturgeons are found only in the Northern Hemisphere; some live in freshwater, and others are marine forms that ascend into freshwater to breed. Commercially important for their rich flesh and as a source of the best caviar, they have been severely depleted by intensive fisheries in much of their range. Dams, river and lake pollution, and siltation have also taken their toll on sturgeon. With intensive watershed-wide conservation measures, they appear to have been making a successful comeback in some areas such as New York's Hudson River, but a new threat has emerged: A growing demand for caviar,

combined with overexploitation, pollution, and alteration of sturgeon habitat around the Caspian Sea, has created a North American sturgeon fishery that once again threatens the survival of these ancient fish.

The two surviving species of paddlefishes, Polyodontidae (Greek *poly* = many and *odon* = tooth), are closely related to the sturgeons but have a still greater reduction of dermal ossification. Their most outstanding feature is a greatly elongate and flattened rostrum, which extends nearly one-third of their 2-meter length (see Figure 6–8c). The rostrum is richly innervated with ampullary organs that are believed to detect minute electric fields. Contrary to the common notion that the paddle is used to stir food from muddy river bottoms, the American paddlefish is a planktivore that feeds by swimming with its prodigious mouth agape, straining crustaceans and small fishes from the water using modified gill rakers as filters. The two species of paddlefishes have a disjunct zoogeographic distribution similar to that of alligators, and they also differ in feeding habits. One is found in the Chang (Yangtze) River valley of China, where it feeds on fishes; it is on the verge of extinction. The planktivorous species is in the Mississippi River drainage of the United States and, like sturgeon, is now being heavily fished for caviar. Fossil paddlefishes are known from western North America.

◼ Primitive Neopterygians

The two extant genera of primitive neopterygians are currently limited to North America and represent widely divergent types. The Lepisosteiformes (Greek *lepis* = scale and *osteo* = bone) is composed of seven species of gars (*Lepisosteus*). They are medium- to large-size (1 to 4 meters) predators of warm, temperate fresh and brackish (estuarine) waters. The elongate body, jaws, and teeth are specialized features, but their interlocking multilayered scales are similar to those of many Paleozoic and Mesozoic actinopterygians (see Figure 6–8d). Gars feed on other fishes taken unaware when the seemingly lethargic and excellently camouflaged gar dashes alongside them and, with a sideways flip of the body, grasps prey with its needlelike teeth. Alligators are the only natural predators able to cope with the thick armor of an adult gar.

The single species of Amiiformes (the bowfin, *Amia calva*; Greek *amia* = a kind of fish, see Figure 6–8e), lives in the same areas as gars. The head skeleton shows modifications of the jaws as a suction device. *Amia*, which are 0.5 to 1 meter long, prey on almost any organism smaller than themselves. Scales of the bowfin are comparatively thin and made up of a single layer of bone as in teleost fishes; however, the asymmetric caudal fin is very similar to the heterocercal caudal fin of more primitive fishes. The interrelationships of gars, *Amia*, and the teleosts have long been controversial.

■ Teleosteans

Most extant fishes are teleosts. They share many characters of caudal and cranial structure and are grouped into four clades of varying size and diversity.

Osteoglossomorpha The Osteoglossomorpha (Greek *osteo* = bone, *gloss* = tongue, and *morph* = form), which appeared in Late Jurassic seas, are now restricted to about 220 species, mostly in tropical fresh waters. *Osteoglossum* (**Figure 6–9a**) is a 1-meter-long predator from the Amazon, familiar to tropical fish enthusiasts as the arawana. *Arapaima* is an even larger Amazonian predator, perhaps the largest strictly freshwater fish. Before intense fishing reduced the populations, they were known to reach a length of at least 3 meters and perhaps as much as 4.5 meters. *Mormyrus* (**Figure 6–9b**), one of the African elephant fishes, is representative of the small African bottom feeders that use weak electric discharges to communicate with other members of their species. As dissimilar as they may

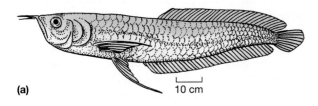

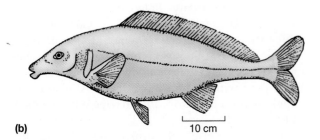

▲ **FIGURE 6–9** Extant osteoglossomorphs. (a) *Osteoglossum,* the arawana, from South America. (b) *Mormyrus,* an elephant-nose from Africa.

seem, the osteoglossomorph fishes are united by unique bony characters of the mouth and by the mechanics of their jaws.

Elopomorpha The Elopomorpha (Greek *elop* = a kind of fish) had appeared by the Late Jurassic period. A specialized leptocephalous (Greek *lepto* = small and *cephal* = head) larva is a unique character of elopomorphs (**Figure 6–10a and b**). These larvae spend a long time adrift, usually at the ocean surface, and are widely dispersed by currents. Elopomorphs include about 35 species of tarpons (Megalopidae), ladyfish (Elopidae), and bonefish (Albulidae), and more than 760 species of true eels (Anguilliformes and Saccopharyngiformes).

Most elopomorphs are eel-like and marine, but some species are tolerant of fresh waters. The common American eel, *Anguilla rostrata*, has one of the most unusual life histories of any fish. After growing to sexual maturity (which takes as long as 10 to 12 years) in rivers, lakes, and even ponds, the **catadromous** eels enter the sea. The North Atlantic eels migrate to the Sargasso Sea, an area in the central North Atlantic between the Azores and the West Indies. Here they are thought to spawn and die, presumably at great depth. The eggs and newly hatched leptocephalous larvae float to the surface and drift in the currents. Larval life continues until the larvae reach continental margins, where they transform into miniature eels and ascend rivers to feed and mature. European eels (*Anguilla anguilla*) spawn in the same region of the Atlantic as their American kin but may choose a somewhat more northeasterly part of the Sargasso Sea and perhaps a different—but also deep—depth at which to spawn. Their leptocephalus larvae remain in the clockwise currents of the North Atlantic (principally the Gulf Stream) and ride to Northern Europe as well as North Africa, the Mediterranean, and even the Black Sea before entering river mouths to migrate upstream and mature. During their long ride, they encounter cooler temperatures than do American eels, so they grow more slowly. Rate of growth and rate of development are separate phenomena, however, and cold temperatures slow the rate of growth more than the rate of development. As a result, adult European eels have more vertebrae than do American eels.

Clupeomorpha Most of the more than 360 species of Clupeomorpha (Latin *clupus* = a kind of fish) are specialized for feeding on minute plankton gathered by a specialized mouth and gill-straining apparatus.

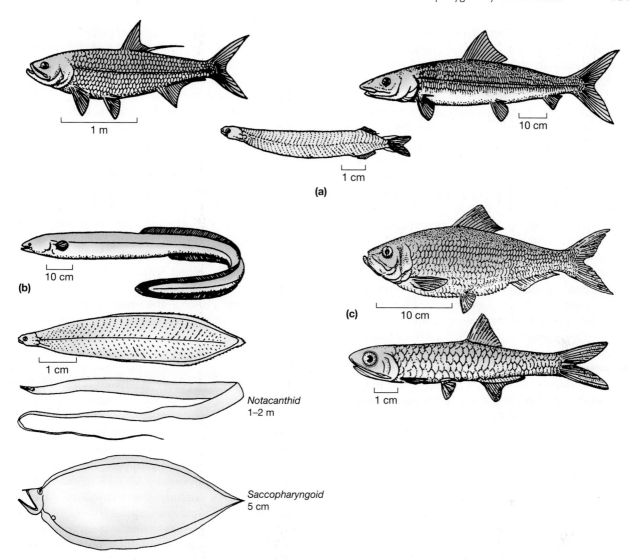

▲ **FIGURE 6–10** Extant teleosts of isolated phylogenetic position. (a) Elopomorpha, represented by a tarpon (*left*), a bonefish (*right*), and a typical fork-tailed leptocephalous larva (*below*). (b) Anguilliform elopormorphs, represented by the common eel, *Anguilla rostrata* (*above*), its leptocephalous larva (*immediately below*), and two other very different eel leptocephali. (c) Clupeomorpha, represented by a herring (*above*) and an anchovy (*below*).

They are silvery, mostly marine schooling fishes of great commercial importance. Common examples are herrings, shad, sardines, and anchovies (**Figure 6–10c**). Several clupeomorphs are **anadromous**, and the springtime migrations of American shad (*Alosa sapidissima*) from the North Atlantic into rivers in eastern North America once involved millions of individuals. The enormous shad runs of the recent past have been greatly depleted by dams and pollution of aquatic environments.

Euteleostei The vast majority of extant teleosts belong to the fourth clade, the Euteleostei (Greek *eu* = good), which evolved before the Late Cretaceous. With so many thousands of species, it is impossible to give more than scant information about them. For more about these fishes, refer to Barton (2007), Helfman et al. (2008), Moyle and Cech (2003) Nelson (2006), and the beautifully illustrated Paxton and Eschmeyer (1998). Here we will consider the stem euteleostian stock as represented today by the specialized ostariophysans and the generalized salmoniforms, but how these two monophyletic groups relate to each other is a matter of much dispute (Stiassny et al. 1996).

Ostariophysi   The Ostariophysi (Greek *ost* = bone and *physa* = bladder), the predominant fishes of the world's fresh waters, represent perhaps 25 to 30 percent of all extant fishes and about 80 percent of the fish species in freshwater. As a group, ostariophysans display diverse traits. For example, many ostariophysans have protrusible jaws and are adept at obtaining food in a variety of ways. In addition, pharyngeal teeth act as second jaws. Many forms have fin spines or special armor for protection, and the skin typically contains glands that produce substances used in olfactory communication. Although they have diverse reproductive habits, most lay sticky eggs or otherwise guard the eggs, preventing their loss downstream.

Ostariophysans have two distinctive derived characters. Their name refers to small bones that connect the swim bladder with the inner ear (**Figure 6–11**). Using the swim bladder as an amplifier and the chain of bones as conductors, this **Weberian apparatus** greatly enhances hearing sensitivity of these fishes. Sound (pressure) waves impinging on the fish cause the swim bladder to vibrate. One of the bones, the tripus, is in contact with the swim bladder; as the bladder vibrates, the tripus pivots on its articulation with the vertebra. This motion is transmitted by ligaments to two other bones, the intercalarium and scaphum. Movement of the scaphum compresses an extension of the membranous labyrinth (inner ear) against a fourth bone, the claustrum, resulting in stimulation of the auditory region of the inner ear. The ostariophysans are more sensitive to sounds and have a broader frequency range of detection than other fishes.

The second derived character uniting the ostariophysans is the presence of a fright or alarm substance in the skin. Chemical signals (pheromones) are released into the water when the skin is damaged, and they produce a fright reaction in nearby members of their own species and other ostariophysan species. The fright reaction may cause fish to rush for cover or form a tighter school.

Although all Ostariophysi have an alarm substance in the skin and a Weberian apparatus (or a rudimentary precursor of it), in other respects they are a diverse group containing some 6500 species. Ostariophysans include the characins (piranhas, neon tetras, and other familiar aquarium fishes) of tropical America and Africa, the carps and minnows (in freshwater worldwide except Central and South America,

Antarctica, and Australia), the catfishes (worldwide in freshwater except Antarctica and in many shallow marine areas as well), and the highly derived electric knifefishes of Central and South America.

The other stem group of euteleostians, the esocid and salmonid fishes (**Figure 6–12a**), include important commercial and game fishes. These fishes have often been lumped into a taxon, the "Protacanthopterygii," but the basis for this classification was often no more than shared ancestral euteleostean characters that are not valid for determining phylogenetic relationships. A new usage of the term includes only salmonids and some deep-water relatives and may be monophyletic. The salmonids include the anadromous salmon, which usually spend their adult lives at sea and make epic journeys to inland waters to breed, as well as the closely related trout, which usually live entirely in freshwater. Among the most primitive extant euteleosteans, at the base of the radiation of all other more derived teleosts, may be the esocids. These temperate freshwater fishes of the Northern Hemisphere include game species such as pickerel, pikes, and muskellunges and their relatives. The Southern Hemisphere galaxiids are also primitive euteleosts that live in habitats similar to those occupied by salmonids in the Northern Hemisphere, although they are in general much smaller fishes.

Another, more diverse group of euteleosts is not recognized as monophyletic. The 800 or more species in this category include the marine mesopelagic lanternfishes and marine hatchetfishes, the lizard fishes, and related fishes—mostly from mesopelagic or bathypelagic waters and thus not well known except to specialists (see Figure 6–21b, c, and e).

Mobile jaws and protective, lightweight spines in the median fins have evolved in several groups of euteleosts. About 1200 species of fishes, including cods and anglerfishes, are grouped as the Paracanthopterygii, although their similarities may represent convergence (**Figure 6–12b**). However, the Acanthopterygii (true spiny-rayed fishes) that dominate the open ocean surface and shallow marine waters of the world do appear to form a monophyletic lineage within the euteleosts. Among the acanthopterygians, the atherinomorphs have protrusible jaws and specializations of form and behavior that suit them to shallow marine and freshwater habitats. This group includes the silversides, grunions, flying fishes, and halfbeaks, as well as the egg-laying and live-bearing cyprinodonts (**Figure 6–12c**). Killifish are examples of oviparous cyprinodonts, and viviparous forms include the guppies, mollies, and swordtails commonly maintained in home aquaria.

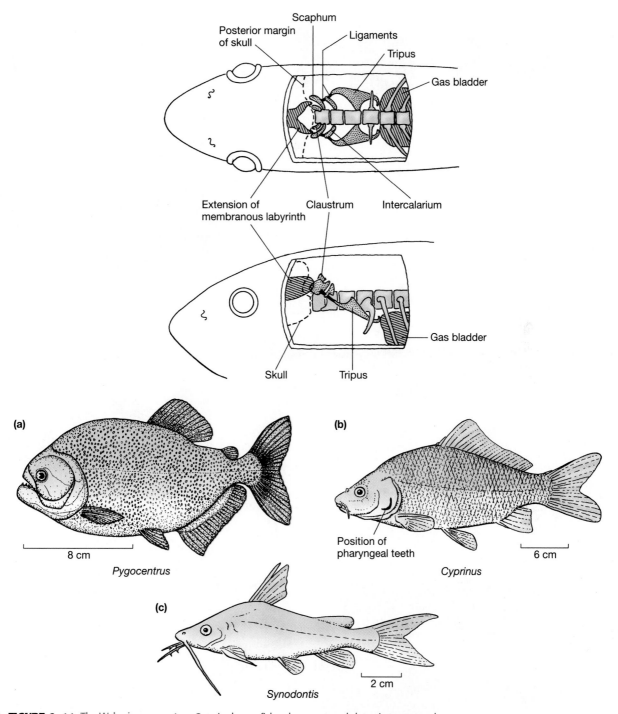

▲ **FIGURE 6–11** The Weberian apparatus. Ostariophysan fishes have a sound-detection system, the Weberian apparatus, which is a modification of the swim bladder and the first few vertebrae and their processes. Typical ostariophysans include (a) characins (here a piranha), (b) minnows, and (c) catfishes.

Most species of acanthopterygians—and the largest order of fishes—are Perciformes, with over 9300 extant species. A few of the well-known members are snooks, sea basses, sunfishes, perches, darters, dolphins (mahi mahi), snappers, grunts, porgies, drums, cichlids, barracudas, tunas, billfish, and most of the fishes found on coral reefs.

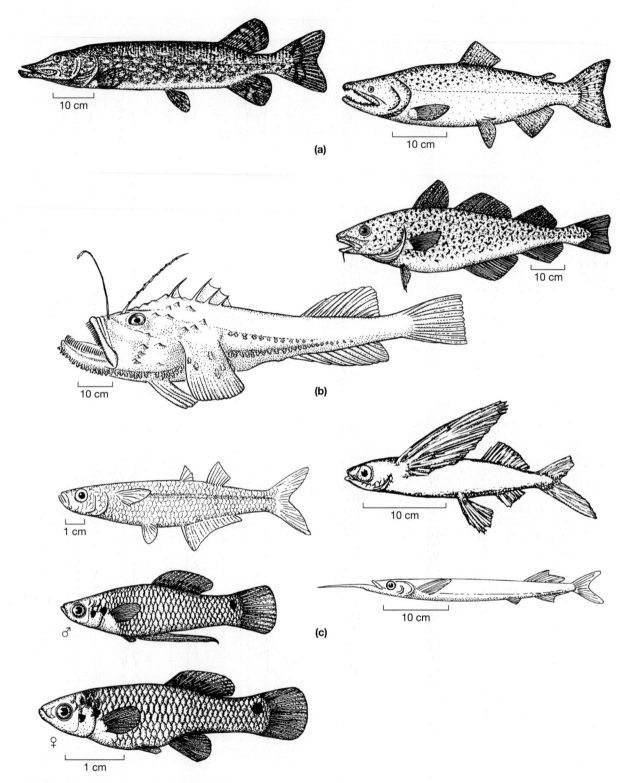

▲ **FIGURE 6–12** Euteleosts. (a) Primitive euteleosts represented by the pike (*left*) and the salmon (*right*). (b) Paracanthopterygians represented by the cod (*right*) and the goosefish angler (*left*). (c) Atherinomorph fishes represented by (*clockwise from upper left*) an Atlantic silverside (*Menidia*), a flying fish, a halfbeak, and live-bearing killifish, the male of which has a modified anal fin used in internal fertilization.

6.5 Locomotion in Water

Perhaps the single most recognizable characteristic of the enormous diversity of fishes is their mode of locomotion. Fish swimming is immediately recognizable, aesthetically pleasing, and—when first considered—rather mysterious, at least when compared to the locomotion of most land animals. Fish swimming results from anterior-to-posterior sequential contrac-

tions of the muscle segments along one side of the body and simultaneous relaxation of those of the opposite side. Thus a portion of the body momentarily bends, the bend is moved posteriorly, and a fish oscillates from side to side as it swims. These lateral undulations are most visible in elongate fishes, such as lampreys and eels (**Figure 6–13**). Most of the power for swimming comes from muscles in the posterior region of the fish.

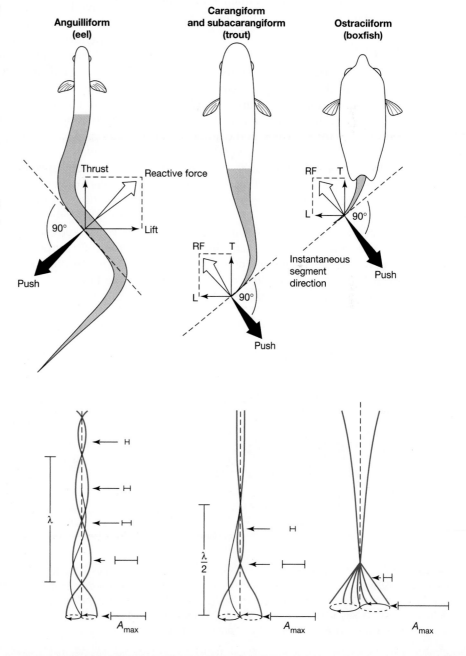

▶ **FIGURE 6–13** Basic movements of swimming fishes. Outlines of some major swimming types (*top*); body regions that undulate are shaded. The lift (lateral) component of the reactive force produced by one undulation's push on the water is canceled by that of the next, oppositely directed undulation so the fish swims in a straight line. The thrust (forward) component from each undulation is in the same direction and thus is additive, and the fish moves through the water. Waveforms created by undulations of points along the body and tail (*bottom*). A_{max} represents the maximum lateral displacement of any point. Note that A_{max} increases posteriorly; λ is the wavelength of the undulatory wave.

In 1926, Charles Breder classified the undulatory motions of fishes into three types:

- Anguilliform—Typical of highly flexible fishes capable of bending into more than half a sinusoidal wavelength. Named for the locomotion seen in the true eels, the Anguilliformes.
- Carangiform—Undulations limited mostly to the caudal region, the body bending into less than half a wavelength. Named for *Caranx*, a genus of fast-swimming marine fishes known as jacks and trevallies.
- Ostraciiform—The body is inflexible; undulation is limited to the caudal fin. Named for the boxfishes, trunkfishes, and cowfishes (family Ostraciidae) whose fused scales form a rigid box around the body, preventing undulations.

Although these forms of locomotion were named for groups of fishes that exemplify them, many other types of fishes use these swimming modes. For example, hagfishes, lampreys, most sharks, sturgeons, arawanas, many catfishes, and countless elongate spiny-rayed fishes are not eels, yet they swim in an anguilliform mode. Nonetheless, Breder's categories are useful for understanding locomotion in water and are central to most modern studies of fish locomotion.

Many specializations of body form, surface structure, fins, and muscle arrangement increase the efficiency of the different modes of swimming. A swimming fish must overcome the downward pull of gravity and the drag of water (**Figure 6–14**). To overcome gravity, fishes generate vertical lift. Teleosts most often do this by generating buoyancy with a gas-filled swim bladder unlike sharks, which use the upwardly angled pectoral fins to produce lift.

■ Overcoming Drag—The Generation of Thrust

In general, fishes swim forward by pushing backward on the water. For every active force, there is an opposite reactive force (Newton's third law of motion). Undulations produce an active force directed backward, and also a lateral force. The overall reactive force is directed forward because the forces to the sides cancel each other. Fish have developed a tremendous range of swimming modes, some emphasizing the body and others relying on the fins to produce thrust (**Figure 6–15**).

Anguilliform and carangiform swimmers increase speed by increasing the frequency of their body undulations. Increasing the frequency of body undulations applies more power (force per unit time) to the water. Different fishes achieve very different maximum speeds—for example, eels are slow and tunas are very fast. An eel's long body limits speed because it induces drag from the friction of water on the elongate surface of the fish. Fishes like tunas that swim rapidly are shorter and less flexible. Force from the contraction of anterior muscle segments is transferred through ligaments to the caudal peduncle and the tail. Morphological specializations of this swimming mode reach their zenith in fishes like tunas, in

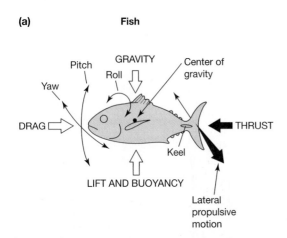

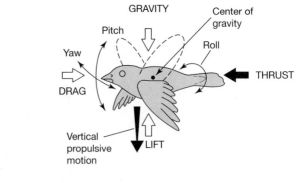

▲ **FIGURE 6–14** Comparison of swimming and a flying. (a) By moving its caudal fin, a fish produces lateral movement far from the center of gravity. The reactive force of the water to this movement causes the head to yaw (move in the horizontal plane) in the same direction as the tail. (b) In a bird, the major propulsive stroke of the wings is downward. Because both the propulsive stroke (thrust) and the upward reactive force (lift) act near the center of gravity, the bird does not pitch (move in the vertical plane).

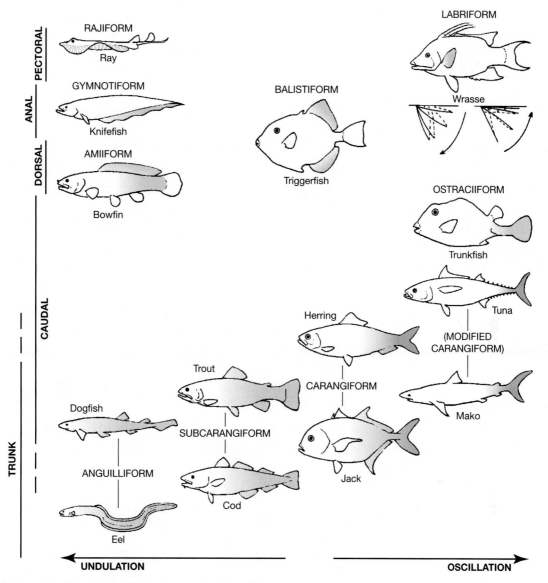

▲ **FIGURE 6–15** Location of swimming movements in various fishes. Shaded areas of body undulate or move in swimming. Names such as carangiform describe the major types of locomotion found in fishes; they are not a phylogenetically based identification of all fishes using a given mode.

which the caudal peduncle is slender and the tail greatly expanded vertically.

Other fishes seldom flex the body to swim but instead undulate the median fins to provide thrust. Using the dorsal fin is amiiform swimming, using the anal fin is gymnotiform, and using both fins is balistiform swimming. Usually, several complete waves are observed along the fin, and very fine adjustment in the direction of motion can be produced. Many fishes—for example, surf perches and many coral reef fishes (such as surgeonfishes, wrasses, and parrotfishes)—generally do not oscil-

late the body or median fins; rather, they row with the pectoral fins to produce movement (known as labriform swimming).

■ Improving Thrust: Minimizing Drag

A swimming fish experiences drag of two forms: viscous drag from friction between the fish's body and the water, and inertial drag from pressure differences created by the fish's displacement of water. Viscous drag is relatively constant over a range of speeds, but inertial drag is low at slow speeds and increases

▶ **FIGURE 6–16** Effect of body shape on drag. (a) Streamlined profiles with width (*d*) equal to approximately one-fourth of length (*l*) minimize drag. The examples are for solid, smooth test objects with thickest section about two-fifths of the distance from the tip. (b) Width-to-length ratios (*d/l*) for several swimming verte-brates. Like the test objects, these vertebrates tend to be circular in cross section. Note that the ratio is near 0.25, and the general body shape approximates a fusiform shape.

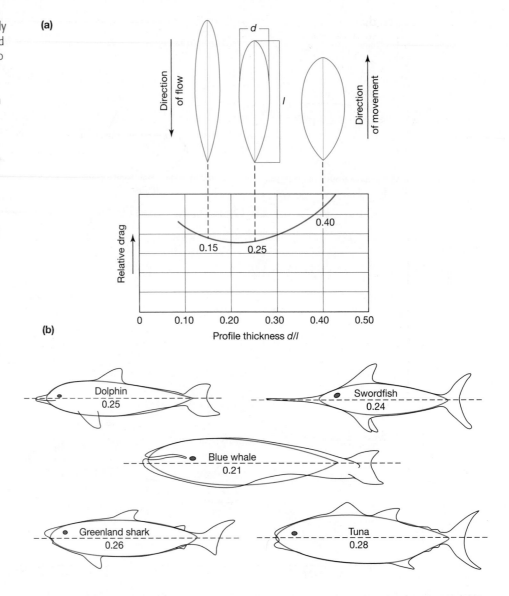

rapidly with increasing speed. Viscous drag is affected by surface smoothness, whereas inertial drag is influenced by body shape. A thin body has high viscous drag because it has a large surface area relative to its muscle mass. A thick body induces high inertial drag because it displaces a large volume of water as it moves forward. Streamlined (teardrop) shapes produce minimum inertial drag when their maximum width is about one-fourth of their length and is situated about one-third of the body length from the leading tip (**Figure 6–16**). The shape of many rapidly swimming vertebrates closely approximates these dimensions. Usually, fast-swimming fishes have small scales or are scaleless, with smooth body contours lowering viscous drag. (Of course, many slow-swimming fishes also are scaleless, which shows that universal general-

izations are difficult to make.) Mucus also contributes to the reduction of viscous drag.

Fishes that swim by undulating only the caudal peduncle and caudal fin are usually called modified carangiform swimmers. Scombroids (mackerels, tunas) and many pelagic sharks have a caudal peduncle that is narrow dorsoventrally but is relatively wide from side to side. The peduncle of carangiform swimmers is often studded laterally with bony plates called scutes. These structures present a knife-edge profile to the water as the peduncle undulates from side to side; they also contribute to the reduction of drag on the laterally sweeping peduncle. The importance of these seemingly minor morphological changes is illustrated by the tail stalk of cetaceans (whales and porpoises), which is also narrow and has a double knife-edge profile. But the tail stalk of cetaceans is narrow from side to side, whereas the caudal peduncle of scombroid fishes is narrow from top to bottom. The difference reflects the plane of undulation—up and down for cetaceans and side to side for fishes. These strikingly similar specializations of modified carangiform swimmers—sharks, scombroids, and cetaceans—produce efficient conversion of muscle contractions into forward motion.

The tail creates turbulent vortices (whirlpools of swirling water) in a fish's wake. These vortices may be a source of inertial drag, or they may be modified to produce thrust (Stix 1994, Triantafyllou and Triantafyllou 1995). The total drag created by the caudal fin depends on its shape. When the aspect ratio of the fin (dorsal-to-ventral length divided by the anterior-to-posterior width) is large, the amount of thrust produced relative to drag is high. The stiff sickle-shaped fin of scombroids and of certain sharks (mako, great white) results in a high aspect ratio and efficient forward motion. Even the cross section of the forks of these caudal fins assumes a streamlined teardrop shape, further reducing drag. Many species with these specializations swim continuously.

The caudal fins of trouts, minnows, and perches are not stiff and seldom have high aspect ratios. These fish bend the body more than carangiform swimmers, in a swimming mode called subcarangiform. Subcarangiform swimmers spread or compress the caudal fin to modify thrust and stiffen or relax portions of the fin to produce vertical movements of the posterior part of the body. The caudal peduncle of these subcarangiform swimmers is laterally compressed and deep because the peduncle contributes a substantial part of the total force of propulsion. These fishes often swim in bursts, usually accelerating rapidly from a standstill. This startle response is controlled by a pair of giant neurons (the Mauthner cells, found in lampreys, teleosts, and many amphibians) that extend the full length of the spinal cord and send branches to the motor nerves serving the myotomes. The large diameter of the Mauthner axons allows an impulse to move rapidly, and the Mauthner system is used in escape reactions.

Perhaps the ultimate in efficient "swimming" has been demonstrated in trout and is probably widespread in fishes living in flowing waters and perhaps those living in dense schools. Trout slip into the vortices that predictably form behind objects such as rocks or woody debris in streams and rivers and, presumably using their sensitive lateral line receptors, adjust their positions so that they are held in place with greatly reduced muscle exertion as long as they relax and allow their bodies to flap in the current like a flag in the breeze (Liao et al. 2003).

Current understanding of these varied adaptations is the result of extensive study of vertebrate swimming (especially that of bony fishes) over the past several decades. Nevertheless, we still do not fully understand how the extraordinary propulsion efficiencies, accelerations, and maneuverability of fishes are achieved. Naval architects have not been able to design boats that perform nearly as well as fishes.

6.6 Actinopterygian Reproduction

Reproductive modes of actinopterygians show greater diversity than is known in any other vertebrate taxon. Despite this diversity, the vast majority of ray-finned fishes are oviparous. Within oviparous teleosts, freshwater and marine species show contrasting specializations. Some marine species of fishes and most freshwater species lay adhesive eggs on rocks or plants or in gravel or sand, and often one or both parents guard the eggs and sometimes the young for varying periods after they hatch. Nests vary from depressions in sand or gravel to elaborate constructions of woven plant material held together by parental secretions. Species of marine fishes that construct nests are smaller than species that are pelagic spawners, perhaps because small species are better able to find secure nest sites than are large species. Eggs may be laid near, on, or within other organisms, such as stinging anemones, mussels, crabs, sponges, and tunicates that protect the eggs. Alternatively the parents may carry the eggs—species are known that carry eggs on their fins; under their lips; on specialized protuberances, skin patches, and pouches; and even in their mouths or gill cavities.

■ Reproductive Characteristics of Freshwater Teleosts

Freshwater teleosts generally produce and care for a relatively small number of large, yolk-rich **demersal** eggs (i.e., eggs that are buried in gravel, placed in a nest, or attached to the surface of a rock or plant). Attachment is important because flowing water could easily carry a pelagic (i.e., free-floating) egg away from habitats suitable for development. The eggs of freshwater teleosts usually hatch into young (called fry) that soon have body forms and behaviors similar to those of adults.

■ Reproductive Characteristics of Marine Teleosts

Demersal eggs may be ancestral for actinopterygians, and producing pelagic eggs and larvae is probably a derived characteristic of euteleosts. Most marine teleosts release large numbers of small, buoyant, transparent eggs into the water. These eggs are fertilized externally and left to develop and hatch while drifting in the open sea. The larvae are also small and usually have little yolk reserve. They begin feeding on microplankton soon after hatching. Marine larvae are generally very different in appearance from their parents, and many larvae have been described for which the adult forms are unknown. Such larvae are often specialized for life in the oceanic plankton, feeding and growing while adrift at sea for weeks or months, depending on the species. The larvae eventually settle into the juvenile or adult habitats appropriate for their species. It is not yet generally understood whether arrival at the appropriate adult habitat (deep-sea floor, coral reef, or river mouth) is an active or passive process on the part of larvae. However, the arrival coincides with metamorphosis from larval to juvenile morphology in a matter of hours to days. Although juveniles are usually identifiable to species, relatively few premetamorphic larvae have successfully been reared in captivity to resolve unknown larval relationships. Studies of bones called otoliths are contributing to our understanding of the early life history of fishes in the wild (**Box 6–2**).

The strategy of producing planktonic eggs and larvae exposed to a prolonged and risky pelagic existence appears to be wasteful of gametes. Nevertheless, complex life cycles of this sort are the principal mode of reproduction of marine fishes. One advantage that fishes may achieve by spawning pelagically is reduction of some types of predation on fertilized eggs. Predators that would capture the eggs may be abundant in the parental habitat but relatively absent from the pelagic realm. Pelagic spawning fishes often migrate to areas of strong currents to spawn or spawn in synchrony with maximum monthly or annual tidal currents, thus assuring rapid offshore dispersal of their eggs. These behaviors would help to ensure that the eggs are quickly carried away from potential predators.

A second advantage of pelagic spawning may involve the high biological productivity of the sunlit surface of the pelagic environment. Microplankton (bacteria, algae, protozoans, rotifers, and minute crustaceans) are abundant where sufficient nutrients reach sunlit waters. If energy is limiting to the parent fishes, it could be advantageous to invest the minimum possible amount of energy by producing eggs that hatch into specialized larvae that feed on pelagic food items too small for the adults to eat.

A final hypothesis involves species-level selection. Producing floating, current-borne eggs and larvae increases the chances of colonizing all patches of appropriate adult habitat in a large area. A widely dispersed species is not vulnerable to local environmental changes that could extinguish a species with a restricted geographic distribution. Perhaps the predominance of pelagic spawning species in the marine environment reflects millions of years of extinctions of species with reproductive behaviors that did not disperse their young as effectively.

6.7 The Adaptable Fish–Teleost Communities in Contrasting Environments

Given the great numbers of both species and individuals of extant fishes, especially the teleosts, it is little wonder that they inhabit an enormous diversity of watery environments. Examining two of the most distinctive, the deep sea and coral reefs, allows us to better understand the amazing adaptability of the teleost fishes.

■ Deep-Sea Fishes

Of all regions on Earth, the deep sea is the least studied. Two major life zones exist in the sea: the pelagic (subdivided into the epipelagic, mesopelagic, and benthopelagic), where organisms live a free-floating or swimming existence, and the benthic, where organisms associate with the bottom either near shore or at great depths (**Figure 6–18**). Sunlight is totally extinguished at a depth of 1000 meters in the clearest oceans and at much shallower depths in coastal seas where concentrations of sediments are higher than they are in the open ocean. As a result, about 75 percent of the ocean is perpetually dark, illuminated only by the flashes and glow of bioluminescent organisms. A distinctive and bizarre array of deep-sea fishes, with representatives from no fewer than five orders, has evolved in those zones.

Fishes decrease in abundance, size, and species diversity at greater depths. These trends are not surprising, for all animals ultimately depend on plant photosynthesis, which is limited to the epipelagic regions (from the surface to 100 meters). Below the

BOX 6–2   What a Fish's Ears Tell About Its Life

Tracking and observing minute fish eggs or translucent larvae in the open sea or turbid rivers presents insurmountable problems. However, there is an indirect method of reconstructing life-history details of an individual fish. A characteristic of bony fishes is the presence of up to three compact, mineralized structures suspended within each inner ear, the otoliths. These "ear stones" are especially well developed in the majority of teleosts, where they are often curiously shaped, fitting into the spaces of the membranous labyrinth very exactly and growing in proportion to the growth of the fish. They are important in hearing, orientation, and locomotion, and they are formed in most teleosts during late embryonic stages.

Otoliths grow by the accretion of mineralized layers deposited on the surface resulting in concentric lamellae, resembling the layers of an onion (**Figure 6–17**). The smallest layer that can be distinguished usually reflects a day's growth. The relative width, density, and interruptions of the layers show the environmental conditions the individual encountered daily, from hatching, including variations in temperature and food capture. For many species, the growth record is complete and uninterrupted for the first several years of life. Seasonal and annual patterns of depositions are often present as well. In such seasonal environments and also as an individual fish's growth slows with age, the daily record may become incomplete or intermittent. Minute quantities of elements and even their distinguishable isotopes characteristic of the environment in which the fish has spent each day are incorporated in the bands, further enhancing the information provided by the otoliths.

Thus a day-by-day record of the individual is written in its otoliths. This may include movement or migrations and enable biologists to determine spawning areas or "home waters." It is even possible to imprint a permanent code in the otoliths by subjecting young fish to a series of carefully controlled temperature increases and decreases: a bar code–like pattern results in the otolith microstructure. This unique pattern allows fishery managers to use this method to mark juvenile fishes before they are released. Years later, when the fishes are recaptured as adults, the code embedded in the otoliths shows the brood to which they belong. Otolith chemistry is also purposely altered as a means to mass mark fish broods using such compounds as strontium chloride and fluorochromes. By injecting gravid females with such substances, researchers have successfully induced transgenerational otolith marks (e.g., marks in the offspring's otoliths) in viviparous surfperches and rockfishes. This may lead to better understanding of juvenile recruitment in the wild.

New techniques of otolith analysis are also providing very important information on the longevity of many species. More traditional aging methods often significantly underestimate fish age. Current information is revealing that it is not uncommon for many fish to attain ages of 50 years, and some may double and triple that figure. While still fraught with problems of interpretation, the study of otolith microstructure is significantly advancing our understanding of reproduction and early life history of fishes.

▶ **FIGURE 6–17** Fine structure of an otolith. Scanning electron micrograph of an otolith from a juvenile French grunt, Haemulon flavolineatum. The central area represents the focus of the otolith from which growth proceeds. The alternating dark and light rings are daily growth increments. Note that the width of the growth increments varies, signifying day-to-day variation in the rate of growth.

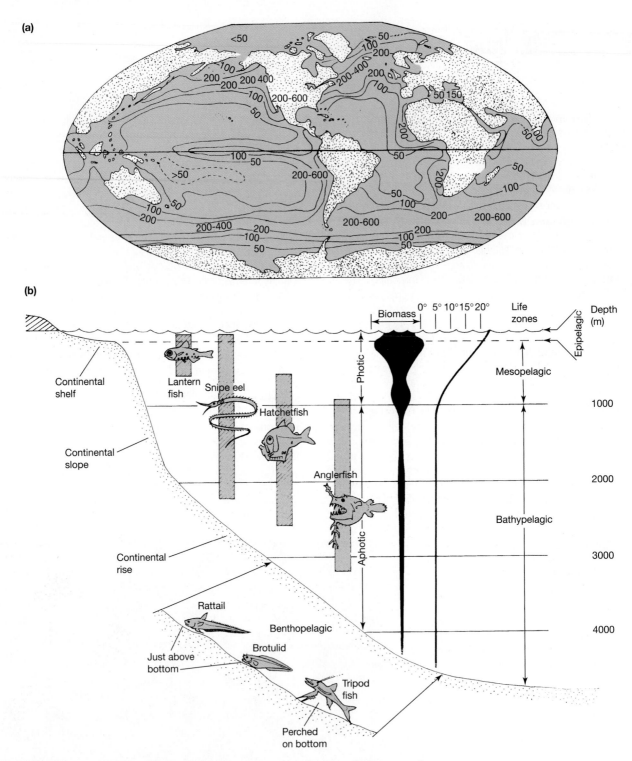

▲ **FIGURE 6–18** Life zones of the ocean depths. (a) Annual productivity at the ocean surface. Numbers are grams of carbon produced per square meter per year. Rich assemblages of deep-sea fishes occur where highly productive waters overlie deep waters. (b) Schematic cross section of the life zones within the deep sea. Temperature and biomass are shown, as are the vertical ranges of several fish species, some of which migrate daily.

epipelagic, animals must depend on a rain of detritus from the surface. The amount of food diminishes with increasing depth because it is consumed during descent. Sampling confirms this decrease in food: surface plankton can reach biomass levels of 500 milligrams wet mass per cubic meter (mg/m³), but at 1000 meters plankton biomass is only 25 mg/m³. At 3000 to 4000 meters, the value falls to 5 mg/m³, and at 10,000 meters there is only 0.5 mg of plankton per cubic meter of water.

Fish diversity parallels this decrease: about 800 species of deep-sea fishes are estimated to occupy the mesopelagic zone (from 100 to 1000 meters below the surface), and only 150 species inhabit the bathypelagic regions (where species live off the bottom but at depths below 1000 meters). Deep seas lying under areas of high surface productivity contain more and larger fish species than do regions underlying less productive surface waters. High productivity occurs in areas of upwelling, where currents recycle nutrients previously removed by the sinking of detritus. Deep-sea fishes tend to be most diverse and abundant in these places.

In tropical waters, photosynthesis continues throughout the year. Away from the tropics it is more cyclic, following seasonal changes in light, temperature, and sometimes currents. Diversity and abundance of deep-sea fishes decrease away from the tropics. Over 300 species of mesopelagic and bathypelagic fishes occur in the vicinity of tropical Bermuda, whereas only 50 species have been described in the entire Antarctic region. The high productivity of Antarctic waters is restricted to a few months of each year, and the detritus that sinks through the rest of the year is not sufficient to nourish a diverse assemblage of deep-sea fishes.

We emphasize that availability of food (energy) is the most formidable environmental problem that deep-sea fishes encounter; indeed, many of their specific characteristics may have been selected by intermittent or continuous food scarcity. Another variable factor is pressure, which changes with depth, whereas low temperature and the absence of light are constants.

Mesopelagic Fishes In general, mesopelagic fishes and invertebrates migrate vertically, ascending toward the surface at dusk to enter areas with more food and descending again near dawn. They apparently follow light-intensity levels as they move up and down, probably remaining within zones where predators have difficulty seeing them (**Figure 6–19**). Sonar signals readily reflect off the swim bladders of

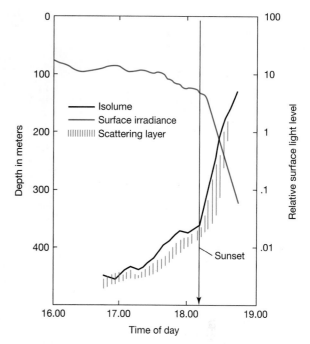

▲ **FIGURE 6–19** Upward migration of mesopelagic fishes. Changes in depth of the deep scattering layer from day to night reveal the vertical movement of fishes. The upward movement closely tracks the changing light intensity (isolume) as the intensity of sunlight at the ocean surface (surface irradiance) decreases.

mesopelagic fishes, and aggregations of fishes and vertically migrating invertebrates produce a sonar reflection called the deep scattering layer. At sunset the intensity of light at the surface decreases, as does the light penetrating the sea. Plotting the depth of a single light intensity (called an isolume) against time illustrates the progressively shallower depth at which a given light intensity occurs during dusk. The light intensity chosen for this example falls between that of starlight and full moonlight. It corresponds closely with the upper boundary of the deep scattering layer, which rises rapidly with the upward migration of mesopelagic fishes and invertebrates.

Vertical migration has both benefits and costs. By rising at dusk, mesopelagic fishes enter a region of higher productivity, where food is more concentrated. However, they also increase their exposure to predators, although many shallower water predators are visual hunters and may be less of a threat at night. Furthermore, ascending mesopelagic fishes are exposed to temperature increases that can exceed 10°C. The energy costs of maintenance at these higher temperatures can double or even triple. In contrast, daytime descent into cooler waters lowers metabolism, which conserves energy and reduces the

▶ **FIGURE 6–20** Reduction of bone in deep-sea fishes. X-rays show the reduced bone density of mesopelagic and bathypelagic fishes compared to a surface dweller. *Top:* The surface-dwelling herring (Clupeidae). *Middle:* A mesopelagic, vertically migrating lanternfish (Myctophidae). *Bottom:* A bathypelagic, deep-living bristlemouth (Gonostomatidae).

chance of predation because fewer predators exist at increased depth.

Bathypelagic Fishes It is less certain that bathypelagic fishes undertake daily vertical migrations. There is little metabolic economy from vertical migration within this lightless zone, because temperatures are uniform (about 5°C). Further, the cost and time of migration over the several thousand meters from the bathypelagic region to the surface would probably outweigh the energy gained from invading the rich surface waters. Instead of migrating, bathypelagic fishes are specialized to live less active lives than those of their mesopelagic counterparts. Deep-sea fishes have less dense bone and less skeletal muscle than do fishes from shallower depths (**Figure 6–20**). Surface fishes have strong, ossified skeletons and some have large red muscles in the axial region that are especially adapted for continuous cruising. Mesopelagic fishes, which swim mostly during vertical migration, have a more delicate skeleton and less axial red muscle. In bathypelagic fishes, the axial skeleton and the mass of muscles are greatly reduced, and locomotion is limited.

The eye size and light sensitivity of pelagic deep-sea fishes correlate with depth. Mesopelagic fishes have larger eyes than do surface dwelling pelagic fishes. The retinas of mesopelagic fishes contain a high concentration of visual pigment, the photosensitive chemical that absorbs light in the process of vision. The visual pigments of deep-sea fishes are most efficient in absorbing blue light, which is the wavelength of light that is transmitted farthest through clear water.

Many deep-sea fishes and invertebrates are emblazoned with startling designs formed by **photophores**, organs that emit blue light, and distinctive bioluminescent patterns characterize the males and females of many bathypelagic fishes. Tiny, light-producing photophores are arranged on their bodies in species- and even sex-specific patterns (Figure 6–21e). The light is produced by symbiotic species of *Photobacterium* and previously unknown groups of bacteria related to *Vibrio*. Some photophores probably act as signals to other members of their species in the darkness of the deep sea, where mates may be difficult to find. Others, such as those in the modified fin-ray lures of anglerfishes, probably attract prey. Female anglerfishes have a bioluminescent lure that looks species-specific. Although it is used to lure prey, the bait probably also attracts males. Because light does not travel far in water, visual detection of other individuals much beyond 40 or 50 meters is not possible, and other senses, such as scent trails, must be used. Female anglerfish secrete a pheromone, and males usually have enlarged olfactory organs. Sensing the pheromone during searching movements, males swim upstream to an intimate encounter.

The jaws and teeth of deep-sea fishes are often enormous in proportion to the rest of the body. Many bathypelagic fishes can be described as a large mouth accompanied by a stomach. If a fish rarely encounters potential prey, it is important to have a mouth large enough to engulf nearly anything it does meet—and a gut that can extend to accommodate a meal (**Figure 6–21**). Increasing the chances of encountering prey is also important to survival in the deep sea. Rather than searching for prey through the blackness of the depths, the ceratioid anglerfishes dangle bioluminescent bait in front of them. The bioluminescent lure is believed to mimic the movements of zooplankton and to lure fishes and larger crustaceans to the angler's mouth. Prey is sucked in with a sudden opening of the mouth, snared in the teeth, and then swallowed. The jaws of many anglerfishes expand, and the stomach stretches to accommodate prey larger than the predator. Thus deep-sea fishes, like most teleosts, show major specializations in locomotor and feeding structures. Unlike surface teleosts, however, deep-sea teleosts have structures that minimize the costs of foraging and maximize the capture of prey.

Most features of the biology of deep-sea fishes are directly related to the scarcity of food, but in some cases the relationship is indirect. For example, most species are small (the average length is less than 5 centimeters), and individuals of a species are not abundant. In so vast a habitat, the number of anglerfishes is very small. The density of females in the most common species is typically less than one female per cubic mile. Imagine trying to find another human under similar circumstances! Yet to reproduce, each fish must locate a mate and recognize it as its own species.

The life history of ceratioid anglerfishes dramatizes how selection adapts a vertebrate to its habitat. The adults typically spend their lives in lightless regions below 1000 meters. Fertilized eggs, however, rise to the surface, where they hatch into larvae. The larvae remain mostly in the upper 30 meters, where they grow, and later descend to the lightless region. Descent is accompanied by metamorphic changes that differentiate females and males. During metamorphosis, young females descend to great depths, where they feed and grow slowly, reaching maturity after several years.

Female anglerfishes feed throughout their lives, whereas males feed only during the larval stage.

Metamorphic changes in males prepare them for a different future; their function is reproduction, literally by lifelong matrimony. The body elongates and axial red muscles develop. The males cease eating and begin an extended period of swimming, apparently fueled by energy stored in the liver. The olfactory organs of males enlarge at metamorphosis, and the eyes continue to grow. These changes suggest that adolescent males spend a brief but active free-swimming period concentrated on finding a female. The journey is precarious, for males must search vast, dark regions for a single female while running a gauntlet of other deep-sea predators. In the young adults, there is an unbalanced sex ratio—often more than 30 males for every female. Apparently, at least 29 of those males will not locate a virgin female.

Having found a female, a male does not want to lose her. He ensures a permanent union by attaching himself as a parasite to the female, biting into her flesh and attaching himself firmly. Preparation for this encounter begins during metamorphosis when the male's teeth degenerate and strong toothlike bones develop at the tips of the jaws. A male remains attached to the female for life; and, in this parasitic state, he grows and his testes mature. Monogamy prevails in this pairing, for females usually have only one attached male. This lifestyle is unknown among other vertebrates, but it has been successful for anglerfish—some 200 ceratioid species exist.

Fishes in Coral Reef Communities— Specialization and Coexistence

A coral reef is one of the most spectacular displays of animal life on Earth. Such concentrations of invertebrate and vertebrate animals occur nowhere else, and the vertebrate component of a coral reef community is drawn almost exclusively from a single taxon, the acanthopterygian teleosts. The unparalleled diversity in feeding modes and ecological niches that characterizes actinopterygians requires precise control of jaw actions and body positions. This precision was achieved through strong evolutionary pressures on the interactions between feeding and locomotion. Nowhere are these interactions more obvious than in coral reef fishes. Over 600 species may be found on a single Indo-Pacific reef. The most primitive spiny-rayed fishes in coral reefs are predators of invertebrates (e.g., squirrelfishes and cardinalfishes).

To avoid predation, many reef invertebrates became nocturnal, limiting their activity to night and

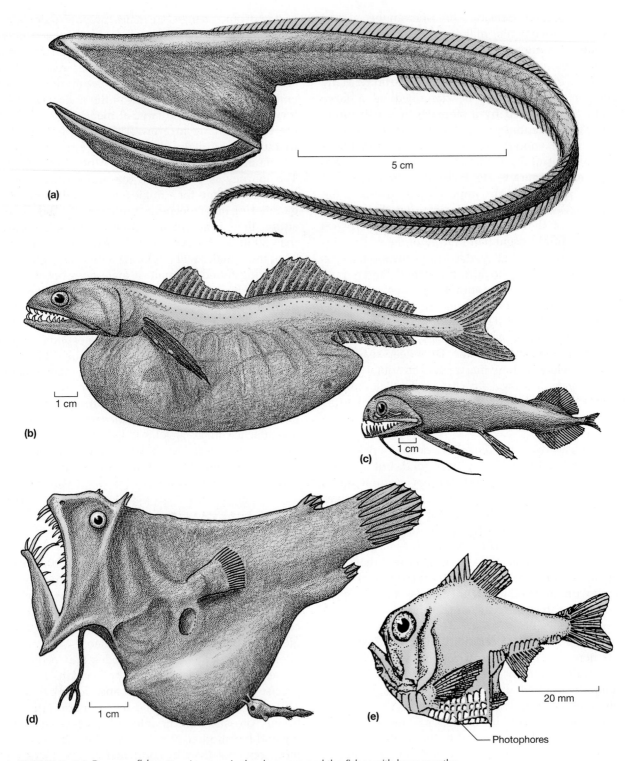

▲ **FIGURE 6–21** Deep-sea fishes. Prey is scarce in the deep sea, and the fishes with large mouths and distensible guts are able to eat any prey items they encounter. Some deep sea fishes lure prey with luminescent organs. (a) Pelican eel, *Eurypharynx pelecanoides*. (b) Deep-sea perch, *Chiasmodus niger,* its belly distended by a fish bigger than itself. (c) Stomiatoid, *Aristostomias grimaldii*. (d) Female anglerfish, *Liophryne argyresca*, with a parasitic male attached to her belly. (e) Hatchetfish, *Sternoptyx diaphana,* with photophores on the ventrolateral region.

remaining concealed during the day. In response to the nocturnal activity of their prey, early acanthopterygians evolved the capacity to locate prey at night. To this day, their descendants disperse over the reef at night to feed; but, during the daylight hours, they congregate in holes and crevices in the reef. The large, sensitive eyes of these nocturnal predators are effective at low light intensities. They use the irregular contours of the reef to conceal their approach and rely on a large protrusible mouth and suction to capture prey.

The evolution of fishes specialized to take food items hidden in the complex reef surface was a major advance among reef acanthopterygians. Some species rely on suction, whereas others use a forceps action of their protrusible jaws. This mode of predation demands sensory specializations, the most important being high visual acuity that can be achieved only in the bright light of day. In addition, delicate positioning is required to direct the jaws. These selection pressures produced fishes capable of maneuvering through the complex three-dimensional reef in search of food. So accurate are their locomotion, visual surveillance, and memory for hiding places and escape routes that these fishes can be diurnal, moving about and feeding in daylight. The refined feeding specializations of diurnal reef fishes allow them to extract small, nocturnally active invertebrates from their daytime hiding places or snip off coral polyps, bits of sponges, and other exposed reef organisms.

Released from heavy predation during daylight because of their ability to find safety in the labyrinthine structure of the reef, many reef fishes have evolved gaudy patterns and colors. These visual signals communicate a wide range of information: species, age, sex, social status, and reproductive condition.

A coral reef has day and night shifts—at dusk the colorful diurnal fishes, dependent as they are on bright light for their acute vision, seek nighttime refuge in the reef, and the nocturnal fishes, with poor vision in bright light, leave these hiding places and replace the diurnal fishes in the water column (**Figure 6–22**). The timing of the shift is controlled by light intensity, and the precision with which each species leaves or enters the protective cover of the reef day after day indicates that this is a strongly selected behavioral and ecological characteristic. Space, time, and the food resources available on a reef are partitioned through this activity pattern.

Many reef fishes belong to a small number of families and genera with many coexisting species. The cardinalfishes, damselfishes, angelfishes, butterflyfishes, wrasses, and parrotfishes are examples. The great number of closely related species is thought to be the result of the repeated partial isolation of sections of the tropical ocean. (For example, the tropical Atlantic was isolated from the tropical Pacific around 3 million years ago by the uplift of the Isthmus of Panama.) The fossil record shows that many genera of reef fish have existed for over 50 million years. During this extensive time, sea levels have varied, volcanic islands have grown and sunk, and continents and archipelagos have drifted with the spreading of the seafloor. Populations of reef fishes, almost all of which have pelagic eggs and/or larvae, have sometimes found themselves isolated in a corner of the vast tropical sea—and at other times, close enough for their larvae to drift and mingle with those of other populations. Periods of isolation probably led to the simultaneous formation of species in multiple isolated reef systems. The result has been a reef fish fauna of great variety, with new species entering the fauna after each period of isolation.

Currently, the fauna is in a period of rather low isolation. The geographic ranges of many species extend from the east coast of Africa to Hawaii, and a few of these same species reach the Pacific Coast of the Americas. Since many groups of reef fish species are closely related, they share most of their adaptations through inheritance of shared derived characters. One would expect the species in these groups to experience intense competition from other species within their group. Such competition should ultimately lead either to elimination of most of the closely related species or to driving related species to exploit different resources despite their original similarity. Surprisingly, neither of these results of competition seems to have occurred extensively. Instead, predation and events such as storms that reduce populations appear to have prevented competitive interactions from running their full course.

6.8 Conservation of Fishes

Fishes face a host of problems. Like all extant organisms, fishes suffer from changes in their habitats that stem directly or indirectly from human activities. In addition, species that have a larval stage occupy distinctly different habitats during their lives and are vulnerable to changes in either habitat. Unlike most other vertebrates, many fishes are enormously important commercially as food and major industries are based on capturing them. Other species of fish are

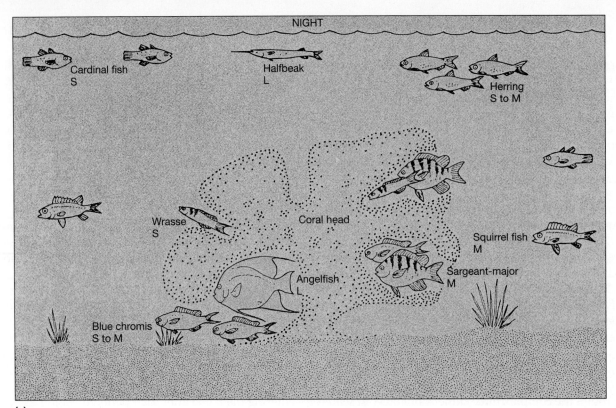

(a)

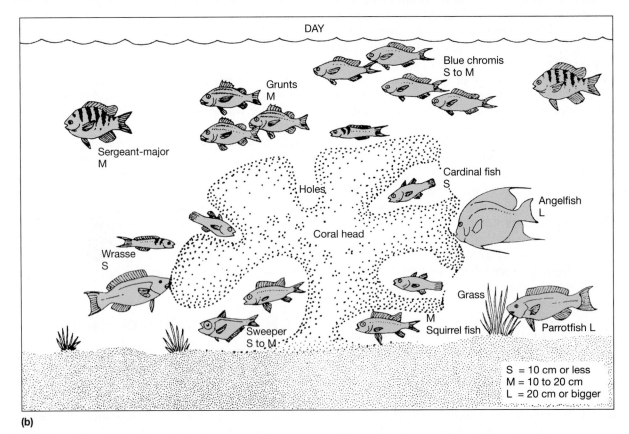

S = 10 cm or less
M = 10 to 20 cm
L = 20 cm or bigger

(b)

▲ **FIGURE 6–22** A Caribbean coral reef. Fish activity and diversity change from day to night as diurnal species retreat into the reef and nocturnal species emerge. (a) at midnight and (b) at midday. Most reef fishes are Perciformes; of the fishes illustrated here, only the herring, halfbeak, and squirrel fish are not perciforms.

captured for the pet trade, sometimes using collection methods that destroy their habitats. As a result of these factors, many species of fishes that were once common have been brought to the verge of extinction, just as we are discovering important new uses for some species (**Box 6–3**).

■ Conservation Concerns for Freshwater Fishes

Nearly 40 percent of fish species live in the world's fresh waters, and all of them are threatened by alteration and pollution of lakes, rivers, and streams.

BOX 6–3   ## A Lot Is Fishy in Genetic Research

During the past decade, research laboratories around the world concerned with human development, genetics, and disease have been overrun by a fish barely 6 centimeters long at its largest. Native to a wide variety of habitats, the zebrafish, *Danio (Brachydanio) rerio* (Ostariophysi: Cyprinidae) ranges from the southern tip of India, around India's east coast, west through the foothill streams of the Himalayas and into Myanmar (Burma) in Southeast Asia. It can be found schooling in small and large rivers, in stagnant waters, flooded rice fields, and canals but is also abundant in rivulets in the foothills of mountains. Discovered about 1797 by a Scottish surgeon working for the East India Company, the zebrafish was not formally described until 1822 nor exported live to Europe until 1905. It immediately became a favorite of the fledgling aquarium hobby in Europe and North America and remains so to this day (**Figure 6–23**). It is the next species that most tropical fish fanciers attempt to keep after mastering the guppy, and it was a favorite of the late fish hobbyist and geneticist, George Streisinger.

Streisinger recognized the unique potential of the zebrafish as a laboratory model for genetic and development research. The zebrafish's natural ability to thrive under a wide variety of conditions in nature make captive husbandry easy. Its small adult size and schooling behavior allows the zebrafish to be maintained in large numbers in small, economical spaces. In this regard, it is superior to the principal other vertebrate models: African clawed frogs, chickens, and mice. Zebrafish are prodigious breeders, mating year-round due to their tropical heritage. Fertilization is external: Eggs and sperm are released into the water. Under ideal conditions females produce 100 to 200 eggs every week from the age of 10 to 12 weeks through their normal life span of about two years. (The largest numbers of eggs are produced by females from 7 to 18 months of age.) Males readily spawn from 10 weeks of age until death. Such fecundity and life span allows genetic manipulations such as back-crossing to parents.

Zebrafish are behaviorally promiscuous, demonstrating little mate selectivity and having only brief courtship behaviors,

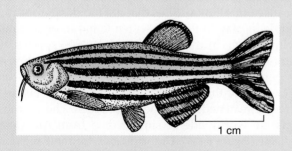

(a)

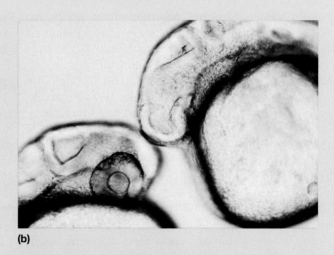

(b)

▲ **FIGURE 6–23** Zebrafish (*Danio rerio*). (a) A wild type adult male. Females are very similar but heavier bodied. (b) Head regions of two 24-hour-old zebrafish: wild type on the left, the mutant "masterblind," which fails to develop eyes on the right.

making it possible for the researcher to decide parental combinations. It is possible to collect eggs and sperm without damage to the adults. If the sperm are sterilized by ultraviolet irradiation, haploid embryos containing only genetic material from the female may be obtained. Pulses of hydrostatic pressure or brief exposure to higher-than-normal temperature can turn very early haploid embryos into diploid ones by interfering with cell division but not with chromosome duplication. Thus it is possible to obtain diploids containing only the mother's genes by fairly simple procedures. Mutations can also be easily produced by exposing one or both parents to chemicals or radiation. Zebrafish genes can be manipulated by injecting one-celled embryos with large doses of foreign DNA. If this DNA has been tagged with fluorescent proteins derived from jellyfish or corals, it will fluoresce in cells in which it is being expressed.

After fertilization, the characteristic of zebrafish most important to developmental biologists becomes apparent: unlike the eggs of many other freshwater fishes, zebrafish eggs are clear (unpigmented) and not adhesive. The eggs of zebrafish are heavier than water; in nature they are spawned into dense aquatic vegetation or over gravel bottoms and depend on sinking into crevices and between fronds and on being crystal-clear to avoid detection by predators. Although many fishes have clear eggs, most of these species are marine and have planktonic, neutrally buoyant eggs requiring the extremely narrow and constant conditions of the open sea to properly develop. Zebrafish development takes place in a hardy crystalline egg in full view of the researcher, providing an unparalleled opportunity to watch the embryonic development of vertebrate tissues and organs. Even the newly hatched young have few pigment cells, so direct observation of internal development can be continued for several weeks.

Zhiyuan Gong at the National University of Singapore has inserted genes from invertebrates into the genome of zebrafish to produce fish that respond to pollutants by changing color. Inducible gene promoters activate the fluorescent color genes in the presence of specific chemicals in the environment—a promoter inducible by estrogen responds to the presence of estrogen in the water, and a second promoter induced by stress responds to the presence of heavy metals and other toxins (National University of Singapore 2002). Samples of water can be pumped into tanks with zebrafish to see if they respond by changing color. This biological assay is substantially faster and less expensive than chemical analyses (Cornwell 1999). The appeal of glowing zebrafish is not limited to environmental research; a zebrafish in which the gene for a fluorescent red color is expressed continuously was introduced to the aquarium trade in the United States in 2004.

Draining, damming, canalization, and diversion of rivers create habitats that no longer sustain indigenous fishes. Perhaps the most imperiled fish species in the world is the one with the smallest range of any vertebrate. The tiny Devil's Hole pupfish (*Cyprinodon diabolis*, an atherinomorph killifish) lives in a single spring-fed pool in a specially designated portion of Death Valley National Monument in southern Nevada. While the long, slitlike pool is approximately 17 by 3 meters at the surface, the pupfish spend their entire lives over a single shallow algae-covered shelf only 18 square meters in area. Although no other species is so extremely limited in range, many other species of fishes in the region are limited to a single spring or pool. Human activity in the region is threatening the Devil's Hole pupfish and the other isolated fishes. Local ranchers and farmers, and cities as far away as Las Vegas, are pumping water from the underground aquifers that feed the pools. The water in the aquifers is fossil water—it has been there since the end of the Pleistocene glaciations—and it is not replaced by rain and snow melt. Pumping fossil water is like pumping oil; the level drops, and eventually the entire reserve is gone. Pool levels and spring flows are dropping; and, unless the aquifers are protected, the pupfish's habitat will dry up, and the species will become extinct. Despite efforts at conservation, the population dropped to just 49 individuals in April of 2006 (Helfman 2007).

In addition to the loss and physical degradation of fish habitat, fresh waters in much of the world are polluted by silt and toxic chemicals of human origin. This is especially true of the Western nations and urbanized regions elsewhere. The United States has had, in recent years, over 2400 instances *annually* of beaches and flowing waterways closed to human use because of pollution. Sites that are too dangerous for people to play in them are often lethal to organisms trying to live in them! Of the nearly 800 species of native freshwater fishes in the United States, almost 20 percent are considered imperiled. As much as 85 percent of the fish fauna of some states is endangered, threatened, or of special concern. These problems are worldwide as similar data from Australia indicate (**Figure 6–24**). Unfortunately for most regions of the world, we have too little sound data on freshwater biota to identify regions of critical concern.

Conservation Concerns for Marine Fishes

The reproductive mode of the vast majority of marine teleosts—eggs that are shed into the environment

(a)

(b)

▲ **FIGURE 6–24** Endangered species of fishes. The maps show the number of known native freshwater fish species in each of the contiguous states of the United States and the states and territories of Australia. The numbers in parentheses show how many species from that state or territory are considered by fisheries professionals to be endangered, threatened, or of special concern.

with little further parental investment—make management of commercial fisheries extraordinarily complex. The eggs tend to be small relative to the size of the adult fish, and they are produced in prodigious numbers. Once the pelagic eggs are launched, many factors determine how many of them survive. Some will always be lost to predation by filter-feeding fishes, but weather-related events are less predictable. Excessively hot or cold water can kill eggs and larvae outright or disrupt their development. Changes in ocean currents produced by the action of wind and global events such as the El Niño/La Niña phenomenon (warming and cooling of the central Pacific) affect the abundance of nutrients and the amount of microplankton. Because so many variables are involved, the number of individuals breeding in any given year (breeding stock size) bears no clear relationship to the number of individuals in the next generation. Thus, a breeding season rich with spawning adults may produce few or no offspring that survive to breed in subsequent seasons if environmental factors prevent the survival of eggs, larvae, or juveniles. Conversely, under exceptionally favorable circumstances, a few breeding adults could produce a very large number of offspring that survive to maturity.

The low predictability of future stock size based on current stock size has been a major stumbling block to effective fisheries management. Because so much of a population's future size depends on the environment experienced by eggs and larvae—conditions not usually obvious to fishermen or scientists—it is difficult to demonstrate the effects of overfishing in its early stages or the direct results of conservation efforts.

The problems inherent in fisheries management have resulted in destruction of commercial fish populations and entire marine food webs by overfishing. Many of the world's richest fisheries are on the verge of collapse. The Georges Bank, which lies east of Cape Cod, is an example of what overfishing can do. For years, conservation organizations called for a reduction in catches of cod, yellowtail flounder, and haddock, but their concerns were not heeded. Many bottom-dwelling fishes are also caught unintentionally by vessels fishing for other species. These unintentional captures are called "by-catch," and the by-catch of a trawler can consist of thousands of animals. Not surprisingly, populations of both the targeted species and several fish species in the by-catch crashed dramatically in the 1990s. By October 1994, the situation was so bad that a government and industry group, the New England Fishery Management Council, directed its staff to devise measures that would reduce the catch of those species essentially to zero. A major portion of Georges Bank remains closed to all fishing by methods that catch these bottom-dwelling species.

Thousands of fishing operations have been affected by draconian measures, such as the complete prohibition of fishing. Some fishers have moved their boats to other heavily fished areas, such as the mid-Atlantic coast and the Gulf of Mexico, and the depletion process will repeat itself in these areas. Many smaller fishing operations, some of which have been family operations for generations, have gone out of business. The U.S. government has bought out many other fishers, and their rights to fish have been retired. These measures and education of the public on the problems of fisheries have brought about significant change in the attitudes and perspectives of fishers and their customers alike. After a decade of complete protection, some of the species endangered by overfishing are again to be found on Georges Bank—the 2003 year class of haddock was the largest on record. But other species—cod and yellowtail flounder, for example—show much weaker signs of recovery.

■ Conservation Concerns for Coral Reef Fishes

The evolution and maintenance of a coral reef's rich fish fauna depends on many factors, including healthy adjacent mangrove forests and sea grass beds that provide nursery areas for reef fish and primary production that is imported to the reef by currents or by fishes migrating between habitats. The complex three-dimensional structure of the reef is very important. When this structure collapses, as in severe storm damage or when the reef is mined for limestone to make concrete, dynamited to capture fish for food or treated with cyanide to capture fish for the aquarium trade, fish populations and diversity plummet. Live reef-building corals across the entire Caribbean basin have dropped from 50 percent coverage of the sea floor in reef areas to just 10 percent coverage in the last three decades. In the last decade, an ominous phenomenon has been observed worldwide. Coral reefs are showing signs of physiological stress and dying en masse. The cause appears to be unusually high sea surface temperatures worldwide. The coral animals that build the reef succumb to prolonged exposure to higher-than-normal temperatures, leaving their dead skeletons exposed to organisms that erode limestone and storms that shatter the three-dimensional structure of the reef. The heat-stressed corals lose the symbiotic algae that normally live with them and become light in color, leading to the term *coral bleaching* to describe the phenomenon.

A two-month period of exceptionally high temperatures in late summer of 1998 left less than 5 percent of inshore reefs of Belize on Central America's Caribbean coast covered by live coral. In the Southern Hemisphere summer of 1998–1999, heat-induced coral stress was documented on the inner reefs along the entire length of Australia's Great Barrier Reef. In January 2001, unusually warm waters resulted in the worst coral bleaching event ever recorded on the Great Barrier Reef. Coral bleaching, documented worldwide, is a sign of severe coral stress and may lead to coral die-off if it lasts more than a few days.

Although it is still too recent a phenomenon to predict the consequences, if reefs around the globe physically collapse, the greatest vertebrate diversity on Earth will also dwindle. Reefs grow slowly. Even if conditions permit regeneration, reefs will be greatly changed over wide geographic areas for decades or longer. If warmer waters prevail into the future, the new reefs that might form in newly warm waters to the north and south of the current occurrence of corals will take centuries to become mature and three-dimensionally complex. Without new habitats, reef organisms face the possibility of massive extinction over the next few years. Will the disappearance of coral reefs and their marvelous vertebrate fauna be the planet's canary in the mine—an early warning of global warming?

Summary

At their first appearance in the fossil record, osteichthyans, the largest vertebrate taxon, are separable into distinct lineages: the Sarcopterygii (fleshy-finned fishes, including actinistians, lungfishes, and tetrapods) and the Actinopterygii (ray-finned fishes). Extant sarcopterygian fishes offer exciting glimpses of adaptations evolved in Paleozoic environments. Actinopterygian fishes were distinct as early as the Silurian period. Actinopterygians inhabit the 73 percent of the Earth's surface that is covered by water, and they are the most numerous and species-rich lineages of vertebrates. Several levels of development in food-gathering and locomotory structures characterize actinopterygian evolution. The radiations of these levels are represented today by a small number of species that are relicts of groups that once had many more species: cladistians (bichirs and reedfishes), chondrosteans (sturgeons and paddlefishes), and the primitive neopterygians (gars and *Amia*). The most derived level of bony fishes, teleosteans, may number close to 27,000 extant species—with two groups, ostariophysans in freshwater and acanthopterygians in seawater, constituting a large proportion of these species. Examining how teleost fish communities are adapted to very specialized habitats like the deep sea and coral reefs helps us to understand how evolution has acted upon the basic teleost body plan, but looking at how humans have changed fish habitats explains why so many fishes are now in danger of extinction.

Additional Readings

Allan, J. D., and A. S. Flecker. 1993. Biodiversity conservation in running water. *Bioscience* 43:32–43.

Barton, M., 2007. *Bond's Biology of Fishes*. 3rd ed. Belmont, CA: Thomson Brooks/Cole.

Bemis, W. E. et al. (eds.). 1987. *The Biology and Evolution of Lungfishes*. New York, NY: Liss.

Birstein, V. J. et al. (eds.). 1997. Sturgeon biodiversity and conservation. *Environmental Biology of Fishes* 48(1–4):9–435.

Breder, C. M. 1926. The locomotion of fishes *Zoologica* 4: 159–256.

Brothers, E. B. 1990. Otolith marking. *American Fisheries Society Symposium* 7:183–202.

Fricke, H. et al. 1987. Locomotion of the coelacanth *Latimeria chalumnae* in its natural environment. *Nature* 324:331–333.

Grande L., and W. E. Bemis. 1997. A comprehensive phylogenetic study of amiid fishes (Amiidae) based on comparative skeletal anatomy. *Journal of Vertebrate Paleontology* (Memoir no. 4): 17.

Grunwals, D. J., and J. S. Eisen. 2002. Headwaters of the zebrafish—Emergence of a new model vertebrate. *Nature Reviews/Genetics* 3: 717–724.

Helfman G. S. 2007. *Fish Conservation*. Washington, DC: Island Press.

Helfman, G. S. et al. 2008. *The Diversity of Fishes*. 2nd ed. Malden, MA: Blackwell Science.

Kingsmill, S. 1993. Ear stones speak volumes to fish researchers. *Science* 260:1233–1234.

Lauder, G. V. 1994. Caudal fin locomotion by teleost fishes: Function of the homocercal tail. *American Zoologist* 34:13A, abst. no. 66.

Lauder, G. V., and K. F. Liem. 1983. The evolution and interrelationships of the actinopterygian fishes. *Bulletin of the Museum of Comparative Zoology* 150:95–197.

Lauder, G. V., and J. H. Long (eds.). 1996. Aquatic locomotion: New approaches to invertebrate and vertebrate biomechanics. *American Zoologist* 36:535–709.

Liao, J. C. et al. 2003. Fish exploiting vortices decrease muscle activity. *Science* 302:1566–1569.

McCune, A. R. 1996. Biogeographic and stratigraphic evidence for rapid speciation in semionotid fishes. *Paleobiology* 22(1):34–48.

McCune, A. R. 1997. How fast is speciation: Molecular, geologic and phylogenetic evidence from adaptive radiations of fishes. In T. Givnish and K. Sytsma (eds.), *Molecular Evolution and Adaptive Radiation*. Cambridge, UK: Cambridge University Press.

McCune, A. R., and N. R. Lovejoy. 1998. The relative rate of sympatric and allopatric speciation in fishes: Tests using DNA sequence divergence between sister species and among clades. In D. Howard and S. Berlocher (eds.), *Endless Forms: Species and Speciation*. Oxford, UK: Oxford University Press.

McPhee, J. 2002. *The Founding Fish*. Farrar, Straus, and Giroux.

Mehta, R. S., and P. C. Wainwright. 2007. Raptorial jaws in the throat help moray eels swallow large prey. *Nature* 449:79–82

Moyle, P. B., and J. J. Cech, Jr. 2003. *Fishes: An Introduction to Ichthyology*, 5th ed. San Francisco, CA: Benjamin Cummings.

National University of Singapore. 2002. Zebrafish as pollution indicators. <http://www.nus.edu.sg/corporate/research/gallery/research12.htm>

Nelson, J. S. 2006. *Fishes of the World*, 4th ed. New York, NY: Wiley.

Nilsson, G. E. 1996. Brain and body oxygen requirements of *Gnathonemus petersii*, a fish with an exceptionally large brain. *Journal of Experimental Biology* 199:603–607.

Northcutt, R. G. 2002. Understanding vertebrate brain evolution. *Integrative and Comparative Biology* 42:743–756.

Pauly, D. et al. 1998. Fishing down marine food webs. *Science* 279:860–863.

Paxton, J. R., and W. N. Eschmeyer (eds.). 1998. *Encyclopedia of Fishes*. San Diego, CA: Academic Press.

Stiassny, M. L. 1996. An overview of freshwater biodiversity: With some lessons from African fishes. *Fisheries* 21(9):7–13.

Stiassny, M. L. et al. 1996. *Interrelationships of Fishes*. San Diego, CA: Academic Press.

Stix, G. 1994. Robotuna. *Scientific American* 270(1):142.

Thompson, K. S. 1992. *Living Fossil: The Story of the Coelacanth*. New York, NY: Norton.

Triantafyllou, M. S., and G. S. Triantafyllou. 1995. An efficient swimming machine. *Scientific American* 274(3):64–70.

Warren, M. L., Jr., and B. M. Burr. 1994. Status of fresh water fishes of the United States: Overview of an imperiled fauna. *Fisheries* 19:6–18.

Weinberg, S. 1999. *A Fish Caught in Time: The Search for the Coelacanth*. London, UK: Fourth Estate.

CHAPTER 7

Geography and Ecology of the Paleozoic Era

The Paleozoic world was very different from the one we know—the continents were in different places, climates were different, and initially there was little structurally complex life on land. By the Early Devonian period, terrestrial environments supported a substantial diversity of plants and invertebrates, setting the stage for the first terrestrial vertebrates (tetrapods) in the Late Devonian. Early plants were represented by primitive groups such as horsetails, mosses, and ferns. The evolution of plants resulted in the production of soils that trapped carbon dioxide. Removal of carbon dioxide from the atmosphere had a profound effect on the Earth's climate, resulting in a reverse greenhouse effect—global cooling. Extensive glaciation occurred during the Late Carboniferous and Early Permian periods, when the tetrapods lived in equatorial regions. Terrestrial ecosystems became more complex in the Carboniferous and Permian; some modern types of plants, such as conifers, appeared; tetrapods diversified; and the first flying insects took to the air. Fluctuations of atmospheric oxygen and carbon dioxide during this time may underlie many events and patterns in vertebrate evolution. Extinctions occurred in terrestrial ecosystems at the end of the Early Permian, and a major extinction (the largest in Earth's history) occurred both on land and in the oceans at the end of the Paleozoic era.

7.1 Earth History, Changing Environments, and Vertebrate Evolution

It is important to realize that the world of today is very different from the world of times past. Our particular pattern of global climates, including such features as ice at the poles and the directions of major winds and water currents, results from the present-day positions of the continents. The world today is, in general, cold and dry in comparison with many past times. It is also unusual because the continents are widely separated from one another, and the main continental landmass is in the Northern Hemisphere. Neither of these conditions existed for most of vertebrate evolution.

■ The Earth's Time Scale and the Early History of the Continents

Vertebrates are known from the portion of Earth's history called the **Phanerozoic eon** (Greek *phanero* = visible and *zoo* = animal). The Phanerozoic began 542 million years ago and contains the Paleozoic (Greek *paleo* = ancient), Mesozoic (Greek *meso* = middle), and Cenozoic (Greek *ceno* = recent) eras. Our own portion of time, the Recent epoch, lies within the Cenozoic era. Each era contains a number of periods, and each period in turn contains a number of subdivisions (see the inside front cover of this textbook). At least 99 percent of described fossil species occur in the Phanerozoic, although the oldest known fossils are from around 3.5 billion years ago, and the origin of life is estimated to be around 4 billion years ago.

The time before the Phanerozoic is often loosely referred to as the Precambrian, because the Cambrian is the first period in the Paleozoic era and thus marks the beginning of the Phanerozoic. However, the Precambrian actually represents seven-eighths of the entire history of the Earth! Precambrian time is better perceived as two eons, comparable to the Phanerozoic eon: the Archean (Greek *archeo* = first or beginning), commencing with the Earth's formation around 4.5 billion years ago (the earlier part of this time period is sometimes referred to as the Hadean), and the Proterozoic (Greek *protero* = former), which began around 2.4 billion years ago. The start of the Proterozoic is marked by the appearance of the large continental blocks seen in today's world. (The pre-Proterozoic world would have looked rather like the South Pacific does now—many little volcanic islands separated by large tracts of ocean.)

Although life dates from the Archean, organisms more complex than bacteria are not known until the Proterozoic. The evolution of eukaryotic organisms, which depend on oxygen for respiration, followed shortly after the first appearance of atmospheric oxygen in the middle Proterozoic (around 2.2 billion years ago). Multicellular organisms are first known near the end of the Proterozoic, about 1 billion years ago. By the start of the Phanerozoic, a major biotic shift occurred—the evolution of forms capable of secreting calcareous (calcium-containing) skeletal parts.

■ Continental Drift—History of Ideas and Effects on Global Climate

The Earth's climate results from the interaction of sunlight, temperature, rainfall, evaporation, and wind during the annual passage of the Earth in its orbit around the Sun. Knowledge of ancient climates helps us to understand the conditions under which plants and animals evolved because climate profoundly affects the kinds of plants and animals that occupy an area. We discussed in Chapter 1 how the positions of the continents could influence global climates and oceanic circulation. The climate of any given area is affected by its latitude (i.e., how far north or south of the equator it is, reflecting the amount of solar energy it receives); its proximity to an ocean (because water buffers temperature change

and provides moisture); and the presence of barriers like mountains, which influence the movement of air masses and thus the amount of rain received.

Our understanding of the dynamic nature of the Earth and the variable nature of the Earth's climate over time is fairly recent. The notion of mobile continents, or **continental drift**, was formally proposed by Alfred Wegener in 1924. However, it was not until the late 1960s, following new oceanographic research demonstrating the spreading of the seafloor as a plausible mechanism for continental movement, that the theory of **plate tectonics** became established. (Plate tectonics is essentially the same as continental drift but focuses on the tectonic plates below the continents.)

The surface rocks of the continents float on the denser underlying mantle rock, much as an ice cube floats in water. Heat in the Earth's core produces slow convective currents in the mantle. Upwelling plumes of molten basalt rise toward the Earth's surface, forming a chain of midoceanic ridges around the globe where they reach the top of the lithosphere (the rocky shell of the Earth) and then spread horizontally (**Figure 7–1**). The youngest seafloor crust is found in the centers of the ridges; moving away from the axis of the ridge, the seafloor becomes older. Subduction zones are regions where the lithosphere sinks back down into the mantle. The ocean floor is continuously renewed by this cycle of upwelling at midoceanic ridges and sinking back into the mantle at subduction zones, and rocks older than 200 million years do not occur anywhere on the ocean floor.

Movements of the tectonic plates are responsible for the sequence of fragmentation, coalescence, and refragmentation of the continents that has occurred during the Earth's history. Plants and animals were carried along as continents slowly drifted, collided, and separated. When continents moved toward the poles, they carried organisms into cooler climates. As once separate continents collided, terrestrial floras and faunas that had evolved in isolation mixed, and populations of marine organisms were separated. A recent (in geological terms) example of this phenomenon is the joining of North and South America around 2.5 million years ago. The faunas and floras of the two continents mingled, which is why we now have armadillos (of South American origin) in Texas and deer (of North American origin) in Argentina. In contrast, the marine organisms originally found in the sea between North and South America were separated, and the populations on the Atlantic and Pacific sides of the land bridge became increasingly different from each other with the passage of time.

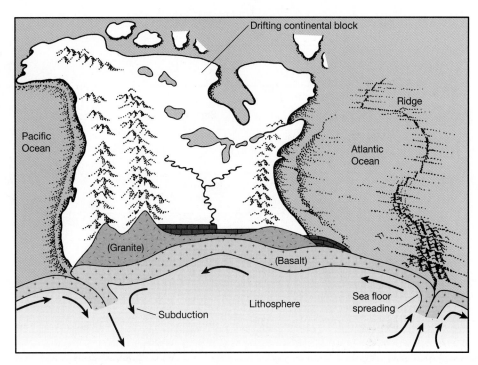

▶ **FIGURE 7–1** Generalized geological structure of the North American continent. The continental blocks (granite) float on a basaltic crust (the bottom part of the picture shows a cross-sectional view of the continent). Arrows show the movements of crustal elements and the interactions with the mantle that produce continental drift.

The position of continents affects the flow of ocean currents; and, because ocean currents transport enormous quantities of heat, changes in their flow affect climates worldwide. For example, the breakup and northern movements of the continents in the late Mesozoic and Cenozoic eventually led to the isolation of the Arctic Ocean, with the formation of Arctic ice by the start of the Pliocene epoch, around 5 million years ago. The presence of this ice cap influences global climatic conditions in a variety of ways, and the world today is colder and drier than it was before the Pliocene.

Arctic waters are a key part of the dynamic conveyor belt of global oceanic currents that influence climates today. This system of currents includes the Gulf Stream, which transports warm water from the equatorial Atlantic and the Gulf of Mexico across the North Atlantic Ocean to Europe. The warming of the Arctic and melting of the Arctic ice cap that are now occurring are likely to have profound effects on the Earth's climate. Some researchers have proposed that, as the Earth as a whole becomes warmer, parts of Western Europe may actually become *colder* because oceanic currents might be disrupted. Without the Gulf Stream, England would probably have the same cold climate as Newfoundland, which is at the same latitude (reviewed in Kunzig 1996). This example shows how very labile the Earth's climate is and how dependent it is on the configuration of

continental masses that influence ice cover and oceanic currents. The British people can only hope that global warming won't make England's climate a test case in demonstrating the validity of this particular hypothesis.

7.2 Continental Geography of the Paleozoic

The world of the early Paleozoic contained at least six major continental blocks (**Figure 7–2**). **Laurentia** included most of present-day North America, plus Greenland, Scotland, and part of northwestern Asia. Four smaller blocks contained other parts of what are now the Northern Hemisphere: Baltica—Scandinavia and much of central Europe; Kazakhstania—central southern Asia; Siberia—northeastern Asia; and China—Mongolia, North China, and Indochina. **Gondwana** (also known as Gondwanaland) included most of what is now the Southern Hemisphere (South America, Africa, Antarctica, and Australia), as well as India, Tibet, South China, Iran, Saudi Arabia, Turkey, southern Europe, and part of the southeastern United States.

In the Late Cambrian, Gondwana and Laurentia straddled the equator; Siberia, Kazakhstania, and China were slightly to the south of the equator; and

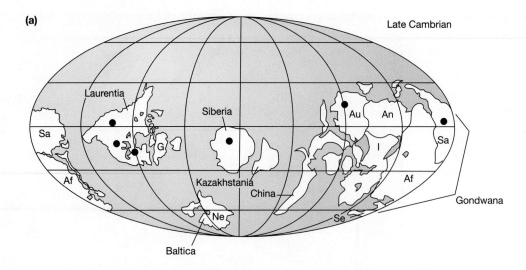

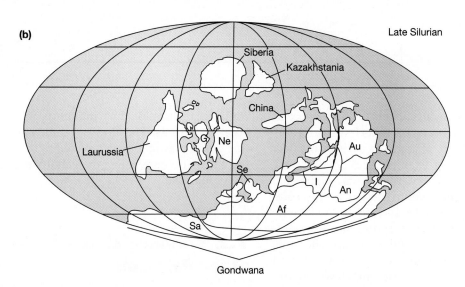

▲ **FIGURE 7–2** Location of continental blocks. (a) in the Late Cambrian and (b) in the Late Silurian periods. The black dots in (a) indicate fossil localities where Ordovician vertebrates have been found. Positions of modern continents are indicated as follows: Af = Africa, An = Antarctica, Au = Australia, G = Greenland, I = India, Ne = Northern Europe, Sa = South America, and Se = Southern Europe.

Baltica was far to the south (see Figure 7–2a). Note that the position of the modern continents within Gondwana was different from today. For example, Africa and South America appear to be upside down. Over the next 100 million years, Gondwana drifted south and rotated clockwise. By the Late Silurian, the eastern portion of Gondwana was over the South Pole, and Africa and South America were in positions similar to those they occupy today (see Figure 7–2b). Laurentia was still in approximately the same position, although it had rotated slightly counter-

clockwise. Baltica had moved north and collided with Laurentia to form a united block called Laurussia. Kazakhstania, Siberia, and China had also moved north and were now in the Northern Hemisphere.

The most dramatic radiation of animal life had occurred by the start of the Cambrian, with the so-called "Cambrian explosion" of animals with hard (preservable) parts, many lineages of which were extinct by the end of the period. A further radiation of marine animals occurred in the Ordovician period,

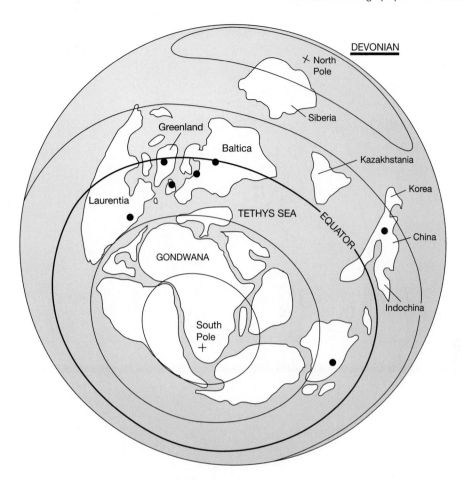

▲ **FIGURE 7–3** Location of continental blocks in the Late Devonian. The black dots indicate fossil localities where Devonian tetrapods have been found. Laurentia, Greenland, and Baltica lie on the equator. An arm of the Tethys Sea extends westward between Gondwana and the northern continents.

and many of the groups that were to dominate the ecosystem of the rest of the Paleozoic appeared and radiated at this time. Although vertebrates date from the Early Cambrian, their major early diversification apparently took place during the Ordovician.

From the Devonian through the Permian, the continents were drifting together (**Figure 7–3**). The continental blocks that correspond to parts of modern North America, Greenland, and Western Europe had come into proximity along the equator. With the later addition of Siberia, these blocks formed a northern supercontinent known as **Laurasia**. Most of Gondwana was in the far south, overlying the South Pole, but its northern edge was separated from the southern part of Laurentia only by a narrow extension of the Tethys Sea. This western arm of the Tethys Sea did not close completely until the Late Carboniferous, when Africa moved northward to contact the east coast of North America (**Figure 7–4**).

During the Carboniferous, the process of coalescence continued, and by the Permian most of the continental surface was united in a single continent, **Pangaea** (sometimes spelled Pangea). At its maximum extent, the land area of Pangaea covered 36 percent of the Earth's surface, compared with 31 percent for the present arrangement of continents. This supercontinent persisted for 160 million years, from the mid-Carboniferous to the mid-Jurassic period, and profoundly influenced the evolution of terrestrial plants and animals.

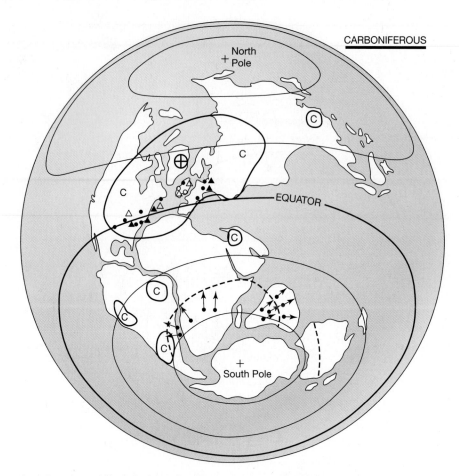

▲ **FIGURE 7–4** Location of continental blocks in the Carboniferous. This map illustrates an early stage of Pangaea. The location and extent of continental glaciation in the Late Carboniferous is shown by the dashed lines and arrows radiating out from the South Pole. The extent of the forests that formed today's coal beds is marked by the heavy lines enclosing each letter "C." The small circles and triangles mark the locations of Carboniferous tetrapods. The triangles are the smaller forms (lepospondyls), and the circles are the larger forms ("labyrinthodonts"). Filled symbols represent Early Carboniferous localities; open symbols represent Late Carboniferous ones. The circle enclosing a cross marks the major Late Devonian tetrapod locality.

7.3 Paleozoic Climates

During the early Paleozoic, sea levels were at or near an all-time high for the Phanerozoic, and atmospheric carbon dioxide levels were also apparently very high. The oxygen level fluctuated but in general was lower than the current level of 21 percent. The high levels of carbon dioxide would have resulted in a greenhouse effect, with the land experiencing very hot and dry climates during the Cambrian and much of the Ordovician, quite unsuitable conditions for the invasion of the terrestrial realm by early plants. However, there was a major glaciation in the Late Ordovician that—combined with falling levels of atmospheric

carbon dioxide—would have created cooler and moister conditions in at least some places on the globe, setting the scene for the development of the Silurian terrestrial ecosystems.

Atmospheric oxygen rose during the Silurian, to reach a maximum of around 25 percent in the Early Devonian, but by the mid-Devonian it had plummeted to around 12 percent. A relatively equable climate continued through the mid-Devonian; however, glaciation was evident again in the Late Devonian, and ice sheets covered much of Gondwana from the mid-Carboniferous until the mid-Permian. The waxing and waning of the glaciers created oscillations in sea level, which resulted in the cyclic formation of coal deposits, especially in eastern North America and Western Europe. The climate over Pangaea was fairly uniform in the Early Carboniferous, but in the Late Carboniferous and Early Permian it was highly differentiated as the result of glaciation, with significant regional differences in the flora. Most vertebrates were found in equatorial regions during this time.

The spread of land plants in the Devonian, and their resultant modification of soils, would have profoundly affected the Earth's atmosphere and climate, resulting in climatic cooling. The formation of soils would speed the breakdown of the underlying rocks as plant roots penetrated them and organic secretions and the decomposition of dead plant material dissolved minerals in the rock. The evidence points to a sharp decrease (by around 90 percent) in atmospheric carbon dioxide during the Late Devonian, possibly related to the spread of land plants (Berner et al. 2007). The organic acids released by plant roots cause a chemical reaction with the silicate rocks on land, called weathering, which traps atmospheric carbon dioxide as carbonate minerals. Carbon in this form eventually gets washed away from the terrestrial soils and buried on the ocean floor, and it is thus removed from the atmospheric pool. Atmospheric levels of carbon dioxide reached an extreme low during the Late Carboniferous and Early Permian, resembling the levels of today's world. (Even with the increase in carbon dioxide that has occurred during the past century, atmospheric concentrations are much lower now than they were in most of the Phanerozoic.) In contrast, oxygen levels were high during this time, reaching present-day levels by the start of the Carboniferous and peaking at around 30 percent in the later Permian. The reverse greenhouse effect of this low atmospheric carbon dioxide probably caused the extensive Permo-Carboniferous glaciations.

7.4 Paleozoic Terrestrial Ecosystems

Photosynthesizing bacteria (cyanobacteria) probably existed in wet terrestrial habitats from their origin in the Archean, and algae, lichens, and fungi have probably occurred on land since the late Proterozoic. Fossilized soils from the Ordovician have mottled patterns that seem to indicate the presence of bacterial mats, and traces of erosion suggest that some of the soil surface was covered by algae, but there is no evidence of rooted plants. Land plants appear to represent a single terrestrial invasion from a particular group of green algae. The first major radiation of

plants onto land probably took place in the Middle to Late Ordovician, as shown by fossilized plant spores and some fragments of whole plants (liverworts), although there are no fossils showing complete land plants until the Late Silurian. These pioneer plants included bryophytes, represented now by mosses, liverworts, and hornworts. The landscape would have looked bleak by our standards—mostly barren, with a few kinds of low-growing vegetation limited to moist areas. Behrensmeyer et al. (1992) and Willis and McElwain (2002) review the evolution of terrestrial ecosystems through the Paleozoic.

As was the case with the evolution of land vertebrates, land plants had to cope with the transition from water to land. The earliest land plants were small and simple, bearing a resemblance to the spore-producing phase of living primitive plants such as mosses. Adaptive responses to life on land included the evolution of an impenetrable outer surface to prevent water loss, water-conducting internal tubes, and spore-bearing organs for reproduction.

The diversity of terrestrial life increased during the Silurian, but plants were only a couple of centimeters in height and were concentrated along river floodplains. Terrestrial fungi were also known among these plant assemblages—as were small arthropods that would have fed on these fungi—and some larger, probably predatory, arthropods. Thus by the latest Silurian, there was a minimal terrestrial food web of primary producers (plants), decomposers (fungi), secondary consumers (fungus-eating arthropods), and predators (millipedes and scorpions).

Terrestrial ecosystems increased in complexity through the Early and Middle Devonian, with the appearance of more derived vascular plants at the start of the period, but food webs remained simple. Today plants form the base of the terrestrial food chain, but there is no evidence that Devonian invertebrates fed on living plants. Instead, they were probably detritivores, consuming dead plant material and fungi. This in turn would recycle the plant nutrients to the soil. Millipedes and scorpions were abundant, and springtails and mites were also present.

In the Early Devonian the land would still have appeared fairly barren. However, the diversity of plant species was greater than it had been in the Silurian, and increased heights were possible for the vascular plants (which could transport water from the site of uptake to other locations). By the Middle Devonian, these plants probably attained heights of 2 meters, and the canopy they created would have modified microclimatic conditions on the ground. Treelike forms evolved independently among several ancient plant lineages, and by the Middle Devonian, there were stratified forest communities consisting of plants of different heights. However, these plants were not related to modern trees and were not really like modern trees in their structure. It would not be possible to make furniture out of Devonian trees; they had narrow trunks and would not have provided enough woody tissue.

Although the terrestrial environment of the mid-Devonian was by now quite complex, it differed in many respects from the modern ecosystems to which we are accustomed. In the first place, there were no terrestrial vertebrates—the earliest of those appeared in the Late Devonian. Furthermore, plant life was limited to wet places—low-lying areas and the margins of streams, rivers, and lakes. Flying and plant-eating insects were absent. Terrestrial animal communities in the mid-Devonian apparently were based on detritivores such as millipedes, springtails, and mites. Those animals in turn were preyed on by scorpions, pseudoscorpions, and spiders.

Terrestrial ecosystems became increasingly complex during the remainder of the Paleozoic. Plants with large leaves first appeared in the Late Devonian. The evolution of large leaves probably coincided with the drop in levels of atmospheric carbon dioxide because leaves with more stomata (pores) admit carbon dioxide more readily. The Late Devonian saw the spread of forests of the progymnosperm (primitive seed plant) *Archaeopteris*, large trees with trunks up to a meter in diameter and reaching heights of at least 10 meters. Giant horsetails (*Calamites*), relatives of the living horsetails that grow in moist areas today, reached heights of several meters. There were also many species of giant clubmosses (lycophytes), a few of which survive today as small ground plants. Other areas were apparently covered by bushlike plants, vines, and low-growing ground cover.

The diversity and habitat specificity of Late Devonian floras continued to expand in the Carboniferous. Most of the preserved habitats represent swamp environments, and vegetation buried in these swamps formed today's coal beds. (The word *carboniferous* means "coal-bearing.") It is a bit of a mystery why such coal beds were laid down during this time in Earth's history but not at later times. It may well be the case that the specialized communities of fungi and bacteria that attack lignin (a tough structural component of plants) had not yet evolved, so dead plants could accumulate without decomposition (Robinson 1990).

Most of the major groups of plants evolved during this time, although the flowering seed plants (**angiosperms**) that dominate the world today were as yet unknown. Seed ferns (the extinct pteridosperms) and ferns (which survive today) lived in well-drained areas, and swamps were dominated by clubmosses, with horsetails, ferns, and seed ferns also present. During the Early Carboniferous, there were forests consisting of trees of varying heights, giving stratification to the canopy, and vinelike plants hung from their branches. The terrestrial vegetation would have looked superficially as it does today, although the actual types of plants present were completely different.

Global drying resulted in changes in the plant communities during the Late Carboniferous and Permian. Seed plants such as conifers became an important component of the flora, replacing the more primitive spore-bearing plants. Seed plants were initially predominant in high latitudes; and, by the end of the Paleozoic, they were the major group of plants worldwide. Ferns also became more predominant in a variety of habitats at this time.

Terrestrial invertebrates diversified during the Carboniferous. Millipedes, arachnids, and insects were common. Flying insects are known from late in the Early Carboniferous. Detritivores continued to be an important part of the food web, but insect herbivory appears to have been well established by the end of the Carboniferous. Fossil leaves from the Late Carboniferous have ragged holes, seeds and wood are penetrated by tunnels, and pollen is found in the guts and feces of fossilized insects. There is speculation that insect flight and herbivory may have evolved hand in hand, allowing the insects to forage throughout the canopy. High levels of atmospheric oxygen at this time would have made flight easier to evolve, in part because flight requires high levels of activity and thus consumes large quantities of food and oxygen and also because an atmosphere with increased oxygen is somewhat denser (Dudley 2002). Increased oxygen levels also enabled insects to attain much larger body sizes than seen today. Huge dragonflies (one species with a wingspan of 63 centimeters) flew through the air, and the extinct predatory arthropleurids were nearly 2 meters long.

Late Paleozoic arthropod communities contained large predatory forms, including scorpions and spiders, but species diversity was lower than today. Most spiders appear to have been burrow-building forms rather than web builders; the earliest evidence of spider silk is from the mid-Devonian. New arthropod types appearing in the Permian include hemipterans (bugs), beetles, and forms

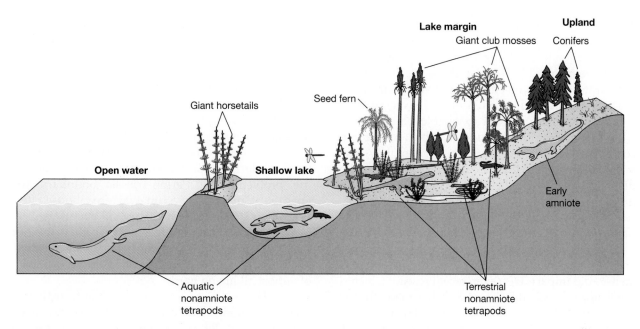

▲ **FIGURE 7–5** Aquatic and terrestrial plants and vertebrates in the Carboniferous. A reconstuction of a scene from a Late Carboniferous lake in Europe and its surroundings.

resembling mosquitoes but not closely related to true mosquitoes.

Terrestrial vertebrates appeared in the Late Devonian and diversified during the Carboniferous. The first **amniotes** were mid-Carboniferous in age; and, by the Late Carboniferous, amniotes had split into two major lineages—one leading to the mammals (synapsids) and the other to modern reptiles and birds (sauropsids). The Carboniferous was dominated by a diversity of semiaquatic primitive tetrapods; however, by the Permian, the more terrestrially adapted amniotes were common, and many vertebrate communities represented upland habitats (**Figure 7–5**).

By the Early Permian several vertebrate lineages had given rise to small insectivorous predators, rather like modern salamanders and lizards. Larger vertebrates (up to 1.5 meters long) were probably predators of these small species, and still larger predators topped the food web. An important development in the Permian was the appearance of herbivorous vertebrates: for the first time, vertebrates were able to exploit the primary production of terrestrial plants directly. By the end of the Permian, the structure and function of terrestrial ecosystems were essentially modern, although the kinds of plants and animals in those ecosystems were almost entirely different from the ones we know today.

7.5 Paleozoic Extinctions

There was a major extinction event among marine invertebrates in the Ordovician, but the record of vertebrates from that time is too poor to know if this event affected them as well. The next major extinction, during the Late Devonian, had severe effects on marine vertebrates. Thirty-five families of fish (76 percent of the existing families) became extinct, including all of the remaining ostracoderms, most of the placoderms (with complete extinction of this group by the end of the Devonian), and many of the lobe-finned fishes.

Major extinctions occurred at the end of the Paleozoic in both terrestrial and marine environments. These were the most severe extinctions of the Phanerozoic, affecting approximately 57 percent of marine invertebrate families and 95 percent of all marine species, including 12 families of fishes (Erwin 1993). There were also significant casualties on land: Twenty-seven families of tetrapods (49 percent) became extinct, with especially heavy losses among the synapsids ("mammal-like reptiles") (Benton 1989). However, in addition to the calamitous end-Permian extinction, there is evidence that trouble was brewing for vertebrates for a number of million years before this event. The background extinction rate was higher in the Late Permian than it had been earlier, suggesting that the environment was becoming less suitable for vertebrates. Levels of atmospheric oxygen were falling during this time, and that may have resulted in hypoxic stress for the amniote vertebrates that evolved and radiated during the Early Permian under conditions of oxygen abundance. Furthermore, lower oxygen levels in general would also limit the altitude at which vertebrates could live, restricting them to smaller, low-lying areas of land, and thus reducing available habitat diversity. There would also have been limitations of migration over mountain ranges, resulting in isolation of populations. All these influences would have made extinctions more likely (Huey and Ward 2005)

Theories abound for the reason for the massive end-Permian extinction, including the possibility of an asteroid impact. However, recent research, including isotopic geochemical evidence, has made the picture clearer (Benton and Twitchett 2003). The age of the Permo-Triassic boundary has now been precisely dated to 251 million years ago; this in turn coincides with massive volcanic eruptions in Siberia concentrated in a time period of less than a million years. Enough molten lava was released to cover an area the size of about half of the United States. These events, together with the release of massive amounts of gases into the atmosphere, would have resulted in global warming of around 6°C. Such warming would not only have had a direct effect on organisms but would have affected oceanic circulation, resulting in stagnation and low oxygen levels, with profound effects on marine organisms.

To make things worse, the global warming also resulted in the melting of underwater frozen gas hydrates, causing the release of massive amounts of methane into the atmosphere. Methane is a greenhouse gas, and this feedback loop in turn would have made the world even warmer, compounding the problem. The Permo-Triassic boundary was apparently marked by a runaway greenhouse effect, with a positive feedback loop of events causing increasing global warming. This disrupted the normal global environmental mechanisms for hundreds of thousands of years and resulted in the extinction of almost all life on Earth.

Another smaller, though significant, extinction occurred among land vertebrates at the end of the Early Permian. Fifteen families of tetrapods became extinct, including many non-amniote tetrapods ("amphibians" in the broad sense) and pelycosaurs (early "mammal-like reptiles"). These extinctions may have been related to climatic changes associated with the end of the Permo-Carboniferous period of glaciation and perhaps also to the accompanying changes in the atmosphere, with a decrease in the levels of oxygen and an increase in the levels of carbon dioxide.

Additional Readings

Beerling, D. J. et al. 2001. Evolution of leaf-form in land plants linked to atmospheric CO_2 decline in the Late Palaeozoic era. *Nature* 410:352–354.

Behrensmeyer, A. K. et al. (eds.). 1992. *Terrestrial Ecosystems Through Time*. Chicago: University of Chicago Press.

Benton, M. J. 1989. Patterns of evolution and extinction in vertebrates. In K. C. Allen and D. E. G. Briggs (eds.), *Evolution and the Fossil Record*. London, U.K.: Bellhaven Press.

Benton, M. J., and R. J. Twitchett. 2003. How to kill (almost) all life: The end-Permian extinction. *Trends in Ecology and Evolution* 18:358–365.

Berner, R. A. et al. 2007. Oxygen and evolution. *Science* 316:557–558.

Cox, C. B., and P. D. Moore. 1993. *Biogeography: An Ecological and Evolutionary Approach*, 5th ed. Oxford, U.K.: Blackwell Scientific.

DiMichele, W. A. et al. 2001. Response of Late Carboniferous and Early Permian plant communities to climate change. *Annual Reviews of Earth and Planetary Science* 29:461–487.

Dudley, R. 2002. The evolutionary physiology of animal flight: Paleobiological and present perspectives. *Annual Reviews of Physiology* 62:135–155.

Erwin, D. H. 1993. *The Great Paleozoic Crisis*. New York, NY: Columbia University Press.

Hallam, A. 1994. *An Outline of Phanerozoic Biogeography*. Oxford, U.K.: Oxford University Press.

Huey, R. B., and P. D. Ward. 2005. Hypoxia, global warming, and terrestrial Late Permian extinctions. *Science* 308:398–401.

Kenrick, P., and P. R. Crane. 1991. The origin and early evolution of plants on land. *Nature* 389:33–39.

Kunzig, R. 1996. In deep water. *Discover* 11(2):86–96.

Retallack, G. J. 1997. Early forest soils and their role in Devonian climate change. *Science* 276:583–585.

Robinson, J. M. 1990. Lignin, land plants and fungi; Biochemical evolution affecting Phanerozoic oxygen balance. *Geology* 15:607–610.

Rothschild. L. J., and A. M. Lister (eds.). 2003. *Evolution on Planet Earth: The Impact of the Physical Environment*. London, UK: Academic Press.

Willis, K. J., and J. C. McElwain. 2002. *The Evolution of Plants*. Oxford, UK: Oxford University Press.

CHAPTER 8

Living on Land

The spread of plants and then invertebrates across the land in the late Paleozoic era provided a new habitat for vertebrates and imposed new selective forces. The demands of terrestrial life are quite different from those in an aquatic environment because water and air have different physical properties. Air is less viscous and less dense than water, so streamlining is a minor factor for tetrapods, whereas a skeleton that supports the body against the pull of gravity is essential. Respiration, too, is different because of the properties of air and water. Gills don't work in air because the gill filaments collapse on each other without the support of water. When that happens, the surface area available for gas exchange is greatly reduced. Terrestrial animals need a structure that won't collapse; and, because the density and viscosity of air are low, tetrapods can use a tidal flow of air in and out of a saclike lung. Heat capacity and heat conductivity are additional differences between water and air that are important to terrestrial animals. Terrestrial habitats can have large temperature differences over a very short distance, and even small tetrapods can maintain body temperatures that are substantially different from air temperatures. As a result, terrestrial animals have far more opportunity than aquatic organisms for regulation of their body temperatures.

8.1 Support and Locomotion on Land

Perhaps the most important difference between water and land is the effect of gravity on support and locomotion. Gravity has little significance for a fish living in water because the bodies of vertebrates are approximately the same density as water, and hence fish are essentially weightless in water. Gravity is a very important factor on land, however, and the skeleton of a tetrapod must be able to support the body.

Water and air also require different forms of locomotion. Because water is dense, fish swim merely by passing a sine wave along the body—the sides of the body and the fins push backward against the water, and the fish moves forward. Pushing backward against the air doesn't move a tetrapod forward unless it has wings. Most tetrapods (including fliers when they are on the ground) use their legs and feet to transmit a backward force to the substrate. Thus, both the skeletons and modes of locomotion of tetrapods are different from those of fishes.

The skeleton is composed of bone, which must be rigid enough to resist the force of gravity and the forces exerted as an animal starts, stops, and turns. The remodeling capacity of bone is of great importance for a terrestrial animal as the internal structure of bone adjusts continuously to the changing demands of an animal's life. In humans, for example, intense physical activity results in an increase in bone mass, whereas inactivity (prolonged bed rest or being in space) results in loss of bone mass. In addition, remodeling allows broken bones to mend, and the skeletons of terrestrial animals experience greater stress than those of aquatic animals and are more likely to break.

■ Bone

Amniotes have bone that is arranged in concentric layers around blood vessels forming cylindrical units called Haversian systems (**Figure 8–1a**). The structure

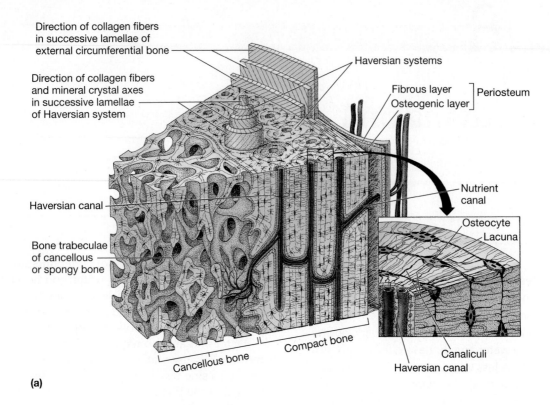

Direction of collagen fibers in successive lamellae of external circumferential bone

Direction of collagen fibers and mineral crystal axes in successive lamellae of Haversian system

Haversian systems

Fibrous layer ⎱ Periosteum
Osteogenic layer ⎰

Nutrient canal

Haversian canal

Osteocyte
Lacuna

Bone trabeculae of cancellous or spongy bone

Canaliculi
Haversian canal

Cancellous bone   Compact bone

Cancellous bone

Compact bone

(a)

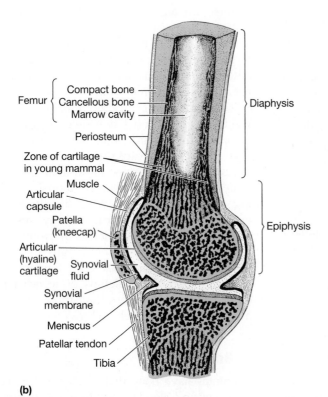

Femur ⎰ Compact bone
       ⎱ Cancellous bone
         Marrow cavity

Diaphysis

Periosteum

Zone of cartilage in young mammal

Muscle

Articular capsule

Patella (kneecap)

Articular (hyaline) cartilage

Synovial fluid

Epiphysis

Synovial membrane

Meniscus

Patellar tendon

Tibia

(b)

▲ **FIGURE 8–1** Structure of bone and joints. (a) Bone from a section of the shaft of a long bone of a mammal, showing Haversian systems. (b) Section through a human knee joint, showing the gross internal structure of a long bone and the structure of a joint capsule. Note that the patella (kneecap) is an example of a sesamoid bone (a bone formed within a tendon).

of a bone is not uniform. If it were, animals would be very heavy. The external layers of a bone are formed of dense, compact or lamellar bone, but the internal layers are lighter, spongy (cancellous) bone. The joints at the ends of bones are covered by a smooth layer of articular cartilage that reduces friction as the joint moves. (Arthritis occurs when this cartilage is damaged or worn.) The bone within the joint is composed of cancellous bone rather than dense bone, and the entire joint is enclosed in a joint capsule, containing synovial fluid for lubrication (**Figure 8–1b**).

■ The Axial System: Vertebrae and Ribs

The vertebrae and ribs of fishes stiffen the body so it will bend when muscles contract, rather than shortening. In tetrapods, the axial skeleton is modified for support on land. Processes called zygapophyses (singular zygapophysis) on the vertebrae of tetrapods (**Figure 8–2**) interlock and resist twisting (torsion) and bending (compression), allowing the spine to act like a suspension bridge to support the weight of the viscera on land (**Figure 8–3**). (Tetrapods that have

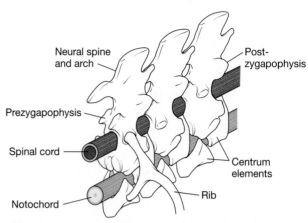

Generalized primitive tetrapod condition

▲ FIGURE 8–2 Gnathostome vertebrae and ribs. In aquatic vertebrates, the vertebral column primarily stiffens the body so that it bends instead of telescoping when muscles on one side of the body contract. Terrestrial vertebrates require more rigidity. Articulating surfaces (zygapophyses) on adjacent vertebrae of tetrapods allow the vertebral column to resist gravity. (In more derived tetrapods, the central elements form a single solid structure, obliterating the notochord in the adult form.)

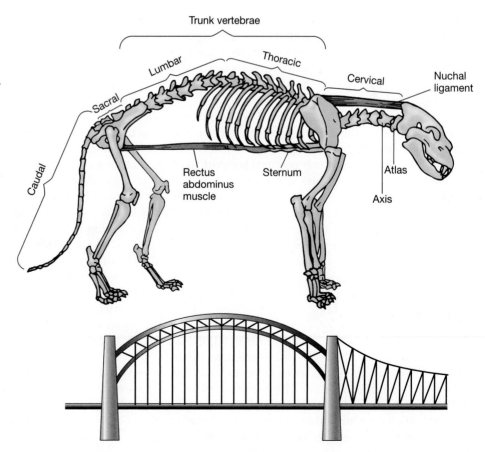

► FIGURE 8–3 Skeleton of a cat. The muscles and ligaments of postural support are shown in comparison with the support elements of a suspension bridge.

permanently returned to the water, such as whales and many of the extinct Mesozoic marine reptiles, have lost the zygapophyses.)

Bony fishes use the opercular bones to protect the gills, and they play a role in gill ventilation. They also rigidly connect the head to the pectoral girdle so a fish cannot turn its head—instead it must pivot its entire body. Tetrapods have lost the connection between the opercular bones and pectoral girdle, and, as a result, tetrapods have a flexible neck region and can move the head separately from the rest of the body. The cervical (neck) vertebrae (see Figure 8–3) allow the head to turn from side to side and up and down relative to the trunk, and the muscles that support and move the head are attached to processes on the cervical vertebrae. The two most anterior cervical vertebrae are the atlas and axis, and they are most highly differentiated in mammals.

The trunk vertebrae are in the middle region of the body and bear the ribs. In mammals, the trunk vertebrae are differentiated into two regions: the thoracic vertebrae (those that bear ribs) and the lumbar vertebrae (those that have lost ribs).

The sacral vertebrae, which are derived from the trunk vertebrae, fuse with the pelvic girdle and allow the hind limbs to transfer force to the appendicular skeleton. Early tetrapods and extant amphibians have a single sacral vertebra, mammals have three to five, and some dinosaurs had a dozen or more. The caudal vertebrae, found in the tail, are usually simpler in structure than the trunk vertebrae.

The ribs of early tetrapods were fairly stout and more prominently developed than those of fishes. They may have stiffened the trunk in animals that had not yet developed much postural support from the axial musculature (**Figure 8–4** [number 7]). The trunk ribs are the most prominent ones in tetrapods in general, and many primitive tetrapods have small ribs extending throughout the entire vertebral column. Modern amphibians have almost entirely lost their ribs; in mammals, ribs are confined to the thoracic vertebrae.

■ Axial Muscles

The axial muscles assume two new roles in tetrapods: postural support of the body and ventilation of the lungs. These functions are more complex than the side-to-side bending produced by the axial muscles of fishes, and the axial muscles of tetrapods are highly differentiated in structure and function. Muscles are important for maintaining posture on land because the body is not supported by water; without muscular action, the skeleton would buckle and collapse. Likewise, the method of ventilating the lungs is different if the chest is surrounded by air rather than by water.

The axial muscles still participate in locomotion in primitive tetrapods, producing the lateral bending of the backbone seen during movement by many amphibians and reptiles. However, in birds and mammals, limb movements have replaced lateral flexion of the trunk by axial muscles. The trunks of birds are rigid, but dorsoventral flexion is an important component of mammalian locomotion. In secondarily aquatic mammals (e.g., whales), the axial muscles again assume a major role in locomotion.

In fishes and modern amphibians, the epaxial muscles form an undifferentiated single mass. This was probably their condition in the earliest tetrapods (see Figure 8–4 [number 11]). The epaxial muscles of extant amniotes are distinctly differentiated into three major components, and their primary role is now postural rather than locomotory.

The hypaxial muscles form two layers in bony fishes (the external and internal obliques), but in tetrapods a third inner layer is added, the transversus abdominus (see Figure 8–4 [number 12]). This muscle is responsible for exhalation of air from the lungs of modern amphibians (which, unlike amniotes, do not use their ribs to breathe), and it may have been essential for respiration on land by early tetrapods. Air-breathing fishes use the pressure of the water column on the body to force air from the lungs, but land-dwelling tetrapods need muscular action. The hypaxial muscles of all tetrapods show a more complex pattern of differentiation into layers than is seen in fishes. In unspecialized amphibians, such as salamanders, both epaxial and hypaxial muscles contribute to the bending motions of the trunk while walking on land, much as they do in fishes swimming in water.

The rectus abdominus, which runs along the ventral surface from the pectoral girdle to the pelvic girdle, is another new hypaxial muscle in tetrapods, and its role appears to be primarily postural. (This is the muscle responsible for the six-pack stomach of human bodybuilders.) The costal muscles in the rib cage of amniotes are formed by all three layers of the hypaxial muscles and are responsible for inhalation as well as for exhalation. The use of the ribs and their associated musculature as devices to ventilate the lungs was probably an amniote innovation.

Thus, in tetrapods, the axial skeleton and its muscles assume very different roles from their original

▶**FIGURE 8–4** Morphological and physiological differences among fishes, primitive tetrapods, and amniotes. Numbers indicate the various systems that are referred to in text, and comparing the three drawings shows where the change occurred—for example, the length of the snout (character 3) changed between fishes and primitive tetrapods. **1.** Mode of reproduction. **2.** Presence of midline fins. **3.** Length of snout. **4.** Length of neck. **5.** Form of lungs and trachea. **6.** Interlocking of vertebral column. **7.** Form of ribs. **8.** Attachment of pelvic girdle to vertebral column. **9.** Form of the limbs. **10.** Form of the ankle joint. **11.** Differentiation of epaxial muscles. **12.** Differentiation of hypaxial muscles. **13.** Presence of urinary bladder. (Note that the kidneys of fishes and non-amniotic tetrapods are in fact elongated structures lying along the dorsal body wall. The kidneys have been portrayed in all the animals as a mammalian bean-shaped form, for familiarity and convenience.) **14.** Form of the acousticolateralis system and middle ear.

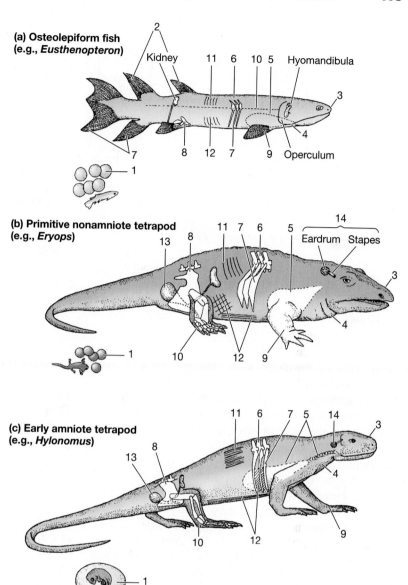

functions in aquatic vertebrates. The skeleton now participates in postural support and ventilation of the lungs, as well as in locomotion, and some of these functions are incompatible. For example, the side-to-side bending of the trunk that occurs when a lizard runs means that it has difficulty using its ribs for lung ventilation, creating a conflict between locomotion and respiration. More derived tetrapods such as mammals and birds have addressed this conflict by a change in posture from sprawling limbs to limbs that are held more directly underneath the body. These tetrapods are propelled by limb movements rather than by trunk bending.

■ The Appendicular Skeleton

The appendicular skeleton includes the limbs and limb girdles. In the primitive gnathostome condition, illustrated by sharks, the pectoral girdle (supporting the front fins) is a simple cartilaginous bar, called the coracoid bar, with a small ascending scapular process. In bony fishes, the pectoral girdle (the scapulocoracoid) is attached to the opercular bones that form the posterior portion of the dermal skull roof. The pelvic girdle in both kinds of fishes is represented by the puboishiatic plate, which has no connection with the vertebral column but merely anchors the

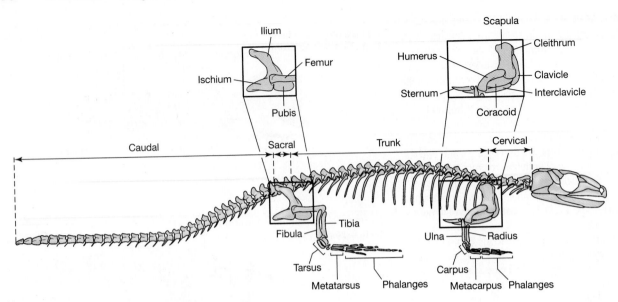

▲ **FIGURE 8–5** Generalized tetrapod skeleton. This is the primitive amniote *Hylonomus*.

hind fins in the body wall. Neither arrangement works well on land.

The tetrapod limb is derived from the fin of fishes. The basic structure of a fin consists of fanlike basal elements supporting one or more ranks of cylindrical radials, which usually articulate with raylike structures that support most of the surface of the fin web. The tetrapod limb is made up of the limb girdle and five segments that articulate end to end. All tetrapods have jointed limbs (with a forwardly pointing knee and a backwardly pointing elbow), wrist/ankle joints, and hands and feet with digits (see Figure 8–4 [number 9]). The feet of primitive tetrapods are used mainly as holdfasts, providing frictional contact with the ground. Propulsion is mainly generated by the axial musculature of the body. In contrast, the feet of amniotes play a more complex role in locomotion. They are used as levers to propel the animal, and the ankle forms a distinct hinge joint (mesotarsal joint; see Figure 8–4 [number 10]). Some non-amniotes (e.g., frogs) parallel amniotes in this condition.

The basic form of the tetrapod skeleton is illustrated in **Figure 8–5**. The pelvic girdle is fused directly to modified sacral vertebrae, and the hind limbs are the primary propulsive mechanism. The pelvic girdle contains three paired bones on each side (a total of six bones): ilium (plural ilia), pubis (plural pubes), and ischium (plural ischia). The ilia on each side connect the pelvic limbs to the vertebral column, forming an attachment at the sacrum via one or more modified ribs (see Figure 8–4 [number 8]).

Support of the forelimb is only a minor role of the pectoral girdle of fishes, which mainly anchors the muscles that move the gills and lower jaw. In bony fishes, the pectoral girdle and forelimb are attached to the back of the head via the series of opercular and gular bones (**Figure 8–6**). In tetrapods these bones are lost, and the pectoral girdle is freed from the dermal skull roof. The main endochondral bones are the scapula and the coracoid; the humerus (upper arm bone) articulates with the pectoral girdle where these two bones meet. However, some dermal postopercular bones (the anocleithrum, the cleithrum, the clavicle, and the interclavicle) become incorporated into the pectoral girdle. The cleithrum and anocleithrum run along the anterior border of the scapula and are seen only in extinct primitive tetrapods. The clavicle (the collar bone of humans) connects the scapula to the sternum or to the interclavicle. The interclavicle is a single midline element lying ventral to the sternum. It has been lost in birds and in most mammals but is still present in the monotremes.

The pectoral girdle does not articulate directly with the vertebral column. (Only in pterosaurs [extinct flying reptiles] is there an equivalent of a sacrum in the anterior vertebral column, a structure called the notarium.) In all other vertebrates the connection between the pectoral girdle and the vertebral column consists of muscles and connective tissue that hold the pectoral girdle to the sternum and the ribs. The sternum is a midventral structure, formed from endochondral bone and usually segmented, which links the lower ends of right and left thoracic ribs in

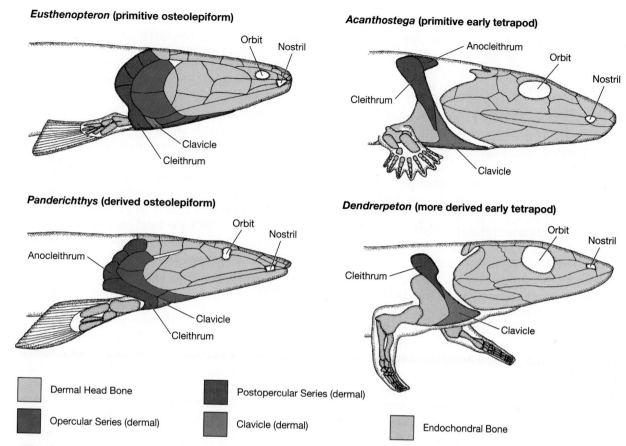

Eusthenopteron (primitive osteolepiform)

Orbit
Nostril
Clavicle
Cleithrum

Acanthostega (primitive early tetrapod)

Anocleithrum
Orbit
Nostril
Cleithrum
Clavicle

Panderichthys (derived osteolepiform)

Orbit
Nostril
Anocleithrum
Clavicle
Cleithrum

Dendrerpeton (more derived early tetrapod)

Orbit
Nostril
Cleithrum
Clavicle

Dermal Head Bone

Opercular Series (dermal)

Postopercular Series (dermal)

Clavicle (dermal)

Endochondral Bone

▲ **FIGURE 8–6** Development of the neck in tetrapods. The bones of the skull are connected to the pectoral girdle by the opercular bones in the primitive osteolepiform fish *Eusthenopteron* and the derived osteolepiform *Panderichthyes*. The opercular bones have been lost in the primitive tetrapod *Acanthostega* and the more derived tetrapod *Dendrerpeton*.

amniotes. The sternum is extensively ossified only in birds and mammals. Bones in the shoulder girdle of frogs and salamanders (called sternal elements) may not be homologous with the amniote sternum.

The midline fins of fishes (the dorsal and anal fins) help to reduce roll, but they have no function on land and are not present in terrestrial animals. The pectoral and pelvic fins are used for hydrodynamic lift, steering, and braking but not usually for propulsion except in rays and some coral reef fishes. The original form of vertebrate appendicular muscles, as seen today in sharks, is fairly simple: a set of muscles to lift the fins up and draw them outward and backward (pectoral and pelvic levators) and another set of muscles to move the fin down, in, and forward (pectoral and pelvic depressors).

The pectoral and pelvic fins of fishes become the limbs of tetrapods, and the limbs become increasingly important for locomotion in derived tetrapods.

The appendicular muscles of tetrapods are more complicated and differentiated than those of fishes, but the ancestral pattern of a major levator and a major depressor is retained. In our arms the levator corresponds to the deltoids (running from the scapula over the shoulder down to the upper arm), and the depressor corresponds to the pectoralis (running from the upper arm to the chest). However, we have many additional muscles in the shoulder region alone—not to mention the ones that move the elbow, the wrist, and the multitude of complex muscles that move the fingers in the hand.

■ Locomotion on Land

A fishlike mode of swimming works only in a dense medium, and energy must be expended to overcome the drag produced by moving though a dense fluid. Overcoming drag from the surrounding medium is not a problem in air, but friction must be generated between the feet and the ground for propulsion. (Walking on ice is difficult because there is very little friction between your foot and the substrate.) Locomotion on land is energetically more expensive than in water. Travel for a given distance requires the most energy for a walker, less for a flier, and least for a swimmer. The high energy expense of terrestrial locomotion is another challenge that tetrapods have to face.

Figure 8–7 shows the modes of locomotion used by tetrapods in a phylogenetic perspective. The basic form of tetrapod limb movement consists of moving diagonal pairs of legs together. The right front and left hind move as one unit and the left front and right hind move as another in a type of gait known as the walking-trot. Even though humans are bipedal, relying entirely on the hind legs for locomotion, we retain this primitive coupling of the limbs in walking, swinging the right arm forward when striding with the left leg and vice versa. This type of coupled, diagonally paired limb movement is probably a primitive feature for gnathostomes because sharks also move their fins in this fashion when bottom-walking over the substrate.

The primitive mode of tetrapod locomotion, still seen today in salamanders, combined axial flexion of the body with limbs moving in a walking-trot and the feet acting primarily as holdfasts on the substrate rather than propelling the body. (That is, the force used to move the animal came from the trunk muscles rather than from the limb muscles.) Lizards retain a modified version of this mode of locomotion, although their limbs are more important for propulsion. The swimming, walking, and jumping modes of locomotion by frogs rely on limb muscles only and are highly specialized.

A new specialization of amniotes is the walk, which is different from the more primitive walking-trot gait. In walking, each leg moves independently in succession, usually with three feet on the ground at any one time. However, it is possible for mammals to employ a speeded-up walk in which only one foot or two feet are on the ground at any time, as seen in the amble of elephants and some horses (such as the Icelandic horses and the South American Paso Finos). All amniotes can also use a form of the trot for faster movement by moving diagonal pairs of limbs together, as in the primitive tetrapod condition.

Derived amniotes, such as mammals and archosaurs (birds, dinosaurs, and crocodiles), are specialized compared to the primitive, sprawling tetrapod condition. These amniotes have an upright posture and hold their limbs more nearly underneath the body.

While archosaurs tended toward bipedalism, mammals devised some new modes of locomotion with the evolution of the dorsoventral flexion of the vertebral column. The characteristic new fast gait of mammals is the bound, which involves jumping off the hind legs and landing on the forelegs, with the flexion of the back contributing to the length of the stride.

Larger mammals must move cautiously to avoid injury (**Box 8–1**), and the bound is modified into the gallop, as seen, for example, in horses. (The canter gait of horses is essentially a slow version of the gallop.) Here the period of suspension in the air with all four feet off the ground is in the bunched-up recovery phase in which the hind legs move forward for the next stride, not with legs stretched out in midleap as in the bound as erroneously illustrated in old British hunting prints. Galloping also involves less bending of the back than does bounding. If you watch domestic pets, you can see that cats usually bound and dogs usually gallop, although many animals larger than dogs (such as cheetahs and antelope) use the bound.

The trot is another gait typical of large mammals, used at medium speeds between the walk and the bound. Basically the trot is diagonal pairs of limbs acting in sequence. But unlike the walking-trot, the trot is a distinct jump from one pair of legs to the other with a period in which all four legs are off the ground (see Figure 8–7m). A more specialized type of mammalian locomotion is the ricochet, or bipedal hopping, probably derived from the bound (see Figure 8–7n). Although kangaroos are famous for this type of gait, several rodents (e.g., kangaroo rats, jerboas, and spring hares) have also evolved this locomotory mode independently of one another. Human locomotion—bipedal striding with an upright trunk (see Figure 8–7p)—is unique among vertebrates, although penguins waddle with an upright trunk.

8.2 Eating on Land

The difference between water and air profoundly affects feeding by tetrapods. In water, most food items are nearly weightless and can be sucked into the mouth and moved within the mouth by creating currents of water. Aquatic vertebrates, from the tiniest tadpoles to the largest whales, can use suction feeding to capture food items suspended in the water column. In contrast, terrestrial animals use their jaws and teeth to seize food items and their tongues and cheeks to manipulate items in the mouth.

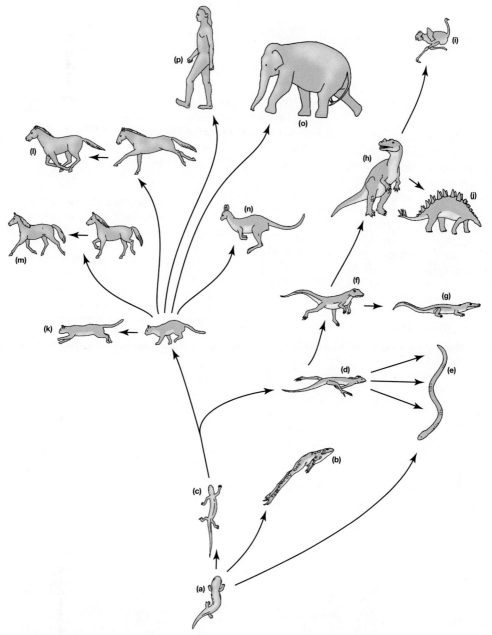

▲ **FIGURE 8–7** Phylogenetic view of tetrapod terrestrial stance and locomotion. (a) Primitive tetrapod condition, retained today in salamanders: movement mainly via axial movements of the body, limbs moved in diagonal pairs (basic walking-trot gait). (b) Derived jumping form of locomotion in the frog. (c) Primitive amniote condition, seen in many extant lizards: limbs used more for propulsion, with development of the walk gait (limbs moved one at a time/independently). (d) Diapsid amniote condition with hind limbs longer than forelimbs; tendency for bipedal running, seen in some extant lizards. (e) Derived limbless condition with snakelike locomotion. Evolved convergently several times among primitive tetrapods (e.g., several types of lepospondyls), lissamphibians (caecilians and salamanders), and lepidosaurs (many lizards, snakes, amphisbaenians). (f) Primitive archosaur condition, with upright posture and tendency to bipedalism. (g) Secondary return to sprawling posture and quadrupedalism in crocodilians. (h) Obligate bipedality in early dinosaurs and (i) birds. (j) Return to quadrupedality several times within dinosaurs. (k) Primitive mammalian condition: upright posture and the use of the bound as a fast gait with dorsoventral flexion of the vertebral column (all mammals use the walk as a slow gait). (l) Condition in larger mammals where the bound is turned into the gallop. (m) The true trot, as seen in larger mammals. (n) The ricochet, a derived hopping gait of kangaroos and some rodents. (o) The amble, a speeded-up walk gait seen as the fast gait of elephants and in some horses. (p) The human condition of upright bipedality.

BOX 8–1   Size and Scaling in Terrestrial Vertebrates

Body size is one of the most important things to know about an organism. Because all structures are subject to the laws of physics, the absolute size of an animal profoundly affects its anatomy and physiology. The design of the skeleto-muscular system is especially sensitive to absolute body size in tetrapods because of the effects of gravity on land.

The study of scaling, or of how shape changes with size, is also known as **allometry** (Greek *allo* = different and *metric* = measure). If the features of an animal show no relative changes with increasing body size (i.e., if a larger animal appeared just like a photo enlargement of a smaller one), all of its component parts would be scaled with **isometry** (Greek *iso* = the same). However, animals are not built this way, and very few body components scale isometrically.

Most body components scale allometrically. For example, although large animals have eyes that are bigger than the eyes of small animals, their eyes are smaller in relation to the size of their heads than are the eyes of small animals (compare the eye of a great dane with the eye of a chihauhua). Hence, eye size scales with negative allometry—that is, less than a one-to-one relationship to body size. On the other hand, the limb bones of large animals are not only absolutely larger than those of small animals, but they are also proportionally thicker. Hence, limb diameter scales with positive allometry.

Underlying all scaling relationships is the issue of how the surface area of an object relates to its volume: when linear dimensions double (i.e., a twofold change), the surface area increases as the square of the change in linear dimensions (a fourfold change) and the volume increases as the cube of the change in linear dimensions (an eightfold change). An animal that is twice as tall as another is eight times as heavy.

Differences in surface area relative to volume have a profound effect on animal function. For example, big animals take much longer to heat up or cool down than do small ones because there is less surface area for exchange relative to the volume of the animal. This relationship affects considerations of body temperature and metabolic rate. Absolute size also affects how animals survive falls from a height. A small animal has a proportionally larger amount of surface area to cushion its fall relative to its weight when it lands (compare the impact on a soft floor of wearing a shoe with a stiletto heel [a small surface area to support one's weight] with a sneaker [a much larger surface area]). A mouse falling out of a second-story window would be stunned but otherwise little hurt, whereas a human would sustain broken bones and a horse would be killed outright (don't try this at home).

It is the cross-sectional area of the bones that actually supports an animal's weight. If an animal increased in size isometrically, its weight would rise as the cube of its linear dimensions, but the cross section of its bones would increase only as the square of its linear dimensions. Thus the bones must increase their surface area by a disproportional amount to keep up with increases in weight. Bone diameter scales with positive allometry: Bigger animals have proportionally thicker limb bones than smaller ones. The skeleton of a bigger animal can easily be distinguished from a smaller one, even when they are drawn to the same size, by virtue of the proportionally thicker bones (**Figure 8–8**).

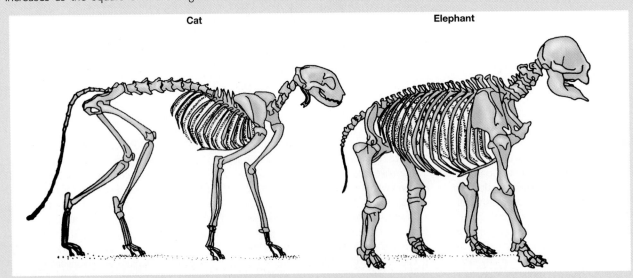

Cat   Elephant

▲ **FIGURE 8–8** Who is bigger: Comparison of skeletons of animals of different sizes. Even though they are drawn to the same size, it is instantly apparent that the animal on the right is larger than the one on the left because of the proportions of the bones, especially the limbs.

However, animals do not simply change the proportions of their limb bones to the extent that might be predicted using simple allometric laws. If they did, their skeletons would become disproportionately heavy. The skeleton of a cat makes up about 5 percent of its total body mass; if an elephant were built to the same structural proportions as a cat, its skeleton would be 78 percent of its total mass. In fact, the skeleton of an elephant is about 13 percent of its mass, almost three times that of a cat—but still within the realm of biological feasibility.

The skeletal mass of terrestrial vertebrates does scale with positive allometry but not in proportion to the stress it experiences. Consequently, the skeleton of larger animals is proportionally more fragile than that of smaller ones. So how do larger terrestrial animals cope with living in the real world? Figure 8–8 and **Figure 8–9** depict some differences between large and small mammals. Larger animals have a different pos-ture than small ones. The primitive posture for mammals is to stand with flexed joints. A bone withstands compressive forces (forces exerted parallel to the axis of the bone) much better than shearing forces (those exerted at an angle to the long axis). The larger the animal, the less flexed are its limb joints, resulting in a more pillarlike, directly weight-bearing stance that reduces the effect of shearing on the limb bones. The relatively shorter limb bones of larger animals also reduce shearing stresses. Larger animals have backs that are shorter and straighter than smaller ones. Finally, larger animals move in a different fashion than smaller ones: They engage in less leaping behavior, galloping instead of bounding, and moving their limbs through smaller angles of excursion (Figure 8–9). The straight-legged gait of elephants ensures that most of the forces on their legs will be compressive rather than shearing. An animal as large as an elephant is actually too big to trot or gallop, and its fastest gait is an amble or speeded-up walk.

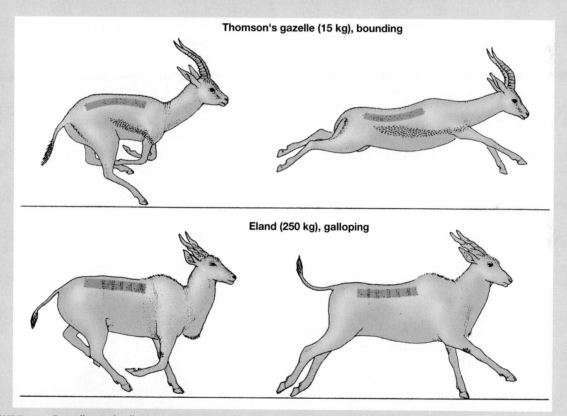

Thomson's gazelle (15 kg), bounding

Eland (250 kg), galloping

▲ **FIGURE 8–9** Bounding and galloping. Small species of antelopes, such as the Thomson's gazelle, bounds, whereas larger species, represented by the eland, gallop.

The skull of early tetrapods is much like that of primitive bony fishes, with an extensive dermatocra-nium that is retained in most extant tetrapods. (Most of the bones of the human skull represent the legacy of the bony fish dermatocranium.) However, the gill skeleton, the operculum, and most of the opercular bones connecting the operculum to the pectoral gir-dle are lost in all but the very earliest known tetrapods. The skull of bony fishes has a short snout, and movements of the jaws and buccal region cause water to be sucked into the mouth for gill ventilation and feeding. In fish skulls, the bony palate can move relative to the braincase to aid in suction, but in many tetrapods (such as crocodilians and mammals) the palate has become fused to the braincase.

Early tetrapods had wide, flat skulls, and the snout had lengthened so that most of the tooth row was now in front of the eye. Their flat heads and long snouts combined the functions of feeding and breathing, as do the heads of extant amphibians, which use the hyoid apparatus (the lower part of the hyoid arch) to ventilate the lungs. The oral cavity is expanded, sucking air into the mouth, and then the floor of the mouth is raised, squeezing the air into the lungs. This method of lung ventilation is called a positive pressure mechanism, or buccal pumping. The same expansion of the buccal cavity is used for suction feeding in water. Suction feeding is not an option on land, however, because air is much less dense than the food particles. (You can suck up the noodles in soup along with the liquid, but you cannot suck up the same noodles placed on the side of the plate.) On land, the head must be moved over the prey, and in tetrapods the skull is greatly lengthened.

The tongue of jawed fishes is small and bony, whereas the tongue of tetrapods is large and muscular. (The tongues of lampreys and hagfish are also muscular, but their tongues are not homologous with the tongues of tetrapods. The tongues of lampreys and hagfishes are innervated by cranial nerve V, whereas the tongues of tetrapods are innervated by cranial nerve XII). The tetrapod tongue works in concert with the hyoid apparatus and is probably a key innovation for feeding on land. Most tetrapods use the tongue to manipulate food in the mouth and transport it to the pharynx. Most terrestrial salamanders and lizards have sticky tongues that help to capture prey and transport it into the mouth—a phenomenon called prehension. In addition, some tetrapods—such as frogs, salamanders, and true chameleon lizards—can project their tongue to capture prey. (The mechanism of tongue projection is different in each group, and these are examples of convergent evolution.)

Salivary glands are known only in terrestrial vertebrates, probably because lubrication is required to swallow food on land. Saliva also contains enzymes that begin the chemical digestion of food while it is still in the mouth. Some insectivorous mammals, two species of lizards, and several lineages of snakes have elaborated salivary secretions into venoms that kill prey.

With the loss of the gills in tetrapods, much of the associated branchiomeric musculature is also lost, but the gill levators are a prominent exception. In fishes, these muscles are combined into a single unit, the cucullaris, and this muscle in tetrapods becomes the trapezius, which runs from the top of the neck and shoulders to the shoulder girdle. In mammals, this muscle helps to rotate and stabilize the scapulae (shoulder blades) in locomotion, and we use it when we shrug our shoulders. Understanding the original homologies of the trapezius muscle explains an interesting fact about human spinal injuries. Because the trapezius is an old branchiomeric muscle, it is innervated directly from the brain by cranial nerves (cranial nerve XI, which is actually part of nerve X), not from the nerves exiting from the spinal cord in the neck. Thus people who are paralyzed from the neck down by a spinal injury can still shrug their shoulders. Small muscles in the throat—for example, those powering the larynx and the vocal cords—are other remnants of the branchiomeric muscles associated with the gill arches. Ingenious biomedical engineering allows quadriplegic individuals to use this remaining muscle function to control prosthetic devices.

The major branchiomeric muscles that are retained in tetrapods are associated with the mandibular and hyoid arches and are solely involved in feeding (**Figure 8-10**). The adductor mandibulae remains the major jaw-closing muscle, and it becomes increasingly complex in more derived tetrapods. The hyoid musculature forms two new important muscles in tetrapods. One is the depressor mandibulae, running from the back of the jaw to the skull and helping the hypobranchials to open the mouth. The other is the sphincter colli that surrounds the neck and aids in swallowing food.

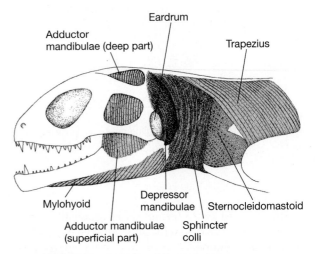

▲ **FIGURE 8–10** Head and neck musculature. This is the generalized tetrapod condition as seen in a tuatara (*Sphenodon*).

8.3 Reproduction on Land

Sauropsids and synapsids, the tetrapods that are most specialized for terrestrial life, are amniotes, and that observation suggests that the amniotic egg has some special advantage for animals that are reproducing on land. More than one mechanism may be involved, and the presence of an egg shell in amniotic eggs may be a key element. An egg shell provides support for the egg, and the shell may be the reason that all the large tetrapods are amniotes. In addition, a shell restricts water movement into or out of the egg, and the shell may allow amniotes to deposit eggs in sites that are not suitable for non-amniotes.

8.4 Breathing Air

In some respects air is an easier medium for respiration than is water. The low density and viscosity of air make tidal ventilation of a saclike lung energetically feasible, and the high oxygen content of air reduces the volume of fluid that must be pumped to meet an animal's metabolic requirements.

The lung is a primitive feature of bony fishes, and lungs were not evolved for breathing on land. For many years it was assumed lungs evolved in fishes living in stagnant, oxygen-depleted water where gulping oxygen-rich air would supplement oxygen uptake by the gills. However, although some lungfishes are found in stagnant, anoxic environments, other air-breathing fishes (e.g., the bowfin) are active animals found in oxygen-rich habitats. Colleen Farmer has suggested an alternative explanation for the evolution of lungs. She proposed that air breathing evolved in well-aerated waters in active fishes in which the additional oxygen is needed primarily to supply the heart muscle itself rather than the body tissues (Farmer 1997).

In contrast to the positive-pressure buccal pump that non-amniotic tetrapods used to inflate the lungs, amniotes use a negative-pressure aspiration pump. Expansion of the rib cage by the intercostal hypaxial muscles creates a negative pressure (i.e., below atmospheric pressure) in the abdominal cavity and sucks air into the lungs. Air is expelled by compressing the abdominal cavity, primarily through elastic return of the rib cage to a resting position and contraction of the elastic lungs, as well as by contraction of the transversus abdominus muscle.

The lungs of many extant amphibians are simple sacs with few internal divisions. They have only a short chamber leading directly into the lungs. In contrast, amniotes have lungs that are subdivided,

sometimes in very complex ways, to increase the surface area for gas exchange. They also have a long trachea (windpipe) strengthened by cartilaginous rings that branches into a series of bronchi in each lung (see Figure 8–4 [number 5]).

The form of lung subdivisions is somewhat different in mammals and other amniotes, suggesting independent evolutionary origins. The combination of a trachea and negative-pressure aspiration allows many amniotes to develop longer necks than those seen in modern amphibians or in extinct non-amniotic tetrapods. Amniotes also possess a larynx (derived from pharyngeal arch elements) at the junction of the pharynx and the trachea that is used for sound production.

8.5 Pumping Blood Uphill

Blood is weightless in water, and the heart needs to overcome only fluid resistance to move blood around the body. Circulation is more difficult for a terrestrial animal because blood tends to pool in low spots, such as the limbs, and it must be forced through the veins and back up to the heart by pumping more blood into the arteries. Thus, tetrapods have blood pressures high enough to push blood upward through the veins against the pull of gravity and valves in the limb veins to resist backflow.

The walls of blood vessels are somewhat leaky, and high blood pressure forces some of the blood plasma (the liquid part of blood) out of the vessels and into the intercellular spaces of the body tissues. This fluid is recovered and returned to the circulatory system by the lymphatic system of tetrapods. The lymphatic system is a one-way system of blind-ended, veinlike vessels that parallel the veins and allow fluid in the tissues to drain into the venous system at the base of the neck. (A lymphatic system is also well developed in teleost fishes, but it is of critical importance on land, where the cardiovascular system is subject to the forces of gravity.) In tetrapods, lymph is kept moving by the contraction of muscles and tissues, and valves in the tubes prevent backflow. Lymph nodes, concentrations of lymphatic tissues, are found in mammals and some birds at intervals along the lymphatic channels. Lymphatic tissue is also involved in the immune system; white blood cells (macrophages) travel through lymph vessels, and the lymph nodes can intercept foreign or unwanted material, such as migrating cancer cells.

With the loss of the gills and the evolution of a distinct neck, the heart has moved posteriorly in

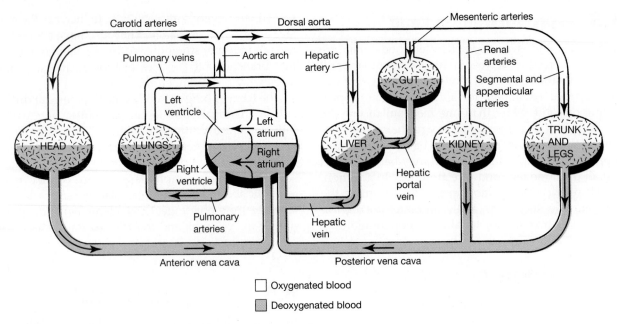

▲ **FIGURE 8–11** Double-circuit cardiovascular system in a tetrapod.

tetrapods. In fishes, the heart lies in the gill region in front of the shoulder girdle, whereas in tetrapods it lies behind the shoulder girdle in the thorax. The sinus venosus and conus arteriosus are reduced or absent in the hearts of tetrapods. With the advent of lungs, vertebrates evolved a double circulation in which the pulmonary circuit supplies the lungs with deoxygenated blood and the systemic circuit supplies oxygenated blood to the body.

The atrium of the heart is divided into left and right atria in lungfish and tetrapods, and the ventricle is divided either by a fixed barrier or by the formation of transiently separate chambers as the heart contracts. The right side of the heart receives deoxygenated blood returning from the body via the systemic veins, and the left side of the heart receives oxygenated blood returning from the lungs via the pulmonary veins. The double circulation of tetrapods can be pictured as a figure eight with the heart at the intersection of the loops (**Figure 8–11**).

The aortic arches have undergone considerable change in association with the loss of the gills in tetrapods. Arches two and five are lost in adult tetrapods (although arch number five is retained in salamanders). Three major arches are retained: the third (carotid arch) going to the head, the fourth (systemic arch) going to the body, and the sixth (pulmonary arch) going to the lungs (**Figure 8–12**). Modern amphibians retain the fishlike condition, in which the aortic arches do not arise directly from the

heart. This condition, with retention of a prominent conus arteriosus and a ventral arterial trunk (the ventral aorta of fishes and the truncus arteriosus of amphibians), was probably found among the earliest tetrapods. In amniotes, the pulmonary artery receives blood from the right ventricle and the right systemic and carotid arches receive blood from the left ventricle, although details of the heart anatomy suggest that this condition evolved independently in mammals and in other amniotes.

In modern amphibians, the skin is of prime importance in the exchange of oxygen and carbon dioxide. In frogs, the pulmonary arch is actually a pulmocutaneous arch, with a major cutaneous artery branching off the pulmonary artery to supply the skin. The cutaneous vein, now carrying oxygenated blood, feeds back into the systemic system via the subclavian vein and hence into the right atrium. A similar, but less well-differentiated, system exists in other amphibians. Thus, oxygenated blood enters the amphibian ventricle from both the left atrium (supplied by the pulmonary vein) and from the right atrium (supplied by veins that return blood from the skin). This type of heart in modern amphibians, with the absence of any ventricular division, may be a derived condition associated with using the skin as well as the lungs for respiration.

A ventricular septum of some sort is present in all amniotes, but the form is different in the various lineages of amniotes. A transient ventricular septum is

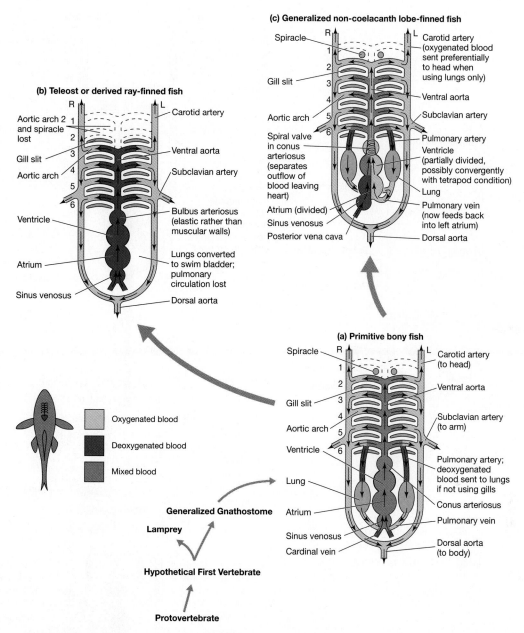

(c) Generalized non-coelacanth lobe-finned fish

Spiracle — Carotid artery (oxygenated blood sent preferentially to head when using lungs only)

Gill slit — Ventral aorta

Aortic arch — Subclavian artery

Spiral valve in conus arteriosus (separates outflow of blood leaving heart) — Pulmonary artery

Ventricle (partially divided, possibly convergently with tetrapod condition)

Atrium (divided) — Lung

Sinus venosus — Pulmonary vein (now feeds back into left atrium)

Posterior vena cava — Dorsal aorta

(b) Teleost or derived ray-finned fish

Aortic arch 2 and spiracle lost — Carotid artery

Gill slit — Ventral aorta

Aortic arch — Subclavian artery

Ventricle — Bulbus arteriosus (elastic rather than muscular walls)

Atrium — Lungs converted to swim bladder; pulmonary circulation lost

Sinus venosus — Dorsal aorta

Oxygenated blood
Deoxygenated blood
Mixed blood

Generalized Gnathostome

Lamprey

Hypothetical First Vertebrate

Protovertebrate

(a) Primitive bony fish

Spiracle — Carotid artery (to head)

Gill slit — Ventral aorta

Aortic arch — Subclavian artery (to arm)

Ventricle — Pulmonary artery; deoxygenated blood sent to lungs if not using gills

Lung — Conus arteriosus

Atrium — Pulmonary vein

Sinus venosus
Cardinal vein — Dorsal aorta (to body)

▲ **FIGURE 8–12** Form of the heart and aortic arches in jawed fishes and primitive tetrapods. (a) The primitive bony fish condition is to have lungs and some sort of pulmonary circuit, as shown in the extant fish with lungs, *Polypterus*. (b) Teleosts convert the lung into a swimbladder and lose the pulmonary circuit; they also lose aortic arch number 2. (c) The generalized sarcopterygian condition above the level of coelacanths: Some living lungfishes have reduced the number of aortic arches from this condition. Here the pulmonary artery feeds back into the left atrium directly, and the ventricle is partially divided. (d) The condition in the early tetrapod (above the *Acanthostega/Ichthyostega* level) is similar to that in lungfishes; however, the internal gills have been lost, and the second aortic arch may also have been lost by this stage (as seen in all tetrapods). (e) In the frog, arch 5 and the connection of the dorsal aorta between arches 3 and 4 (the ductus carotidus) have been lost, but both these features are retained in salamanders, so this condition cannot have been inherited from an early tetrapod ancestor. A derived condition in modern amphibians is to have a cutaneous artery leading from the pulmonary artery taking blood to the skin to be oxygenated; the blood returns via the cutaneous vein that feeds into the subclavian vein. The ventricular septum is absent in modern amphibians. This may be a secondary condition that is related to having oxygenated blood returning from the skin to the right side of the heart. (f) The proposed early amniote condition is similar to that of the frog with the exception of the absence of a cutaneous circuit. The ventricular septum would have been simple, and there would have been no division of the vessels leaving the heart because the more derived conditions in extant amniotes have been evolved independently. (R = right, L = left)

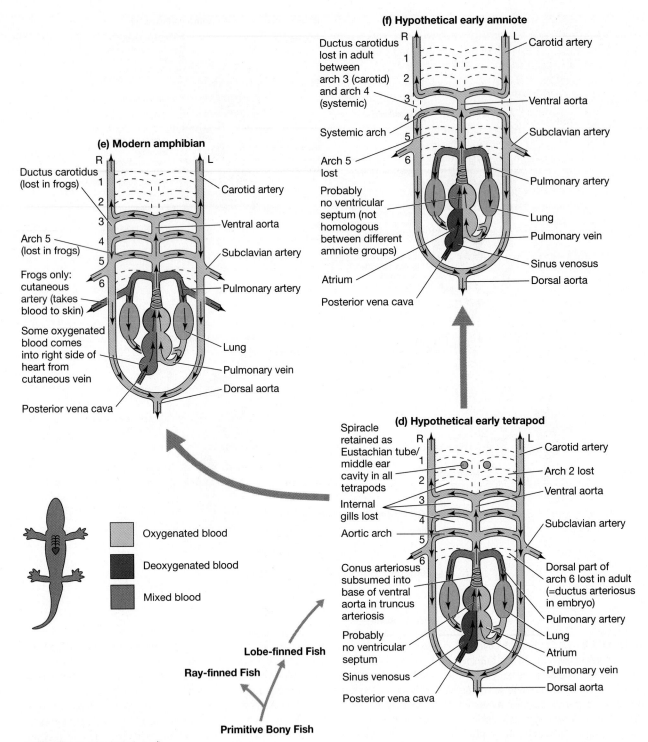

(f) Hypothetical early amniote

Ductus carotidus lost in adult between arch 3 (carotid) and arch 4 (systemic)

Systemic arch

Arch 5 lost

Probably no ventricular septum (not homologous between different amniote groups)

Atrium

Posterior vena cava

Carotid artery

Ventral aorta

Subclavian artery

Pulmonary artery

Lung

Pulmonary vein

Sinus venosus

Dorsal aorta

(e) Modern amphibian

Ductus carotidus (lost in frogs)

Arch 5 (lost in frogs)

Frogs only: cutaneous artery (takes blood to skin)

Some oxygenated blood comes into right side of heart from cutaneous vein

Posterior vena cava

Carotid artery

Ventral aorta

Subclavian artery

Pulmonary artery

Lung

Pulmonary vein

Dorsal aorta

Oxygenated blood

Deoxygenated blood

Mixed blood

Lobe-finned Fish

Ray-finned Fish

Primitive Bony Fish

(d) Hypothetical early tetrapod

Spiracle retained as Eustachian tube/ middle ear cavity in all tetrapods

Internal gills lost

Aortic arch

Conus arteriosus subsumed into base of ventral aorta in truncus arteriosis

Probably no ventricular septum

Sinus venosus

Posterior vena cava

Carotid artery

Arch 2 lost

Ventral aorta

Subclavian artery

Dorsal part of arch 6 lost in adult (=ductus arteriosus in embryo)

Pulmonary artery

Lung

Atrium

Pulmonary vein

Dorsal aorta

▲ **FIGURE 8–12** (Continued)

formed during ventricular contraction in turtles and lizards, whereas a permanent septum is present in mammals, crocodiles, and birds. This pattern of occurrence suggests that a permanent ventricular septum evolved independently in the sauropsid and synapsid lineages.

The heavy workload and divided ventricle of birds and mammals introduce another complication: how to supply oxygen to the heart muscle. Modern amphibians and nonavian reptiles have lower blood pressures than mammals and birds, their hearts don't work as hard, and their ventricles allow some mixing of oxygenated and deoxygenated blood. The hearts of these animals lack coronary arteries, presumably because enough oxygen diffuses into the cardiac muscle from the blood in the lumen of the ventricle. In contrast, the ventricular muscles of mammals and birds are thicker and must work harder than those of amphibians and reptiles to generate higher blood pressures. In addition, these animals have a permanent ventricular septum, so the right ventricle contains only deoxygenated blood. Both birds and mammals have coronary arteries that supply oxygenated blood to the muscles of both ventricles, and these arteries apparently evolved separately in the two lineages.

8.6 Sensory Systems in Air

Some of the sensory modes that are exquisitely sensitive in water are useless in air, whereas other modes work better in air than in water. Air is not dense enough to stimulate the mechanical receptors of the lateral line system, for example, and does not conduct electricity well enough to support electrosensation. Chemical systems work well on land, however, at least for molecules small enough to be suspended in air, and air offers advantages for both vision and hearing.

■ Vision

The sense of vision is actually easier to use on land than in the water because light is transferred through air with less disturbance than through water. Air is rarely murky in the way that water can frequently be, and, as a result, vision is more useful as a distance sense in air than in water. In air, the cornea (the transparent covering of the front of the eye) participates in focusing light on the retina, and tetrapods have flatter lenses than do fishes and focus an image on the retina differently. Fishes focus light by moving the position of the lens within the eye, while tetrapods focus by changing the shape of the lens. (Snakes are an odd exception here; they move the lens to focus the eye.) In air, the eye's surface must be protected and kept moist and free of particles. New features in tetrapods include eyelids, glands that lubricate the eye and keep it moist (including tear-producing lacrimal glands), and a nasolacrimal duct to drain the tears from the eyes into the nose.

■ Hearing

Sound perception is very different in air than in water. The density of animal tissue is nearly the same as the density of water, and sound waves pass freely from water into animal tissue. Because water is dense, movement of water molecules directly stimulates the hair cells of the lateral line system. Air is not dense enough to move hair cells, and the lateral line system is lost in all tetrapods except for larval or permanently aquatic amphibians. The inner ear assumes the function of hearing airborne sounds, with the transmission of sound waves through a bone (or a chain of bones) in a middle ear, whereas in water sound must be channeled via other routes to the middle ear (**Box 8–2**, including **Figure 8–13** and **Figure 8–14**).

Considerably more energy is needed to set the fluids of the inner ear in motion than most airborne sounds impart, and the middle ear is a sound amplifier. It receives the relatively low energy of sound waves on its outer membrane, the tympanum (eardrum), and these vibrations are transmitted by the bones of the middle ear to the oval window of the otic capsule in the skull. The area of the tympanum is much larger than that of the oval window, and the difference in area plus the lever system of the bones that connect the tympanum to the oval window amplifies sound waves. In-and-out movement of the oval window produces waves of compression in the fluids of the inner ear, and these waves stimulate the hair cells in the organ of Corti. This organ discriminates the frequency and intensity of vibrations it receives and transmits this information to the central nervous system. The organ of Corti lies within a flask-shaped structure, the lagena (Greek *lagenos* = flask) (**Figure 8–15**). The lagena is larger in derived tetrapods, and in mammals it is called the cochlea.

The middle ear is not an airtight cavity—if it were, the movement of the eardrum would be resisted by pressure changes in the middle ear. The Eustachian (auditory) tube, derived from the spiracle of fishes, connects the mouth with the middle ear. Air flows in or out of the middle ear as air pressure changes. (These tubes sometimes become blocked. When that happens, changes in external air pressure can produce a painful sensation in addition to reduced auditory sensitivity. Anyone who has traveled in an airplane while they had a bad cold knows about this.)

The middle ear of tetrapods has evolved convergently several times, although in each case the stapes (the old fish hyomandibula, often called the columella in nonmammalian tetrapods) transmits vibrations between the tympanum and oval window. Modern amphibians have an organization of the inner ear that is different from that of amniotes, indicating an independent evolution of hearing, and a middle ear has been evolved independently in mammals and other

BOX 8–2   Echolocation in Air and Water

Bats and cetaceans rely on hearing as their primary distance sense for navigation and location of prey. An examination of **echolocation** illustrates how a sensory system has been modified in media with different physical properties. The most thoroughly studied echolocating mammals are the microchiropteran bats and the toothed cetaceans (porpoises and dolphins). Bats hunt for insects at night, and porpoises hunt for fish and other marine animals in murky water. Both kinds of animals use **ultrasound** (i.e., frequencies above 20 kilohertz), but they differ in their production and reception of echolocation signals because they operate in different media. In some ways bats are less specialized than cetaceans because air is the original medium for the tetrapod ear.

Bats produce a stream of ultrasound from the larynx, which is enlarged but not greatly modified from the general mammalian condition. The sounds are emitted through the mouth or the nose, which often has highly complex folds and flaps to focus the sound waves, giving microchiropteran bats their typically gargoylelike faces. The external ears of microchiropteran bats are also large and complexly shaped to receive the echoes. Figure 8–13 shows the echolocation calls of a little brown bat (*Myotis lucifugus*) as it located and captured an insect. While the bat was searching, it emitted about 15 sound pulses per second. The frequency dropped from about 85 kilohertz at the start of a pulse to about 35 kilohertz at the end. The rate of calling increased and the frequency of the sound decreased as the bat detected an echo from an insect and turned toward it. In the final tenth of a second before capture, the rate of calling increased to more than 100 pulses per second, and the frequency dropped to 25 kilohertz to 30 kilohertz.

Because water is denser than air, sound travels faster in water than it does in air, and that means that the sound wave for any given frequency is longer in water than in air. That relationship creates a problem for porpoises because objects reflect only wavelengths that are shorter than the target object. Cetaceans must emit sounds with extremely short wavelengths to locate fish; and, to do that in water, they must use extremely high frequency sounds—much higher frequencies than a bat. The echolocation calls of bottlenosed dolphins (*Tursiops truncatus*) extend up to 220 kilohertz.

Cetaceans do not produce sound in the larynx, probably because they dive with only a limited amount of air in the lungs and it is wasteful to bubble any of the air out the mouth to produce sounds. Instead, sound is produced by moving air

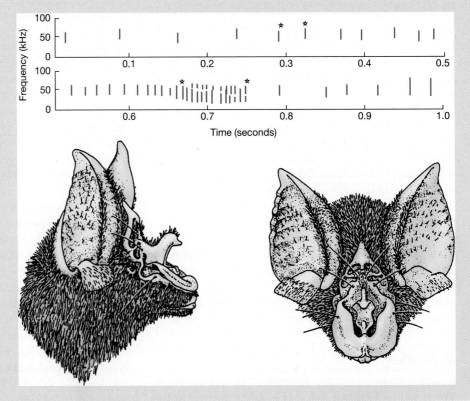

▶ **FIGURE 8–13** Echolocation by a little brown bat (*Myotis lucifugus*). The graphs show the sound pulses emitted by a bat as it captured an insect. Frequency in kilohertz is plotted against time during a 1-second record. The upper record shows searching calls during the first 0.5 second. The two stars indicate loud pulses near the time of detection of an insect. The lower record shows the next 0.5 second as the bat homed in on the insect, captured it, and then resumed its searching calls. The two stars in the lower record show the onset and completion of the highest rate of calling just before capture of prey.

back and forth between air sacs in the nasal passages, and the sound beam is reflected off the front of the skull and focused by an oil-filled body on the forehead called the melon that acts as an acoustic lens (Figure 8–14). If you watch an echolocating cetacean closely you can see the melon change shape as it moves the sound beam around.

The density of water creates two problems for cetaceans in hearing the echoes of their echolocation calls. First, the middle ear of mammals is designed to transmit sound from an air-air interface at the eardrum through the air-filled middle ear to the fluid-filled inner ear. Because air and water have such different densities, sound bounces back from an air-water interface instead of passing from one medium to the other. When you put the air-filled mammalian middle ear in water, sound bounces off the eardrum rather than being transmitted through

the middle ear to the inner ear. Second, because the density of water and animal body tissue is very similar, sound readily enters everywhere else and bounces around inside the body. (That is why you can hear only a diffuse roaring sound when your ears are underwater.) To solve these problems, cetaceans receive sound through an acoustic window of thin bone toward the rear of the lower jaw. Sound waves pass through the bone and into a fat body inside the mandible that extends back to the inner ear. The inner ear is isolated from the rest of the skull with sound-absorbing tissues everywhere except at the point of contact with the mandibular fat body. Cetaceans are so good at perceiving objects via echolocation that they can even tell the difference between objects of similar size but different shapes.

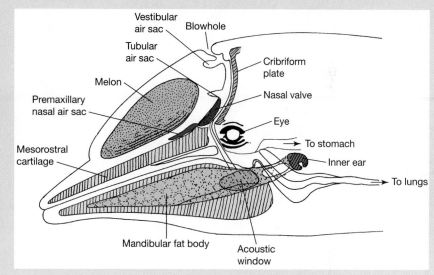

▶ **FIGURE 8–14** Sound production and reception by the bottlenosed dolphin (*Tursiops truncatus*). Sound is produced by moving air past the nasal valve. The sound reflects off the bony cribriform plate at the front of the skull and is focused into beam by the oil in the melon. Some sound may also be guided by the mesorostral cartilage. Echoes are channeled to the otherwise isolated and fused middle and inner ear via the mandibular fat body.

amniotes. Even within nonmammalian amniotes, differences in anatomy suggest that evolution of the middle ear occurred independently in turtles, lizards, and archosaurs.

■ Olfaction

The olfactory receptors responsible for the sense of smell are located in the olfactory epithelium in the nasal passages of tetrapods, and air passes over the olfactory epithelium with each breath. The receptors can be extraordinarily sensitive, and some chemicals can be detected at concentrations below 1 part in 1 million trillion (10^{15}) parts of air. Among tetrapods, mammals probably have the greatest olfactory sensitivity, and the area of the olfactory epithelium in mammals is increased by the presence of scrolls of

thin bone called the turbinates (**Figure 8–16**). Primates, including humans, have a relatively poor sense of smell because our snouts are too short to accommodate large turbinates and an extensive olfactory epithelium.

Tetrapods have an additional chemosensory system located in a unique organ of olfaction in the anterior roof of the mouth—the vomeronasal organ or Jacobson's organ. When snakes flick their tongues in and out of their mouth they are capturing molecules in the air and transferring them to this organ. Many male ungulates (hoofed mammals) sniff or taste the urine of a female, a behavior that permits them to determine the stage of her reproductive cycle. This sniffing is usually followed by flehmen, a behavior in which the male curls the upper lip and often holds his head high, probably inhaling molecules of

(a) Frog

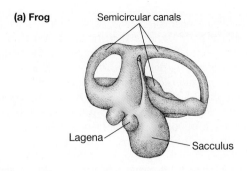

(b) Lizard

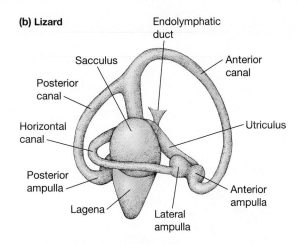

▲ **FIGURE 8–15** The vestibular apparatus of tetrapods. (a) Generalized amphibian condition. Note the small lagena for hearing airborne sounds. (b) Generalized nonmammalian amniote condition with a larger lagena.

pheromones into the vomeronasal organ. Primates, with their relatively flat faces, were thought to have lost their vomeronasal organs, but some recent work suggests the presence of a remnant of this structure in humans that is used for pheromone detection.

Proprioception—Where Are Your Parts?

Aquatic vertebrates don't have long appendages, and their appendages have relatively little range of movement in relation to the body. Their heads are attached to their pectoral girdles, and their fins move either from side to side or forward and back. That is not true of terrestrial animals, which have necks and limbs that can move in three dimensions with respect to the body. It is important for a terrestrial vertebrate to know where all the parts of the body are, and proprioception provides that information. (It is the proprioceptors in your arm that enable you to touch your finger to your nose when your eyes are closed.) Proprioceptors include muscle spindles, which detect the amount of stretch in the muscle, and ten-

Generalized mammalian condition

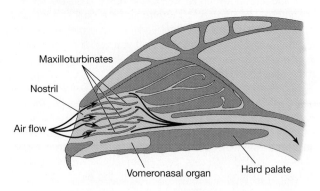

Ungulate condition

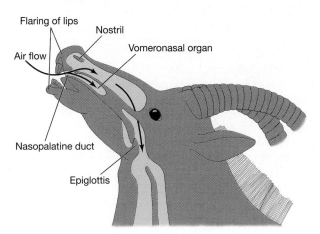

▲ **FIGURE 8–16** Olfactory system of a mammal. (a) Generalized mammal, showing the positions of the turbinates and vomeronasal organ. (b) The flehmen behavior of an ungulate.

don organs, which convey information about the position of the joints. Muscle spindles are found only in the limbs of tetrapods, and they are important for determining posture and balance on land.

8.7 Conserving Water in a Dry Environment

Bony fishes are covered with scales, a remnant of the ancient dermal exoskeleton. Although most modern fishes have thin proteinaceous scales that do not contain mineralized tissues, the immediate ancestors of tetrapods were covered in heavy dermal scales containing layers of enamel, dentine, and bone like the scales of gars. These scales were mostly lost in the earliest tetrapods: only scales on the belly remained, and these had lost the enamel and dentine layers. Presumably they would have afforded some protection from abrasion on land.

In terrestrial environments, water is evaporated from the body surface and respiratory system as water vapor and lost through the kidneys as liquid water. The permeability of the skin of terrestrial vertebrates depends on its structure and varies from very high in most extant amphibians to very low in most amniotes. It is difficult to say what the skin of the earliest tetrapods would have been like. The very thin, glandular skin of modern amphibians appears to be a recent specialization that is associated with using the skin for gas exchange. Probably the early tetrapods had scaly skins that protected their bodies from abrasion and limited the rate of evaporative water loss. The epidermal cells of vertebrates synthesize keratin (Greek *keratin* = horn), which is an insoluble protein that ultimately fills those cells. The outer layer of the skin of vertebrates is composed of layers of keratinized epidermal cells, forming the stratum corneum (Latin *cornu* = horn). The stratum corneum is only a few cell layers deep in fishes and amphibians but many layers deep in the skins of amniotes. The keratinized cells resist physical wear; the presence of an insoluble protein has some waterproofing effect, but lipids in the skin are the main agents that limit evaporative water loss.

The sauropsid and synapsid lineages developed different solutions to the problem of minimizing water loss through the kidney, and the urinary systems of the two groups are compared in Chapter 11. What they have in common, however, is a urinary bladder—a saclike structure that receives urine from the kidney and voids it to the outside (see Figure 8–4 [number 13]). A bladder is a new feature of tetrapods, although some bony fishes have a bladderlike extension of the kidney duct, and a bladder was probably a primitive character of tetrapods. It may have served as a water recovery device in early tetrapods, as it still does in extant lissamphibians and some sauropsids. Amniotes have a new duct draining the kidney—the ureter—derived from the base of the archinephric duct. In most vertebrates the urinary, reproductive, and digestive systems reach the outside through a single common opening, the cloaca (**Figure 8–17**). Only in therian mammals (marsupials and placentals) is the cloaca replaced by separate openings for the urogenital and digestive systems. The penis is a conduit for urine only in therian mammals—in all other amniotes it is purely an intromittent organ, used to introduce sperm into the reproductive tract of the

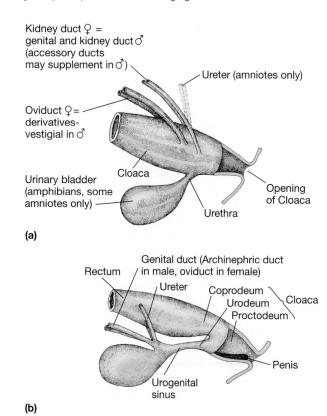

(a)

(b)

▲ **FIGURE 8–17** Anatomy of urogenital ducts in tetrapods. (a) Generalized jawed vertebrate condition, including nonamniotic tetrapods and some primitive amniotes. (b) More derived amniote condition (except for therian mammals), as illustrated by a male monotreme. The urogenital sinus is the name given to the urethra past the point where the genital duct has joined the system. In this example, the entire structure between the bladder and the cloaca is the urogenital sinus, but this is variable in different species of amniotes.

female so the egg can be fertilized before it is encased in a shell.

8.8 Controlling Body Temperature in a Changing Environment

An animal on land is in a physical environment that varies over small distances and can change rapidly. Temperature, in particular, varies dramatically in time and space in terrestrial environments, and the changes in environmental temperature have a direct impact on terrestrial animals, especially small ones, because they gain and lose heat rapidly.

This difference in the aquatic and terrestrial environments results from differences in the physical properties of water and air (see Box 4–1). The heat capacity of water is high, as is its ability to conduct heat. As a result, water temperatures are relatively stable and do not vary much from one place to another in a pond or stream. An aquatic animal has little capacity to change its body temperature by moving or to maintain a body temperature different from water temperature. As soon as an animal moves out onto the shore, however, it encounters a patchwork of warm and cool spots with temperatures that may differ by several degrees within a few centimeters. Terrestrial animals can select favorable temperatures within this thermal mosaic, and the low heat conductivity of air means that they can maintain body temperatures that are different from air temperature.

In fact, thermoregulation (regulating body temperature) is essential for most tetrapods because they encounter temperatures that are hot enough to kill them or cold enough to incapacitate them. In general, tetrapods maintain body temperatures that are higher than air temperature (some exceptions to this generalization are discussed in subsequent chapters), and to do this they need a source of heat. The heat used to raise the body temperature to levels permitting normal activity can come from the chemical reactions of metabolism (endothermy) or from basking in the sun or being in contact with a warm object such as a rock (ectothermy). The convergent evolution of endothermy in the synapsid and sauropsid lineages will be discussed in Chapter 11; here we will focus on ectothermy, which is the ancestral form of thermoregulation by tetrapods.

Ectothermy is the method of thermoregulation used by nearly all non-amniotes and by turtles, lepidosaurs, and crocodilians. Despite the evolutionary antiquity of ectothermy, it is a complex and effective way to control body temperature. Ectothermal thermoregulation is based on balancing the movement of heat between an organism and its environment. A study of thermoregulation of lizards by Raymond Cowles and Charles Bogert (1944) showed that they can maintain stable body temperatures with considerable precision and that many species have body temperatures very similar to those of birds and mammals.

■ Ectothermal Thermoregulation

A brief discussion of the pathways by which thermal energy moves between a living organism and its environment is necessary to understand the thermoregulatory mechanisms employed by terrestrial ectotherms. An organism can gain or lose energy by several pathways: solar radiation, thermal (infrared) radiation, convection, conduction, evaporation, and metabolic heat production. Adjusting the flow through various pathways allows an animal to warm up, cool down, or maintain a stable body temperature.

Figure 8–18 illustrates pathways of thermal energy exchange. Heat comes from both internal and external sources, and flow in some of the pathways can be either into or out of an organism.

- For most organisms, the sun is the primary source of heat, and solar energy always results in heat gain. Solar radiation reaches an animal directly when it is standing in a sunny spot. In addition, solar energy is reflected from clouds and dust particles in the atmosphere and from other objects in the environment, and reflected solar radiation reaches the animal by these paths. The wavelength distribution of solar radiation is the portion of the solar spectrum that penetrates the Earth's atmosphere. About half this energy is contained in the visible wavelengths (400 to 700 nanometers), and most of the rest is in the infrared region of the spectrum (i.e., wavelengths longer that 700 nanometers).

  Energy exchange in the infrared is an important part of the radiative heat balance. All objects, animate or inanimate, radiate energy at wavelengths determined by their absolute temperatures. Objects in the temperature range of animals and the Earth's surface (roughly −20°C to + 50°C) radiate in the infrared portion of the spectrum. Animals continuously radiate heat to the environment and receive infrared radiation from the environment. Thus, infrared radiation can lead to either gain or loss of heat, depending on the relative temperature of the animal's body surface and the environmental surfaces, as well as on the radiation characteristics of the surfaces themselves. In Figure 8–18, the lizard is cooler than the sunlit rock in front of it and receives more energy from the rock than it loses to the rock. However, the lizard is warmer than the shaded rock behind it and has a net loss of energy in that exchange. The radiative temperature of the clear sky is about 20°C, so the lizard loses energy by radiation to the sky.

- Heat is exchanged between objects in the environment and the air via convection—the transfer of heat between an animal and a fluid. Convection can result in either gain or loss of heat. If the air temperature is lower than an animal's surface

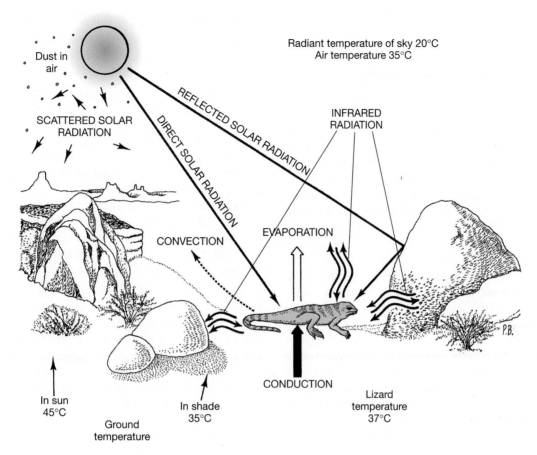

▲ FIGURE 8–18 Energy exchange. Energy is exchanged between a terrestrial organism and its environment by several pathways. These are illustrated in simplified form by a lizard resting on the floor of a desert arroyo. Small adjustments of posture or position can change the magnitude of the various routes of energy exchange and give a lizard considerable control over its body temperature.

temperature, convection leads to heat loss—in other words, it's a cooling breeze. If the air is warmer than the animal, however, convection results in heat gain. In still air, convective currents formed by local heating produce convective heat exchange. When the air is moving—that is, when there is a breeze—forced convection replaces natural convection, and the rate of heat exchange is greatly increased. In the example shown, the lizard is warmer than the air and loses heat by convection.

- Heat exchange occurs where the body and the substrate are in contact by conduction—the transfer of energy between an animal and a solid material. Conductive heat exchange resembles convection in that its direction depends on the relative temperatures of the animal and environment. It can be modified by changing the surface area of

the animal in contact with the substrate and by changing the rate of blood flow in the parts of the animal's body that are in contact with the substrate. In this example, the lizard gains heat by conduction from the warm ground.

- Evaporation of water occurs from the body surface and from the pulmonary system. Each gram of water evaporated represents a loss of about 2450 joules (the exact value changes slightly with temperature). Evaporation of water transfers heat from the animal to the environment and thus represents a loss of heat. (The inverse situation, condensation of water vapor on an animal, would produce heat gain, but it rarely occurs under natural conditions.)

- Metabolic heat production is the final pathway by which an animal can gain heat. Among ectotherms, metabolic heat gain is usually trivial in relation to the heat gained directly or indirectly from solar energy.

▉ Endothermal Thermoregulation

Endotherms (birds and mammals for the purpose of this discussion) exchange energy with the environment by the same routes as ectotherms. Everyone has had the experience of getting hot in the sun (direct solar radiation) and starting to sweat (evaporation). When you were in that situation you probably moved into the shade, and that is the same behavioral thermoregulatory response that a lizard would exhibit.

What is different about endotherms is the magnitude of their metabolic heat production. Endotherms have metabolic rates that are seven to ten times higher than those of ectotherms of the same body size. During cellular metabolism, chemical bonds are broken, and the energy in those bonds is captured in the bonds of other molecules, such as adenosine triphosphate (ATP). Metabolism is an energetically inefficient process, however, and only a portion of the energy released when a bond is broken is captured—the rest is released as heat. This wasted energy from metabolism is the heat that endotherms use to maintain their body temperatures at stable levels.

▉ Ectothermy versus Endothermy

Neither ectothermy nor endothermy can be considered the better mode of thermoregulation because each one has advantages and disadvantages:

- By producing heat internally, endotherms gain a considerable independence from environmental temperatures. Endotherms can live successfully in cold climates and can be nocturnal in situations that would not be possible for ectotherms. These benefits come at the cost of high energy (food) requirements, however.
- Ectotherms save energy by relying on solar heating, and an ectotherm eats less than an endotherm of the same body size. Because of that difference, ectotherms can live in places that do not provide enough energy to sustain an endotherm.

The examples of ectothermy and endothermy described here represent the extreme ends of a spectrum of thermoregulatory patterns that includes intermediate conditions. Fishes that maintain the body core at a temperature higher than that of the surrounding sea are examples of ectotherms with significant metabolic heat production, and some birds and mammals allow their body temperatures to fall during the night and then bask in the sun to warm up in the morning.

Summary

Life on land differs from life in water in a host of ways because the physical properties of water and air are so different. Aquatic vertebrates are nearly neutrally buoyant in water, so gravity is a small factor in their lives. The vertebral column of a fish need only resist lengthwise compression as trunk muscles contract. In contrast, the skeleton of a tetrapod has to support the animal's weight. Zygapophyses in the vertebral column of tetrapods transmit forces from one vertebra to the next, resisting the pull of gravity. The heads of tetrapods are not connected to their pectoral girdles like the heads of fishes, and the vertebral column shows regional modification in the shapes of vertebrae that are associated with functional specializations. The limbs of a tetrapod lift the body off the ground and push against the substrate as the animal moves. Extensive modifications of the limbs are associated with locomotor specializations, and several lineages have independently lost their limbs.

Air is easier to breathe than water because it is less dense and has a higher concentration of oxygen. Tidal respiration is feasible for air-breathing vertebrates, and movement of the ribs in amniotes creates a negative pressure in the abdominal cavity that can draw air into the lungs through a long neck. Suction feeding is ineffective in air, however, and tetrapods use their mobile heads to seize prey and muscular tongues to manipulate it in the mouth.

Even the cardiovascular system feels the effect of gravity because venous blood must be forced upward as it returns from regions of the body that are lower than the heart. Tetrapods have high blood pressures that are created by thick-walled muscular hearts. Separation of the oxygenated (arterial) and deoxygenated (venous) blood streams in the heart allows tetrapods to maintain higher blood pressures in the systemic portion of the circulatory system than in the lungs. When the separation of blood streams is produced by a septum that permanently divides the ventricle, the right side of the heart receives only deoxygenated blood, and coronary arteries are needed to bring oxygen to the heart muscle itself.

The difference in the function of sensory systems in air and water is profound. The cornea of the eye participates in focusing light on the retina of terrestrial vertebrates, and an image is focused by changing the shape of the lens rather than by moving the lens, as is the case in fishes. Air is not dense enough to activate the hair cells of a lateral line system. The hair cells of terrestrial vertebrates are found in the ear (in the organ of Corti), and hearing airborne sounds requires a lever system that amplifies sound-pressure waves as they are transmitted from air to the fluid in the inner ear. Somewhat surprisingly, the structural details of the ears of tetrapods show that hearing evolved independently in different lineages of terrestrial vertebrates. Chemosensation is as important to terrestrial vertebrates as it is to aquatic forms, but the receptor cells are internal. The vomeronasal system is a chemosensory system unique to tetrapods that is intimately involved with reproductive behaviors. Unlike fishes, tetrapods have mobile necks and long limbs that can move in three dimensions relative to the body. Locating the parts of the body in space is the function of the proprioceptive system.

Physical abrasion and evaporation of water through the skin are potential problems for terrestrial vertebrates, and the skin of the earliest tetrapods was probably covered by a stratum corneum containing keratinized epidermal cells that resisted abrasion and lipids that reduced water loss. Temperature varies far more on land than in water, both from one spot to another and from hour to hour. Aquatic animals have few options for selecting favorable temperatures, and most have little ability to maintain body temperatures that are different from water temperatures. In contrast, terrestrial animals can exploit a mosaic of temperatures created by patches of sunlight and shade, and they can have body temperatures that are very different from air temperatures.

Additional Readings

Akamatsu, T, et al. 2005. Biosonar behavior of free-ranging porpoises. *Philosophical Transactions of the Royal Society B: Biological Sciences*. 272:797–801.

Carroll, R. L. et al. 2005. Thermal physiology and the origin of terrestrialty in vertebrates. *Zoological Journal of the Linnean Society*. 143-345-358.

Clack, J. A. 2002. *Gaining Ground: The Origin and Evolution of Tetrapods*. Bloomington, IN: Indiana University Press.

Cowles, R. B., and C. M. Bogert. 1944. A preliminary study of the thermal requirements of desert reptiles. *Bulletin of the American Museum of Natural History* 83:261–296.

Domenici, P. et al. 2007. Environmental constraints upon locomotion and predator–prey interactions in aquatic organisms: an introduction. *Philosophical Transactions of the Royal Society B: Biological Sciences* 362:1929–1936.

Farmer, C. 1997. Did the lungs and intracardiac shunt evolve to oxygenate the heart in vertebrates? *Paleobiology* 23:358–372.

Harley, J. E. et al. 2003. Bottlenose porpoises perceive object features through echolocation. *Nature* 424:667–669.

Hillenius, W. J. 1992. The evolution of nasal turbinates and mammalian endothermy. *Paleobiology* 18:17–29.

Johnson, N. et al. 2008. Echolocation behavior adapted to prey in foraging Blainville's beaked whale (*Mesoplodon densirostris*). *Proceedings of the Royal Society B: Biological Sciences*. 275:133–139.

Jones, G. and M. W. Holdereid. 2007. Bat echolocation calls: adaptation and convergent evolution. *Proceedings of the Royal Society B: Biological Sciences*. 274:905–912.

Miller, P. O. et al. 2004. Sperm whale behaviour indicated the use of echolocation click buzzes 'creaks' in prey capture. *Proceedings of the Royal Society B: Biological Sciences*. 271:2239–2247.

Reilly, S. M. et al. 2006. Tuataras and salamanders show that walking and running mechanics are ancient features of tetrapod locomotion. *Proceedings of the Royal Society B: Biological Sciences*. 273:1563–1568.

Ulanosky, N. et al. 2004. Dynamics of jamming avoidance in echolocating bats. *Proceedings of the Royal Society B: Biological Sciences*. 271:1467–1475.

CHAPTER 9

Origin and Radiation of Tetrapods

By the Late Devonian period, the stage was set for the appearance of terrestrial vertebrates, whose origin can be found among the lobe-finned fishes. New fossil evidence shows that the earliest tetrapods were actually aquatic animals and that many of the anatomical changes that were later useful for life on the land were first evolved primarily in the water. The early tetrapods underwent a rapid radiation in the Carboniferous period: many were probably amphibious, but some lineages became secondarily fully aquatic, while others became increasingly specialized for terrestrial life. However, only one of the terrestrial lineages of Paleozoic tetrapods made the next major transition in vertebrate history, developing the embryonic membranes that define the amniotic vertebrates. Amniote diversification shows an initial early split between the synapsids, the lineage that includes mammals, and the sauropsids, the lineage that includes reptiles and birds.

9.1 Tetrapod Origins

Our understanding of the origin of tetrapods is advancing rapidly. The earliest known tetrapods are from the Late Devonian, some 360 million years ago. Until fairly recently, the genus *Ichthyostega* (originally found in East Greenland in 1932) was the only well-known representative of the earliest tetrapods. In the past couple of decades, however, we have discovered new material from this fossil site, including both skulls and skeletons of a different genus, *Acanthostega*, which was a more fishlike animal, and a new undescribed taxon has also been identified. Fragmentary fossil material of other Late Devonian tetrapods has also been found in Latvia, Scotland, Australia, Asia, and North America (see **Table 9.1**).

Analysis of the new specimens has focused on derived characters, and this perspective has emphasized the sequence in which the characteristics of tetrapods were acquired. The gap between fishes and tetrapods has narrowed, and the earliest tetrapods now appear to have been much more fishlike than we had previously realized. That information provides a basis for hypotheses about the ecology of animals at the transition between aquatic and terrestrial life.

The next stage in the history of tetrapods was their radiation into different lineages and different ecological types during the late Paleozoic and Mesozoic eras (**Figure 9–1**). By the Early Carboniferous, tetrapods had split into two lineages. One of these lineages is the batrachomorphs, which includes the temnospondyls—the largest and longest-lasting group of primitive, extinct non-amniotic tetrapods. Although the heyday of temnospondyls was in the Early Permian period, some lineages extended into the Cretaceous period, and at least some of the living amphibians are derived from temnospondyls. The second lineage is the reptilomorphs, which includes amniotes (reptiles, mammals, and birds) and their non-amniotic relatives, the anthracosaurs. Anthracosaurs were also at their peak in the Early Permian, but their diversity waned during the Late Permian; only one taxon survived into the earliest Triassic period.

TABLE 9.1 Major groups of Paleozoic non-amniotic tetrapods

Stem Tetrapods

Late Devonian taxa: e.g., *Acanthostega* and *Ichthyostega* from Greenland (see Figure 9–8a,b); *Metaxygnathus* (Australia, 1977—lower jaw); *Tulerpeton* (Russia, 1984—skeleton only); *Ventustega* (Latvia, 1994—fragments); *Hynerpeton* (Pennsylvania, North America, 1994—shoulder girdle and partial lower jaw); *Obruchevichthys* (Latvia, 1995—lower jaw); *Elginerpeton* (Scotland, 1995—skeletal fragments); *Dusignathus* (Pennsylvania, North America, 2000—lower jaw); *Sinostega* (China, 2002—lower jaw); unnamed form (Pennsylvania, North America, 2004—humerus); unnamed form (Belgium, 2004—lower jaw).

Enigmatic late Early Carboniferous taxa: e.g., *Pederpes* and *Crassigyrinus* (see Figure 9–8c) from Europe and *Whatcheeria* from North America.

Colosteidae: Aquatic late Early Carboniferous forms, possibly secondarily so, known from North America and Europe, with elongate, flattened bodies, small limbs, and lateral line grooves (e.g., *Greererpeton, Pholidogaster, Colosteus*).

Baphetidae (formerly Loxomattidae): Late Early and early Late Carboniferous forms from North America and Europe, with crocodile-like skulls and distinctive keyhole-shaped orbits (e.g., *Eucritta, Megalocephalus*).

Batrachomorphs

Temnospondyli: The most diverse, longest-lived group, ranging worldwide from the late Early Carboniferous to the Early Cretaceous. Possessed large heads with akinetic skulls. Paleozoic forms (e.g., *Eryops, Cacops*) were terrestrial or semiaquatic (see Figure 9–8d); Mesozoic forms (e.g., *Cyclotosaurus, Trematosaurus, Gerrothorax*) were all secondarily fully aquatic (see Figure 9–8e-g).

Reptilomorphs

Anthracosauroidea: The other diverse, long-lived group, although to a lesser extent than the temnospondyls. Known from the late Early Carboniferous to the earliest Triassic of North America and Europe. Anthracosauroids had deeper skulls than temnospondyls with prominent tabular horns, and they retained cranial kinesis. Some forms (e.g., *Gephyrostegus*) were terrestrial (see Figure 9–9b). Others, grouped together as embolomeres, were secondarily aquatic (e.g., *Pholiderpeton, Archeria;* see Figure 9–9a).

Seymouriamorpha: Known from the Permian only. Early Permian forms known from North America (e.g., *Seymouria;* see Figure 9–9c) were large and fully terrestrial. Later Permian forms known from Europe and China (discosauriscids and kotlassiids) were secondarily fully aquatic.

Diadectomorpha: Known from Late Carboniferous and the Early Permian of North America and Europe. Large, fully terrestrial forms, now considered to be the sister group of amniotes. Diadectidae (e.g., *Diadectes;* see Figure 9–9d) had laterally expanded cheek teeth suggestive of a herbivorous diet. Limnoscelidae and Tseajaiidae had sharper, pointed teeth and were probably carnivorous.

Lepospondyls

Microsauria: Distinguished by a single bone in the temporal series termed the tabular. Many (e.g., the tuditanomorphs) were terrestrial and rather lizardlike, with deep skulls and elongate bodies, but only had four toes (*microsaur* = small reptile; e.g., *Pantylus*, see Figure 9–10a). Some forms (microbranchomorphs) were probably aquatic. Known from Late Carboniferous and the Early Permian of North America and Europe.

Aïstopoda: Limbless forms, lacking limb girdles, with elongate bodies (up to 200 trunk vertebrae) and rather snakelike skulls that may have allowed them to swallow large prey items (e.g., *Lethicus, Ophiderpeton;* see Figure 9–10c). They may have been aquatic or have lived in leaf litter. Known from the Middle to Late Carboniferous of North America and Europe.

Adelogyrinidae: Limbless, long-trunked forms, but retaining the dermal shoulder girdle. Known from the Early Carboniferous of Europe.

Lysorophia: Elongate forms with greatly reduced limbs. Known from the Late Carboniferous and the Early Permian of North America.

Nectridia: Also rather elongate, but with a long tail rather than a long trunk. Distinguished by having fan-shaped neural and hemal arches in the vertebral column. Limbs small and poorly ossified, indicative of an aquatic mode of life. Some nectridians (keraterpetontids) had broad flattened skulls with enlarged tabular bones (e.g., *Diplocaulus;* see Figure 9–10b). Known from Late Carboniferous and Early Permian of North America, Europe, and North Africa.

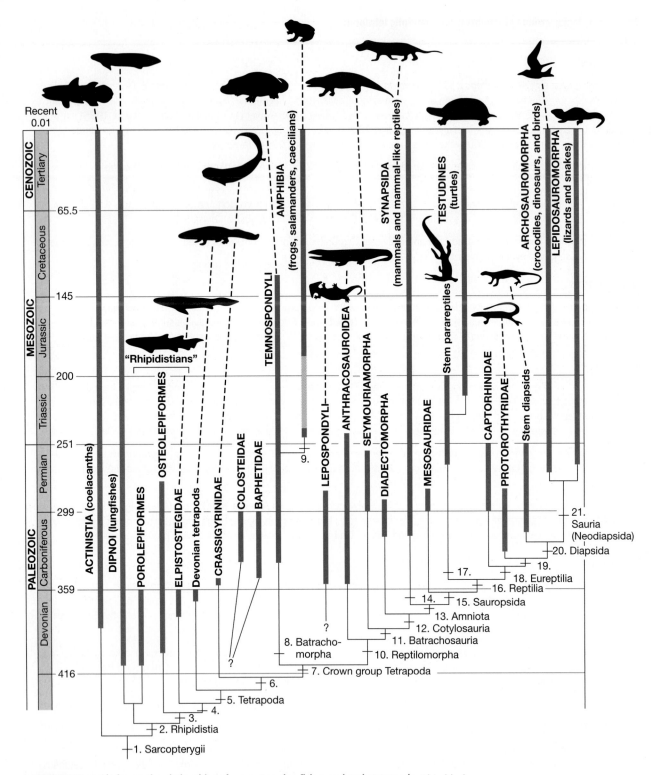

▲ **FIGURE 9–1** Phylogenetic relationships of sarcopterygian fishes and early tetrapods. Thin black lines show interrelationships only and are not indicative of the times of divergence of or the unrecorded presence of taxa in the fossil record. Lightly shaded bars indicate ranges of time when the taxon is known to be present but is unrecorded in the fossil record. Numbers indicate derived characters that distinguish the lineages. (An alternative interpretation of amphibian relationships is described in the text.)

Legend: 1. Sarcopterygii—fleshy pectoral and pelvic fins with a single basal element, muscular lobes at the base of those fins, true enamel on teeth, plus features of jaws and limb girdles. **2.** Rhipidistia—heart with separated pulmonary and systemic circulations. **3.** True choana (internal nostril), labyrinthine folding of tooth enamel, and details of limb skeleton. **4.** Tetrapodomorpha—flattened head with elongate snout, external nares situated on the margin of the mouth, orbits with eye ridges and on top of skull, body flattened, humerus with enlarged (deltoid) ridge, absence of dorsal and anal fins, enlarged ribs.

5. Tetrapoda—limbs with carpals, tarsals, and digits, vertebrae with zygapophyses, large ornamented interclavicle, iliac blade of pelvis attached to vertebral column, loss of contact between dermal skull and pectoral girdle. ("Devonian tetrapods" is a paraphyletic assemblage of Late Devonian genera, including [in order of more primitive to more derived] *Acanthostega*, *Ichthyostega*, and *Tulerpeton*, plus others known from fragmentary material.) **6.** Absence of anocleithrum (dermal bone in shoulder girdle), five or fewer digits. **7.** Crown group Tetrapoda—presence of occipital condyles (projections on skull for articulation with vertebral column), notochord excluded from braincase in adult. **8.** Batrachomorpha—skull roof attached to braincase via the exoccipital bones at back of skull, loss of primitive mobility within the skull (kinesis), only four fingers in hand. **9.** Lissamphibia—pedicellate teeth: (Lissamphibia includes the Salientia [frogs], the Caudata [salamanders], and the Apoda [caecilians].) **10.** Reptilomorpha—several skull characters, plus vertebrae with the pleurocentrum (posterior central element) as the predominant element. **11.** Batrachosauria—intercentrum (the anterior central vertebral element) reduced in size, enlarged caninelike tooth in maxilla (upper jaw). **12.** Cotylosauria—sacrum with more than one vertebra, robust claws on feet, more derived atlas-axis complex, plus other skull characters. **13.** Amniota—loss of labyrinthodont teeth, hemispherical and well-ossified occipital condyles, frontal bone contacts orbit in skull, transverse pterygoid flange present (reflects differentiation of pterygoideus muscle), three ossifications in scapulocoracoid (shoulder girdle), distinct astragalus bone in ankle. **14.** Synapsida—presence of lower temporal fenestra. **15.** Sauropsida—single centrale bone in ankle, maxilla separated from quadratojugal in skull, single coronoid bone in lower jaw. **16.** Reptilia—suborbital foramen in palate, tabular bone in skull small or absent, large post-temporal fenestra in skull. **17.** Parareptilia—loss of caniniform maxillary teeth, posterior emargination of skull, quadratojugal bone in skull expanded dorsally, expanded iliac blade in pelvis. ("Stem Parareptiles" includes the Late Permian Millerettidae and Pareiasauridae and the Late Permian and Triassic Procolophonidae. Opinions vary as to whether Testudines [turtles] are derived from pareiasaurs or from procolophonids—or even if they might be included with the diapsids.) **18.** Eureptilia—supratemporal bone in skull is small, parietal and squamosal bones in skull broadly in contact, tabular bone in skull not in contact with opisthotic, horizontal ventral margin of postorbital portion of skull, ontogenetic fusion of atlas pleurocentrum and axis intercentrum. **19.** Postorbital region of skull short, anterior pleurocentra keeled ventrally, limbs long and slender, hands and feet long and slender, metapodials overlap proximally. **20.** Diapsida—upper and lower temporal fenestrae present, exoccipitals not in contact on occipital condyle, ridge-and-groove tibia-astralagal joint. (Stem Diapsids is a paraphyletic assemblage of Late Carboniferous and Permian diapsids including [in order of more primitive to more derived] Araeoscelidia, Coelurosauravidae, and Younginiformes.) **21.** Sauria (Neodiapsida)—dorsal origin of temporal musculature, quadrate exposed laterally, tabular bone lost, unossified dorsal process of stapes, loss of caniniform region in maxillary tooth row, sacral ribs oriented laterally, ontogenetic fusion of caudal ribs, modified ilium, short and stout fifth metatarsal, small proximal carpals and tarsals.

By the start of the Cenozoic era, the only remaining non-amniotic tetrapods were the lineages of amphibians that we see today: frogs, salamanders, and caecilians. Amniotes have been the dominant tetrapods since the late Paleozoic. They have radiated into many of the terrestrial life zones that were previously occupied by non-amniotes and have developed feeding and locomotor specializations that had not previously been seen among tetrapods. Figure 9–1 shows a detailed phylogeny of primitive tetrapods, and **Figure 9–2** shows a simplified cladogram.

■ Fish-Tetrapod Relationships

Tetrapods are clearly related to the sarcopterygian (lobe-finned) fishes, which survive today as lungfishes, and the coelacanth. The discovery of lungfishes seemed to provide an ideal model of a prototetrapod—what more could one ask for than an air-breathing fish? However, lungfishes are very specialized animals, and many of their apparent similarities to tetrapods (such as the internal nostril, or choana) appear to have been evolved independently. The coelacanth lacks the specializations of lungfishes and, for a while after its discovery in 1938, was hailed as a surviving member of the group ancestral to tetrapods. However, most scientists now consider that the lungfishes are more closely related to tetrapods than is the coelacanth because coelacanths are primitive in a number of ways and also have many of their own specializations (e.g., they have converted the original bony fish lung into a swimbladder).

Both lungfishes and coelacanths have an extensive Paleozoic fossil record, along with a third group, called the "rhipidistians." Traditionally the coelacanths and the rhipidistians were grouped together as

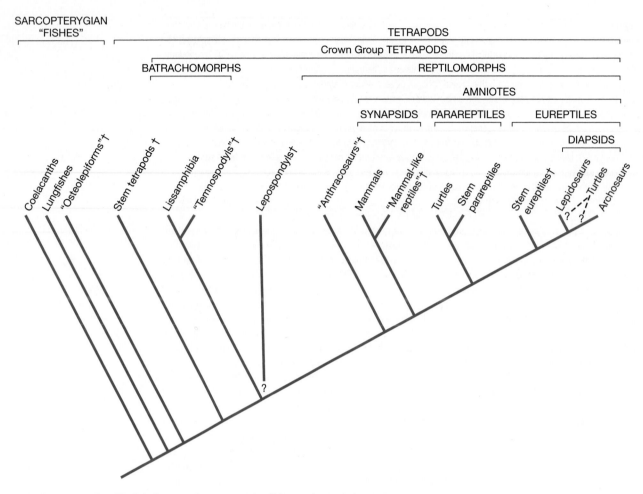

▲ **FIGURE 9–2** Simplified cladogram of sarcopterygian fishes and tetrapods. Quotation marks indicate paraphyletic groups. A dagger (†) indicates an extinct taxon. Note the alternative phylogenetic position of turtles.

the "crossopterygians," a group considered ancestral to tetrapods. We now consider the term crossopterygian invalid because it uses primitive characters to group together coelacanths and other sarcopterygian fishes lacking the specializations of lungfishes. The term *rhipidistian* in this context also has fallen into disuse because it, too, represents paraphyletic grouping, including some fishes that were more closely related to lungfishes (porolepiforms and rhizodontiforms) and others that were more closely related to tetrapods (osteolepiforms and elpistostegids). The term *Rhipidistia* is now used for the monophyletic grouping of lungfishes and tetrapods (see Figure 9–1).

The osteolepiforms were cylindrical-bodied, large-headed fishes with thick scales and apparently were shallow water predators. Many osteolepiforms resembled early tetrapods in having paired crescentic vertebrae and teeth with labyrinthine infolding of enamel (**Figure 9–3**). The most likely sister group of

tetrapods is a newly defined lineage of Late Devonian osteolepiforms called the Elpistostegidae (informally referred to as tetrapodomorph fishes), including the genera *Panderichthys*, *Livoniana*, and *Elpistostege*. Other osteolepiforms now form the sister group of elpistostegids plus tetrapods (see Figure 9–1). Note that both these other osteolepiforms and the elpistostegids are paraphyletic groupings, as some osteolepiforms are more closely related to the elpistostegids than others, and in turn some elpistostegids are more closely related than others to tetrapods.

Elpistostegids were more derived than osteolepiforms for shallow water life in that they had eyes on the top of their heads (in a crocodile-like fashion), had lost their dorsal and anal fins, and greatly reduced their tail fin. Their bodies and heads were dorsoventrally flattened, with ventrally projecting ribs, and their snouts were long, very much like the condition in the earliest tetrapods. Elpistostegids

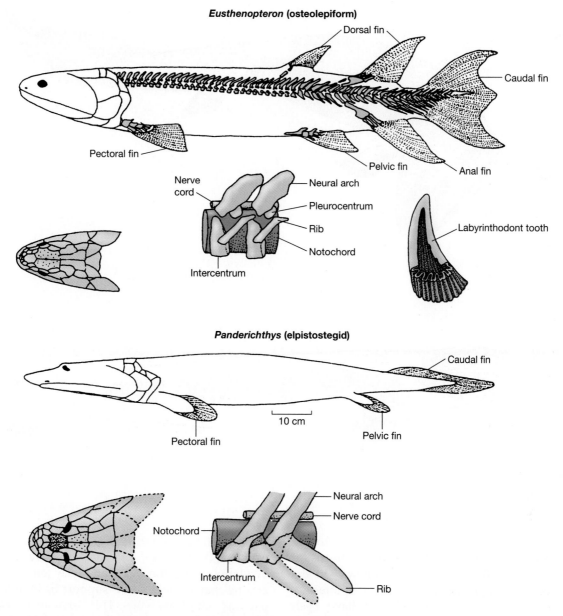

▲ **FIGURE 9–3** A Devonian osteolepiform and elpistostegid. The osteolepiform *Eusthenopteron* has a cylindrical body, a short snout, and four unpaired fins in addition to the paired pectoral and pelvic appendages. The elpistostegid *Panderichthys* has a dorsoventrally flattened body with a long, broad snout and eyes on top of the head. The dorsal and anal fins have been lost, and the caudal fin has been reduced in size. In the vertebral column of *Eusthenopteron* the ribs are short and probably extended dorsally. The ribs are larger in *Panderichthys* and project laterally and ventrally. In the skull roof of *Eusthenopteron* the area anterior to the parietals (sparsely stippled) is occupied by a single, median element (densely stippled). In the skull roof of *Panderichthys* there is a single pair of large frontal bones (densely stippled) immediately anterior to the parietals, as in tetrapods.

share with early tetrapods a derived form of the humerus (the upper arm bone) indicating powerful forelimbs that would be capable of propping the front end of the animal out of the water (Shubin et al. 2004).

A spectacular new find of a derived elpistostegid, from Ellesmere Island in the Canadian Arctic, has filled a morphological gap between the previously known most-derived sarcopterygian fishes and the first tetrapods (Daeshler et al. 2006, Shubin et al.

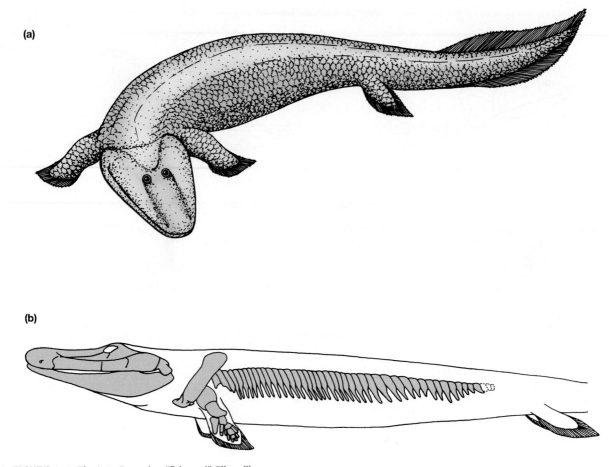

(a)

(b)

▲ **FIGURE 9–4** The Late Devonian "fishapod" *Tiktaalik*.

2006). This animal, *Tiktaalik* (pronounced with the accent on the second syllable; the name means "a large freshwater fish seen in the shallows" in the local Inuktitut language), is sufficient of an intermediate between fish and tetrapods to have been dubbed a "fishapod" (**Figure 9–4**). The fact that it retained fin rays means that it was definitely on the fish side of the transition, but it also possessed many derived, tetrapod-like features.

Tiktaalik is found in the early Late Devonian, around 385 million years ago, approximately 20 million years older than *Acanthostega* and *Ichthyostega* and 10 million years older than the oldest tetrapod fragments from Scotland and Latvia, but 2 to 3 million years younger than the elpistostegid *Panderichthys*. Perhaps the most tetrapod-like feature is the loss of the bony operculum; this change, which means that *Tiktaalik* would be able to raise its head above its body, was related to feeding on prey items outside of the water, snapping at them

with its long snout rather than sucking them in with water as other fishes do. The ribs were large and overlapping, like those of many early tetrapods, suggesting that *Tiktaalik* could support its body at least partway out of the water. This behavior is also suggested by the structure of its pectoral fin, which could bend in the middle to prop up the body in air. The pectoral fin also had an elaboration of the distal elements, not precisely homologous to those of the tetrapod wrist and hand but certainly providing an example of the kind of structures that might later be elaborated into a tetrapod limb. (Note that this was not unique to *Tiktaalik*; other "rhipidistians," like the more distantly related rhizodontid *Sauripterus*, were also apparently "experimenting" at this time with the development of fingerlike bones at the end of the fin.)

Features of *Tiktaalik* that remained definitely fishlike include well-developed gills, poorly ossified vertebrae, and a long body. The pelvic region is still unknown, although related fish had hindlimbs apparently used to anchor themselves in the substrate.

■ The Earliest Tetrapods

The new specimens of the Late Devonian *Acanthostega* and *Ichthyostega* from East Greenland have changed our views about early tetrapod biology. These fossils show that early tetrapods were primarily aquatic rather than terrestrial (Coates and Clack 1995). In addition, one of the most widespread features of tetrapods, the pentadactyl (five-fingered) limb, turns out not to be an ancestral character. Very recently more complete material of *Ventastega*, a tetrapod from Latvia previously known only from fragments, has shown it to be intermediate in its anatomy between *Acanthostega* and *Tiktaalik* (Ahlberg et al. 2008).

The evidence for an aquatic way of life for early tetrapods comes partly from the presence of a groove on the ventral surface of the ceratobranchials, part of the branchial apparatus supporting the gills of fishes. In fishes, this groove carries blood to the gills, and the presence of a similar groove on the ceratobranchials of *Acanthostega* and *Ichthyostega* strongly suggests that these tetrapods also had internal fishlike gills, which are different from the external gills found in the larvae of modern amphibians and in some adult salamanders. Additional evidence of internal gills in *Acanthostega* is provided by a flange, called the postbranchical lamina, on the anterior margin of the cleithrum, which is a bone in the shoulder girdle. In fishes, this ridge supports the posterior wall of the opercular chamber.

The picture of the earliest tetrapods that emerges from these features is of animals with fishlike internal gills that were capable of fishlike aquatic respiration. These animals probably also had lungs because lungs are present in lungfishes, the closest living relatives of tetrapods, and are probably a primitive osteichthyian feature.

Another unexpected feature of Devonian tetrapods is polydactyly—that is, having more than five toes. *Acanthostega* had eight toes on its front and hind feet, and *Ichthyostega* had seven toes on its hind foot (its fore foot is unknown) (**Figure 9–5**). Additionally, *Tulerpeton* (known from Russia) had six toes. These discoveries confound long-standing explanations of the supposed homologies of bones in the fins of sarcopterygian fins with those in tetrapod hands and feet, but they correspond beautifully with predictions based on the embryology of limb development (**Box 9–1**).

BOX 9–1   Early Feet

How a tetrapod limb could evolve from the fin of a sarcopterygian fish has been hotly debated for more than a century. In the nineteenth century, Gegenbaur tried to equate specific bones in fish fins with those in tetrapod limbs and suggested that extension and additional segmentation of the radials seen in the limb of the osteolepiform fish *Eusthenopteron* could produce a limb with digits like those seen in tetrapods (Figure 9–5a,b). Similar speculations have been applied to the fin skeletons of lungfishes (Figure 9–5c,d).

A new evolutionary perspective has been supplied by studies of the embryonic development. Recent work detailing the involvement of *Hox* genes in tetrapod limb formation is reviewed in Shubin et al. (1997) and Coates and Cohn (1998). *Hox* genes determine anteroposterior patterning in the vertebrate embryo. All jawed vertebrates have four different *Hox* gene clusters, labeled *Hox A–Hox D*. Limb development in vertebrates involves the 5' *Hox* genes that are expressed in the posterior part of these four *Hox* gene clusters, but similarities in gene expression between fishes and tetrapods extends only to the proximal (nearest to the body) part of the limb. The hands and feet are a new feature of tetrapods and were originally thought to be a tetrapod novelty, as nothing similar had been seen in extant fishes. Tetrapod limb buds have a phase of 5' *Hox D* expression that occurs at the end of the development of digits and affects the tips of the digits. This pattern of genetic expression is not apparent in teleost development, and thus the tetrapod condition was seen as being an evolutionary novelty. However, telosts are highly derived actinopterygian (ray-finned) fishes that have greatly reduced their fin structure. Examination of a more basal actinopterygian, the paddlefish *Polyodon*, revealed a similar expression of 5' *Hox D* genes in the fin to that seen in tetrapod hands and feet (Davis et al. 2007). Thus the origin of tetrapod digits is now seen as a result of a change in patterns of gene regulation rather than the evolution of new genes, with the absence of these genes in teleosts representing a secondary loss.

The limbs of all tetrapods, both forelimbs and hindlimbs, have a common sequence of events during development. In the forelimb, the humerus branches to form the radius (anteriorly) and the ulna (posteriorly). The developmental axis runs through the ulna. Subsequent development on the anterior side of the limb occurs (1) by segmentation of the radius, which produces the radiale (shown in Figure 9–5e) and sometimes two additional segments, and (2) by both branching and segmentation of the ulna. A preaxial branch of the ulna produces the proximal bones in the wrist—the intermedium and centrales. The carpals (distal wrist bones), metacarpals (bones of the palm), and digits (fingers) result from postaxial branching.

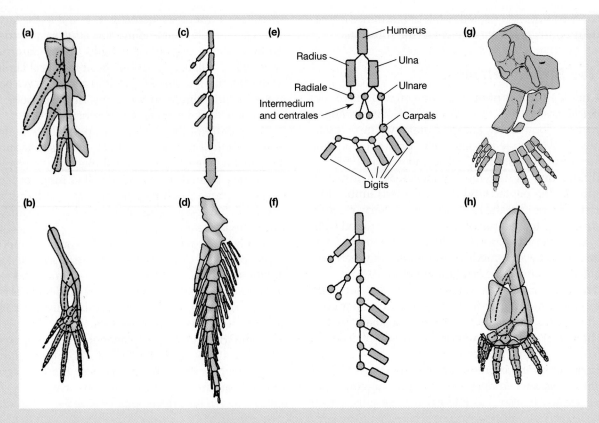

▲ **FIGURE 9–5** Three hypotheses of the origin of tetrapod limbs. In every case, the head of the animal is to the left. Thus, the preaxial direction (i.e., anterior to the axis of the limb) is to the left. (a) The pectoral fin skeleton of *Eusthenopteron*. The limb has a longitudinal axis (solid line) and preaxial radials. (b) Gegenbaur's nineteenth-century hypothesis of the origin of the vertebrate fore foot and fingers from preaxial radials. (c,d) The pectoral fin skeleton of the Australian lungfish, *Neoceratodus*. (c) The fin axis and the preaxial radials that appear early in embryonic development by branching from the axis. The radials of *Eusthenopteron* may have developed in the same way. (d) The postaxial radials seen in the adult fin of *Neoceratodus* develop by condensation of tissue, not by branching. (e,f) A diagram of the forelimb skeleton of a mouse during early development. (e) The proximal (nearest to the shoulder) parts of the limb skeleton consist of an axis with preaxial branches (radials) as seen in *Eusthenopteron* and *Neoceratodus*; compare the upper part of (e) with (a) and (c). The digits, however, are formed as postaxial branches. (f) The embryonic limb skeleton of a mouse, straightened for comparison with *Eusthenopteron* (a) and *Neoceratodus* (c). (g) The forelimb of *Acanthostega* showing the polydactylous condition of eight digits. (h) The hindlimb of *Ichthyostega* showing the inferred position of the axis (solid line) and radials. Note that the tibia and fibula and the first two rows of tarsal bones in the foot are formed by preaxial radials, whereas the third row of tarsals, the metatarsals, and the digits originate as postaxial branches.

For pentadactyl tetrapods, the formation of digits starts with digit 4 and concludes with digit 1 (which is the thumb of humans). Digit 5 (our little finger) forms at different times in different lineages, and one of the interesting features of this developmental process is the ease with which more or fewer than five digits can develop. If the process of segmentation and branching continues, a polydactylous foot is produced with the extra digits forming beyond the thumb. If the developmental process is shortened, fewer than five digits are produced, and the thumb is the first digit to be lost.

Changes in the timing of development do not have to produce an all-or-none addition or loss of a digit; a reduction in size is common. Dogs, for example, have four well-developed digits (numbers 2 through 5), plus a vestigial digit (number 1) called a dewclaw. Many dogs are born without external dewclaws (a carpal may be present internally), and some breeds of dogs are required by their breed standards to have double dewclaws. This pattern of increase or reduction in the number of digits results from a change in the timing of development during evolution. Reduction in the number of digits in evolutionary lineages of birds and mammals is frequently associated with specialization for high-speed running—ostriches and some artiodactyls (e.g., antelope) have two digits, and some perissodactyls (horses) have a single digit. The embryonic process by which toes are formed explains why variation in the number of digits is so widespread.

These new discoveries leave us with a paradoxical situation: Animals with well-developed limbs and other structural features that suggest they were capable of locomotion on land appear to have retained gills that would function only in water. How does a land animal evolve in water?

■ Evolution of Tetrapod Characters in an Aquatic Habitat

Tantalizingly incomplete as the skeletal evidence is, it is massive compared to the information we have about the ecology of elpistostegids and early tetrapods. It is not possible even to be certain if the evolution of tetrapods occurred in purely freshwater habitats. Eurasia and Gondwana, the two landmasses from which the earliest tetrapods are known, are believed to have been separated by marine environments in the Devonian. Tetrapods may have evolved in brackish or saline lagoons or even in marine habitats.

How Does a Land Animal Evolve in Water? It is clear from the fossil record that tetrapod characteristics did not evolve because they would someday be useful to animals that would live on land; rather, they evolved because they were advantageous for animals that were still living in water.

Elpistostegids were large fish—as much as a meter long—with heavy bodies, long snouts, and large teeth. They presumably could breathe air by swimming to the surface and gulping or by propping themselves on their pectoral fins in shallow water to lift their heads to the surface. Their lobed fins may have been useful in slow, careful stalking behavior in dense plants on the bottom of a lagoon or other body of water. A group of living fishes, the frogfishes, provides a model for the usefulness of a tetrapod-like limb in water. The pectoral fins of frogfishes are modified into structures that look remarkably like the limbs of tetrapods (**Figure 9–6**) and are used to walk over the substrate. An analysis of frogfish locomotion revealed that they employ two gaits that are used by tetrapods—a walk and a slow gallop.

What Were the Advantages of Terrestrial Activity? The classic theory of the evolution of tetrapods proposed that the Devonian was a time of seasonal droughts. Fishes trapped in shrinking ponds during the dry season are doomed unless the next rainy season begins before the pond is completely dry. We know that the

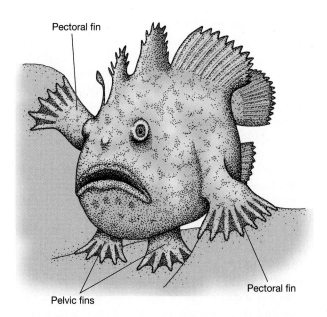

▲ **FIGURE 9–6** The frogfish *Antennarius pictus.* The small pelvic fins are in an anterior position but are not fused. When the animal walks, the left and right pelvic fins make contact with the substrate independently, allowing the gait of the fish to be compared with the gaits of tetrapods.

living African and South American lungfishes cope with this situation by estivating in the mud of their dry pond until the rains return. Perhaps the limblike fins of Devonian sarcopterygians allowed them to crawl from a drying pond and move overland to larger ponds that still held water. Could millions of years of selection of the fishes best able to escape death by finding their way to permanent water produce a lineage showing increasing ability on land?

This hypothesis, popular several decades ago, has been criticized on several grounds and almost certainly does not represent what actually happened in evolution. After all, a fish that succeeds in moving from a drying pond to one that still holds water has enabled itself to go on leading the life of a fish. Juvenile *Ichthyostega* and *Acanthostega* might have congregated in shallow water (as do juveniles of living fishes and amphibians) to escape predation from deeper water fishes and then ventured out onto land. Because the earliest tetrapods were relatively large animals, the smaller body size of a juvenile would have greatly simplified the difficulties of support, locomotion, and respiration in the transition from an aquatic to a terrestrial habitat. Other hypotheses for coming onto land include searching for food, dispersal of juveniles, laying eggs in moist environments, and basking in the sun to elevate body temperature. (Note that these hypotheses are not mutually exclusive.) Specialized behavioral and morphological adaptations to life on land can be found among several species of living teleost fishes, such as the mudskippers, climbing perches, and walking catfishes, which make extensive excursions out of the water—even climbing trees and capturing food on land.

Many of the anatomical changes seen in the transition between fish and tetrapods can be interpreted as adaptations to life in a shallow-water habitat. For example, limbs with digits could be useful in navigating bottom vegetation, the development of ankles and wrists may increase manipulative ability, and the attachment of the pelvic girdle to the vertebral column may provide support for a hindlimb-propelled predatory lunge under water. The development of a distinct neck, with loss of the opercular bones and the later gain of a specialized articulation between the skull and the vertebral column (not yet present in the earliest tetrapods), may be related to lifting the snout out of water to breathe air or to snap at prey items. A longer, flatter snout would also be of use here—similar modifications are seen in some living fish that gulp air, and these features were also seen in the derived fish *Tiktaalik*.

Note also that other problems exist in the transition from water to land apart from breathing air and supporting the body's weight. One particular issue is that of kidney function. The gills of fishes have other functions besides gas exchange—they also are the site of monovalent ion (sodium and chloride) regulation and of nitrogen loss (in the form of ammonia). However, gills do not work in air; and, even if they were strengthened so that they did not collapse in air, they would be a liability to a terrestrial animal as they would inadvertently serve as a site for water loss. With the loss of the gills in post-Devonian tetrapods, the kidney now assumes these functions, excreting nitrogen in the form of urea. The kidney also acts to regulate acid-base balance as the blood is now likely to become more acidic with the relative difficulty of getting rid of carbon dioxide on land (see Janis and Farmer 1999).

9.2 Radiation and Diversity of Non-amniotic Paleozoic Tetrapods

For over 200 million years, from the Late Devonian to the Early Cretaceous, non-amniotic tetrapods (excluding the groups of modern amphibians) radiated into a great variety of terrestrial and aquatic forms. Non-amniotic tetrapods are often called amphibians, but that term is now reserved for the extant non-amniotic tetrapods, the amphibians (frogs, salamanders, and caecilians). It is misleading to think of the primitive Paleozoic tetrapods as being amphibians for several reasons. First, many of them were much larger than any living amphibians and would have been more crocodile-like in appearance and habits. Second, they lacked the specializations of modern amphibians. For example, many forms had dermal scales, making it unlikely that they relied on their skin for respiration, as do modern amphibians. Finally, and most important, many of them were actually more closely related to amniotes than to amphibians.

Table 9–1 lists the different types of Paleozoic non-amniotic tetrapods, and Figure 9–1 illustrates a current consensus of their interrelationships. Note that some authors prefer to restrict the term *tetrapod* to the crown group, which encompasses only extant taxa and extinct taxa that fall within the range of characters seen in the extant taxa. Thus, under this scheme in the cladogram portrayed in Figure 9–1, the taxonomic term *Tetrapoda* would be shifted to node 7.

■ Devonian Tetrapods

Ichthyostega and *Acanthostega* from Greenland have been known since the 1930s, but only in the past couple of decades have other Devonian tetrapods been described—and are known from less complete material. These taxa, the skeletal material that they are known from, and their year of discovery are summarized in Table 9–1. These animals ranged from about 0.5 to 1.2 meters in length, and they differ enough from each other to show that by the end of the Devonian, some 7 million years after their first appearance, tetrapods had diversified into several niches. *Tulerpeton* and *Hynerpeton* appear to have been more terrestrial than *Acanthostega* or *Ichthyostega*. *Acanthostega* was more aquatic than *Ichthyostega* (**Figure 9–7**), with short ribs and wrists and ankles that would have been incapable of bearing weight on land.

We have already seen how new discoveries of *Acanthostega* have transformed our ideas about the earliest tetrapods, and new studies of *Ichthyostega* have revealed that this animal is anything but the standard "first tetrapod" that is often portrayed. While *Acanthostega* appears to be genuinely primitive, *Ichthyostega* is specialized in its own unique fashion and has modifications that appear to be for

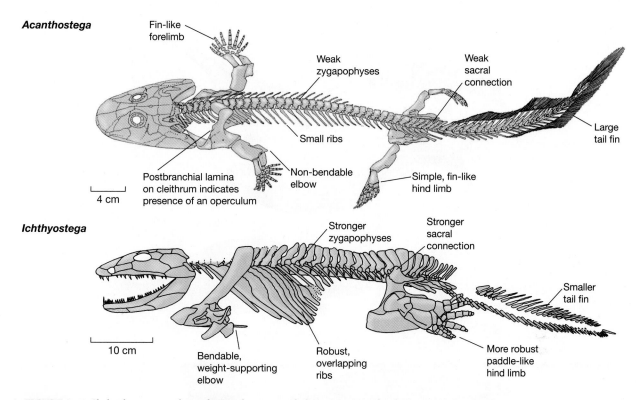

▲ FIGURE 9–7 Skeletal reconstructions of Devonian tetrapods from East Greenland. The notations illustrate the more aquatic nature of *Acanthostega* (The fore foot of *Ichthyostega* is unknown.)

both aquatic and terrestrial life (Ahlberg et al. 2005). Unlike most other tetrapods, and in fact rather resembling the mammals, its trunk vertebrae were regionally specialized so that it had distinct "thoracic" and "lumbar" portions. The ribs were longer and more overlapping than previously realized and were confined to the thoracic region (**Figure 9–8**). These ribs would have allowed the body to be elevated above the ground but would have prohibited the type of fishlike lateral flexion that was retained in *Acanthostega* and is maintained in many tetrapods even today. The lumbar region is interpreted as allowing mammal-like dorsoventral flexion; and, with the rigid ribs and the strange form of the backbone, *Ichthyostega* may have inch-wormed its way along, using first both forelimbs together and then both hindlimbs, rather as sea lions hump their way along the beach. However, *Ichthyostega* also had elongated ribs at the base of the tail, which may have provided anchorage for muscles that powered the tail during swimming, a hindlimb that is shaped more like a seal's flipper than a leg for walking on land, and an ear region that seems to indicate specialization for underwater hearing analogous to that seen today in certain teleost fishes.

▣ Carboniferous-Mesozoic Non-amniotic Tetrapods

General Patterns of Radiation The groups listed in Table 9–1 are all well established; the difficulty is trying to understand how these different groups are related to one another and to the modern groups of tetrapods, the amphibians, and amniotes. A major problem is that we are missing a critical piece of the geological record of tetrapod history. Although fossils are known from the Late Devonian, the subsequent record is an almost complete blank for 20 to 30 million years, with virtually no further fossils known until the later part of the Early Carboniferous. However, some more taxa from this time period have been found recently. *Pederpes* was about the same size as *Acanthostega*, with feet that appeared to be adapted for terrestrial life and functionally five-toed (with just the remnants of an extra finger on the hand) (Clack 2002b). In the beginning of the Early Carboniferous, we find the long-bodied, rather snakelike lepospondyl *Lethicus*, which had lost its limbs completely. However, by the late Early Carboniferous we find many tetrapod taxa, ranging from a few centimeters

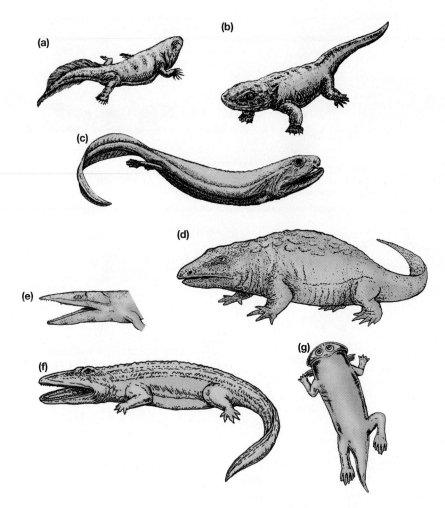

▲ **FIGURE 9–8** Stem tetrapods and temnospondyls. (a) *Acanthostega*, an aquatic Late Devonian stem tetrapod. (b) *Ichthyostega*, an aquatic Late Devonian stem tetrapod. (c) *Crassigyrinus*, an aquatic Early Carboniferous stem tetrapod. (d) *Eryops*, a semiterrestrial Early Permian eryopoid temnospondyl. (e) *Trematosaurus*, an aquatic (marine) Early Triassic trematosaurid temnospondyl. (f) *Cyclotosaurus*, an aquatic Middle Triassic capitosaurid temnospondyl. (g) *Gerrothorax*, an aquatic Late Triassic plagiosaurid temnospondyl. *Eryops* was around 2 meters long (the size of a medium-size crocodile). The other animals are drawn approximately to scale.

to a few meters in length and displaying a great diversity of feeding and locomotor types.

This gap in the fossil record may bias our understanding of how the groups known from the later Carboniferous are related to one another. The interrelationship of early tetrapods shown in Figure 9–1 broadly agrees with the 1994 review by Ahlberg and Milner; but, since that time, a plethora of new phylogenies has been suggested, most offering dramatically different opinions. The discovery of more Early Carboniferous tetrapod fossils is crucially needed to help confirm or refute the various competing hypotheses. For the purposes of this volume, we will treat the phylogeny in Figure 9–1 as a working hypothesis.

The Paleozoic tetrapods were originally divided into groups called "labyrinthodonts" and "lepospondyls" (see Table 9–1). Labyrinthodonts were mainly larger forms (large lizard- to crocodile-size, with a skull at least 5 centimeters long; see Figure 9–8 and **Figure 9–9**), with a multipartite vertebral centrum, and teeth with complexly infolded enamel (labyrinthodont teeth). Lepospondyls were small forms (small lizard- or salamander-size, with a skull less than 5 centimeters long; **Figure 9–10**) with a single, spool-shaped vertebral centrum and without the

▲ **FIGURE 9–9** "Anthracosaur" (non-amniotic reptilomorph) tetrapods. (a) *Pholiderpeton*, an aquatic Late Carboniferous embolomere. (b) *Gephyrostegus*, a terrestrial Late Carboniferous anthracosauroid. (c) *Seymouria*, a terrestrial Early Permian seymouriamorph. (d) *Diadectes*, a terrestrial Early Permian diadectomorph. (e) *Westlothiana*, an Early Carboniferous form, probably closely related to amniotes. *Seymouria* is around 1 meter long (the size of a golden retriever). The other animals are drawn approximately to scale (*Diadectes* should be a little larger and *Westlothiana* a little smaller).

labyrinthine form of enamel. The various groups of lepospondyls are probably related to each other, despite the fact that the distinguishing features just mentioned probably relate merely to smaller body size, but how they are related to other early tetrapods remains a point of controversy. The large early tetrapods were a diverse taxonomic grouping; some were stem tetrapods, some were closely related to modern amphibians, and others were more closely related to amniotes. For this reason the term *labyrinthodont* is no longer formally employed but is still in common use in popular texts.

Current consensus recognizes two main groups of Paleozoic large non-amniotic tetrapods: the "temnospondyls" (paraphyletic because they probably contain the origin of at least some modern amphibians; see Figure 9–8) and the "anthracosaurs" (the paraphyletic grouping of non-amniotic reptilomorphs: anthracosauroids, seymouriamorphs and diadectids; see Figure 9–9). Temnospondyls were in general more aquatic, characterized by flat, immobile skulls and a reduction of the hand to four fingers. Anthracosaurs were in general more terrestrial, are characterized by domed skulls retaining some kinetic ability, and had a five-fingered hand. The affinities of the lepospondyls (see Figure 9–10) are open to question. Figure 9–1 shows them as a monophyletic group of uncertain phylogenetic position, in

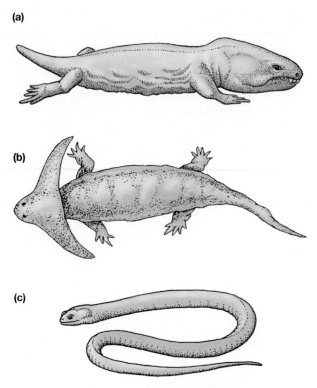

(a)

(b)

(c)

▲ **FIGURE 9–10** Lepospondyl tetrapods. (a) *Pantylus*, a terrestrial Early Permian microsaur. (b) *Diplocaulus*, an aquatic Early Permian nectridian. (c) *Ophiderpeton*, an aquatic (or possibly terrestrial burrowing form) Late Carboniferous aïstopod. *Pantylus* is around 20 centimeters long (the size of a hamster). The other animals are drawn approximately to scale.

between temnospondyls and anthracosaurs; however, lepospondyls may be more closely related to the living amphibians or even to the amniotes.

The origins of the modern amphibians (Lissamphibia) are also in a state of controversy and debate. Some current hypotheses (e.g., Ruta et al. 2003) propose that they are a monophyletic group derived from within the temnospondyls (as shown in Figure 9–1). The most recent hypothesis was proposed by Robert Carroll (Carroll 2007), who has made an exhaustive study of the features of modern amphibians and their possible fossil relatives, including study of larval forms. He concluded that frogs and salamanders are derived from within the dissorophoid temnospondyls but from different families: frogs from the amphibamids (which include many highly terrestrial forms) and salamanders from the branchiosaurids (which include many forms with persistent larval characters as adults). In contrast, caecilians appear to be rather different animals entirely and most closely related to certain microsaurs (lepospondyls). In this new arrangement,

the amphibians are no longer a monophyletic group with a single origin from the Paleozoic non-amniotic tetrapods.

Non-amniotic tetrapods reached their peak of generic diversity in the Late Carboniferous and Early Permian, when they consisted of fully aquatic, semi-aquatic, and terrestrial forms found over central and western Europe and eastern North America. Most lineages were extinct by the mid-Permian period. The only groups to survive past the earliest Mesozoic were the ancestors of the modern groups of amphibians and the fully aquatic types of temnospondyls. Temnospondyls were extinct in most of the world by the end of the Triassic but persisted into the Early Cretaceous in Australia. Although all living amphibian groups had their origins in the Mesozoic, the generic diversity of non-amniotic tetrapods did not return to Permian levels until the mid-Cenozoic era.

Ecological and Adaptive Trends One of the most striking aspects of early tetrapods is the number and diversity of forms that returned to a fully aquatic mode of life. Among living amphibians, many salamanders and some frogs are fully aquatic as adults, and this was also true of a diversity of Paleozoic forms (see Table 9–1). Many of these forms independently acquired an elongated body with the reduction or loss of the limbs, a morphology that may also be associated with burrowing, as seen in the living caecelian amphibians.

Some of the most bizarre aquatic forms were found among the lepospondyls. Some nectridians had broad, flattened skulls with enlarged tabular bones (see Figure 9–10b). These tabular horns were up to five times the width of the anterior part of the skull, and skin imprints show that they were covered by a flap of skin extending back to the shoulder (not shown in the figure). These horns may have acted as a hydrofoil to help in underwater locomotion, and they may have supported highly vascularized skin to help in underwater respiration.

The temnospondyls were the only group of non-amniotic tetrapods (aside from the amphibians) to survive the Paleozoic, and all of the Mesozoic forms were large, flattened, fully aquatic predators, such as capitosaurids and plagiosaurids (see Figure 9–8f,g). Some trematosaurids (Figure 9–8e) evolved the elongated snout characteristic of specialized fish eaters and are found in marine beds, making them the only known fully marine non-amniotic tetrapods. How did these animals osmoregulate in the marine environment? Even if the adults had evolved a reptilelike impermeable skin, the larvae still would have had gills. Perhaps trematosaurids retained high levels of urea to raise their internal osmotic pressure, as some modern estuarine frogs do (see Section 4.4).

In contrast to the temnospondyls, the anthracosaurs appear to have been predominantly terrestrial as adults, and many have been mistaken for early reptiles (especially animals such as *Seymouria* and *Diadectes*). Terrestriality also evolved convergently among other early tetrapods, predominately in the microsaurs (see Figure 9–10), and the dissorophid temnospondyls that may have been ancestral to frogs. These animals acquired skeletal adaptations such as longer, more slender limb bones.

9.3 Amniotes

Amniotes include most of the tetrapods alive today. Their name refers to the amniotic egg, which is one of the most obvious features distinguishing living amniotes from living amphibians. Amniotes appeared somewhat later in the fossil record than the earliest tetrapods of the Devonian, but they seem to have been well established by the time of the later radiation of non-amniotic tetrapods, although they were a minor part of the Carboniferous fauna. Their initial major radiation occurred in the Permian.

The first known candidates for the status of amniote—or near amniote—are from the Early Carboniferous of Scotland, and they are only 20 million years younger than the earliest known tetrapods. These include the mouse-size *Casineria* and the slightly younger salamander-size *Westlothiana* (see Figure 9–9e), both discovered in the 1990s. They appear to have been small, agile animals, most likely with insectivorous diets, rather resembling present-day lizards. Whether or not these animals turn out to be true early amniotes, they are certainly representative of what the first amniotes would have been like—small and more terrestrial than other early tetrapods.

Amniotes soon began to radiate into many of the life zones originally occupied by non-amniotic tetrapods, and all of these animals at this time were carnivorous (including fish- and invertebrate-eaters). No living adult amphibian is herbivorous, and there is little evidence in the fossil record to suggest that Paleozoic non-amniotic tetrapods were herbivores. (The Permian *Diadectes*, the probable sister taxon to amniotes, is the only exception to this generalization.) A key event in the radiation of early tetrapods may thus have been the great diversification of insects in the Late Carboniferous, probably in response to the increasing quantity and diversity of terrestrial vegetation. Probably for the first time in evolutionary history, the food supply was adequate to support a diverse fauna of fully terrestrial vertebrate predators; but, from the start of the Mesozoic onward, terrestrial habitats were dominated by a series of radiations of amniote tetrapods (**Table 9.2** and **Figure 9–11**).

■ Derived Features of Amniotes

Traditionally, amniotes have been distinguished by the amniotic egg (sometimes called the *cleidoic egg*; Greek *cleido* = closed or locked) and a waterproof skin. The amniotic egg is characteristic of turtles, lepidosaurs (lizards and their relatives), crocodilians, birds, and monotremes (egg-laying mammals). Further, embryonic membranes that contribute to the placenta of therian mammals (marsupials and placentals) are homologous to certain membranes in the egg. The amniotic egg is assumed to have been the reproductive mode of Mesozoic diapsids, and fossil dinosaur eggs are relatively common. In many other ways, however, amniotes represent a more derived kind of tetrapod than either the living amphibians or the Paleozoic non-amniotic tetrapods.

Skin permeability varies widely among living amphibians and amniotes. Although amniotes have a thicker skin than amphibians and a keratinized epidermis, it is the presence of lipids in the skin that makes the skin relatively impermeable to water. Compared to amphibians, amniotes have a greater variety of skin elaborations—scales, hair, and feathers—all formed from keratin. The lack of similar structures in living amphibians may be related to their use of the skin in respiration. Another important derived amniote feature is costal (rib) ventilation of the lungs. Because amniotes rely on the lungs for gas exchange, the skin need not be moist, and cutaneous water loss is reduced.

Costal ventilation has other consequences. It allows an animal to have a long neck because movement of the ribs can produce a pressure differential large enough to draw air down a long, thin tube, such as the trachea in the necks of amniotes. In addition, some of the muscles involved in buccal pumping insert onto the shoulder girdle, and this arrangement also may limit the development of a long neck.

A longer neck provides space for elaboration of the nerves supplying the forelimb. Nerves supplying the forelimb leave the spinal cord in the neck and

▲ **FIGURE 9–11** Diversity of Paleozoic amniotes. Early amniotes varied in size from a few centimeters long to a couple of meters, and their ecological roles were equally diverse. (a) *Hylonomus*, a protorothyridid (lizard-size), represents the typical lizardlike body form of many early amniotes. (b) *Haptodus*, a synapsid (dog-size). (c) *Mesosaurus*, a mesosaur (cat-size). (d) *Captorhinus*, a captorhinid (lizard-size). (e) *Petrolacosaurus kansensis*, a stem diapsid (araeoscelidan; lizard-size). (f) *Procolophon*, a procolophonid (dog-size). (g) *Pareiasaurus*, a pareiasaur (cow-size). (See Table 9–2 for more details).

join together in a nerve complex called the brachial plexus. (There is a similar sacral plexus for nerves supplying the hindlimb.) The brachial plexus in living amphibians is simple; only two nerves are involved, in contrast to the plexus of amniotes, which has at least five nerves. Thus, amniotes have more complex innervation of the forelimb, which improves control of the limb and the ability for manipulation. This example shows how anatomical features may be linked together in evolution in unexpected ways because the animals evolve as an integrated whole. Who would suspect that our distant ancestors' using their ribs to ventilate the lungs could be linked to our ability to use our hands for tasks such as writing?

■ Structure of the Amniotic Egg

An amniotic egg is a remarkable example of biological complexity (**Figure 9–12**). The shell, which may be leathery and flexible (as in many lizards and turtles and in monotreme mammals) or calcified and rigid (as in other lizards, turtles, crocodiles, and birds), provides mechanical protection while being porous enough to allow movement of respiratory gases and water vapor. The albumin (egg white) gives further protection against mechanical damage and provides a reservoir of water and protein. The large yolk is the energy supply for the developing embryo. At the beginning of embryonic development, the embryo is represented by a few cells resting on top of the yolk.

TABLE 9.2 Major groups of Paleozoic amniotes

Synapsida

Synapsids, or "mammal-like reptiles" are discussed in Chapter 18. Early synapsids (Figure 9–11b) were somewhat larger than early eureptiles, and their larger heads and teeth suggest a more specialized carnivorous habit.

Sauropsida

Mesosaurs: The first secondarily aquatic amniotes (Figure 9–11c). Known from freshwater deposits in the Early Permian of South Africa and South America, they provide one of the classic pieces of evidence for continental drift because these continents were united in Gondwana at that time. Swimming adaptations include large, probably webbed, hind feet, a laterally flattened tail, and heavily ossified ribs that may have acted as ballast in diving. The long jaws and slender teeth may have been used to strain small crustaceans from the surrounding water.

Parareptilia

Millerettids: Rather like the eureptile protorothyridids shown in Figure 9–11a. Known from the Late Permian of South Africa.

Procolophonids: Medium size, with peglike teeth that were laterally expanded in later members of the group, apparently specialized for crushing or grinding, suggestive of herbivory (Figure 9–11f). Known worldwide (except Australia) from the Late Permian to Late Triassic.

Pareiasaurs. Large size, approaching 3 meters long (Figure 9–11g). Known from the Late Permian of Europe, Asia, and Africa. Their teeth were laterally compressed and leaf-shaped, like the teeth of herbivorous lizards. Pareiasaurs were evidently the dominant terrestrial herbivores of the later Permian.

Eureptilia

Protorothyridids: Small, relatively short-legged, rather lizardlike forms, probably insectivorous in habits (Figure 9–11a). Known from mid-Carboniferous to Early Permian in North America and Europe.

Captorhinids: Tetrapods with more robust skulls and flatter teeth than protorothyridids and early diapsids and may have had more of an omnivorous diet that required crushing (Figure 9–11d). Known from the Early and mid-Permian of North America and Europe and the Late Permian of East Africa.

Araeoscelidans (stem diapsids): Early diapsids with shorter bodies and longer legs than protorothyridids and probably also insectivorous (Figure 9–11e). Araeoscelidans were known from the Late Carboniferous and Early Permian of North America and Europe.

As development proceeds, these multiply, and endodermal and mesodermal tissue surrounds the yolk, enclosing it in a yolk sac that is part of the developing gut. Blood vessels differentiate rapidly in the mesodermal tissue surrounding the yolk sac and transport food and gases to the embryo. By the end of development, only a small amount of yolk remains, and this is absorbed before or shortly after hatching.

While all vertebrates have an extraembryonic membrane (or membranes) enclosing the yolk sac, amniotes have three additional extraembryonic membranes—the **chorion, amnion,** and **allantois.** The chorion and amnion develop from outgrowths of the body wall at the edges of the developing embryo and spread outward and around the embryo until they meet. At their junction, the membranes merge and leave an outer membrane (the chorion), which sur-

rounds the entire contents of the egg, and an inner membrane (the amnion), which surrounds just the embryo itself. The allantoic membrane develops as an outgrowth of the hindgut posterior to the yolk sac and lies within the chorion. The allantois appears to have evolved as a storage place for nitrogenous wastes produced by the metabolism of the embryo, and the urinary bladder of the adult grows out from its base. The allantois also serves as a respiratory organ during later development because it is vascularized and can transport oxygen from the surface of the egg to the embryo and carbon dioxide from the embryo to the surface. The allantois is left behind when the embryo emerges, so the nitrogenous wastes stored in it do not have to be reprocessed. The embryo in an amniotic egg bypasses the larval stage typical of amphibian embryos and does not form

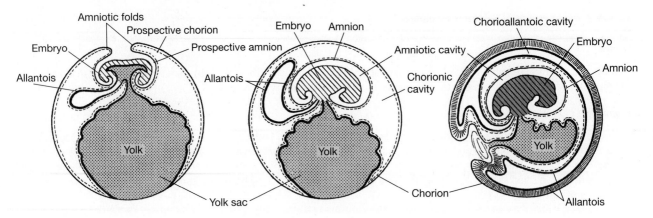

▲ **FIGURE 9–12** Distinctive features of the amniotic egg. Progressive stages in development are illustrated from left to right.

external gills at any stage in development. All traces of the lateral line are also lost. This loss of the larval form relates to one significant disadvantage of the amniotic egg—it can no longer be laid in water because the gill-less embryo would drown. Marine amniotes must either come ashore to lay eggs, such as sea turtles and penguins, or be viviparous, such as sea snakes and marine mammals.

How and why might the amniotic egg have evolved? It is not essential for development on land because many species of living amphibians, some fishes, and many invertebrates lay non-amniotic eggs that develop quite successfully on land. Both amniotic and non-amniotic eggs must be laid in relatively moist conditions to avoid desiccation, and both types of eggs are usually buried in the soil or deposited under objects such as rocks and logs. (Birds, which represent a highly derived condition, are an exception to this generalization.)

Various plausible explanations for the development of the amniote egg have been proposed. For example, the extraembryonic membranes may improve within-egg respiratory capacities, and the shell may provide mechanical support on land; together these features would allow the evolution of a large egg that produced a large hatchling that in turn grew into a large adult because egg size is related to adult size. However, the truth is that we do not really understand what evolutionary forces would have led to the first amniote eggs, even though this kind of egg was doubtless important in the later evolution of amniotes (see Skulan 2000).

How could we tell if a fossil animal laid an amniotic egg when features such as extraembryonic membranes are not preserved in the fossil record? We can estimate the latest point of origin of the amniote egg from the tetrapod phylogeny (see Figure 9–1). The synapsids (mammals and their extinct relatives) branched off from other amniotes very early, and all other fossil animals that we consider to be amniotes are more closely related to sauropsids (living reptiles and birds) than to mammals. Because the egg membranes of mammals are homologous with those of other living amniotes, all tetrapods higher than node 13 in the cladogram must have inherited an amniotic egg from the common ancestor of mammals and other amniotes.

A more difficult question is whether any fossil tetrapods lower down in the phylogeny might have laid an amniotic egg. We know that this was not true of seymouriamorphs because larval forms with external gills and lateral lines are known. That leaves the diadectomorphs as the only possible candidates for the earliest amniotic vertebrates. Diadectomorphs such as *Diadectes* have teeth that look like teeth of herbivorous lizards, and only amniote tetrapods are herbivorous (see the following discussion). In addition, they were much larger than any terrestrial non-amniotes, and those two characters may suggest that they laid amniotic eggs.

■ Patterns of Amniote Temporal Fenestration

Amniotes have traditionally been subdivided by the number of holes in their head—that is, on the basis of **temporal fenestration** (Latin *fenestra* = window). The major configurations that give names to different lineages of amniotes are **anapsid** (Greek *an* = without and *apsid* = junction), seen in primitive amniotes and in turtles; **synapsid** (single arch, Greek *syn* = joined), seen in mammals and their ancestors; and **diapsid** (double arch, Greek *di* = two), seen in birds and other reptiles. The term *arch* refers to the temporal bars lying below and between the holes. **Figure 9–13** illustrates the different patterns in the different groups.

Note that the phylogenetic pattern of fenestrae suggests that temporal openings arose independently in the synapsid and diapsid lineages because early sauropsids lack holes entirely. Turtles have

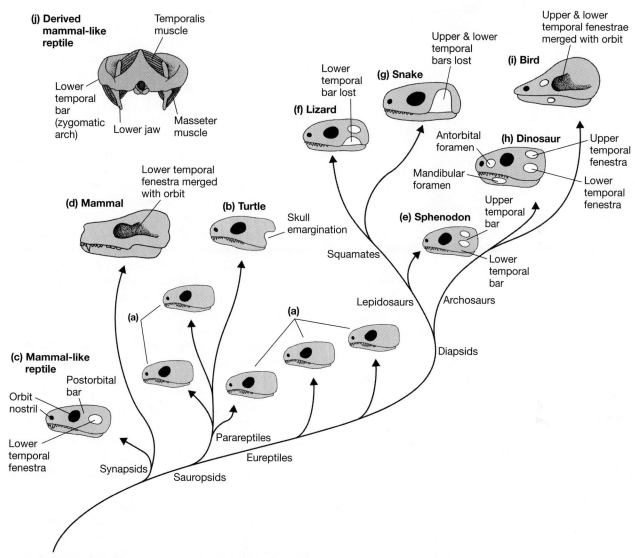

▲ **FIGURE 9–13** Patterns in amniote skull fenestration. (a) Primitive anapsid condition, as seen in the common ancestor of all amniotes and in basal members of the parareptiles and eureptiles. (b) Modified anapsid condition with emargination of the posterior portion of the skull, as seen in turtles. (Turtles may be secondarily anapsid—see text.) (c) Primitive synapsid condition, with lower temporal fenestra only. (d) Derived mammalian synapsid condition, in which the orbit has become merged with the temporal opening and dermal bone has grown down from the skull roof to surround the braincase. (e) Primitive diapsid condition, as seen today in the reptile *Sphenodon*; both upper and lower temporal fenestrae are present. (f) Lizardlike condition, typical of most squamates, where lower temporal bar has been lost. (g) Snake condition; upper temporal bar has been lost in addition to the lower bar. (h) Primitive archosaur diapsid condition, as seen in thecodonts and most dinosaurs; an antorbital foramen and a mandibular foramen have been added to the basic diapsid pattern (note that the antorbital foramen is secondarily reduced or lost in crocodiles). (i) Derived avian archosaur condition, convergent with the condition in mammals—the orbit has become merged with the temporal openings, and the braincase is enclosed in dermal bone. (j) Posterior view through the skull of a synapsid (a cynodont mammal-like reptile) showing how the temporal fenestra allows muscles to insert on the outside of the skull roof. The temporalis and masseter muscles are divisions of the original amniote adductor muscle complex.

(a) Nonamniote tetrapod

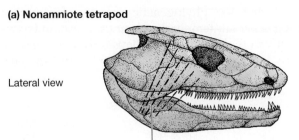

Lateral view

Undifferentiated adductor muscle mass

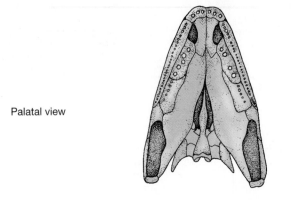

Palatal view

(b) Amniote (captorhinid)

Adductor mandibularis muscle

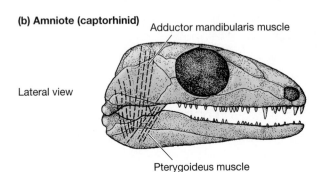

Lateral view

Pterygoideus muscle

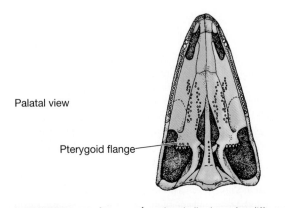

Palatal view

Pterygoid flange

▲ **FIGURE 9–14** Jaw muscles. The skulls show the differences in jaw muscles of non-amniotic and amniotic tetrapods.

traditionally been classified with the other anapsids, as shown in Figure 9–1 and **Figure 9–14**, but molecular data suggests that their origins may lie within the diapsid radiation (see Figure 9–2), which implies that they secondarily have roofed in their skull from an originally fenestrated condition.

Even though the skull of living mammals is highly modified from the primitive synapsid condition, you can still feel these skull features in yourselves. If you put your hands on either side of your eyes, you can feel your cheekbone (the zygomatic arch)—that is the temporal bar that lies below your synapsid skull opening. Then, if you clench your jaw, you can feel the muscles bulging above the arch. They are passing through the temporal opening, running from their origin from the top of the skull to the insertion on your lower jaw.

What is the function of these holes? As you just discovered, they provide room for muscles to bulge. Amniotes have larger and more complex jaw muscles than non-amniotes, and the notion of room for bulging was originally the preferred evolutionary explanation. However, Frazzetta (1968) pointed out that the *initial* evolutionary reason for developing these holes must have been something different. Only a large hole will allow enough room for a bulging muscle, so what could be the evolutionary advantage of the initial, small hole? And why does no non-amniote ever develop temporal fenestration?

Frazzetta suggested that the key to the evolution of the temporal fenestrae lies in changes in the complexity and orientation of the jaw-closing (adductor) muscles. The large, flat skull of non-amniotes, which may be related to their buccal-pumping mode of respiration (in which the skull is acting as a pair of bellows), does not permit a change in the orientation of the jaw muscles from the basic fish condition (see Figure 9–14a). A muscle originating from the skull roof, like the amniote adductor mandibularis, would be too short to allow the jaw to open wide because muscles can only be stretched for one third of their resting length. With the evolution of costal ventilation, the head size and shape is no longer important: amniotes were now able to evolve smaller, more domed skulls, allowing differentiation of the simple fishlike jaw adductors into the adductor mandibularis and the pterygoideus (see Figure 9–14b). The pterygoideus originates from a distinct pterygoid flange on the base of the skull, which is a characteristic feature of amniotes, revealing that a change in jaw musculature has occurred.

The advantage of this change in musculature would be a change in the feeding abilities. Fish and non-amniotic tetrapods can only close their jaws with a single snap (inertial feeding), whereas amniotes can snap the jaws closed and also apply pressure with the teeth when the jaws are closed (static pressure feeding). This difference may have allowed more complex types of feeding in amniotes, such as the ability of herbivores to nip off vegetation with their front teeth. Occlusion between the upper and lower teeth is seen for the first time in *Diadectes* and the amniotes (Reisz 2006). Dental occlusion is probably also related to the ability to be herbivorous, as herbivores require more oral processing to break down tough food. The change in jaw musculature also may be the reason why amniotes have lost the labyrinthodont teeth of large early non-amniotic tetrapods, as they would have more fine control over their jaw movements and thus would not need such heavily reinforced teeth to resist inertial snapping.

However, there could be a biomechanical problem with the origin of the adductor mandibularis from the underside of the skull roof. These muscles would have a 90-degree orientation to the bone of attachment, resulting in considerable stress on the thin covering of the bone (the periosteum). A way of compensating for this would be to leave portions of the dermal skull roof partially open in development in the areas where three bones meet and to have the muscle originate in part from the connective tissue covering the fenestra (**Figure 9–15**). In this way, a small hole would still reflect an important structural change to reduce stresses on the skull. Later evolutionary developments would then include the enlargement of this hole, with the muscles running through the hole to attach to the outside of the skull roof as they do in extant amniotes. Perhaps differences in muscle actions, relating to different feeding styles, encouraged temporal fenestration to take a different form in the synapsids and diapsids groups.

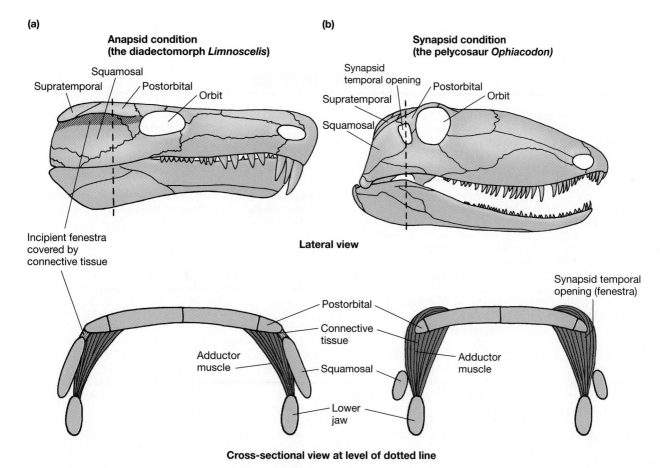

▲ **FIGURE 9–15** Hypothetical origin of the temporal fenestra in amniotes.

Summary

The origin of tetrapods from elpistostegid lobe-finned fishes in the Devonian is inferred from similarities in the bones of the skull and braincase, vertebral structure, and limb skeleton. Paleozoic tetrapods comprise about a dozen distinct lineages of uncertain relationships. One current view divides them into three major groups: batrachomorphs, including the predominantly aquatic temnospondyls; reptilomorphs, including the predominantly terrestrial anthracosaurs; and the lepospondyls. Temnospondyls radiated extensively in the Late Carboniferous and Permian, and several lineages extended through the Triassic into the Early Cretaceous. At least some of the modern amphibians—salamanders and frogs—may be derived from the temnospondyl lineage. The anthracosaurs radiated during the Carboniferous and became extinct by the earliest Triassic. Amniotes may be derived from this group. The lepospondyls were small forms of uncertain phylogenetic affiliation and may contain the ancestry of the modern caecilians.

The amniotic egg, with its distinctive extraembryonic membranes, is a shared derived character that distinguishes the amniotes (turtles, lepidosaurs, crocodilians, birds, and mammals) from the non-amniotes (fishes and amphibians). The earliest amniotes were small animals, and their appearance coincided with a major radiation of terrestrial insects in the Carboniferous. By the end of the Carboniferous, amniotes had begun to radiate into most of the terrestrial life zones that had been occupied by non-amniotes, and only the relatively aquatic groups of non-amniotic tetrapods maintained much diversity through the Triassic.

The major groups of amniotes can be distinguished by different patterns of temporal fenestration—holes in the dermal skull roof that reflect increasing complexity of jaw musculature. The major division of amniotes is into synapsids (mammals and their relatives) and sauropsids (reptiles and birds).

Additional Readings

Ahlberg, P. E. and A. R. Milner. 1994. The origin and early diversification of tetrapods. *Nature* 368:507–514.

Ahlberg, P. E. et al. 2005. The axial skeleton of the Devonian tetrapod *Ichthyostega*. *Nature* 437:137–140.

Ahlberg, P. E. et al. 2008. *Ventastega curonica* and the origin of tetrapod morphology. *Nature* 437:1199–1204.

Boisvert, C. A. 2005. The pelvic fin and girdle of *Panderichthys* and the original of tetrapod locomotion. *Nature* 438:1145–1147.

Carroll, R. L. 2001. The origin and early radiation of terrestrial vertebrates. *Journal of Paleontology* 75:1202–1213.

Carroll, R. L. 2007. The Palaeozoic ancestry of salamanders, frogs and caecilians. *Zoological Journal of the Linnean Society* 150 (Suppl. 1):1–140.

Clack, J. A. 2002a. *Gaining Ground: The Origin and Evolution of Tetrapods*. Bloomington, IN: Indiana University Press.

Clack, J. A. 2002b. An early tetrapod from "Romer's Gap." *Nature* 418:72–76.

Clack, J. A. 2006. The emergence of early tetrapods. *Palaeogeography, Palaeoclimatology, Palaeoecology* 232:167–189.

Coates, M. I., and J. A. Clack. 1990. Polydactyly in the earliest known tetrapod limbs. *Nature* 347:66–69.

Coates, M. I., and J. A. Clack. 1991. Fishlike gills and breathing in the earliest known tetrapod. *Nature* 352:234–235.

Coates, M. I., and J. A. Clack. 1995. Romer's gap: Tetrapod origins and terrestriality. *Bulletin de Musée National Histoire Naturelle, Paris, 4ᵉ series* 17:373–388.

Coates, M. I., and M. J. Cohn. 1998. Fins, limbs, and tails: Outgrowths and axial patterning in vertebrate evolution. *BioEssays* 20:371–381.

Daeschler E. B. et al. 2006. A Devonian tetrapod-like fish and the evolution of the tetrapod body plan. *Nature* 440:757–763.

Davis, M. C. et al. 2007. An autopodiallike pattern of *Hox* expression in the fins of a basal actinopterygian fish. *Nature* 447:473–476.

Frazzetta, T. H. 1968. Adaptive problems and possibilities in the temporal fenestration of tetrapod skulls. *Journal of Morphology* 125:145–158.

Janis, C. M., and C. Farmer. 1999. Proposed habits of early tetrapods. Gills, kidneys, and the water-land transition. *Zoological Journal of the Linnean Society* 126:117–126.

Janis, C. M., and J. C. Keller. 2001. Modes of ventilation in early tetrapods: Costal aspiration as a key feature of amniotes. *Acta Palaeontologica Polonica* 46:137–170.

Laurin, M. et al. 2000. Early tetrapod evolution. *Trends in Ecology and Evolutionary Biology* 15:118–123.

Long, J. A., and M. S. Gordon. 2004. The greatest step in vertebrate history: A paleobiological review of the fish-tetrapod transition. *Physiological and Biochemical Zoology* 77:700–719.

Reisz, R. R. 1997. The origin and early evolutionary history of amniotes. *Trends in Ecology and Evolution* 12:218–222.

Reisz, R. R. 2006. Origin of dental occlusion in tetrapods: signal for vertebrate evolution? *Journal of Experimental Zoology (Mol Dev Evol)* 306B:261–277.

Ruta, M. et al. 2003. Early tetrapod relationships revisited. *Biological Reviews* 78:251–345.

Shubin, N. H. et al. 1997. Fossils, genes and the evolution of animal limbs. *Nature* 388:639–648.

Shubin, N. H. et al. 2004. The early evolution of the tetrapod humerus. *Science* 304:90–93.

Shubin, N. H. et al. 2006. The pectoral fin of *Tiktaalik roseae* and the origin of the tetrapod limb. *Nature* 440:764–771.

Skulan, J. 2000. Has the importance of the amniote egg been overstated? *Zoological Journal of the Linnean Society* 130:235–261.

Sumida, S. S., and K. L. M. Martin (eds.). 1997. *Amniote Origins: Completing the Transition to the Land*. San Diego, CA: Academic.

CHAPTER 10

Salamanders, Anurans, and Caecilians

The three lineages of extant amphibians (salamanders, frogs, and caecilians) have very different body forms, but they are identified as a monophyletic evolutionary lineage by several shared derived characters. Some of these characters—especially the moist, permeable skin—have channeled the evolution of the three lineages in similar directions. Frogs are the most successful amphibians, and it is tempting to think that the variety of locomotor modes permitted by their specialized morphology may be related to their success: frogs can jump with simultaneous movements of the hind legs, swim with either simultaneous or alternating leg movements, and walk or climb with alternating leg movements. In contrast, salamanders retain the ancestral tetrapod locomotor pattern of lateral undulations combined with alternating limb movements in a walking-trot.

The range of reproductive specializations of amphibians is nearly as great as that of fishes, a remarkable fact considering that there are more than five times as many species of fishes as amphibians. The ancestral reproductive mode of amphibians probably consisted of laying large numbers of eggs that hatched into aquatic larvae, and many amphibians still reproduce this way. An aquatic larva gives a terrestrial species access to resources that would not otherwise be available to it. Modifications of the ancestral reproductive mode include bypassing the larval stage, viviparity, and parental care of eggs and young, including females that feed their tadpoles.

The permeable skin of amphibians is central to many aspects of their lives. The skin is a major site of respiratory gas exchange and must be kept moist. Evaporation of water from the skin limits the activity of most amphibians to relatively moist microenvironments. The skin contains glands that produce substances used in courtship, as well as other glands, which produce toxic substances that deter predators. Many amphibians advertise their toxicity with bright warning colors, and some nontoxic species deceive predators by mimicking the warning colors of toxic forms.

10.1 Amphibians

The extant amphibians, or Amphibia, are tetrapods with moist, scaleless skins. The group includes three distinct lineages: anurans (frogs), urodeles (salamanders), and gymnophionans (caecilians or apodans). Most amphibians have four well-developed limbs, although a few salamanders and all caecilians are limbless. Frogs lack tails (hence the name *anura*, which means "without a tail"), whereas most salamanders have long tails. The tails of caecilians are short, as are those of other groups of elongate, burrowing animals.

At first glance, the three lineages of amphibians appear to be very different kinds of animals: frogs have long hindlimbs and short, stiff bodies that don't bend when they walk; salamanders have forelimbs and hindlimbs of equal size and move with lateral

TABLE 10.1 Shared derived characters of amphibians

1. *Structure of the skin and the importance of cutaneous gas exchange.* All amphibians have mucus glands that keep the skin moist. A substantial part of an amphibian's exchange of oxygen and carbon dioxide with the environment takes place through the skin. All amphibians also have poison (granular) glands in the skin.

2. *Papilla amphibiorum.* All amphibians have a special sensory area, the papilla amphibiorum, in the wall of the sacculus of the inner ear. The papilla amphibiorum is sensitive to frequencies below 1000 hertz (Hz; cycles per second), and a second sensory area, the papilla basilaris, detects sound frequencies above 1000 Hz.

3. *Operculum-columella complex.* Most amphibians have two bones that are involved in transmitting sounds to the inner ear. The columella, which is derived from the hyoid arch, is present in salamanders, caecilians, and most frogs. The operculum develops in association with the fenestra ovalis of the inner ear. The columella and operculum are fused in anurans, caecilians, and some salamanders.

4. *Green rods.* Salamanders and frogs have a distinct type of retinal cell, the green rod. Caecilians apparently lack green rods; however, the eyes of caecilians are extremely reduced, and these cells may have been lost.

5. *Pedicellate teeth.* Nearly all modern amphibians have teeth in which the crown and base (pedicel) are composed of dentine and are separated by a narrow zone of uncalcified dentine or fibrous connective tissue. A few amphibians lack pedicellate teeth, and the boundary between the crown and base is obscured in some other genera. Pedicellate teeth also occur in some actinopterygian fishes, which are not thought to be related to amphibians.

6. *Structure of the levator bulbi muscle.* This muscle is a thin sheet in the floor of the orbit that is innervated by the fifth cranial nerve. It causes the eyes to bulge outward, thereby enlarging the buccal cavity. This muscle is present in salamanders and anurans and in modified form in caecilians.

undulations; and caecilians are limbless and employ snakelike locomotion. These obvious differences are all related to locomotor specializations, however, and closer examination shows that amphibians have many derived characters in common, indicating that they form a monophyletic evolutionary lineage (Table 10.1). We will see that many of these shared characters play important roles in the functional biology of amphibians. Perhaps the most important derived character of extant amphibians is a moist, permeable skin. The name applied to the lineage, *Lissamphibia*, refers to the texture of the skin (Greek *liss* = smooth). Many of the Paleozoic non-amniote tetrapods had dermal armor in the form of bony scutes in the skin; a permeable, unadorned skin is a derived character shared by amphibians.

All living adult amphibians are carnivorous, and relatively little morphological specialization is associated with different dietary habits within each group. Amphibians eat almost anything they are able to catch and swallow. The tongue of aquatic forms is broad, flat, and relatively immobile, but some terrestrial amphibians can protrude the tongue from the mouth to capture prey. The size of the head is an important determinant of the maximum size of prey that can be taken, and species of salamanders that occur in the same habitat frequently have markedly different head sizes, suggesting that this may be a feature that reduces competition for food. Frogs in the tropical American genera *Lepidobatrachus* and *Ceratophrys*, which feed largely on other frogs, have such large heads that they are practically walking mouths.

The anuran body form probably evolved from a more salamander-like starting point. Both jumping and swimming have been suggested as the mode of locomotion that made the change advantageous. Salamanders and caecilians swim as fishes do—by passing a sine wave down the body. Anurans have inflexible bodies and swim with simultaneous thrusts of the hind legs. Some paleontologists have proposed that the anuran body form evolved because of the advantages of that mode of swimming. An alternative hypothesis traces the anuran body form to the advantage gained by an animal that could rest near the edge of a body of water and escape aquatic or terrestrial predators with a rapid leap followed by locomotion on either land or water. The stem anuran *Triadobatrachus* may be an example of that body form.

The oldest fossils that may represent modern amphibians are isolated vertebrae of Permian age that appear to include both salamander and anuran types. The oldest true frogs are from the Early Jurassic period, and salamanders and caecilians also are known from the Jurassic. Clearly, the modern orders of amphibians have had separate evolutionary histories for a long time. The continued presence of such common characteristics as a permeable skin, after at least 250 million years of independent evolution, suggests that the shared characteristics are central to the lives of modern amphibians. In other characters, such as reproduction, locomotion, and defense, amphibians show tremendous diversity.

■ What's in a Name?

Cladograms display the branching sequences of evolutionary lineages so clearly that it can be difficult to remember that they are hypotheses, and, like other hypotheses, cladograms are tested repeatedly as new information becomes available. Sometimes new data show that the branching sequences in a cladogram are not correct; when that happens, the cladogram is redrawn. Because the names of evolutionary lineages are based on their branching sequences, these rearrangements may change the names of species, genera, or families.

A recent reanalysis of the phylogeny of amphibians has resulted in new generic names and family allocations for a large number of species. Darrel Frost of the American Museum of Natural History and a large number of colleagues have developed an online catalog of amphibians that presents current phylogenetic information (http://www.research.amnh.org/herpetology/amphibia/index.php). Because new information about the phylogenetic relationships of amphibians is published nearly daily, keeping this list up to date is a continuing project.

This catalog proposes new generic names for many familiar species of amphibians, some of which are discussed in this book. For example, the leopard frog that has been known to generations of biology students as *Rana pipiens* is *Lithobates pipiens* in the online catalog, and name of the American bullfrog has been changed from *Rana catesbeiana* to *Lithobates catesbeianus*.

The new names reflect advances in our understanding of the evolution of amphibians. Prior to the rearrangement, more than 20 species of pond and meadow frogs in North America were placed in the genus *Rana*, but recent studies have shown that this group is not a single evolutionary lineage. The species of *Rana* in eastern North America are derived from a lineage that radiated in North, Central, and South America, whereas the species in western North America are part of a radiation of frogs that extends from the temperate parts of Eurasia through Indochina to western North America. The generic names in the online catalog—*Lithobates* for the eastern species and *Rana* for the western species—recognize the separation of the lineages.

Changing the names of species that are studied in a wide variety of biological specialties can cause con-fusion. Furthermore, the names of species do not change in databases when new names are proposed, so the old names must still be used for literature searches—at this stage, a search for *Rana pipiens* will produce thousands of hits, whereas a search for *Lithobates pipiens* will produce only a few hundred. To minimize the confusion that can result from unfamiliar names, we will use both the new name and the former name the first time a species is mentioned.

■ Salamanders—Caudata or Urodela

The salamanders have the most generalized body form and locomotion of the living amphibians. Salamanders are elongate, and all but a very few species of completely aquatic salamanders have four functional limbs (**Figure 10–1**). Their walking-trot gait is probably similar to that employed by the earliest tetrapods. It combines the lateral bending characteristic of fish locomotion with leg movements. The eight families, containing approximately 556 species, are almost entirely limited to the Northern Hemisphere; their southernmost occurrence is in northern South America (**Table 10.2**). North and Central America have the greatest diversity of salamanders—more species of salamanders are found in Tennessee than in all of Europe and Asia combined. Paedomorphosis is widespread among salamanders, and several families of aquatic salamanders consist solely of paedomorphic forms. Paedomorphs retain larval characteristics, including larval tooth and bone patterns, the absence of eyelids, a functional lateral line system, and (in some cases) external gills.

The largest living salamanders are the Japanese and Chinese giant salamanders (*Andrias*), which reach lengths of 1 meter or more. The related North American hellbenders (*Cryptobranchus*) grow to 60 centimeters. All are members of the Cryptobranchidae and are paedomorphic and permanently aquatic. As their name indicates (Greek *crypto* = hidden and *branchus* = a gill), they do not retain external gills, although they do have other larval characteristics. Another group of large aquatic salamanders are the mudpuppies (*Necturus*), which consist of paedomorphic species that retain external gills. Mudpuppies occur in lakes and streams in eastern North America. The congo eels (three species of aquatic salamanders in the genus *Amphiuma*) live in the lower Mississippi Valley and coastal plain of the United States. They have well-developed lungs and can estivate in the mud of dried ponds for up to two years.

Several lineages of salamanders have adapted to life in caves. The constant temperature and moisture

▶ **FIGURE 10–1** Diversity of salamanders. The body forms of salamanders reflect differences in their life histories and habitats. Aquatic salamanders may retain gills as adults as in (a) the North American mudpuppy (*Necturus maculosus*) and (b) the North American siren (*Siren lacertian*). Others have folds of skin that are used for gas exchange or rely on lungs and the body surface as in (c) the North American hellbender (*Cryptobranchus alleganiensis*) and (d) the North American Congo eel (*Amphiuma means*). Specialized cave-dwelling salamanders such as (e) the Texas blind salamander (*Eurycea* [*Typhlomolge*] *rathbuni*) and (f) the European olm (*Proteus anguinus*), are white and lack eyes. Terrestrial salamanders usually have sturdy legs like (g) the North American tiger salamander (*Ambystoma tigrinum*) and its aquatic larva, (h) the European fire salamander (*Salamandra salamandra*) and (i) the North American slimy salamander (*Plethodon glutinosus*).

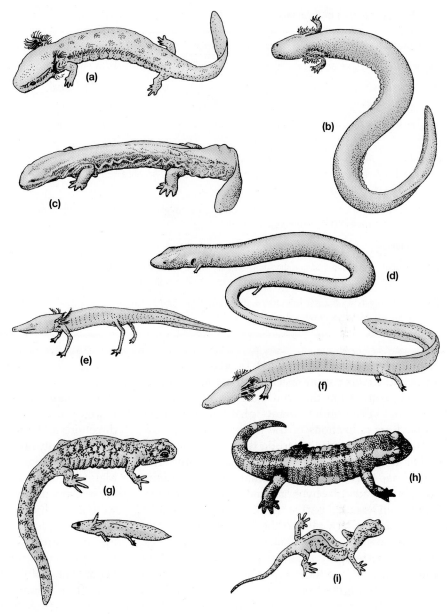

of caves make them good salamander habitats, and cave-dwelling invertebrates supply food. The brook salamanders (*Eurycea*, Plethodontidae) include species that form a continuum from those with fully metamorphosed adults inhabiting the twilight zone near cave mouths to fully paedomorphic forms in the depths of caves or sinkholes. The Texas blind salamander, *Eurycea* [formerly *Typhlomolge*] *rathbuni*, is a highly specialized cave dweller—blind, white, with external gills, extremely long legs, and a flattened snout used to probe underneath pebbles for food. The unrelated European olm (*Proteus*) is another cave salamander that has converged on the same body form.

Terrestrial salamanders like the North American mole salamanders (*Ambystoma*) and the European salamanders (*Salamandra*) have aquatic larvae that lose their gills at metamorphosis. The most fully terrestrial salamanders, the lungless plethodontids (such as the slimy salamander, *Plethodon glutinosus*), include species in which the young hatch from eggs as miniatures of the adult and there is no aquatic larval stage.

Feeding Specializations of Plethodontid Salamanders

Lungs seem an unlikely organ for a terrestrial vertebrate to abandon, but among salamanders the evolutionary loss of lungs has been a successful tactic. The

TABLE 10.2 Families of Salamanders

Ambystomatidae: Small to large terrestrial salamanders with aquatic larvae (35 species in North America; total length up to 30 centimeters).

Amphiumidae: Elongate aquatic salamanders lacking gills (3 species in North America; total length about 1 meter).

Cryptobranchidae: Very large paedomorphic aquatic salamanders with external fertilization of the eggs (1 species in North America and 2 species in Asia; total length 1 to 1.5 meters).

Hynobiidae: Terrestrial or aquatic salamanders with external fertilization of the eggs and aquatic larvae (50 species in Asia; total length up to 30 centimeters).

Plethodontidae: Aquatic or terrestrial salamanders, some with aquatic larvae, others lay eggs on land and omit the larval stage (about 369 species in North, Central, and northern South America, plus 7 species in Europe and 1 in Asia; total length from 3 to 30 centimeters).

Proteidae: Paedomorphic aquatic salamanders with external gills (5 species in North America [*Necturus*] and 1 in Europe [*Proteus*]; total length up to 30 centimeters).

Rhyacotritonidae: Semiaquatic salamanders with aquatic larvae (4 species in North America; total length less than 10 centimeters).

Salamandridae: Terrestrial and aquatic salamanders with aquatic larvae (68 species in Europe, Asia, and extreme northwestern Africa, 6 species in North America; total length up to 20 centimeters).

Sirenidae: Elongate aquatic salamanders with external gills and lacking the pelvic girdle and hindlimbs (4 species in North America, total length 15 to 75 centimeters).

Plethodontidae is characterized by the absence of lungs and contains more species and has a wider geographic distribution than any other lineage of salamanders. Furthermore, many plethodontids have evolved specializations of the hyobranchial apparatus that allow them to protrude the tongue a considerable distance from the mouth to capture prey. This ability has not evolved in salamanders with lungs, probably because the hyobranchial apparatus in these forms is an essential part of the respiratory system.

Salamanders lack ribs, so they cannot expand and contract the rib cage to move air in and out of the lungs. Instead, they employ a buccal pump that forces air from the mouth into the lungs. A sturdy hyobranchial apparatus in the floor of the mouth and throat is an essential part of this pumping system, whereas tongue protrusion requires that parts of the hyobranchial apparatus be elongated and lightened. The modification of the hyobranchial apparatus that allows tongue protrusion is not compatible with buccal pump respiration. Reliance on the skin instead of the lungs for gas exchange may have been a necessary first step in the evolution of tongue protrusion by plethodontids.

The modifications of the respiratory system and hyobranchial apparatus that allow tongue protrusion appear to be linked with several other characteristics of the biology of plethodontids (Roth and Wake 1985, Wake and Marks 1993). These associations can be seen most clearly in the bolitoglossine plethodontids, which have the most specialized tongue-projection mechanisms (**Figure 10–2**). Bolitoglossine plethodontids

(Greek *bola* = dart and *glossa* = tongue) can project the tongue a distance equivalent to their head plus trunk length and can pick off moving prey. This ability requires fine visual discrimination of distance and direction, and the eyes of bolitoglossines are placed more frontally on the head than the eyes of less specialized plethodontids. Furthermore, the eyes of bolitoglossines have a large number of nerves that travel to the ipsilateral (same side) visual centers of the brain as well as the strong contralateral (opposite side) visual projection that is typical of salamanders. Because of this neuroanatomy, bolitoglossines have a complete dual projection of the binocular visual fields to both hemispheres of the brain. They can estimate their distance from a prey object very exactly and rapidly.

Tongue projection is reflected in many different aspects of the life-history characteristics of plethodontid salamanders, including their reproductive modes. Aquatic larval salamanders employ suction feeding, opening the mouth and expanding the throat to create a current of water that carries the prey item with it. The hyobranchial apparatus is an essential part of this feeding mechanism, and the first ceratobranchial becomes well developed during the larval period. In contrast, enlargement of the second ceratobranchial is associated with the tongue-projection mechanism of adult plethodontids. Furthermore, larval salamanders have laterally placed eyes, and the optic nerves project mostly to the contralateral side of the brain. Thus the morphological specializations that make aquatic plethodontid

▶ **FIGURE 10–2** A European bolitoglossine salamander, *Hydromantes*. This species captures prey by trapping the prey on the sticky tip of its tongue, which can be projected from the mouth.

larvae successful are different from the specializations of adults that allow tongue projection, and this situation creates a conflict between the selective forces that act on juveniles and adults.

The bolitoglossines do not have aquatic larvae, and the morphological specializations of adult bolitoglossines appear during embryonic development. In contrast, hemidactyline plethodontids do have aquatic larvae that use suction feeding. As adults, hemidactylines have considerable ability to project the tongues, but they retain the large first ceratobranchial that appears in the larvae. This is a mechanically less efficient arrangement than the large second ceratobranchial of bolitoglossines, and the ability of hemidactylines to project their tongues is correspondingly less than that of bolitoglossines. Thus, the development of a specialized feeding mechanism by plethodontid salamanders has gone hand in hand with such diverse aspects of their biology as respiratory physiology and life history and demonstrates that organisms evolve as whole functioning units, not as collections of independent characters.

Social Behavior of Plethodontid Salamanders Plethodontid salamanders can be recognized externally by the nasolabial groove that extends ventrally from each external naris (nostril opening) to the lip of the upper jaw (**Figure 10–3**). These grooves are an important part of the chemosensory system of plethodontids. As a plethodontid salamander moves about, it repeatedly presses its snout against the substrate. Fluid is drawn into the grooves and moves upward

to the external nares, into the nasal chambers, and over the chemoreceptors of the vomeronasal organ.

Studies of plethodontid salamanders have contributed greatly to our understanding of the roles of competition and predation in shaping the structure of ecological communities. Much recent work in this area has focused on experimental manipulations of animals in the field or laboratory. Because plethodontid salamanders have small home ranges and often remain in a restricted area for their entire lives, they are excellent species to use for these studies.

Males of many plethodontid salamanders defend all-purpose territories that are used for feeding and reproduction. Studies of these salamanders have revealed patterns of social behavior and foraging that seem remarkably complex for animals with skulls the size of a match head and brains little larger than the head of a pin. Robert Jaeger and his colleagues have studied the territorial behavior of the red-backed salamander, *Plethodon cinereus*, a common

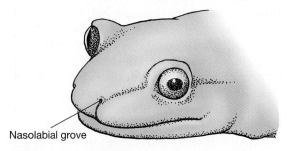

Nasolabial grove

▲ **FIGURE 10–3** Nasolabial grooves of a plethodontid salamander.

species in woodlands of eastern North America. Male red-backed salamanders readily establish territories in cages in the laboratory. A resident male salamander marks the substrate of its cage with pheromones. A salamander can distinguish between substrates it has marked and those marked by another male salamander or by a female salamander. Male salamanders can also distinguish between the familiar scent of a neighboring male salamander and the scent of a male they have not previously encountered, and they react differently to those scents.

In laboratory experiments, red-backed salamanders select their prey in a way that maximizes their energy intake: when equal numbers of large and small fruit flies are released in the cages, the salamanders first capture the large flies. This is the most profitable foraging behavior for the salamanders because it provides the maximum energy intake per capture. In a series of experiments, Jaeger and his colleagues showed that territorial behavior and fighting can interfere with the ability of salamanders to select the most profitable prey (Jaeger et al. 1983). These experiments used surrogate salamanders that were made of a roll of moist filter paper the same length and diameter as a salamander. The surrogates were placed in the cages of resident salamanders to produce three experimental conditions: a control surrogate, a familiar surrogate, and an unfamiliar surrogate. In the control experiment, male red-backed salamanders were exposed to a surrogate that was only moistened filter paper; it did not carry any salamander pheromone. For both of the other groups, the surrogate was rolled across the substrate of the cage of a different male salamander to absorb the scent of that salamander before being placed in the cage of a resident male.

The experiments lasted seven days; the first six days were conditioning periods, and the test itself occurred on the seventh day. For the first six days, the resident salamanders in both experimental groups were given surrogates bearing the scent of another male salamander. The residents thus had the opportunity to become familiar with the scent of that male. On the seventh day, however, the familiar and unfamiliar surrogate groups were treated differently. The familiar surrogate group once again received a surrogate salamander bearing the scent of the same individual it had been exposed to for the previous six days, whereas the resident salamanders in the unfamiliar surrogate group received a surrogate bearing the scent of a different salamander, one to which they had never been exposed before. After a five-minute pause, a mixture of large and small fruit flies was placed in each cage, and the behavior of the resident salamander was recorded.

The salamanders in the familiar surrogate group showed little response to the now familiar scent of the other male salamander. They fed as usual, capturing large fruit flies. In contrast, the salamanders that were exposed to the scent of an unfamiliar surrogate began to give threatening and submissive displays, and their rate of prey capture decreased as a result of the time they spent displaying. In addition, salamanders exposed to unfamiliar surrogates did not concentrate on catching large fruit flies, so the average energy intake per capture also declined. The combined effects of the reduced time spent feeding and the failure to concentrate on the most profitable prey items caused an overall 50 percent decrease in the rate of energy intake for the salamanders exposed to the scent of an unfamiliar male.

The ability of male salamanders to recognize the scent of another male after a week of habituation in the laboratory cages suggests that they would show the same behavior in the woods. That is, a male salamander could learn to recognize and ignore the scent of a male in the adjacent territory, while still being able to recognize and attack a strange intruder. Learning not to respond to the presence of a neighbor may allow a salamander to forage more effectively, and it may also help to avoid injuries that can occur during territorial encounters.

Resident male red-backed salamanders challenge strange intruders, and the encounters involve aggressive and submissive displays and biting. Bites on the body (**Figure 10–4a**) can drive another male away but are not likely to do permanent damage. A bite to the tail (**Figure 10–4b**) may cause the bitten salamander to autotomize (break off) its tail. Salamanders store fat in their tails, and this injury may delay reproduction for a year while the tail is regenerated. Most bites are directed at the snout of an opponent (**Figure 10–4c**) and may damage the nasolabial grooves. Because the nasolabial grooves are used for olfaction, these injuries can reduce a salamander's success in finding prey. Twelve salamanders that had been bitten on the snout were able to capture an average of only 5.8 fruit flies in a two hour period compared with an average of 18.6 flies for 12 salamanders that had not been bitten. In a sample of 144 red-backed salamanders from the Shenandoah National Forest, 11.8 percent had been bitten on the nasolabial grooves, and these animals weighed less than the unbitten animals, presumably because their foraging success had been reduced (Jaeger 1981).

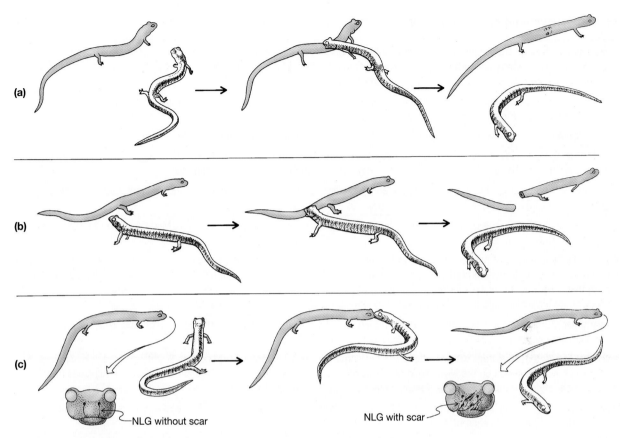

▲ **FIGURE 10–4** Aggressive behaviors of the red-backed salamander, *Plethodon cinereus*. The resident salamander is dark, and the intruder is colored in these drawings. (a) Resident bites the intruder on the body. (b) Bitten on the tail by the resident, the intruder autotomizes its tail to escape. (c) Resident bites intruder on the snout, injuring the nasolabial grooves. (NLG = nasolabial groove)

The possibility of serious damage to an important sensory system during territorial defense provides an additional advantage for a red-backed salamander in being able to distinguish neighbors (which are always there and are not worth attacking) from intruders (which represent a threat and should be attacked). The phenomenon of being able to recognize territorial neighbors has been called *dear enemy recognition* and may be generally advantageous because it minimizes the time and energy that territorial individuals expend on territorial defense and also minimizes the risk of injury during territorial encounters. Similar dear enemy recognition has been described among territorial birds that show more aggressive behavior on hearing the songs of strangers than they do when hearing the songs of neighbors.

■ Frogs and Toads–Anura

In contrast to the limited number of species of salamanders and their restricted geographic distribution, the anurans (Greek *an* = without and *uro* = tail) include 45 families with nearly 5400 species and occur on all the continents except Antarctica (**Table 10.3**). Specialization of the body for jumping is the most conspicuous skeletal feature of anurans (**Figure 10–5**). The hindlimbs and muscles form a lever system that can catapult an anuran into the air, and numerous morphological specializations are associated with this type of locomotion; the hind legs are elongate, and the tibia and fibula are fused. A powerful pelvis strongly fastened to the vertebral column is clearly necessary, as is stiffening of the vertebral column. The ilium is elongate and reaches far anteriorly, and the posterior vertebrae are fused into a solid rod, the **urostyle**. The pelvis and urostyle render the posterior half of the trunk rigid. The vertebral column is short, with only five to nine presacral vertebrae, and these are strongly braced by zygapophyses that restrict lateral bending. The strong forelimbs and flexible pectoral girdle absorb the impact of landing. The eyes are large and are placed well forward on the head, giving binocular vision. Specializations of the

TABLE 10.3 Noteworthy families of anurans

Brachycephalidae: Mainly small terrestrial frogs from the New World tropics. More than half the 803 species are in the genus *Eleutherodactylus*, which was formerly placed in the Leptodactylidae. Many brachycephalids have direct development, and 1 species of *Euhyas* is known to be viviparous; 2 to 6 centimeters.

Bufonidae: Mainly terrestrial frogs; most have aquatic larvae, but some species of *Nectophrynoides* are viviparous (495 species in North, Central, and South America, Africa, Europe, and Asia; 2 to 25 centimeters).

Centrolenidae: Arboreal frogs with aquatic larvae that live in streams (144 species in Central and South America; 5 centimeters).

Dendrobatidae: Small terrestrial frogs, many of which are brightly colored and extremely toxic; terrestrial eggs hatch into tadpoles that are transported to water by an adult (163 species in Central and South America; 2 to 6 centimeters).

Hylidae: Mostly arboreal frogs, but a few species are aquatic or terrestrial (830 species in North, Central, and South America, the West Indies, Europe, Asia, and the Australo-Papuan region; 2 to 15 centimeters).

Hyperoliidae: Small to medium-size mostly arboreal frogs with aquatic larvae (199 species in Africa, Madagascar, and the Seychelles Islands; 1.5 to 8 centimeters).

Leiopelmatidae: Two genera of primitive frogs up to 5 centimeters. *Ascaphus* (2 species) are found in torrential streams in coastal mountains from British Columbia to northern California (North America). Fertilization is internal and the extension of the cloaca of the male that is used to introduce sperm into the cloaca of the female gives the genus its common name—tailed frogs. *Leiopelma* (4 species) are terrestrial frogs, the only frogs native to New Zealand. Eggs are laid on land and cared for by the male.

Leptodactylidae: Aquatic and terrestrial frogs with aquatic larvae (91 species in southern North America, Central, and northern South America and the West Indies; 5 to 25 centimeters).

Mantellidae: Mostly terrestrial frogs with aquatic larvae, but some species have direct development. Brightly colored, toxic species of *Mantella* have converged on the New World dart-poison frogs (Dendrobatidae) in ecology and behavior (165 species from Madagascar and Mayotte Island in the Comoros Islands chain; 3 to 10 centimeters).

Megophryidae: Forest floor and stream edge frogs (136 species from Pakistan and western China east to the Philippines and islands on the Sunda Shelf; 5 to 10 centimeters).

Microhylidae: Terrestrial or arboreal frogs; many have aquatic larvae, but some species have nonfeeding tadpoles and others lay eggs on land and omit the tadpole stage (419 species in North, Central, and South America, sub-Saharan Africa, India, and Korea to northern Australia; 5 to 10 centimeters).

Pipidae: Specialized aquatic frogs; *Xenopus, Hymenochirus, Pseudohymenochirus*, and some species of *Pipa* have aquatic larvae; other species of *Pipa* have eggs that develop directly into juvenile frogs (31 species in South America and sub-Saharan Africa; 2 to 15 centimeters).

Ranidae: Aquatic or terrestrial frogs; most have aquatic tadpoles (319 species in North, Central, and northern South America, Europe, Asia, and Africa; the genus *Rana* contains 44 species from temperate Eurasia to Indochina and western North America; the genus *Lithobates* includes 49 species from eastern North America through Central America to southern Brazil; 10 to 30 centimeters).

Rhacophoridae: Mostly arboreal frogs; some species have filter-feeding aquatic larvae, and others lay eggs in holes in trees and have larvae that do not feed (278 species in sub-Saharan Africa and southern Asia; 2 to 15 centimeters).

locomotor system can be used to distinguish many different kinds of anurans (**Figure 10–6**).

The hindlimbs generate the power to propel the frog into the air, and this high level of power production results from structural and biochemical features of the limb muscles. The internal architecture of the semimembranosus muscle and its origin on the ischium and insertion below the knee allow it to operate at the length that produces maximum force during the entire period of contraction. The muscle shortens faster and generates more power than muscles from most other animals. Furthermore, the intracellular physiological processes of muscle contraction continue at the maximum level throughout contraction rather than declining, as is the case in muscles of most vertebrates.

The diversity of anurans exceeds the number of common names that can be used to distinguish various ecological specialties (**Figure 10–7**). Animals called frogs usually have long legs and move by jumping, and this body form is found in many lineages. Semi-aquatic frogs are moderately streamlined and have webbed feet. Stout-bodied terrestrial anurans that make short hops instead of long leaps are often called toads. They usually have blunt heads, heavy bodies, relatively short legs, and little webbing between the toes. Members of the Bufonidae represent this body form, and very similar body forms are found in other families, including the spadefoot toads of western North America (Scaphiopodidae) and the horned frogs of South America (Ceratophryidae). Spadefoot

(a)

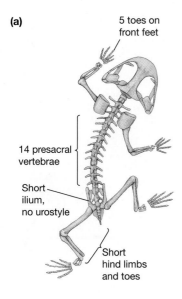

5 toes on front feet

14 presacral vertebrae

Short ilium, no urostyle

Short hind limbs and toes

(b)

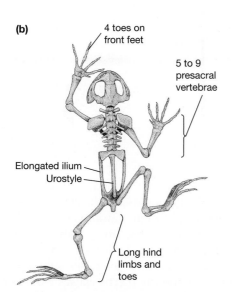

4 toes on front feet

5 to 9 presacral vertebrae

Elongated ilium
Urostyle

Long hind limbs and toes

▲ **FIGURE 10–5** *Triadobatrachus* and a modern anuran. The Triassic fossil *Triadobatrachus* (a) is considered the sister group of modern anurans (b). Derived characters of anurans visible in this comparison include shortening of the body, elongation of the ilia, and fusion of the posterior vertebrae to form a urostyle.

toads take their name from a keratinized structure on the hind foot that they use for digging backward into the soil with rapid movements of their hind legs. The horned frogs have extremely large heads and mouths. They feed on small vertebrates, including birds and mammals, but particularly on other frogs. The tadpoles of horned frogs also are carnivorous and feed on other tadpoles. Many frogs that burrow headfirst have pointed heads, stout bodies, and short legs.

Arboreal frogs usually have large heads and eyes and often slim waists and long legs. Arboreal frogs in many different families move by quadrupedal walking and climbing as much as by leaping. Many arboreal species of hylids and rhacophorids have enlarged toe discs and are called tree frogs. The surfaces of the toe pads consist of an epidermal layer with peglike projections separated by spaces or canals (**Figure 10–8**). Mucus glands distributed over the discs secrete a viscous solution of long-chain, high-molecular-weight polymers in water. Arboreal species of frogs use a mechanism known as wet adhesion to stick to smooth surfaces. (This is the same mechanism by which a wet scrap of paper sticks to glass.) The watery mucus secreted by the glands on the toe discs forms a layer of fluid between the disc and the surface and establishes a meniscus at the interface between air and fluid at the edges of the toes. As long as no air bubble enters the fluid layer, a combination of surface tension (capillarity) and viscosity holds the toe pad and surface together.

Frogs can adhere to vertical surfaces and even to the undersides of leaves. Cuban tree frogs (*Osteopilus septentrionalis*) can cling to a sheet of smooth plastic as it is rotated past the vertical; the frogs do not begin to slip until the rotation reaches an average of 151 degrees—that is, 61 degrees past vertical. Adhesion and detachment of the pads alternate as a frog walks across a leaf. As a frog moves forward, its pads are peeled loose, starting at the rear, as the toes are lifted. The two most specialized families of tree frogs, the Hylidae and Rhacophoridae, have a cartilage (the intercalary cartilage) that lies between the last two bones in the toes. The intercalary cartilage may promote adhesion by increasing the angle through which the toe can move before peeling begins.

Tree frogs are not able to rest facing downward because in that orientation the frog's weight causes the toe pads to peel off the surface. Frogs invariably orient their bodies facing upward or across a slope, and they rotate their feet if necessary to keep the toes pointed upward. When a frog must descend a vertical surface, it moves backward. This orientation keeps the toes facing upward. During backward locomotion, toes are peeled loose from the tip backward by a pair of tendons that insert on the dorsal surface of the terminal bone of the toe.

Toe discs have evolved independently in several lineages of frogs and show substantial convergence in structure. Expanded toe discs are not limited exclusively to arboreal frogs; many terrestrial species that move across fallen leaves on the forest floor also have toe discs.

Several aspects of the natural history of anurans appear to be related to their different modes of

▶ **FIGURE 10–6** The relation of body form and locomotor mode among anurans. Short forelimbs and hindlimbs are associated with burrowing, whereas long forelimbs and hindlimbs are found in species that climb and leap. Hindlimbs that are distinctly longer than the forelimbs usually indicate that a species is a jumper if the amount of webbing on the hind feet is limited or a swimmer if the hind feet are fully webbed.

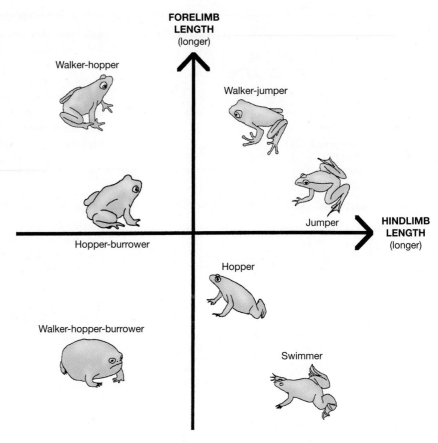

locomotion. In particular, short-legged species that move by hopping are frequently wide-ranging predators that cover large areas as they search for food. This behavior makes them conspicuous to their own predators, and their short legs prevent them from fleeing rapidly enough to escape. Many of these anurans have potent defensive chemicals that are released from glands in the skin when they are attacked. Species of frogs that move by jumping, in contrast to those that hop, are usually sedentary predators that wait in ambush for prey to pass their hiding places. These species are usually cryptically colored, and they often lack chemical defenses. If they are discovered, they rely on a series of rapid leaps to escape. Anurans that forage widely encounter different kinds of prey from those that wait in one spot, and differences in dietary habits may be associated with differences in locomotor mode.

Aquatic anurans use suction to engulf food in water, but most species of semiaquatic and terrestrial anurans have sticky tongues that can be flipped out to trap prey and carry it back to the mouth (**Figure 10–9**). Most anurans use a catapult-like mechanism

to project the tongue. As the mouth is opened, contraction of the genioglossus muscles causes the front of the tongue to stiffen. Simultaneously, contraction of a short muscle at the front of the jaws (the submentalis) provides a fulcrum, and the stiffened tongue rotates forward over the submentalis and flips out of the mouth. Inertia causes the rear portion of the tongue to elongate as it emerges; and, because the tongue has rotated, its dorsal surface slams down on the prey. The tongue is drawn back into the mouth by the hyoglossus muscle, which originates on the hyoid apparatus and inserts within the tongue.

■ Caecilians—Gymnophiona

The third group of living amphibians is the least known and does not even have an English common name (**Figure 10–10**). These are the caecilians, three families with about 173 species of legless burrowing or aquatic amphibians that occur in tropical habitats around the world (**Table 10.4**). The eyes of caecilians are greatly reduced and covered by skin or even by bone. Some species lack eyes entirely but, the retinas

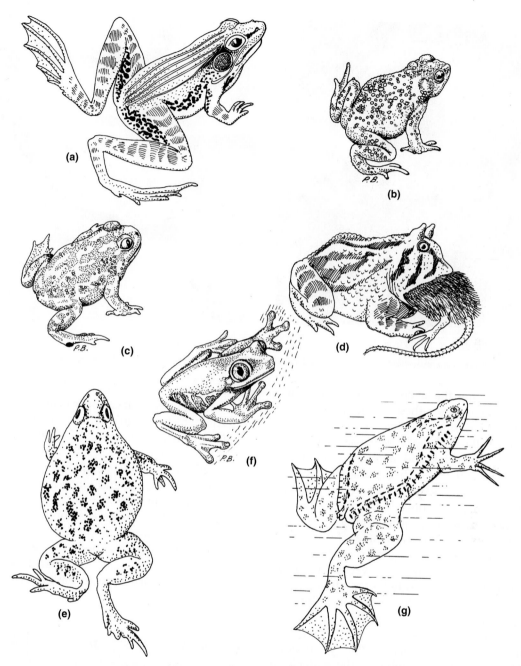

▲ **FIGURE 10–7** Anuran body forms.
Body shape and limb length reflect specializations for different habitats and different methods of locomotion. (a) Semiaquatic frogs that both jump and swim, such as the African ridged frog (Ptychadena oxyrhyncus), have streamlined heads and bodies and long limbs with webbed hind feet, whereas terrestrial walkers and hoppers like (b) the spotted toad (Anaxyrus [Bufo] punctatus), (c) the western spadefoot toad (Spea hammondii), and (d) the Argentine horned frog (Ceratophrys ornata) have blunt heads, stout bodies, short limbs, and hind feet with little or no webbing. Frogs that burrow into the ground head-first, like (e), the African shovel-nosed frog (Hemisus marmoratus), have short limbs and pointed snouts, whereas arboreal frogs, like (f) the red-eyed leaf frog (Agalychnis callidryas), have long limbs and broad snouts. Specialized aquatic frogs, like (g) the African clawed frog (Xenopus laevis), have a smooth surface contour and a well-developed system of lateral lines.

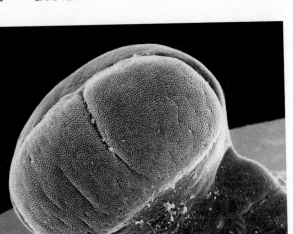

▲ **FIGURE 10–8** Toe discs of a hylid frog. *Left:* A single toe pad. *Right:* Detail of the polygonal plates.

of other species have the layered organization that is typical of vertebrates, and these species appear to be able to detect light. Conspicuous dermal folds (annuli) encircle the bodies of caecilians. The primary annuli overlie vertebrae and myotomal septa and reflect body segmentation. Many species of caecilians have dermal scales in pockets in the annuli; scales are not known in the other groups of living amphibians. A second unique feature of caecilians is a pair of protrusible tentacles, one on each side of the snout between the eye and nostril. Some structures that are associated with the eyes of other vertebrates have become associated with the tentacles of caecilians. One of the eye muscles, the retractor bulbi, has become the retractor muscle for the tentacle; the levator bulbi moves the tentacle sheath; and the Harderian gland (which moistens the eye in other tetrapods) lubricates the channel of the tentacle of caecilians. The tentacle is probably a sensory organ that allows chemical substances to be transported from the animal's surroundings to the vomeronasal organ on the roof of the mouth. The eye of caecilians in the African family Scolecomorphidae is attached to the side of the tentacle near its base. When the tentacle is protruded, the eye is carried along with it, moving out of the tentacular aperture beyond the roofing bones of the skull.

The earliest caecilian known is *Eocaecilia*, an Early Jurassic fossil from the Kayenta formation of western North America. It has a combination of ancestral and derived characters and is the sister taxon of extant caecilians. *Eocaecilia* has a fossa for a chemosensory tentacle, which is a unique derived character of caecilians, but it also has four legs, whereas all living caecilians are legless.

Caecilians feed on small or elongate prey—termites, earthworms, and larval and adult insects—and the tentacle may allow them to detect the presence of prey when they are underground. Females of some species of caecilians brood their eggs, whereas other species give birth to fully formed young. The embryos of terrestrial species have long, filamentous gills, and the embryos of aquatic species have saclike gills.

10.2 Diversity of Life Histories of Amphibians

Of all the characteristics of amphibians, none is more remarkable than the variety they display in modes of reproduction and parental care. It is astonishing that the range of reproductive modes among the approximately 6000 species of amphibians far exceeds that of any other group of vertebrates except for fishes, which outnumber amphibian species by five to one.

Most species of amphibians lay eggs. The eggs may be deposited in water or on land, and they may hatch into aquatic larvae or into miniatures of the terrestrial adults. The adults of some species of frogs

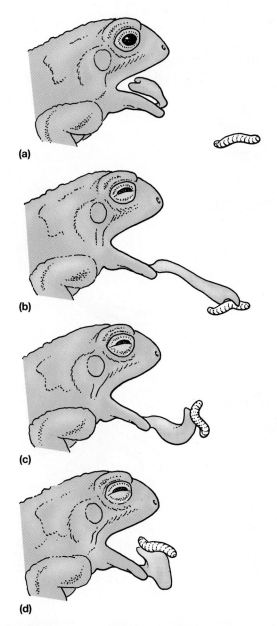

(a)

(b)

(c)

(d)

▲ **FIGURE 10–9** Prey capture by a toad. The tongue is attached at the front of the lower jaw and pivots around a stiffened muscle as it is flipped out. The tip of the tongue (the portion that is at the rear of the mouth when the tongue is retracted) has glands that excrete a sticky mucus that adheres to the prey as the tongue is retracted into the mouth.

carry eggs attached to the surface of their bodies. Others carry their eggs in pockets in the skin of the back or flanks, in the vocal sacs, or even in the stomach. In still other species, the females retain the eggs in the oviducts and give birth to metamorphosed young. Many amphibians have no parental care of their eggs or young; but in many other species, a par-

ent remains with the eggs and sometimes with the hatchlings or transports tadpoles from the nest to water. In a few species, an adult even feeds the tadpoles.

Amphibians have two characteristics that make their population ecology hard to study. First, fluctuation in size appears to be a normal feature of amphibian populations. Many species of amphibians lay hundreds of eggs, and the vast majority of these eggs never reach maturity. In a good year, however, survival may be unusually high and a large number of individuals may be added to the population. Conversely, in a year of drought, the entire reproductive output of a population may die and no individuals will be added to the population in that year. Thus, year-to-year variation in recruitment creates natural fluctuations in populations that can obscure long-term trends in population size. In addition, many species of amphibians live in metapopulations in which individual animals move among local populations that are often centered on breeding sites. In the shifting existence of a metapopulation, breeding populations may disappear from some sites while a healthy metapopulation continues to exist and breed at other sites. A limited study might conclude that a species was vanishing, whereas a broader analysis would show that the total population of the species had not changed. *Measuring and Monitoring Biological Diversity: Standard Methods for Amphibians* (Heyer et al. 1993) was compiled to provide standard methods for studies of amphibian populations so that data from different studies could be compared and combined.

■ Caecilians

The reproductive adaptations of caecilians are as specialized as their body form and ecology. A male intromittent organ that is protruded from the cloaca accomplishes internal fertilization. Some species of caecilians lay eggs, and the female may coil around the eggs, remaining with them until they hatch (see Figure 10–10b). Viviparity is widespread, however, and about 75 percent of the species are viviparous and matrotrophic. At birth young caecilians are 30 to 60 percent of their mother's body length. A female *Typhlonectes* 500 millimeters long may give birth to nine babies, each 200 millimeters long. The initial growth of the fetuses is supported by yolk contained in the egg at the time of fertilization, but this yolk is exhausted long before embryonic development is complete. *Typhlonectes* fetuses have absorbed all of the yolk in the eggs by the time they are 30 millimeters long. Thus, the energy they need to grow to

▶ **FIGURE 10–10** Body form and reproduction of caecilians. (a) The mountain caecilian, *Gymnopis syntrema*, is a small terrestrial species from Guatemala. The body is slim, and the eyes are covered with bone. (b) A female sticky caecilian, *Ichthyophis glutinosus,* coiled around her eggs. This is a terrestrial species from Sri Lanka. The embryo of (c) the Uluguru black caecilian, *Scolecomorphus uluguruensis,* has branched gills, whereas (d) the Rio Cauca caecilian, *Typhlonectes natans,* has saclike gills.

(a)

(b)

(c)

(d)

200 millimeters (a 6.6-fold increase in length) must be supplied by the mother. The energetic demands of producing nine babies, each one increasing its length 6.6 times and reaching 40 percent of the mother's length at birth, must be considerable.

The fetuses obtain this energy by scraping material from the walls of the oviducts with specialized embryonic teeth. The epithelium of the oviduct proliferates and forms thick beds surrounded by ramifications of connective tissue and capillaries. As the

TABLE 10.4 Families of caecilians

Caeciliidae: Terrestrial and aquatic caecilians with both oviparous and viviparous species; no aquatic larval stage (120 species in Central and South America, Africa, India, and the Seychelles Islands; 10 centimeters to 1.5 meters).

Ichthyophiidae: Terrestrial caecilians with aquatic larvae (44 species in the Philippines, India, Thailand, southern China, and the Malayan Archipelago; up to 50 centimeters).

Rhinatrematidae: Terrestrial caecilians believed to have aquatic larvae (9 species in northern South America; up to 30 centimeters total length).

fetuses exhaust their yolk supply, these beds begin to secrete a thick, white, creamy substance that has been called uterine milk. When their yolk supply has been exhausted, the fetuses emerge from their egg membranes, uncurl, and align themselves length-wise in the oviducts. The fetuses apparently bite the walls of the oviduct, stimulating secretion and strip-ping some epithelial cells and muscle fibers that they swallow with the uterine milk. Small fetuses are reg-ularly spaced along the oviducts. Large fetuses have their heads spaced at intervals, and the body of one fetus overlaps the head of the next. This spacing probably gives all the fetuses access to the secretory areas on the walls of the oviducts. Gas exchange appears to be achieved by close contact between the fetal gills and the walls of the oviducts. All the terres-trial species of caecilians have fetuses with a pair of triple-branched filamentous gills. In preserved speci-mens, the fetuses frequently have one gill extending forward beyond the head and the other stretched along the body. In the aquatic genus *Typhlonectes*, the gills are saclike but are usually positioned in the same way. Both the gills and the walls of the oviducts are highly vascularized, and it seems likely that exchange of gases, and possibly of small molecules such as metabolic substrates and waste products, takes place across the adjacent gill and oviduct. The gills are absorbed before birth, and cutaneous gas exchange may be important for fetuses late in devel-opment.

A remarkable method of feeding the young has recently been described for a caecilian from Kenya, *Boulengerula taitana* (Kupfer et al. 2006). Females give birth to young that are in a relatively undeveloped (altricial) stage, and the young remain with their mother in subterranean nests. The cells in the outer layer of the skin (the stratum corneum) of brooding females are thickened and contain vesicles filled with lipids. The young have a specialized fetal dentition that allows them to peel off single layers of skin cells from the surface of their mother's body. The same feeding mechanism may have evolved in a caecilian from South America, *Siphonops annulatus*. The altri-cial young have the same sort of dentition as young *B. taitana*, and the skin of brooding female *S. annula-tus* has the same pale, milky appearance as that of brooding *B. taitana*. Newborn young of other species of caecilians from Africa and South America also have dentition that may permit them to strip lipid-rich skin from their mother. This mechanism of feed-ing young may represent an evolutionary step inter-mediate between producing young that obtain all their nutrition from the egg yolk and the viviparous condition in which embryos strip cells from the walls of the oviduct.

■ Salamanders

Most groups of salamanders use internal fertiliza-tion, but the Cryptobranchidae, Hynobiidae, and probably the Sirenidae retain external fertilization. Internal fertilization in salamanders is accomplished not by an intromittent organ but by the transfer of a packet of sperm (the **spermatophore**) from the male to the female (**Figure 10–11**). The form of the sper-matophore differs in various species of salamanders, but all consist of a sperm cap on a gelatinous base. The base is a cast of the interior of the male's cloaca, and in some species it reproduces the ridges and furrows in accurate detail. Males of the Asian sala-mandrid *Euproctus* deposit a spermatophore on the body of a female and then, holding her with their

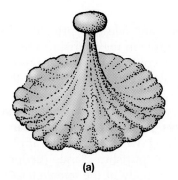
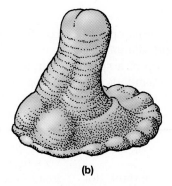

(a)   (b)   (c)

▲ **FIGURE 10–11** Spermatophores. Male salamanders deposit a spetmatorphore that contains a capsule of sperm supported on a gelatinous base. (a) Red-spotted newt, Notophthalmus viridescens. (b) Dusky salamander, Desmognathus fuscus. (c) Two-lined salamander, Eurycea bislineata.

▶ **FIGURE 10–12** Transfer of pheromones by male salamanders during courtship.
(a) A male rough-skinned newt (*Taricha granulosa*) rubbing his chin on the female's snout. (b) A female Jordan's salamander (*Plethodon jordani*) following a male in the tail walk behavior that precedes deposition of a spermatophore. (c) A male two-lined salamander (*Eurycea bislineata*) using enlarged teeth to scrape the top of the female's head. (d) A male smooth newt (*Lissotriton* [*Triturus*] *vulgaris*) on the right, using his tail to waft pheromones toward the female.

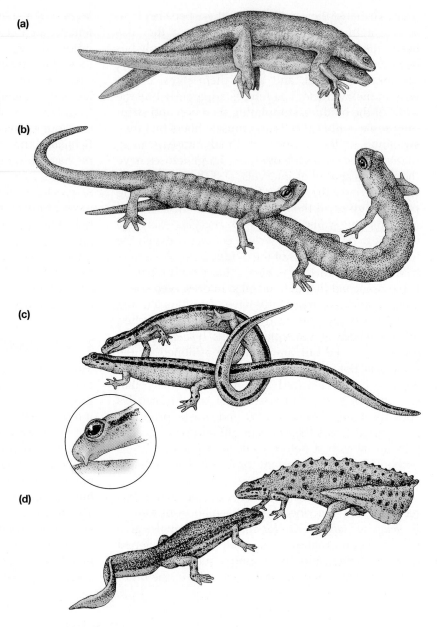

tail or jaws, use their feet to insert the spermatophore into her cloaca. Females of the hynobiid salamander *Ranodon sibiricus* deposit egg sacs on top of a spermatophore. In derived species of salamanders, the male deposits a spermatophore on the substrate, and the female picks up the cap with her cloaca. The sperm are released as the cap dissolves, and fertilization occurs in the oviducts.

Courtship Courtship patterns are important for species recognition, and they show great interspecific variation. Males of some species have elaborate secondary sexual characters that are used during courtship. Pheromones are released primarily by males and play a large role in the courtship of salamanders; they probably contribute to species recognition and may stimulate endocrine activity that increases the receptivity of females.

Pheromone delivery by most salamanders that breed on land involves physical contact between a male and female, during which the male applies secretions of specialized courtship glands (hedonic glands) to the nostrils or body of the female (**Figure 10–12**). Several modes of pheromone delivery have been described. Males of many plethodontids (e.g., *Plethodon jordani*) have a large gland beneath the chin (the mental gland), and secretions of the gland are applied to the nostrils of the female with a slapping motion. The anterior teeth of males of many species of *Desmognathus* and *Eurycea* (both members of the Plethodontidae) hypertrophy during the breeding season. A male of these species spreads secretion from his mental gland onto the female's skin, and then abrades the skin with his teeth, inoculating the female with the pheromone. Males of two small species of *Desmognathus* use specialized mandibular teeth to bite and stimulate the female. Male salamandrids (Salamandridae) rub the female's snout with hedonic glands on their cheeks (the red-spotted newt, *Notophthalmus viridescens*), chin (the rough-skinned newt, *Taricha granulosa*), or cloaca (the Spanish newt, *Pleurodeles waltl*).

The males of many species of newts perform elaborate courtship displays in which the male vibrates its tail to create a stream of water that wafts pheromones secreted by a gland in his cloaca toward the female. Two trends are apparent within evolutionary lineages of newts: an increase in diversity of the sexual displays performed by the male and an increase in the importance of positive feedback from the female. The behaviors seen in *Mesotriton* [*Triturus*] *alpestris* may represent the ancestral condition. This species shows little sexual dimorphism (**Figure 10–13c**), and the male's display consists only of fanning (a display in which the tail is folded back against the flank nearest the female and the tail tip is vibrated rapidly). The male's behavior is nearly independent of response by the female—a male *M. alpestris* may perform his entire courtship sequence and deposit a spermatophore without active response by the female he is courting.

A group of large newts, including *Triturus cristatus* and *Ommatotriton* [*Triturus*] *vittatus* (**Figure 10–13a,b**), are highly sexually dimorphic, and males defend display sites. Their displays are relatively static and lack the rapid fanning movements of the tail that characterize other groups of newts. A male of these species does not deposit a spermatophore unless the female he is courting touches his tail with her snout.

A group of small-bodied newts includes *Lissotriton* [*Triturus*] *vulgaris* and *Lissotriton* [*Triturus*] *boscai* (**Figure 10–13d,e**). These species show less sexual dimorphism than the large species and have a more diverse array of behaviors, including a nearly static lateral display, whipping the tail violently against the female's body, fanning with the tail tip, and other displays (with names like wiggle and flamenco) that occur in some species in the group.

Response by the female is an essential component of courtship for these species—a male will not move on from the static display that begins courtship to the next phase unless the female approaches him repeatedly, and he will not deposit a spermatophore until the female touches his tail.

These trends toward greater sexual dimorphism, more diverse displays, and more active involvement of the female in courtship may reflect sexual selection by females within the derived groups. Halliday (1990) has suggested that in the ancestral condition there was a single male display, and females mated with the males that performed it most vigorously. That kind of selection by females would produce a population of males that all display vigorously, and males that added new components to their courtship might be more attractive than their rivals to females.

Eggs and Larvae In most cases, salamanders that breed in water lay their eggs in water. The eggs may be laid singly or in a mass of transparent gelatinous material. The eggs hatch into gilled aquatic larvae that, except in aquatic species, transform into terrestrial adults. Some families, including the lungless salamanders (Plethodontidae), have a number of species that have dispensed in part or entirely with an aquatic larval stage. The dusky salamander, *Desmognathus fuscus*, lays its eggs beneath a rock or log near water, and the female remains with them until after they have hatched. The larvae have small gills at hatching and may either take up an aquatic existence or move directly to terrestrial life. The red-backed salamander, *Plethodon cinereus*, lays its eggs in a hollow space in a rotten log or beneath a rock. The embryos have gills, but these are reabsorbed before hatching so the hatchlings are miniatures of the adults.

Viviparity Only a few species of salamanders in the genera *Salamandra* and *Lyciasalamandra* are viviparous. The European alpine salamander (*Salamandra atra*) gives birth to one or two fully developed young, each about one-third the adult body length, after a gestation period that lasts from 2 to 4 years. Initially the clutch contains 20 to 30 eggs, but only one or two of these eggs are fertilized and develop into embryos. When the energy in their yolk sacs is exhausted, these embryos consume the unfertilized eggs; and, when that source of energy is gone, they scrape the reproductive tract of the female with specialized teeth. *Lyciasalamandra antalyana* [formerly *Mertensiella luschiana antalyana*], a species of salamandrid found in a small area of Turkey, has a

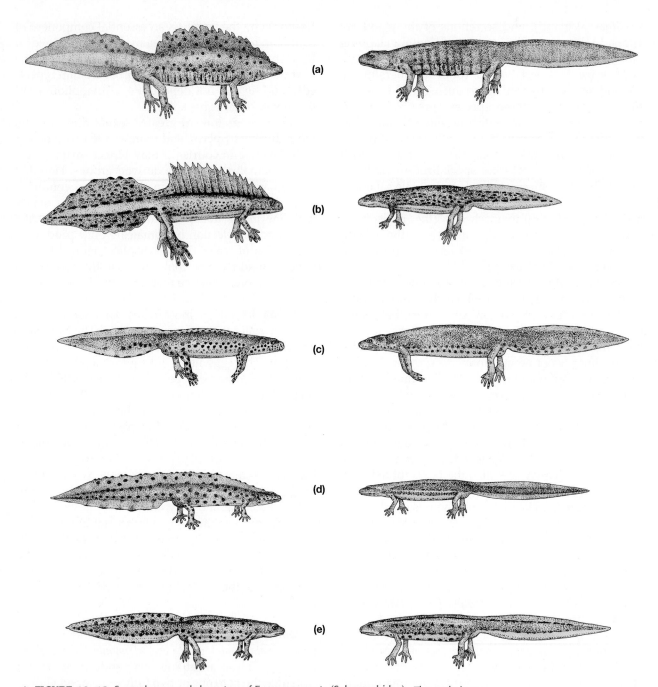

▲ **FIGURE 10–13** Secondary sexual characters of European newts (Salamandridae). The male is on the left, the female is on the right. (a) The great crested newt, *Triturus cristatus*. (b) The banded newt, *Ommatotriton* [*Triturus*] *vittatus*. (c) The alpine newt, *Mesotriton* [*Triturus*] *alpestris*. (d) The smooth newt, *Lissotriton* [*Triturus*] *vulgaris*. (e) Bosca's newt, *Lissotriton* [*Triturus*] *boscai*.

reproductive mode similar to that of *Salamandra atra*, but the gestation period is shorter, lasting about a year.

Females in some populations of the European fire salamander (*Salamandra salamandra*) produce 20 or more small larvae, each about one-twentieth the length of an adult. The embryos probably get all the energy needed for growth and development from egg yolk. The larvae are released in water and have an aquatic stage that lasts about three months. In other populations of this species, the eggs are retained in the oviducts and, when all of the unfertilized eggs have been consumed, some of the embryos cannibalize other embryos. The surviving embryos pass through metamorphosis in the oviducts of the female.

Paedomorphosis Paedomorphosis is the rule in families like the Cryptobranchidae and Proteidae, and it characterizes most cave dwellers. It also appears as a variant in the life history of species of salamanders that usually metamorphose and can be a short-term response to changing conditions in aquatic or terrestrial habitats. The life history of a species of salamanders from eastern North America provides an example of the flexibility of paedomorphosis.

The small-mouthed salamander, *Ambystoma talpoideum,* is the only species of mole salamander in eastern North America that displays paedomorphosis, although a number of species of *Ambystoma* in the western United States and in Mexico are paedomorphic. Small-mouthed salamanders breed in the autumn and winter; during the following summer, some larvae metamorphose to become terrestrial juveniles. These animals become sexually mature by autumn and return to the ponds to breed when they are about a year old. Ponds in South Carolina also contain paedomorphic larvae that remain in the ponds throughout the summer and mature and breed in the winter. Some of these paedomorphs metamorphose after breeding, whereas others do not metamorphose and remain in the ponds as permanently paedomorphic adults.

▤ Anurans

Anurans are the most familiar amphibians, largely because of the vocalizations associated with their reproductive behavior. It is not even necessary to get outside a city to hear them. In springtime, a weed-choked drainage ditch beside a highway or a trash-filled marsh at the edge of a shopping center parking lot is likely to attract a few toads or tree frogs that have not yet succumbed to human usurpation of their habitat.

The mating systems of anurans can be divided roughly into **explosive breeding**, in which the breeding season is very short (sometimes only a few days), and **prolonged breeding**, with breeding seasons that may extend for several months. Explosive breeders include many species of toads and other anurans that breed in temporary aquatic habitats, such as vernal ponds or pools that form after rainstorms in the desert. Because these bodies of water do not last very long, breeding congregations of anurans usually form as soon as the site is available. Males and females arrive at the breeding sites nearly simultaneously and often in very large numbers. The number of males and females present is approximately equal because the entire population breeds in a short time. Time is the main constraint on how many females a male is able to court, and mating success is usually approximately the same for all the males in a chorus.

In species with prolonged breeding seasons, the males usually arrive at the breeding sites first. Males of some species, such as green frogs (*Lithobates* [*Rana*] *clamitans*), establish territories in which they spend several months, defending the spot against the approach of other males. The males of other species move between daytime retreats and nocturnal calling sites on a daily basis. Females come to the breeding site to breed and leave when they have finished. Only a few females arrive every day, and the number of males at the breeding site is greater than the number of females every night. Mating success may be skewed, with many of the males not mating at all and a few males mating several times. Males of anuran species with prolonged breeding seasons compete to attract females, usually by vocalizing. The characteristics of a male frog's vocalization (pitch, length, or repetition rate) might provide information that a female frog could use to evaluate his quality as a potential mate. This is an active area of study in anuran behavior.

Vocalizations Anuran calls are diverse; they vary from species to species, and most species have two or three different sorts of calls used in different situations. The most familiar calls are the ones usually referred to as mating calls, although a less specific term such as **advertisement calls** is preferable. These calls range from the high-pitched *peep* of a spring peeper to the nasal *waaah* of a spadefoot toad or the bass *jug-o-rum* of a bullfrog. The characteristics of a call identify the species and sex of the calling individual. Many species of anurans are territorial, and males of at least one species, the North American bullfrog (*Lithobates catesbeianus* [formerly *Rana catesbeiana*]) recognize one another individually by voice.

An advertisement call is a conservative evolutionary character, and among related species there is often considerable similarity in advertisement calls. Superimposed on the basic similarity are the effects of morphological factors, such as body size, as well as ecological factors that stem from characteristics of the habitat. Most toads have an advertisement call that consists of a train of repeated pulses, and the pitch of the call varies with body size, extending downward from 5200 hertz for the oak toad (*Anaxyrus [Bufo] quercicus*), which is only 2 or 3 centimeters long; to 1800 hertz for the American toad (*Anaxyrus [Bufo] americanus*), about 6 centimeters; and down to 600 hertz for the giant toad (*Chaunus [Bufo] marinus*), with a body length of nearly 20 centimeters. A Bornean frog *Metaphrynella sundana* calls from cavities in trees, and males of this species adjust the frequency of their calls to match the resonant frequency of the hole from which they are calling (Lardner and bin Lakin 2002). The concave-eared torrent frog (*Wurana [Amolops] tormotus*), which lives in the Huangshan Hot Springs in Anhui Province, China, produces birdlike calls that include ultrasonic frequencies (i.e., frequencies above 20 kilohertz). The hot spring environment is noisy with background sound frequencies extending upward to 22 kilohertz, and it seems likely that ultrasonic components of the torrent frogs' vocalizations prevent them from being masked by background noise (Feng et al. 2006).

Female frogs are responsive to the advertisement calls of males of their species for a brief period when their eggs are ready to be laid. The hormones associated with ovulation are thought to sensitize specific cells in the auditory pathway that respond to the species-specific characteristics of the male's call. Mixed choruses of anurans are common in the mating season; a dozen species may breed simultaneously in one pond. A female's response to her own species' mating call is a mechanism for species recognition in that situation.

Costs and Benefits of Vocalization   The vocalizations of male frogs are costly in two senses. The actual energy that goes into call production can be very large, and the variations in calling pattern that accompany social interactions among male frogs in a breeding chorus can increase the cost per call (**Box 10–1**). Another cost of vocalization for a male frog is the risk of predation. A critical function of vocalization

BOX 10–1   The Energy Cost of Vocalization by Frogs

The vocalizations of frogs, like most acoustic signals of tetrapods, are produced when air from the lungs is forced over the vocal cords, causing them to vibrate. Contraction of trunk muscles provides the pressure in the lungs that propels the air across the vocal cords, and these contractions require metabolic energy. Measurement of the actual energy expenditure by frogs during calling is technically difficult because a frog must be placed in an airtight metabolism chamber to measure the amount of oxygen it consumes, and that procedure can frighten the frog and prevent it from calling. Ted Taigen and Kent Wells (1985) at the University of Connecticut overcame that difficulty in studies of the gray tree frog, *Hyla versicolor*, by taking the metabolism chambers to the breeding ponds (**Figure 10–14**). Calling male frogs were placed in the chambers early in the

▶ **FIGURE 10–14**   Measuring oxygen consumption of a calling frog.   A male gray tree frog (*Hyla versicolor*) is placed in a metabolism chamber beside a breeding pond. A microphone in the chamber records the frog's calls, and a thermocouple measures the temperature inside the chamber. Gas samples are drawn from the tube for measurements of oxygen consumption.

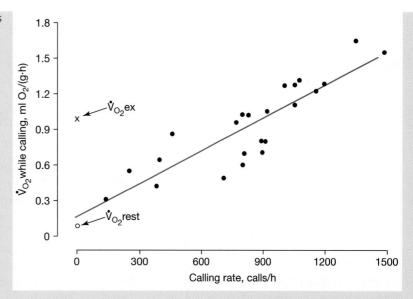

▶ **FIGURE 10–15** Energetic cost of calling. Rates of oxygen consumption of frogs calling inside metabolism chambers (on the vertical axis) is recorded as a function of the rate of calling (on the horizontal axis). The energy expended by a calling frog increases linearly with the number of times it calls per hour. The rates of oxygen consumption for frogs at rest (o) and during maximum exercise (x) are shown for comparison.

evening and then left undisturbed. With the stimulus of the chorus around them, frogs would call in the chambers. Their vocalizations were recorded with microphones attached to each chamber, and the amount of oxygen they used during calling was determined from the decline in the concentration of oxygen in the chamber over time.

The rates at which individual frogs consumed oxygen were directly proportional to their rates of vocalization (**Figure 10–15**). At the lowest calling rate, 150 calls per hour, oxygen consumption was barely above resting levels. However, at the highest calling rates, 1500 calls per hour, the frogs were consuming oxygen even more rapidly than they did during high levels of locomotor activity. Examination of the trunk muscles of the male frogs, which hypertrophy enormously during the breeding season, revealed biochemical specializations that appear to permit this high level of oxygen consumption during vocalization (Marsh and Taigen 1987).

The advertisement call of the gray tree frog is a trill that lasts from 0.7 to 1.7 second. During their studies, Wells and Taigen found that gray tree frogs gave short calls when they were in small choruses, and lengthened their calls when many other males were calling near them (Wells and Taigen 1986). It has subsequently been shown that long calls are more attractive to female frogs than short calls (Klump and Gerhardt 1987), but the long calls require more energy. A long call requires about twice as much energy as a short call, and the rate of oxygen consumption during calling increases as the length of the calls increases. That relationship suggests that, all else being equal, a male gray tree frog that increased its call duration to be more attractive to female frogs would pay a price for its attractiveness with a higher rate of energy expenditure.

Indirect evidence suggests that the energy cost of calling might limit the time a male gray tree frog could spend in a breeding chorus. The tree frogs call for only two to four hours each night, and stores of glycogen (the metabolic substrate

used by calling frogs) decreased by 50 percent in that time. Wells and Taigen were able to simulate the effects of different chorus sizes by playing tape recordings of vocalizations to frogs. The frogs matched their own calls to the recorded calls they heard—short responses to short calls, medium to medium calls, and long responses to long calls. As the length of their calls increased, the frogs reduced the rate at which they called. The reduction in rate of calling approximately balanced the increased length of each call, so the overall calling effort (the number of seconds of vocalization per hour) was nearly independent of call duration. Thus, it appears that male gray tree frogs compensate for the higher energy cost of long calls by giving fewer of them. However, that compromise may not entirely eliminate the problem of high energy costs for frogs giving long calls. Even though the calling effort was approximately the same, males giving long calls at slow rates spent fewer hours per night calling than did frogs that produced short calls at higher rates.

The high energy cost of calling offers an explanation for the pattern of short and long calls produced by male gray tree frogs. The length of time that an isolated male can call may be the most important determinant of his success in attracting a female, and the trade-off between rate of calling and call duration suggests that male frogs are performing at or near their physiological limits. Giving short calls and calling for several hours every night may be the best strategy if a male has that option available. In a large chorus, however, competition with other males is intense and giving a longer and more attractive call may be important, even if the male can call for only a short time.

From the perspective of a female frog, however, there could be additional significance to call length if the ability to produce high-cost calls identified the most fit males. The Good Genes hypothesis predicts that the characteristics of males that are favored by females, such as long calls, identify genetically superior males. That hypothesis was studied in gray tree frogs by comparing the growth rates of tadpoles sired by males producing long calls to those of tadpoles from males with short calls (Welch et al. 1998). In most cases, tadpoles sired by long-call males grew faster than those from short-call males, especially when the food available to the tadpoles was limited.

There was no difference in larval survival to metamorphosis, but after the froglets had metamorphosed, the ones sired by long-call males grew faster than the ones from short-call males. This study suggests that a female gray tree frog can increase the fitness of her offspring by mating with a male who is giving long calls.

is to permit a female frog to locate a male, but female frogs are not the only animals that can use vocalizations as a cue to find male frogs; predators of frogs also find that calling males are easy to locate.

The túngara frog (*Engystomops* [*Physalaemus*] *pustulosus*) is a small terrestrial anuran that occurs in Central America (**Figure 10–16**). Túngara frogs breed in small pools, and breeding assemblies range from a single male to choruses of several hundred males. The advertisement call of a male túngara frog is a strange noise, a whine that sounds as if it would be more at home in an arcade of video games than in the tropical night. The whine starts at a frequency of 900 hertz and sweeps downward to 400 hertz in about 400 milliseconds (**Figure 10–17**). The whine may be produced by itself, or it may be followed by one or several *chucks*. When a male túngara frog is calling alone in a pond, it usually gives only the whine portion of the call; however, as additional males join a chorus, more and more of the frogs produce calls that include chucks. Male túngara frogs calling in a breeding pond added chucks to their calls when they heard playbacks of calls of other males. That observation suggested that it was the presence of other calling males that stimulated frogs to make their calls more complex by adding chucks to the end of the whine.

What advantage would a male frog in a chorus gain from using a whine-chuck call instead of a whine? Perhaps the complex call is more attractive to female frogs than the simple call. Michael Ryan and Stanley Rand tested that hypothesis by placing female túngara frogs in a test arena with a speaker at each side. One speaker broadcast a whine call, and the second speaker broadcast a whine-chuck. When female frogs were released individually in the center of the arena, 14 of the 15 frogs tested moved toward the speaker broadcasting the whine-chuck call (Ryan 1985).

If female frogs are attracted to whine-chuck calls in preference to whine calls, why do male frogs give whine-chuck calls only when other males are present? Why not always give the most attractive call possible? One possibility is that whine-chuck calls require more energy than whines, and males save energy by using whine-chucks only when competition with other males makes the energy expenditure necessary. However, measurements of the energy expenditure of calling male túngara frogs showed that energy cost was not related to the number of chucks. Another possibility is that male frogs giving whine-chuck calls are more vulnerable to predators than frogs giving only whine calls. Túngara frogs in breeding choruses are preyed upon by frog-eating

▲ **FIGURE 10–16** Male túngara frog, *Engystomops* [*Physalaemus*] *pustulosus*. As the frog calls, air is forced from the lungs (*left*) into the vocal sacs (*right*).

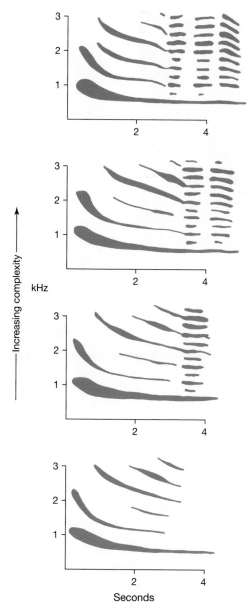

▲ **FIGURE 10–17** Sonographs of the advertisement call of *Engystomops pustulosus*. A sonograph is a graphic representation of a sound: time is shown on the horizontal axis and frequency on the vertical axis. The calls increase in complexity from bottom (a whine only) to top (a whine followed by three chucks).

bats, *Trachops cirrhosus*, and the bats locate the frogs by homing on their vocalizations.

In a series of playback experiments, Ryan and Merlin Tuttle placed pairs of speakers in the forest and broadcast vocalizations of túngara frogs. One speaker played a recording of a whine and the other a recording of a whine-chuck. The bats responded as if the speakers were frogs: They flew toward the speakers and even landed on them. In five experiments at different sites, the bats approached speakers broadcasting whine-chuck calls twice as frequently as those playing simple whines (168 approaches versus 81). Thus, female frogs are not alone in finding whine-chuck calls more attractive than simple whines—an important predator of frogs also responds more strongly to the complex calls. Predation can be a serious risk for male túngara frogs. Ryan and his colleagues measured the rates of predation in choruses of different sizes. The major predators were frog-eating bats, a species of opossum (*Philander opossum*), and a larger species of frog (*Leptodactylus pentadactylus*); the bats were the most important predators of the túngara frogs. Large choruses of frogs did not attract more bats than small choruses, and consequently the risk of predation for an individual frog was less in a large chorus than in a small one. Predation was an astonishing 19 percent of the frogs per night in the smallest chorus and a substantial 1.5 percent per night even in the largest chorus. When a male frog shifts from a simple whine to a whine-chuck call, it increases its chances of attracting a female, but it simultaneously increases its risk of attracting a predator. In small choruses, the competition from other males for females is relatively small, and the risk of predation is relatively large. Under those conditions it is apparently advantageous for a male túngara frog to give simple whines. However, as chorus size increases, competition with other males also increases while the risk of predation falls. In that situation, the advantage of giving a complex call apparently outweighs the risks.

■ Modes of Reproduction

Fertilization is external in most anurans; the male uses his fore legs to clasp the female in the pectoral region (axillary amplexus) or pelvic region (inguinal amplexus). Amplexus may be maintained for several hours or even days before the female lays eggs. Males of the tailed frog of the Pacific Northwest (*Ascaphus truei*) have an extension of the cloaca (the "tail" that gives them their name) that is used to introduce sperm into the cloaca of the female. Internal fertilization has been demonstrated for the Puerto Rican coquí (*Eleutherodactylus coquí*) and may be widespread among frogs that lay eggs on land. Fertilization must also be internal for the few species of viviparous anurans.

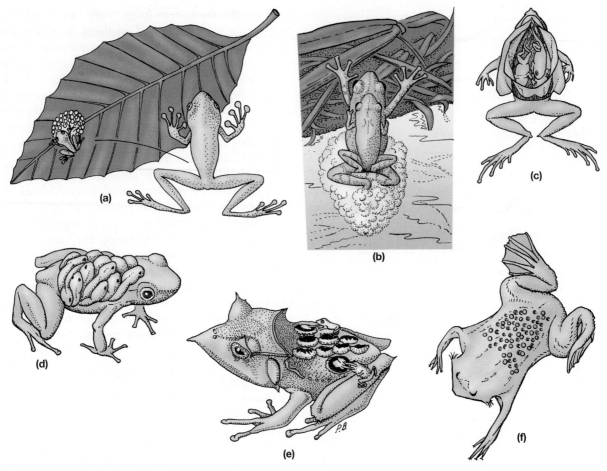

▲ **FIGURE 10–18** Reproductive modes and parental care among anurans.
(a) The female northern glass frog, *Hyalinobatrachium* [*Centrolenella*] *fleischmanni,* attaches the eggs to the underside of a leaf overhanging a stream, and the male frog remains with the eggs. When the eggs hatch, the tadpoles drop into the water. (b) A male túngara frog (*Engystomops pustulosus*) takes the eggs and egg jelly with his hind legs as the female releases them and beats the jelly into a foam nest that contains the eggs and floats on the surface of the water. When the tadpoles hatch, they release an enzyme that dissolves the foam and allows them to drop into the water. (c) The female Darwin's frog, *Rhinoderma darwinii,* deposits eggs on the ground; the male frog picks them up and transfers them to his vocal sacs, where they develop through metamorphosis. (d) The female rocket frog, *Colostethus inguinalis,* deposits eggs in a nest site within the territory of a male frog, who remains with the eggs until they hatch into tadpoles. The newly hatched tadpoles squirm onto the male's back and are carried to water. (e) As a female Spix's horned frog (*Hemiphractus scutatus*) releases her eggs, the male pushes them onto her back and into the thickened skin. They develop in capsules that form around them, passing through metamorphosis and hatching as tiny frogs. (f) A female Surinam toad, *Pipa pipa,* with eggs on her back. The eggs of this species hatch as tadpoles, whereas other species of *Pipa* retain the eggs through metamorphosis, and they hatch as tiny frogs.

Anurans show even greater diversity in their modes of reproduction than salamanders (**Figure 10–18**). Similar reproductive habits have clearly evolved independently in different lineages. Large eggs produce large offspring that probably have a better chance of surviving than smaller ones, but large eggs also require more time to hatch and are exposed to predators for a longer period. Thus, the evolution of large eggs and hatchlings has often been accompanied by the simultaneous evolution of behaviors that protect the eggs, and sometimes the tadpoles as well, from predation. A study of Amazon rain forest frogs revealed a positive relationship between the intensity of predation on frogs' eggs in a pond and the proportion of frog species in the area that laid eggs in terrestrial situations (Magnusson and Hero 1991). Many arboreal frogs lay their eggs on the leaves of trees overhanging water. The eggs undergo their embryonic development out of the reach of aquatic egg predators, and when the tadpoles hatch they drop into the water and take up an aquatic existence. Other frogs, such as *Engystomops pustulosus*, achieve the same result by constructing foam nests that float on the water surface. The female emits mucus during amplexus, which the pair of frogs beat into foam with their hind legs. The eggs are laid in the foam mass, and, when the tadpoles hatch, they drop through the foam into the water.

Although these methods reduce egg mortality, the tadpoles are subjected to predation and competition. Some anurans avoid both problems by finding or constructing breeding sites free from competitors and predators. Some frogs, for example, lay their eggs in the water that accumulates in bromeliads—epiphytic tropical plants that grow in trees and are morphologically specialized to collect rainwater. A large bromeliad may hold several liters of water, and the frogs pass through egg and larval stages in that protected microhabitat. Many tropical frogs lay eggs on land near water. The eggs or tadpoles may be released from the nest sites when pond levels rise after a rainstorm. Other frogs construct pools in the mud banks beside streams. These volcano-shaped structures are filled with water by rain or seepage and provide a favorable environment for the eggs and tadpoles. Some frogs have eliminated the tadpole stage entirely. These frogs lay large eggs on land that develop directly into little frogs. This reproductive mode, called direct development, is characteristic of about 20 percent of all anuran species.

Parental Care Adults of many species of frogs guard the eggs. In some cases it is the male that protects the eggs; in others it is the female. In most cases it is not clearly known which sex is involved because external sex identification is difficult with many anurans. Many of the frogs that lay their eggs over water remain with them. Some species sit beside the eggs; others rest on top of them. Most of the terrestrial frogs that lay direct-developing eggs remain with the eggs and will attack an animal that approaches the nest. Removing the guarding frog frequently results in the eggs being eaten by predators or desiccating and dying before hatching. Male African bullfrogs (*Pyxicephalus adspersus*) guard their eggs and then continue to guard the tadpoles after they hatch. The male frog moves with the school of tadpoles and will even dig a channel to allow the tadpoles to swim from one pool in a marsh to an adjacent one. Tadpoles of several species of the tropical American frog genus *Leptodactylus* follow their mother around the pond. These species of *Leptodactylus* are large and aggressive, and the adult frogs are able to deter many potential predators.

Some of the dart-poison frogs of the American tropics deposit their eggs on the ground, and one of the parents remains with the eggs until they hatch into tadpoles. The tadpoles adhere to the adult and are transported to water. Females of the Panamanian frog *Colostethus inguinalis* carry their tadpoles for more than a week, and the tadpoles increase in size during this period. The largest tadpoles being carried by females had small amounts of plant material in their stomachs, suggesting that they had begun to feed while they were still being transported by their mother. Females of another Central American dart-poison frog, *Oophaga* [*Dendrobates*] *pumilio*, release their tadpoles in small pools of water in the leaf axils of plants and then return at intervals to the pools to deposit unfertilized eggs that the tadpoles eat.

Other anurans, instead of remaining with the eggs, carry the eggs with them. The male of the European midwife toad (*Alytes obstetricians*) gathers the egg strings about his hind legs as the female lays them. He carries them with him until they are ready to hatch, at which time he releases the tadpoles into water. The male of the terrestrial Darwin's frog (*Rhinoderma darwinii*) of Chile snaps up the eggs the female lays and carries them in his vocal pouches, which extend back to the pelvic region. The embryos pass through metamorphosis in the vocal sacs and emerge as fully developed froglets. Male frogs are not alone in caring for eggs. The females of a group of tree frogs carry the eggs on their back, in an open oval depression, a closed pouch, or individual pockets. The eggs develop into miniature frogs before they leave their mother's back. A similar specialization is seen in the completely aquatic Surinam toad, *Pipa pipa*. In the breeding season, the skin of the female's back thickens and softens. During egg laying, the male and female in amplexus swim in vertical loops in the water. On the upward part of the loop, the female is above the male and releases a few eggs, which fall onto his ventral surface. He fertilizes them and, on the downward loop, presses them against the female's back. They sink into the soft skin, and a cover forms over each egg, enclosing it in a small capsule. Tadpoles of the two species of the Australian frog genus *Rheobatrachus* are carried in the stomach of the female frog. The female swallows eggs or newly hatched larvae and retains them in her stomach through metamorphosis. This behavior was first described in *Rheobatrachus silus* and is accompanied by extensive morphological and physiological modifications of the stomach. These changes include distension of the proximal portion of the stomach, separation of individual muscle cells from the surrounding connective tissue, and inhibition of hydrochloric acid secretion, perhaps by prostaglandin released by the tadpoles. In January 1984, a second species of gastric-brooding frog, *R. vitellinus*, was discovered in Queensland. Strangely, this species lacks the extensive structural changes in the stomach that characterize the gastric brooding of *R. silus*. The striking differences between the two species suggest the surprising possibility that this bizarre reproductive mode might have evolved independently. Both species of *Rheobatrachus* disappeared within a few years of their discovery and are thought to be extinct, victims of a worldwide decline of amphibian populations.

Females of the Jamaican brachycephalid frog *Euhyas* [*Eleutherodactylus*] *cundalli* and males of two microhylid frogs from New Guinea (*Liophryne schlaginhaufeni* and *Sphenophryne cornuta*) carry froglets on their backs. *Liophryne* males carried the froglets for periods extending from three to nine nights, and a few froglets jumped off each night (Bickford 2002).

Viviparity Only a few species of anurans are viviparous (Wake 1993). Species in the African bufonid genus *Nectophrynoides* show a spectrum of reproductive modes. One species deposits eggs that are fertilized externally and hatch into aquatic tadpoles, two species produce young that are nourished by yolk, and other species have embryos that feed on secretions from the walls of the oviduct. The golden coquí of Puerto Rico (*Eleutherodactylus jasperi*) also gives birth to fully formed young, but in this case the energy and nutrients come from the yolk of the egg. (The golden coquí has not been seen since 1981 and is presumed to be extinct.)

■ The Ecology of Tadpoles

Although many species of frogs have evolved reproductive modes that bypass an aquatic larval stage, a life history that includes a tadpole has certain advantages. A tadpole is a completely different animal from an adult anuran, both morphologically and ecologically.

Tadpoles are as diverse in their morphological and ecological specializations as adult frogs, and they occupy nearly as great a range of habitats (**Figure 10–19**). Tadpoles that live in still water usually have

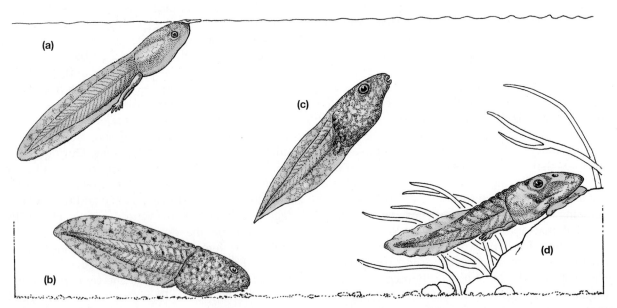

▲ **FIGURE 10–19** Body forms of tadpoles. (a) The tadpole of the Kwangshien spadefoot toad, *Xenophrys* [*Megophrys*] *minor*, is a surface feeder. The mouthparts unfold into a platter over which water and particles of food on the surface are drawn into the mouth. (b) The tadpole of the red-legged frog, *Rana aurora*, scrapes food from rocks and other submerged objects. (c) The tadpole of the red-eyed leaf frog, *Agalychnis callidryas*, is a midwater suspension feeder that filters particles of food from the water column. This species shows the large fins and protruding eyes that are typical of midwater tadpoles. It maintains its position in the water column with rapid undulations of the end of its tail, which is thin and nearly transparent. (d) A stream-dwelling tadpole (an unidentified species of Australasian tree frog, *Litoria*) that adheres to rocks in swiftly moving water with a suckerlike mouth while scraping algae and bacteria from the rocks. The low fins and powerful tail are characteristic of tadpoles living in swift water.

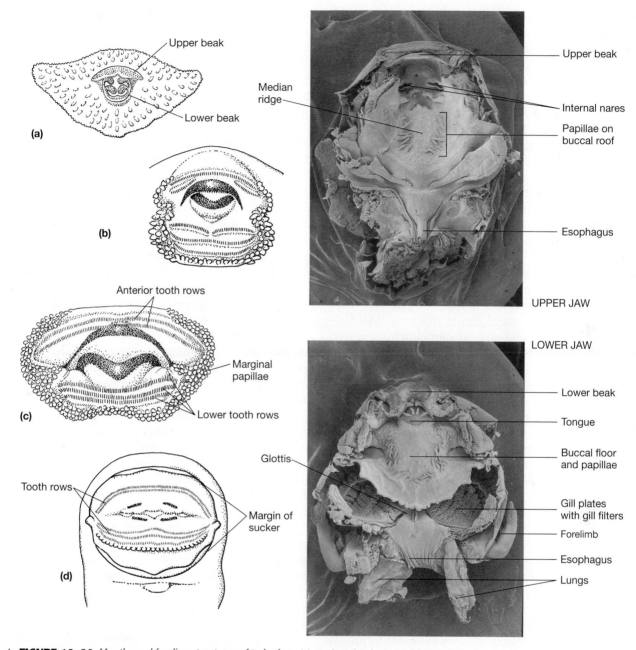

▲ **FIGURE 10–20** Mouths and feeding structures of tadpoles. (a) Surface feeder, *Xenophrys minor*. (b) Surface scraper, *Rana aurora*. (c) Midwater feeder, *Agalychnis callidryas*. (d) Stream-dweller, *Litoria*. (e) Scanning electron micrograph of the inside of the mouth and buccal region of a tadpole of the island spiny-chested frog, *Alsodes monticola*.

ovoid bodies and tails with fins that are as large as the muscular part of the tail, whereas tadpoles that live in fast-flowing water have more streamlined bodies and smaller tail fins. Semiterrestrial tadpoles wiggle through mud and leaves and climb on damp rock faces; they are often dorsoventrally flattened and have little or no tail fin, and many tadpoles that live in bromeliads have a similar body form. Direct-developing tadpoles have large yolk supplies and reduced mouthparts and tail fins. The mouthparts of tadpoles also show variation that is related to diet (**Figure 10–20**). Filter-feeding tadpoles that hover in midwater lack keratinized mouth parts, whereas species that graze from surfaces have small beaks that are often surrounded by rows of denticles. Predatory tadpoles have larger beaks that can bite pieces from other tadpoles. Funnel-mouthed, surface-feeding tadpoles have greatly expanded mouth parts that skim material from the surface of the water.

Tadpoles of most species of anurans are filter-feeding herbivores, whereas all adult anurans are carnivores that catch prey individually. Because of these differences, tadpoles can exploit resources that are not available to adult anurans. This advantage may be a factor that has led many species of frogs to retain the ancestral pattern of life history in which an aquatic larva matures into a terrestrial adult. Many aquatic habitats experience annual flushes of primary production, when nutrients washed into a pool by rain or melting snow stimulate the rapid growth of algae. The energy and nutrients in this algal bloom are transient resources that are available for a brief time to the organisms able to exploit them.

Tadpoles are excellent eating machines. All tadpoles extract suspended food particles from water, and feeding and ventilation of the gills are related activities. The stream of water that moves through the mouth and nares to ventilate the gills also carries particles of food. As the stream of water passes through the branchial basket, small food particles are trapped in mucus secreted by epithelial cells. The mucus, along with the particles, is moved from the gill filters to the ciliary grooves on the margins of the roof of the pharynx and then transported posteriorly to the esophagus.

Although all tadpoles filter food particles from a stream of water that passes across the gills, the method used to put the food particles into suspension differs among species. Some tadpoles filter floating plankton from the water. Tadpoles of this type are represented in several families of anurans, especially the Pipidae and Hylidae, and usually hover in the water column. Midwater-feeding tadpoles are out in the open, where they are vulnerable to predators, and they show various characteristics that may reduce the risk of predation. Tadpoles of the African clawed frog, for example, are nearly transparent and may be hard for predators to see. Some midwater tadpoles form schools that, like schools of fishes, may confuse a predator by presenting so many potential prey items that it has difficulty concentrating its attack on one individual.

Many tadpoles are bottom-feeders that scrape bacteria and algae off the surfaces of rocks or the leaves of plants. The rasping action of their keratinized mouthparts frees the material and allows it to be whirled into suspension in the water stream entering the mouth, and then filtered out by the branchial apparatus. Some bottom-feeding tadpoles, such as toads and spadefoot toads, form dense aggregations that create currents to lift particles of food into suspension in the water. These aggregations may be groups of siblings. Tadpoles of American toads (*Anaxyrus americanus*) and cascade frogs (*Rana cascadae*) are able to distinguish siblings from nonsiblings, and they associate preferentially with siblings. They probably recognize siblings by scent. Toad tadpoles can distinguish full siblings (both parents the same) from maternal half-siblings (only the mother the same), and they can distinguish maternal half-siblings from paternal half-siblings.

Some tadpoles are carnivorous and feed on other tadpoles. Predatory tadpoles have large mouths with a sharp, keratinized beak. Predatory individuals appear among the tadpoles of some species of anurans that are normally herbivorous. Some species of spadefoot toads in western North America are famous for this phenomenon. Spadefoot tadpoles are normally herbivorous; but, when tadpoles of the southern spadefoot toad, *Spea multiplicata*, eat freshwater shrimp that occur in some breeding ponds, they are transformed into the carnivorous morph. These carnivorous tadpoles have large heads and jaws and a powerful beak that allow them to bite off bits of flesh that are whirled into suspension and then filtered from the water stream. In addition to eating shrimp, they prey on other tadpoles.

In an Amazonian rain forest, tadpoles are by far the most important predators of frog eggs. In fact, egg predation decreases as the density of fish increases, apparently because the fish eat tadpoles that would otherwise eat frog eggs (Magnusson and Hero 1991). Carnivorous tadpoles are also found among some species of frogs that deposit their eggs or larvae in bromeliads. These relatively small reservoirs of water may have little food for tadpoles. It seems possible that the first tadpole to be placed in a bromeliad pool may feed largely on other frog eggs—either unfertilized eggs deliberately deposited by the mother of the tadpole, as is the case for the dart-poison frog *Oophaga pumilio,* or fertilized eggs subsequently deposited by unsuspecting female frogs.

The feeding mechanisms that make tadpoles such effective collectors of food particles suspended in the water allow them to grow rapidly, but that growth contains the seeds of its own termination.

As tadpoles grow bigger, they become less effective at gathering food because of the changing relationship between the size of food-gathering surfaces and the size of their bodies. The branchial surfaces that trap food particles are two-dimensional. Consequently, the food-collecting apparatus of a tadpole increases in size approximately as the square of the linear dimensions of the tadpole. However, the food the tadpole collects must nourish its entire body, and the volume of the body increases in proportion to the cube of the linear dimensions of the tadpole. The result of that relationship is a decreasing effectiveness of food collection as a tadpole grows; the body it must nourish increases in size faster than does its food-collecting apparatus.

10.3 Amphibian Metamorphosis

The transition from tadpole to frog involves a complete metamorphosis in which tadpole structures are broken down and their chemical constituents are rebuilt into the structures of adult frogs. In the early twentieth century, Friedrich Gudersnatch discovered the importance of thyroid hormones for amphibian metamorphosis quite by accident when he induced rapid precocious metamorphosis in tadpoles by feeding them extracts of beef thyroid glands. Some details of the interaction of neurosecretions and endocrine gland hormones have been worked out, but no fully integrated explanation of the mechanisms of hormonal control of amphibian metamorphosis is yet possible (Hayes 1997).

Anuran larval development is generally divided into three periods: (1) during premetamorphosis, tadpoles increase in size with little change in form; (2) in prometamorphosis, the hind legs appear and growth of the body continues at a slower rate; and (3) during metamorphic climax, the fore legs emerge and the tail regresses. These changes are stimulated by the actions of thyroxine, and production and release of thyroxine is controlled by a product of the pituitary gland, thyroid-stimulating hormone (TSH).

The metamorphosis of a tadpole to a frog involves readily visible changes in almost every part of the body. The tail is absorbed and recycled into the production of adult structures. The small tadpole mouth that accommodated algae broadens into the huge mouth of an adult frog. The long tadpole gut, characteristic of herbivorous vertebrates, changes to the short gut of a carnivorous animal. The action of

TABLE 10.5 Some morphological and physiological changes induced by thyroid hormones during amphibian metamorphosis

Body form and structure
  Formation of dermal glands
  Restructuring of mouth and head
  Intestinal regression and reorganization
  Calcification of skeleton

Appendages
  Degeneration of skin and muscle of tail
  Growth of skin and muscle of limbs

Nervous system and sense organs
  Increase in rhodopsin in retina
  Growth of extrinsic eye muscles
  Formation of nictitating membrane of the eye
  Growth of cerebellum
  Growth of preoptic nucleus of the hypothalamus

Respiratory and circulatory systems
  Degeneration of the gill arches and gills
  Degeneration of the operculum that covers the gills
  Development of lungs
  Shift from larval to adult hemoglobin

Organs
  Pronephric resorption in the kidney
  Induction of urea-cycle enzymes in the liver
  Reduction and restructuring of the pancreas

thyroxine on larval tissues is both specific and local. In other words, it has a different effect in different tissues, and that effect is produced by the presence of thyroxine in the tissue; it does not depend on induction by neighboring tissues. The particular effect of thyroxine in a given tissue is genetically determined, and virtually every tissue of the body is involved (**Table 10.5**).

Metamorphic climax begins with the appearance of the forelimbs and ends with the disappearance of the tail. This is the most rapid part of metamorphosis, occupying only a few days after a larval period that lasts for weeks or months. One reason for the rapidity of metamorphic climax may lie in the vulnerability of larvae to predators during this period. A larva with legs and a tail is neither a good tadpole nor a good frog: the legs inhibit swimming, and the tail interferes with jumping. As a result, predators are more successful at catching anurans

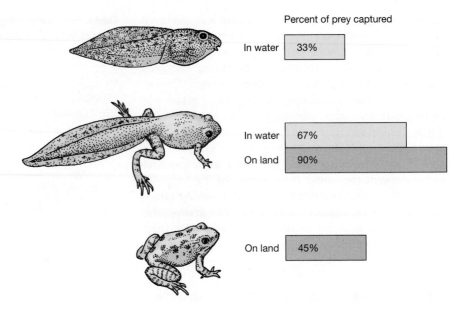

▶ **FIGURE 10–21** Predation on tadpoles and frogs. Metamorphosing chorus frogs (*Pseudacris triseriata*) cannot swim as well as tadpoles nor hop as well as adults. As a result, metamorphosing frogs are more vulnerable to garter snakes than are tadpoles or fully transformed frogs.

Percent of prey captured

In water 33%

In water 67%
On land 90%

On land 45%

during metamorphic climax than they are in prometamorphosis or following the completion of metamorphosis. Metamorphosing chorus frogs (*Pseudacris triseriata*) were most vulnerable to garter snakes when they had developed legs and still retained a tail. Both tadpoles (with a tail and no legs) and metamorphosed frogs (with legs and no tail) were more successful than the metamorphosing individuals at escaping from snakes (**Figure 10–21**). In water, the snakes captured 33 percent of the tadpoles offered, compared with 67 percent of the transforming frogs. On land, the snakes captured 45 percent of the fully transformed frogs that were offered and 90 percent of the transforming ones. Life-history theory predicts that selection will act to shorten the periods in the lifetime of a species when it is most vulnerable to predation, and the speed of metamorphic climax may be a manifestation of that phenomenon.

10.4 Exchange of Water and Gases

Amphibians have a glandular skin that lacks external scales and is highly permeable to gases and water. Both the permeability and glandularity of the skin have been of major importance in shaping the ecology and evolution of amphibians. Mucus glands are distributed over the entire body surface and secrete mucopolysaccharide compounds. The primary function of the mucus is to keep the skin moist and per-

meable. For an amphibian, a dry skin means reduction in permeability. That, in turn, reduces oxygen uptake and the ability of the animal to use evaporative cooling to maintain its body temperature within equable limits. Experimentally produced interference with mucus gland secretion can lead to lethal overheating of frogs undergoing normal basking activity.

Both water and gases pass readily through amphibian skin. In biological systems, permeability to water is inseparable from permeability to gases, and amphibians depend on cutaneous respiration for a significant part of their gas exchange. Although the skin permits passive movement of water and gases, it controls the movement of other compounds. Sodium is actively transported from the outer surface to the inner, and urea is retained by the skin. These characteristics are important in the regulation of osmolal concentration and in facilitating uptake of water by terrestrial species.

■ Blood Flow in Larvae and Adults

Larval amphibians rely on their gills and skin for gas exchange, whereas adults of species that complete full metamorphosis lose their gills and develop lungs. Lungs develop at different larval stages in different families of amphibians; as the lungs develop, they are increasingly used for respiration. Late in their development, tadpoles and partly metamorphosed froglets can be seen swimming to the surface to gulp air. As the gills lose their respiratory function, the carotid arches also change their roles. Arches 1 and 2 are lost early in embryonic development. In

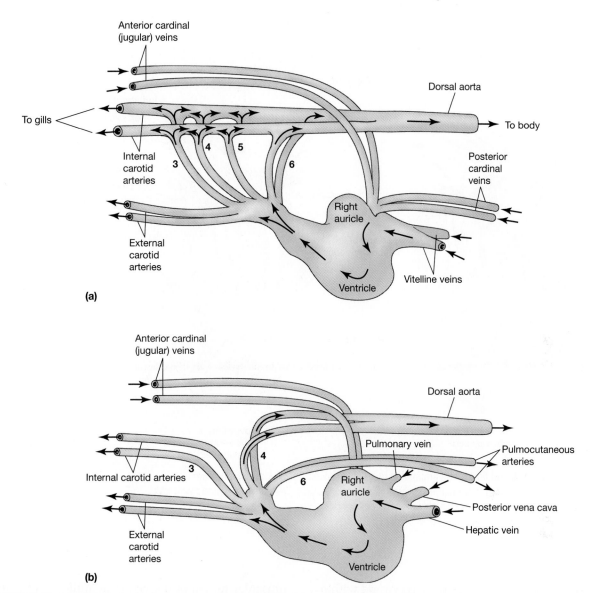

▲ **FIGURE 10–22** Changes in circulation at metamorphosis. Blood flow through the aortic arches of (a) a larval amphibian and (b) an adult without gills. The head is to the left. Arches 3, 4, and 5 carry blood to the gills of larvae, and arch 6 flows into the dorsal aorta. In adults, arch 3 carries blood to the brain via the internal carotid arteries, and arch 6 sends blood to the pulmocutaneous arteries. The posterior cardinal veins return blood from the posterior body, and the vitelline veins carry blood from the intestine.

tadpoles, arches 3 through 5 supply blood to the gills and thence to the internal carotid arteries that carry the blood to the head (**Figure 10–22a**). Arch 6 carries blood to the dorsal aorta via a connection called the ductus arteriosus. At metamorphosis, arch 3 becomes the supply vessel for the internal carotid arteries (**Figure 10–22b**). Initially, arches 4 and 5 supply blood to the dorsal aorta; however, arch 5 is usually lost in anurans, so arch 4 becomes the main route by which

blood from the heart enters the aorta. Arch 6 primarily supplies blood to the lungs and skin via the pulmocutaneous arteries.

■ Cutaneous Respiration

All amphibians rely on the skin surface for gas exchange, especially for the release of carbon dioxide. The balance between cutaneous and pulmonary

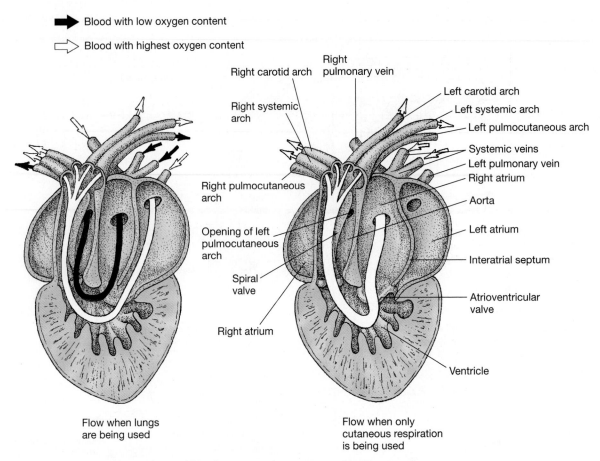

➡ Blood with low oxygen content

⇨ Blood with highest oxygen content

Right carotid arch

Right pulmonary vein

Right systemic arch

Left carotid arch

Left systemic arch

Left pulmocutaneous arch

Systemic veins

Left pulmonary vein

Right atrium

Aorta

Left atrium

Interatrial septum

Right pulmocutaneous arch

Atrioventricular valve

Opening of left pulmocutaneous arch

Spiral valve

Ventricle

Right atrium

Flow when lungs are being used

Flow when only cutaneous respiration is being used

▲ **FIGURE 10–23** Blood flow in the amphibian heart. *Left:* pattern of flow when lungs are being ventilated. *Right:* flow when only cutaneous respiration is taking place. Dark arrows = blood with low oxygen content; light arrows = most highly oxygenated blood.

uptake of oxygen varies among species, and within a species it depends on body temperature and the animal's rate of activity. Amphibians show increasing reliance on the lungs for oxygen uptake as temperature and activity increase.

The patterns of blood flow within the hearts of adult amphibians reflect the use of two respiratory surfaces. The following description is based on the anuran heart (**Figure 10–23**). The atrium of the heart may be divided anatomically into right and left chambers by a septum, or patterns of blood flow may keep oxygenated and deoxygenated streams of blood separate within an undivided chamber. Blood from the systemic veins flows into the right side of the heart, and blood from the lungs flows into the left side. The ventricle shows a variable subdivision that correlates with the physiological importance of pulmonary respiration to the species. The spongy

muscular lumen of the ventricle minimizes the mixing of right- and left-side blood, and the position within the ventricle of a particular parcel of blood appears to determine its fate on leaving the contracting ventricle. The short conus arteriosus contains a spiral valve of tissue that differentially guides the blood from the left and right sides of the ventricle to the aortic arches. The anatomical relationships within the heart are such that oxygen-rich blood returning to the heart from the pulmonary veins enters the left atrium, which injects it on the left side of the common ventricle. Contraction of the ventricle tends to eject blood in laminar streams that spiral out of the pumping chamber, carrying the left-side blood into the ventral portion of the spirally divided conus. This half of the conus is the one from which the carotid (head-supplying) and systemic aortic arches arise.

Thus, when the lungs are actively ventilated, oxygen-rich blood returning from them to the heart is selectively distributed to the tissues of the head and body. Oxygen-poor venous blood entering the right atrium is directed into the dorsal half of the spiral valve in the conus. It goes to the pulmocutaneous arch, destined for oxygenation in the lungs. However, when the skin is the primary site of gaseous exchange, as it is when a frog is underwater, the highest oxygen content is in the systemic veins that drain the skin. The lungs may actually be net users of oxygen, and, because of vascular constriction, little blood passes through the pulmonary circuit. Because the ventricle is undivided and the majority of the blood is arriving from the systemic circuit, the ventral section of the conus receives blood from an overflow of the right side of the ventricle. The scant left atrial supply to the ventricle also flows through the ventral conus. Thus the most oxygenated blood coming from the heart flows to the tissues of the head and body during this shift in primary respiratory surface, a phenomenon possible only because of the undivided ventricle. Variability of the cardiovascular output in amphibians is an essential part of their ability to use alternative respiratory surfaces effectively.

Permeability to Water

The internal osmolal pressure of amphibians is approximately two-thirds that of most other vertebrates. The primary reason for the dilute body fluids of amphibians is low sodium content—approximately 100 milliequivalents compared with 150 milliequivalents in other vertebrates. Amphibians can tolerate a doubling of the normal sodium concentration, whereas an increase from 150 milliequivalents to 170 milliequivalents is the maximum humans can tolerate.

Amphibians are most abundant in moist habitats, especially temperate and tropical forests, but a surprisingly large number of species live in dry regions. Anurans have been by far the most successful amphibian invaders of arid habitats. All but the harshest deserts have substantial anuran populations, and different families have converged on similar specializations. Avoiding the harsh conditions of the ground surface is the most common mechanism by which amphibians have managed to invade deserts and other arid habitats. Anurans and salamanders in deserts may spend nine or ten months of the year in moist retreat sites, sometimes more than a meter underground, emerging only during the rainy season and compressing feeding, growth, and reproduction into just a few weeks.

Many species of arboreal frogs have skins that are less permeable to water than the skin of terrestrial frogs, and a remarkable specialization is seen in a few tree frogs. The African rhacophorid *Chiromantis xerampelina* and the South American hylid *Phyllomedusa sauvagii* lose water through the skin at a rate only one-tenth that of most frogs. *Phyllomedusa* has been shown to achieve this low rate of evaporative water loss by using its legs to spread the lipid-containing secretions of dermal glands over its body surface in a complex sequence of wiping movements, but the basis for the impermeability of *Chiromantis* is not yet understood. These two frogs are also unusual because they excrete nitrogenous wastes as precipitated salts of uric acid (as do lizards and birds) rather than as urea. This method of disposing of nitrogen provides still more water conservation.

Behavioral Control of Evaporative Water Loss

For animals with skins as permeable as those of most amphibians, the main difference between rain forests and deserts may be how frequently they encounter a water shortage. The Puerto Rican coquí (*Eleutherodactylus coqui*) lives in wet tropical forests; nonetheless, it has elaborate behaviors that reduce evaporative water loss during its periods of activity. Male coquís emerge from their daytime retreat sites at dusk and move 1 or 2 meters to calling sites on leaves in the understory vegetation. They remain at their calling sites until shortly before dawn, when they return to their daytime retreats. The activities of the frogs vary from night to night, depending on whether it rained during the afternoon. On nights after a rainstorm, when the forest is wet, the coquís begin to vocalize soon after dusk and continue until about midnight, when they fall silent for several hours. They resume calling briefly just before dawn. When they are calling, coquís extend their legs and raise themselves off the surface of the leaf (**Figure 10–24a**). In this position they lose water by evaporation from the entire body surface.

On dry nights, the behavior of the frogs is quite different. The males move from their retreat sites to their calling stations, but they call only sporadically. Most of the time they rest in a water-conserving posture in which the body and chin are flattened against the leaf surface, the eyes are closed, and the limbs are pressed against the body (**Figure 10–24c**). A frog in this posture exposes only half its body surface to the air, thereby reducing its rate of evaporative water

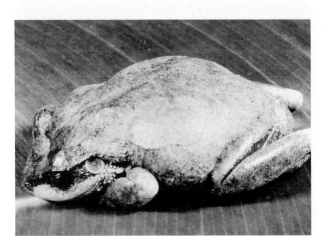

▲ **FIGURE 10–24** A male Puerto Rican coquí, *Eleutherodactylus coquí*. (a) During vocalization, nearly all the body surface is exposed to evaporation. (b) In the alert posture in which frogs wait to catch prey, most of the body surface is exposed. (c) In the water-conserving posture adopted on dry nights, half the body surface is protected from exposure.

loss. The effectiveness of the postural adjustments is illustrated by the water losses of frogs in the forest at El Verde, Puerto Rico, on dry nights. Frogs in one test group were placed individually in small wire-mesh cages that were placed on leaf surfaces. A second group was composed of unrestrained frogs sitting on leaves. The caged frogs spent most of the night climbing around the cages trying to get out. This activity, like vocalization, exposed the entire body surface to the air, and the caged frogs had an evaporative water loss that averaged 27.5 percent of their initial body mass. In contrast, the unrestrained frogs adopted water-conserving postures and lost an average of only 8 percent of their initial body mass by evaporation (Pough et al. 1983).

Experiments showed that the jumping ability of coquís was not affected by an evaporative loss of as much as 10 percent of the initial body mass, but a loss of 20 percent or more substantially decreased the

distance frogs could jump. Thus, coquís can use behavior to limit their evaporative water losses on dry nights to levels that do not affect their ability to escape from predators or to capture prey. Without those behaviors, however, they would probably lose enough water by evaporation to affect their survival (Beuchat et al. 1984).

■ Uptake and Storage of Water

The mechanisms that amphibians use for obtaining water in terrestrial environments have received less attention than those for retaining it. Amphibians do not drink water. Because of the permeability of their skins, species that live in aquatic habitats face a continuous osmotic influx of water that they must balance by producing urine. The impressive adaptations of terrestrial amphibians are ones that facilitate rehydration from limited sources of water. One such specialization is the **pelvic patch**. This is an area of highly vascularized skin in the pelvic region that is responsible for a very large portion of an anuran's cutaneous water absorption. Toads that are dehydrated and completely immersed in water rehydrate only slightly faster than those placed in water just deep enough to wet the pelvic area. In arid regions, water may be available only as a thin layer of moisture on a rock or as wet soil. The pelvic patch allows an anuran to absorb this water.

The urinary bladder plays an important role in the water relations of terrestrial amphibians, especially anurans. Amphibian kidneys produce urine that is hyposmolal to the blood, so the urine in the bladder is dilute. Amphibians can reabsorb water from urine to replace water they lose by evaporation, and terrestrial amphibians have larger bladders than aquatic species. Storage capacities of 20 to 30 percent of the body mass of the animal are common for terrestrial anurans, and some species have still larger bladders: the Australian desert frogs *Notaden nichollsi* and *Neobatrachus wilsmorei* can store urine equivalent to about 50 percent of their body mass, and a bladder volume of 78.9 percent of body mass has been reported for the Australian frog *Heleioporus eyrei*.

Behavior is as important in facilitating water uptake as it is in reducing water loss. Leopard frogs, *Lithobates pipiens,* spend the summer activity season in grassy meadows where they have no access to ponds. The frogs spend the day in retreats they create by pushing vegetation aside to expose moist soil. In the retreats, the frogs rest with the pelvic patch in contact with the ground, and tests have shown that the frogs are able to absorb water from the soil. On nights when dew forms, many frogs move from their retreats and spend some hours in the early morning sitting on dew-covered grass before returning to their retreats. Leopard frogs show a daily pattern of water gain and loss during a period of several days when no rain falls: in the morning the frogs are sleek and glistening with moisture, and they have urine in their bladders. That observation indicates that in the morning the frogs have enough water to form urine. By evening, the frogs have dry skins and little urine in the bladder, suggesting that as they lost water by evaporation during the day they had reabsorbed water from the urine to maintain the water content of their tissues.

By the following morning the frogs have absorbed more water from dew and are sleek and well hydrated again. Net gains and losses of water are shown by daily fluctuations in body masses of the frogs; in the mornings they are as much as 4 or 5 percent heavier than their overall average mass, and in the evenings they are lighter than the average by a similar amount. Thus, these terrestrial frogs are able to balance their water budgets by absorbing water from moist soil and from dew to replace the water they lose by evaporation and in urine. As a result, they are independent of sources of water like ponds or streams and are able to colonize meadows and woods far from any permanent sources of water.

10.5 Poison Glands and Other Defense Mechanisms

The mucus that covers the skin of an amphibian has a variety of properties. In at least some species, it has antibacterial activity, and a potent antibiotic that may have medical applications has been isolated from the skin of the African clawed frog (*Xenopus laevis*). It is mucus that makes some amphibians slippery and hard for a predator to hold. Other species have mucus that is extremely adhesive. Many salamanders, for example, have a concentration of mucus glands on the dorsal surface of the tail. When one of these salamanders is attacked by a predator, it bends its tail forward and buffets its attacker. The sticky mucus causes debris to adhere to the predator's snout or beak, and with luck the attacker soon concentrates on cleaning itself, losing interest in the salamander. When the California slender salamander (*Batrachoseps attenuatus*) is seized by a garter snake, the salamander curls its tail around the snake's head and neck. This behavior makes the salamander hard for the snake to swallow and also spreads sticky secretions on the snake's body. A small snake can find its body glued into a coil from which it is unable to escape.

Although secretions of the mucus glands of some species of amphibians are irritating or toxic to predators, an amphibian's primary chemical defense system is located in the poison glands (**Figure 10–25**). These glands are concentrated on the dorsal surfaces of the animal, and defense postures of both anurans and salamanders present the glandular areas to potential predators.

A great diversity of pharmacologically active substances has been found in the skins of amphibians. Some of these substances are extremely toxic; others are less toxic but capable of producing unpleasant sensations when a predator bites an amphibian. Biogenic amines such as serotonin and histamine, peptides such as bradykinin, and hemolytic proteins

▶ **FIGURE 10–25 Amphibian skin.** This cross section of skin is from the base of the tail of a red-backed salamander, *Plethodon cinereus*. Three types of glands can be seen—hedonic glands produce pheromones used in social interactions with other individuals of the species, poison glands produce toxins that deter predators, and mucus glands secrete a mucopolysaccharide that helps to keep the skin moist and may have antibacterial properties.

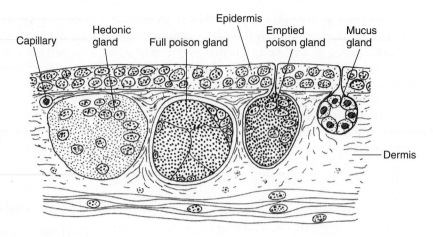

have been found in frogs and salamanders belonging to many families. Many of these substances, such as bufotoxin, epibatadine, leptodactyline, and physalaemin are named for the animals in which they were discovered. Cutaneous alkaloids are abundant and diverse among the dart-poison frogs, the family Dendrobatidae, of the New World tropics. Most of these frogs are brightly colored and move about on the ground surface in daylight, making no attempt at concealment. More than 200 new alkaloids have been described from dendrobatids. Most of the alkaloids found in the skins of dart-poison frogs are similar to those found in ants, beetles, and millipedes that live in the leaf litter with the frogs, suggesting that frogs obtain alkaloids from their prey.

That hypothesis was tested by John Daly and his colleagues by raising juvenile Panamanian dart-poison frogs, *Dendrobates auratus,* under three dietary conditions (Daly et al. 2000). Dart-poison frogs live in the leaf litter on the forest floor, and leaf litter was used as the cage bedding for all experiments. Frogs with the most restricted diet were kept in a glass cage with carefully sealed joints and were fed only fruit flies. (Even these measures did not prevent a few tiny myrmicine ants from entering the cage during the experiment.) To ensure that the frogs in the glass cage had no natural prey, the leaf litter in their cage was frozen for two weeks before the experiment to kill the ants and other arthropods that live in it. Frogs receiving the other two diets were kept outdoors in screened cages with a mesh size large enough to allow fruit flies, ants, and other small arthropods to enter the cages. The frogs in one cage lived in leaf litter that had been frozen and ate primarily fruit flies that were attracted by bananas and fruit fly medium in the cage. Fruit flies were present in such abundance in this cage that they made up the bulk of the

frogs' diet, but myrmicine ants also got into the cage. Frogs in the other screened cage lived in freshly gathered leaf litter that was replaced weekly; these frogs ate the ants and other arthropods that came in with the leaf litter. At the end of the experiment, analyses of the skins of the frogs showed that one frog from the glass cage had no detectable alkaloids and a second had a trace amount of an alkaloid found in myrmicine ants—probably this frog had eaten the ants that had entered the cage despite the sealed joints. Frogs raised in the screened cage that was baited with bananas to attract fruit flies contained four alkaloids, all characteristic of myrmicine ants. Frogs in the cage that received fresh leaf litter had at least 16 different alkaloids in their skins that can be traced to millipedes, beetles, and ants. Wild-caught frogs from the area of the experiment had still more alkaloids in their skins—more than 40 different compounds. This experiment shows clearly that the frogs obtain toxic alkaloids from the prey that they eat, and that the more varied a frog's diet is, the more different alkaloids it contains. Furthermore, some species of *Dendrobates* can modify an alkaloid from insects, converting it to a more toxic form by adding a hydroxyl group to the molecule (Daly et al. 2003).

The name *dart-poison frog* refers to the use by South American Indians of the toxins of some of these frogs to poison the tips of the blowgun darts used for hunting. The use of frogs in this manner appears to be limited to three species of *Phyllobates* that occur in western Colombia; plant poisons like curare are used to poison blowgun darts in other parts of South America. A unique alkaloid, batrachotoxin, occurs in the genus *Phyllobates*. Batrachotoxin is a potent neurotoxin that prevents the closing of sodium channels in nerve and muscle cells, leading to irreversible depolarization and

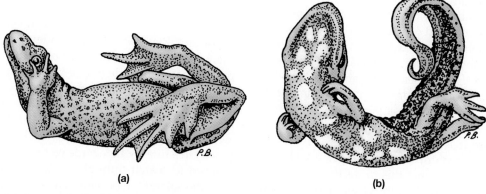

(a)   (b)

▲ **FIGURE 10–26** Aposematic displays by amphibians. Brightly colored species of amphibians have displays that present colors as warnings that predators can learn to associate with the animals' toxic properties. (a) The European fire-bellied toad (*Bombina bombina*) has a cryptically colored dorsal surface and a brightly colored underside that is displayed in the *unken* reflex when the animal is attacked. (b) The Hong Kong newt (*Paramesotriton hongkongensis*) has a brownish dorsal surface and a mottled red and black ventral surface that is revealed by its aposematic display.

producing cardiac arrhythmias, fibrillation, and cardiac failure.

The bright yellow *Phyllobates terribilis* is the largest and most toxic species in the genus. The Emberá Choco Indians of Colombia use *Phyllobates terribilis* as a source of poison for their blowgun darts. The dart points are rubbed several times across the back of a frog and set aside to dry. The Indians handle the frogs carefully, holding them with leaves—a wise precaution because batrachotoxin is exceedingly poisonous. A single frog may contain up to 1900 micrograms of batrachotoxin, and less than 200 micrograms is probably a lethal dose for a human if it enters the body through a cut. Batrachotoxin is also toxic when it is eaten. In fact, the investigators inadvertently caused the death of a dog and a chicken in the Indian village in which they were living when the animals got into garbage that included plastic bags in which the frogs had been carried. Cooking destroys the poison and makes prey killed by darts anointed with the skin secretions of *Phyllobates terribilis* safe to eat.

Many amphibians advertise their distasteful properties with conspicuous **aposematic** (warning) colors and behaviors. A predator that makes the mistake of seizing one is likely to spit it out because it is distasteful. The toxins in the skin may also induce vomiting that reinforces the unpleasant taste for the predator. Subsequently, the predator will remember its unpleasant experience and avoid the distinctly marked animal that produced it. Some toxic amphibians combine a cryptic dorsal color with an aposematic ventral pattern. Normally, the cryptic color conceals them from predators, but if they are attacked, they adopt a posture that displays the brightly colored ventral surface (**Figure 10–26**).

Some salamanders have a morphological specialization that enhances the defensive effects of their chemical secretions. The European salamander *Pleurodeles waltl* and two genera of Asian salamanders (*Echinotriton* and *Tylotriton*) have ribs that pierce the body wall when a predator seizes the salamander. You can imagine the shock for a predator that bites a salamander and finds its tongue and palate impaled by a dozen or more bony spikes! Even worse, the ribs penetrate poison glands as they emerge through the body wall, and each rib carries a drop of poison into the wound.

Many anurans make long leaps to escape a predator, and others feign death. Some cryptically colored frogs extend their legs stiffly when they play dead. In this posture they look so much like the leaf litter on the ground that they may be hard for a visually oriented predator such as a bird to see. Very large frogs attack potential predators. They increase their apparent size by inflating the lungs and hop toward the predator, often croaking loudly. That alone can be an unnerving experience, and some of the carnivorous species such as the horned frogs of South America (*Ceratophrys*), which have recurved teeth on the maxillae and toothlike serrations on the mandibles, also can inflict a painful bite.

Red efts (the terrestrial life stage of the red-spotted newt, *Notophthalmus viridescens*) are classic examples of aposematic animals (see the color insert). They are bright orange and are active during the day, making no effort to conceal themselves. Efts contain tetrodotoxin, which is a potent neurotoxin. Touching an eft to your lips produces an immediate unpleasant numbness and tingling sensation, and the behavior of animals that normally prey on salamanders indicates that it affects them the same way. As a result, an eft that is attacked by a predator is likely to be rejected before it is injured. After one or two such experiences, a predator will no longer attack efts. Support for the belief that this protection may operate in nature is provided by the observation that 4 of 11 wild-caught blue jays (*Cyanocitta cristata*) refused to attack the first red eft they were offered in a laboratory feeding trial (Tilley et al. 1982). That behavior suggests that those four birds had learned to avoid red efts before they were captured. The remaining seven birds attacked at least one eft but dropped it immediately. After one or two experiences of this sort, the birds made retching movements at the sight of an eft and refused to attack.

Of course, aposematic colors and patterns work to deter predation only if a predator can see the aposematic signal. Nocturnal animals may have difficulty being conspicuous if they rely on visual signals. One species of dendrobatid frog apparently deters predators with a foul odor. The aptly named skunk frog (*Aromobates nocturnus*) from the cloud forests of the Venezuelan Andes is an inconspicuous frog, about 5 centimeters long with a dark olive color. These frogs emit a foul, skunklike odor when they are handled.

10.6 Mimicry

The existence of unpalatable animals that deter predators with aposematic colors and behaviors offers the opportunity for other species that lack noxious qualities to take advantage of predators that have learned by experience to avoid the aposematic species. In this phenomenon, known as **mimicry**, the mimic (a species that lacks noxious properties) resembles a noxious model, and that resemblance causes a third species—the dupe—to mistake the mimic for the model. Some of the best-known cases of mimicry among vertebrates involve salamanders. One that has been investigated involves two color morphs of the common red-backed salamander, *Plethodon cinereus*.

Red-backed salamanders normally have dark pigment on the sides of the body; but, in some regions an erythristic (Greek *erythr* = red) color morph is found that lacks the dark pigmentation and has red-orange on the sides as well as on the back. These erythristic morphs resemble red efts and could be mimics of efts. Red-backed salamanders are palatable to many predators, and mimicry of the noxious red efts might confer some degree of protection on individuals of the erythristic morph. That hypothesis was tested in a series of experiments (Brodie and Brodie 1980). Salamanders were put in leaf-filled trays from which they could not escape, and the trays were placed in a forest where birds were foraging. The birds learned to search through the leaves in the trays to find the salamanders. This is a very lifelike situation for a test of mimicry because some species of birds are important predators of salamanders. For example, red-backed salamanders and dusky salamanders (*Desmognathus ochrophaeus*) made up 25 percent of the prey items fed to their young by hermit thrushes in western New York.

Three species of salamanders were used in the experiments, and the number of each species was adjusted to represent a hypothetical community of salamanders containing 40 percent dusky salamanders, 30 percent red efts, 24 percent striped red-backed salamanders, and 6 percent erythristic red-backed salamanders. The dusky salamanders are palatable to birds and are light brown; they do not resemble either efts or red-backed salamanders, and they served as a control in the experiment. The striped red-backed salamanders represent a second control: The hypothesis of mimicry of red efts by erythristic salamanders leads to the prediction that the striped salamanders, which do not look like efts, will be eaten by birds, whereas the erythristic salamanders, which are as palatable as the striped ones but which do look like the noxious efts, will not be eaten.

A predetermined number of each kind of salamander was put in the trays, and birds were allowed to forage for two hours. At the end of that time the salamanders that remained were counted. As expected, only 1 percent of the efts had been taken by birds, whereas 44 percent to 60 percent of the palatable salamanders had disappeared. As predicted, the birds ate fewer of the erythristic form of the red-backed salamanders than they ate of the striped form.

These results show that the erythristic morph of the red-backed salamander does obtain some protection from avian predators as a result of its resemblance to the red eft. In this case the resemblance is visual, but mimicry can exist in any sensory mode to which a dupe is sensitive. Olfactory mimicry by amphibians might be effective against predators such as shrews and snakes, which rely on scent to find and identify prey. This possibility has scarcely been considered, but careful investigations may yield fascinating new examples.

10.7 Why Are Amphibians Vanishing?

Biologists from many countries met in England in 1989 at the First World Congress of Herpetology. In a week of formal presentations of scientific studies and in casual conversations at meals and in pubs, the participants discovered that an alarmingly large proportion of them knew of populations of amphibians that had once been abundant and now were rare or even entirely gone. Events that had appeared to be isolated instances began to look like parts of a global pattern. As a result of that discovery, David Wake, of the University of California at Berkeley, persuaded the National Academy of Sciences to convene a meeting of biologists concerned about vanishing amphibians. Biologists from all over the world met at the West Coast center of the Academy in February 1990. All reported that populations of amphibians in their countries were disappearing, and often there was no apparent reason. Following that meeting, an international effort to identify the causes of amphibian declines was initiated by the Declining Amphibian Populations Task Force of the Species Survival Commission of the World Conservation Union (IUCN). In 1998 another meeting was convened, this time in Washington, DC, that brought together authorities in disciplines ranging from herpetology and population biology through toxicology and infectious diseases to climate change and science policy. The conference concluded that "there is compelling evidence that, over the last 15 years, unusual and substantial declines have occurred in abundance and numbers of populations in globally distributed geographic regions" (Wake 1998). Rapid declines occurred during the 1950s and 1960s and are continuing; at least 9 species, and perhaps as many as 122, have become extinct since 1980 (Mendelson et al. 2006).

The current high rate of declines and extinctions of amphibians contrasts dramatically with the fossil record of the Pleistocene, in which amphibian species persisted over millions of years while birds and mammals disappeared (Stuart et al. 2007). For example, the amphibian fauna of a Middle Pleistocene fossil site in Italy consists exclusively of extant species that lived with species of mammals that are either extinct (saber-toothed tigers) or no longer found in Italy (apes, elephants, rhinoceroses, and leopards). Clearly, something new is happening that affects amphibians more severely than other terrestrial vertebrates.

But what is this new factor that is so lethal to amphibians? The most frightening aspect of the problem is that in many cases we have little idea *why* a species has disappeared from places in which it was formerly abundant. Probably multiple causes are involved, and some of these factors may interact to produce effects that are more severe than simply the sum of their individual effects.

In some cases, local events appear to provide an explanation for the decline of populations of amphibians. Habitat changes produced by logging are usually destructive to amphibians, for example, because frogs and salamanders depend on cool, moist microhabitats on the forest floor. When the forest canopy is removed, sunlight reaches the ground and conditions become too hot and dry for amphibians, and when forest remnants are separated from streams by cultivated fields, stream-breeding species of frogs cannot reach their breeding sites. Chemical contamination—including the use of pesticides—has been implicated in some disappearances. Atrazine, one of the most widely used agricultural herbicides, drains from fields into bodies of water where frogs breed. Exposure of tadpoles of *Lithobates pipiens* to concentrations of 0.1 parts per billion produced feminization: 36 percent of the male frogs showed retarded gonadal development, and 29 percent formed oocytes in their testicles (Hayes et al. 2002). Mining and extraction of oil also cause damage that can extend over large areas. The rock removed from mines often releases acid or toxic chemicals; cyanide that is used to extract gold from ore poisons surface water; oil wells spread toxic hydrocarbons, nitrates and nitrites drain from farmland and reach levels that cause deformities and death of larval amphibians—the list of abuses is nearly endless.

Some local causes of amphibian mortality are not only obvious, they are positively undignified. Federal land in the western United States is leased for grazing, and cattle drink from the ponds that are breeding sites for anurans. As the ponds shrink during the summer, they leave a band of mud that cattle cross when they come to drink. The deep hoof prints the cattle make can be death traps for newly metamorphosed frogs and toads that tumble in and cannot climb out. Even worse, a few juvenile anurans that have the bad luck to pass behind a cow at exactly the wrong moment are trapped and suffocated beneath a pile of fresh manure!

The global scope of the problem suggests that we should look for global explanations (Collins and Storfer 2003). Four factors that have probably contributed to some of these declines are global warming, acid precipitation, increased ultraviolet radiation, and disease.

■ Global Warming

Populations of amphibians at high altitudes appear to be declining even faster than most species, and global warming may be one of the reasons for this pattern (Stuart et al. 2004, Pounds et al. 2006). As the climate has become warmer and drier in the mountains of Costa Rica, lowland species of birds have moved upward and high altitude species of amphibians have declined or disappeared. An analysis of the effect of global warming on mountain forests in Queensland, Australia, predicts that an increase of just 1°C in average annual temperature would diminish the core habitat of high altitude species by 65 percent, and an increase of 3.5°C would completely eliminate the habitats of nearly half of the 65 species of vertebrates that are endemic to the mountains (Williams 2003).

■ Acid Precipitation

Rain, snow, and fog over large parts of the world, especially the Northern Hemisphere, is at least a hundredfold more acidic than it would be if the water were in equilibrium with carbon dioxide in the air. The extra acidity is produced by nitric and sulfuric acids that form when water vapor combines with oxides of nitrogen and sulfur released by combustion of fossil fuels. Water in the breeding ponds of many amphibians in the Northern Hemisphere has become more acidic in the past 50 years, and this acidity has both direct and indirect effects on amphibian eggs and larvae. Embryos of many species of frogs and salamanders are killed or damaged at pH 5 or less. Larvae that hatch may be smaller than normal and sometimes have strange lumps or kinks in their bodies. Spotted salamander larvae grow slowly in acid water because their prey-capture efforts are clumsy and they eat less than do larvae at higher pH (Preest 1993).

■ Ultraviolet Radiation

Destruction of ozone in the stratosphere by chemical pollutants is allowing more intense ultraviolet radiation to reach the Earth's surface. Ozone holes in the stratosphere develop annually at both poles, and they are spreading into lower latitudes in both hemispheres. The role of increased ultraviolet radiation in amphibian declines is controversial and probably complex. Ultraviolet light, especially the 280- to 320-nanometer UV-B band, kills amphibian eggs and embryos (Blaustein and Belden 2003). Only 50 to 60 percent of the eggs of the Cascade frog (*Rana cascadae*) and the western toad (*Anaxyrus* [*Bufo*] *boreas*) in ponds in the Cascade Mountains of Oregon survived when they were exposed to incident sunlight; but, when a filter that blocked UV-B was placed over the eggs, survival climbed to 70 percent to 85 percent.

Some species of amphibians and some breeding sites do not appear to be affected, however, and the damage caused by UV-B radiation may depend on the interaction of several variables (Licht 2003). For example, the presence of overhanging vegetation or turbidity (suspended solids in the water of breeding ponds) may block UV-B before it reaches the eggs of amphibians. Similarly, some dissolved organic chemicals absorb ultraviolet light and may protect amphibian eggs. Species of amphibians differ in their sensitivity to UV-B, and resistance may be correlated with the amount of the enzyme photolyase (an enzyme that repairs UV-induced damage to DNA) in the eggs of different species.

■ Disease

Recently attention has focused on the role of disease in the global decline of amphibians, and especially on two organisms that infect amphibians, iridoviruses and chytrid fungi (Jancovich et al. 1997, Morgan et al. 2007). Disease-causing organisms and amphibians have coexisted as long as amphibians have existed, and some amphibian species have established apparently stable relationships with their pathogens. For example, iridoviruses infect some pond-breeding salamanders and appear to have shaped the biology of populations. Tiger salamanders (*Ambystoma tigrinum*) in Arizona have two larval forms, the normal morph and a cannibalistic morph with an enlarged head and jaws. Some populations consist entirely of the normal morph, whereas in others some individuals develop into cannibalistic larvae that eat individuals of the normal morph. A survey of the occurrence of normal and cannibalistic forms revealed that cannibals are absent from populations that are infected by iridoviruses but occur regularly in populations without the viruses (Pfennig et al. 1991). Apparently in virus-free populations it is safe to eat other larvae, but, when viruses are present, a cannibal risks infection when it eats other larval salamanders.

The correspondence between the presence of iridoviruses and the life-history structure of salamander populations suggests that this host-pathogen association is ancient and that the salamander hosts and their iridovirus pathogens have evolved a stable relationship. This generalization probably applies to other amphibians that serve as hosts for iridoviruses, and it appears that iridoviruses cause fluctuations in amphibian populations but are not usually responsible for extinctions.

A chytrid fungus, *Batrachochytrium dendrobatidis*, is responsible for the disappearances of populations at sites from the Americas to Australia and New Zealand. Chytrids were first linked to the worldwide decline of amphibians in 1998, when they were identified as the probable cause of deaths of frogs in Australia and Central America. The following year, chytrids were identified in dead frogs at the U.S. National Zoo. The chytrid that infects amphibians has motile reproductive zoospores that live in water and can penetrate the skin of an amphibian, causing a disease called chytridiomycosis. When a zoospore enters an amphibian, it matures to form a spherical reproductive body, the zoosporangium, and branching structures that extend through the skin. These structures apparently interfere with respiration and control of water movement, and adult frogs die. Chytrid infections do not appear to be lethal to tadpoles, although infected tadpoles grow slowly and are smaller than normal when they metamorphose and die soon after metamorphosis (Parris 2004). The fungus grows best between 17° and 25 °C, and it is associated especially with the disappearances of mountain populations.

African clawed frogs (*Xenopus laevis* and other species in the genus *Xenopus*) are the source of the fungus. These species are resistant to the infection, but studies of preserved specimens of *Xenopus* in museums in South Africa showed that the fungus was present in that country as early as 1938 (Weldon et al. 2004). The worldwide spread of chytridiomycosis can be traced to the discovery in 1934 that female *Xenopus* provide a more convenient assay for human pregnancy than the rabbit test (Pollack 1949). The Second World War made access to clawed frogs difficult, but after the end of the war enormous numbers of clawed frogs were exported from South Africa. Some of these frogs were released or escaped from laboratories and came into contact with native frogs. Soon chytrid infections began to appear in native frogs. Although chytridiomycosis was not recog-

nized at the time, retrospective studies have revealed chytrid infections in frogs in North America in 1961, in South America in 1977, in Australia in 1978, and in Europe in 1997.

Amphibian declines have been studied especially well in Central and South America and eastern Australia, and chytrid infections and amphibian declines are found in both areas. In Central America, a wave of infection was detected in Costa Rica in 1987 and has been moving southward through Panamá (**Figure 10–27**). Additional waves of infection have spread along mountain chains in northern South America from Peru to Venezuela (Lips et al. 2008). In Australia, a wave of chytrid infection and amphibian population declines is moving northward through Queensland and the Northern Territories (Berger et al. 1998, Schloegel et al. 2006).

Once they have been infected, native species of frogs can spread the fungus. Susceptibility to chytridiomycosis varies among species, and wild individuals of some species can appear to be healthy, even though they are carrying mild infections. Hundreds of thousands of wild-caught amphibians are shipped annually in the food and pet trades, and these shipments are a new route for the spread of chytrids.

Zoos, aquariums, and botanical gardens are in the forefront of attempts to curb amphibian declines (Reid and Zippel 2008). The Amphibian Ark program was initiated as a response to the spread of chytrid infections in Central America (http://www. amphibia-nark.org/index. htm). In March 2005, as the southward advance of chytrid fungi threatened the rich amphibian fauna of El Valle de Antón, Panamá, a team of scientists collected founder populations of 35 species of amphibians and transported them to Zoo Atlanta and the Atlanta Botanical Garden. In addition, a collaborative program with the Houston Zoo and El Nispero Zoo in El Valle established the El Valle Amphibian Conservation Center where additional breeding populations are being maintained. The chytridiomycosis epidemic reached El Valle de Antón only six months after the frogs were sent to Atlanta.

Captive breeding of frogs rescued from advancing epidemics is a first step, but it is far from easy. Many species of frogs are difficult to breed in captivity—they require specific combinations of temperature, photoperiod, and rainfall to bring them into breeding condition, and these conditions are difficult to simulate in captivity. Furthermore, stringent measures are needed to ensure that chytrids are not inadvertently introduced into a breeding facility when new frogs are added or into a natural population when captive-bred frogs are released (Young et al. 2007).

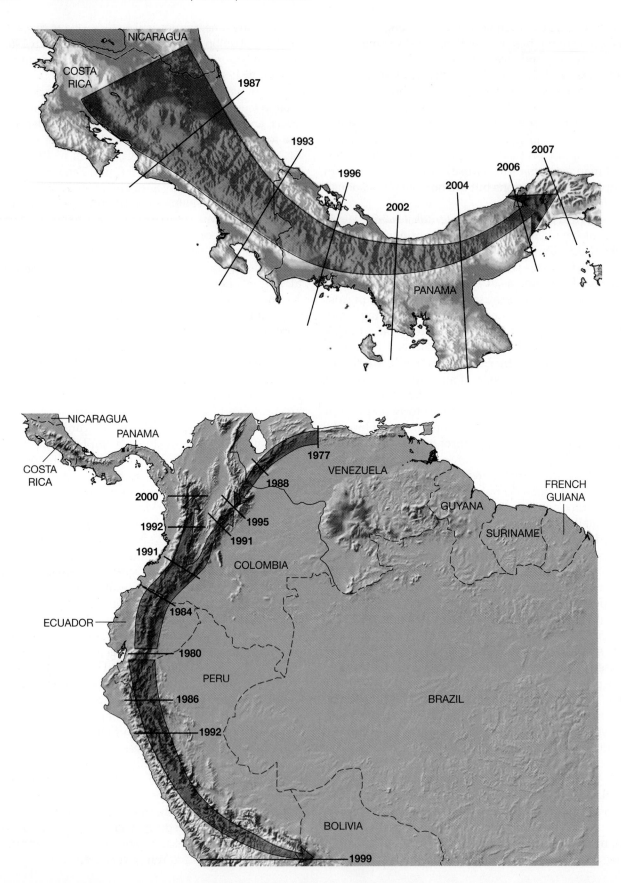

◀ **FIGURE 10–27** Disappearing amphibians. Waves of amphibian population disappearances have flowed across Central and South America as a result of infections by *Batrachochytrium dendrobatidis*. (a) In Central America disappearances were first noted in Monteverde, Costa Rica, in 1987, and by 2007 the wave had swept through most of Panamá and was approaching the border of Colombia. (b) In South America the first population decline occurred near Caracas, Venezuela, in 1977 and a wave of infection spread westward into Colombia along the Cordillera Oriental. The oldest record of *Batrachochytrium dendrobatidis* in South America is from 1980 in Cañar, Ecuador, and that infection spread northward into Colombia and Venezuela along both the Cordillera Occidental and Cordillera Oriental. A third wave spread from Cañar southward into Peru along the Cordillera Occidental and the Cordillera Oriental. An additional introduction of *Batrachochytrium dendrobatidis* into the Atlantic coastal forest of Brazil in 1981 is not shown. (Source: Lips et al. 2008 and personal communications from Karen R. Lips and Douglas C. Woodhams.)

Environmental changes may be contributing to the spread and lethality of chytridiomycosis. Amphibians have immune systems that normally protect them from pathogens in the environment, but stress interferes with the immune system of vertebrates (Carey et al. 1999). Amphibians that are stressed by pollution, acid precipitation, or ultraviolet radiation may be susceptible to diseases or parasites that they would ordinarily be able to resist. Interacting mechanisms of that sort are hard to identify, and the National Science Foundation is supporting a multidisciplinary study that combines the efforts of participants ranging from field biologists through experts in infectious disease to molecular biologists. This group is engaged in a multiyear program of field and laboratory studies to test hypotheses about the roles of iridoviruses and chytrid fungi in amphibian declines and the interactions of these factors with environmental changes and evolutionary changes in the disease organisms. Issues of conservation biology quickly transcend specific fields of study, and our best chance of finding solutions to the acute problems that affect many kinds of animals lies in pooling the techniques and perspectives of a broad spectrum of biological specialties.

Summary

Locomotor adaptations distinguish the lineages of amphibians. Salamanders (Urodela) usually have short, sturdy legs that are used with lateral undulation of the body in walking. Aquatic salamanders use lateral undulations of the body and tail to swim, and some specialized aquatic species are elongate and have very small legs. Frogs and toads (Anura) are characterized by specializations of the pelvis and hind limbs that permit both legs to be used simultaneously to deliver a powerful thrust used both for jumping and for swimming. Many anurans walk quadrupedally when they move slowly, and some are agile climbers. The caecilians (Gymnophiona) are legless tropical amphibians; some are burrowers, and others are aquatic.

The diversity of reproductive modes of amphibians exceeds that of any other group of vertebrates except the fishes. Fertilization is internal in derived salamanders, but most frogs rely on external fertilization. All caecilians have internal fertilization. Many species of amphibians have aquatic larvae. Tadpoles, the aquatic larvae of anurans, are specialized for life in still or flowing water, and some species of frogs deposit their tadpoles in very specific sites, such as the pools of water that accumulate in the leaf axils of bromeliads or other plants. Specializations of tadpoles are entirely different from specializations of frogs, and metamorphosis causes changes in all parts of the body. Direct development that omits the larval stage is also widespread among anurans and is often combined with parental care of the eggs. Viviparity occurs in all three orders.

In many respects the biology of amphibians is determined by properties of their skin. Hedonic glands are key elements in reproductive behaviors, poison glands protect the animals against predators, and mucus glands keep the skin moist, facilitating gas exchange. Above all, the permeability of the skin to water limits most amphibians to microhabitats in which they can control water gain and loss. That sounds like a severe restriction, but, in the proper microhabitat, amphibians can utilize the permeability of their skin to achieve a remarkable degree of independence of standing water. Thus the picture that is sometimes presented of amphibians as animals barely hanging on as a sort of evolutionary oversight is misleading. Only a detailed examination of all facets of their biology can produce an accurate picture of amphibians as organisms.

An examination of that sort reinforces the view that the skin is a dominant structural characteristic of amphibians. This is true not only for the limitations and opportunities presented by the skin's permeability to water and gases but also as a result of the intertwined functions of the skin glands in defensive and reproductive behaviors. The structure and function of the skin may be primary characteristics that have shaped the evolution and ecology of amphibians, and it may also be responsible for some aspects of their susceptibility to pollution. All over the world, populations of amphibians are disappearing at an alarming rate. Some of these extinctions may be caused by regional or global effects of human activities that are likely to affect other organisms as well.

Additional Readings

Alford, R. A., and S. J. Richards. 1999. Global amphibian declines: A problem in applied ecology. *Annual Review of Ecology and Systematics* 30:133–165.

Anderson, J. S. et al. 2008. A stem batrachian from the Early Permian of Texas and the origin of frogs and salamanders. *Nature* 435:515–518.

Berger, L. et al. 1998. Chytridiomycosis causes amphibian mortality associated with population declines in the rainforests of Australia and Central America. *Proceedings of the National Academy of Sciences USA.* 95:9031–9036.

Beuchat, C. A. et al. 1984. Response to simultaneous dehydration and thermal stress in three species of Puerto Rican frogs. *Journal of Comparative Physiology B.* 154:579–585.

Bickford, D. 2002. Male parenting of New Guinea froglets. *Nature* 418:601–602.

Blaustein, A. R. and L. K. Belden. 2003. Amphibian defenses against ultraviolet-B radiation. *Evolution and Development* 5:89–97.

Blaustein, A. R. and A. Dobson. 2006. A message from the frogs. *Nature* 439:143–144.

Brodie, E. D., Jr., and E. D. Brodie III. 1980. Differential avoidance of mimetic salamanders by free-ranging birds. *Science* 208:181–182.

Carey, C. et al. 1999. Amphibian declines: An immunological perspective. *Developmental and Comparative Immunology.* 23:459–472.

Collins, J. P., and A. Storfer. 2003. Global amphibian declines: Sorting the hypotheses. *Diversity and Distributions* 9:89–98.

Daly, J. W. et al. 2000. Arthropod-frog connection: Decahydroquinoline and pyrrolizidine alkaloids common to microsympatric myrmicine ants and dendrobatid frogs. *Journal of Chemical Ecology* 26:73–85.

Daly, J. W. et al. 2003. Evidence of an enantioselective pumilotoxin 7-hydroxylase in dendrobatid poison frogs of the genus *Dendrobates. Proceedings of the National Academy of Sciences, USA* 100:11092–11097.

Feng, A. S. et al. 2006. Ultrasonic communication in frogs. *Nature* 440:333–336.

Frost, D. 2007. Amphibian Species of the World: An Online Reference. Version 5.0 (1 February, 2007). Electronic Database accessible at <http://research.amnh.org/herpetology/ amphibia/index.php>. American Museum of Natural History, New York.

Hayes, T. B. (ed.). 1997. Amphibian metamorphosis: An integrative approach. *American Zoologist* 37:121–207.

Hayes, T. et al. 2002. Feminization of male frogs in the wild. *Nature* 419:895–896.

Heyer, W. R. et al. (ed.). 1993. *Measuring and Monitoring Biological Diversity: Standard Methods for Amphibians.* Washington, DC: Smithsonian Institution Press.

Jaeger, R. G. 1981. Dear enemy recognition and the costs of aggression between salamanders. *American Naturalist* 117: 962–974.

Jaeger, R. G. et al. 1983. Foraging tactics of a terrestrial salamander: Costs of territorial defense. *Animal Behaviour* 31:191–198.

Jancovich, J. K. et al. 1997. Isolation of a lethal virus from the endangered tiger salamander *Ambystoma tigrinum stebbinsi. Diseases of Aquatic Organisms* 31:161–167.

Kiesecker, J. M. et al. 2001. Complex causes of amphibian population declines. *Nature* 410:681–684.

Klump, G. M. and H. C. Gerhardt. 1987. Use of non-arbitrary acoustic criteria in mate choice by female gray tree frogs. *Nature* 326286–288.

Kupfer, A. et al. 2006. Parental investment by skin-feeding in a caecilian amphibian. *Nature* 440:926–929.

Lardner, B., and M. bin Lakin. 2002. Tree-hole frogs exploit resonance effects. *Nature* 420:475.

Licht, L. E. 2003. Shedding light on ultraviolet radiation and amphibian embryos. *Bioscience* 53:551–561.

Lips, K. R. et al. 2006. Emerging infectious disease and the loss of biodiversity in a Neotropical amphibian community. *Proceedings of the National Academy of Sciences USA* 103:3165–3170.

Lips, K. R. et al. 2008. Riding the wave: Reconciling the roles of disease and climate change in amphibian declines. *PLoS Biology* 6(3):e72. doi:10.1371/journal.pbio.0060072

Magnusson, W. E. and J-M. Hero. 1991. Predation and the evolution of complex oviposition behaviour in Amazon rainforest frogs. *Oecologia* 86:310–318.

Marsh, R. L. and T. L. Taigen, 1987. Properties enhancing aerobic capacity of calling muscles in gray tree frogs, *Hyla versicolor. American Journal of Physiology* 252:R786–R793.

McDiarmid, R. W., and R. Altig. 1999. *Tadpoles: The Biology of Anuran Larvae.* Chicago, IL: University of Chicago Press.

Mendelson, J. R. III et al. 2006. Confronting amphibian declines and extinctions. *Science* 313:48.

Min, M. S. et al. 2005. Discovery of the first Asian plethodonid salamander. *Nature* 435:87–90.

Morgan, J. A. T. et al. 2007. Enigmatic amphibian declines and emerging infectious disease: Population genetics of the frog-killing fungus *Batrachochytrium dendrobatidis. Proceedings of the National Academy of Sciences USA* 104:13845–13850.

Parris, M. J. 2004. Hybrid response to pathogen infection in interspecific crosses between two amphibian species (Anura: Ranidae). *Evolutionary Ecology Research* 6:457–471.

Pfennig, D. W. et al. 1991. Pathogens as a factor limiting the spread of cannibalism in tiger salamanders. *Oecologia* 88:161–166.

Pollack, S. S. 1949. The *Xenopus* pregnancy test. *Canadian Medical Association Journal* 60:159–161.

Pough, F. H. 1983. Behavioral modification of evaporative water loss by a Puerto Rican frog. *Ecology* 64:244–252.

Pounds, J. A. et al. 2006. Widespread amphibian extinctions from epidemic disease driven by global warming. *Nature* 439:161–167.

Preest, M. R. 1993. Mechanism of growth rate reduction in acid-exposed larval salamanders, *Ambystoma maculatum*. *Physiological Zoology* 66:686–707.

Reid, G. M. and K. C. Zippel. 2008. Can zoos and aquariums ensure the survival of amphibians in the 21st century? *International Zoo Yearbook* 42:1–6.

Roth, G., and D. B. Wake. 1985. Trends in the functional morphology and sensorimotor control of feeding behavior in salamanders: An example of the role of internal dynamics in evolution. *Acta Biotheoretica* 34:175–192.

Ryan, M. J. 1985. *The Túngara Frog: A Study in Sexual Selection and Communication*. Chicago, IL: University of Chicago Press.

Schloegel, L. M. et al. 2006. The decline of the sharp-snouted day frog (*Taudactylus acutirostris*): The first documented case of extinction by in infection in a free-ranging wildlife species? *EcoHealth* 3:35–40.

Stuart, S. N. et al. 2004. Status and trends of amphibian declines and extinctions worldwide. *Science* 306:1783–1786.

Stuart, S. N. et al. 2007. The past and future of extant amphibians. *Science* 308:49–50.

Taigen, T. L., and K. D. Wells. 1985. Energetics of vocalization by an anuran amphibian (*Hyla versicolor*). *Journal of Comparative Physiology* B155:163–170.

Tilley, S. G. et al. 1982. Erythrism and mimicry in the salamander *Plethodon cinereus. Herpetologica* 38:409–417.

Wake, D. B. 1998. Action on amphibians. *Trends in Ecology and Evolution* 13:379–380.

Wake, D. B., and S. B. Marks. 1993. Development and evolution of plethodontid salamanders: A review of prior studies and a prospectus for future research. *Herpetologica* 49: 194–203.

Wake, M. H. 1993. Evolution of oviductal gestation in amphibians. *Journal of Experimental Zoology* 266:394–413.

Welch, A. M. et al. 1998. Call duration as an indicator of genetic quality in male gray tree frogs. *Science* 280:1928–1930.

Weldon, C. et al. 2004. Origin of the Amphibian Chytrid Fungus. *Emerging Infectious Diseases*, 10:2100–2105 <www.cdc.gov/eid>

Wells, K. D., and T. L. Taigen. 1986. The effect of social interactions on calling energetics in the gray treefrog (*Hyla versicolor*). *Behavioral Ecology and Sociobiology* 19:9–18.

Williams, S. E. et al. 2003. Climate change in Australian tropical rainforests: An impending environmental catastrophe. *Proceedings of the Royal Society* B 270:1887–1892.

Young, S. et al. 2007. Amphibian chytridiomycosis: Strategies for captive management and conservation. *International Zoo Yearbook* 41:85–95.

PART III

Sauropsida: Turtles, Lepidosaurs, and Archosaurs

An early division of amniotes produced the two evolutionary lineages that include the vast majority of extant terrestrial vertebrates, the Synapsida and Sauropsida. Mammals are synapsids, and turtles, tuatara, lizards, snakes, crocodilians, and birds are sauropsids, as are all extinct reptiles except for the so-called mammal-like reptiles.

The lineages can be distinguished in the fossil record by the mid-Carboniferous period, and they show remarkable similarities and differences in the solutions they found to the challenges of life on land. Both approaches were successful. We tend to think of mammals as the preeminent terrestrial vertebrates, but that opinion reflects our own position in the synapsid lineage. The extant species of sauropsids greatly outnumber mammals, and sauropsids have exploited virtually all of the terrestrial adaptive zones occupied by mammals plus many that mammals have never penetrated, such as the gigantic body size achieved by some dinosaurs and the elongate body form of snakes.

CHAPTER 11

Synapsids and Sauropsids: Two Approaches to Terrestrial Life

The terrestrial environment provided opportunities for new ways of life that amniotes have exploited. The amniotic egg may be a critical element of the success of synapsids and sauropsids because amniotic eggs are larger than non-amniotic eggs and produce larger hatchlings that grow into larger adults. Early in their evolutionary history, amniotes split into the two evolutionary lineages that dominate terrestrial habitats today, the Sauropsida and the Synapsida. Extant sauropsids include turtles, the scaly reptiles (tuatara, lizards, and snakes), crocodilians, and birds, whereas mammals are the only extant synapsids. Both lineages underwent great radiations in the Paleozoic and Mesozoic eras that include animals that are now extinct and have no modern equivalents—the dinosaurs and pterosaurs were sauropsids, and the pelycosaurs and therapsids were synapsids.

At the time the sauropsid and synapsid lineages separated, amniotes had evolved few derived characters associated with terrestrial life. As a result, the sauropsid and synapsid lineages independently developed most of the advanced characters that are necessary for terrestrial life, such as respiratory and excretory systems that conserve water and locomotor systems that are compatible with high rates of lung ventilation. Both lineages developed fast-moving predators that could pursue fleeing prey (as well as fleet-footed prey that could run away from predators), and both lineages include species capable of powered flight. Both lineages had members that became endothermal, evolving high metabolic rates and insula-

tion to retain metabolic heat in the body, and both lineages evolved extensive parental care and complex social behavior.

Despite the parallel evolutionary trends in synapsids and sauropsids, differences in the way they carry out basic functions show that they evolved those derived characters independently:

- A terrestrial animal that runs for long distances must eliminate the conflict between respiratory movements and locomotion that is characteristic of primitive amniotes. Derived sauropsids became bipedal and retained expansion and contraction of the rib cage as the primary method of creating the pressure differences that move air in and out of the lungs. In contrast, derived synapsids shifted the primary site of respiratory movements from the rib cage to the diaphragm.
- The high rates of oxygen consumption that are needed to sustain rapid locomotion require respiratory systems that can take up oxygen and release carbon dioxide rapidly and still conserve water. In derived sauropsids, these functions are accomplished with a one-way flow of air through the lung (a through-flow lung), whereas airflow in the lungs of synapsids is in and out (a tidal-flow lung).
- High rates of oxygen consumption during activity produce large amounts of heat, and a layer of insulation allows an animal to retain the heat and raise its body temperature (endothermy). Derived sauropsids have feathers for insulation and derived synapsids have hair.

- A terrestrial animal requires an excretory system that eliminates nitrogenous wastes while conserving water. Sauropsids do this by having an insoluble waste product (uric acid), kidneys that cannot produce concentrated urine, and glands that secrete salt, whereas synapsids excrete a highly soluble waste product (urea) through kidneys that can produce very concentrated urine and lack salt-secreting glands.

These differences in structural and functional characters of sauropsids and synapsids show that there is more than one way to succeed as a terrestrial amniotic vertebrate.

11.1 Taking Advantage of the Opportunity for Sustained Locomotion

Running involves much more than just moving the legs rapidly. If an animal expects to run for very far, the muscles used to move the limbs require a steady supply of oxygen, and that is where the ancestral form of vertebrate locomotion encounters a problem. Early tetrapods moved with lateral undulations of the trunk, as salamanders and lizards do today. The axial muscles provide the power in this form of locomotion, bending the body from side to side. The limbs and feet are used in alternate pairs (i.e., left front and right rear, left rear and right front) to provide purchase on the substrate as the trunk muscles move the animal.

This is an effective form of locomotion (and animals as large as crocodilians use it to move astonishingly fast), but it works only for short dashes. The problem with this ancestral locomotor mode is that the axial muscles are responsible for two essential functions—bending the trunk unilaterally for locomotion and compressing the rib cage bilaterally to ventilate the lungs—and these activities cannot happen simultaneously. **Figure 11–1** illustrates the problem: the side-to-side bending of the lizard's rib cage compresses one lung as it expands the other, so air flows from one lung into the other, interfering with airflow in and out of the mouth (Carrier 1987).

Short sprints are feasible for animals that use lateral bending of the trunk for locomotion because the energy for a sprint is supplied initially by a reservoir of high-energy phosphate compounds (such as adenosine triphosphate [ATP] and creatine phosphate) that are present in the muscle cells. When

▲ **FIGURE 11–1** Lung ventilation and locomotion. The effect of axial bending on lung volume of a running lizard (top view) and a galloping dog (side view). The bending axis of the lizard's thorax is between the right and left lungs. As the lizard bends laterally, the lung on the concave (*left*) side is compressed and air pressure in that lung increases (shown by +), while air pressure on the convex side decreases (shown by –). Air moves between the lungs (*arrow*), but little or no air moves in or out of the animal. In contrast, the bending axis of a galloping mammal's thorax is dorsal to the lungs. As the vertebral column bends, volume of the thoracic cavity decreases; pressure in both lungs rises (shown by +), pushing air out of the lungs (*arrow*). When the vertebral column straightens, volume of the thoracic cavity increases, pressure in the lungs falls (shown by –), and air is pulled into the lungs (*arrow*).

those compounds are used up, the muscles switch to anaerobic metabolism, which draws on glycogen stored in the cells and does not require oxygen. The problem arises when rapid locomotion must be sustained beyond a minute or two because, at that point, the supply of glycogen in the muscles has been consumed. Because of this conflict between running and breathing, lizards that retain the ancestral modes of locomotion and ventilation are limited to short bursts of activity.

Sustained locomotion requires a way to separate respiration from locomotion. Synapsids and sauropsids both developed modes of locomotion that allow the trunk to be held rigid and use limbs as a major source of propulsion, but the ways they did this are quite different.

Locomotion and Respiration of Synapsids

Early nonmammalian synapsids retained the short limbs, sprawling posture, and long tail that are ancestral characters of amniotes and are still seen in extant lizards and crocodilians. Later nonmammalian synapsids, a group called therapsids, had adopted an upright posture with limbs held more underneath the trunk as they are in extant mammals (**Figure 11–2**). Limbs in this position can move fore and aft without bending the trunk.

(a) Pelycosaur *(Mycterosaurus)*

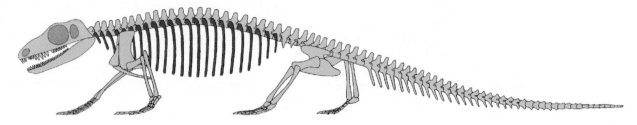

(b) Noncynodont therapsid *(Titanophoneus)*

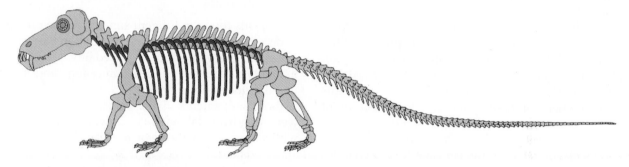

(c) Primitive cynodont therapsid *(Thrinaxodon)*

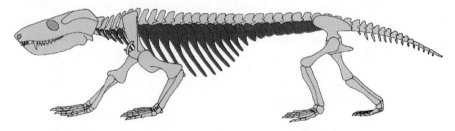

(d) Derived cynodont therapsid *(Massetognathus)*

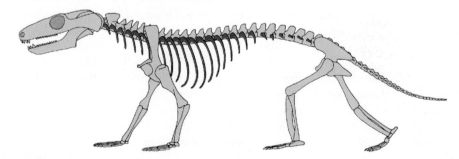

▲ **FIGURE 11–2** Changes in the anatomy of synapsids. Early synapsids such as *Mycterosaurus* (top) retained the ancestral conditions of ribs on all thoracic vertebrae, short legs, and long tails, whereas later synapsids like *Massetognathus* (bottom) had lost ribs from the posterior vertebrae and had longer legs and shorter tails. These changes probably coincided with the development of a diaphragm for respiration and fore-and-aft movement of the legs during locomotion.

A second innovation in the synapsid lineage also contributed to resolving the conflict between locomotion and respiration. Ancestrally, contraction of the trunk muscles produced the reduced pressure within the trunk that draws air into the lungs for inspiration, but this situation changed with the development of a diaphragm by some derived therapsids. The diaphragm is a sheet of muscle that separates the body cavity into an anterior portion (the pulmonary cavity) and a posterior portion (the abdominal cavity). The diaphragm is convex anteriorly (i.e., it bulges toward the head) when it is relaxed and flattens when it contracts. This flattening increases the volume of the pulmonary cavity, creating a negative pressure that draws air into the lungs. Simultaneous contraction of the hypaxial muscles pulls the ribs forward and outward, expanding the rib cage—you can feel this change when you take a deep breath. Relaxation of the diaphragm permits it to resume its domed shape, and relaxation of the hypaxial muscles allows elastic recoil of the rib cage. These changes raise the pressure in the pulmonary cavity, causing air to be exhaled from the lungs.

Movements of the diaphragm do not conflict with locomotion, and in fact the bounding gait of therian mammals carries the resolution of the conflicting demands of locomotion and respiration a step further (Bramble and Carrier 1983, Boggs 2002). The inertial backward and forward movements of the viscera (especially the liver) with each bounding stride work with the diaphragm to force air in and out of the lungs (see Figure 11–1). Thus, in derived mammals, respiration and locomotion work together in a synergistic fashion rather than conflicting.

Humans have little direct experience of this basic mammalian condition because our bipedal locomotion has separated locomotion and ventilation, but locomotion and respiration interact positively in many quadrupedal mammals, and that relationship explains some features of locomotion. Gait and respiration are coupled in most quadrupedal mammals; that is, an animal inhales and exhales in synchrony with limb movements. This coupling of breathing and limb movement depends on matching the respiratory rate to a multiple of the stride interval. The most efficient respiratory rate corresponds to the frequency of the movements of the respiratory system, so the most efficient stride frequency is a multiple of this resonant frequency. If you ride a galloping horse, you will notice that the horse takes a breath with each stride because that frequency matches the reso-

nant frequency of the horse's respiratory system. A galloping horse increases its speed by increasing its stride length rather than its step frequency, thus maintaining the match between stride frequency and the resonant frequency of the respiratory system. Large mammals have lower resonant frequencies and correspondingly lower stride frequencies than do small mammals.

■ Locomotion and Respiration of Sauropsids

Sauropsids found a different solution to the problem of decoupling locomotion and respiration that culminated in the evolution of bipedality, as seen in derived archosaurs such as dinosaurs and birds. Bipedal locomotion involves only the hindlimbs without movements of the trunk. Early sauropsids were quadrupedal animals, however, that moved with lateral undulations of the trunk, just as early synapsids did.

Instead of developing a diaphragm, sauropsids appear to have incorporated pelvic movements and the ventral ribs (gastralia, which are formed by dermal bone) in lung ventilation (Carrier and Farmer 2000a). Extant crocodilians provide a model for understanding the respiratory mechanism used by stem archosaurs. Crocodilians move with lateral undulations of the trunk, but lung ventilation is not limited by locomotor movements—quite the contrary, in fact, alligators hyperventilate during locomotion. Examination of alligators shows that they use three methods of changing the volume of the trunk to move air in and out of the lungs: movement of the ribs, movement of the liver, and rotation of the pubic bones. These movements are shown in **Figure 11–3**.

Birds also use movements of the pelvis to ventilate the lungs, although the muscles and mechanisms of bird respiration are different from those used by the alligator. Nonetheless, the use of pelvic movements for lung ventilation by both crocodilians and birds raises the possibility that this character may have been present in the common ancestor of these lineages. If it is an ancestral character of the archosaur lineage, dinosaurs might have had a similar mechanism of lung ventilation. If that is the case, it may explain some puzzling features of their anatomy, such as the retention of gastralia and the extension of the pubic and ischial bones in the pelvic girdle.

Gastralia are an ancestral character of amniotes that consist of V-shaped bony rods located in the

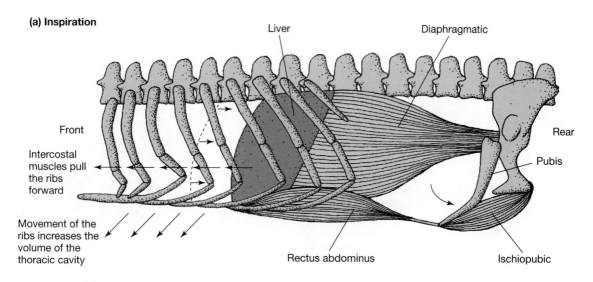

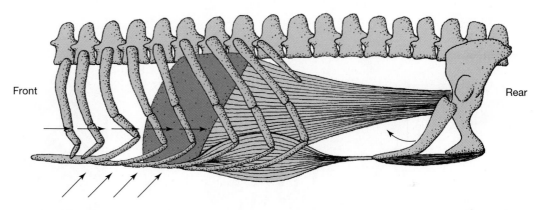

▲ **FIGURE 11–3** Lung ventilation by the alligator. During inspiration (*top*), contraction of the intercostal muscles (not shown) moves the ribs anteriorly, contraction of the diaphragmatic muscle pulls the liver posteriorly, and contraction of the ischiopubic muscle rotates the pubic bones ventrally, increasing the volume of the thoracic cavity. During expiration (*bottom*), the rectus abdominus and transversus abdominus muscles rotate the pubic bones dorsally. This movement forces the viscera anteriorly as the diaphragmatic and intercostal muscles relax, reducing the volume of the thorax and forcing air out of the lungs.

ventral body wall with the apex of the V pointing forward. They appear as bony ventral armor in Paleozoic tetrapods and persist in extant crocodilians, the Mesozoic flying reptiles known as pterosaurs, and in the carnivorous dinosaurs (theropods) that gave rise to birds, although they have been lost in extant birds. Dinosaurs may have used the gastralia in combination with the ribs and with extended pelvic bones to change the volume of the thoracic cavity by a mechanism called cuirassal breathing, a term that is derived from *cuirass*—a type of armor that protects the chest (Carrier and Farmer 2000b) (**Figure 11–4**).

■ Evolution of Lung Ventilation and Locomotor Stamina

The conflicting demands placed on the hypaxial trunk muscles by their dual roles in locomotion and respiration probably limited the ability of early amniotes to occupy many of the adaptive zones that are potentially available to a terrestrial vertebrate. If respiration nearly ceases when an animal moves, as is the case for modern lizards, both speed and distance of movement are limited. Separating locomotion and respiration allows tetrapods to move far and fast, and that separation was achieved in both the synapsid

(a) Inspiration

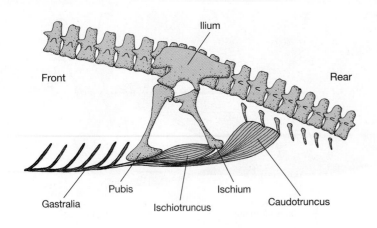

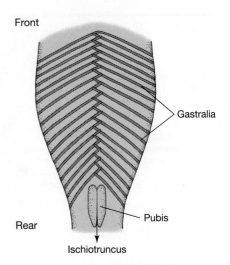

(b) Expiration

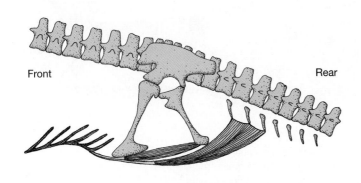

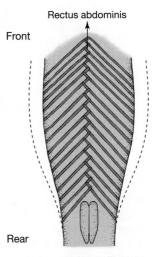

▲ **FIGURE 11–4** Proposed mechanism of cuirassal breathing by a nonavian sauropsid. This reconstruction is based on carnivorous theropod dinosaur *Allosaurus*. As the dinosaur inhales (*top*), the ischiotruncus and caudotruncus muscles pull the gastralia posteriorly, pushing the body wall laterally and ventrally. The expanded area on the ventral end of the elongate pubis may have been a guide that oriented the pull of the muscles. The pubis extends anteriorly and the ischium posteriorly, and the distal ends of those bones are widely separated. As a result, the ischiotruncus muscle is long, which is mechanically significant because muscles can contract by about one-third of their resting length. Expiration (*bottom*) is accomplished by contracting the rectus abdominus muscle, which pulls the gastralia anteriorly, narrowing the V's and pulling the body wall inward and upward.

and sauropsid lineages (**Figure 11–5**). The synapsid solution—loss of the gastralia and the ribs in the lumbar portion of the trunk and development of a muscular diaphragm—appeared early in the development of the lineage and is found in synapsids from the Early Triassic period through extant mammals. In contrast, sauropsids devised a variety of solutions. Archosaurs retained the gastralia and used them for cuirassal breathing, whereas lizards emphasized rib movements and increased flexibility of the trunk resulting from the loss of the gastralia. The contrast between the single solution adopted by synapsids and the multiple solutions of sauropsids probably reflects the diversity of body form in the two lineages. Synapsids remained quadrupedal, and terrestrial through the Mesozoic, while sauropsids became enormously diverse, with species ranging from the size of modern lizards to dinosaurs 30 meters or

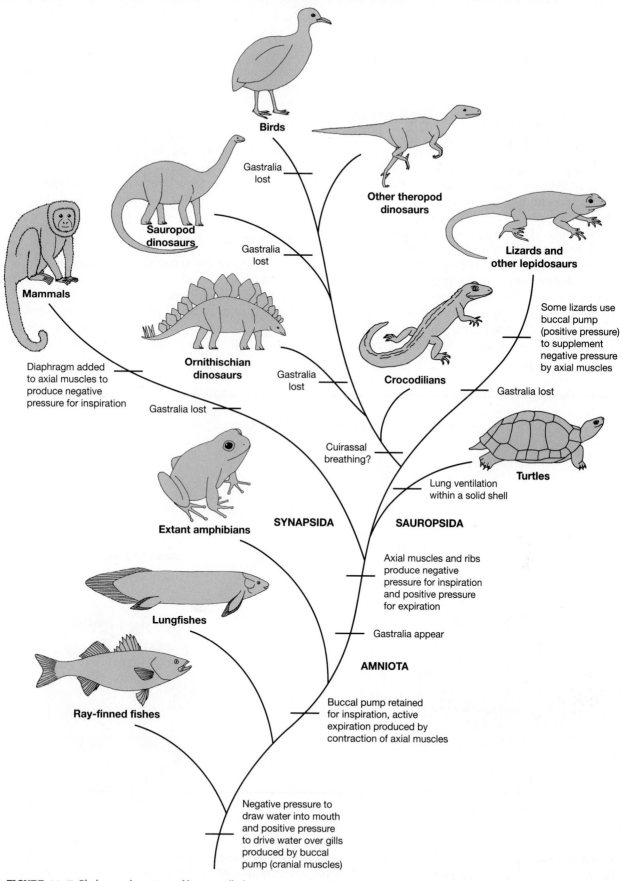

▲ **FIGURE 11–5** Phylogenetic pattern of lung ventilation among tetrapods.

longer, as well as quadrupedal, bipedal, flying, and secondarily aquatic species.

11.2 Increasing Gas Exchange: The Trachea and Lungs

Ventilating the lungs by moving the ribs is a primitive feature of amniotes, and a trachea is probably ancestral as well. The limited speed and endurance of early tetrapods minimized the importance of high rates of gas exchange. Simple lungs—basically internal sacs in which inhaled air could exchange oxygen and carbon dioxide with blood in capillaries of the lung wall—were probably sufficient for these animals. Rates of oxygen consumption would have increased as sustained locomotion appeared, and more surface area would have been needed in the lungs for gas exchange. Complex lungs appeared in both the synapsid and

sauropsid lineages but did so in very different ways: even the most derived synapsids merely increased the effectiveness of the ancestral pattern of in-and-out tidal flow of air to the lungs by elaborating the air passages and gas-exchange surfaces, whereas derived sauropsids (birds, for example) developed a through-flow system for ventilating the lungs. The appearance of faveolar lungs in derived sauropsids appears to coincide with an increase in atmospheric oxygen concentration during the middle of the Mesozoic.

■ Alveolar Lungs

The design of the synapsid respiratory system is an elaboration of the saclike lungs of the earliest tetrapods (**Figure 11–6**). Air passes from the trachea through a series of progressively smaller passages—beginning with the primary bronchi and extending through 50 or more branch points—and ultimately

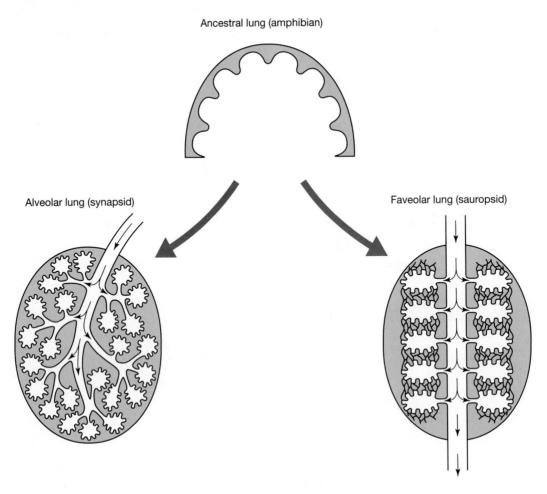

Ancestral lung (amphibian)

Alveolar lung (synapsid)   Faveolar lung (sauropsid)

▲ **FIGURE 11–6** Alveoli and faveoli. Alveoli in the lungs of synapsids are dead-ends; air enters and leaves by the same route. The faveoli of sauropsids are part of a flow-through system by which air enters at one side and leaves from the other.

reaches the respiratory bronchioles and alveolar sacs. The individual alveoli within the alveolar sacs are the sites of gas exchange. Alveoli are tiny (about 0.2 millimeters in diameter) and thin-walled. Blood in the capillaries of the alveolar walls is separated from air in the lumen of the alveolus by only 0.2 micrometers of tissue. This very short diffusion distance is critically important because an individual red blood cell passes through an alveolus in less than a second, and in that time it must release carbon dioxide and take up oxygen. The alveoli expand and contract as the lungs are ventilated, and elastic recoil of alveoli in the mammalian lung helps to expel air. The alveoli are so tiny that they would collapse if were it not for the presence of a substance secreted by alveolar cells that reduces the surface tension of water. The total surface of the alveoli is enormous—in humans it is 70 square meters, equivalent to the floor space of a large room.

■ Faveolar Lungs

The synapsid pattern of branching airways ending in sacs where gas exchange occurs is not the only way to increase the internal surface area of a lung. The alternative that sauropsids have adopted is faveoli, small chambers that open from a common space. Faveolar lungs (which are also called septate lungs) can be quite simple—for example, most lizards have only shallow chambers on the walls of saclike lungs. Monitor lizards (*Varanus*), in contrast are active predators that sustain relatively high levels of oxygen consumption, and they have lungs with extensive faveolar subdivisions.

The faveolar lungs of most sauropsids employ a tidal flow of air, but in birds faveolar lungs are combined with a complex system of air sacs that create a through-flow passage of air in the lung. The respiratory system of birds is unique among extant vertebrates (Maina 2005). Two groups of air sacs, anterior and posterior, occupy much of the dorsal part of the body and extend into cavities (called pneumatic spaces) in many of the bones (**Figure 11–7**). The air sacs are poorly vascularized and do not participate in gas exchange, but they are large—about nine times the volume of the lung—and are reservoirs that store air during parts of the respiratory cycle to create a through-flow lung in which air flows in only one direction.

The trachea branches into a pair of primary bronchi, one passing through each lung. Secondary bronchi branch from each primary bronchus and several parabronchi open from each secondary bronchus. Millions of short interconnected chambers called air capillaries branch from each parabronchus. The air

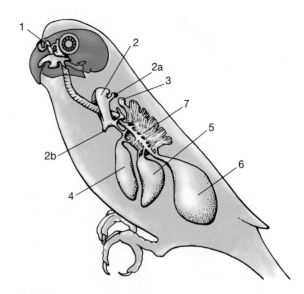

▲ **FIGURE 11–7** The lung and air sac system of the budgerigar. **1**. Infraorbital sinus; **2**. clavicular air sac; **2a**. axillary diverticulum to the humerus; **2b**. sternal diverticulum; **3**. cervical air sac; **4**. cranial thoracic air sac; **5**. caudal thoracic air sac; **6**. abdominal air sacs; **7**. parabronchial lung (Only the left side is shown.).

capillaries intertwine closely with vascular capillaries that carry blood. Airflow and blood flow pass in opposite directions, although they are not exactly parallel because the air and blood capillaries follow winding paths. This arrangement is called a crosscurrent exchange system.

An opposing flow of blood and air in the bird is possible only because air flows through the parabronchial lung in the same direction during both inspiration and exhalation (**Figure 11–8**). Movements of the sternum and pelvis contribute to the changes in pressure that draw air into the air sacs (**Figure 11–9**). During inspiration the sternum rotates ventrally and the pelvic girdle is elevated, increasing the volume of the thorax. On expiration, the pelvis and tail rotate downward, decreasing the volume of the thorax. Two respiratory cycles are required to move a unit of air through the lung: On the first inspiration, the volume of the thorax increases, drawing fresh air through the trachea and primary bronchi into the posterior air sacs. On the first expiration, the volume of the thorax decreases, forcing the air from the posterior sacs into the parabronchial lung. The second inspiration draws that unit of air into the anterior air sacs, and the second expiration sends it out through the trachea.

The through-flow passage of air in the bird lung has several benefits. The crosscurrent flow of air in the air capillaries and blood in the blood capillaries

First cycle

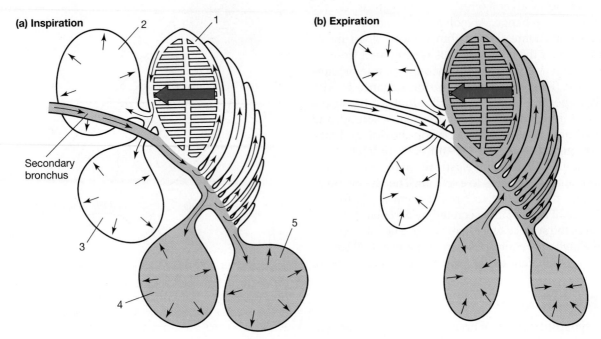

Second cycle

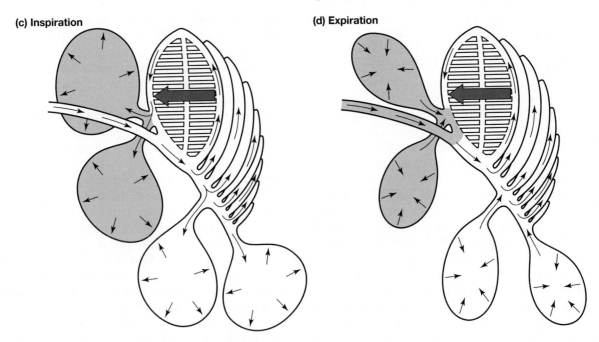

▲ **FIGURE 11–8** Pattern of airflow during inspiration and expiration by a bird. Note that air flows in only one direction through the parabronchial lung. **1.** Parabronchial lung; **2.** clavicular air sac; **3.** cranial thoracic air sac; **4** caudal thoracic air sac; **5** abdominal air sacs.

(a) Inspiration

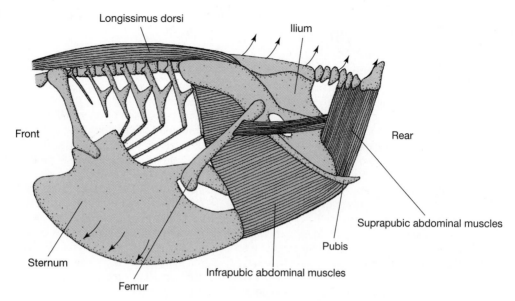

(b) Expiration

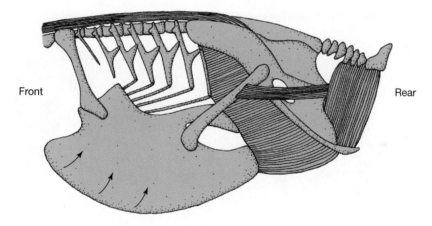

▲ **FIGURE 11–9** Respiratory movements of extant birds. In the pigeon, inspiration is produced by a ventral rotation of the sternum while the longissimus dorsi muscle pulls on the ilium and lifts the pelvis; both of these movements increase the volume of the thorax. On expiration, the sternum returns to its resting position, and contraction of the suprapubic and infrapubic abdominal muscles pull the pelvis and tail downward, reducing the volume of the thorax.

allows efficient exchange of gases, like the countercurrent flow of blood and water in the gills of fishes described in Chapter 4. Because of this efficiency, birds can breathe at very high altitudes (**Box 11–1**). In addition, through-flow passage of air through the lungs minimizes the anatomic dead space in the respiratory system. (The anatomic dead space is the portion of the respiratory system in which air is only pumped back and forth with each breath, not replaced by fresh air.) Any respiratory system in which air enters and leaves via the same opening has some anatomic dead space, and

a tidal flow system like that of mammals leaves some stale air in the lungs with each breath. In humans, the anatomic dead space in the trachea and bronchi is about 150 milliliters, which is about 30 percent of the 500 milliliters of air inhaled in a normal breath. The flow-through lung of a bird reduces the amount of stale air that remains in the lungs and air sacs, allowing birds such as flamingos to have very long necks. (You can demonstrate the problem of combining a long neck with a tidal airflow by trying to breathe through a length of hose. Don't do this for more than a few breaths

BOX 11–1 High-Flying Birds

Birds regularly reach altitudes higher than human mountain climbers can ascend without using auxiliary breathing apparatus, both as residents and during migration. For example, radar tracking of migrating birds shows that they sometimes fly as high as 6500 meters, the alpine chough lives at altitudes around 8200 meters on Mount Everest, and bar-headed geese pass directly over the summit of the Himalayas at altitudes of 9200 meters during their migrations. The ability of birds to sustain activity at high altitudes is a result of the morphological characteristics of their pulmonary systems.

To fully appreciate the feats of these high-flying birds, we need to consider what factors are in play in the Earth's atmosphere. At the surface of the Earth, the atmosphere is most dense because the entire weight of the atmosphere is pressing down on it. At higher altitudes, the atmosphere becomes less and less dense.

At sea level, atmospheric pressure is 760 millimeters of mercury (mm Hg; or 760 torr in International System units). The composition of dry air by volume is 79.02 percent nitrogen and other inert gases, 20.94 percent oxygen, and 0.04 percent carbon dioxide. These gases contribute to the total atmospheric pressure in proportion to their abundance, so the contribution of oxygen is 20.94 percent of 760 torr, or 159.14 torr. The pressure exerted by an individual gas is called the partial pressure of that gas. The rate and direction of diffusion of gas between the air in the lungs and the blood in the pulmonary capillaries is determined by the difference in the partial pressures of the gas in the blood and in the lungs. Oxygen diffuses from air in the lungs into blood in the pulmonary capillaries because oxygen has a higher partial pressure in the air than in the blood, whereas carbon dioxide

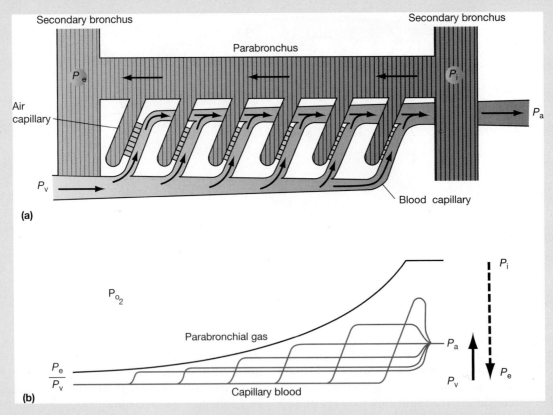

▲ **FIGURE 11–10** Gas exchange in a crosscurrent lung. Air flows from right to left in this diagram, and blood flows from left to right. Darker shading indicates a higher concentration of oxygen. ($P_e =$ oxygen pressure in the air exiting the parabronchus; $P_v =$ oxygen pressure in the mixed venous blood entering the blood capillaries; $P_a =$ oxygen pressure in the blood leaving the blood capillaries; $P_i =$ oxygen pressure in the air entering the parabronchus.) *Top:* General pattern of air and blood flow through the parabronchial lung. *Bottom:* Diagrammatic representation of crosscurrent gas exchange.

diffuses in the opposite direction because its partial pressure is higher in blood than in air.

At higher altitudes, the atmospheric pressure is lower. At 7700 meters, the atmospheric pressure is only 282 torr, and the partial pressure of oxygen in dry air is about 59 torr. Because of the low atmospheric pressure at this altitude, the driving force for diffusion of oxygen into the blood is small. (The actual pressure differential is reduced even below this figure because air in the lungs is saturated with water, and water vapor contributes to the total pressure in the lungs. The vapor pressure of water at 37°C is 47 torr. Thus, the partial pressure of oxygen in the lungs is 20.94 percent of [282 torr minus 47 torr], which is 49 torr.)

How does all this affect breathing? In mammals, it makes breathing at high altitudes difficult. The tidal ventilation pattern of the lungs of mammals means that the partial pressure of oxygen in the pulmonary capillaries can never be higher than the partial pressure of oxygen in the expired air. The best that a tidal ventilation system can accomplish is to equilibrate the partial pressures of oxygen in the pulmonary air and in the pulmonary circulation. In fact, failure to achieve complete mixing of the gas within the pulmonary system means that oxygen

exchange falls short even of this equilibration, and blood leaves the lungs with a partial pressure of oxygen slightly lower than the partial pressure of oxygen in the exhaled air.

In birds, breathing is a different process. The crosscurrent blood flow system in their parabronchial lungs ensures that the gases in the air capillaries repeatedly encounter a new supply of deoxygenated blood (**Figure 11–10**). When blood enters the system (on the left side of the diagram), it has the low oxygen pressure of mixed venous blood. The blood entering the leftmost capillary is exposed to air that has already had much of its oxygen removed farther upstream. Nonetheless, the low oxygen pressure of the mixed venous blood ensures that even in this part of the parabronchus, oxygen uptake can occur. Blood flowing through capillaries farther to the right in the diagram is exposed to higher partial pressures of oxygen in the parabronchial gas and takes up correspondingly more oxygen. The oxygen pressure of the blood that flows out of the lungs is the result of mixing of blood from all the capillaries. The oxygen pressure of the mixed arterial blood is higher than the partial pressure of oxygen in the exhaled air. As a result, birds are more effective than mammals at oxygenating their blood at high altitudes.

because the hose increases the dead space of your respiratory system and prevents fresh air from reaching your lungs.)

■ The Lungs of Dinosaurs

The trachea, lungs, and air sacs are made of soft tissue that does not fossilize, but the fossilized bones of saurischian dinosaurs—the lineage of dinosaurs that includes birds—have cavities and openings that indicate the presence of pneumatic spaces and air sacs in life (Britt 1997). This condition is called pneumaticity, and the most spectacular examples of these spaces are found among the huge, secondarily quadrupedal, long-necked sauropod dinosaurs where the vertebrae have grooves showing the presence of four large air sacs. In *Diplodocus* and related species, which had exceptionally long necks, these grooves extend the entire length of the trunk and onto the anterior vertebrae of the tail. These extremely long-necked animals might have required air sacs and one-way flow of air through the lungs to compensate for the large anatomic dead space in the trachea.

Theropod saurischian dinosaurs, the forms most closely related to birds also had pneumatic vertebrae. Even early theropods appear to have had cervical

and abdominal air sacs and a thoracic skeleton that formed an air pump, and through-flow ventilation of the lungs was probably a general characteristic of theropods (O'Connor and Claessens 2005, Codd et al. 2007)

11.3 Transporting Oxygen to the Muscles: Structure of the Heart

Changes in the mechanics of lung ventilation resolved the conflict between locomotion and breathing, and internal divisions of the lungs increased the capacity for gas exchange. These features were essential steps toward occupying adaptive zones that require sustained locomotion, but another element is necessary—oxygen must be transported rapidly from the lungs to the muscles and carbon dioxide from the muscles to the lungs to sustain high levels of cellular metabolism. A powerful heart can produce enough pressure to move blood rapidly, but there is a complication: although high blood pressure is needed in the systemic circulation to drive blood from the heart to the limbs, high blood pressure would be bad for the lungs. Lungs are delicate structures because of the very short diffusion distances between blood and air that are needed for rapid gas exchange, and high blood pressure in the lungs forces plasma out of the thin-walled capillaries into the air spaces. When these spaces are partly filled with fluid instead of air—as in pneumonia, for example—gas exchange is reduced. Thus, amniotes must maintain different blood pressures in the systemic and pulmonary systems while they are pumping blood at high speed. The solution that derived synapsids and sauropsids found to that problem was separation of the ventricle into systemic and pulmonary sides with a permanent septum, and differences in the hearts of the two lineages indicate that this solution was reached independently in each lineage.

The primitive amniote heart probably lacked a ventricular septum. The flow of blood through the ventricle was probably directed by a spongelike internal structure and perhaps by a spiral valve in a conus arteriosus, as in extant lungfishes and lissamphibians. Turtles and lizards do not have a permanent septum in the ventricle formed by tissue. Instead, during ventricular contraction the wall of the ventricle presses against a muscular ridge in the interior of the ventricle, keeping oxygenated and deoxygenated blood separated. This anatomy, which is probably derived and more complex than the primitive amniote condition, plays an important functional role in the lives of turtles and lizards because it allows them to shunt blood between the systemic and pulmonary circuits in response to changing conditions.

Differences in resistance to flow in the pulmonary and systemic circuits are important in controlling the movement of blood through the hearts of turtles and lizards, and their blood pressures and rates of blood flow are low compared to those of birds and mammals. It may be that increasing blood pressure and rates of flow made a permanent division necessary for derived synapsids (mammals) and derived sauropsids (crocodilians and birds) (**Figure 11–11**). However, each heartbeat must send the same volume of blood to the lungs of mammals and birds as it does to the body. Because the volume of blood in the pulmonary circuit is much smaller than the volume in the systemic circuit, this restriction may limit blood flow to the body. Additionally, blood can no longer be shunted from the (oxygenated) left ventricle to the (deoxygenated) right ventricle, and ventricular coronary vessels must be developed to oxygenate the heart muscle. The muscles in the right ventricle receive oxygenated blood via the coronary arteries, which branch off from the aorta. (These are the vessels in which blockage causes a heart attack.) All amniotes have some coronary supply to the ventricle, but an extensive ventricular coronary system has been evolved independently in mammals and derived sauropsids (archosaurs). Ventricular coronary vessels also evolved convergently in sharks and in some teleosts.

When the ventricle is permanently divided, the relative resistance to blood flow in the systemic and pulmonary circuits no longer determines where blood goes when it leaves the ventricle; instead, blood can flow only into the vessels that exit from each side of the ventricle: the right ventricle leads to the pulmonary circuit and the left ventricle to the systemic circuit. Synapsids and sauropsids have both reached this stage, but they must have done it independently because the relationship of the systemic arches to the left ventricle differs in the two lineages. Mammals retain the left systemic arch as the primary route for blood flow from the left ventricle, whereas birds retain the right arch. Portions of the old right systemic arch remain in adult mammals as the right brachiocephalic artery that gives rise to the right carotid artery (or both carotids in some mammals) and the right subclavian artery. This situation contrasts with the usual sauropsid condition, in which all of the carotids and subclavians branch from the right systemic arch.

Why have birds and mammals each lost one of the systemic arches (or, in the case of mammals, the bottom part of the right systemic arch)? Developmental studies show that synapsid and sauropsid embryos both start off with two arches that are subsequently reduced to a single one. The independent reduction to a single arch in each lineage suggests that this design is somehow better than two arches, although two arches appear to be entirely functional for less derived sauropsids. Perhaps the advantage of a single arch is related to the high blood pressures and high rates of blood flow in the aortic arches of mammals and birds. One vessel with a large diameter creates less friction between flowing blood and the wall of the vessel than two smaller vessels carrying the same volume of blood. In addition, turbulence may develop where the two arches meet, and that would reduce flow. Thus, a single arch may be the best conduit for blood leaving the heart at high pressure.

11.4 Taking Advantage of Wasted Energy: Endothermy

Resolving the conflicts between locomotion and ventilation and modifying the lungs and heart to supply oxygen to muscles did more than just increase the endurance of synapsids and sauropsids—it produced a lot of heat. The synthesis and consumption of chemical energy in compounds like ATP is not very efficient, and a substantial amount of energy is lost as heat. This is why you get hot when you exercise vigorously, and the increase in body temperature during exercise can be substantial; it is overheating rather than exhaustion that forces a cheetah to end its pursuit of a gazelle within a minute of starting its sprint.

The body forms of derived synapsids and derived theropod dinosaurs clearly indicate that increasing activity was developing in both lineages. Both synapsids (therapsids and mammals) and sauropsids (derived archosaurs, including dinosaurs and birds) evolved a posture in which the limbs are held more or less vertically beneath the trunk. Synapsids remained quadrupedal, but archosaurian sauropsids initially became bipedal, although secondarily quadrupedal forms—quadrupedal dinosaurs, for example—appeared later. Synapsids appear to have reorganized their hindlimb musculature more extensively than birds and dinosaurs in association with the new limb position. The increasing locomotor activity and endurance suggested by the changes in body forms would have been accompanied by increased metabolic rates that would have generated substantial amounts of heat during activity, and that heat could have been a critical step toward endothermy.

Endothermy and ectothermy are both effective methods of temperature regulation, but there is a barrier to the evolutionary transition from ectothermy (which is the primitive condition for amniotes) to endothermy. Endothermy requires two characteristics that ectotherms lack—a high metabolic rate and insulation that retains heat in the body. The difficulty in moving from ectothermy to endothermy is that neither a high metabolic rate nor insulation *by itself* is sufficient for endothermy. *Both* characters must be present for an animal to maintain a high and stable body temperature.

A small endotherm, such as a mouse or a sparrow, has a metabolic rate that is about 10 times that of an ectotherm of the same body size, such as a lizard. The heat produced by metabolism in the endotherm is retained by insulation (hair for the mouse, feathers for the bird) and raises body temperature. A lizard lacks insulation, so metabolic heat would be lost, but heat from sunlight is rapidly absorbed. Adding a layer of insulation to the lizard does not allow it to be endothermal because it still lacks a high metabolic rate, but it does prevent the lizard from absorbing heat from sunlight. Raymond Cowles demonstrated that fact in 1958 when he made small fur coats for lizards and measured their rates of warming and cooling. The potential benefit of a fur coat for a lizard is, of course, that it will keep the lizard warm as the environment cools off. However, the well-dressed lizards in Cowles's experiments never achieved that benefit because when they were wearing fur coats they were unable to get warm in the first place.

So the evolution of endothermy involves a catch-22—insulation provides no advantage to an animal without a high metabolic rate, and the heat produced by a high metabolic rate is lost unless the animal has insulation. This paradox confounded discussions of the evolution of endothermy for decades as authors offered scenarios in

▶ **FIGURE 11–11** Diagrammatic view of the form of the heart and aortic arches in synapsids and sauropsids. (a) Early amniote condition: a conus arteriosus with a spiral valve and a truncus arteriosus are retained, and a ventricular septum is lacking. This condition, basically like that of living amphibians, is proposed here to account for the differences in these structures between synapsids and sauropsids. (b). Hypothetical early synapsid condition. Mammal ancestors cannot have had the sauropsid pattern of dual systemic arches, as it would be impossible with the sauropsid condition to retain the left arch only, as seen in mammals. Here a separation of the truncus arteriosus into separate pulmonary and (single) systemic trunks is proposed. Some degree of shunting between pulmonary and systemic circuits and within the heart may have been possible with an incomplete ventricular septum. The sinus venosus is shown as retained because a small sinus venosus is still present in monotremes. (c). Mammal (therian). The ventricular septum is complete and the lower portion of the right systemic arch has been lost. (d) Generalized sauropsid condition, seen in turtles and lepidosaurs. The truncus arteriosus is divided into three parts: a pulmonary arch, and two separate systemic arches (the left arch exits from the right side of the heart and the right arch exits from the left side). The ventricular septum is incomplete, although it may be complexly subdivided. This system allows blood to be shunted within the heart and between pulmonary and systemic circuits. (e) Crocodile. The ventricular septum (possibly a separate development in archosaurs) is now complete, but shunting between left and right systemic arches is still possible via the foramen of Panizza. (f). Bird. The entire left systemic arch is now lost, and the sinus venosus has been subsumed into the right atrium. (R = right; L = left)

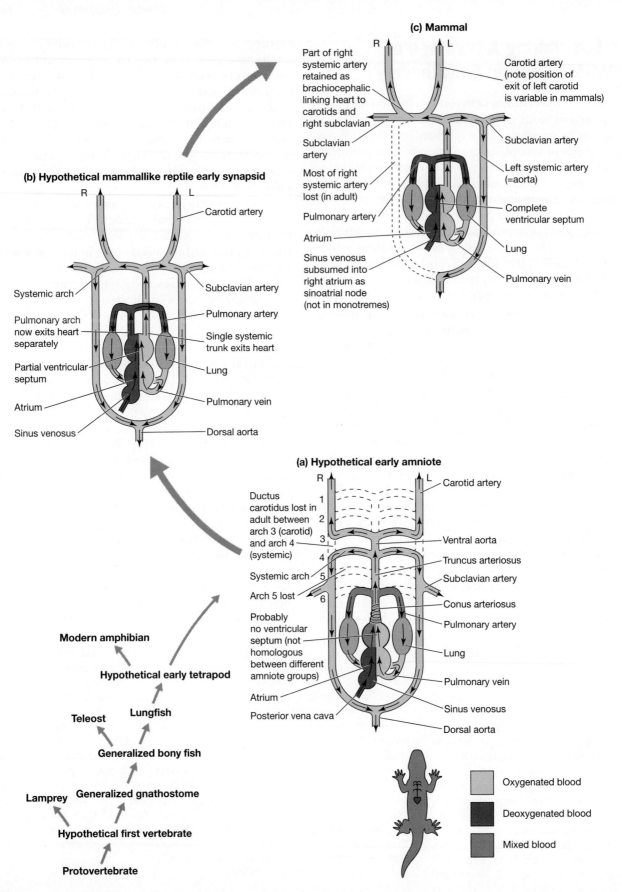

(c) Mammal

Part of right systemic artery retained as brachiocephalic linking heart to carotids and right subclavian

Carotid artery (note position of exit of left carotid is variable in mammals)

Subclavian artery

Subclavian artery

Most of right systemic artery lost (in adult)

Left systemic artery (=aorta)

Pulmonary artery

Complete ventricular septum

Atrium

Lung

Sinus venosus subsumed into right atrium as sinoatrial node (not in monotremes)

Pulmonary vein

(b) Hypothetical mammallike reptile early synapsid

Carotid artery

Systemic arch

Subclavian artery

Pulmonary arch now exits heart separately

Pulmonary artery

Single systemic trunk exits heart

Partial ventricular septum

Lung

Atrium

Pulmonary vein

Sinus venosus

Dorsal aorta

(a) Hypothetical early amniote

Carotid artery

Ductus carotidus lost in adult between arch 3 (carotid) and arch 4 (systemic)

1
2
3
4

Ventral aorta

Truncus arteriosus

Systemic arch

5

Subclavian artery

Arch 5 lost

6

Conus arteriosus

Pulmonary artery

Probably no ventricular septum (not homologous between different amniote groups)

Lung

Pulmonary vein

Atrium

Sinus venosus

Posterior vena cava

Dorsal aorta

Modern amphibian

Hypothetical early tetrapod

Teleost   Lungfish

Generalized bony fish

Generalized gnathostome

Lamprey

Hypothetical first vertebrate

Protovertebrate

Oxygenated blood

Deoxygenated blood

Mixed blood

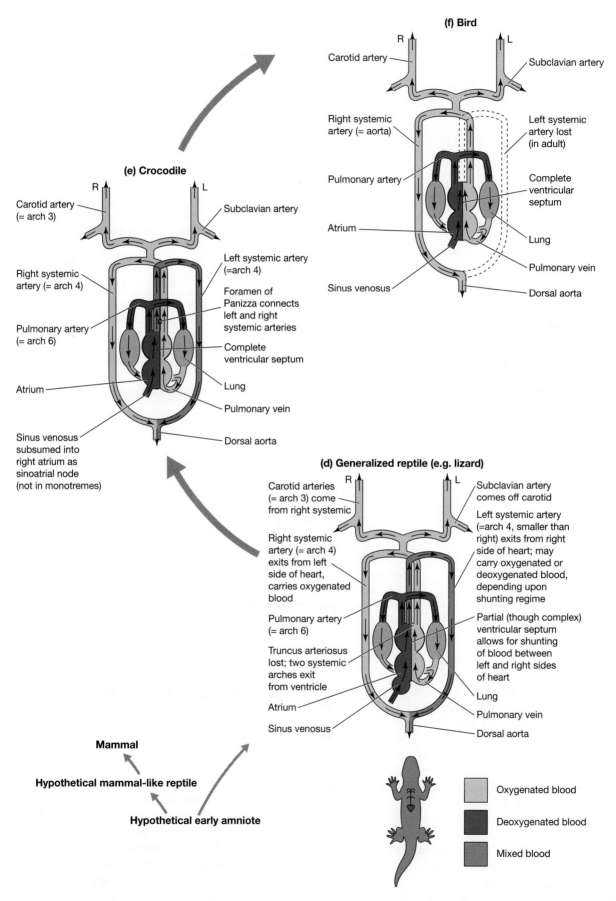

(f) Bird

Carotid artery

Subclavian artery

Right systemic artery (= aorta)

Left systemic artery lost (in adult)

Pulmonary artery

Complete ventricular septum

Atrium

Lung

Pulmonary vein

Sinus venosus

Dorsal aorta

(e) Crocodile

Carotid artery (= arch 3)

Subclavian artery

Right systemic artery (= arch 4)

Left systemic artery (=arch 4)

Foramen of Panizza connects left and right systemic arteries

Pulmonary artery (= arch 6)

Complete ventricular septum

Atrium

Lung

Pulmonary vein

Sinus venosus subsumed into right atrium as sinoatrial node (not in monotremes)

Dorsal aorta

(d) Generalized reptile (e.g. lizard)

Carotid arteries (= arch 3) come from right systemic

Subclavian artery comes off carotid

Right systemic artery (= arch 4) exits from left side of heart, carries oxygenated blood

Left systemic artery (=arch 4, smaller than right) exits from right side of heart; may carry oxygenated or deoxygenated blood, depending upon shunting regime

Pulmonary artery (= arch 6)

Truncus arteriosus lost; two systemic arches exit from ventricle

Partial (though complex) ventricular septum allows for shunting of blood between left and right sides of heart

Atrium

Lung

Sinus venosus

Pulmonary vein

Dorsal aorta

Mammal

Hypothetical mammal-like reptile

Hypothetical early amniote

Oxygenated blood

Deoxygenated blood

Mixed blood

TABLE 11.1 Changes associated with the development of endothermy in synapsids and sauropsids

Physiological Issue	Anatomical Correlates	
	Synapsids (mammals)	Sauropsids (birds)
	Derived members of both lineages adopted upright posture so the trunk does not bend laterally as the limbs move.	
Need to resolve conflict between locomotion and lung ventilation that results from the primitive tetrapod method of locomotion with axial flexion of trunk.	Primitive quadrupedal posture was retained with derived changes in the hindlimb muscles (e.g., gluteal muscles rather than caudofemoral muscles are used to retract the hindlimb.) Changes in the limbs are visible in the fossil record.	Derived upright posture was developed in correlation with bipedality, and primitive pattern of hindlimb muscles was retained. Changes in the limbs are visible in the fossil record.
Need more oxygen to support the high metabolic rates associated with sustained locomotion.	Develop diaphragm to aid in lung ventilation. Can be inferred in fossils from loss of lumbar ribs. Develop secondary palate to eat and breathe at the same time. Preserved in fossils.	Develop flow through lung with one-way passage of air. Can be inferred in fossils from cavities and openings in bones that reveal the presence of air sacs in some dinosaurs and perhaps from their very long necks.
Need to warm and humidify large volumes of air on inspiration and recover water and heat on expiration.	Turbinate bones in the nasal passages that provide a large, moist surface. Turbinates are bony, and traces can be seen in fossils.	Narrow nasal passages suggest that cartilaginous turbinates were a late development.
Need more food to fuel high rates of metabolism.	Develop complex teeth with precise occlusion to reduce particle size of food. Preserved in fossils. Increased volume of jaw musculature to masticate food. Can infer presence of large muscles from skull features of fossils.	Muscular gizzard used to reduce particle size of food. Not preserved in fossils, but gizzard stones have been found in association with the fossils of dinosaurs.
Need to retain heat produced by metabolism within the body.	Development of hair and perhaps subcutaneous fat deposits (blubber) No fossils yet from pre-Cretaceous sediments that show such fine detail.	Development of feathers. Visible in fossils preserved in very fine-grained sediments.

which incipient insulation could initially have functioned in ectothermal thermoregulation before it became effective in retaining metabolic heat. These proposals were not very convincing and were not generally accepted. The relationship of locomotor activity to the evolution of endothermy was not appreciated until the late twentieth century. Clearly endothermy evolved in a stepwise process in which the appearance of one new feature created conditions in which another new feature could be advantageous; the complication is deciphering the sequence in which these changes occurred (Kemp 2006).

■ Endothermy in Synapsids

The evolution of a high metabolic rate via increased capacity for locomotor activity involved changes in many parts of the body of synapsids. Changes in the skeleton can be seen in fossils, but changes in the soft tissues and in physiological characteristics are not fossilized. These changes must be inferred from the parts of an animal that do fossilize. **Table 11.1** lists several characteristics of derived amniotes that are associated with the evolution of endothermy.

The changes in locomotion and respiration that we have described set the stage for additional demands on the physiology of synapsids, and one of the most important is associated with high rates of respiration. The air an endothermal animal inhales is dry and cooler than the temperature deep inside an animal's body (the core body temperature). Lungs are delicate tissues that would dry out if they were ventilated with air that was not saturated with water vapor at the core body temperature, so inhaled air must be warmed and humidified before it reaches the lungs. Furthermore, when the air has been saturated with water vapor and warmed to the core body temperature, merely exhaling the air would represent a loss of water and heat that the animal can't afford. Thus, animals need a way to warm and humidify air they inhale and then to recover water and heat when they exhale—that is, a recycling mechanism for water and heat.

Recycling heat and water is not difficult for animals with low rates of metabolism and lung ventilation. Extant lizards do this, and the moist walls of their tubular nasal passages provide enough surface area to meet their needs. As metabolic rate increases, however, the rate of ventilation increases, and that places greater demands on the recycling system. An animal that inhales and exhales air rapidly needs a larger surface area than just the walls of simple tubular nasal passages.

This additional surface area is provided in extant mammals by an array of thin sheets of bone or cartilage in the nasal passages that are covered in life with moist tissue (**Figure 11–12**). These are the turbinates, which are also called conchae, because they form spiral curves that look like the interiors of some shells. Mammals have two kinds of turbinates: olfactory and respiratory. The olfactory turbinates support the olfactory epithelium that contains the sensory cells used for olfaction. They are located above and behind the nasal passages, out of the direct flow of air. (This is why you sniff when you are trying to smell something—the abrupt inhalation draws air over the olfactory turbinates.)

The respiratory turbinates protrude directly into the main pathway of respiratory airflow, and air passes over them with each inspiration and expiration. Inhaled air is warmed and humidified, and the turbinates are cooled by the combination of cool outside air and the cooling effect of water evaporating from their surface. Then the warm, moist air leaving the lungs is cooled as it passes back over the turbinates, and some of the water vapor in the air condenses. The warm air and condensation of water rewarm the turbinates, preparing them for the next inhalation. Thus, the turbinates recycle both water and heat as air flows in and out of the lungs. A recycling system of this sort is probably essential, because without it the loss of heat and water would be too large for an endotherm to sustain (Ruben et al. 2003). Respiratory turbinates are found in all extant mammals, and traces of the ridges that support them can be seen in the nasal passages of derived therapsids and early mammals.

High metabolic rates require high rates of food intake and rapid digestion. The changes in body form seen in therapsids gave them greater capacity

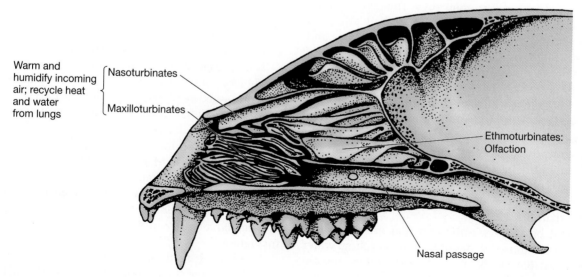

Warm and humidify incoming air; recycle heat and water from lungs

Nasoturbinates

Maxilloturbinates

Ethmoturbinates: Olfaction

Nasal passage

▲ **FIGURE 11–12** Longitudinal section through the snout of a raccoon.

for sustained locomotion, allowing them to forage over larger areas or to pursue prey. In addition, the teeth became more complex with precise occlusion of the surfaces of teeth in the upper and lower jaw. These changes in dentition allowed therapsids to process food more completely, slicing chunks of meat into small pieces that would provide a large surface area for the digestive enzyme to attack. In addition, the jaw muscles became larger, increasing the force that could be applied by the teeth.

In addition to metabolic heat production, an endotherm needs a layer of insulation to retain heat in the body. Most extant terrestrial mammals have hairy pelts that provide insulation, although some bare-skinned mammals such as pigs have very sparse hair and rely on a subcutaneous layer of fat for insulation. Hair does not provide good insulation for animals that spend all their time in the water, and specialized aquatic mammals such as whales and porpoises have thick layers of blubber for insulation. Neither hair nor blubber fossilizes well, and we have no direct evidence of insulation in therapsids.

▇ Endothermy in Sauropsids

Sauropsids faced the same challenges to the evolution of endothermy as synapsids, and they found similar solutions, though with some differences in detail. The capacity to sustain locomotor activity was probably the starting point for the evolution of endothermy by sauropsids, just as it was for synapsids. The body forms of dromeosaurs (derived theropod dinosaurs that are close to the transition to birds) strongly suggest that they were fleet-footed predators that pursued their prey. The faveolar lung is effective in gas exchange, and derived dromeosaurs probably had a system of air sacs and flow-through ventilation of the lungs. The teeth of carnivorous dinosaurs never attained the complex occluding surfaces that we see in mammals, but they were capable of slicing through flesh. Birds, of course, lack teeth and process food in a muscular gizzard that can exert tremendous pressure. Many birds purposely swallow gravel that lodges in the gizzard and helps to pulverize food particles. Some dinosaurs also had stones in their gizzards, although the gizzard stones of dinosaurs are large pebbles rather than fine gravel.

Extant birds have turbinates that are at least as effective as those of mammals in recycling water and heat (Geist 2000), but they often remain cartilaginous and do not have bony contacts with the nasal passages that could be seen in fossils.

Turbinates have not been identified in dinosaurs, and computed tomography (CT) scans of the fossilized skulls of advanced theropod dinosaurs and primitive birds suggest that the nasal passages were narrow. Thus, pre-avian sauropsids and early birds probably did not have turbinates (Ruben et al 2003), but inhaled air may have passed over an extensive moist surface area in the nasal passages, allowing both humidification of the air and olfactory sensing (Witmer 2001).

Derived theropod dinosaurs were probably endothermal. Not only do their body forms suggest that they were similar in their ecology and behavior to flightless birds such as ostriches, but the discovery in the late 1990s of dromeosaurs with feathers (Qiang et al. 1998, 2001) shows that insulation evolved before flight (**Figure 11–13**). These fossils were found at a site in Liaoning Province in northeastern China that has produced spectacular material of Early Cretaceous age (Stokstad 2001). *Caudipteryx*, the first of the feathered dinosaurs to be discovered, and *Protoarchaeopteryx* have both downlike feathers like the feathers that provide insulation to extant birds and vaned feathers like the flight feathers of extant birds—although both dromeosaurs had relatively short arms and neither could fly. *Protoarchaeopteryx* has downlike feathers on the body and tail and a fan of symmetrical vaned feathers on the tail. *Caudipteryx* has vaned feathers attached to the second finger of the hand, where remiges (primary feathers) are found in modern birds, and a tuft of vaned feathers on the tail. The entire body of *Sinosauropteryx*, another dromeosaur from Liaoning Province, is covered with downy filaments interpreted as protofeathers. On the tail the filaments making up the structures radiate from a single point, and on the arms the filaments radiate from a central stem, as in the feathers of modern birds.

The presence of two kinds of feathers in *Caudipteryx* and *Protoarchaeopteryx*, downlike feathers and vaned feathers, suggests that two selective processes may have been operating simultaneously. Down insulates birds, just as it insulates humans when it is stuffed into a down vest, so the presence of down is consistent with the hypothesis that the metabolic rates of dromeosaurs were high enough to produce heat that could be retained in the body by insulation. The filamentous body covering of *Sinosauropteryx* could also have been a layer of insulation.

The vaned feathers on the wings and tail of *Caudipteryx* and *Protoarchaeopteryx* would not have been useful for insulation. In modern birds vaned

▲ **FIGURE 11–13** *Caudipteryx zoui.* This ia a feathered dinosaur from Liaoning Province, China. The body was covered with downy feathers, and there were long feathers on the arms and tail. The snout was short with a few teeth.

feathers provide lift, propulsion, and steering during flight, but neither of the fossil species was capable of flight. *Protoarchaeopteryx* had arms only half the length of modern birds, and *Caudipteryx* had even shorter arms. Furthermore, the vaned feathers of both species were symmetrical. Modern flying birds have asymmetric feathers, and that asymmetry allows the feathers to produce lift and thrust. Probably the vaned wing and tail feathers of *Caudipteryx* and *Protoarchaeopteryx* were colorful and used for social displays. Possibly feathers were initially used in social displays and were subsequently modified for insulation and then for flight.

■ Metabolism and Thermoregulation: Was It Aerobic Capacity or Something Else?

Synapsids and sauropsids both present pictures of lineages of animals that were becoming more capable of sustained locomotion and endothermy, but there are some gaps—either in the evidence or in our interpretation of the evidence. Body form, the presence of turbinates, and mechanisms that improve processing of food are all seen among derived synapsids, but we have no direct evidence of insulation, either hair or subcutaneous fat. The lack of fossil traces of hair or blubber is not surprising because neither one would fossilize except under unusual conditions.

Sauropsids are more puzzling because, in addition to evidence of locomotor capacity provided by body form and of food-processing capacity shown by gizzard stones, we also have fossil evidence of downlike feathers that should provide insulation for an endothermal animal. But contradicting that interpretation is the absence of turbinates, which we think are an essential element of endothermy. Where does this leave us?

Although the anatomical evidence of high levels of activity and metabolism is persuasive for derived therapsids and suggestive for derived dromeosaurs, a physiological element of the evolution of endothermy remains speculative. *Must* animals that are capable of high levels of metabolism during activity *also* have high metabolic rates when they are inactive? That may be true—at least for amniotes—because measurements of many different species show that in most cases the maximum metabolic rates they achieve during activity are about 10 times their resting metabolic rates. That observation is the basis for the aerobic capacity hypothesis of the evolution of endothermy, which proposes that resting rates of metabolism (and hence of heat production) increased in step with maximum rates (Bennett and Ruben 1979). In other words, selection for increased rates of metabolism during activity also led to increased resting rates and ultimately to resting rates that produced enough heat for endothermal thermoregulation.

The difficulty with this hypothesis lies in explaining *why* active and resting metabolic rates are so often linked. Perhaps it has something to do with the cost of maintaining the lungs, heart, circulatory system, and mitochondria that are needed to produce ATP at high rates during activity, but the mechanistic basis for the relationship is not clear. Furthermore, the relationship between active and resting metabolic rates is found when comparisons are made among species; however, when individuals within a species are tested, that relationship does not necessarily appear.

Is failure to find the relationship among individuals important? On one hand, that is the level at which the aerobic capacity hypothesis proposes that selection on incipiently endothermal forms would have acted, so failure to find a relationship between resting and active metabolic rates in an experiment could falsify the hypothesis. But biological experiments are rarely simple to interpret, and in this instance small variations in the physical condition and motivation of individual animals in a study could obscure a relationship that was actually present. With selection acting over tens of millions of years, a tiny heritable link is all that is needed for the aerobic capacity hypothesis to work, and that link could be much too small to demonstrate in a laboratory experiment with animals that are already specialized ectotherms or endotherms.

Although the aerobic capacity hypothesis is a plausible explanation for the evolution of endothermy, the absence of a clear mechanistic basis for the relationship between active and resting metabolic rates has fostered alternative explanations. Colleen Farmer, for example, has proposed that the evolution of endothermy derived primarily from the advantage that homeothermy (maintaining a stable body temperature) confers during reproduction (Farmer 2000, 2001). She suggests that if parental thermoregulation accelerated the growth of embryos and juveniles, selection might have acted first to produce endothermy during embryonic development (i.e., while the parents were brooding their eggs) and subsequently to prolong endothermy to include the period while the parents were caring for the young in a nest. (This argument applies to both synapsids and sauropsids because Mesozoic synapsids are believed to have laid eggs, as do extant monotreme mammals.) The parental care hypothesis is a subject of vigorous discussion (Angilletta et al. 2002, Farmer 2003).

▨ Insulation: Hair, Feathers, Skin, and Fat

The importance of hair or feathers in providing a layer of insulation that retains metabolic heat in the body is obvious, but why do derived synapsids have hair while derived sauropsids have feathers? Keratin, a protein synthesized by epidermal cells, forms both hair and feathers. Thick epidermal keratin is a derived feature of amniotes, and X-ray diffraction distinguishes two types of keratin—alpha and beta—that differ in molecular structure. Alpha keratins are present in the epithelial tissues of all vertebrates. Alpha keratin forms hair and the derivatives of hair that are found in mammals, such as claws and fingernails, hooves, and the horn of a rhinoceros. Beta keratin is a derived feature of the sauropsid lineage, and sauropsids have both alpha and beta keratin, either in layers (alpha-beta-alpha-beta) or in regions.

Alpha keratin is found in the hinge regions of scales, for example, and beta keratin on the scale surfaces. Feathers consist almost entirely of beta keratin feather proteins, which are different in size and structure from the beta keratins in the skin.

The synapsid lineage probably never had a scale-covered skin like that of sauropsids, and that may be why mammals retain a soft skin with a variety of mucus glands. A glandular skin like that seen in mammals may not be an option with the harder skin of sauropsids, and the relative absence of skin glands in sauropsids has ramifications in other aspects of their lives. Sweating, for example, occurs only in mammals because sauropsids do not have sweat glands in their skin, and it is unlikely that a sauropsid could have evolved lactation.

Differences in the skin of synapsids and sauropsids may have shaped their reproductive and social behavior. Mammary glands are modified skin glands, and lactation may have had its evolutionary origin in the form of skin glands that provided water for eggs of early synapsids that were developing in underground burrows with the mother in attendance (Oftedal 2002a,b).

Parchment-shelled eggs (i.e., eggs like those of turtles and many lizards that have flexible shells) lose water by evaporation even in the humid air of a burrow, but they are also able to take up water across the shell. Turtle and lizard eggs do this from the moist soil in which the nests are constructed, and a female synapsid might have been able to provide water to her eggs by releasing fluid onto them. Subsequent elaboration of this process could include the addition of nutrients to the glandular secretions and eventually the nippleless mammary patch of milk-secreting glands found in extant monotremes.

Only female mammals have functional mammary glands, and parental care is largely provided by the mother. The primary social bond of mammals is between mother and infant, and male parental care is rare. In contrast, both mother and father can be equally effective in providing parental care among sauropsids, and adult crocodilians and birds commonly (although not universally) form pair bonds. Both parents often participate in brooding the eggs and feeding and caring for the young.

Fat storage is another feature that has both thermoregulatory and reproductive consequences. The phylogenetic pattern of fat storage is a bit more complex than of keratin. Most fishes store fat as lipid droplets in the liver and muscles. (Cod liver oil is a commercial product that takes advantage of fat storage in the liver of a fish, and the omega-3 and omega-6 fatty acids that appear to be helpful in protecting humans against coronary heart disease are found in the muscles of oily fishes such as salmon.) Adipose tissue, where fat is stored in specialized cells called adipocytes, is a feature of tetrapods. Amphibians and all sauropsids except birds store fat only in discrete fat bodies in the abdomen and at the base of the tail.

Mammals and birds store fat in many locations—subcutaneously and in association with muscles, the heart, the gut, and around other organs where it is available for rapid energy release (Pond 1998). The fat bodies in the posterior abdominal and caudal region of birds are easy to see when you are preparing a chicken or turkey for roasting, and intramuscular and subcutaneous fat is especially well developed in aquatic birds like ducks and geese. No birds, not even penguins, have subcutaneous fat deposits as extensive as the blubber layer of marine mammals.

Fat is an important energy store that female mammals draw on during lactation, and female mammals in good condition have a higher proportion of body fat than males. For humans, fat stores equivalent to 15 percent of body mass are considered normal for men, and 20 percent is normal for women. The primary locations of fat storage differ between the sexes. The main site of fat deposition for men is in the abdomen, creating the familiar beer belly of overweight males. Females have important fat deposits in the thighs and buttocks; this is the human female pattern of fat storage that has been represented in art at least as far back as the Venus figurines made in Europe some 27,000 years ago.

11.5 Getting Rid of Wastes: The Kidneys and Bladder

High metabolic rates are beneficial for locomotor endurance and thermoregulation, but they have costs associated with them. As we have noted, high rates of energy use require correspondingly high rates of feeding and digestion, and we will consider additional examples of the complex interactions among energy requirements, foraging behavior, and feeding and digestion in subsequent chapters. High rates of food intake have another consequence—high rates of nitrogenous waste production—and synapsids and sauropsids found different ways to reconcile the conflicting demands of excreting nitrogen while retaining water.

Metabolism of protein produces ammonia, NH_3. Ammonia is quite toxic, but it is very soluble in water and diffuses rapidly because it is a small molecule (**Table 11.2**). Aquatic non-amniotes (bony fishes and aquatic amphibians) excrete a large proportion of their nitrogenous waste as ammonia, and ammonia is a nitrogenous waste product of terrestrial amniotes as well—human sweat contains small amounts of ammonia.

Ammonia can be converted to urea, $CO(NH_2)_2$, which is a less toxic substance than ammonia and is even more soluble than ammonia. Because it is both soluble and relatively nontoxic, urea can be accumulated within the body and released in a concentrated solution in urine, thereby conserving water. Urea synthesis is an ancestral character of amniotes and probably of all gnathostomes.

TABLE 11.2 Characteristics of the major nitrogenous waste products of vertebrates

Compound	Chemical Formula	Molecular Weight	Solubility in Water $(g \cdot L^{-1})$	Toxicity	Metabolic Cost of Synthesis	Water Conservation Efficiency*
Ammonia	NH_3	17	890	High	None	1
Urea	$CO(NH_2)_2$	60	1190	Moderate	Low	2
Uric Acid	$C_5H_4O_3N_4$	168	0.065	Low	High	4
Sodium Urate	$C_5H_4O_3N_4Na_2$	212	0.88	Low	High	4
Potassium Urate	$C_5H_4O_3N_4K_2$	244	2.32	Low	High	4

*The efficiency of water conservation is expressed as the number of nitrogen (N) atoms per osmotically active particle; higher ratios mean more nitrogen is excreted per osmotic unit.

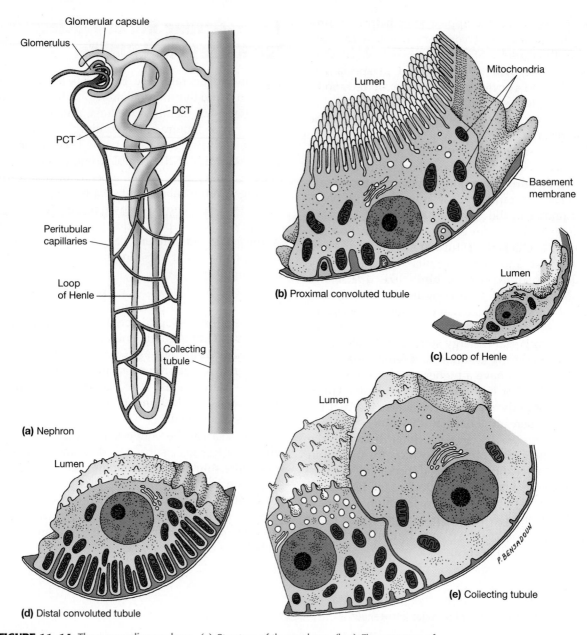

▲ **FIGURE 11–14** The mammalian nephron. (a) Structure of the nephron. (b-e) Fine structure of the cells lining the walls of the nephron. (PCT = proximal convoluted tubule; DCT = distal convoluted tubule)

A complex metabolic pathway converts several nitrogen-containing compounds into uric acid, $C_5H_4O_3N_4$. Unlike ammonia and urea, uric acid is insoluble, and it readily combines with sodium and potassium ions to precipitate as sodium or potassium urate.

Synapsids and sauropsids have taken different paths in dealing with nitrogenous wastes: synapsids retained the ancestral pattern of excreting urea and developed a kidney that is extraordinarily effective in producing concentrated urine, whereas sauropsids developed the capacity to synthesize and excrete uric acid and recover the water that is released when it precipitates.

Nitrogen Excretion by Synapsids: The Mammalian Kidney

The mammalian kidney is a highly derived organ composed of millions of nephrons, the basic units of kidney structure that are recognizable in all vertebrates (**Figure 11–14**). Each nephron consists of a glomerulus that filters the blood and a long tube in which the chemical composition of the filtrate is altered. A portion of this tube, the loop of Henle, is a derived character of mammals that is largely responsible for the ability of mammals to produce concentrated urine. The mammalian kidney is capable of producing urine more concentrated than that of any non-amniote—and in most cases, more concentrated than that of sauropsids as well (**Table 11.3**). Understanding how the mammalian kidney works is important for understanding how mammals can thrive in places that are seasonally or chronically short of water.

Urine is concentrated by removing water from the ultrafiltrate that is produced in the glomerulus when water and small molecules are forced out of the capillaries. Because cells are unable to transport water directly, they use osmotic gradients to manipulate the movement of water molecules. In addition, the cells lining the nephron actively reabsorb substances important to the body's economy from the ultrafiltrate and secrete toxic substances into it. The cells lining the nephron differ in permeability, molecular and ion transport activity, and reaction to the hormonal and osmotic environments in the surrounding body fluids.

The cells of the proximal convoluted tubule (PCT) have an enormous surface area produced by long, closely spaced microvilli, and the cells contain many mitochondria. These structural features reflect the function of the PCT in actively moving sodium from the lumen of the tubule to the peritubular space and capillaries; passive movement of chloride and water follows sodium transport to the peritubular space to neutralize electric charge (**Figure 11–15**). Farther down the nephron, the cells of the thin segment of the loop of Henle are waferlike and contain fewer mitochondria. The descending limb of the loop of Henle permits passive flow of sodium and water, and the ascending limb actively removes sodium from the ultrafiltrate. Finally, cells of the collecting tubule appear to be of two kinds. Most seem to be suited to the relatively impermeable state characteristic of periods of sufficient body water. Other cells are mitochondria-rich and have a greater surface area. They are probably the cells that respond to the presence of antidiuretic

TABLE 11.3 Maximum urine concentrations of some synapsids and sauropsids

Species	Maximum Observed Urine Concentration (mmol · kg^{-1})	Approximate Urine: Plasma Concentration Ratio
Synapsids		
Human (*Homo sapiens*)	1430	4
Bottlenose porpoise (*Tursiops truncatus*)	2658	7.5
Hill kangaroo (*Macropus robustus*)	2730	7.5
Camel (*Camelus dromedarius*)	2800	8
White rat (*Rattus norvegicus*)	2900	8.9
Marsupial mouse (*Dasycercus eristicauda*)	3231	10
Cat (*Felis domesticus*)	3250	9.9
Desert woodrat (*Neotoma lepida*)	4250	12
Vampire bat (*Desmodus rotundus*)	6250	20
Kangaroo rat (*Dipodomys merriami*)	6382	18
Australian hopping mouse (*Notomys alexis*)	9370	22
Sauropsids		
American alligator (*Alligator mississippiensis*)	312	0.95
Desert iguana (*Dipsosaurus dorsalis*)	300	0.95
Desert tortoise (*Gopherus agassizii*)	622	1.8
Pelican (*Pelecanus erythrorhynchos*)	700	2
House sparrow (*Passer domesticus*)	826	2.4
House finch (*Carpodacus mexicanus*)	850	2.4
Savannah sparrow (*Passerculus sandvicensis*)	2000	5.8

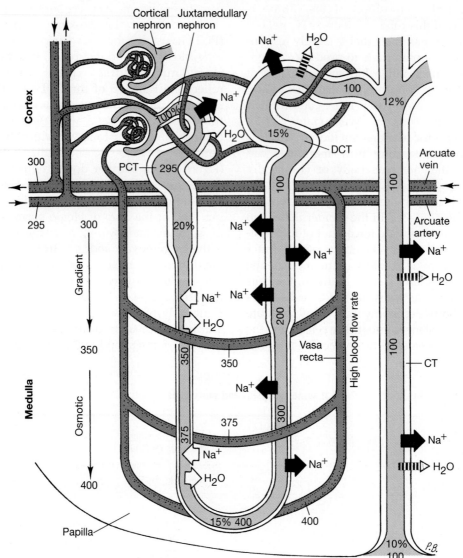

▲ **FIGURE 11–15** Function of the mammalian kidney. The mammalian kidney produces dilute urine when the body is hydrated and concentrated urine when the body is dehydrated. Black arrows indicate active transport, and white arrows indicate passive flow. The numbers represent the approximate osmolality of the fluids in the indicated regions. Percentages are the volumes of the forming urine relative to the volume of the initial ultrafiltrate. (a) When blood osmolality drops below normal concentration (about 300 mmol · kg⁻¹), excess body water is excreted. (b) When osmolality rises above normal, water is conserved. (ADH = antidiuretic hormone; CT = collecting tubule; DCT = distal convoluted tubule; H_2O = water; Na⁺ = sodium ion; PCT = proximal convoluted tubule)

hormone (ADH) from the pituitary gland, triggered by insufficient body fluid. Under the influence of ADH, the collecting tubule actively exchanges ions, pumps urea, and becomes permeable to water, which flows from the lumen of the tubule into the concentrated peritubular fluids.

The nephron's activity is a six-step process, each step localized in a region with special cell characteristics and distinctive variations in the osmotic environment. The first step is production of an ultrafiltrate at the glomerulus. The ultrafiltrate is isosmolal with blood plasma and resembles whole blood after

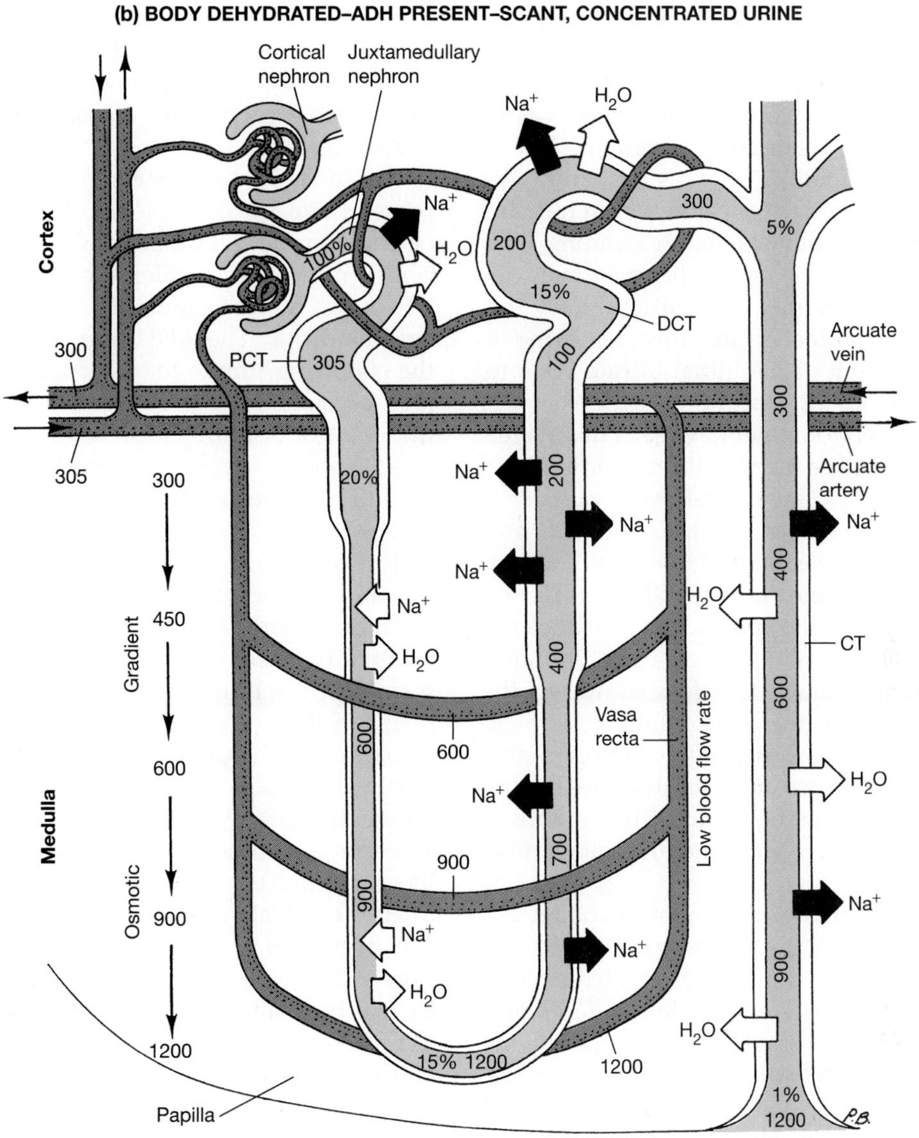

(b) BODY DEHYDRATED–ADH PRESENT–SCANT, CONCENTRATED URINE

▲ **FIGURE 11–15** Continued

the removal of (1) cellular elements, (2) substances with a molecular weight of 70,000 or greater (primarily proteins), and (3) substances with molecular weights between 15,000 and 70,000, depending on the shapes of the molecules. Humans produce about 120 milliliters of ultrafiltrate per minute—that is 170 liters (45 gallons) of glomerular filtrate per day! We excrete only about 1.5 liters of urine because the kidney reabsorbs more than 99 percent of the ultrafiltrate produced by the glomerulus.

The second step in the production of the urine is the action of the PCT in decreasing the volume of the ultrafiltrate. The PCT cells have greatly enlarged surfaces that actively transport sodium from the lumen of the tubules to the exterior of the nephron. Chloride and water move passively through the PCT cells in response to the removal of sodium. By this process, about two-thirds of the salt is reabsorbed in the PCT, and the volume of the ultrafiltrate is reduced by the same amount. Although the urine is still very nearly isosmolal with blood at this stage, the substances contributing to the osmolality of the urine after it has passed through the PCT are at different concentrations than in the blood.

The next alteration occurs in the descending limb of the loop of Henle. The thin, smooth-surfaced cells freely permit diffusion of sodium and water. Because the descending limb of the loop passes through tissues of increasing osmolality as it plunges into the medulla, water is lost from the urine and it becomes more concentrated. In humans the osmolality of the fluid in the descending limb may reach 1200 mmol · kg^{-1}, and other mammals can achieve considerably higher concentrations. By this mechanism, the volume of the forming urine is reduced to 25 percent of the initial filtrate volume, but it is still large. In an adult human, for example, 25 liters to 40 liters of fluid per day reach this stage.

The fourth step takes place in the ascending limb of the loop of Henle, which has cells with numerous large, densely packed mitochondria. The ATP produced by these organelles is used to remove sodium from the forming urine. Because these cells are impermeable to water, the volume of urine does not decrease as sodium is removed; and, because sodium was removed without loss of water, the urine is hyposmolal to the body fluids as it enters the next segment of the nephron. Although this sodium-pumping, water-impermeable, ascending limb does not concentrate or reduce the volume of the forming urine, it sets the stage for these important processes.

The very last portion of the nephron changes in physiological character, but the cells closely resemble those of the ascending loop of Henle. This region, the distal convoluted tubule (DCT), is permeable to water. The osmolality surrounding the DCT is that of the body fluids, and water in the entering hyposmolal fluid flows outward and equilibrates osmotically. This process reduces the fluid volume to as little as 5 percent of the original ultrafiltrate.

The final touch in the formation of a small volume of highly concentrated mammalian urine occurs in the collecting tubules. Like the descending limb of the loop of Henle, the collecting tubules course through tissues of increasing osmolality, which withdraw water from the urine. The significant phenomenon associated with the collecting tubule, and to a lesser extent with the DCT, is its conditional permeability to water. During excess fluid intake, the collecting tubule demonstrates low water permeability: only half of the water entering it may be reabsorbed and the remainder excreted. In this way, copious, dilute urine can be produced. When a mammal is dehydrated, the collecting tubules and the DCT become very permeable to

water, and the final urine volume may be less than 1 percent of the original ultrafiltrate volume. In certain desert rodents, so little water is contained in the urine that it crystallizes almost immediately upon urination.

Antidiuretic hormone (also known as vasopressin) is a polypeptide produced by specialized neurons in the hypothalamus, stored in the posterior pituitary, and released into the circulation whenever blood osmolality is elevated or blood volume drops. The ADH increases permeability of the collecting tubule to water and facilitates water reabsorption to produce a scant, concentrated urine. Alcohol inhibits the release of ADH, inducing a copious urine flow, and this can result in dehydrated misery the morning after a drinking binge.

The key to concentrated urine production clearly depends on the passage of the loops of Henle and collecting tubules through tissues with increasing osmolality. These osmotic gradients are formed and maintained within the mammalian kidney as a result of its structure (**Figure 11–16**), which sets it apart from the kidneys of other vertebrates. Particularly important are the structural arrangements within the kidney medulla of the descending and ascending segments of the loop of Henle and its blood supply, the vasa recta. These elements create a series of parallel tubes with flow passing in opposite directions in adjacent vessels (countercurrent flow). As a result, sodium secreted from the ascending limb of the loop of Henle diffuses into the medullary tissues to increase their osmolality, and this excess salt is distributed by the countercurrent flow to create a steep osmotic gradient within the medulla. The final concentration of a mammal's urine is determined by the amount of sodium accumulated in the fluids of the medulla. Physiological alterations in the concentration in the medulla result primarily from the effect of ADH on the rate of blood flushing the medulla. When ADH is present, blood flow into the medulla is retarded and salt accumulates to create a steep osmotic gradient. Another hormone, aldosterone, from the adrenal gland increases the rate of sodium secretion into the medulla to promote an increase in medullary salt concentration.

In addition to these physiological means of concentrating urine, a variety of mammals have morphological alterations of the medulla. Most mammals have two types of nephrons: those with a cortical glomerulus and abbreviated loops of Henle that do not penetrate far into the medulla, and those with juxtamedullary

(a)

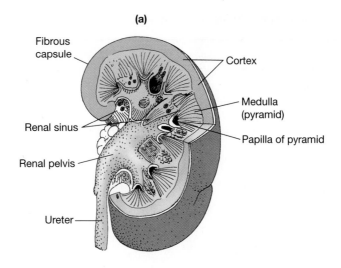

(b)

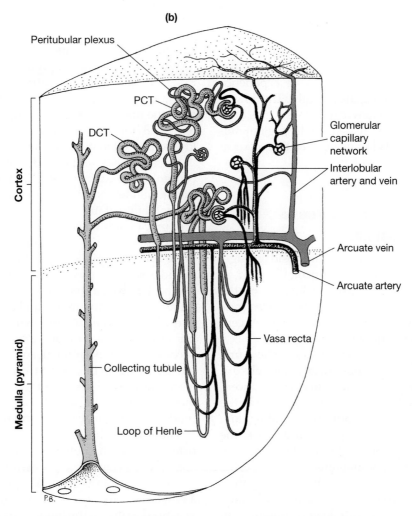

▲ **FIGURE 11–16** Gross morphology of the mammalian kidney. (a) Structural divisions of the kidney and proximal end of the ureter. (b) Enlarged diagram of a section extending from the outer cortical surface of the apex of a renal pyramid, the renal papilla. (DCT = distal convoluted tubule; PCT = proximal convoluted tubule)

TABLE 11.4 Distribution of nitrogenous end products among sauropsids

Group	Total Urinary Nitrogen (%)		
	Ammonia	Urea	Salts of Uric Acid
Lepidosaurs			
Tuatara	3–4	10–28	65–80
Lizards and snakes	Small	0–8	90–98
Archosaurs	0	1	2
Crocodilians	25	0–5	70
Birds	6–17	5–10	60–82
Turtles	0	1	2
Aquatic	4–44	45–95	1–24
Desert	3–8	15–50	20–50

glomeruli, deep within the cortex, with loops that penetrate as far as the papilla of the renal pyramid. Obviously, the longer, deeper loops of Henle experience large osmotic gradients along their lengths. The flow of blood to these two populations of nephrons seems to be independently controlled. Juxtamedullary glomeruli are more active in regulating water excretion; cortical glomeruli function in ion regulation.

■ Nitrogen Excretion by Sauropsids: Renal and Extrarenal Routes

All extant representatives of the sauropsid lineage, including turtles and birds, are uricotelic—that is, they excrete nitrogenous wastes primarily in the form of uric acid. Except for freshwater turtles, uric acid and its salts account for 80 percent to 90 percent of urinary nitrogen in sauropsids (Table 11.4).

The strategy for water conservation when uric acid is the primary nitrogenous waste is entirely different from that required when urea is produced. Because urea is so soluble, concentrating urine in the kidney can conserve water, but concentrating uric acid would cause it to precipitate and block the nephrons. The kidneys of lepidosaurs lack the long loops of Henle that allow mammals to reduce the volume of urine and increase its concentration (Figure 11–17). Urine from the kidneys of lepidosaurs has the same (or even a slightly lower) osmotic concentration as the blood plasma. The kidneys of birds have two types of nephrons: short loop nephrons like those of lepidosaurs and long loop nephrons that

extend down into the medullary cone. The long loop nephrons allow birds to produce urine that is two to three times more concentrated than the plasma. These ratios are lower than those of mammals, and even the highest urine to plasma ratio recorded for a bird—5.8 for the savannah sparrow—is relatively low compared to mammalian ratios.

If sauropsids depended solely on the urine-concentrating capacity of their kidneys, they would excrete all their body water in urine. This is where the low solubility of uric acid becomes advantageous: uric acid precipitates when it enters the cloaca or bladder. The dissolved uric acid combines with ions in the urine and precipitates as a light-colored mass that includes sodium, potassium, and ammonium salts of uric acid, as well as other ions held by complex physical forces. This mixture is familiar to anyone who has parked a car beneath a tree where birds roost. When uric acid and ions precipitate from solution, the urine becomes less concentrated and water is reabsorbed into the blood. In this respect, excretion of nitrogenous wastes as uric acid is even more economical of water than is excretion of urea because the water used to produce urine is reabsorbed and reused.

Water is not the only substance that is reabsorbed from the cloaca, however. Many sauropsids also reabsorb sodium and potassium ions and return them to the bloodstream. At first glance, that seems a remarkably inefficient thing to do. After all, energy was used to create the blood pressure that forced the ions through the walls of the glomerulus into the urine in the first place, and now more energy is being used in the cloaca or bladder to operate the

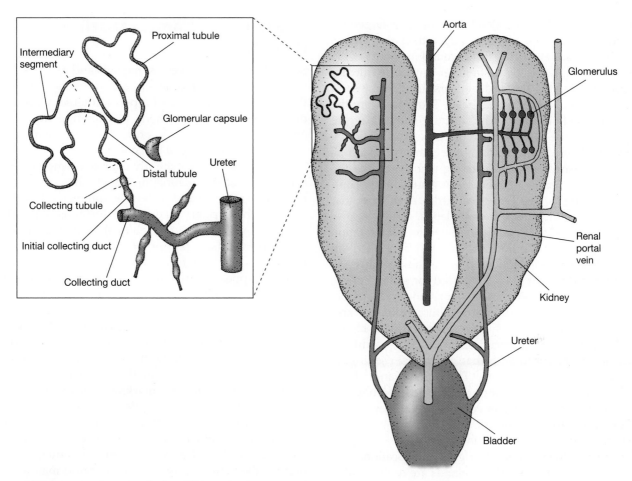

▲ **FIGURE 11–17** Structure of a lizard kidney. The left side shows a nephron in detail (*inset*) and its position within the kidney. The right side shows the relationship of glomeruli and nephrons within a segment of the kidney.

active transport system that returns the ions to the blood. The animal has used two energy-consuming processes, and it is back where it started—with an excess of sodium and potassium ions in the blood. Why do that?

The solution to the paradox lies in a water-conserving mechanism that is present in many sauropsids: salt-secreting glands that provide an extrarenal (i.e., in addition to the kidney) pathway that disposes of salt with less water than urine. Nasal salt glands are widespread among lizards, and in all cases it is the lateral nasal glands that excrete salt. The secretions of the glands empty into the nasal passages, and a lizard expels them by sneezing or by shaking its head. In birds, also, the lateral nasal gland has become specialized for salt excretion. The glands are situated in or around the orbit, usually above the eye. Marine birds (pelicans, albatrosses, penguins) have well-developed salt glands, as do many freshwater birds

(ducks, loons, grebes), shorebirds (plovers, sand-pipers), storks, flamingos, carnivorous birds (hawks, eagles, vultures), upland game birds (pheasants, quail, grouse), the ostrich, and the roadrunner. Depressions in the supraorbital region of the skull of the extinct aquatic birds *Hesperornis* and *Ichthyornis* suggest that salt glands were present in these forms as well.

In sea snakes and the marine elephant trunk snakes, the posterior sublingual gland secretes a salty fluid into the tongue sheath, from which it is expelled when the tongue is extended. In some species of homalopsine snakes (a group of marine snakes from the Indoaustralian region), the premaxillary gland in the front of the upper jaw secretes salt. Salt-secreting glands on the dorsal surface of the tongue have been identified in several species of crocodiles, in a caiman, and in the American alligator.

Finally, in sea turtles and in the diamondback terrapin (a North American species of turtle that inhabits estuaries), the lacrimal glands are greatly enlarged (in some species each gland is larger than the turtle's brain) and secrete a salty fluid around the orbits of the eyes. Photographs of nesting sea turtles frequently show clear paths streaked by tears through the sand that adheres to the turtle's head. Those tears are the secretions of the salt glands. The huge glands leave an imprint on the skull, and the oldest sea turtle known, *Santanachelys gaffneyi* from the Early Cretaceous period, clearly had salt-secreting glands. Unlike the situation in lizards, however, salt glands are uncommon among turtles. Terrestrial turtles, even those that live in deserts, do not have salt glands.

The diversity of glands involved in salt excretion among sauropsids indicates that this specialization has evolved independently in various groups. At least five different glands are used for salt secretion, indicating that a salt gland is not an ancestral character for the group. Furthermore, although lizards and birds are not very closely related, both use the lateral nasal gland for salt secretion, whereas crocodilians (which are closer to birds than to lizards) use lingual glands for salt secretion. Thus salt glands have evolved repeatedly among sauropsids, perhaps in response to the water-conserving opportunities offered by excretion of uric acid, which is a derived character of the sauropsid lineage.

Despite their different origins and locations, the functional properties of salt glands are quite similar. They secrete fluid containing primarily sodium or potassium cations and chloride or bicarbonate anions in high concentrations (**Table 11.5**). Sodium is the predominant cation in the salt-gland secretions of marine vertebrates, and potassium is present in the secretions of terrestrial lizards, especially herbivorous species such as the desert iguana. Chloride is the major anion, and herbivorous lizards may also excrete bicarbonate ions.

The total osmolal concentration of the salt gland secretion may reach 2000 mmol · kg^{-1}, which is more than six times the concentration of urine that can be produced by the kidney. This efficiency of excretion is the explanation of the paradox of active uptake of salt from the urine. As ions are actively reabsorbed, water follows passively, so an animal recovers both water and ions from the urine. The ions can then be excreted via the salt gland at much higher concentrations, with a proportional reduction in the amount of water needed to dispose of the salt. Thus, by investing energy in recovering ions from urine, sauropsids with salt glands can conserve water by excreting ions through the more efficient extrarenal route.

■ Correlates of Nitrogen Excretion Methods

Some differences between synapsids and sauropsids, such as the retention or loss of a urinary bladder and the presence or absence of a penis, interact with the differences in their methods of excreting nitrogen and have consequences in areas as diverse as reproductive mode and social behavior.

Many fishes have a slightly enlarged region at the caudal end of the archinephric ducts that are called urinary bladders, and extant amphibians have a large, bilobed urinary bladder that forms as an evagination from the cloaca. Thus a bladder appears to be an ancestral character for amniotes, but it is not retained in all extant forms. It is present in mammals, where it appears to be only a storage area for urine because urine does not undergo any modification after it enters the bladder. The functional significance of the mammalian bladder is unclear, and it has facetiously been called an organ of social convenience. That may be an accurate characterization, because urine plays a large role in the social behavior of mammals, especially in marking territories. Male therian mammals use the penis for urination as well as for sperm transport, an arrangement that would seem bizarre to other amniotes that use the intromittent organ only to transport sperm. This dual function for the penis is possible only because mammals have retained the bladder, and that anatomy allows the gonadal ducts to connect to the urethra (the duct draining the bladder).

Despite the extensive processing that the urine of sauropsids undergoes after it leaves the kidney, many sauropsids lack a urinary bladder entirely and process the urine in the cloaca; others have an ephemeral bladder that is lost shortly after hatching; some have a functional bladder throughout life. The universal absence of a bladder among birds is probably associated with flight—one of many weight-reducing characteristics of birds. Most turtles have bladders, and some aquatic species use them as accessory respiratory structures, pumping water in and out of the bladder via the cloaca. Terrestrial turtles use urine in the bladder as a water reserve and rely on it to survive for long periods without drinking.

Uricotely may be a necessary element of oviparity as it occurs in birds. Unlike the eggs of many other sauropsids, the egg of a bird does not absorb water during incubation. On the contrary, a bird egg must *lose* enough water during the development of the embryo to create an air space at the wide end of the

TABLE 11.5 Salt gland secretions from sauropsids

Species and Condition	Ion Concentration (mmole · kg^{-1})		
	Na^+	K^+	Cl^-
Turtles			
Loggerhead sea turtle (*Caretta caretta*), seawater	732–878	18–31	810–992
Diamondback terrapin (*Malaclemys terrapin*), seawater	322–908	26–40	N/R
Lizards			
Desert iguana (*Dipsosaurus dorsalis*), estimated field conditions	180	1700	1000
Fringe-toed lizard (*Uma scoparia*), estimated field conditions	639	734	465
Snakes			
Sea snake (*Pelamis platurus*), salt loaded	620	28	635
Homalopsine snake (*Cerberus rhynchops*), salt loaded	414	56	N/R
Crocodilian			
Saltwater crocodile (*Crocodylus porosus*), natural diet	663	21	632
Birds			
Blackfooted albatross (*Diomeda nigripes*), salt loaded	800–900	N/R	N/R
Herring gull (*Larus argentatus*), salt loaded	718	24	N/R

N/R = nor reported

egg from which the hatchling draws its first breath before it pips the shell. The insolubility of uric acid allows the nitrogenous wastes produced by the embryo to precipitate inside the allantois so they do not raise the osmotic concentration inside the egg and are left behind when the bird hatches.

11.6 Sensing and Making Sense of the World: Eyes, Ears, Tongues, Noses, and Brains

In many respects synapsids and sauropsids perceive the world around them very differently because the lineages are quite different in some of their sensory capacities. Most synapsids are exquisitely sensitive to odors but have relatively poor vision. (Primates in general—and humans in particular—are an exception to that generalization; indeed, primates and especially humans have lost many of the genes associated with olfactory reception [Gilad et al. 2003].) Sauropsids have the reverse combination of characters—most have good vision, and many have a rather poor sense of smell. These sensory capacities are reflected in the social behaviors of the groups: scent-marking is a com-

mon element of territorial behavior of mammals but not of birds, whereas territorial displays that emphasize color and pattern are common among lizards and birds but rare among mammals.

■ Vision

The vertebrate retina contains two types of cells that respond to light—rods and cones. Rod cells are sensitive to a wide range of wavelengths, and the electrical responses of many rod cells are transmitted to a single bipolar cell that sums their inputs. Because of these characteristics, rod cells are sensitive to low light levels but do not produce high visual acuity. The population of cone cells in the retina of a vertebrate includes subgroups of cells that contain pigments that are sensitive to different wavelengths of light. Blue light stimulates one set of cells, red light another, and so on. The variety of cone cells found in vertebrate eyes is extensive and includes cells with pigments that are sensitive to wavelengths from deep red to ultraviolet light. Only a few cone cells transmit their responses to a bipolar cell, so cone cells require higher levels of illumination than rod cells; however, they are capable of producing sharper images than do rod cells.

Fishes (teleosts, at least), amphibians, and sauropsids use retinal cone cells to perceive colors. Thus the capacity for good color vision is probably primitive for amniotes. Mammals are commonly supposed to see in black and white, a notion that is becoming less familiar now that black-and-white movies are being colorized! Extant small, nocturnal mammals, such as insectivores, have predominately rod cells in their retinas, but all mammals, including monotremes, have at least some cone cells. At some point in the evolution of the synapsid lineage, the ability to perceive color was evidently reduced, possibly in connection with the adoption of a nocturnal lifestyle by early mammals. However, most mammals perceive some color and have monochromatic (one visual color pigment) or dichromatic (two pigments) vision. The ability to distinguish colors is well documented in dogs, cats, and horses, but these animals—and mammals in general—probably perceive the world like a person with red-green color blindness. Our ability to perceive a richly colored world with subtle variations in hues is the result of a type of trichromatic vision that we share only with the other anthropoid primates (apes and monkeys). We anthropoids perceive three primary colors (red, green, and blue), each associated with one type of cone. The spectral sensitivities of the three types of cone overlap, and intermediate hues are produced by graded responses from the cones in response to mixtures of wavelengths in the light entering the eye. The excellent color vision of anthropoid primates is unique among synapsids and probably relates to the increased reliance of anthropoids on vision and reduced reliance on the sense of smell.

Cones and color vision were retained in sauropsids, and many species have colored oil droplets in the cone cells. (Colored oil droplets are also found in the eyes of turtles, monotremes, and marsupials and are probably a primitive amniote character.) The oils range in color from reddish through orange to yellow, and they may improve visual acuity by filtering out the very short wavelengths of light that are not focused effectively by the vertebrate eye. You can easily demonstrate the difficulty of focusing blue light by comparing the sharpness of the image produced by a blue Christmas tree bulb compared to a red one.

Chemosensation: Gustation and Olfaction

Tasting and smelling are forms of chemosensation mediated by receptor cells that respond to the presence of chemicals with specific characteristics. Tasting and smelling appear distinct from our perspective as terrestrial animals, but the distinction is blurred among aquatic non-amniotes where taste buds may be widely distinguished across the body surface.

Taste buds are clusters of cells derived from the embryonic endoderm that open to the exterior by pores. Chemicals must be in contact with the sensory cells in a taste bud to elicit a response, and they produce a relatively narrow range of responses. Amniotes have taste buds in the oral cavity (especially on the tongue) and in the pharynx. The taste buds of mammals are broadly distributed over the tongue and oral cavity, whereas in sauropsids they are on the back of the tongue and palate. We perceive chemicals as combinations of salt, sweet, sour, and bitter sensations.

Olfactory cells are derived from neural crest tissue and are distributed in the epithelium of the olfactory turbinates that are adjacent to the nasal passages. Chemicals that stimulate olfactory cells are called odorants, and some mammals are exquisitely sensitive to olfactory stimuli. Even humans, who have poor olfactory sensitivity by mammalian standards, may be able to detect 10,000 different odors. The chemical structures of odorant molecules allow them to bind only to specific receptor proteins in the membranes of olfactory cells. Genetically defined families of receptor proteins respond to difference categories of odorants, but it is not clear how we distinguish so many odors. Olfaction and gustation interact to produce the sensation we call taste, and the process can be highly stylized as in the sniffing, sipping, and swirling ritual of wine tasting. In general, olfaction is better developed among synapsids than sauropsids.

Hearing

We saw in Chapter 8 that a hearing ear was not a feature of the earliest tetrapods but that it evolved independently among amniotes and non-amniotes. The lagena (called the cochlea in mammals) is a structure in the inner ear that is devoted to hearing, and it seems to be an ancestral feature for all amniotes. In contrast, the middle ear appears to have evolved independently several times. Mammals have a completely enclosed middle ear involving the stapes (the old hyomandibula) and two additional bones, whereas all sauropsids have a single-boned middle ear, consisting only of the stapes. But even within sauropsids, details of the ear structure suggest that the final condition evolved independently in turtles, lepidosaurs, and archosaurs (Clack 2002).

■ Brains

All amniotes have a relatively enlarged forebrain in comparison with amphibians, especially with respect to the telencephalon. Both birds and mammals have larger brains in proportion to their body size than do more primitive amniotes, but the increase in size has been achieved differently in the two groups (**Figure 11–18**). In both there is a distinct cerebrum, formed by the expansion of the dorsal portion of the telencephalon, the so-called dorsal pallium. The dorsal pallium has two components in the original amniote condition, which have different fates in the sauropsid and synapsid lineages. Sauropsids enlarge one of these portions into what is called the dorsal ventricular ridge, while the other portion (the lemnopallium) remains small. In synapsids, the reverse occurs, with the cerebrum formed from the lemnopallium (Butler

and Hodos 1996). In addition, the mammalian dorsal pallium forms a distinctive six-layered structure, now known as the neocortex. This structure becomes highly convoluted in larger-brained mammals, resulting in the type of wrinkled surface that is so apparent in the human cerebrum.

The brains of mammals and birds are larger in proportion to their body size than are the brains of nonavian reptiles (**Figure 11–19**). For a 1-kilogram animal, an average-size mammalian brain would weigh 9.9 grams, a bird's brain 6.7 grams, and a nonavian reptile's brain only 0.7 grams. Thus both birds and mammals are "brainy" compared to nonavian reptiles, but their cerebrums are not strictly homologous. We often tend to think of birds as being less intelligent than mammals (the epithet "bird-brained" is a common one, after all), but new studies show that at least some birds are capable of highly complex behavior and that birds might attain conscious experience (Butler et al. 2005). Differences also exist in the way visual information is

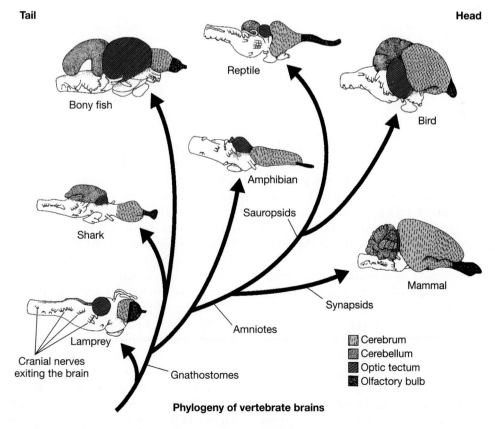

▲ **FIGURE 11–18** Phylogenetic pattern of brain enlargement among vertebrates.

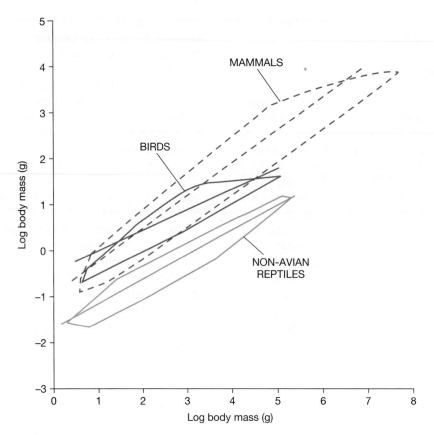

▲ **FIGURE 11–19** Body size and brain size. The relationship between body size and brain size. The polygons enclose the values for 301 genera of mammals, 174 species of birds, and 62 species of non-avian reptiles.

processed. Birds and other sauropsids retain the primitive vertebrate condition of relying mainly on the optic lobes of the midbrain for visual processing. In mammals, the optic lobes are small and are used mainly for optical reflexes such as tracking motion, while it is a new portion of the cerebrum (the visual cortex) that actually interprets the information.

Evaluating how brainy an animal is can be problematic. For a start, the size of brains scales with negative allometry; that is, larger animals have proportionally smaller brains for their size than do small animals. The reason for this negative scaling is not known for certain but may be related to the fact that bigger bodies don't need absolutely more brain tissue to operate. For example, if 200 nerve cells in the brain were needed to control the right hind leg of a mouse, there is no reason to suppose that more cells would be needed to control the right hind leg of

an elephant. Of course, larger animals have brains that are absolutely larger than smaller ones; they are just not quite as large as one would predict for their size, all other things being equal. Comparisons of brain size often neglect differences in body size. For example, television programs often state that the brains of dolphins are just as big as human brains to emphasize how intelligent dolphins are, but most dolphins are much bigger than humans so they would be expected to have *larger* brains.

The size of brains is usually expressed in relation to body size. One such estimate, pioneered by Harry Jerison (1973), is the encephalization quotient (EQ), a measure of the actual brain size in relation to the expected brain size for an animal of that body mass. In other words, the actual mass of the brain is divided by the mass predicted by equations such as those in Figure 11–19. An EQ of 1.0 means that the brain is exactly the size expected, whereas values less than 1

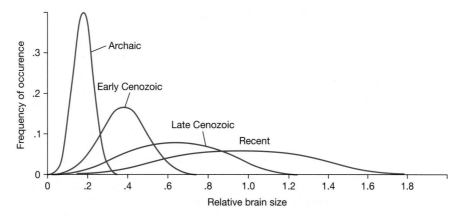

▲ **FIGURE 11–20** Brain gain. Changes in the encephalization quotient of Northern Hemisphere ungulates during the Cenozoic era.

indicate brains that are smaller than expected and values greater than 1 are brains larger than expected.

Many mammals (e.g., rodents, shrews, marsupials) have EQ values less than 1, whereas others (including dogs, cats, horses, elephants, primates, and whales) have values greater than 1. Great apes have values of about 3, and humans have a value of about 8.

Although we think of the tendency to evolve a large brain as a natural outcome of mammalian evolution, the situation is not that simple. Pack-hunting carnivores are bigger-brained than solitary ones, for example, but many small-brained animals display complex behavior and substantial capacity for learning. **Figure 11–20** shows the changes in EQs of hooved herbivorous mammals (ungulates) over time—graphs for other mammalian orders would have similar shapes. Although average brain size has increased and the largest brains known belong to extant species, some living ungulates have brains as small as some of the early Cenozoic forms. In other words, rather than seeing an *overall* increase in brain size, we find that the *range* of brain sizes has increased and small-brained species coexist with much larger-brained forms.

Summary

Synapsids (represented now by mammals) and sauropsids (represented by turtles, tuatara, lizards and snakes, crocodilians, and birds) dominate the terrestrial vertebrate fauna, and both include lineages specialized for flight, burrowing, and secondarily aquatic life. The two lineages have been separate at least from the late Paleozoic, and they have faced the same evolutionary challenges. The solutions they developed to the problems of life on land are similar in some respects and different in others. In some cases ancestral characters were retained, and in others new, derived characters appeared. For example, synapsids retained the ancestral pattern of tidal lung ventilation and urea excretion, while some sauropsids derived different methods. Both lineages developed new forms of locomotion, endothermal thermoregulation, and more complex hearts and brains, achieving functionally equivalent conditions by different routes.

More important than the differences themselves are the consequences of some of the differences in terms of the adaptive zones occupied by synapsids and sauropsids. Sauropsids have specialized in diurnal activity, with excellent color vision and extensive use of color, pattern, and movement in social displays. Synapsids may have passed through a stage of being largely nocturnal, during which color vision was limited and scent and hearing were the primary senses.

Most important is that the different solutions that synapsids and sauropsids have found to the challenges of terrestrial life emphasize that there is more than one way to succeed as a terrestrial amniote.

Additional Readings

Angilletta, M. J. et al. 2002. The evolution of thermal physiology in ectotherms. *Journal of Thermal Biology* 27:249–268.

Baumel, J. J. et al. 1990. The ventilatory movements of the avian pelvis and tail: Function of the muscles of the tail region of the pigeon (*Columba livia*). *Journal of Experimental Biology* 151:263–277.

Bennett, A. F., and J. A. Ruben. 1979. Endothermy and activity in vertebrates. *Science* 206:649–654.

Boggs, D. F. 2002. Interaction between locomotion and ventilation in tetrapods. *Comparative Biochemistry and Physiology* 133A:279–288.

Bramble, D. M., and D. R. Carrier. 1983. Running and breathing in mammals. *Science* 219:251–256.

Britt, B. B. 1997. Postcranial pneumaticity. In *Encyclopedia of Dinosaurs*, ed. P. J. Currie and K. Padian, 590–593. San Diego, CA: Academic.

Butler, A. B., and W. Hodos. 1996. *Comparative Vertebrate Neuroanatomy*. New York, NY: Wiley-Liss.

Butler, A. B. et al. 2005. Evolution of the neural basis of consciousness: A bird-mammal comparison. *BioEssays* 27:923–936.

Carrier, D. R. 1987. The evolution of locomotor stamina in tetrapods: Circumventing a mechanical constraint. *Paleobiology* 13:326–341.

Carrier, D. R., and C. G. Farmer. 2000a. The integration of ventilation and locomotion in archosaurs. *American Zoologist* 40:87–100.

Carrier, D. R., and C. G. Farmer. 2000b. The evolution of pelvic aspiration in archosaurs. *Paleobiology* 26:271–293.

Clack, J. A. 2002. *Gaining Ground: The Origin and Evolution of Tetrapods*. Bloomington, IN: Indiana University Press.

Codd, J. R. et al. 2007. Avian-like breathing mechanics in maniraptoran dinosaurs. *Proceedings of the Royal Society B* 275:157–161.

Cowles, R. B. 1958. Possible origin of dermal temperature regulation. *Evolution* 12:347–357.

Farmer, C. 1997. Did lungs and intracardiac shunt evolve to oxygenate the heart in vertebrates? *Paleobiology* 23:358–372.

Farmer, C. G. 2000. Parental care: The key to understanding endothermy and other convergent features in birds and mammals. *American Naturalist* 155:326–334.

Farmer, C. G. 2001. Parental care: A new perspective on the origin of endothermy. In *New Perspectives in the Origin and Early Evolution of Birds*, Proceedings of the International Symposium in Honor of John H. Ostrom, ed. J. A. Gauthier, 389–412. New Haven, CT: Yale University, Peabody Museum of Natural History.

Farmer, C. G. 2003. Reproduction: The adaptive significance of endothermy. *American Naturalist* 162:826–840.

Farmer, C. G., and D. R. Carrier. 2000. Pelvic aspiration in the American alligator (*Alligator mississippiensis*). *Journal of Experimental Biology* 203:1679–1687.

Geist, N. R. 2000. Nasal respiratory turbinate function in birds. *Physiological Biochemistry and Zoology* 73:581–589.

Gilad, Y. et al. 2003. Human specific loss of olfactory receptor genes. *Proceedings of the National Academy of Sciences USA* 100:3324–3327.

Jacobs, G. H. 1993. The distribution and nature of colour vision among the mammals. *Biological Reviews* 68:413–471.

Jerison, H. J. 1973. *Evolution of the Brain and Intelligence*. New York, NY: Academic Press.

Karten, H. J. 1991. Homology and evolutionary origins of the "neocortex." *Brain, Behavior and Evolution* 38:264–272.

Kemp, T. A. 2006. The origin of mammalian endothermy: A paradigm for the evolution of complex biological structure. *Zoological Journal of the Linnean Society* 147:473–488.

Maina, J. N. 2006. Development, structure, and function of a novel respiratory organ, the lung-air sac system of birds: To go where no other vertebrate has gone. *Biological Reviews* 89:541–579.

Oftedal, O. T. 2002a. The mammary gland and its origin during synapsid evolution. *Journal of Mammary Gland Biology and Neoplasia* 7:225–252.

Oftedal, O. T. 2002b. The origin of lactation as a water source for parchment-shelled eggs. *Journal of Mammary Gland Biology and Neoplasia* 7:253–266.

O'Connor, P. M., and L. P. A. M. Claessens. 2005. Basic avian pulmonary design and flow-through ventilation in non-avian theropod dinosaurs. *Nature* 436:253–256.

Pond, C. M. 1998. *The Fats of Life*. Cambridge, UK: Cambridge University Press.

Qiang, J. et al. 1998. Two feathered dinosaurs from northeastern China. *Nature* 395:753–761.

Qiang, J. et al. 2001. The distribution of integumentary structures in a feathered dinosaur. *Nature* 410:1084–1088.

Ritter, D. 1995. Epaxial muscle function during locomotion in a lizard (*Varanus salvator*) and the proposal of a key innovation in the vertebrate skeletal and musculoskeletal system. *Journal of Experimental Biology* 198:2477–2490.

Ruben, J. A. et al. 2003. Respiratory and reproductive paleophysiology of dinosaurs and early birds. *Physiological and Biochemical Zoology* 76:141–164.

Stokstad, E. 2001. Exquisite Chinese fossils add new pages to book of life. *Science* 291:232–236.

Witmer, L. M. 2001. Nostril position in dinosaurs and other vertebrates and its significance for nasal function. *Science* 293:850–853.

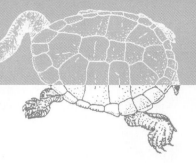

CHAPTER 12

Turtles

Turtles provide a contrast to amphibians in the relative lack of diversity in their life histories. All turtles lay eggs, and none exhibit parental care of the hatchlings. Turtles show morphological specializations associated with terrestrial, freshwater, and marine habitats, and marine turtles make long-distance migrations rivaling those of birds. Probably turtles and birds use many of the same navigation mechanisms to find their way. Most turtles are long-lived animals with relatively poor capacities for rapid population growth. Many, especially sea turtles and large tortoises, are endangered by human activities. Some efforts to conserve turtles have apparently been frustrated by a feature of the embryonic development of some species of turtles—the sex of an individual is determined by the temperature to which it is exposed in the nest. This experience emphasizes the critical importance of information about the basic biology of animals to successful conservation and management.

12.1 Everyone Recognizes a Turtle

Turtles found a successful approach to life in the Triassic period and have scarcely changed since. The shell, which is the key to their success, has also limited the group's diversity. For obvious reasons, flying or gliding turtles have never existed, and even arboreality is only slightly developed. Perhaps the most distinctive turtles are an extinct group of very large terrestrial turtles with horns and frills on the head and a clublike expansion on the end of the tail, the Meiolaniinae (**Figure 12–1**). Meiolanines lived in South America in the Cretaceous period and Eocene epoch, and in Australia and New Caledonia from the Miocene through the Pleistocene epochs.

Shell morphology reflects the ecology of turtle species: The most terrestrial forms, the tortoises of the family Testudinidae, have high domed shells and elephant-like feet. Many smaller species of tortoises have flat, spadelike front feet that they use for burrowing (**Figure 12–2**). The gopher tortoises of North America are an example—their front legs are flattened into scoops, and the dome of the shell is reduced. The Bolson tortoise of northern Mexico constructs burrows a meter or more deep and several meters long in the hard desert soil. These tortoises bask at the mouths of their burrows; when a predator appears, they throw themselves down the steep entrance tunnels of the burrows to escape, just as an aquatic turtle dives off a log. The pancake tortoise of Africa is a radical departure from the usual tortoise morphology. The shell is flat and flexible because its ossification is much reduced. This turtle lives in rocky foothill regions and scrambles over the rocks with nearly as much agility as a lizard. When threatened by a predator, the pancake tortoise crawls into a rock crevice and uses its legs to wedge itself in place. The flexible shell presses against the overhanging rock and creates so much friction that it is almost impossible to pull the tortoise out.

Other terrestrial turtles have moderately domed carapaces (upper shells), like the box turtles of the

▲ **FIGURE 12–1** An extinct horned turtle. *Meiolania platyceps* is from Lord Howe Island, New South Wales, Australia. This late Pleistocene species grew to a shell length of 2.5 m, and the head was about 60 cm wide.

family Emydidae. This is one of several kinds of turtles that have evolved flexible regions in the **plastron** (lower shell), which allow the front and rear lobes to be pulled upward to close the openings of the shell. Aquatic turtles have low carapaces that offer little resistance to movement through water. The Emydidae and Bataguridae contain a large number of pond turtles, including the painted turtles and the red-eared turtles often seen in pet stores and anatomy and physiology laboratory courses.

The snapping turtles (family Chelydridae) and the mud and musk turtles (family Kinosternidae) prowl along the bottoms of ponds and slow rivers and are not particularly streamlined. The mud turtle has a hinged plastron, but the musk and snapping turtles have very reduced plastrons. They rely on strong jaws for protection. A reduction in the size of the plastron makes these species more agile than most turtles, and musk turtles may climb several feet into trees, probably to bask. If a turtle falls on your head while you are canoeing, it is probably a musk turtle.

The soft-shelled turtles (family Trionychidae) are fast swimmers. The ossification of the shell is greatly reduced, lightening the animal, and the feet are large with extensive webbing. Soft-shelled turtles lie in ambush partly buried in debris on the bottom of the pond. Their long necks allow them to reach considerable distances to seize the invertebrates and small fishes on which they feed.

Extant turtles can be placed in 13 families with about 300 species (**Table 12.1**). The two lineages of living turtles can be traced through fossils to the Mesozoic. The **cryptodires** (Greek *crypto* = hidden, *dire* = neck) retract the head into the shell by bending the neck in a vertical S shape. The **pleurodires** (Greek *pleuro* = side) retract the head by bending the neck horizontally. All the turtles discussed so far have been cryptodires, and these are the dominant group of turtles. Cryptodires are the only turtles now found in most of the Northern Hemisphere, and there are aquatic and terrestrial cryptodires in South America and terrestrial ones in Africa. Only Australia has no cryptodires. Pleurodires are now found only in the Southern Hemisphere, although they had worldwide distribution in the late Mesozoic and early Cenozoic eras. *Stupendemys*, a pleurodire from the Pliocene epoch of Venezuela, had a carapace more than 2 meters long. All the living pleurodires are at least semi-aquatic, but some fossil pleurodires had high, domed shells that suggest they may have been terrestrial. The most terrestrial of the living pleurodires are probably the African pond turtles, which readily move overland from one pond to another.

The snake-necked pleurodiran turtles (family Chelidae) are found in South America, Australia, and New Guinea. As their name implies, they have long, slender necks. In some species, the length of the neck is considerably greater than that of the body. These forms feed on fishes that they catch with a sudden dart of the head. Other snake-necked turtles have much shorter necks. Some of these feed on mollusks and have enlarged palatal surfaces used to crush shells. The same specialization for feeding on mollusks is seen in certain cryptodiran turtles.

An unusual feeding method among turtles is found in a pleurodire, the matamata of South America. Large matamatas reach shell lengths of 40 centimeters. They are bizarre-looking animals, with a broad heads and flat shells and flaps of skin projecting from the sides of the head and the broad neck. To these are added trailing bits of adhering algae. The effect is exceedingly cryptic. It is hard to recognize a matamata as a turtle even in an aquarium, and it is practically impossible to see one against the mud and debris of a river bottom. In addition to obscuring the shape of the turtle, the flaps of skin on the head are sensitive to minute vibrations in water caused by the passage of a fish. When it senses the presence of prey, the matamata abruptly opens its mouth and expands its throat. Water rushes in, carrying the prey with it, and the matamata closes its mouth, expels the water, and swallows the prey. The matamata lacks the horny beak that other turtles use for seizing prey or biting off pieces of plants.

▲ FIGURE 12–2 Body forms of turtles. (a) Tortoise, *Testudo,* Testudinidae. (b) Pancake tortoise, *Malacochersus,* Testudinidae. (c) Terrestrial box turtle, *Terrapene,* Emydidae. (d) Pond turtle, *Trachemys,* Emydidae. (e) Soft-shelled turtle, *Apalone,* Trionychidae. (f) Mud turtle, *Kinosternon,* Kinosternidae. (g) Alligator snapping turtle, *Macroclemys,* Chelydridae. (h) African pond turtle, *Pelusios,* Pelomedusidae. (i) Australian snake-necked turtle, *Chelodina,* Chelidae. (j) South American matamata, *Chelys,* Chelidae. (k) Loggerhead sea turtle, *Caretta,* Cheloniidae. (l) Leatherback sea turtle, *Dermochelys,* Dermochelyidae (The turtles are not drawn to scale.)

Marine turtles are cryptodires. The families Cheloniidae and Dermochelyidae show more extensive specialization for aquatic life than any freshwater turtle; for example, the forelimbs are modified as flippers in both families. The largest of the sea turtles of the family Cheloniidae is the loggerhead, which once reached weights exceeding 400 kilograms. The largest marine turtle, the leatherback, reaches shell lengths of more than 2 meters and weights in excess of 600 kilograms. In this species the dermal ossification has been reduced to bony platelets embedded in dense connective tissue, which gives the leatherback its name. This is a pelagic turtle that ranges far from land, and it has a wider geographic distribution than any other ectothermal amniote. Leatherback turtles penetrate far into cool temperate seas and have been recorded in the Atlantic from Newfoundland to Argentina

TABLE 12.1 Families of turtles

Cryptodira

Testudinidae: About 51 species of small (15 centimeters) to very large (1 meter) terrestrial turtles with a worldwide distribution in temperate and tropical regions. Although tortoises are clumsy swimmers, they float well and withstand long periods without food or water. These characteristics have allowed tortoises swept to sea by flooding rivers to populate oceanic islands, such as the Galápagos Islands.

Bataguridae: About 69 species of small (12 centimeters) to large (75 centimeters) freshwater, semiaquatic, and terrestrial turtles, primarily Asian with 1 genus in Central America.

Emydidae: About 41 species of small (12 centimeters) to large (60 centimeters) freshwater, semiaquatic, and terrestrial turtles, mostly in North America, 1 genus in Central and South America, and 1 in Europe, Asia, and north Africa.

Trionychidae: About 30 species of small (25 centimeters) to very large (130 centimeters) freshwater turtles with flattened bodies and reduced ossification of the shell from North America, Africa, and Asia.

Carettochelyidae: A single species of large (70 centimeters) freshwater turtle, *Carettochelys insculpta*, that lacks epidermal scutes on the shell and has paddlelike forelimbs. From New Guinea and extreme northern Australia.

Dermatemydidae: A single species of large (65 centimeters) freshwater turtle, *Dermatemys mawii,* from Mexico and Central America.

Kinosternidae: About 22 species of small (11 centimeters) to medium (40 centimeters) bottom-dwelling freshwater turtles from North America and South America.

Cheloniidae: 6 species of large (70 centimeters) to very large (150 centimeters) sea turtles with bony shells covered with epidermal scutes and paddlelike forelimbs, found worldwide in tropical and temperate oceans.

Dermochelyidae: A single species, *Dermochelys coriacea*, the largest extant turtle (up to 240 centimeters). This is a marine turtle in which the shell is reduced to thousands of small bones embedded in a leathery skin. It occurs in oceans worldwide, and its range extends north and south into seas too cold for other sea turtles,

Chelydridae: 2 species of large (*Chelydra,* 50 centimeters) to very large (*Macroclemys,* 70 centimeters and 80 kilograms) freshwater turtles in North and Central America and a small (*Platysternon,* 18 centimeters), agile turtle living in mountain streams from southeast China to Burma and Thailand.

Pleurodira

Chelidae: About 52 species of small (15 centimeters) to large (50 centimeters) aquatic turtles from South America, Australia, and New Guinea.

Pelomedusidae: 19 species of small aquatic turtles from Africa, Madagascar, and the Seychelles Islands. All extant pelomedusids inhabit freshwater, but some extinct species may have been marine.

Podocnemidae: 8 species of aquatic turtles found in northern South America and Madagascar. *Podocnemis expansa*, which occurs in the Amazon and Orinoco Rivers, is the largest extant pleurodire; females reach shell lengths of 90 centimeters. The extinct *Stupendemys* (from Late Tertiary of Venezuela) was over 2 meters long.

and in the Pacific from Japan to Tasmania. Leatherback turtles dive to depths of more than 1000 meters. One dive that drove the depth recorder off scale is estimated to have reached 1200 meters, which exceeds the deepest dive recorded for a sperm whale (1140 meters). Leatherback turtles feed largely on jellyfish, whereas the smaller hawksbill sea turtles eat sponges that are defended by spicules of silica (glass) as well as a variety of chemicals (including alkaloids and terpenes) that are toxic to most vertebrates. Green turtles (*Chelonia mydas*) are the only herbivorous marine turtles.

12.2 But What Is a Turtle? Phylogenetic Relationships of Turtles

Turtles show a combination of ancestral features and highly specialized characters that are not shared with any other group of vertebrates, and their phylogenetic affinities are not fully understood. The turtle lineage probably originated among the early amniotes of the Late Carboniferous period. Like those animals, turtles have anapsid skulls, but the shells and postcranial skeletons of turtles are unique. One view emphasizes the importance of the anapsid skull and places the affinities of turtles among the parareptiles (see Figure 9–1; Laurin and Reisz 1995; Lee 1993, 1995). A radically different opinion regards the anapsid skull of turtles as being a secondarily derived feature and places turtles among the diapsids (Rieppel and de Braga 1996, Rieppel 2000, Hill 2005). The case for a diapsid origin dominates, but a lively debate is in progress and the issue is not fully resolved (**Figure 12–3**).

The earliest turtles are found in Late Triassic deposits in Germany, Thailand, and Argentina. These animals had nearly all the specialized characteristics of derived turtles and shed no light on the phylogenetic affinities of the group. *Proganochelys*, from Triassic deposits in Germany, was nearly a meter long (larger than most living turtles) and had a high, arched shell. The marginal teeth had been lost, and the maxilla, premaxilla, and dentary bones were probably covered with a horny beak, just as they are in derived turtles. The skull of *Proganochelys* retained the supratemporal and lacrimal bones and the lacrimal duct, and the palate had rows of denticles; all these structures have been lost by derived turtles. The plastron of *Proganochelys* also contained some bones that have been lost by derived turtles, and the vertebrae of the neck lack specializations that would have allowed the head to be retracted into the shell. *Palaeochersis* (Late Triassic period of Argentina) and *Australochelys* (Early Jurassic period of South Africa) are probably the sister group of later turtles, including Cryptodira and Pleurodira.

Turtles with neck vertebrae specialized for retraction are not known before the Cretaceous, but differences in the skulls and shells allow the pleurodiran and cryptodiran lineages to be traced back to the Late Triassic (shells of pleurodires) and Late Jurassic (skulls of cryptodires). The otic capsules of all turtles

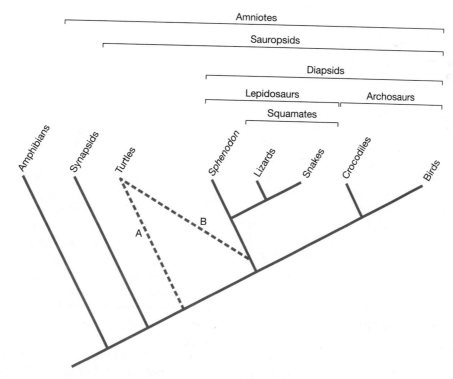

▲ **FIGURE 12–3** Simplified cladogram of tetrapods. The two competing hypotheses of the phylogenetic position of turtles are shown: an origin from parareptiles (A) or from within diapsid reptiles (B).

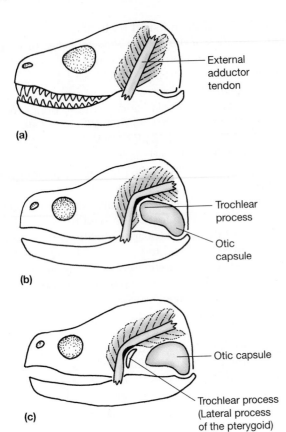

(a)

(b)

(c)

— External adductor tendon

— Trochlear process

— Otic capsule

— Otic capsule

— Trochlear process (Lateral process of the pterygoid)

▲ **FIGURE 12–4** Position of the external adductor tendon. (a) In the ancestral (parareptilian) condition. (b) In cryptodiran turtles. (c) In pleurodiran turtles.

beyond the proganochelids are enlarged, and the jaw adductor muscles bend posteriorly over the otic capsule (**Figure 12–4**). The muscles pass over a pulleylike structure, called the trochlear process. In cryptodires the trochlear process is formed by the anterior surface of the otic capsule itself, whereas in pleurodires it is formed by a lateral process of the pterygoid. Fusion of the pelvic girdle to the carapace and plastron further distinguishes pleurodires from cryptodires, which have a bony connection attaching the shell to the girdle. The beginnings of these changes are seen in *Australochelys*.

12.3 Turtle Structure and Functions

Turtles are among the most bizarre vertebrates. Covered in bone, with the limbs inside the ribs and with horny beaks instead of teeth—if turtles had become extinct at the end of the Mesozoic, they would rival dinosaurs in their novelty. However, because they survived, they are regarded as commonplace and are often used in comparative anatomy courses to represent primitive amniotes (inappropriately, because they are so specialized).

■ Shell and Skeleton

The shell is the most distinctive feature of a turtle (**Figure 12–5**). The carapace is composed of dermal bone that typically grows from 59 separate centers of ossification. Eight plates along the dorsal midline form the neural series and are fused to the neural arches of the vertebrae. Lateral to the neural bones are eight paired costal bones, which are fused to the broadened ribs. The ribs of turtles are unique among tetrapods in being external to the girdles. Eleven pairs of peripheral bones, plus two unpaired bones in the dorsal midline, form the margin of the carapace. The plastron is formed largely from dermal ossifications, but the entoplastron is derived from the interclavicle; the paired epiplastra anterior to it are derived from the clavicles. Processes from the hyoplastron and hypoplastron fuse with the first and fifth pleurals, forming a rigid connection between the plastron and carapace.

The bones of the carapace are covered by horny scutes of epidermal origin that do not coincide in number or position with the underlying bones. The carapace has a row of five central scutes, bordered on each side by four lateral scutes. Ten to twelve marginal scutes on each side turn under the edge of the carapace. The plastron is covered by a series of six paired scutes.

Flexible areas, called hinges, are present in the shells of many turtles. The most familiar examples are the North American and Asian box turtles (*Terrapene* and *Cuora*), in which a hinge between the hyoplastral and hypoplastral bones allows the anterior and posterior lobes of the plastron to be raised to close off the front and rear openings of the shell. Mud turtles (*Kinosternon*) have two hinges in the plastron; the anterior hinge runs between the epiplastra and the entoplastron (which is triangular in kinosternid turtles rather than diamond shaped), and the posterior hinge is between the hypoplastron and xiphiplastron. Some species of tortoises have plastral hinges; in *Testudo* the hinge lies between the hyoplastron and xiphiplastra, as it does in *Kinosternon*; but, in another genus of tortoise (*Pyxis*), the hinge is anterior and involves a break across the entoplastron. The African forest tortoises (*Kinixys*)

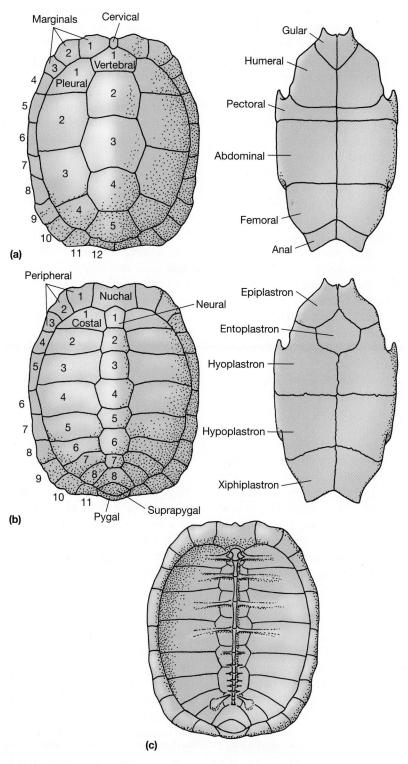

▲ **FIGURE 12–5** Shell and vertebral column of a turtle. (a) Epidermal scutes of the carapace (*left*) and plastron (*right*). (b) Dermal bones of the carapace (*left*) and plastron (*right*). (c) Vertebral column of a turtle, seen from the inside of the carapace. Note that anteriorly the ribs articulate with two vertebral centra.

have a hinge on the posterior part of the carapace. The margins of the epidermal shields and the dermal bones of the carapace are aligned, and the hinge runs between the second and third pleural scutes and the fourth and fifth costals. The presence of hinges is sexually dimorphic in some species of tortoises. The erratic phylogenetic occurrence of kinetic shells and differences among related species indicate that shell kinesis has evolved many times in turtles.

Asymmetry of the paired epidermal scutes is quite common among turtles, and modifications of the bony structure of the shell are seen in some families. Soft-shelled turtles lack peripheral ossifications and epidermal scutes. The distal ends of the broadened ribs are embedded in flexible connective tissue, and the carapace and plastron are covered with skin. The New Guinea river turtle (*Carretochelys*) is also covered by skin instead of scutes, but in this species the peripheral bones are present and the edge of the shell is stiff. The leatherback sea turtle (*Dermochelys*) has a carapace formed of cartilage with thousands of small polygonal bones embedded in it, and the plastral bones are reduced to a thin rim around the edge of the plastron. The neural and costal ossifications of the pancake tortoise (*Malacochersus*) are greatly reduced, but the epidermal plates are well developed.

Extant turtles have only 10 vertebrae in the trunk and 8 in the neck. The centra of the trunk vertebrae are elongated and lie beneath the dermal bones in the dorsal midline of the shell. The centra are constricted in their centers and fused to each other. The neural arches in the anterior two-thirds of the trunk lie between the centra as a result of anterior displacement, and the spinal nerves exit near the middle of the preceding centrum. The ribs are also shifted anteriorly; they articulate with the anterior part of the neurocentral boundary; and, in the anterior part of the trunk, where the shift is most pronounced, the ribs extend onto the preceding vertebra.

Cryptodires have two sacral vertebrae (the 19th and 20th vertebrae) with broadened ribs that meet the ilia of the pelvis. Pleurodires have the pelvic girdle firmly fused to the dermal carapace by the ilia dorsally and by the pubic and ischial bones ventrally, and the sacral region of the vertebral column is less distinct. The ribs on the 17th, 18th, 19th, and sometimes the 20th vertebrae are fused to the centra and end on the ilia or the iliocarapacial junction.

The cervical vertebrae of cryptodires have articulations that permit the S-shaped bend used to retract the head into the shell. Specialized articulating surfaces between vertebrae, called ginglymes, permit vertical rotation. This type of rotation, ginglymoidy, is peculiar to cryptodires, but the anatomical details vary within the group. In most families, the hinge is formed by two successive ginglymoidal joints between the 6th and 7th and the 7th and 8th cervical vertebrae. The lateral bending of the necks of pleurodiran turtles is accomplished by ball-and-socket or cylindrical joints between adjacent cervical vertebrae.

■ The Heart

The circulatory systems of tetrapods can be viewed as consisting of two circuits: The systemic circuit carries oxygenated blood from the heart to the head, trunk, and appendages, whereas the pulmonary circuit carries deoxygenated blood from the heart to the lungs. The blood pressure in the systemic circuit is higher than the pressure in the pulmonary circuit, and the two circuits operate in series. That is, blood flows from the heart through the lungs, back to the heart, and then to the body. The morphology of the hearts of derived synapsids and sauropsids (mammals, crocodilians, and birds) makes this sequential flow obligatory, but the hearts of turtles and lepidosaurs have the ability to shift blood between the pulmonary and systemic circuits.

This flexibility in the route of blood flow can be accomplished because the ventricular chambers in the hearts of turtles and lepidosaurs are in anatomical continuity, instead of being completely divided by a septum like the ventricles of birds and mammals. The flow of blood is controlled partly by the relative resistance to flow in the pulmonary and systemic circuits. The pattern of blood flow can best be explained by considering the morphology of the heart and how intracardiac pressure changes during a heartbeat. **Figure 12–6** shows a schematic view of the heart of a turtle. The left and right atria are completely separate, and three subcompartments can be distinguished in the ventricle. A muscular ridge in the core of the heart divides the ventricle into two spaces, the cavum pulmonale and the cavum venosum. The muscular ridge is not fused to the wall of the ventricle, and thus the cavum pulmonale and the cavum venosum are only partly separated. A third subcompartment of the ventricle, the cavum arteriosum, is located dorsal to the cavum pulmonale and cavum venosum. The cavum arteriosum communicates with the cavum venosum through an intraventricular canal. The pulmonary artery opens from

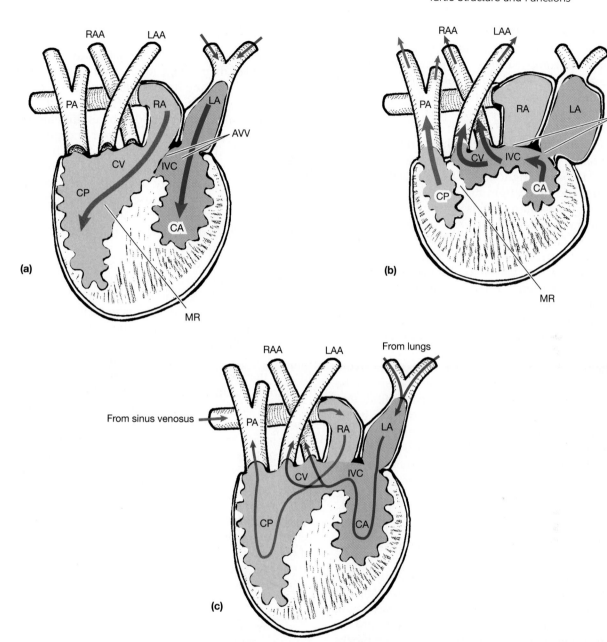

▲ **FIGURE 12–6** Blood flow in the heart of a turtle. (a) As the atria contract, oxygenated blood (colored arrows) from the left atrium (LA) enters the cavum arteriosum (CA), while deoxygenated blood (*gray arrows*) from the right atrium (RA) first enters the cavum venosum (CV) and then crosses the muscular ridge (MR) and enters the cavum pulmonale (CP). The atrioventricular valve (AVV) blocks the intraventricular canal (IVC) and prevents mixing of oxygenated and deoxygenated blood. (b) As the ventricle contracts, the deoxygenated blood in the cavum pulmonale is expelled through the pulmonary arteries (PA); the AVV closes, no longer obstructing the IVC; and the oxy- genated blood in the cavum arteriosum is forced into the cavum venosum and expelled through the right and left aortic arches (RAA and LAA). The adpression of the wall of the ventricle to the muscular ridge prevents mix- ing of deoxygenated and oxygenated blood. (c) Summary of the pattern of blood flow through the heart of a turtle.

the cavum pulmonale, and the left and right aortic arches open from the cavum venosum.

The right atrium receives deoxygenated blood from the body via the sinus venosus and empties into the cavum venosum, and the left atrium receives oxygenated blood from the lungs and empties into the cavum arteriosum. The atria are separated from the ventricle by flaplike atrioventricular valves that open as the atria contract and then close as the ven- tricle contracts, preventing blood from being forced back into the atria. The anatomical arrangement of the connections between the atria, their valves, and the three subcompartments of the ventricle is crucial because it is those connections that allow pressure differentials to direct the flow of blood and to prevent mixing of oxygenated and deoxygenated blood.

When the atria contract, the atrioventricular valves open, allowing blood to flow into the ventricle. Blood from the right atrium flows into the cavum venosum, and blood from the left atrium flows into the cavum arteriosum. At this stage in the heartbeat, the large median flaps of the valve between the right atrium and the cavum venosum are pressed against the opening of the intraventricular canal, sealing it off from the cavum venosum. As a result, the oxygenated blood from the left atrium is confined to the cavum arteriosum. Deoxygenated blood from the right atrium fills the cavum venosum and then continues over the muscular ridge into the cavum pulmonale.

When the ventricle contracts, blood pressure inside the heart increases. Ejection of blood from the heart into the pulmonary circuit precedes flow into the systemic circuit because resistance is lower in the pulmonary circuit. As deoxygenated blood flows out of the cavum pulmonale into the pulmonary artery, the displacement of blood from the cavum venosum across the muscular ridge into the cavum pulmonale continues. As the ventricle shortens during contraction, the muscular ridge comes into contact with the wall of the ventricle and closes off the passage for blood between the cavum venosum and cavum pulmonale.

Simultaneously, blood pressure inside the ventricle increases, and the flaps of the right atrioventricular valve are forced into the closed position, preventing backflow of blood from the cavum venosum into the atrium. When the valve closes, it no longer blocks the intraventricular canal. Oxygenated blood from the cavum arteriosum can now flow through the intraventricular canal and into the cavum venosum. At this stage in the heartbeat, the wall of the ventricle is pressed firmly against the muscular ridge, separating the oxygenated blood in the cavum venosum from the deoxygenated blood in the cavum pulmonale.

As pressure in the ventricle continues to rise, oxygenated blood in the cavum venosum is ejected into the aortic arches. This system effectively prevents mixing of oxygenated and deoxygenated blood in the heart, despite the absence of a permanent morphological separation of the two circuits.

Respiration

As we have seen, primitive amniotes probably used movements of the rib cage to draw air into the lungs and to force it out, and lizards still employ that mechanism. The fusion of the ribs of turtles with their rigid shells makes that method of breathing impossible. Only the openings at the anterior and posterior ends of the shell contain flexible tissues. The lungs of a turtle, which are large, are attached to the carapace dorsally and laterally. Ventrally, the lungs are attached to a sheet of nonmuscular connective tissue that is itself attached to the viscera (**Figure 12–7**). The weight of the viscera keeps this diaphragmatic sheet stretched downward.

Turtles produce changes in pressure in the lungs by contracting muscles that force the viscera upward, compressing the lungs and expelling air, followed by contracting other muscles that increase the volume of the visceral cavity, allowing the viscera to settle downward. Because the viscera are attached to the diaphragmatic sheet, which in turn is attached to the lungs, the downward movement of the viscera expands the lungs, drawing in air. In turtles, both inhalation and exhalation require muscular activity. The viscera are forced upward against the lungs by the contraction of the transverse abdominus muscle posteriorly and the pectoralis muscle anteriorly. The transverse abdominus inserts on the cup-shaped connective tissue (the posterior limiting membrane) that closes off the posterior opening of the visceral cavity. Contraction of the transverse abdominus flattens the cup inward, thereby reducing the volume of the visceral cavity. The pectoralis draws the shoulder girdle back into the shell, further reducing the volume of the visceral cavity.

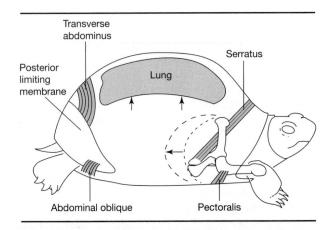

▲ **FIGURE 12–7** Schematic view of the lungs and respiratory movements of a tortoise.

The inspiratory muscles are the abdominal oblique, which originates near the posterior margin of the plastron and inserts on the external side of the posterior limiting membrane, and the serratus, which originates near the anterior edge of the carapace and inserts on the pectoral girdle. Contraction of the abdominal oblique pulls the posterior limiting membrane outward, and contraction of the serratus rotates the pectoral girdle outward. Both of these movements increase the volume of the visceral cavity, allowing the viscera to settle back downward and causing the lungs to expand. The in-and-out movements of the forelimbs and the soft tissue at the rear of the shell during breathing are conspicuous.

Box turtles (*Terrapene carolina*) show no interaction between locomotion and lung ventilation (Landberg et al. 2003). In this respect, they differ from lizards (and presumably from primitive amniotes) in not having a conflict between locomotion and lung ventilation and from derived sauropsids and synapsids in not coupling step frequency to respiratory frequency. The absence of interaction between locomotion and breathing in box turtles was unexpected because female green sea turtles (*Chelonia mydas*) appear to alternate periods of locomotion with periods of breathing when they emerge from the sea to lay eggs. Furthermore, when box turtles are not walking, their front limbs move in and out in tempo with their respiration, as would be expected from the involvement of the pectoralis and serratus muscles in lung ventilation. Green sea turtles are highly specialized for swimming, and on the beach they crawl with bilateral movements of the fore limbs. In contrast, box turtles are terrestrial and use both fore- and hindlimbs in a diagonal sequence during walking, so a pressure increase produced by the movement of one limb is balanced by a pressure decrease from the other limb. The balanced effect of symmetrical limb movements probably explains the absence of an effect of locomotion on lung ventilation, and it could be the key to the evolutionary origin of the turtle shell. The ventilatory movements of turtles may be another solution to the conflict between locomotion and lung ventilation, replacing contraction of thoracic muscles by contraction of abdominal muscles to change the volume of the thorax. When movement of the ribs was no longer required to ventilate the lungs, it would be possible for the ancestors to develop a rigid shell to which the ribs were fused.

The basic problems of respiring within a rigid shell are the same for most turtles, but the mechanisms show some variation. For example, aquatic turtles can use the hydrostatic pressure of water to help move air in and out of the lungs. In addition, many aquatic turtles are able to absorb oxygen and release carbon dioxide to the water. The pharynx and cloaca appear to be the major sites of aquatic gas exchange. In 1860, in *Contributions to the Natural History of the U.S.A.*, Louis Agassiz pointed out that the pharynx of soft-shelled turtles contains fringelike processes and suggested that these structures are used for underwater respiration. Subsequent study has shown that, soft-shelled turtles use movements of the hyoid apparatus to draw water in and out of the pharynx when they are confined under water, and that pharyngeal respiration accounts for most of the oxygen absorbed from the water. The Australian turtle *Rheodytes leukops* uses cloacal respiration. Its cloacal orifice is as much as 30 millimeters in diameter, and the turtle holds it open. Large bursae (sacs) open from the wall of the cloaca, and the bursae have a well-vascularized lining with numerous projections (villi). The turtle pumps water in and out of the bursae at rates of 15 to 60 times per minute. Captive turtles rarely surface to breathe, and experiments have shown that the rate of oxygen uptake through the cloacal bursae is very high.

■ Patterns of Circulation and Respiration

The morphological complexity of the hearts of turtles and of squamates allows them to adjust blood flow through the pulmonary and systemic circuits to meet short-term changes in respiratory requirements. The key to these adjustments is changing pressures in the systemic and pulmonary circuits.

Recall that in the turtle heart, deoxygenated blood from the right atrium normally flows from the cavum venosum across the muscular ridge and into the cavum pulmonale. The blood pressure inside the heart increases as the ventricle contracts, and blood is first ejected into the pulmonary artery because the resistance to flow in the pulmonary circuit is normally less than the resistance in the systemic circuit. However, the resistance to blood flow in the pulmonary circuit can be increased by muscles that narrow the diameter of blood vessels. When this happens, the delicate balance of pressure in the heart that maintained the separation of oxygenated and deoxygenated blood is changed. When the resistance of the pulmonary circuit is essentially the same as that of the systemic circuit, blood flows out of the cavum pulmonale and cavum venosum at the same time, and some deoxygenated blood bypasses the lungs and flows into the systemic circuit (**Figure 12–8**). This process is called a right-to-left intracardiac

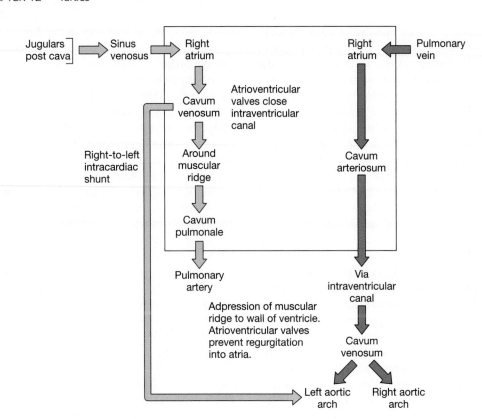

▲ **FIGURE 12–8** Right-to-left shunt of blood in the heart of a turtle. Light colored arrows show deoxygenated blood, and dark colored arrows show oxygenated blood. The box encloses the cycle of events during normal blood flow (compare with Figure 12.6). When resistance in the pulmonary circuit increases, some deoxygenated blood from the cavum venosum flows into the left aortic arch instead of into the pulmonary arteries.

shunt. *Right-to-left* refers to the shift of deoxygenated blood from the pulmonary circuit into the systemic circuit, and *intracardiac* means that it occurs in the heart rather than by flow between the major blood vessels.

Why would it be useful to divert deoxygenated blood from the lungs into the systemic circulation? The ability to make this shunt is not unique to turtles—it occurs also among lizards and snakes and in crocodilians. The heart morphology of lizards is very like that of turtles, and the same mechanism of changing pressures in the pulmonary and systemic circuits is used to achieve an intracardiac shunt. Crocodilians have hearts in which the ventricle is permanently divided into right and left halves by a septum, and they employ a different mechanism to achieve a right-to-left shunt.

The most general function for blood shunts may lie in the ability they provide to match patterns of lung ventilation and pulmonary gas flow (Wang et al. 1997). Lizards, snakes, crocodilians, and turtles normally breathe intermittently, and periods of lung ventilation alternate with periods of **apnea** (no breathing). A mathematical model indicates that a combination of right-to-left and left-to-right shunts could stabilize oxygen concentration in blood during alternating periods of apnea and breathing.

Another function of intracardiac shunts may be to reduce blood flow to the lungs during breath-holding to permit more effective use of the oxygen stored in the lungs. Diving is one situation in which reptiles hold their breath. Many reptiles are excellent divers, and even terrestrial and arboreal forms such as green iguanas may dive into water to escape predators, but that is not the only situation in which defensive behaviors interfere with breathing. Turtles are particularly prone to periods of breath-holding because their method of lung ventilation means they cannot breathe when they withdraw their heads and legs into their shells.

Yet another function of blood shunts may be to enhance digestion: a right-to-left shunt that retains carbon dioxide-rich blood in the body could enhance secretion of gastric acid in the digestive system. That hypothesis was tested by blocking the shunt in one group of American alligators (*Alligator missisippiensis*) and performing a sham operation on a control group of alligators (Farmer et al, 2008). As predicted by the hypothesis, the group in which the shunt was blocked had lower rates of gastric acid secretion and digested bone more slowly than did the control group.

12.4 Ecology and Behavior of Turtles

Turtles are long-lived animals. Even small species like the painted turtle (*Chrysemys picta*) do not mature until they are seven or eight years old, and they may live to be 14 or older. Larger species of turtles live longer. Estimates of centuries for the life spans of tortoises are exaggerated, but large tortoises and sea turtles may live at least as long as humans, and even box turtles may live over 50 years. These longevities make the life histories of turtles hard to study. Furthermore, a long lifetime is usually associated with a low replacement rate of individuals in the population, and species with those characteristics are at risk of extinction when hunting or habitat destruction reduces their numbers. Conservation efforts for sea turtles and tortoises are especially important areas of concern.

■ Temperature Regulation and Body Size of Turtles

Turtles can achieve a considerable degree of stability in body temperature by regulating their exchange of heat energy with the environment. Turtles basking on a log in a pond are a familiar sight in many parts of the world because few pond turtles are large enough to maintain body temperatures higher than the temperature of the water surrounding them. Emerging from the water to bask is the only way most pond turtles can raise their body temperatures to speed digestion, growth, and the production of eggs. In addition, basking may help aquatic turtles to rid themselves of algae and leeches. Exposure to ultraviolet light may activate vitamin D, which is involved in controlling calcium deposition in their bones and shell. A few turtles are quite arboreal; these turtles have small plastrons that allow considerable freedom of movement for the limbs. The big-headed turtle (*Platysternon*

megacephalum) from Southeast Asia lives in fast-flowing streams at high altitudes and is said to climb on rocks and trees to bask. In North America, musk turtles (*Sternotherus*) bask on overhanging branches and drop into the water when they are disturbed.

Small terrestrial turtles, such as box turtles and small species of tortoises, can thermoregulate by moving between sunlight and shade. Small tortoises warm and cool quite rapidly, and they appear to behave very much like other small reptiles in selecting suitable microclimates for thermoregulation. Familiarity with a home range may assist this type of thermoregulation. A study conducted in Italy compared the thermoregulation of resident Hermann's tortoises (animals living in their own home ranges) with individuals that were brought to the study site and tested before they had learned their way around (Chelazzi and Calzolai 1986). The resident tortoises warmed faster and maintained more stable shell temperatures than did the strangers.

Turtles are unusual among reptiles in having a substantial number of species that reach large body sizes. The bulk of a large tortoise provides considerable thermal inertia, and large species like the Galápagos and Aldabra tortoises heat and cool slowly. The giant tortoises of Aldabra Atoll (*Geochelone gigantea*), which weigh 60 kilograms or more, allow their body temperatures to rise to 32°C to 33°C on sunny days and cool to 28°C to 30°C overnight.

Large body size slows the rate of heating and cooling, but it can make temperature regulation more difficult. A small turtle can find shade beside a bush or even a clump of grass, but a giant tortoise needs a bigger object—a tree, for example. In open, sunny habitats, overheating can be a problem for giant tortoises. The difficulty is particularly acute for some tortoises on Grande Terre, an island in the Indian Ocean (Swingland and Frazier 1979, Swingland and Lessells 1979). During the rainy season each year, some of the turtles on the island move from the center of the island to the coast. This movement has direct benefits because the migrant turtles gain access to a seasonal flush of plant growth on the coast. As a result of the extra food, migrant females are able to lay more eggs than females that remain inland. There are risks to migrating, however, because shade is limited on the coast and the rainy season is the hottest time of the year. Tortoises on the coast must limit their activity to the vicinity of patches of shade, which may be no more than a single tree in the midst of a grassy plain. During the morning tortoises forage on the plain; but, as their body temperatures rise, they move back toward the shade of the tree. As the day grows hotter, tortoises try to get into the deepest shade, and the biggest individuals do this most successfully. As the big tortoises (which are mostly males) push their way into the shade, they force smaller individuals (most of which are females) out into the sunlight, and some of these tortoises die of overheating.

Marine turtles are large enough to achieve a considerable degree of endothermy (Spotila and Standora 1985, Paladino et al. 1990). A body temperature of 37°C was recorded by telemetry from a green turtle swimming in water that was 20°C. The leatherback turtle is the largest living turtle; adults may weigh up to 1000 kilograms. Leatherbacks range far from warm equatorial regions and in the summer can be found off the coasts of New England and Nova Scotia in water as cool as 8°C to 15°C. Body temperatures of these turtles appear to be 18°C or more above water temperatures, and a countercurrent arrangement of blood vessels in the flippers may contribute to retaining heat produced by muscular activity.

▓ Social Behavior and Courtship

Tactile, visual, and olfactory signals are employed by turtles during social interactions. Many pond turtles have distinctive stripes of color on their heads, necks, and forelimbs and on their hindlimbs and tail. These patterns are used by herpetologists to distinguish the species, and they may be species-isolating mechanisms for the turtles as well. During the mating season, male pond turtles swim in pursuit of other turtles, and the color and pattern on the posterior limbs may enable males to identify females of their own species. At a later stage of courtship, when the male turtle swims backward in front of the female and vibrates his claws against the sides of her head, both sexes can see the patterns on their partner's head, neck, and forelimbs (**Figure 12–9**).

Among terrestrial turtles, the behavior of tortoises is best known. Many tortoises vocalize during courtship; the sounds they produce have been described as grunts, moans, and bellows. The frequencies of the

(a)

(b)

▲ **FIGURE 12–9** Social behavior of turtles. (a) Male painted turtle (*Chrysemys picta*) courting a female by vibrating the elongated claws of his forefeet against the sides of her head. (b) The head-raising dominance posture of a Galápagos tortoise, *Geochelone*. (This behavior can sometimes be elicited by crouching in front of a male tortoise and raising your arm.)

calls that have been measured range from 500 hertz to 2500 hertz. Some tortoises have glands that become enlarged during the breeding season and appear to produce pheromones. The secretion of the subdentary gland found on the underside of the jaw of tortoises in the North American genus *Gopherus* appears to identify both the species and the sex of an individual. During courtship, males and females of the Florida gopher tortoise rub their subdentary gland across one or both forelimbs, and then extend the limbs toward the other individual, which may sniff at them. Males also sniff the cloacal region of other tortoises, and male tortoises of some species trail females for days during the breeding season. Fecal pellets may be territorial markers; fresh fecal pellets from a dominant male tortoise have been reported to cause dispersal of conspecifics.

Tactile signals used by tortoises include biting, ramming, and hooking. These behaviors are used primarily by males, and they are employed against other males and also against females. Bites are usually directed at the head or limbs, whereas ramming and hooking are directed against the shell. The epiplastral region is used for ramming, and in some species the epiplastral bones of males are elongated and project forward beneath the neck. A tortoise about to ram another individual raises itself on its legs, rocks backward, and then plunges forward, hitting the shell of the other individual with a thump that can be heard from a distance of 100 meters in large species. During hooking, the epiplastral projections are placed under the shell of an adversary, and the aggressor lifts the front end of its shell and walks forward. The combination of lifting and pushing hustles the adversary along and may even overturn it.

Movements of the head appear to act as social signals for tortoises, and elevating the head is a signal of dominance in some species. Herds of tortoises have social hierarchies that are determined largely by aggressive encounters. Ramming, biting, and hooking are employed in these encounters, and the larger individual is usually the winner—although experience may play some role. These social hierarchies are expressed in the priority of different individuals in access to food or forage areas, mates, and resting sites. Dominance relationships also appear to be involved in determining the sequence in which individual tortoises move from one place to another. The social structure of a herd of tortoises can be a nuisance for zookeepers trying to move the animals from an outdoor pen into an enclosure for the night because the tortoises resist moving out of their proper rank sequence.

Nesting Behavior

All turtles are oviparous. Female turtles use their hindlimbs to excavate a nest in sand or soil and deposit a clutch that ranges from four or five eggs for small species to more than 100 eggs for the largest sea turtles. Turtles in the families Cheloniidae, Dermochelyidae, and Chelydridae lay eggs with soft, flexible shells, as do most species in the families Bataguridae, Emydidae, and Pelomedusidae. The eggs of turtles in the families Carettochelyidae, Chelidae, Kinosternidae, Testudinidae, and Trionychidae have rigid shells. Embryonic development typically requires 40 to 60 days; and, in general, soft-shelled eggs develop more rapidly than hard-shelled eggs. Some species of turtles lay their eggs in late summer or fall, and the eggs have a diapause (a period of arrested embryonic development) during the winter and resume development when temperatures rise in the spring. The Australian pleurodire *Chelodina rugosa* lays eggs underwater in temporary ponds, but embryonic development does not begin until the ponds dry out and the eggs are exposed to air (Kennett et al. 1998).

Environmental Effects on Egg Development

Temperature, wetness, and the concentrations of oxygen and carbon dioxide can have profound effects on the embryonic development of turtles. The temperature of a nest affects the rate of embryonic development, and excessively high or low temperatures can be lethal. The discovery that the sex of some reptiles is determined by the temperature they experienced during embryonic development has important implications for understanding patterns of life history, as well as conservation of these species. Temperature-dependent sex determination is widespread among turtles, apparently universal among crocodilians, and is known for tuatara and a few species of lizards. The effect of temperature on sex determination is correlated with sexual size dimorphism of adults— high incubation temperatures produce the larger sex, which is usually females for turtles. The switch from one sex to the other occurs within a span of 3°C or 4°C (**Figure 12–10**). Male crocodilians, tuatara, and lizards are usually larger than females, and in these groups high temperatures during embryonic development produce males.

Temperature-dependent sex determination would make sense if temperature during embryonic development had a different effect on the lifetime reproductive

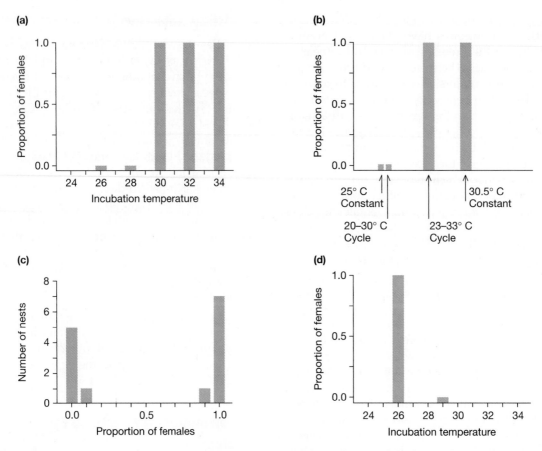

▲ **FIGURE 12–10** Temperature-dependent sex determination. (a) Eggs of the European pond turtle *Emys orbicularis* hatch into males when they are incubated at 26°C or 28°C and into females at 30°C or above. (b) The North American map turtle *Graptemys ouachitensis* shows the same pattern. A temperature that cycles between 20°C and 30°C produces males, whereas a temperature cycle of 23°C to 33°C produces females. (c) Natural nests of map turtles produce predominantly males or females, depending on the nest temperature. (d) Eggs of the lizard *Agama agama* also show temperature-dependent sex determination, but the male- and female-determining temperatures are the opposite of those in turtles—for the lizard, low temperatures produce females, and high temperatures produce males.

success of males versus females. For example, if males benefit from being large more than females do, evolution should coordinate the effects of temperature on sex determination so that the same temperatures that promote large body size also produce males. This plausible hypothesis has been tested by measuring the lifetime reproductive success of male and female jacky lizards (*Amphibolurus muricatus*) that were produced from eggs incubated at different embryonic temperatures (Warner and Shine 2008). Sure enough, the fitness of each sex was highest at the embryonic temperatures that produces that sex.

Temperatures of natural nests are not completely stable, of course. There is some daily temperature variation superimposed on a seasonal cycle of chang-

ing environmental temperatures. The middle third of embryonic development is the critical period for sex determination; the sex of the embryos depends on the temperatures they experience during those few weeks. When eggs are exposed to a daily temperature cycle, the high point of the cycle is most critical for sex determination.

Because of the narrowness of the thermal windows involved in sex determination and the variation that exists in environmental temperatures, both sexes are produced under field conditions, but not necessarily in the same nests. A nest site may be cooler in late summer when it is shaded by vegetation than in early spring when it is exposed to sunlight. Thus, eggs laid early in the season would produce females, whereas eggs deposited in the same place later in the year would produce males. Temperature can also differ between the top and bottom of a nest. For example, temperatures averaged from 33°C to 35°C in the top center of dry nests of American alligators in marshes, and males subsequently hatched from eggs in this area. At the bottom and sides of the same nests, average temperatures were from 30°C to 32°C, and eggs from those regions hatched into females.

Unexpectedly, a viviparous lizard, the Australian skink *Eulamprus tympanum*, has been found to have temperature-dependent sex determination; this is the first example of the phenomenon to be described in a viviparous species (Robert and Thompson 2001.) In this species, warm temperatures produce males; and, in laboratory experiments, the proportion of males increased from 55 percent males for lizards maintained at a constant temperature of 25°C to 100 percent males at 32°C.

In the wild, female lizards can regulate their body temperatures during gestation, and the combination of thermoregulation and temperature-dependent sex determination would allow females to determine the sex of their offspring (Wapstra et al. 2004). If a population contained an excess of males, for example, producing female offspring would be advantageous. A viviparous skink from Tasmania does just that, adjusting the proportion of males and females in a litter in response to the ratio of males and females in the breeding population, and temperature-dependent sex determination may be the mechanism by which this is accomplished. The sex ratio in the population of *Eulamprus* studied was nearly 1:1. When the sex ratio is even, some males mate with more than one female and other males do not mate at all, but every female in the population is likely to mate and produce young. In this situation female offspring have higher fitness, and if they can adjust the ratio of males to females in their young, they should produce more female offspring than male offspring. Sure enough, wild females from this population produced nearly 50% more female babies than male babies. Temperature-dependent sex determination has important implications for efforts to conserve endangered species. Nests of the American alligator in marshes are cooler than those built on levees; females hatch from marsh nests and males from nests built on levees. The sex ratio of hatchlings from 8000 eggs collected from natural nests was five males to one female, reflecting the relative abundance of levee and marsh nests. It is not clear whether this large imbalance of the sexes represents a normal condition for alligators or if it reflects a recent change in the availability of nest sites as a result of construction of levees.

Some conservation efforts have been confounded by temperature-dependent sex determination in sea turtles. A number of programs have depended on collecting eggs from natural nests and incubating them under controlled conditions. Unfortunately, these unnaturally uniform conditions of incubation can result in producing hatchlings of only one sex.

The amount of moisture in the soil surrounding a turtle nest is another important variable during embryonic development of the eggs. The wetness of a nest interacts with temperature in sex determination, as well as influencing the rate of embryonic development and the size and vigor of the hatchlings produced. Dry substrates induced the development of some female painted turtles at low temperatures (26.5°C and 27°C) that would normally have produced only males. The wetness of the substrate did not affect the sex of turtles from eggs incubated at 30.5°C and 32°C: all the hatchlings from these eggs were females, as would be expected on the basis of temperature-dependent sex determination alone.

Wet incubation conditions produce larger hatchlings than do dry conditions, apparently because water is needed for metabolism of the yolk. When water is limited, turtles hatch early and at smaller body sizes, and their guts contain a quantity of yolk that was not used during embryonic development. Hatchlings from nests under wetter conditions are larger and contain less unmetabolized yolk. The large hatchlings that emerge from moist nests are able to run and swim faster than hatchlings from drier nests, and, as a result, they may be more successful at escaping predators and catching food.

■ Hatching and the Behavior of Baby Turtles

Turtles are self-sufficient at hatching, but in some instances interactions among the young may be essential to allow them to escape the nest. Sea-turtle nests are quite deep; the eggs may be buried 50 centimeters beneath the sand, and the hatchling turtles must struggle upward through the sand to the surface. After several weeks of incubation, the eggs all hatch within a period of a few hours, and a hundred or so baby turtles find themselves in a small chamber at the bottom of the nest hole. Spontaneous activity by a few individuals sets the whole group into motion, crawling over and under one another. The turtles at the top of the pile loosen sand from the roof of the chamber as they scramble about, and the sand filters down through the mass of baby turtles to the bottom of the chamber.

Periods of a few minutes of frantic activity are interspersed by periods of rest, possibly because the turtles' exertions reduce the concentration of oxygen in the nest and they must wait for more oxygen to diffuse into the nest from the surrounding sand. Gradually, the entire group of turtles moves upward through the sand as a unit until it reaches the surface. As the baby turtles approach the surface, high sand temperatures probably inhibit further activity, and they wait a few centimeters below the surface until night falls, when a decline in temperature triggers emergence. All the babies emerge from a nest in a very brief period, and all the babies in different nests that are ready to emerge on a given night leave their nests at almost the same time, probably because their behavior is cued by temperature. The result is the sudden appearance of hundreds or even thousands of baby turtles on the beach, each one crawling toward the ocean as fast as it can.

Simultaneous emergence is an important feature of the reproduction of sea turtles because the babies suffer severe mortality crossing the few meters of beach and surf. Terrestrial predators—crabs, foxes, raccoons, and other predators—gather at the turtles' breeding beaches at hatching time and await their appearance. Some of the predators come from distant places to prey on the baby turtles. In the surf, sharks and bony fishes patrol the beach. Few, if any, baby turtles would get past that gauntlet were it not for the simultaneous emergence that brings all the babies out at once and temporarily swamps the predators.

Turtles exhibit no parental care, and the long period of embryonic development renders their nests vulnerable to predators. Females of many species of turtles scrape the ground in a wide area around the nest when they have finished burying their eggs. This behavior may make it harder for predators to identify the exact location of the nest. Major breeding sites of sea turtles are often on islands that lack mammalian predators that could excavate the nests. Another important feature of a nesting beach is provision of suitable conditions for the hatchling turtles. Newly hatched sea turtles are small animals; they weigh 25 grams to 50 grams, which is less than 0.05 percent of the body mass of an adult sea turtle. The enormous disparity in body size of hatchling and adult turtles is probably accompanied by equally great differences in their ecological requirements and their swimming abilities. Many of the major sea-turtle nesting areas are upcurrent from the feeding grounds, and that location may allow currents to carry the baby turtles from the breeding beaches to the feeding grounds.

We know even less about the biology of baby sea turtles than we do about the adults. Where the turtles go in the period following hatching has been a long-standing puzzle in the life cycle of sea turtles. For example, green turtles hatch in the late summer at Tortuguero on the Caribbean coast of Costa Rica. The turtles disappear from sight as soon as they are at sea, and they are not seen again until they weigh 4 or 5 kilograms. Apparently, they spend the intervening period floating in ocean currents. Material drifting on the surface of the sea accumulates in areas where currents converge, forming drift lines of flotsam that include sargassum (a brown algae) and the vertebrate and invertebrate fauna associated with it. These drift lines are probably important resources for juvenile sea turtles.

■ Navigation and Migration

Pond turtles and terrestrial turtles usually lay their eggs in nests that they construct within their home ranges. The mechanisms of orientation that they use to find nesting areas are probably the same ones they use to find their way among foraging and resting areas. Familiarity with local landmarks is an effective method of navigation for these turtles, and they may also use the sun for orientation. Sea turtles have a more difficult time, partly because the open ocean lacks conspicuous landmarks but also because feeding and nesting areas are often separated by hundreds or thousands of kilometers. Most sea turtles are carnivorous. The leatherback turtle feeds on jellyfishes, loggerhead and Ridley sea turtles eat crabs and other benthic invertebrates, and the hawksbill turtle uses its beak to scrape encrusting organisms (sponges, tunicates, bryozoans, mollusks, and algae) from reefs. Juvenile green turtles are carnivorous, but the adults feed on vegetation, particularly turtle grass (*Thalassia testudinium*), which grows in shallow water on protected shorelines in the tropics. The areas that provide food for sea turtles often lack the characteristics needed for successful nesting, and many sea turtles move long distances between their feeding grounds and their breeding areas.

The ability of sea turtles to navigate over thousands of kilometers of ocean and find their way to nesting beaches that may be no more than tiny coves on a small island is astonishing. The migrations of sea turtles, especially the green turtle, have been studied for decades. Turtles captured at breeding sites in the Caribbean and Atlantic Oceans have been

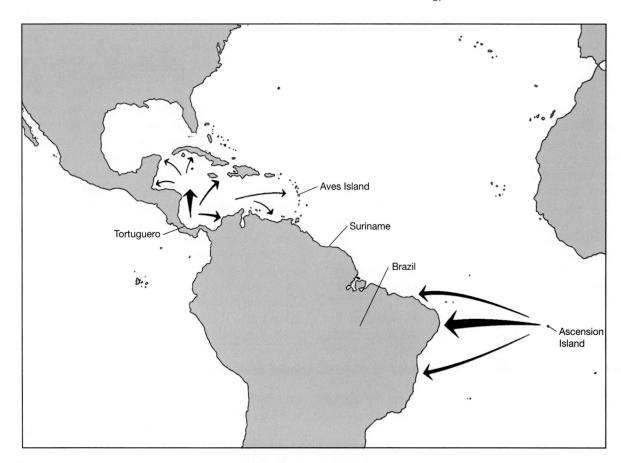

▲ **FIGURE 12–11** Migratory movements of green turtles (*Chelonia mydas*). The population that nests on beaches in the Caribbean is drawn from feeding grounds in the Caribbean and Gulf of Mexico. The turtles that nest on Ascension Island feed along the coast of northern South America.

individually marked with metal tags since 1956, and tag returns from turtle catchers and fishing boats have allowed the major patterns of population movements to be established (**Figure 12–11**).

Four major nesting sites of green turtles have been identified in the Caribbean and South Atlantic: one at Tortuguero on the coast of Costa Rica, one on Aves Island in the eastern Caribbean, one on the coast of Suriname, and one on Ascension Island between South America and Africa. Male and female green turtles congregate at these nesting grounds during the nesting season. The male turtles remain offshore, where they court and mate with females, and the female turtles come ashore to lay eggs on the beaches. A typical female green turtle at Tortuguero produces three clutches of eggs about 12 days apart. About a third of the female turtles in the Tortuguero population nest in alternate years, and the remaining two-thirds of the turtles follow a three-year breeding cycle. The coast at Tortuguero lacks the beds of turtle

grass on which green turtles feed, and the turtles come to Tortuguero only for nesting. In the intervals between breeding periods, the turtles disperse around the Caribbean. The largest part of the Tortuguero population spreads northward along the coast of Central America. The Miskito Bank off the northern coast of Nicaragua appears to be the main feeding ground for the Tortuguero colony. A smaller number of turtles from Tortuguero swim south along the coast of Panama, Colombia, and Venezuela. Female green turtles return to their natal beaches to nest, and the precision with which they home is astonishing. Female green turtles at Tortuguero return to the same kilometer of beach to deposit each of the three clutches of eggs they lay in a breeding season.

Probably the most striking example of the ability of sea turtles to home to their nesting beaches is provided by the green turtle colony that has its feeding grounds on the coast of Brazil and nests on Ascension Island, a small volcanic peak that emerges from the ocean. The island is 2200 kilometers east of Brazil and less than 20 kilometers in diameter—a tiny target in the vastness of the South Atlantic.

How do turtles find their way across thousands of kilometers of ocean? Other animals use a variety of cues for navigation, and sea turtles probably do so as well. Chemosensory information may be one important component of their navigation. The South Atlantic equatorial current flows westward, washing past Ascension Island and continuing toward Brazil. Newly hatched green turtles may drift with the current from their natal beaches to the adult feeding grounds, and the odor plume of the island may help to guide female turtles back to the island to nest. That is, a female turtle leaving the coast of Brazil may swim upstream in the South Atlantic equatorial current (i.e., up the odor gradient) to locate Ascension Island.

It is impractical to locate female turtles off the coast of Brazil as they are about to begin their journey to the island, but it is easy to find turtles that have completed nesting at Ascension Island and are ready to start back to Brazil. Five female green turtles were tracked on their return trip using the Argos satellite system (Luschi et al. 1998). The turtles traveled 1777 to 2342 kilometers and reached Brazil in 33 to 74 days.

For the first 500 kilometers of the journey, they followed a west-south-west heading that carried them slightly south of a direct route toward the bulge of Brazil (**Figure 12–12**). At this stage they were following the route of the South Atlantic equatorial current. Perhaps the turtles were simply being carried off course by the current, but they may have been using the same guidance system that they rely on for their outward journey—that is, staying within the plume of the island's scent. Even though the current carries them slightly south of a direct route to Brazil, they may save energy and move faster by initially staying in the current.

If they remained in the current for the entire trip, they would be carried too far south, so the turtles make a midcourse correction. The new course heads west-north-west on a nearly direct route to the bulge on the coast of Brazil. The shift in direction might be triggered by the waning strength of the scent plume. The turtles spend more than 90 percent of the journey underwater, suggesting that they may be sampling the plume in three dimensions.

A study of navigation by hatchling loggerhead turtles showed that they used at least three cues for orientation: light, wave direction, and magnetism (Lohmann 1991). These stimuli play sequential roles in the turtles' behavior. When they emerge from their nests, the hatchlings crawl toward the brightest light they see. Normally the sky at night is lighter over the ocean than over land, and this behavior brings them to the water's edge. (Shopping centers, streetlights, even porch lights on beachfront houses can confuse these and other species of sea turtles and lead them inland, where they are crushed on roads or die of overheating the next day.)

In the ocean, the loggerhead hatchlings swim into the waves. This response moves them away from shore and ultimately into the Gulf Stream. They drift with the current along the coast of the United States and then eastward across the Atlantic. Off the coast of Portugal, the Gulf Stream divides into two branches. One turns north toward England, and the other swings south past the bulge of Africa and even-

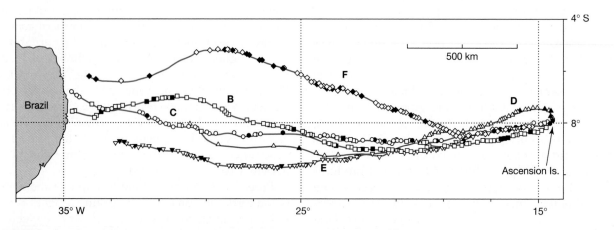

▲ **FIGURE 12–12** Turtle tracks. The paths followed by five green turtles (*Chelonia mydas*) were tracked by satellite as they returned from Ascension Island to Brazil.

tually back westward across the Atlantic. It's essential for the baby turtles to turn right at Portugal; if they fail to make that turn, they are swept past England into the chilly North Atlantic, where they perish. If they do turn southward off the coast of Portugal, they are eventually carried back to the coast of tropical America—a round-trip that takes five to seven years.

Magnetic orientation appears to tell the turtles when to turn right to catch the current that will carry them to the South Atlantic. We usually think of the Earth's magnetic field as providing two-dimensional information—north-south and east-west—but it's more complicated than that. The field loops out of the north and south magnetic poles of the Earth. At the equator, the field is essentially parallel to the Earth's surface (in other words, it forms an angle of 0 degrees), and at the poles it intersects the surface at an angle of 90 degrees. Thus, the three-dimensional orientation of the Earth's magnetic field provides directional information (which way is magnetic north?) and information about latitude (what is the angle at which the magnetic field intersects the Earth's surface?).

When loggerheads in a pool on land that had no waves were exposed to an artificial magnetic field at the 57-degree angle of intersection with the Earth's surface that is characteristic of Florida, they swam toward artificial east—even when the magnetic field was changed 180 degrees, so that the direction they thought was east was actually west. That is, they were able to use a compass sense to determine direction. But that wasn't all they could do. When the angle of intersection of the artificial magnetic field was increased to 60 degrees, as if they were further north than they really were, the turtles turned south. A 60-degree angle of intersection corresponds to the latitude where the Gulf Stream forks off the coast of Portugal—and where the turtles must turn south to reach the South Atlantic. Thus, it appears that loggerhead turtles can use magnetic sensitivity to recognize both direction and latitude.

12.5 Conservation of Turtles

Slow rates of growth and delayed maturity are characteristics that predispose a species to the risk of extinction when changing conditions increase the mortality of adults or drastically reduce recruitment of juveniles into the population. The plight of large tortoises and sea turtles is particularly severe, partly because these species are among the largest and slowest growing of turtles but also because other aspects of their biology expose them to additional risk (**Figure 12–13**). The conservation of tortoises and sea turtles is a subject of active international concern and has led to the founding of a new journal, *Chelonian Conservation and Biology*.

The largest living tortoises are found on the Galápagos and Aldabra Islands. The relative isolation of these small and (for humans) inhospitable landmasses has probably been an important factor in the survival of tortoises. Human colonization of the islands has brought with it domestic animals such as goats and donkeys, which compete with tortoises for the limited quantities of vegetation to be found in these arid habitats, and dogs, cats, and rats that prey on tortoise eggs and on baby tortoises.

The limited geographic range of a tortoise that occurs only on a single island makes it vulnerable to extinction. In 1985 and again in 1994 brush fires on the island of Isabela in the Galápagos Archipelago threatened the 20 surviving individuals of *Geochelone guntheri* and emphasized the advantage of moving some or all of the turtles to the breeding facility operated by the Charles Darwin Research Station on Santa Cruz Island. This station has a successful record of breeding and releasing another species of Galápagos tortoise, *Geochelone nigra hoodensis*, which is native to Española Island. In the early 1960s, only 14 individuals of this form could be located. All were adults and apparently had not bred successfully for many years. All of the tortoises were moved to the research station, and the first babies were produced in 1971. On March 24, 2000, the one-thousandth captive-bred tortoise was released on Española. This success story shows that carefully controlled captive breeding and release programs can be an effective method of conservation for endangered species of turtles. However, these programs also have inherent risks (**Box 12–1**).

Entire turtle faunas are threatened in some areas. Nearly all species of turtles in Southeast Asia are now at risk because of economic and political changes in the region (van Dijk et al. 2000). Turtles have traditionally been used in China for food and for their supposed medicinal benefits. A two-day survey in just two Chinese food markets found an estimated 10,000 turtles for sale. If the turnover time is estimated conservatively to be a week, those two markets would consume more than a quarter of a million turtles annually. When that rate is extrapolated to all of the markets in China, the estimate rises to 12 million turtles sold annually in China alone (Altherr and Freyer 2000).

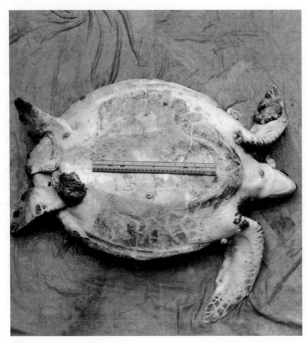

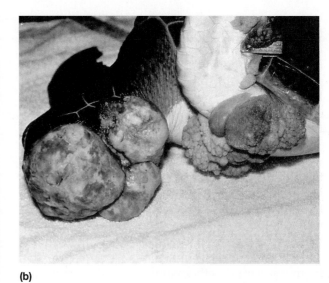

(b)

(a)

▲ **FIGURE 12–13** A green turtle with cutaneous papillomas. (a) These tumors, which are probably caused by a virus, have been found on most species of sea turtles and in most parts of the world (Herbst 1994), and new occurrences are reported with depressing frequency (e.g., Formia et al. 2007). The tumors grow to more than 30 centimeters in diameter and can appear on any skin-covered surface. (b) They are especially common on the conjuctiva of the eyes and may grow over the cornea. Tumors were not recorded on green turtles in the Indian River Lagoon, Florida, until 1982, and by 1994 approximately 50 percent of the green turtles were affected. The first record of papilloma on green turtles in Kaneohe Bay, Hawaii, was in 1958. Since 1989 the incidence has ranged from 49 percent to 92 percent. The tumors can be lethal, and their increased frequency is an ominous development for species that were already endangered.

BOX 12-1   Sick Turtles

The desert tortoise (*Gopherus agassizi*) is one of the largest terrestrial turtles in North America. Its geographic range includes the southwestern corner of Utah, the southwestern third of Arizona, and adjacent parts of Nevada and California, and it extends southward into Mexico.

Populations of desert tortoises have declined since the 1950s as human activity has intruded on the desert habitat. Between 1979 and 1989, most tortoise populations in the Mohave and Colorado deserts decreased by 30 to 70 percent. The situation has become even more serious with the appearance of upper respiratory tract disease (URTD), which attacks desert tortoises, often with fatal results. Infected turtles first have a runny nose, which becomes progressively worse until the turtles exude foam from their nostrils, wheeze when they breathe, cease feeding, become listless, and ultimately die. In 1988, tortoises in the Desert Tortoise Natural Area in Kern County, California, first showed symptoms of URTD (Jacobson et al. 1991). In 1989, 627 dead tortoises were found, and 43 percent of the live tortoises in the Natural Area showed symptoms of URTD.

A large variety of bacteria was cultured from the nasal passages of the sick turtles, including *Mycoplasma,* which has subsequently been shown to be the cause of the disease (Jacobson et al. 1995). Desert tortoises are popular pets in the desert southwest, and a high proportion of pet turtles are infected by *Mycoplasma.* The infection may have been introduced to the Desert Tortoise Natural Area when pet tortoises were released, and its spread may have been accelerated by the poor physical condition of the wild tortoises that resulted from habitat degradation and a prolonged drought. *Mycoplasma* infections are notoriously difficult to cure. Captive tortoises can be treated with a combination of antibiotics, but there is no practical treatment for wild tortoises.

Upper respiratory tract disease has now been reported in a population of the gopher tortoise (*Gopherus polyphemus*) on Sanibel Island off the coast of Florida (**Figure 12–14**). Again, captive tortoises appear to have introduced the infection into a wild population: until 1978, tortoises used in tortoise races in Fort Myers were released on Sanibel Island, and infected tortoises from the races may have carried *Mycoplasma* with them.

These examples emphasize the risk of releasing animals that have been held in captivity into wild populations. Captive breeding programs must take extraordinary measures to ensure that the animals to be released are quarantined in a facility that is isolated from other animals. A breeding colony should be self-contained; once it is established, no outside animals should be introduced, and no equipment or containers should be moved in or out. Even the clothing of animal caretakers can carry pathogens, and a dressing room must be provided so caretakers can wash and change their clothes when they enter or leave. These precautions are time-consuming and expensive, but neglecting them can be disastrous.

▲ **FIGURE 12–14** A gopher tortoise with a runny nose. Nasal discharge and swollen eyes are signs that this tortoise is infected with the *Mycoplasma* that causes upper respiratory tract disease.

As long-lived animals with low reproductive rates, turtles have exactly the wrong characteristics to withstand heavy predation. Very little is known about the natural history of Chinese turtles. In fact, some species such as *Cuora mccordi* are known scientifically only from specimens purchased in markets—wild populations have never been described. These species may never be known in the wild; specimens have not been seen in markets for several years, and the species may be extinct.

As populations of turtles in China have been depleted, turtles have increasingly been imported from other countries in Southeast Asia. China is a huge country, and movement within the country is restricted (especially for foreigners), so complete data are not available. Records for Hong Kong show that importation of turtles for food rose from 139,200 kilograms in 1977 to 1,800,024 kilograms in just the first 10 months of 1994. Vietnam is estimated to send between 1600 and 16,000 kilograms of turtles to China *every day!* No effective control of this trade is in place. Species protected by national laws and international conventions are included, and turtles are collected from nature reserves.

Although China is considered the biggest black hole for turtles (Behler 2000), it is by no means alone. Madagascar is home to endangered species of many kinds of animals, including tortoises. Although these species are protected by international treaty, they are smuggled out of the country by the score for sale as pets in Japan, Europe, and North America, where they fetch high prices. In the United States, our own protected species are collected and sold illegally as pets, and the most endangered species command the highest prices. In addition, all species of turtles face threats ranging from death on roads and loss of habitat to being shot for target practice as they bask on logs. It is no exaggeration to say that "turtles are in terrible trouble" (Rhodin 2000). The conservation crisis for turtles is as severe as that facing amphibians.

Summary

The earliest turtles known, fossils from the Triassic, have nearly all the features of derived turtles. The first Triassic forms were not able to withdraw their heads into the shell, but this ability appeared in the two major lineages of living turtles, which were established by the Late Triassic. The cryptodiran turtles retract the head with a vertical flexion of the neck vertebrae, whereas the pleurodires use a sideward bend.

Turtles are among the most morphologically specialized vertebrates. The shell is formed of dermal bone that is fused to the vertebral column and ribs. In most turtles, the dermal shell is overlain by a horny layer of epidermal scales. The limb girdles are inside the rib cage. Breathing presents special difficulties for an animal that is encased in a rigid shell: Exhalation is accomplished by muscles that squeeze the viscera against the lungs, and inhalation is accomplished by muscles that increase the volume of the visceral cavity, thereby allowing the lungs to expand. The heart of turtles (and of lepidosaurs as well) is able to shift blood between the pulmonary and systemic circuits in response to the changing requirements of gas exchange and thermoregulation.

The social behavior of turtles includes visual, tactile, and olfactory signals used in courtship. Dominance hierarchies shape the feeding, resting, and mating behaviors of some of the large species of tortoises. All species of turtles lay eggs, and none provides parental care to their young. Coordinated activity by hatchling sea turtles may be necessary to enable them to dig themselves out of the nest, and simultaneous emergence of baby sea turtles from their nests helps them to evade predators as they rush down the beach into the ocean. Sea turtles migrate tens, hundreds, and even thousands of kilometers between their feeding areas and their nesting beaches and use a large variety of cues for navigation.

The life history of many turtles makes them vulnerable to extinction. Slow rates of growth and long periods required to reach maturity are characteristic of turtles in general and of large species of turtles in particular. Turtles cannot withstand commercial exploitation of the sort that has occurred historically on many oceanic islands and is currently in progress in parts of Asia.

Additional Readings

Altherr, A., and D. Freyer. 2000. Asian turtles are threatened by extinction. *Turtle and Tortoise Newsletter* No. 1:7–11.

Behler, J. 2000. Letter from the IUCN tortoise and freshwater turtle specialist group. *Turtle and Tortoise Newsletter*, No. 1:4–5.

Burke, V. J. et al. 1993. Conservation of turtles: The chelonian dilemma. In *Proceedings of the 13th Annual Symposium on Sea Turtle Biology and Conservation*. Jekyll Island, GA: U.S. Department of Commerce, National Oceanic and Atmospheric Administration, 35–38.

Chelazzi, G., and R. Calzolai. 1986. Thermal benefits from familiarity with the environment in a reptile. *Oecologia* 68:557–558.

Congdon, J. D. et al. 1994. Demographics of common snapping turtles (*Chelydra serpentina*): Implications for conservation and management of long-lived organisms. *American Zoologist* 34:397–408.

Farmer, C. G. et al. 2008. The right-to-left shunt of crocodilians serves digestion. *Physiological and Biochemical Zoology* 81:125-127.

Formia, A. et al. 2007. Fibropapillomatosis confirmed in Chelonia myda in the Gulf of Guinea, West Africa. *Marine Turtle Newsletter* 116: 20–22.

Herbst, L. H. 1994. Fibropapillomatosis of marine turtles. *Annual Review of Fish Diseases* 4:389–425.

Hicks, J. W. et al. 1996. The mechanism of cardiac shunting in reptiles: A new synthesis. *Journal of Experimental Biology* 199:1435–1446.

Hill, R. V. 2005. Integration of morphological data sets for phylogenetic analysis of Amniota: The importance of integumentary characters and increased taxonomic sampling. *Systematic Biology* 54:530–547.

Jacobson, E. R. et al. 1995. Mycoplasmosis and the desert tortoise (*Gopherus agassizii*) in Las Vegas Valley, Nevada. *Chelonian Conservation and Biology* 1:279–284.

Jacobson, E. R. et al. 1991. Chronic upper respiratory disease of free-ranging desert tortoises (*Xerobates agassizi*). *Journal of Wildlife Diseases* 27:296–316.

Kennett, R. et al. 1998. Underwater nesting by the Australian freshwater turtle *Chelodina rugosa*: Effects of prolonged immersion and eggshell thickness on incubation period, egg survivorship, and hatchling size. *Canadian Journal of Zoology* 76:1–5.

Klemens, M. W. (ed.) 2000. *Turtle Conservation*. Washington, DC: Smithsonian Institution Press.

Landberg, T. et al. 2003. Lung ventilation during treadmill locomotion in a terrestrial turtle. *Journal of Experimental Biology* 206:3391–3404.

Laurin, M. and R. R. Reisz, 1995. A re-evaluation of early amniote phylogeny. *Zoological Journal of the Linnean Society* 113:165–223.

Lee, M. S. Y. 1993. The origin of the turtle body plan: Bridging a famous morphological gap. *Science* 261:1716–1720.

Lee, M. S. Y. 1995. Historical burden in systematics and the interrelationships of "Parareptiles." *Biological Reviews of the Cambridge Philosophical Society*. 70:459–547.

Lohmann, K. J. 1991. Magnetic orientation by hatchling loggerhead sea turtles. *Journal of Experimental Biology* 155: 37–49.

Luschi, P. et al. 1998. The navigational feats of green sea turtles migrating from Ascension Island investigated by satellite telemetry. *Proceedings of the Royal Society (London) Series B*. 265:2279–2284.

Merlen, G. 1999. *Restoring the Tortoise Dynasty: The Decline and Recovery of the Galápagos Giant Tortoise*. Quito, Ecuador: Charles Darwin Foundation.

Packard, G. C., and M. J. Packard. 1988. Physiological ecology of reptile eggs. In C. Gans and R. B. Huey (eds.), *Biology of the Reptilia*, Vol. 16. Philadelphia, PA: Alan Liss, 523–605.

Paladino, F. V. et al. 1990. Metabolism of leatherback turtles, gigantothermy, and thermoregulation of dinosaurs. *Nature* 344:858–860.

Plotkin, P. T. (ed.) 2007. *Biology and Conservation of Ridley Sea Turtles*. Johns Hopkins University Press, Baltimore, MD.

Rhodin, A. G. J. 2000. Publishers Editorial: Turtle Survival Crisis. *Turtle and Tortoise Newsletter* 1:2–3.

Rieppel, O. 2000. Turtles as diapsid reptiles. *Zoologica Scripta* 29:199–212.

Rieppel, O., and M. de Braga. 1996. Turtles as diapsid reptiles. *Nature* 384:453–455.

Robert, K. A., and M. B. Thompson. 2001. Viviparous lizard selects sex of embryos. *Nature* 412:698–699.

Spotila, J. R. 2004. *Sea Turtles: A Compete Guide to Their Biology, Behavior, and Conservation.* Johns Hopkins University Press, Baltimore, MD.

Spotila, J. R. and E. A. Standora. 1985. Environmental constraints on the thermal energetics of sea turtles. *Copeia* 1985:694–702.

Swingland, I. R., and J. G. Frazier. 1979. The conflict between feeding and overheating in the Aldabran giant tortoise. In *A Handbook on Biotelemetry and Radio Tracking*, ed. by C. J. Amlaner, Jr., and D. W. MacDonald. Oxford, UK: Pergamon, 611–615.

Swingland, I. R., and C. M. Lessells. 1979. The natural regulation of giant tortoise populations on Aldabra Atoll. Movement polymorphism, reproductive success and mortality. *Journal of Animal Ecology* 48:639–654.

van Dijk, P. P. et al. (eds). 2000. Asian turtle trade. *Chelonian Research Monographs*, No. 2, Lunenburg, MA: Chelonian Research Foundation.

Wang, T. et al. 1997. The role of cardiac shunts in the regulation of arterial blood gases. *American Zoologist* 37:12–22.

Wapstra, E. et al. 2004. Maternal basking behaviour determines offspring sex in a viviparous reptile. *Proceedings of the Royal Society of London B* (supplement) 271:S230–S232.

Warner, D. A. and R. Shine. 2008. The adaptive significance of temperature-dependent sex determination in a reptile. *Nature* 451:566–568.

CHAPTER 13

The Lepidosaurs: Tuatara, Lizards, and Snakes

Some aspects of the biology of lepidosaurs may give us an impression of the ancestral way of life of amniotes, although many of the structural characteristics of lizards and snakes are derived. One derived characteristic of lizards is determinate growth. That is, increase in body size stops when the growth centers of the long bones ossify. This mechanism sets an upper limit to the size of individuals of a species and may be related to the specialization of most lizards as predators of small animals, such as insects. The predatory behavior of lizards ranges from sitting in one place and ambushing prey to seeking food by traversing a home range in an active, purposeful way. Broad aspects of the biology of lizards are correlated with these foraging modes, including morphology, exercise physiology, reproductive mode, defense against predators, and social behavior. The anatomical specializations of snakes are associated with their elongate body form and include modifications of the jaws and skull that allow them to subdue and swallow large prey.

13.1 The Lepidosaurs

Lepidosaurs are the largest group of nonavian reptiles, containing more than 4800 species of lizards and 2900 species of snakes in addition to the two species of tuatara (Table 13.1). Lepidosaurs are predominantly terrestrial tetrapods with some secondarily aquatic species, especially among snakes. The skin of lepidosaurs is covered by scales and relatively impermeable to water. The outer layer of the epidermis is shed at intervals. Tuatara and most lizards have four limbs; however, reduction or complete loss of limbs is widespread among lizards, and all snakes are limbless. Lepidosaurs have a transverse cloacal slit rather than the longitudinal slit that characterizes other tetrapods.

Lepidosaurs are the sister lineage of archosaurs (crocodilians and birds). Within the Lepidosauria, the Sphenodontidae (tuatara) is the sister group of Squamata (lizards and snakes). Within the squamates, lizards can be distinguished from snakes in colloquial terms but not phylogenetically because snakes are derived from lizards. Thus "lizards" is a paraphyletic group because it does not include all the descendants of a common ancestor. Nonetheless, lizards and snakes are distinct in many aspects of their ecology and behavior, and a colloquial separation is useful in discussing them.

13.2 Radiation of Sphenodontids and the Biology of Tuatara

The sphenodontids were a diverse group in the Mesozoic era, including terrestrial, arboreal, and marine forms and both insectivores and herbivores. Triassic forms were small, with body lengths of only 15 to 35 centimeters, but during the Jurassic and Cretaceous periods, some sphenodontids reached lengths of 1.5 meters. Most Triassic sphenodontids

TABLE 13.1 Lineages of extant lepidosaurs

Sphenodontidae: The sister group of squamates. Two species of tuatara (*Sphenodon*), now restricted to islands off the coast of New Zealand.

Iguania

Iguanidae: A group that includes more than 900 species of lizards, primarily found in the New World but with representatives on the islands of Fiji and Madagascar. The family Iguanidae includes eight lineages; see Table 13.2.

"Agamidae": About 381 species of small to large (10 centimeters to 1 meter) lizards that extend through the Middle East into Africa and along the Indo-Australian archipelago into Australia.

Chamaeleonidae: About 161 species of primarily arboreal lizards but including a few grassland and terrestrial species, found in Africa and Madagascar and extending into southern Spain and along the west coast of the Mediterranean. Chameleons have the laterally flattened bodies characteristic of many arboreal lizards, and additional specializations including zygodactyl feet (i.e., feet with two groups of toes that oppose each other when grasping), prehensile tails, eyes that swivel to provide a 360-degree field of view, and a projectile tongue. The leaf chameleons (*Brookesia, Rhampholeon*) are as small as 25 millimeters, whereas some species of *Chamaeleo* grow to more than 60 centimeters. (Note that the family name and the genus are spelled with an *ae*, but the common name *chameleon* is spelled with an *e*.)

Gekkota

Gekkonidae: About 1076 species of geckos with modified scales (setae) on the bottoms of the toes that allow them to climb vertical surfaces and even to hang by a single toe. The geographic distribution of gekkonids includes every continent except Antarctica. Extant species range in size from very small (30 millimeters) to medium size (30 centimeters). Eublepharines are about 25 species of small- to medium-size terrestrial geckos, lacking the modifications of the toes that allow gekkonids to climb. Their eyes have moveable lids; geographic distribution extends from the southwestern United States through Central America and includes species in Africa and Asia.

Pygopodidae: About 36 species of elongate, nearly limbless terrestrial lizards found only in Australia.

Dibamidae: About 19 species of small- to medium-size (5–25 centimeters) limbless, burrowing lizards. *Dibamus* is found in Indo-Malaysia, whereas the single species of *Anelytropsis* is from Mexico.

Scincomorpha

Cordylidae: 54 species of small- to medium-size terrestrial or rock-dwelling lizards from sub-Saharan Africa. *Cordylus* is a rock dweller and has the dorsoventrally flattened body typical of lizards that seek shelter in crevices. It is heavily armored, and many species have exceedingly sharp spines along the body margins and the tail. Skull kinesis is discussed in the text in the context of feeding mechanisms employed by lizards, but *Cordylus* uses skull kinesis in a different way: by contracting its jaw muscles, *Cordylus* raises the braincase relative to the lower jaw, and it uses this mechanism to wedge itself into crevices. *Platysaurus* is a small rock dweller without the heavy body armor of *Cordylus*, and *Chamaesaura* is an elongate, nearly limbless lizard. Most cordylids are viviparous, with the exception of *Platysaurus*, which lays eggs.

Gerrhosauridae: The 32 species of 10- to 70-centimeter lizards are found in sub-Saharan Africa and on Madagascar. This lineage shows parallels to the cordylids: *Gerrhosaurus* has heavy body armor, and *Tetradactylus* is elongate and has reduced limbs. *Angolosaurus skoogi* lives in shifting sand dunes in the Namib Desert, where it eats seeds that blow in from outside the dunes and buries itself in sand to avoid extreme temperatures.

Gymnophthalmidae: About 193 species of small (less than 6 centimeters) lizards that live in the leaf litter of neotropical forests. Limb reduction is widespread in this lineage.

Teiidae: About 121 species of active terrestrial lizards ranging from near the Canadian border in North America to central Argentina, and including the West Indies. Teiids range from small insectivorous species of *Ameiva*, *Cnemidophorus*, and *Kentropyx* to species such as the tegus (*Tupinambis*) and the caiman lizard (*Dracena*), which can grow to a meter or more.

Lacertidae: About 279 species of small- to medium-size terrestrial lizards from the Old World with a range of body forms very like those seen among teiids. Lacertids are found over all of Europe, Africa, and Asia. Like the teiids, lacertids are mostly terrestrial and include some large predators as well as a great number of smaller species.

Scincidae: Some 1305 species make this one of the most species-rich lineages of lepidosaurs. Skinks occur on all continents except Antarctica. Nearly all are very small- to medium-size (5–20 centimeters) terrestrial lizards, and many show limb reduction. Most are insectivorous, but some of the large (40 centimeters) species of the Australian blue-tongued skinks (*Tiliqua*) consume plant material, and the Solomon Islands arboreal skink (*Corucia zebrata*) has specializations of the gut that promote fermentative digestion of plant material by symbiotic microorganisms.

Xantusiidae: 25 species of tiny to small (3–10 centimeters), secretive lizards from southwestern North America, Central America, and Cuba. All have the eyelids fused into a transparent scale that covers the eye and gives the group its common name, spectacled lizards.

(Continued)

TABLE 13.1 *Continued*

Diploglossa

Anguidae: About 112 species of anguids occur in North and South America and in Europe, the Middle East, and southern China. All anguids have body armor and a fold along the side of the body that allows the trunk to expand and contract as they breathe. Most are terrestrial, foraging in leaf litter, and four genera are legless. *Anguis* (Europe), *Ophisaurus* (North America, Europe, and Asia), and *Ophioides* (South America) are limbless surface dwellers that forage in leaf litter and dense vegetation. *Anguis* is small, but *Ophisaurus* grows to a trunk length of 0.5 meters, with a tail at least as long as its trunk.

Anniellidae: 2 species of *Anniella* (California and Baja California), small (20–30 centimeters) limbless lizards that spend the day underground and emerge at night to hunt on the surface.

Xenosauridae: This small family includes 6 species of small (10–15 centimeters) lizards in the genus *Xenosaurus*, that live in the moist leaf litter of high-altitude cloud forests in Mexico.

Shinisauridae: A single species of *Shinisaurus*, a semiaquatic lizard from China; about 25 centimeters long; lives along stream banks and forages in the water, catching fishes and aquatic invertebrates.

Platynota

Helodermatidae: The 2 species of helodermatids are the only poisonous lizards. These are large (25- to 40-centimeter trunk length), heavy-bodied lizards found in the southwestern United States and Mexico

Lanthanotidae: The single species of *Lanthanotus*, the Bornean earless monitor, is an elongate lizard with a trunk 15–20 centimeters long. It is secretive and semiaquatic, spending the day in a burrow and foraging at night on both land and water.

Varanidae: The 59 species in this group range in size from *Varanus brevicauda*, which has a total length of only 10 centimeters, to the Komodo dragon, *V. komodoensis*, which can exceed 3 meters in length and may weigh 75 kilograms. All monitors are active predators, and the larger species patrol a large home range searching for food in holes and beneath rocks and logs. Some species are moderately arboreal. Varanids occur in Africa, Asia, and the East Indies, but about half the species are limited to Australia.

Amphisbaenia

Four families of elongate, legless (except for *Bipes*) burrowing squamates ranging in length from about 10 to 80 centimeters.

Amphisbaenidae: 154 species of round-headed, spade-snouted, and keel-headed amphisbaenians found in the West Indies, South America, sub-Saharan Africa, and around the Mediterranean Sea.

Trogonophiidae: 6 species of round-headed amphisbaenians that use an oscillating movement of the head in digging. Trogonophiids are found in North Africa and the Middle East.

Rhineuridae: 1 species, a spade-snouted amphisbaenian, found in Florida.

Bipedidae: 4 species of round-headed amphisbaenians (*Bipes*) found in Mexico. This lineage is unique among amphisbaenians in retaining well-developed forelimbs that are used to help penetrate the soil surface. Once underground, *Bipes* use the head for burrowing.

Serpentes (see Table 13.3).

had teeth that were fused to the top edges of the jawbones (**acrodont**) like the teeth of extant tuatara (*Sphenodon*), but others had teeth attached to the inner sides of the jawbones (**pleurodont**) like those of some lizards.

The two species of *Sphenodon*, known as tuatara, are the only extant sphenodontids (**Figure 13–1**). (*Tuatara* is a Maori word meaning "spines on the back," and no *s* is added to form the plural.) Tuatara formerly inhabited the north and south islands of New Zealand, but the advent of humans and their associates (cats, dogs, rats, sheep, and goats) exterminated tuatara on the mainland. Now populations are found on only about 30 small islands off the coast.

Tuatara have been fully protected in New Zealand since 1895, but only one species, *Sphenodon punctatus*, was recognized. In fact, there is a second species of tuatara, *S. guntheri*, which was described in 1877. Because *S. guntheri* is much less common than *S. punctatus*, it was overlooked when laws protecting tuatara were written. As a result, *S. guntheri*

(a)   (b)

▲ **FIGURE 13–1** Tuatara. The two species of tuatara are the sister group of the other extant lepidosaurs. (a) *Sphenodon punctatus,* the common tuatara, is more abundant with natural populations on about 30 small islands off the coast of New Zealand and introduced populations on an additional three islands. (b) In contrast, *Sphenodon guntheri,* the Brothers tuatara, occurs naturally only on North Brother Island, but has been released on at least two other islands.

did not receive the special protection that it needed (Daugherty et al. 1990). The only surviving natural population of *S. guntheri* is a group of fewer than 300 adults living on 1.7 hectares of scrub on the top of North Brother Island. These animals were regarded as not very important from a conservation perspective compared with the large populations of *S. punctatus* on some of the other islands. Probably only the presence of a lighthouse on North Brother Island saved these tuatara: it was staffed until 1990 by resident keepers who deterred illegal landings and poaching. The East Island population of *S. guntheri* (the only other natural population of the species) became extinct during this century. This example illustrates the crucial role that taxonomy plays in conservation—a species must be recognized before it can be protected.

Adult tuatara are about 60 centimeters long. They are nocturnal; and, in the cool, foggy nights that characterize their island habitats, they cannot raise their body temperatures during activity by basking in sunlight as lizards do. Body temperatures from 6°C to 16°C have been reported for active tuatara, and these are low compared with most lizards. During the day, tuatara do bask, and they raise their body temperatures to 28°C or higher. Tuatara feed largely on invertebrates, with an occasional frog, lizard, or seabird for variety. The jaws and teeth of tuatara produce a shearing effect during chewing: The upper jaw contains two rows of teeth, one on the maxilla and the other on the palatine bones. The teeth of the lower jaw fit between the two rows of upper teeth, and the lower jaw closes with an initial vertical movement, followed by an anterior sliding movement. As the lower jaw slides, the food item is bent or sheared between the triangular cusps on the teeth of the upper and lower jaws.

Tuatara live in burrows that they may share with nesting seabirds. The burrows are spaced at intervals of 2 to 3 meters in dense colonies, and both male and female tuatara are territorial. They use vocalizations, behavioral displays, and color change in their social interactions.

The ecology of tuatara rests to a large extent on exploitation of the resources provided by colonies of seabirds. Tuatara feed on the birds, which are most vulnerable to predation at night. In addition, the quantities of guano produced by the birds, scraps of the food they bring to their nestlings, and the bodies of dead nestlings attract huge numbers of arthropods that are eaten by tuatara. These arthropods are largely nocturnal and must be hunted when they are active. Thus the nocturnal activity of tuatara and the low body temperatures resulting from being active at night are probably specializations that stem from the association of tuatara with colonies of nesting seabirds. This pattern of behavior and thermoregulation probably does not represent the ancestral condition even for sphenodontids, and there is no reason to interpret it as being ancestral for lepidosaurs or diapsids.

13.3 Radiation of Squamates

Determinate growth may be the most significant derived character of squamates. Growth occurs as cells proliferate in the cartilaginous epiphyseal plates at the ends of long bones, continues while the epiphyseal plates are composed of cartilage, and stops completely when the epiphyses fuse to the shafts of the bones, obliterating the cartilaginous plates. Determinate growth of this sort is characteristic of squamates (and also of birds and mammals). Crocodilians and turtles continue to grow all through their lives, although the growth rates of adults are much slower than those of juveniles. Development of determinate growth in squamates may initially have been associated with the insectivorous diet that researchers believe was characteristic of early lepidosaurs. Generalized lizard-size animals can readily capture insects, whereas large insect-eating vertebrates, such as mammalian anteaters, require morphological or ecological specializations to capture tiny prey.

The fossil record of lizards is largely incomplete through the middle of the Mesozoic, but Late Jurassic deposits in China and Europe include members of most lineages of extant lizards. The major groups of lizards had probably diverged by the end of the Jurassic.

The phylogenetic relationships of squamates are well understood in general, but disagreement surrounds some details. Two major lineages are recognized, Iguania and the Scleroglossa. Iguanians include the sister lineages "Agamidae" (which is not considered monophyletic) and Chamaeleonidae, as well as the large family Iguanidae (**Table 13.2**). Scleroglossans include two large families (geckos and skinks) and several smaller families. The amphisbaenians (specialized burrowing lizards) and Serpentes (snakes) are nested within the Scleroglossa. Amphisbaenians and snakes have morphological and ecological specializations that make it convenient to discuss them individually, but their evolutionary status as lineages within the scleroglossan squamates must be remembered.

▪ Lizards

The approximately 4800 species of lizards range in size from diminutive geckos less than 3 centimeters long to the Komodo monitor lizard, which is 3 meters long at maturity and weighs some 75 kilograms. A reconstruction of the skeleton of a fossil monitor lizard,

TABLE 13.2 Lineages (subfamilies) of iguanid lizards. Some classifications consider these lineages to be families

Corytophaninae: 9 species of small- to medium-size (10–20 centimeters) arboreal neotropical lizards. They have long tails, laterally flattened bodies, and some (*Corytophanes, Laemanctus*) have crests on their heads that may make it hard for predators to recognize them as lizards when they are seen as silhouettes against a patch of light.

Crotaphytinae: The collared and leopard lizards; 10 species of medium-size (10–15 centimeters) North American desert lizards that prey on other lizards.

Hoplocercinae: About 11 species of medium-size (up to 16 centimeters) terrestrial lizards with a geographic range extending from Panama to Brazil.

Iguaninae: 36 species of large (to >1.5 meters) terrestrial and arboreal herbivorous lizards. The neotropical green iguana (*Iguana iguana*) is the most familiar member of the family, which also includes the black and ground iguanas of Central America and the West Indies, the Galápagos marine and land iguanas, and the Fijian iguanas. The North American chuckwalla (*Sauromalus*) is an iguanine. This rock-dwelling lizard is dorsoventrally flattened and seeks shelter from predators in rock crevices, where it inflates its lungs to wedge itself in place.

Oplurinae: 7 species of medium-size (20–40 centimeters), rock-dwelling and arboreal lizards from Madagascar.

Phrynosomatinae: About 125 species of North and Central American lizards. The group takes its name from the horned lizards (*Phrynosoma*), but more than half the species are in the genus *Sceloporus* (spiny swifts), which range from southern Canada to Panama. Most phrynosomatines are terrestrial, and some are specialized to enter rock crevices (dorsoventrally flattened) or to live on loose sand (fringes of scales on the toes; valvular nostrils that exclude sand; earless or with skin folds and elongated scales that prevent sand grains from entering the ears). Some species of *Sceloporus* and *Urosaurus* are arboreal.

Polychrotinae: About 353 species of primarily small (10 centimeters) South American lizards, most of them in the closely related genera *Anolis* and *Norops*. *Anolis carolinensis*, the green anole, has a geographic range that extends northward to North Carolina. Most species of polychrotines live in trees, bushes, or grass clumps.

Tropidurinae: About 309 species of small to medium-size (10–20 centimeters) South American lizards that occupy regions from sea level to the high Andes, including deserts, rain forests, and grasslands.

Megalania prisca, from the Pleistocene epoch of Australia, is 5.5 meters long, and in life the lizard may have weighed more than 1000 kilograms. About 80 percent of extant lizards weigh less than 20 grams as adults and are insectivorous. Spiny swifts and japalures (**Figure 13–2a,b**) are examples of these small, generalized insectivores. Other small lizards have specialized diets: the North American horned lizards and the Australian spiny devil (**Figure 13–2e,f**) feed on ants. Most geckos (**Figure 13–2g**) are nocturnal, and many species are closely associated with human habitations.

Lizards are adaptable animals that have occupied habitats ranging from swamp to desert and even above the timberline in some mountains. Many species are arboreal. The most specialized of these are frequently laterally flattened, and they often have peculiar projections from the skull and back that help to obscure their outline. The Old World chameleons (Chamaeleonidae) are the most specialized arboreal lizards (**Figure 13–2h**). Their **zygodactylous** (Greek *zygo* = joined and *dactyl* = digit) feet grasp branches firmly, and additional security is provided by a prehensile tail. The tongue and hyoid apparatus are specialized, and the tongue can be projected forward more than a body's length to capture insects that adhere to its sticky tip. This feeding mechanism requires good eyesight, especially the ability to gauge distances accurately so that the correct trajectory can be employed. The chameleon's eyes are elevated in small cones and can move independently. When the lizard is at rest, the eyes swivel back and forth, viewing its surroundings. When it spots an insect, the lizard fixes both eyes on it and cautiously stalks to within shooting range.

Most large lizards are herbivores. Many iguanas (subfamily Iguaninae) are arboreal inhabitants of the tropics of Central and South America. Large terrestrial iguanas occur on islands in the West Indies and the Galápagos Islands, probably because the absence of predators has allowed them to spend a large part of their time on the ground. Smaller terrestrial herbivores like the black iguanas (**Figure 13–2c**) live on the mainland of Mexico and Central America, and still smaller relatives such as the chuckwallas and desert iguanas range as far north as the western United States. Many species of lizards live on beaches, but few extant species actually enter the water. The marine iguana of the Galápagos Islands is an exception. The feeding habits of the marine iguana are unique. It feeds on seaweed, diving 10 meters or more to browse on algae growing below the tide mark.

An exception to the rule of herbivorous diets for large lizards is found in the monitor lizards (family Varanidae). Varanids (**Figure 13–2j**) are active preda-

tors that feed on a variety of vertebrate and invertebrate animals, including birds and mammals. They circumvent the conflict between locomotion and lung ventilation that constrains the activity of other lizards by using a positive pressure gular pump to assist the axial muscles and are able to sustain high levels of activity (Owerkowicz et al. 1999). The Komodo monitor lizard is capable of killing adult water buffalo, but its normal prey is deer and feral goats. Large monitor lizards were widely distributed on the islands between Australia and Indonesia during the Pleistocene and may have preyed on pygmy elephants that also lived on the islands.

The hunting methods of the Komodo monitor are very similar to those employed by mammalian carnivores, showing that a simple brain is capable of complex behavior and learning. In the late morning, a Komodo monitor waits in ambush beside the trails deer use to move down from the hilltops, where they rest during the morning, to the valleys, where they sleep during the afternoon. The lizards, familiar with trails used by the deer, often wait where several deer trails converge. If no deer pass the lizard's ambush, it moves into the valleys, systematically stalking the thickets where deer are likely to be found. This purposeful hunting behavior, which demonstrates familiarity with the behavior of prey and with local geography, is in strong contrast to the opportunistic seizure of prey that characterizes the behavior of many lizards, but it is very like the hunting behavior of some snakes (Greene 1997).

The effectiveness of monitor lizards as predators is reflected in reciprocal geographic distributions of monitors and small mammalian carnivores in the Indo-Australian Archipelago, Australia, and New Guinea. Wallace's Line is a zoogeographic boundary that extends between the islands of Borneo (on the west) and Sulawesi (to the east). It traces the location of a deepwater trench, and areas to the west of the line have never had a land connection with areas to the east. Because of this ancient separation, Wallace's Line marks the easternmost boundary of the occurrence of many species from the Asian mainland and the westernmost boundary of species from Australia and New Guinea. In particular, small carnivorous placental mammals are found west of the line and small marsupial carnivores to the east. Large species of monitors occur on both sides of the line, but small species are found only on the eastern side. West of the line, the ecological niche for small carnivores is filled by mammals (such as small cats, civets, mongooses, and weasels), but east of the line—especially in Australia and New Guinea—most of the small carnivores are monitor lizards (Sweet and Pianka 2003).

Limb reduction has evolved more than 60 times among lizards, and every continent has one or more families with legless, or nearly legless, species (**Figure 13–2i**). Leglessness in lizards is usually associated with life in dense grass or shrubbery in which a slim, elongate body can maneuver more easily than a short one with functional legs. Some legless lizards crawl into small openings among rocks and under logs, and a few are subterranean.

The amphisbaenians are extremely **fossorial** lizards (Latin *fossor* = a digger), which have specializations that are different from those of other squamates. The earliest amphisbaenian known is a fossil from the Late Cretaceous. Most amphisbaenians are legless, but the three species in the Mexican genus *Bipes* have well-developed forelegs that they use to assist entry into the soil but not for burrowing underground. Suprisingly, a phylogenetic analysis indicates that external limbs were lost independently three times during the evolution of amphisbaenians (Kearny and Stuart 2004).

The skin of amphisbaenians is distinctive. The **annuli** (rings; singular *annulus*) that pass around the circumference of the body are readily apparent from external examination, and dissection shows that the integument is nearly free of connections to the trunk. Thus, it forms a tube within which the amphisbaenian's body can slide forward or backward. The separation of trunk and skin is employed during locomotion through tunnels. Integumentary muscles run longitudinally from annulus to annulus. The skin over this area of muscular contraction is then telescoped and buckles outward, anchoring that part of the amphisbaenian against the walls of its tunnel. Next, contraction of muscles that pass anteriorly from the vertebrae and ribs to the skin slide the trunk forward within the tube of integument. Amphisbaenians can move backward along their tunnels with the same mechanism by contracting muscles that pass posteriorly from the ribs to the skin. (The name amphisbaenian is derived from

▶ **FIGURE 13–2** Variation in body form among lizards. Small, generalized insectivores: (a) spiny swift, *Sceloporus* (Phrynosomatinae, Iguanidae); (b) japalure, *Calotes* ("Agamidae"). Herbivores: (c) black iguana, *Ctenosaura* (Iguaninae, Iguanidae); (d) mastigure, *Uromastyx* ("Agamidae"). Ant specialists: (e) horned lizard, *Phrynosoma* (Phrynosomatinae, Iguanidae); (f) spiny devil, *Moloch* ("Agamidae"). Nocturnal lizard: (g) Tokay gecko, *Gekko* (Gekkonidae). Arboreal lizard: (h) African chameleon, *Chamaeleo* (Chamaeleonidae). Legless lizard: (i) North American glass lizard, *Ophisaurus* (Anguidae). Large predator: (j) monitor lizard, *Varanus* (Varanidae).

Greek roots [*amphi* = double, *baen* = walk] in reference to the ability of amphisbaenians to move forward and backward with equal facility.) A similar type of rectilinear (straight-line) locomotion is used by some heavy-bodied snakes. However, the telescoping ability of the skin of snakes is generally restricted to the lateroventral portions of the body, whereas in amphisbaenians the skin is loose around the entire circumference of the body.

The dental structure is also distinctive: Amphisbaenians possess a single median tooth in the upper jaw—a feature unique to this group of vertebrates. The median tooth is part of a specialized dental battery that makes amphisbaenians formidable predators, capable of subduing a wide variety of invertebrates and small vertebrates. The upper tooth fits into the space between two teeth in the lower jaw and forms a set of nippers that can bite out a piece of tissue from a prey item too large for the mouth to engulf as a whole.

The skulls of amphisbaenians are used for tunneling, and they are rigidly constructed. The burrowing habits of amphisbaenians make them difficult to study, but three major functional categories can be recognized. Some species have blunt heads; the rest have either vertically keeled or horizontally spade-shaped snouts. Blunt-snouted forms burrow by ramming

▲ **FIGURE 13–2** *Continued*

their heads into the soil to compact it (**Figure 13–3b**). Sometimes an oscillatory rotation of the head with its heavily keratinized scales is used to shave material from the face of the tunnel. Shovel-snouted amphisbaenians ram the end of the tunnel, then lift the head to compact soil into the roof (**Figure 13–3c**). Wedge-snouted forms ram the snout into the end of the tunnel and then use the snout or the side of the neck to compress the material into the walls of the tunnel (**Figure 13–3d-f**). In parts of Africa, representatives of the three types occur together and share the subsoil habitat. The unspecialized blunt-headed forms live near the surface where the soil is relatively easy to tunnel through, and the specialized forms live in deeper, more compact soil. The geographic range of unspecialized forms is greater than that of specialized ones, and in areas where only a single species of amphisbaenian occurs, it is a blunt-headed species.

The ecological relationship between unspecialized and specialized burrowers is puzzling. One would expect that the specialized forms with their more elaborate methods of burrowing would replace the unspecialized ones, but this has not happened. The explanation may lie in the conflicting selective forces

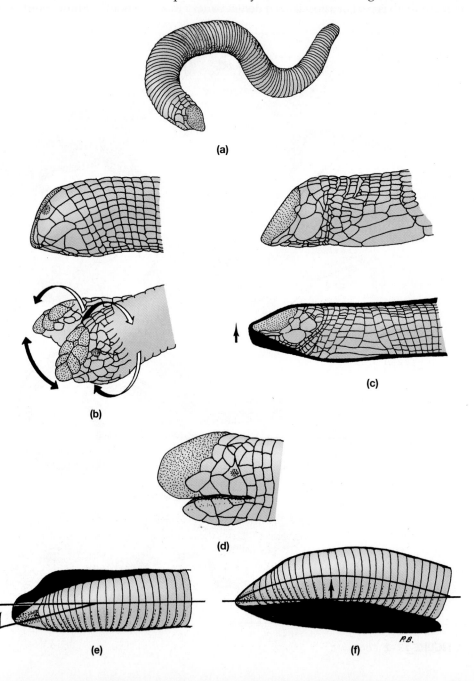

▶ **FIGURE 13–3** Body forms of amphisbaenians. (a) *Monopeltis* from Africa. Variation in snout shape: (b) blunt snouts (represented by *Agamodon*, from Africa); (c) shovel snouts (represented by *Rhineura* from Florida); or (d) wedge snouts (represented by *Anops*, Brazil). Widening the tunnel (e) with movements of the head in loose soil, or (f) with the anterior part of the body in dense soil.

(a)

(b)

(c)

(d)

(e)

(f)

on the snout. On one hand, it is important to have a snout that will burrow through soil; but, on the other hand, it is important to have a mouth capable of tackling a wide variety of prey. The specializations of the snout that make it an effective structure for burrowing appear to reduce its effectiveness for feeding. The blunt-headed amphisbaenians may be able to eat a wider variety of prey than the specialized forms can. Thus, in loose soil where it is easy to burrow, the blunt-headed forms may have an advantage. Only in soil too compact for a blunt-headed form to penetrate might the specialized forms find the balance of selective forces shifted in their favor.

■ Snakes

The 2900 species of snakes range in size from diminutive burrowing species, which feed on termites and grow to only 10 centimeters, to the large constrictors, which approach 10 meters in length (**Table 13.3**). The phylogenetic affinities of snakes are hotly debated. Clearly they are nested within the squamates, and they appear to have evolved in a terrestrial environment

TABLE 13.3 Lineages of extant snakes (Serpentes)

Scolecophidia

Anomalepididae: 16 species of small- (20 centimeters) to medium-size (75 centimeters) fossorial snakes from Central and South America.

Leptotyphlopidae: About 93 species of tiny (10 centimeters) to small fossorial snakes found in Africa, southwestern Asia, South and Central America, and southwestern North America.

Typhlopidae: About 233 species of small- to medium-size fossorial snakes with reduced eyes. Typhlopids are found on all continents except North America and Antarctica.

Alethinophidia

Aniliidae: A single species of fossorial snake from northern South America.

Anomochelidae: 2 species of small (40 centimeters) snakes from Malaysia, Sumatra, and Borneo.

Boidae: Three subfamilies of boas are usually recognized.

Pythoninae: 26 species, many of which are large (2 meters) to enormous (10 meters); found in Asia, Africa, and Australia.

Boinae: 33 species of mainly large to enormous terrestrial (*Boa*), semiaquatic (*Eunectes*), and arboreal (*Corallus*, *Epicrates*) snakes found from western North America through subtropical South America and the West Indies.

Erycinae: 15 species of medium-size semifossorial snakes. *Eryx* and *Gongylophis* are found from Africa and India to central Asia; *Charina* is found in northwestern North America.

Boyleriidae: 2 species of medium-size boalike snakes known only from Round Island, near Mauritius in the Indian Ocean.

Cylindrophiidae: 10 species of medium-size (to 1 meter) stout, blunt-headed burrowing snakes with shiny scales from Sri Lanka through Malaysia to Indonesia.

Loxocemidae: A single medium-size species of semifossorial snake from southern Mexico and Central America.

Tropidophiidae: 27 species of small, terrestrial boalike snakes from Central America, northern South America, and the West Indies.

Uropeltidae: About 47 species of small- to medium-size fossorial snakes (20–70 centimeters) from India and Sri Lanka.

Xenopeltidae: Two species of medium-size fossorial snakes found from southern China to Borneo and the Celebes. Xenophidiidae: 2 species of nocturnal snakes from tropical forests found in Borneo and Malaysia.

Achrochordidae: 3 species of medium-size to large (3 meters) aquatic snakes from southern Asia, the East Indies, and northern Australia.

Colubroidea

Atractaspidae: About 66 species of small- to medium-size African and Asian snakes. The atractaspids are secretive, living in leaf litter and spending time underground. They have elongated fangs on the maxillae, sometimes preceded by several small teeth.

Colubridae: More than 1827 species of tiny to very large snakes, found on all continents except Antarctica. Many colubrids have glands that secrete venom that kills prey, but they lack hollow teeth specialized for injecting venom.

Elapidae: About 300 species of venomous snakes with hollow fangs near the front of relatively immobile maxillae. Elapids occur on all continents except Antarctica and are most diverse in Australia. The sea snakes are elapids.

Viperidae: About 259 species of medium-size to large (2 meters) venomous snakes in which the maxillae are rotated about their attachment to the prefrontals, allowing the fangs to rest horizontally when the mouth is closed. True vipers (about 60 species) are found in Eurasia and Africa; pit vipers are found in the New World and in Asia. Viperids are absent from Australia and Antarctica.

(Vidal and Hedges 2004). (The Scolecophidia probably represent the primitive condition for snakes. This group includes three families of small burrowing snakes with shiny scales and reduced eyes. Traces of the pelvic girdle remain in most species, but the braincase is snakelike. Their anatomical and ecological characters appear consistent with a long-standing hypothesis that snakes evolved from a subterranean lineage of lizards with greatly reduced eyes. The hypothesis that the eyes of extant surface-dwelling snakes were redeveloped after nearly disappearing during a fossorial

stage in their evolution could explain some puzzling differences in eyes of snakes and lizards.

Burrowing snakes in the families Aniliidae and Uropeltidae use their heads to dig through soil, and the bones of their skulls are solidly united. The sole xenopeltid, the sunbeam snake of Southeast Asia, is a ground-dwelling species that takes its common name from its highly iridescent scales. Boa constrictors (Boinae) are mostly New World snakes, whereas pythons (Pythoninae) are found in the Old World. The anaconda, a semiaquatic species of boa from

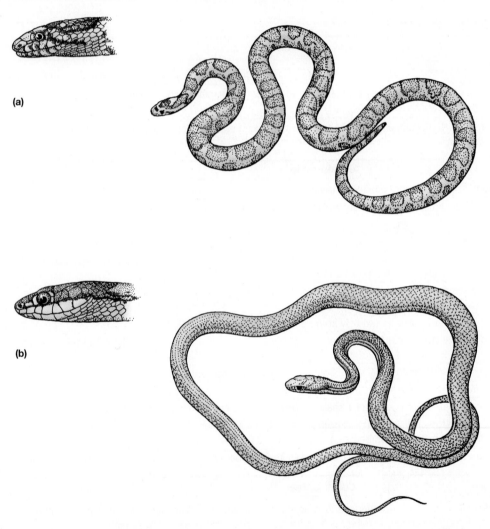

▲ **FIGURE 13–4** Body forms of snakes. Slow-moving constrictors such as the milk snake, (a) *Lampropeltis*, are relatively short and stout. Active, visually oriented snakes such as racers, (b) *Masticophis*, are longer and faster moving. Arboreal snakes such as the parrot snake, (c) *Leptophis*, are still more elongate and can follow their prey out among the twigs at the ends of branches. Burrowing snakes such as the blind snakes, (d) *Typhlops*, have small rounded or pointed heads with little distinction between head and neck, short tails, and smooth, often shiny scales; their eyes are greatly reduced in size. Vipers, especially the African vipers like the puff adder, (e) *Bitis arietans*, have large heads and stout bodies that accommodate large prey. Sea snakes, such as (f) *Laticauda*, have a tail that is flattened from side to side and valves that close the nostrils when they dive.

South America, is considered the largest extant species of snake—it probably approaches a length of 10 meters—and the reticulated python of Southeast Asia is nearly as large. Not all boas and pythons are large, however; some secretive and fossorial species are considerably less than 1 meter long as adults. The wart snakes in the family Acrochordidae are entirely aquatic; they lack the enlarged ventral scales that characterize most terrestrial snakes, and they have difficulty moving on land.

The Colubroidea includes most of the extant species of snakes, and the family Colubridae alone contains two-thirds of the extant species. The diversity of the group makes characterization difficult.

Colubroids have lost all traces of the pelvic girdle, they have only a single carotid artery, and the skull is very kinetic. Many colubroid snakes are venomous, and snakes in the families Elapidae and Viperidae have hollow fangs at the front of the mouth that inject extremely toxic venom into their prey.

The body form of even a generalized snake such as the milk snake (**Figure 13–4a**) is so specialized that little further morphological specialization is associated with different habits or habitats. King snakes and milk snakes are constrictors, and they crawl slowly, poking their heads under leaf litter and into holes that might shelter prey. Chemosensation is an important means of detecting prey for these snakes. Snakes

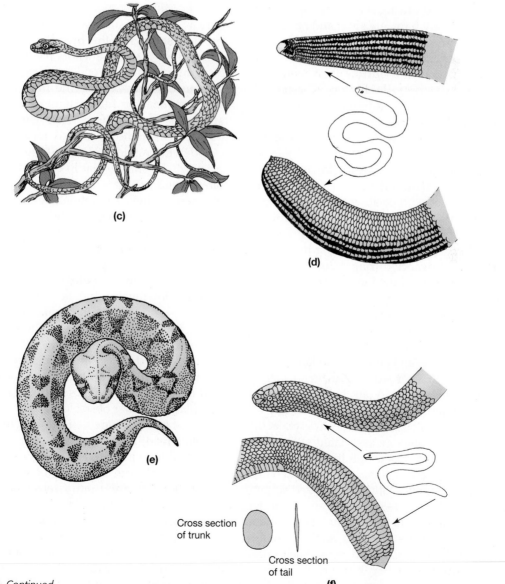

(c)

(d)

(e)

Cross section of trunk

Cross section of tail

(f)

▲ **FIGURE 13–4** *Continued*

have forked tongues, with widely separated tips that can move independently. When the tongue is projected, the tips are waved in the air or touched to the ground. The tongue then is retracted, and chemical stimuli are transferred to the paired vomeronasal organs. The forked shape of the tongue of snakes (which is seen also among the Amphisbaenia, Lacertiformes, and Varanoidea) may allow them to detect gradients of chemical stimuli and localize objects.

Nonconstrictors such as the whip snakes and racers (**Figure 13–4b**) move quickly and are visually oriented. They forage by crawling rapidly, frequently raising the head to look around. Many arboreal snakes are extremely elongated and frequently have large eyes (**Figure 13–4c**). Their length distributes their weight and allows them to crawl over even small twigs without breaking them. Burrowing snakes, at the opposite extreme of snake body form, are short and have blunt heads and very small eyes (**Figure 13–4d**). The head shape assists in penetrating soil, and a short body and tail create less friction in a burrow than would the same mass in an elongate body. Vipers, especially forms like the African puff adder (**Figure 13–4e**), are heavy-bodied with broad heads.

The sea snakes (**Figure 13–4f**) are derived from terrestrial elapids. Sea snakes are characterized by extreme morphological specialization for aquatic life: The tail is laterally flattened into an oar, the large ventral scales are reduced or absent in most species, and the nostrils are located dorsally on the snout and have valves that exclude water. The lung extends back to the cloaca and apparently has a hydrostatic role in adjusting buoyancy during diving, as well as having a respiratory function. Oxygen uptake through the skin during diving has been demonstrated in sea snakes. *Laticauda* are less specialized than other sea snakes and may represent a separate radiation into the marine habitat. They retain enlarged ventral scales and emerge onto land to bask and to lay eggs. The other sea snakes are so specialized for marine life that they are helpless on land, and these species are viviparous.

The locomotor specializations of snakes reflect differences in their morphology associated with different predatory modes (discussed in the following section) and the properties of the substrates on which they move. In lateral undulation (also called serpentine locomotion; **Figure 13–5a**), the body is thrown into a series of curves. The curves may be irregular, as shown in the illustration of a snake crawling across a board dotted with fixed pegs. Each curve presses backward; the pegs against which the snake is exerting force are shown in solid color. The lines numbered 1 to 7 are at 3-inch intervals, and the position of the snake at intervals of 1 second is shown.

Rectilinear locomotion (**Figure 13–5b**) is used primarily by heavy-bodied snakes. Alternate sections of the ventral integument are lifted off the ground and pulled forward by muscles that originate on the ribs and insert on the ventral scales. The intervening sections of the body rest on the ground and support the snake's body. Waves of contraction pass from anterior to posterior, and the snake moves in a straight line. Rectilinear locomotion is slow, but it is effective even when there are no surface irregularities strong enough to resist the sideward force exerted by serpentine locomotion. Because the snake moves slowly and in a straight line, it is inconspicuous, and rectilinear locomotion is used by some snakes when stalking prey.

Concertina locomotion (**Figure 13–5c**) is used in narrow passages such as rodent burrows that do not provide space for the broad curves of serpentine locomotion. A snake anchors the posterior part of its body by pressing several loops against the walls of the burrow and extends the front part of its body. When the snake is fully extended, it forms new loops anteriorly and anchors itself with these while it draws the rear end of its body forward.

Sidewinding locomotion (**Figure 13–5d**) is used primarily by snakes that live in deserts where wind-blown sand provides a substrate that slips away during serpentine locomotion. A sidewinding snake raises its body in loops, resting its weight on two or three points that are the only body parts in contact with the ground. The loops are swung forward through the air and placed on the ground, the points of contact moving smoothly along the body. Force is exerted downward; the lateral component of the force is so small that the snake does not slip sideward. This downward force is shown by imprints of the ventral scales in the tracks. Because the snake's body is extended nearly perpendicular to its line of travel, sidewinding is an effective means of locomotion only for small snakes that live in habitats with few plants or other obstacles.

Snake skeletons are delicate structures that do not fossilize readily. In most cases, we have only vertebrae, and little information has been gained from the fossil record about the origin of snakes. The earliest fossils known are from Cretaceous deposits and seem to be related to boas. Colubrid snakes are first known from the Oligocene epoch, and elapids and viperids appeared during the Miocene epoch.

The specializations of snakes compared with legless lizards appear to reflect two selective pressures—locomotion and predation. Elongation of the

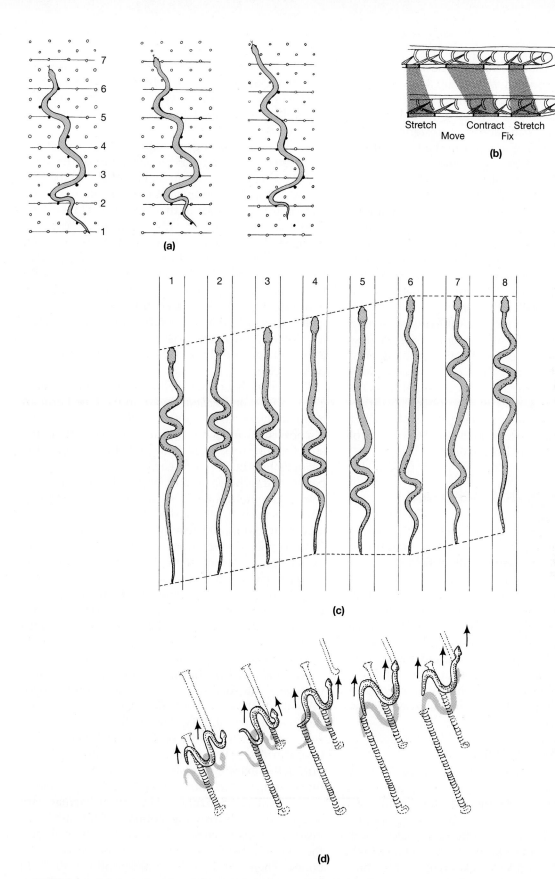

▲ **FIGURE 13–5** Locomotion of snakes. (a) Lateral undulation. (b) Rectilinear. (c) Concertina. (d) Sidewinding.

body is characteristic of snakes. The reduction in body diameter associated with elongation has been accompanied by some rearrangement of the internal anatomy of snakes. The left lung is reduced or entirely absent, the gallbladder is posterior to the liver, the right kidney is anterior to the left, and the gonads may show similar displacement.

Legless lizards face problems in swallowing prey. The primary difficulty is not the loss of limbs because few lizards use the legs to seize or manipulate food. The difficulty stems from the elongation that is such a widespread characteristic of legless forms. As the body lengthens, the mass is redistributed into a tube with a smaller diameter. As the mouth gets smaller, the maximum diameter of the prey that can be swallowed also decreases, and an elongate animal is faced with the difficulty of feeding a large body through a small mouth. Most legless lizards are limited to eating relatively small prey, whereas snakes have morphological specializations that permit them to engulf prey considerably larger than the body diameter (see Section 13.4). This difference may be one element in the great evolutionary success of snakes in contrast to the limited success of legless lizards and amphisbaenians.

13.4 Ecology and Behavior of Squamates

The past quarter century has seen an enormous increase in the number and quality of field studies of the ecology and behavior of snakes and lizards. Studies of lizards have been particularly fruitful, in large measure because many species of lizards are conspicuous and active during the day. These diurnal species dominate the literature—much less is known about species with cryptic habits. The discussions in this section rely on studies of particular species, and it is important to remember that no single species or family is representative of lizards or snakes as a group.

■ Foraging and Feeding

The methods that snakes and lizards use to find, capture, subdue, and swallow prey are diverse, and they are important in determining the interactions among species in a community. Astonishing specializations have evolved: blunt-headed snakes with long lower jaws that can reach into a shell to winkle out a snail, nearly toothless snakes that swallow bird eggs intact and then slice them open with sharp ventral processes (hypapophyses) on the neck vertebrae, and chameleons that project their tongues to capture insects or small vertebrates on the sticky tips are only a sample of the diversity of feeding specializations of squamates.

Many of the feeding specializations of squamates are related to changes in the structure of the skull and jaws. The most conspicuous of these is the loss of the lower temporal bar and the quadratojugal bone that formed part of that bar (**Figure 13–6**). This modification is part of a suite of structural changes in the skull that contribute to the development of a considerable degree of kinesis. The living tuatara show the ancestral condition for squamates, with the quadratojugal linking the jugal and the quadrate bones to form a complete lower temporal arch. (This fully diapsid condition is not characteristic of all sphenodontids, however; some of the Mesozoic forms did not have a complete lower temporal arch.)

Early lizards are not well known. The fossil genera *Paliguana* and *Palaeagama* from the Late Permian and Early Triassic periods of South Africa are probably not true lizards, but they do show changes in the structure of the skull that probably parallel the changes that occurred in early squamates. The gap between the quadrate and jugal widened, and the complexly interdigitating suture between the frontal and parietal bones on the roof of the skull became straighter and more like a hinge. Additional areas of flexion evolved at the front and rear of the skull and in the lower jaw of some lizards. These changes were accompanied by the development of a flexible connection at the articulation of the quadrate bone with the squamosal, which provided some mobility to the quadrate. This condition, known as streptostyly, increases the force the pterygoideus muscle can exert when the jaws are nearly closed.

In snakes, the flexibility of the skull was increased still further by loss of the second temporal bar, which was formed by a connection between the postorbital and squamosal bones. A further increase in the flexibility of the joints between other bones in the palate and the roof of the skull produced the extreme flexibility of snake skulls. The third group of squamates has a completely different sort of skull specialization. The amphisbaenians are small, legless, burrowing animals. They use their heads as rams to construct tunnels in the soil, and the skull is heavy with rigid joints between the bones. Their specialized dentition allows them to bite small pieces out of large prey.

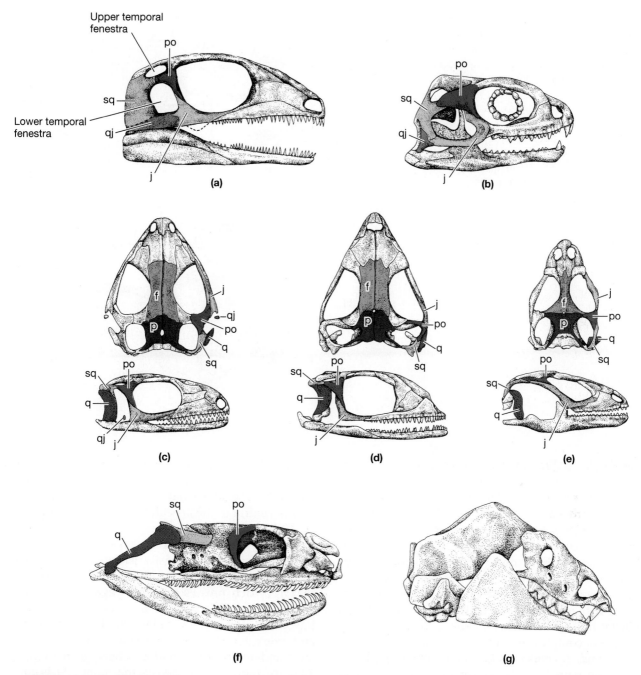

▲ **FIGURE 13–6** Modifications of the diapsid skull among lepidosauromorphs. Fully diapsid forms like the Permian *Petrolacosaurus* (a) retain two complete arches of bone that define the upper and lower temporal fenestrae. This condition is seen in living tuatara, *Sphenodon* (b). Lizards have achieved a kinetic skull by developing a gap between the quadrate and quadratojugal and by simplifying the suture between the frontal and parietal bones, as shown by the modern collared lizard *Crotaphytus* (e). Probable transitional stages allowing increasing skull kinesis that occurred in a nonsquamate lineage are illustrated by *Paliguana* (c) and *Palaeagama* (d). In snakes (f), skull kinesis is further increased by loss of the upper temporal arch. Amphisbaenians (g), which use their heads for burrowing through soil, have specialized akinetic skulls. (f = frontal; j = jugal; p = parietal; po = postorbital; q = quadrate; qj = quadratojugal; sq = squamosal)

▶ **FIGURE 13–7** Skull of a snake. (a) Lateral and (b) ventral views. A snake skull contains eight movable links: (1) braincase; (2) supratemporal; (3) prefrontal; (4) palatine; (5) pterygoid; (6) pterygoquadrate ligament; (7) quadrate; (8) quadratosupratemporal tie. (ang = angular; art = articular; boc = basioccipital; bsp = basisphenoid; col = columella; den = dentary; ecptg = ectopterygoid; fro = frontal; max = maxilla; nas = nasal; pal = palatine; par = parietal; pmax = premaxilla; po = postorbital; prf = prefrontal; ptg = pterygoid; q = quadrate; spl = splenial; sur = surangular; sut = supratemporal; vom = vomer).

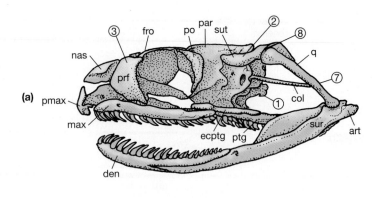

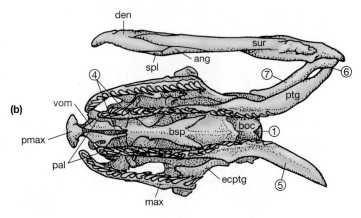

Feeding Specializations of Snakes

The entire skull of advanced snakes is much more flexible than the skull of a lizard. In popular literature, snakes are sometimes described as "unhinging" their jaws during feeding. That's careless writing and rather silly—unhinged jaws would merely flap back and forth. What those authors are trying to say is that snakes have extremely kinetic skulls that allow extensive movement of the jaws. A snake skull contains eight links, with joints between them that permit rotation (**Figure 13–7**). This number of links gives a staggering degree of complexity to the movements of the snake skull; and, to make things more complicated, the links are paired—each side of the head acts independently. Furthermore, the pterygoquadrate ligament and quadrato-supratemporal ties are flexible. When they are under tension, they are rigid; but, when they are relaxed, they permit sideward movement as well as rotation. All of this results in a considerable degree of three-dimensional movement in a snake's skull.

The mandibles of lizards are joined at the front of the mouth by a rigid bony connection, but in snakes the mandibles are attached only by muscles and skin so they can spread sideward and move forward or back independently. Loosely connected mandibles and flexible skin in the chin and throat allow the jaw tips to spread, so that the widest part of the prey passes ventral to the articulation of the jaw with the skull.

Swallowing movements take place slowly enough to be observed easily (**Figure 13–8**). A snake usually swallows prey headfirst, perhaps because that approach presses the limbs against the body, out of the snake's way. Small prey may be swallowed tail first or even sideward. The mandibular and pterygoid teeth of one side of the head are anchored in the prey, and the head is rotated to advance the opposite jaw as the mandible is protracted and grips the prey ventrally. As this process is repeated, the snake draws the prey item into its mouth. Once the prey has reached the esophagus, it is forced toward the stomach by contraction of the snake's neck muscles. Usually the neck is bent sharply to push the prey along.

Most species of snakes seize prey and swallow it as it struggles. The risk of damage to the snake during this process is a real one, and various features of snake anatomy seem to give some protection from struggling prey. The frontal and parietal bones of a snake's

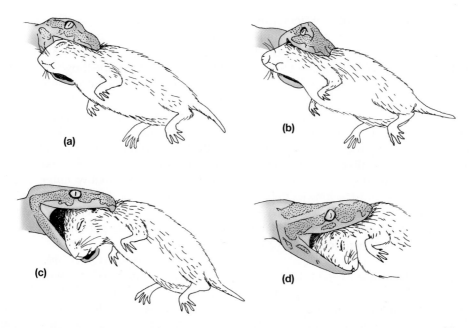

(a)

(b)

(c)

(d)

▲ **FIGURE 13–8** Jaw movements of a snake during feeding. Snakes use a combination of head movements and protraction and retraction of the jaws to swallow prey. (a) Prey grasped by left and right jaws at the beginning of the swallowing process. (b) The upper and lower jaws on the right side have been protracted, disengaging the teeth from the prey. (c) The head is rotated counterclockwise, moving the right upper and lower jaws over the prey. The recurved teeth slide over the prey like the runners of a sled. (d) The upper and lower jaw on the right side are retracted, embedding the teeth in the prey and drawing it into the mouth. Notice that the entire head of the prey has been engulfed by this movement. Next the left upper and lower jaws will be advanced over the prey by clockwise rotation of the head. The swallowing process continues with alternating left and right movements until the entire body of the prey has passed through the snake's jaws.

skull extend downward, entirely enclosing the brain and shielding it from the protesting kicks of prey being swallowed. Possibly the kinds of prey that can be attacked by snakes without a specialized feeding mechanism are limited by the snake's ability to swallow the prey without being injured in the process.

Constriction and venom are predatory specializations that permit a snake to tackle large prey with little risk of injury to itself. Constriction is characteristic of the boas and pythons as well as some colubrid snakes. Despite travelers' tales of animals crushed to jelly by a python's coils, the process of constriction involves very little pressure. A constrictor seizes prey with its jaws and throws one or more coils of its body about the prey. The loops of the snake's body press against adjacent loops, and friction prevents the prey from forcing the loops open. Each time the prey exhales, the snake takes up the slack by tightening the loops slightly. Two hypotheses have been proposed to explain the cause of death from constriction. The traditional view holds that prey suffocates because it cannot expand its thorax to inhale. Another possibility is that the increased internal pressure interferes with, and eventually stops, the heart (Hardy 1994).

Snakes that constrict their prey must be able to throw the body into several loops of small diameter to wrap around the prey. Constrictors achieve these small loops by having short vertebrae and short trunk muscles that span only a few vertebrae from the point of origin to the point of insertion. Contraction of these muscles produces sharp bends in the trunk that allow constrictors to press tightly against their prey. However, the trunk muscles of snakes are also used for locomotion, and the short muscles of constrictors produce several small-radius curves along the length of the snake's body. That morphology limits the speed with which constrictors can move because rapid locomotion by snakes is accomplished by throwing the body into two or three broad loops. This is the pattern seen in fast-moving species such as whip snakes, racers, and mambas. The muscles that produce these broad loops are long, spanning many vertebrae, and the vertebrae are longer than those of constrictors.

In North America, fast-moving snakes (colubroids) first appear in the fossil record during the Miocene, a time when grasslands were expanding. Constrictors, largely erycines, predominated in the snake fauna of the early Miocene, but by the end of that epoch the snake fauna was composed primarily of colubroids. Fast-moving colubroid snakes may have had an advantage over slow-moving boids in the more open habitats that developed during the Miocene, and that radiation of colubroids may have involved a complex interaction between locomotion and feeding. Rodents were probably the most abundant prey available to snakes, and rodents are dangerous animals for a snake to swallow while they are alive and able to bite and scratch. Constriction provided a relatively safe way for boids to kill rodents, but the long vertebrae and long trunk muscles that allowed colubroids to move rapidly through the open habitats of the later Miocene would have prevented them from using constriction to kill their prey.

Early colubroids may have used venom to immobilize prey. Duvernoy's gland, found in the upper jaw of many extant colubrid snakes, is homologous to the venom glands of viperids and elapids and produces a toxic secretion that immobilizes prey. (Some extant colubrids have venom that is dangerously toxic, even to animals as large as humans.) Thus, the evolution of venom that could kill prey may have been a key feature that allowed Miocene colubroid snakes to dispense with constriction and become morphologically specialized for rapid locomotion in open habitats. The presence of Duvernoy's gland appears to be an ancestral character for colubroid snakes, as this hypothesis predicts. Some colubrids, including the rat snakes (*Elaphe*), gopher snakes (*Pituophis*), and king snakes (*Lampropeltis*), have lost the venom-producing capacity of the Duvernoy's gland, and these are the groups in which constriction has been secondarily developed as a method of killing prey.

In this context, the front-fanged venomous snakes (Elapidae and Viperidae) are not a new development, but instead represent alternative specializations of an ancestral venom delivery system. Given the ancestral nature of venom for colubroid snakes, you might expect that different specializations for venom delivery would be represented in the extant snake fauna, as indeed they are. A variety of snakes have enlarged teeth (fangs) on the maxillae. Three categories of venomous snakes are recognized (**Figure 13–9**): opisthoglyphous, proteroglyphous, and solenoglyphous. A fourth category, the aglyphous snakes, is reserved for snakes with no fangs. This classification is descriptive and represents convergent evolution by different phylogenetic lineages.

Opisthoglyphous (Greek *opistho* = behind and *glyph* = knife) snakes have one or more enlarged teeth near the rear of the maxilla, with smaller teeth in front. In some forms, the fangs are solid; in others there is a groove on the surface of the fang that may help to conduct saliva into the wound. Several African and Asian opisthoglyphs can deliver a dangerous or even lethal bite to large animals, including humans, but their primary prey is lizards or birds, which are often held in the mouth until they stop struggling and are then swallowed.

Proteroglyphous snakes (Greek *prot* = first) include the cobras, mambas, coral snakes, and sea snakes in the Elapidae. The hollow fangs of the proteroglyphous snakes are located at the front of the maxilla, and there are often several small, solid teeth behind the fangs. The fangs are permanently erect and relatively short.

Solenoglyphous (Greek *solen* = pipe) snakes include the pit vipers of the New World and the true vipers of the Old World. In these snakes, the hollow fangs are the only teeth on the maxillae, which rotate so that the fangs are folded against the roof of the mouth when the jaws are closed. This folding mechanism permits solenoglyphous snakes to have long fangs that inject venom deep into the tissues of the prey. The venom, a complex mixture of enzymes and other substances (**Table 13.4**), first kills the prey and then speeds its digestion after it has been swallowed.

Snakes that can inject a disabling dose of venom into their prey have evolved a very safe prey-catching method. A constrictor is in contact with its prey while it is dying and runs some risk of injury from the prey's struggles. A solenoglyphous snake needs only to inject venom and allow the prey to run off to die. Later the snake can follow the scent trail of the prey to find its corpse. This is the prey-capture pattern of most vipers, and experiments have shown that a viper can distinguish the scent trail of a mouse it has bitten from trails left by uninjured mice.

Several features of the body form of vipers allow them to eat larger prey in relation to their own body size than can most nonvenomous snakes. Many vipers, including rattlesnakes, the jumping viper, the African puff adder, and the Gaboon viper, are very stout snakes. The triangular head shape that is usually associated with vipers is a result of the outward extension of the rear of the skull, especially the quadrate bones. The wide-spreading quadrates allow bulky objects to pass through the mouth, and even a large meal makes little bulge in the stout body and thus does not interfere with locomotion. Vipers have specialized as relatively sedentary predators that wait in ambush and can prey even on quite large animals. The other family of terrestrial venomous snakes, the elapids—cobras,

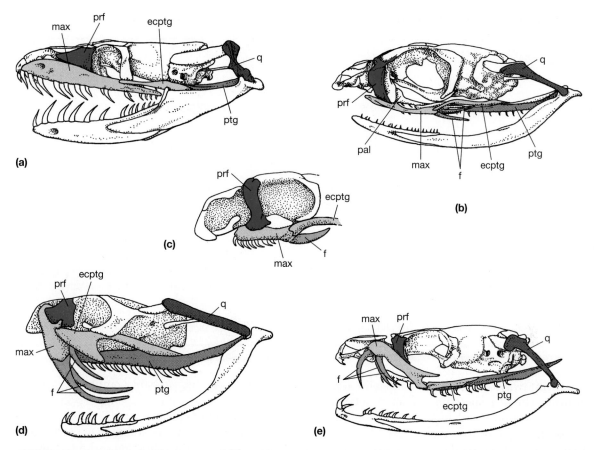

▲ **FIGURE 13–9** Dentition of snakes. (a) Aglyphous (without fangs), African python, *Python sebae*. (b,c) Opisthoglyphous (fangs in the rear of the maxilla), African boomslang, *Dispholidus typus*, and Central American false viper, *Xenodon rhabdocephalus*. (d) Solenoglyphous (fangs on a rotating maxilla), African puff adder, *Bitis arietans*. (e) Proteroglyphous (permanently erect fangs at the front of the maxilla), African green mamba, *Dendroaspis jamesoni*. The fangs of solenoglyphs (d) are erected by an anterior movement of the pterygoid that is transmitted through the ectopterygoid and palatine to the maxilla, causing it to rotate about its articulation with the prefrontal, thereby erecting the fang. Some opisthoglyphs, especially *Xenodon* (c), have the same mechanism of fang erection. (ecptg = ectopterygoid; f = fang; max = maxilla; pal = palatine; prf = prefrontal; ptg = pterygoid; q = quadrate).

mambas, and their relatives—are primarily slim-bodied snakes that actively search for prey.

Foraging Behavior and Energetics of Lizards The activity patterns of lizards span a range from extremely sedentary species that spend hours in one place to species that are in nearly constant motion. Field observations of the tropidurid lizard *Leiocephalus schreibersi* and the teiid *Ameiva chrysolaema* in the Dominican Republic revealed two extremes of behavior. *Leiocephalus* rested on an elevated perch from sunrise to sunset and was motionless for more than 99 percent of the day. Its only movements consisted of short, rapid dashes to capture insects or to chase away other lizards. These periods of activity never lasted longer than 2 seconds, and the frequency of movements averaged 9.6 per hour. In contrast, *Ameiva* were active for only 4 or 5 hours in the middle of the day, but they were moving more than 70 percent of that time, and their velocity averaged one body length every 2 to 5 seconds.

The same difference in behavior was seen in a laboratory test of spontaneous activity: *Ameiva* was more than 20 times as active as *Leiocephalus*. In fact, the teiids were as active in exploring their surroundings as small mammals tested in the same apparatus. A xantusiid lizard tested in the laboratory apparatus had a pattern of spontaneous activity that fell

TABLE 13.4 Components of the venoms of squamates

Compound	Occurrence	Effect
Proteinases	All venomous squamates, especially vipers	Tissue destruction
Hyaluronidase	All venomous squamates	Increases tissue permeability, hastening the spread of other constituents of venom through the tissues
L-amino acid oxidase	All venomous squamates	Attacks a wide variety of substrates and causes great tissue destruction
Cholinesterase	High in terrestrial elapids; may be present in sea snakes; low in vipers	Unknown; it is not responsible for the neurotoxic effects of elapid venom
Phospholipases	All venomous squamates	Destroys cell membranes
Phosphatases	All venomous squamates	Breaks down high-energy compounds such as ATP, preventing cells from repairing damage
Basic polypeptides	Terrestrial elapids and sea snakes	Blocks neuromuscular transmission

approximately midway between that of the teiid and the tropidurid. Thus, a spectrum of spontaneous locomotor activity is apparent in lizards, extending from species that are nearly motionless through species that move at intermediate rates to species that are as active as mammals.

For convenience, the extremes of the spectrum are frequently called sit-and-wait predators and widely foraging predators, respectively, and the intermediate condition has been called a cruising forager. Other field studies have shown that this spectrum of locomotor behaviors is widespread in lizard faunas. In North America, for example, spiny swifts (*Sceloporus*) are sit-and-wait predators, many skinks (*Eumeces*) appear to be cruising foragers, and whiptail lizards (*Cnemidophorus*) are widely foraging predators. The ancestral locomotor pattern for lizards may have been that of a cruising forager, and both sit-and-wait predation and active foraging may represent derived conditions. (A spectrum of foraging modes is not unique to lizards; it probably applies to nearly all kinds of mobile animals, including fishes, mammals, birds, frogs, insects, and zooplankton.)

The ecological, morphological, and behavioral characteristics that are correlated with the foraging modes of different species of lizards appear to define many aspects of the biology of these animals. For example, sit-and-wait predators and widely foraging predators consume different kinds of prey and fall victim to different kinds of predators. They have dif-

ferent social systems, probably emphasize different sensory modes, and differ in some aspects of their reproduction and life history.

These generalizations are summarized in **Table 13.5** and are discussed in the following sections. However, a weakness of this analysis must be emphasized: Sit-and-wait species of lizards (at least, the ones that have been studied most) are primarily iguanians, whereas widely foraging species are mostly scleroglossans. That phylogenetic split raises the question of whether the differences we see between sit-and-wait and widely foraging lizards are really the consequences of the differences in foraging behavior, or if they are ancestral characteristics of iguanian versus scleroglossan lizards. If the latter is true, the association with different foraging modes may be misleading. In either case, however, the model presented in Table 13.5 provides a useful integration of a large quantity of information about the biology of lizards; it represents a hypothesis that may be modified as more information becomes available.

Lizards with different foraging modes use different methods to detect prey. Sit-and-wait lizards normally remain in one spot from which they can survey a broad area. These motionless lizards detect the movement of an insect visually and capture it with a quick dash from their observation site. Sit-and-wait lizards may be most successful in detecting and capturing relatively large insects like beetles and grasshoppers. Active foragers spend most of their time on the ground

TABLE 13.5 Ecological and behavioral characteristics associated with the foraging modes of lizards. Foraging modes are presented as a continuum from sit-and-wait predators to widely foraging predators. In most cases, data are available only for species at the extremes of the continuum. (See the text for details.)

	Foraging Mode		
Character	Sit-and-Wait	Cruising Forager	Widely Foraging
Foraging behavior			
Movements/hour	Few	Intermediate	Many
Speed of movement	Low	Intermediate	Fast
Sensory modes	Vision	Vision and olfaction	Vision and olfaction
Exploratory behavior	Low	Intermediate	High
Types of prey	Mobile, large	Intermediate	Sedentary, often small
Predators			
Risk of predation	Low	?	Higher
Types of predators	Widely foraging	?	Sit-and-wait and widely foraging
Body form			
Trunk	Stocky	Intermediate?	Elongate
Tail	Often short	?	Often long
Physiological characteristics			
Endurance	Limited	?	High
Sprint speed	High	?	Intermediate to low
Aerobic metabolic capacity	Low	?	High
Anaerobic metabolic capacity	High	?	Low
Heart mass	Small	?	Large
Hematocrit	Low	?	High
Energetics			
Daily energy expenditure	Low	?	Higher
Daily energy intake	Low	?	Higher
Social behavior			
Size of home range	Small	Intermediate	Large
Social system	Territorial	?	Not territorial
Reproduction			
Mass of clutch (eggs or embryos) relative to mass of adult	High	?	Low

surface, moving steadily and poking their snouts under fallen leaves and into crevices in the ground. These lizards apparently rely largely on chemical cues to detect insects, and they probably seek out local concentrations of patchily distributed prey such as termites. Widely foraging species of lizards appear to eat more small insects than do lizards that are sit-and-wait predators. Thus, the different foraging behaviors of lizards lead to differences in their diets, even when the two kinds of lizards occur in the same habitat.

The different foraging modes also have different consequences for the exposure of lizards to their own predators. A lizard that spends 99 percent of its time resting motionless is relatively inconspicuous, whereas a lizard that spends most of its time moving is more easily seen. Sit-and-wait lizards are probably most likely to be discovered and captured by predators that are active searchers, whereas widely foraging lizards are likely to be caught by sit-and-wait predators. Because of this difference, foraging modes may alternate at successive levels in the food chain: Insects that move about may be captured by lizards that are sit-and-wait predators, and those lizards may be eaten by widely foraging predators. Insects that are sedentary are more likely to be discovered by a widely foraging lizard, and that lizard may be picked off by a sit-and-wait predator.

The body forms of sit-and-wait lizard predators may reflect selective pressures different from those that act on widely foraging species. Sit-and-wait lizards are often stout bodied, short tailed, and cryptically colored. Many of these species have dorsal patterns formed by blotches of different colors that probably obscure the outlines of the body as the lizard rests motionless on a rock or tree trunk. Widely foraging species of lizards are usually slim and elongate with long tails, and they often have patterns of stripes that may produce optical illusions as they move. However, one predator-avoidance mechanism, the ability to break off the tail when it is seized by a predator (**autotomy**), does not differ among lizards with different foraging modes (**Box 13-1**).

What physiological characteristics are necessary to support different foraging modes? The energy requirements of a dash that lasts for only a second or two are quite different from those of locomotion that is sustained nearly continuously for several hours. Sit-and-wait and widely foraging species of lizards differ in their relative emphasis on the two metabolic pathways that provide adenosine triphosphate (ATP) for activity and in how long that activity can be sustained. Sit-and-wait lizards move in brief spurts, and they rely largely on anaerobic metabolism to sustain their movements. Anaerobic metabolism uses glycogen stored in the muscles as a metabolic substrate and produces lactic acid as its end product. It is a way to synthesize ATP quickly (because the glycogen is already in the muscles), but it is not good for sustained activity, because the glycogen is soon exhausted and lactic acid inhibits cellular metabolism. Lizards that rely on anaerobic metabolism can make brief sprints but become exhausted when they are forced to run continuously. In contrast, aerobic metabolism uses glucose that is carried to the muscles by the circulatory system as a metabolic substrate, and it produces carbon dioxide and water as end products. Aerobic exercise can continue for long periods because the circulatory system brings more glucose and carries carbon dioxide away. As a result, widely foraging species can sustain activity for long periods without exhaustion.

The differences in exercise physiology are associated with differences in the oxygen transport systems of the lizards: Widely foraging species of lizards have larger hearts and more red blood cells in their blood than do sit-and-wait species. As a result, each beat of the heart pumps more blood, and that blood carries more oxygen to the tissues of a widely foraging species of lizard than a sit-and-wait species.

Sustained locomotion is probably not important to a sit-and-wait lizard that makes short dashes to capture prey or to escape from predators, but sprint speed might be vitally important in both these activities. Speed may be relatively unimportant to a widely foraging lizard that moves slowly, methodically looks

BOX 13-1   Caudal Autotomy—Your Tail or Your Life

Autotomy (self-amputation) of appendages is a common predator-escape mechanism among invertebrates and vertebrates. The tail is the only appendage that vertebrates are known to autotomize, and the capacity for caudal autotomy is developed to some degree among salamanders, tuatara, lizards, and a few amphisbaenians, snakes, and rodents (Arnold 1988). In most cases, autotomy is followed by regeneration of a new tail.

The caudal autotomy of squamates occurs at distinctive fracture planes that are found in all but the four to nine anteriormost caudal vertebrae. The caudal muscles are segmental, and pointed processes from adjacent segments interdigitate. The caudal arteries have sphincter muscles just anterior to each fracture site, and the veins have valves. Autotomy appears to be an active process that requires contraction of the caudal muscles, bending the tail sharply to one side and initiating separation. The vertebral centrum ruptures, and the processes of the caudal muscles separate. The arterial sphincter muscles contract and venous valves close, preventing loss of blood. An autotomized tail twitches rapidly for several minutes, and its violent writhing can distract the attention of a predator while the lizard itself scurries to safety (**Figure 13-10**). The tails of some juvenile skinks are bright blue; and lizards with these colorful tails were more effective at using autotomy to escape from predatory snakes than were lizards with tails that had been painted black. Of course, an auto-

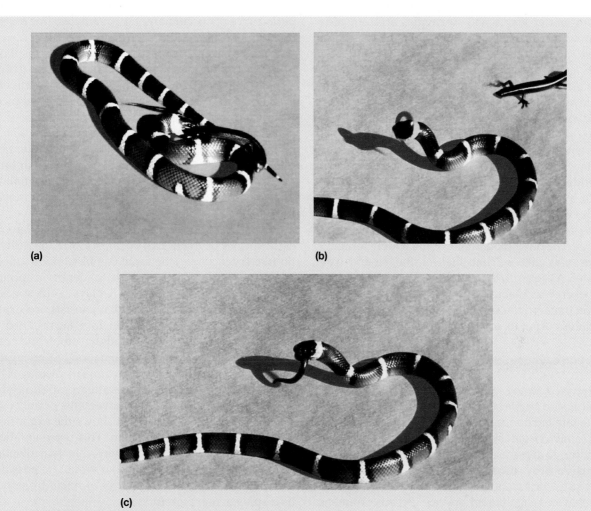

(a)   **(b)**

(c)

▲ **FIGURE 13–10** Autotomy as a defensive mechanism. The freshly autotomized tail of a skink writhes and jerks, engaging the attention of a predatory king snake while the lizard escapes. In this sequence of photographs, a king snake seizes a skink by the tail (a). The skink autotomizes its tail and runs off (b), leaving the snake struggling to swallow the wriggling tail (c).

tomized tail receives no blood flow and its muscular activity is sustained by anaerobic metabolism. The anaerobic metabolic capacity of lizard tail muscles appears to be substantially greater than that of limb muscles.

The point of autotomy is normally as far posterior on the tail as possible. When the tail of a lizard is seized with forceps, autotomy usually occurs through the plane of the vertebra immediately anterior to the point at which the tail was being held, thereby minimizing the amount of tail lost. When the tail is regenerated, the vertebrae are replaced by a rod of cartilage that does not contain fracture planes. Consequently, future autotomy must occur anterior to the regenerated portion of the tail. Some geckos adjust the point of autotomy according to their body temperature: autotomy occurs closer to the body when the lizard is cold than when it is warm. When a lizard is cold, it cannot run as fast as when it's warm, and perhaps the longer segment of tail left behind by a cold lizard occupies the predator's attention long enough to allow the lizard to reach safety.

Leaving your tail in the grasp of a predator is certainly better than being eaten, but it is not free of costs. Because the tail acts as a signal of status among some species of lizards, a lizard that autotomizes a large part of its tail may fall to a lower rank in the dominance hierarchy. Losing the tail also affects the energy balance of a lizard; for example, the rate of growth of juvenile lizards that have autotomized their tails is reduced while their tails are being regenerated. Many lizards store fat in the tail, and the females mobilize this energy while they are depositing yolk in eggs. Sixty percent of the total fat storage was located in the tails of female geckos (*Coleonyx brevis*). Autotomy of the tail by gravid female geckos resulted in their producing smaller clutches of eggs than lizards with tails. Some lizards, especially the small North American skink *Scincella lateralis*, eat their autotomized tails if they can, thereby recovering the lost energy.

for prey under leaves and in cavities and can hide under a bush to confuse a predator. As you might predict from these considerations, sit-and-wait lizards generally have high sprint speeds and low endurance, whereas widely foraging species usually have lower sprint speeds and greater endurance.

The continuous locomotion of widely foraging species of lizards is energetically expensive. Measurements of energy expenditure of lizards in the Kalahari showed that the daily energy expenditure of a widely foraging species averaged 150 percent of that of a sit-and-wait species. However, the energy that the widely foraging species invested in foraging was more than repaid by its greater success in finding prey. The daily food intake of the widely foraging species was 200 percent that of the sit-and-wait predator. As a result, the widely foraging species had more energy available to use for growth and reproduction than did the sit-and-wait species, despite the additional cost of its mode of foraging.

■ Social Behavior

Squamates employ a variety of visual, auditory, chemical, and tactile signals in the behaviors they use to maintain territories and to choose mates. Iguanians use mainly visual signals, whereas scleroglossans (including snakes) use pheromones extensively. The various sensory modalities employed by animals have biased the amount of information we have about the behaviors of different species. Because humans are primarily visually oriented, we perceive the visual displays of other animals quite readily. The auditory sensitivity of humans is also acute, and we can detect and recognize vocal signals that are used by other species. However, the olfactory sensitivity of humans is low and we lack a well-developed vomeronasal system, so we are unable to perceive most chemical signals used by squamates. One result of our sensory biases has been a concentration of behavioral studies on organisms that use visual signals. Because of this focus, the extensive repertoires of visual displays of iguanian lizards figure largely in the literature of behavioral ecology but much less is known about the chemical and tactile signals that are probably important for other lizards and for snakes.

The social behaviors of squamates appear to be limited in comparison with those of crocodilians, but many species show dominance hierarchies or territoriality. The signals used in agonistic encounters between individuals are often similar to those used for species and sex recognition during courtship. Parental attendance at a nest during the incubation period of eggs occurs among squamates, but extended parental care of the young is unknown.

Iguanian lizards employ primarily visual displays during social interactions. The polychrotine genus *Anolis* includes some 400 species of small- to medium-size lizards that occur primarily in tropical America. Male *Anolis* have gular fans, areas of skin beneath the chin that can be distended by the hyoid apparatus during visual displays (see color insert). The brightly colored scales and skin of the gular fans of many species of *Anolis* are conspicuous signaling devices, and they are used in conjunction with movements of head and body.

Figure 13–11 shows the colors of the gular fans of eight species of *Anolis* that occur in Costa Rica. Since no two species have the same combination of colors on their gular fans, it is possible to identify a species solely by seeing the colors it displays. Each species also has a behavioral display that consists of raising the body by straightening the fore legs (called a push-up), bobbing the head, and extending and contracting the gular fan. The combination of these three sorts of movements allows a complex display. The three movements can be represented graphically by an upper line that shows the movements of the body and head and a lower line that shows the expansion and contraction of the gular fan. This representation is called a display action pattern. No two display action patterns are the same, so it would be possible to identify any of the eight species of *Anolis* by seeing its display action pattern.

The behaviors that territorial lizards use for species and sex recognition during courtship are very much like those employed in territorial defense—push-ups, head bobs, and displays of the gular fan. A territorial male lizard is likely to challenge any lizard in its territory, and the progress of the interaction depends on the response it receives. An aggressive response indicates that the intruder is a male and stimulates the territorial male to defend its territory, whereas a passive response from the intruder identifies a female and stimulates the territorial male to initiate courtship. These behaviors are illustrated by the displays of a male *Anolis carolinensis* shown in **Figures 13–12** and **13–13**. The first response of a territorial lizard to an intruder is the assertion-challenge display shown in **Figure 13–12a**. The dewlap is extended, and the lizard bobs at the intruder. The nuchal (neck) and dorsal crests are also slightly raised, and a black spot appears behind the eye. The next stage depends on the intruder's sex and its response to the territorial male's challenge (**Figure 13–13**). If the intruder is a male and does not retreat

(a) SIMPLE   **(b)** COMPOUND   **(c)** COMPLEX

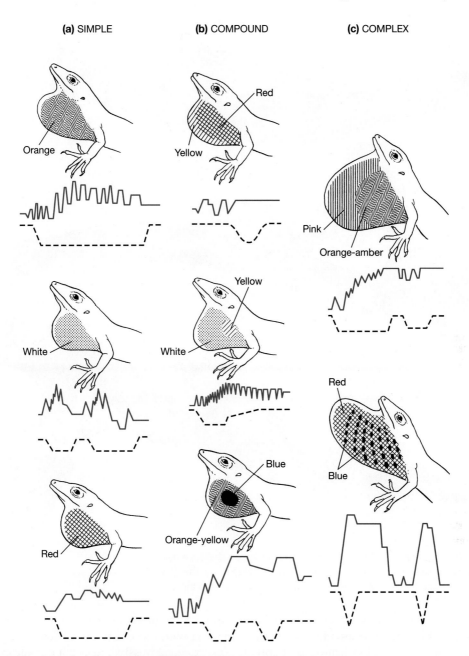

▲ **FIGURE 13–11** Species-typical displays of *Anolis* lizards. Eight species of *Anolis* from Costa Rica can be separated into three groups based on the size and color pattern of their gular fans. *Simple* fans are unicolored (a), *compound* fans are bicolored (b), and *complex* fans are bicolored and very large (c). Display action patterns for each species are graphed beneath the lizard drawings. The horizontal axis is time (the duration of these displays is about 10 seconds), and the vertical axis is vertical height. Solid line shows movements of the head; dashed line indicates extension of the gular fan.

from the initial challenge, both males become more aggressive. During aggressive posturing (**Figure 13–12b**), the males orient laterally to each other, the nuchal and dorsal crests are fully erected, the body is compressed laterally, the black spot behind the eye darkens, and the throat is swelled. All these postural changes make the lizards appear larger and presumably more formidable to the opponent. If the intruder is a receptive female, the territorial male initiates courtship (**Figure 13–12c**).

The differences in color and movement that characterize the dewlaps and display action patterns of

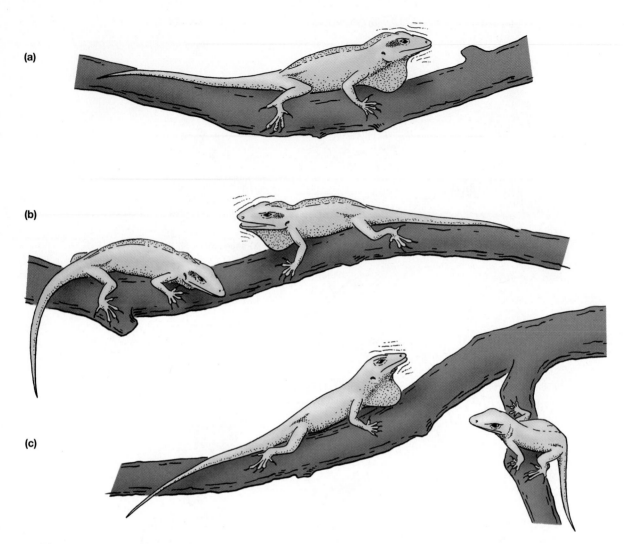

▲ **FIGURE 13–12** Displays by a male *Anolis carolinensis*. (a) Assertion-challenge display. (b) Aggressive posturing between males. (c) Courtship. Note the extension of the dewlap, the species-typical head bob, and the absence of the dorsal and nuchal crests and the eyespot.

Anolis are conspicuous to human observers, but do the lizards also rely on them for species identification? Indirect evidence suggests that the lizards probably do use gular fan color and display action patterns for species identification. For example, examination of communities of *Anolis* containing many species shows that the differences in colors of gular fans and in display action patterns are greatest for those species that encounter one another most frequently.

Studies of common species of lizard in western North America, the side-blotched lizard *Uta stansburiana*, have revealed a complex association between gular color and territorial and reproductive behavior. Male lizards have one of three colors in the gular region—blue, orange, or yellow. Males with blue throats are territorial, maintaining territories that over-

lap with the home ranges of one or more females and mating with those females. Males with orange throats are more aggressive than blue males; they do not maintain territories themselves, but displace blue-throated males from their territories and mate with the females. Yellow-throated males try to sneak into the territories of blue-throated males and steal a mating before they are chased away by the territorial male (Calsbeek et al. 2002). The throat color of a male is determined by the level of testosterone in its blood and is fixed during early development (Sinervo et al. 2000). The fitness relations among the forms—that is, which mating tactic is most effective—resembles a game of rock, paper, scissors. That is, yellow sneakers beat aggressive orange males, who beat territorial blue males, who beat yellow sneakers. The fitness of any

▶ **FIGURE 13–13** Territorial behavior. Normal sequence of behaviors for a territorial male *Anolis carolinensis* confronting an intruding male or female anole. A territorial male challenges any intruder, and the response of the intruder determines the subsequent behavior of the territorial male.

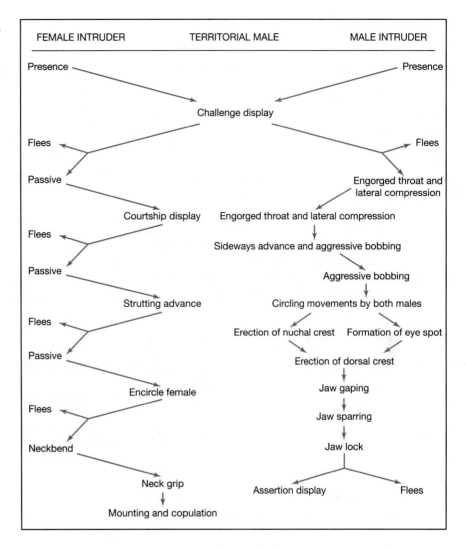

one of the color forms relative to the others depends on the proportions of the three forms in the population and also on the spatial distribution of the forms within a habitat. Blue-throated males that are neighbors of other blue-throated males have three times the fitness of blue-throated males with orange- or yellow-throated neighbors, and orange-throated males do best when they are not near another orange-throated male (Sinervo and Clobert 2003).

Pheromonal communication probably occurs in several lineages of lizards, primarily scleroglossans, although chemical cues may be more important for some iguanians than has been realized. Territorial male *Sceloporus* and other phrynosomatine lizards rub secretions from glands located on the posterior side of the femoral region onto objects in their territories. These secretions contain protein and sometimes lipids. Exploratory behavior by lizards, including iguanians, involves touching the tongue to the sub-

strate, and the vomeronasal organ may detect pheromones in the femoral gland secretions. In addition, the secretions absorb light strongly in the ultraviolet portion of the spectrum. Some lizards are sensitive to ultraviolet light, and the femoral gland secretions rubbed onto rocks and branches may be both visual and olfactory signals.

Territoriality, the relative importance of vision compared with olfaction, and foraging behavior appear to be broadly correlated among lizards. The elevated perches from which sit-and-wait predators survey their home ranges allow them to see both intruders and prey, and they dash from the perch to repel an intruder or to catch an insect. In contrast, widely foraging lizards are almost entirely nonterritorial, and olfaction is as important as vision in their foraging behavior. These lizards spend most of their time on the ground, where their field of vision is limited and they probably have little opportunity to detect intruders.

■ Reproduction

Squamates show a range of reproductive modes from **oviparity** (development occurs outside the female's body and is supported entirely by the yolk—i.e., lecithotrophy) to **viviparity** (eggs are retained in the oviducts and development is supported by transfer of nutrients from the mother to the fetuses—matrotrophy). Intermediate conditions include retention of the eggs for a time after they have been fertilized and the production of **precocial** young that were nourished primarily by material in the yolk. Oviparity is assumed to be the ancestral condition, and viviparity has evolved at least 45 times among lizards and 35 times among snakes. Viviparous squamates have specialized chorioallantoic placentae; in the Brazilian skink, *Mabuya heathi*, more than 99 percent of the mass of the fetus results from transport of nutrients across the placenta.

Viviparity is usually a high-investment reproductive strategy. Females of viviparous squamates generally produce relatively small numbers of large young, although there are exceptions to that generalization. Viviparity is not evenly distributed among lineages of squamates: nearly half the origins of viviparity in the group have occurred in the family Scincidae, whereas it is unknown in teiid lizards and occurs in only two genera of lacertids. Viviparity has advantages and disadvantages as a mode of reproduction. The most commonly cited benefit is the opportunity it provides for a female snake or lizard to use her own thermoregulatory behavior to control the temperature of the embryos during development. This hypothesis is appealing in an ecological context because a relatively short period of retention of the eggs by the female might substantially reduce the total amount of time required for development, especially in a cold climate.

Viviparity potentially lowers reproductive output because a female who is retaining one clutch of eggs cannot produce another. Lizards in warm habitats may produce more than one clutch of eggs in a season, but that is not possible for a viviparous species because development takes too long. In a cold climate, lizards are not able to produce more than one clutch of eggs in a breeding season anyway, and viviparity would not reduce the annual reproductive output of a female lizard. Phylogenetic analyses of the origins of viviparity suggest that it has evolved most often in cold climates, as this hypothesis predicts, but other origins appear to have taken place in warm climates; more than one situation favoring viviparity among squamates appears likely.

Viviparity has other costs. The agility of a female lizard is substantially reduced when her embryos are large. Experiments have shown that pregnant female lizards cannot run as fast as nonpregnant females and that snakes find it easier to capture pregnant lizards than nonpregnant ones. Females of some species of lizards become secretive when they are pregnant, perhaps in response to their vulnerability to predation. They reduce their activity and spend more time in hiding places. This behavioral adjustment may contribute to the reduction in body temperature seen in pregnant females of some species of lizards, and it probably reduces their rate of prey capture as well.

In general, large species of squamates produce more eggs or fetuses than do small species; and, within a single species, large individuals often have more offspring in a clutch than do small individuals. Both phylogenetic and ecological constraints play a role in determining the number of young produced, however. All geckos have a clutch size of either one or two eggs, and all *Anolis* produce only one egg at a time. Lizards with stout bodies usually have clutches that are a greater percentage of the mother's body mass than do lizards with slim bodies. The division between stout and slim bodies approximately parallels the division between sit-and-wait predators and widely foraging predators. It is tempting to infer that a lizard that moves about in search of prey finds a bulky clutch of eggs more hindrance than a lizard that spends 99 percent of its time resting motionless. However, because some of the divisions among modes of predatory behavior, body form, and relative clutch mass also correspond to the phylogenetic division between iguanian and scleroglossan lizards, it is not possible to decide which characteristics are ancestral and which may be derived.

■ Parthenogenesis

All-female (**parthenogenetic**) species of squamates have been identified in six families of lizards and one snake. The phenomenon is particularly widespread in the teiids (especially *Aspidoscelis*, [*Cnemidophorus*]) and lacertids (*Lacerta*) and occurs in several species of geckos. Parthenogenetic species are known or suspected to occur among chameleons, agamids, xantusiids, and typhlopids. However, parthenogenesis is probably more widespread among squamates than this list indicates because parthenogenetic species are not conspicuously different from bisexual species. Parthenogenetic species are usually detected when a study undertaken for an entirely different purpose reveals that a species contains no males. Confirmation of parthenogenesis can be obtained by obtaining fertile eggs from females raised in isolation or by making reciprocal skin grafts between individuals. Individuals of bisexual species usually reject tissues transplanted from another individual because genetic differences between them lead to immune reactions. Parthenogenetic species, however, produce progeny that are genetically identical to the mother, so no immune reaction occurs and grafted tissue is retained.

The chromosomes of lizards have allowed the events that produced some parthenogenetic species to be deciphered. Many parthenogenetic species appear

to have had their origin as interspecific hybrids. These hybrids are diploid ($2n$), with one set of chromosomes from each parental species. For example, the diploid parthenogenetic whiptail lizard, *Aspidoscelis tesselatus*, is the product of hybridization between the bisexual diploid species *A. tigris* and *A. septemvittatus* (**Figure 13–14**). Some parthenogenetic species are triploids ($3n$). These forms are usually the result of a backcross of a diploid parthenogenetic individual to a male of one of its bisexual parental species or, less commonly, the result of hybridization of a diploid parthenogenetic species with a male of a bisexual species different from

(a)

(b)

(c)

(d)

(e)

▲ **FIGURE 13–14** Unisexual lizards. The photographs show the species of whiptail lizards involved in a sequence of crosses leading to the formation of diploid and triploid unisexual species of *Aspidoscelis*. Hybridization of the bisexual diploid species (a) *A. tigris* and (b) *A. septemvittatus* produced a unisexual diploid form with half of its genetic complement derived from each of the parental species (an allodiploid). This parthenogenetic form is called *A. tesselatus* (c). Hybridization between a diploid *A. tesselatus* and a male of the bisexual species *A. sexlineatus* (d) produced a unisexual triploid form with its genetic complement derived from three different parental species (an allotriploid). This parthenogenetic triploid form is also called *A. tesselatus* (e). Thus *A. tesselatus* consists of clones of both diploid and triploid lineages, although taxonomists soon will probably treat these two forms as separate species.

its parental species. A parthenogenetic triploid form of *A. tesselatus* is apparently the result of a cross between the parthenogenetic diploid *A. tesselatus* and the bisexual diploid species *A. sexlineatus*.

It is common to find the two bisexual parental species and a parthenogenetic species living in overlapping habitats. Parthenogenetic species of *Aspidoscelis* often occur in habitats like the floodplains of rivers that are subject to frequent disruption. Disturbance of the habitat may bring together closely related bisexual species, fostering the hybridization that is the first step in establishing a parthenogenetic species. Once a parthenogenetic species has become established, its reproductive potential is twice that of a bisexual species because every individual of a parthenogenetic species is capable of producing young. Thus, when a flood or other disaster wipes out most of the lizards, a parthenogenetic species can repopulate a habitat faster than a bisexual species.

■ Parental Care

Parental care has been recorded for more than 100 species of squamates. A few species of snakes and a larger number of lizards remain with the eggs or nest site. Some female skinks remove dead eggs from the clutch. Some species of pythons brood their eggs: The female coils tightly around the eggs, and, in some species, muscular contractions of the female's body produce sufficient heat to raise the temperature of the eggs to about 30°C, which is substantially above air temperature. One unconfirmed report exists of baby pythons returning at night to their empty eggshells, where their mother coiled around them and kept them warm. Little interaction between adult and juvenile squamates has been documented. In captivity, female prehensile-tailed skinks (*Corucia zebrata*) have been reported to nudge their young toward the food dish, as if teaching them to eat. Prehensile-tailed skinks, which occur only on the Solomon Islands, are herbivorous and viviparous.

Free-ranging baby green iguanas have a tenuous social cohesion that persists for several months after they hatch. The small iguanas move away from the nesting area in groups that may include individuals from several different nests. One lizard may lead the way, looking back as if to see that others are following. The same individual may return later and recruit another group of juveniles. During the first three weeks after they hatch, juvenile iguanas move up into the forest canopy and are seen in close association with adults. During this time, the hatchlings probably ingest feces from the adults, thereby inoculating their guts with the symbiotic microbes that facilitate digestion of plant material. After their fourth week of life, the hatchlings move down from the forest canopy into low vegetation, where they continue to be found in loosely knit groups of two to six or more individuals that move, feed, and sleep together. This association might provide some protection from predators; and, if the hatchlings continue to eat fecal material, it is another opportunity to ensure that each lizard has received its full complement of gut microorganisms.

13.5 Behavioral Control of Body Temperatures by Ectotherms

The behavioral mechanisms involved in ectothermal temperature regulation are quite straightforward and are employed by insects, birds, and mammals (including humans), as well as by ectothermal vertebrates. Lizards, especially desert species, are particularly good at behavioral thermoregulation. Movement back and forth between sunlight and shade is the most obvious thermoregulatory mechanism they use. Early in the morning or on a cool day, lizards bask in the sun, whereas in the middle of a hot day they retreat to shade and make only brief excursions into the sun. Sheltered or exposed sites may be sought out. In the morning, when a lizard is attempting to raise its body temperature, it is likely to be in a spot protected from the wind. Later in the day, when it is getting too hot, the lizard may climb into a bush or onto a rock outcrop where it is exposed to the breeze and its convective heat loss is increased.

An animal can alter the amount of solar radiation it absorbs by changing its orientation to the sun, its body contour, and its skin color. Lizards use all of these mechanisms. An animal oriented perpendicular to the sun's rays intercepts the maximum amount of solar radiation, and one oriented parallel to the sun's rays intercepts minimum radiation. Lizards adjust their orientation to control heat gained by direct solar radiation. Many lizards can spread or fold their ribs to change the shape of the trunk. When the body is oriented perpendicular to the sun's rays and the ribs are spread, the surface area exposed to the sun is maximized and heat gain is increased. Compressing the ribs decreases the surface exposed to the sun and can be combined with orientation parallel to the rays to minimize heat gain. Horned lizards (*Phrynosoma*) provide a good example of this type of control. If the surface area that a horned lizard exposes to the sun directly overhead when the lizard sits flat on the ground with its ribs held in a resting position is considered to be 100 percent, the maximum surface area the lizard can expose by orientation and change in body contour is 173 percent, and the minimum is 28 percent. That is, the lizard can change its radiant heat gain more than six-fold solely by changing its position and body shape.

Color change can further increase a lizard's control of radiative exchange (see the color insert). Objects look dark because they are absorbing energy in the visible part of the solar spectrum, and the radiant energy they absorb warms them. The lightness or darkness of a lizard affects the amount of solar radiation it absorbs, and lizards can darken or lighten by moving dark pigment in their skin. Melanophores are cells that contain the pigment melanin. They are shaped rather like mushrooms, with a broad upper portion connected by a stalk to a lower section. When melanin granules are dispersed into the upper part of the cell, close to the skin surface, the skin appears dark; when the granules are drawn away from the surface into the lower section of the cell, the skin appears light. Lizards heat 10 percent to 75 percent faster when they are dark than they do when they are light.

Combining these mechanisms gives lizards a remarkable independence from air temperature. Lizards occur above the timberline in many mountain ranges, and they are capable of maintaining body temperatures 30°C or more above air temperature during their periods of activity on sunny days. While air temperatures are near freezing, these lizards scamper about with body temperatures as high as those of species that inhabit lowland deserts.

The repertoire of thermoregulatory mechanisms seen in lizards is greater than that of many other ectothermal vertebrates. Turtles, for example, cannot change their body contour or color, and their behavioral thermoregulation is limited to movements between sunlight and shade and in and out of water. Crocodilians are very like turtles, although young individuals make minor changes in body contour and color. Most snakes cannot change color, but rattlesnakes and boas lighten and darken as they warm and cool.

Activity Temperature Ranges

The extensive repertoire of thermoregulatory mechanisms employed by ectotherms allows many species of lizards and snakes to keep body temperature within a range of a few degrees during the active part of their day. Many species of lizards have body temperatures between 33°C and 38°C while they are active (the **activity temperature range**), and snakes often have body temperatures between 28°C and 34°C. This is the region of temperature in which an ectotherm carries out its full repertoire of activities—feeding, courtship, territorial defense, and so on.

These activity temperature ranges have been the focus of much research: Field observations show that thermoregulatory activities may occupy a considerable portion of an animal's time. Less obvious, but just as important, are the constraints that the need for thermoregulation sets on other aspects of the behavior and ecology of ectotherms. For example, some species of lizards and snakes are excluded from certain habitats because it is impossible to thermoregulate. In temperate regions, the activity season lasts only during the months when it is warm and sunny enough to permit thermoregulation; at other times of the year, snakes and lizards hibernate. Even during the activity season, time spent on thermoregulation may not be available for other activities. Avery (1976) proposed that lizards in temperate regions show less extensive social behavior than do tropical lizards because thermoregulatory behavior in cool climates requires so much time.

◾ Effects of Nutritional Status, Reproductive Status, and Bacterial Infections on Temperature Regulation

Several internal states of ectotherms influence body temperature. A thermophilic (Greek *thermo* = heat and *philo* = loving) response after feeding is widespread: Individuals with food in the gut maintain higher body temperatures than do individuals without food. A higher body temperature accelerates digestion and increases digestive efficiency and water uptake, so a warm animal digests its food more rapidly and assimilates a higher proportion of the energy and water present in the food. Conversely, fasting animals regulate their body temperatures at low levels that reduce their metabolic rates and conserve their stored energy.

Pregnancy affects thermoregulation by lizards. The rate of embryonic development is strongly affected by temperature, and a major advantage of viviparity is thought to be the opportunity it gives the mother to control the temperature of embryos during development. The body temperatures of female lizards during pregnancy may be different from the temperatures they would normally maintain. For example, pregnant female spiny swifts (*Sceloporus jarrovi*) had an average body temperature of 32.0°C, whereas male lizards in the same habitat had an average body temperature of 34.5°C (Beuchat 1986). The female lizards changed their thermoregulatory behavior after they had given birth, and the average body temperature of female lizards after they had given birth to their young was 34.5°C, like that of the males.

The low body temperatures of pregnant lizards were unexpected because giving birth as early in the year as possible would be advantageous for the lizards, and a high body temperature would speed the development of the embryos. That line of reasoning suggests that female lizards should maintain higher-than-normal body temperatures during pregnancy, or at least they should not reduce their body temperatures. Contrary to that prediction, however, the body temperatures maintained by pregnant lizards of several different species appear to converge toward 32°C, whether the normal body temperature for the species is higher or lower. Perhaps the body temperature of pregnant female lizards is a compromise between the thermal requirements of the mother and the best temperature for embryonic development (Beuchat and Ellner 1987).

Behavioral fever is another common response of ectotherms. Individuals infected by bacteria change their thermoregulatory behavior and maintain body temperatures several degrees higher than those of uninfected animals. These behavioral fevers have been demonstrated in arthropods, fishes, frogs, salamanders, turtles, and lizards. The release of prostaglandin E_1, which acts on thermoregulatory centers of the anterior hypothalamus, appears to be the immediate cause of both the behavioral fevers of ectotherms and the physiological fevers of endotherms. Survival is enhanced by fever because bacterial growth is limited by a reduction in the availability of iron at higher temperatures.

Physiological Control of the Rate of Change of Body Temperature

A new dimension was added to studies of ectothermal thermoregulation in the 1960s by the discovery that ectotherms can use physiological mechanisms to adjust their rate of temperature change. The original observations showed that several different kinds of large lizards were able to heat faster than they cooled when exposed to a 20°C difference between body and ambient temperatures. Subsequent studies by other investigators extended these observations to turtles

and snakes. From the animal's viewpoint, heating rapidly and cooling slowly prolongs the time it can spend in the normal activity range.

The basis of this control of heating and cooling rates lies in changes in peripheral circulation. Heating the skin of a lizard causes a localized expansion of dermal blood vessels (vasodilation) in the warm area. Dilation of the blood vessels, in turn, increases the blood flow through them, and the blood is warmed in the skin and carries the heat into the core of the body. Thus, in the morning, when a cold lizard orients its body perpendicular to the sun's rays and the sunlight warms its back, local vasodilation in that region speeds heat transfer to the rest of the body.

The same mechanism can be used to avoid overheating. The Galápagos marine iguana is a good example. Marine iguanas live on the bare lava flows on the coasts of the Galápagos Islands. In midday, beneath the equatorial sun, the black lava becomes extremely hot—uncomfortably if not lethally hot for a lizard. Retreat to the shade of the scanty vegetation or into rock cracks would be one way the iguanas could avoid overheating; however, the males are territorial, and those behaviors would mean abandoning their territories and probably having to fight for them again later in the day. Instead, a male iguana stays where it is, using physiological control of circulation and the cool breeze blowing off the ocean to form a heat shunt that absorbs solar energy on the dorsal surface, carries it through the body, and dumps it out the ventral surface.

The process is as follows: In the morning the lizard is chilled from the preceding night and basks to bring its body temperature to the normal activity range. When its temperature reaches this level, the lizard uses postural adjustments to slow the increase in body temperature, finally facing directly into the sun to minimize its heat load. In this posture, the forepart of the body is held off the ground (**Figure 13–15**). The ventral surface is exposed to the cool wind blowing off the ocean, and its body shades a patch of lava under the animal. This lava is soon cooled by the wind. Local vasodilation is produced by warming the blood vessels; it does not matter whether the heat comes from the outside (from the sun) or from inside (from warm blood). Warm blood circulating from the core of the body to the ventral skin warms it and produces vasodilation, increasing blood flow to the ventral surface. The lizard's ventral skin is cooler than the rest of its body because it is shaded and cooled by the wind. The warm blood heats the ventral skin, which loses heat by radiation

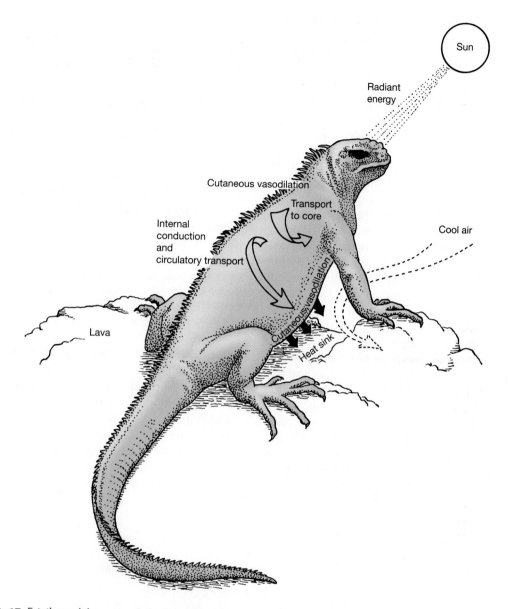

▲ **FIGURE 13–15** Ectothermal thermoregulation. The Galápagos marine iguana uses a combination of behavioral and physiological thermoregulatory mechanisms to shunt heat absorbed by its dorsal surface out its ventral surface.

to the cool lava in the shadow created by the lizard's body and by convection to the breeze. In this way, the same cardiovascular mechanism that earlier in the day allowed the lizard to warm rapidly is converted to a regulated heat shunt that rapidly transports solar energy from the dorsal to the ventral surface and keeps the lizard from overheating. In combination with postural adjustments and other behavioral mechanisms, such as the choice of a site where the breeze is strong, these physiological adjustments allow a male iguana to remain on station in its territory all day.

Organismal Performance and Temperature

Minimizing variation in body temperature greatly simplifies the coordination of biochemical and physiological processes. An organism's body tissues are the site of a tremendous variety of biochemical reactions, proceeding simultaneously and depending on one another to provide the proper quantity of the proper substrates at the proper time for reaction sequences to occur. Each reaction has a different sensitivity to temperature, and regulation is greatly facilitated when temperature variation is limited. Thus, coordination of internal processes may be a major benefit of thermoregulation for ectotherms. If the temperature stability that a snake or lizard achieves by thermoregulation is important to its physiology and biochemistry, you might expect that the internal economy of an animal functions best within its activity temperature range, and that is often the case. Examples of physiological processes that work best at temperatures within the activity range can be found at the molecular, tissue, system, and whole-animal levels of organization.

The wandering garter snake (*Thamnophis elegans vagrans*) provides examples of the effects of body temperature on a variety of physiological and behavioral functions (Stevenson et al. 1985). Wandering garter snakes are diurnal, semiaquatic inhabitants of lakeshores and stream banks in western North America. They hunt for prey on land and in water and feed primarily on fishes and amphibians. Chemosensation, accomplished by flicking the tongue, is an important mode of prey detection for snakes. Scent molecules are transferred from the tips of the forked tongue to the epithelium of the vomeronasal organ in the roof of the mouth. Garter snakes spend the night in shelters, where their body temperatures fall to ambient levels (4°C to 18°C), and emerge in the morning to bask. During activity on sunny days, the snakes maintain body temperatures between 28°C and 32°C.

Stevenson and his associates measured the effect of temperature on the speed of crawling and swimming, the frequency of tongue flicks, the rate of digestion, and the rate of oxygen consumption of the snakes (**Figure 13–16**). Crawling, swimming, and tongue flicking are elements of the foraging behavior of garter snakes, and the rates of digestion and oxygen consumption are involved in energy utilization. The ability of garter snakes to crawl and swim was severely limited at the low temperatures they experience during the night when they are inactive. At 5°C snakes often refused to crawl, and at 10°C they were able to crawl only 0.1 meter per second and could swim only 0.25 meter per second. The speed of both types of locomotion increased at higher temperatures. Swimming speed peaked near 0.6 meter per second at 25°C and 30°C, and crawling speed increased to an average of 0.8 meter per second at 35°C. The rate of tongue flicking increased from less than 0.5 flick per second at 10°C to about 1.5 flicks per second at 30°C. The rate of digestion increased slowly from 10°C to 20°C and more than doubled between 20°C and 25°C. It did not increase further at 30°C and dropped slightly at higher temperatures. The rate of oxygen consumption rose as temperature increased from 20°C to 35°C, which was the highest temperature tested because higher body temperatures would have been injurious.

All five measures of performance by garter snakes increased with increasing temperature, but the

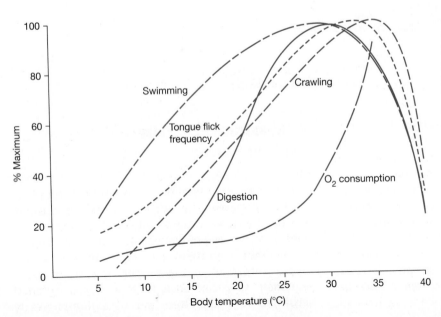

◀ **FIGURE 13–16** Effect of temperature on performance. The ability of a wandering garter snake, *Thamnophis elegans vagrans*, to perform many activities essential to survival depends on its body temperature. The vertical axis shows the percentage of maximum performance achieved at each temperature.

responses to temperature were not identical. For example, swimming speed did not increase substantially above 20°C, whereas crawling speed continued to increase up to 35°C. The rate of digestion peaked at 25°C to 30°C and then declined, but the rate of oxygen consumption increased steadily to 35°C. More striking than the differences among the functions, however, is the apparent convergence of maximum performance for all the functions on temperatures between 28.5°C and 35°C. This range of temperatures is close to the body temperatures of active snakes in the field on sunny days (28°C to 32°C). Anywhere within that range of body temperatures, snakes would be able to crawl, swim, and tongue-flick at rates that are at least 95 percent of their maximum rates.

The relationship between the body temperatures of active garter snakes and the temperature sensitivity of various behavioral and physiological functions reported by Stevenson and his colleagues is probably common for ectotherms. That is, in most cases, the body temperatures they maintain during activity are the temperatures that maximize organismal performance.

13.6 Temperature and Ecology of Squamates

As we discussed in Section 8.8, squamates, especially lizards, are capable of very precise thermoregulation, and the thermal environment can be a very important feature of their ecology. Microhabitats at which particular body temperatures can be maintained may be one of the dimensions that define the ecological niches of lizards. For example, the five most common species of *Anolis* on Cuba partition the habitat in several ways (**Figure 13–17**). First, they divide the habitat along the continuum, from sunny to shady: two species (*A. lucius* and *A. allogus*) occur in deep shade in forests, one (*A. homolechis*) in partial shade in clearings and at the forest edge, and two (*A. allisoni* and *A. sagrei*) in full sunlight. Within habitats in the sunlight-shade continuum, the lizards are separated by the substrates they use as perch sites. In the forest, *A. lucius* perches on large trees up to 4 meters above the ground, whereas *A. allogus* rests on small trees within 2 meters of the ground. *A. homolechis*, which does not share its habitat with another common species of *Anolis*, perches on both large and small trees. In open habitats, *A. allisoni* perches more than 2 meters above the ground on tree trunks and

houses, and *A. sagrei* perches below 2 meters on bushes and fence posts. Within this temperature and spatial distribution, the lizards have developed further specializations. Species that live on tree trunks near the ground have long hindlimbs and long tails, whereas species that live on twigs high in the canopy have short hindlimbs and short tails. The body proportions of the different forms appear to be related to locomotion on broad versus narrow surfaces (Irschick and Losos 1998, Irschick and Garland 2001). These relationships among thermal ecology, habitat, and body form have evolved repeatedly on different islands in the Caribbean (Losos 2001, Losos et al. 2003).

Some species of lizards do not thermoregulate. Lizards that live beneath the tree canopy in tropical forests often have body temperatures very close to air temperature (that is, they are thermally passive), whereas species that live in open habitats thermoregulate more precisely. The relative ease of thermoregulation in different habitats may be an important factor in determining whether a species of lizard thermoregulates or allows its temperature to vary with ambient temperature.

The distribution of sunny areas is one factor that determines the ease of thermoregulation. Sunlight penetrates the canopy of a forest in small patches that move across the forest floor as the sun moves across the sky. These patches of sunlight are the only sources of solar radiation for lizards that live at or near the forest floor, and the patches may be too sparsely distributed or too transient to be used for thermoregulation. In open habitats, sunlight is readily available, and thermoregulation is easier. The difference in thermoregulatory behavior of lizards in open and shaded habitats can be seen even in comparisons of different populations within a species. For example, *Anolis sagrei* occurs in both open and forest habitats on Abaco Island in the Caribbean. Lizards in open habitats bask and maintain body temperatures between 32°C and 35°C from about 8:30 in the morning through about 5:00 in the afternoon. Lizards in the forest do not bask, and their body temperatures vary from a low of 24°C to a high of 28°C over the same period.

The task of integrating thermoregulatory behavior with foraging is relatively simple for sit-and-wait foragers such as *Anolis*. These lizards can readily change their balance of heat gain and loss by making small movements in and out of shade or between calm and breezy perch sites, while continuing to scan their surroundings for prey. Widely foraging species may have more difficulty integrating thermoregulation

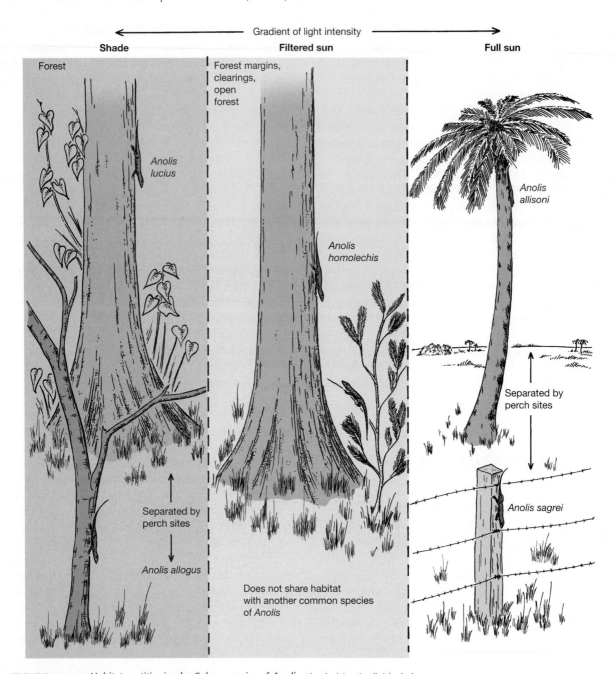

▲ **FIGURE 13–17** Habitat partitioning by Cuban species of *Anolis*. The habitat is divided along a gradient from shaded forest to open pastures and a vertical gradient in perch height.

and predation. They are continuously moving between sunlight and shade and in and out of the wind, and their body temperatures are affected by their foraging activity. These lizards sometimes have to stop foraging to thermoregulate, resuming foraging only when they have warmed or cooled enough to return to their activity temperature range.

Body size is yet another variable that can affect thermoregulation. An example of the interaction of body size, thermoregulation, and foraging behavior is provided by three species of teiid lizards (*Ameiva*) in Costa Rica. *Ameiva* are widely foraging predators that move through the habitat, pushing their snouts beneath fallen leaves and into holes. Three species of *Ameiva* occur together on the Osa Peninsula of Costa Rica in a habitat that extends from full sunlight (a roadside) to deep shade (forest). The largest of the three species, *A. leptophrys*, has an average body mass of 83 grams; the middle species, *A. festiva*, weighs approximately 32 grams; and the smallest, *A. quadrilineata*, on average weighs 10 grams (**Figure 13–18**). The three species forage in different parts of the habitat: *A. quadrilineata* spends most of its time in the short vegetation at the edge of the road, *A. festiva* is found on the bank beside the road, and *A. leptophrys* forages primarily beneath the forest canopy (**Figure 13–19**). The different foraging sites of the three species may reflect differences in thermoregulation that result from the variation in body size.

The thermoregulatory behavior of the three species is the same: A lizard basks in the sunlight until its body temperature rises to 39°C to 40°C, and then moves through the mosaic of sunlight and shade as it searches for food. The body temperature of the lizard drops as it forages, and a lizard ceases foraging and resumes basking when its body temperature has fallen to 35°C. Thus the time that a lizard can forage depends on how long it takes for its body temperature to cool from 39°C or 40°C to 35°C. The rate of cooling for *Ameiva* in the shade is inversely proportional to the body size of the three

species: *A. quadrilineata* cools in 4 minutes, *A. festiva* in 6 minutes, and *A. leptophrys* in 11 minutes (**Figure 13–20**). That relationship appears to explain part of the microhabitat separation of the three species: The smallest species, *A. quadrilineata*, cools so rapidly that it may not be able to forage effectively in shady microhabitats, whereas *A. leptophrys* cools slowly and can forage in the shade beneath the forest canopy. The species of intermediate body size, *A. festiva*, uses the habitat with an intermediate amount of shade.

The slow rate of cooling of *A. leptophrys* may explain why it is able to forage in the shade. However, field observations indicate that its foraging is actually restricted to shade; it emerges from the forest only to bask. Does some environmental factor prevent *A. leptophrys* from foraging in sunny areas? The answer to that question may lie in the way the body temperatures of the three species increase when they are in open microhabitats. Body size profoundly affects the equilibrium temperature of an organism in the sunlight. A lizard warms by absorbing solar radiation; as it gets warmer, its heat loss by

(a)

(b)

(c)

▲ **FIGURE 13–18** Three sympatric species of *Ameiva* from Costa Rica. (a) *Ameiva leptophrys*, which has an average adult mass of 83 grams. (b) *Ameiva festiva*, average adult mass 32 grams. (c) *Ameiva quadrilineata*, average adult mass 10 grams.

▶ **FIGURE 13–19** Foraging sites of three species of *Ameiva* in Costa Rica. The histograms show the number of individuals of the three species seen in each of six locations: (A) a small clearing in the forest; (B) immediately inside the forest edge; (C) outside the edge of the forest; (D) midway between the edge of the forest and open area; (E) low vegetation beside a road; and (F) low vegetation in a large open area without trees.

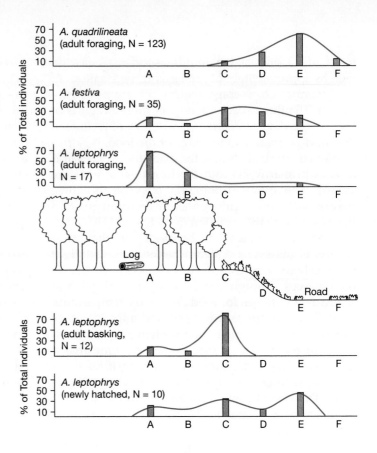

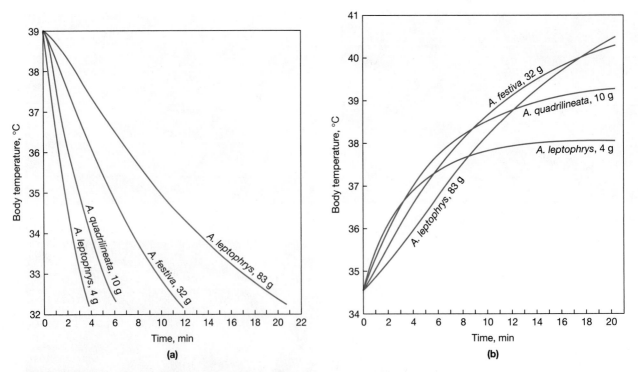

▲ **FIGURE 13–20** Cooling and heating rates of the three sympatric species of *Ameiva*. The largest species, *A. leptophrys*, heats and cools more slowly than the smaller species. If it remained in the sunlight, its body temperature would rise above 40°C. The smaller species heat and cool more rapidly than the large species and reach temperature equilibrium at lower body temperatures.

convection, evaporation, and reradiation also increases. When the rate of heat loss equals the rate of heat gain, the body temperature does not increase further. Large lizards reach that equilibrium at higher body temperatures than do small ones. Computer simulations of the heating rates of the three *Ameiva* in sunlight showed that *A. quadrilineata* and *A. festiva* would reach equilibrium at body temperatures of 37°C to 40°C, but *A. leptophrys* would continue to heat until its body temperature reached a lethal 45°C (see **Figure 13–20b**). This analysis suggests that *A. leptophrys* would die of heat stress if it spent more than a few minutes in a sunny microhabitat, but that the two smaller species of *Ameiva* would not have that problem.

Thus, as a result of the biophysics of heat exchange, the large body size of *A. leptophrys* apparently allows it to forage in shaded habitats (because it cools slowly) but prevents it from foraging in sunny habitats (because it would overheat). Field observations of the foraging behavior of hatchling *A. leptophrys* emphasize the importance of heat exchange in the foraging behavior of these lizards. Hatchling *A. leptophrys* forage in open habitats like *A. festiva* and *A. quadrilineata* rather than under the forest canopy like adult *A. leptophrys* (see Figure 13–19). That is, the juveniles of the large species of *Ameiva* behave like adults of the smaller species, probably because of the importance of body size and heat exchange in determining the microhabitats in which lizards can thermoregulate.

The difference in the use of various microhabitats by these three species of lizards looks, at first glance, like an example of habitat partitioning in response to interspecific competition for food. That is, because all three species eat the same sort of prey, they could be expected to concentrate their foraging efforts in different microhabitats to reduce competition. However, this analysis of the thermal requirements of the lizards suggests that interspecific competition for food is, at most, a secondary factor.

If competition for food were important, we would not expect to find hatchling *A. leptophrys* foraging in the same microhabitat as adult *A. quadrilineata*, because the similarity in size of the two lizards would intensify competition for food. The hypothesis that competition for food determines the microhabitat distribution of the animals predicts that the forms most similar in body size should be widely separated in the habitat. In contrast, the hypothesis that energy exchange with the environment is critical in determining where a lizard can forage predicts that species of similar size will live in the same habitat, and that is approximately the pattern seen. Apparently, the physical environment (radiant energy) is more important than the biological environment (interspecific competition for food) in determining the microhabitat distributions of these lizards. That conclusion reflects the broad-scale ecological significance of the morphological and physiological differences between ectotherms and endotherms, a theme that is developed in the next chapter.

Summary

The extant lepidosaurs include the squamates (lizards and snakes) and their sister group, the Sphenodontidae. The lepidosaurs, with more than 8000 species, form the second largest group of extant tetrapods. The two species of tuatara of the New Zealand region are the sole extant sphenodontids. They are lizardlike animals, about 60 centimeters long, with a dentition and jaw mechanism that give a shearing bite. Sphenodontids were diverse in the Mesozoic and included terrestrial insectivorous and herbivorous species as well as a marine form.

Lizards range in size from tiny geckos less than 3 centimeters long to the Komodo monitor lizard, which reaches a length of 3 meters. The Iguania is composed mainly of stout-bodied lizards with sturdy legs and considerable diversity of body form. Most Scleroglossa are elongate, and leglessness has developed independently many times within this lineage.

Differences in ecology and behavior parallel the phylogenetic divisions. Many iguanians are sit-and-wait predators that maintain territories and detect prey and intruders by vision. Iguanians often employ colors and patterns in visual displays during courtship and territorial defense. Many scleroglossan lizards are widely foraging predators that detect prey by olfaction and do not maintain territories. Pheromones are important in the social behaviors of many of these lizards.

Amphisbaenians are specialized burrowing lizards. Their skulls are solid structures that they use for tunneling through soil. Many amphisbaenians have blunt heads, and others have vertically keeled or horizontally spade-shaped snouts. The dentition of amphisbaenians appears to be specialized for nipping small pieces from prey too large to be swallowed whole. The skin of amphisbaenians is loosely attached to the trunk, and amphisbaenians slide backward or forward inside the tube of their skin as they move through tunnels with concertina locomotion.

Snakes are derived from scleroglossan lizards. Repackaging the body mass of a vertebrate into a serpentine form has been accompanied by specializations of the mechanisms of locomotion (serpentine, rectilinear, concertina, and sidewinding), prey capture (constriction and the use of venom), and swallowing (a highly kinetic skull).

Many squamates have complex social behaviors associated with territoriality and courtship, but parental care is only slightly developed. Viviparity has evolved 80 or more times among squamates. Thermoregulation is another important behavior of squamates, and various activities are influenced by body temperature. The ecological niches of some lizards may be defined in part by the microhabitats needed to maintain particular body temperatures.

Additional Readings

Andrews, R. M., and B. R. Rose. 1994. Evolution of viviparity: Constraints on egg retention. *Physiological Zoology* 67:1006–1024.

Arnold, E. N. 1988. Caudal autotomy as a defense. In C. Gans and R. B. Huey (eds.), *Biology of the Reptilia*, Vol. 16. New York: Alan Liss, 235–273.

Avery, R. A. 1976. Thermoregulation, metabolism, and social behaviour in Lacertidae. In A. d'A. Bellairs and C. B. Cox (eds.), *Morphology and Biology of Reptiles*. London, UK: Academic Press, 245–259.

Beuchat, C. A. 1986. Reproductive influence on the thermoregulatory behavior of a live-bearing lizard. *Copeia* 1986: 971–979.

Beuchat, C. A., and S. Ellner. 1987. A quantitative test of life history theory: Thermoregulation by a viviparous lizard. *Ecological Monographs* 57:45–60.

Calsbeek, R. et al. 2002. Sexual selection and alternative mating behaviours generate demographic stochasticity in small populations. *Proceedings of the Royal Society of London* B 269:157–164.

Coates, M. and M. Ruta. 2000. Nice snakes, shame about the legs. *Trends in Ecology and Evolution* 15:503–507.

Daugherty, C. H. et al. 1990. Neglected taxonomy and continuing extinctions of tuatara (*Sphenodon*). *Nature* 347:177–179.

Greene, H. 1997. *Snakes*. Berkeley, CA: University of California Press.

Greene, H. W., and D. Cundall. 2000. Limbless tetrapods and snakes with legs. *Science* 287:1939–1941.

Hardy, D. L., Sr. 1994. A reevaluation of suffocation as the cause of death during constriction. *Herpetological Review* 25:45–47.

Huey, R. B. and A. F. Bennett. 1986. A comparative approach to field and laboratory studies in evolutionary biology. Feder, M. E. and G. V. Lauder (eds.), *Predator-Prey Relationships*, Chicago, IL: University of Chicago Press, 82–98.

Irschick, D. J., and T. Garland, Jr. 2001. Integrating function and ecology in studies of adaptation: Investigations of locomotor capacity as a model system. *Annual Review of Ecology and Systematics* 32:367–396.

Irschick, D. J., and J. B. Losos. 1998. A comparative analysis of the ecological significance of locomotor performance in Caribbean *Anolis* lizards. *Evolution* 52:219–226.

Kearney. M. and B. L. Stuart. 2004. Repeated evolution of limblessness and digging heads in worm lizards revealed by DNA from old bones. *Proceedings of the Royal Society of London* B 271:1677-1683.

Losos, J. B. 2001. Evolution: A lizard's tale. *Scientific American* (March 2001):64–69.

Losos, J. B. et al. 2003. Niche lability in the evolution of a Caribbean lizard community. *Nature* 424:542–545.

Owerkowicz, T. et al. 1999. Contribution of gular pumping to lung ventilation in monitor lizards. *Science* 284:1661–1663.

Rieppel, O., and M. Kearney. 2001. The origin of snakes: Limits of a scientific debate. *Biologist* 48:110–114.

Sinervo, B., and J. Clobert. 2003. Morphs, dispersal behavior, genetic similarity, and the evolution of cooperation. *Science* 300:1949–1951.

Sinervo, B. et al. 2000. Testosterone, endurance, and Darwinian fitness: Natural and sexual selection on the physiological basis of alternative male behaviors in side-blotched lizards. *Hormones and Behavior* 38:222–233.

Stevenson, R. D. et al. 1985. The thermal dependence of locomotion, tongue flicking, digestion and oxygen consumption in the wandering garter snake. *Physiological Zoology* 58:46–57.

Sweet, S. S., and E. R. Pianka. 2003. The lizard kings. *Natural History* 112(9):40–45 (November 2003).

Vidal, N., and S. B. Hedges. 2004. Molecular evidence for a terrestrial origin of snakes. *Proceedings of the Royal Society of London* B (Supplement) 271:S226–S229.

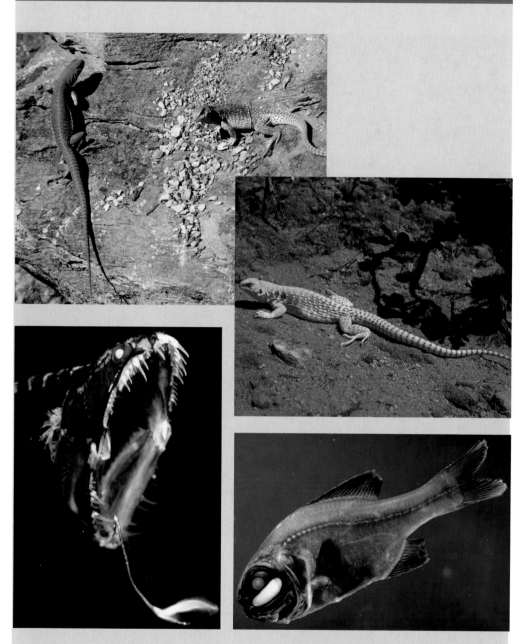

Color change is a temperature-regulating mechanism used by lizards such as the desert iguana (*Dipsosaurus dorsalis*). When they first emerge in the morning, desert iguanas are dark (upper left). By the time it has reached its activity temperature, a lizard has turned light (upper right). This color change reduces heat gained from the sun by 23 percent.

Luminescent bacteria in the light organs of fishes emit light as a by-product of their metabolism. (Lower left) A black dragonfish, *Idiacanthus*. A long barbel on the chin bears a luminous lure that is believed to attract prey close enough to be engulfed by the enormous jaws lined with sharp teeth. (Lower right) The flashlight fish, *Photoblepharon*, has a light-emitting organ under each eye. The fish can cover the organ with a pigmented shutter to conceal the light, or open the shutter to reveal it. It uses the light organ in social interactions with other flashlight fish, and in a blink-and-run defense to startle and confuse predators.

Three species of salamanders form a mimicry complex in eastern North America. The red eft (*Notophthalmus viridescens*, top left) and red salamander (*Pseudotriton ruber*, top right) have skin toxins that deter predators. The red-backed salamander (*Plethodon cinereus*, middle left) is not protected by toxins, but predators confuse the erythristic form of that species (middle right) with the toxic species. The experiment described in the text used the mountain dusky salamander (*Desmognathus ochrophaeus*, bottom left) as a palatable control.

The gular fans of lizards are used in social displays. Color, size, and shape identify the species and sex of an individual. (All the lizards in these photographs are males.) (Top left) Carolina anole, *Anolis carolinensis*, from Florida. (Top right) knight anole, *Anolis equestris*, from Cuba. (Bottom left) *Anolis grahami* from Jamaica. (Bottom right) *Anolis chrysolepis* from Brazil.

Amur tiger

Indo-Chinese tiger

Malaysian tiger

South China tiger

Bengal tiger

Sumatran tiger

Less than a century ago there were nine subspecies of tigers that ranged from Turkey across Asia to the Pacific Ocean, but only six of those forms now survive. They are genetically distinct and differ greatly in body size—an adult male Amur tiger weighs 300 kilograms, whereas adult males of the two island forms, the Sumatran and Malaysian tigers, weigh only 120 kilograms. The genetic relationships of the subspecies are shown in Figure 25–13 on page 682.

CHAPTER 14

Ectothermy: A Low-Cost Approach to Life

Ectothermy is an ancestral character of vertebrates; but, like many ancestral characters, it is just as effective as its derived counterpart, endothermy. Furthermore, the mechanisms of ectothermal thermoregulation are as complex and specialized as those of endothermy. Here we consider the consequences of ectothermy in shaping broader aspects of the lifestyle of fishes, amphibians, and reptiles. The general conclusion from this examination is that success in difficult environments is as likely to reflect the ancestral features of a group as its derived characters.

14.1 Vertebrates and Their Environments

Vertebrates manage to live in the most unlikely places. Amphibians live in deserts where rain falls only a few times a year and several years may pass with no rainfall at all. Lizards live on mountains at altitudes above 4000 meters, where the temperature falls below freezing nearly every night of the year and does not rise much above freezing during the day.

Of course, vertebrates do not seek out only inhospitable places to live—birds, lizards, mammals, and even amphibians can be found on the beaches at Malibu (sometimes running between the feet of surfers), and fishes cruise the shore. However, even this apparently benign environment is harsh for some animals. Examining the ways that vertebrates live in extreme environments has provided much information about

how they function as organisms; that is, how morphology, physiology, ecology, and behavior interact.

In some cases, elegant adaptations allow specialized vertebrates to colonize demanding habitats. More common and more impressive than these specializations, however, is the realization of how minor are the modifications of the ancestral vertebrate body plan that allow animals to endure environmental temperatures from –70°C to +70°C or water conditions ranging from complete immersion in water to complete independence of liquid water. No obvious differences distinguish animals from vastly different habitats—an Arctic fox looks very much like a desert fox, and a lizard from the Andes Mountains looks like one from the Atacama Desert. The adaptability of vertebrates lies in the combination of minor modifications of their ecology, behavior, morphology, and physiology. A view that integrates these elements shows the startling beauty of organismal function of vertebrates.

14.2 Dealing with Dryness— Ectotherms in Deserts

Deserts are produced by various combinations of topography, air movements, and ocean currents and are found from the poles to the equator. But whatever their cause, deserts have in common a scarcity of liquid water. A desert is defined as a region in which the potential loss of water (via evaporation and transpiration of water by plants) exceeds the input of water via precipitation. Dryness is at the root of many features of deserts that make them difficult places for vertebrates to live. The dry air characteristic of most deserts seldom contains enough moisture to form clouds that would block solar radiation during the day or radiative cooling at night. As a result, the daily temperature excursion in deserts is large compared with that of more humid areas. Scarcity of water is reflected by sparse plant life. Without plants to eat, deserts have low densities of insects for small vertebrates to eat and correspondingly sparse populations of small vertebrates that would be prey for larger vertebrates. Food shortages are chronic and are worsened by seasonal shortages and by unpredictable years of drought when the usual pattern of rainfall does not develop.

Not all deserts are hot; indeed, some are distinctly cold—most of Antarctica and the region of Canada around Hudson Bay and the Arctic Ocean are deserts. The low-latitude deserts north and south of the equator are hot deserts, however, and it is the combination of heat and dryness in low-latitude deserts that creates the most difficult problems.

The scarcity of rain that contributes to the low primary production of deserts also means that sources of liquid water for drinking are usually unavailable to small animals that cannot travel long distances. These animals obtain water from the plants or animals they eat, but plants and insects have sodium and potassium concentrations that differ from those of vertebrates. In particular, potassium is found in higher concentrations in both plants and insects than it is in vertebrate tissues, and excreting the excess potassium can be difficult if water is too scarce to waste in the production of large quantities of urine.

The low metabolic rates of ectotherms alleviate some of the difficulty caused by scarcity of food and water, but many desert ectotherms must temporarily extend the limits within which they regulate body temperatures or body fluid concentrations, become inactive for large portions of the year, or adopt a combination of these responses.

■ Terrestrial Ectotherms

Terrestrial habitats in deserts are often harsh—hot by day and cold at night. Solar radiation is intense, and air does not conduct heat rapidly. As a result, a sunlit patch of ground can be lethally hot, whereas a shaded area just a few centimeters away can be substantially cooler. Underground retreats offer protection from both heat and cold. The annual temperature extremes at the surface of the ground in the Mohave Desert extend from a low that is below freezing to a maximum above 50°C; but, just 1 meter below the surface of the ground, the annual temperature range is only from 10°C to 25°C. Desert animals rely on the temperature differences between sunlight and shade and between the surface and underground burrows to escape both hot and cold.

The Desert Tortoise The largest ectothermal vertebrates in the deserts of North America are tortoises. The Bolson tortoise (*Gopherus flavomarginatus*) of northern Mexico probably once reached a shell length of a meter, although predation by humans has apparently prevented any tortoise in recent times from living long enough to grow that large. The desert tortoise (*G. agassizii*) of the southwestern United States is smaller than the Bolson tortoise, but it is still an impressively large turtle (**Figure 14–1**).

(a)

(b)

▲ **FIGURE 14–1** The desert tortoise, *Gopherus agassizii.* (a) An adult tortoise. (b) A tortoise approaching its burrow.

Adults can reach shell lengths approaching 50 centimeters and may weigh 5 kilograms or more. A study of the annual water, salt, and energy budgets of desert tortoises in Nevada shows the difficulties they face in that desert habitat (Nagy and Medica 1986).

Tortoises in the Mohave Desert construct shallow burrows that they use as daily retreat sites during the summer and deeper burrows for hibernation in winter. The tortoises in the study area emerged from hibernation in spring, and aboveground activity extended through the summer until they began hibernation again in November. Doubly labeled water was used to measure the energy expenditure of free-ranging tortoises (**Box 14–1**).

BOX 14-1   Doubly Labeled Water

The technique of doubly labeling water is widely employed in studies of the energy consumption of wild vertebrates, because it is the only method of measuring metabolism without restraining the animal in some way. The method measures carbon dioxide production, which in turn can be used to estimate oxygen consumption. *Doubly labeled* refers to water that carries isotopic forms of oxygen and hydrogen: the oxygen atom has been replaced with its stable isotope oxygen-18 (^{18}O) and one hydrogen atom (H) has been replaced by its radioactive form, tritium (^{3}H). (Doubly labeled water is prepared by mixing appropriate quantities of water containing the hydrogen isotope [^{3}HHO] with water containing the oxygen isotope [$H_2{}^{18}O$]. When this mixture is diluted by the body water of an animal, the concentrations of the isotopes are so low that few individual water molecules contain both isotopes.) A measured amount of doubly labeled water is injected into an animal and allowed to equilibrate with the body water for several hours (**Figure 14–2a**), and then a small blood sample is withdrawn. The concentrations of tritium and oxygen-18 in the blood are measured, and the total volume of body water can be calculated from the dilution of the doubly labeled water that was injected.

After that first blood sample has been withdrawn, the animal is released and recaptured at intervals of several days or weeks. Each time the animal is recaptured, a blood sample is taken and the concentrations of tritium and oxygen-18 are measured. The calculation of the amount of carbon dioxide produced by the animal is based on the difference in the rates of loss of tritium and oxygen-18. Tritium, which behaves chemically like hydrogen, is lost as water—that is, as ^{3}HHO—whereas oxygen-18 is lost both as water ($H_2{}^{18}O$) and as carbon dioxide ($C^{18}OO$). Thus, the decline in the concentration of tritium in the blood of the animal is a measure of the rate of water loss, and the decline in the concentration of oxygen-18 is a measure of the rates of loss of carbon dioxide and water (**Figure 14–2b**). The difference between the decreases in concentrations of tritium and oxygen-18, therefore, is the rate of loss of carbon dioxide, and this is proportional to the rate of oxygen consumption. A full description of the use of doubly labeled water for metabolic studies and a summary of measurements of metabolic rates of free ranging animals can be found in Nagy (1983) and Nagy et al. (1999).

▶ **FIGURE 14–2** Doubly labeled water. (a) The reaction of carbon dioxide and water to produce carbonic acid, and the subsequent reconversion of carbonic acid to water and carbon dioxide produce an equilibrium of ^{18}O between H_2O and CO_2 when water labeled with the isotope is injected into a vertebrate. (b) Differential washout of hydrogen and oxygen isotopes in the body water of an animal that has been injected with doubly labeled water.

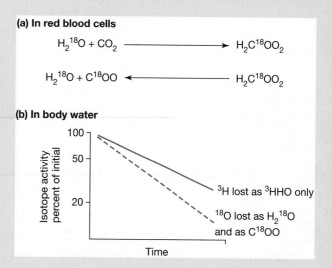

(a) In red blood cells

$$H_2{}^{18}O + CO_2 \longrightarrow H_2C^{18}OO_2$$

$$H_2{}^{18}O + C^{18}OO \longleftarrow H_2C^{18}OO_2$$

(b) In body water

Isotope activity percent of initial

^{3}H lost as ^{3}HHO only

^{18}O lost as $H_2{}^{18}O$ and as $C^{18}OO$

Time

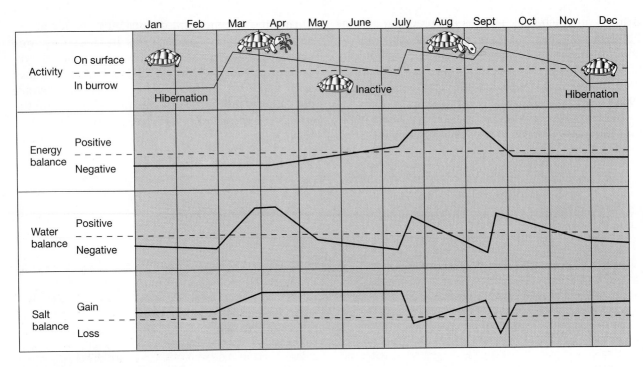

▲ **FIGURE 14–3** Annual cycle of desert tortoises.

Figure 14–3 shows the annual cycle of time spent above ground and in burrows by the tortoises and the annual cycles of energy, water, and salt balance. A positive balance means that the animal shows a net gain, whereas a negative balance represents a net loss. Positive energy and water balances indicate that conditions are good for the tortoises, but a positive salt balance means that ions are accumulating in the body fluids faster than they can be excreted. That situation indicates a relaxation of homeostasis and is probably stressful for the tortoises. The figure shows that the animals were often in negative balance for water or energy, and they accumulated salt during much of the year. Examination of the behavior and dietary habits of the tortoises through the year shows what was happening.

After they emerged from hibernation in the spring, the tortoises were active for about 3 hours every fourth day; the rest of the time they spent in their burrows. From March through May the tortoises were eating annual plants that had sprouted after the winter rains. They obtained large amounts of water and potassium from this diet, and their water and salt balances were positive. Desert tortoises lack salt glands and their kidneys cannot produce concentrated urine, so they have no way to excrete the salt without losing a substantial amount of water in the process. Instead they retain the salt,

and the osmolality of the tortoises' body fluids increased by 20 percent during the spring. This increased concentration shows that they were osmotically stressed as a result of the high concentrations of potassium in their food. Furthermore, the energy content of the plants was not great enough to balance the metabolic energy expenditure of the tortoises, and they were in negative energy balance. During this period, the tortoises were using stored energy by metabolizing their body tissues.

As ambient temperatures increased from late May through early July, the tortoises shortened their daily activity periods to about 1 hour every sixth day. The rest of the time the tortoises spent in shallow burrows. The annual plants died, and the tortoises shifted to eating grass and achieved positive energy balances. They stored this extra energy as new body tissue. The dry grass contained little water, however, and the tortoises were in negative water balance. The osmolal concentrations of their body fluids remained at the high levels they had reached earlier in the year.

In mid-July, thunderstorms dropped rain on the study site, and most of the tortoises emerged from their burrows. They drank water from natural basins, and some of the tortoises constructed basins by scratching shallow depressions in the ground that could then trap rainwater. The tortoises drank large quantities of water (nearly 20 percent of their body mass) and voided the contents of their urinary bladders. The osmolal concentrations of their body fluids and urine decreased as they moved into positive water balance and excreted the excess salts they had accumulated when water was scarce. The behavior of the tortoises changed after the rain: they ate every two or three days and often spent their periods of inactivity above ground instead of in their burrows.

August was dry, and the tortoises lost body water and accumulated salts as they fed on dry grass. They were in positive energy balance, however, and their body tissue mass increased. More thunderstorms in September allowed the tortoises to drink again and to excrete the excess salts they had been accumulating. Seedlings sprouted after the rain, and in late September the tortoises started to eat them.

In October and November the tortoises continued to feed on freshly sprouted green vegetation; however, low temperatures reduced their activity, and they were in slightly negative energy balance. Salts accumulated and the osmolal concentrations of the body fluids increased slightly. In November, the tortoises entered hibernation. Hibernating tortoises had low metabolic rates and lost water and body tissue mass slowly. When they emerged from hibernation the following spring, they weighed only a little less than they had in the fall. Over the entire year, the tortoises increased their body tissues by more than 25 percent and balanced their water and salt budgets, but they did this by tolerating severe imbalances in their energy, water, and salt relations for periods that extended for several months at a time.

The Chuckwalla The chuckwalla (*Sauromalus obesus*) is an herbivorous iguanine lizard that lives in the rocky foothills of desert mountain ranges (**Figure 14–4**). The annual cycle of the chuckwallas, like that of the desert tortoises, is shaped by the availability of water. The lizards face many of the same problems that the tortoises encounter, but their responses are different. The lizards have nasal glands that allow them to excrete salt at high concentrations, and they do not drink rainwater but instead depend on water they obtain from the plants they eat.

Two categories of water are available to an animal from the food it eats: free water and metabolic water. Free water corresponds to the water content of the food—that is, molecules of water (H_2O) that are absorbed across the wall of the intestine. Metabolic water is a by-product of the chemical reactions of metabolism. Protons are combined with oxygen during aerobic metabolism, yielding a molecule of water for every two protons. The amount of metabolic water produced can be substantial; more than a gram of water is released by metabolism of a gram of fat (**Table 14.1**). For animals like the chuckwalla that do not drink liquid water, free water and metabolic water are the only routes of water gain that can replace the water lost by evaporation and excretion.

Chuckwallas were studied at Black Mountain in the Mohave Desert of California (Nagy 1972, 1973). They spent the winter hibernating in rock crevices and emerged from hibernation in April. Individual lizards spent about 8 hours a day on the surface in

▶ **FIGURE 14–4** Chuckwalla, *Sauromalus obesus*.

TABLE 14.1 Quantity of water produced by metabolism of different substrates

Compound	Grams of Water/Gram of Compound
Carbohydrate	0.556
Fat	1.071
Protein	0.396 when urea is the end product; 0.499 when uric acid is the end product

April and early May (**Figure 14–5**). By the middle of May, air temperatures were rising above 40°C, and the chuckwallas retreated into rock crevices for about 2 hours during the hottest part of the day, emerging again in the afternoon. At this time of year, annual plants that sprouted after the winter rains supplied both water and nourishment, and the chuckwallas gained weight rapidly. The average increase in body mass between April and mid-May was 18 percent (**Figure 14–6**). The water content of the chuckwallas increased faster than the total body mass, indicating that they were storing excess water.

By early June, the annual plants had withered, and the chuckwallas were feeding on perennial plants that contained less water and more ions than the annual plants. Both the body masses and the water contents of the lizards declined. The activity of the lizards decreased in June and July: Individual lizards emerged in the morning or in the afternoon but not at both times. In late June, the chuckwallas reduced their feeding activity; in July, they stopped

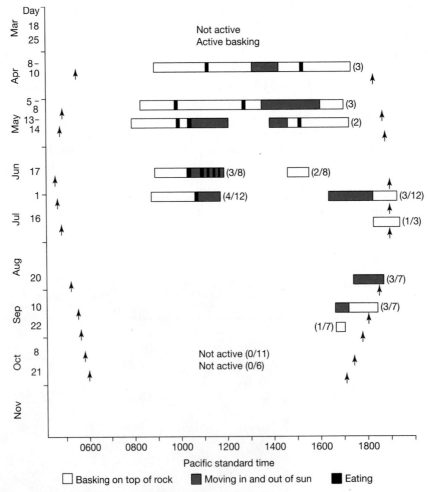

▲ **FIGURE 14–5** Daily behavior patterns in chuckwallas. Arrows indicate sunrise and sunset. Numbers in parentheses for April and May indicate the number of animals whose behavior was recorded. Thereafter the fraction in parentheses indicates the number of lizards active and observed out of the number known to be present.

▶**FIGURE 14–6** Seasonal changes in body composition of chuckwallas. The total body water is composed of the extracellular fluid (ECF; blood plasma, urine, and water in lymph sacs) and the intracellular fluid (ICF; the water inside cells).

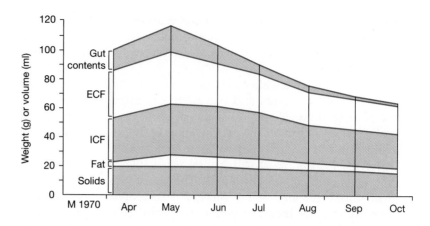

eating altogether. They spent most of the day in the rock crevices, emerging only in late afternoon to bask for an hour or so every second or third day. From late May through autumn, the chuckwallas lost water and body mass steadily, and in October they weighed an average of 37 percent *less* than they had in April when they emerged from hibernation.

The water budget of a chuckwalla weighing 200 grams is shown in **Table 14.2**. In early May, the annual plants it is eating contain more than 2.5 grams of free water per gram of dry plant material, and the lizard shows a positive water balance, gaining about 0.8 gram of water per day. By late May, when the plants have withered, their free water content has dropped to just under a gram of water per gram of dry plant matter, and the chuckwalla is losing about 0.8 gram of water per day. The rate of

water loss falls to 0.3 gram per day when the lizard stops eating.

Evaporation from the respiratory surfaces and from the skin accounts for about 61 percent of the total water loss of a chuckwalla. When the lizards stop eating, they also become inactive and spend most of the day in rock crevices. The body temperatures of inactive chuckwallas are lower than the temperatures of lizards that remain active. Because of their low body temperatures, the inactive chuckwallas have lower rates of metabolism. They breathe more slowly and lose less water from their respiratory passages. Also, the humidity is higher in the rock crevices than it is on the surface of the desert, and this reduction in the humidity gradient between the animal and the air further reduces evaporation. Most of the remaining water loss by a chuckwalla occurs in the feces (31 percent) and urine (8 percent). When a lizard stops

TABLE 14.2 Seasonal changes in the water balance of a 200-gram chuckwalla

	Early May	Late May	September
Food intake (g dry mass/day)	2.60	2.86	0.00
Water content of food (g/g dry mass)	2.53	0.96	—
Water gain (g/day)			
Free water	6.56	2.74	0.0
Metabolic water	0.68	0.68	0.20
Total water gain	7.24	3.41	0.20
Water loss (g/day)	6.41	4.26	0.52
Net water flux (g/day)	+0.81	−0.84	−0.32

eating, it also stops producing feces and reduces the amount of urine it must excrete. The combination of these effects reduces the daily water loss of a chuckwalla by almost 90 percent.

The food plants were always hyperosmolal to the body fluids of the lizards and had high concentrations of potassium. Despite this dietary salt load, osmotic concentrations of the chuckwallas' body fluids did not show the variation seen in tortoises, because the lizards' nasal salt glands were able to excrete ions at high concentrations. The concentration of potassium ions in the salt-gland secretions was nearly 10 times their concentration in urine. The formation of potassium salts of uric acid was the second major route of potassium excretion by the lizards and was nearly as important in the overall salt balance as nasal secretion. The chuckwallas would not have been able to balance their salt budgets without the two extrarenal routes of ion excretion, but with them they were able to maintain stable osmolal concentrations.

Both the chuckwallas and tortoises illustrate the interaction of behavior and physiology in responding to the characteristics of their desert habitats. The tortoises lack salt-secreting glands and store the salt they ingest, tolerating increased body fluid concentrations until a rainstorm allows them to drink water and excrete the excess salt. The chuckwallas were able to stabilize their body fluid concentrations by using their nasal glands to excrete excess salt, but they did not take advantage of rainfall to replenish their water stores. Instead they became inactive, reducing their rates of water loss by almost 90 percent and relying on energy stores and metabolic water production to see them through the period of drought.

Conditions for the chuckwallas were poor at the Black Mountain site during Nagy's study. Only 5 centimeters of rain had fallen during the preceding winter, and that low rainfall probably contributed to the early withering of the annual plants that forced the chuckwallas to cease activity early in the summer. Unpredictable rainfall is a characteristic of deserts, however, and the animals living there must be able to adjust to the consequences. Rainfall records from the weather station closest to Black Mountain showed that in five of the previous ten years, the annual total rainfall was about 5 centimeters. Thus the year of the study was not unusually harsh; conditions are sometimes even worse—only 2 centimeters of rain fell during the winter after the study. However, conditions in the desert are sometimes good. Fifteen centimeters of rain fell in the winter of 1968, and vegetation remained green and lush all through the following summer and fall. Chuckwallas and tortoises live for decades, and their responses to the boom-or-bust conditions of their harsh environments must be viewed in the context of their long life spans. A temporary relaxation of the limits of homeostasis in bad years is an effective trade-off for survival that allows the animals to exploit the abundant resources of good years.

Desert Amphibians Permeable skins and high rates of water loss are characteristics that would seem to make amphibians unlikely inhabitants of deserts, but certain species are abundant in desert habitats. Most remarkably, these animals succeed in living in the desert *because* of their permeable skins, not despite them. Anurans are the most common desert amphibians; tiger salamanders are found in the deserts of North America, and several species of plethodontid salamanders occupy seasonally dry habitats in California.

The spadefoot toads are the most thoroughly studied desert anurans (**Figure 14–7**). They inhabit the desert regions of North America—including the edges of the Algodones Sand Dunes in southern California, where the average annual precipitation is only 6 centimeters and in some years no rain falls at all. An analysis of the mechanisms that allow an amphibian to exist in a habitat like that must include consideration of both water loss and gain. The skin of desert amphibians is as permeable to water as that of species from moist regions. A desert anuran must control its water loss behaviorally by its choice of sheltered microhabitats free from solar radiation and wind movement. Different species of anurans utilize different microhabitats—a hollow in the bank of a desert wash, the burrow of a ground squirrel or kangaroo rat, or a burrow the anuran excavates for itself. All these places are cooler and wetter than exposed ground.

Desert anurans spend extended periods underground, emerging on the surface only when conditions are favorable. Spadefoot toads construct burrows about 60 centimeters deep, filling the shaft with dirt and leaving a small chamber at the bottom, which they occupy. In southern Arizona, the spadefoots construct these burrows in September, at the end of the summer rainy season, and remain in them until the rains resume the following July.

At the end of the rainy season when the spadefoots first bury themselves, the soil is relatively moist. The water tension created by the normal osmolal concentration of a spadefoot's body fluids establishes a gradient favoring movement of water from the soil into the toad. In this situation, a buried spadefoot can absorb water from the soil just as the

▶ **FIGURE 14–7** A desert spadefoot toad, *Spea multiplicata*.

roots of plants do. With a supply of water available, a spadefoot toad can afford to release urine to dispose of its nitrogenous wastes.

As time passes, the soil moisture content decreases, and the soil moisture potential (the driving force for movement of water) becomes more negative until it equals the water potential of the spadefoot. At this point, there is no longer a gradient allowing movement of water into the toad. When its source of new water is cut off, a spadefoot stops excreting urine and instead retains urea in its body, increasing the osmotic pressure of its body fluids. Osmotic concentrations as high as 600 millimoles · $[kg\ H_2O]^{-1}$ have been recorded in spadefoot toads emerging from burial at the end of the dry season. The low water potential produced by the high osmolal concentration of the spadefoot's body fluids may reduce the water gradient between the animal and the air in its underground chamber so that evaporative water loss is reduced. Sufficiently high internal concentrations should create potentials that would allow spadefoot toads to absorb water from even very dry soil.

The ability to continue to draw water from soil enables a spadefoot toad to remain buried for nine or ten months without access to liquid water. In this situation, its permeable skin is not a handicap to the spadefoot—it is an essential feature of the toad's biology. If the spadefoot had an impermeable skin, or if it formed an impermeable cocoon as some other amphibians do, water would not be able to move from the soil into the animal. Instead, the spadefoot would have to depend on the water contained in its body when it was buried. Under those circumstances, spadefoot toads would probably not be able to invade

the desert because their initial water content would not see them through a nine-month dry season.

A different response to arid conditions is seen in a few tree frogs. The African rhacophorid *Chiromantis xerampelina* and the South American hylid *Phyllomedusa sauvagei* lose water through the skin at a rate only one-tenth that of most frogs. *Phyllomedusa* has been shown to achieve this low rate of evaporative water loss by using its legs to spread the lipid-containing secretions of dermal glands over its body surface in a complex sequence of wiping movements. These two frogs are unusual also because they excrete uric acid instead of urea, and this characteristic reduces their urinary water loss.

■ Aquatic Ectotherms in the Desert

An aquatic habitat in a desert sounds like a contradiction. In fact, however, several different kinds of aquatic habitats are found in deserts, and some of them have distinctive faunas of fishes or amphibians. Temporary pools that are formed by heavy rains and last for only a few weeks are the breeding sites for most species of desert amphibians. The ephemeral nature of these habitats puts a premium on rapid development. Spadefoot toads, for example, can grow from an egg to metamorphosis in two or three weeks. Fishes require permanent water, and desert fishes are found in rivers, springs, and desert lakes.

Aquatic habitats in the desert have the same temperature extremes as terrestrial habitats, but there are some differences that are important to the animals living in them. Water has a much higher heat capacity than does air, so water temperature changes more slowly than does air temperature. The temperature variation from day to night is usually smaller in a pool of water than on land, but pools of water in the desert can reach lethally hot temperatures by the end of summer. Water conducts heat faster than does air, and the high heat capacity and conductance of water ensure that the body temperatures of small aquatic organisms are always close to water temperature. Aquatic organisms must regulate their body temperature by moving between areas of different water temperature.

Tadpoles The ability of aquatic animals to select the most favorable temperatures available is shown by observations of tadpoles of the foothill yellow-legged frog, *Rana boylii* (Brattstrom 1962). The tadpoles were in a small cove with a maximum depth of about 40 centimeters and a shallow area that was only 10 centimeters deep, and they moved about in this area as temperature changed (**Figure 14–8**). At night (2100 hours), all the tadpoles were in the deepest part of the cove where the temperature was warmest, and they remained there until morning. During the morning, sunlight warmed the water in the pool, and the shallow area warmed fastest. Between 0900 and 1000 hours, the tadpoles moved from the deep water into the shallow area, and by 1100 hours all of the tadpoles were in the shallows. By midday, the shallow water had become too warm, and all of the tadpoles had moved back into the deep part of the pool, which remained cool. The shallow parts of the cove cooled rapidly in the late afternoon. Water temperature was essentially the same everywhere, and the tadpoles were distributed all over the cove. As temperature continued to fall, the tadpoles moved back into the deepest part of the pool for the night.

Desert Fishes Permanent aquatic habitats in the desert may have populations of fishes, but relatively few species can tolerate the high temperatures and high salinities that characterize many of these bodies of water. Pupfish (Cyprinodontidae), minnows (Cyprinidae), and cichlids (Cichlidae) are the groups most often found in deserts. Desert lakes and springs rarely contain more than five species of fishes, and many habitats are occupied by a single species.

The problems that desert fishes encounter depend on the habitat in which they occur. Springs, such as those at Ash Meadows in Death Valley, may be relatively large or very small. Big Spring is 15 meters in diameter and 9 meters deep, whereas Mexican Spring is less than 2 meters in diameter and only 2 to 5 centimeters deep. The water in a spring usually has the same temperature and salinity year-round because the water is emerging from underground aquifers, but some springs are 30°C or hotter, while others are near 20°C.

Ponds and lakes often show more seasonal variation in temperature and salinity than do springs.

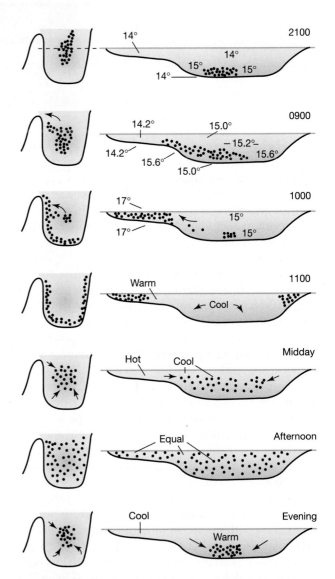

▲ **FIGURE 14–8** Temperature selection by tadpoles. Tadpoles of *Rana boylii* change positions in a pool of water in response to temperature changes during the day. The left column shows the view looking down on the pool, and the right column shows a cross section of the pool at the position indicated by the dashed line in the top drawing. Temperatures shown are in degrees Celsius.)

Some of these bodies of water are the remnants of enormous lakes that filled the desert valleys during the Pleistocene. Pyramid Lake in Nevada, for example, is a remnant of Pleistocene Lake Lahonton, which once covered more than 20,000 square kilometers in California and Nevada. Former shorelines of Lake Lahonton are marked by three terraces that are now 34, 100, and 163 meters above the present lake level. Other remnants of Lake Lahonton include Honey Lake in California and Walker Lake and the Carson and Humboldt Sinks in Nevada. Similar Pleistocene lake terraces can be found in the Sahara Desert and in the deserts of the Middle East. As the Pleistocene lakes shrank, once widespread populations of fishes were isolated in the remnant bodies of water. (The Salton Sea of California has a different origin from the other lakes of the Southwest. It lies in the basin of Pleistocene Lake LaConte, but the Salton Sea was formed in 1905 when the Colorado River broke through the banks of an irrigation system and formed a lake 25 meters deep and more than 1300 square kilometers in area.)

High temperature, salinity, and variation in oxygen concentration make desert lakes difficult environments for fishes. Pupfish (species of *Cyprinodon*) are widespread in desert lakes in North America and cope well with both heat and salinity. At Quitobaquito Springs in Organ Pipe National Monument, for example, desert pupfishes (*Cyprinodon macularis*) are found in shallow water at temperatures of 40°C to 41°C (which is only 2 or 3°C below the lethal temperature for the fish) rather than in water of 30°C that is a few meters away. Other species of *Cyprinodon* have been found voluntarily inhabiting water with temperatures of 43°C to 44°C.

Desert lakes are often characterized by high salinities as well as by high temperatures. Juvenile *Cyprinodon* can tolerate salt concentrations at least up to 90 grams of dissolved salt per liter of water, which is approximately three times the concentration of seawater. Adult pupfishes are slightly less tolerant of salinity than are juveniles, and eggs will not develop in salinities greater than 70 grams of salt per liter. A mosaic of salinities is often available to pupfishes because the density of water increases as the salinity increases, and water masses with different salinities tend not to mix. Thus, a fish can select areas of different salinity within a habitat just as it can select areas with different temperatures.

Streams are a third habitat for desert fishes. Streamflow varies through the year, and in dry seasons surface water in intermittent streams is reduced to isolated pools separated by dry areas of the stream bed. In the rainy season, however, the stream may flow continuously and even flood. Populations of fishes in desert streams typically wax and wane in response to changes in water flow. Salt Creek in Death Valley is inhabited by Devil's Hole pupfishes (*Cyprinodon salinus*). The population increases 100-fold as water from winter rains in the Mohave Desert flows down the stream in the winter and spring, and then crashes in summer and autumn as the stream dries. Some desert fishes, such as pupfishes and the longfin dace (*Agosia chrysogaster*), penetrate far into temporary aquatic habitats during periods of heavy rain and runoff. When desert streams are close to drying up, evaporation may consume the entire flow of water during the day, leaving only a damp streambed. Most species of fishes die under these conditions, but longfin dace survive beneath water-saturated mats of algae. Streamflow resumes at night when the temperature falls, and the dace emerge from beneath the algae, swimming about and feeding in a few millimeters of water. The fishes can survive in these conditions for several weeks; and, if rain refills the stream, they are able to return to areas of permanent water.

14.3 Coping with Cold—Ectotherms in Subzero Conditions

Temperatures drop below freezing in the habitats of many vertebrates on a seasonal basis, and some animals at high altitudes may experience freezing temperatures on a daily basis for a substantial part of the year. Birds and mammals respond to cold by increasing metabolic heat production and insulation, but ectotherms do not have those options. Instead, ectotherms show one of two responses—they avoid freezing by supercooling or synthesizing antifreeze compounds, or they tolerate freezing and thawing by using mechanisms that prevent damage to cells and tissues.

■ Cold Fishes

The temperature at which water freezes is affected by its osmolal concentration: Pure water freezes at 0°C, and increasing the osmolal concentration lowers the freezing point. Body fluid concentrations of marine fishes are 300 to 400 millimoles · [kg H_2O]$^{-1}$, whereas seawater has a concentration near 1000 millimoles · [kg H_2O]$^{-1}$. The osmolal concentrations of the body fluids of marine fishes correspond to freezing points of –0.6°C to –0.8°C, and the freezing point of seawater is –1.86°C. The temperature of Arctic and Antarctic seas falls to –1.8°C in winter, yet the fishes swim in this water without freezing.

A classic study of freezing avoidance of fishes was conducted by P. F. Scholander and his colleagues in Hebron Fjord in Labrador (Scholander et al. 1957). In summer, the temperature of the surface water at Hebron Fjord is above freezing, but the water at the bottom of the fjord is –1.73°C (**Figure 14–9**). In winter, the surface temperature of the water also falls to –1.73°C, like the bottom temperature. Several species of fishes live in the fjord; some are bottom dwellers, whereas others live near the surface. These two zones present different problems to the fishes. The temperature near the bottom of the fjord is always

below freezing, but ice is not present because ice is lighter than water and remains at the surface. Surface-dwelling fishes live in water temperatures that rise well above freezing in the summer and drop below freezing in winter, and they are also in the presence of ice.

The body fluids of bottom-dwelling fish in Hebron Fjord have freezing points of –0.8°C year-round. Because the body temperatures of these fishes are –1.73°C, the fishes are supercooled; that is, the water in their bodies is in the liquid state even though it is below its freezing point. When water freezes, the water molecules become oriented in a crystal lattice. The process of crystallization is accelerated by nucleating agents that hold water molecules in the proper spatial orientation for freezing. In the absence of

▶ **FIGURE 14–9** Water temperatures and distribution of fishes in Hebron Sound. The water temperature and occurrence of ice are shown for summer (a) and winter (b). The vertical axis shows the water depth in meters and the horizontal axis shows the freezing point depression in degrees centigrade. In summer (a) the temperature in the upper 10 meters of water rises to about 5 °C, but in winter (b) it falls to − 1.73 °C. Deeper water remains at − 1.73 °C year-round. The fish that live in deep water have freezing points of − 0.8 °C in summer and winter, whereas the fish living in shallow water decrease their freezing points from about − 0.8 °C in summer to about − 1.5 °C in winter.

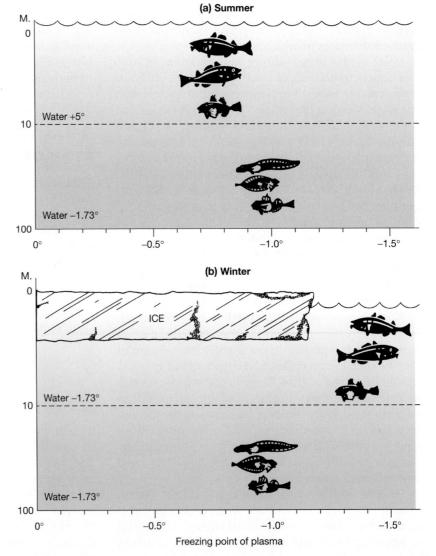

nucleating agents, pure water can remain liquid at –20°C. In the laboratory, the fishes from the bottom of Hebron Fjord can be supercooled to –1.73°C without freezing, but if they are touched with a piece of ice, which serves as a nucleating agent, they freeze immediately. At the bottom of the fjord there is no ice, and the bottom-dwelling fishes exist year-round in a supercooled state.

What about the fishes in the surface waters? They do encounter ice in winter when the water temperature is below the osmotically determined freezing point of their body fluids, and supercooling would not be effective in that situation. Instead, the surface-dwelling fishes synthesize antifreeze substances in winter that lower the freezing point of their body fluids to approximately the freezing point of the seawater in which they swim.

Antifreeze compounds are widely developed among vertebrates (and also among invertebrates and plants). Marine fishes have two categories of organic molecules that protect against freezing: glycoproteins with molecular weights of 2600 to 33,000 and polypeptides and small proteins with molecular weights of 3300 to 13,000 (Hew et al. 1986, Davies et al. 1988). These compounds are extremely effective in preventing freezing. For example, the blood plasma of the Antarctic fish *Trematomus borchgrevinki* contains a glycoprotein that is several hundred times more effective than salt (sodium chloride) in lowering the freezing point. The glycoprotein is adsorbed onto the surface of ice crystals and hinders their growth by preventing water molecules from assuming the proper orientation to join the ice-crystal lattice.

Chilly Lizards and Turtles

Some terrestrial ectotherms rely on supercooling. Mountain-dwelling lizards, such as Yarrow's spiny lizard (*Sceloporus jarrovi*), which lives at altitudes up to 3000 meters in western North America, are exposed to temperatures below freezing on cold nights, but sunny days permit thermoregulation and activity. These animals have body fluid concentrations that correspond to freezing points of –0.6°C, but they withstand substantially lower temperatures before they freeze. For example, spiny lizards supercooled to an average temperature of –5.5°C before they froze. At –3°C, the lizards had not frozen after 30 hours. The lizards spent the nights in rock crevices that were 5°C to 6°C warmer than the air temperature, and the combination of this protection and their ability to supercool was usually sufficient to allow

the lizards to survive. However, a few individuals at the highest altitudes were found frozen in their rock crevices during most winters.

Newly hatched painted turtles (*Chrysemys picta*) also bet their lives on their ability to supercool (Packard and Packard 2001). Painted turtles lay their eggs in nests that the female turtle excavates in the soil. A turtle can dig down only a few centimeters, so the nests are not very deep. Unlike the hatchlings of other species of aquatic turtles, baby painted turtles remain in the nest from the time they hatch in late summer or early autumn until the following spring. In northern parts of the geographic range of the species, baby turtles are exposed to temperatures that fall to –10°C or less, well below the –0.7°C freezing point of their body fluids.

The soil of the nest freezes, so the turtles are in intimate contact with ice crystals. The baby turtles are in a riskier situation than the spiny lizards, because the lizards retreat to dry crevices in rocks whereas the turtles are in moist soil where ice crystal can form. The key to the turtles' survival is avoiding contact with ice crystals that could initiate freezing of their body tissues. The outer layer of the turtles' skin appears to form a barrier that resists penetration by ice crystals. This layer of skin in painted turtles consists of a layer of alpha keratin with a layer of lipid on its inner surface (Willard et al. 2000). Hatchlings of other species of turtles lack the lipid layer, and ice crystals readily penetrate the skin of those species. Recordings of survival of hatchlings in nests in north-central Nebraska reveal the effectiveness of the protection provided by the skin of painted turtles. All of the turtles survived in nests that cooled to temperatures as low as –8°C, and more than half the turtles survived minimum temperatures down to –9.7°C. Mortality increased to 60 percent to 100 percent below that temperature, but seven of eight turtles survived in one nest that cooled to –12.7°C.

Frozen Frogs

Terrestrial amphibians that spend the winter in hibernation show at least two categories of responses to low temperatures. One group, which includes salamanders, toads, and aquatic frogs, buries deeply in the soil or hibernates in the mud at the bottom of ponds. These animals apparently are not exposed to temperatures below the freezing point of their body fluids. As far as we know, they have no antifreeze substances and no capacity to tolerate freezing. However, other amphibians apparently hibernate close to the soil surface, and these animals are exposed to temperatures below their freezing points. Unlike fishes and lizards, these amphibians freeze at low temperatures but are not killed by freezing (**Figure 14–10**). These species can remain frozen at –3°C for several weeks, and they tolerate repeated bouts of freezing and thawing without damage. However, temperatures below –10°C are lethal.

Tolerance of freezing refers to the formation of ice crystals in the extracellular body fluids; freezing of the fluids inside the cells is apparently lethal. Thus, freeze tolerance involves mechanisms that control the distribution of ice, water, and solutes in the bodies of animals (Storey 1986). The ice content of frozen frogs is usually in the range of 34 percent to 48 percent. Freezing of more than 65 percent of the body water appears to cause irreversible damage, probably because too much water has been removed from the cells.

Freeze-tolerant frogs accumulate low–molecular weight substances in the cells that prevent intracellular ice formation. Wood frogs, spring peepers, and chorus frogs use glucose as a cryoprotectant, whereas gray tree frogs use glycerol. Glycogen in the liver appears to be the source of the glucose and glycerol. The accumulation of these substances is apparently stimulated by freezing and is initiated within minutes of the formation of ice crystals. This mechanism of triggering the synthesis of cryoprotectant substances has not been observed for any other vertebrates or for insects.

Frozen frogs are, of course, motionless. Breathing stops, the heartbeat is exceedingly slow and irregular or may cease entirely, and blood does not circulate through frozen tissues. Nonetheless, the cells are not frozen; they have a low level of metabolic activity that is maintained by anaerobic metabolism. The glycogen content of frozen muscle and kidney cells decreases, and concentrations of lactic acid and alanine (two end products of anaerobic metabolism) increase.

The ecological significance of freeze tolerance in some species of amphibians is unclear. The four species of frogs so far identified as being freeze tolerant all breed relatively early in the spring, and shallow hibernation may be associated with early emergence in the spring. Being among the first individuals to arrive at the breeding pond may increase the chances for a male of obtaining a mate, and it gives larvae the longest possible time for development and metamorphosis; however, it also entails risks. Frequently, frogs and salamanders move across snowbanks to reach the breeding ponds, and they enter ponds that are still partly covered with ice. A cold snap can lead to the entire surface of the pond freezing again, trapping some animals under the ice and others in shallow retreats under logs and rocks around the pond. Freeze tolerance may be important to these animals even after their winter hibernation is over.

14.4 The Role of Ectothermal Tetrapods in Terrestrial Ecosystems

Life as an animal is costly. In thermodynamic terms, an animal lives by breaking chemical bonds that were formed by a plant (if the animal is an herbivore) or by another animal (if it's a carnivore) and using

(a)

(b)

▲ **FIGURE 14–10** Wood frog (*Lithobates* [*Rana*] *sylvatica*). (a) At normal temperature. (b) Frozen.

the energy from those bonds to sustain its own activities. Vertebrates are particularly expensive animals because vertebrates generally are larger and more mobile than invertebrates. Big animals require more energy (i.e., food) than small ones, and active animals use more energy than sedentary ones.

In addition to body size and activity, an animal's method of temperature regulation (ectothermy, endothermy, or a combination of the two mechanisms) is a key factor in determining how much energy it uses and therefore how much food it requires. Because ectotherms rely on external sources of energy to raise their body temperatures to the level needed for activity and endotherms use heat generated internally by metabolism, ectotherms use substantially less energy than do endotherms. The metabolic rates (i.e., rates at which energy is used) of terrestrial ectotherms are only 10 percent to 14 percent of the metabolic rates of birds and mammals of the same body size. The lower energy requirements of ectotherms mean that they need less food than would an endotherm of the same body size.

Body size is another major difference between ectotherms and endotherms that relates directly to their mode of temperature regulation and affects their roles in terrestrial ecosystems. Ectotherms are smaller than endotherms, partly because the energetic cost of

endothermy is very high at small body sizes. As body mass decreases, the mass-specific cost of living (energy per gram) for an endotherm increases rapidly, becoming nearly infinite at very small body sizes (**Figure 14–11**). This is a finite world, and infinite energy requirements are just not feasible. Thus energy requirements, among other factors, apparently set a lower limit to the body size possible for an endotherm.

The mass-specific energy requirements of ectotherms also increase at small body sizes; but, because the energy requirements of ectotherms are about one-tenth those of endotherms of the same body size, an ectotherm can be about an order of magnitude smaller than an endotherm. A mouse-size mammal weighs about 20 grams, and few adult birds and mammals have body masses less than 10 grams. The very smallest species of birds and mammals weigh about 3 grams, but many ectotherms are only one-tenth that size (0.3 grams). Amphibians are especially small—20 percent of the species of salamanders and 17 percent of the species of anurans have adult body masses less than 1 gram, and 65 percent of salamanders and 50 percent of anurans are smaller than 5 grams. Squamates are generally larger than amphibians, but 18 percent of the species of lizards and 2 percent of snakes weigh less than 1 gram. Even the largest amphibians and squamates weigh substantially less than 100 kilograms, whereas more

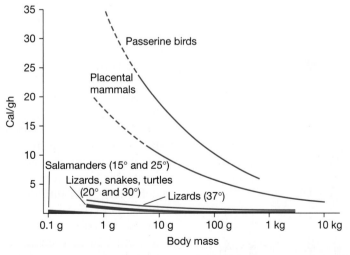

▲ **FIGURE 14–11** Energy cost. Resting metabolic rate (cal · [g · h]$^{-1}$) is shown as a function of body size for terrestrial vertebrates. Metabolic rates for salamanders are shown at 15°C and 25°C as the lower and upper limits of the darkened area, and data for lizards, snakes, and turtles combined are shown at 20°C and 30°C. Data from lizards alone are shown for 37°C. The curve for anurans falls within the lizards-snakes-turtles area, and the relationship for nonpasserine birds is similar to that for placental mammals. Dotted portions of the lines for birds and mammals show hypothetical extensions into body sizes below the minimum for adults of most species of birds and mammals.

than 5 percent of the species of extant mammals weigh 100 kilograms or more (**Figure 14–12**).

Body shape is another aspect of vertebrate body form in which ectothermy allows more flexibility than does endothermy. An animal exchanges heat with the environment through its body surface, and the surface area of the body in relation to the mass of the body (the surface/mass ratio) is one factor that determines how rapidly heat is gained or lost. Small animals have higher surface/mass ratios than do large ones, and that is why endothermy becomes increasingly expensive at progressively smaller body sizes. Metabolic rates of small endotherms must be high enough to balance the high rates of heat loss across their large body surface areas.

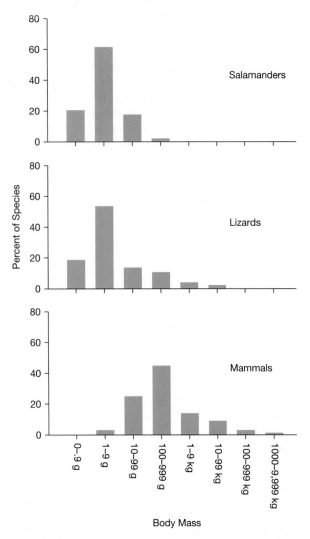

▲ **FIGURE 14–12** Adult body masses of salamanders, lizards, and mammals.

Similarly, body shapes that increase surface/mass ratios have an energy cost that makes them disadvantageous for endotherms. There are no highly elongate endotherms, whereas elongate body forms are widespread among fishes (true eels, moray eels, pipefishes, barracudas, and many more), amphibians (all caecilians and most salamanders, especially the limbless aquatic forms), and reptiles (many lizards and all snakes). Dorsoventral or lateral flattening is another shape that increases surface/mass ratio. There are no flat endotherms, but flat fishes are common (dorsoventral flattening—skates, rays, flounders; lateral flattening—many coral-reef and freshwater fishes). Some reptiles are also flat (dorsoventral flattening—aquatic turtles, especially soft-shelled turtles, horned lizards; lateral flattening—many arboreal lizards, especially chameleons). Small body sizes and specialized body forms allow ectotherms to fill ecological niches that are not available to endotherms.

The amount of energy required by ectotherms and endotherms is not the only important difference between them; equally significant is what they do with that energy once they have it. Endotherms expend more than 90 percent of the energy they take in to produce heat to maintain their high body temperatures. Less than 10 percent—often as little as 1 percent—of the energy a bird or mammal assimilates is available for net conversion (that is, increasing the species' biomass by growth of an individual or production of young). Ectotherms do not rely on metabolic heat. The solar energy they use to warm their bodies is free in the sense that it is not drawn from their food. Thus, most of the energy they ingest is converted into the biomass of their species. Values of net conversion for amphibians and reptiles are between 30 percent and 90 percent (**Table 14.3**).

Because of the difference in how energy is used, a given amount of chemical energy invested in an ectotherm produces a much larger biomass return than it would have from an endotherm. A study of salamanders in the Hubbard Brook Experimental Forest in New Hampshire showed that, although their energy consumption was only 20 percent that of the birds or small mammals in the watershed, their conversion efficiency was so great that the annual increment of salamander biomass was equal to that of birds or small mammals. Similar comparisons can be made between lizards and rodents in deserts.

Small amphibians and squamates occupy key positions in the energy flow through an ecosystem. Because these animals are so small, they can capture tiny insects and arachnids that are too small to be

TABLE 14.3 Efficiency of biomass conversion by ectotherms and endotherms. These are net conversion efficiencies calculated as (energy converted/energy assimilated) × 100

Ectotherm Species	Efficiency (%)	Endotherm Species	Efficiency (%)
Red-backed salamander (Plethodon cinereus)	48	Kangaroo rat (Dipodomys merriami)	0.8
Mountain salamander (Desmognathus ochrophaeus)	76–98	Field mouse (Peromyscus polionotus)	1.8
Panamanian anole (Anolis limifrons)	23–28	Meadow vole (Microtus pennsylvanicus)	3.0
Side-blotched lizard (Uta stansburiana)	18–25	Red squirrel (Tamiasciurius hudsonicus)	1.3
Hognose snake (Heterodon platirhinos)	81	Least weasel (Mustela rixosa)	2.3
Python (Python curtus)	6–33	Savanna sparrow (Passericulus sandwichensis)	1.1
Common adder (Vipera berus)	49	Marsh wren (Telmatodytes palustris)	0.5
Average of 12 species	50	Average of 19 species	1.4

Source: Pough 1980, 1983.

eaten by birds and mammals. Because they are ectotherms, they are efficient at converting the energy in the food they eat into their own body tissues. As a result, the small ectothermal vertebrates in terrestrial ecosystems can be viewed as repackaging energy into a form that avian and mammalian predators can exploit. In other words, when a shrew or a bird searches for a meal in the Hubbard Brook Forest, the most abundant vertebrate prey it will find is salamanders. In this context, frogs, salamanders, lizards, and snakes occupy a position in terrestrial ecosystems that is important both quantitatively (in the sense that they constitute a substantial energy resource) and qualitatively (in that ectotherms exploit food resources that are not available to endotherms).

In a very real sense, small ectotherms can be thought of as living in a different world from that of endotherms. As we saw in the case of the three species of *Ameiva* lizards in Costa Rica (see Section 13.6), interactions with the physical world may be more important in shaping the ecology and behavior of small ectotherms than are biological interactions such as competition. In some cases, these small vertebrates may have their primary predatory and competitive interactions with insects and arachnids rather than with other vertebrates. For example, orb-web spiders and *Anolis* lizards on some Caribbean islands are linked by both predation (adult lizards eat spiders, and spiders may eat hatchling lizards) and competition (lizards and spiders eat many of the same kinds of insects). The competitive relationship between these distantly related animals was demonstrated by experiments. When lizards were removed from experimental plots, the abundance of insect prey increased, and the spiders consumed more prey and survived longer than in control plots with lizards present (Schoener and Spiller 1987).

Ectothermy and endothermy thus represent fundamentally different approaches to the life of a terrestrial vertebrate. An appreciation of ectotherms and endotherms as animals requires understanding the functional consequences of the differences between them. Ectothermy is an ancestral character of vertebrates, but it is a very effective way of life in modern ecosystems.

Summary

All living organisms are mosaics of ancestral and derived characters, interacting to make an organism a functional entity. Ectothermy is an ancestral character that is entirely functional in the modern world. Ectotherms do not use chemical energy from the food they eat to maintain high body temperatures. The results of that ancestral vertebrate characteristic are far-reaching for extant ectothermal vertebrates, and ectotherms and endotherms represent quite different approaches to vertebrate life.

Because of their low energy requirements, ectotherms can colonize habitats in which energy is in short supply. Ectotherms are able to extend some of their limits of homeostasis to tolerate high or low body temperatures and high or low body-water contents when doing so allows them to survive in difficult conditions.

When food is available, ectotherms are efficient at converting the energy it contains into their own body tissues for growth or reproduction. Net conversion efficiencies of ectotherms average 50 percent of the energy assimilated compared with an average of 1.4 percent for endotherms.

Ectotherms can be smaller than endotherms because their mass-specific energy requirements are low, and many ectotherms weigh less than a gram, whereas most endotherms weigh more than 10 grams. Because of this difference in body size, many small ectotherms—such as salamanders, frogs, and lizards—eat prey that is too small to be consumed by endotherms. The efficiency of energy conversion by ectotherms and their small body sizes lead to a distinctive role in modern ecosystems, one that is in many respects quite different from that of terrestrial ectotherms. Understanding these differences is an important part of understanding the organismal biology of terrestrial ectothermal vertebrates.

Additional Readings

Brattstrom, B. H. 1962. Thermal control of aggregation behavior in tadpoles. *Herpetologica* 18:38–46.

Davic, R. H., and H. H. Welsh, Jr. 2005. On the ecological role of salamanders. *Annual Review of Ecology and Systematics* 35:405–434.

Davies, P. L. et al. 1988. Fish antifreeze proteins: Physiology and evolutionary biology. *Canadian Journal of Zoology* 66:2611–2617.

Hew, C. L. et al. 1986. Molecular biology of antifreeze. In H. C. Heller et al. (eds.), *Living in the Cold: Physiological and Biochemical Adaptation*. New York, NY: Elsevier, 117–123.

Nagy, K. A. 1972. Water and electrolyte budgets of a free-living desert lizard, *Sauromalus obsesus. Journal of Comparative Physiology* 79:39–62.

Nagy, K. A. 1973. Behavior, diet and reproduction in a desert lizard, *Sauromalus obsesus. Copeia* 1973:93–102.

Nagy, K. A. 1983. The doubly labeled water (^{3}HH^{18}O) method: A guide to its use. *UCLA Publication 12–1417*. Los Angeles, CA: University of California.

Nagy, K. A., and P. A. Medica. 1986. Physiological ecology of desert tortoises in southern Nevada. *Herpetologica* 42:73–92.

Nagy, K. A. et. al. 1999. Energetics of free-ranging mammals, reptiles, and birds. *Annual Review of Nutrition* 19:247–277.

Packard G. C., and M. J. Packard. 2001. The overwintering strategy of hatchling painted turtles, or how to survive in the cold without freezing. *BioScience* 51:199–207.

Pough, F. H. 1980. The advantages of ectothermy for tetrapods. *The American Naturalist* 115:92–112.

Pough, F. H. 1983. Amphibians and reptiles as low-energy systems. In W. P. Aspey and S. I. Lustick (eds.), *Behavioral Energetics: Vertebrate Costs of Survival*. Columbus, OH: Ohio State University Press, 141–188.

Schoener, T. W., and D. A. Spiller. 1987. Effect of lizards on spider populations: Manipulative reconstruction of a natural experiment. *Science* 236:949–952.

Scholander, P. F. et al. 1957. Supercooling and osmoregulation in Arctic fish. *Journal of Cellular and Comparative Physiology* 49:5–24.

Storey, K. B. 1986. Freeze tolerance in vertebrates: Biochemical adaptation of terrestrially hibernating frogs. In H. C. Heller, X. J. Musacchia, and L. C. H. Wang (eds.), *Living in the Cold: Physiological and Biochemical Adaptations*. New York, NY: Elsevier, 131–138.

Willard. R. et al. 2000. The role of the integument as a barrier to penetration of ice into overwintering hatchlings of the painted turtles (*Chrysemys picta*). *Journal of Morphology* 246:150–159.

CHAPTER 15

Geography and Ecology of the Mesozoic Era

By the early Mesozoic era, the Earth's entire land surface had coalesced into a single continent, Pangaea, that stretched from pole to pole. Early Mesozoic (Triassic) faunas and floras showed some regional differentiation due to climate but would have had no oceanic barriers to dispersal. With the breakup of Pangaea in the later Mesozoic (Jurassic and Cretaceous periods), floras and faunas were geographically isolated and became distinct in different parts of the world.

Many new types of insects appeared in the Mesozoic, including social insects such as bees, ants, and termites. The appearance and rapid radiation of the **angiosperms** (flowering seed plants) during the Cretaceous was an important floral change. A major turnover occurred in terrestrial vertebrates at the end of the Triassic, when the large-animal fauna, including a diverse assemblage of mammal-like reptiles, was replaced by dinosaurs. Jurassic dinosaurs added a new ecological form to the ecosystem—gargantuan herbivorous sauropods. More advanced herbivorous dinosaurs, with jaws and teeth specialized for processing tough vegetation, appeared in the Cretaceous.

Mammals, birds, and modern types of amphibians and reptiles appeared in the Mesozoic. The Mesozoic was also the time of a great radiation of marine reptiles—ichthyosaurs, plesiosaurs, mosasaurs, placodonts, and others, none of which survived the end of the Cretaceous. The ecological niches of these Mesozoic marine tetrapods were reinvaded in the Cenozoic era by marine mammals such as whales and seals.

Extinctions were prominent at the end of the Triassic period, which established the dinosaur-dominated faunas of the later Mesozoic and at the end of the Cretaceous. Faunal changes during the Triassic may be related to low levels of atmospheric oxygen. The end-Cretaceous extinctions are famous as the ones that resulted in the demise of the dinosaurs, but in actuality many types of organisms were affected: invertebrates and plants, as well as nondinosaurian vertebrates. Evidence for an impact with an extraterrestrial body (a meteor or asteriod) at the end of the Cretaceous has strengthened the argument for these extinctions being sudden rather than gradual in nature, but problems remain with the resolution of the fossil record and the pattern of the kinds of animals that survived versus the ones that became extinct.

15.1 Mesozoic Continental Geography

By the middle of the Triassic, the entire land area of Earth was concentrated in the supercontinent Pangaea, which straddled the equator. The southern part of Pangaea, areas that now form Antarctica and Australia, was close to the South Pole, and the northern part of modern Eurasia was within the Arctic Circle (**Figure 15–1a**).

The fragmentation of Pangaea began in the Jurassic with the separation of Laurasia (northern) and

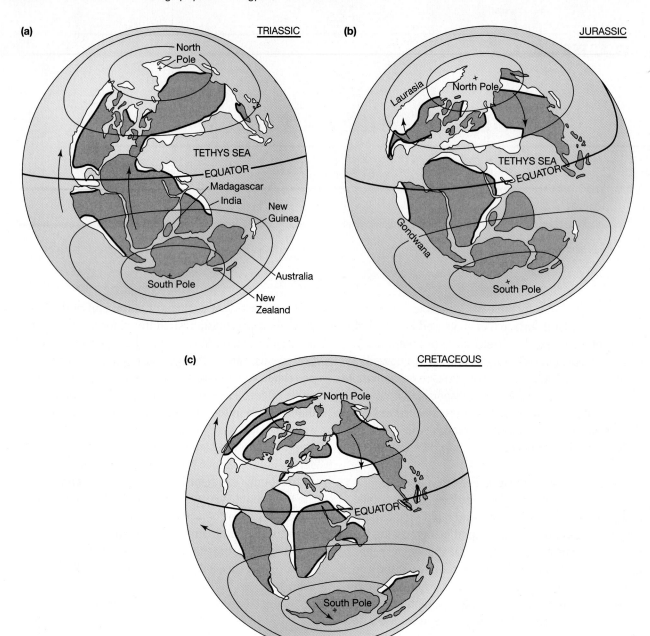

▲ **FIGURE 15–1** Location of continental blocks in the Mesozoic. Continental land areas are shaded, and epicontinental seas are the unshaded areas at the edges of the modern continents. Arrows indicate the direction of continental rotation.

Gondwana (southern) by a westward extension of the Tethys Sea. Laurasia rotated away from the other continents, ripping North America from its connection with South America and increasing the size of the newly formed Atlantic Ocean (**Figure 15–1b**). Epicontinental seas (areas of sea that covered what is now continenal land area) were more extensive than they had been in the Triassic.

Separation of the continents and rotation of the northern continents continued through the Mesozoic. Epicontinental seas became even more extensive. By the Late Cretaceous, the continents were approaching their current positions, although India was still close to Africa, and Australia, New Zealand, and New Guinea were still well south of their present-day positions (**Figure 15–1c**).

15.2 Mesozoic Terrestrial Ecosystems

The Mesozoic was marked by a series of large-scale changes in flora and fauna beginning with the Permo-Triassic (Permian–Triassic) extinction event, the repercussions of which extended over a period of about 25 million years, and culminating in the Cretaceous–Tertiary transition. Terrestrial ecosystems had achieved an essentially modern food web by the end of the Permian. Plants grew in communities that contained mixtures of species, herbivorous insects and vertebrates consumed these plants, and carnivorous invertebrates and vertebrates preyed on the herbivores. The kinds of plants and animals in early Mesozoic ecosystems were quite different from those in modern ecosystems, however.

The Permo-Triassic extinctions left a fairly impoverished terrestrial fauna. Insects lost a considerable amount of diversity, especially among plant-sucking forms, but new herbivorous forms appeared rapidly in the Triassic, including stick insects. Early Triassic vertebrate faunas were dominated by the herbivorous mammal-like reptile *Lystrosaurus,* as well as carnivorous mammal-like reptiles and early archosaurs ("thecodonts"), diapsid reptiles that would eventually give rise to the dinosaurs. Later in the Triassic, the climate in the places where tetrapods were found shifted from warm and moist to hot and dry, and the archosaurs increased in diversity and faunal predominance. Triassic faunas still contained a diversity of mammal-like reptiles, including dicynodonts (her-bivorous), cynodonts (both carnivorous and herbivorous), and rhyncosaurs (large herbivorous diapsid reptiles) (**Figure 15–2**).

Triassic vegetation included familiar modern groups of gymnosperms, such as conifers (pines and other cone-bearing trees), ginkgophytes (relatives of the living ginkgo), and cycads and seed ferns. Other plants included ferns, tree ferns, and horsetails. While many of these plants exist today, they are not nearly as common and widespread as they were in the Mesozoic. Although Pangaea was a single continent in the Triassic, Triassic floras and faunas show regional characters that probably reflect differences in patterns of rainfall and seasonal temperature extremes in areas far from the sea. Conifers ranged from small bushes to tall forest trees, and many appear to have lived in relatively dry habitats. Cycads, too, appear to have lived in dry regions.

Triassic herbivorous vertebrates ranged in body mass between 10 and 1000 kilograms. This size distribution suggests that the major vegetational landscape may have been open woodland because dense forest supports primarily small-bodied herbivores. All of these herbivores (rhyncosaurs, dicynodonts, gomphodont cynodonts, and aetosaur archosaurs) were generalized browsers that would have foraged within a meter of the ground. Higher-level browsers were not present until the evolution of prosauropod dinosaurs in the latest Triassic.

There were virtually no arboreal herbivorous vertebrates, although some sphenodontid reptiles and early mammals (haramyids) from the end of the Triassic might have been arboreal herbivores. Only

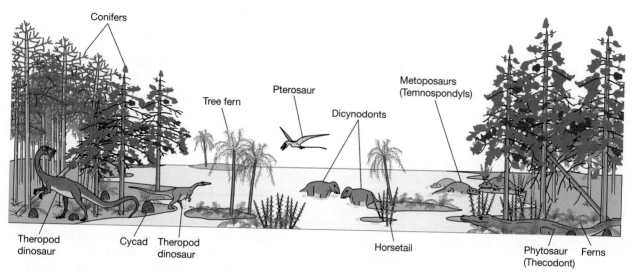

▲ **FIGURE 15–2** Reconstruction of a scene from the early Late Triassic of New Mexico.

when mammals occupied this habitat in the Cenozoic was there a significant radiation of canopy-dwelling herbivorous vertebrates. However, there were plenty of herbivorous insects in the Triassic, and along with them were several arboreal insectivorous diapsid reptiles, including gliding forms.

During the Late Triassic there was a shift in vegetational communities, especially in Gondwana where the floras (dominated by the seed fern *Dicroidium*) were replaced by ones dominated by conifers. This floral turnover was matched by a major faunal turnover; the assemblage of large Triassic tetrapods became extinct. Both dinosaurs and mammals made a first appearance at this time, as did other modern groups such as sphenodontids (the sister group of lizards), turtles, and crocodiles. Groups that are now extinct such as pterosaurs (flying reptiles) also appeared, along with various marine reptiles such as ichthyosaurs and plesiosaurs.

We saw in Chapter 7 that low levels of atmospheric oxygen in the Late Permian period may have contributed to heightened levels of extinction before the mass extinction at the end of the Paleozoic era. Oxygen levels continued to drop for much of the Triassic, reaching a low of 15 percent by the middle of the period. These low levels may explain why the recovery of postextinction Triassic faunas was so slow. Low oxygen levels may also explain an evolutionary quandary in the Triassic. The previously successful "mammal-like reptiles" were now progressively overshadowed by the rise of a new group, the archosaurs ("thecodonts" at this point in time). Many explanations have been advanced for this, including so-called improved posture (i.e., more upright) in archosaurs or changes in vegetational habitats. But archosaurs also may have had an advantage in terms of respiration in low levels of oxygen (Berner et al. 2007). The lungs of the mammal-like reptiles were probably modified in the mammalian direction to form an alveolar lung, which is tidally ventilated. Work by Colleen Farmer suggests that the lungs of crocodiles, like those of birds, have a bias to the flow so that the air moves through the lung in primarily one direction, rather than simply in and out. Crocodiles and birds "bracket" all of the archosaurs (i.e., all extinct types known fall either on the crocodile line or the bird line), and thus some degree of unidirectional flow through the lung was probably a basal archosaurian feature. Unidirectional flow coupled with cross- or countercurrent vascular alignment (see Section 11.2) may have given archosaurs a competitive edge in the hypoxic conditions of the Triassic.

Another proposal, not necessarily mutually exclusive with the one above, for archosaur success in the Triassic has been proposed by Nick Geist. The reproductive tract of archosaurs differs from that of other amniotes, enabling them to produce eggs with a rigid calcareous shell (with a uniquely ordered arrangement of calcium carbonate) and a high albumen content. Such eggs would be resistant to desiccation (and also to microbial infection and predation) and may have given archosaurs an advantage in the widespread Triassic drying (Jones and Geist 2008).

The conifer- and fern-dominated vegetation of the Late Triassic continued into the Jurassic, though the seed ferns were then reduced in diversity. Several new kinds of insects appeared in the Jurassic, including beetles, thrips, and a variety of bugs (hemipterans). An abrupt change in terrestrial ecosystems occured at the start of the Jurassic when the dinosaurs became the dominant element of the large vertebrate fauna. Small carnivorous dinosaurs were a rare element of the fauna in the Late Triassic, but the Jurassic saw radiations of larger carnivores (allosaurids) and several kinds of large herbivores. The latter included the high-browsing prosauropods and sauropods and the low-browsing stegosaurs. There was also a continuation of smaller carnivorous therapod dinosaurs, now matched by small herbivorous ornithopods. Lizards and the modern groups of amphibians first appeared in the Jurassic, and the earliest bird, *Archaeopteryx*, is known from the Late Jurassic.

Sauropods were gigantic animals, with body masses ranging from 10,000 to 50,000 kilograms or even more, and they may have influenced patterns of vegetational growth and structure as elephants do today. It is not clear how the vegetation of the Jurassic was able to support such a great biomass of herbivores. However, atmospheric carbon dioxide levels were exceptionally high during the Jurassic, and this may have allowed high levels of plant productivity. Similarly the levels of atmospheric oxygen, which rose slightly (back up to around 18 percent) at the end of the Triassic, then plummeted to an all-time Phanerozoic low (around 11 percent) at the start of the Jurassic, and did not really recover from this until the Cretaceous (when oxygen levels rose to around 18 percent by the end of the period). This sustained environmental hypoxia would have continued to provide archosaurs with a competitive edge.

We have a fantastic window into the world of the Early Cretaceous from the Jehol Biota in northeastern China (Zhou et al. 2003), which was formed in deposits that preserved soft tissue. Not only is an entire ecosystem preserved—plants, invertebrates, fish, and tetrapods—but many of the new spectacular finds of feathered dinosaurs, early birds, and early mammals.

The vegetation of the Early Cretaceous was similar to that of the Late Jurassic. However, by the Late Cretaceous, the pattern of global vegetation was quite different. Flowering plants (angiosperms) appear in the fossil record in the Early Cretaceous and were the first plants to be pollinated by insects. Insects of the Early Cretaceous include butterflies, aphids, short-horned grasshoppers, and gall wasps. In the Late Cretaceous, a variety of social insects made their first appearance, including termites, ants, and hive-forming bees.

Angiosperms were initially located near the equator; but, by the middle of the Cretaceous, they were well established in the middle latitudes. Angiosperms formed 50 percent to 80 percent of the plants in many fossil assemblages by the end of the Cretaceous, but their role in the vegetational landscape was different from that of today. The large trees were still conifers. The growth forms of angiosperms included small trees and low-growing shrubs and herbs that replaced ferns and cycads as the ground cover. Grasses, which are such a dominant feature of landscapes today, would not have been a predominant feature of the landscape until well into the Cenozoic, although we now have evidence that grasses had evolved by the end of the Cretaceous (Prasad et al. 2005).

Several types of herbivorous dinosaurs appeared in the Cretaceous. The most diverse of these were the hadrosaurs (duck-billed dinosaurs) and the ceratopsians (horned dinosaurs)—both low-level feeders with complex, shearing cheek teeth, probably for dealing with tough vegetation. Both kinds of dinosaurs probably lived in migratory or nomadic herds that may have numbered in the thousands of individuals. It has been suggested that the feeding habits of these dinosaurs altered the nature of the flora and that their appearance was causally related to the spread of angiosperms. However, careful analysis of the fossil record reveals that dinosaurs did not really "invent flowers" (Barrett and Willis 2001).

Other later Cretaceous faunal changes include the evolution of mammals with complex molars by the Early Cretaceous. By the mid-Cretaceous the first mammals belonging to the three modern groups appeared—monotremes, marsupials, and placentals. There was a considerable radiation of birds during the Cretaceous, but few of these forms were closely related to the modern birds that diversified in the Cenozoic. Snakes and modern types of crocodiles are also first known from the Cretaceous.

Thus, although the Jurassic and Cretaceous are usually thought of as the Age of Dinosaurs, there was a considerable diversity of other types of vertebrates present, albeit playing the role of the smaller members of the fauna. The largest mammal was about the size of a small dog, although most were the size of mice and shrews, and the lizards and other reptiles had a size range similar to that of extant species. The very smallest dinosaur would have been about the size of the largest mammal, but, of course, most were considerably larger. Some of the smaller vertebrates might have preyed on dinosaur eggs or served as food for juvenile dinosaurs. On the other hand, we have evidence that the largest known mammal of the time, *Repenomamus*, actually ate juvenile dinosaurs, as one of these mammals has been found with a tiny dinosaur (a psittacosaur ceratopsian) in its stomach (Hu et al. 2005).

Although dinosaurs are often depicted as inhabiting a strange, extinct world, in fact the dinosaur ecosystem was not so different from our present day one. The main difference is that the ecological roles for larger vertebrates in the later Mesozoic ecosystems were taken by dinosaurs rather than mammals. All modern tetrapod groups—frogs, salamanders, lizards, snakes, turtles, crocodilians, birds, and mammals—arose in the Late Triassic or Jurassic. The fauna included far more species of vertebrates that were not dinosaurs than species of dinosaurs, even though dinosaurs may have comprised much of the biomass (**Figure 15–3**).

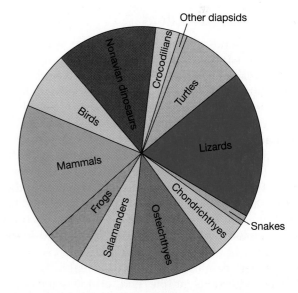

▲ **FIGURE 15–3** The relative abundance of genera at the Late Cretaceous Lance locality in Montana. This locality appears to represent a wooded, swampy habitat with large streams and some ponds.

Mesozoic Climates

Mesozoic climates were equable worldwide, and there were no polar ice caps at any time during the era. Large temnospondyls (aquatic non-amniote tetrapods) are found in Triassic deposits from Australia (where they lasted into the Cretaceous), Antarctica, Greenland, and Spitzbergen, and coal deposits in both the Northern and Southern Hemispheres point to moist climates. In contrast, low and middle latitudes were probably dry—either seasonally or year-round—until the Late Cretaceous and early Cenozoic, when coal deposits of Mesozoic age in middle latitudes indicate the presence of swamps and suggest that the climate had become wetter. These dry lower latitudes had a type of vegetation different from the equatorial vegetation of today, with the absence of wet tropical rain forests. The types of vertebrates found in the Late Cretaceous Arctic, such as champsosaurs (large aquatic crocodile-like reptiles), suggests a frost-free environment with a mean annual temperature exceeding 14°C (Tarduno et al. 1998).

The Cretaceous plant record also suggests a highly equable world with temperate forests extending into the polar regions, and the Arctic faunas contained large aquatic reptiles (champsosaurs) which would have required a temperate or subtropical climate. A reduction in the diversity of plants at higher latitudes at the end of the Cretaceous signaled an episode of worldwide cooling, however.

Mesozoic Extinctions

The first round of Mesozoic extinctions was at the end of the Triassic, and it had considerable impact on both terrestrial vertebrates and marine invertebrates. The fauna of the Triassic, which was essentially a holdover from the late Paleozoic, was replaced by forms that would dominate the Mesozoic, such as the dinosaurs; 18 families of tetrapods became extinct at this time. The Triassic extinctions coincided with the initial breakup of Pangaea, but there is now some evidence for an impact with an asteroid or meteor at this time (Olsen et al. 2002). (The type of evidence for such an event is discussed below.) Lesser extinctions of tetrapods occurred in the Early Triassic and in the Late Jurassic, and both of these terrestrial extinction events were paralleled in the marine realm.

The extinctions at the end of the Cretaceous (known as the K-T boundary) include the demise of the dinosaurs even though the magnitude of the effect on the global fauna was nowhere near as large

as the late Paleozoic (end Permian) extinctions. Many hypotheses have been proposed to explain why dinosaurs became extinct, ranging from the extraterrestrial (their gonads were zapped by radiation from an exploding supernova) to the ridiculous (constipation caused by angiosperms; see Benton 1990). It must be remembered that dinosaurs were not the only animals that suffered extinctions at this time. Thirty-six of the tetrapod families (40 percent) were extinct by the end of the Cretaceous, including not only all nonavian dinosaurs but also all flying reptiles (pterosaurs) and marine reptiles (ichthyosaurs, plesiosaurs, mosasaurs, and others). Birds and mammals suffered lesser extinctions, and insects were among the few groups that survived the Cretaceous relatively intact. There were also extensive extinctions among plants and marine invertebrates. Any explanation for the demise of the dinosaurs must also account for the disappearances of a large variety of other organisms, both on land and in the sea, as well as for the survival of many other lineages.

There has long been debate as to whether the end-Cretaceous extinctions were sudden or gradual in nature. This is difficult to gauge from the fossil record because fossils are not usually found in the type of uninterrupted sequence that would be necessary for that type of resolution. The absence of an animal near the K-T boundary might indicate nonpreservation at that point in time rather than extinction.

The debate about gradual or sudden Late Cretaceous extinctions has been particularly virulent when applied to dinosaurs (see Fastovsky and Wieshampel 2005). Several issues complicate this debate. For a start, the terrestrial record of the latest Cretaceous is primarily known from North America; patterns of dinosaur evolution observed on this continent might not apply to the rest of the world. Second, while it has been noted that dinosaur diversity in the latest Cretacous is rather lower than the immediately preceding time period, dinosaur diversity waxed and waned throughout the late Mesozoic, and in North America the time preceding the latest Cretaceous actually contains the most diverse dinosaur faunas known. So was the observed decline in diversity a few million years before the end of the Cretaceous indicative of a long-term decline to extinction or merely part of the normal ups and downs of dinosaur diversity through time? The answer is impossible to determine.

In the past couple of decades, evidence has mounted for the impact of the Earth with a large extraterrestrial body, an asteroid or meteor, at the K-T boundary. A narrow band of rock containing a high concentration of iridium (known as an iridium anomaly) occurs in many parts of the world in sediments that were deposited at the rock that marks the K-T boundary layer. Iridium is normally an extremely rare element in sedimentary rocks, but it occurs at somewhat higher concentrations in the core of the Earth and in some extraterrestrial objects. The worldwide deposition of iridium may indicate an impact that occurred with such force as to pulverize the extraterrestrial body and distribute its remains as atmospheric dust, later to fall to the ground and coat the entire globe. The dust cloud raised by such an impact might have limited the sunlight reaching the Earth's surface for several years, resulting in greatly reduced plant photosynthesis and ecosystem collapse. Other potentially catastrophic events were also happening on Earth around this time: huge volcanoes in India, known as the Deccan Traps, were erupting during the period from 67 million years ago to 65 million years ago. Recent work has quantified the amount of sulfur and chlorine gas that would have been released from these eruptions, which would have had a huge effect on Earth's atmosphere, with climatic effects that may have caused extinction of organisms (Self et al. 2008).

Evidence for an impact by an extraterrestrial object has been found in the form of a crater of the appropriate age (64.4 ± 0.5 million years), the Chicxulub crater off the Yucatán coast of Mexico. This crater contains many of the geological signals found in the sites of known meteor impacts of historical times. Thus we have excellent evidence that an impact did occur at approximately the same time as the end-Cretaceous extinctions, but there is still debate as to what would have been the exact effects of that impact and whether other prior factors influenced patterns of organismal diversity so that the impact was a final blow rather than a sudden catastrophe.

However, where vertebrates are concerned, there are problems in figuring out how such a catastrophe could have resulted in the observed patterns of those animals surviving and those becoming extinct. For the most part, the difference in survival seems to have depended on body size. Almost all vertebrates greater than 10 kilograms became extinct, which included all of the nonavian dinosaurs at that time. Many species of smaller vertebrates (including birds and mammals) also became extinct, but of course some survived to reradiate in the Cenozoic. Smaller animals have a greater capacity to reproduce quickly and rebuild their populations. (This is why many endangered mammals today are large species, such as rhinos and big cats.) Thus the relationship between body size, reproduction rate, and the capacity to repopulate can explain some of the pattern of winners and losers in the K-T extinctions. However, we do not know exactly why large crocodiles and turtles survived while other large vertebrates became extinct at this time.

Another problem with the pattern of survivors of the Cretaceous extinctions lies in the survival of the modern groups of amphibians. Frogs, salamanders, and caecilians all survived the K-T boundary, yet modern amphibians today appear to be extremely sensitive to environmental changes, and species are becoming extinct at an alarming rate. It is hard to believe that the type of environmental changes supposed to have followed an impact, including widespread acid rain falling over the Earth's surface, could have spared the groups of modern amphibians. Again, further fossil record studies might throw light on these problems, but the vital evidence is not always present.

In summary, the Cretaceous extinctions affected many other organisms besides dinosaurs, and the patterns of those extinctions make sense in some cases but are problematical in others. We are now fairly certain that an extraterrestrial object hit the Earth at the end of the Mesozoic, but debate continues about whether that event can explain why some lineages became extinct whereas others survived.

Additional Readings

Barrett, P. M., and K. J. Willis. 2001. Did dinosaurs invent flowers? Dinosaur-angiosperm coevolution revisited. *Biological Reviews* 76:411–447.

Benton, M. J. 1989. Patterns of evolution and extinction in vertebrates. In K. C. Allen and D. E. G. Briggs (eds.), *Evolution and the Fossil Record*. London, UK: Belhaven Press, 218–241.

Benton, M. J. 1990. Scientific methodologies in collision: The history of the study of the extinction of the dinosaurs. *Evolutionary Biology* 24:371–400.

Berner, R. A. et al. 2007. Oxygen and evolution. *Science* 316:557–558.

Farmer, C. G., and D. R. Carrier. 2000. Pelvic aspiration in the American alligator (*Alligator mississippiensis*). *Journal of Experimental Biology* 203:1679–1687.

Fastovsky, D. E., and D. B. Weishampel. 2005. *The Evolution and Extinction of Dinosaurs*, 2nd ed. Cambridge, UK: Cambridge University Press.

Hallam, A. 1994. *An Outline of Phanerozoic Biogeography.* Oxford, UK: Oxford University Press.

Hu, Y. et al. 2005. Large Mesozoic mammals fed on young dinosaurs. *Nature* 433:149–152.

Jones, T. D., and N. R. Geist. 2008. Reproductive biology of dinosaurs. In M. K. Brett-Surman et al. (eds.), *The Complete Dinosaur*, 2nd ed. Indiana University Press.

Olsen, P. E. et al. 2002. Ascent of dinosaurs linked to an iridium anomaly at the Triassic–Jurassic boundary. *Science* 296:1305–1307.

Prasad, V. et al. 2005. Dinosaur coprolites and the early evolution of grasses and grazers. *Science* 310:1177–1180.

Self, S. et al. 2008. Sulfur and chlorine in Late Cretaceous Deccan magmas and eruptive gas release. *Science* 319:1654–1657.

Tarduno, J. A. et al. 1998. Late Cretaceous Arctic volcanism: tectonic and climatic connections. *American Geophysical Union, Spring Meeting Abstracts,* Washington, DC.

Willis, K. J., and J. C. McElwain. 2002. *The Evolution of Plants.* Oxford, UK: Oxford University Press.

Wing, S. L., and H.-D. Sues (rapporteurs). 1992. Mesozoic and early Cenozoic terrestrial ecosystems. In A. K. Behrensmeyer et al. (eds.), *Terrestrial Ecosystems through Time.* Chicago. IL: University of Chicago Press, 327–416.

Zhou, Z. et al. 2003. An exceptionally preserved Lower Cretaceous ecosystem. *Nature* 421:807–814.

CHAPTER 16

Mesozoic Diapsids: Dinosaurs, Crocodilians, Birds, and Others

The Diapsida is the most diverse lineage of amniotic vertebrates. The huge nonavian dinosaurs of the Mesozoic era are the most spectacular diapsids, but the lineage also includes most species of extant terrestrial vertebrates. Crocodilians and birds are diapsids, as are lizards, snakes, and turtles. A variety of extinct forms—such as pterosaurs, ichthyosaurs, plesiosaurs, and placodonts—fills the roster of diapsids.

The dinosaur fauna of the Mesozoic was unlike anything that has existed before or since. Many dinosaurs were so enormous that it is difficult to re-create the details of the lives they led because we have no living models of truly large terrestrial vertebrates. Even elephants are only as large as a medium-sized dinosaur.

Birds are the most derived extant diapsids. Their feathers and high metabolic rates are essential elements of their ability to fly, but those features evolved separately. Feathers evolved before flight, illustrating the maxim that the current function of a character is not necessarily the same as its evolutionary origin.

16.1 Mesozoic Fauna

The Mesozoic era, frequently called the Age of Reptiles, extended for more than 180 million years from the close of the Paleozoic 251 million years ago to the beginning of the Cenozoic only 65.5 million years ago. Through this vast period evolved a worldwide fauna that diversified and radiated into most of the adaptive zones occupied by all the terrestrial vertebrates living today and some that no longer exist (e.g., the enormous herbivorous and carnivorous tetrapods called dinosaurs). Although the dinosaurs are the most familiar representatives of the Age of Reptiles, they are only one of many groups.

Inevitably, dealing with such a huge number of animals is complicated and confusing, not only on first acquaintance but even after prolonged study. Parallel and convergent evolution were widespread in Mesozoic tetrapods. Long-snouted fish eaters evolved repeatedly, as did heavily armored quadrupeds and highly specialized marine forms. A trend toward bipedalism was general, and a secondary reversion to quadrupedal locomotion occurred in many forms. Knowledge of phylogenetic relationships is in a state of flux, and the scheme outlined in **Figures 16–1** and **16–2** will undoubtedly need revision as additional material is analyzed.

This chapter begins with a brief review of the phylogenetic relationships of Mesozoic tetrapods and some aspects of their functional morphology and major evolutionary trends. Crocodilians and birds are the living animals most closely related to the Mesozoic forms, and the phylogenetic relationship of crocodilians and birds allows us to make inferences about the ecology and behavior of extinct lineages. Superbly preserved fossils from fine-grained rocks in China are revealing structural details that we have never before been able to see in fossils, and new fossil localities on the African continent, Madagascar, and South America have expanded our understanding of the worldwide diversity of Mesozoic diapsids (Zhou et al. 2003, Wang et al. 2005, Irmis et al. 2007).

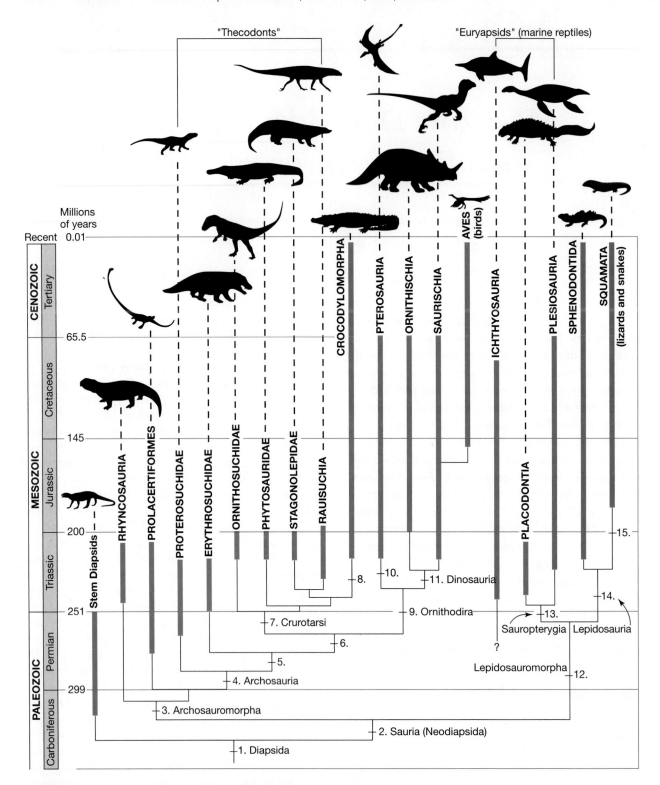

▲ **FIGURE 16–1** Phylogenetic relationships of the Diapsida. This diagram shows the probable relationships among the major groups of diapsids. (Turtles, which may be diapsids, are not included.) Dotted lines show interrelationships only; they do not indicate times of divergence nor the unrecorded presence of taxa in the fossil record. The numbers indicate derived characters that distinguish the lineages.

Figure 16–1 Legend: 1. Diapsida—skull with upper and lower temporal fenestrae, upper temporal arch formed by triradiate postorbital and triradiate squamosal, suborbital fenestra, ossified sternum, complex ankle joint between tibia and astragalus, first metatarsal less than half the length of the fourth metatarsal. **2.** Sauria (Neodiapsida)—anterior process of squamosal narrow, squamosal mainly restricted to top of skull, tabular absent, stapes slender, cleithrum absent, fifth metatarsal hooked, trunk ribs mostly single-headed. **3.** Archosauromorpha—cervical ribs with two heads, various features of limbs including concave-convex articulation between astragalus and calcaneum. **4.** Archosauria—presence of an antorbital fenestra, orbit shaped like an inverted triangle, teeth laterally compressed with serrations. **5.** Pubis and ilium elongated, fourth trochanter on femur. **6.** Crown group Archosauria— parietal foramen absent, no palatal teeth on pterygoid, palatine, or vomer. **7.** Crurotarsi—ankle (tarsus) in which the astragalus forms a distinct peg that fits into a deep socket on the calcaneum, and characters of the cervical ribs and the humerus. **8.** Crocodylomorpha—secondary palate present and includes at least the maxillae. **9.** Ornithodira—anterior cervical vertebrae longer than mid-dorsals, interclavicles absent, clavicles reduced or absent, tibia longer than femur, calcaneal tuber rudimentary or absent, metatarsals bunched together and 2–4 elongated. **10.** Pterosauria—hand with three short fingers and elongate fourth finger supporting wing membrane, pteroid bone in wrist, short trunk, short pelvis with prepubic bones. **11.** Dinosauria—S-shaped swanlike neck, forelimb less than half the length of hindlimb, hand digit 4 reduced, and other characteristics of the palate, pectoral and pelvic girdles, hand, hindlimb, and foot. **12.** Lepidosauromorpha—postfrontal enters border of upper temporal fenestra, supratemporal absent, teeth absent on lateral pterygoid flanges, characteristics of the vertebrae, ribs, and sternal plates. **13.** Sauropterygia—elongation of postorbital region of skull, enlargement of upper temporal fenestra, elongate and robust mandibular symphysis, curved humerus, equal length of radius and ulna. **14.** Lepidosauria—determinant growth with epiphyses on the articulating surfaces of the long bones, postparietal and tabular absent, fused astragalus and calcaneum, and other characteristics of the skull, pelvis, and feet. **15.** Squamata—loss of lower temporal bar (including loss of quadratojugal), highly kinetic skull with reduction or loss of squamosal, nasals reduced, plus other characteristics of the palate and skull roof, vertebrae, ribs, pectoral girdle, and humerus.

16.2 Phylogenetic Relationships Among Diapsids

Our understanding of the phylogenetic relationships of several groups of Mesozoic tetrapods has changed in the past two decades. Many of these forms (crocodilians, pterosaurs, dinosaurs, squamates, and rhynchosaurs) had skulls with two temporal openings (a diapsid skull), and the Diapsida is considered a monophyletic lineage that includes most major groups of Mesozoic tetrapods, as well as the living crocodilians, birds, turtles, tuatara, and squamates.

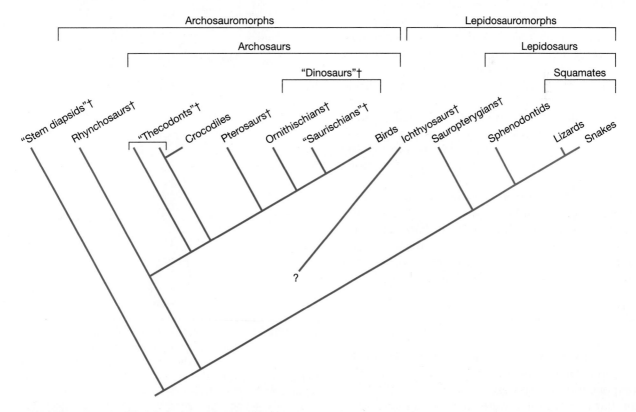

▲ **FIGURE 16–2** Simplified cladogram of the Diapsida. Quotation marks indicate paraphyletic groups and daggers indicate extinct lineages.

The name *diapsid* means "two arches" and refers to the presence of an upper and a lower fenestra in the temporal region of the skull. More distinctive than the openings themselves are the bones that form the arches that border the openings. The upper temporal arch is composed of a three-pronged postorbital bone and a three-pronged squamosal. The lower arch is formed by the jugal and quadratojugal bones. The lower arch has been lost repeatedly in the radiation of diapsids, and the upper arch is also missing in some forms. Extant lizards and snakes clearly show the importance of those modifications of the skull in permitting increased skull kinesis during feeding, and the same significance may attach to loss of the arches in some extinct forms. In addition to the two temporal fenestrae, derived diapsids have a fenestra on each side of the head anterior to the eye, and the presence of this opening modifies the relationships among the bones of the palate and the side of the skull.

The earliest diapsid known is *Petrolacosaurus*, from Late Carboniferous deposits in Kansas. It is a moderately small animal, 60 to 70 centimeters from snout to tail tip, with a long neck, large eyes, and long limbs. It gives the impression of having been an agile terrestrial animal that may have fed on large insects and other arthropods.

The derived diapsids can be split into two groups, the Archosauromorpha (Greek *archo* = ruling, *saur* = reptile, and *morph* = form) and the Lepidosauromorpha (Greek *lepi* = scale). The archosauromorphs include crocodilians and birds, the extinct pterosaurs, dinosaurs, and several Late Permian and Triassic forms. The lepidosauromorphs include the tuatara (Sphenodontidae) and squamates, as well as their extinct relatives. In addition, four groups of specialized marine tetrapods (the placodonts, plesiosaurs, ichthyosaurs, and *Hupehsuchus*) are tentatively considered to be lepidosauromorphs. The skulls of these animals have a dorsal temporal opening but lack a lower temporal fenestra, and the postorbital and squamosal bones do not have the three-pronged shape characteristic of diapsids. However, these patterns are within the range of modifications of the basic diapsid skull that is seen among other members of the clade.

16.3 Archosauria

The archosaurs are the animals most frequently associated with the great radiation of tetrapods in the Mesozoic. Dinosaurs and pterosaurs are distinctive components of many Mesozoic faunas, and other less familiar archosaurs were also abundant. The archosaurs are distinguished by the presence of an antorbital (in front of the eye) fenestra. The skull is deep, the orbit of the eye is shaped like an inverted triangle rather than being circular, and the teeth are laterally compressed (**Figure 16–3a**). A trend toward bipedalism was widespread but not universal among archosaurs, and the ventral side of the femur shaft had a distinctive area with a rough surface, the fourth trochanter (**Figure 16–3b**). The powerful caudofemoral muscle originated on the base of the tail and inserted on the trochanter. When this muscle contracted, it retracted (i.e., pulled back on) the thigh, propelling the animal forward.

■ Crurotarsi

The archosaur stock gave rise to two lineages of aquatic fish eaters, the phytosaurs and crocodilians, that were similar in appearance and probably ecologically

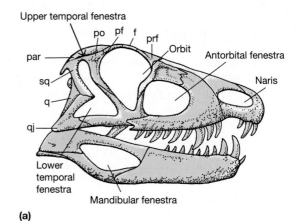

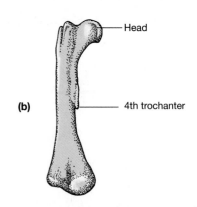

▲ **FIGURE 16–3** Morphological features of archosaurs. (a) Skull of *Ornithosuchus* showing the characteristic features of archosaurs: two temporal arches, an orbit shaped like an inverted triangle, and an antorbital fenestra. (b) Femur of *Thescelosaurus* showing the fourth trochanter. (f = frontal; par = parietal; pf = postfrontal; po = postorbital; prf = prefrontal; q = quadrate; qj = quadratojugal; sq = squamosal)

similar as well. The phytosaurs were the earlier radiation; and, during the Late Triassic period, they were abundant and important elements of the shoreline fauna. The nostrils of crocodilians are at the tip of the snout, and a secondary palate separates the nasal passages from the mouth. A flap of tissue arising from the base of the tongue can form a watertight seal between the mouth and throat. Thus, a crocodilian can breathe without inhaling water while only its nostrils are exposed. Phytosaur nostrils were located on an elevation just anterior to the eyes, and the nasal passages opened into the rear of the mouth, which was probably closed off by a soft-tissue seal like the one found in crocodilians. Both of these arrangements allow animals to float at the water surface with only the nostrils and eyes exposed as they wait for their prey. True crocodilians appeared in the Triassic and seem to have replaced phytosaurs by the end of that period. In most respects, crocodilians conform closely to the skeletal structure of archosaurs, but the skull and pelvis are specialized. The increasing involvement of the premaxilla, the maxillae, and the pterygoids in the secondary palate can be traced from Mesozoic crocodilians to modern forms.

Modern crocodilians are semiaquatic animals, but Triassic crocodilians were terrestrial. They were thin, slender animals about the size of a large cat and give the impression of having been active hunters that probably preyed on smaller diapsids. Extant crocodilians have a ventricular septum in the heart that separates the left (systemic) and right (pulmonary) blood flows. This morphology may be a legacy of the high metabolic rates of the early terrestrial crocodilians (Seymour et al. 2004).

The Cretaceous period was the high point in crocodilian evolution in terms of geographic distribution and ecological diversity. *Simosuchus clarki* (Greek *simo* = flat-nosed and *suchus* = crocodile), a fossil crocodilian recently discovered in Madagascar, was about a meter long. Unlike all other known crocodilians, *Simosuchus* had a short, blunt snout and teeth with multiple cusps, and it may have been herbivorous. The largest crocodilians were as big as the great carnivorous dinosaurs. *Sarcosuchus* (Greek *sarco* = flesh) from Africa and *Deinosuchus* (Greek *deino* = terrible) from Texas were Cretaceous crocodiles 12 to 15 meters long that might have preyed on marine turtles and possibly even on dinosaurs (Schwimmer 2002).

Enormous crocodilians persisted long after dinosaurs disappeared. A skull of the Miocene crocodile *Purussaurus brasiliensis* found in the Amazon Basin in 1986 is 1.5 meters long. If that animal had the same proportions as an alligator, it would have had a total length of 11 to 12 meters and have stood 2.5 meters tall—that is, as high as the ceiling in most houses. An isolated lower jaw in the paleontology museum at the Universidade Federal do Acre is 30 centimeters longer than the jaw of the complete skull and may have come from an animal 13 to 14 meters long. These crocodilians would have been as large as *Tyrannosaurus rex*.

A heavy, laterally flattened tail propels a swimming crocodilian, and the legs are held against the sides of the body. In the Late Jurassic period, a lineage of specialized marine crocodiles enjoyed brief success. These metriorhynchids had long skulls with pointed snouts. They lacked the dermal body armor typical of most crocodilians. They had a lobed tail with the vertebral column turned downward into the lower lobe and the upper lobe supported by stiff tissue, and the feet were modified into paddles. These marine crocodiles were just one among several lineages of diapsids that radiated in the sea (**BOX 16–1**).

Biology of Extant Crocodilians In many respects, crocodilians are the living archosaurs most like Mesozoic forms, and we will consider them first to set a context for understanding dinosaurs. Only 23 species of crocodiles now survive. Most are found in the tropics or subtropics, but three species have ranges that extend into the temperate zone. Systematists divide living crocodilians into three lineages (Brochu 2003): the Alligatoridae, Crocodylidae, and the Gavialidae. The Alligatoridae includes the two species of living alligators and the caimans (**Figure 16–5**). Except for the Chinese alligator, the Alligatoridae is solely a New World group. The American alligator is found in the Gulf coast states, and several species of caimans range from Mexico to South America and through the Caribbean. Alligators and caimans are freshwater forms, whereas the Crocodylidae includes species such as the saltwater crocodile that inhabits estuaries, mangrove swamps, and the lower regions of large rivers. This species occurs widely in the Indo-Pacific region and penetrates the Indo-Australian archipelago to northern Australia. In the New World, the American crocodile is quite at home in the sea and occurs in coastal regions from the southern tip of Florida through the Caribbean to northern South America.

The saltwater crocodile is probably the largest living species of crocodilian. Until recently, adults may have reached lengths of 7 meters. Crocodilians grow slowly once they reach maturity, and it takes a long time to attain large size. In the face of intensive hunting in the past two centuries, few crocodilians now

BOX 16–1 Marine Diapsids of the Mesozoic Era

Archosaurs dominated the land and skies of the Mesozoic era, but lepidosauromorphs were the diapsids that became secondarily aquatic and exploited the resources of the oceans, especially in coastal regions and in the shallow epicontinental seas that spread across both North America and Eurasia during the late Mesozoic. Although all of the marine forms were carnivorous, they specialized on prey ranging from sessile mollusks to fish and free-swimming cephalopod mollusks.

Placodonts The Triassic placodonts were stocky, short-legged mollusk eaters specialized for crushing hard-shelled food rather than for rapid pursuit of prey **(Figure 16–4a)**. *Placodus* had large, flat maxillary teeth and a heavy palate with enormous teeth on the palatine bones. The anterior teeth projected forward and might have been used to pull mussels or oysters off rocks. Some placodonts—such as *Henodus* and *Placochelys*—converged on the body form of turtles, with horny beaks and bony body coverings.

Plesiosaurs The plesiosaurs appeared in the Late Triassic and persisted to the Cretaceous period. Primitive forms had slightly elongated necks, and their heads were proportional to their body size **(Figure 16–4b)**. Two ecological specializations are represented among derived plesiosaurs: pliosauroids had long skulls (more than 3 meters in some forms) and short necks (about 13 cervical vertebrae), whereas plesiosauroids had small skulls and exceedingly long necks (32 to 76 vertebrae). Both types had heavy, rigid trunks and appear to have rowed through the water with limbs that acted like oars and may also have served as hydrofoils, increasing the efficiency of swimming. Hyperphalangy, the addition of extra bones lengthening the fingers and toes, increased the size of the paddles, and some plesiosaurs had as many as 17 phalanges per digit. In both types of plesiosaurs, the nostrils were located high on the head just in front of the eyes.

The pliosauroids developed an increasingly streamlined body form during their evolution as the neck became shorter and the paddles larger, whereas plesiosauroids became less streamlined as their necks lengthened and the paddles became smaller in proportion to body size. Pliosauroids were probably speedy swimmers that might have captured swimming cephalopod mollusks and fish by pursuing them the way seals and sea lions their prey. Plesiosauroids may have been ambush hunters, although it is not clear how they would have captured prey because their necks appear to have been quite rigid.

Ichthyosaurs Ichthyosaurs were the most specialized of the aquatic tetrapods of the Mesozoic **(Figure 16–4c)**. In many aspects of their body form, they resemble tunas, sharks, and porpoises (Motani 2005). Ichthyosaurs had a dorsal fin that was supported only by stiff tissue, not by bone, and the upper lobe of the caudal fin similarly lacked skeletal support. (The vertebral column of derived ichthyosaurs bent sharply downward into the ventral lobe of the caudal fin.) We know of the presence of these soft tissues because many ichthyosaur fossils in fine-grained sediments near Holzmoden in southern Germany contain an outline of the entire body preserved as a carbonaceous film.

Ichthyosaurs had both forelimbs and hindlimbs (unlike whales and dolphins, which retain only the forelimbs). The limbs of ichthyosaurs were modified into paddles by both hyperphalangy (as in plesiosaurs) and by hyperdactyly (the addition of extra fingers and toes). Fossil ichthyosaurs with embryos in the body cavity indicate that these animals gave birth to fully formed young. One fossil appears to be an individual that died in the process of giving birth, with a young ichthyosaur emerging tail-first as do baby porpoises.

The stomach contents of ichthyosaurs, preserved in some specimens, include cephalopods, fishes, and an occasional pterosaur. Ichthyosaurs had large heads with long, pointed jaws that were armed with sharp teeth in most forms, although a few ichthyosaurs were toothless. Ichthyosaurs had very large eyeballs that were supported by a ring of sclerotic bones. *Ophthalmosaurus,* which had larger eyes than any other vertebrate, is believed to have hunted at great depth—500 meters

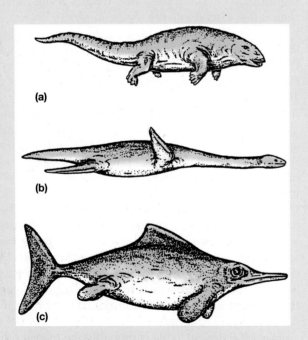

▲ **FIGURE 16–4** Mesozoic aquatic reptiles. (a) The placodont *Placodus* from the Middle Triassic (approx. 1 meter long). (b) Late Jurassic plesiosaur *Cryptoclidus* (approx. 3 meters long). (c) Late Jurassic ichthyosaur *Opthalamosaurus* (approx. 2.5 meters long).

or more—and detected light emitted by the photophores of its prey. Deep-diving animals risk caisson disease (the bends) if an emergency such as the need to avoid a predator forces them to rise rapidly to the surface. The two ichthyosaurs with the largest eyes (suggesting that they were the deepest-diving forms) were also the two that showed the highest incidence of caisson disease (Motani et al. 1998).

Triassic ichthyosaurs were elongate and poorly streamlined and may have used anguilliform locomotion. The greater streamlining of later forms may have been associated with the development of carangiform locomotion and rapid pursuit of prey like that of extant tunas. The Jurassic was the high point of ichthyosaur diversity. They were less abundant in the Early Cretaceous, and only a single genus remained in the Late Cretaceous. Ichthyosaurs became extinct before the end of the period.

attain the sizes they are genetically capable of reaching. Not all crocodilians are giants; several diminutive species live in small bodies of water. The dwarf caiman of South America and the dwarf crocodile of Africa are about a meter long as adults and live in swift-flowing streams.

The third family of crocodilians, the Gavialidae, contains only a single species—the gharial, which once lived in large rivers from northern India to Burma. This species has the narrowest snout of any crocodilian; the mandibular symphysis (the fusion between the mandibles at the anterior end of the lower jaw) extends back to the level of the 23rd or 24th tooth in the lower jaw. A very narrow snout of this sort is a specialization for feeding on fish that are caught with a sudden sideward jerk of the head.

The muscle that opens a crocodilian's mouth (the depressor mandibulae) runs from the rear of the skull to a retroarticular process (i.e., an extension of the bone beyond its articulation) on the mandible. The depressor mandibulae is a short muscle with little mechanical advantage. A person can readily hold a crocodilian's mouth closed, as viewers can see on television nature shows almost daily. The jaw-closing muscles, in contrast, are very powerful. Broad-snouted crocodilians feed on adult turtles that they crush in their jaws. The snout shapes of crocodilians are so closely linked to their feeding specializations that similar shapes have evolved repeatedly in different lineages from the Mesozoic onward (Brochu 2001).

Extant crocodilians are basically aquatic, although they have well-developed limbs and some species

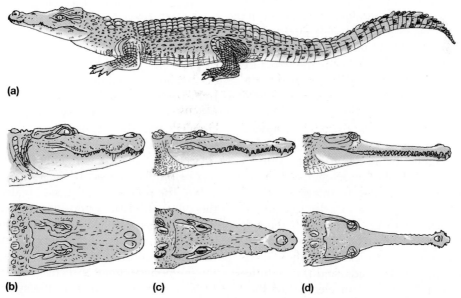

(a)

(b) (c) (d)

▲ **FIGURE 16–5** Crocodilians. Modern crocodilians differ little from one another and from Late Mesozoic forms. The greatest interspecific variation in living crocodilians is seen in the head shape. Alligators and caimans are broad-snouted forms with varied diets. Crocodiles include a range of snout widths. The widest crocodile snouts are almost as broad as those of most alligators and caimans, and these species of crocodilians have varied diets that include turtles, fishes, and terrestrial animals. Other crocodiles have very narrow snouts, and these species are primarily fish eaters. (a) Cuban crocodile. (b) Chinese alligator. (c) American crocodile. (d) Gharial.

make extensive overland movements. Crocodilians can gallop, moving the limbs from their normal laterally extended posture to a nearly vertical position beneath the body. Unlike some of the more terrestrial Mesozoic forms, extant crocodilians hunt in water. The upper and lower jaws are covered with small bulges that are exquisitely sensitive pressure receptors. In complete darkness, American alligators can lunge toward the point of impact of a single drop of water falling on a water surface (Soares 2002).

Some crocodilians, such as the Nile crocodile, wait in ambush at the water's edge and attack large mammals like antelope and zebra when they come to drink. In tropical areas of Australia, warnings are posted beside rivers and lakes to alert people to the very real danger of attack by crocodiles. After seizing an animal, a crocodilian drags it under water to drown. When the prey is dead, the crocodilian bites off large pieces and swallows them whole. Sometimes crocodilians wedge a dead animal into a tangle of submerged branches or roots to hold it as the crocodilian pulls chunks loose. Alternatively, crocodilians can use the inertia of a large prey item to pull off pieces: the crocodilian bites the prey and then rotates rapidly around its own long axis, tearing loose the portion it is holding. Some crocodilians leave large prey items to decompose for a few days until they can be dismembered easily.

Living crocodilians are ectotherms, and small individuals bask in the sunlight to raise their body temperatures. A basking crocodilian can increase its rate of heating by using a right-to-left intracardiac blood shunt to increase blood flow in the peripheral circulation, just as lizards do. However, the structure of the crocodilian heart is different from that of the squamate and turtle heart, and the intracardiac blood shunt is achieved in a different way.

Unlike turtles and squamates, crocodilians have a four-chambered heart—that is, the ventricle is divided into left and right halves by a septum. This feature of crocodilian hearts leads to the conjecture that modern ectothermal crocodilians are descendants of endothermal ancestors. It is the absence of a septum in the hearts of squamates and turtles that permits them to use pressure differentials to shift blood from the pulmonary [right] side of the ventricle across the muscular ridge to the systemic [left] side, and this right-to-left shift of blood is an important element of their physiological control of body temperature. Extant crocodilians use a right-to-left shift of blood to speed heating, but the mechanism is different from that of squamates and turtles. In the crocodilian heart, the right aortic arch opens from the left ventricle and

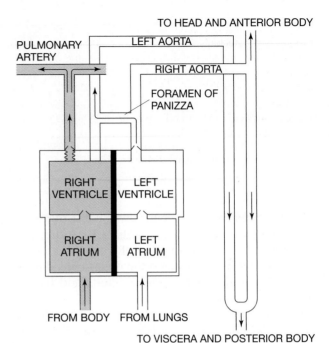

▲ **FIGURE 16–6** The relationship of the heart and major vessels of a crocodilian. The right aortic arch opens from the left ventricle and receives oxygenated blood, which flows to both the anterior and posterior parts of the body. The left aortic arch opens from the right ventricle. When pressure in the right ventricle equals or exceeds pressure in the left, the right atrioventricular valve opens and deoxygenated blood flows into the left aorta, which carries blood only to the posterior part of the body. When pressure in the left ventricle exceeds pressure in the right ventricle, the right atrioventricular valve is held shut, and oxygenated blood flows via the foramen of Panizza into the left aortic arch. (White = oxygenated blood; shaded = deoxygenated blood.)

receives oxygenated blood (**Figure 16–6**). The left aortic arch and the pulmonary artery both open from the right ventricle. The flow of blood is controlled by the resistance to flow in the systemic versus pulmonary circuits, and the pressures change depending on what the alligator is doing. Small, rounded nodules project into the outflow tract to the pulmonary artery just before blood reaches the ventricular pulmonary valve. The nodules on one side match to nodules on the other side like opposing knuckles and form a valve that is controlled by the hormones that activate beta-adrenoceptors (epinephrine and norepinephrine). Activation of the valve increases resistance in the pulmonary circuit and creates a right-to-left shunt of deoxygenated blood into the systemic circulation. Although beta-receptors are found in cardiac muscle of vertebrates where they are involved in regulating pressure in the coronary arteries, this is believed to be the first example of an actively controlled heart valve (Franklin and Axelsson 2000).

When an alligator is at rest, blood pressure is approximately the same in the right and left ventricles. In this situation, deoxygenated blood does flow from the right ventricle into the left aortic arch and then posteriorly to the viscera. Deoxygenated blood contains hydrogen ions that are produced when carbon dioxide combines with the bicarbonate buffering system of the blood. The hydrogen ions that enter the left aortic arch may be used for the secretion of hydrochloric acid in the stomach during digestion. Note that the *right* aortic arch supplies blood to the head, so even in this situation the brain receives only oxygenated blood.

A different pattern of blood flow occurs when the alligator is active. In this condition, more blood must be delivered to the muscles; and, to accomplish that, the pressure in the left ventricle rises above that of the right ventricle. The left and right aortic arches are connected via the foramen of Panizza. When pressure in the left aortic arch exceeds that in the right, blood flows through this passage from the left aortic arch into the right (a left-to-right shunt). The increased pressure in the right aortic arch holds the ventricular valve closed, preventing entry of deoxygenated blood from the right ventricle. Thus, during activity both aortic arches receive oxygenated blood.

A third pattern of blood flow probably occurs when an alligator dives and is holding its breath. In this situation, blood vessels in the pulmonary circuit are constricted and the pressure in the right ventricle rises to match the pressure in the left ventricle. Under these conditions a right-to-left shunt forms and a substantial volume of deoxygenated blood flows into the left aortic arch. The same right-to-left shunt probably occurs when a cold crocodilian is basking in the sun. Crocodilians (at least small individuals—no one has tested this with adult crocodilians!) can increase their rate of heating by increasing blood flow through the limbs. The legs have a large surface area in relation to their volume, and blood flowing through the legs rapidly transfers heat to the core of the body. Increasing the resistance to flow in the pulmonary circuit to produce a right-to-left shunt of blood in the heart is probably how crocodilians increase systemic blood flow to warm rapidly in the sun.

Crocodilians and Birds as Models for Dinosaurs

A strength of the cladistic method of determining evolutionary relationships is the emphasis it places on shared derived characteristics of related organisms. Usually these are morphological characters, and they are used to draw inferences about phylogenetic relationships, but the process can be used in other ways. For example, if a phylogeny can be established by using morphological characters, other characteristics—ecological, behavioral, and physiological—can be superimposed on the phylogeny, and their evolution can be interpreted in a phylogenetic context.

Morphological features common to crocodilians and birds place them in the archosaur lineage, and this relationship can be combined with information about the social behavior of the two groups to answer questions about the probable behavior of dinosaurs. For example, what sorts of parental care and vocal behavior did dinosaurs display? This question can be addressed by analyzing the phylogenetic history of parental care and vocalization in the archosaur lineage.

The extensive parental care provided by many birds has long been known, partly because birds are generally conspicuous animals that are relatively easy to study. Vocalizations are important in the parental care of birds. The chicks cluck within the eggs as they near hatching, chirp as they beg for food, and shriek when they feel threatened. Adult birds of some species give warning calls that cause their young to freeze and remain motionless until the danger has passed.

Parental care and vocal communication with their young by crocodilians is less well known than that of birds, but it appears to be as extensive. All crocodilians probably protect their nests, and elaborate parental care has been described for some species. Vocal communication between juveniles and adults begins before eggs hatch and continues after the babies are out of the nest (Lang 1986).

Baby crocodilians begin to vocalize before they have fully emerged from the eggs, and these vocalizations are loud enough to be heard some distance away. The calls of the babies stimulate one or both parents to excavate the nest, using their feet and jaws to pull away vegetation or soil (**Figure 16–7**). For example, the female American alligator bites chunks of vegetation out of her nest to release the young when they start to vocalize. Then she picks up the babies in her mouth and carries them—one or two at a time—to water, where she releases them. This process is repeated until all the hatchlings have been carried from the nest to the water. The parents of some species of crocodilians gently break the eggshells with their teeth to help the young escape. The sight of a crocodile, with jaws that could crush the leg of a zebra, delicately cracking the shell of an egg little larger than a hen's egg and releasing the hatchling unharmed is truly remarkable.

Young crocodilians stay near their mother for a considerable period—two years for the American

(a)

(b)

(c)

▲ **FIGURE 16–7** Parental care by the mugger crocodile, *Crocodylus palustris*. The numbered tag on the head allows individuals to be recognized. (a) Male parent picking up a hatchling. (b,c) Male parent releasing the hatchling in the water, 9 meters away, where the mother is waiting.

alligator, three years for the spectacled caiman of South America—and may feed on small pieces of food the female drops in the process of eating. Like many birds, baby crocodilians are capable of catching their own food shortly after they hatch and are not dependent on their parents for nutrition.

Adult crocodilians, like birds, are vocal archosaurs (Lang 1989). Male crocodilians emit a variety of vocalizations during courtship and territorial displays and also slap their heads and tails against the water. Vocal displays are especially important for crocodilians like American alligators that live in dense swamps, because males' territories are often out of sight of other males and females. The roar of a male alligator resounds through the swamp and announces his presence to other alligators up to 200 meters away. Female alligators also roar, but only males produce a subaudible vocalization (i.e., in a frequency below the range of human hearing) that causes drops of water to dance on the water surface and travels for long distances underwater.

Frightened crocodilian hatchlings emit a distress squeak that stimulates adult male and female crocodilians to come to their defense. In addition to summoning the parents, these vocalizations may attract unrelated adults. When staff members at a crocodile farm in Papua New Guinea rescued a hatchling New Guinea crocodile that had strayed from the pond, its distress call brought 20 adult crocodiles surging toward the hatchling (and the staff members!). The dominant male head-slapped the water repeatedly, and then charged into the chain-link fence where the staff members were standing while the females swam about, gave deep guttural calls, and head-slapped the water.

Nesting behaviors and parental care of crocodilians overlap those of many birds. For example, the bush turkeys (megapodes) of Australia bury their eggs in piles of soil and vegetation in craters they excavate in the ground and release their young at the end of incubation. The young disperse as soon as they emerge from the nest. The young of many birds—including familiar species such as ducks, chickens, and quail—are well developed at hatching (**precocial**) and able to find their own food. In these birds, as in crocodilians, the important function of parental care appears to be protecting the nests and young.

The most parsimonious explanation of the presence of well-developed vocalizations, elaborate nest construction and nest guarding, and care of young in crocodilians and birds is that these behaviors were present in the common ancestor of these two groups. In other words, parental care of young and vocalization for social communication appear to be ancestral characters of the archosaur lineage, at least at the level of crocodilians. If that is the case, dinosaurs would have inherited parental care and vocal communication as a part of their ancestral behavioral repertoire, and we would expect them to have exhibited the behaviors seen in birds and crocodilians. Behavior is difficult to decipher from the fossil record, but evidence is accumulating to suggest that dinosaurs did indeed engage in parental care, vocal communication, and other forms of complex social behaviors.

16.4 Dinosaurs

By far the most generally known of the archosaurs are the Ornithischia and Saurischia. These groups are linked in popular terminology as dinosaurs, but they are independent radiations and differ in the specializations they developed. The groups had a common ancestor that was bipedal, and both evolved some secondarily quadrupedal forms. Initially most of our information about dinosaurs came from North America and Europe; but, in the past 20 years, new discoveries in South America, Russia, China, Africa, and Madagascar have revealed details of a worldwide dinosaur fauna in the Mesozoic. The pattern that is emerging suggests that early in the Mesozoic, when the continents were still broadly connected, dinosaur faunas were quite similar. By the Cretaceous, the continents had moved almost to their current positions, and were more widely separated than they had been earlier in the Cretaceous. This geographic separation allowed regional differences in dinosaur faunas to develop and the Late Cretaceous was a high point in dinosaur diversity (e.g., Rich et al. 2002, Sereno et al. 1999, Sampson et al. 2001, Xu et al. 2002).

Many morphological trends that can be traced in archosaur evolution appear to be associated with increased locomotor efficiency. The two most important developments were movement of the legs under the body and a widespread tendency toward bipedalism (Hutchinson 2006). Early archosauromorphs had a sprawling posture like that of many living amphibians and squamates. The humerus and femur were held out horizontally from the body, and the elbow and knee were bent at a right angle. Derived archosaurs have legs that are held vertically beneath the body.

Among early tetrapods, muscles originating on the pubis and inserting on the femur protract the leg (move it forward), muscles originating on the ischium adduct the femur (move it toward the midline of the body), and muscles originating on the tail retract the femur (move it posteriorly). The ancestral tetrapod pelvis, little changed from *Ichthyostega* through early archosauromorphs, was platelike (**Figure 16–8a**). The ilium articulated with one or two sacral vertebrae, and the pubis and ischium did not extend far anterior or posterior to the socket for articulation with the femur (acetabulum). The pubofemoral and ischiofemoral muscles extended ventrally from the pelvis to insert on the femur. (The downward force of their contraction was countered by iliofemoral muscles that ran from the ilium to the dorsal surface of the femur.) As long as the femur projected horizontally from the body, this system was effective. The pubofemoral and caudofemoral muscles were long enough to swing the femur through a large arc relative to the ground. As the legs were held more nearly under the body, the pubofemoral muscles became less effective. As the femur rotated closer to the pubis, the sites of muscle origin and insertion moved closer together, and the muscles themselves became shorter. A muscle's maximum contraction is about 30 percent of its resting length; thus the shorter muscles would have been unable to swing the femur through an arc large enough for effective locomotion had there not been changes in the pelvis associated with the evolution of bipedalism.

The bipedal ornithischian and saurischian dinosaurs carried the legs completely under the body and show associated changes in pelvic structure. The two groups attained the same mechanically advantageous result in different ways. In quadrupedal saurischians, the pubis and ischium both became elongated and the pubis was rotated anteriorly, so that the pubofemoral muscles ran back from the pubis to the femur and were able to protract it (**Figure 16–8b**). The pubis of early ornithischians did not project anteriorly. Instead, the ilium was elongated

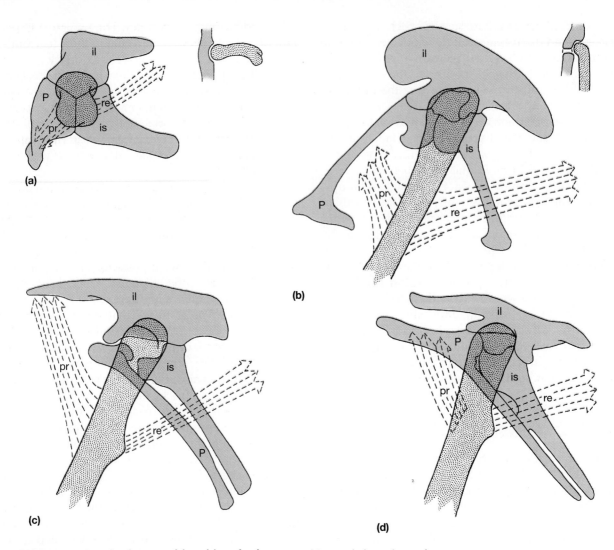

▲ **FIGURE 16–8** Functional aspects of the pelvises of archosaurs. Pelvic morphology of an early archosaur (a, *Euparkeria*), a saurischian theropod dinosaur (b, *Ceratosaurus*), and two ornithischian dinosaurs (c, *Scelidosaurus*; d, *Thescelosaurus*). The presumed action of femoral protractor muscles (pr) and retractors (re) is shown by dotted arrows. Insets show an anterior view of the articulation of the femur with the pelvis. The prominent expansion of the end of the pubis of the theropod *Ceratosaurus* is probably associated with cuirassal ventilation of the lungs. (P = pubis; il = ilium; is = ischium)

anteriorly, and it appears likely that the femoral protractors originated on the anterior part of the ilium, from which they ran posteriorly to the femur. This condition is seen in the pelvis of ornithischians such as *Scelidosaurus* (**Figure 16–8c**) and appears to be maintained in the ankylosaurs, a group of derived quadrupedal ornithischians. Other ornithischians developed an anterior projection of the pubis that ran parallel to and projected beyond the anterior part of the ilium (**Figure 16–8d**). This development occurred in both bipedal and quadrupedal lineages and provided a still more anterior origin for protractor muscles.

The trend toward bipedalism was important in opening new adaptive zones to archosaurs. A fully quadrupedal animal uses its forelimbs for walking, and any changes in limb morphology must be compatible with that function. As animals become increasingly bipedal, the importance of the forelimbs for locomotion decreases and the scope of specialized functions that can develop increases. Many of the smaller carnivorous dinosaurs that were fully bipedal used their forelimbs to seize prey. Specialization of forelimbs as wings occurred twice among diapsids, once in the evolution of birds and once in pterosaurs.

Bipedal animals have hind legs that are considerably longer than their front legs, and the degree of disproportion between hind legs and fore legs is assumed to reflect the extent of bipedalism in a given species. The quadrupedal dinosaurs had longer hind legs than front legs, indicating that they had evolved from bipedal ancestors and were secondarily quadrupedal.

16.5 The Ornithischian Dinosaurs

Differences in the structure of the pelvis of saurischian and ornithischian dinosaurs indicate an early separation of the two groups. However, ornithischians and saurischians show similarities in body form that probably reflect the mechanical problems of being very large terrestrial animals. Ornithischians were herbivorous and radiated into considerably more diverse morphological forms than did the herbivorous sauropod saurischians. Many ornithischians probably had cheeks and horny beaks rather than teeth at the front of the mouth. The larger ornithischians were not as bipedal as some large theropod saurischians, and the forelimbs were never so greatly reduced.

Three groups of ornithischian dinosaurs can be distinguished (**Figures 16–9** and **16–10**).

- Thyreophora—The armored dinosaurs. Quadrupedal forms including stegosaurs (forms with a double row of plates or spines on the back and tail) and ankylosaurs (heavily armored forms, some with clublike tails).
- Ornithopoda—Bipedal or quadrupedal forms, including the duck-billed dinosaurs.
- Marginocephalia—The pachycephalosaurs (bipedal dinosaurs with enormously thick skulls) and ceratopsians (the quadrupedal horned dinosaurs).

■ Thyreophora–Stegosaurs and Ankylosaurs

The stegosaurs were a group of quadrupedal herbivorous ornithischians that were most abundant in the Late Jurassic, although some forms persisted to the end of the Cretaceous. *Stegosaurus*, a large form from the Jurassic of western North America, is the most familiar of these dinosaurs. It was up to 6 meters long, and its front legs were much shorter than its hind legs (**Figure 16–11**). A double series of leaf-shaped plates were probably set alternately on the left and right sides of the vertebral column. Two pairs of spikes on the tail made it a formidable weapon.

The function of the plates of *Stegosaurus* has been a matter of contention for decades. Originally they were assumed to have provided protection from predators, and some reconstructions have shown the plates lying flat against the sides of the body as shields. A defensive function is not very convincing, however. Whether the plates were erect or flat, they left large areas on the sides of the body and the belly unprotected. Another idea is that the plates were used for heat exchange. Examination shows that the plates were extensively vascularized and could have carried a large flow of blood to be warmed or cooled according to the needs of the animal. *Kentrosaurus*, the African counterpart of *Stegosaurus*, had much smaller dorsal plates than *Stegosaurus*; the plates on *Kentrosaurus* extended only from the neck to the middle of the trunk. Posteriorly a double row of spikes extended down the tail. These spikes appear to have had a primarily defensive function rather than a thermoregulatory one. It is frustrating not to be able to compare the thermoregulatory behaviors of the two kinds of stegosaurs in a controlled experiment.

The short front legs of stegosaurs kept their heads close to the ground, and their heavy bodies do not give the impression that they were able to stand upright on their hind legs to feed on trees as ornithopods and perhaps sauropods did. Stegosaurs may have browsed on ferns, cycads, and other low-growing plants. The skull was surprisingly small for such a large animal and had the familiar horny beak at the front of the jaws. The teeth show none of the specializations seen in hadrosaurs or ceratopsians, and the coronoid process of the lower jaw is not well developed. Unlike hadrosaurs and ceratopsians, which appear to have been able to grind or cut plant material into small pieces that could be digested efficiently, stegosaurs may have eaten large quantities of food without much chewing and used **gastroliths** (Greek *gastro* = stomach and *lith* = stone) in a muscular gizzard to pulverize plant material.

The ankylosaurs are a group of heavily armored dinosaurs that are found in Jurassic and Cretaceous deposits in North America and Eurasia. Ankylosaurs were quadrupedal ornithischians that ranged from 2 to 6 meters in length. They had short legs and broad bodies, with **osteoderms** (bones embedded in the skin) that were fused together on the neck, back, hips, and tail to produce large shieldlike pieces. Bony plates also covered the skull and jaws, and even the eyelids of *Euoplocephalus* had bony armor. Ankylosaurs had short tails, and some species had a lump of bone at the end of the tail that could apparently be swung like a club. The posteriormost caudal vertebrae of these

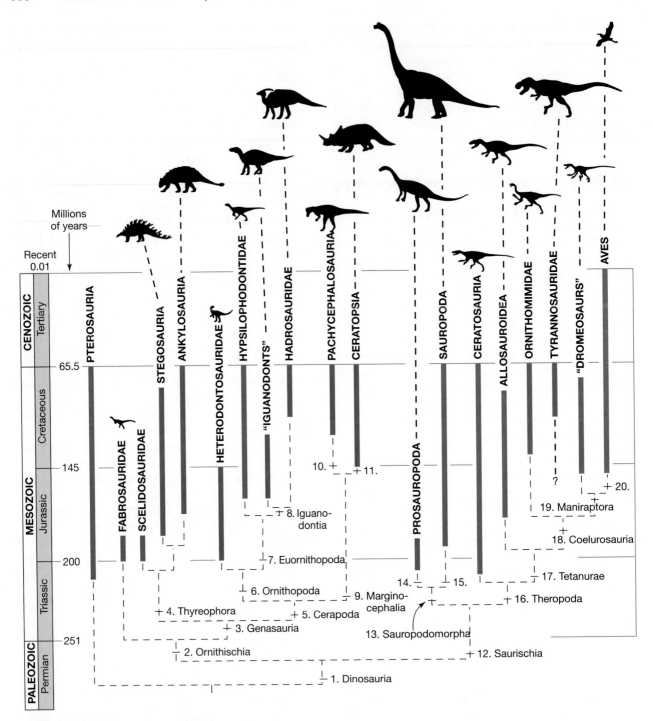

▲ **FIGURE 16–9** Phylogenetic relationships of the Dinosauria. This diagram shows the probable relationships among the major groups of dinosaurs, including birds. Dotted lines show interrelationships only; they do not indicate times of divergence or the unrecorded presence of taxa in the fossil record. The numbers indicate derived characters that distinguish the lineages. Quotation marks indicate paraphyletic groups.

Legend: 1. Dinosauria—S-shaped swanlike neck, forelimb less than half the length of hindlimb, hand digit 4 reduced, and other characteristics of the palate, pectoral and pelvic girdles, hand, hindlimb, and foot. **2.** Ornithischia—cheek teeth with low subtriangular crowns, reduced antorbital opening, predentary bone, toothless and roughened tip of snout, jaw joint set below level of upper tooth row, at least five sacral vertebrae, ossified tendons above sacral region, pelvis with pubis directed backward, small prepubic process on pubis. **3.** Genasauria—muscular cheeks, reduction in size of mandibular foramen.

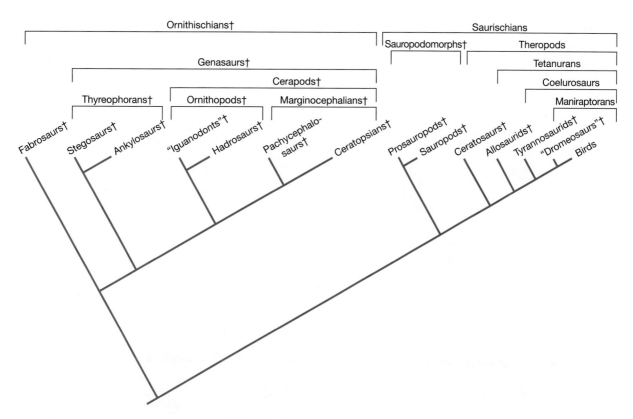

▲ **FIGURE 16–10** Simplified cladogram of the Dinosauria. Quotation marks indicate paraphyletic groups and daggers indicate extinct lineages.

club-tailed forms have elongated neural and hemal arches that touch or overlap the arches on adjacent vertebrae, and ossified tendons running down both sides of the vertebrae. Contraction of the muscles that inserted on these tendons probably pulled the posterior caudal vertebrae together to form a stiff handle for swinging the club head at the end of the tail. The tail of these animals resembles nothing so much as an

4. Thyreophora—characters of the orbit of the eye, rows of keeled scutes on the dorsal body surface. **5.** Cerapoda—five or fewer premaxillary teeth, a diastema between premaxillary and maxillary teeth, characters of the pelvis. **6.** Ornithopoda—premaxillary tooth row offset ventrally compared to maxillary tooth row, jaw joint set well below level of tooth rows by ventral extension of quadrate. **7.** Euornithopoda—absence of a bony prominence in cheek region. **8.** Iguanodontia—premaxillary teeth absent, external naris enlarged, wrist bones fused, also characters of the tooth surfaces and tooth enamel, lower jaw and skull. **9.** Marginocephalia—a shelf formed by the parietals and squamosals extends over the occiput, characters of the snout and pelvis. **10.** Pachycephalosauria—thickened skull roof (frontals and parietals), other characters of the skull, dorsal vertebrae, forelimbs, and pelvis. **11.** Ceratopsia—head triangular in dorsal view; tall, narrow anterior beak; jugals flare beyond the skull roof; deep, transversely arched palate, immobile mandibular symphysis. **12.** Saurischia—construction of the snout including subnarial foramen, extension of the temporal musculature onto the frontal bones, elongation of the neck, modifications of the articulations between vertebrae, and modifications of the hand (including a large thumb). **13.** Sauropodomorpha—relatively small skull, anterior end of premaxilla deflected, teeth with serrated crowns, at least 10 cervical vertebrae forming an elongated neck. **14.** Prosauropoda—elongation of claw on digit 1 of hand. **15.** Sauropoda—four or more sacral vertebrae, straight femur with lesser trochanter reduced or absent. **16.** Theropoda—articulation in middle of the lower jaw, construction of bones of the skull roof and palate, fenestra in the maxilla, characters of the vertebrae and neural arches, and no transverse processes posterior to a transition point from mobile to fixed articulations between vertebrae in the middle of the tail, hand with elongated digits 1–3 armed with highly recurved claws, fibula and tibia closely adpressed, foot long and narrow with fifth metatarsal reduced to a splint, thin-walled (hollow) long bones. **17.** Tetanurae—large fenestra posteriorly located in the maxilla, large fanglike teeth absent from dentary, maxillary tooth row ends anterior to orbit of eye, transition point in the tail is farther anterior than in other theropods, expanded distal portion of pubis, characters of the foot. **18.** Coelurosauria—fenestra in roof of mouth, characters of cervical vertebrae and ribs, furcula (wishbone) formed by fused clavicles, fused bony sternal plates, elongate forelimb and hand, characters of the foot. **19.** Maniraptora—prefrontals reduced or absent, characters of the vertebrae, transition point in tail vertebrae close to base of tail, characteristics of the feet and pelvis. **20.** Aves—progressive loss of teeth on maxilla and dentary, well-developed bill, feathers, characteristics of skull, jaws, vertebrae, and axial and appendicular skeleton.

▶ **FIGURE 16–11** Quadrupedal ornithischians. (a) *Stegosaurus* (Late Jurassic, up to 9 meters) and (b) *Kentrosaurus* (Late Jurassic, up to 5 meters) were stegosaurs. (c) *Euoplocephalus*, an ankylosaur (Late Cretaceous, up to 7 meters). (d) *Styracosaurus*, a ceratopsian (Late Cretaceous, up to 5.5 meters).

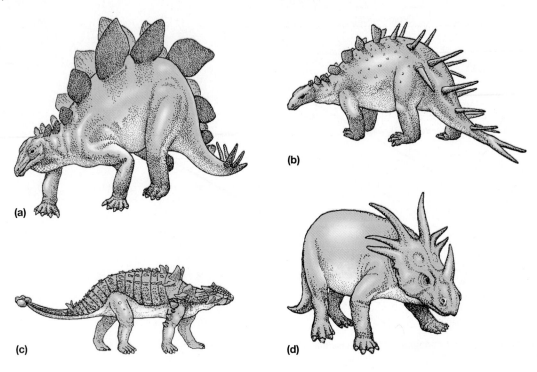

enormous medieval mace. Other species had spines projecting from the back and sides of the body, and ankylosaurs must have been difficult animals for tyrannosaurids to attack. The brains of ankylosaurs appear to have had large olfactory stalks leading to complex nasal passages that probably had sheets of bone supporting an epithelium with chemosensory cells. If this interpretation is correct, ankylosaurs may have had a keen sense of smell.

■ Ornithopods

Ornithopods from the Early Jurassic, the heterodontosaurids and related groups, were mostly small (1 to 3 meters long). The hind limbs are substantially longer than the forelimbs (**Figure 16–12**), and that difference has long been interpreted as indicating that they were bipedal. Unlike the bipedal saurischians, however, ornithopods retained five toes on the front feet, and a recent examination of the semicircular

canals of ornithopods suggests that they may have been quadrupedal (Sipla et al. 2004, Stokstad 2004).

The first dinosaur fossil to be recognized as such was an ornithopod, *Iguanodon*, found in Cretaceous sediments in England. Specimens have subsequently been found in Europe and Mongolia, and related forms have been discovered in Africa and Australia. *Iguanodon* reached lengths of 10 meters, although most specimens are smaller. Iguanodontids from the Early Cretaceous had large heads and elongated snouts that ended in broad, toothless beaks. Their teeth, which were in the rear of the jaws, were laterally flattened and had serrated edges, very like the teeth of living herbivorous lizards like *Iguana*.

The first digit on each front foot of derived ornithopods was modified as a spine that projected upward. These spines show a striking resemblance to spines on the front feet of some frogs that are used as defensive weapons and during intraspecific encounters. *Ouranosaurus*, an ornithopod known from the early

▲ **FIGURE 16–12** Bipedal ornithischians. (a) *Iguanodon* (Early Cretaceous, up to 9 meters). (b) *Pachycephalosaurus* (Late Cretaceous, up to 4.5 meters). (c) *Ouranosaurus* (Early Cretaceous, up to 7 meters). (d) *Hadrosaurus* (Late Cretaceous, up to 10 meters).

Middle Cretaceous of Africa, had a large sail that was supported by elongated neural spines on the vertebrae of the trunk and tail.

Hadrosaurs The derived ornithopods include several specialized forms of hadrosaurs (duck-billed dinosaurs). Hadrosaurs were the last group of ornithopods to evolve, appearing in the middle of the Cretaceous. As their name implies, some duck-billed dinosaurs had flat snouts with a ducklike bill.

These were large animals, some reaching lengths of over 10 meters and weights greater than 10,000 kilograms. The anterior part of the jaws was toothless, but a remarkable battery of teeth occupied the rear of the jaws. On each side of the upper and lower jaws were four tooth rows, each containing about 40 teeth packed closely side by side to form a massive tooth plate. Several sets of replacement teeth lay beneath those in use, so a hadrosaur had several thousand teeth in its mouth, of which several hundred were in use simultaneously. Fossilized stomach contents of hadrosaurs consist of pine needles and twigs, seeds, and fruits of terrestrial plants.

The rise of the hadrosaurs was approximately coincident with a change in the terrestrial flora during the Middle and Late Cretaceous. The conifers, bennettitaleans, and seed ferns (gymnosperms) that had spread during the Triassic now were replaced by flowering plants (angiosperms). Simultaneous with the burgeoning of the angiosperms and hadrosaurian dinosaurs was a decline in the enormous sauropod dinosaurs, such as *Diplodocus* and *Brachiosaurus*. Those lineages were most diverse in the Late Jurassic and Early Cretaceous, and only a few forms persisted after the middle of the Cretaceous.

The coincidence in time of the rise of angiosperms and the radiation of large herbivorous dinosaurs has fueled speculation about a cause-and-effect relationship between these events. Did angiosperms become the dominant plants because they were more resistant to browsing by dinosaurs than were gymnosperms? A detailed analysis failed to support this intriguing hypothesis (Barrett and Willis 2001). A close examination of the timing of the radiations of dinosaurs and angiosperms on different continents shows that they were not closely linked. Indeed, dinosaurs were apparently absent from northern Gondwana, where the earliest angiosperms are found. It seems more likely that rising levels of carbon dioxide in the atmosphere during the Mesozoic, and the global warming that resulted, were the primary factors in the rise of angiosperms.

Three kinds of hadrosaurs are distinguished: flat-headed, solid-crested, and hollow-crested (**Figure 16–13**). In the flat-headed forms (hadrosaurines), the nasal bones are not especially enlarged, although the nasal region may have been covered by folds of flesh. In the solid-crested forms (saurolophines), the nasal and frontal bones grew upward, meeting in a spike that projected over the skull roof. In the hollow-crested forms (lambeosaurines), a similar projection was formed by the premaxillary and nasal bones. In *Corythosaurus,* those bones formed a helmetlike crest that covered the top of the skull; whereas in *Parasaurolophus,* a long, curved structure extended over the shoulders. Although the crests of the saurolophines contained only bone, the nasal passages ran through the crests of the lambeosaurines. Inspired air traveled a circuitous route from the external nares through the crest to the internal nares, which were located in the palate just anterior to the eyes.

Perhaps these bizarre structures were associated with species-specific visual displays and vocalizations. The crests might have supported a frill attached to the neck, which could have been used in behavioral displays analogous to the displays of many living lizards that have similar frills. Possibly, in the noncrested forms, the nasal regions were covered by extensive folds of fleshy tissue that could be inflated by closing the nasal valves. Analogous structures can be found in the inflatable proboscises of elephant seals and hooded seals. The inflated structures of seals are resonating chambers used to produce vocalizations. The size and shape of the nasal cavities of lambeosaurine hadrosaurs suggest that adults produced low-frequency sounds, but juveniles would have had higher-pitched vocalizations.

■ Marginocephalia

The Marginocephalia, two lineages of highly specialized Late Cretaceous herbivores, were the last groups of ornithischians to appear. By this time the continents were close to their present locations, and both of the northern continents were separated into eastern and western parts by inland seas that extended in North America from what is now the Arctic Ocean to the present Gulf of Mexico and in Eurasia from the Arctic to the Indian Ocean. Thus, the geography of the Late Mesozoic did not allow the extensive overland movements that had been possible earlier in the Mesozoic, and lineages of dinosaurs that radiated late in the Cretaceous have limited geographic distributions compared to the lineages that radiated earlier.

Pachycephalosaurs The pachycephalosaurs, which are among the most bizarre ornithischians, occurred in western North America and eastern Asia. In the Late Cretaceous, these regions may have had a land connection across what is now the Bering Sea. A distribution limited to those two areas would be consistent with our understanding of the geography of the Late Cretaceous, but fossil material that may represent pachycephalosaurs has also been found at one locality in western Europe and on Madagascar. Occurrences of this sort may result from dispersal and regional extinction combined with the movements of continents (Upchurch et al. 2002).

The body and limbs of pachycephalosaurs were similar to those of other ornithopods, but an enormous bony dome on the head thickens the skull roof. The bone is as much as 25 centimeters thick in a skull only 60 centimeters long. The angle of the occipital condyle indicates that the head was held so that the axis of the neck extended directly through the dome.

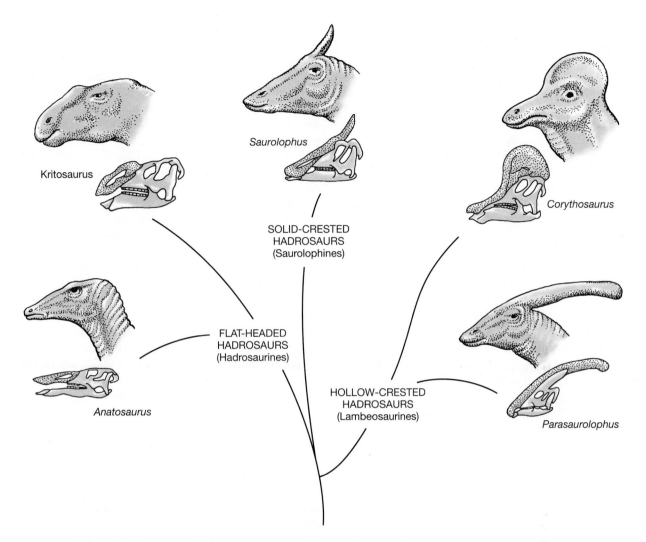

Kritosaurus

Saurolophus

Corythosaurus

SOLID-CRESTED
HADROSAURS
(Saurolophines)

Anatosaurus

FLAT-HEADED
HADROSAURS
(Hadrosaurines)

HOLLOW-CRESTED
HADROSAURS
(Lambeosaurines)

Parasaurolophus

▲ **FIGURE 16–13** Hadrosaurs. The bizarre development of the nasal and maxillary bones of some hadrosaurs gave their heads a superficially antelope-like appearance. In the flat-headed and solid-crested forms, the nasal passages ran directly from the external nares to the mouth. In the hollow-crested forms, the premaxillary and nasal bones contributed to the formation of the crests, and the nasal passages were diverted up and back through the crests before they reached the internal nares.

The trunk vertebrae have articulations and ossified tendons that appear to have stiffened the vertebral column and resisted twisting. The pelvis was attached to at least six, and possibly to eight, vertebrae.

The thickened skull roof and the features of the postcranial skeleton have led some paleontologists to suggest that pachycephalosaurs used their heads like battering rams, perhaps for defense against carnivorous dinosaurs or for intraspecific combat. An analogy has been drawn with goats and especially mountain sheep, in which males and females use head-to-head butting in social interactions. However, sheep and goats have horns that absorb some of the impact, and they have protective air sacs at the front of the brain. Pachycephalosaurs had neither of these specializa-

tions, although stretchable ligaments in the neck may have helped to absorb the shock of impact. The Galápagos marine iguana may be a better model for the behavior of pachycephalosaurs. These lizards have blunt heads with spikes very like miniature versions of the heads of pachycephalosaurs. Male marine iguanas press their heads together, twisting and wrestling during territorial disputes. Perhaps pachycephalosaurs used their bony heads in the same way.

Ceratopsians The most diverse marginocephalians, the horned dinosaurs or ceratopsians, appeared in the Late Jurassic or Early Cretaceous. Ceratopsians are found in western North America and eastern Asia, but they were apparently excluded from the rest of the Northern Hemisphere by shallow epicontinental seas that covered the central parts of both North America and Eurasia in the late Mesozoic.

The distinctive features of ceratopsians are found in the frill over the neck, which is formed by an enlargement of the parietal and squamosal bones, a parrotlike beak, and a battery of shearing teeth in each jaw. The earliest ceratopsians were the bipedal psittacosaurs from Asia. These bipedal dinosaurs had no trace of a frill, but they did have a horny beak that covered a rostral bone at the front of the upper jaw. (The rostral bone is a distinctive feature of ceratopsian dinosaurs.) Early quadrupedal ceratopsians, *Liaoceratops*, *Leptoceratops*, and *Protoceratops*, had modest frills that extended backward over the neck and formed the origin for powerful jaw-closing muscles that extended anteriorly through slits at the rear of the skull and inserted on the coronoid process of the lower jaw. The teeth were arranged in batteries in each jaw, somewhat like those of hadrosaurs but with an important difference. The teeth of ceratopsians formed a series of knifelike edges rather than a solid surface like hadrosaur teeth. The feeding method of ceratopsians seems likely to have consisted of shearing vegetation into short lengths rather than crushing it, as hadrosaurs did.

Early ceratopsians had simple frills, unadorned by spikes, and lacked nasal horns. Derived ceratopsians had both of these elaborations. Two groups can be distinguished: In the short-frilled ceratopsians (*Monoclonius*, *Styracosaurus*, and others), the frill extended backward over the neck, whereas in the long-frilled forms (*Chasmosaurus*, *Pentaceratops*, and others) the frill extended past the shoulders. Both short- and long-frilled ceratopsians had nasal and brow horns developed to varying degrees. Probably the initial stages in the evolution of the frill involved jaw mechanics and the importance of strong jaw muscles. Even in *Protoceratops*, however, males had larger frills than females; this sexual dimorphism suggests that frills played a role in the social behavior of ceratopsians. Furthermore, the nasal and brow horns would have been formidable weapons for defense and for intraspecific combat. An analogy to the horns of antelope or the antlers of deer, which function in both defense and social behavior, seems appropriate for ceratopsians.

■ Social Behavior of Ornithischians

The morphological diversity of the ornithischian dinosaurs suggests that their behavior and ecology were equally diverse. Social interactions based on visual displays and vocalizations may have been well developed among hadrosaurs, and pachycephalosaurs may have engaged in shoving contests. Individual interactions of these sorts may have been integrated into group behaviors. Fossilized eggs of dinosaurs provide information about nesting behaviors (**BOX 16–2**). Evidence of parental care may be revealed by a nest of 15 baby hadrosaurs (*Maiasaura*) in the Late Cretaceous Two Medicine Formation in Montana. The babies were about a meter long—approximately twice the size of other hatchlings found in the same area—indicating that the group had remained together after they hatched. The teeth of the baby hadrosaurs showed that they had been feeding; some teeth were worn down to one-quarter of their original length. The object presumed to be a nest was a mound 3 meters in diameter and 1.5 meters high, with a saucer-shaped depression in the center. Such a large structure would have made the babies conspicuous to predators, and it seems likely that a parent remained with the young. (*Maiasaura* can be translated as "good mother reptile.") The morphology of the inner ears of lambeosaurs suggests that adults would have been able to hear the high-pitched vocalizations of juveniles, strengthening the inference of parental care. The association between adults and young appears to have lasted for a considerable time. Fossils suggest that *Maiasauria* and the lambeosaur *Hypacrosaurus* grew to one-quarter of adult size before they left the nesting grounds, and a species of hypsilophodontid found at the same site grew to half its adult size. Both vocal communication and prolonged association of parent dinosaurs with their young are plausible in light of the behaviors known for crocodilians and birds.

Additional fossils in the same formation suggest that the area contained nest sites of other species of hadrosaurs and of ceratopsians as well. Eggshells and baby dinosaurs are abundant in the Two Medicine Formation but rare in adjacent sediments. A similar concentration of conspicuous nests, eggs, and juveniles of the small ceratopsian *Protoceratops* discovered in Mongolia also suggests parental care in this species.

16.6 The Saurischian Dinosaurs

Two groups of saurischian dinosaurs are distinguished, the Sauropodomorpha (Greek *saur* = reptile, *pod* = foot, and *morph* = form) and the Theropoda (Greek *thero* = wild beast). Sauropodomorphs, all of

BOX 16–2   Dinosaur Eggs and Nests

More than 200 sites have yielded fossils of dinosaur eggs, primarily from Late Cretaceous deposits but including a few as old as the Triassic (Carpenter 1999). Most of these fossils are fragments of eggshells, but intact eggs containing embryos have been discovered (**Figure 16–14**).

Concentrations of nests and eggs ascribed to sauropods in Cretaceous deposits in southern France and Patagonia suggest that these animals had well-defined nesting grounds. Eggs thought to be those of the large sauropod *Hypselosaurus priscus* are found in association with fossilized vegetation similar to that used by alligators to construct their nests. The orientation of the nests suggests that each female dinosaur probably deposited about 50 eggs. The eggs had an average volume of 1.9 liters, about 40 times the volume of a chicken egg. Fifty

might have been filled with rotting vegetation to provide both heat and moisture for the eggs. (This method of egg incubation is used by many of the living crocodilians and by bush turkeys.) Nests of *Protoceratops* fall in this second category—the 30 to 35 eggs in each nest are arranged in concentric circles with their blunt ends up. Eggs of *Orodromeus makelai*, a hypsilophodontid, are also oriented vertically with the blunt end up but are arranged in a spiral within the circular nest.

Nesting and parental care by nonavian dinosaurs was probably similar to that of crocodilians (Horner 2000). At least some of the smaller species of dinosaurs, both ornithischians and saurischians, probably remained with the nest (**Figure 16–15**). The fossil of a theropod dinosaur that apparently died while attending a nest of eggs was discov-

(a)

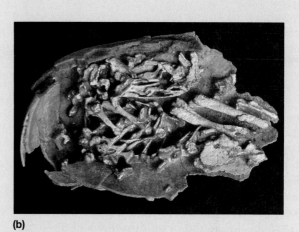
(b)

▲ **FIGURE 16–14** Dinosaur nests and eggs. Fossilized nest of a small coelurosaur (*left*). The fossilized skeleton of an embryo of a nonavian coelurosaur (*right*). This is the first embryo of a carnivorous dinosaur ever found. (*Source:* Image no. K17088. Courtesy Department of Library Services, American Museum of Natural History.)

of these eggs together would weigh about 100 kilograms, or 1 percent of the estimated body mass of the mother. Crocodilians and large turtles have egg outputs that vary from 1 percent to 10 percent of the adult body mass, so an estimate of 1 percent for *Hypselosaurus* seems reasonable. The eggs might have been deposited in small groups instead of all together because 50 eggs in one clutch would have consumed oxygen faster than it could diffuse through the walls of the nest (Seymour 1979).

Two patterns of egg laying can be distinguished (Mikhailov 1997). Sauropod dinosaurs laid eggs in nests dug into the soil, much like those of extant turtles. In contrast, ornithischian and theropod dinosaurs laid eggs in an excavation that

ered in the Gobi Desert in 1923, but its significance was not recognized until 70 years later. The eggs, which were about 12 centimeters long and 6 centimeters in diameter, were thought to have been deposited by the small ceratopsian *Protoceratops andrewsi* because adults of that species were by far the most abundant dinosaurs at the site. The theropod was assumed to have been robbing the nest and was given the name *Oviraptor philoceratops,* which means "egg seizer, lover of ceratops." In 1993, paleontologists from the American Museum of Natural History, the Mongolian Academy of Sciences, and the Mongolian Museum of Natural History discovered a fossilized embryo in an egg identical to the supposed *Protoceratops* eggs. To their surprise, the

(continued)

BOX 16–2   Continued

embryo was an *Oviraptor* nearly ready to hatch. This discovery suggests that the adult *Oviraptor* probably died while resting on its own nest sheltering its eggs from the sandstorm.

▲ **FIGURE 16–15** An adult oviraptor brooding a nest of eggs. This reconstruction is based on a fossil of an oviraptor found in the Gobi Desert. The adult was apparently brooding its eggs when it was buried by a giant sandstorm. The forearms are spread over the eggs in the same posture used by some ground-nesting extant birds. If this oviraptor had vaned feathers on its forearms like those of *Caudipteryx*, the feathers would cover the eggs.

which are now extinct, were primarily quadrupedal herbivores, whereas theropods, which include the extant birds, are bipedal carnivores. Ten shared derived characters unite saurischians (Gauthier 1986); the most obvious is elongation of the mobile, S-shaped neck. This character distinguishes birds among living amniotes. Other birdlike characters of saurischians are found in modifications of the hand, skull, and postcranial skeleton.

■ Sauropodomorph Dinosaurs

The earliest sauropodomorph dinosaurs were the prosauropods, a group that was abundant and diverse in the Late Triassic and Early Jurassic. Three types of prosauropods are known, differing in size and tooth structure. The anchisaurids ranged in size from *Anchisaurus* (2.5 meters) to *Plateosaurus* (6 meters). The anchisaurids had long necks and small heads (**Figure 16–16**), and the teeth of the best-known forms had large serrations. Modern herbivorous lizards (iguanas) have teeth with very much the same form, and anchisaurids were probably herbivorous. Supporting this view is the presence of gastroliths associated with some fossil prosauropods. Prosauropods appear to have had cheeks that retained food in the mouth as it was processed by the teeth. The earliest prosauropods were small, lightly built, and bipedal. Later forms were larger and heavier. Their body proportions suggest that they could stand vertically on their hind legs, but they probably used a quadrupedal posture most of the time.

Prosauropods such as *Plateosaurus* had 10 cervical vertebrae, 15 trunk vertebrae, 3 sacral vertebrae, and about 46 caudal vertebrae. The long necks of all the prosauropods suggest that they were able to browse on plant material at heights up to several meters above the ground. The ability to reach tall plants might have been a significant advantage during the shift from the low-growing *Dicroidium* flora to the taller bennettitaleans and conifers that occurred in the Late Triassic.

The derived sauropods of the Jurassic and Cretaceous were enormous quadrupedal herbivores. Most fossils consist of fragmentary material, and nearly complete skeletons are known for only about five of the nearly 90 genera that have been named. The sauropods were the largest terrestrial vertebrates that have ever existed. The largest of them may have exceeded 30 meters in length and weighed 35,000 to 40,000 kilograms. (For comparison, a large African elephant is about 5 meters long and weighs 4000 to 5000 kilograms.)

Various authors recognize different subgroups of derived sauropods. The six groups listed here are recognized by most authors.

▲ **FIGURE 16–16** Sauropodomorph dinosaurs. (a) *Plateosaurus* (Late Triassic, up to 12 meters). (b) *Camarasaurus* (Late Jurassic, up to 18 meters). (c) *Diplodocus* (Late Jurassic, up to 27 meters).

Vulcanodontidae: This group is represented by a single form, *Vulcanodon*, from the Late Triassic or Early Jurassic of Zimbabwe. Only the limbs and part of the limb girdles have been found, and they combine characters of prosauropods and sauropods.

Cetiosauridae: These were generalized sauropods from the Middle Jurassic with hindlimbs that were only slightly longer than the forelimbs and simple vertebral spines in the neck and trunk. The neck and tail were only moderately long. *Shunosaurus* is a primitive Middle Jurassic cetiosaurid from China that is represented by several complete skeletons. Its vertebral count was 12 cervical, 13 trunk, 4 sacral, and 44 tail vertebrae. Other cetiosaurids include *Cetiosaurus* from England, *Patagosaurus* from Argentina, *Barapasaurus* from India, and *Bellusaurus* from China.

Brachiosauridae: These Middle to Late Jurassic forms had large teeth, greatly elongated neck ver-

tebrae, and relatively short tails. The fore legs were substantially longer than the hindlimbs. There were 13 vertebrae in the neck, 11 or 12 in the trunk, and 5 in the sacrum; the total number of tail vertebrae is unknown. The trunk vertebrae had deep indentations in the sides (pleurocoels) that probably accommodated air sacs in life (**Figure 16–17**). Brachiosaurs were enormous animals; and, because they were heavily built, they may have weighed more than the diplodocids, even though the diplodocids were longer. *Brachiosaurus* from Tanzania and Colorado is a member of this group. *Bothriospondylus* is known from England and perhaps from Madagascar.

Camarasauridae: These Late Jurassic sauropods had large teeth at the front of a short muzzle. The hindlimbs were long. They had relatively short necks and trunks (12 vertebrae in each), a sacrum

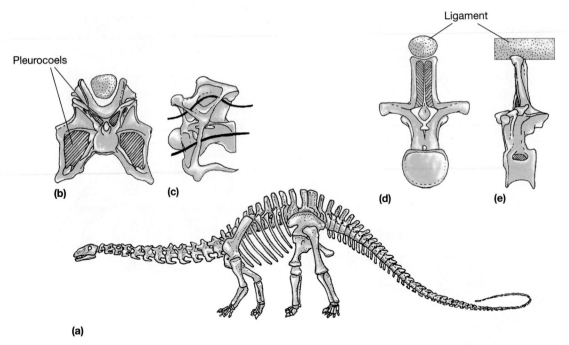

▲ **FIGURE 16–17** Structural features of sauropods. The skeletons of large sauropods (a) combined lightness with strength. Vertebrae from the neck region seen from the rear (b) and side (c) show pleurocoels that are thought to have contained air sacs. (The black ribbons in [c] indicate the position of the pleurocoels.) The neural arches of the dorsal region (d, anterior view; e, lateral view) show a U-shaped depression that accommodated a large ligament the supported the neck and head.

that attached to 5 vertebrae, and moderately long tails (53 caudal vertebrae). Both the neck and trunk vertebrae had pleurocoels, and the neural spines of the vertebrae had a deep U-shaped cleft in the center that probably accommodated a heavy tendon (see Figure 16–17). *Camarasaurus* is the most common North American sauropod, and similar forms are known from Europe (*Aragosaurus*) and China (*Euhelopus*).

Diplodocidae: *Diplodocus* and *Apatosaurus* (formerly known as *Brontosaurus*) from the Late Jurassic are two of the best-known sauropods from North America. Diplodocids had long necks (15 vertebrae), short trunks (10 vertebrae), and extremely long tails (more than 80 vertebrae) ending in a whiplash. The hemal arches that lie beneath the vertebral centra in the tail extended anteriorly and posteriorly and probably stiffened the tail, and it may have been used to beat off predators or in combat with other members of their own species. The vertebral spines had a deep cleft in the top, and the vertebrae had extensive pleurocoels. Diplodocids were enormously long (perhaps 40 meters), but much of this was the long neck and tail, so they were not as heavy as the

brachiosaurs. *Barosaurus* (from North America and East Africa) and *Cetiosauriscus* (from England—not to be confused with *Cetiosaurus*, which is a cetiosaurid also from England) are diplodocids, as are three forms known from fragmentary remains in North America—*Amphicoelias*, *Supersaurus*, and *Seismosaurus*.

Titanosauridae: These were the last surviving sauropods and had a worldwide distribution in the Middle and Late Cretaceous. Their forelimbs and hindlimbs were about the same length, and the cervical and trunk vertebrae had pleurocoels. Titanosaurids have been known primarily as fragmentary remains from India, Europe, Africa, and South America. *Rapetosaurus*, a fossil from Madagascar described by Kristina Rogers and Catherine Forster (2001), is the first nearly complete titanosaurid. It had 16 cervical vertebrae and 11 trunk vertebrae with deep pleurocoels.

Camarasauroids and brachiosaurids had compact skulls with stout jaws and large, chisel-shaped teeth. The teeth of *Camarasaurus* and *Brachiosaurus* show evidence of heavy wear, suggesting that they fed on abrasive material. The skulls of diplodocids and titanosaurids were elongate, teeth were limited to the front of the mouth, and the modest development of the lower jawbones suggests that the jaw muscles were not particularly powerful.

Sauropods dominated terrestrial habitats in the Late Jurassic and Early Cretaceous, evolving with a flora of conifers, ginkgos, ferns, cycads, and horsetails. Later in the Cretaceous, sauropods were replaced in North America and western Asia by derived ornithischians. Sauropods persisted on the southern continents, however, and titanosaurids from South America apparently reinvaded the southern part of North America at the end of the Cretaceous.

■ Biology of Sauropods

Both diplodocoids and camarasauroids were enormously heavy, and their vertebrae show features that helped the spinal column to withstand the stresses to which it was subjected. The vertebrae themselves were massive, and the neural arches were well developed. Strong ligaments transmitted forces from one arch to adjacent ones to help equalize stress. The head and tail were cantilevered from the body, supported by a heavy spinal ligament. The feet of these forms were elephant-like, and fossilized tracks indicate that the hind legs bore about two-thirds of the body weight. Some trackways show no tail marks, suggesting that the tails were carried in the air, not dragged along the ground in the manner shown in older illustrations of these dinosaurs.

Habitats of Sauropods From the earliest discovery of sauropod fossils, paleontologists doubted that such massive animals could have walked on land, instead believing they must have been limited to a semi-aquatic life in swamps. Mechanical analysis of sauropod skeletons does not support that conclusion, however. The skeletons of the large sauropods clearly reveal selective forces favoring a combination of strength with light weight. The arches of the vertebrae acted like flying buttresses on a large building, while the neural spines of diplodocoids held a massive and possibly elastic ligament that helped to support the head and tail. In cross section, the trunk was deep, shaped like the body of a terrestrial animal such as an elephant rather than rounded like that of the aquatic hippopotamus. The tails of sauropods are not laterally flattened like tails used for swimming. Instead they are round in cross section and, in diplodocoids, terminate in a long, thin whiplash. These structures look like counterweights and defensive weapons.

Fossil trackways of sauropods indicate that the legs were held under the body; the tracks of the left and right feet are only a single foot-width apart. Analysis of the limb bones suggests that they were held straight in an elephant-like pose and moved fore and aft parallel to the midline of the body. This is what would be expected on mechanical grounds because no other leg morphology is possible for a very large animal. Bone is far less resistant to bending forces exerted across its long axis than it is to compressional forces exerted parallel to the axis. As an animal's body increases in size, its mass grows as the cube of its linear dimensions, but the cross-sectional area of the bones increases as the square of their linear dimensions. The strength of bone is roughly proportional to its cross-sectional area. As a result, when the body size of an animal increases, the strength of the skeleton increases more slowly than the stress to which it is subjected. One evolutionary response to this relationship is a disproportionate increase in the diameter of bones—an elephant skeleton is proportionally larger than a mouse skeleton. Another response is to transform bending forces to compression forces by bringing the legs more directly under the body. In a large animal, such as an elephant or a sauropod dinosaur, not only are the legs held under the body but the knee joint tends to be locked as the animal walks. This morphology produces the straight-legged locomotion familiar in elephants; sauropods probably walked with an elephant-like gait, holding their heads and tails in the air.

Diets of Sauropods What did sauropods eat? The teeth of sauropods are sometimes described as being small and weak. Certainly they were small in proportion to the size of the body, as was the entire skull. In absolute terms, however, they were neither small nor weak. They were larger than the teeth of today's browsing mammals, and there is no reason to believe that plant material was tougher in the Mesozoic than it is today. There were no flat (molariform) teeth to crush the ingested plant material. This function may have been served by gastroliths, and the breakdown of plant material may have been aided by symbiotic microorganisms as in extant herbivorous reptiles, birds, and mammals.

The fossilized stomach contents of a sauropod dinosaur, found in Jurassic sediments in Utah, includes pieces of twigs and small branches about 2.5 centimeters long and 1 centimeter in diameter. The fragmented and shredded character of the woody material indicates that even without molariform teeth, the sauropod could crush its food. This discovery appears to confirm the view of sauropod ecology that was developed by studying the skeleton and analyzing plant fossils found in association with sauropod fossils. Sauropods probably occupied open country with an undergrowth of ferns and cycads and an upper story of conifers. They were preyed on by the large theropod carnivores and sought escape in flight or defended themselves by whipping their tails.

Sauropod Necks: Food or Sex? The long necks of sauropods are a puzzle. The conventional view is that they grazed from treetops, perhaps even standing on their hind legs by using the tail as a counterweight, and many reconstructions show them with their necks extended upward in a giraffelike posture. It may be significant that Mesozoic conifers bore branches only near the tops of the trees, far out of reach of any but a very long-necked dinosaur. On the other hand, two lines of reasoning argue against the idea that sauropods had giraffelike feeding habits. An analysis of the joints between vertebrae in the necks of *Diplodocus* and *Apatosaurus* suggests that the necks were less flexible than had been assumed (Stevens and Parrish 1999, 2005). If this interpretation is correct, sauropods may have swept the head horizontally and vertically through limited arcs, covering a large volume of feeding space without having to move because their necks were so long. Both *Diplodocus* and *Apatosaurus* could lower their heads to ground level, and an analysis of fossilized dung from Late Cretaceous sauropods revealed the presence of silica crystals (phytoliths) from at least five families of grasses (Prasad et al. 2005).

Another mechanical problem that sauropods would have faced while feeding from treetops was the difficulty of pumping blood to a head that might have been as much as 20 meters above the ground and 6 or 7 meters above the level of the heart (Seymour and Lillywhite 2000). Blood is mostly water, and water is heavy. When their heads were raised to browse on trees, the tallest sauropods would have required ventricular blood pressures exceeding 500 millimeters of mercury to overcome the pressure of a 7-meter column of blood between the heart and the brain. A column of blood extending to a head 20 meters above the ground could have produced blood pressures as great as 1000 millimeters of mercury in the vessels of the legs and feet of a large sauropod. Pressures that high would have forced water across the walls of the capillaries, causing the legs and feet to swell. These problems would not have occurred if sauropods fed with their heads close to the level of their hearts.

An alternative explanation has been suggested for the long necks of sauropods—that they were sexually selected characters—and again, giraffes provide an analogy. The classic hypothesis that the long necks of giraffes allow them to avoid competition with other grazing species does not withstand scrutiny (Simmons and Scheepers 1996). Even during droughts, when competition for food should be most severe, giraffes usually feed from low shrubs rather than from trees, and they consume food faster when their necks are bent than when they are raised.

In contrast to feeding, sexual competition between male giraffes provides a convincing explanation for their long necks. Dominance hierarchies among male giraffes are decided by fights in which males club each other by swinging their long necks and large heads, and injury or even death can result from the battering they receive in these combats. Male giraffes with large necks are dominant and gain access to females.

A similar function has been suggested for the long necks of sauropod dinosaurs (Senter 2007), and this proposal is consistent with the relatively limited ability of sauropods to lift their heads vertically. Male sauropods may have competed for access to females by pummeling each other with blows from basketball sized heads on necks that were several meters long.

Social Behavior of Sauropods Sauropods lacked frills and other sexually dimorphic display structures of the sort seen among ornithischian dinosaurs, but that does not mean that social behavior was absent. After all, modern crocodilians are not sexually dimorphic or ornamented with frills, yet they have elaborate social behaviors.

Fossil trackways reveal a few details from which glimpses of sauropod behavior can be reconstructed (Thulborn 1990). A famous trackway found in Texas—parts of which are now on display at the University of Texas at Austin and in the American Museum of Natural History in New York City— shows the footprints of a sauropod that was apparently being trailed by a large theropod, which was a few steps behind and slightly to the left. The theropod tracks duplicate several small changes in direction by the sauropod, and the rhythm of the theropod's stride was adjusted to match that of the sauropod. Mammalian predators such as lions make similar adjustments to match the stride of their prey before they attack. A drag mark made by the sauropod's right rear foot and two consecutive marks of the theropod's right foot (i.e., a hop) may even mark the point of an attack (Thomas and Farlow 1997).

Evidence of possible herd behavior by sauropods may be revealed by a series of tracks found in Early Cretaceous sediments at Davenport Ranch in Texas. These reveal the passage of 23 apatosaur-like dinosaurs some 120 million years ago. A group of individuals moving in the same direction at the same time would be remarkable for most living diapsids, but the apatosaur tracks suggest that this is what happened. Furthermore, the tracks may show that the herd moved in a structured fashion with the young animals in the middle, surrounded by adults.

■ Theropodomorph Dinosaurs

Theropod dinosaurs included three general types of animals: large, probably slow-moving predators that attacked large prey using their jaws as weapons (ceratosaurs, allosaurs, and tyrannosaurs), fast-moving predators that seized small prey with their forelimbs (ornithomimids), and fast-moving predators that used a huge claw on the hind foot to attack prey larger than themselves (dromeosaurids) (**Figure 16–18**).

Current phylogenetic arrangements place large theropods in two evolutionary lineages. The ceratosaurs are the earliest theropods and are the sister lineage of the Tetanurae (see Figure 16–9). *Dilophosaurus*, from the Early to Middle Jurassic of North America, is named for the paired bony crests on its head (Greek *di* = double, *loph* = crest, and *saurus* = lizard). Its jaws were slender and appear too weak to withstand the strain of attacking large prey. Although it was large (6 meters long), *Dilophosaurus* may have been a scavenger. *Ceratosaurus* of the Late Jurassic was also about 6 meters long but had a heavier skull and jaws than *Dilophosaurus*. The head was large in proportion to the body, and the long teeth were fearsome weapons.

▲ **FIGURE 16–18** Theropod dinosaurs. (a) *Coelophysis* (Late Triassic, up to 3 meters). (b) *Ornithomimus* (Late Cretaceous, up to 3.5 meters). (c) *Tyrannosaurus* (Late Cretaceous, up to 12 meters). (d) *Deinonychus* (Early Cretaceous, up to 4 meters).

Its front feet had four fingers with large claws. Ceratosaurs have a space in the upper jaw between the premaxilla (the bone at the front of the jaw) and the maxilla, which lies behind it: a tooth from the lower jaw fits into this space. (To be precise, all ceratosaurs *except Ceratosaurus* have this derived character, and it is possible that *Ceratosaurus* is not a ceratosaur. If that turns out to be correct, the group will need a new name.)

Tyrannosaurs, the second lineage of large theropods, are the sister lineage of the coelurosaurs. Two fossils from the beginning of the radiation of tyrannosaurs, *Dilong paradoxus* and *Guanlong wucaii*, confirm the relationship of this lineage to coelurosaurs. These early tyrannosaurids were small, lightly built dinosaurs with long arms and legs. The first modifications of the ancestral pattern are seen in the skull, which became heavier and better suited to killing prey.

Derived tyrannosaurids were large: *Allosaurus*, a Jurassic form, was 12 meters long, and Late Cretaceous tyrannosaurids, such as *Tarbosaurus* and *Tyrannosaurus*, were as much as 15 meters long and 6 meters high. As tyrannosaurs grew larger and concentrated their weaponry in their jaws, the size of the head increased relative to the body, and the neck shortened. The head was lightened by the elaboration of antorbital and mandibular fenestrae, reducing the skull to a series of bony arches and providing maximum strength for a given weight. The forelimbs became shorter, and the number of digits was reduced; *Allosaurus* had only three claws on its front feet, and *Tyrannosaurus* had only two small fingers on each hand.

The teeth of large therapods were as much as 15 centimeters long, dagger-shaped with serrated edges and driven by powerful jaw muscles. Marks from the teeth of predatory dinosaurs are sometimes found on fossilized dinosaur bones, and these records of prehistoric predation provide a way to estimate the force of a dinosaur's bite. The pelvis of a horned dinosaur (*Triceratops*) found in Montana bears dozens of bite marks from a *Tyrannosaurus rex*, some as deep as 11.5 millimeters. A fossilized *Tyrannosaurus* tooth was used to make an indentation that deep in the pelvis of a cow (Erickson et al. 1996). The force required to make the marks on the *Triceratops* pelvis were estimated to range from 6410 Newtons to 14,400 Newtons. These values exceed the force that can be exerted by several extant predators (dog, wolf, lion, shark). Interestingly, an alligator was the only predator tested that could deliver a bite as powerful as that of the *Tyrannosaurus*, and the jaws and teeth of alligators have many of the same structural characters as the jaws and teeth of *Tyrannosaurus*. A mechanical analysis of the skull of another large theropod, *Allosaurus*, suggests that this species used a slash-and-tear bite that could have killed the prey through loss of blood rather than with a crushing bite (Rayfield et al. 2001). The largest extant lizard, the Komodo dragon, attacks large prey (deer and even water buffalo) in this manner.

Other experimental studies that used fossilized tyrannosaur teeth to bite meat showed that the serrations increased the cutting effect only slightly, but they trapped and retained meat fibers. This debris would have supported the growth of bacteria, and a tyrannosaur bite would almost surely have become infected. Perhaps tyrannosaurs did not necessarily kill large prey such as sauropods in the initial attack but relied on infection to weaken the victim and make it susceptible to a subsequent attack. Bacteria on the teeth and claws of the Komodo dragon are thought to play exactly this role.

A coprolite (fossilized dung) the size of a loaf of bread from Saskatchewan, Canada, is believed to have been deposited by a *Tyrannosaurus rex*. It contains bone fragments from a juvenile ornithischian, possibly the head frill of a *Triceratops* (Chin et al. 1998). The shattered bone in the coprolite suggests that tyrannosaurs repeatedly bit down on food in the mouth before they swallowed it. This feeding behavior is different from that of extant crocodilians, which swallow large mouthfuls of food without processing it.

Tyrannosaurids were fearsome animals, but is the pursuit of a speeding jeep by a *Tyrannosaurus rex* portrayed in *Jurassic Park* fiction in more than one sense? How fast could *Tyrannosaurus rex* run? Two recent studies have used different methods to address that question, and they came to opposite conclusions. John Hutchison and Mariano Garcia (2002) compared the hindlimb muscles of crocodilians and chickens and found that the leg muscles of *Tyrannosaurus* would have to comprise an impossible 80-plus percent of the total body mass for *Tyrannosaurus* to be a swift runner. In contrast, William Sellers and Philip Manning (2007) used a biomechanical modeling method and concluded that the maximum running speed of adult *Tyrannosaurus rex* was just under 30 km · h^{-1}. That's not fast enough to catch a jeep, but it is about the same as a fleeing human.

Small Theropods The coelurosaurs are a mainly Cretaceous group that includes birds and all the theropods more closely related to birds than to allosaurs. Unlike the large theropods, which used teeth as their primary weapons, coelurosaurs used their hands or feet to capture prey. Many characters of living birds can be found in coelurosaurs (Gauthier 1986). The most interesting of these from the perspective of the origin of birds include a fused bony sternum and a furcula (wishbone) formed by fusion of the clavicles. The widespread occurrence of a furcula among non-flying relatives of birds shows that the original function of the structure did not involve flight. Thus, the important role the furcula plays in flight by extant birds has evolved secondarily.

Small theropods were also found in another lineage of tetanurans, the Ornithomimidae ("bird mimics"). Despite their name, the ornithomimids are not closely related to birds, but they had evolved into very birdlike forms. *Ornithomimus* was ostrichlike in size, shape, and probably in ecology as well (Figure 16–18b). It had a small skull on a long neck, and its toothless jaws were covered with a horny bill. The forelimbs were long, and only three digits were developed on the hands. The inner digit was opposable, and the wrist was flexible, making the hand an effective organ for capturing small prey. Like ostriches, *Ornithomimus* was probably omnivorous and fed on fruits, insects, small vertebrates, and eggs. Quite possibly it lived in groups, as do ostriches, and its long legs suggest that it inhabited open regions rather than forests.

Apparently not all ornithomimids preyed on small animals. A fossil from the Gobi Desert, *Deino-cheirus* (terrible hand), had fingers more than 60 centimeters long that appear to have been used for grasping and dismembering large prey. The proportions of the hands and arms are like those of coelurosaurs. If this theropod had the same body proportions as coelurosaurs, it may have been more than 7.5 meters tall—exceeding *Tyrannosaurus rex*, previously the tallest theropod known.

Dromeosaurids *Deinonychus* was unearthed by an expedition from Yale University in Early Cretaceous sediments in Montana (see Figure 16–18d). It is a small theropod, a little over 2 meters long. Its distinctive features are the claw on the second toe of the hind foot and the tail. In other theropods, the hind feet are clearly specialized for bipedal locomotion and are very similar to bird feet—the third toe is the largest, the second and fourth are smaller, and the fifth has sometimes disappeared entirely. The first toe is turned backward, as in birds, to provide support behind the axis of the leg. The second toe of dromeosaurs, especially the claw on that toe, is enlarged (**Figure 16–19**). In its normal position, the

▶ **FIGURE 16–19** The foot of *Deinonychus,* showing the enlarged claw.

claw was apparently held off the ground, and it could be bent upward even farther.

It seems likely that dromeosaurs used these claws in hunting, disemboweling prey with a kick. The structure of the tail was equally remarkable. The caudal vertebrae were surrounded by bony rods that were extensions of the prezygapophyses (dorsally) and hemal arches (ventrally) that ran forward about 10 vertebrae from their place of origin. Contraction of muscles at the base of the tail would be transmitted through these bony rods, drawing the vertebrae together and making the tail a rigid structure that could be used as a counterbalance or swung like a heavy stick. Possibly the tail was part of the armament of *Deinonychus*, used to knock prey to the ground where it could be kicked, and it may have served as a counterweight for balance as *Deinonychus* made sharp turns. Dromeosaurs probably relied on fleetness of foot to capture active prey. The discovery of five *Deinonychus* skeletons in close association with the skeleton of *Tenontosaurus*, an ornithischian three times their size, might indicate that *Deinonychus* hunted in packs. Deinonychosaurs probably used their clawed fore feet to seize prey and then slashed at it with the sicklelike claws on the hind feet. This tactic appears to be illustrated by a remarkable discovery in Mongolia of a dromeosaur called *Velociraptor*. It was preserved in combat with a *Protoceratops*, its hands grasping the head of its prey and its enormous claw embedded in the midsection of the *Protoceratops*.

The discovery of *Deinonychus* stimulated a reexamination of fossils of several other genera of small theropod dinosaurs from the Cretaceous, including *Dromeosaurus* and *Velociraptor*. All these forms have an enlarged claw on the second toe of the hind foot, and they are now grouped with *Deinonychus* and birds in the Maniraptora. *Deinonychus*-like claws 35 centimeters long were discovered in Early Cretaceous sediments in Utah during autumn 1991. The claws were probably from a previously unknown theropod (nicknamed "super-slasher" by paleontologists) that was nearly as large as a *Tyrannosaurus rex* and had the speed, agility, and predatory behavior of *Deinonychus*.

16.7 The Origin of Birds: Feathers and Flight

Two lineages of diapsids have exploited the ecological and evolutionary opportunities of powered flight. The birds are the best known flying diapsids, but they were latecomers. By the time birds appeared, an entirely different lineage of diapsids—the pterosaurs (Greek *ptero* = wing)—had diversified into an enormous array of flying reptiles. Pterosaurs appeared in the Triassic and persisted through the Cretaceous; so, for nearly 100 million years, two different groups of flying diapsids lived side-by-side (**BOX 16–3**).

BOX 16–3 Pterosaurs: The First Flying Reptiles

The archosaurs gave rise to two independent radiations of fliers. Birds are more familiar, but pterosaurs came first, appearing in the Late Triassic, some 50 million years earlier than birds, and persisting into the Cretaceous. Pterosaurs radiated into a wide variety of body forms and ecological types, from the sparrow-size *Pterodactylus* to *Quetzalcoatlus*, which had a wingspan of 13 meters (**Figure 16–20**).

The wing of pterosaurs was formed by skin and was entirely different from the feathered wings of birds. The fourth finger of pterosaurs was elongate and supported a membrane of skin anchored to the side of the body and perhaps to the hind leg. A small splintlike bone was attached to the front edge of the finger and probably supported a membrane that ran forward to the neck. The early rhamphorhynchoid pterosaurs had a long tail with an expanded portion on the end that was presumably used for steering, but later pterodactyloids lacked a tail.

The mechanical demands of flight are reflected in the structure of flying vertebrates, and it is not surprising that pterosaurs and birds show a high degree of parallel evolution. The long bones of pterosaurs were hollow, as they are in birds and many other archosaurs, reducing weight with little loss of strength. The sternum, to which the powerful flight muscles attach, was well developed in pterosaurs, although it lacked the keel seen in birds. (A keeled sternum is not essential for flight—bats have a flat sternum.) The eyes were large, and casts of the brain cavities of pterosaurs show that parts of the brain associated with vision were large and olfactory areas were small, as they are in birds. The cerebellum, which is concerned with balance and coordination of movement, was large in proportion to other parts of the brain, and this is another way in which pterosaurs were similar to birds (Witmer et al. 2003).

Some pterosaurs lost their teeth and evolved a birdlike beak. Others had sharp, conical teeth in blunt skulls reminiscent of those of bats. Some pterosaurs with elongate skulls

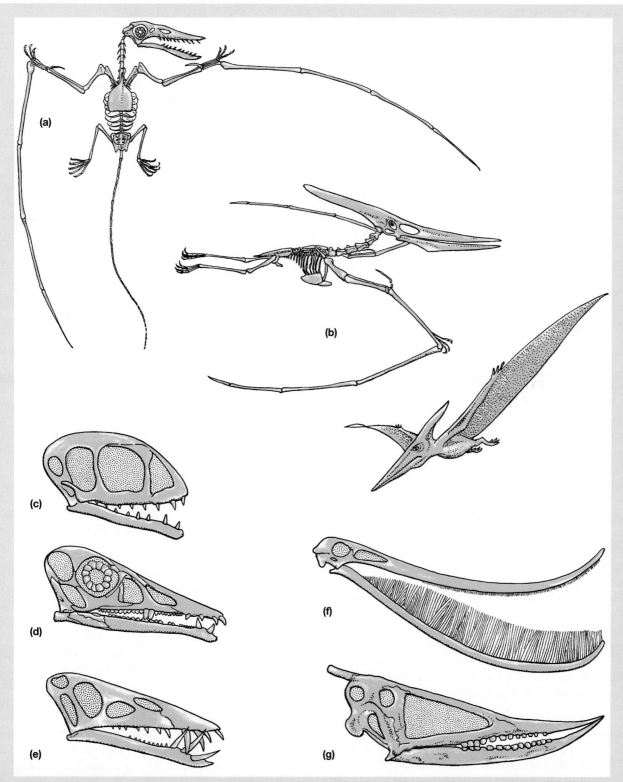

▲ **FIGURE 16–20** Pterosaurs. (a) *Rhamphorhynchus* from the Jurassic. (b) *Pteranodon* from the Cretaceous. The skulls of pterosaurs suggest dietary specializations: (c) *Anurognathus* may have been insectivorous. (d) *Eudimorphodon* may have eaten small vertebrates. (e) *Dorygnathus* may have been a fish eater. (f) *Pterodaustro* had a comblike array of teeth that may have been used to sieve plankton. (g) *Dsungaripterus* may have pulled mollusks from rocks with a horny beak and then crushed them with its molariform teeth.

(continued)

BOX 16–3   Continued

and stout, sharp teeth may have caught fish or small tetrapods. *Pterodaustro* had an enormously long snout lined with a comblike array of fine teeth that may have been used for sieving small aquatic organisms from the water. *Dsungaripterus* had long jaws that met at the tips like a pair of forceps. The tips of the jaws were probably covered with a horny beak, and blunt teeth occupied the rear of the jaw. These animals may have plucked snails from rocks with their beaks and then crushed them with their broad teeth.

The flight capacities of pterosaurs have long been debated, and most hypotheses about their ecology have been based on the assumption that they were weak fliers. That assumption has led to suggestions of restrictions of activities and habitats of pterosaurs that seem unlikely for a group of animals that was clearly diverse and successful through much of the Mesozoic. An aerodynamic analysis suggests that the flying abilities of pterosaurs have been underestimated (Hazlehurst and

Rayner 1992). This view suggests that small pterosaurs were slow, maneuverable fliers like bats. The large pterosaurs appear to have been specialized for soaring like frigate birds and some vultures.

Birds appeared in the Late Jurassic, so birds and pterosaurs coexisted during much of the Cretaceous. Did the two kinds of flying reptiles occupy different adaptive zones, or did competition with birds contribute to the extinction of pterosaurs? Pterosaurs and birds may have occupied different biomes, at least initially, with pterosaurs primarily in coastal areas and birds inland (Wang et al. 2005). Subsequently, however, competition between the two lineages may have occurred. The derived lineages of pterosaurs show a progressive increase in body size through time, and small- and medium-size species of pterosaurs disappeared during the Jurassic and Cretaceous, just when small- and medium-size birds were appearing (Hone and Benton 2007).

■ Shared Derived Characters of Theropods and Birds

Birds are derived theropod dinosaurs, and few evolutionary transitions are as clearly recorded in the fossil record as the appearance of birds. More than a century ago, Thomas Henry Huxley was an ardent advocate of that relationship, writing that birds are nothing more than "glorified reptiles." Huxley, in fact, was so impressed by their many similarities that he placed birds and reptiles together in his classification as the class Sauropsida. For most of the next century, traditional systematics, with its emphasis on strict hierarchical categories, obscured that evolutionary relationship by placing reptiles and birds at the same taxonomic level (class Reptilia and class Aves). Cladistic systematics emphasizes monophyletic evolutionary lineages, and now birds are again seen as the most derived theropod dinosaurs. The similarities of birds and theropods include the following derived characters:

- Hollow, pneumatic bones.
- Elongate, mobile S-shaped neck.
- A foot with three toes pointed forward and one extending backward (called a tridactyl foot).
- Digitigrade posture (i.e., with the toes bearing the weight of the body).
- Ankle joint forms between tarsal bones (an intertarsal joint) rather than between the tarsals and tibia + fibula.
- Feather precursors or true feathers.
- Reduced genome size.

These characters can be traced far back into the theropod lineage. When we look specifically at derived theropods, such as coelurosaurs, still more shared derived characters emerge:

- A furcula (wishbone) formed by fusion of the clavicles.
- A fused bony sternum.
- A birdlike egg-brooding posture resting on the clutch with the forearms (wings) spread to cover the eggs.
- A birdlike sleeping posture with the head tucked under the forearm (wing).

A remarkable biochemical discovery has recently added a new line of evidence supporting the theropod-bird connection—the discovery of protein preserved in an unfossilized state in the fossil of a *Tyrannosaurus rex*. Finding organic material in a fossil is an extraordinary occurrence because organic materials are normally replaced by minerals during fossilization. Thus, fossils are rocks that retain the exact shape of the original bone but contain none of the chemicals that were in the bone during the life of the animal. Very rarely, however, areas of soft tissues are left unmineralized. If the organic material has not decayed, it may be possible to compare the chemical characteristics of tissues from the fossil with tissues from extant species. In this case, a sample of collagen from a 68 million year old *Tyrannosaurus rex* skeleton yielded peptides that were compared to peptides from a variety of extant vertebrates. The closest match was to peptides from a bird, just as would be predicted on the basis of our understanding of the phylogenetic relationships of theropods and birds (Asara et al. 2007, Organ et al. 2008).

Mosaic Evolution of Avian Anatomical Characters

Major evolutionary changes, such as the transition from nonavian dinosaurs to birds, do not occur all at once. Instead, one character changes, then another, and then another in a stepwise fashion so the transitional forms present mosaics of ancestral and derived characters. The mosaic nature of evolution is superbly illustrated by the transition from nonavian dinosaurs to birds (**Figure 16–21**).

The dromeosaurs, a derived group of coelurosaurs that includes *Velociraptor* of *Jurassic Park* fame, had several birdlike characters, including a wrist structure that permitted them to flex the wrists sideways while rotating them. This motion probably allowed coelurosaurs to use their hands to seize prey, and it is recognized in the name Maniraptora (Latin *manus* = hand and *rapt* = seize), which is the lineage that includes dromeosaurs

and birds. Birds use this wrist motion to produce a flow of air over the primary feathers of the wings to create lift during flapping flight.

Some still more derived dromeosaurs—such as *Unenlagia*, which was a 2-meter-long terrestrial predator—have a change in the shoulder joint that allowed greater freedom of movement of the arms. The glenoid fossa (where the humerus articulates with the pectoral girdle) is oriented laterally rather than ventrally in these animals, allowing the arms to be lifted upward and backward and to strike downward and forward to seize prey. This anatomical change, which probably made dromeosaurs more effective as terrestrial predators, is the origin of the up-and-back/down-and-forward arm motion that birds use to flap their wings in flight.

Feathered Dinosaurs Feathers are the most dramatic birdlike characters of dromeosaurs (**Figure 16–22**). Nonavian dinosaurs with feathers were first described in the late 1990s, and feathers have subsequently been described in a dozen species of coelurosaurs, mostly

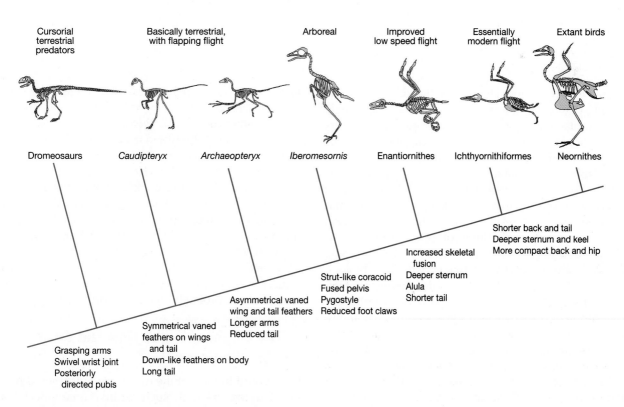

▲ **FIGURE 16–21** Mosaic evolution of the derived characters of birds.

▶ **FIGURE 16–22** *Sinornithosaurus.* This feathered dromeosaur was found in the Liaoning Cretaceous fossil beds of China.

from the Liaoning fossil beds (Norell and Xu 2005). The simplest kind of feathers found in dinosaurs are single hollow filaments from 1 centimeter to 5 centimeters long. These filaments bear little superficial resemblance to the derived feathers seen in birds, and some paleontologists initially doubted that they are really feathers. Those doubts were relieved by a fossil of *Shuvuuia deserti* in which the filaments retained enough keratin for chemical tests; results showed that the filaments were composed of the form of beta keratin that is unique to feathers (Schweitzer et al. 1999). Combining the skills and perspectives of two very different biological specialties—paleontology and developmental biology—has given rise to a new specialty, evolutionary development (usually called Evo-Devo), which allows interpretation of the genetic basis of evolutionary changes such as the appearance of feathers (**BOX 16–4**).

Vaned feathers—that is, feathers with flat surfaces on both sides of a central shaft like the feathers of extant birds—are preserved with fossils of *Caudipteryx* and *Protoarchaeopteryx*. *Caudipteryx* had vaned feathers attached to the second finger of the hand, where remiges (primary flight feathers) are found in modern birds, and a tuft of vaned feathers on the tail. *Protoarchaeopteryx* has downlike feathers on the body and tail, as well as a fan of symmetrical vaned feathers on the tail.

The first vaned feathers had flat surfaces that were symmetrical about the central shaft. This shape is effective in feathers that are used in social displays or for insulation, and symmetrical vaned feathers serve these functions in extant birds. Flight, however, requires asymmetric vaned feathers on the wings because the twisting of the feathers contributes thrust. Thus, the appearance of asymmetric vaned feathers in *Archaeopteryx* probably signals the origin of flight.

■ The Origin of Flight: How Did Birds Get Off the Ground?

One would think that the discovery of feathered non-avian dinosaurs with such birdlike anatomical features as a furcula, a keeled sternum, and arms capable of flapping movement would lead to a clear understanding of the origin of flight, but this is not the case. Nonavian dinosaurs could have used feathers in combination with body postures and movements in social interactions, just as extant birds do, and this hypothesis could explain the origin of feathers. At a later stage, well-developed feathers, like those on the forelimbs of *Caudipteryx*, might have been used to cover eggs in a nest, shielding them from the sun during the day and trapping warmth in the nest at night, and still later a covering of feathers on the body could have retained heat produced by metabolic processes.

If feathers initially evolved as social signals and as insulation, how did the ancestors of birds change the functions of their forearms and feathers to make them wings and airfoils? What were the selective advantages for the evolution of wings and flight? How did the transition from terrestrial feathered dinosaurs like *Caudipteryx* to fliers like *Archaeopteryx* occur? Two competing hypotheses have existed for a century—the arboreal ("from the trees down") theory and the terrestrial ("from the ground up") theory (**Figure 16–24**).

From the Trees Down The arboreal theory long dominated the field. According to this view, the ancestors of *Archaeopteryx* were tree climbers that jumped from branch to branch and from tree to tree much as some squirrels, lizards, and monkeys do. Under selective pressures favoring increased distance and accuracy of

---

BOX 16–4   The Evolution of Feathers: Evo-Devo and Fossils

A merger of evolutionary biology with our rapidly growing understanding of the genetic mechanisms that control growth and development has given rise to a new biological specialty—evolutionary developmental biology—which is known as Evo-Devo for short. The evolution of feathers has been illuminated by Evo-Devo studies of the actions of two genes that are crucial for the development of vertebrate limbs: A gene called *sonic hedgehog* produces a protein that induces cell proliferation, and *bone morphogenetic protein 2* regulates cell proliferation and promotes cell differentiation. Both of these genes are involved in the growth of feathers, turning on and off to control the tempo and pattern of growth and differentiation (Prum 1999). An Evo-Devo perspective based on the action of the genes during the evolution of feathers can be compared with the growing body of information about the occurrence of feathers in dinosaur fossils (**Figure 16–23**).

The embryonic development of a feather begins with a placode formed when the epidermis thickens over a condensation of cells in the underlying dermis. The same process initiates formation of a scale, so it would be an ancestral character for feathered archosaurs. Stage 1 represents a new evolutionary and developmental feature as the placode forms an elongated tube that is the feather germ. Proliferation of cells around the base of the feather germ creates a follicle, the organ that produces feathers in extant birds. In Stage 2, the follicle becomes differentiated to produce two layers; the outer layer is the feather sheath that protects the growing feather. As the genes controlling proliferation and differentiation are turned on and off, the inner layer produces a series of separate structures that will be the barbs. *Sinornithosaurus* (see Figure 16–22) illustrates this stage. Two changes occurred in Stage 3: The barbs grew together at one end to form the rachis, and they branched to produce barbules. The sequence in which these changes occurred has not been determined, but in combination they would produce feather with an open pennaceous vane. In Stage 4, the development of distal barbules with hooks and proximal barbules with grooves would allow adjacent barbs to attach to each other, producing a closed pennaceous vane. In Stage 5, the sheetlike surface of a pennaceous vane adds new functional properties that would give rise to additional developmental modifications, such as asymmetric flight feathers.

Fossilized feathers are imprints in the rock surrounding fossil bones, as in the fossils of *Archaeopteryx*. Only fine-grained sediments preserve traces of delicate structures like feathers, and the position of the feathers on the body of the animal can be determined only if the carcass was nearly intact when it was fossilized. Fossils do not reveal all of the details of the stepwise evolution of feathers that are predicted by the Evo-Devo model, but the two approaches show a substantial level of agreement. Fossils of *Sinosauropteryx* are surrounded by traces that many paleontologists interpret as single filaments, corresponding to Stage 1 of the Evo-Devo model, and *Caudipteryx* has symmetrical vaned feathers on its arms and tail and tufts of barbs on its body. *Archaeopteryx* is the earliest occurrence of asymmetric vaned fathers, and these feathers are also found on later forms.

(continued)

BOX 16–4 Continued

Predictions of the Evo-Devo model

Evidence of feathers in fossils

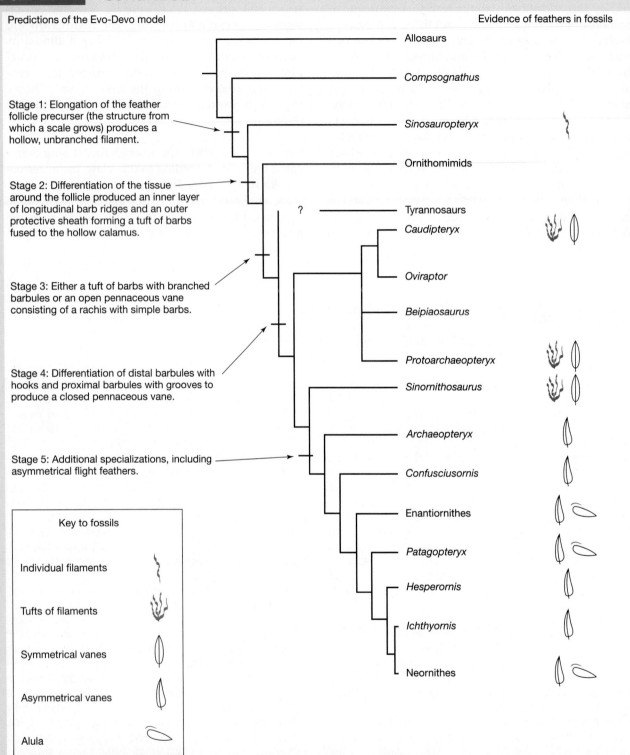

Stage 1: Elongation of the feather follicle precurser (the structure from which a scale grows) produces a hollow, unbranched filament.

Stage 2: Differentiation of the tissue around the follicle produced an inner layer of longitudinal barb ridges and an outer protective sheath forming a tuft of barbs fused to the hollow calamus.

Stage 3: Either a tuft of barbs with branched barbules or an open pennaceous vane consisting of a rachis with simple barbs.

Stage 4: Differentiation of distal barbules with hooks and proximal barbules with grooves to produce a closed pennaceous vane.

Stage 5: Additional specializations, including asymmetrical flight feathers.

Allosaurs

Compsognathus

Sinosauropteryx

Ornithomimids

Tyrannosaurs

Caudipteryx

Oviraptor

Beipiaosaurus

Protoarchaeopteryx

Sinornithosaurus

Archaeopteryx

Confusciusornis

Enantiornithes

Patagopteryx

Hesperornis

Ichthyornis

Neornithes

Key to fossils

Individual filaments

Tufts of filaments

Symmetrical vanes

Asymmetrical vanes

Alula

▲ **FIGURE 16–23** Two perspectives on the evolution of feathers. *Left:* The sequence of changes predicted by evolutionary development studies of the action of the genes controlling feather development in extant birds. *Right:* Feathers that have been found with fossilized nonavian dinosaurs and birds. The earliest fossilized featherlike structures are probably single filaments, as predicted by Stage 1 of the Evo-Devo model. The next clear fossil evidence of feathers includes both downy tufts and symmetrical vane feathers that correspond to Stage 3 of the model. The structural characteristics of Stage 4 feathers are beyond the level of resolution in fossils currently available, but Stage 5 feathers can be identified in the fossil record. The appearance of the **alula** (a structure that reduces turbulence in the flow of air across the wing feathers) is a further specialization of flight of the sort envisioned by the Evo-Devo model.

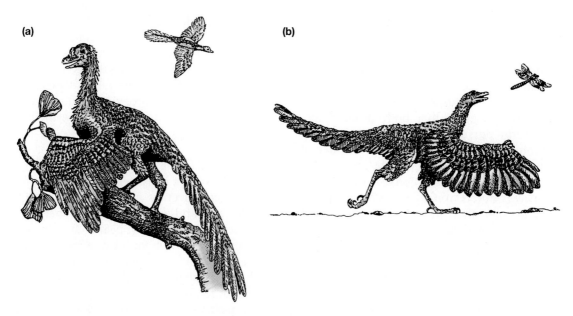

▲ **FIGURE 16–24** Two reconstructions of *Archaeopteryx*. (a) The from-the-trees-down hypothesis, showing *Archaeopteryx* climbing in a tree and (in the distance) gliding. (b) The from-the-ground-up hypothesis showing *Archaeopteryx* as a cursorial hunter about to use its forelimb to capture an insect.

travel between trees, structures that provided some surface area for lift would be advantageous. *Draco* are arboreal lizards from the East Indies with extremely elongate ribs that support wings of skin on the sides of the trunk. They use their wings to glide from tree to tree. A flight starts with a dive from an elevated perch. The lizard descends at an angle of about 45 degrees, and then levels out and uses the kinetic energy developed during the dive to glide nearly horizontally. A brief upward glide immediately precedes landing on another perch. Glides as long as 60 meters have been recorded with a loss of altitude of less than 2 meters. By this hypothesis, the evolution of flying forms would have passed from gliding stages through intermediate stages—such as *Archaeopteryx*, in which gliding was aided by weak flapping flight—to fully airborne flapping fliers. The arboreal hypothesis received a boost from the description of a new dromeosaur from China, *Microraptor gui*, which, with its vaned feathers on both the forelimbs and the hindlimbs and a tuft of feathers on the end of the tail, may have been a glider (Xu et al. 2003).

From the Ground Up On the other hand, given that the dromeosaur lineage consisted of bipedal, cursorial, terrestrial predators, is it plausible to invoke gliding as the evolution of avian flight? The from-the-ground-up hypothesis proposed by John Ostrom (1974) postulates that flapping flight evolved directly from ground-dwelling, bipedal runners.

According to the first version of the cursorial hypothesis, the proavian ancestors of birds were fast bipedal runners that used their wings as planes to increase lift and lighten the load for running. In a later development, the wings were flapped as the animal ran to provide additional forward propulsion, much as a chicken flaps across the barnyard to escape from a dog. Finally, the pectoral muscles and flight feathers became sufficiently developed for full-powered flight through the air.

The cursorial hypothesis in its original form failed as an explanation for the origin of flight because it would not work. Flapping would not have been an effective way for proavian dinosaurs to increase running speed because maximum traction on the ground is needed to achieve acceleration, and this traction is provided by solid contact between the feet and the substrate. Extending feathered arms to create lift would have taken some of the body's weight off the feet and *reduced* traction, slowing the animal rather than adding speed.

A specimen of *Archaeopteryx* that had been misidentified as a nonavian coelurosaur for over 100 years revealed some previously unknown anatomical details and led to a modification of the cursorial theory. Some elements of the hand are extremely well preserved in this specimen and show the actual horny claws on digits 1 and 3. These claws look like the talons of a predatory bird. Further evidence supporting the cursorial theory comes from the seventh

Archaeopteryx to be discovered; this specimen shows a hyperextensible second toe (like the clawed toes of *Deinonychus* and *Velociraptor*), although in *Archaeopteryx* this toe lacks a large killer claw.

The similarities in the hand, metacarpus, forearm, humerus, pectoral apparatus, and feet of *Archaeopteryx* and dromeosaurs suggest that both dromeosaurs and *Archaeopteryx* pursued their prey and seized it with their forelimbs. The forelimb and shoulder of *Archaeopteryx* have been only slightly structurally modified from the skeletal condition of dromeosaurs, and *Archaeopteryx* differs from all other birds in lacking several features that are critical for powered flight—fused carpometacarpus, restricted wrist and elbow joints, modified coracoids, and a platelike sternum with keel (Jenkins 1993). In fact, the only skeletal features of *Archaeopteryx* that suggest flight are the well-developed furcula (wishbone) and the lateral position of the glenoid fossa that permits up-and-back and down-and-forward motion by the arms. Both of these features are present in proavian dromeosaurs, and *Archaeopteryx* appears to have been as well adapted for predation as for flight. From these considerations, Ostrom postulated that the incipient wings of the proavian ancestors of *Archaeopteryx* evolved first as snares to trap insects or other prey against the ground or to bat them out of the air, so they could be grasped by the claws and teeth. The wings subsequently became modified into flapping appendages capable of subduing larger prey.

Further refinements of the ground-up hypothesis have been proposed. For example, a wing could have assisted horizontal jumps after prey. By spreading or moving its forelimbs, a proavian dromeosaur could not only control pitch, roll, and yaw while leaping to catch a flying insect but could also maintain its balance when it landed (Rayner 1988). In addition, the observation that chuckar partridges beat their wings when they run up steep slopes led Kenneth Dial (2003) to suggest that Wing Assisted Incline Running (WAIR) may have played a role in the origin of flight. In laboratory experiments, Dial found that partridges flapped their wings when they ran up slopes steeper than 45 degrees. Even newly hatched chicks could ascend 50-degree slopes using WAIR, 4-day-old chicks (which had longer wing feathers) were able to climb 60-degree slopes, and adult birds could ascend vertical (90-degree) slopes. The contribution of the wings to the birds' climbing ability was tested by trimming the flight feathers. During the first week after hatching the feathers are very short and trimming them did not affect climbing performance;

however, for older birds, removing half the feather surface area reduced the maximum angle of climb by 10 degrees to 20 degrees. When the flight feathers were entirely removed, the birds were unable to climb slopes steeper than 60 degrees. Perhaps *Caudipteryx* and other proavian dromeosaurs initially used their winged arms to increase their climbing ability.

Body Size and Flight At present, a terrestrial origin of flight seems most consistent with the multiple lines of evidence currently available (Padian 2001). The complex arm and wrist movements that birds use to generate lift and thrust appear to be consistent with prey-seizing movements by terrestrial predators rather than with grasping and pulling movements used to clamber through tree branches. Indeed, feathered forelimbs like those of *Caudipteryx* and *Archaeopteryx* seem poorly suited to arboreal life.

The evolution of fight was accompanied by a marked reduction in body size (Turner et al. 2007). A flying bird is supported by the lift produced by its wings. Wing loading, the weight of the bird divided by the surface area of the wings, is a critical component of flight. Wing loadings in the range of 2.5 gm · cm^{-2} are typical of extant birds that run with flapping wings to take off and are plausible values for the earliest birds.

A wing is basically two dimensional, and the surface area of a wing increases as the *square* of its linear dimensions. The weight of a bird, however, is proportional to its volume, which increases as the *cube* of its linear dimensions. As a result, large birds require proportionally larger wings than do small birds to achieve the same wing loading. The conspicuous reduction in body size that occurred during the transition from dromeosaurs to birds is probably the result of the importance of wing loading in the evolution of flight.

16.8 *Archaeopteryx* and Other Birds

Archaeopteryx is the earliest-known bird, and birds (Aviale) are defined as *Archaeopteryx* plus living birds (Neornithes) and all descendants of their most recent common ancestor. Fossils of *Archaeopteryx* from fine-grained sediments show imprints of feathers that were much better differentiated than the feathers of *Caudipteryx* and *Protoarchaeopteryx*.

In addition to a presumed covering of body-contour feathers, *Archaeopteryx* had wing feathers that were differentiated into an outer series of primaries

▶ **FIGURE 16–25** A fossil of *Archaeopteryx lithographica.* Details of the feathers were preserved in fine-grained limestone.

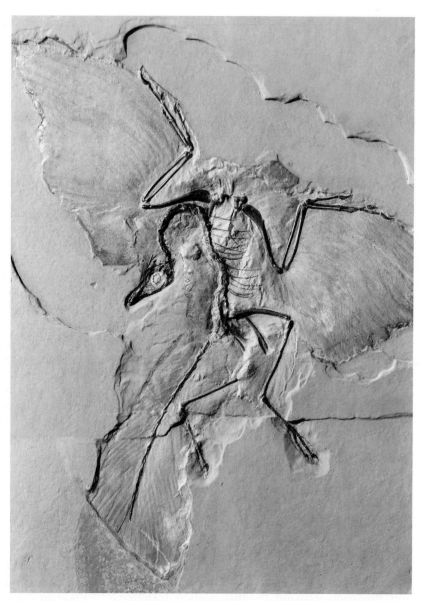

on the hand bones and an inner series of secondaries along the outer arm (see Figure 16–24). This arrangement of flight feathers is essentially the same as that seen in extant birds. Furthermore, the flight feathers on the wings of *Archaeopteryx* have asymmetrical vanes like those of flying birds, suggesting that they had been shaped by aerodynamic forces associated with flapping flight. The rectrices (tail feathers) of *Archaeopteryx* are arranged in 15 pairs along the sides of the 6th through 20th caudal vertebrae.

■ How Well Could *Archaeopteryx* Fly?

Archaeopteryx was probably a late-surviving relict that was contemporaneous with more typically avian birds; and, if the fine-grained limestone in the Solnhofen quarry had not faithfully preserved the imprint of its feathers, *Archaeopteryx* could readily be mistaken for a small dinosaur (**Figure 16–25**). However, several lines of evidence suggest that *Archaeopteryx* was capable of flight: Its skeletal proportions were similar to those of some extant flying birds, the number of primaries and secondaries were identical to those of extant birds, the asymmetry of its flight feathers is like that seen in extant flying birds, and the furcula was large. The seventh specimen of *Archaeopteryx* reveals a feature not visible in the fossils previously known—a rectangular sternum that was probably associated with strong flight muscles. On balance, these characteristics are consistent with the view that *Archaeopteryx* was a flying bird (Martin 1983; Rayner 1988).

Unlike the proavian ancestors of birds, *Archaeopteryx* had a wing loading comparable to extant birds that run when they take flight, and this is probably what *Archaeopteryx* did. Once an *Archaeopteryx* was airborne, calculations of its metabolic capacity suggest that it could fly at least 1.5 kilometers at a velocity of 40 kilometers per hour (Ruben 1991, 1993).

An animal that could take off from the ground and fly rapidly for several hundred meters would be able to escape predators. Many living birds, including cursorial predators such as the North American roadrunner and the African secretary bird, use flight in exactly this way. Thus, *Archaeopteryx* can plausibly be interpreted as a ground-dwelling predator that leaped into the air to seize flying insects, flew rapidly for short distances to escape from its own predators, and made running landings on the ground.

Summary

The major groups of tetrapods in the Mesozoic were members of the diapsid (two arches) lineage. This group is distinguished particularly by the presence of two fenestrae in the temporal region of the skull that are defined by arches of bone. The archosauromorph lineage of diapsids contains the most familiar of the Mesozoic tetrapods, the dinosaurs. Two major groups of dinosaurs are distinguished, the Ornithischia and Saurischia.

The ornithischian dinosaurs were herbivorous, and many had horny beaks on the snout and batteries of specialized teeth in the rear of the jaw. The ornithopods (duck-billed dinosaurs) and pachycephalosaurs (thick-headed dinosaurs) were bipedal, and the stegosaurs (plated dinosaurs), ceratopsians (horned dinosaurs), and ankylosaurians (armored dinosaurs) were quadrupedal.

The saurischians included the sauropod dinosaurs—enormous herbivorous quadrupedal forms like *Apatosaurus* (formerly *Brontosaurus*), *Diplodocus*, and *Brachiosaurus*—and the theropods, which were bipedal carnivores. Large theropods (of which *Tyrannosaurus rex* is the most familiar example) probably preyed on large sauropods. Other theropods were smaller: Ornithomimids were probably very like ostriches, and some had horny beaks and lacked teeth. Dromeosaurs were fast-running predators. Ornithomimids probably seized small prey with hands that had three fingers armed with claws, whereas dromeosaurs probably were able to prey on dinosaurs larger than themselves. They may have hunted in packs and used the enormous claw on the second toe to slash their prey. Birds had evolved by the Jurassic: *Archaeopteryx*, the earliest-known bird, is very like small theropods, and the sequence of appearance of the derived characters of birds can be traced in non-avian dromeosaurs.

The phylogenetic relationship of crocodilians and birds allows us to draw inferences about some aspects of the biology of dinosaurs. Characters that are shared by crocodilians and birds are probably ancestral for dinosaurs. Social behavior, vocalization, and parental care are the norm for crocodilians and birds, and increasing evidence suggests that dinosaurs, too, had elaborate social behavior and vocalizations, and that at least some species cared for the young.

Additional Readings

Asara, J. M. et al. 2007. Protein sequences from mastodon and *Tyrannosaurus rex* revealed by mass spectrometry. *Science* 316:280–285.

Barrett, P. M., and K. J. Willis. 2001. Did dinosaurs invent flowers? Dinosaur-angiosperm coevolution revisited. *Biological Reviews* 76:411–447.

Brochu, C. A. 2001. Crocodylian snouts in space and time: Phylogenetic approaches toward adaptive radiation. *American Zoologist* 41:564–585.

Brochu, C. 2003. Phylogenetic approaches toward crocodylian history. *Annual Review of Earth and Planet Science* 31:357–397.

Carpenter, K. 1999. *Eggs, Nest, and Baby Dinosaurs: A Look at Dinosaur Reproduction*. Bloomington, IN: Indiana University Press.

Chiappe, L. M. 2007. *Glorified Dinosaurs*. Hoboken, NJ: Wiley.

Chiappe, L. M. et al. 1998. Sauropod dinosaur embryos from the Late Cretaceous of Patagonia. *Nature* 396:258–261.

Chiappe, L. M., and G. J. Dyke. 2002. The Mesozoic radiation of birds. *Annual Review of Ecology and Systematics* 33:91–124.

Chiappe, L. M., and L. M. Witmer (eds.) 2002. *Mesozoic Birds: Above the Heads of Dinosaurs.* Berkeley, CA: University of California Press.

Chin, K. et al. 1998. A king-size theropod coprolite. *Nature* 393:680–682.

Dial, K. P. 2003. Wing-assisted incline running and the evolution of flight. *Science* 299:402–404.

Erickson, G. M. et al. 1996. Bite-force estimation for *Tyrannosaurus rex* from tooth marks on bones. *Nature* 382:706–708.

Franklin, C. E., and M. Axelsson. 2000. An actively controlled heart valve. *Nature* 406:847–848.

Gauthier, J. 1986. Saurischian monophyly and the origin of birds. In K. Padian (ed.), *The Origin of Birds and the Evolution of Flight, Memoirs of the California Academy of Sciences,* no. 8, pp. 1–55.

Hazlehurst, G. A., and J. M. V. Rayner. 1992. Flight characteristics of Triassic and Jurassic Pterosauria: An appraisal based on wing shape. *Paleobiology* 18:447–463.

Hone, D. W. E., and M. J. Benton. 2007. Cope's rule in the Pterosauria, and differing perceptions of Cope's rule at different taxonomic levels. *Journal of Evolutionary Biology* 20:1164–1170.

Horner, J. R. 2000. Dinosaur reproduction and parenting. *Annual Review of Earth and Planet Science* 28:19–45.

Hutchinson, J. R. 2006. The evolution of locomotion in archosaurs. *Comptes Rendus Palevol* 5:519–530.

Hutchison, J. R., and M. Garcia. 2002. *Tyrannosaurus* was not a fast runner. *Nature* 415:1018–1021. (See also Erratum *Nature* 447:349.)

Irmis, R. B. et al. 2007. A Late Triassic dinosauromorph assemblage from New Mexico and the rise of dinosaurs. *Science* 317:358–361.

Jenkins, F. A., Jr. 1993. The evolution of the avian shoulder joint. *American Journal of Science* 293A:253–267.

Kellner, A. W., and D. A. Campos. 2002. The function of the cranial crest and jaws of a unique pterosaur from the Early Cretaceous of Brazil. *Science* 297:389–392.

Kirkland, J. I. 1994. Predation of dinosaur nests by terrestrial crocodilians. In K. Carpenter et al. (eds.), *Dinosaur Eggs and Babies,* Cambridge, UK: Cambridge University Press, 124–133.

Lang, J. W. 1986. Male parental care in mugger crocodiles. *National Geographic Research* 2:519–525.

Lang, J. W. 1989. Social behavior. In C. A. Ross (ed.), *Crocodiles and Alligators,* New York, NY: Facts on File, 102–117.

Martin, L. D. 1983. The evolution of birds and of avian flight. In R. F. Johnston (ed.), *Current Ornithology* vol 1. New York: Plenum, 105–129.

Massare, J. A. 1988. Swimming capabilities of Mesozic marine reptiles: Implications for methods of predation. *Paleobiology* 14:187–205.

Mikhailov, K. E. 1997. Eggs, eggshells, and nests. In P. J. Currie and K. Padian (eds.), *Encyclopedia of Dinosaurs.* San Diego, CA: Academic Press, 205–209.

Motani, R. 2005. Evolution of fish-shaped reptiles (Reptilia: Ichthyopterygia) in their physical environments and constraints. *Annual Review of Earth and Planetary Science* 33:395–420.

Motani, R. et al. 1998. Large eyeballs in diving ichthyosaurs. *Nature* 402:747.

Norell, M. A., and J. A. Clarke. 2001. Fossil that fills a critical gap in avian evolution. *Nature* 409:181–184.

Norell, M. A. and X. Xu 2005. Feathered dinosaurs. *Annual Review of Earth and Planetary Sciences* 33:277–299.

Organ, C. L. et al. 2008. Molecular phylogenetics of mastodon and *Tyrannosaurus rex. Science* 320:499.

Ostrom, J. H. 1974. *Archaeopteryx* and the evolution of flight. *Quarterly Review of Biology* 49:27–47.

Padian, K. 2001. Cross-testing adaptive hypotheses: Phylogenetic analysis and the origin of bird flight. *American Zoologist* 41:598–607.

Prasad, V. et al. 2005. Dinosaur coprolites and the early evolution of grasses and grazers. *Science* 310:1177–1180.

Prum, R. O. 1999. Development and evolutionary origin of feathers. *Journal of Experimental Zoology* 285:291–306.

Rayfield, E. J. et al. 2001. Cranial design and function in a large theropod dinosaur. *Nature* 409:1033–1037.

Rayner, J. M. V. 1988. The evolution of vertebrate flight. *Biological Journal of the Linnean Society* 34:269–287.

Rich, T. H. et al. 2002. Polar dinosaurs. *Science* 295:979–980.

Rogers, K. C., and C. A. Forster. 2001. The last of the dinosaur titans: A new sauropod from Madagascar. *Nature* 412:530–534.

Ruben, J. 1991. Reptilian physiology and the flight capacity of *Archaeopteryx. Evolution* 45:1–17.

Ruben, J. 1993. Powered flight in *Archaeopteryx:* Response to Speakman. *Evolution* 47:935–938.

Sampson, S. D. et al. 2001. Bizarre predatory dinosaur from the Late Cretaceous of Madagascar. *Nature* 2001:504–506.

Schweitzer, M. H. et al. 1999. Beta-keratin specific immunological reactivity in feather-like structures of the Cretaceous Alvarezsaurid *Shuuvia deserti. Journal of Experimental Zoology* 285:146–157.

Schwimmer, D. R. 2002. *King of the Crocodylians: The Paleobiology of* Deinosuchus. Bloomington, IN: Indiana University Press.

Sellers, W. I., and P. L. Manning. 2007. Estimating dinosaur maximum running speeds using evolutionary robotics. *Proceedings of the Royal Society of London,* Series B. 274:2711–2716.

Senter, P. 2007. Necks for sex: Sexual selection as an explanation for sauropod dinosaur neck elongation. *Journal of Zoology* 271:45–53.

Sereno, P. C. et al. 1999. Cretaceous sauropods from the Sahara and the uneven rate of skeletal evolution among dinosaurs. *Science* 286:1342–1347.

Sereno, P. C. et al. 2001. The giant crocodyliform *Sarcosuchus* from the Cretaceous of Africa. *Science* 294:1516–1519.

Seymour, R. S. 1979. Dinosaur eggs: Gas conductance through the shell, water loss during incubation and clutch size. *Paleobiology* 5:1–11.

Seymour, R. S. and H. B. Lillywhite. 2000. Hearts, neck posture and metabolic intensity of sauropod dinosaurs. *Proceedings of the Royal Society of London,* Series B. 267:1883–1887.

Seymour, R. S. et al. 2004. Evidence for endothermic ancestors of crocodiles at the stem of archosaur evolution. *Physiological and Biochemical Zoology* 77:1051–1067,

Simmons, R. and L. Scheepers. 1996. Winning by a neck: Sexual selection in the evolution of giraffe. *American Naturalist* 148:771–786.

Sipla, J. et al. 2004. The semicircular canals of dinosaurs: Tracking major transitions in locomotion. *Journal of Vertebrate Paleontology* 24:113A.

Soares, D. 2002. An ancient sensory organ in crocodilians. *Nature* 417:241–242.

Stevens, K. A., and J. M. Parrish. 1999. Neck posture and feeding habits of two Jurassic sauropod dinosaurs. *Science* 284:798–800.

Stevens, K. A., and J. M. Parrish. 2005. Digital reconstructions of sauropod dinosaurs and implications for feeding. In Curry Rogers, K. A. and Wilson, J. A. (eds.), *Sauropods: Evolution and Paleobiology*, University of California Press, Berkeley, CA, 178–200.

Stokstad, E. 2004. Head games show whether dinos went on two legs or four. *Science* 306:1466–1467.

Thomas, D. A., and J. O. Farlow. 1997. Tracking a dinosaur attack. *Scientific American* 277(6):74–79.

Thulborn, T. 1990. *Dinosaur Tracks*. London, UK: Chapman & Hall.

Turner, A. H. et al. 2007. A basal dromaeosaurid and size evolution preceding avian flight. *Science* 317:1378–1381.

Upchurch, P. et al. 2002. An analysis of dinosaurian biogeography: Evidence for the existence of vicariance and dispersal patterns caused by geological events. *Proceedings of the Royal Society of London*. Series B. 269:613–621.

Wang, X. et al. 2005. Pterosaur diversity and faunal turnover in Cretaceous terrestrial ecosystems in China. *Nature* 437:875–879.

Witmer, L. M. et al. 2003. Neuroanatomy of flying reptiles and implications for flight, posture, and behavior. *Nature* 425:950–953.

Xu, X. et al. 2002. An unusual oviraptoraurian dinosaur from China. *Nature* 419:291–293.

Xu, X. et al. 2003. Four-winged dinosaurs from China. *Nature* 421:335–340.

Xu, X., and M. A. Norell. 2004. A new troodontid dinosaur from China with avian-like sleeping posture. *Nature* 431:838–841.

Zhou, Z. et al. 2003. An exceptionally preserved Lower Cretaceous ecosystem. *Nature* 421:807–814.

CHAPTER 17

Avian Specializations

Flight is a central characteristic of birds. At the structural level, the mechanical requirements of flight shape many aspects of the anatomy of birds. In terms of ecology and behavior, flight provides options for birds that terrestrial animals lack. The ability of birds to make long-distance movements is displayed most dramatically in their migrations. Even small species like hummingbirds travel thousands of kilometers between their summer and winter ranges. Birds use a variety of methods to navigate on these trips, including orienting by the sun and the stars and probably also by using a magnetic sense.

Wing movements are used during swimming by wing-propelled aquatic birds, such as penguins, whereas others (ducks are familiar examples) use their feet to swim. And not all birds fly—birds are derived from terrestrial dromeosaurs, and some species of birds (among them ostriches, emus, and the kiwi, as well as several lineages of extinct predatory birds) are secondarily flightless. Many more species (both herbivorous birds like grouse and pheasants and predators like roadrunners and bustards) spend most of their time on the ground and fly only short distances to escape from predators.

A second characteristic of birds is diurnality; most species are active only by day. In addition, most birds have excellent vision, and colors and movements play important roles in their lives. Because humans, too, are diurnal and sensitive to color and movement, birds have been popular subjects for behavioral and ecological field studies, and important areas of modern biology draw heavily on studies of birds for data that can be generalized to other vertebrates.

17.1 Early Birds and Extant Birds

Even the limited flying abilities that we think *Archaeopteryx* possessed allowed it to move into an adaptive zone that had never been exploited by vertebrates. An animal that could run, launch itself into the air, and then land and run again had new opportunities to attack prey or to escape from predators, and the birds that appeared after *Archaeopteryx* quite rapidly developed the range of specializations that we now associate with birds.

Archaeopteryx was shaped like a bipedal dinosaur; it had a long body with its center of gravity near the hind legs, a shallow trunk (because it lacked the deep sternum of extant birds), a long bony tail, front legs that retained claws, and a full set of teeth. It was not very birdlike in appearance; and, if you saw an *Archaeopteryx* while you were out walking, your immediate reaction would not be "There's a bird." However, by the Early Cretaceous period a large number of anatomical changes had taken place, so you'd have no doubt that you were looking at a bird if you saw a *Sinornis*, or *Confuciusornis* (**Figure 17–1**).

The most obvious changes are in the general body form—the center of gravity has shifted forward toward the wings as in modern birds, the bony tail is greatly shortened, and fused vertebrae at the end of the tail form a pygostyle as they do in modern birds.

(a)

(b)

▲ **FIGURE 17–1** Early Cretacous enantiornithine birds. (a) *Sinornis* is a sparrow-size bird from Early Cretaceous lake bed deposits in China. (b) *Confuciusornis* is a crow-size bird, also from the Early Cretaceous of China.

Less conspicuous changes are the strutlike coracoids (bones that help the shoulder girdle resist the pressures exerted on the chest by the wing muscles), a reduction in the size of the claws on the feet (making them better suited to perching in trees), a larger sternum (giving more area for the origin of flight muscles), and a wrist that can bend back sharply to tuck the wing against the body.

The Mesozoic era saw two independent radiations of birds, and between them they produced a worldwide avian fauna that included most of the ecological types of birds we know today (**Figure 17–2**). The birds in the earliest radiation are known as Enantiornithes ("opposite birds") because the metatarsals are fused differently from those of modern birds. In addition, the Enantornithes retained teeth. Enantornithines were the dominant birds of the Cretaceous, and they ranged from small to large: *Sinornis*, the smallest enantiornithine discovered so far, was the

size of a sparrow and the largest, *Enantiornis* from Argentina, was the size of a turkey vulture. Most enantiornithines were small- to medium-sized and probably lived in trees, but some had long legs and appear to have been wading birds, while still others had powerful claws like modern hawks.

A second radiation of birds, the Ornithurae, appeared later in the Cretaceous and radiated into a wide variety of ecological types. Early ornithurines were small, finchlike arboreal species; but, by the Late Cretaceous, the lineage had expanded to waders, perchers, and secondarily flightless foot-propelled swimmers and divers such as *Hesperornis* and *Ichthyornis* (**Figure 17–3**).

Modern birds, the Neornithes, probably began to diversify during the last part of the Cretaceous. The time of origin of modern birds is currently the subject of debate because molecular and biogeographic studies suggest that the origins of modern orders of

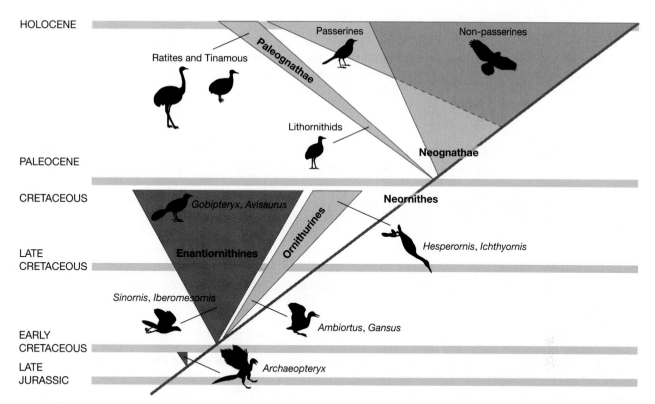

▲ **FIGURE 17–2** Mesozoic and modern birds. The Enantornithes was the first lineage to radiate and were the dominant group of birds during the Cretaceous period. A second radiation, the primitive ornithurines, completed the Mesozoic avian fauna. Neither of these lineages survived the end of the Mesozoic era, and the extant birds (Neornithes) are modern ornithurines. The shaded areas indicate the changes in diversity of the lineages.

birds can be traced into the Cretaceous, possibly even as long as 90 to 100 million years ago (Cracraft 1986, 2002). We have fossils of modern birds (Neornithes) from the Cretaceous (Clarke et al. 2005), but this record is sparse compared to the fossil records of enantornithines and ornithurines. That contrast suggests that Neornithes were a minor part of the Mesozoic bird fauna (Feduccia 2003, Fountaine et al. 2004).

▶ **FIGURE 17–3** Hesperornithiformes. By the Late Cretaceous a lineage of birds had already become specialized as flightless fish-eaters. (a) Reconstruction of *Hesperornis*. (b) Skull of a related species, *Parahesperornis*; note the teeth in the maxilla and dentary bones.

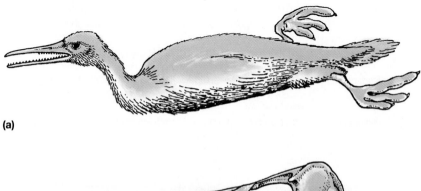

(a)

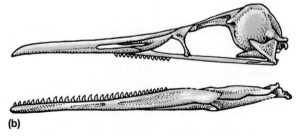

(b)

With about 9672 extant species, birds are a complex group of animals, and phylogenetic relationships within the Neoaves are especially controversial (Hackett et al. 2008). **Table 17.1** describes the characteristics of the major lineages of extant birds. The passerine lineage, with about 5739 species, has been particularly successful (Raikow and Bledsoe 2000, Barker et al. 2004). Passerines are characterized by modifications of the feet and legs that allow the toes to hold tightly to a perch even when the bird is

TABLE 17.1 Major lineages of extant birds (Neornithes). Except as noted, the lineages have representatives on all continents except Antarctica. For additional information see the Tree of Life Web Project (http:/ /tolweb. org/ Neornithes). The table lists the approximate number of species and general characteristics for the major taxa.

Palaeognathae (Ratites)
Struthiorniformes: ostrich, emu, rhea cassowary. 10 species of flightless birds, all from the Southern Hemisphere.

Neognathae
Galoanserae: ducks, geese, swans
Anseriformes: ducks, geese, and relatives. More than 150 species of semiaquatic birds.
Galliformes: Powl, quail, megapodes. More than 200 species of ground-dwelling, herbivorous birds.
Neoaves: most extant birds
Piciformes: woodpeckers and relatives. More than 200 species of mostly arboreal insectivorous birds.
Coraciiformes: kingfishers, kookaburras, and relatives. More than 100 species of mostly arboreal, carnivorous, or insectivorous birds.
Cuculiformes: cuckoos. About 150 species of arboreal and ground-dwelling birds that include both herbivorous and predatory forms.
Psittaciformes: parrots. More than 300 species of arboreal birds, mostly fruit and seed eaters.
Columbiformes: pigeons. More than 300 species of mostly rock dwellers, primarily seed eaters.
Apodiformes: hummingbirds and swifts. More than 400 species of arboreal birds with specialized flight. Swifts feed on insects they capture in flight, and hummingbirds are specialized nectar feeders that also capture flying insects.
Strigiformes: owls. About 200 species of nocturnal predatory birds that hunt on the wing.
Gruiformes: cranes, rails, and coots. More than 300 species of mostly herbivorous birds that are associated with freshwater habitats.
Ciconiiformes: storks, herons, ibises, pelicans, and their relatives. About 100 species of mostly predatory birds found near water. All except pelicans are long-legged waders.
Falconiformes: hawks, eagles, and vultures. More than 50 species of diurnal birds of prey or scavengers.
Charadriiformes: stilts, plover, oystercatchers, and their relatives. Fewer than 100 species of long-billed shorebirds that probe for invertebrates buried in mud or sand.
Phoenicopteriformes: flamingos. Five species of highly specialized aquatic filter feeders in tropical regions.
Procellariiformes: albatrosses, petrels, and their relatives. More than 100 species of seabirds; many of them are pelagic and fly thousands of kilometers before returning to land.

TABLE 17.1 *Continued*

Sphenisciformes: penguins. Seventeen species of wing-propelled divers that feed on fish. Largely confined to the cold Southern Hemisphere, including Antarctica, but one species lives in the Galápagos Islands.

Passeriformes: perching birds. Nearly 6000 species of birds with feet that are specialized for holding onto perches.

Suboscines: ovenbirds, antbirds, tyrant flycatchers, and their relatives. Six families with about 1000 species. Suboscines have limited control of the syrinx muscles and do not produce complex vocalizations.

Oscines: songbirds. About 70 families containing about 5000 species. Oscines have great control of their syrinx muscles and produce complex birdsongs.

---

asleep, and the passerines are commonly called the perching birds.

17.2 The Structure of Birds

In many respects, birds are variable: Beaks and feet are specialized for different modes of feeding and locomotion, the morphology of the intestinal tract is related to dietary habits, and wing shapes reflect flight characteristics. Despite that variation, however, the morphology of birds is more uniform than that of mammals. Much of this uniformity is a result of the specialization of birds for flight.

Consider body size as an example: Flight imposes a maximum body size on birds. The muscle power required for takeoff increases by a factor of 2.25 for each doubling of body mass. That is, if species B weighs twice as much as species A, it will require 2.25 times as much power to fly at its minimum speed. If the proportion of the total body mass allocated to flight muscles is constant, the muscles of a large bird must work harder than the muscles of a small bird. In fact, the situation is still more complicated because the power output is a function of both muscular force and wing-beat frequency, and large birds have lower wing-beat frequencies than small birds. As a result, if species B weighs twice as much as species A, it will develop only 1.59 times as much power from its flight muscles, although it needs 2.25 times as much power to fly. Therefore, large birds require longer takeoff runs than do small birds, and a bird could ultimately reach a body mass at which any further increase in size would move it into a realm in which its leg and flight muscles were not able to provide enough power to take off.

Taking off is particularly difficult for large birds, which have to run and flap their wings to reach the speed needed for liftoff. The great kori bustard (10 kilograms), the largest extant flying bird, requires such a long takeoff run that an observer wonders if it is ever going to get airborne. The largest flying bird known is *Argentavis*, a fossil from the upper Miocene epoch of Argentina. This gigantic predator, which is related to present-day storks and vultures, had a wingspan that is estimated at 7 meters (equivalent to the wingspan of a single-engine airplane) and a body mass of about 70 kilograms. *Argentavis* is thought to have taken flight by launching itself from cliffs and to have spent much of the time it was airborne gliding on updrafts that formed as air masses from the Pacific met the western slopes of the Andes Mountains (Chatterjee et al. 2007).

Flightless birds are spared the mechanical constraints associated with producing power for flight, but they still do not approach the body sizes of mammals. The largest extant bird is the flightless ostrich, which weighs about 150 kilograms, and the largest bird known, one of the extinct elephant birds, weighed an estimated 450 kilograms. In contrast, the largest terrestrial mammal, the African elephant, weighs up to 5000 kilograms.

The structural uniformity of birds is seen even more clearly if their body shapes are compared with those of other sauropsids. There are no quadrupedal birds, for example, nor any with horns or bony armor. Even those species of birds that have become secondarily flightless retain the basic body form of birds.

■ Feathers and Flight

Feathers develop from follicles in the skin, generally arranged in tracts or **pterylae**, which are separated by patches of unfeathered skin, the **apteria** (**Figure 17–4**).

▶ **FIGURE 17–4** Feather tracts of a typical songbird.

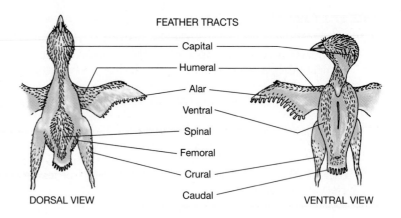

FEATHER TRACTS

Capital
Humeral
Alar
Ventral
Spinal
Femoral
Crural
Caudal

DORSAL VIEW   VENTRAL VIEW

Some species—such as ratites, penguins, and mousebirds—lack pterylae, and the feathers are uniformly distributed over the skin.

For all their structural complexity, the chemical composition of feathers is remarkably simple and uniform. More than 90 percent of a feather consists of a specific type of beta keratin, a protein related to the keratin that forms the scales of lepidosaurs. About 1 percent of a feather consists of lipids, about 8 percent is water, and the remaining fraction consists of small amounts of other proteins and pigments, such as melanin. The colors of feathers are produced by structural characters and pigments.

■ Types of Feathers

Feathers are anchored in the skin by a short, tubular base, the **calamus**, which remains firmly implanted within the follicle until molt occurs (**Figure 17–5**). A long, tapered **rachis** extends from the calamus and bears closely spaced side branches called **barbs**. Barbules branch from the barbs, and proximal and distal barbules branch from opposite sides of the barbules. The ends of the distal barbules bear hooks that insert in grooves in the proximal barbules of the adjacent barb. The hooks and grooves act like Velcro to hold adjacent barbs together, forming a flexible vane.

A body-contour feather has several regions that reflect differences in structure. Near the base of the rachis, the barbs and barbules are flexible, and the barbules lack hooks. This portion of a feather has a soft, loose, fluffy texture called plumulaceous or downy. It gives the plumage of a bird its properties of thermal insulation. Farther from the base, the barbs form a tight surface called the vane, which has a pennaceous (sheetlike) texture. This is the part of the feather that is exposed on the exterior surface of the plumage, where it serves as an airfoil, protects the downy undercoat,

sheds water, and reflects or absorbs solar radiation. The barbules are the structures that maintain the closed pennaceous character of the feather vanes. They are arranged in such a way that any physical disruption to the vane is easily corrected by preening behavior, in which the bird realigns the barbules by drawing its slightly separated bill over them.

Ornithologists usually distinguish five types of feathers: (1) contour feathers, including typical body feathers and the flight feathers (remiges and rectrices); (2) semiplumes; (3) down feathers of several sorts; (4) bristles; and (5) filoplumes.

The **remiges** (wing feathers, singular *remex*) and **rectrices** (tail feathers, singular *rectrix*) are large, stiff, mostly pennaceous contour feathers that are modified for flight. For example, the distal portions of the outer primaries of many species of birds are abruptly tapered or notched so that, when the wings are spread, the tips of these primaries are separated by conspicuous gaps or slots (**Figure 17–6**). This condition reduces the drag on the wing and, in association with the marked asymmetry of the outer and inner vanes, allows the feather tips to twist as the wings are flapped and to act somewhat as individual propeller blades.

Semiplumes are feathers intermediate in structure between contour feathers and down feathers. They combine a large rachis with entirely plumulaceous vanes and can be distinguished from down feathers in that the rachis is longer than the longest barb (**Figure 17–7a**). Semiplumes are mostly hidden beneath the contour feathers. They provide thermal insulation and help to fill out the contour of a bird's body.

Down feathers of various types are entirely plumulaceous feathers in which the rachis is shorter than the longest barb or entirely absent. Down feathers provide insulation for adult birds of all species. In addition, natal down, which is structurally simpler than adult down, provides an insulating covering on

▶ **FIGURE 17–5**

Typical vaned feathers.
(a) A wing quill, showing its main structural features.
(b) A body-contour feather. The inset and electron micrograph (c) show details of the interlocking mechanism of the proximal and distal barbules.

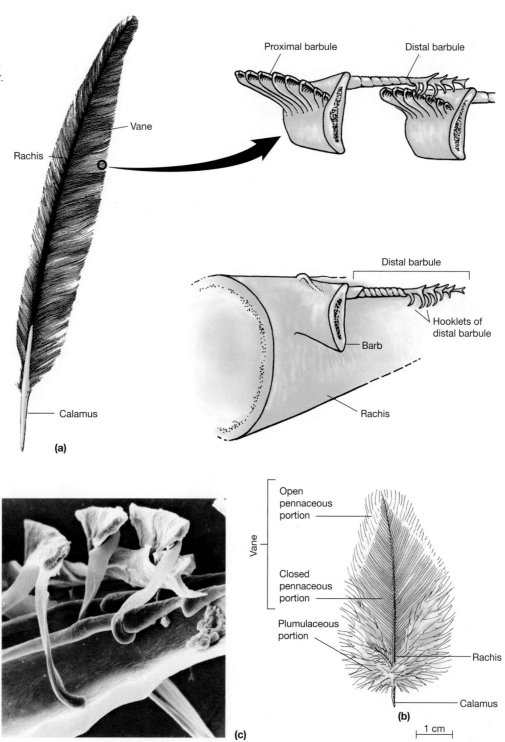

many birds at hatching or shortly thereafter. Natal downs usually precede the development of the first contour feathers, and down feathers are associated with apteria. Definitive downs develop as part of the full body plumage, which consists of down and contour feathers. Uropygial gland downs are associated with the large sebaceous gland found at the base of the tail in most birds. The papilla of the gland usually bears a tuft of modified, brushlike down feathers that aid in transferring the oily secretion from the gland to the bill to provide waterproof dressing to the plumage.

▲ **FIGURE 17–6** A red-tailed hawk. Slotting is visible in the outer primaries.

Powder down feathers, which are difficult to classify by structural type, break down to produce an extremely fine white powder composed of granules of keratin. The powder, which is shed into the general plumage, is nonwettable and is therefore assumed to provide another kind of waterproof dressing for the contour feathers. All birds have powder down, but it is best developed in herons.

Bristles are specialized feathers with a stiff rachis and barbs only on the proximal portion or none at all (**Figure 17–7b**). Bristles occur most commonly around the base of the bill, around the eyes, as eyelashes, and on the head or even on the toes of some birds. The distal rachis of most bristles is colored dark brown or black by melanin granules. The melanin not only colors the bristles but also adds to their strength, resistance to wear, and resistance to photochemical damage. Bristles and structurally intermediate feathers called semibristles screen out foreign particles from the nostrils and eyes of many birds; they also act as tactile sense organs and possibly as aids in the aerial capture of flying insects, as, for example, do the long bristles at the edges of the jaws in nightjars and flycatchers.

Filoplumes are fine, hairlike feathers with a few short barbs or barbules at the tip (**Figure 17–7c**). In some birds, such as cormorants and bulbuls, the filoplumes grow out over the contour feathers and contribute to the external appearance of the plumage, but usually they are not exposed. Filoplumes are sensory structures that aid in the operation of other feathers. Filoplumes have numerous free nerve endings in their follicle walls, and these nerves connect to pressure and vibration receptors around the follicles and transmit information about the position and movement of the contour feathers. This sensory system probably plays a role in keeping the contour feathers in place and adjusting them properly for flight, insulation, bathing, or display.

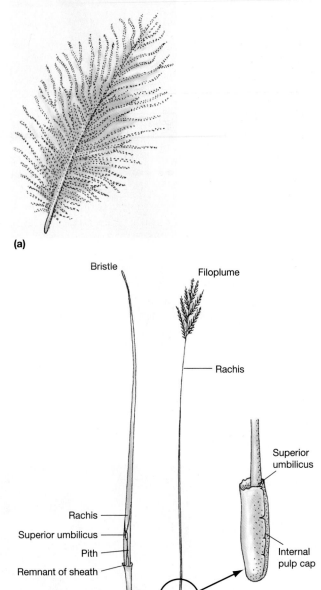

▲ **FIGURE 17–7** Types of feathers. (a) Semiplume. (b) Bristle. (c) Filoplume.

■ Streamlining and Weight Reduction

Birds are the only vertebrates that move fast enough in air for wind resistance and streamlining to be important factors in their lives. Many passerine birds are probably able to fly 50 kilometers per hour or even faster when they must, although their normal cruising speeds are lower. Ducks and geese can fly at 80 or 90 kilometers per hour, and peregrine falcons reach speeds as high as 200 kilometers per hour when they dive on prey. Fast-flying birds have many of the same structural characters as those seen in fast-flying aircraft. Contour feathers make smooth junctions between the wings and the body and often between the head and body as well, eliminating sources of turbulence that would increase wind resistance. The feet are tucked close to the body during flight, further improving streamlining.

At the opposite extreme, some birds are slow fliers. Many of the long-legged, long-necked wading birds, such as spoonbills and flamingos, fall in this category. Their long legs trail behind them as they fly, and their necks are extended. They are far from streamlined, although they may be strong fliers.

Characteristics of some of the organs of birds reduce body mass. For example, birds lack urinary bladders, and most species have only one ovary (the left). The gonads of both male and female birds are usually small; they hypertrophy during the breeding season and regress when breeding has finished.

■ Skeleton

Structural modifications can be seen in several aspects of the skeleton of birds. The avian skeleton is not lighter in relation to the total body mass of a bird than is the skeleton of a mammal of similar size, but the distribution of mass is different. Many bones are air filled (pneumatic, **Figure 17–8**), and the skull is especially light; however, the leg bones of birds are heavier than those of mammals. Thus, the total mass of the skeleton of a bird is similar to that of a mammal, but more of a bird's mass is concentrated in its hindlimbs.

Except for the specializations associated with flight, the skeleton of a bird is very much like that of a small dromeosaur (**Figure 17–9**). The pelvic girdle of birds is elongated, and the ischium and ilium have broadened into thin sheets that are firmly united with a **synsacrum**, which is formed by the fusion of

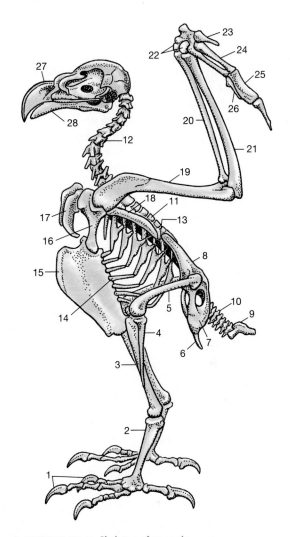

▲ **FIGURE 17–9** Skeleton of an eagle.

Legend: 1. Toes. **2.** Tarsometatarsus, formed by fusion of the distal tarsals to the metatarsals. **3.** Tibiotarsus, formed by fusion of the proximal tarsal bones (the calcaneum and astragalus) to the end of the tibia. **4.** Fibula. **5.** Femur. **6.** Pubis. **7.** Ischium. **8.** Ilium. **9** Pygostyle. **10.** Caudal vertebrae. **11.** Thoracic vertebrae. **12.** Cervical vertebrae. **13.** Uncinate process of rib. **14.** Sternal rib. **15.** Sternum. **16.** Coracoid. **17.** Furcula. **18.** Scapula. **19.** Humerus. **20.** Radius. **21.** Ulna. **22.** Carpal bones. **23.** First digit. **24.** Metacarpal. **25.** Second digit. **26.** Third digit. **27.** Upper jaw. **28.** Lower jaw.

▲ **FIGURE 17–8** Bird bone. The hollow core and reinforcing struts in a long bone from a bird.

10 to 23 vertebrae. The long tail of ancestral diapsids has been shortened in birds to about five free caudal vertebrae and a **pygostyle** formed by the fusion of the remaining vertebrae. The pygostyle supports the tail feathers (rectrices). The thoracic vertebrae are joined by strong ligaments that are often ossified. The relatively immobile thoracic vertebrae, the synsacrum, and the pygostyle combine with the elongated, rooflike pelvis to produce a nearly rigid vertebral column. Flexion is possible only in the neck, at

the joint between the thoracic vertebrae and the synsacrum, and at the base of the tail.

The center of gravity is beneath the wings, and the sternum is greatly enlarged compared with other vertebrates. The sternum of flying birds bears a keel from which the pectoralis and supracoracoideus muscles originate. Strong fliers have well-developed keels and large flight muscles. The scapula extends posteriorly above the ribs and is supported by the coracoid, which is fused ventrally to the sternum. Additional bracing is provided by the clavicles, which, in most birds, are fused at their distal ends to form the **furcula** (wishbone).

The hind foot of birds is greatly elongated, and the ankle joint is within the tarsals (a mesotarsal joint) as it was in Mesozoic theropods. The fifth toe is lost, and the metatarsals of the remaining toes are fused with the distal tarsals to form a bone called the **tarsometatarsus**. In other words, birds walk with their toes (tarsals) flat on the ground and the tarsometatarsus projecting upward at an angle. The actual knee joint is concealed within the contour feathers on the body, and what looks like the lower part of a bird's leg is actually the tarsometatarsus. That is why birds appear to have knees that bend backward; the true knee lies between the femur and lower leg (i.e., between the thigh and the drumstick). The outer (distal) end of the tibia fuses with the uppermost (proximal) tarsal bones, forming a composite bone called the **tibiotarsus** that forms most of the lower leg (drumstick). The fibula is reduced to a splint of bone.

■ Muscles

Power-producing features are equally important components of the ability of birds to fly. The pectoral muscles of a strong flier may account for 20 percent of the total body mass. The power output per unit mass of the pectoralis major of a turtledove during level flight has been estimated to be 10 to 20 times that of most mammalian muscles. Birds have large hearts and high rates of blood flow and complex lungs that use crosscurrent flows of air and blood to maximize gas exchange and to dissipate the heat produced by high levels of muscular activity during flight.

The relative size of the leg and flight muscles of birds is related to their primary mode of locomotion. Flight muscles comprise 25 percent to 35 percent of the total body mass of strong fliers such as hummingbirds and swallows. These species have small legs; the leg muscles account for as little as 2 percent of the body mass. Predatory birds such as hawks and owls use their legs to capture prey. In these species, the flight muscles make up about 20 percent of the body mass and the limb muscles 10 percent. Swimming birds—ducks and grebes, for example—have an even division between limb and flight muscles; the combined mass of these muscles may be 30 percent to 60 percent of the total body mass. Birds such as rails, which are primarily terrestrial and run to escape from predators, have limb muscles that are larger than their flight muscles.

Muscle fiber types and metabolic pathways also distinguish running birds from fliers. The familiar distinction between the light meat and dark meat of a chicken reflects those differences. Fowl, especially domestic breeds, rarely fly, but they are capable of walking and running for long periods. The dark color of the leg muscles reveals the presence of myoglobin in the tissues and indicates a high capacity for aerobic metabolism in the limb muscles of these birds. The white muscles of the breast lack myoglobin and have little capacity for aerobic metabolism. The flights of fowl (including wild species such as pheasants, grouse, and quail) are of brief duration and are used primarily to evade predators. The bird uses an explosive takeoff, fueled by anaerobic metabolic pathways, followed by a long glide back to the ground. Birds that are capable of strong, sustained flight have dark breast muscles with high aerobic metabolic capacities.

17.3 The Avian Wing

Unlike the fixed wings of an airplane, the wings of a bird function both as an airfoil (lifting surface) and as a propeller for forward motion. The primaries, inserted on the hand bones, do most of the propelling when a bird flaps its wings down, and the secondaries along the arm provide lift (**Figure 17–10**). Removing flight feathers from the wings of doves and pigeons shows that the ability to fly is greatly reduced when only a few of the primaries are pulled out, but a bird can still fly when as much as 55 percent of the total area of the secondaries has been removed.

A bird can alter the area and shape of its wings and their position with respect to the body. These changes in area and shape cause corresponding changes in velocity and lift that allow a bird to maneuver, change direction, land, and take off. The ability of swifts to modify their flight by changes in wing shapes is particularly noteworthy (Lentink et al. 2007),

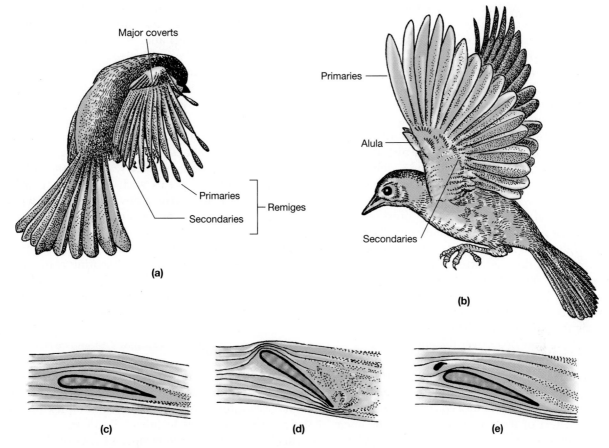

▲ **FIGURE 17–10** The wing in flight. (a,b) Drawings from high-speed photographs show the twisting and opening of the primaries during flapping flight. (c-e) Airflow around a cambered airfoil. At a low angle of attack (c), the air streams smoothly over the upper surface of the wing and creates lift. When the angle of attack becomes steep (d), air passing over the wing becomes turbulent, decreasing lift enough to produce a stall. A wing slot formed by the alula (e) helps to prevent turbulence by directing a flow of rapidly moving air close to the upper surface of the wing.

Obviously, the aerodynamic properties of a bird's wing in flight—even in nonflapping flight—are vastly more complex than those of a fixed wing on an airplane or glider. Nevertheless, it is instructive to consider a bird's wing in terms of the basic performance of a fixed airfoil. Although a bird's wing actually moves forward through the air, it is easier to think of the wing as stationary with the air flowing past. The flow of air produces a force, which is usually called the reaction. It can be resolved into two components: the **lift**, which is a vertical force equal to or greater than the weight of the bird, and the **drag**, which is a backward force opposed to the bird's forward motion and to the movement of its wings through the air.

When the leading end of a symmetrically streamlined body cleaves the air, it thrusts the air equally upward and downward, reducing the air pressure equally on the dorsal and ventral surfaces. No lift results from such a condition. There are two ways to modify this system to generate lift. One is to increase the **angle of attack** of the airfoil, and the other is to bend its surface. Either change increases lift at the cost of increasing drag.

When the contour of the dorsal surface of the wing is convex and the ventral surface is concave (a **cambered airfoil**), the air pressure against the two surfaces is unequal because the air has to move farther and faster over the dorsal convex surface relative to the ventral concave surface (Figure 17–10c). The result is a lower air pressure above the wing than beneath it. This difference in pressure is called lift, and when the lift equals or exceeds the bird's body weight, the bird becomes airborne. The camber of the wing varies in birds with different flight characteristics; it also changes along the length of the wing. Camber is greatest close to the body and decreases toward the wing tip. This change in camber is one reason that the inner part of the wing generates greater lift than the outer part.

If the leading edge of the wing is tilted up so that the angle of attack is increased (Figure 17–10d), the result is increased lift—up to an angle of about 15 degrees, the stalling angle. This lift results more from a decrease in pressure over the dorsal surface than from an increase in pressure below the airfoil. If the smooth flow of air over the wing becomes disrupted, the airflow begins to separate from the wing because of the increased air turbulence over the wing. The wing is then stalled. Stalling can be prevented or delayed by the use of slots or auxiliary airfoils on the leading edge of the main wing. The slots help to restore a smooth flow of air over the wing at high angles of attack and at slow speeds. The bird's **alula** has this effect, particularly during landing or takeoff (Figure 17–10e). Also, the primaries act as a series of independent, overlapping airfoils, each tending to smooth out the flow of air over the one behind.

Another characteristic of an airfoil has to do with wing-tip vortexes—eddies of air resulting from outward flow of air from under the wing and inward flow from over it. This is induced drag. One way to reduce the effect of these wing-tip eddies and their drag is to lengthen the wing, so that the tip vortex disturbances are widely separated and there is proportionately more wing area where the air can flow smoothly. Another solution is to taper the wing, reducing its area at the wing tip where induced drag is greatest. The ratio of length to width is called the **aspect ratio**. Long, narrow wings have high aspect ratios and high lift-to-drag (L/D) ratios. High-performance sailplanes and albatrosses, for example, have aspect ratios of 18:1 and L/D ratios in the range of about 40:1.

Wing loading is another important consideration. This is the mass of the bird divided by the wing area. The lighter the loading, the less power is needed to sustain flight. Small birds usually have lighter wing loading than do large birds, but wing loading is also related to specializations for powered versus soaring flight. The comparisons in **Table 17.2** illustrate both of these trends. Small species such as hummingbirds, barn swallows, and mourning doves have lighter wing loading than do large species such as the peregrine, golden eagle, and mute swan; yet the 3-gram hummingbird, a powerful flier, has a heavier wing loading than the more buoyant, sometimes soaring barn swallow, which is more than five times heavier. Similarly, the rapid-stroking peregrine has a heavier wing loading than the larger, often soaring golden eagle.

■ Flapping Flight

Flapping flight is remarkable for its automatic, unlearned performance. A young bird on its initial flight uses a form of locomotion so complex that it defies precise analysis in physical and aerodynamic terms. The nestlings of some species of birds develop in confined spaces, such as burrows in the ground or cavities in tree trunks, in which it is impossible for them to spread their wings to practice flapping before they leave the nest. Despite this seeming handicap, many of them are capable of flying considerable distances on their first flights. Diving petrels

TABLE 17.2 Wing loading of some representative birds. In general, wing loading increases with body size, but the highest wing loading recorded is found in the thick-billed murre, a sea bird that swims and dives. The murre's wing loading is believed to approach the maximum value possible for a flying bird.

Species	Body Mass (g)	Wing Area (cm²)	Wing loading (g/cm²)
Ruby-throated hummingbird	3	12.4	0.24
Barn swallow	17	118.5	0.14
Mourning dove	130	357	0.36
Thick-billed murre	1033	397	2.6
Peregrine falcon	1222	1342	0.91
Golden eagle	4664	6520	0.72
Mute swan	11,602	6808	1.70

may fly as far as 10 kilometers the first time out of their burrows. On the other hand, young birds reared in open nests—especially large birds such as albatrosses, storks, vultures, and eagles—frequently flap their wings vigorously in the wind for several days before flying. Such flapping may help to develop muscles, but it is unlikely that these birds are learning to fly. However, a bird's flying abilities do improve with practice for a period after it leaves the nest. Landing is especially challenging, and inexperienced young birds may miss the branch they were aiming for and tumble through the tree, grabbing frantically for anything within reach.

There are so many variables involved in flapping flight that it is difficult to understand exactly how it works. A beating wing is both flexible and porous, and it yields to air pressure. Its shape, wing loading, camber, angle relative to the body, and the position of the individual feathers all change remarkably as a wing moves through its cycle of locomotion. This is a formidable list of variables, far more complex than those involved in the analysis of the fixed wing of an airplane, and it is no wonder that the aerodynamics of flapping flight are not yet fully understood. However, the general properties of a flapping wing can be described.

We can begin by considering the flapping cycle of a small bird in flight. A bird cannot continue to fly straight and level unless it can develop a force or thrust to balance the drag operating against forward momentum. The downward stroke of the wings produces this thrust. The wingtips (primary feathers) are the site where thrust is generated, whereas the inner wings (secondaries) generate lift. It is easiest to consider the forces operating on the inner and outer wing separately.

The forces on the wing tips derive from two motions that have to be added together. The tips are moving forward with the bird, but at the same time, they are moving downward relative to the bird. The wing tip would have a very large angle of attack and would stall if it were not flexible. As it is, the forces on the tip cause the individual primaries to twist as the wing is flapped downward (see Figure 17–10a) and to produce the forces diagrammed in **Figure 17–11**.

The forces acting on the two parts of the wing combine to produce the conditions for equilibrium flight. The positions of the inner and outer wings during the upstroke (dotted lines) and downstroke reveal that vertical motion is applied mostly at the wing tips (Figure 17–11a). Thus, the inner wing acts as if the bird were gliding and generates the forces shown in Figure 17–11b, whereas the outer wing generates the

force shown in Figure 17–11c. Most of the lift to counter the pull of gravity (M) is generated by the inner wing and body. Tilting the wing tip during the downstroke (see Figure 17–11c) produces a resultant force (R) that is directed forward. The movement of the wing tip relative to the air is affected by the forward motion of the bird through the air (Figure 17–11d). As a result of this motion and the tilting of the wing tips during the downstroke, the flow of air across the primaries is different from the flow across the secondaries and the body. When flight speed through the air is constant, the forces acting on the inner wing and the body and on the outer wing combine to produce a set of summed vectors in which thrust exceeds total drag and lift at least equals the body mass (Figure 17–11e).

As the wings move downward and forward on the downstroke the trailing edges of the primaries bend upward under air pressure, and each feather acts as an individual propeller biting into the air and generating thrust. Contraction of the pectoralis major, the large breast muscle, produces the forceful downstroke during level flapping flight. During this downbeat, the thrust is greater than the total drag, and the bird accelerates. In small birds, the return stroke, which is upward and backward, provides little or no thrust. It is mainly a passive recovery stroke, and the bird slows down during this part of the wingbeat cycle.

For large birds with slow wing actions, the upstroke lasts too long to spend in a state of deceleration, and a similar situation exists when any bird takes off—in these situations the bird needs to generate thrust on the upstroke as well as on the downstroke. Thrust on the upstroke is produced by bending the wings slightly at the wrists and elbow and by rotating the humerus upward and backward. This movement causes the upper surfaces of the twisted primaries to push against the air and to produce thrust as their lower surfaces did in the downstroke. In this type of flight, the wing tip describes a rough figure eight through the air. As speed increases, the figure-eight pattern is restricted to the wing tips.

A powered upstroke results mainly from contraction of the supracoracoideus, a deep muscle underlying the pectoralis major and attached directly to the keel of the sternum. It inserts on the dorsal head of the humerus by passing through the foramen triosseum, formed where the coracoid, furcula, and scapula join (**Figure 17–12b**). In most species of birds, the supracoracoideus is a relatively small, pale muscle with low myoglobin content, easily fatigued. In species that rely on a powered upstroke—for fast,

▶ **FIGURE 17–11** Physical aspects of bird flight. The forces acting on the inner and outer wings and the body of a bird during flapping flight are shown. M = gravitational force; R = resultant force. See text for explanation.

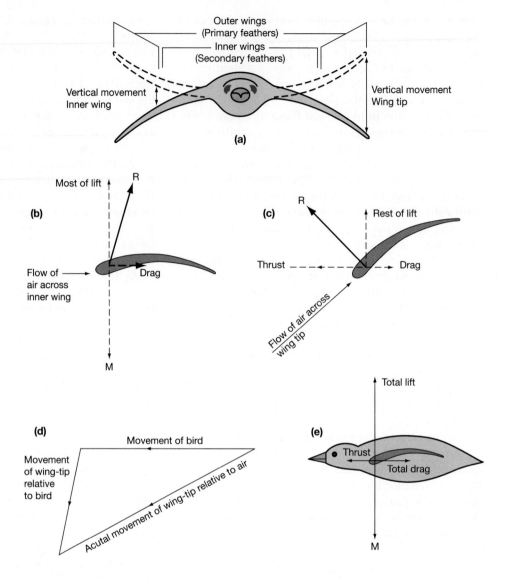

steep takeoffs, for hovering, or for fast aerial pursuit—the supracoracoideus is relatively larger. The ratio of weights of the pectoralis major and the supracoracoideus is a good indication of a bird's reliance on a powered upstroke; such ratios vary from 3:1 to 20:1. The total weight of the flight muscles shows the extent to which a bird depends on powered flight. Strong fliers such as pigeons and falcons have breast muscles comprising more than 20 percent of body weight, whereas in some owls, which have very light wing loading, the flight muscles make up only 10 percent of total weight.

■ Wing Proportions

To a small degree, a bird can change the surface area of the wing by changing the position of its feathers,

but most change in wing proportions occurs over evolutionary time. Wings may be large or small in relation to body size, resulting in light wing loading or heavy wing loading. They may be long and pointed, short and rounded, highly cambered or relatively flat, and the width and degree of slotting are additional important characteristics. Depending on whether a bird is primarily a powered flier or a soaring form, the various segments of the wing (hand, forearm, upper arm) are lengthened to different degrees. Hummingbirds have very fast, powerful wing beats, requiring maximum propulsive force from the primaries. The hand bones of hummingbirds are longer than the forearm and upper arm combined. Most of the flight surface is formed by the primaries, and hummingbirds have only six or seven secondaries. Frigate birds are marine

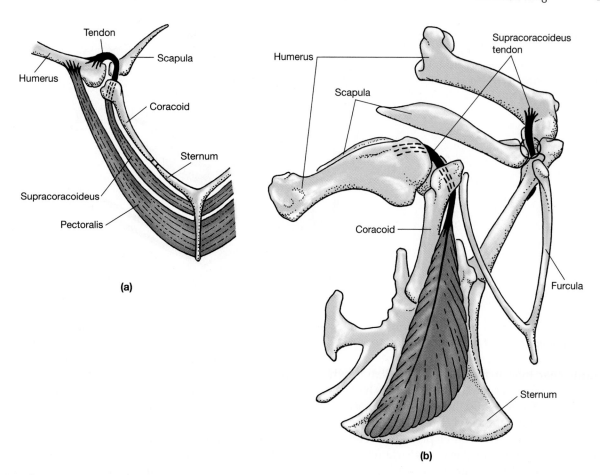

▲ **FIGURE 17–12** Major flight muscles of birds. (a) Cross section through the sternum of a bird showing the relationships of the pectoralis major and supracoracoideus muscles. (b) Frontal and lateral view of the sternum and pectoral girdle of a bird showing the insertion of the supracoracoideus tendon through the foramen triosseum onto the dorsal head of the humerus. The foramen is formed by the articulation of the furcula, coracoid, and scapula.

species with long, narrow wings specialized for powered flight as well as for gliding and soaring; they have the lowest wing loading of any extant bird. All three segments of the forelimb are about equal in length. The soaring albatrosses have carried lengthening of the wing to the extreme limit found in birds: the humerus or upper arm is the longest segment, and there may be as many as 32 secondaries in the inner wing (**Figure 17–13**).

■ Wing Structure and Flight Characteristics

Ornithologists recognize four structural and functional types of wings (**Figure 17–14**). Seabirds, particularly those such as albatrosses and shearwaters that rely on dynamic soaring, have long, narrow, flat wings lacking slots in the outer primaries. Dynamic

soaring is possible only where there is a pronounced vertical wind gradient, with the lower 15 or so meters of air being slowed by friction against the ocean surface. Furthermore, dynamic soaring is feasible only in regions where winds are strong and persistent, such as in the latitudes of the Roaring Forties (between latitudes 40 and 50 degrees south, in which there are strong westerly winds). This is where most albatrosses and shearwaters are found. Starting from the top of the wind gradient, an albatross glides downwind with great increase in ground speed (kinetic energy). Then, as it nears the surface, it turns and gains altitude while gliding into the wind. Because the bird flies into wind of increasing speed as it rises, its loss of airspeed is not as great as its loss of ground speed. Consequently it does not stall until it has mounted back to the top of the wind gradient, where the air velocity becomes stable. At that point,

▶ **FIGURE 17–13** Wing proportions. Comparison of the relative lengths of the proximal, middle, and distal elements of the wing bones of a hummingbird (*top*), frigate bird (*middle*), and albatross (*bottom*) drawn to the same size.

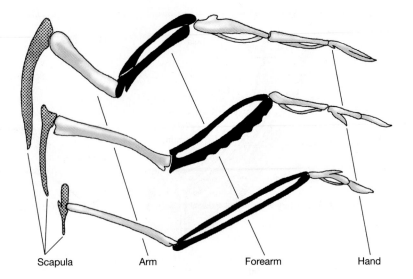

Scapula   Arm   Forearm   Hand

the bird has converted much of its kinetic energy to potential energy, and it turns downwind to repeat the cycle.

Birds that live in forests and woodlands where they must maneuver around obstructions have elliptical wings. These wings have a low aspect ratio, tend to be highly cambered, and usually have a high degree of slotting in the outer primaries. These features are generally associated with rapid flapping, slow flight, and a high degree of maneuverability.

Birds such as swifts, that are aerial foragers, make long migrations, or have a heavy wing loading that is related to some other aspect of their lives, such as diving, have high aspect ratio wings. These wings have a flat profile (little camber) and often lack slots in the outer primaries. In flight, they show the swept-back attitude of jet-fighter plane wings. All fast-flying birds have converged on this form.

The slotted high-lift wing is a fourth type. It is associated with static soaring typified by vultures, eagles, storks, and some other large birds. This wing has an intermediate aspect ratio between the elliptical wing and the high aspect ratio wing, a deep camber, and marked slotting in the primaries. When the bird is in flight, the tips of the primaries turn upward under the influence of air pressure and body weight (see Figure 17–6). Static soarers remain airborne mainly by seeking out and gliding in air masses that are rising at a rate faster than the bird's sinking speed. Hence, a light wing loading and maneuverability (slow forward speed and small turning radius) are advantageous. Broad wings provide the light wing loading, and the highly developed slotting enhances maneuverability by responding to changes in wind currents with changes in the positions of individual feathers instead of movements of the entire wing.

▶ **FIGURE 17–14** Comparison of four basic types of bird wings.
(a) Dynamic soaring. (b) Elliptical.
(c) High aspect ratio. (d) High lift.

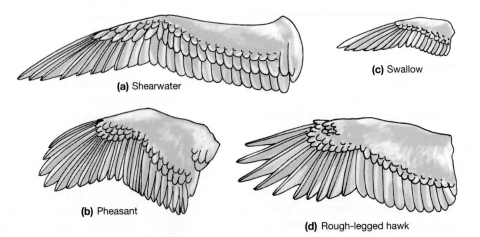

(a) Shearwater

(c) Swallow

(b) Pheasant

(d) Rough-legged hawk

In regions where topographic features and meteorological factors provide currents of rising air, static soaring is an energetically cheap mode of flight. By soaring rather than flapping, a large bird the size of a stork can decrease by a factor of 20 or more the energy required for flight per unit of time, whereas the saving is only one-tenth as much for a small bird such as a warbler. Probably the gigantic *Argentavis* of the Miocene could not have lived in a region without updrafts of this sort because the energy cost of continuous flapping flight would have been prohibitive for a 70-kilogram bird. It is little wonder, then, that most large land birds perform their annual migrations by soaring and gliding as much of the time as possible, and some condors and vultures cover hundreds of kilometers each day soaring in search of food.

Flying Speed and the Cost of Flight

A flying bird increases its speed by increasing the amplitude of its wing beats, but the frequency of wing beats remains nearly constant at all speeds during level flight. Large birds have slower wing-beat frequencies than do small birds, and strong fliers usually have slower beat frequencies than weak fliers. The furcula (wishbone) acts as a spring and a timing device. The frequency of respiration of many birds during flight appears to have a constant relationship to the wing-beat frequency. Some birds, especially those that have low wing-beat frequencies, breathe once per wing-beat cycle. In contrast, birds with high wing-beat frequencies have breathing cycles that span several wing beats. In general, inspiration appears to occur during or at the end of a wing upstroke, with expiration at the end of a downstroke.

Initial studies suggest that the most efficient flying speed may depend on anatomical characteristics of a bird's wings and wing muscles. Aerodynamic theory predicts that the mechanical power required for flight should show a U-shaped relationship to flying speed, with the bottom of the U being the most efficient flight speed. That relationship was observed for two fast-flying species of birds with high aspect ratio wings: a dove (*Streptopelia risoria*) and a cockatiel (*Nymphicus hollandicus*), both of which had sharply U-shaped curves with distinct minima at 5 meters per second to 7 meters per second (Tobalske et al. 2003). In contrast, a black-billed magpie (*Pica pica*), which has short, broad elliptical wings, had a much flatter curve with a power output that varied little between 4 meters per second and 12 meters per second.

Birds often fly in formation—V-shaped skeins of migrating geese are familiar spring and autumn sights in the Northern Hemisphere, and coastal residents can see pelicans flying in formation year-round. The heart rates of free-flying great white pelicans (*Pelecanus onocrotalus*) were monitored to compare energy use during individual flight and flight in formation (Weimerskirsch et al. 2001). Pelicans have a distinctive flight pattern that alternates several wing beats with a glide that lasts 1 to 2 seconds. When pelicans fly in formation, they either match the leader wing beat for wing beat or flap in regular succession from the leader backward through the formation. Heart rate was lowest, averaging about 150 beats per minute, when a pelican was gliding. Pelicans flying individually had average heart rates of 180 beats per minute to 190 beats per minute, but when they flew in formation the average heart rate was only 165 beats per minute. A bird flying in the wake of one or more other birds benefits from the upwash of air that is generated by the wings of the other birds. For pelicans this effect could produce an energy saving of 1.7 to 3.4 percent, which is similar to the energy saving calculated for geese flying in formation. In addition, pelicans spent more time gliding when they were in formation than when they flew individually. When the effect of additional gliding time is included, the estimate of energy savings rises to 11.4 percent to 14 percent.

17.4 The Hindlimbs

The ability to perch in trees was an early feature of bird evolution. *Archaeopteryx* was apparently no better adapted for perching than most coelurosaurs; however, in the Cretaceous, both the entantiform and ornithourine lineages included perching birds. On the ground birds hop, walk, or run, and in water they swim using either the hindlimbs or the wings.

Perching

The most specialized avian foot for perching on branches is one in which all four toes are free and mobile and of moderate length. Three toes extend forward and are opposable to one toe that extends backward in the same plane. This is an **anisodactyl** foot. It allows a firm grip and is highly developed in the passerine birds. The zygodactylous condition, with two toes extending forward that are opposable to two toes extending backward, is characteristic of birds such as parrots and woodpeckers, which climb or perch on vertical surfaces.

The tendons that flex the toes of a perching bird can lock the foot in a tight grip so that the bird does not fall off its perch when it relaxes or goes to sleep. The plantar tendons, which insert on the individual phalanges of the toes, tighten when the legs bend and curl the toes around the perch. Furthermore, the tendons lying underneath the toe bones have hundreds of minute, rigid, hobnail-like projections that mesh with ridges on the inside surface of the surrounding tendon sheath. The projections and ridges lock the tendons in place in the sheaths and help to hold the toes in their grip around the branch. As long as the legs are flexed, the toes are locked around the perch and the bird cannot fall off. Indeed, the bird must extend its legs by standing up before its toes will uncurl.

■ Hopping, Walking, and Running

When they are on the ground, birds hop (make a succession of jumps in which both legs move together), walk (move the legs alternately with at least one foot in the contact with the ground at all times) or run (move the legs alternately with both feet off the ground at some times). Modifications of these basic gaits include climbing, wading in shallow water or on insubstantial surfaces such as lily pads, and supporting heavy bodies.

Hopping is a special form of locomotion found mostly in perching, arboreal birds. It is most highly developed in the passerines, and only a few non-passerine birds regularly hop. Many passerines cannot walk, and hopping is their only mode of terrestrial locomotion. Groups of passerines that have a relatively terrestrial mode of existence (larks, pipits, starlings, and grackles, for example) are able to walk as well as hop. The separation between walking and hopping passerines cuts across families. For example, in the family Corvidae, the ravens, crows, and rooks are walkers, whereas the jays and magpies are hoppers.

Running is a modification of walking in which both feet are off the ground at the same time for a portion of each step cycle. In general, vertebrates that are fast runners have long, thin legs and small feet. The significance of long legs is obvious—they allow an animal to cover more distance with each stride. The reduction in leg and foot mass is a bit more subtle and involves the physics of momentum. When the foot is in contact with the ground, it is motionless; after it pushes off from the ground, the foot and the lower part of the leg must be accelerated to a speed faster than the trunk of the running animal. Reducing the mass of the foot and leg reduces the energy needed for this change in momentum and allows the animal to run faster.

Reducing the number and length of the toes makes the foot smaller. No bird has reduced the length and number of toes in contact with the ground to the extent that the hoofed mammals have; but, the large, fast-running ostrich has only two toes on each foot, and many other species have only the three forward-directed toes in contact with the ground (**Figure 17–15**).

Weight-bearing characteristics, such as large, heavy leg bones arranged in vertical columns as supports of great body mass, are well known in large mammals such as elephants. No surviving birds show specializations of this kind, although these features are seen in some of the large, flightless terrestrial birds of earlier times. The extinct elephant birds (Aepyornithiformes) of Madagascar and moas (Dinornithiformes) of New Zealand were herbivores that evolved on

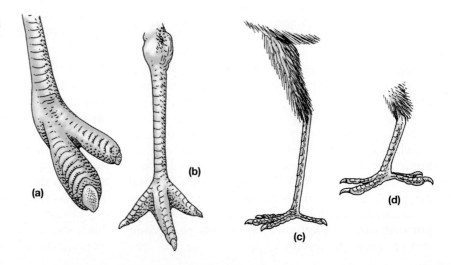

▶ **FIGURE 17–15** Avian feet with various specializations for terrestrial locomotion. (a) Ostrich, with only two toes. (b) Rhea, with three toes. (c) Secretary bird, with a typical avian foot. (d) Roadrunner, with zygodactyl foot. (Not drawn to scale.)

(a)

(b)

(c)

(d)

▶ **FIGURE 17–16** Large, flightless, terrestrial birds: (a) the extinct elephant bird of Madagascar; (b) the cursorial ostrich.

1 m

1 m

oceanic islands in the absence of large carnivorous mammals and survived there until contact with humans in the post-Pleistocene period (**Figure 17–16**).

Climbing

Birds climb on tree trunks or other vertical surfaces by using their feet, tails, beaks, and, rarely, their forelimbs. Several distantly related groups of birds have independently acquired specializations for climbing and foraging on vertical tree trunks. Species such as woodpeckers and woodcreepers, which use their tails as supports, begin foraging near the base of a tree trunk and work their way vertically upward, head first, clinging to the bark with strong feet on short legs. The tail is used as a prop to brace the body against the powerful pecking exertions of head and neck, and the pygostyle and free caudal vertebrae in these species are much enlarged and support strong, stiff tail feathers. A similar modification of the tail is found in certain swifts that perch on cave walls and inside chimneys.

Nuthatches and similarly modified birds climb on trunks and rock walls in both head-upward and head-downward directions while foraging. In these species, which do not use their tails for support, the claw on the backward-directed toe (the hallux) is larger than those on the forward-directed toes and is strongly curved.

Swimming on the Surface

Although no birds have become fully aquatic like the ichthyosaurs and cetaceans, nearly 400 species of birds are specialized for swimming. Nearly half of these aquatic species also dive and swim underwater.

Modifications of the hindlimbs are the most obvious avian specializations for swimming. Other changes include a wide body that increases stability in water; dense plumage that provides buoyancy and insulation; a large preen gland, producing oil that waterproofs the plumage; and structural modifications of the body feathers that retard penetration of water to the skin. The legs are at the rear of a bird's body, where the mass of leg muscles interferes least with streamlining and where the best control of steering can be achieved.

The feet of aquatic birds are either webbed or lobed (**Figure 17–17**). Webbing between the three forward toes (palmate webbing) has been independently acquired at least four times in the course of avian evolution. Totipalmate webbing of all four toes is found in pelicans and their relatives. The hydrodynamic forces acting on the foot of a swimming bird are complex. At the beginning of the power stroke, when the foot is in its forward-most position, the web is nearly parallel to the water surface and moving downward. At this stage, the effect of foot movement is to lift the bird and propel it forward. Later in the stroke, however, when the web passes through the point of being perpendicular to the water surface, hydrodynamic drag is produced on the forward-facing (dorsal) side of the web by the vortex that develops as water flows around the web. At this stage, the drag is the major force propelling the bird forward.

Lobes on the toes have evolved convergently in several phylogenetic lineages of aquatic birds. There are two different types of lobed feet. Grebes are unique in that the lobes on the outer sides of the toes are rigid and do not fold back as the foot moves forward. Instead of folding the lobes, grebes rotate

▶ **FIGURE 17–17** Webbed and lobed feet of some aquatic birds. (a) Duck, showing partial webbing. (b) Cormorant, showing totipalmate webbing. The lobed foot of a grebe, showing how it is rotated during a stroke: (c) position of toes during backward power stroke in side view; (d) front view; (e) side view of the rotated foot during the forward recovery stroke.

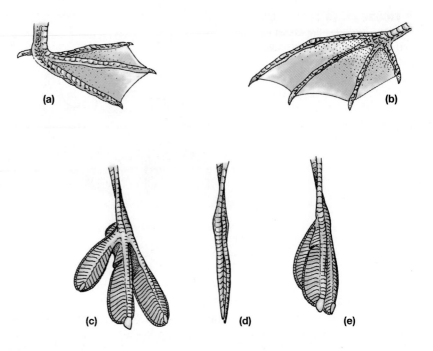

the third and fourth toes 90 degrees at the beginning of the recovery stroke and move the second toe behind the tarsometatarsus. In this position, the sides of the toes slice through the water like knife blades, minimizing resistance. A simpler mechanism for the recovery stroke occurs in all the other lobe-footed swimmers, where the lobes are flaps that fold back against the toes during forward movement through water and flare open to present a maximum surface on the backward stroke.

Diving and Swimming Underwater

The transition from a surface-swimming bird to a subsurface swimmer has occurred in two fundamentally different ways: either by further specialization of a hindlimb already adapted for swimming or by modification of the wing for use as a flipper under water. Highly specialized foot-propelled divers have evolved independently in grebes, cormorants, loons, and the extinct Hesperornithidae. All these families, except the loons, include some flightless forms (**Figure 17–18**). Wing-propelled divers have evolved in the Procellariiformes (the diving petrels), the Sphenisciformes (penguins), and the Charadriiformes (auks and related forms). Only among the waterfowl are there both foot-propelled and wing-propelled diving ducks, but none of these species is as highly modified for diving as specialists like the loons or auks. The water ouzels or dippers (*Cinclus*) are passerine birds that dive and swim underwater with great facility using their small, round wings, but they lack any other morphological specializations.

17.5 Feeding and Digestion

With the specialization of the forelimbs as wings, which largely prevents any substantial role in prey capture, birds have concentrated their predatory mechanisms in their beaks and feet. Modifications of the beak, tongue, and intestines are often associated with dietary specializations.

Beaks and Tongues

The presence of a horny beak in place of teeth is not unique to birds—turtles have beaks, and so did the rhynchosaurs, many dinosaurs, pterosaurs, and the dicynodonts. However, the diversity of beaks among birds is remarkable. The range of morphological specializations of beaks defies complete description, but some categories can be recognized (**Figure 17–19**).

Insectivorous birds such as warblers—which find their food on leaf surfaces—usually have short, thin, pointed bills that are adept at seizing insects. Aerial sweepers such as swifts, swallows, and nighthawks—which catch their prey on the wing—have short, weak beaks and a wide gape.

► **FIGURE 17–18** Parallel evolution of swimming and diving birds. Petrels and penguins (Procellariiformes and Sphenisciformes) occupied this niche in the Southern Hemisphere. In the Northern Hemisphere gulls and auks (Charadriiformes) developed parallel specializations.

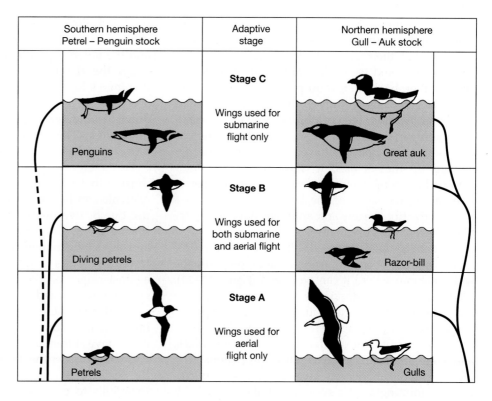

Southern hemisphere Petrel – Penguin stock	Adaptive stage	Northern hemisphere Gull – Auk stock
Penguins	**Stage C** Wings used for submarine flight only	Great auk
Diving petrels	**Stage B** Wings used for both submarine and aerial flight	Razor-bill
Petrels	**Stage A** Wings used for aerial flight only	Gulls

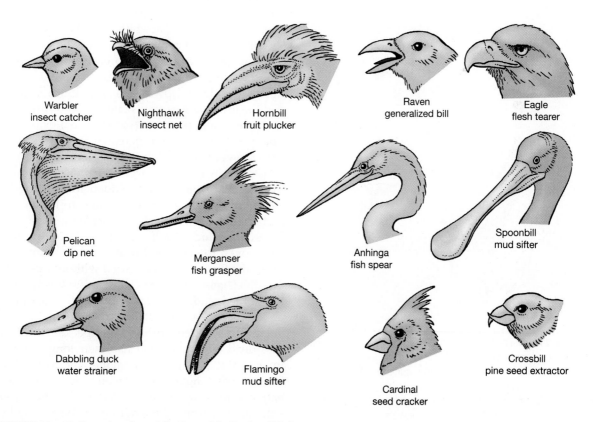

Warbler
insect catcher

Nighthawk
insect net

Hornbill
fruit plucker

Raven
generalized bill

Eagle
flesh tearer

Pelican
dip net

Merganser
fish grasper

Anhinga
fish spear

Spoonbill
mud sifter

Dabbling duck
water strainer

Flamingo
mud sifter

Cardinal
seed cracker

Crossbill
pine seed extractor

▲ **FIGURE 17–19** Examples of specializations of the beaks of birds.

Many carnivorous birds, such as gulls, ravens, crows, and roadrunners, use their heavy pointed beaks to kill their prey. On the other hand, most hawks, owls, and eagles kill prey with their talons and use their beaks to tear off pieces small enough to swallow. Falcons stun prey with the impact of their dive, and then bite the neck of the prey to disarticulate the cervical vertebrae. Fish-eating birds such as cormorants and pelicans have beaks with a sharply hooked tip that is used to seize fish, while mergansers have long, narrow bills with a series of serrations along the sides of the beak, in addition to a hook at the tip. Darters and anhingas have harpoon-like bills that they use to impale fish.

Spoonbills have flattened bills with broad tips, which they use to create currents that lift prey into the water column where it can be seized (**Figure 17–20**). Many aquatic birds strain small crustaceans or plankton from water or mud with bills that incorporate some sort of filtering apparatus. Dabbling ducks have bills with horny lamellae that form crosswise ridges, and their tongues also have horny projections. The tongue and bill are densely invested with sensory corpuscles, forming a filter system that allows ducks to scoop up a billfull of water and mud, filter out the

prey, and allow the debris to escape. Flamingos have a similar system. The bill of a flamingo is sharply bent, and the anterior part is held in a horizontal position when the flamingo lowers its head to feed. The flamingo's lower jaw is smaller than the upper jaw, and it is the upper jaw that vibrates rapidly up and down during feeding while the lower jaw remains motionless. (This reversal of the usual vertebrate pattern is made possible by the kinetic skull of birds.)

Seeds are usually protected by hard coverings (husks) that must be removed before the nutritious contents can be eaten. Specialized seed-eating birds use one of two methods to husk seeds before swallowing them. One group holds the seed in its beak and slices it by making fore-and-aft movements of the lower jaw. Birds in the second group hold the seed against ridges on the palate and crack the husk by exerting an upward pressure with their robust lower jaw. After the husk has been opened, both kinds of birds use their tongues to remove the contents. Other birds have different specializations for eating seeds: Crossbills extract the seeds of conifers from between the scales of the cones, using the diverging tips of their bills to pry the scales apart and a prehensile tongue to capture the seed inside. Woodpeckers, nuthatches, and chickadees may wedge a nut or acorn into a hole in the bark of a tree and then hammer at it with their sharp bills until it cracks.

The skulls of most birds consist of four bony units that can move in relation to each other. This skull kinesis is important in some aspects of feeding. The upper jaw flexes upward as the mouth is opened, and the lower jaw expands laterally at its articulation with the skull (**Figure 17–21a**). The flexion of the upper and lower jaws increases the bird's gape in both the vertical and horizontal planes and probably assists in swallowing large items. Many birds use their beaks to search for hidden food. Long-billed shorebirds probe in mud and sand to locate worms and crustaceans. These birds display a form of skull kinesis in which the flexible zone in the upper jaw has moved toward the tip of the beak, allowing the tip of the upper jaw to be lifted without opening the mouth (**Figure 17–21b**). This mechanism enables long-billed waders to grasp prey under the mud.

Tongues are an important part of the food-gathering apparatus of many birds. Woodpeckers drill holes into dead trees and then use their long tongues to investigate passageways made by wood-boring insects. The tongue of the green woodpecker, which extracts ants from their tunnels in the ground, extends four times the length of its beak. The hyoid bones that support the tongue are elongated and housed in a sheath of muscles that passes around the

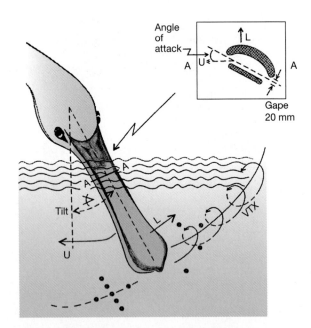

▲ **FIGURE 17–20** Feeding mechanism of the spoonbill. When a spoonbill sweeps its beak through the water, the curved upper surface and flat lower surface create vortex currents (shown by spiraling arrows marked VTX) that lift prey from the bottom. In this drawing, the bill is being swept from right to left (indicated by the arrow marked U). The sweeping motion generates a lift (L). The line A–A indicates the position of the cross section of the beak (shown in the inset).

▶ **FIGURE 17–21** Avian skull and jaw kinesis. (a) A yawning herring gull (*Larus argentatus*) shows the kinetic movements of the skull and jaws that occur during swallowing. White arrows show the positions of outward flexion; black arrows show inward flexion. (b) Long-billed wading birds that probe for worms and crustaceans in soft substrates can raise the tips of the upper bill without opening their mouth.

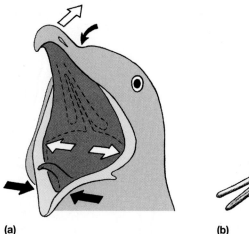

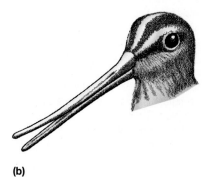

(a)   (b)

outside of the skull and rests in the nasal cavity (**Figure 17–22**). When the muscles of the sheath contract, the hyoid bones are pushed around the back of the skull and the tongue is projected from the bird's

Elongate hyoid bones are enclosed by circular muscles. Contraction of the muscles pushes the bones forward and causes the tongue to project.

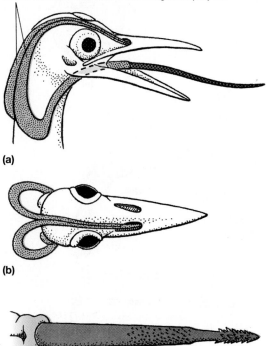

(a)

(b)

(c)

▲ **FIGURE 17–22** Tongue projection mechanism of a woodpecker. The tongue itself is about the length of the bill, and it can be extended well beyond the tip of the beak by muscles that squeeze the posterior ends of the elongated hyoid apparatus, forcing the tongue forward. The detail of the tongue shows the barbs on the tip that impale prey.

mouth. The tip of a woodpecker tongue has barbs that impale insects and allow them to be pulled from their tunnels. Nectar-eating birds, such as hummingbirds and sunbirds, also have long tongues and a hyoid apparatus that wraps around the back of the skull. The tip of the tongue of nectar-eating birds is divided into a spray of hair-thin projections, and capillary force causes nectar to adhere to the tongue.

■ The Digestive System

The digestive systems of birds show some differences from those of other vertebrates. The absence of teeth prevents birds from doing much processing of food in the mouth, and the gastric apparatus takes over some of that role.

Esophagus and Crop Birds often gather more food than they can process in a short period, and the excess is held in the esophagus. Many birds have a **crop,** an enlarged portion of the esophagus that is specialized for temporary storage of food. The crop of some birds is a simple expansion of the esophagus, whereas in others it is a unilobed or bilobed structure (**Figure 17–23**). An additional function of the crop is transportation of food for nestlings. As the foraging adult gathers food, it stores it in the crop. When the adult returns to the nest, it regurgitates the material from its crop and feeds it to the young. In doves and pigeons, the crop of both sexes produces a nutritive fluid (crop milk) that is fed to the young. The milk is produced by fat-laden cells that detach from the squamous epithelium of the crop and are suspended in an aqueous fluid. Crop milk is rich in lipids and proteins but contains no sugar. Its chemical composition is similar to that of mammalian milk, although it differs in containing intact cells.

The hoatzin (*Opisthocomus hoazin*), a South American bird, is the only avian species known to employ

▶ **FIGURE 17–23** Anterior digestive tract of birds. (a) The relationship among the parts. The relative sizes of the proventriculus and gizzard vary in relation to diet. Carnivorous and fish-eating birds, like the great cormorant, have a relatively small crop (b) and gizzard (c), whereas seed eaters and omnivores like the peafowl have a large crop (d) and muscular gizzard (e). (f) The digestive tract of the hoatzin. The muscular crop and the posterior part of the esophagus are greatly enlarged. Cornified ridges on the inner surface of the crop probably grind its contents, reducing the particle size.

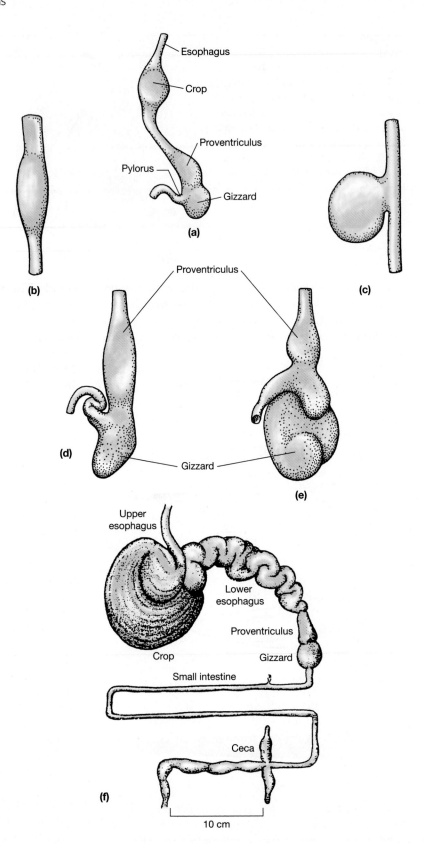

foregut fermentation. Hoatzins are herbivorous; green leaves make up more than 80 percent of the diet. More than a century ago, naturalists observed that hoatzins smell like fresh cow manure, and a study of the crop and lower esophagus (Figure 17–23f) revealed that volatile fatty acids are the source of the odor. Volatile fatty acids produced by the process of fermentation are absorbed by the gut. Bacteria and protozoa like those found in the rumen of cows break down the plant cell walls, and bacterial extracts from the hoatzin's crop are as effective as those from a cow's rumen in digesting plant material.

The Stomach The form of the stomachs of birds is related to their dietary habits. Carnivorous and piscivorous (fish-eating) birds need expansible storage areas to accommodate large volumes of soft food, whereas birds that eat insects or seeds require a muscular organ that can contribute to the mechanical breakdown of food. Usually, the gastric apparatus of birds consists of two relatively distinct chambers, an anterior glandular stomach (**proventriculus**) and a posterior muscular stomach (**gizzard**; see Figure 17–23a). The proventriculus contains glands that secrete acid and digestive enzymes. The proventriculus is especially large in species that swallow large items such as intact fruit (Figure 17–23d,e).

The gizzard has several functions, including food storage while the chemical digestion that was begun in the proventriculus continues, but its most important function is the mechanical processing of food. The thick, muscular walls of the gizzard squeeze the contents, and small stones that are held in the gizzards of many birds help to grind the food. In this sense, the gizzard is performing the same function as a mammal's teeth. The pressure that can be exerted on food in the gizzard is intense. A turkey's gizzard can grind two dozen walnuts in as little as four hours, and it can crack hickory nuts that require 50 kilograms to 150 kilograms of pressure to break.

Intestine, Ceca, and Cloaca The small intestine is the principal site of chemical digestion, as enzymes from the pancreas and intestine break down the food into small molecules that can be absorbed across the intestinal wall. The mucosa of the small intestine is modified into a series of folds, lamellae, and villi that increase its surface area. The large intestine is relatively short, usually less than 10 percent of the length of the small intestine. Passage of food through the intestines of birds is quite rapid: Transit times for carnivorous and fruit-eating species are in the range of a few minutes to a few hours. Passage of food is slower in herbivores and may require a full day. Birds generally have a pair of ceca at the junction of the small and large intestines (see Figure 17–23f). The ceca are small in carnivorous, insectivorous, and seed-eating species, but they are large in herbivorous and omnivorous species such as cranes, fowl, ducks, geese, and the ostrich. Symbiotic microorganisms in the ceca apparently ferment plant material.

Birds respond to seasonal changes in diet with changes in the morphology of the gut. Many species of birds feed on insects and other animal prey during the summer and switch to plant food (such as berries) during the winter. Plant material takes longer to digest than animal food, so it must pass through the gut more slowly. To accommodate differences in passage time, the intestine changes length in response to changes in diet. **Figure 17–24** illustrates this seasonal change for starlings (*Sternus vulgaris*), which feed mainly on insects from March through June and add progressively more plant material to their diet from late summer through winter. The length of the intestine shows a corresponding cycle, increasing in length by about 20 percent during fall and winter and decreasing by the same amount during spring and early summer. In addition to the anatomical changes in the intestine, the digestive enzymes change to match the reduction in protein and fat and increase in simple sugars in fruit compared to animal food.

The cloaca temporarily stores waste products while water is being reabsorbed. The precipitation of uric acid in the form of urates (i.e., the potassium and sodium salts of uric acid) frees water from the urine, and this water is returned to the bloodstream. Species of birds that have salt-secreting glands can accomplish further conservation of water by reabsorbing some of the ions that are in solution in the cloaca and excreting them in more concentrated solutions through the salt glands (See Section 11.5). The mixture of white urates and dark fecal material that is voided by birds is familiar to anyone who has washed an automobile.

17.6 Sensory Systems

A flying bird moves rapidly through three-dimensional space and requires a continuous flow of sensory information about its position and the presence of obstacles in its path. Vision is the sense best suited to provide this sort of information on a rapidly renewed basis, and birds have a well-developed visual system that remains active when the bird is asleep (**Box 17–1**). The importance of vision is reflected in the brain: The optic lobes are large, and the midbrain is an

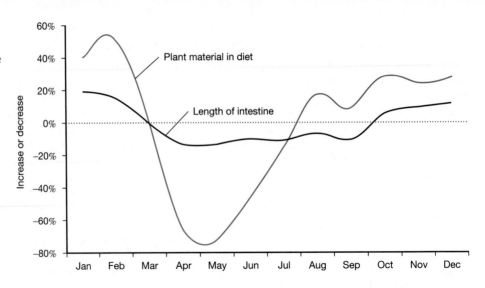

► **FIGURE 17–24** Seasonal changes in diet and intestinal length of the starling (*Sternus vulgaris*). The length of the intestine decreases from winter through spring as plant material becomes a less important part of the diet, then increases in autumn as the amount of plant material increases.

BOX 17–1   Chaucer the Ornithologist

The Prologue to *The Canterbury Tales*, which was written in Middle English by Geoffrey Chaucer between 1387 and 1400, evokes images of spring—April showers, soft breezes, sprouting leaves, burgeoning flowers,

> *And smale foweles maken melodye,*
>
> *That slepen al the nyght with open eye*
>
> (And small birds are singing
>
> that sleep all night with their eyes open).

Chaucer used the birds as a metaphor for the feeling of well-being that pervades even the most timid creatures on a lovely spring day, but it turns out that he was biologically correct—birds on the lookout for predators can sleep with one eye open and one hemisphere of the brain awake (Rattenborg et al. 1999).

This phenomenon is called unihemispheric slow wave sleep (USWS) and had previously been known only in aquatic mammals that must sleep and surface to breathe simultaneously. Sleep is more efficient when both hemispheres of the brain sleep,

and birds do this when they feel safe; however, in potentially dangerous situations, they use USWS and keep watch with one eye. Furthermore, the eye that remains open is the one facing in the direction from which a predator is most likely to approach.

These conclusions are based on a study of mallard ducks (*Anas platyrhynchos*) sleeping in rows of four. The end birds in each row had one side of their body exposed and the other side protected by the presence of another bird, whereas the two middle birds in each row were protected on both sides. The ducks at the ends of the rows used USWS sleep more than twice as much as the birds in the center. Furthermore, the end birds had the eye facing away from the center of the group open most of the time, whereas the birds in the central positions showed no preference for which eye was open (**Figure 17–25**). Electroencephalographic recordings of electrical activity showed that the hemisphere receiving input from the open eye was more active than the sleeping hemisphere, demonstrating that information was passing from the open eye to the awake hemisphere.

► **FIGURE 17–25** One-eyed sleep. Ducks at the ends of a row have one eye facing outward (the direction from which a predator is most likely to approach) and the other eye facing toward the center of the row. Ducks in the two central positions in the row have another duck on both their outward and central sides. The ducks in the end positions keep the eye that faces away from the center of the group open almost five times as long as the ducks in the central positions.

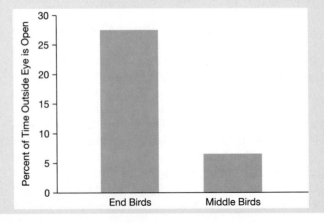

important area for processing visual and auditory information. Olfaction is relatively unimportant for most birds, and the olfactory lobes are small. The cerebellum, which coordinates body movements, is large. The cerebrum is less developed in birds than it is in mammals and is dominated by the corpus striatum.

■ Vision

The eyes of birds are large—so large that the brain is displaced dorsally and caudally, and in many species the eyeballs meet in the midline of the skull. The eyes of some hawks, eagles, and owls are as large as the eyes of humans. In its basic structure, the eye of a bird is like that of any other vertebrate, but the shape varies from a flattened sphere to something approaching a tube (**Figure 17–26**). An analysis of the optical characteristics of birds' eyes suggests that these differences are primarily the result of fitting large eyes into small skulls. The eyes of a starling are small enough to be contained within the skull, whereas the eyes of an owl bulge out of the skull. An owl would require an enormous, unwieldy head to accommodate a flat eye like that of a starling. The tubular shape of the owl's eye allows it to fit into a reasonably sized skull.

The pecten is a conspicuous structure in the eye of birds (see Figure 17–26a). It is formed by blood capillaries surrounded by pigmented tissue and covered by a membrane; it lacks muscles and nerves. The pecten arises from the retina at the point where the nerve fibers from the ganglion cells of the retina join to form the optic nerve. In some species of birds the pecten is small, but in other species the pecten extends so far into the vitreous humor of the eye that it almost touches the lens. The function of the pecten

remains uncertain after 200 years of debate. Proposed functions include reduction of glare, a mirror to reflect objects above the bird, production of a stroboscopic effect, and a visual reference point—but none of these seems very likely. The large blood supply flowing to the pecten suggests that it may provide nutrition for the retinal cells and help to remove metabolic waste products that accumulate in the vitreous humor.

Oil droplets are found in the cone cells of the avian retina, as they are in other sauropsids. The droplets range in color from red through orange and yellow to green. The oil droplets are filters, absorbing some wavelengths of light and transmitting others. The function of the oil droplets is unclear; it is certainly complex, because the various colors of droplets are combined with different kinds of photoreceptor cells and different visual pigments. Birds like gulls, terns, gannets, and kingfishers that must see through the surface of water have a preponderance of red droplets. Aerial hawkers of insects (swifts, swallows) have predominantly yellow droplets.

■ Hearing

In birds, as in other sauropsids, the columella (stapes) and its cartilaginous extension, the extracolumella, transmit vibrations of the tympanum to the oval window of the inner ear. The cochlea of birds appears to be specialized for fine distinctions of the frequency and temporal pattern of sound. The cochlea of a bird is about one-tenth the length of the cochlea of a mammal, but it has about 10 times as many hair cells per unit of length. The space above the basilar membrane (the scala vestibuli) is nearly filled in birds by a folded, glandular tissue. This structure may dampen

▶ **FIGURE 17–26**
Variation in the shape of the eye of birds. (a) Flat, typical of most birds. (b) Globular, found in most falcons. (c) Tubular, characteristic of owls and some eagles.

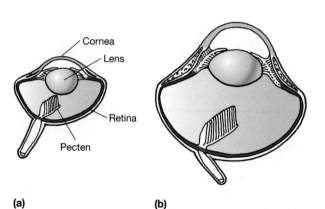

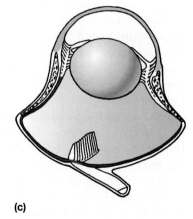

Cornea
Lens
Retina
Pecten

(a)   (b)   (c)

sound waves, allowing the ear of a bird to respond very rapidly to changes in sounds.

Localization of sounds in space can be difficult for small animals such as birds. Large animals localize the source of sounds by comparing the time of arrival, intensity, or phase of a sound in their left and right ears, but none of these methods is very effective when the distance between the ears is small. The pneumatic construction of the skulls of birds may allow them to use sound that is transmitted through the air-filled passages between the middle ears on the two sides of the head to increase their directional sensitivity. If this is true, internally transmitted sound would pass from the middle ear on one side to the middle ear on the other side and reach the *inner* surface of the contralateral tympanum. There it would interact with the sound arriving on the external surface of the tympanum via the external auditory meatus. The vibration of each tympanum would be the product of the combination of pressure and phase of the internal and external sources of sound energy, and the magnitude of the cochlear response would be proportional to the difference in pressure across the tympanic membrane.

The sensitivity of the auditory system of birds is approximately the same as that of humans, despite the small size of birds' ears. Most birds have tympanic membranes that are large in relation to the head size. A large tympanic membrane enhances auditory sensitivity, and owls (which have especially sensitive hearing) have the largest tympani relative to their head size among birds. Sound pressures are amplified during transmission from the tympanum to the oval window of the cochlea because the area of the oval window is smaller than the area of the tympanum. The reduction ratio for birds ranges from 11:1 to 40:1. High ratios suggest sensitive hearing, and the highest values are found in owls; songbirds have intermediate ratios (20:1 to 30:1). (The ratio is 36:1 for cats and 21:1 for humans.) The inward movement of the tympanum as sound waves strike it is opposed by air pressure within the middle ear, and birds show a variety of features that reduce the resistance of the middle ear. The middle ear is continuous with the dorsal, rostral, and caudal air cavities in the pneumatic skulls of birds. In addition to potentially allowing sound waves to be transmitted to the contralateral ear, these interconnections increase the volume of the middle ear and reduce its stiffness, thereby allowing the tympanum to respond to faint sounds.

Owls are acoustically the most sensitive of birds. At frequencies up to 10 kilohertz (10,000 cycles per second), the auditory sensitivity of an owl is as great as that of a cat. Owls have large tympanic membranes, large cochleae, and well-developed auditory centers in the brain. Some owls are diurnal, others crepuscular (active at dawn and dusk), and some are entirely nocturnal. In an experimental test of their capacities for acoustic orientation, barn owls were able to seize mice in total darkness. If the mice towed a piece of paper across the floor behind them, the owls struck the rustling paper instead of the mouse, showing that sound was the cue they were using.

A distinctive feature of many owls is the facial ruff, which is formed by stiff feathers (**Figure 17–27a**). The ruff acts as a parabolic sound reflector, focusing sounds with frequencies above 5 kilohertz on the external auditory meatus and amplifying them by 10 decibels. The ruffs of some owls are asymmetric, and that asymmetry appears to enhance the ability of owls to locate prey because barn owls made large errors in finding targets when their ruffs were removed.

Asymmetry of the aural system of owls goes beyond the feathered ruff. The skull itself is markedly asymmetric in many owls (**Figure 17–27b**), and these are the species with the greatest auditory sensitivity. The asymmetry ends at the external auditory meatus; the middle and inner ears of owls are bilaterally symmetric. The asymmetry of the external ear openings of owls assists with localization of prey in the horizontal and vertical axes. Minute differences in the time and intensity at which sounds are received by the two ears indicate the direction of the source. The brains of owls integrate time and intensity information with extraordinary sensitivity to produce a map of their environment that integrates auditory and visual information.

■ Olfaction

The sense of smell is well developed in some birds but poorly developed in most species. The size of the olfactory bulbs is a rough indication of the sensitivity of the olfactory system. Relatively large bulbs are found in ground-nesting and colonial-nesting species, species that are associated with water, and carnivorous and piscivorous species of birds. Some birds use scent to locate prey. The kiwi, for example, has nostrils at the tip of its long bill and finds earthworms underground by smelling them. Turkey vultures follow airborne odors of carrion to the vicinity of a carcass, which they then locate by sight. Shearwaters, fulmars, albatrosses, and petrels were attracted to sponges that had been soaked in fish oil and placed on floating buoys, and they could find the sponges even at night.

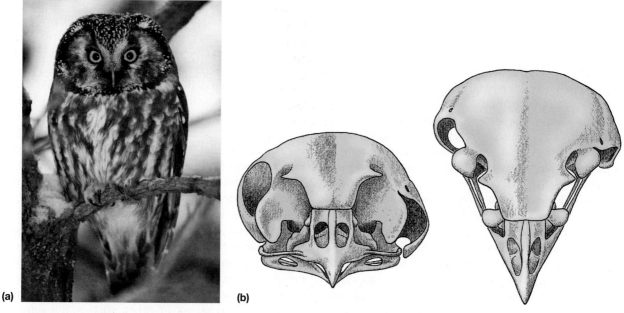

▲ **FIGURE 17–27** Anatomical features of owls that increase their auditory sensitivity. (a) A boreal owl (*Aegolius funereus*) showing its facial ruff. (b) The skull of a boreal owl. The pronounced asymmetry in the position of the ear opening (the external auditory meatus) is characteristic of owls. This asymmetry assists in localization of sound.

Olfaction probably plays a role in the orientation and navigation abilities of some birds. The tubenosed seabirds (albatrosses, shearwaters, fulmars, and petrels) nest on islands; and, when they return from foraging at sea, they approach the islands from downwind. Homing pigeons use olfaction (as well as other mechanisms) to navigate (Ioalè et al. 1990). The well-developed nasal bulbs of colonial-nesting species of birds suggest the possibility that olfaction is used for social functions such as recognition of individuals, but this has never been demonstrated.

17.7 Mating Systems, Reproduction, and Parental Care

The activities associated with reproduction are among the most complex and conspicuous behaviors of birds. Much of our understanding of the evolution and function of the mating systems of vertebrates is derived from studies of birds. Classic work in avian ethology, such as Konrad Lorenz's studies of imprinting and Niko Tinbergen's demonstration of innate responses of birds to specific visual stimuli, has formed a basis for current studies of behavioral ecology. Recent work has focused on the sources of variation in mating systems.

■ Social Behavior and Mating Systems

Vision and hearing are the major sensory modes of birds, as they are of humans. One result of this correspondence has been the important role played by birds in behavioral studies. Most birds are active during the day and are thus relatively easy to observe. A tremendous amount of information has been accumulated about the behavior of birds under natural conditions. This background has contributed to the design of experimental studies in the field and in the laboratory.

Colors and Patterns Three types of pigments are widespread in birds. Dark colors are produced by melanin—eumelanins produce black, gray, and dark brown, whereas phaeomelanins are responsible for reddish brown and tan shades. Carotenoid pigments are responsible for most red, orange, and yellow colors. Birds obtain these pigments from their diet, and in some cases the intensity of color can be used to gauge the fitness of a prospective mate. Porphyrins are metal-containing compounds chemically similar to the pigments in hemoglobin and liver bile. Ultraviolet light causes porphyrins to emit a red fluorescence. Porphyrins are destroyed by prolonged exposure to sunlight, so they are most conspicuous in new plumage.

Structural colors result from tiny particles of melanin in the cells on the surface of feather barbs, which reflect specific wavelengths of light. Blue is produced by very small particles that reflect the shortest wavelengths, whereas some greens are produced by slightly larger particles of melanin. Structural colors can be combined with pigments—green parakeets combine a structural blue with a yellow carotenoid, and blue parakeets have a gene that blocks formation of the carotenoid. That gene is a simple recessive, so two blue parakeets can produce only blue offspring.

Iridescent colors, such as those on the heads and throats of hummingbirds and in the eye-shaped patterns on a peacock's tail, result from interference of light waves reflected from the outer and inner surfaces of hollow structures. The hue of the iridescent color depends on the distance between the reflecting surfaces, and the intensity of the color depends on how many layers of reflective structures the feathers contain. Hummingbirds, for example, have 7 to 15 layers of hollow melanin platelets in the barbules of the feathers. Perception of interference colors depends on the angle of view—color is visible when light is being reflected toward the eye of a viewer, but the feathers appear black when viewed from a different angle. As a result, iridescent colors flash on and off as a bird changes its position.

Vocalizations and Visual Displays Birds use colors, postures, and vocalizations for species, sex, and individual identification. Studies of birdsong have contributed greatly to our understanding of communication by vertebrates, and important general concepts such as species specificity in signals and innate predisposition to learning were first developed in studies of birdsong. Studies of the neural basis of song are leading to a close integration of behavior and neurobiology.

The term *birdsong* has a specific meaning that is distinct from a birdcall. The song is usually the longest and most complex vocalization produced by a bird. In many species, songs are produced only by mature males and only during the breeding season. Song is a learned behavior that is controlled by a series of song control regions (SCRs) in the brain. During the period of song learning, which occurs early in life, new connections form between a part of the SCR that is associated with song learning and a region of the brain that controls the vocal muscles. Thus, song learning and song production are closely linked in male birds.

The SCRs are under hormonal control; and, in many species of birds, the SCRs of males are larger than those of females and have more and larger neurons and longer dendritic processes. The vocal behavior of female birds varies greatly across taxonomic groups: in some species females produce only simple calls, whereas in other species the females engage with males in complex song duets. The SCRs of females of the latter species are very similar in size to those of males (**Table 17.3**). The function of the SCR in female birds of species in which females do not vocalize has been unclear, but recent experiments suggest that it plays a role in species recognition. When the SCR of female canaries was inactivated, the birds no longer distinguished the vocalizations of male canaries from those of sparrows.

A birdsong consists of a series of notes with intervals of silence between them. Changes in frequency (frequency modulation) are conspicuous components of the songs of many birds, and the avian ear may be very good at detecting rapid changes in frequency. Birds often have more than one song type, and some species may have repertoires of several hundred songs.

Birdsongs identify the particular species of bird that is singing, and they often show regional dialects.

TABLE 17.3 Sexual dimorphism in the song control regions (SCR) of the brains of birds. The average ratio of the volumes of the five SCRs in males compared to females (male:female) parallels the difference in the sizes of the song repertoires of males and females.

	Zebra Finch	Canary	Chat	Bay Wren	Buff-Breasted Wren
SCR volume ratio	4.0:1.0	3.1:1.0	2.3:1.0	1.3:1.0	1.3:1.0
Song repertoire	Males only	Males very much greater than females	Males much greater than females	Males the same as females	Males the same as females

These dialects are transmitted from generation to generation as young birds learn the songs of their parents and neighbors. In the indigo bunting, one of the best-studied species, song dialects that were characteristic of small areas persisted up to 15 years, which is substantially longer than the life of an individual bird. Bird songs also show individual variation that allows birds to recognize the songs of residents of adjacent territories and to distinguish the songs of these neighbors from those of intruders. Male hooded warblers remember the songs of neighboring males—and recognize them as individuals when they return to their breeding sites in North America after spending the winter in Central America.

The songs of male birds identify their species, sex, and occupancy of a territory. When a recording of a male's song is played back through a speaker that has been placed in the territory of another male, the territorial male responds with vocalizations and aggressive displays, and he may even attack the speaker. These behaviors repel intruders, and even hearing the song of a territorial male keeps intruders at a distance—broadcasting recorded songs in a territory from which the territorial male has been removed delays occupation of the vacant territory by a new male.

Visual displays are frequently associated with songs; for example, a particular body posture that displays colored feathers may accompany singing. Male birds are often more brightly colored than females and have feathers that have become modified as the result of sexual selection. In this process, females mate preferentially with males that have certain physical characteristics. Because of that response by females, those physical characteristics contribute to the reproductive fitness of males, even though they may have no useful function in any other aspect of the animal's ecology or behavior. The colorful areas on the wings of male ducks, the red epaulets on the wings of male red-winged blackbirds, the red crowns on kinglets, and the elaborate tails of male peacocks are familiar examples of specialized areas of plumage that are involved in sexual behavior and display.

Truth in Advertising The bright colors and other adornments of male birds may be an indication of good nutritional status, of resistance to parasites, or of the ability to evade predators. If these visual signals really do correlate with the quality of a male, they could provide a basis for females to evaluate the merits of several potential mates. This hypothesis, which is referred to as "truth in advertising," has been tested for several species of birds and seems to

apply in most cases. The tail of a male peacock (*Pavo cristatus*) is a classic example of a sexually selected trait. Female peafowl mate preferentially with males that have long tails with many eyespots. Cutting out some of the eyespots from a male peacock's tail reduces its success in attracting females. A study of peafowl showed that chicks sired by males whose tails have many eyespots grew faster than chicks sired by males with smaller tails and had higher survival under seminatural conditions (Petrie 1994).

It was long assumed that preferred males produce better chicks because those males are genetically superior, but a study of mallard ducks (*Anas platyrhynchos*) has suggested an additional possibility. When female mallards were paired with either attractive or less attractive males, both the male and female offspring of attractive males had higher viability, and the viability of the hatchlings was positively correlated with the size of the eggs they hatched from. That is, large eggs produced better hatchlings than did small eggs, and egg size is controlled by the female. Apparently the female ducks that were paired with attractive males invested more energy in their eggs than did females paired with less attractive males. Thus, at least some of the advantage enjoyed by offspring of attractive males may result from behavior by the female.

Mating Systems and Parental Investment The mating systems of vertebrates are believed to reflect the distribution of food, breeding sites, and potential mates. The key factor in the situation is whether one sex can control resources the other sex needs. For example, a male could potentially increase his opportunities to mate by defending resources that females need. If he can exclude other males, he can mate with all the females in the area. However, a male's ability to control access to resources depends on how the resources are distributed in the habitat.

- If food and nest sites are more or less evenly distributed, it is unlikely that a male could control a large enough area to monopolize many females. Under those conditions, all males will have access to the resources and to the females.
- On the other hand, if resources are clumped in space with barren areas between the patches, the females will be forced to aggregate in the resource patches, and it will be possible for a male to monopolize several females by defending a patch. Males that are able to defend good patches should attract more females than males defending patches of lower quality.

Monogamy, Extra-Pair Copulations, and Polygamy

Social vertebrates exhibit one of two broad categories of mating systems—monogamy or polygamy. **Monogamy** (Greek *mono* = one and *gamy* = marriage) refers to a pair bond between a single male and a single female. The pairing may last for part of a breeding season, an entire season, or for a lifetime. **Polygamy** (Greek *poly* = many) refers to a situation in which an individual has more than one mate in a breeding season.

Polygamy can be exhibited by males, females, or both sexes. In **polygyny** (Greek *gyn* = female), a male mates with more than one female; in **polyandry** (Greek *andr* = male), a female mates with two or more males. **Promiscuity** is a mixture of polygamy and polyandry in which both males and females mate with several different individuals.

Monogamy has been considered the most widespread social system of birds. Both parents in monogamous mating systems usually participate in caring for the young. Monogamy does not necessarily mean fidelity to a mate, however. Genetic studies of monogamous birds have shown that extra-pair copulation (mating with a bird other than the partner) is common. Thus, some of the eggs in a nest may have been fertilized by a male other than the partner of the female who deposited them, and the male partner may have fertilized eggs of other females that are being incubated in their nests. As evidence of extra-pair copulation has accumulated, the term **social monogamy** has been introduced for species in which a male and female share responsibility for a clutch of eggs but do not demonstrate fidelity. **Genetic monogamy** describes a social system in which a male and female share parental responsibilities and do not have extra-pair copulations.

Social Monogamy Perhaps the most important incentive for pair formation for many species of birds is the need for attendance by both parents to raise a brood to fledging (i.e., leaving the nest). Dramatic examples include situations in which continuous nest attendance by one parent is necessary to protect the eggs or chicks from predators while the other parent forages for food. This situation is commonly observed in seabirds, which nest in dense colonies that sometimes include mixtures of two or more species. In the absence of an attending parent, neighbors raid the nest and kill the eggs or chicks. The male and female alternate periods of nest attendance and foraging, and some species engage in elaborate displays when the parents switch duties (**Figure 17–28**).

Extra-Pair Copulations In the 1960s, David Lack concluded from his field work that more than 90 percent of bird species are monogamous, but there were hints that appearances might be deceptive. For example, an attempt to control reproduction by red-winged blackbirds (*Agelaius phoeniceus*) by vasectomizing males failed when the female partners of vasectomized males laid eggs that hatched. Since the 1990s, DNA analysis has ended the myth of avian genetic monogamy by allowing the paternity of nestlings to be determined. A male that was not the socially bonded partner of the nesting female sired at least some of her babies in half of the species of birds studied. Indeed, 10 percent to 20 percent—and sometimes up to 40 percent—of the babies in a nest turn out to have been sired by extra-pair males.

Both males and female birds can benefit from certain aspects of extra-pair copulations:

- Either sex could increase the fitness of its offspring by mating with an individual of better genetic quality than the one it is paired with.
- Either sex could benefit from increasing the heterozygosity of its offspring by mating with more than one partner, as well as increasing the chance of producing genetically compatible combinations of maternal and paternal genes.

In addition, there are potential benefits of extra-pair copulations that are specific to just one sex. For males, the advantages of extra-pair copulations are probably both genetic and practical:

- A male may be able to sire more offspring in a season by mating with multiple females than it could with a single female.
- A male benefits from the parental care provided to his offspring by the male partner of an extra-pair female with which he has mated.
- Spreading a male's reproductive investment among several nests reduces the risk of losing his entire reproductive effort if a nest is raided by a predator.

From the perspective of a female also, extra-pair copulations offer potential benefits:

- Some males are more fertile than others, and mating with several males reduces the risk that some of a female's eggs will not be fertilized. (This is the situation that the female blackbirds avoided in the study described above.)

▶ **FIGURE 17–28** Nest exchange display of northern gannets (*Morus bassanus*). The birds engage in an elaborate ritual when one member of the pair returns to the nest after foraging.

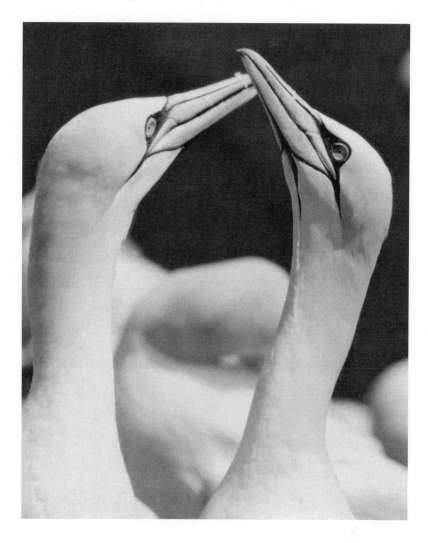

- A female may be able to increase the fitness of her male offspring by mating with a male who has especially attractive secondary sexual characteristics. (This phenomenon is called the *"sexy son" hypothesis.*)
- A female may lay the egg in the nest of the male who fertilized it, thus benefiting from the parental care provided by the female that is pair-bonded to that male. (This phenomenon has been called *quasi nest parasitism.*)

Polygyny When an individual male can control or gain access to several females, the male can increase his reproductive success by mating with more than one female. In resource defense polygyny, males control access to females by monopolizing critical resources—such as nest sites or food—that have patchy distributions. A male that stakes out a territory in a high-quality patch can attract many females. For this system to work, a female must benefit from mating with a male that already has one mate. That is, the reproductive fitness of a female must be greater as a secondary mate on a high-quality territory than it would be as a primary mate on a territory of lower quality. Red-winged blackbirds are a familiar example of resource defense polygyny. Male blackbirds arrive at their marshy breeding areas before females and compete for territories. When the females arrive, they have a choice among a variety of territories of different quality, each defended by a male blackbird. A female should choose to mate polygynously if the difference in quality of the territories is large enough so that she will raise more young than she would by mating monogamously with a male on a poorer territory.

In male dominance polygyny, males are not defending a resource that females require. Instead, males compete for females by establishing patterns of dominance or by demonstrating their quality through displays. Aggregations of many males in a

▶ **FIGURE 17–29** Male greater prairie chicken (*Tympanuchus cupido*). During the breeding season, male prairie chickens congregate in traditional lek sites. Each male occupies a small territory from 13 to 100 square meters in area in the lek, and within this territory performs a courtship display that includes elaborate postures. Two colorful sacs that are outgrowths from the esophagus are filled with air and project through the breast feathers. Air is expelled from these sacs with a popping sound.

small area are called leks (**Figure 17–29**). Females visit the leks and copulate with a single male. The central sites in a lek appear to be the most favored; and the males that occupy the most central sites obtain most of the matings.

Turning the Tables: Male Incubation and Polyandry In polyandrous mating systems, it is the females that control a limited resource, and they use that control to gain access to multiple males. This pattern of breeding among birds seems to be typical of situations in which the cost of each reproductive effort for the female is low (because food is abundant and a clutch contains only a few small eggs) and the probability of successful fledging is small. Spotted sandpipers (*Actitis macularia*) provide an example: Male spotted sandpipers form territories, and females move among the territories of several males. When a female spotted sandpiper mates with a male, she remains in his territory until she has laid three eggs. After that she moves away to breed with other males, leaving parental care to the male she leaves behind. A female that has moved away and bred with other males may return and breed again with the original male if their first clutch is destroyed. Predation on sandpiper nests is high, and a male stands a good chance of losing the clutch of eggs it is incubating. The resource that female sandpipers control is replacement clutches for these males, and the value of that resource allows females to mate with multiple males.

■ Eggs and Nests

Elaborate and diverse behaviors are associated with egg laying and parental care. Nest preparation by birds runs the gamut from nothing more than a fairy tern's selection of a branch on which to balance its egg to weaver birds' construction of multiroom communal nests to be used by generation after generation. Incubation provides heat for the development of eggs, and the presence of a parent is a deterrent to many predators. However, some birds leave their eggs for periods of days while they forage. Brood parasites, such as cuckoos, deposit their eggs in the nests of other species of birds and play no role in brooding or rearing their young.

Oviparity Although birds have a great diversity of mating strategies, their mode of reproduction is limited to laying eggs. No other group of vertebrates that contains such a large number of species is exclusively oviparous. Why is this true of birds?

Constraints imposed on birds by their specializations for flight are often invoked to explain their failure to evolve viviparity. Those arguments are not particularly convincing, however, considering that bats have successfully combined flight and viviparity. Furthermore, flightlessness has evolved in at least 15 families of birds, but none of these flightless species has evolved viviparity.

Oviparity is presumed to be the ancestral reproductive mode for archosaurs, and it is retained by both extant groups of archosaurs—crocodilians and birds. However, viviparity evolved in the marine reptiles of the Mesozoic and it has evolved nearly 100 times in the other major lineage of extant sauropsids, the lizards and snakes (lepidosaurs), so the capacity for viviparity is clearly present in sauropsids—why not in birds?

A key element in the evolution of viviparity among lizards and snakes appears to be the retention of eggs in the oviducts of the female for some period before they are deposited. This situation occurs when the benefits of egg retention outweigh its costs. For example, the high incidence of viviparity among snakes and lizards in cold climates may be related to the ability of a female ectotherm to speed embryonic development by thermoregulation. A lizard that basks in the sun can raise the temperature of eggs retained in her body; but, after depositing the eggs in a nest, she has no control over their temperature and rate of development. Birds are endotherms and can control egg temperature by brooding the eggs. Thus, the reasoning goes, egg retention provides no thermoregulatory advantage for birds so they never took that first step toward viviparity.

Egg Biology The inorganic part of eggshells contains about 98 percent crystalline calcite, $CaCO_3$,

and the embryo obtains about 80 percent of its calcium from the eggshell. An organic matrix of protein and mucopolysaccharides is distributed through the shell and may serve as a support structure for the growth of calcite crystals. Eggshell formation begins in the isthmus of the oviduct (**Figure 17–30**). Two shell membranes are secreted to enclose the yolk and albumen, and carbohydrate and water are added to the albumen by a process that involves active transport of sodium across the wall of the oviduct followed by osmotic flow of water. The increased volume of the egg contents at this stage appears to stretch the egg membranes taut. Organic granules are attached to the egg membrane, and these **mammillary bodies** appear to be the sites of the first formation of calcite crystals (**Figure 17–31**). Some crystals grow downward from the mammillary bodies and fuse to the egg membranes, and other crystals grow away from the membrane to form cones. The cones grow vertically and expand horizontally, fusing with crystals from adjacent cones to form the palisade layer.

An eggshell is penetrated by an array of pores that allow oxygen to diffuse into the egg and carbon dioxide and water to diffuse out (**Figure 17–32**). Pores occur at the junction of three calcite cones, but only 1 percent or less of those junctions form pores; the rest

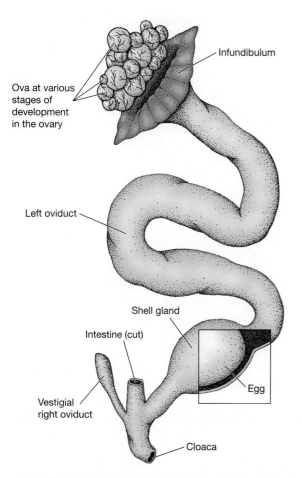

▲ **FIGURE 17–30** Oviduct of a bird. Ova released from the ovary enter the infundibulum of the oviduct. Fertilization occurs at the upper end of the oviduct, then albumen and shell membranes are secreted, and finally the egg is enclosed in calcareous shell.

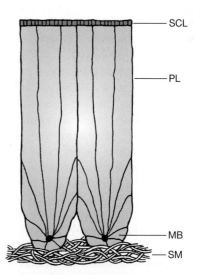

▲ **FIGURE 17–31** Diagram of the crystal structure of an avian eggshell. Crystallization begins at the mammillary bodies (MB). Crystals grow into the outer shell membrane (SM) and upward to form the palisade layer (PL). Changes in the chemical composition of the fluid surrounding the growing eggshell are probably responsible for the change in crystal form in the surface crystalline layer (SCL).

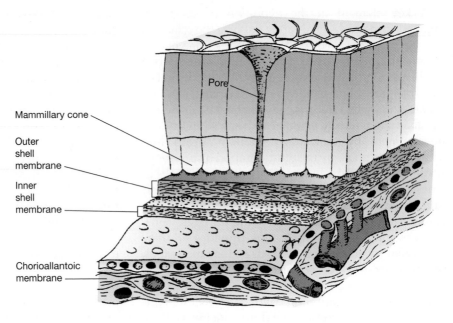

▶ **FIGURE 17–32** A diagram of the structure of an eggshell. Pore canals penetrate the calcified region, allowing oxygen to enter and carbon dioxide and water to leave the egg. The gases are transported to and from the embryo via blood vessels in the chorioallantoic membrane.

Pore

Mammillary cone

Outer shell membrane

Inner shell membrane

Chorioallantoic membrane

are fused shut. Pores occupy about 0.02 percent of the surface of an eggshell. The morphology of the pores varies in different species of birds: some pores are straight tubes, whereas others are branched, and the pore openings on the eggshell's surface may be blocked to varying degrees with organic or crystalline material—all of these factors affect the rate of evaporation of water from the egg.

Bird eggs must lose water to hatch, because the loss of water creates an air cell at the blunt end of the egg. The embryo penetrates the membranes of this air cell with its beak one or two days before hatching begins, and ventilation of the lungs begins to replace the chorioallantoic membrane in gas exchange. Pipping, the formation of the first cracks on the surface of the eggshell, follows about half a day after penetration of the air cell, and actual emergence begins half a day later. Shortly before hatching, the chick develops a horny projection on its upper jaw. This structure is called the egg tooth, and it is used in conjunction with a hypertrophied muscle on the back of the neck (the hatching muscle) to thrust vigorously against the shell. The egg tooth and hatching muscle disappear soon after the chick has hatched.

In those species of birds that delay the start of incubation until all the eggs have been laid, an entire clutch nears hatching simultaneously. Hatching may be synchronized by clicking sounds that accompany breathing within the egg, and both acceleration and retardation of individual eggs may be involved. A low-frequency sound produced early in respiration,

before the clicking phase is reached, appears to retard the start of clicking by advanced embryos. That is, the advanced embryos do not begin clicking while other embryos are still producing low-frequency sounds. Subsequently, vocalizations from advanced embryos appear to accelerate hatching by late embryos. Both effects were demonstrated in experiments with bobwhite quail eggs: pairing two eggs, one of which had started incubation 24 hours before the other, accelerated hatching of the late-starting egg by 14 hours and delayed hatching of the early-starting egg by 7 hours.

Sex Determination Birds have genetic sex determination (GSD) and heterogametic sex chromosomes, as do mammals, except that in birds the female is the heterogametic sex and the male is homogametic. To avoid confusion with mammals, the sex chromosomes of female birds are designated ZW and male birds are ZZ. The presence of the W sex chromosome causes the primordial gonad to secrete estrogen, which stimulates the left gonad to develop as an ovary and the left Müllerian duct system to develop into an oviduct and shell gland. (Only the left ovary and oviduct normally develop in birds.) In the absence of estrogen (i.e., when the genotype is ZZ), a male develops. (This is the opposite of the sex determination process in mammals in which a male-determining gene on the Y chromosome causes an XY individual to develop as a male.)

The evolutionary origin of GSD in birds is unclear because all crocodilians have temperature dependent sex determination (TSD) and lack heterogametic sex chromosome. Because most birds maintain relatively stable egg temperatures by incubating their eggs during embryonic development, TSD would not be an effective sex-determining mechanism, but it not possible to say whether GSD is a derived character of birds or TSD is a derived character of crocodilians. Moving farther back in the sauropsid lineage only complicates the question because some families of lepidosauromorphs have TSD and others have GSD. For example, in the lizard families Iguanidae, Scincidae, and Teiidae, males are heterogametic; but in the Anguidae and Varanidae, females are heterogametic, and in the Gekkonidae both male and female heterogamety have been reported. Including turtles merely adds to the confusion because most families have TSD; in Chelidae and Kinosternidae, males are heterogametic; and, in Bataguridae, both male and female heterogamety is known.

Sex-Biased Broods To add to the complications of GSD in birds, female birds can exert some control over the sex of their offspring. Many birds lay one egg per day until all the eggs in a clutch have been produced. Eggs hatch in approximately the sequence in which they were laid, and female birds may be able to adjust the sex of offspring in relation to the laying sequence of their eggs. A study of house finches (*Carpodacus mexicanus*) found that the eggs laid first by birds in Montana hatched into females, whereas first-laid eggs in an Alabama population hatched into males. Sex-specific growth and survival characteristics appear to be the basis for this difference because babies that hatch first grow faster and reach larger adult sizes than those that hatch later. In Montana large females survive better than smaller ones, whereas in Alabama large males are favored. The sex-biased hatching order increases chick survival by 10 percent to 20 percent.

Nests, Incubation, and Parental Care Construction of nests, incubation of the eggs, and care of the young are important aspects of avian reproduction. Like other aspects of avian biology, these elements show tremendous variation.

Nests protect the eggs not only from such physical stresses as heat, cold, and rain but also from predators. Bird nests range from shallow holes in the ground to enormous structures that represent the combined efforts of hundreds of individuals over many generations (**Figure 17–33**). The nests of passerines are usually cup-shaped structures composed of plant materials that are woven together. Swifts use sticky secretions from buccal glands to cement material together to form nests, and grebes, which are marsh-dwelling birds, build floating nests from the buoyant stems of aquatic plants.

Most birds nest individually, but some lineages are exceptions. Only 16 percent of passerines nest in colonies, but 98 percent of seabirds are colonial nesters. Nesting colonies of some species of penguins, petrels, gannets, gulls, terns, and auks contain hundreds of thousands of individuals. Colonies are smaller in most other groups of birds; colonies of herons, storks, doves, swifts, and passerines contain 10 to 50 nests. Colonial nesting offers both advantages and disadvantages. A colony is a concentration of potential prey that may attract predators, but the density of nesting birds may provide some protection to colony members. In many colonies, the nests are located two neck lengths apart, and an intruder is menaced from all sides by snapping beaks. Centrally placed nests may be better protected against predators than nests on the periphery of the colony.

Clutch Size How many eggs should a bird produce? That question has bedeviled ornithologists and ecologists at least since David Lack's pioneering work in the 1940s proposed that there would be an optimal number of eggs for a species of bird that reflected the amount of food the parent birds could bring to their nestlings. The size of a bird at fledging depends on how much it is fed in the nest, and larger fledglings (up to some limit) are more likely to survive than smaller ones. Thus the optimal clutch size should equal the maximum number of babies the parents can bring to the fledgling size, which maximizes their chances of survival.

Lack's reasoning is plausible, but observed clutch sizes are often larger or smaller than would be predicted by the food limitation hypothesis alone, and other factors may be involved. For example, the lifetime reproductive success of an individual bird may be more important than the success of an individual clutch. Large clutches place more physiological strain on females than do small clutches, and both parents are probably more exposed to predation when they are foraging to feed a large brood rather than a smaller one. A bird might maximize the total number of offspring it produces during its lifetime by producing clutches that have fewer babies than would be permitted by the availability of food.

Predation might play a role in determining optimal clutch size. Large clutches are probably more likely to be detected by predators than small clutches

(a)

(b)

(c)

(d)

▲ **FIGURE 17–33** **Diversity of bird nests.** Some nests are no more than shallow depressions; other birds build elaborate structures. The piping plover (a) like many shorebirds, lays its eggs in a depression scraped in the soil. The bald eagle (b) constructs an elaborate nest that is used year after year. Coots (c) build floating nests, and the Australian mallee fowl (d) scrapes together a pile of sand and buries its eggs. Heat from the sun warms the eggs, and the male mallee fowl adds and removes sand to keep the temperature stable.

because they are noisier, smellier, and the parent birds must make more trips to and from the nest. Thus, birds that nest in exposed sites might minimize the risk of predation by producing small clutches, whereas birds that nest in protected sites might be able to have large clutches.

Incubation All birds control the temperature of their eggs during development, and an adult bird resting on top of its eggs is a familiar sight. This form of brooding is the most widespread method of incuba-

tion, but it is not the only one. The megapodes, a group of birds found in the Indo-Australian region, are an interesting exception. Megapodes are known as mound birds because they bury their eggs in sand or soil and rely on heat from sunlight or rotting vegetation for incubation. The male remains at the nest, monitoring its temperature and adding or removing material to keep the temperature constant. Sex determination is temperature dependent in at least one megapode, the Australian brush turkey (Göth and Booth 2004, Göth 2007). The sex ratio is biased toward females in warm nests and toward males in cool nests. Megapode chicks are fully feathered when they hatch and can fly when they emerge from the nest. In this respect they resemble Early Cretaceous birds and the birdlike dinosaurs (Pennisi 2004, Zhou and Zhang 2004)

Some species of birds begin incubation as soon as the first egg is laid, and others wait until the clutch is complete. Though starting incubation immediately may protect the eggs, it means that the first eggs in the clutch hatch while the eggs that were deposited later are still developing. This forces the parents to divide their time between incubation and gathering food for the hatchlings. Furthermore, the eggs that hatch last produce young that are smaller than their older nestmates, and these young probably have less chance of surviving to fledge. Most passerines, as well as ducks, geese, and fowl, do not begin incubation until the next-to-last or last egg has been laid.

Prolactin, secreted by the pituitary gland, suppresses ovulation and induces brooding behavior, at least in those species of birds that wait until a clutch is complete to begin incubation. The insulating properties of feathers that are so important a feature of the thermoregulation of birds become a handicap during brooding, when the parent must transfer metabolic heat from its own body to the eggs. Prolactin plus estrogen or androgen stimulates the formation of brood patches in female and male birds, respectively. These brood patches are areas of bare skin on the ventral surface of a bird. The feathers are lost from the brood patch, and blood vessels proliferate in the dermis, which may double in thickness and give the skin a spongy texture. Not all birds develop brood patches, and in some species only the female has a brood patch, although the male may share in incubating the eggs. Ducks and geese create brood patches by plucking the down feathers from their breasts; they use the feathers to line their nests. Some penguins lay a single egg that they hold on top of their feet and cover with a fold of skin from the belly, thus enveloping the egg.

The temperature of eggs during brooding is usually maintained within the range 33°C to 37°C, although some eggs can withstand periods of cooling when the parent is off the nest. Tube-nosed seabirds (Procellariiformes) are known for the long periods that adults spend away from the nest during foraging. Fork-tailed storm petrels (*Oceanodroma furcata*) lay a single egg in a burrow or rock crevice. Both parents participate in incubation; but, the adults forage over vast distances, and both parents may be absent from the nest for several days at a time. For storm petrels in Alaska, parents averaged 11 days of absence during an incubation period of 50 days. Eggs were exposed to ambient temperatures of 10°C while the parents were away. Experimental studies showed that storm petrel eggs were not damaged by being cooled to 10°C every four days. The pattern of development of chilled eggs was like that of eggs incubated continuously at 34°C, except that each day of chilling added about one day to the total time required for the eggs to hatch.

Incubation periods are as short as 10 to 12 days for some species of birds and as long as 60 to 80 days for others. In general, large species of birds have longer incubation periods than small species, but ecological factors also contribute to determining the length of the incubation period. A high risk of predation may favor rapid development of the eggs. Among tropical tanagers, species that build open-topped nests near the ground are probably more vulnerable to predators than are species that build similar nests farther off the ground. The incubation periods of species that nest near the ground are short (11 to 13 days) compared to those of species that build nests at greater heights (14 to 20 days). Species of tropical tanagers with roofed-over nests have still longer incubation periods (17 to 24 days).

Parental Care It is a plausible inference—both from phylogenetic considerations and from fossils of coelurosaurs—that the ancestral form of reproduction in the avian lineage includes depositing eggs in a well-defined nest site, one or both parents attending the nest, young that are able to feed themselves at hatching (precocial young), and a period of association between the young and one or both parents after hatching. All of the crocodilians that have been studied conform to this pattern, and evidence is increasing that at least some dinosaurs remained with their nests and young.

Extant birds follow these ancestral patterns, but not all species produce precocial young. Instead, hatchling birds show a spectrum of maturity that extends from **precocial y**oung that are feathered and self-sufficient from the moment of hatching to **altricial** forms that are naked and entirely dependent on their parents for food and thermoregulation (**Figure 17–34** and **Table 17.4**). The distinction between precocial and altricial birds includes differences in the amount of yolk originally in the eggs, the relative development of organs and muscles at hatching, and the rates of growth after hatching (**Table 17.5**).

After they hatch, altricial young are guarded and fed by one or both parents. Adults of some species of

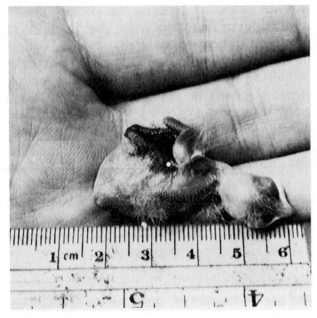

(a)   (b)

▲ **FIGURE 17–34** Altricial and precocial chicks. Altricial chicks (a) such as that of the tree swallow (*Tachycinta bicolor*) are entirely naked when they hatch and unable even to stand up. Precocial species (b) such as the snowy plover (*Charadius alexandrinus*) have chicks that are covered with down when they hatch and can stand erect and even walk. The plover in this photograph has just hatched; it retains the egg tooth on the tip of its bill, whereas the tree swallow chick is five days old. The dark color on the leg of the swallow was applied by a researcher to identify individual hatchlings for a study of parental care.

birds carry food to nestlings in their beaks, but many species swallow food and later regurgitate it to feed the young. Hatchling altricial birds respond to any disturbance that might signal the arrival of a parent at the nest by gaping their mouths widely. The sight of an open mouth appears to stimulate a parent bird to feed it, and the young of many altricial birds have brightly colored mouth linings. Ploceid finches have covered nests, and the mouths of the nestlings of some species are said to have luminous spots that have been likened to beacons showing the parents where to deposit food in the gloom of the nest.

The duration of parental care is variable. The young of small passerines leave the nest about two weeks after hatching and are cared for by their parents for an additional one to three weeks. Larger

TABLE 17.4 Maturity of birds at hatching

Precocial: Eyes open, covered with feathers or down, leave nest after one or two days

1. Independent of parents: megapodes
2. Follow parents but find their own food: ducks, shorebirds
3. Follow parents and are shown food: quail, chickens
4. Follow parents and are fed by them: grebes, rails

Semiprecocial: Eyes open, covered with down, able to walk but remain at nest and are fed by parents: gulls, terns

Semialtricial: Covered with down, unable to leave nest, fed by parents

1. Eyes open: herons, hawks
2. Eyes closed: owls

Altricial: Eyes closed, little or no down, unable to leave nest, fed by parents: passerines

---

TABLE 17.5 Comparison of birds with altricial and precocial chicks

Eggs	
Amount of yolk in eggs	precocial > altricial
Amount of yolk remaining at hatching	precocial > altricial
Chicks	
Size of eyes and brain	precocial > altricial
Development of muscles	precocial > altricial
Size of gut	altricial > precocial
Rate of growth after hatching	altricial > precocial

species of birds, such as the tawny owl, spend a month in the nest and receive parental care for an additional three months after they have fledged, and young wandering albatrosses require a year to become independent of their parents.

17.8 Migration and Navigation

The mobility that characterizes vertebrates is perhaps most clearly demonstrated in their movements over enormous distances. These displacements, which may cover half the globe, require both endurance and the ability to navigate. Other vertebrates migrate, some of them over enormous distances, but migration is best known among birds.

■ Migratory Movements of Birds

Migration is a widespread phenomenon among birds. About 40 percent of the bird species in the Palearctic are migratory, and an estimated total of some 5 billion birds migrate from the Palearctic every year. Migrations often involve movements over thousands of kilometers, especially in the case of birds nesting in northern latitudes, some marine mammals, sea turtles, and fishes. Short-tailed shearwaters (*Puffinus tenuirostris*), for example, make an annual migration between the North Pacific and their breeding range in southern Australia that requires a round-trip of more than 30,000 kilometers (**Figure 17–35**).

Many birds return each year to the same migratory stopover sites, just as they may return to the same breeding and wintering sites year after year. Migrating birds may be concentrated at high densities at certain points along their traditional migratory routes. For example, species that follow a coastal route may be funneled to small points of land—such as Cape

May, New Jersey—from which they must initiate long overwater flights. At these stopovers, migrating birds must find food and water to replenish their stores before venturing over the sea, and they must avoid the predators congregating at these sites. Development of coastal areas for human use has destroyed many important resting and refueling stations for migratory birds. The destruction of coastal wetlands has caused serious problems for migratory birds on a worldwide basis. Loss of migratory stopover sites may remove a critical resource from a population at a particularly stressful stage in its life cycle.

Advantages of Migration The high energy costs of migration must be offset by energy gained as a result of moving to a different habitat. The normal food sources for some species of birds are unavailable in the winter, and the benefits of migration for those species are starkly clear. Other species may save energy mainly by avoiding the temperature stress of northern winters. In other cases, the main advantage of migration may come from breeding in high latitudes in the summer where resources are abundant and long days provide more time to forage than the birds would have if they remained closer to the equator.

Physiological Preparation for Migration Migration is the result of a complex sequence of events that integrate the physiology and behavior of birds. Fat is the principal energy store for migratory birds, and birds undergo a period of heavy feeding and premigratory fattening (**Zugdisposition,** migratory preparation) in which fat deposits in the body cavity and subcutaneous tissue increase tenfold, ultimately reaching 20 percent to 50 percent of the nonfat body mass. Fat is metabolized rapidly when migration begins, and many birds migrate at night and eat

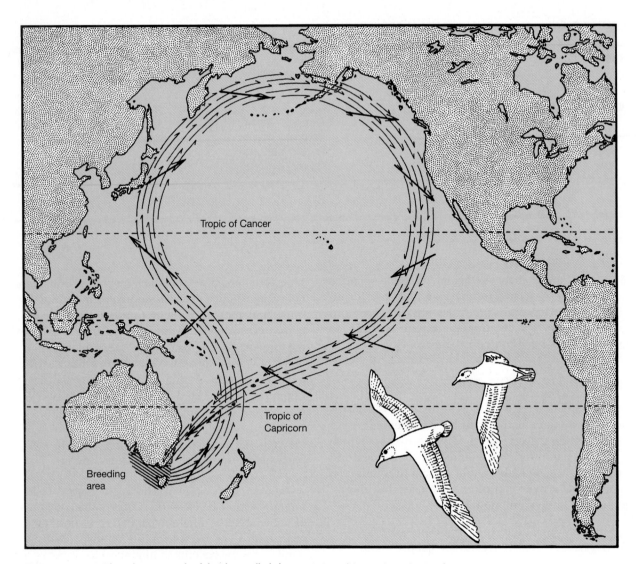

▲ **FIGURE 17–35** The migratory path of the short-tailed shearwater. As this species migrates from its Australian breeding area to its northern range, it takes advantage of prevailing winds in the Pacific region to reduce the energetic cost of migration.

during the day. Even diurnal migrants divide the day into periods of migratory flight (usually early in the day) and periods of feeding. In addition, pauses of several days to replenish fat stores are a normal part of migration. *Zugdisposition* is followed by **Zugstimmung** (migratory mood), in which the bird undertakes and maintains migratory flight. In caged birds, which are prevented from migrating, this condition results in the well-known phenomenon of **Zugunruhe** (migratory restlessness).

Timing of Migration Preparation for migration must be integrated with environmental conditions, and this coordination appears to be accomplished by the interaction of internal rhythms with an external stimulus.

Day length is the most important cue for *Zugdisposition* and *Zugstimmung* for birds in northern temperate regions. Northward migration in spring is induced by increasing day length (**Figure 17–36**). The direction in which migratory birds orient during *Zugunruhe* depends on their physiological condition. In one experiment, the photoperiod was manipulated to bring one group of indigo buntings into their autumn migratory condition at the same time that a second group of birds was in its spring migratory condition. When the birds were tested under an artificial planetarium sky, the birds in the spring migratory condition oriented primarily in a northeasterly direction (Figure 17–36b), whereas birds in the autumn migratory condition oriented in a southerly direction

▶ **FIGURE 17–36** Orientation of migratory birds during *Zugunruhe*. (a) The birds were tested in circular cages that allow a view of the sky. The bird stands on an ink pad, and each time it hops onto the sloping wall it leaves a mark on the blotting paper that lines the cage. (b) In spring, the birds oriented toward the north and (c) in autumn toward the south.

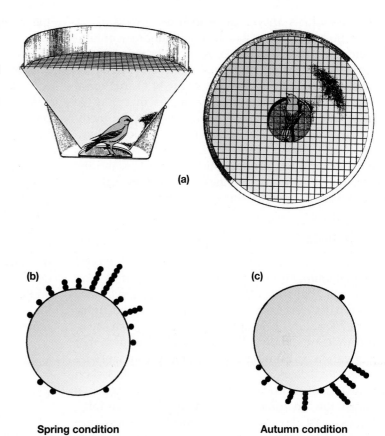

(a)

(b) **Spring condition**

(c) **Autumn condition**

(Figure 17–36c). Dark circles show the mean nightly headings, pooled for several observations, for each of six birds in the spring migratory condition and five birds in the autumn condition.

Underlying the responses of birds to changes in day length is an endogenous (internal) rhythm. This circannual (about a year) cycle can be demonstrated by keeping birds under constant conditions. Fat deposition and migratory restlessness coincide in most species and alternate with gonadal development and molt, as they do in wild birds. When the rhythms are free running (that is, when they are not cued by external stimuli), they vary between 7 and 15 months. In other words, the birds' internal clocks continue to run, but in the absence of the cue normally provided by changing day length, the internal rhythms drift away from precise correspondence with the seasons.

Synchronizing Migration Migratory birds that are pair-bonded to their mates face an additional complication in timing migration—the members of the pair must synchronize their internal rhythms so they migrate at the same time. If that degree of coordination seems complicated for species of birds that migrate together and live as a pair at both ends of

their migratory journey, think of the problem that confronts the Icelandic black-tailed godwit (*Limosia limosia islandica*), a species in which the males and females migrate separately and spend the winter in locations that are hundreds of kilometers apart!

Icelandic godwits breed in Iceland during the summer and migrate south to spend the winter in England, Ireland, France, or Spain (Gunnarsson et al. 2004). Males and females depart from Iceland separately and winter in different locations: the average distance between the members of 14 pairs during the winter was 955 km, and the most distant pair was separated by almost 2000 km. Returning godwits start to appear in Iceland in mid-April, and all of the birds have returned by mid-May. Although the return dates for the population as a whole extend over a period of 30 days, the average interval between the arrivals of the members of a pair was only 3 days.

Mate fidelity appears to be important to the lifetime reproductive success of long-lived migratory birds, and a change of partners is often followed by a reduction in reproductive success. Synchrony in arrival dates appears to contribute to the maintenance of pair bonds; the only divorces recorded during Gunnarsson's study occurred in two of the three pairs in which the individuals arrived more than eight days apart. How do a male and a female living hundreds of kilometers apart coordinate their return trips to Iceland so well that they arrive nearly simultaneously? Unfortunately, the study offers no clear answer to that question but it does suggest some factors that may be important. It is possible, for example, that the members of a pair winter in areas with similar food abundance so that they are in similar nutritional states as the time for the return migration approaches, and the availability of food during the stopovers they make on the return trip may fine-tune their timing.

■ Orientation and Navigation

The homing pigeon has become a favorite experimental animal for studies of navigation. For as long as people have raised and raced pigeons, we have known that birds released in unfamiliar territory vanish from sight flying in a straight line, usually in the direction of home. Experiments have shown that navigation by homing pigeons (and presumably by other vertebrates as well) is complex and is based on a variety of sensory cues. On sunny days, pigeons vanish toward home and return rapidly to their lofts. On overcast days, vanishing bearings are less precise, and birds more often get lost. These observations led to the idea that pigeons use the sun as a compass.

Of course, the sun's position in the sky changes from dawn to dusk. That means a bird must know what time of day it is in order to use the sun to tell direction, and its time-keeping ability requires some sort of internal clock. If that hypothesis is correct, it should be possible to fool a bird by shifting its clock forward or backward. For example, if lights are turned on in the pigeon loft six hours before sunrise every morning for about five days, the birds will become accustomed to that artificial sunrise. At any time during the day, they will act as if the time is six hours later than it actually is. When these birds are released they will be directed by the sun, but their internal clocks will be wrong by six hours. That error should cause the birds to fly off on courses that are displaced by 90 degrees from the correct course for home, and that is what clock-shifted pigeons do on sunny days (**Figure 17–37 a** and **b**). Under cloudy skies, however, clock-shifted pigeons head straight for home despite the 6-hour error in their internal clocks (**Figure 17–37c** and **d**). Clearly, pigeons have more than one way to navigate: when they can see

the sun they use a sun compass, but when the sun is not visible they use another mechanism that is not affected by the clock-shift.

Probably this second mechanism is an ability to sense the Earth's magnetic field and use it as a compass. On sunny days, attaching small magnets to a pigeon's head does not affect its ability to navigate, but on cloudy days pigeons with magnets cannot return to their home lofts.

Polarized and ultraviolet light are additional cues that pigeons probably use to determine direction, and they can also navigate by recognizing airborne odors and familiar visual landmarks. In addition, pigeons can detect extremely low-frequency sounds (infrasound), well below the frequencies that humans can hear. Those sounds are generated by ocean waves and air masses moving over mountains and can signal a general direction over thousands of kilometers, but their use as cues for navigation remains obscure.

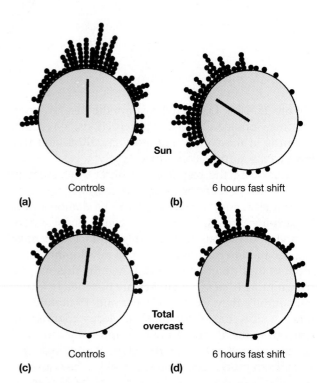

▲ **FIGURE 17–37** Orientation of clock-shifted pigeons under sunny and cloudy skies. Each dot in these plots shows the direction in which a pigeon vanished from sight when it was released in the center of the large circle. The home loft is straight up in each diagram. The solid bar extending outward from the center of each circle shows the average direction chosen by the birds.

Results of this sort are being obtained with other vertebrates as well and lead to the general conclusion that a great deal of redundancy is built into navigation systems. Apparently, there is a hierarchy of useful cues. For example, a bird relies on the sun and polarized light to navigate on clear days, but it can switch to magnetic direction sensing on heavily overcast days. In both conditions, the bird can use local odors and recognition of landmarks as it approaches home.

Many birds migrate only at night. Under these conditions a magnetic sense of direction might be important. Several species of nocturnally migrating birds use star patterns for navigation. Apparently, each bird fixes on the pattern of particular stars and uses their motion in the night sky to determine a compass direction. As in sun compass navigation, an internal clock is required for this sort of celestial navigation, and artificially changing the time setting of the internal clock produces predictable changes in the direction in which a bird orients.

Despite numerous studies, the complexities of navigation mechanisms of vertebrates are far from being fully understood, and much controversy surrounds some hypotheses. The built-in redundancy of the systems makes it difficult to devise experiments that isolate one mechanism. When it is deprived of the use of one sensory mode, an animal is likely to have several others it can use instead. This redundancy and the remarkable sophistication with which many vertebrates navigate show the importance of migration in their lives.

Summary

Flight is the distinctive mode of avian locomotion and many elements of the anatomy of birds are shaped by the demands of flight. Flapping flight is a more complicated process than flight with fixed wings like those of aircraft, but it can be understood in aerodynamic terms. Feathers compose the aerodynamic surfaces responsible for lift and propulsion during flight, and they also provide streamlining. There are many variations and specializations within the four basic types of wings, but, in general, high speed and elliptical wings are used in flapping flight and high aspect ratio and high-lift wings are used for soaring and gliding.

The hollow bones of birds, an ancestral character of the coleurosaur lineage, combine lightness and strength. Some parts of the skeletons of birds are light in relation to the sizes of their bodies, and some of this lightness has been achieved by modification of the skull. Several skull bones that are separate in other diapsids are fused in birds, and teeth are absent. The function of teeth in processing food has been taken over by the muscular gizzard, which is part of the stomach of birds. The gizzard is well developed in birds that eat hard items such as seeds; it may contain stones that probably assist in grinding food.

The basic design of the bird foot is very similar to the feet of coleurosaurs. Modifications of the feet are related to the lifestyle of a species—many flightless birds have reduced the number of toes, and aquatic birds have developed lobes on the toes or webs between the toes to increase the surface area acting on the water.

All birds are oviparous, perhaps because egg retention is the first step in the evolution of viviparity, and the specializations of the avian way of life do not make egg retention advantageous for birds. Female birds can exert a degree of control over the number of eggs they lay and the sex of the hatchlings. Most birds have extended periods of parental care, and during this period young birds learn species-specific behaviors such as song.

Many of the complex social behaviors of birds are associated with reproduction, and birds have contributed greatly to our understanding of the relationship between ecological factors and the mating systems of vertebrates. Social monogamy, with both parents caring for the young, is the most common mating system for birds, but extra-pair copulations are a common feature of monogamous mating systems and provide opportunities for both males and females to increase their fitness.

Migration is the most dramatic manifestation of the mobility of birds, and some species travel tens of thousands of kilometers in a year. Migrating birds use a variety of cues for navigation, including the position of the sun, stars, polarized light, and the Earth's magnetic field.

Additional Readings

Badyaev, A. V. et al. 2002. Sex-biased hatching order and adaptive population divergence in a passerine bird. *Science* 295:316–318.

Barinaga, M. 2002. Sight, sound converge in owl's mental map. *Science* 297:1462–1463.

Barker, F. K. et al. 2004. Phylogeny and diversification of the largest avian radiation. *Proceedings of the National Academy of Sciences USA* 101:11040–11045.

Bennett, P. M., and I. P. F. Owens. 2002. *Evolutionary Ecology of Birds: Life Histories, Mating Systems, and Extinction*. Oxford, UK: Oxford University Press.

Blount, J. D. et al. 2003. Carotenoid modulation of immune function and sexual attractiveness in zebra finches. *Science* 300:125–127.

Chatterjee, S. et al. 2007. The aerodynamics of *Argentavis*, the world's largest flying bird, from the Miocene of Argentina. *Proceedings of the National Academy of Sciences (United States)* 104:12398–12403.

Chiappe. L. M., and G. J. Dyke. 2002. The Mesozoic radiation of birds. *Annual Review of Ecology and Systematics* 33:91–124.

Clarke, J. A. et al. 2005. Definitive fossil evidence for the extant avian radiation in the Cretaceous. *Nature* 433:305–308.

Cracraft, J. A. 1986. The origin and early diversification of birds. *Paleobiology* 12:383–399.

Cracraft, J. A. 2002. Gondwana genesis. *Natural History* December 2001/January 2002, pages 64–73.

Faivre, B. 2003. Immune activation rapidly mirrored in a secondary sexual trait. *Science* 300:103–125.

Feduccia, A. 2003. "Big Bang" for Tertiary birds? *Trends in Ecology and Evolution* 18:172–176.

Fountaine, T. M. R. et al. 2004. *The quality of the fossil record of Mesozoic birds. Proceedings of the Royal Society B.* 272:289–294.

Gil, D. et al. 1999. Male attractiveness and differential testosterone investment in zebra finch eggs. *Science* 286:126–128.

Göth, A. 2007. Incubation temperatures and sex ratios in Australian brush-turkey (*Alectura lathami*) mounds. *Austral Ecology* 32:378–385.

Göth, A., and D. T. Booth. 2004. Temperature-dependent sex ratio in a bird. *Biology Letters*. Published online <doi:10.1098/rsbl.2004.0247>.

Gunnarsson, T. G. et al. 2004. Arrival synchrony in migratory birds. *Nature* 431:646.

Hackett, S. J. et al. 2008. A phylogenomic study of birds reveals their evolutionary history. *Science* 320:1763–1768.

Hansell, M. 2000. *Bird Nests and Construction Behavior.* Cambridge, UK: Cambridge University Press.

Ioalè, P. et al. 1990. Homing pigeons do extract directional information from olfactory stimuli. *Behavioral Ecology and Sociobiology* 26:301–306.

Johansson, L. C., and R. Å. Norberg. 2003. Delta-wing function of webbed feet gives hydrodynamic life for swimming propulsion in birds. *Nature* 424:65–68.

Johansson, L. C., and U. M. Norberg. 2001. Lift-based paddling in diving grebe. *Journal of Experimental Biology* 204:1687–1696.

Lentink, D. et al. 2007. How swifts control their glide performance with morphing wings, *Nature* 446:1082–1085.

Lovejoy, T. E. and J. Elphick. 2007. *Atlas of Bird Migrations: Tracing the Great Journeys of the World's Birds.* Buffalo, NY: Firefly Books.

Martin, T. E. et al. 2000. Parental care and clutch sizes in North and South American birds. *Science* 287: 1482–1485.

Norberg, R. Å. 1977. Occurrence and independent evolution of bilateral ear asymmetry in owls and implications for owl taxonomy. *Philosophical Transactions of the Royal Society of London B* 280:375–408.

Norberg, R. Å. 1978. Skull asymmetry, ear structure and function, and auditory localization in Tengmalm's owl, *Aeogilius funereus* (Linné). *Philosophical Transactions of the Royal Society of London B* 282:325–410.

Padian, K. et al. 2001. Feathered dinosaurs and the origin of flight. In *Mesozoic Vertebrate Life*, D. H. Tanker and K. Carpenter (eds.), Bloomington, IN: Indiana University Press, 117–135..

Pennisi, E. 2003. Colorful males may flaunt their health. *Science* 300:29–30.

Pennisi, E. 2004. Newly hatched dinosaur babies hit the ground running. *Science* 305:1396.

Petrie, M. 1994. Improved growth and survival of offspring of peacocks with more elaborate trains. *Nature* 371:598–599.

Petrie, M. et al. 2001. Sex differences in avian yolk hormone levels. *Nature* 412:498–499.

Raikow, R. J., and A. H. Bledsoe. 2000. Phylogeny and evolution of passerine birds. *BioScience* 50:487–499.

Rattenborg, N. C. et al. 1999. Half-awake to the risk of predation. *Nature* 397:397–398.

Tobalske, B. W. et al. 2003. Comparative power curves in bird flight. *Nature* 421:363–366.

Weihs, D., and G. Katzir. 1994. Bill sweeping in the spoonbill, *Platalea leucordia*: Evidence for a hydrodynamic function. *Animal Behaviour* 47:649–654.

Wiemerskirsch, H. et al. 2001. Energy saving in flight formation. *Nature* 413:697–698.

Zhou, Z., and F. Zhang. 2004. A precocial avian embryo from the lower Cretaceous of China. *Science* 306:653.

PART IV

Synapsida: The Mammals

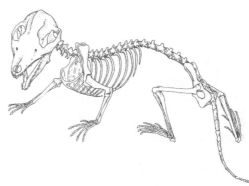

Because we *are* mammals, humans tend to think of mammals as the dominant kinds of vertebrates. That perspective does not withstand examination—there are barely more than half as many extant species of mammals as of birds, for example, and ray-finned fishes include as many extant species as all of the tetrapods combined—but it is pervasive. In some contexts, mammals are exceptionally successful derived vertebrates. The size range of mammals—from shrews to whales—is impressive, as is the development of flight by bats and echolocation by bats and cetaceans (whales and dolphins).

Mammals probably have more complex social systems than any other kinds of vertebrates, and many features of their biology are related to interactions with other individuals of their species, ranging from the often prolonged dependence of young on their mother to lifelong alliances between individuals that affect their social status and reproductive success in a group.

The anatomical and physiological characteristics of mammals—respiration with a diaphragm, hair that provides insulation, high metabolic rates, teeth with complex surfaces that process food efficiently—make them successful in a wide variety of habitats. The progressive appearance of these characteristics can be traced clearly through the stem groups in the mammalian lineage.

Humans differ from other vertebrates in the extent to which they have come to dominate all of the habitats of Earth and in their effects—direct and indirect—on other vertebrates. In this portion of the book we explore the evolution of mammals, the adaptive zones opened to them by their distinctive characteristics, the origin and radiation of humans, and the impact of humans on other species.

CHAPTER 18

The Synapsida and the Evolution of Mammals

We must backtrack to the end of the Paleozoic era to find the origin of the final lineage of vertebrates, the synapsids. The synapsids actually had their first radiation (pelycosaurs) and a significant portion of their second radiation (therapsids) in the Paleozoic, before the radiations of the diapsids we have already discussed. The third radiation of the synapsid lineage (mammals) did not reach its peak until the Cenozoic era. Nonetheless, through the late Paleozoic and early Mesozoic, the synapsid lineage was becoming increasingly mammal-like, and mammals and dinosaurs both appeared on the scene in the Late Triassic period.

The three groups of extant mammals—monotremes, marsupials, and placentals—had evolved by the late Mesozoic era, and they were accompanied by several groups of mammals that are now extinct.

18.1 The Origin of Synapsids

Synapsids include mammals and their extinct predecessors, commonly called "mammal-like reptiles" (**Figure 18–1** and **Figure 18–2**), and we considered in Chapter 11 the ways in which synapsids differed in various aspects of their biology from the other major group of amniotes, the sauropsids. Synapsids are distinguished from other amniotes by the presence of a lower temporal fenestra, plus a few other skull features. Changes in the structure of the skull and skeleton of nonmammalian synapsids, and their probable relation to metabolic status and the evolution of the mammalian condition, are described later.

The term *synapsid* is often used incorrectly to refer to just the extinct nonmammalian forms, but it in fact includes all amniotes descended from a common ancestor with the synapsid type of temporal fenestration. *Mammal-like reptile* is an appealing term for the ancestors of mammals, yet it is a misleading one, and is now seldom used in technical publications. As we saw in Chapters 9 and 11, mammals are not the descendants of any animals closely related to modern reptiles, and to think of early synapsids as some sort of large, peculiar lizardlike beasts would indeed be inaccurate. Moreover, since mammals originated within the group of animals called "mammal-like reptiles," it is a paraphyletic assemblage rather than a true evolutionary group. (This is the same problem we faced in Chapter 16 with "dinosaurs," which is also a paraphyletic assemblage unless birds are included, and we adopted the term *nonavian dinosaur*.) In this chapter, we will use the term nonmammalian synapsid.

The synapsid lineage was the first group of amniotes to radiate widely in terrestrial habitats. During the Late Carboniferous and the entire Permian periods, synapsids were the most abundant terrestrial vertebrates; and, from the Early Permian into the Triassic, they were the top carnivores in the food web. Most synapsids were medium- to large-size animals, weighing between 10 kilograms and 200 kilograms, with a few weighing as much as a half ton or more (e.g., the dinocephalian therapsid *Moschops*). Most of the synapsid lineages disappeared by the Late Triassic period. The surviving forms (represented only by mammals past the Early Cretaceous

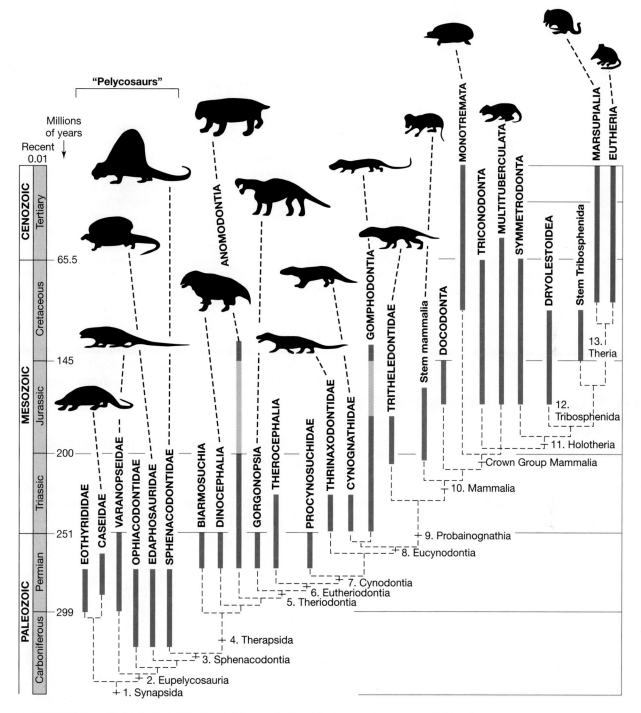

▲ **FIGURE 18–1** Phylogenetic relationships of the synapsida. This diagram shows the probable relationships among the major groups of synapsids. Note that some researchers (e.g., Kemp 2005) would favor the tritylodontids (here included with the Gomphodontia) as the sister taxon to Mammalia. The dotted lines indicate interrelationships only and are not indicative of times of divergence or of the presence of the taxon unrecorded in the fossil record. The numbers indicate derived characters that distinguish the lineages.

◀ **FIGURE 18–1** *Continued*

Legend: 1. Synapsida—lower temporal fenestra present. **2.** Eupelycosauria—snout deeper than it is wide, frontal bone forming a large portion of the margin of the orbit. **3.** Sphenacodontia—a reflected lamina on the angular bone, retroarticular process of the articular bone turned downward, high coronoid eminence on dentary, narrowing of scapular blade. **4.** Therapsida—temporal fenestra enlarged, upper canine plus the bone containing it (the maxilla) enlarged, mobile scapula with loss of pelycosaur screw-shaped glenoid, ossified sternum, limb bones more slender, limbs held more underneath body (indicated by inturned heads of femur and humerus), femur with greater trochanter (for insertion of gluteal muscles), shorter feet. **5.** Theriodontia—coronoid process on dentary, flatter skull with wider snout. **6.** Eutheriodontia—temporal fossa completely open dorsally, differentiation of the vertebral column into distinct lumbar and thoracic regions.

7. Cynodontia—postcanine teeth with anterior and posterior accessory cusps and small cusps on the inner side, partial bony secondary palate, masseteric fossa on dentary and bowing out of zygomatic arch (evidence for the presence of a masseter muscle), large sagittal crest on top of skull, double occipital condyle, lumbar ribs reduced or lost, coracoid reduced, ilium expanded forwards, pubis reduced, femur with inturned head, distinct calcaneal heel. **8.** Eucynodontia—dentary greatly enlarged, postdentary bones reduced, phalangeal formula of 2-3-3-3-3. **9.** Probainognathia (also included but not shown here are the families Probainognathidae and Chiniquodontidae)—pineal foramen absent, posteriorly enlongated secondary palate, at least incipient contact between dentary and squamosal bones. Tritheledontidae and Mammalia share the derived

features of prismatic enamel and unilateral action of the lower jaw. **10.** Mammalia—dentary-squamosal jaw articulation, double-rooted postcanine teeth, and specializations of the portion of the skull housing the inner ear. ("Stem Mammalia" is a paraphyletic assemblage of Late Triassic and Early Jurassic genera, including [listed from more primitive to more derived] *Adelobasileus, Sinoconodon, Megazostrodon,* and *Morganucodon.*) **11.** Holotheria—reversed triangles molar pattern. **12.** Tribosphenida—tribosphenic molars. ("Stem Tribosphenida" is a paraphyletic assemblage of Early Cretaceous genera, including [listed from more primitive to more derived] *Aegialodon, Pappotherium,* and *Holoclemensia,* plus others. The somewhat more primitive *Vincelestes* might also be included in this grouping.) **13.** Theria—details of braincase structure, and many features of the soft anatomy.

period) were considerably smaller, mostly under a kilogram in body mass (i.e., rat-size or smaller).

18.2 Diversity of Nonmammalian Synapsids

The two major groups of nonmammalian synapsids were the pelycosaurs and the therapsids; **Table 18.1** lists the major subgroups within each. Pelycosaurs, the more primitive of the two groups, were mainly found in the paleoequatorial latitudes of the Northern Hemisphere (Laurasia) and were predominantly known from the Early Permian. The more

derived therapsids were found mainly in the Southern Hemisphere (Gondwana). They range in age from the Late Permian to the Early Cretaceous but were predominantly known from the Late Permian to Early Triassic.

■ Pelycosaurs—Primitive Nonmammalian Synapsids

Pelycosaurs are known as the sailbacks of the late Paleozoic, although only a minority of them actually had sails. The best-known pelycosaur is probably *Dimetrodon* (**Figure 18–3a**), an animal frequently mislabeled as a dinosaur in children's books.

▶ **FIGURE 18–2** Simplified cladogram of synapsids. Quotation marks indicate paraphyletic groups. Extinct lineages are indicated by a dagger (†).

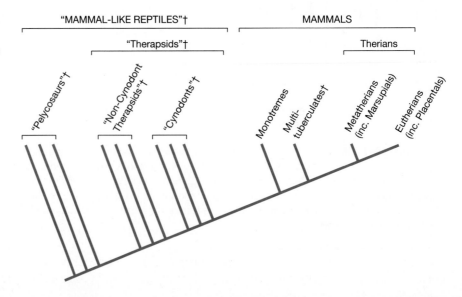

TABLE 18.1 Major groups of nonmammalian synapsids

Note that *large*, *medium*, and *small* refer to size within a particular group. A medium-size noncynodont therapsid is much larger than a medium-size cynodont.

Pelycosaurs

Eothyrididae: The most primitive pelycosaurs, small (cat-size) and probably insectivorous. Early Permian of North America (e.g., *Eothyris*).

Caseidae: Large (pig-size) herbivorous forms. Middle Permian of North America and Europe (e.g., *Caseia, Cotylorhynchus* [see Figure 18–3b]).

Varanopseidae: Generalized, medium-size forms. Early Permian of North America (e.g., *Varanops*) and Late Permian of South Africa and South America.

Ophiacodontidae: The earliest known pelycosaurs (but not the most primitive). Medium-size with long, slender heads, reflecting semi-aquatic fish-eating habits. Late Carboniferous to Early Permian of North America and Europe (e.g., *Ophiacodon* [see Figure 18–4a]).

Edaphosauridae: Large-size herbivores, some with a sail. Early Permian of North America (e.g., *Edaphosaurus*).

Sphenacodontidae: Large-size carnivores, some with a sail. Early Permian of North America and Europe (e.g., *Haptodus, Dimetrodon* [see Figures 18–3a and 18–4b]).

Noncynodont Therapsids

Biarmosuchia: The most primitive therapsids, medium-size (dog-size) carnivores. Late Permian of Eastern Europe (e.g., *Biarmosuchus*).

Dinocephalia: Medium- to large-size (cow-size) carnivores and herbivores. Some large herbivores had thickened skulls, possibly for head-butting in intraspecific combat. Late Permian of Eastern Europe and South Africa (e.g., *Titanophoneus, Moschops* [see Figure 18–3c]).

Anomondontia: Small- (rabbit-size) to large-size herbivores, the most diverse of the therapsids. Includes the dicynodonts, which retained only the upper canines and substituted the rest of the dentition with a turtlelike horny beak. Included burrowing and semiaquatic forms. Late Permian to Late Triassic worldwide (including Antarctica) and Early Cretaceous of Australia (e.g., *Dicynodon* [see Figure 18–4c], *Lystrosaurus* [see Figure 18–5]).

Gorgonopsia: Medium- to large-size carnivores. Late Permian of Eastern Europe and South Africa (e.g., *Scymnognathus* [see Figure 18–4d], *Lycaenops* [see Figure 18–3d]).

Therocephalia: Small- to medium-size carnivores and insectivores. Paralleled cynodonts in acquisition of secondary palate, complex postcanine teeth, and evidence for nasal turbinate bones. Late Permian to mid-Triassic of Eastern Europe and South and East Africa (e.g., *Pristerognathus*).

Cynodont Therapsids

Procynosuchidae: The most primitive cynodonts, medium-size (rabbit-size) carnivores and insectivores. Late Permian of Eastern Europe and South and East Africa (e.g., *Procynosuchus*).

Thrinaxodontidae: Medium-size (rabbit-size) carnivores and insectivores. Early Triassic of Eastern Europe, South Africa, South America, and Antarctica (e.g., *Thrinaxodon* [see Figures 18–7b and 18–9]).

Cynognathidae: Medium- to large-size (dog-size) carnivores. Early Triassic of South Africa and South America (e.g., *Cynognathus* [see Figure 18–4e]).

Gomphodontia: Small- (mouse-size) to medium-size (dog-size) herbivores. Includes the large diademodontids (e.g., *Diademodon*, Early Triassic of Africa and East Asia) and transversodontids (e.g., *Transversodon*, Middle Triassic of South and East Africa, South America, East Asia) and the small-size tritylodontids (e.g., *Tritylodon, Oligokyphus* [see Figure 18–4f]). Late Triassic to Middle Jurassic of North America, Europe, East Asia and South Africa, plus Early Cretaceous of Russia. Tritylodonts were rodentlike forms that paralleled the mammalian condition of the postcranial skeleton. Some researchers consider them as separate from the other gomphodonts and as the sister group to mammals.

Chiniquodontidae and Probainognathidae: Small-size carnivores and insectivores, closely related to Tritheledontidae (not included in Figure 18–1). Middle to Late Triassic of North and South America (e.g., *Probelesodon* [see Figures 18–3e and 18–7c]).

Tritheledontidae: Small-size carnivores and insectivores. Approached mammalian condition in form of jaw joint and postcranial skeleton. Late Triassic of South America to Early Jurassic of South Africa (e.g., *Diarthrognathus*).

Pelycosaurs contain the ancestors of the more derived synapsids, including the mammals; thus they represent a paraphyletic assemblage. However, they were a unique group while they lived in the Late Carboniferous and the Permian, with morphological and ecological features that set them apart from later,

▲ FIGURE 18–3 Diversity of nonmammalian synapsids. (a) The sphenacosaurid pelycosaur *Dimetrodon* (about the size of a Saint Bernard dog). (b) The caseid pelycosaur *Cotylorhyncus* (similar size to *Dimetrodon* or a little smaller). (c) The dinocephalian therapsid *Moschops* (about the size of a cow). (d) The gorgonopsid therapsid *Lycaenops* (about the size of a Labrador retriever). (e) The cynodont *Probelesodon* (about the size of Jack Russell terrier). (The species are shown approximately to scale.)

more derived therapsids. Pelycosaurs were basically generalized amniotes, albeit with some of their own specializations, and none showed any evidence of increased locomotor capacity or metabolic rate.

Most pelycosaurs were generalized carnivores. Some (the ophiacodontids) had long snouts and multiple pointed teeth and were semiaquatic fish-eaters. The caseids (**Figure 18–3b**) and edaphosaurids were herbivores, as we can determine by their blunt, peg-like teeth. Both forms had expanded rib cages, indicating that they had the large guts typical of herbi-

vores, and heads that look surprisingly small in comparison to their barrel-shaped bodies.

The most derived pelycosaurs were the sphenacodonts, such as *Dimetrodon*—mainly large, carnivorous forms with large, sharp teeth (**Figure 18–4b**). An enlarged caninelike tooth in the maxillary bone is a key feature linking sphenacodonts to more derived synapsids. Additional derived features include an arched palate, which was the first step toward the development of a separation of the mouth and nasal passages seen in some therapsids

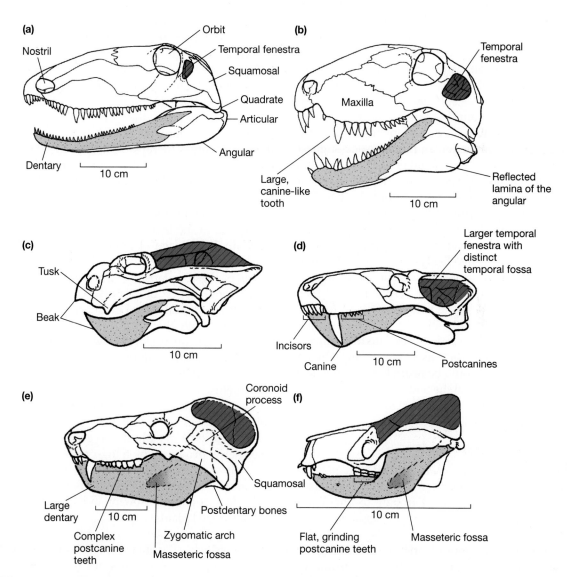

▲ **FIGURE 18–4** Skulls of nonmammalian synapsids. (a) The ophiacodontid pelycosaur *Ophiacodon*. (b) The sphenacodontid pelycosaur *Dimetrodon*. (c) The dicynodont therapsid *Dicynodon*. (d) The gorgonopsid therapsid *Scymnognathus*. (e) The cynodont *Cynognathus*. (f) The tritylodontid cynodont *Oligokyphus*. Light colored stippling = the dentary bone; dark-colored cross-hatching = the temporal fossa. (The opening in the skull is the temporal fenestra, and the temporal fossa is the depression on the skull exposed by the temporal fenestra where the jaw muscles originate.)

and in mammals. Sphenacodonts also had a little flange on the angular bone in the lower jaw that jutted out and backward, creating a space between it and the main body of the jaw, called the reflected lamina (see Figure 18–4b). This structure was probably originally evolved for the insertion of an enlarged and more elaborated pterygoideus muscle, wrapping around the lower edge of the jaw from its main insertion on the inside of the jaw. However, it became more prominent in therapsids and eventually assumed an important role in the evolution of the mammalian middle ear.

Elongation of the neural spines of the trunk into the well-known pelycosaur sail was a remarkable feature of some edaphosaurids and sphenacodontids. These sails must have been evolved independently in the two groups, because they are absent from primitive members of both groups and differ in structure when they are present— sphenacodontid spines were smooth, and edaphosaurid spines had horizontal projections. Marks of blood vessels on the spines and spines that had been broken and then healed suggest that connective tissue and a web of skin covered the spines. There is no evidence of any degree of sexual dimorphism in these sails, so it is unlikely that they were devices evolved primarily for display.

Such a large increase in the surface area of an animal would affect its heat exchange with the environment, and it seems likely that the sails were temperature-regulating devices. In the morning, a *Dimetrodon* could orient its body perpendicular to the sun's rays and allow a large volume of blood to flow through the sail, where the blood would be warmed by the sun and the heat carried back into the animal's body. When a *Dimetrodon* was warm enough, blood flow through the sail could be restricted, and the heat would be retained within the body.

■ Therapsids—More Derived Nonmammalian Synapsids

A flourishing fauna of more derived synapsids— grouped under the general name of therapsids— extended from the Middle Permian to the Early Cretaceous. (Note that the term *therapsid* properly includes not only these beasts but also mammals.) Therapsids all had modifications suggesting an increase in metabolic rate over the more primitive pelycosaurs; thus they are often portrayed as having hair. They were all fairly heavy-bodied, large-headed, stumpy-legged forms. The image of this body form, combined with incipient hairiness, prompted the cartoonist Larry Gonick (1990) to declare them "too ugly to survive."

Therapsid Diversity Therapsids appeared rather suddenly in the fossil record at around 267 million years ago, in the Late Permian, when nine different lineages were apparent. However, this apparent sudden and rapid diversification may have been preceded by a slower radiation, as yet undiscovered, in the Early Permian (Kemp 2006a).

Like the earlier pelycosaurs, therapsids radiated into herbivorous and carnivorous forms. Some of the herbivorous therapsids were large, heavy-bodied animals and may have congregated in herds as ungulates (hoofed mammals) do today. Other herbivorous forms were small and superficially rather like rodents. The carnivorous therapsids included large, ferocious-looking animals that may have played an ecological role similar to that of big cats today, smaller ones that may have been more foxlike, and rabbit-size forms that were probably insectivorous. One of the numerous lineages of therapsids— the cynodonts—was the group that gave rise to mammals. Although cynodonts were highly derived therapsids, other therapsid lineages could be considered equally derived in their own fashions, including the dicynodonts, which were specialized herbivores.

Tom Kemp (2006a) has proposed a paleobiological scenario whereby the pelycosaurs were replaced by therapsids in the later Permian. This hypothesis fits with our knowledge of global climates and vegetational zones in the Permian and with where we find fossils of these synapsids. Although it is difficult to prove these types of hypotheses, in an absolute sense, further fossil record evidence can serve to later support or refute them.

Pelycosaurs are mainly known from the equatorial ever-wet (i.e., high rainfall year round) zone in the Early Permian, with their food base rooted in aquatic ecosystems. Kemp proposed that therapsids originally evolved in a tropical summer-wet zone, and subsequently evolved higher metabolic rates during a period when atmospheric oxygen levels were high. The energy produced by a high metabolic rate could have allowed them to regulate their body temperature and meet their water demands, enabling them to thrive in a drier environment, in which the food base was terrestrial plants.

Sea levels rose at the end of the Early Permian, greatly reducing the area of tropical ever-wet habitat, and forcing the pelycosaurs out of existence. However, at this time the therapsids, having already become adapted to more challenging environments, were now able to move into more temperate zones as tropical habitats disappeared. Late Permian therapsid fossils are found in the cool temperate biome in Gondwana and the winter-wet biome in Pangaea.

The therapsids were the dominant large land mammals in the Late Permian, sharing center stage only with the large herbivorous parareptilian pareiasaurs. The transition from the Permian to the Triassic was a period of great extinction. From the diversity of Permian therapsids, only the tusked dicynodonts (specialized herbivores) and the therocephalians and cynodonts (both derived carnivores) survived and diversified in the Early Triassic. But other vertebrate groups also diversified in the Triassic, most notably the diapsid reptiles that gave rise to the dinosaurs. Therapsids became an increasingly minor component of the terrestrial fauna during the Triassic and were near extinction at its end. Only a few lineages of nonmammalian therapsids survived past the end of the Triassic, when (apart from the mammals) they are known from a few relict forms only.

Derived Features of Therapsids Primitive therapsids had various new features suggesting that they had a higher metabolic rate than the pelycosaurs. A larger temporal fenestra provided space for the origin of large external adductor muscles on the skull roof, the upper canines were longer, and the dentition now showed a distinct differentiation into incisors, canines,

and postcanine teeth. The choanae (internal nostrils) were enlarged, and a trough in the roof of the mouth (possibly covered by a soft tissue secondary palate in life) indicates the evolution of a dedicated airway passage separate from the rest of the oral cavity. The entire skull was more rigid, and the head was capable of increased dorsoventral flexion on the neck (i.e., a nodding action), and the neck as a whole was more flexible, allowing the head to be turned left and right. The pectoral and pelvic girdles were less massive than those of pelycosaurs, the limbs were more slender, and the shoulder joint appears to have allowed more freedom in movement of the forelimb, which would have enabled a longer stride. Therapsid bones were more vascularized than those of pelycosaurs, possibly indicative of more mammal-like growth rates (Kemp 2006a).

More derived therapsids include the dicynodonts (Anomodontia) and the theriodonts (gorgonopsids, therocephalians, and cynodonts). Dicynodonts dominated terrestrial diversity in the Late Permian and diversified into several different ecological types, including long-bodied burrowers and stout, probably semiaquatic forms such as *Lystrosaurus* (**Figure 18–5**). A distinctive feature of dicynodonts was the extreme specialization of the skull for an herbivorous diet. In

▲ **FIGURE 18–5** A reconstruction of the Early Triassic dicynodont therapsid *Lystrosaurus*.

most forms, all of the marginal teeth were lost, except for the upper canines, which were retained as a pair of tusks; the jaws were covered with a horny beak, like that of turtles, which would have provided a replaceable continuous cutting surface for shearing vegetation. The structure of the jaw articulation permitted extensive fore-and-aft movement of the lower jaw, shredding the food between the two cutting plates. Dicynodonts were thought to have become extinct at the end of the Triassic, but a relict form has recently been described from the Early Cretaceous of Australia (Thulborn and Turner 2002).

Theriodonts were the major predators of the Late Permian and Early Triassic. They were characterized by the development of the coronoid process of the dentary, a vertical flange near the back of the lower jaw that provided additional area for the insertion of the jaw-closing muscles, as well as a lever arm for their action. The temporal fossa was correspondingly enlarged for the origin of these muscles.

Cynodonts appeared in the Late Permian and had their heyday in the Triassic; nonmammalian cynodonts were mostly extinct by the end of that period. However, a few species from the tritylodont and tritheledont lineages persisted into the Jurassic period, and a tritylodont has recently been found in the Early Cretaceous of Russia. Soon after their first appearance, cynodonts split into two major lineages: the Cynognathia and the Probainognathia. The Cynognathia included some large carnivorous forms, such as *Cynognathus*, and a variety of herbivorous forms, with expanded postcanine teeth with blunt cusps, including the larger gomphodonts and the smaller tritylodonts. The members of the Probainognathia were in general smaller, less specialized carnivorous and insectivorous forms, such as the tritheledontids; however, this was the lineage that gave rise to the mammals.

The evolution of cynodonts was in general characterized by a reduction of body size. In fact, several Middle Triassic lineages of cynodonts were independently evolving smaller body size and, along with this, attaining various "mammal-like" features related to this miniaturization (Kemp 2005). Some early cynodonts were the size of large dogs; but, by the Middle Triassic, the carnivorous cynodonts were only about the size of a rabbit. The earliest mammals of the latest Triassic were less than 100 millimeters, about the size of a shrew, while the contemporaneous tritylodontid cynodonts were a only little larger, around the size of a mouse.

Cynodonts had a variety of derived features, making them more mammal-like than other therap-

sids: All cynodonts had multicusped cheek teeth (that is, with small accessory cusps anterior and posterior to the main cusp). An enlarged infraorbital foramen—the hole under the eye through which the sensory nerves from the snout pass back to the brain—suggests a highly innervated face, perhaps indicative of a mobile, sensitive muzzle with lips and whiskers. Cynodonts and therocephalians also had evidence of turbinates, scroll-like bones in the nasal passages that warm and humidify the incoming air and help to prevent respiratory water and heat loss.

18.3 Evolutionary Trends in Synapsids

The synapsid lineage crossed a physiological boundary as the animals moved from ectothermy to endothermy, and this change was accompanied by changes in ecology and behavior. Physiology, ecology, and behavior do not fossilize directly, but some of the changes that were occurring in metabolism, ventilation, and locomotion can be traced indirectly through changes in the skull and the postcranial skeleton. However, in tracing the evolution of a feature like endothermy, we must not forget that evolution is not goal-oriented; presumably, each small change was advantageous for the animal that had it at that time, and only in retrospect can we detect a pattern leading to a more modern condition. The nonmammalian synapsids were well-adapted animals in their own right, not merely evolutionary stages on the road to full mammalhood.

Tom Kemp (2006b) presents a comprehensive review of hypotheses concerning the reasons for and the timing of the evolution of endothermy in synapsids. He considers endothermy to have been evolved over a long period of time, first starting with the therapsids and continuing through to mammals, in a series of small steps that he terms "correlated progression." While the most important physiological transition was between pelycosaurs and therapsids, another major one took place at the transition to the cynodont therapsids.

■ Skeletal Modifications and Their Relationship with Metabolic Rate

While metabolic rate does not directly show on the skeleton, features associated with a higher metabolic rate may indeed be apparent. Animals with high metabolic rates require greater quantities of food and oxygen per day, so any changes suggesting improvements in the rate of feeding or respiration may indicate metabolic rate increase. Only animals with high metabolic rates are capable of sustained activity (on land at least—locomotion is cheaper in the water, which is why fishes can be continuously active despite being ectotherms). Thus indicators of greater levels of activity may also reflect higher metabolic rates.

Numbers of the features listed here refer to those in **Figure 18–6**, which illustrates such changes.

1. *Size of the temporal fenestra*—A larger fenestra indicates a greater volume of jaw musculature, and hence implies more food eaten per day. A small opening in pelycosaurs becomes an increasingly larger opening in more derived therapsids. This change, plus an increasing tendency to enclose the braincase with dermal bone, results in a distinct temporal fossa for the origin of a larger volume of jaw musculature (see also Figure 18–4). (The *temporal fenestra* is the opening in the dermal skull roof, while the *temporal fossa* is the depression or area on the skull roof now enclosing the braincase that is exposed by this opening). The external adductor (jaw-closing) muscles pass from their origin on the temporal fossa, through the temporal fenestra, to insert onto the lower jaw. In mammals, and also in a few derived cynodonts, the fenestra is enlarged further with the loss of the postorbital bar so that the orbit is now confluent with the temporal fenestra.

2. *Condition of the lower temporal bar*—A bar of bone bowed out from the skull in the region of the orbit indicates the presence of a **masseter muscle,** originating from this bony bar and inserting on the lower jaw, again suggesting more effective food processing (**Figure 18–7**). The temporal bar originally lay very close to the upper border of the lower jaw, leaving no room for muscle insertion on the outside of the lower jaw. In cynodonts and mammals, the bar is bowed outward, forming the **zygomatic arch** and indicating the presence of a masseter muscle. A corresponding masseteric fossa on the dentary in these animals also indicates the presence of this muscle.

3. *Lower jaw and jaw joint*—Changes reflect the increased compromise between food processing and hearing in synapsids, as explained later. The lower jaw in pelycosaurs resembles the general amniote condition. The tooth-bearing portion, the dentary, takes up only about half of the jaw. By the level of cynodonts, the dentary has greatly expanded, and the postdentary bones have been reduced in size (see also Figure 18–4). In mammals, the dentary now forms a new jaw joint with the skull.

4. *Teeth*—Greater specialization of the dentition reflects an increased emphasis on food processing. The teeth of pelycosaurs are **homodont;** that is, they are virtually all the same size and shape, with no evidence of regionalization of function. In more derived synapsids, the teeth become increasingly **heterodont;** that is, differentiated in size, form, and function (see Figure 18–4), although actual mastication with precise occlusion of the cheek teeth is a mammalian feature. Mammals have reduced the number of tooth replacements to two (**diphyodonty**), have the postcanine teeth further differentiated into premolars (replaced) and molars (not replaced), and have the lower teeth set closer together than the uppers—indicative of chewing with a rotary jaw motion. Mammals also have double-rooted molars and prismatic enamel on their teeth.

5. *Development of a bony secondary palate*—A secondary palate separates the nasal passages from the mouth and allows breathing and eating at the same time. It also reinforces the skull against stresses from increased amounts of food processing. No bony secondary palate is apparent in pelycosaurs. An incipient, incomplete one is present in some noncynodont therapsids, although as previously mentioned this may have been completed by soft tissue, and a complete one is present in derived cynodonts and in mammals (dicynodont therapsids evolved a bony secondary palate convergently). A ridge inside of the nasal passage, suggesting the presence of nasal turbinates, is found in cynodonts and therocephalians. Mammals have merged the originally double nasal opening into a single median one, probably reflecting an increase in size of the nasal passages and a higher rate of ventilation.

6. *Presence of a parietal foramen*—A hole in the skull for the pineal eye reflects control of temperature regulation by behavioral means. It is present in pelycosaurs and in most therapsids but lost in derived cynodonts and mammals.

7. *Position of the limbs*—Limbs placed more underneath the body (upright posture) are reflective of a higher level of activity, resolving the conflict between running and breathing while also increasing agility and capacity for acceleration. Pelycosaurs have the sprawling limb posture

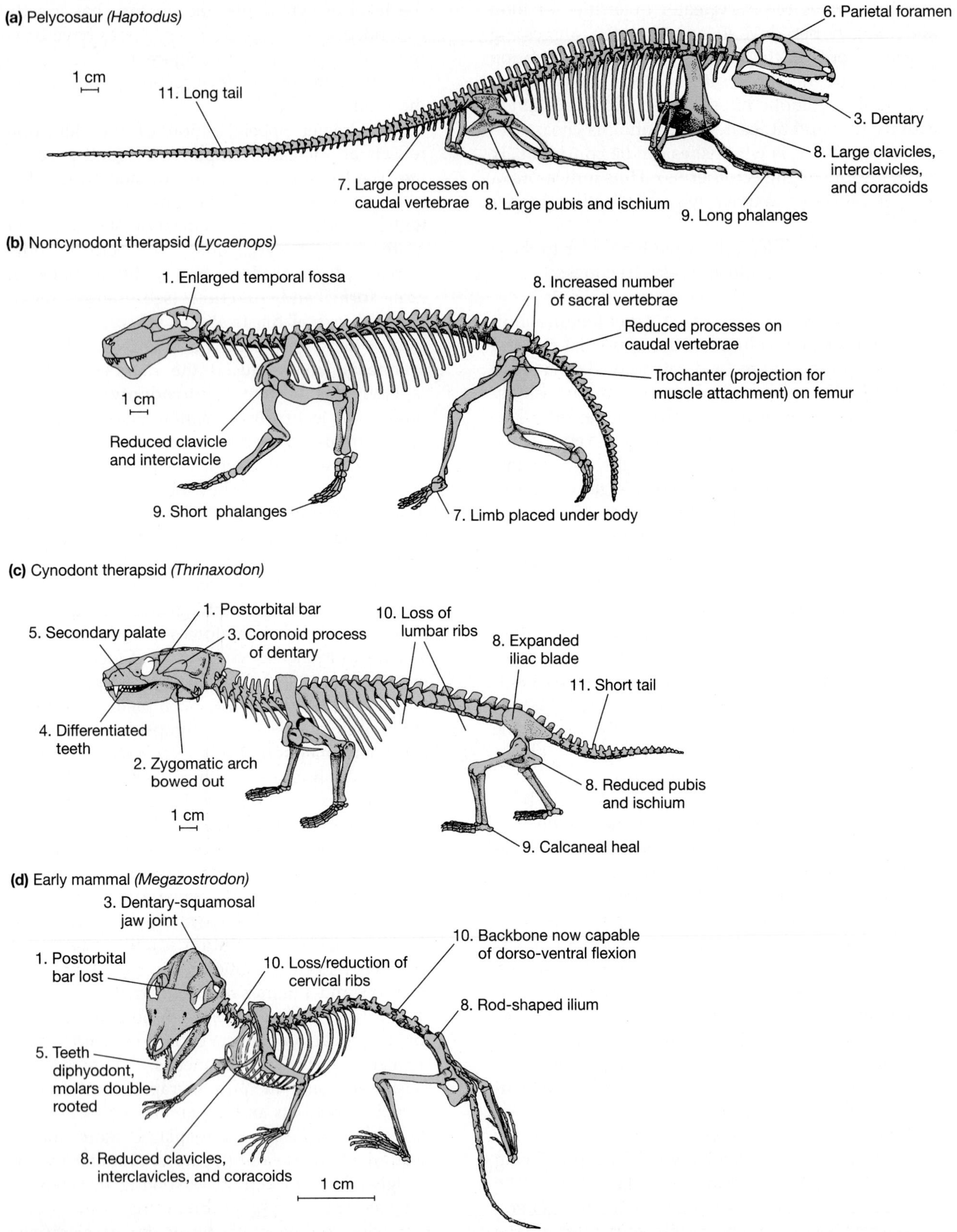

(a) Pelycosaur (*Haptodus*)

6. Parietal foramen

1 cm

11. Long tail

3. Dentary

8. Large clavicles, interclavicles, and coracoids

7. Large processes on caudal vertebrae

8. Large pubis and ischium

9. Long phalanges

(b) Noncynodont therapsid (*Lycaenops*)

1. Enlarged temporal fossa

8. Increased number of sacral vertebrae

Reduced processes on caudal vertebrae

Trochanter (projection for muscle attachment) on femur

1 cm

Reduced clavicle and interclavicle

9. Short phalanges

7. Limb placed under body

(c) Cynodont therapsid (*Thrinaxodon*)

1. Postorbital bar

10. Loss of lumbar ribs

5. Secondary palate

3. Coronoid process of dentary

8. Expanded iliac blade

11. Short tail

4. Differentiated teeth

2. Zygomatic arch bowed out

1 cm

8. Reduced pubis and ischium

9. Calcaneal heal

(d) Early mammal (*Megazostrodon*)

3. Dentary-squamosal jaw joint

10. Backbone now capable of dorso-ventral flexion

1. Postorbital bar lost

10. Loss/reduction of cervical ribs

8. Rod-shaped ilium

5. Teeth diphyodont, molars double-rooted

8. Reduced clavicles, interclavicles, and coracoids

1 cm

▲ **FIGURE 18–6** Skeletal features of synapsids. Numbers correspond to the numbered list in the text.

▶ **FIGURE 18–7** Cynodont skulls and jaw musculature. The squamosal forms the posterior portion of the zygomatic arch.

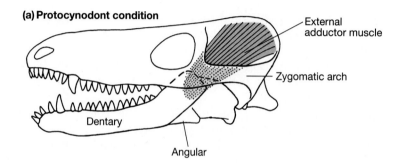

(a) Protocynodont condition

External adductor muscle

Zygomatic arch

Dentary

Angular

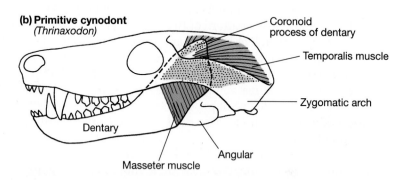

(b) Primitive cynodont
(Thrinaxodon)

Coronoid process of dentary

Temporalis muscle

Zygomatic arch

Dentary

Angular

Masseter muscle

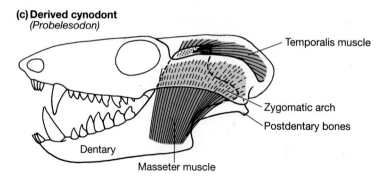

(c) Derived cynodont
(Probelesodon)

Temporalis muscle

Zygomatic arch

Postdentary bones

Dentary

Masseter muscle

typical of primitive amniotes. All therapsids show some degree of development of an upright posture. Changes in the therapsid shoulder girdle would now allow for a greater fore-and-aft movement of the forelimb. Also in therapsids, the expanded iliac blade and the development of the greater trochanter of the femur are considered evidence of a switch to a gluteal type of hindlimb musculature typical of mammals. In pelycosaurs, large processes on the caudal vertebrae indicate the retention of the more primitive amniote method of limb retraction using the caudofemoralis muscle, although this muscle was probably also retained in more primitive therapsids. Evidence from the structure of the knee and ankle joints suggests that therapsids were capable of a dual-gait locomotion, with a slow sprawling gait and a faster more upright gait, as seen in modern crocodiles (Kemp 2005).

8. *Shape of the limb girdles*—The primitive amniote condition reflects a sprawling posture, with large ventral components to the limb girdles. With a more upright posture, more of the weight passes directly through the limbs, and the supportive undercarriage of the limb girdles can be reduced. Therapsids have more lightly built limb girdles than pelycosaurs, with the reduction of the ventral elements and the expansion of the dorsal ones. Therapsids also show an increase in the number of sacral vertebrae from the two or three of pelycosaurs to four or more. Mammals have a very reduced pubis and a rod-shaped ilium, probably reflecting a change in muscle positioning and muscle forces generated with the change

in the vertebral column to allow dorsoventral flexion (see item 10). All mammals except placentals have epipubic bones on the front of the pelvis, a feature also shared with derived small cynodonts such as tritheledontids and tritylodontids (Kemp 2005).

9. *Shape of the feet*—Long toes indicate that the feet are used as holdfasts and are typical of a sprawling gait. Short toes indicate that the feet are used as levers, in a more upright posture. Pelycosaurs have long fingers and toes. All therapsids and mammals have shorter feet; derived cynodonts and mammals have a pattern of two segments on the first digit (thumb or big toe), and three on the other four fingers or toes (you can count this on your own hands). A distinct calcaneal heel is seen in cynodonts and mammals, providing a lever arm for a greater degree of push-off from the gastrocnemius (calf) muscle. Mammals also have an opposable big toe.

10. *Form of the vertebral column*—The loss of the lumbar ribs suggests the presence of a muscular diaphragm, indicating a higher rate of respiration. Extant mammals use the diaphragm in addition to the ribs to inhale air into the lungs. Experiments in which the diaphragm was immobilized showed that it is especially important in obtaining additional oxygen during activity (Ruben et al. 1987).

     Lumbar ribs are present in pelycosaurs and noncynodont therapsids, are absent from mammals, and are reduced or absent in cynodonts. With the complete loss of the lumbar ribs, mammals have evolved distinctive differences between the thoracic and lumbar vertebrae, indicative of the mammalian mode of dorsoventral flexion, although the distinctive bounding gait of modern mammals is probably limited to therians (marsupials and placentals). Most therapsids have restricted the number of neck vertebrae to seven, a characteristic feature of modern mammals. Mammals have also reduced or lost the ribs on the cervical vertebrae.

11. *Tail*—A long, heavy tail is the primitive amniote condition and is retained in pelycosaurs. It indicates that locomotion was conducted primarily by axial movements. A shorter tail, as in most therapsids and mammals, indicates a more upright posture in which limb propulsion is more important than axial flexion. The reduction of the large ventral processes on the caudal vertebrae is another character showing a switch in limb retractor muscles (see item 7).

■ Evolution of Jaws and Ears

In the original synapsid condition a tooth-bearing dentary bone forms the anterior half of the jaw, with a variety of bones (known collectively as postdentary bones) forming the posterior half. (This is the primitive condition that is present in other tetrapods and bony fishes.) The jaw articulated with the skull via the articular bone in the lower jaw and the quadrate bone in the skull. Within cynodonts we can see a progressive enlargement in the size of the dentary and a decrease in the size of the postdentary bones. This trend was probably related to the increase in the volume of jaw adductor musculature, which inserted onto the dentary (see Figure 18–7). In the most derived cynodonts, a condylar process of the dentary grew backward and eventually contacted the squamosal bone of the skull. In mammals and some very derived cynodonts, this contact between the dentary and the squamosal formed a new jaw joint, the dentary-squamosal jaw joint. In these derived cynodonts, and in the earliest mammals, this new jaw joint coexisted with the old one, but in later mammals the dentary-squamosal jaw joint is the sole one. The bones forming the old jaw joint are now part of the middle ear (**Box 18–1**).

The bare facts of this transition have been known for a century or so, but the evolutionary interpretation of these facts has changed over time. Originally it was assumed that a lizardlike middle ear, with the stapes alone forming the auditory ossicle (the bone that transmits vibrations from the tympanum to the oval window), was the primitive condition for all tetrapods and hence the one possessed by the early synapsids. We now have good evidence that an enclosed middle ear evolved separately in modern amphibians and amniotes and probably at least three times convergently within amniotes. However, before anatomists had this information, they assumed that, with the formation of a new jaw joint, an originally single-boned middle ear was transformed into a three-boned one using some leftover jawbones. (This evolutionary scenario also carried the tacit assumption that the mammalian condition of both ear bones and jaw articulation was inherently superior to that of other tetrapods.)

On reconsideration, there are some problems with this story. Even if it were true that the mammalian type of jaw joint *was* somehow superior, how could researchers explain the millions of years of cynodont evolution during which the dentary was enlarging prior to contacting the skull? (Recall that evolution has no foresight.) Why is the original jaw articulation

BOX 18–1 Evolution of the Mammalian Middle Ear

More than a century ago, embryological studies demonstrated that the malleus and incus of the middle ear of mammals are homologous with the articular and quadrate bones that formed the ancestral jaw joint of other gnathostomes. More recently, the transition has been traced in fossils, from its beginning in basal synapsids through therapsids to early mammals. Nonmammalian synapsids have a jaw joint formed from the quadrate (in the skull) and the articular (in the lower jaw), as in other jawed vertebrates. The first indication of a mammalian type of middle ear is seen in the sphenacodontid pelycosaurs. These animals have a structure called the "reflected lamina" on the angular bone of the lower jaw, which is believed, at least in later synapsids, to have housed a tympanum (eardrum). In primitive tetrapods and many early amniotes, the stapes (the old hyomandibula) played a role in bracing the dermal skull roof to the braincase, in addition to retaining its original contact with the quadrate bone as in the primitive gnathostome condition. In therapsids this bracing role is lost, and the stapes is reduced in size: the quadrate-stapes contact in therapsids resembles the incus-stapes articulation of mammals, suggesting that they were using their jaw joint to hear with as well as to form the jaw hinge.

The transition from the nonmammalian to the mammalian condition can be visualized by comparing the posterior half of the skull of *Thrinaxodon*, a fairly primitive cynodont, with that of *Didelphis*, the Virginia opossum (**Figure 18–8**). The reflected lamina was probably the principal support of the tympanic membrane, and vibrations were transmitted to the stapes via the articular and quadrate. The mammalian jaw joint is a new structure, formed by the dentary of the lower jaw and squamosal bone of the skull. The tympanum and middle-ear ossicles lie behind the jaw joint and are much reduced in size, but they retain the same relation to each other as they had in cynodonts. The lower jaw of a fetal mammal viewed from the medial side shows that the angular (tympanic), articular (malleus), and quadrate (incus) develop in the same positions they have in the cynodont skull (Figure 18–8c). The homologue of the angular, the tympanic bone, supports the tympanum of mammals. The retroarticular process of the articular bone is the manubrium of the malleus, and the ancestral jaw joint persists as the articulation between the malleus and incus.

▶ **FIGURE 18–8** Anatomy of the back of the skull and the middle ear bones of synapsids. (a) *Thrinaxodon*, a cynodont. (b) *Didelphis*, the Virginia opossum. (c) Embryonic mammal.

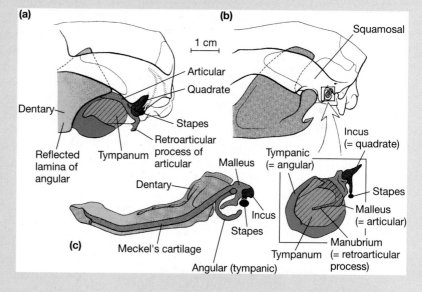

an inherently weak one? It appears to work well enough in other vertebrates; no one has ever accused *Tyrannosaurus rex* of having a weak jaw! And why disrupt a perfectly functional middle ear to insert some extra bones that happened to be available? Even if a three-boned middle ear were ultimately superior, there would be a period of adjustment to a new condition that would probably be less effective than the previous condition.

These questions were addressed by Edgar Allin (1975). He proposed that the chain of bones that make up the middle ear of mammals had been used for hearing all along in synapsids (at least since the level of derived pelycosaurs). Using the same set of bones for two different functions, as a jaw joint and as a hearing device, might seem like a rather clumsy, makeshift arrangement. However, this mode of hearing would probably be adequate for the early synapsids, which, as ectotherms with low food intakes, would not have been using their jaw joints in complex chewing movements. A problem would arise in more derived synapsids because their higher metabolic rates would require more food, more jaw musculature, more use of the jaw in feeding, and thus greater stress on the jaw joint that would be incompatible with the role of these bones in hearing. Thus, Allin interpreted the evolutionary history of the jaw in cynodonts as representing a conflict between feeding and hearing.

An initial evolutionary step would be to enlarge the dentary bone and transfer all the jaw muscle insertions to this bone. This change would isolate the auditory postdentary bones somewhat from the feeding apparatus, allowing them to become smaller as befits auditory ossicles. Note that the reason earlier researchers had interpreted the jaw of synapsids as weak is that the postdentary bones remained loose and wobbly; further, they did not fully fuse with one another and with the dentary as in many other tetrapods. However, given Allin's interpretation, there was a very good reason for these bones *not* to fuse: if they had done so, they would have compromised their role as vibrating auditory ossicles. **Figure 18–9** illustrates the possible appearance of a cynodont with this type of ear, with an eardrum held in the lower jaw.

The dentary increased in size in progressively more derived cynodonts. Finally, the condylar process of the dentary was large enough to contact the squamosal in the skull and to form a new jaw joint, freeing the original jaw joint from its old function and allowing it to be devoted entirely to hearing. A classic example of an evolutionary intermediate is provided by the tritheledontid *Diarthrognathus*. This animal gets its name from its double jaw joint (Greek: *di* = two, *arthro* = joint, and *gnath* = jaw). It has both the ancestral articular-quadrate joint and, next to it, the mammalian dentary-squamosal articulation. The postdentary bones are still retained in a groove in the lower jaw in the earliest mammals.

Other researchers have extended and refined Allin's hypothesis. A. W. Crompton emphasized the role of the masseter muscle that first appeared in cynodonts. In mammals, the masseter can move the lower jaw laterally and can be thought of as holding the lower jaw in a supportive sling (Crompton 1995). Crompton interpreted the evolution of the masseter muscle as helping to relieve stresses at the jaw joint by this slinglike action and thereby helping to resolve the conflict between jaw use and hearing.

Dennis Bramble (1978) interpreted a preadaptive role for the condylar process of the dentary that grows back over the postdentary bones and eventually forms the new jaw joint. This process increases in size in cynodonts, but its initial function prior to the formation of a new jaw joint had never been deciphered. Bramble's biomechanical analysis of jaw function interpreted the initial function of this structure as preventing dislocation between the dentary and the postdentary bones during biting.

Then, in 1996, Tim Rowe proposed that the final separation of the middle ear bones from the lower jaw in true mammals is correlated with an increase in the size of the mammalian forebrain. He proposed that the expansion of the skull during embryonic development to accommodate the expansion of the brain would dislocate these bones from their original position and thus provide the initial condition for their subsequent enclosure in a discrete middle-ear cavity distinct from the lower jaw.

Note that while the earliest mammals had evolved the dentary-squamosal jaw joint, they had not yet lost the connection of the postdentary bones with the lower jaw. This condition can be seen in fossils because even when these little, loose bones fall out and are not preserved, the distinctive trough in which they lay remains on the internal side of the lower jaw. In fact, the jaw and ear region of an adult Jurassic mammal would resemble the condition seen in fetal mammals today, as shown in Figure 18–8c. A newly discovered jaw of an Early Cretaceous fossil monotreme shows the retention of this trough, leading to the conclusion (long suspected on other grounds) that the final enclosure of the auditory ossicles into a middle ear occurred convergently in monotremes and therian mammals (Rich et al. 2005). Examination of a diversity of Mesozoic mammals belonging to extinct lineages shows that at least one other lineage also lost this trough, implying that an enclosed middle ear evolved convergently at least three times (Martin and Luo 2005).

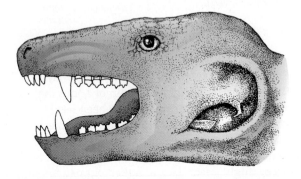

▲ **FIGURE 18–9** The primitive cynodont *Thrinaxodon*. This reconstruction shows the probable appearance of the ear region.

18.4 The First Mammals

Extant mammals are characterized by two salient features: hair and mammary glands. Neither of these features is directly preserved in the fossil record, although we may be able to infer the point in synapsid evolution at which these features were acquired. Another feature typical of most living mammals is viviparity—giving birth to young rather than laying eggs. However, we know that this was not a feature of the earliest mammals because some living primitive mammals, the monotremes (the platypus and echidna of Australia and New Guinea), still lay eggs. How, then, can we define where in synapsid evolutionary history we should start to apply the term *mammal*?

Traditionally, the acquisition of a dentary-squamosal jaw joint was used to define the first fossil mammal; but, as we have seen, a few derived cynodonts have a dentary-squamosal joint while retaining the old quadrate-articular one, as was also the condition in the earliest mammals. In practice, it has always been pretty easy to determine an early mammal. They were tiny, a couple of orders of magnitude smaller than cynodonts (a body mass of less than 100 grams—shrew size). In contrast, the smallest cynodonts would have weighed about a kilogram—the size of a rat—except for some smaller tritylodontids, which were around the size of a mouse but lacked the dentary-squamosal jaw articulation. But did this transition to a very small body size really signify a biological shift into a different type of adaptive zone that we can clearly signify as mammalian?

The point of mammalian transition is traditionally marked not only by the jaw joint but also by some derived features of the skull, reflecting enlargement of the brain and inner ear regions, as well as by postcanine teeth with divided roots. Some workers (e.g., Rowe 1988) would prefer to limit the group *Mammalia* to those animals bracketed by the interrelationships of surviving mammals (i.e., to the "Crown Group Mammalia" in Figure 18–1). However, we will argue later that the traditional division between mammals and cynodonts represents biological innovations, including the evolution of lactation and suckling.

We have already argued that at least some cynodonts had higher metabolic rates than the other therapsids; and, if they were indeed endothermal, it is possible they had at least some hair for insulation. Some bony evidence allows inferences about when an insulating coat of fur first evolved. Most tetrapods have a gland associated with the eye, called the **Harderian gland**. In modern mammals, this gland secretes an oily substance that travels down the nasolacrimal duct to the nose, where it can be wiped off on the paws. This secretion is then used for preening the fur, providing a lipid barrier that repels water, probably an essential feature for the insulating fur of a small mammal.

Willem Hillenius (2000) has shown that mammals have a characteristic anatomy of the nasolacrimal duct, possibly associated with this new function of the Harderian gland. This anatomy is present in *Morganucodon*, one of the earliest-known mammals, but not in any of the cynodonts. This suggests that a hairy coat was a feature of the first mammals but was absent in cynodonts. Cynodonts may have had mammal-like metabolic rates and may have had some type of hair (perhaps sensory whiskers). Perhaps the need to evolve fur as insulation was not really pressing until the very small size of the earliest mammals, which would have been more subject to heat loss due to the relatively greater body surface area.

■ Features of the Earliest Mammals

The oldest well-known mammals from the earliest Jurassic include *Morganucodon* (also known as *Eozostrodon*) from Wales (**Figure 18–10**), *Megazostrodon* from South Africa (see Figure 18–6d), and *Sinoconodon* from China. A somewhat older possible mammal is *Adelobasileus* (about 225 million years old), known only from an isolated braincase from the Late Triassic of Texas. Most of our information about mammals comes from their teeth. At their small size, the fragile mammal bones are not easily preserved, but the harder, enamel-containing teeth are more likely to fossilize. Fortunately, mammalian teeth turn out to be very informative about their owners' lifestyle.

▲ **FIGURE 18–10** Reconstruction of the Early Jurassic mammal *Morganucodon*.

Most vertebrates have multiply-replacing sets of teeth and are hence termed **polyphyodont**. In contrast, the general mammalian condition is to be **diphyodont**; that is, they have only two sets of teeth (like our milk teeth and our permanent teeth), and the molars are not replaced at all but instead erupt fairly late in life. This seems to have been the condition in the earliest known mammals, with the exception of the very primitive *Sinoconodon*. Mammals also have molars with precise occlusion that is produced by an interlocking arrangement of the upper and lower teeth. We can determine this in fossils because the teeth show distinct wear facets, produced from abrasion between the teeth and the food, showing that the teeth always met in the same manner. Precise occlusion makes it possible for the cusps on the teeth to cut up food very thoroughly, creating a large surface area for digestive enzymes to act on and thereby promoting rapid digestion. Only mammals **masticate** (thoroughly chew) their food in this fashion. The cheek teeth of mammals also are set in an alternating fashion, and the lower teeth are closer together than the uppers. In addition, the jaws are moved in a rotary fashion—in contrast with the simple up-and-down movement of the jaws in most other tetrapods—which means that mammals can chew only on one side of the jaw at any one time (**Figure 18–11**).

Fossils of early mammals have areas of wear on their teeth, indicating precise occlusion, and they also have narrow-set lower jaws that would require sideways movement of the lower jaw to occlude the teeth. Thus we can infer that they used the basic mammalian pattern of jaw movement. Mammalian teeth must be durable because mammals have only two sets of teeth per lifetime, and teeth with prismatic, wear-resistant enamel are a distinctive mammalian feature (also shared with the derived tritheledontid cynodonts), as are double-rooted molars.

We can also deduce some features of the soft tissue anatomy of the earliest mammals by comparing the monotremes with more derived living mammals, the **therians** (marsupials and placentals). Monotremes have a more primitive ear than therians. The cochlea in their inner ear is not highly coiled (as also seen in the preserved braincases of the earliest mammals), and they lack a pinna (external ear). Thus it is likely that the earliest mammals were also pinna-less, even though this makes them look rather nonmammalian in reconstruction (see Figure 18–10). Although monotremes produce milk, their mammary glands lack nipples, so these were also probably lacking in the earliest mammals. Monotremes, although endothermal, have a metabolic rate lower than that of therians and are not good at evaporative cooling; early mammals were probably similar.

However, monotremes are not good examples for deductions about the behavior and ecology of the earliest mammals. Monotremes are relatively large mammals, and they are specialized in their habits (Chapter 20). The earliest mammals were small and, judging by their teeth, insectivorous. Like small, insectivorous marsupials and placentals today, they were probably nocturnal and solitary in their behavior, with the mother-infant bond as the only strong social bond. We do know from preserved braincases that early mammals had large olfactory lobes, indicating the importance of the sense of smell, and proportionally larger cerebral hemispheres than even the most derived cynodonts. Early mammals may have been capable of much more sophisticated sensory processing than were cynodonts.

■ The Evolution of Lactation and Suckling

Researchers have two main questions about the evolution of the typically mammalian features of lactation and suckling. First, when did providing the young with milk originate, and did it coincide with the point that we consider as the cynodont-mammal transition? Second, how did milk and the mammary gland itself evolve?

Lactation   Caroline Pond (1977) has argued that precise occlusion and diphyodonty indicate the presence of lactation in early mammals. She points out that precise occlusion would not work with the polyphyodont condition in cynodonts because only a fully-erupted tooth occludes properly with its counterpart in the other half of the jaw. This would be especially true for the tall, pointy teeth of early mammals (see Figure 18–11), for which a small degree of malocclusion could break the tooth. Precise monitoring of occlusion is less important in mammals with flatter teeth, such as humans. Some herbivorous cynodonts did independently evolve a form of occlusion, but this was with flat teeth that did not require precise alignment for their function.

Thus diphyodonty must have preceded precise occlusion in evolution. However, an animal could reduce its number of sets of teeth only if it was fed milk during its early life. If the newborn were fed only on liquid food, the jaw could grow while it had no need of teeth, and permanent teeth could erupt in a near adult-size jaw. Thus, if an animal has precise occlusion and diphyodonty, it must first have evolved lactation.

Pond's hypothesis allows us to infer that the earliest mammals lactated, but had they inherited this

▶ **FIGURE 18–11** Occlusion and molar form in cynodonts and early mammals. (a) Cross-sectional view through the muzzle of a cynodont. Lower teeth and upper teeth are the same distance apart (isognathy), and jaw movement is a simple vertical up-and-down motion (arrows indicate direction). (b) Cross-sectional view through the muzzle of a mammal. Lower teeth are closer together than upper ones because of a narrower lower jaw (anisognathy); jaw movement is rotary (arrows indicate direction), with chewing on only one side of the jaw at any one time. (c) Side view of the jaws of a cynodont (*Thrinaxodon*). The postcanine teeth all look similar (and are replaced continually) and are arranged so each tooth in the upper jaw lines up with one in the lower jaw. (d). Side view of the jaws of a mammal (*Morganucodon*). Postcanine teeth are now divided into simple premolars (replaced once) and more complex molars (not replaced); upper and lower molars are offset so that they interdigitate on occlusion (each tooth in the upper jaw meshes with two in the lower jaw and vice versa). (e) Schematic upper molar of *Morganucodon* in occlusal ("tooth's eye") and lateral views, representing the original mammalian pattern (similar to that of cynodonts). The three main cusps are set in a straight line. (f) Schematic upper molar of *Kuehneotherium* in occlusal and lateral view, representing the more derived holotherian pattern. Arrangement of the cusps is triangular, with the major cusp set more medially (in the uppers, more laterally in the lowers), to effect the "reversed triangles" pattern of occlusion.

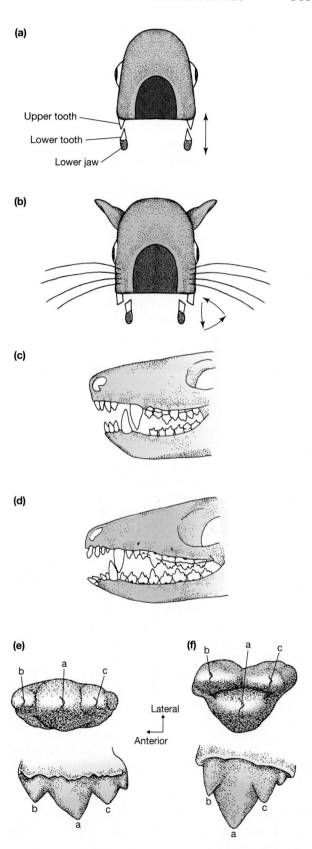

feature from the cynodonts? In the next section we discuss the evolution of suckling and conclude that cynodont babies would not have been able to suckle. This does not mean that cynodonts did not have parental care. Obviously, parental care would have to be in place before lactation could evolve. Parental care would also be essential for an endotherm in order to ensure the proper thermal environment for juveniles and the developing young.

Thus we have some fairly strong evidence to suggest that lactation appeared with the origin of mammals, but how did the whole system evolve in the first place? Mammary glands appear to be rather closely linked to the type of glands associated with hair (sebaceous glands). Thus they probably had the primitive ability to secrete small quantities of organic materials, as do sebaceous glands in mammals today. One suggestion for the evolution of lactation is that these glands originally secreted pheromones signaling the offspring to recognize their mother and to stay with her.

When we think of the functions of milk we usually think of its nutritional value, but milk also has an important immune system role in all mammals, passing on innate immunity to the newly-birthed (or newly-hatched) young. Milk contains proteins that are related to the lysozyme enzymes that attack bacteria—even human milk has antimicrobial properties—and it has been proposed that the original use of milk might have been for protection of the eggs in a nest against microorganisms (Blackburn et al. 1989). New molecular evidence suggests that mammary glands and milk production arose as part of the immune system and that milk was especially involved in controlling inflammatory response. Once a secretion of this type had evolved, any evolutionary change to a more copious, more nutritive secretion accidentally ingested by the young could only have been of benefit. In fact, it has been shown that gene sharing and gene duplication events led to antimicrobial enzymes in milk adopting the role of secreting fat droplets, producing whey protein and sugar, and accumulating water (Vorbach et al. 2006). This protomilk would initially have supplemented the egg yolk and then later supplanted it.

What is the evolutionary advantage of lactation for mammals? Lactation allows the production of offspring to be separated from seasonal food supply. Unlike birds, which must lay eggs only when there is the appropriate food supply for the fledglings (spring and summer), mammals can store food as fat and convert it into milk at a later date. Provision of food in this manner by the mother alone also means

that she does not have to be dependent on paternal care to rear her young. Finally, lactation makes viviparity less strenuous on the mother because the young can be born at a relatively undeveloped stage and cared for outside of the uterus.

Suckling The ability to suckle is a unique mammalian feature. Mammals can form fleshy seals against the bony hard palate with the tongue and with the epiglottis, effectively isolating the functions of breathing and swallowing (**Figure 18–12**). Mammals use these seals to suckle on the nipple while breathing through the nose. Adult humans have lost the more posterior seal that allows this action (because the larynx shifts more ventrally in early childhood). This makes us more liable to choke on our food and also enables us to breathe voluntarily through the mouth as well through the nose. We retain the more anterior seal into adulthood: this is what stops us from swallowing water when we gargle.

This mammalian pharyngeal anatomy is also important for our mode of swallowing a discrete, chewed bolus of food rather than shoving large items wholesale down the gullet, like a snake swallowing a mouse. Changes in the bony anatomy of the palate and surrounding areas indicate that these functions came into use only with the most derived cynodonts, suggesting that this type of swallowing and the capacity for suckling are fundamentally mammalian attributes.

Facial muscles are another characteristic feature of mammals that is absent from other vertebrates (**Figure 18–13**). These muscles make possible our varieties of facial expressions, but they were probably first evolved in the context of mobile lips and cheeks that enable the young to suckle. The facial muscles are thought to be homologous with the neck constrictor muscles (constrictor colli, also called the sphincter colli) of other amniotes because both types of muscles are innervated by the facial nerve (cranial nerve VII). This muscle aids in the transport of food down the esophagus and, with the mammalian mode of swallowing a discrete bolus of food, this muscle could now be co-opted for a different function. There is some evidence that the acquisition of facial muscles may have occurred somewhat differently in various mammal lineages because monotremes extend a different portion of the constrictor colli onto the face during development than do therians. Monotremes also lack mobile lips, although this could be a secondary feature related to the development of a beak.

There is a great deal of difference in the elaboration of the facial muscles in different mammals. Most mammals do not have highly expressive faces. However, it is not just primates that are capable of facial

▶ **FIGURE 18–12** Longitudinal-section views of the oral and pharyngeal regions. Seal 1 prevents substances from entering the pharynx by appressing the back of the tongue against the soft palate. Seal 2 (lacking in post-infant humans) appresses the epiglottis against the back of the soft palate: this allows air to enter the trachea but blocks the entrance of material from the oropharynx. However, liquids can pass from the oropharynx, around the trachea, and into the esophagus while this seal is in place.

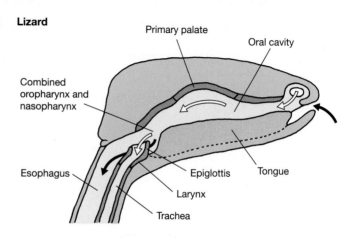

Lizard

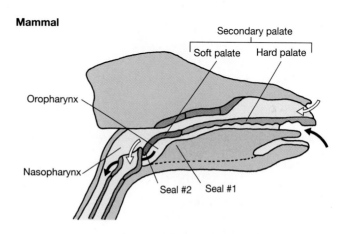

Mammal

expressions. Horses use their lips in feeding and are capable of quite a wide variety of expressions, whereas cows use their tongues and are poker-faced. This is why Mr. Ed (the television talking horse) seems plausible to us, whereas a cow could never play that role. All mammals with well-developed facial muscles display similar expressions for similar emotions; for example, an angry human snarls in a similar way to an angry dog or even an angry horse. This similarity of expression is surprising because different mammals must have elaborated their facial muscles independently from ancestors with little capacity for facial expression, as seen in small mammals today.

■ The Radiation of Mesozoic Mammals

The Cenozoic is often called the Age of Mammals, but this description overlooks a tremendous amount of mammalian evolution that occurred during the Mesozoic. While it is true that mammals were small and basically shrewlike until the end of the Cretaceous, the radiation of Mesozoic mammals nonetheless represents two-thirds of mammalian history. Mesozoic mammals were diverse taxonom-ically but were fairly homogenous in body form. They were all fairly small: many were shrew-size, and only one has been found that was much bigger than a present-day opossum (weighing 2 to 3 kilograms). Their teeth reveal mainly insectivorous or carnivorous diets with some omnivory, but these mammals were too small to be herbivores with a highly fibrous diet.

Although it is unlikely there was ever direct competition between mammals and dinosaurs (scenarios of mammals causing dinosaur extinction by eating their eggs are works of fiction), it is true that mammals did not diversify into larger-bodied forms with more varied diets until the dinosaurs' extinction. The presence of dinosaurs must in some way have been preventing the radiation of mammals into a broader variety of adaptive niches.

There were two major periods of mammalian diversification during the Mesozoic. The first, spanning the Jurassic to Early Cretaceous, produced an early radiation of forms that in the main did not survive past the Mesozoic: morganu-codonts, docodonts, triconodonts, symmetrodonts, dryolestids, and the like. Although these early

505

Lizard (reptile)—no muscles of facial expression

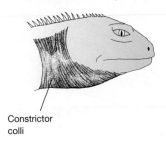

Constrictor
colli

Rodent (mammal)—moderate development of muscles
of facial expression

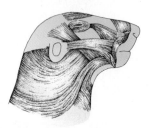

Primate (mammal)—extensive development of muscles
of facial expression

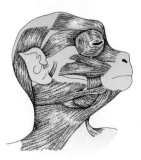

▲ **FIGURE 18–13** Muscles of facial expression.

mammals are fascinating to students of early mammalian history, we will not discuss them in more detail here. (For an extensive review, see Kielan-Jaworoswka et al. 2004 and also Kemp 2005.) We used to think that all Mesozoic mammals were small and fairly generalized in body form. How-ever, recent discoveries, primarily in China, have yielded new, exceptionally well-preserved early mammals, as dramatic as remarkable new Chinese birds and dinosaurs, although none is larger than the size of a medium-sized dog (i.e., around 12 kilograms to 14 kilograms). **Figure 18–14** shows a number of other recently discovered forms, whose skeletons are indicative of swimming, digging, and gliding habits, and also a relatively large (about the size of a pitbull dog) predatory form, *Repenomamus*, famous for being preserved with a juvenile dinosaur in its stomach (Hu et al. 2005).

The second radiation, which got under way in the Cretaceous, was composed of more derived mammals, including the first therians and the rodentlike multituberculates (now extinct but which survived into the Cenozoic), and is discussed more extensively in Chapter 20. Changes in other aspects of the terrestrial ecosystem are also apparent at this time. The Early Cretaceous marks the time of the initial radiation of the angiosperms, the flowering plants. Among other tetrapods, the Cretaceous dinosaur faunas were distinctively different from earlier ones, and snakes made their first appearance, possibly in association with a diversity of small mammals on which they fed.

The diversity of Mesozoic mammals appears to represent two major groups: one in Laurasia (the boreosphenidans—Greek *boreo* = northern and *sphen* = a wedge) that contains the origin of therians and the extinct multituberculates, and one in Gondwana (the australosphenidans—Latin *austral* = southern) that contains the origin of monotremes (Luo et al. 2002). The earliest eutherian (the group that includes placentals and also their extinct stem-group relatives) has recently been described from the Early Cretaceous of China, estimated as around 125 million years old (Ji et al. 2002). This specimen is so well preserved that it even shows impressions of fur and appears to have been a generalized but somewhat climbing-adapted form about the size of a hamster. Rather appropriately, it has been named *Eomaia* (Greek *eo* = dawn and *maia* = mother). The same deposits have also yielded an animal called *Sinodelphis*, which may represent the first metatherian (marsupials and their extinct stem-group relatives) (Luo et al. 2003).

Summary

The synapsid lineage is characterized by a single lower temporal fenestra on each side of the skull. The first synapsids were the pelycosaurs of the Late Carboniferous and Early Permian, including the familiar sailbacks. Many pelycosaurs were large animals for their time, up to the size of a large pig. The

▲ **FIGURE 18–14** Newly discovered Mesozoic mammals. (a) *Repenomamus*, a large (for a Mesozoic mammal!) predator from the Early Cretaceous of China (about the size of a corgi dog). (b) *Volaticotherium*, a gliding form from the Late Jurassic or earliest Cretaceous of China (about the size of a flying squirrel). (c) *Castorocauda*, a swimming form from the Middle Jurassic of China (about the size of a rat). (d) *Fruitafossa*, a digging form from the Late Jurassic of North America (about the size of a chipmunk).

most derived pelycosaurs, the sphenacodontids, had several mammal-like features, including the beginnings of a three-boned middle ear.

The earliest therapsids were derived in having an upright posture with the feet and limbs acting as levers (rather than as holdfasts) during locomotion and a greater volume of jaw musculature. The most derived therapsids, the Triassic cynodonts, had many features suggesting that they were endothermal, including turbinate bones in the nose, a secondary palate, and a reduction of the lumbar rib cage, indicating the presence of a diaphragm. Many features of the cynodont skull that changed over its history can be understood in the context of an evolutionary conflict between chewing and hearing, because some of the bones of the mammalian middle ear formed the original synapsid jaw joint.

The first true mammals are known from the earliest Jurassic, with a possible candidate from the Late Triassic. Most of these early mammals had teeth that precisely interlocked to chew food and were replaced only once. Evolution of mammals from cynodonts was accompanied by a substantial reduction in body size—early mammals were the size of shrews, about 100 millimeters long, and probably weighed less than 50 grams. We can deduce quite a lot about the probable biology of these early mammals, not just from their skeletal remains but also from considering the biology of the living monotremes and primitive therians. From patterns of tooth replacement and the anatomy of the nasolacrimal duct, we can infer that both lactation and a fur coat evolved with the earliest mammals. Milk itself may originally have been evolved for its antimicrobial and immunity properties, with its nutritional value being added later.

Mesozoic mammals were mainly small insectivorous or omnivorous forms, although we know of greater diversity now than we did a few years ago. The modern groups of mammals—monotremes and therians—can trace their origin back to the Early Cretaceous, a time of evolutionary turnover not only in mammals but also in other tetrapods as well as in plants. Monotremes and therians may respectively represent the survivors of an initial separate southern and northern radiation of mammalian lineages. Mammals did not diversify into larger, more specialized forms until the Cenozoic, after the extinction of the dinosaurs.

Additional Readings

Allin, E. F. 1975. Evolution of the mammalian middle ear. *Journal of Morphology* 147:403–437.

Blackburn, D. G. 1991. Evolutionary origins of the mammary gland. *Mammal Review* 21:81–96.

Blackburn, D. G. et al. 1989. The origins of lactation and the evolution of milk: A review with new hypotheses. *Mammal Review* 19:1–26.

Bramble, D. M. 1978. Origin of the mammalian feeding complex: Models and mechanisms. *Paleobiology* 4:271–301.

Crompton, A. W. 1995. Masticatory function in nonmammalian cynodonts and early mammals. In J. J. Thomason (ed.), *Functional Morphology in Vertebrate Paleontology*. Cambridge, UK: Cambridge University Press, 55–75.

Gonick, L. 1990. *The Cartoon History of the Universe*, volumes 1–7. New York, NY: Doubleday.

Hillenius, W. J. 2000. The septomaxilla of nonmammalian synapsids: Soft tissue correlates and a new functional interpretation. *Journal of Morphology* 245:207–229.

Hotton, N. III, et al. (eds.). 1986. *The Ecology and Biology of Mammallike Reptiles*, Washington, DC: Smithsonian Institution Press.

Hu, Y. et al. 2005. Large Mesozoic mammals fed on young dinosaurs. *Nature* 433:149–152.

Ji, Q. et al. 2002. The earliest-known eutherian mammal. *Nature* 416:816–822.

Kemp, T. S. 2005. *The Origin and Evolution of Mammals*. Oxford, UK: Oxford University Press.

Kemp, T. S. 2006a. The origin and early radiation of the therapsid mammallike reptiles: A palaeobiological hypothesis. *Journal of Evolutionary Biology* 19:1231–1247.

Kemp, T. S. 2006b. The origin of mammalian endothermy: A paradigm for the evolution of complex biological structure. *Zoological Journal of the Linnean Society* 147:473–488.

Kielan-Jaworoswka, Z. et al. 2004. *Mammals from the Age of Dinosaurs*. New York: Columbia University Press.

King, G. 1990. *The Dicynodonts: A Study in Paleobiology*. London: Chapman & Hall.

Luo, Z.-X., et al. 2002. Dual origin of tribosphenic molars. *Nature* 409:53–57.

Luo, Z.-X., et al. 2003. An Early Cretaceous tribosphenic mammal and metatherian evolution. *Science* 302:1934–40.

Martin, T., and Z.-X. Luo. 2005. Homoplasy in the mammalian ear. *Science* 307:861–862.

Pond, C. M. 1977. The significance of lactation in the evolution of mammals. *Evolution* 31:187–199.

Rich, T. H. et al. 2005. Independent origins of middle ear bones in monotremes and therians. *Science* 307:910–914.

Rowe, T. 1988. Definition, diagnosis, and origin of Mammalia. *Journal of Vertebrate Paleontology* 8:241–264.

Rowe, T. 1996. Coevolution of the mammalian middle ear and neocortex. *Science* 273:651–654.

Ruben, J. A. 1995. The evolution of endothermy in mammals and birds: From physiology to fossils. *Annual Review of Physiology* 57:69–95.

Ruben, J. A. et al. 1987. Selective factors in the origin of the mammalian diaphragm. *Paleobiology* 13:54–59.

Ruben, J. A., and T. D. Jones. 2000. Selective factors associated with the origin of fur and feathers. *American Zoologist* 40:585–596.

Thulborn, T., and S. Turner. 2002. The last dicynodont: An Australian Cretaceous relict. *Proceedings of the Royal Society, London, B* 270:985–993.

Vorbach, C. et al. 2006. Evolution of the mammary gland from the innate immune system? *BioEssays* 28:606–616.

19

Geography and Ecology
of the Cenozoic Era

The role of Earth's history in shaping the evolution of vertebrates is difficult to overestimate. The positions of continents have affected climates and the ability of vertebrates to migrate from region to region. The geographical continuity of Pangaea in the late Paleozoic and early Mesozoic eras allowed tetrapods to migrate freely across continents, and the faunas were fairly similar in composition across the globe. By the late Mesozoic, however, Pangaea no longer existed as a single entity. Epicontinental seas extended across the centers of North America and Eurasia, and the southern continents were separating from the northern continents and from one another. This isolation of different continental blocks resulted in the isolation of their tetrapod faunas and limited the possibility for migration. As a result, tetrapod faunas became progressively more different on different continents. Distinct regional differences became apparent in the dinosaur faunas of the late Mesozoic, and they have been a prominent feature of Cenozoic mammalian faunas.

Continental drift in the late Mesozoic and early Cenozoic moved the northern continents that had formed the old Laurasia (North America and Eurasia) from their early Mesozoic near-equatorial position into higher latitudes. The greater latitudinal distribution of continents, plus changes in the patterns of ocean currents that were the result of continental movements, caused cooling trends in high northern and southern latitudes during the late Cenozoic era. With the formation of the Arctic ice cap some 5 million years ago, this cooling led to a series of ice ages that began in the Pleistocene and continued to the present. We are currently living in a relatively ice-free interglacial period. Both the fragmentation of the landmasses and the changes in climate during the Cenozoic have been important factors in the evolution of mammals.

19.1 Cenozoic Continental Geography

The breakup of Pangaea in the Jurassic period began with the movement of North America that opened the ancestral Atlantic Ocean. Rifts formed in Gondwana, and India then moved northward on its separate oceanic plate (**Figure 19–1**), eventually to collide with Eurasia. The collision of the Indian and Eurasian plates in the mid-Cenozoic produced the Himalayas, the highest mountain range in today's world.

South America, Antarctica, and Australia separated from Africa during the Cretaceous but maintained connections into the early Cenozoic. In the middle to late Eocene, Australia separated from Antarctica and, like India, drifted northward. (Note that New Guinea, today an island north of Australia that is separated from the mainland continent by a shallow sea, is actually a part of the Australian continental block. It was in direct land contact with Australia for much of Earth's history—this is why the faunas and of the two regions are so similar.) Intermittent land connections between South America and

▲ **FIGURE 19–1** Continental positions in the late Paleocene and early Eocene epochs. An epicontinental sea, the Turgai Straits (crosshatching), extended across Eurasia. Dashed arrows show the direction of continental drift, and solid arrows indicate major land bridges mentioned in the text.

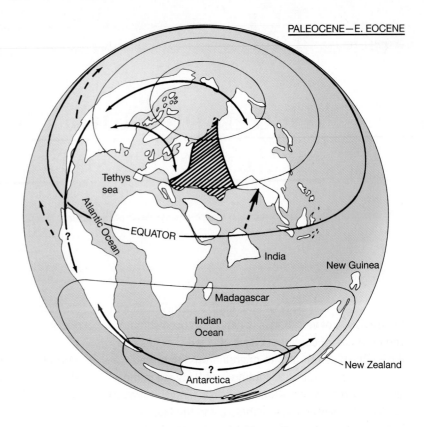

PALEOCENE—E. EOCENE

Antarctica were retained until the middle Cenozoic via the Scotia Island arc, and mammalian faunas from Antarctica in the late Eocene bear a strong resemblance to those from South America. New Zealand separated from Australia sometime in the middle Mesozoic and has a diversity of endemic tetrapods—most notably the primitive diapsid reptile *Sphenodon* (tuatara). Until recently, it was thought that the only endemic mammal was a bat, which evidently reached the island by flying from Australia rather than being part of the original fauna when New Zealand broke away from Australia. However, a recent find of Miocene age shows the presence of a primitive mouse-sized mammal, evidently a relic of some primitive Mesozoic stock (Worthy et al. 2006). This mammal lineage became extinct some time before the present day. Other nonflying mammals that are found in New Zealand today, such as possums, deer, and hedgehogs, were brought in by humans.

In the Northern Hemisphere, a land bridge between Alaska and Siberia—the trans-Bering Bridge (situated where the Bering Straits are today)—broadly connected North America to Asia at high but relatively ice-free latitudes. Intermittent connections also persisted between eastern North America and Europe via Greenland and Scandinavia from the Cretaceous to the early Eocene. However, migrations

of mammals between North America and Europe via this route appear to have been prominent only during the early Eocene. The connection apparently broke during the Eocene, and later migrations between North America and the Old World occurred via the Bering land bridge linking Siberia and Alaska. Other tectonic movements also influenced Cenozoic climates. For example, the uplift of mountain ranges such as the Rockies, the Andes, and the Himalayas in the middle Cenozoic resulted in alternations of global and local rainfall patterns, resulting in the spread of grasslands—a new habitat type—in the higher latitudes.

19.2 Cenozoic Terrestrial Ecosystems

The evolution of animals and plants during the Cenozoic is closely connected with climatic changes, which in turn are related to changes in the positions of continents. The key to understanding changes in climatic conditions lies in knowing that the major landmasses were moving *away* from the equatorial region and toward the poles. The passage of landmasses over the polar region—Antarctica in the early Cenozoic and Greenland in the later Cenozoic—

allowed the formation of polar ice caps, culminating in the periods of glaciation (ice ages) of the Pleistocene. The story of Cenozoic terrestrial ecosystems is also the story of the temperate regions of the higher latitudes, which became cooler and drier with accompanying changes in vegetation, such as the replacement of lush tropical-like forests with woodland and grassland starting around 45 million years ago. (The term *tropical-like* indicates a forest that had a multilayered structure like modern tropical forests but with a somewhat different assortment of tree species.)

The radiation of modern types of mammals is a prominent feature of the Cenozoic, following their origin in the latest Triassic period and their persistence through the Mesozoic at small body size and low levels of morphological diversity. Thus the Cenozoic is commonly known as the Age of Mammals, even though the time elapsed since the start of the Cenozoic represents only about a third of the time spanned by the total history of mammals (and even though the most diverse vertebrates of the Cenozoic, in terms of numbers of species, are the teleost fishes). The radiation of larger mammals is almost certainly related to the extinction of the dinosaurs, which left the world free of large tetrapods and provided a window of opportunity for other groups.

In the warm world of the early Cenozoic, tropical-like forests were found in high latitudes—even extending into the Arctic Circle. These forests would have rarely, if ever, experienced frost, and the average polar temperature has been estimated at around 12°C (Jahren 2007). Apparently the vegetation of these polar regions could withstand three months of continuous light and three months of continuous darkness. The vegetation of polar regions in the early Cenozoic was an exceptionally broad-leaved type, unknown in the world today. The broad leaves presumably allowed the trees to obtain the maximum amount of sunlight possible during the short summer and under the dim conditions of winter. A fossil assemblage from the early Eocene of Ellesmere Island (within the Canadian Arctic Circle) shows the presence of reptiles such as turtles and crocodiles, as well as of mammals resembling—although not closely related to—the tree-dwelling primates and flying lemurs of present-day Southeast Asia. **Figure 19–2** illustrates the subtropical conditions seen in Europe in the early Eocene. Fossil plants from the early Eocene of Patagonia, at the southern tip of South America, show that tropical forests also existed in the high latitudes in the Southern Hemisphere.

Many of the early Cenozoic lineages of mammals are now extinct. These groups are often called archaic mammals, a rather pejorative term that really refers to our own perspective from the comfort of the Recent epoch. Archaic mammals were mainly small- to medium-size generalized forms with a predominance of arboreal types. Larger, specialized predators, and herbivores with teeth suggesting a high-fiber herbivorous diet, did not appear until the late Paleocene. Members of most present-day orders did not make their first appearance until the Eocene, but there are some notable exceptions, most importantly the orders Carnivora (dogs, cats, and others) and Xenarthra (sloths, armadillos, and related forms) that are first known from the early Paleocene of South America, and Rodentia (first known from the late Paleocene of Asia and North America).

It appears that browsing by herbivorous dinosaurs kept forests at bay during the Cretaceous, much as large herbivorous mammals such as elephants maintain savanna habitats today. Because forests dominated terrestrial habitats after the dinosaurs' extinction, most of the niches available for early Cenozoic mammals were arboreal ones. Tropical forests were established in North America a mere million years or so after the start of the Cenozoic. However, it was not until the climatic changes of the Eocene resulted in more open canopied forests capable of supporting denser undergrowth vegetation that larger mammals began to radiate into the diversity of terrestrial niches they occupy today.

Modern types of birds also diversified at the start of the Cenozoic, including large terrestrial birds—both carnivorous forms (now all extinct—see Figure 19–2) and herbivores like the present-day ostrich. Many of the present-day groups of lizards had their origins in the latest Cretaceous or early Cenozoic, while the modern groups of turtles and amphibians were largely established in the Cretaceous, and snakes are largely a Cenozoic radiation. Modern types of freshwater crocodiles were the only ones to survive into the later Cenozoic, and there was a modest radiation of terrestrial crocodiles in the Southern Hemisphere during the Paleocene and Eocene. Champsosaurs (crocodile-like in form but only very distantly related to crocodilians) were also prominent freshwater aquatic predators in the early Cenozoic. Among the insects, modern butterflies and moths first appeared in the middle Eocene.

The later Cenozoic world was generally cooler and drier than that of the early Cenozoic. Changes in terrestrial ecosystems reflected these climatic changes. Temperate forests and woodlands replaced

◀ **FIGURE 19–2** A reconstruction of a scene from the early Eocene of Europe The trees in this tropical-like rainforest include sequoia, pine, birch, palmetto, swamp cypress, and tree ferns. In the foreground are cycads and magnolia. The birds are ibises, with a *Gastrornis* (an extinct flightless predatory bird) in the background. The hippolike mammals are *Coryphodon*, belonging to the extinct ungulate-like order Dinocerata. An early primate (*Cantius*, an omomyid) climbs up a liana vine. Crouching among the ferns is the oxyaenid *Palaeonictis*, a catlike predator belonging to the extinct order Creodonta. Its potential prey are the early artiodactyls, *Diacodexis*, in the foreground.

the tropical-like vegetation of the higher latitudes, and tropical forests were confined to equatorial regions. Extensive grasslands first appeared in the Miocene epoch in the northern latitudes, forming swaths of savanna (grassland with scattered trees) across North America and central Asia. However, the types of mammals present in the South American faunas suggest that grasslands appeared earlier in the high southern latitudes, perhaps as early as the end of the Eocene. In the late Pliocene, 2 to 3 million years ago, savannas appeared in more tropical areas such as East Africa, and the more temperate grasslands turned into treeless prairie or steppe. This vegetational change coincided with the emergence of our own genus *Homo*, a hominid well adapted to life on the tropical savannas. New types of vegetation also appeared in the Plio-Pleistocene: tundra (treeless shrubland) and taiga (boreal evergreen forests) in the Arctic regions and deserts in the tropical and temperate regions.

The radiation of mammals in the late Cenozoic reflected these vegetational changes. Large grazing mammals such as horses, antelope, rhinoceroses, and elephants evolved along with the emerging grasslands—and with them the carnivores that preyed on them, such as large cats and dogs. Some small mammals also diversified, most notably modern types of rodents, such as rats and mice. This diversification of small mammals may explain the concurrent late Cenozoic diversification of modern types of snakes. Late Cenozoic lizards included some very large varanoids, including not only the largest lizard known today, the Komodo dragon, but also a lion-size predator, *Megalania*, in the Pleistocene of Australia. Many modern types of birds first appeared in the late Cenozoic, including the passerines (songbirds) and modern types of birds of prey such as eagles, hawks, and vultures. The more open habitats of the late Cenozoic also favored the diversification of social insects that live in grasslands, such as ants and termites.

19.3 Cenozoic Climates

The Cenozoic was, in general, a cooler and dryer time than the Mesozoic, although the early part (until the mid-Eocene, around 50 million years ago) was a period of relative warmth. The later Cenozoic was characterized by the buildup of ice at the poles, culminating in the Ice Age at the start of the Pleistocene, 2 million years ago.

■ Tertiary (Paleocene–Pliocene) Climates

The warm and humid conditions typical of the Jurassic and Early to Middle Cretaceous changed toward the end of the Cretaceous period, when widespread moderate cooling took place. Yet the world of the early Cenozoic still reflected the hothouse world of the Mesozoic. There was an increase in atmospheric carbon dioxide at the start of the Eocene, which would have resulted in warming via a greenhouse effect. A rise in atmospheric oxygen levels in the Eocene, from slightly below present-day values (around 17 percent) to present-day values (21 percent) or above has been invoked as a possibly causal mechanism in the evolution of larger-sized placental mammals at this point (Falkowsi et al. 2005).

The peak of climatic warming in the higher latitudes occurred at the start of the Eocene, 55 million years ago. There was a transitory spike in temperature at this time (**Figure 19–3**), lasting around 100,000 years, known as the Paleocene-Eocene thermal maximum. The probable cause was the release of methane, a greenhouse gas, into the atmosphere from shallowly buried sediments on the ocean continental shelf. This warming seems to have had profound effects on mammalian evolution, and many of the modern orders of mammals appeared at this time (Gingerich 2006). A more stable and prolonged period of warm temperatures followed, peaking at around the early-middle Eocene boundary 50 million years ago (see Figure 19–3). From this high point, the higher latitude regions started to cool, with a rather precipitous drop in mean annual temperature (and in atmospheric CO_2 levels) in the latest Eocene and earliest Oligocene epochs, plunging the Earth into the start of the colder world of the later Cenozoic (Zachos et al. 2008).

What caused this dramatic change in the Earth's temperatures? A primary reason was probably a reverse greenhouse effect related to sharply falling

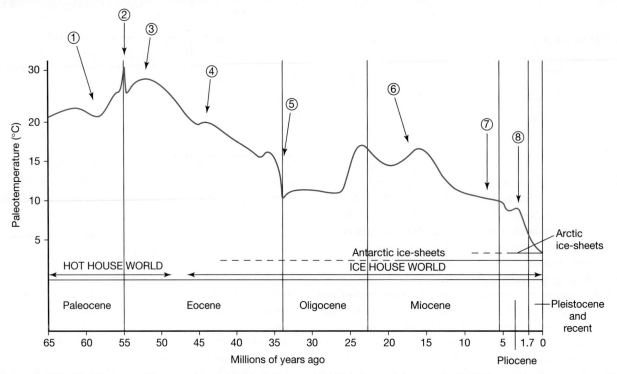

▲ **FIGURE 19–3** Mean annual paleotemperatures in the northern higher latitudes during the Cenozoic. The bars indicating the presence of ice sheets are dashed when representing minimal (partial or ephemeral) ice coverage and solid when representing maximal (full-scale and permanent) ice coverage.
1. Tropical forests within confines of Arctic Circle. Archaic mammals predominate.
2. Paleocene-Eocene thermal maximum. Transitory (100,000 years) time of high temperatures, many modern types of mammals appear.

3. Temperatures reach a Cenozoic maximum. Modern types of mammals diversify, and archaic mammals decline.
4. Cooling brings winter frosts to higher latitudes. Tropical forests replaced by woodlands in higher latitudes, and polar forests disappear; more tropical types of vertebrates (e.g., primates, many kinds of reptiles) now disappear from higher latitudes, and remaining archaic mammals become extinct.
5. Eocene-Oligocene temperature plunge results in extinctions and an overall cooler world.

6. Miocene warming brings grassland to higher latitudes. Subsequent cooling and drying results in spread of grasslands and loss of woodlands. In higher latitudes, grazing mammals predominate, browsers decline in numbers, all crocodiles and large turtles now disappear.
7. C_4 grasses now expand.
8. Significant further cooling at 2.5 million years ago plunges the world into an ice age. (Modified from Zachos et al. [2008] and based on deep-sea oxygen isotope data from foraminifera.)

levels of atmospheric carbon dioxide, which plummeted at this time from around four times today's levels to levels resembling that of the present day. Additionally, cold polar-bottom water massed over the poles when Australia broke away from Antarctica and Greenland broke away from Norway. Ocean circulation carried the cold water toward the equator, cooling the temperate latitudes. The Antarctic ice cap probably formed by the end of the Eocene, although the Arctic ice cap did not form until some 30 million years later, in the latest Miocene.

Following a rather cool early Oligocene, temperatures rose again in the higher latitudes in the late Oligocene, dipped a little in the earliest Miocene, and reached a second peak in the middle Miocene. The mid-Cenozoic warming may be due to the opening of Drake's Passage between Antarctica and South America, which isolated the cold polar water around Antarctica. However, expansion of the Antarctic ice cap in the late Miocene once again brought cooling to the higher latitudes, a trend that has persisted to the present day with occasional remissions. The closing of the Isthmus of Panama that joined North and South America in the Pliocene epoch also profoundly affected global climate. Warm waters could no longer circle the equator, and it is during this time that the Gulf Stream was formed, carrying warm water from the subtropical American regions to Western Europe. Elsewhere in the world, disruption of oceanic circulation resulted in cooling and led to the formation of the Arctic ice cap.

The Miocene world was in general drier than in the Eocene, and the combination of warmth and dryness promoted the spread of grasslands. At the end of the Miocene, around 7 million years ago, there was a shift in the type of photosynthetic pathway used by grasses, as registered by carbon isotopes in both fossil soils and the enamel of the teeth of grazing mammals. This was the shift from the C_3 pathway, seen today in dicotyledonous plants and grasses in cooler regions, to the C_4 pathway, seen today in grasses in subtropical and tropical regions. The C_4 pathway is more efficient under conditions of both high temperature and low carbon dioxide.

Note that today's world is a very cool and dry place in comparison with most of the Cenozoic (indeed, with most of Earth's history), and it is also quite varied in terms of different types of habitats and climatic zones. At the start of the Cenozoic, the world was covered with tropical-like forests and with the types of animals that live in such habitats. We still have forests of this type, now confined to the equatorial regions, but have added other types of environments at other latitudes (temperate woodland and grasslands) and also have tropical grasslands, plus very new types of habitats, such as deserts and arctic tundra. Thus the numbers of different types of animals are enormously increased to reflect this great diversity of habitat types. In addition, the division of the world into separate continents has meant that many faunas and floras have evolved in comparative isolation.

■ The Pleistocene Ice Ages

The extensive episodic continental glaciers that characterize the Pleistocene epoch were events that had been absent from the world since the Paleozoic. These ice ages had an important influence not only on Cenozoic mammal evolution in general—almost all of today's species have their origin in the Pleistocene—but also on our own evolution and even on our present civilizations. We still live in a world with abundant polar ice, but right now we inhabit a warmer, interglacial, period.

For example, currently the volume of ice on Earth constitutes about 26 million cubic kilometers. During glacial episodes in the Pleistocene, there was as much as 77 million cubic kilometers of ice, perhaps even more. An enormous volume of water still is locked up in glaciers and polar ice caps. The melting of the glaciers of the last glacial episode at the end of the Pleistocene, around 10 thousand years ago, caused the sea level to rise by about 140 meters (almost half the height of the Empire State Building) comparable to its present relatively stable condition. If the present-day glaciers were to melt, sea level

would rise at least another 50 meters, submerging most of the world's coastal cities. Entire countries such as Bangladesh and some Pacific Island nations would be completely submerged. No wonder that the present-day global warming caused by human activities is the focus of so much concern!

Today glaciers cover 10 percent of the Earth's land surface, mostly in polar regions but also on high mountains. At times in the Pleistocene, an ice mass that was probably between 3 and 4 kilometers thick covered as much as 30 percent of the land and extended southward in North America to 38°N latitude (southern Illinois; **Figure 19–4**). A similar ice sheet covered northern Europe. However, much of Alaska, Siberia, and Beringia (Pleistocene land that is now underwater between Alaska and northeast Asia) were free of ice and housed a biome known as the Mammoth Steppe or the steppe-tundra that is unknown today. This steppe-tundra habitat was obviously much more productive than present-day high-latitude habitats because it contained a mammalian fauna that rivaled the diversity of modern African savanna faunas. This fauna combined mammals now absent from higher latitudes, such as lions and rhinoceroses, with animals that persist in Arctic latitudes today, such as reindeer and musk ox. Fossil pollen also shows that types of sage and grass plants existed in the steppe-tundra of Beringia that are absent from tundra habitats today.

These continental glaciers advanced and retreated several times during the Pleistocene. (The Southern Hemisphere was less affected because at that time the southern continental landmasses were farther from the poles than the northern ones, as they are today.) There were four major episodes of glaciation, but we now know that many (perhaps 20 or more) minor ones were interspersed among these major ones.

Continental glaciation had a greater effect on world climates than just ice covering the high latitudes. Popular books depict mammoths struggling to free themselves from ice, but glaciers advance slowly enough for animals to migrate toward the equator—although problems may occur if routes are blocked by mountain ranges or seaways. Eurasian animals fleeing colder climates had the advantage of broad connections between temperate and tropical zones, both in Asia and in Africa. In contrast, North American animals would have to traverse the narrow Isthmus of Panama to reach the more tropical areas of South America. This geographic bottleneck may have limited the migration of certain types of mammals.

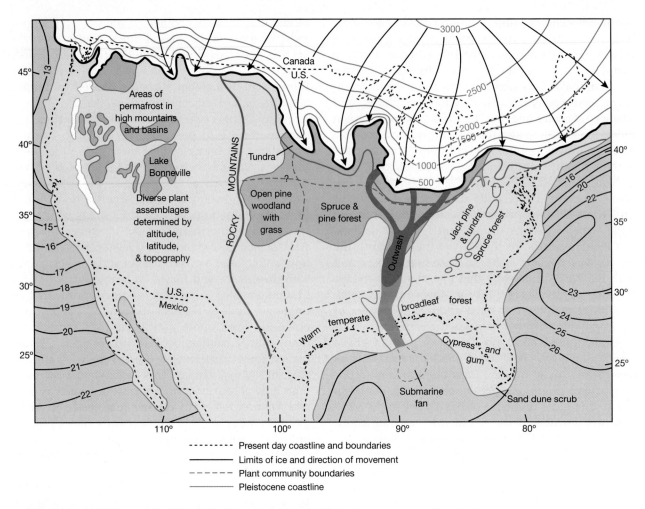

▲ **FIGURE 19–4** Pleistocene glaciation in North America about 18,000 years ago. Contours show the depth of ice sheets, and arrows indicate the direction of movement. Dominant vegetation is shown in the east; in the west, mountainous terrain produced complex variation in vegetation. Lake Bonneville and other large pluvial lakes are shown in the west. Sea surface temperatures (°C) are based on analysis of microfossils from deep sea cores.

Drying of the ice-free portions of the Earth due to the volume of water tied up in glaciers was at least as important for terrestrial ecosystems as the glaciers themselves. Many of the equatorial areas that today are covered by lowland rain forests were much drier then, even arid. Today's relatively mild interglacial period is apparently colder and drier than other interglacial periods in the Pleistocene. For example, during other interglacial periods hippopotamuses were found in what is now the Sahara Desert and in England.

What caused these episodes of glaciation? A long-standing theory suggests that the amount of solar radiation impinging on the Earth varies enough to affect the Earth's climate. In the 1930s, the Yugoslavian astronomer Milutin Milankovitch proposed that

episodes of glaciation are initiated by the combination of several small variations of the passage of the Earth's orbit around the sun, and the position of the Earth relative to the sun. Three cycles interact here, each with its own characteristic periodicity (time elapsed between extremes of the cycle): (1) the 100,000-year cycle of the Earth's elliptical orbit around the sun; (2) the 41,000-year cycle of the tilt of the Earth's rotational axis; and (3) the 23,000-year cycle in the precession (wobble) of the Earth's rotational axis.

Each of these orbital properties produces different effects. Change in tilt and precession modify the distribution of sunlight with respect to season and latitude, but not total global **insolation** (incoming solar radiation), whereas changes in the Earth's orbit result in minute changes in global insolation. Normally these properties are cycling out of phase, like discordant keys played on a piano, but every so often they line up together like notes making a chord. Milankovitch suggested that the critical factor leading to a glacial episode is a change in the amount of summer insolation at high latitudes. It appears that glacial episodes get their start not from the world as a whole getting colder year-round, but from cool summers that prevent the melting of winter ice. In contrast, the winters during glacial periods may have been warmer than those of the present day.

It is important to realize that these Milankovitch cycles must have been in existence throughout Earth's history. However, it was only after the formation of the Arctic ice cap in the Pleistocene (or possibly as early as the Pliocene) that enough polar ice existed to plunge the Northern Hemisphere into an ice age.

19.4 Cenozoic Extinctions

The best-known extinction of the Cenozoic is probably the one at the end of the Pleistocene, although this was by no means the extinction of greatest overall magnitude. The Pleistocene extinction appears dramatic because of the extinction of the megafauna (mammals over 20 kilograms in body mass). This included many very large mammals that are not only extinct but are not represented by any similar types today, such as glyptodonts (giant armored mammals related to armadillos) and ground sloths in North and South America, and diprotodontids (the largest of all marsupials, some as large as a hippopotamus) in Australia. It also included many larger and exotic forms of more familiar mammals, such as the saber-toothed cats of Holarctica and Africa; the Irish elk, cave bears, and woolly rhinoceroses of Eurasia; the mammoths of Holarctica and Africa; and kangaroos, wombats, and echidnas in Australia that were much larger than the extant forms. Large terrestrial birds also suffered in these extinctions—including herbivores such as the moas of New Zealand and the elephant bird of Madagascar and carnivores such as the New World phorusrachiforms and the Australian dromornithids.

There is much debate about the cause of these extinctions. The main extinctions occurred at the end of the last glacial period, between 13,000 and 11,500 years ago. Surprisingly enough, animals appear to be more vulnerable to extinction when the climate changes from glacial to interglacial rather than the other way around, probably because warming occurs faster. Thus climatic change would be an obvious

explanation. However, many scientists have noted that it is only the last glacial period—rather than any of the previous ones—that brought extinctions of such magnitude. This observation suggests that part, if not all, of the blame for megafaunal extinctions should be placed on the spread of modern humans and modern hunting techniques, which were concurrent with that time period.

Many scientists today are adamant that human activity, rather than climatic change, must be the root cause of the Pleistocene extinctions (e.g., Burney and Flannery 2005). This is the overkill hypothesis, and the survival of mammoths until only a few thousand years ago on human-free Wrangel Island, off the coast of Siberia, appears to support this view. However, Steve Wroe and colleagues (Wroe et al. 2004) point out that our actual knowledge about the effect of humans on mammal extinctions is drawn from historical examples of island faunas. Hunting and/or habitat disturbance may have been the cause of the comparatively recent (within the past few thousand years) extinctions of the giant lemurs on Madagascar or the moas (giant herbivorous birds) in New Zealand, but there are problems with extending such scenarios to larger landmasses, such as North America or Australia.

Other researchers argue for climate change being the key factor in megafaunal extinctions. The extinctions in North America did not follow a north-to-south pattern, as would be expected with the invasion of humans from Beringia (Beck 1996). Furthermore, horses in Alaska underwent a rapid decrease in body size shortly before becoming extinct—and before human arrival—and that pattern is consistent with the hypothesis of climatic change being the extinction agent (Guthrie 2006). Human hunters and climate change are not mutually exclusive hypotheses, of course, and the two forces may have acted synergistically, with hunting pressure providing the last straw that drove already unstable populations to extinction. Koch and Barnosky (2006) provide a comprehensive review of current knowledge of the Pleistocene extinctions and hypotheses about their causes.

About 30 percent of mammalian genera became extinct at the end of the Pleistocene. That is approximately the magnitude of the other two major Cenozoic extinctions (late Eocene and late Miocene). However, the preceding two extinctions differ in several critical ways from that of the Pleistocene. The late Eocene extinctions were associated with the dramatic fall in higher-latitude temperatures. Higher-latitude forests turned to temperate woodlands, with the accompanying disappearance of mammals adapted to these tropical-like forests. This included not only a diversity of archaic mammals but also some early, more modern types, such as higher-latitude primates and early horses. The early Cenozoic diversity of amphibians and reptiles in higher latitudes was also greatly reduced during the late Eocene.

The late Miocene extinctions were associated again not only with falling higher-latitude temperatures but also with global drying. The major extinctions were of browsing mammals (including a variety of large browsing horses), which suffered habitat loss as the savanna woodlands turned into open grasslands and prairie. North America was especially hard hit by the climatic events of the late Miocene because of its relatively high latitudinal position and the fact that animals could not migrate to more tropical areas in South America before the formation of the Isthmus of Panama in the Pliocene.

Most significantly for the overkill hypotheses, mammals of all body sizes (not just large ones) were affected in both the Eocene and Miocene. Other organisms, both terrestrial and marine, also experienced profound extinctions. The late Pleistocene extinction affected primarily large mammals and birds, which are the species mostly likely to be viewed as prey or competitors by human hunters.

Additional Readings

Beck, M. W. 1996. On discerning the cause of late Pleistocene megafaunal extinctions. *Paleobiology* 22:91–103.

Burney, D. A., and T. F. Flannery. 2005. Fifty millennia of catastrophic extinctions after human contact. *Trends in Ecology and Evolutionary Biology* 20:395–401.

Ehleringer, J. R. et al. 1991. Climate change and the evolution of C_4 photosynthesis. *Trends in Ecology and Evolution* 6:95–99.

Falkowski, P. G. et al. 2005. The rise of oxygen over the past 205 million years and the evolution of large placental mammals. *Science* 309:2202–2204.

Gingerich P. D. 2006. Environment and evolution through the Paleocene-Eocene thermal maximum. *Trends in Ecology and Evolution* 21:246–253.

Guthrie, R. D. 1990. *Frozen Fauna of the Mammoth Steppe.* Chicago, IL: University of Chicago Press.

Guthrie, R. D. 2006. New carbon dates link climatic change with human colonization and Pleistocene extinctions. *Nature* 441:207–209.

Jahren, A. H. 2007. The Arctic forest of the middle Eocene. *Annual Review of Earth and Planetary Science*s 35:509–540.

Janis, C. M. 1993. Tertiary mammal evolution in the context of changing climates, vegetation, and tectonic events. *Annual Review of Ecology and Systematics* 24:467–500.

Koch, P. L., and A. D. Barnosky. 2006. Late Quaternary extinctions: state of the debate. *Annual Review of Ecology, Evolution and Systematics* 37:215–250.

Lister, A. M. 2004. The impact of Quaternary Ice Ages on mammalian evolution. *Philosophical Transactions of the Royal Society* 359:221–241.

Martin, P. S., and R. G. Klein. 1984. *Quaternary Extinctions.* Tucson, AZ: University of Arizona Press.

Prothero, D. R. 2006. *After the Dinosaurs: The Age of Mammals.* Bloomington, IN: Indiana University Press.

Worthy, T. H. et al. 2006. Miocene mammal reveals a Mesozoic ghost lineage on insular New Zealand, southwest Pacific. *Proceedings of the National Academy of Sciences* 103:19419–19423.

Wroe, S. et al. 2004. Megafaunal extinction in the Late Quaternary and the global overkill hypothesis. *Alcheringa* 28:291–332.

Zachos, J. et al. 2008. An early Cenozoic perspective on greenhouse warming and carbon-cycle dynamics. *Nature* 451:279–283.

CHAPTER 20

Mammalian Diversity and Characteristics

Cenozoic mammals are a highly diverse group of organisms, adapted to a wide variety of lifestyles and displaying great diversity in body form and ecology. In addition to the familiar features of lactation and hair, mammals have other derived characters that set them apart from other vertebrates. Although most mammals alive today are placentals, an understanding of mammal diversity and specializations requires an understanding of how placentals differ from marsupials, as well as how therians (marsupials and placentals) differ from monotremes. Much of the diversity seen among Cenozoic therian mammals reflects the isolation of different groups of mammals on different continental landmasses; many different types of mammals evolved convergently on different continents. The changing climates of the higher latitudes during the Cenozoic era also resulted in a wider diversity of mammals adapted to new habitats, such as grasslands, and changes in sea level and continental positions resulted in various mammalian intercontinental migrations.

20.1 Major Lineages of Mammals

Mammals are perceived as the dominant terrestrial animals of the Cenozoic, but their species diversity (about 4800 species) is only a little more than half that of birds (about 9000 species) and considerably less than that of lepidosaurian reptiles (about 7750 species). Mammalian species diversity is in fact about the same as that of amphibians (about 4800 species), making mammals and amphibians the smallest lineages of tetrapods (at least in terms of numbers of species).

Mammals do, of course, include the largest living terrestrial and aquatic vertebrates (the blue whale, at around 120 tons, is the largest animal ever known). Mammals have great morphological diversity; no other vertebrate group has forms as different from each other as a whale is from a bat. Even among strictly terrestrial mammals, there is a tremendous morphological difference between, for example, a mole and a giraffe. However, it is prudent to consider that mammals do not rise above other vertebrates when some measures of evolutionary success, such as species diversity, are considered.

Traditionally, the class Mammalia has been divided into three subclasses: **Allotheria** (multituberculates, now extinct), **Prototheria** (monotremes), and **Theria**. This classification does not really take into account the large diversity of Mesozoic mammals (see Chapter 18), but these three groups do reflect basic divisions in body plans among those mammals surviving into the Cenozoic. Therians are subdivided into two infraclasses: Metatheria, including marsupials and their extinct Mesozoic stem-group relatives, and Eutheria, including placentals and their extinct stem-group relatives. The term *placental* is a bit of a misnomer, as marsupials also possess a placenta (see Chapter 21 Section 21.1). In the previous edition of this book, we followed the custom of the time by referring to placentals as *eutherians*. However, more recently, researchers have reverted to the term *placental* for the crown-group of the extant lineages of eutherians placentals that made their first appearance in the Cenozoic and probably originated in the latest part of the Cretaceous period.

Multituberculates and monotremes were originally seen as being very distantly related to therians. Living monotremes retain some primitive anatomical features in addition to their egg-laying habits. Until quite recently, monotremes were considered to be derived from a different group of Late Triassic stem mammals than the therians, and they were once even considered to be an independent radiation from the cynodonts. The teeth of multituberculates are so different from those of other early mammals that they, too, were thought to have a very early offshoot from the main lineage.

Living monotremes are toothless as adults, so the characters of dental anatomy on which much of mammalian phylogeny is based cannot easily be used to evaluate their relationships. However, teeth are found in juvenile platypuses, and teeth are known from fossils. A single jaw of a fossil (*Steropodon*) from the Early Cretaceous of Australia shows fully formed teeth that are clearly triangular. The shape of these teeth suggests that monotremes are at least more closely related to therians than to some of the Mesozoic mammals, although there is also speculation that the triangular teeth of monotremes are convergent with those of therian mammals.

■ Multituberculates—Rodentlike Mammals of the Mesozoic Era

Multituberculates are the best known and the most common of the Mesozoic mammals. They were a very long-lived group, known from the Late Jurassic period to the early Cenozoic (late Eocene epoch). Multituberculates were probably small terrestrial and semiarboreal omnivores like extant rodents. Their very narrow pelvis suggests that they did not lay eggs but may have given birth to poorly developed young.

Some multituberculates were rather squirrel-like: The structure of their ankle bones shows that they could rotate their foot backward to descend trees headfirst, like a squirrel, and their caudal vertebrae indicate a prehensile tail (**Figure 20–1**). These squirrel-like multituberculates also had an enlarged lower posterior premolar that formed a shearing blade, perhaps used to open hard seeds. Other multituberculates were more terrestrial and rather like a groundhog or a wombat. The extinction of multituberculates

▶ **FIGURE 20–1** The Early Cenozoic multituberculate *Ptilodus*.

in the Eocene was probably due to competition with rodents, which first appeared in the late Paleocene epoch.

Multituberculates get their name from their molars, which are broad, multicusped (multituberculed) teeth specialized for grinding rather than shearing. Wear on the teeth indicates that multituberculates moved their lower jaw backward while bringing their teeth into occlusion. The teeth and jaw movements of multituberculates were also similar to those of rodents, except rodents move their lower jaws *forward* into occlusion.

The position of the multituberculates within the Mammalia is controversial. Multituberculates have a rather derived, therian-like shoulder girdle, which originally was considered an independent evolution within this group. Some researchers now consider this feature to link multituberculates even more closely to therians than are monotremes. In contrast, Zofia Kielan-Jaworowska (1997), who has spent a lifetime studying multituberculates, considers them to be very primitive, especially in the nature of their brains (as determined from casts of the inside of the skull). If this interpretation is correct, the derived nature of the shoulder girdle must represent convergence between multituberculates and therians. The phylogenetic position of multituberculates is an area of active debate and controversy; better fossil material of Jurassic multituberculates would greatly help to resolve their relationships.

■ Mammalian Ordinal Diversity

One current hypothesis of interrelationships of living therian mammal orders is shown in **Figure 20–2**. Controversies about the interrelationships among the various placental mammal orders are discussed later.

Monotremes Monotremes are grouped in the infraorder Ornithodelphia (Greek *ornitho* = bird and *delphy* = womb, referring to the single functional oviduct in the platypus and many birds; contrast with the *didelphid* opossums) and in the order Monotremata (Greek *mono* = one and *trema* = hole, referring to the cloaca—the single opening of the excretory and reproductive tracts). (An old term for therians was *Ditremata*, referring to the two separate openings.) There are two families: the Ornithorhynchidae (Greek *rhynchus* = beak) contains the platypus, a semiaquatic animal that feeds on aquatic invertebrates in the streams of eastern Australia, and the Tachyglossidae (Greek *tachy* = fast and *glossa* = tongue) contains two types of echidnas, the short-nosed echidna of Australia (which eats mainly ants and termites) and the long-nosed echidna of New Guinea (which includes earthworms in its diet) (**Figure 20–3** on page 523).

Small teeth belonging to Cretaceous Australian monotremes have recently been discovered, suggesting a previously unsuspected Mesozoic diversity of these animals (Flannery et al. 1995). However, the few known Cenozoic monotreme fossils represent only echidnas or platypuses, and monotreme diversity overall has probably always been fairly limited. That is to say, the platypus and echidna are not the relicts of a once much larger radiation but rather have been its mainstay.

The rather sprawling stance of monotremes, reminiscent of reptiles, may reflect specializations for swimming or digging rather than a truly primitive condition. Monotremes also have their own unique specializations. Both platypuses and echidnas lack teeth as adults and have a leathery (rather than horny) bill or beak. This beak contains receptors that sense electromagnetic signals from the muscles of other animals and are used for sensing prey underwater or in a termite nest. Male platypuses have a spur on the hind leg attached to a venom gland, which is used to poison rivals or predators; however, evidence of a similar spur is seen in some primitive Mesozoic mammals, so this may not be a unique monotreme feature (Hurum et al. 2006).

Monotremes are similar to birds in some ways—for example, the spermatozoa of the platypus are thread-like, as are those of birds, rather than having a distinct head and tail like the spermatozoa of therians. In addition, the platypus genome has more bird characteristics than do the genomes of therian mammals. These characteristics are presumably primitive amniote features that are retained in monotremes but lost in therians (Warren et al. 2008).

The possession of venom is a rare feature in mammals. Other extant venomous mammals include some lipotyphlans (insectivores) such as a few species of shrews and the cat-size *Solenodon* found on the Caribbean island of Hispaniola. These mammals have venom in their saliva, which they inject via a bite with their canines, but their canines are not specialized for this purpose (unlike the condition in snakes). However, an extinct type of insectivorous placental that is not closely related to shrews has been found in the late Paleocene of Alberta, which has a distinct groove in its upper canines that is hypothesized to serve as a venom-delivery system (Fox and Scott 2005).

Marsupials Extant marsupials traditionally have been considered to represent a single order, Marsupialia. More recent studies suggest they can be divided into at least four lineages that are equivalent in morphological and genetic diversity to placental orders. The current scheme of marsupial interrelationships is shown in Figure 20–2 and **Table 20.1** on page 524, and **Figure 20–4** on page 525 illustrates a diversity of living marsupials.

There is a fundamental split in the interrelationships of living marsupials into the Ameridelphia of the New World and the Australidelphia of (mainly) Australia. A third group, now extinct and possibly more primitive than either of these living groups, was the Deltatheroidea of the Late Cretaceous of Asia. Note that the original metatherian/eutherian divergence was most likely in Asia.

The ameridelphian order Didelphimorpha includes didelphoids (opossums plus related extinct forms, such as the Northern Hemisphere marsupials of the early Cenozoic) and some extinct Cretaceous forms. Present-day opossums are quite a diverse group of small- to medium-size marsupials, mainly arboreal or semiarboreal omnivores, including animals such as the herbivorous woolly opossum and the otterlike yapok. A much greater diversity of didelphoids existed in the Tertiary period, including jerboa-like hopping forms and molelike burrowing

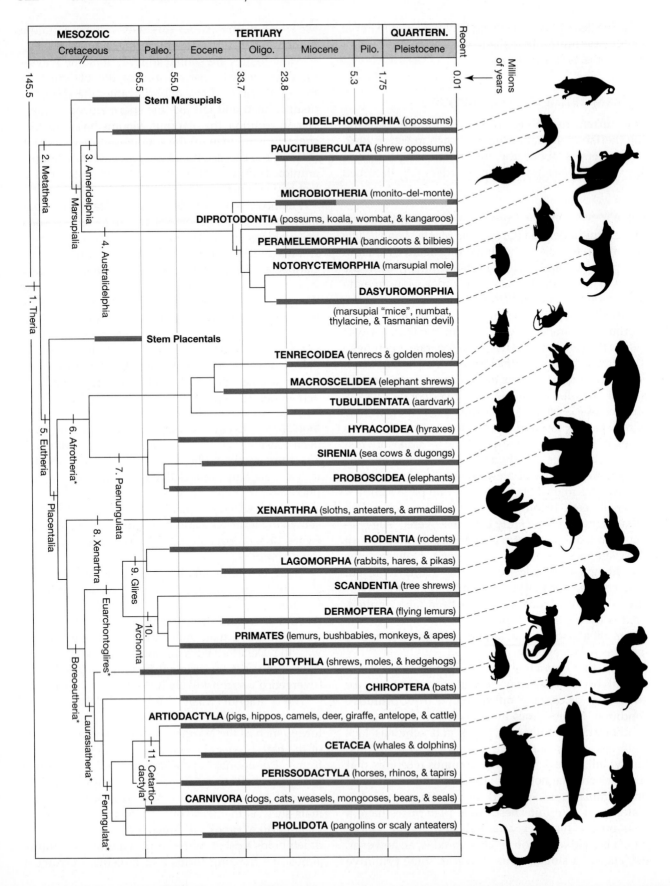

◀ **FIGURE 20–2** Phylogenetic relationships of extant mammalian orders, excluding Monotremata. This diagram shows the probable relationships among living therian mammals. Dotted lines show interrelationships only and are not indicative of the times of divergence nor of the unrecorded presence of taxa in the fossil record. Numbers indicate derived characters that distinguish the lineages. This cladogram is based on a combination of morphological and molecular characters. Higher taxa for which the primary (or only) evidence is molecular are indicated by an asterisk (*). Afrotheria is placed here as the basal placental clade: note, however, that other researchers promote Afrotheria and Xenarthra as sister groups or one or both groups in a more nested position within the other placentals.

Legend: 1. Theria–mammary glands with nipples, viviparity with loss of eggshell, digastric muscle used in jaw opening, anal and urogenital openings separate in adults, spiraled cochlea, scapula with supraspinous fossa, and numerous features of skull and dentition. **2.** Metatheria–dentition essentially **monophyodont** (P3 is only tooth replaced), development of chorioallantoic membrane suppressed, ureters pass medial to Müllerian ducts to enter the bladder, pseudovaginal canal present at parturition, and various detailed features of skull, dentition (including upper molars with wide stylar shelves), and ankle joint. **3.** Ameridelphia– sperm paired in epididymis. **4.** Australidelphia–details of dentition and ankle joint. **5.** Eutheria–egg shell membrane lost, intrauterine gestation prolonged with suppression of estrous cycle, corpus callosum connects cerebral hemispheres, ureters pass lateral to Müllerian ducts to enter the bladder, fusion of Müllerian ducts into a median vagina, penis simple (not bifid at tip), plus details of dentition (including upper molars with narrow stylar shelves). **6.** Afrotheria*–abdominal testes, more than 19 thoracolumbar vertebrae, details of fetal membranes. **7.** Paenungulata–styloglossus tongue muscle bifurcate, details of structure of skull, wrist bones, and placenta. **8.** Xenarthra–details of skull anatomy, sacrum strongly fused to pelvis, tooth development suppressed with loss of anterior teeth and enamel poorly developed or absent. **9.** Glires–enlarged pair of ever-growing upper and lower incisors, which represent the deciduous second incisors of other mammals, plus details of skull anatomy. **10.** Archonta (or Euarchonta)– pendulous penis, plus details of ankle structure. **11.** Cetartiodactyla*–double-trochleated (= "double-pulley") astragalus.

forms. Other ameridelphians are the Paucituberculata (caenolestids or rat opossums), which are small, terrestrial, shrewlike forms. The Borhyaenoids (Eocene-Pliocene) formed a large radiation of carnivorous forms, including ferretlike, doglike, and bearlike forms, and even a Plio-Pleistocene saber-tooth parallel, *Thylacosmilus*.

The one remaining type of South American marsupial is the *monito del monte*, a tiny mouselike animal living in the montane forests of Chile and Argentina, placed in its own distinct order Microbiotheria. Molecular studies have shown this animal to be related to the Australian marsupials, and it may be a distant relict of the stock of originally South American mar-

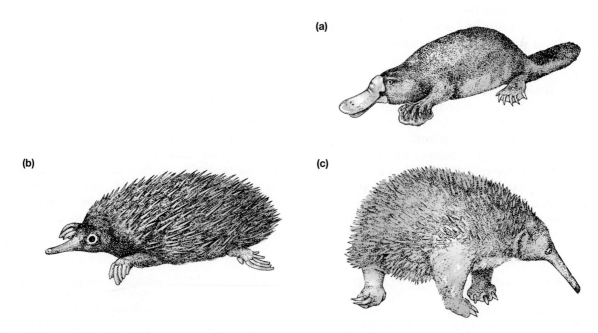

▶ **FIGURE 20–3** Diversity of living monotremes (a) The platypus, *Ornithorhynchus anatinus*. (b) The short-nosed echidna (Australia), *Tachyglossus aculeatus*. (c) The long-nosed echidna (New Guinea), *Zaglossus bruijni*. (The species are drawn approximately to scale; [a] is the size of a large house cat).

TABLE 20.1 Classification of extant marsupial orders and approximate numbers of families and species.
Note that different authorities recognize different numbers of families within an order.

Order	Number of Families/Species	Major Examples and Higher-Level Classification
		Ameridelphia
Didelphimorpha	1/77	Opossums; 20 g to 6 kg; Neotropical region (plus one North American species).
Paucituberculata	1/5	Caenolestids or rat opossums; 15 to 40 g; Neotropical region.
		Australidelphia
Microbiotheria	1/1	The monito del monte; ~25 g; Neotropical region.
Dasyuromorphia	3/60	Marsupial mice, native cats, Tasmanian devil, Tasmanian wolf (thylacine), marsupial anteater (numbat); 5 g to 20 kg; Australian region.
Notoryctemorphia	1/1	Marsupial mole; 50 g; Australian region.
Peramelemorphia	2/21	Bandicoots and bilbies; 100 g to 5 kg; Australian region.
Diprotodontia	9/110	Possums, flying phalangers, cuscuses, honey possum (noolbenger), koala, wombats, potoroos, wallabies, kangaroos; 12 g to 90 kg; Australian region; possums and wallabies introduced into New Zealand by humans.

supials that migrated across Antarctica to Australia in the early Cenozoic.

The Australian Australidelphia fall into three major orders. The Dasyuromorphia is mainly composed of the carnivorous dasyurids: the marsupial cats and marsupial mice (which would be better called marsupial shrews, since they are carnivorous and insectivorous rather than omnivorous). Some larger dasyurids include the rather doglike Tasmanian devil and the Tasmanian (marsupial) wolf or thylacine (the latter sometimes placed in its own family, Thylacinidae). The thylacine has reportedly been extinct since the 1930s, but occasional claims for its continued existence come from purported sightings or footprints. Tasmania, once part of Australia when sea levels were lower, is now an island off its south coast, and these Tasmanian animals were known from mainland Australia before the arrival of humans and their true (placental) dogs, the dingoes. The numbat (the marsupial anteater) is related to the dasyurids (included in Figure 20–2 with Dasyuromorphia), and the tiny marsupial mole *Notoryctes* may be a more distant relation.

The Peramelina includes the peramelids, the bandicoots and bilbies. These animals look rather like rabbits, and Australians celebrate an "Easter Bilby" rather than an "Easter Bunny," but they are insectivorous rather than herbivorous. Peramelids share with the final group of marsupials (the diprotodontians) a condition of the hind feet known as syndactyly, in which the second and third toes are reduced in size and enclosed within the same skin membrane so that they appear to be a single toe. The syndactylous toes are used for grooming.

The largest group of marsupials is the diprotodontians. Diprotodontians get their name from their modified lower incisors, which project forward rather like the incisors of rodents (Greek *di* = two, *proto* = first, and *dont* = tooth). This lineage includes herbivorous and omnivorous forms today, although marsupial "lions" such as the Pleistocene *Thylacoleo* appear to represent an independent evolution of carnivory from an herbivorous ancestry. *Thylacoleo* shows some interesting specializations: coming from an herbivorous ancestry, it had lost its canines but modified the incisors into caninelike teeth.

▶ **FIGURE 20–4** Diversity of living marsupials. (a) The common North American opossum, *Didelphis virginiana* (Didelphidae: Didelphimorpha). (b) The shrew opossum, *Lestoros inca* (Caenolestidae: Paucituberculata). (c) The monito del monte, *Dromiciops australis* (Microbiotheriidae: Microbiotheria). (d) The Tasmanian devil, *Sarcophilus harrisii* (Dasyuridae: Dasyuromorphia). (e) The marsupial mole, *Notoryctes typhlops* (Notoryctidae: Notoryctemorphia). (f) The bilby, or rabbit-eared bandicoot, *Macrotis lagotis* (Thylacomyidae: Peramelemorphia). (g) The honey possum, *Tarsipes rostratus* (Tarsipedidae, Phalangeroidea, Diprotodontia). (h) The koala, *Phascolarctos cinereus* (Phascolarctidae, Phascolartoidea, Diprotodontia). (i) The long-nosed potoroo (rat kangaroo), *Potorous tridactylus* (Macropodidae, Macropodoidea, Diprotodontia). (The species are drawn approximately to scale; [a] is the size of a large house cat).

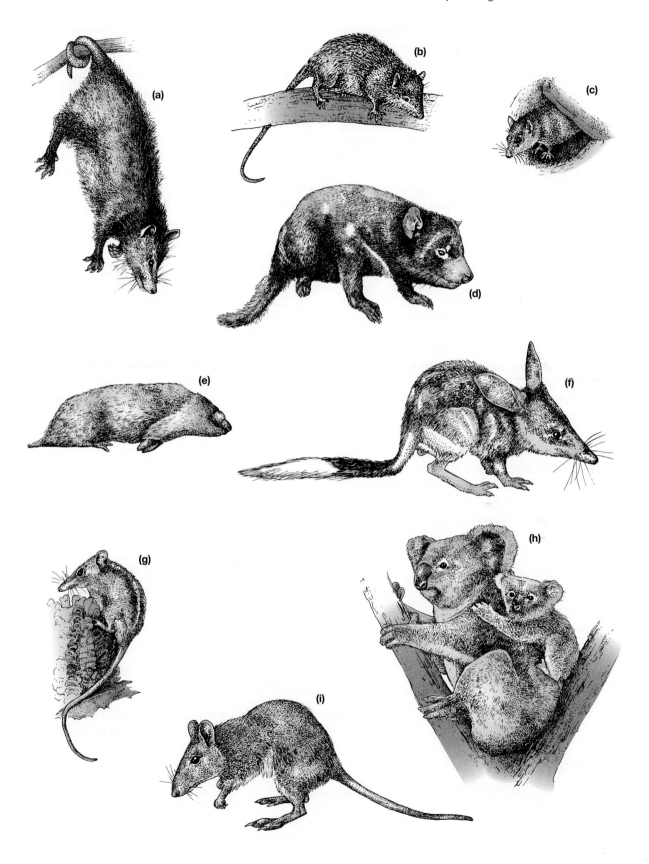

The three major radiations within the diprotodontians are the phalangeroids, phascolarctoids, and macropodoids. Phalangeroids represent an arboreal radiation of rather primatelike animals, also including gliders. They comprise six families, including possums, phalangers, ringtails, cuscuses, and the diminutive honey possum or noolbenger—the only nectar-eating mammal that is not a bat. Phascolarctoids or vombatiformes include the arboreal koala and the terrestrial, burrowing wombats. Extinct phascolarctoids include the bison-size diprotodontids that looked like giant wombats and grazed on the Plio-Pleistocene Australian savannas. Macropodoid include the small, omnivorous rat kangaroos (or potaroos) and the larger, herbivorous true kangaroos (including wallabies and tree kangaroos). Some extinct rat kangaroos were the size of a large dog and may have been carnivorous (Wroe 1999). The largest kangaroos today have a body mass of about 90 kilograms, but larger ones (up to three times that size) existed in the Pleistocene, including a radiation of one-toed, short-faced browsing sthenurine kangaroos, now all extinct. Sthenurines, with their short faces and stout forearms, could be the source of legends of giant rabbits in Australia's interior. Perhaps, as desert forms, they also had large rabbitlike ears for cooling.

Placentals Extant placental mammals can be grouped into a number of distinct taxa, but their interrelationships have been a source of controversy, suggesting that the diversification of these groups from an ancestral stock occurred very rapidly without leaving many morphological clues. The current scheme of the interrelationships among placental mammals is shown in Figure 20–2 and **Table 20.2**, and **Figure 20–5** illustrates a diversity of placental mammals.

Over the past decade, molecular studies have resulted in a very different view of placental interrelationships from phylogenies based on morphological data alone (see Madsen et al. 2001, Springer et al. 2004). Probably the most radical difference is the creation of an endemic grouping of African mammals, the Afrotheria. A grouping of Paenungulata, the rodentlike hyraxes (the conies of the Bible), the aquatic sirenians (manatees and dugongs), and the proboscideans (elephants and extinct relatives), had long been supported by morphology, but the molecular data shows that aardvarks, elephant shrews, tenrecs (Madagascan insectivorous forms) and the golden

TABLE 20.2 Classification of extant placental orders and approximate numbers of families and species. Note that different authorities recognize different numbers of families within an order. The geographic regions (explained in Figure 20–17) represent the distribution of the entire orders (and for the present day only); families within an order often have smaller distributions.

Order	Number of Families/Species	Major Examples and Higher-Level Classification
Afrotheria		
Tenrecoidea	2/44	Tenrecs, otter shrews, and golden moles; 5 g to 2 kg; Ethiopian region.
Macroscelidea	1/15	Elephant shrews; 25 to 500 g; Ethiopian region with one species in Morocco and northern Africa.
Tubulidentata	1/1	Aardvark; 64 kg; Ethiopian region.
Hyracoidea	1/7	Hyraxes (= conies or dassies); 4 kg; Ethiopian region and Asia Minor.
Sirenia	2/4	Dugongs, manatees; 140 to over 1000 kg; coastal waters and estuaries of all tropical and subtropical oceans except the eastern Pacific (in the Atlantic drainage they enter rivers).
Proboscidea	1/2	Elephants and fossil relatives; 4500 to 7000 kg; Ethiopian and Oriental regions.
Xenarthra		
Xenarthra	3/30	Anteaters, sloths, armadillos; 20 g to 33 kg; Neotropical region (plus some armadillos in southern United States of America).

(Continued)

TABLE 20.2 Continued

Order	Number of Families/Species	Major Examples and Higher-Level Classification
		Boreoeutheria
		EUARCHONTOGLIRES
		<u>Glires</u>
Lagomorpha	2/69	Rabbits, hares, pikas; 180 g to 7 kg; worldwide except Antarctica, introduced in Australia by humans.
Rodentia	29/1814	Rats, mice, squirrels, guinea pigs, capybara; 7 g to over 50 kg; worldwide except Antarctica.
		<u>Archonta</u>
Scandentia	1/16	Tree shrews; 400 g; Oriental region.
Primates	9/235	Lemurs, monkeys, apes, humans; 85 g to over 275 kg; primarily Oriental, Ethiopian, and Neotropical regions, humans are now worldwide.
Dermoptera	1/2	Flying lemurs; 1 to 2 kg; Oriental region.
		LAURASIATHERIA
Lipotyphla	4/346	Hedgehogs, moles, shrews; 2 g to 1 kg; worldwide except Australia and Antarctica (although only a single species of shrew is known from South America, a Pleistocene immigrant).
Chiroptera	15/986	Bats; 4 g to 1.4 kg; worldwide (including New Zealand) except Antarctica.
		Ferungulata
Carnivora	12/274	Dogs, bears, raccoons, weasels, hyenas, cats, sea lions, walruses, seals (these last three are often assigned to the suborder Pinnipedia); 70 g to 760 kg, some marine forms over 100 kg; worldwide.
Pholidota	1/7	Pangolins (scaly anteaters); 2 to 33 kg; Ethiopian and Oriental regions.
Perissodactyla	3/17	Odd-toed ungulates: horses, tapirs, rhinoceroses; 150 to 3600 kg; worldwide except Antarctica (horses introduced by humans into North America and Australia).
Artiodactyla	10/213	Even-toed ungulates: swine, hippopotamuses, camelids, deer, giraffe, antelope, sheep, cattle; 2 to 2500 kg; worldwide except Antarctica (introduced into Australia and New Zealand by humans).
Cetacea	9/80	Porpoises, dolphins, sperm whales, baleen whales; 20 to 120,000 kg; worldwide in oceans and in some rivers and lakes in Asia, South America, northern America, and Eurasia. Molecular data place whales within the order Artiodactyla.

mole are also closely related to these larger African mammals. Previously the tenrecs and golden moles were classified within the Insectivora (now termed the Lipotyphla), the elephant shrews were linked with rabbits and rodents, and the relationships of the

aardvark remained obscure. This endemic grouping, which usually appears to fall at the base of the placental phylogeny (as shown in Figure 20–2, although other researchers promote Afrotheria and Xenarthra as sister groups or one or both groups in a

◀ **FIGURE 20–5** Diversity of living placentals. Taxa indicated below with an asterisk (*) belong to the newly identified group Afrotheria. (a) Two-toed sloth, *Choloepus didactylus* (Megalonychidae: Xenarthra); cat-size. (b) Common tenrec, *Tenrec ecaudatus* (Tenrecidae: Tenrecoidea)*; rat-size. (c) Golden-rumped elephant shrew, *Rhynchocyon chrysopygus* (Macroscelididae: Macroscelidae)*; rat-size. (d) Naked mole rat, *Heterocephalus glaber* (Bathyergidae: Rodentia); mouse-size. (e) Flying lemur, *Cynocephalus volans* (Cynocephalidae: Dermoptera); cat-size. (f) The spotted hyena, *Crocuta crocuta* (Hyaenidae: Carnivora); wolf-size. (g) Asiatic tapir, *Tapirus indicus* (Tapiridae: Perissodactyla); pony-size. (h) African water chevrotain, *Hyemoschus aquaticus* (Tragulidae: Artiodactyla); rabbit-size. (i) The rock hyrax, *Procavia capensis* (Procaviidae: Hyracoidea)*; rabbit-size. (The species are not drawn to scale.)

more nested position within the other placentals), is indicative of a long period of independent isolated evolution of mammals on the African continent. This fits in with the patterns of Cenozoic continental movements, as Africa was isolated from Eurasia during the first half of the Cenozoic.

The embryos of elephants show evidence of an aquatic ancestry in the structure of their kidneys (Gaeth et al. 1999), suggesting that the grouping of sirenians, desmostylians (semiaquatic extinct forms), and proboscideans (termed the Tethytheria) may have been an originally aquatic paenungulate stock. Proboscideans would then represent a secondary return to a more terrestrial existence. However, popular speculation that the elephant's trunk was originally a snorkel is mistaken; the skulls of early fossil proboscideans show that these animals lacked trunks.

The next major grouping of placental mammals, long considered primitive on morphological grounds, is the edentates (order Xenarthra): sloths, anteaters, and armadillos. *Edentate* means "without teeth"; and, while only anteaters are completely toothless, all of these animals have simplified their dental pattern. Edentates are South American endemics and have only been found on the North American continent since the late Pliocene epoch. The pangolins or scaly anteaters (order Pholidota) were originally thought to be related to the edentates based on morphological features, but molecular data place them as the sister taxon to the Carnivora.

All other placental orders are placed in the supergrouping Boreoeutheria (meaning "northern mammals"). This grouping can be subdivided into the Euarchontaglires (rodents, primates, and relatives) and the Laurasiatheria (insectivores, bats, carnivores, and ungulates, including whales). Tree shrews were once thought to be primitive primates. This is now no longer believed to be true, but they are grouped with the primates, bats, and dermopterans (flying lemurs) in the Archonta. Bats (Chiroptera) were long considered to belong with primates in the Archonta, but molecular evidence has reassigned them as

closer to the insectivores. The Glires includes the rodents and rabbits.

Insectivores (members of the order Insectivora) were often considered to be the basal stock from which other placentals were derived. However, although modern insectivores such as shrews may superficially resemble ancestral placental mammals, they are not closely related to them. With the loss of tenrecs and golden moles to the Afrotheria, the grouping of true shrews, moles, and hedgehogs, is now called the Lipotyphla.

The largest grouping within the Laurasiatheria is the Ferungulata, mainly comprising carnivores and ungulates. The Carnivora, (informally termed *carnivorans* to distinguish them from other, unrelated carnivorous mammals) are distinguished by having a pair of specialized shearing teeth, or **carnassials**, formed from the upper last (fourth) premolar and the lower first molar. Carnivorans include not only specialized carnivores, such as cats, but also secondarily herbivorous forms, such as the panda bear, and one of the three living groups of secondarily aquatic mammals, the pinnipeds (seals, sea lions, and walruses), that are related to bears.

The extant ungulates (hoofed mammals), the **perissodactyls** (odd-toed ungulates) and **artiodactyls** (even-toed ungulates), were previously placed with some extinct groups in the higher taxon Ungulata, but the molecular data implies that these two orders, although closely related, are not each other's closest relatives. Some archaic ungulates in the early Cenozoic were also omnivorous or even carnivorous; in South America, there were several hoofed orders that are now entirely extinct. These extinct orders are probably related to the living ungulates, but of course they cannot be slotted into a molecular phylogeny.

A surprise in the reexamination of mammalian interrelationships over the past couple of decades has been the realization that whales and dolphins (order Cetacea) are technically ungulates (even though they lack hooves), related to the artiodactyls.

However, this grouping of *Cetartiodactyla* has been known and accepted for longer than the other changes to the traditional ideas resulting from molecular evidence and was even proposed on morphological grounds (from paleontological evidence) by some earlier researchers. There has been more dissent between molecular biologists and morphologists about the molecular placement of the whales *within* the order Artiodactyla as the sister group to hippopotamuses. Recent fossil discoveries (Gingerich et al. 2001) have shown that the earliest whales had a key morphological feature that unites the order Artiodactyla—a form of the astragulus in the ankle called the double-pulley astragalus—but the proposal of a close relationship to hippos is more controversial.

The position of the extinct mesonychids has not been resolved. These terrestrial carnivorous archaic ungulates bear a strong resemblance to early whales and were originally considered as whale ancestors. Chapter 21, Section 21.6 discusses the evolutionary history of whales in more detail.

To people familiar only with living animals, the concept of a whale-hippo link is rather appealing. Both are large, aquatic animals with several features in common, such as the ability to suckle their young underwater. However, the fossil record shows that it is highly unlikely that whales and hippos shared a common *aquatic* ancestry, even if they do prove to be sister taxa. It remains possible that whales and hippos share a very distant sister-group relationship within the Artiodactyla, possibly from an early offshoot of piglike artiodactyls called anthracotheres, but no aquatic hippolike animals are known until about 10 million years ago. Indeed, the living pygmy hippo is a predominantly terrestrial beast.

20.2 Features Shared by All Mammals

■ Lactation

The females of all mammalian species lactate, feeding their young by producing milk. Mammary glands are entirely absent from the males of marsupials, but they are present and potentially functional in male monotremes and placentals. There have been examples of human males producing milk under certain circumstances, and recently a species of fruit bat has been identified in which the male actually produces milk (Francis et al. 1994). It has long been a mystery why male mammals retain mammary

glands but do not lactate (see Daly 1979); indeed, breast cancer affects human males as well as females.

Although all mammals lactate, only therians have nipples so that the young can suck directly from the breast rather than from the mother's fur. Mammary hairs, probably a primitive mammalian feature, are present in monotremes and marsupials, and the mammae develop from areola patches confined to the abdominal region. Placentals lack mammary hairs, and the mammae develop from mammary lines that form along the entire length of the abdomen.

■ Skeletomuscular System

In Chapter 18 we discussed how the mammalian skeletal system was gradually evolved within the synapsid lineage. Here we will focus on those features that are uniquely mammalian in comparison with other extant amniotes. **Epiphyses** on the long bones reflect the mammalian feature of determinate growth. The epiphyses are the ends of the bones that are separated from the shaft (the diaphysis) by a zone of growth cartilage in immature mammals. At maturity, the ossification centers in the epiphyses and the shaft of the bone fuse, and the epiphyses no longer appear as distinct structures.

Cranial Features In the mammalian skull, the dermal bones that originally formed the skull roof have grown down around the brain and completely enclose the braincase (**Figure 20–6a**). The bones that form the lower border of the synapsid temporal opening are bowed out into a zygomatic arch, which we commonly refer to as the cheekbone. This is the bony bar that you can feel just below your eyes.

The dentition of mammals is divided into several types of teeth (a condition called heterodonty): incisors, canines, premolars, and molars. Most mammals have two sets of dentitions in their lifetime. The first set—or the milk teeth—consists of incisors, canines, and premolars only, although the form of these premolars may be like that of the adult molars. The permanent, adult dentition consists of the second set of the original teeth with the addition of the later-erupting molars. (Our last molars are known as wisdom teeth, so called because they erupt at the age—late teens—by which we supposedly have attained wisdom.) Mammals are the only animals that masticate (chew the food) and swallow a discrete bolus of food. Therian mammals have unique types of molars, called **tribosphenic molars** (**Box 20–1**).

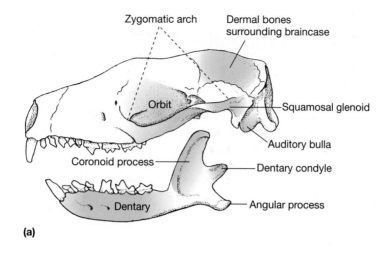

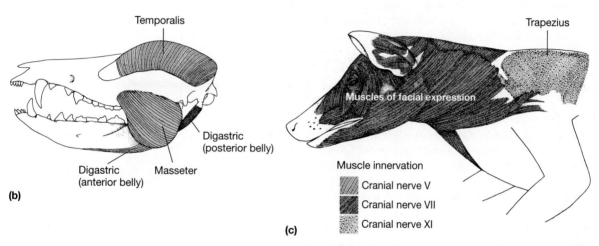

▲ **FIGURE 20–6** Cranial anatomy of mammals. (a) Skull of a generalized mammal (a hedgehog): the condyle on the dentary fits into the glenoid on the squamosal to form the jaw joint. (b) The muscles of mastication. (c) Superficial view of the muscles of facial expression plus shoulder muscles innervated by cranial nerves.

Postcranial Features Unlike the sprawling posture of extant reptiles, mammals have a more upright posture with the limbs positioned underneath the body (**Figure 20–8**). However, the highly upright posture of familiar mammals such as cats, dogs, and horses is a derived one; the semisprawling stance of a mammal such as the opossum probably represents the primitive mammalian condition.

Mammals have an ankle joint that differs from that of other amniotes. The site of movement is not within the bones of the ankle joint (the mesotarsal joint); it is between the tibia and one of the proximal ankle bones, the astragalus (a crurotarsal joint). Along with this new joint, mammals have a projection of the other proximal ankle bone, the calcaneum, to form the calcaneal heel. The heel is the point of

insertion of the gastrocnemius (calf) muscle. In the pelvic girdle, the ilium is rod-shaped and directed forward, and the pubis and ischium are short in contrast with the more platelike pubis and ischium of reptiles. The femur has a distinct trochanter on the proximal lateral side (the greater trochanter) for the attachment of the gluteal muscles, which are now the major retractors of the hindlimbs. It is the gluteals that give mammals their characteristic rounded rear ends.

With very few exceptions, all mammals have seven cervical (neck) vertebrae. (Manatees and one type of tree sloth have six cervical vertebrae, and the giraffe may have eight [Solounias 1999].) They also have a uniquely specialized atlas-axis complex of the first two cervical vertebrae; a mammal can rotate its

BOX 20–1 The Evolution of Tribosphenic Molars

Very early in mammalian history, as early as the earliest Jurassic, there was a difference observable in the types of molar teeth and patterns of occlusion. The primitive type, exemplified by the basal mammal *Morganucodon*, is to have the three main molar cusps in a more or less straight line. In the more derived type, exemplified by the contemporaneous symmetrodont *Kuehneotherium*, the principal, middle cusp is shifted so that the teeth assume a triangular form in occlusal (food's-eye) view (**Figure 20–7a**). The apex of the triangle formed by the upper tooth points inward, while that of the lower tooth points outward, forming an intermeshing relationship called reversed triangle occlusion. The upper triangle is known as the trigon, and the lower one the trigonid. The longer sides of these triangular teeth result in a greater amount of available area for shearing action. The teeth also interdigitate in a more complex fashion than do the teeth of more primitive mammals.

A further dental complication seen in later, more derived mammals is the development of the tribosphenic molar. The group of mammals having tribosphenic molars is called the Tribosphenida and includes the Theria (marsupials and placentals). In this type of molar, a new cusp, the protocone, is added to the trigon in the uppers, which occludes against a basined addition to the lowers called the talonid (**Figure 20–7b**). The tribosphenic molar adds the function of crushing and punching to the original tooth, which acted mainly to cut and shear. The possession of this tooth presumably reflects a greater diversity of dietary items taken.

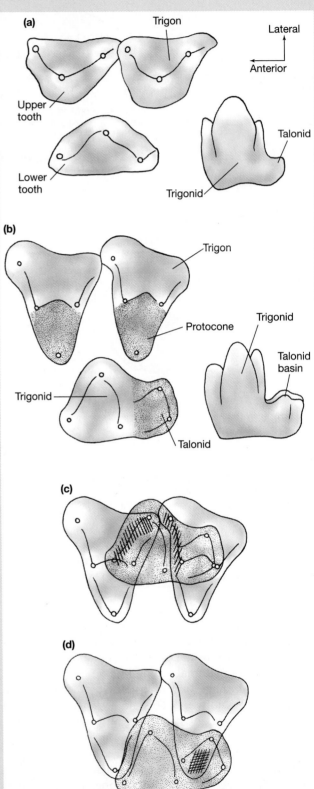

▶ **FIGURE 20–7** Evolution of mammalian molars. (a) Schematic occlusal view of reversed triangle molars of a nontherian holotherian mammal (e.g., *Kuehneotherium*). The lower molar is also illustrated in side view. (b) Similar view of the tribosphenic molars of a therian. The new portions (the protocone in the uppers and the basined talonid in the lowers) are shaded. Parts (c) and (d) show the action of the lower molars in occlusion with the uppers: The lower molar is shaded, and areas of tooth contact are crosshatched. (c) Initial contact between the teeth at the start of occlusion. The trigonid cusps produce a shearing action alongside the cusps at the back of the trigon of the anterior upper tooth and at the front of the trigon of the posterior upper tooth. (d) Mortar-and-pestle action of the molars at the end of the occlusal power stroke. The protocone (= the pestle of the combination) fits into the basin formed between the cusps in the trigonid (the mortar).

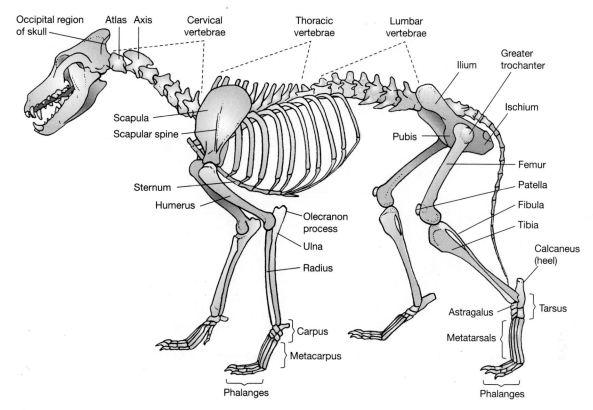

▲ **FIGURE 20–8** Skeleton of an extinct Pleistocene wolf (the dire wolf, *Canis dirus*). The long neural spines of the thoracic vertebrae serve as the attachment area of the nuchal ligament, which runs from the back of the head and helps to hold the head up. This animal is standing on the phalanges in a digitigrade form of foot posture, in contrast with the more primitive plantigrade foot posture of the opossum in Figure 20–14b.

head on its neck in two places: not only in the more general up-and-down fashion (at the joint between the skull and the atlas) but also in the more derived side-to-side fashion (at the joint between the atlas and the axis). Mammals have restricted the ribs to the anterior (thoracic) trunk vertebrae. The lumbar ribs now have zygapophyseal connections that allow only dorsoventral flexion; they also have large transverse processes for the attachment of the longissimus dorsi (one of the epaxial muscles) that produces this movement during locomotion. The capacity to twist the spine in both lateral and dorsoventral directions in mammals may relate to the ability of mammals to lie down on their sides, something that other vertebrates cannot do easily. This ability may have been important in the evolution of suckling because the nipples are on the ventral surface of the trunk.

■ The Integument

In many ways the outside covering of mammals is the key to their unique way of life. We have empha-

sized that endothermy is an energetically expensive process, and much of the mammals' ability to live in harsh climates is attributable to properties of their integument. The variety of mammalian integuments is enormous. Some small rodents have a delicate epidermis only a few cells thick. Human epidermis varies from a few dozen cells thick over much of the body to over a hundred cells thick on the palms and soles. Elephants, rhinoceroses, hippopotamuses, and tapirs were once classified together as pachyderms (Greek: *pachy* = thick and *derm* = skin) because their epidermis is several hundreds of cells thick. The texture of the external surface of the epidermis varies from smooth (in fur-covered skins and the hairless skin of whales and dolphins) to rough, dry, and crinkled (many hairless terrestrial mammals). The tail of opossums and many rodents is covered by epidermal scales similar to those of lizards but lacking the hard beta keratin found in birds and reptiles.

Figure 20–9 illustrates the typical structure of mammalian skin. Note that while mammalian skin is like that of other vertebrates in basic form, with

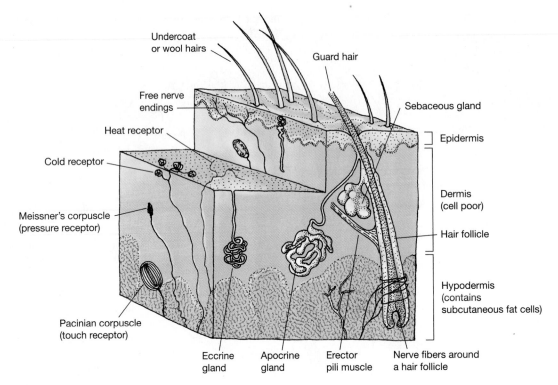

▲ **FIGURE 20–9** Structure of mammalian skin.

epidermal, dermal, and hypodermal layers, there are also unique components. Mammalian skin has hair, lubricant- and oil-producing sebaceous glands, and apocrine and eccrine glands that secrete volatile substances, water, and ions. Typical mammalian structures derived from the keratinous layer of the epidermis include nails, claws, hoofs, and horns. Sensory nerve endings include free nerve endings (probably pain receptors), beaded nerve nets around blood vessels, Meissner's corpuscles (touch receptors, seen in primates and some primatelike marsupials), Pacinian corpuscles (pressure receptors), nerve terminals around hair follicles, and warmth and cold receptors. Vascular plexuses (intertwined blood vessels) of the skin are the basis for countercurrent blood flow.

Hair Hair has a variety of functions including camouflage, communication, and sensation via **vibrissae** (whiskers). Vibrissae grow on the muzzle, around the eyes, or on the lower legs; touch receptors are associated with these specialized hairs. However, the basic function of hair is insulation. Fur consists of closely placed hairs, often produced by multiple hair shafts arising from a single complex root. Its insulating effect depends on its ability to trap air within the fur coat (or pelage), and its insulating ability is proportional to the length of the hairs. The erector pili

muscles that attach midway along the hair shaft pull the hairs erect to deepen the layer of trapped air.

Cold stimulates a general contraction of the erector pili via the sympathetic nerves, as do other stressful conditions such as fear and anger. A curious side effect noticeable in near-naked mammals such as humans is the dimples (goose pimples) on the skin's surface over the insertion of contracted erector pili muscles. Hair erections can serve for communication as well as for thermoregulation; mammals can use them to send a warning of fear or anger (as seen in the display of a puffed-up cat or the raised hackles of a dog).

Prominent features of the hairy covering of extant mammals are its growth, replacement, color, and mobility. A hair is composed of keratin, and it grows from a deep invagination of the germinal layer of the epidermis called the hair follicle. The color of hair depends on the quality and quantity of melanin injected into the forming hair by melanocytes at the base of the hair follicle (see Figure 20–9). The color patterns of mammals are built up by the colors of individual hairs. Because exposed hair is nonliving, it wears and bleaches. Replacement occurs by growth of an individual hair or by **molting**, in which old hairs fall out and are replaced by new hairs. Most mammals have pelage that grows and rests in seasonal phases; molting usually occurs only once or twice a year.

Glandular Structures Secretory structures of the skin develop from the epidermis. There are three major types of skin glands in vertebrates: **eccrine, sebaceous**, and **apocrine** glands. Except for the eccrine glands, skin glands are associated with hair follicles, and the secretion in all of them is under neural and hormonal control. A full component of these skin glands is found in monotremes as well as in therians, and thus they may be assumed to be a basic feature of all mammals.

Eccrine glands produce a secretion that is mainly watery, with little organic content. In most mammals, eccrine glands are restricted to the soles of the feet, prehensile tails, and other areas that contact environmental surfaces, where they improve adhesion or enhance tactile perception. Eccrine glands are found over the body surface only in primates and especially in humans, where they secrete copious amounts of fluid for evaporative cooling. Profuse thermoregulatory sweating has evolved convergently within mammals, because eccrine glands function as sweat glands in humans, whereas ungulates sweat via apocrine glands. The sweat glands of humans do not appear to be a primitive mammalian feature, and most mammals do not thermoregulate by secreting fluid from skin glands for evaporative cooling. For example, dogs pant to keep cool, and kangaroos lick their forearms. In humans, sweat glands may act in conjunction with nearby apocrine glands, contributing to odor production under conditions of stress and excitement.

Sebaceous glands are found over the entire body surface. They produce an oily secretion, sebum, which lubricates and waterproofs the hair and skin. Sheep lanolin, our own greasy hair, and the grease spots that the family dog leaves on the wallpaper where it curls up in the corner are all sebaceous secretions.

Apocrine glands have a restricted distribution in most mammals, and their secretions appear to be used in chemical communication. In humans, apocrine glands are found in the armpit and pubic regions—these are the secretions that we usually try to mask with deodorant. In some other mammals, such as large ungulates, these glands are scattered over the body surface and are used in evaporative cooling.

Many mammals have specialized scent glands that are modified sebaceous or apocrine glands. Sebaceous glands secrete a viscous substance, usually employed to mark objects, whereas apocrine glands produce volatile substances that may be released into the air as well as placed on objects.

Scent marking is used to indicate the identity of the marker and to define territories. Scent glands are usually placed on areas of the body that can be easily applied to objects, such as the face, chin, or feet. Domestic cats often rub their face and chin to mark objects, including their owners. Many carnivorans have anal glands so that scents can be deposited along with the urine and feces, and apocrine anal scent glands are a well-known feature of skunks. There are also apocrine glands in the ear that produce earwax.

Mammary glands have a more complex, branching structure than do other skin glands. They have several basic features in common with apocrine and sebaceous glands: structure, body distribution, and chemical composition of secretion. The evolution of mammary glands may have occurred with the formation of a new type of skin gland that contained properties of both apocrine and sebaceous glands; though they resemble the other two types of glands, mammary glands cannot be fully homologized with either one alone (Blackburn 1991).

Claws, Nails, Hooves, and Horns Some integumentary appendages are involved in locomotion, offense, defense, or display. Claws, nails, and hooves are accumulations of keratin that protect the terminal phalanx of the digits (**Figure 20–10**). Permanently extended claws are the primitive condition; the retractable claws of catlike carnivores are derived. The fingernails of humans and other primates are a simpler structure than either the retractable claw or the hoof but were derived from ancestral claws. The hoof of ungulates is an extensively modified nail covering the entire third phalanx. This morphology gives ungulates a small foot (which is mechanically advantageous for a running animal) that is solid enough to bear the animal's weight. Hyraxes, which are ungulate-like mammals, have an intermediate condition of hooflike nails that do not bear the animals' weight (they retain an extensive foot pad); this may have been the condition in many of the smaller, primitive extinct ungulates.

The horns of ungulates may also be formed, at least in part, from keratin. Rhinoceros horns are formed entirely from matted keratin fibers, while the horns of cows and antelope are formed from a keratin sheath covering a bony core (see Figure 20–10d).

■ Internal Anatomy

Mammals have numerous differences from other amniotes in their internal anatomy and physiology.

(a)

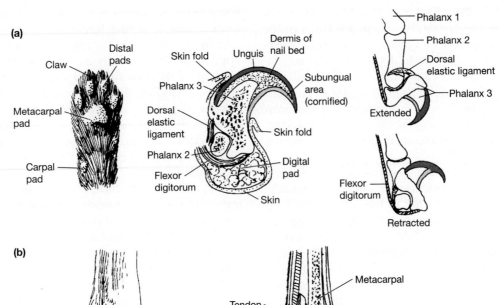

Claw

Distal pads

Metacarpal pad

Carpal pad

Skin fold

Phalanx 3

Unguis

Dermis of nail bed

Subungual area (cornified)

Dorsal elastic ligament

Phalanx 2

Flexor digitorum

Skin fold

Digital pad

Skin

Phalanx 1

Phalanx 2

Dorsal elastic ligament

Phalanx 3

Extended

Flexor digitorum

Retracted

(b)

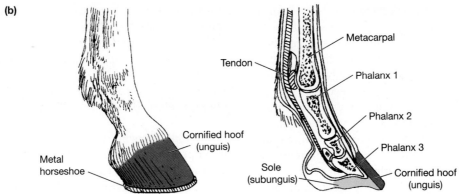

Metal horseshoe

Cornified hoof (unguis)

Tendon

Metacarpal

Phalanx 1

Phalanx 2

Phalanx 3

Sole (subunguis)

Cornified hoof (unguis)

(c)

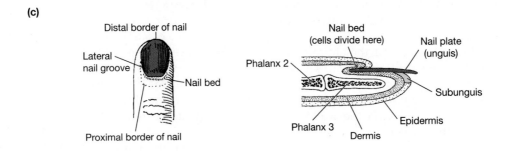

Distal border of nail

Lateral nail groove

Nail bed

Proximal border of nail

Nail bed (cells divide here)

Nail plate (unguis)

Phalanx 2

Subunguis

Phalanx 3

Dermis

Epidermis

(d)

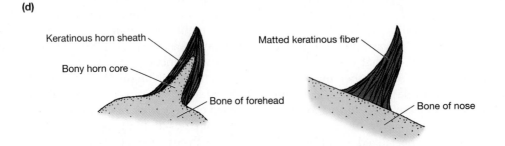

Keratinous horn sheath

Bony horn core

Bone of forehead

Matted keratinous fiber

Bone of nose

◀ **FIGURE 20–10** Skin appendages associated with terminal phalanges. The unguis is the fully keratinized portion of the appendage (claw, nail or hoof), and the subunguis is the less keratinized portion underlying it. (a) Retractable claws. *Left:* Hair and thick epidermal pads associated with the base of the claws (primitive mammalian condition). *Center:* Longtitudinal section of a claw showing its close relationships with the blood vessels, dermis, and bone of the third (terminal) phalanx. *Right:* Claw retraction mechanism characteristic of cats. (b) The hoof of a horse. *Left:* Normal appearance of the hoof of a shod horse. (Horseshoes are devices used to minimize wear of the hoof on unnaturally hard and abrasive surfaces.) *Right:* Longitudinal section of lower front foot showing relationship of phalanges to hoof. (c) The human nail. *Left:* The external appearance of the nail. *Right:* A longtitudinal section of the end of a finger showing the association of the nail with the epidermis and terminal phalanx of the digit. (d) Horns of ungulates. *Left:* the horn of a cow (animal facing toward the left), a bony core covered with a keratin sheath. *Right:* the horn of a rhinoceros (animal facing toward the right), made entirely from keratin fibers.

Some of these relate to their endothermic metabolism; similar systems have evolved convergently in the other endothermic vertebrates, birds. Others are uniquely mammalian and reflect their evolutionary history.

Adipose Tissue Mammalian adipose tissue (white fat) is distributed as subcutaneous fat, as fat associated with various internal organs (e.g., heart, intestines, kidneys), as deposits in skeletal muscles, and as cushioning for joints. Adipose tissue is not simply inert material used only as an energy store when fasting or as an insulating layer of blubber in marine mammals, as has often been assumed. More recent studies have revealed that adipocytes (fat cells) secrete a wide variety of messenger molecules that coordinate important metabolic processes. Some small fat storage sites have specific properties that equip them to interact locally with the immune system and possibly other organs (see Pond 1998).

Mammals also have a unique type of adipose tissue—brown fat. This tissue is specially adapted to generate heat and can break down lipids or glucose to generate energy as heat at a rate up to ten times that of the muscles. Brown fat is especially prominent in newborn mammals, where it is important for general thermoregulation, and in the adults of hibernating species, where it is used to rewarm the body during emergence from hibernation.

Cardiovascular System The mammalian heart has a complete ventricular septum and only a single systemic arch (**Figure 20–11**), although the original double arch is apparent in development. Mammals differ from other vertebrates in the form of their erythrocytes (red blood cells), which lack nuclei in the mature condition. Additionally, while monotremes retain a small sinus venosus as a distinct chamber, therians have incorporated this structure into the

▶ **FIGURE 20–11** Diagrammatic view of the mammalian heart. Black arrows indicate the passage of deoxygenated blood, green arrows the passage of oxygenated blood. The ductus arteriosus (ligamentum arteriosum in the adult condition) is the remains of the dorsal part of the pulmonary arch; it is functional in the amniote fetus, where it is used as a bypass shunt for the lungs, but closes after birth.

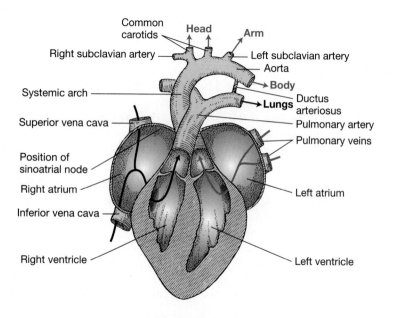

right atrium as the sinoatrial node, which now acts as the heart's pacemaker.

Respiratory System Mammals have large, lobed lungs with a spongy appearance due to the presence of a finely branching system of bronchioles in each lung, terminating in tiny thin-walled, blind-ending chambers (the sites of gas exchange) called **alveoli**. They also have a muscular sheet, the diaphragm, which aids the ribs in inspiration and divides the original pleuroperitoneal cavity into a **peritoneal cavity** surrounding the viscera and paired **pleural cavities** surrounding the lungs.

Urogenital System All mammals retain the bladder and excrete relatively dilute urine. Mammals have entirely lost the renal portal system seen in other vertebrates, which supplies venous blood to the kidney in addition to the arterial blood supplied by the renal artery. Mammals also have a new portion of the kidney tubule called the loop of Henle, correlating with their ability to excrete urine that has a higher concentration of salt than the body fluids.

Therian mammals differ from other vertebrates in various ways. In most vertebrates, the urinary, reproductive, and alimentary systems reach the outside via a single common opening, the cloaca, while in therians the cloaca is replaced by separate openings for the urogenital and alimentary systems. In most (but not all) species of therians, the testes are placed in a scrotum outside of the body in the males (see Section 21.1), and the penis is used for urination as well as for the passage of sperm, with the ducts leaving the testes and the bladder combined into a single **urethra** (**Figure 20–12**). The clitoris in the female is the homolog of the penis but is not used to pass urine.

The glans is the bulbous distal end of the penis. Monotremes and most marsupials have a bifid (forked) glans, whereas placentals have a single glans. Some male placentals have a bone in their penis, the os penis or **baculum**; females may also have a corresponding structure, called the os clitoris. This structure is seen among certain species in nonhuman primates, rodents, insectivores, carnivorans, and chiropterans (more astute students may notice a mnemonic here) but is presumed to be a primitive placental feature and thus homologous among these groups.

The urethra and vagina are joined in a single **urogenital** sinus leading to the outside in most female mammals, but primates and some rodents have the more derived condition of separate openings for the urinary and genital systems. Note that, in the more usual mammalian condition, the clitoris is

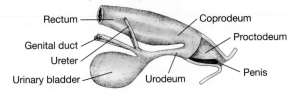

Male monotreme (similar to general amniote condition)

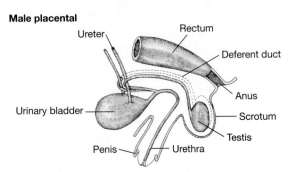

Male placental

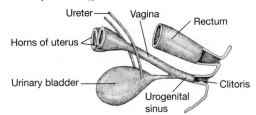

Female placental (generalized)

Female primate

▲ **FIGURE 20–12** Anatomy of the urogenital ducts in mammals. The head is to the left.

within the urogenital sinus, where it receives direct stimulation from the penis during copulation, a rather more practical arrangement than in humans. Primates also are unusual in having a pendulous penis that cannot be retracted into the body. Most male mammals extend the penis from an external sheath (normally the only visible portion) only during urination and copulation, again a seemingly more practical design. The anatomy of primates may be related to the relatively high position of the body wall. In most mammals the body wall encloses a greater portion of the femur (thigh bone) than it does in primates, providing more intra-abdominal space for luxuries such as a retractable penis.

■ Sex Determination and Sex Chromosomes

Mammals always have distinct sexes, and sex determination is by distinctive sex chromosomes, the X and Y chromosomes: females have two X chromosomes, and males have an XY combination. However, the platypus is peculiar in having multiple sex chromosomes: males have 5X and 5Y chromosomes, and the mode of sex determination is unclear (Warren et al. 2008). In therians it appears that a gene located on the Y chromosome in mammals initiates male gonadal development, and female gonadal development results from its absence. Once a gonad has had its primary sex declared as female or male, one of the sex hormones—estrogen or testosterone—is produced. These hormones affect development of the secondary sex characteristics. In humans, the genitalia, breasts, hair patterns, and differential growth patterns are secondary sex characteristics. Horns, antlers, and dimorphic color patterns are familiar differences that we associate with gender in other mammals.

■ Sensory Systems

The sensory systems of mammals differ from those of other tetrapods in various ways. Mammals have exceptionally large brains among vertebrates, and their brains evolved along a pathway somewhat different from that of other amniotes. Mammals are more reliant on hearing and olfaction and less reliant on vision than are most other tetrapods.

The Brain   The enlarged portion of the cerebral hemispheres of mammals, the neocortex or neopallium, forms differently from the enlarged forebrain of derived sauropsids (see Section 11.6). Other unique features of the mammalian brain include an infolded cerebellum and a large representation of the area for cranial nerve VII, which is associated with the facial musculature. In addition to the anterior commissure, a structure that links the hemispheres in all amniotes, placentals have a new nerve tract linking the two cerebral hemispheres—the corpus callosum. Variations in the brains of mammals are related to ecological specializations, and many convergences are apparent (e.g., Brown 2001).

Mammals have divided optic lobes in the midbrain. As their name suggests, the optic lobes are normally concerned with processing nerve impulses from the eyes, although these impulses are interpreted in the brain. However, the optic lobes take on other roles in some species of mammals. In mole rats, an entirely subterranean group of rodents, a specific structure for detecting magnetic fields is found in the larger of the optic lobes, the superior colliculus (Nêmec et al. 2001).

Olfaction   The keen sense of smell of most mammals is probably related to their primarily nocturnal behavior. The olfactory receptors are located in specialized epithelium on the scroll-like nasoturbinal and ethmoturbinal bones in the nose. The olfactory bulb is a prominent portion of the brain in many mammals, but primates have a relatively small olfactory portion of the brain and a poor sense of smell, probably in association with their diurnal and arboreal habits. The sense of smell is also reduced or completely absent in whales.

Vision   Mammals evolved as nocturnal animals, and visual sensitivity (forming images in dim light) was more important than visual acuity (forming sharp images). Mammals have retinas composed primarily of rod cells, which have high sensitivity to light (thus providing good night vision) but are relatively poor at acute vision. Acute vision is possible only in one small region of the retina, the all-cone fovea. In addition to providing high visual acuity, cones are the basis for color vision. However, most mammals can perceive some color and have monochromatic (one visual color pigment) or dichromatic (two pigments) vision. Anthropoid primates are unique among mammals in having good color vision and a brain that is specialized for the visual sensory mode (Barton 2007).

Hearing   The middle ear of mammals is more complex than that of other tetrapods. It contains a chain of three bones (stapes, malleus, and incus) rather than just a single bone. Although mammals have a greater acuity of hearing than other tetrapods, and a chain of ear ossicles seems to amplify the force of vibration on the inner ear, the anatomy of the mammalian middle ear is an evolutionary accident rather than the result of direct selection for increased auditory capacity (see Box 18–1).

Several other features of therian mammals also contribute to increased auditory acuity. These include a long cochlea capable of precise pitch discrimination. The cochlea of mammals is so long that it must be coiled to fit inside the otic capsule (**Figure 20–13**). In addition, the external ear (or **pinna**) helps to determine sound direction. The pinna, in combination with the narrowing of the external auditory meatus of mammals, concentrates sound from the relatively large area encompassed by the external opening of the pinna to the small, thin, tympanic membrane. The pinna is unique to mammals, although it has a feathery analog

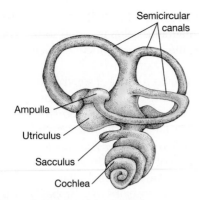

▲ **FIGURE 20–13** Design of the vestibular apparatus in therian mammals. The long, coiled cochlea is equivalent to the lagena of other vertebrates.

in certain owls (see Section 17.6). Most mammals can move their pinnae to detect sound, although anthropoid primates lack this capacity. The auditory sensitivity of a terrestrial mammal is reduced if the pinnae are removed. Aquatic mammals use entirely different systems to hear underwater and have reduced or lost the pinnae. Cetaceans, for example, use the lower jaw to channel sound waves to the inner ear.

20.3 Features that Differ Between Mammal Groups

Differences Between Therians and Nontherians

Therians are distinguished from monotremes by a number of derived features. The most obvious one is giving birth to young rather than laying eggs. The differences in reproduction between different types of mammals are considered in Chapter 21. Therians are also distinguished by the possession of mammae with nipples, a cochlea in the inner ear with at least two and a half coils, an external ear (pinna), and tribosphenic molars.

Therians also have distinctive features of the skull and skeleton. Therians have completely lost the sclerotic bony rings around the eyes. Monotremes retain sclerotic cartilages, although they do not ossify to form a bony ring, as seen in many other amniotes, including nonmammalian synapsids. Almost all therians have lost the septomaxilla, a bone in the skull near the nostril that is present in monotremes. While all mammals have a specialized ankle joint, known as a crurotarsal joint, in therians the astragalus bone in the ankle sits on top of the calcaneus, rather than

being placed side by side with it (**Figure 20–14**). The ankle hinge joint is now restricted to the astragalus and the tibia, whereas in monotremes the fibula-calcaneum joint forms part of the hinge as well. This more derived ankle joint probably made possible new forms of locomotion, such as bounding and hopping.

The therian shoulder girdle is also extremely derived. While monotremes have the derived mammalian form of pelvis, their shoulder girdle is more reminiscent of the typically reptilian condition. Therians have lost the ventral elements of the shoulder girdle, the coracoid and interclavicle bones, although these bones are retained in marsupial newborns (see Section 21.1). Therians also have expanded the scapula with the addition of a scapular spine. This spine was long thought to represent the anterior border of the scapula, but developmental studies now show that most of the scapular spine is a new addition (Sánchez-Villagra and Maier 2003). The clavicle (collarbone) is retained in most therians but is lost in many running-adapted placentals (e.g., dogs and horses).

This reduction of the ventral elements of the shoulder girdle allows the scapula to move as an independent limb segment around its dorsal border, adding to the length of the stride during locomotion. Additionally, certain hypaxial muscles have become modified to hold the limb girdle in a muscular scapular sling (**Figure 20–15**). This muscle arrangement probably also aids with scapular mobility during locomotion and cushions the impact of body weight landing on the front limbs during bounding. The therian mammalian type of bounding gait may require this modification of the shoulder girdle anatomy and musculature (see Fischer et al. 2002).

The shoulder musculature has also been reorganized to reflect the change in scapular function. A muscle that originally protracted the humerus on the shoulder girdle, the supracoracoideus, has been modified into two new muscles, the infraspinatus and supraspinatus, running from the humerus to either side of the scapula spine. This muscular arrangement may help to stabilize the limb on the shoulder girdle during bounding.

Differences Between Marsupials and Placentals

Although there are numerous differences between marsupials and placentals in their reproductive biology, as will be discussed in Chapter 21, there are few major differences in their anatomy.

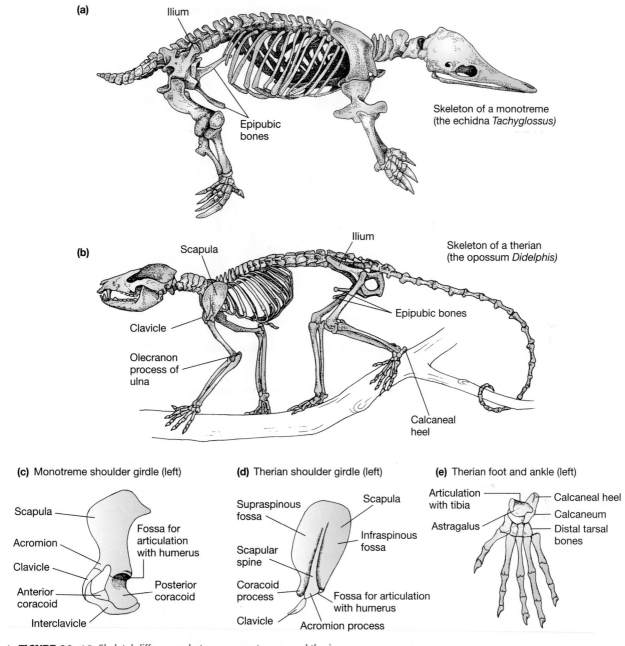

(a)

Ilium

Epipubic bones

Skeleton of a monotreme (the echidna *Tachyglossus*)

(b)

Scapula

Ilium

Skeleton of a therian (the opossum *Didelphis*)

Clavicle

Epipubic bones

Olecranon process of ulna

Calcaneal heel

(c) Monotreme shoulder girdle (left)

Scapula

Acromion

Clavicle

Anterior coracoid

Interclavicle

Fossa for articulation with humerus

Posterior coracoid

(d) Therian shoulder girdle (left)

Supraspinous fossa

Scapula

Infraspinous fossa

Scapular spine

Coracoid process

Clavicle

Fossa for articulation with humerus

Acromion process

(e) Therian foot and ankle (left)

Articulation with tibia

Astragalus

Calcaneal heel

Calcaneum

Distal tarsal bones

▲ **FIGURE 20–14** Skeletal differences between monotremes and therians.

Features in the skull and dentition distinguish marsupials from placentals, although not all of these features apply to all marsupials. The lower jaw of marsupials (with the exception of the koala) has a distinct inturned projection from the angle of the dentary (where the pterygoideus muscle inserts) that is lacking in placentals, and their nasal bones abut the frontal bones with a flared, diamond shape in contrast to the rectangular shape of the placentals' nasals (**Figure 20–16**). Many placentals also have an elaboration of bone around the ear region, the **auditory bulla**, that probably increases auditory acuity. Marsupials usually lack a bulla, or have a small one formed by a different bone from that of placentals. Herbivorous placentals may have a bar of bone behind the orbit called the postorbital bar, but this is never seen in marsupials. During ontogeny, most placentals replace all their teeth except for the molars, whereas marsupials replace only the last premolar.

▶ **FIGURE 20–15** Specializations of shoulder girdle musculature. The rhomboideus and the serratus ventralis in the mammalian scapular sling are new muscles, derived from the hypaxial layers.

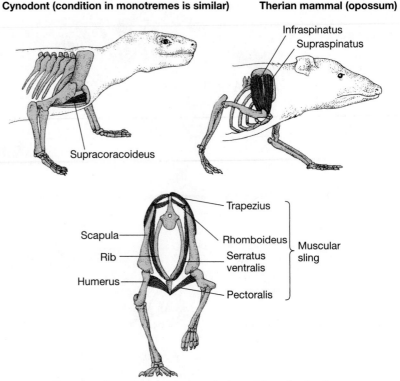

Cynodont (condition in monotremes is similar)

Therian mammal (opossum)

Infraspinatus
Supraspinatus

Supracoracoideus

Trapezius

Scapula

Rhomboideus

Rib

Serratus ventralis

Humerus

Pectoralis

Muscular sling

Scapular sling of therian mammal, viewed from the front

Marsupials also differ from placentals in their dental formula, or numbers of different types of teeth. The primitive dental formula of placentals (i.e., the maximum number of teeth usually seen, as counted on one side of the skull only) is three upper and three lower incisors (a total of 12 incisors, count-ing both sides), one upper and one lower canine (a total of 4), four upper and four lower premolars (a total of 16), and three upper and three lower molars (a total of 12). Many placentals have fewer teeth than this, but only a few mammals with highly special-ized diets, such as armadillos and porpoises, have

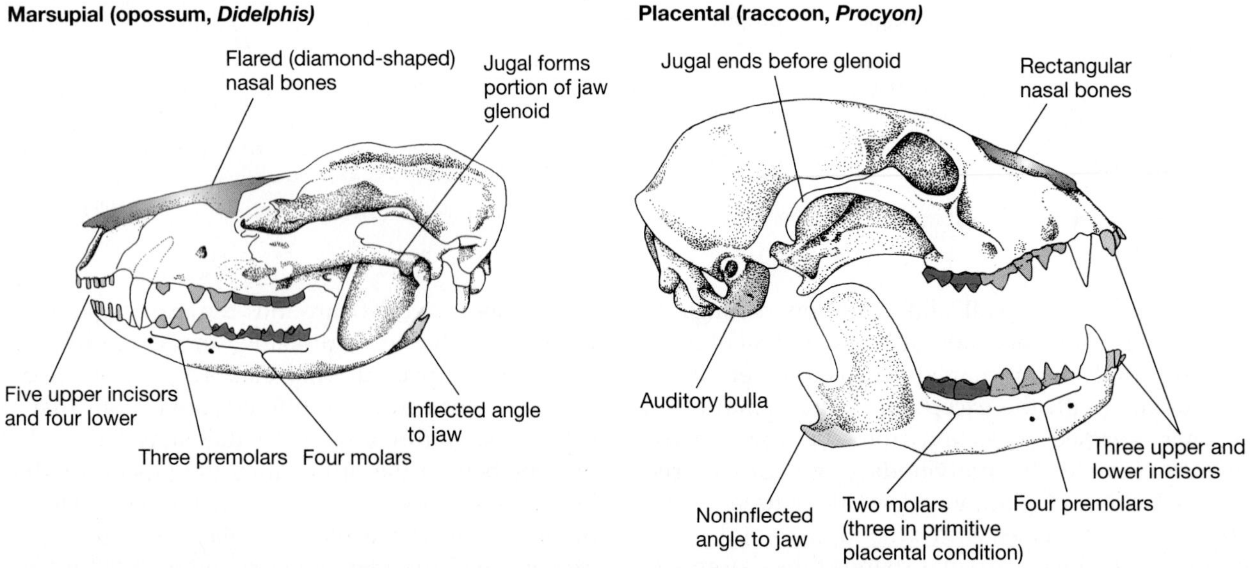

Marsupial (opossum, *Didelphis*)

Flared (diamond-shaped) nasal bones

Jugal forms portion of jaw glenoid

Five upper incisors and four lower

Three premolars   Four molars

Inflected angle to jaw

Placental (raccoon, *Procyon*)

Jugal ends before glenoid

Rectangular nasal bones

Auditory bulla

Noninflected angle to jaw

Two molars (three in primitive placental condition)

Four premolars

Three upper and lower incisors

▲ **FIGURE 20–16** Skull differences between marsupials and placentals.

more teeth than the standard placental formula. Humans have lost a pair of incisors and two pairs of premolars from each side, so we have a total of only 32 teeth instead of the primitive placental component of 44. Marsupials have a greater number of incisors (five uppers and four lowers) and molars (four uppers and lowers) than do placentals and have fewer premolars (three uppers and lowers) (see Figure 20–16).

The postcranial skeleton of marsupials can be distinguished from the skeleton of placentals primarily by the **epipubic bones** that project forward from the pubis. Epipubic bones were once considered a unique feature of marsupials that supported the pouch, but they are now recognized as a primitive mammalian feature that is retained in a few Cretaceous placentals (Novacek et al. 1997). Recent studies of extant marsupials (Reilly and White 2003) have shown that the epipubic bones act in concert with the thigh and abdominal muscles to stiffen the torso and to resist bending of the trunk during locomotion. This feature is probably related to the upright stance of the earliest true mammals and also to the fact that they have a reduced pubis and ischium in comparison with cynodonts.

However, it appears that epipubic bones limit asymmetrical movements of the hind limbs in gaits such as the gallop. This may be the reason why the only quadrupedal cursorial marsupial, the recently extinct Tasmanian wolf, has lost these bones (interestingly enough, this was also the case in the long extinct borhyaenids that took the role of marsupial wolflike predators in South America). The loss of epipubic bones in most placentals may be for the locomotor reasons outlined above or may relate to the fact that these bones are rigid components of the abdominal wall that would interfere with the expansion of the abdomen during pregnancy. (These two hypotheses are not mutually exclusive.)

20.4 Cenozoic Mammal Evolution

During the early Cenozoic, the continents had not moved as far from their equatorial positions as they are today, but the major landmasses were more isolated than at present. Australia had broken free from the other southern continents by the mid-Eocene, but North and South America did not come into contact until the Pliocene. India collided with Asia in the Miocene and formed the Himalayan mountain ranges. Africa made contact with Eurasia in the late Oligocene or early Miocene epochs, closing off the original east-west expanse of the Tethys Sea to form the now enclosed Mediterranean basin

■ Time and Place of Modern Mammal Origins

The more derived mammals with tribosphenic molars (including therians) were probably originally a Northern Hemisphere radiation, perhaps originating after North America and Eurasia separated from Gondwana. The fossil record shows that the evolutionary split between marsupial and placental lineages had occurred by the late Early Cretaceous. This event probably occurred in Asia, as the earliest eutherians and metatherians are both known from China (see Section 18.4). Both metatherians and eutherians are reasonably well known from the Late Cretaceous of Asia and North America; these two continents were in land contact via Beringia at this time. There has been much controversy as to whether the modern placental orders originated in the Northern Hemisphere (as suggested by the pattern of the fossil record) or the Southern Hemisphere (as suggested by the pattern of phylogeny, with both of the basal placental clades, Afrotheria and Xenarthra, being of southern origin). The balance of evidence now indicates a northern origin (Wible et al. 2007).

Multituberculates were primarily Northern Hemisphere residents, but a distinct southern branch, the gondwanatheres, has now been identified from the Cretaceous of South America, Africa, and Madagascar. Monotremes were probably originally Australian in origin, but a fossil monotreme was found in Patagonia, at the tip of South America (Pascual et al. 1992). It is not clear if this animal represents dispersal across Antarctica from Australia or the remnants of an earlier, unrecorded, Gondwanan fauna that spread across the southern continents.

The radiation of extant orders of placentals has long been assumed to be correlated with the disappearance of the dinosaurs, and there is little fossil evidence of members of extant orders until the start of the Cenozoic, 65 million years ago. In contrast, many molecular researchers, using molecular clock estimates of times of divergence of the major orders, would place the origin of modern placental orders much earlier, between 120 and 148 million years ago in the Early Cretaceous (Kumar and Hedges 1998, Bininda-Emonds et al. 2007). This interpretation is not only at odds with the fossil record, however, but it also implies that there is a gap in the fossil record for placentals during the time of the dinosaurs that is longer than their Cenozoic record following dinosaur extinction, and a gap that large seems unlikely (Foote et al. 1999).

The discrepancy between the fossil record evidence and molecular clock inferences about the time of origin of the modern placental orders could perhaps be explained if the orders had diverged genetically that long ago but still looked pretty similar during the Mesozoic, with changes in morphology not occurring until the Cenozoic. An alternative interpretation is that rapid morphological evolution at the start of the Cenozoic resulted in the speeding up of the rate of molecular evolution. If this were true, it could mean that traditional molecular techniques of estimating divergence times would provide a date that was much too early, as more molecular change would have happened during the Cenozoic than would be expected. Additionally, the fossil record for the Cretaceous is actually rather good. Many types of mammals are preserved, but none have the unique features shared by the modern placental orders until perhaps the very end of the Cretaceous (although good fossil evidence is lacking from the southern continents). Mathematical models of the likelihood that placentals were actually present in all but the latest Cretaceous, but just not preserved in the fossil record, show that this hypothesis is highly unlikely (Foote et al. 1999, Benton 1999). Some recent molecular studies (Kitazoe et al. 2007) and new fossil record evidence (Wible et al. 2007) propose a placental radiation commencing in the latest Cretaceous, no more than 85 million years ago. Paleontologists are more comfortable with this date: obviously placental mammals must have started diverging some time before their first fossil record appearance at 65 million years ago, and a fossil record gap of 20 million years is more plausible than one of 60 million years or more.

■ Biogeography of Cenozoic Mammals

At the start of the Cenozoic, all mammals were small and fairly unspecialized. The marsupials appear to have been omnivorous and arboreal, like modern opossums, while the placentals were mostly shrewlike terrestrial primitive insectivorous forms or archaic ungulates.

The radiation of mammals occurred during this fractionation of the continental masses, when different stocks became isolated on different continents. This separation of ancestral stocks from one another by the Earth's physical processes, rather than by their own movements, is called **vicariance.** Some of the difference in the distribution of present-day mammals (e.g., the isolation of the monotremes in Australia and New Guinea) results from vicariance biogeography, that is, animals and plants being carried passively on moving landmasses.

Other patterns of mammal distribution can be explained by **dispersal**, which reflects movements of the animals themselves, usually by the spread of populations rather than the long-distance movements of individual animals. The presence of marsupials in Australia apparently represents dispersal from South America via Antarctica in the early Cenozoic. The subsequent isolation of Australia may have allowed marsupials to diversify on that continent. Dispersal can produce extinctions as well as radiations. For example, when the Turgai Straits that had separated Europe and Asia during the Paleocene and Eocene dried up in the early Oligocene, mammals from Asia flooded into Europe, and some uniquely European mammals (mainly archaic types) became extinct. This episode of extinctions was so dramatic that it is known as *La Grande Coupure* (the Great Separation).

While today marsupials are considered the quintessential Australian mammals, they did not reach that continent until the early Cenozoic. Marsupials reached North America (from Asia) in the Late Cretaceous, where they enjoyed a modest radiation of small forms, and the initial Cenozoic diversification of marsupials occurred primarily in South America. During the Cenozoic, marsupials dispersed not only across the southern continents but also across the northern ones. Generalized, primitive, rather opossum-like marsupials are known from the Paleocene, Eocene, and Oligocene of North America, where they are fairly numerous as fossils, and the Eocene and/or Oligocene of Europe, Asia, and northern Africa, where they were apparently much more rare. Some fragmentary remains of marsupials are known from as late as the mid-Miocene in North America and Europe, but the abundance and diversity of marsupials in the Northern Hemisphere was severely limited after the mid-Eocene. Their mid-Cenozoic extinction in that part of the world may be just part of the extinction of many archaic mammals taking place at that time. There is no need to invoke a complex scenario of competition with placentals. The original stock of North American marsupials has been extinct for at least 12 million years. The present-day North American native marsupial, the common opossum *Didelphis virginiana*, is a Pleistocene immigrant from South America. Opossums are still moving north and were first recorded in Canada in the 1950s.

Today the mammals of the Northern Hemisphere (Holarctica) and of Africa, Madagascar, South America, and Australia are strikingly different from one another. These geographic groupings fall into three major faunal provinces: (1) Laurasian fauna (in Holarctica, Africa, and Madagascar), consisting almost exclusively of placentals; (2) South American, or New World tropical fauna, containing a mixture of placentals and marsupials; and (3) Australian fauna, containing monotremes, marsupials, and a sprinkling of placentals.

However, the fauna of Africa was even more distinct from the fauna of the rest of Holarctica before Africa collided with Eurasia in the Oligocene-Miocene times. For example, hyraxes, which now survive only as small rodentlike forms, assumed ecological roles taken today by antelope and pigs. Similarly, the South American fauna was more distinct from that of Holarctica before South America was connected with North America in the Pliocene, with the role of large carnivores taken by the now-extinct borhyaenoid marsupials (**Figure 20–17**), and the large

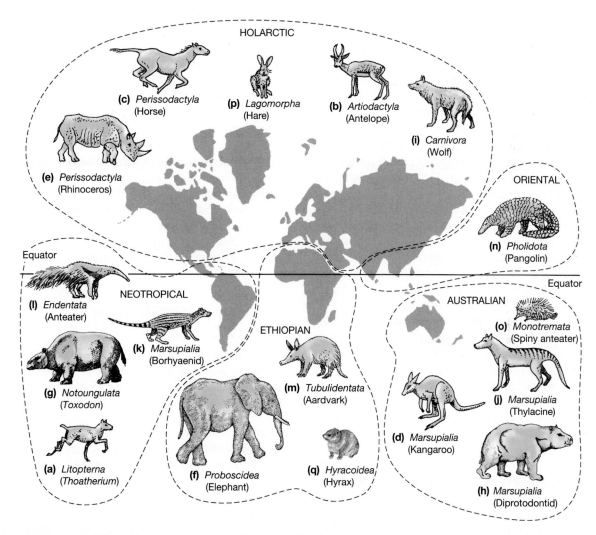

▲ **FIGURE 20–17** Radiation and convergence of mammals evolving in isolation during the Cenozoic era. Mammals are grouped with landmasses on which they probably originated. Major biogeographical regions of the world today are illustrated. Current convention, reflecting the historical patterns of colonization by Europeans, is to call American continents the New World and other continents (including Australia) the Old World. The northern continents, North America and northern Eurasia, can be grouped together as Holarctica; they can be further subdivided into the Nearctic (North America, including Greenland) and the Palaearctic (Europe and northern Asia, including Asia Minor). India and Southeast Asia fall within the tropics and together are called the Oriental region. Africa (including Madagascar) forms the Ethiopian or African region. The Ethiopian and Oriental regions are sometimes grouped together as the Old World Tropics, or Paleotropics. South and Central America form the Neotropical region, or New World Tropics. Australia and associated islands (including New Guinea, Tasmania, and New Zealand) make up the Australian region.

herbivores by the now-extinct native ungulates of uncertain phylogenetic affinities.

Even the North American fauna was much more distinct from that of the rest of Holarctica for most of the Cenozoic than it is today. Much of the present North American fauna (e.g., deer, bison, and rodents such as voles) emigrated from Eurasia in the Pliocene and Pleistocene via the Bering land bridge. Today only Madagascar and Australia retain distinctly different faunas from the rest of the world.

■ Convergent Evolution of Mammalian Ecomorphological Types

The term *ecomorphology* describes the way in which an animal's form (its morphology) is adapted to its activities in its environment (its ecology). The separate evolution of different basic mammalian stocks on different continents illustrates the convergent evolution of ecomorphs in Cenozoic mammals.

Specialized running herbivores (Figure 20–17a-d) include the extinct litopterns of South America, the Holarctic artiodactyl antelope of Eurasian origin, and the perissodactyl horses (most of the evolution of horses took place in North America). These ungulates are strikingly similar to each other in many respects. Even the Australian marsupials such as the kangaroo demonstrate convergences in jaws, teeth, and feeding behavior to placental herbivores. Convergent evolution also took place among large-bodied, slower-moving herbivores (Figure 20–17e-h), such as the Holarctic rhinoceroses (perissodactyls), the elephants (which originated in Africa), the extinct notoungulates of South America, and the extinct marsupial diprotodontids of Australia.

Carnivorous mammals show similar convergences (Figure 20–17i-k). The true wolves of Holarctica, the recently extinct marsupial wolf or thylacine of Australia, and the extinct marsupial borhyaenids of South America have similar body shapes and tooth forms, although only the true wolves have the long legs that distinguish fast-running predators. Similar types of intercontinental convergences, again involving both marsupials and placentals, occurred with a carnivorous ecomorph that does not exist today—the saber-toothed catlike predator.

Mammals specialized for feeding on ants and termites (myrmecophagy; Figure 20–17l-o) include the giant anteater in South America, the aardvark in Africa, the pangolin in tropical Asia and Africa, and the spiny anteater (a monotreme) in Australia. Ants and termites are social insects that employ group defense. They often build impressive earthen nests containing thousands of individuals, some of which are strong-jawed soldiers. The convergent specializations of these unrelated or distantly related mammals include a reduction in the number and size of teeth, changes in jaw and skull shape and strength, and forelimbs modified for digging. Some mammals with a carnivorous ancestry have adopted a myrmecophagous lifestyle and show similar, though less extensive, specializations. These species include the Australian numbat, an African hyena, the aardwolf, and a Madagascan viverrid, the fanalouc (Malagasy mongoose).

Other examples of convergent evolution of ecomorphs abound, such as burrowing moles or mole-like animals and gliders. Gliding ecomorphs include the flying squirrels of the Northern Hemisphere and the marsupial flying phalangers in Australia, as well as the flying lemurs (not true primates) of Southeast Asia (see Figure 20–5e) and a completely different type of flying rodent (the "scaly-tailed squirrel," which is not closely related to northern squirrels) in Africa. Sometimes similar forms evolved under what appear to be much less isolated conditions. Thus hares and rabbits (Figure 20–17p) evolved in Holarctica where other small herbivores, the related rodents, also occurred. The original rodentlike animals of Africa are the hyraxes (Figure 20–17q), which had a much greater diversity of sizes and body forms in the past.

■ Cenozoic Mammals of the Southern Continents

The mammals of Australia, the island of Madagascar, and South America differ from those of the Northern Hemisphere, providing clear examples of the effects of biogeographical isolation.

Cenozoic Isolation of Australia The mammalian fauna of Australia has always been composed almost entirely of marsupials and monotremes. The earliest Australian marsupial fossils are of early Eocene age, and there are abundant remains from the early Miocene onward. South America, Antarctica, and Australia were still close together in the early Cenozoic. Australia was probably populated by marsupials that moved from South America across Antarctica, which was warm and ice-free until about 35 million years ago. Marsupial fossils are known from the Eocene of western Antarctica, and, in fact, the major barrier to dispersal to Australia was probably crossing the midcontinental mountain ridge between western and eastern Antarctica.

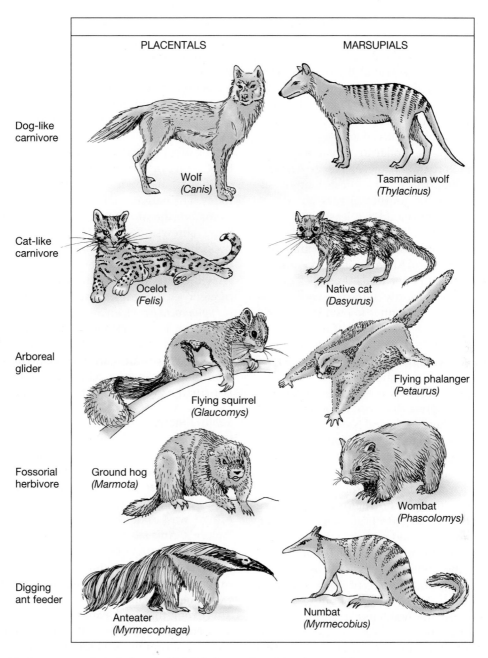

PLACENTALS | MARSUPIALS

Dog-like carnivore — Wolf (*Canis*) | Tasmanian wolf (*Thylacinus*)

Cat-like carnivore — Ocelot (*Felis*) | Native cat (*Dasyurus*)

Arboreal glider — Flying squirrel (*Glaucomys*) | Flying phalanger (*Petaurus*)

Fossorial herbivore — Ground hog (*Marmota*) | Wombat (*Phascolomys*)

Digging ant feeder — Anteater (*Myrmecophaga*) | Numbat (*Myrmecobius*)

▲ **FIGURE 20–18** Convergences in body form and habits between placental and marsupial mammals.

Once marsupials reached Australia, they enjoyed the advantages of long-term isolation, evolving to fill a variety of niches with food habits ranging from complete herbivory to carnivory. **Figure 20–18** illustrates some Australian marsupials and notes their convergences with northern placentals. Note that the possums and phalangers are a diverse radiation of arboreal mammals that are convergent not only with squirrels but also with primates.

A particularly striking example of convergence exists between the aye-aye of Madagascar (a lemur, a type of primitive primate) and the striped possum of the York Peninsula of northeastern Australia. Both pry wood-boring insects out from under tree bark with an elongated finger used as a probe, and both have chisel-like incisors. Certain insectivores (apatemyids) from the early Cenozoic of Holarctica had similar specializations but became extinct in the early

Miocene when woodpeckers evolved—birds that also feed on wood-boring insects. Madagascar and Australia are the only places in the world today that lack woodpeckers, which suggests that the woodpeckers elsewhere outcompeted the wood-boring mammals.

The early Eocene fauna of Australian mammals revealed a remarkable discovery: in addition to several marsupial species, it preserved the molar of a placental mammal that was similar to archaic ungulates known from the early Cenozoic of North America (Godthelp et al. 1992). This fossil suggests that marsupials did not make the journey across Antarctica alone; they were accompanied by at least one type of placental. This placental lineage evidently became extinct because other terrestrial placentals are unknown in Australia until the latest Miocene, around 7 million years ago. This animal rather turns the tables on the supposed marsupial inferiority in the presence of placentals. The only other placentals known in the early Cenozoic of Australia are bats. Their affinities appear to be with the bats of Asia, which suggests that they flew to Australia by moving from island to island along the Indo-Australian Archipelago.

Rodents originally arrived in Australia in the latest Miocene, probably via dispersal along the island chain between Southeast Asia and New Guinea. Australian rodents are an interesting endemic radiation today; they are related to the mouse-rat group of Eurasian rodents but have evolved into unique Australian forms such as the small jerboa-like hopping mice and the large (otter-size) water rats. True mice and rats arrived later, in the Pleistocene. However, these rodents apparently had surprisingly little overall effect on the Australian marsupials. A far greater threat has been the prehistoric invasion by humans and dogs (dingoes), and the recent introduction of domestic mammals such as foxes, rabbits, and cats, which exterminated the mainland marsupial carnivores. Today, numerous marsupial species are threatened or endangered by humans and their introduction of exotic species.

Madagascar Madagascar is an island off the coast of East Africa. It apparently separated from Africa in the mid-Mesozoic, and its present-day mammalian fauna represents subsequent immigration. However, the source of these immigrants, whether from the African mainland or from Asia (the source of the present-day indigenous people of Madagascar), remains in debate. The best-known endemic mammals of Madagascar are the lemurs, a radiation of primitive primates known nowhere else in the world. Diverse as today's lemurs are, their diversity was much greater in the recent past. As recently as a few hundred years ago, much larger (up to gorilla-size) lemurs—both terrestrial and arboreal—paralleled the radiation of the great apes among the anthropoid primates. The extinction of these giant lemurs appears to be related to the arrival of humans a couple of thousand years ago and is probably due to habitat destruction rather than to hunting.

The other native Madagascan mammals belong to the Carnivora and the Afrotheria, and are composed of forms unknown from elsewhere in the world today. The native small omnivores and insectivorous forms are the afrotherian tenrecs, many resembling large shrews. Tenrecs also have diversified into hedgehoglike spiny forms, molelike burrowers, and otterlike swimmers. The carnivorans are viverrids, related to the African and Asian mainland civets and mongooses. The most remarkable of these is the fossa (*Cryptoprocta ferox*), an animal a little larger than a big house cat. In the absence of true cats, the fossa has evolved to become a catlike predator. Although it is more stockily built than true cats, with shorter, more muscular legs that reflect its arboreal habits, its skull is virtually indistinguishable from that of a cat.

South American Mammals and the Great American Interchange From the Late Jurassic until the late Cenozoic, South America was isolated from North America by a seaway between Panama and the northwestern corner of South America. In the Pliocene, about 2.5 million years ago, the Panamanian land bridge was established between North and South America. Animals from the two continents were free to mix for the first time in more than 100 million years (Marshall 1988). Faunal interchange by island dispersal or rafting commenced in the late Miocene as the two American landmasses drew nearer to each other. This event, the Great American Interchange (GAI), is a spectacular example of the effects of dispersal and faunal intermingling between two previously separated landmasses (**Figure 20–19**).

Superficially the mammals moving from North America to South America appear to have been more numerous and to have fared better than those moving in the opposite direction. For many years, the interchange was viewed as an example of the competitive superiority of Northern Hemisphere mammals. However, with reconsideration of the available evidence, a different interpretation is preferred today. Before discussing the GAI in more detail, first we need to consider the diversity of mammals that inhabited South America before the interchange (see MacFadden 2006).

▲ FIGURE 20–19 Mammalian taxa involved in the Great American Interchange. A dagger (†) indicates taxa that are now totally extinct; an asterisk (*) indicates taxa now extinct in that area.

Three major groups of mammals can be distinguished in South America prior to the GAI. In fact, much of the original diversity of the South American mammal fauna was extinct by the time of the GAI, and competition with northern immigrants was not responsible for their demise.

• *Early inhabitants, known from the Paleocene, evolving in situ or originating from North America: marsupials,*

edentates, and archaic ungulates. Edentates were known only from South America prior to the Pliocene. The past diversity of edentates was considerably greater than that of today, including the armadillo-related glyptodonts (cow-size beasts encased in a turtlelike carapace of dermal bone) and ground sloths (ranging from the size of a large dog to the size of a rhinoceros).

The endemic South American ungulates, now all extinct, radiated into several orders. The ones surviving at the time of the GAI were the litopterns and the notoungulates. Litopterns were lightly built, cursorial animals that were small and pony-like or larger and camel-like. Notoungulates were stockier and diverged into small rodentlike forms or large rhinolike forms.

Equally important were the groups of North American mammals that failed to colonize South America at this time, notably insectivores, carnivorans, and rodents. We have no explanation of why these groups did not succeed in populating South America, but their absence is important because it left adaptive zones open, and marsupials radiated into them during the isolation of South America in the early Cenozoic.

- *Late Eocene or early Oligocene colonizers, probably arriving by rafting from Africa.* These include the caviomorph rodents (e.g., guinea pigs, agoutis, and capybaras) and the South American monkeys. The caviomorphs diversified during the Miocene and Pliocene to include the largest rodents that have ever lived. The recently discovered *Phoberomys*, from the late Miocene of Venezuela (Sánchez-Villagra et al. 2003) was the size of a small rhinoceros, and even today the pig-size capybara is extremely large for a rodent.
- *Late Miocene arrivals, arriving from North America via an island arc linking the two continents.* These include raccoons (moving from north to south) and some types of small ground sloths (moving from south to north).

With the establishment of the Panamanian land bridge in the Pliocene some animals from North America moved southward, and some South American forms moved northward. Overall, more species moved from north to south than in the opposite direction, but this difference may be an artifact of the relative sizes of North and South America. North America has a greater land area than South America, and, as a result, it has more mammalian taxa. The greater number of north-to-south movements may only mean that there were more species of mammals to the north when the Panamanian land bridge formed.

On each continent the newcomers and the native fauna appeared to coexist, and for the most part the immigrants enriched the existing fauna rather than displacing it. However, a disparity appeared during the Pleistocene extinctions. Although these extinctions affected the largest mammals on both continents, the southern immigrants in North America

were affected more profoundly than were the North American forms in South America.

Today about half the generic diversity of South American mammals consists of forms with a North American origin, although some notable northern immigrants to South America are now extinct there—for example, gomphotheres, which were related to elephants (Figure 20–19), and horses. The southern species that persist in North America are mostly confined to Central America (e.g., capybaras) and the southern United States (e.g., armadillos). Opossums and porcupines are exceptions—they have extended their geographic ranges to northern North America and have remained successful there.

A key to understanding the apparent greater success of the immigrants to South America lies in understanding the importance of biogeography. The equator and much of tropical America lies within South America. During climatic stress, such as the glacial periods, South America retains more equable habitats than North America, so fewer extinctions would be expected. Vegetational changes associated with climatic changes obviously also played a role in the survival or extinction of species. The Isthmus of Panama probably had savanna-like habitats linking North and South America during the Pliocene, acting as a corridor for the southward movement of horses and deer and the northward movement of glyptodonts and ground sloths.

By the Pleistocene, the Central American corridor was evidently closed to migration of savanna-adapted mammals, possibly because of the development of tropical forests. Mammals that arrived in North America only in the Pleistocene, such as bison and mammoths, never reached South America. The glyptodonts and ground sloths that had migrated northward would also have been unable to return to South America. When the North American savannas disappeared at the end of the Pleistocene, these southern-originating edentate mammals were unable to adapt to the cooler prairies that housed the bison and so became extinct.

A final point of consideration in the GAI is that counts of who moved where—and when—usually consider only species known from the fossil record (as in Figure 20–19). Because tropical habitats rarely preserve fossils, we have little information about the fossil history of Central America. Yet the large diversity of opossum-like marsupials, edentates, monkeys, and caviomorph rodents in Central America today can only have come from South America. Some other Central American mammals that are traditionally considered as northern taxa, such as cats (ocelots and pumas) and ungulates (brocket deer and tapirs), may also have evolved into their present form in South America and reimmigrated back into North America. Because of our present-day political boundaries, we often forget that Central America is geologically part of North America, not South America, even though we group its flora and fauna together with that of South America as the Neotropics. A proper tally of the immigrants from South America to Central America is necessary before we can write the final chapter on the Great American Interchange.

Summary

The major groups of living mammals are monotremes and therians (marsupials and placentals); another important group, the multituberculates, did not survive the early Cenozoic. Placentals are by far the most diverse group of living mammals, and they are found worldwide, while monotremes and marsupials are confined to the Southern Hemisphere (with the exception of the North American opossum). New molecular techniques have altered previous notions of the phylogenetic relationships of placental mammals: most notable is the recognition of an endemic group of African mammals, the Afrotheria.

All mammals share uniquely derived features. Lactation is the most obvious mammalian feature, and all mammals have hair and a variety of skin glands used for hair lubrication and waterproofing, olfactory communication, and thermoregulation. Mammary glands evolved from these types of skin glands. Therians are not only more derived than monotremes in their mode of reproduction (viviparity as opposed to egg-laying) but are more derived in many features of the skull, dentition, skeleton, and soft anatomy. Only a few dental and skeletal differences distinguish marsupials and placentals, most notably the loss of the epipubic bones in placentals.

The diversity of Cenozoic mammals can be best understood in the context of changing patterns of biogeography. Distribution of some mammalian groups reflects vicariance, as in the monotremes of Australia. Other patterns of distribution reflect dispersal, such as the migration of marsupials to Australia from South America. Isolation of the various continents has resulted in unique mammalian faunas in Australia, Madagascar, and South America. The South American fauna was even more different from the rest of the world until the formation of the Isthmus of Panama around 2 million years ago.

Additional Readings

Archer, M. et al. 1985. First Mesozoic mammal from Australia—An early Cretaceous monotreme. *Nature* 318:363–366.

Barton, R. A. 2007. Evolutionary specialization in mammalian cortical structure. *Journal of Evolutionary Biology* 20:1504–1511.

Benton, M. J. 1999. Early origins of modern birds and mammals: Molecules vs. morphology. *BioEssays* 21:1043–1051.

Bininda-Emonds, O. R. P. et al. 2007. The delayed rise of present-day mammals. *Nature* 446:507–511.

Blackburn, D. G. 1991. Evolutionary origins of the mammary gland. *Mammal Review* 21:81–96.

Brown, W. M. 2001. Natural selection of mammalian brain components. *Trends in Ecology and Evolution* 16:471–473.

Cifelli, R. L. 2001. Early mammalian radiations. *Journal of Paleontology* 75:1214–1226.

Daly, M. 1979. Why don't male mammals lactate? *Journal of Theoretical Biology* 78:325–345.

Fischer, M. S. et al. 2002. Basic limb kinematics of small therian mammals. *Journal of Experimental Biology* 205:1315–1338.

Flannery, T. F. et al. 1995. A new family of monotremes from the Cretaceous of Australia. *Nature* 377:418–420.

Foote, M. et al. 1999. Evolutionary and preservational constraints on the origins of major biologic groups: divergence times of eutherian mammals. *Science* 283:1310–1314.

Fox, R. C., and C. S. Scott. 2005. First evidence of a venom delivery system in extinct mammals. *Nature* 435:1091–1093.

Francis, C. M. et al. 1994. Lactation in male fruit bats. *Nature* 367:691–692.

Gaeth, A. P. et al. 1999. The developing renal, reproductive, and respiratory systems of the African elephant suggest an aquatic ancestry. *Proceedings of the National Academy of Sciences* 96:5555–5559.

Gingerich, P. D. et al. 2001. Origin of whales from early artiodactyls: Hands and feet of Eocene Protocetidae from Pakistan. *Science* 293:2239–2242.

Godthelp, H. et al. 1992. Earliest known Australian Tertiary mammal fauna. *Nature* 356:514–516.

Hurum, J. H. et al. 2006. Were mammals originally venomous? *Acta Palaeontologica Polonica* 51:1–11.

Kielan-Jaworowska, Z. 1997. Characters of multituberculates neglected in phylogenetic analyses of early mammals. *Lethaia* 29:249–266.

Kitazoe, V. et al. 2007. Robust time estimation reconciles views of the antiquity of placental mammals. *PLoS ONE* 4:e834.

Kumar, S., and S. B. Hedges. 1998. A molecular timescale for vertebrate evolution. *Nature* 392:917–920.

Long, J. et al. 2002. *Prehistoric Mammals of Australia and New Guinea*. Baltimore, MD: The Johns Hopkins Press.

MacFadden, B. J. 2006. Extinct mammalian biodiversity of the ancient New World tropics. *Trends in Ecology and Evolution* 21:157–165.

Madsen, O. et al. 2001. Parallel adaptive radiations in two major clades of placental mammals. *Nature* 409:610–614.

Marshall, L. G. 1988. Land mammals and the Great American Interchange. *American Scientist* 76:380–388.

Nêmec, P. et al. 2001. Neuroanatomy of magnetoreception: The superior colliculus involved in magnetic orientation in a mammal. *Science* 294:366–368.

Novacek, M. J. et al. 1997. Epipubic bones in eutherian mammals from the Late Cretaceous of Mongolia. *Nature* 389:483–486.

Pascual, R. et al. 1992. First discovery of monotremes in South America. *Nature* 256:704–706.

Pond, C. M. 1998. *The Fats of Life*. Cambridge, UK: Cambridge University Press.

Reilly, S. M., and T. D. White. 2003. Hypaxial motor patterns and the function of epipubic bones in primitive mammals. *Science* 299:400–402.

Rose, K. D. 2006. *The Beginning of the Age of Mammals*. Baltimore: Johns Hopkins University Press.

Rose, K. D., and J. D. Archibald (eds.) 2005. *The Rise of Placental Mammals*. Baltimore: Johns Hopkins University Press.

Sánchez-Villagra, M. R., and W. Maier. 2003. Ontogeny of the scapula in marsupial mammals, with special emphasis on perinatal stages of didelphids and remarks on the origin of the therian scapula. *Journal of Morphology* 258:115–129.

Sánchez-Villagra, M. R. et al. 2003. The anatomy of the world's largest extinct rodent. *Science* 301:1708–1710.

Solounias, N. 1999. The remarkable anatomy of the giraffe's neck. *Journal of Zoology* 247:257–268.

Springer, M.S. et al. 2004. Molecules consolidate the placental mammal tree. *Trends in Ecology and Evolutionary Biology* 19:430–438.

Vaughan, T. A. et al. 2000. *Mammalogy*, 4th. ed. Philadelphia, PA: Saunders.

Vrba, E. S. 1992. Mammals as a key to evolutionary theory. *Journal of Mammalogy* 73:1–28.

Warren, W. C., et al. 2008. Genome analysis of the platypus reveals unique signatures of evolution. *Nature* 453:175–184.

Webb, S. D. 1991. Ecogeography and the Great American Interchange. *Paleobiology* 17:266–280.

Wible. J. R. et al. 2007. Cretaceous eutherians and Laurasian origin for placental mammals near the K/T boundary. *Nature* 447:1003–1006.

Wroe, S. 1999. Killer kangaroos and other murderous marsupials. *Scientific American* 280:58–64.

CHAPTER 21

Mammalian Specializations

Cenozoic mammals are a highly diverse group of organisms, adapted to a wide variety of lifestyles and displaying broad ecological and morphological diversity. The three major types of living mammals—monotremes, marsupials, and placentals—can be distinguished by profound differences in their reproduction. Differences in morphological specializations can also be seen in the skull and teeth for feeding, in the postcranial skeleton for locomotion, and in the brain and sense organs.

21.1 Mammalian Reproduction

The mode of reproduction is the major, and most obvious, difference among the three major groups of extant mammals. While all mammals lactate and care for their young, monotremes are unique among living mammals in laying eggs. The egg-laying mode of monotremes is most likely the primitive mammalian condition because egg laying is the generalized condition for amniotes. With the evolution of viviparity, the uterine glands that add the shell and other egg components are lost. Thus it seems highly likely that it would be difficult or impossible to return to oviparity once a lineage has become dedicated to viviparity. Among the therians, marsupials and placentals have profound differences in their relative lengths of gestation; marsupials are familiar to many people as pouched mammals, the pouch housing the young that would still be inside the uterus of a comparable placental mammal.

All mammals grow from an initial embryonic ball of cells (the **blastocyst**) that forms both the embryo and the **trophoblast.** The trophoblast is a differentiation of extraembryonic tissue specialized for obtaining nutrition in the uterus, for producing hormones to signal the state of pregnancy to the mother, and (in therians) for helping the embryo implant into the uterine wall. All mammals have a trophoblast, but the distinction between the inner cell layers of the blastocyst (forming the embryo) and outer cell layers (forming the trophoblast) is more distinct in placentals than in monotremes and marsupials and develops slightly later during the initial stages of cell division (Selwood and Johnson 2006).

Additionally, all mammals have a glandular uterine epithelium (**endometrium**), which secretes materials that nourish embryos in the uterus, and a **corpus luteum** (Latin *corpus* = body and *lut* = yellowish; plural *corpora lutea*), a hormone-secreting structure formed in the ovary from the follicular cells remaining after the egg is shed. Hormones secreted by the corpus luteum are essential for the establishment and at least the initial maintenance of pregnancy, although only in placentals is there feedback from the placenta that maintains the life of the corpus luteum.

■ Reproductive Mode of Monotremes

The reproductive tract of monotremes retains the primitive amniote condition. The two oviducts remain separate and do not fuse in development except at the base, where they join with the urethra from the bladder to form the **urogenital sinus.** The oviducts swell to form a uterus, in which the fertilized egg is retained (only the left oviduct is functional in the platypus). In all mammals, the eggs are fertilized in the anterior portion of the oviduct, the **fallopian tube,** before they enter the uterus. The ovaries of monotremes are bigger than those of therians, and monotremes provide the embryo with yolk. However, the eggs of monotremes are much smaller at ovulation than are those of reptiles or birds of similar body size. The amount of yolk is not sufficient to sustain the embryo until hatching, and the eggs are retained in the uterus, where they are nourished by maternal secretions and increase in size before the shell is secreted. The eggshell is leathery, like that of some lizards, rather than rigid like the calcareous eggshells of birds.

Monotremes lay one or two eggs, and the young hatch soon after the egg is laid (after only 12 days for the platypus). The young are at an almost embryonic stage when they hatch (**Figure 21–1**), and brooding by the mother continues for a further 16 weeks. The platypus usually lays its eggs in a burrow, whereas echidnas keep their eggs in a ventral pouch that resembles the pouch of marsupials but is probably not homologous with it. All monotremes have a low reproductive rate—no more than once a year.

■ Reproductive Mode of Therians

Placentation All therians have a placenta, which is formed from the extraembryonic membranes of the fetus. Marsupials and placentals are often thought to differ in their type of placentation, and it is sometimes stated that marsupials lack a placenta. However, there is more similarity in placentation between these two types of mammals than is commonly believed.

All marsupials and some placental mammals have an initial **choriovitelline** placenta, developed from the yolk sac (although this structure is vestigial in humans). Placentals also have a later-developing

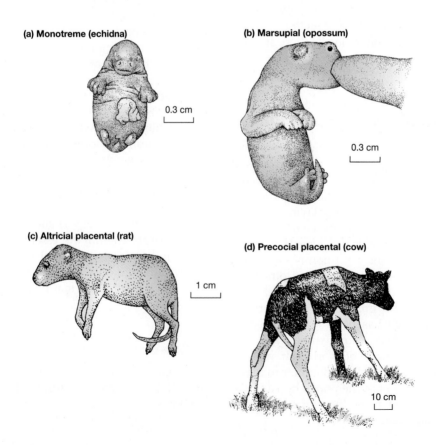

(a) Monotreme (echidna)

0.3 cm

(b) Marsupial (opossum)

0.3 cm

(c) Altricial placental (rat)

1 cm

(d) Precocial placental (cow)

10 cm

▲ **FIGURE 21–1** Mammalian neonates.

▶ **FIGURE 21–2** Two types of mammalian placental structures as seen in a transitional stage of an implanted embryo of a cat. Both a choriovitelline placenta and a chorioallantoic placenta are present at this stage. The chorioallantoic placenta grows outward and takes over the function of the earlier-forming choriovitelline placenta.

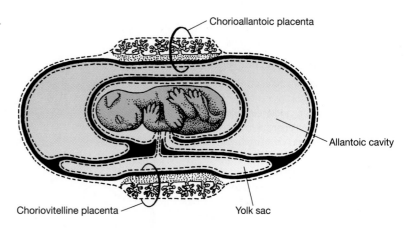

Chorioallantoic placenta

Allantoic cavity

Choriovitelline placenta

Yolk sac

chorioallantoic placenta, developed from the combination of the chorionic and allantoic amniote membranes. Some placentals retain a choriovitelline placenta even after the chorioallantoic placenta has developed (**Figure 21–2**). Six distinct layers of tissues separate fetal and maternal blood in most placentals, but the placenta of some placental mammals (including anthropoid primates and some rodents) penetrates so deeply into the uterine lining that only a layer or two of tissues separates the fetal and maternal blood systems. There is much variety in the form of the mammalian placenta as well as in the types of placentation (see Mossman 1987 for details).

Most marsupials have only a choriovitelline placenta during their short gestation, but some marsupials show a transitory chorioallantoic placenta near the end of gestation (**Figure 21–3**). The chorioallantoic placenta is best developed in bandicoots, where it is vascularized and invasive, but a less developed chorioallantoic placenta is also seen in koalas and wombats. The lack of a chorioallantoic placenta in most marsupials represents suppression of the chorioallantoic membrane during development; the rate of outgrowth of this membrane in development is slower than that seen in all other amniotes, not just other mammals.

Embryonic diapause The ability to maintain the embryo in a state of suspended animation prior to implantation is an important reproductive feature of some therians. This capacity enables the mother to space successive litters and to separate the time of mating and fertilization from the start of gestation. Thus, diapause allows mating and the birth of young to occur at optimal times of the year. Embryonic diapause is particularly well developed in kangaroos and has often been perceived as a derived marsupial feature. However, embryonic diapause occurs in a wide variety of placentals, including carnivorans, rodents, bats, edentates, and at least one artiodactyl, the roe deer (Renfree and Shaw 2000).

Testes and Scrotum Monotremes are like other vertebrates in having testes that are retained within the abdomen. Some therians, both marsupials and placentals, also retain the testes in the abdomen, either in the original position high within the body or partially descended and housed subcutaneously at the base of the abdomen. However, the testes of most therian mammals descend into a scrotum during development.

The value of a scrotum is obscure. The traditional idea is that a scrotum provides a cooler environment for sperm production, but there is no simple correlation between core body temperature and testicular position among mammals. The scrotum probably evolved convergently in marsupials and placentals because the control of scrotal development is different in the two groups. In marsupials testicular descent is under direct genetic control, whereas in placentals it is hormonally determined. Additionally, the scrotum is in front of the penis in marsupials and behind the penis in most placentals, although there are some exceptions (e.g., rabbits have a prepenile scrotum).

▓ Reproduction of Placental Mammals

In all therians, the ureters draining the kidney enter the base of the bladder rather than the cloaca or urogenital sinus, as in most other vertebrates. Backwashing of the urine from the cloaca into the bladder may be a good enough design for most vertebrates; indeed, the bladder is lost in a number of nonmammalian amniotes, including many lepidosaurs and most birds. However, in a viviparous

▶ **FIGURE 21–3** Types of placentation in marsupials and placentals. (a) Egg-laying monotreme. (b) Dasyurid—the allantois reaches the chorion and then retreats from it without forming a placental structure. (c) Bandicoot—a complex chorioallantoic placenta is formed at the close of gestation, and the choriovitelline placenta remains functional until the young are born. (d) Possums and kangaroos—the allantois may grow to a large size but remains enshrouded in the folds of the yolk sac wall. (e) Koala and wombat—the allantois reaches the chorion, forming an apposed chorioallantoic placenta. (f) Placental—choriovitelline placenta is short-lived, and a complex chorioallantoic placenta is the functional one for most of the gestation. Dark blue indicates areas of placentation. (1 = vascular choriovitelline placenta; 2 = novascular choriovitelline placenta; 3 = syncytialized choriovitelline placenta; 4 = apposed (nonvascular) chorioallantoic placenta; 5 = syncytialized (vascular) chorioallantoic placenta; AC = allantoic cavity; EC = extraembryonic coelom; S = shell; YC = yolk-sac cavity)

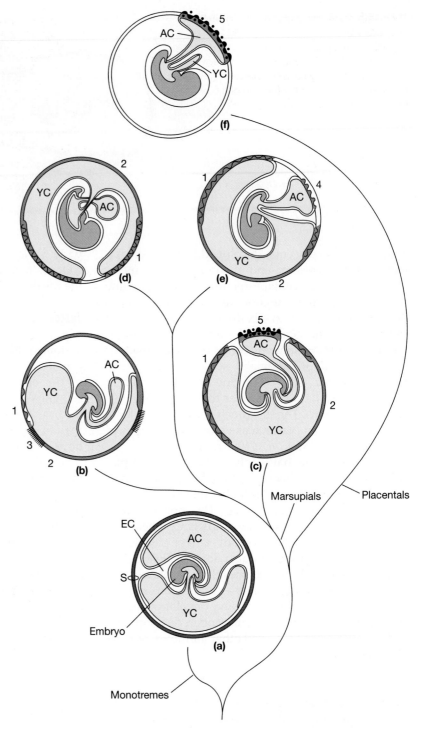

mammal, urine might enter the uterus containing the developing young, and repositioning of the ureters may have occurred at this evolutionary juncture (Renfree 1993).

In placentals, the ureters pass laterally around the developing reproductive ducts to enter the bladder. This arrangement allows the oviducts of females to fuse in the midline anterior to the urogenital sinus for much of their length (**Figure 21–4**). In males, this anatomical arrangement results in the **vasa deferentia** (the male reproductive tracts; Latin *vas* = vessel; singular *vas deferens*) looping around the ureters in their passage from the scrotum to the urogenital sinus. All placentals have a single, midline vagina,

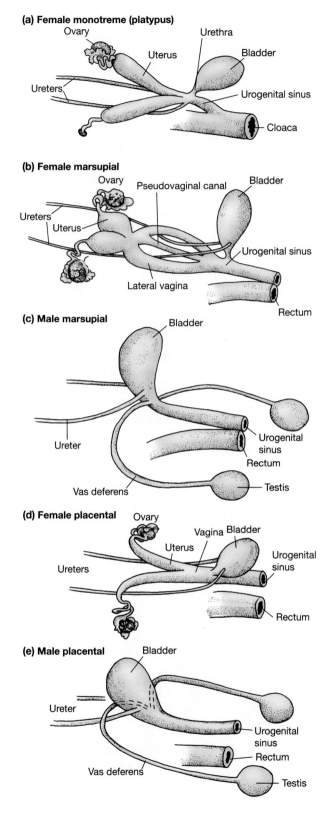

▲ **FIGURE 21–4** Mammalian reproductive tracts. The structures are shown as if the animal were lying on its back with its head to the left.

but a few have a single median uterus as seen in humans. Most placentals have a uterus that is bipartite (divided lengthwise into left and right sides) for some or all of its length, and a bipartite uterus sometimes occurs as a developmental abnormality in humans.

In most placentals, the urogenital sinus and the alimentary canal have separate openings with a distinct external space (the perineum) between them. The perineum is not so apparent in marsupials, and a well-defined separation of female urogenital openings into distinct external urethral and vaginal openings is seen only in primates and some rodents.

Some placentals (e.g., rodents and insectivorans) are born in a highly altricial state in which they are only slightly more developed than some marsupial young; others are born in more developed stages, extending to a highly precocial state (most ungulates) in which the young can run within a few hours of birth. All placentals, however precocial, still require a period of lactation for the transfer of essential antibodies from the mother as well as for nutrition. The period of lactation in most placentals is relatively short in comparison with other mammals—usually shorter than the period of gestation.

In all mammals, larger-bodied species tend to have fewer young per litter than do smaller ones—but the total number of young that the mother produces in her lifetime is probably similar because larger mammals have a longer life span than do smaller ones. In placentals, a distinct difference exists between large carnivores and large herbivores. While large carnivores such as bears and lions have several altricial young per litter, almost all ungulates have only a single, precocial young or twins at the most. Pigs are an exception here, with their large litters, but this condition is apparently derived for pigs, rather than representing the primitive ungulate condition as is often assumed (Gaucher et al. 2004). (Obviously, having several precocial young is not a viable option for any mammal: there would not be enough room inside the mother for them all to develop!)

It is often assumed that herbivores need to have young that are more developed at birth so that they can run with the herd and escape. Alternatively, the dietary habits of herbivores may explain the occurrence of precocial young in this group. Plants have low energy and nutrient contents compared to animal tissue, and an herbivore must consume and process more food than a carnivore to receive the same amount of nutrition. Placental transfer is more efficient than lactation, and a young ungulate might obtain better nutrition by staying in the uterus and allowing its mother to concentrate nutrients and transfer them via the placenta than it would by being born earlier and nursing.

■ Reproduction of Marsupials

In marsupials, the ureters pass medial to the developing reproductive ducts to enter the bladder. This arrangement prevents the oviducts of the females from fusing in the midline, at least posteriorly, and means that the vasa deferentia of the males do not have to loop around the ureters. The female reproductive tract consists of two lateral vaginae that unite anteriorly, from which point the two separate uteri diverge (see Figure 21–4b). The lateral vaginae are for the passage of sperm only. Birth of the young is through a midline structure, the median vagina or **pseudovaginal canal,** which develops at the first parturition.

Hormonal feedback from the embryonic trophoblast to the pituitary and hypothalamus alerts the mother to the state of pregnancy and influences the secretory activity of the uterus. Unlike placentals, marsupials do not maintain the corpus luteum in the ovary, and the young of most species are ejected at the end of the estrous cycle. The length of gestation in marsupials is relatively independent of body size, although the total time taken to rear the young is not. In contrast to placentals, marsupials retain direct evidence of their oviparous ancestry; a transient shell membrane appears, and the eggs still contain a small amount of yolk.

Marsupial embryonic development is very different from that of both monotremes and placentals. It clearly represents a derived condition because both placentals and monotremes are more similar to other amniotes than are marsupials. Marsupial neonates have well-developed forelimbs in comparison with other altricial neonates, and their lungs are relatively large at birth. Development of the jaws, secondary palate, facial muscles, and tongue is advanced, while that of the central nervous system is retarded, so that the newborn marsupial can attach itself to a nipple and begin suckling.

Most, but not all, marsupials enclose the nipples within a pouch—some dasyurids (marsupial mice, etc.) and some didelphids (opossums) lack a pouch. When young marsupials are born, they make their way from the vagina to the pouch where they attach themselves to a nipple to complete their development. They make the journey without assistance from their mother. Marsupial neonates have a novel condition of their shoulder girdle, called a shoulder arch, that allows them to make this trip (**Figure 21–5**). In making this journey, the young marsupials do not lever themselves up with their limbs like a gymnast on the parallel bars. Rather, they wriggle their bodies, using the front claws as holdfasts, and this shoulder arch aids in the functional requirements for the crawl to the nipple. The marsupial shoulder arch includes the coracoid and interclavicle bones, which are portions of the tetrapod shoulder girdle that are present in most tetrapods, including monotremes, but absent from all therians except for their transient

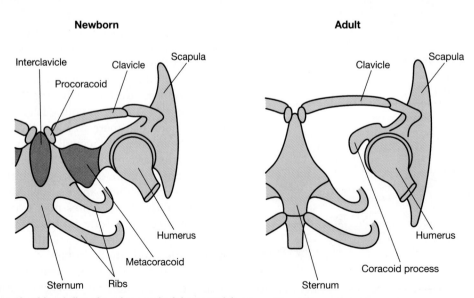

▲ **FIGURE 21–5** Shoulder girdles of newborn and adult marsupials. Viewed from the ventral side with only the left limb shown.

retention in newborn marsupials. Here they provide a strong brace for the front limbs during the crawl to the nipple, but this specialized neonatal anatomy limits the subsequent development of the shoulder girdle in the adults, and marsupials have less variation than placentals in the form of their shoulder girdles (Sears 2004).

In kangaroos, as in most other species, the young climb to the pouch unaided. The mother adopts a distinct sitting birth posture (**Figure 21–6**); she licks a path from the vagina to the pouch but does not otherwise aid the young in its journey. Marsupial neonates are evidently born with an instinct to climb upward. Some dasyurids and didelphids have young even more altricial than kangaroos. These babies are ejected directly into the pouch (or into the mammary area of pouchless species) at birth. The newborn young of these species are passive, unlike newborn kangaroos.

The time that young marsupials spend developing while attached to the nipple greatly exceeds the length of gestation. Lactation also continues for some time after the young have become sufficiently mature to detach from the nipple. This is when we typically see the young-at-foot, hopping in and out of the pouch.

Although the composition of the milk varies little during pregnancy in placentals, there is a marked variation in marsupials and monotremes. The first milk is dilute and protein-rich, while the later milk is more concentrated and richer in fats. Concurrent asynchronous lactation has been observed in some kangaroos; that is, an immature pouch young is attached to one nipple while a more mature, independent pouch young drinks from another nipple, and the mother produces different kinds of milk at the two nipples. The composition of milk is probably determined by how long the young spend suckling on the nipple per day.

■ The Primitive Therian Condition

Highly altricial young are probably the primitive condition for mammals in general and for all therian mammals in particular. Altricial young may be advantageous for small, endothermal vertebrates because they are essentially ectothermal as newborns, with correspondingly low metabolic rates. Parental brooding keeps them warm and, because they are not using metabolic energy to keep themselves warm, the newborns convert a high proportion of the food provided by the parents into growth.

(a) Birth posture of red kangaroo

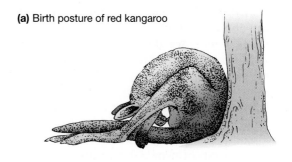

(b) Birth posture of grey kangaroo

(c) Red kangaroo with 3 different young at 3 different developmental stages

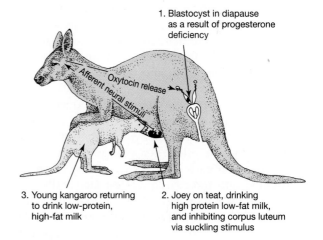

1. Blastocyst in diapause as a result of progesterone deficiency

Oxytocin release

Afferent neural stimuli

3. Young kangaroo returning to drink low-protein, high-fat milk

2. Joey on teat, drinking high protein low-fat milk, and inhibiting corpus luteum via suckling stimulus

▲ **FIGURE 21–6** Birth posture and embryonic diapause in kangaroos.

The reproductive mode of the earliest therians is not clear. The specializations of the reproductive anatomy of marsupials and placentals cannot easily be derived from each other, and it is possible that viviparity evolved independently in the two groups. The early therians of the Cretaceous period were small animals, most weighing less than a kilogram. They are likely to have had the same life-history features as extant small living therians—short life spans, several litters produced in rapid succession or a single large litter, and a short gestation period. Some fetal placentals show evidence of mouth seals—tissue that develops around the lateral margins of the mouth of neonatal marsupials to aid in attachment to the nipples. This feature might suggest that attachment to a nipple is a primitive therian feature. However, a pouch is definitely a derived feature of marsupials. Indeed, a pouch may even have been evolved independently in different lineages within marsupials because its form differs among lineages, and many extant primitive marsupials lack a pouch.

21.2 Some Extreme Placental Mammal Reproductive Specializations

Numerous reproductive specializations exist among living placentals. Some of the most bizarre are those seen in the naked mole rat (*Heterocephalus glaber*) and the spotted hyena (*Crocuta crocuta*). Naked mole rats are found in arid areas in sub-Saharan Africa and live as underground burrowers feeding on plant roots and tubers. They are **eusocial,** a type of social system otherwise seen only in social insects such as ants, termites, and bees. Like these insects, animals within a colony are closely related; but, unlike the insects, they are all diploid in chromosome number.

Mole rats live in colonies of up to 40 individuals with only one breeding female: the queen. The queen's presence suppresses breeding by the other female members of the colony, and the queen produces one to four litters per year with up to two dozen young in each litter. Other colony members are divided into three social castes: the smaller frequent workers that engage in cooperative burrowing and feed the community; the infrequent workers (also small) whose role is similar but who appear to do only about 25 percent as much work as the frequent workers; and the larger, nonworkers that care only for the young. Males and females are equally represented in these castes. All the males in a colony produce sperm, but only the nonworkers are large enough to copulate successfully with the queen. If the queen dies, one of the faster-growing female infrequent workers may become the new queen.

Spotted hyenas, found in the African savannas, are the only hyenas that regularly hunt in packs, and they also have females with masculinized genitalia. The clitoris is so enlarged that it resembles a male penis (complete with the capacity for erection), and the labia are fused to form a structure resembling a scrotum. For many years it was mistakenly thought that these hyenas were hermaphrodites.

Various adaptive hypotheses have been proposed for these features, but it is unlikely that the appearance of the external genitalia themselves is adaptive (although for an adaptive explanation, see Muller and Wrangham 2002). Rather, it appears that high levels of male hormones (androgens such as testosterone) are advantageous for females because they support aggressive behavior, which could be useful if more aggressive females and their offspring obtain a larger share of food during communal predation. According to this hypothesis, masculinization of females' genitalia during development is a side effect of the high levels of male hormones.

Although hyena females use their genitalia for behavioral display, there are clearly disadvantages to this condition. All the functions of the original urogenital sinus must now be transmitted through this penislike structure; thus female hyenas urinate, copulate, and give birth through the enlarged clitoris. Perhaps unsurprisingly, there is a high incidence of mortality among females giving birth for the first time.

21.3 Are Placental Mammals Reproductively Superior to Marsupials?

It used to be considered that the marsupial mode of reproduction was inferior to that of placentals. This opinion was based primarily on the assumption that marsupials had been unable to compete with placentals except in Australia, where they were thought to have evolved in isolation. (We now know that one type of placental did reach Australia along with the marsupials: see Section 20.4). It was said that marsupials would be unable to maintain the young in the uterus past one estrous cycle because the mother was unable recognize the condition of pregnancy due to the lack of hormonal feedback between the developing young and the maternal brain, but we now know that hormonal feedback of this nature does exist in marsupials, even though they lack the extended gestation of placentals.

It has also been argued that the lack of midline fusion of the oviducts in marsupials would make it impossible to carry large young to full term. However, many placentals have uteri that are almost completely separate with midline fusion occurring only at the base. This type of duplex uterus is seen in cows, for example, which give birth to large young. It seems more likely that, rather than the marsupial reproductive mode being an inferior version of the placental condition, marsupials and placentals evolved different but equivalent reproductive strategies. The largest energy investment in reproduction by marsupials comes during the extended period of postgestation lactation, whereas placentals invest most of their energy during intrauterine development.

Marsupial reproduction may be superior to that of placental mammals under some conditions (e.g., Kirsch 1977). For example, kangaroos eject pouch young while fleeing predators (actually, humans chasing them in cars; there have never been native pursuit predators in Australia). It has been suggested that in this situation the marsupial mother, freed from the burden of carrying her young, could escape to breed again, whereas in the case of a pregnant placental both mother and young would perish.

An alternative hypothesis points out that, although marsupial and placental mothers invest similar amounts of resources in their young, marsupials supply energy at slower rates because the time from conception to weaning is half again as long for a marsupial as for a placental. The slower rate of investment could be less stressful for the mother, and it might limit the energy loss if a baby dies. Having invested less energy, a marsupial might have enough stored energy to conceive again immediately, whereas a placental would have to wait until the following year. This feature of marsupial reproduction might be adaptive in arid, unpredictable climates such as those in Australia, where droughts and food shortages may occur frequently.

These proposals and other similar ideas seem plausible, but they have a critical flaw in that they consider only present-day mammals. In seeking evolutionary reasons for the development of marsupial or placental reproductive strategies, we must consider the animals and conditions that existed at the time of divergence, back in the Early Cretaceous period. The ability to dump the pouch young if pursued is unlikely to be adaptive for a small mammal that has only a single litter per lifetime. This is the life-history pattern of many extant dasyurids, and it

is probably the ancestral pattern for marsupials—possibly even for all therians. Likewise, harsh, arid climates were not a feature of the Cretaceous. Australia did not develop its arid interior until the late Cenozoic era, and marsupials did not even reach Australia until the early Cenozoic. Both reproductive strategies probably represent perfectly good means of achieving the same goal for the small, primitive therians of the Early Cretaceous.

Is there any feature of marsupial reproduction that would make their potential adaptive diversity different from placentals at the time in the Cenozoic when mammals became larger and diversified into a greater variety of ecomorphological types? The lack of marsupial species diversity (about 6 percent that of placentals) has been cited as an example of evolutionary inferiority. However, due to accidents of history, marsupials have had less land area to evolve on during the later Cenozoic than have placentals and would thus be expected to have less species diversity.

Only one specialization is probably impossible for a marsupial—fully aquatic life, such as a whale. A fully aquatic marsupial could not carry altricial young in a pouch because they would be unable to breathe air. There is only one semiaquatic marsupial, the South American yapok or water opossum, which seals up its pouch during short underwater forays, but marsupials generally have avoided aquatic situations. It is also hard to imagine a marsupial giving birth under water in the fashion of a whale—the tiny neonates would probably be swept away by currents before they could reach the pouch. There are no marsupial equivalents to bats, but this might just be a matter of evolutionary chance. Gliding has evolved several times convergently among both marsupials and placentals, but flight has evolved only once in placentals. It is also probably the case that a marsupial could not afford to reduce its front feet to nonclasping limbs with few fingers, like the front limbs of hoofed ungulates, as the young would be unable to climb up to the pouch. This is probably why the marsupial equivalents of horses and antelope, the large kangaroos, have specialized only their back legs for locomotion.

Although the marsupial mode of reproduction may not be adaptively inferior overall, it is true that placentals have a faster reproductive rate than marsupials, especially at smaller body sizes. Certainly one reason that feral placental mammals are diversifying at the expense of the native Australian marsupials is because they can reproduce faster and more often.

21.4 Specializations for Feeding

Mammals need large quantities of food, and processing food in the mouth speeds the digestive processes. All mammals masticate their food, and all mammals have a distinct swallowing reflex whereby they ingest a discrete bolus of finely chewed food. The mammalian tongue, important in both oral food processing and in swallowing, has a unique system of intrinsic muscles, and the muscular cheeks of mammals—which keep food in the mouth as it is chewed—are derived from a unique set of facial muscles.

▇ Dentition

Mammalian teeth are shown in **Figure 21–7**. The incisors are used to seize food. The incisors of mammals that gnaw, such as rodents and rabbits, may be enormously enlarged and grow continuously throughout life. Rodent incisors have enamel only on the anterior surface. Because the enamel is the hardest part of the tooth, it wears more slowly than the dentine behind it, producing a self-sharpening chisel edge.

Canines are used to stab prey and are often lost in herbivores, which have no need to subdue their food items, but they are also used in social signals and may be retained in modified form. The tusks of pigs and walruses are modified canines (but the tusks of elephants are modified incisors). Upper canines are generally larger in male primates than in females, even slightly so in humans. Hornless ruminants, such as the mouse deer, may retain large upper canines in the males for fighting and display.

Premolars pierce and slice food, and molars break food into fine particles. Premolars and molars are usually different in form. Primitively, premolars have a single cusp, whereas molars have three or more cusps. Many herbivores use the entire postcanine tooth row for mastication, and the premolars resemble the molars (are molarized). This condition can be seen in ruminant artiodactyls and, to a still greater degree, in horses, in which all the postcanine teeth appear identical.

Omnivorous and fruit-eating mammals have reduced the originally pointed cusps of their molars to rounded, flattened structures suitable for crushing and pulping. They have also added a fourth cusp to the upper molars and increased the size of the talonid basin in the lower molars so that these teeth now appear square (quadritubercular) rather than triangular. These teeth are called **bunodont** (Greek *buno* = a hill or mound and *dont* = tooth) in reference to the rounded cusps. We have bunodont molars, as do most primates and other omnivores, such as pigs and raccoons.

In herbivores, the simple cusps of the bunodont tooth run together into ridges, or **lophs**. Lophed teeth work best when the enamel has been worn off the top of the ridges to expose the underlying dentine. Each ridge then consists of a pair of sharp enamel blades lying on either side of the intervening dentine. When these teeth occlude and the jaws move sideways, the food is grated between multiple sets of flat, shearing blades.

All herbivorous mammals face a similar problem with their teeth—that of dental durability. Durability of the dentition is not an issue for most vertebrates, which continually replace their teeth, but all mammals have inherited the condition of diphyodonty (a single set of replacement teeth) from their original ancestor. Diphodonty was probably essential for precise occlusion in early mammals (see discussion in Section 18.4), but it presents a problem in that the adult dentition must last a lifetime. Herbivores face a particular problem because vegetation is more abrasive than other forms of food. Grazers have especially high tooth wear because grasses contain silica in the cell wall.

Herbivorous mammals have made their dentition more durable in a variety of ways. The most common way to make the molars more durable is to make them high crowned or **hypsodont** (Greek: *hyps* = high). Hypsodont cheek teeth look like regular teeth when seen in the jaw, but the division at the base between the crown and the root is not visible, as with low-crowned (**brachydont,** from the Greek *brachy* = short) teeth. The crowns of hypsodont teeth extend into the depth of the jawbones, and hypsodont species have very deep lower jaws and deep cheek regions. In fact, the eyes of grazers have moved backward to accommodate the roots of the upper teeth in the maxilla.

Hypsodont ungulates also extend the layer of cementum to cover the entire tooth. Cementum is a bonelike material that covers only the root and the base of the crown in most mammals, but in ungulates the high lophs of the teeth must be laid down during tooth development. Without cementum acting as filler, the individual lophs would be tall, free-standing blades once the tooth had erupted and would be likely to fracture. As the chewing surface is worn away, the teeth erupt from the base to provide a continuously renewing occlusal surface, much like the way lead in a mechanical pencil pushes up as it is worn away.

Most hypsodont mammals have a finite amount of tooth crown. When the tooth is worn out, the animals can no longer eat. (Most mammals in the wild will

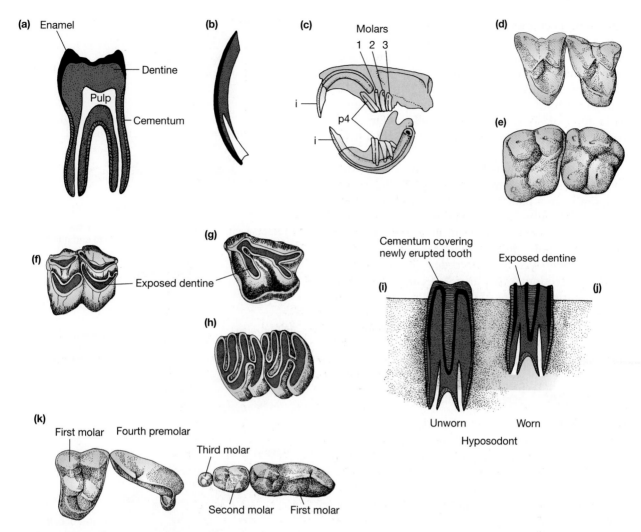

▲ **FIGURE 21–7** The structure of mammalian teeth. (a) Sectioned mammalian molar showing general dental form. (b) Sectioned rodent ever-growing incisor, showing enamel on anterior surface only. (c) Sectioned rodent skull, showing ever-growing incisors and ever growing (hypselodont) cheek teeth. (d) Tritubercular upper molar, as found in the primitive therian mammal condition. (e) Bunodont, quadritubercular upper molars, as found in omnivorous mammals. The teeth of ruminant artiodactyls (f, upper molar of a deer) have crescentic lophs that run in a predominantly anterior to posterior direction across the teeth; these teeth are called **selenodont**. Different lineages of herbivorous mammals evolved lophed teeth convergently. The teeth of perissodactyls (g, upper molar of a rhinoceros) have straight lophs that run predominantly in a lateral to medial direction across the teeth; these teeth are called **lophodont**. Wombats, rodents (h, upper molars of a rodent), warthogs, and elephants have a complex form of molars called multilophed or lamellar. (i) Sectioned unworn hypsodont molar, showing covering of cementum. (j) Worn hypsodont molar, showing sharp enamel ridges interspersed with softer areas of dentine and cementum. (k) Carnivore carnassial shearing teeth (in a coyote) (upper teeth to the left, lowers to the right). (i = incisor; p4 = fourth premolar; 1, 2, and 3 = first, second, and third molars, respectively)

have died of natural causes long before their teeth wear out, but domestic horses surviving into their late twenties and thirties often must be fed soft food because they have virtually no molars left.) However, some mammals have molars in which the roots do not close and the tooth is functionally ever-growing,

or **hypselodont.** For a variety of reasons, this appears to be an option primarily for small mammals (Janis and Fortelius 1988). Hypselodont molars are seen most commonly in rabbits and some rodents.

Elephants do not have ever-growing cheek teeth but instead employ the novel feature of molar progression. Each molar is now not only hypsodont but also greatly enlarged, being the size of the entire original tooth row. Each molar is now erupted and worn in turn: when it becomes worn down, its remaining stub falls out of the front of the jaw, and the molar behind erupts from the back. Elephants have a total of six molars (three milk molars and three permanent molars) in each upper and lower jaw half.

Carnivorous mammals have specialized shearing postcanine teeth. Mammals in the placental order Carnivora have a pair of specialized teeth modified into a set of tightly shearing blades, the carnassials, formed from the last premolar in the upper jaw and the first molar in the lower jaw. In the early Cenozoic, there was another order of carnivorous placentals (creodonts) that also had carnassials, but these were formed from different teeth farther back in the jaws. The marsupial wolf lacked true carnassials; instead, each molar was somewhat specialized into a bladelike shape, but none was significantly larger than the others.

Craniodental Specializations of Mammals

The hedgehog shows the generalized insectivorous condition that can be taken to represent the primitive mammalian mode (**Figure 21–8**). The molars of generalized mammals usually retain the primitive triangular shape and pointed individual cusps that are useful for puncturing insect cuticle.

Anteaters Mammals that specialize on ants and termites are called **myrmecophagous** (Greek *myrme* = ant and *phago* = to eat). Myrmecophagous mammals have elongated jaws and teeth that are reduced or absent. They also have enlarged salivary glands and highly elongated tongues. Reduction of the teeth is also seen in mammals that specialize on nectar (e.g., some bats and the honey possum).

Aquatic Feeders Aquatic fish-eating and squid-eating mammals such as porpoises and dolphins have highly elongate jaws that have lost the anteriormost teeth. The form of their skulls and teeth has become convergent with other piscivorous tetrapods, such as ichthyosaurs and crocodiles.

The baleen whales (mysticetes) have replaced their teeth with sheets of a fibrous, stiff, hornlike epidermal derivative known as **baleen,** which extends downward from the upper jaw. These whales use the baleen to strain planktonic organisms from the water.

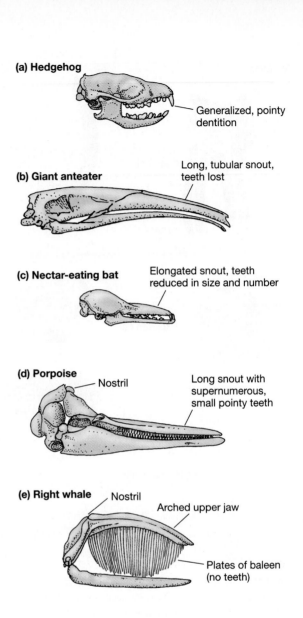

(a) Hedgehog — Generalized, pointy dentition

(b) Giant anteater — Long, tubular snout, teeth lost

(c) Nectar-eating bat — Elongated snout, teeth reduced in size and number

(d) Porpoise — Nostril — Long snout with supernumerous, small pointy teeth

(e) Right whale — Nostril — Arched upper jaw — Plates of baleen (no teeth)

(f) Walrus — Flat teeth — Large tusk

▲ **FIGURE 21–8** Some feeding specializations of the teeth and skulls of mammals.

The walrus feeds on mollusks, crushing their shells with its flat postcanine teeth. (The enormous canine tusks are used mainly for display.)

Differences between Carnivorous and Herbivorous Mammals The basic mode of mammalian mastication is best understood by considering the difference between carnivorous and herbivorous mammals (**Figure 21–9**). All mammals use a combination of masseter, temporalis, and pterygoideus muscles to close the jaws, and the digastric muscle (in therians) to open the jaw. The relative sizes of the muscles and the shape of the skulls reflect the different demands of masticating flesh and vegetation. The temporalis has its greatest mechanical advantage at initiating jaw closure at moderate to large gapes, when the incisors and canines are likely to be used. Thus a large temporalis is typical of carnivores, which use a forceful bite with their canines to kill and subdue prey. The jaw joint is on the same level as the tooth row so the teeth come into contact sequentially as the jaw closes, like the blades of a pair of scissors—a design well suited for teeth that primarily cut and shear. The postglenoid process prevents the strong temporalis muscle from dislocating the lower jaw. Strong muscles run from the high occipital region of head to the cervical vertebrae. These muscles are probably important for resisting struggling prey.

The skulls of herbivorous mammals are modified to grind up tough, resistant food. The protein content of leaves and stems is usually low, and the protein is enclosed by a tough cell wall formed by **cellulose** (a complex carbohydrate). Thus the skulls and teeth of herbivorous mammals must process large quantities of tough, fibrous material. The masseter creates forces at the back of the tooth row and also moves the jaw from side to side.

Herbivorous mammals have a large masseter and small temporalis in comparison to the primitive mammalian condition. This morphology is reflected in the increased size of the angle of the lower jaw (where the masseter inserts) and the reduced size of the coronoid process and temporal fossa (where the temporalis inserts).

The jaw joint has been shifted in herbivores so that it is high on the skull, offset from the tooth row. This offset brings all the teeth in the upper and lower jaws together simultaneously, with a grinding action that shreds plant material between the lophs of upper and lower teeth, much as the offset handle of a cooking spatula allows you to apply the entire blade of the spatula to the bottom of the frying pan while keeping your hand above the pan's rim. (Some herbivorous dinosaurs also had offset jaw joints, but in this case the jaw joint was below the level of the tooth row. It is the offsetting of the jaw joint from the tooth row that is the important mechanical feature, not whether the joint is above or below the row.)

Herbivores also usually have elongated snouts, resulting in a gap between the cheek teeth and the incisors called the **diastema**. The function of the diastema is uncertain. It may allow extra space for the tongue to manipulate food, or it may just be a reflection of the elongation of the jaw for other reasons. A long jaw allows an animal to select food with its incisors without poking its eye on the vegetation.

Many herbivorous placental mammals ossify the cartilaginous partition at the back of the orbit to form a bony postorbital bar. This bar is probably important in absorbing stress from the jaws during the constant chewing of herbivores, thus protecting the braincase. Herbivores usually have a fairly low occipital region because they do not need to have a carnivore-like attachment for muscles that help brace the head on the neck. An exception can be found in pigs, which root with their snouts and have strong muscles linking the back of the head to the neck.

Many rodents have a highly specialized type of food processing. Their upper and lower tooth rows are the same distance apart—unlike the condition in most mammals, in which the lower tooth rows are closer together than the uppers. This derived condition in rodents is combined with a rounded jaw condyle, which allows forward and backward jaw movement, and the insertion of a portion of the masseter muscle far forward on the skull so that the lower jaw can be pulled forward into occlusion. This jaw apparatus allows rodents to chew on both sides of the jaw at once, presumably a highly efficient mode of food processing. Note that this mode of chewing can be achieved only with flattened, lamellar teeth because the high ridges of other types of teeth would prevent this jaw motion.

Herbivores, Microbes, and the Ecology of Digestion

The specialized teeth of herbivores can rupture the cell walls of vegetation and expose the cell contents, but only special enzymes (**cellulases**) can digest the cellulose that forms the cell wall and constitutes a large percentage of the plant material. However, no multicellular animal has the ability to synthesize cellulase. Thus the efficient use of plants as food requires cellulase enzymes produced by microorganisms that live as symbionts in the guts of herbivorous

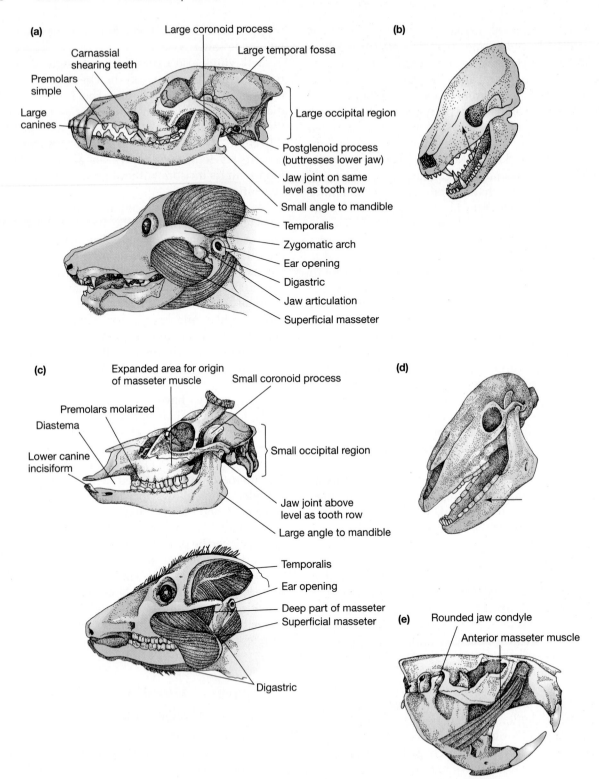

▲ **FIGURE 21–9** Craniodental differences between carnivorous and herbivorous mammals. (a) Carnivore skull and musculature (a dog). (b) Action of carnivore jaws. (c) Herbivore skull and musculature (a deer). (d) Action of herbivore jaws. (e) Rodent skull and musculature (beaver).

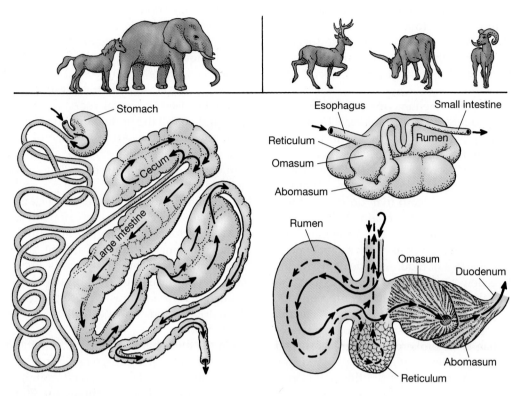

▲ **FIGURE 21–10** Hindgut and ruminant foregut digestive systems. *Left:* The hindgut fermenting system. Fermentation occurs in the enlarged cecum and colon (large intestine). *Right:* The ruminant system, showing the four-chambered stomach.

animals. While all mammals have gut symbiotic microorganisms of some kind, herbivorous mammals have independently evolved specialized chambers within the digestive tract to house symbiotic microorganisms that convert the cellulose and lignin of plant cell walls into digestible nutrients (volatile fatty acids) (**Figure 21–10**).

The many separate evolutions of fermentative digestion among vertebrates have resulted in distinctly different solutions to the problems posed by plants as food. Horses and other perissodactyls are examples of **hindgut (monogastric) fermenters.** These have a simple stomach and have enlarged both the large intestine and the cecum as fermentation chambers. Other hindgut fermenters include elephants, hyraxes, New World howler monkeys, wombats, koalas, rabbits, and many rodents. Some degree of hindgut fermentation is probably a primitive vertebrate character: it occurs among birds, lizards, turtles, and fishes, as well as omnivorous and carnivorous mammals, such as humans and dogs.

Cows and other ruminant artiodactyls are examples of **foregut (ruminant) fermenters,** in which the nonabsorptive forestomach is divided into three chambers

that store and ferment food, followed by a fourth chamber in which digestion occurs. Ruminants are so called because they ruminate, or chew, the cud. Camels resemble other ruminants in many respects but have only three-chambered stomachs (the omasum is lacking). A simpler type of foregut fermentation, without extensive stomach division or cud-chewing, is found in many other mammals, including Old World colobine monkeys, kangaroos, hippos, and some rodents. Interestingly, the gut symbionts of hindgut fermenters and foregut fermenters are somewhat distinctive in each group, regardless of phylogenetic associations (e.g., the sheep's gut flora is more like that of a kangaroo than of a horse) (Ley et al. 2008).

Hindgut fermenters chew their food thoroughly as they eat, fracturing the plant cell walls with their teeth so that the cell contents are released. These cell contents are processed and absorbed in the stomach and small intestine. The cellulose of the plant cell wall is not digested until it reaches the cecum and large intestine, where it is attacked by the symbiotic microorganisms. Cellulose is fermented to form substances known as volatile fatty acids, which are absorbed through the walls of the gut.

Some small hindgut fermenters, such as rabbits and rodents, ferment the food largely in the cecum and do not absorb much of the initial products of fermentation. Instead they rely on **coprophagy,** re-eating the first set of feces that are produced and recycling the nutrients.

Ruminant foregut fermenters do not need to chew their food as thoroughly on initial mastication, partly because the plant cell walls will be chemically disrupted in the stomach and partly because boluses of food (the cud) will be returned to the mouth and chewed again. Ruminants have less extensive modifications of the skull and teeth than do hindgut fermenters. The food is initially retained in the two front chambers of the stomach, the rumen and the reticulum. Here microorganisms break down the cellulose, and the boluses of food are repeatedly regurgitated and rechewed. Food cannot pass from the reticulum into the third and fourth compartments, the omasum and the abomasum (the true stomach), until the particles are very small. Most or all of the cellulose has been broken down and absorbed before the food reaches the abomasum, and the digestive process of ruminants from this point on is like that of most other mammals.

The consequences of being a hindgut versus a foregut fermenter are profound, and each of the two kinds of fermentative digestion has distinct advantages and disadvantages. If the teeth of a hindgut fermenter have broken down the plant cell walls effectively, then the cell contents (proteins, lipids, sugars) have been released, digested in the stomach, and are ready to be absorbed when they reach the small intestine. The cellulose in the cell walls is still intact, however, and it represents a substantial part of the energy in the food that is not digested until it reaches the cecum and large intestine. Nutrient uptake from these portions of the intestine is not as effective as uptake in the small intestine, so hindgut fermenters lose some of the energy of the food in the feces.

In contrast, foregut fermentation can be extremely efficient because the microorganisms have broken down the cellulose *before* it reaches the small intestine. The symbiotic microorganisms in the rumen ferment *all* of the chemical compounds in the food—including lipids, sugars, and proteins that the mammalian system has no problem digesting—and they use these materials to produce more microorganisms. That sounds as if it would be a loss of energy for the ruminant, but this is not the case because surplus microorganisms subsequently pass into the abomasum where they are digested, so the ruminant ultimately gets those nutrients.

In fact, this protein fermentation actually ends up being beneficial to the ruminant via a process called nitrogen cycling. Microorganisms ferment the protein into ammonia, which is then taken via the circulation to the liver, where it is converted to urea. This urea is transported by the circulatory system to the rumen, where it is used by the microorganisms for their own growth. Thus all the protein that the ruminant digests is microbial protein. An advantage of this system is that the microorganisms make all of the essential amino acids needed in the diet. Thus, a ruminant can be more limited in its selection of plant species than a hindgut fermenter, which must find all of its essential amino acids by eating a variety of plant sources.

An additional advantage of foregut fermentation is the role microorganisms in the rumen play in detoxifying chemical compounds that would be harmful to a vertebrate. A hindgut fermenter receives no such benefit and must absorb plant toxins into its bloodstream and transport them to its liver for detoxification.

On the other hand, a hindgut fermentation system processes material rapidly, whereas ruminants process food more slowly. Food moves through the gut of a horse in 30 to 45 hours, compared to 70 to 100 hours for a cow. Hindgut fermentation works well with food that has relatively high concentrations of fiber because a large volume of food can be processed rapidly. The system is not efficient at extracting energy from the cellulose; but, by processing a large volume of food rapidly, a horse can obtain a large quantity of energy from the cell contents in a short time.

In contrast, a ruminant foregut system is slow because food cannot pass out of the rumen until it has been ground into very fine particles. Ruminants do not do well on diets containing high levels of fiber because this slows the passage rate of the food even further—the animal can literally starve to death with a stomach full of food. Ruminants are very efficient at extracting maximum amounts of energy from the cellulose in food of moderate fiber content, but they cannot process highly fibrous food.

Another possible limitation of the ruminant mode of digestion is body size restriction. Ruminants past and present are abundant in the body size range from around 5 kg (small deer or antelope) to 1000 kg (giraffe), but there are no rabbit-size ruminants and no rhinoceros- or elephant-size ones. The slow rate of ruminant food passage probably limits the size range at both ends. Rabbit-size ruminants would have relatively greater metabolic demands (a simple effect of metabolic scaling, and the necessary long retention of food in the rumen would mean that they couldn't eat enough per day to survive. However, the reason why there were no rhino-size ruminants was enigmatic until recent work by Marcus Clauss and colleagues (Clauss et al. 2003). Clauss noted that in addition to the fact that ruminants have longer passage times of their digesta than nonruminants, larger animals also have longer passage times than do smaller ones. A cow, with a passage rate of up to 100 hours, is on the large size for a ruminant; but such lengthy passage times pose problems. After about 100 hours, methanogenic bacteria in the gut start to attack the food, and much of the energy in the food is harvested by the bacteria, not by the ruminant mammal. Elephants and rhinos do not approach passage times of this length even at their very large body sizes because they are hindgut fermenters, but a ruminant much larger than a cow would have passage times longer than 100 hours and would not be able to maintain digestive efficiency. (Note that one of the reasons why giraffes can attain greater body weights than bison (a relative of cows, but somewhat larger) is because they eat browse (the shoots, twigs, and leaves of shrubs and trees), a more digestible food than grass that requires less gut retention time.

These differences in digestive physiology are reflected in the ecology of foregut and hindgut fermenters. Hindgut fermenters can survive on very low-quality food such as straw, as long as it is available in large quantities. Consequently, the feral horses in the American West can live on land too poor for cattle to graze; these ruminants are unable to process the low-quality food fast enough to subsist. In contrast, hindgut fermenters cannot survive so well in areas where the absolute quantity of food is the limiting factor.

Ruminants are the main herbivores in places like the Arctic and deserts, where the food is of moderately good quality but severely limited in quantity; such food best supports an animal that can make the most efficient use of its dietary intake. Ruminants also have an advantage in desert conditions because of their nitrogen cycling. Because this cycle uses some of the waste urea from other sources, less urea needs to be excreted by the kidneys—thus less water needs to be used to produce the urine. Ruminants are better able to go without water for a few days than are hindgut fermenters, and ruminants are more typically found in arid habitats. Ruminants may have become the dominant herbivores in the later Cenozoic because the increased seasonality resulted in a habitat in which it was easier for a selective feeder to make a living.

21.5 Specializations for Locomotion

A tree shrew illustrates the bounding and scrambling that is the basic mode of mammalian locomotion (**Figure 21–11**). The limbs and back are flexed during locomotion, and the basic therian gait appears to be highly dependent on this flexed spine and limbs, which bend at the elbow-ankle, shoulder-knee, and top of scapula-hip joints in a three-part zigzag fashion. The independent movement of the scapula (a new feature in therians) is critical to this arrangement: the major pivot points of the limbs during locomotion are formed by the dorsal border of the scapula (rotating around its own axis) in the forelimb and by the hip joint (the articulation of the femur with the pelvic girdle) in the hindlimb. This morphology is probably an adaptation to locomotion at small body size on an irregular ground surface (Fischer et al. 2002). In contrast, larger mammals, such as the horse, move with a stiffer back and straighter legs, and they gallop rather than bound.

■ Specialized Forms of Locomotion

Larger animals experience the world differently from smaller ones because of physical size and scaling effects, and larger mammals are usually modified for more specialized forms of locomotion. **Figure 21–12** contrasts the specializations of running (**cursorial**) mammals and digging (**fossorial**) mammals.

Cursorial Limb Morphology   The number of strides it takes to travel a given distance determines the cost of locomotion, and a long-legged mammal will cover a given distance in fewer strides than a shorter-legged one. Long legs also provide a long out-lever arm for the major locomotor muscles, such as the triceps in the forelimb and the gastrocnemius in the hindlimb. This arrangement favors speed of motion rather than power.

Elongation is limited mostly to the lower portions of the limb: the radius and ulna in the forelimb, the tibia and fibula in the hindlimb, and the metapodials (a collective term used to describe the metacarpal and metatarsal bones). The humerus and femur are not elongated, nor are the phalanges.

Muscles are limited to the proximal portion of the limb, reducing the mass in the lower limb. There is almost no muscle in a horse's leg below the wrist (the horse's so-called knee joint) or the ankle (the hock). That anatomy makes sense in mechanical

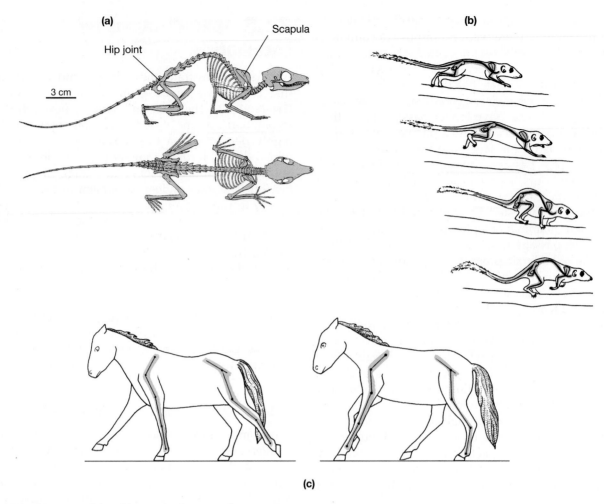

▲ **FIGURE 21–11** Gait and locomotion in mammals. (a) Skeleton of tree shrew (*Tupaia glis*) in typical posture with flexed limbs and quadrupedal stance. (b) Sequential phases of the bounding run in a tree shrew. Note the relatively flexed limbs and mobile back. (c) Sequential phases of the gallop in a horse. Note the relatively straight limb angles and immobile back.

terms because the foot is motionless at the start of each stride (as it pushes against the ground); and, when it leaves the ground, it must be accelerated from zero velocity to a speed greater than the body speed as it moves forward ready for the next contact with the ground. The lighter the foot, the less effort is needed.

The force of muscular contraction by the muscles in the upper limb is transmitted to the lower limb via long elastic tendons. These tendons are stretched with each stride, storing and then releasing elastic energy and contributing to locomotor efficiency.

The leg tendons of a hopping kangaroo are an obvious example of elastic storage, as the animal bounces on landing as if using a pogo stick. However, all cursorial animals rely on energy storage in tendons for gaits faster than a walk (**Figure 21–13**).

Even humans rely on elastic energy storage in tendons, especially in the Achilles tendon that attaches the gastrocnemius (calf) muscle to the calcaneal heel. People who have damaged their Achilles tendons (a common sports injury) find running difficult or impossible. Lengthening these tendons to increase the amount of stretch and recoil may be part of the evolutionary reason for limb elongation and changes in foot posture.

Other cursorial modifications restrict the motion of the limb to a fore-and-aft plane so that most of the thrust on the ground contributes to forward movement. The clavicle is reduced or lost, the wrist and ankle bones allow motion only in a fore-and-aft plane, and the forelimb cannot be supinated. (Note how easily you can turn your hand at the wrist so that your palm faces upward; a dog can't turn its

FOOT POSTURE

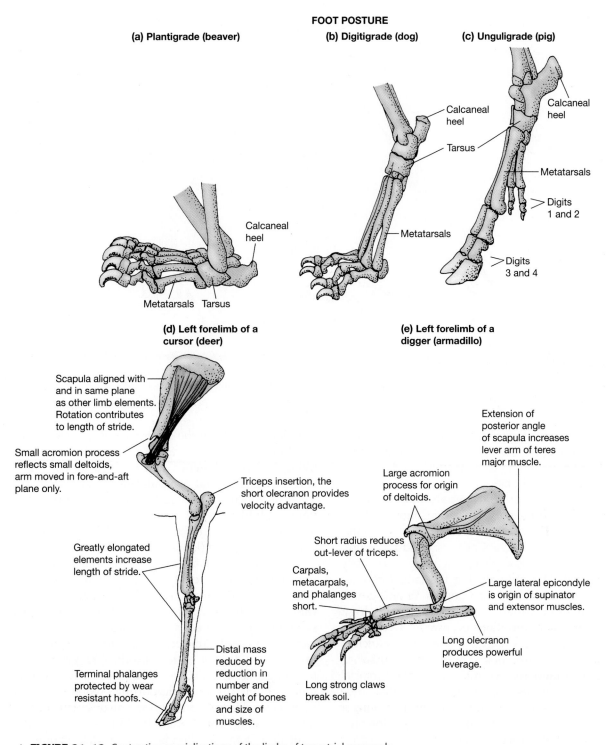

(a) Plantigrade (beaver)

(b) Digitigrade (dog)

(c) Unguligrade (pig)

Calcaneal heel

Tarsus

Metatarsals

Calcaneal heel

Tarsus

Metatarsals

Digits 1 and 2

Digits 3 and 4

Metatarsals　Tarsus

(d) Left forelimb of a cursor (deer)

Scapula aligned with and in same plane as other limb elements. Rotation contributes to length of stride.

Small acromion process reflects small deltoids, arm moved in fore-and-aft plane only.

Triceps insertion, the short olecranon provides velocity advantage.

Greatly elongated elements increase length of stride.

Distal mass reduced by reduction in number and weight of bones and size of muscles.

Terminal phalanges protected by wear resistant hoofs.

(e) Left forelimb of a digger (armadillo)

Extension of posterior angle of scapula increases lever arm of teres major muscle.

Large acromion process for origin of deltoids.

Short radius reduces out-lever of triceps.

Carpals, metacarpals, and phalanges short.

Large lateral epicondyle is origin of supinator and extensor muscles.

Long olecranon produces powerful leverage.

Long strong claws break soil.

▲ **FIGURE 21–12**　Contrasting specializations of the limbs of terrestrial mammals.

forepaw that much, and a horse has almost no ability to rotate this joint.)

The number of digits may also be reduced, perhaps to reduce the weight of the foot so that it can be accelerated and stopped more easily. Carnivores often lose digit 1 but otherwise compress the digits together rather than reducing their number. Artiodactyls reduce or lose digits 1, 2, and 5, becoming

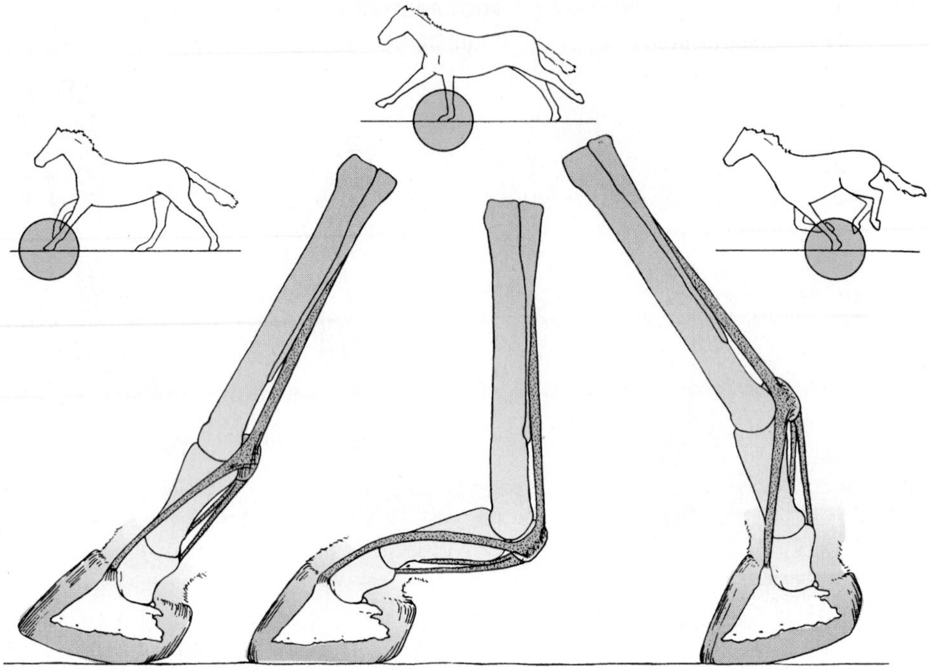

▲ **FIGURE 21–13** Springing action of the elastic tendons in the foot of a horse. The tendon is stretched as the horse's body moves forward over the leg and shortens as the foot leaves the ground, providing additional propulsive force.

effectively four-toed like a pig or two-toed like a deer. Perissodactyls lose digits 1 and 5 and reduce digits 2 and 4, becoming three-toed like a rhinoceros or single-toed like a horse.

The evolution of cursorial specializations in mammals is an interesting story. No early Cenozoic (Paleocene) mammal showed cursorial specializations, and early members of the Carnivora and the ungulate orders of the Eocene epoch were also fairly unspecialized. However, during the later Cenozoic we find these mammals becoming larger and developing a more cursorial form of limb morphology. Such cursorial specializations occurred convergently among many different ungulate lineages. As ancient ungulates and carnivorans were not cursorial, whereas their modern descendants do have cursorial specializations, it has long been assumed that these specializations, especially the evolution of longer legs, must have arisen in the context of predator-prey relationships. Longer legs would have given a carnivore a little more speed to pursue the herbivore, resulting in selection for ungulates with longer legs to make a faster escape.

This idea of a coevolutionary arms race between predator and prey is appealing, but the fossil record does not support it. If coevolution were the driving force, ungulates and carnivores would have evolved their longer limbs at the same time, in lockstep fashion, but this is not the case. For example, ungulates living in North America had evolved longer limbs and other cursorial specializations by the early Miocene epoch, some 20 million years ago, but carnivores built like present-day pursuit predators do not appear in the fossil record until the Pliocene, around 5 million years ago (Janis and Wilhelm 1993).

Why did ungulates evolve cursorial specializations if not to flee predators? Probably because all of the limb modifications that make a mammal a faster runner also make it more efficient at slower gaits, such as a trot. The early Miocene is the time when habitats in North America started to turn from productive woodlands to less productive grasslands, meaning that ungulates would have to forage further each day for food. Supporting evidence for this hypothesis comes from looking at ungulate evolution on other continents. In Eurasia, where the transition from woodlands to grasslands happened later in the Miocene, the cursorial specializations of ungulates also appeared later, contemporaneous with the change in vegetation. In South America, the vegetation changed earlier—in the Oligocene epoch—and cursorial ungulates appeared at that time. Yet carnivores with a pursuit type of cursorial morphology first appeared on all continents only

5 million years ago—15 to 25 million years later than their cursorial prey. Thus the evolution of cursorial limb adaptations in ungulates appears to be related to endurance in foraging for food because there were no carnivores able to pursue them as they evolved this condition. However, modifications for endurance at slow gaits provide speed at fast gaits, and speed became valuable when cursorial predators evolved.

A final unresolved issue is why ungulates appear to be more specialized for cursorial locomotion than carnivores. A horse has more elongated legs, a more derived foot stance, greater reduction of toes, and more restricted movement in the limb joint than a dog, but horses are not significantly faster runners than dogs. (If they were, foxhunting on horseback with hounds would be impossible because the hounds would rapidly be left behind.) The answer may lie in the digestive physiology of ungulates. The gut and gut contents of an ungulate may comprise up to 40 percent of its total body mass (more in a ruminant than in a horse). Thus, for any given body weight, an ungulate has proportionally less muscle mass than a carnivore of the same size because so much of its mass is composed of the gut. It may be that ungulates must have more skeletal modifications than carnivores to compensate for the difference in relative muscle mass in proportion to total body weight.

Fossorial Limb Morphology Mammals do several types of digging. The most common is scratch digging at the surface—that is, the type of digging that a dog uses to bury a bone. Animals that are truly fossorial (burrowing under the ground surface) have a variety of anatomical specializations. No mammal is limbless and elongated like a burrowing lizard or caecilian, although mammals that follow their prey down burrows (such as weasels and ferrets) have elongated bodies and short legs.

A limb specialized for digging is almost exactly the opposite of a limb specialized for running. Running limbs maximize speed at the expense of power, whereas digging limbs maximize power at the expense of speed. Fossorial mammals achieve this mechanical design in the forelimb with a long olecranon process and a relatively short forearm.

Some subterranean diggers are called rapid scratch diggers. These animals, such as the African golden mole and the Australian marsupial mole, dig with both fore and hind feet and move through sandy soil by pushing the grains aside and back without constructing an open burrow. True moles,

living in more compact soil, rotate their forelimb to the earth rather than simply retracting it, a form of digging termed *rotation thrust digging*. Moles burrow just below the surface of the ground, seeking worms and insect larvae in the roots of plants, and push the soil upward as they tunnel. Finally, many rodents (gophers are an example) use their ever-growing incisors rather than their limbs to dislodge soil, which is termed *chisel-tooth digging*. Gophers have a pronounced diastema and can pull their lips together behind the incisors to prevent soil from entering the mouth. Gophers push the soil they excavate out onto the surface of the ground, forming the mounds that are a familiar feature of the landscape wherever gophers occur.

Digging mammals retain all five digits, tipped with stout claws for breaking the substrate. They also have large projections on their limb bones for attachment of strong muscles, such as the enlarged acromion process on the scapula. Scratch diggers at the surface, such as anteaters, have a very stout pelvis, with many vertebrae involved in the sacrum, for bracing the hindlimb while digging with the forelimb. However, underground burrowers such as moles do not have this type of strengthened pelvis and sacrum.

21.6 Evolution of Aquatic Mammals

Semiaquatic mammals are not very different from terrestrial mammals except for somewhat more paddlelike limbs and a denser fur coat, and lineages of semiaquatic mammals have evolved numerous times. Among extant mammals, we can see examples in monotremes (the platypus), marsupials (the yapok, or water opossum), and within placentals in the orders Lipotyphla (water shrews, desmans), Tenrecoidea (otter tenrec), Rodentia (beavers, coypu, muskrat, Australian water rat), Carnivora (otters, mink, polar bear), and Artiodactyla (hippopotamus). The fossil record adds hippopotamus-like rhinoceroses, several independent evolutions of otterlike animals, and even a semiaquatic sloth.

Specialization for fully aquatic life is a different matter, however, and fully aquatic mammals—primarily marine forms that never or rarely come out onto land—have evolved only three times: in the orders Cetacea (whales, porpoises, and dolphins), Sirenia (dugongs and manatees), and Carnivora (seals, sea lions, and walruses) (**BOX 21–1**).

| BOX 21–1 | Comparison of Different Types of Fully Aquatic Mammals |

The cetaceans (whales, porpoises, and dolphins), sirenians (manatees and dugongs), and pinnipeds (seals, sea lions, and walruses) are lineages of specialized aquatic mammals (**Figure 21–14**). Cetaceans and sirenians cannot come on land, but pinnipeds are more amphibious. Cetaceans and pinnipeds are carnivores, but sirenians are herbivores. All of them use blubber (a thick layer of subcutaneous fat) for insulation instead of hair.

Pinnipeds come on land to court, mate, and give birth; and, with their long, flexible necks, they are not as streamlined as whales and sirenians. They also retain zygapophyses in their trunk vertebrae, and their teeth and jaws are little modified.

Pinnipeds appear to have lost the hindlimbs, but what seems to be a tail is actually a modified pair of hindlimbs that have been turned backward. True seals are unable to change the position of the hindlimbs and are clumsy on land, but sea lions and walruses can turn the hind legs forward and move on land with vertical flexions of the vertebral column in an effective, although ungainly, humping fashion. Sea lions additionally have a derived type of paraxial swimming in which the fore-limbs are used in synchrony for underwater flying, creating lift in the water in concert with dorsoventral movements of the back and hind legs.

▼ **FIGURE 21–14** Diversity of marine mammals. (a) Toothed whale: the bottlenose dolphin, *Tursiops truncatus* (Cetacea: Odontoceti: Delphinidae). (b) Baleen whale: Northern right whale, *Eubalaena glacialis* (Cetacea: Mysticeti: Balaenidae). (c) Sea cow: dugong, *Dugong dugon* (Sirenia: Dugongidae). (d) Sea lion: the Cape fur seal, *Arctocephalus pusilus* (Carnivora: Otariidae). (e) True seal: the harbor seal, *Phoca vitulina* (Carnivora, Phocidae). The drawings are approximately to scale except for the baleen whale, which is about one quarter of its normal size in comparison to the other animals.

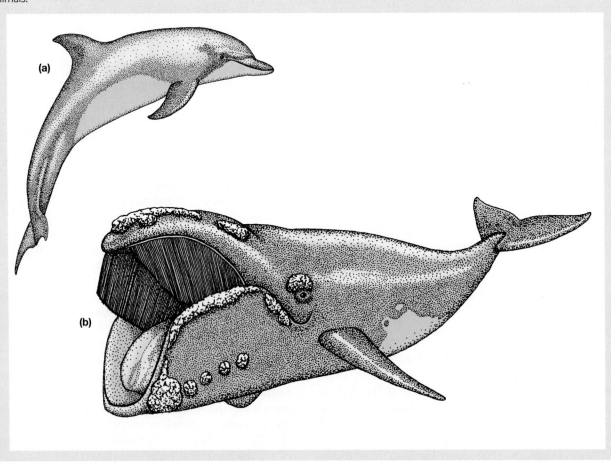

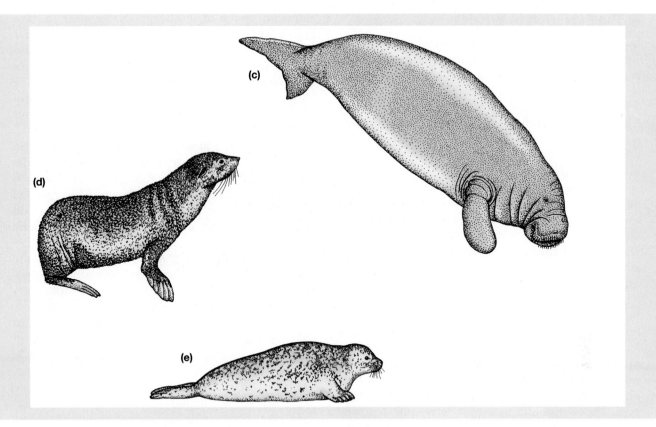

■ Morphological Adaptations for Life in Water

Most semiaquatic mammals use the limbs to swim (paraxial swimming), as we do ourselves. This type of swimming is fairly inefficient in terms of the drag forces created in the water. Most fully aquatic mammals use undulations of the body (axial swimming) via dorsoventral flexion, rather than laterally like fishes and aquatic sauropsids. This swimming motion is a modification of the flexion of the vertebral column that is used by terrestrial mammals. Fully aquatic mammals have short paddlelike limbs, with a short proximal portion and elongated phalanges, and these limbs are used like fish fins for braking and steering, not for propulsion. Modern cetaceans and sirenians have lost the hindlimb entirely. **Figure 21–15** illustrates the aquatic modifications of a generalized whale.

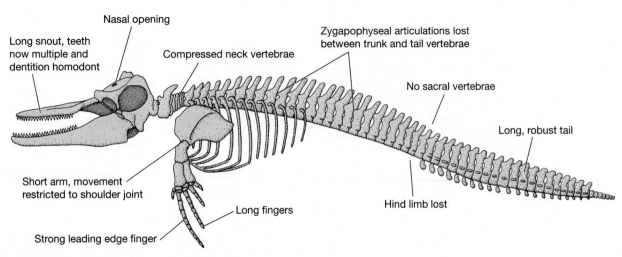

▲ **FIGURE 21–15** Specializations for aquatic locomotion in whales.

◼ The Evolution of Whales

In the past decade, an almost perfect fossil record sequence of early cetacean evolution has been assembled, starting in the Eocene of Pakistan along the shores of the ancient Tethys Sea. This sequence shows a progression among early whales from more terrestrial forms with a full set of legs to more aquatic forms with reduced and modified limbs (**Figure 21–16**). The relationship of whales to other mammals is discussed in Section 20.1.

These fossil whales all belong to the extinct suborder Archaeoceti, which was largely extinct by the end of the Eocene. The modern suborders of whales, Odontoceti (toothed whales) and Mysticeti (baleen whales), probably arose from a common ancestor among the archaeocetes and first appeared in the latest Eocene or early Oligocene.

The earliest archaeocetes are in the family Pakicetidae from the late early Eocene. These were fairly small animals in comparison to modern whales (coyote- to wolf-size), with skulls and teeth suggesting adaptations for aquatic predation. They had orbits situated on the top of their head, like a hippopotamus; later whales had more laterally placed eyes. They are found mainly in terrestrial deposits, but their postcranial remains suggest that they were amphibious, using quadrupedal paddling. Their middle ears were not modified for underwater hearing, and the oxygen isotope ratio in their bones indicates that they drank fresh water.

The Ambulocetidae were somewhat later forms from the early middle Eocene. They were about the size of a sea lion and had limbs with enormous feet, possibly a specialization in this group for a mode of locomotion involving dorsoventral flexion and paddling with the hind feet pointed backward like those of a seal. However, while their skeletons were more modified for aquatic life than the pakicetids, their fingers were not embedded in a webbed flipper, and both fingers and toes had a little hoof at the tips. Their fossil remains are found in coastal environments, but their bone isotopes show that they still drank fresh water. The robust skull and teeth of ambulocetids suggest that they were specialized crocodile-like ambush predators on large prey.

The Remingtoncetidae is a more derived group. They were seal-size and had relatively robust hind limbs that were probably still capable of bearing their weight on land. Unlike the ambulocetids, they had delicate skulls with long, narrow snouts and very small eyes. They may have fed on fish that they detected by sensing vibrations, catching them with a sideward sweep of the head. They appear to have been nearshore dwellers like ambulocetids but were probably more aquatic because their middle ear and semicircular canals show modifications like those of modern whales (Spoor et al. 2002). In addition, their bone isotopes show that they did not drink fresh water; they must have obtained all of their water from their food as modern whales do.

Animals that appear to be still more whalelike in appearance and behavior appeared in the late middle Eocene. These were the Protocetidae, which had more reduced hindlimbs than earlier whales. Most retained a connection of the hindlimb to the sacrum but would probably have been clumsy on land, if indeed they came ashore at all. The earlier protocetids probably still relied on their hindlimbs for paddling, but the later ones may have had an axial swimming locomotion, with oscillations of the lumbar spine. Protocetids were found in offshore marine habitats, and they were the first whales to be found outside of the Indo-Pakistan region, as far afield as the coasts of North America and Africa. They may have had a lifestyle like modern seals: fully aquatic, but not obligatorily so.

Finally, in the later Eocene, the Basilosauridae appeared. These whales had lost the hindlimb-sacral connection and had greatly reduced hindlimbs; their neck was short; their forelimbs were flipperlike; and the morphology of their tail vertebrae suggests that they had a tail fluke like modern whales. All these features suggest that they were obligatorily aquatic. The basilosaurines, known from the Northern Hemisphere, were long-bodied (up to 16 meters long), with greatly elongated trunk vertebrae and small heads. The hindlimbs were tiny and may have been used as copulatory guides. The durodontines were more dolphinlike in appearance, known from both Northern and Southern Hemispheres, and contained the ancestry of the modern whales.

The initial radiation of modern whales more or less coincides with the extinction of the archaeocetes at the Eocene-Oligocene boundary, a time when higher-latitude temperatures fell dramatically. Toothed whales and baleen whales are distinct radiations, even though the earliest baleen whales retained teeth. Modern whales differ from archaeocetes in having a telescoped skull, in which the nostrils have been moved to the top of the head where they form the blowhole, although the mode of skull telescoping is different in the two suborders.

The modern whale radiation is probably related to changes in oceanic circulation that resulted in the increased productivity of the oceans—resulting in the novel feeding strategies of echolocation-assisted

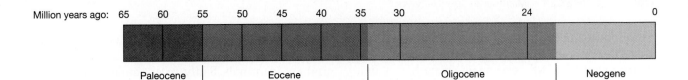

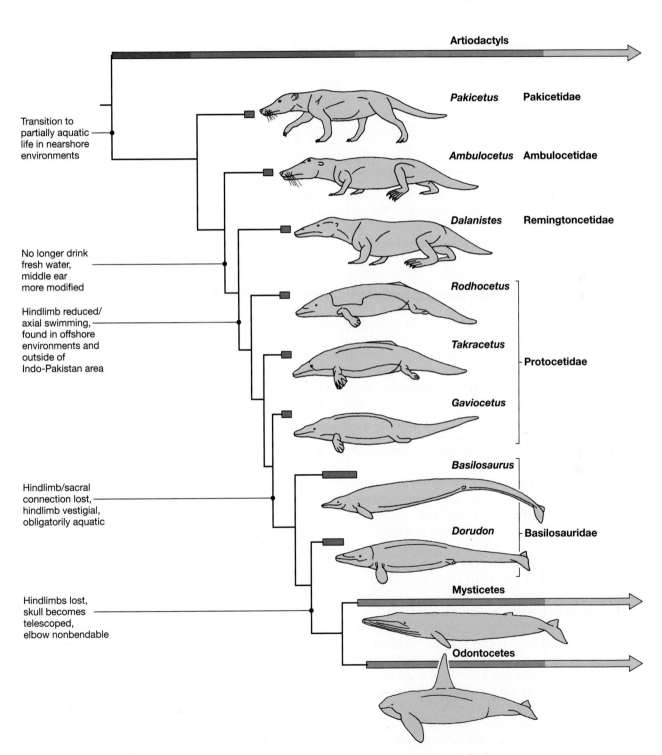

Million years ago: 65 60 55 50 45 40 35 30 24 0

Paleocene | Eocene | Oligocene | Neogene

Artiodactyls

Pakicetus **Pakicetidae**

Ambulocetus **Ambulocetidae**

Dalanistes **Remingtoncetidae**

Rodhocetus

Takracetus **Protocetidae**

Gaviocetus

Basilosaurus

Dorudon **Basilosauridae**

Mysticetes

Odontocetes

Transition to partially aquatic life in nearshore environments

No longer drink fresh water, middle ear more modified

Hindlimb reduced/axial swimming, found in offshore environments and outside of Indo-Pakistan area

Hindlimb/sacral connection lost, hindlimb vestigial, obligatorily aquatic

Hindlimbs lost, skull becomes telescoped, elbow nonbendable

▲ **FIGURE 21–16** A phylogeny of whales showing the sequence of appearance of aquatic specializations.

predation (toothed whales) and filter feeding (baleen whales). Another pulse of evolutionary radiation occurred in the late Miocene, again concurrent with a decline in higher-latitude temperatures and changes in oceanic circulation. At this time, many existing families became extinct, and some modern forms (such as dolphins and porpoises) made their first appearance.

Thus the fossil record shows that whales have long been an extremely diverse and successful group of mammals. Sadly, recent molecular analyses suggest that North Atlantic baleen whales were up to 20 times as abundant prior to the start of human commercial exploitation than they are today (Roman and Palumbi 2003).

Summary

The major groups of living mammals can be distinguished by their mode of reproduction. Therians (marsupials and placentals) are more derived than monotremes, giving birth to live young (viviparity) as opposed to laying eggs. Marsupials give birth to very immature young that complete their development attached to the nipples—usually, but not always—enclosed in a pouch. Placentals give birth to more mature young and have a shorter period of lactation than marsupials. The marsupial mode of reproduction has often been considered inferior to the placental one, but the differences may reflect only their separate evolutionary histories, with neither method being inherently superior to the other.

Cenozoic mammals have diversified into a variety of feeding types, reflected in different anatomies of their skulls and dentitions. The skulls and teeth of herbivores are in general more specialized than those of omnivores and carnivores because fibrous plant material is difficult to chew and abrades the teeth. Vegetation is also more difficult to digest than other diets, and many herbivores have evolved a symbiotic association with microorganisms in their guts, which ferment the plant fiber and aid in its chemical breakdown.

The teeth of herbivorous mammals are particularly specialized. In order to shred tough plant material, the original cusps have been run together into cutting blades, or lophs. Also, herbivorous mammals have to ensure that their teeth last a lifetime, a problem all mammals share since they replace their teeth only once, and the molars are not replaced at all. The most common way is to make the teeth high-crowned, or hypsodont, although different solutions exist in some mammals, such as elephants. Carnivorous mammals usually have a pair of specialized bladelike cutting teeth, the carnassials.

The primitive mammalian mode of locomotion is probably some sort of bounding. With the radiation of larger mammals in the Cenozoic, more specialized types of locomotion evolved. Cursorial (running) mammals have elongated their legs and changed their foot posture, restricting the range of limb motion and sometimes reducing the number of digits. Although such specializations bestow the ability for fast running and long limbs are assumed to have evolved in a coevolutionary fashion in predator and prey, ungulates in fact evolved their longer legs many millions of years before carnivores did. An alternative explanation for the evolution of cursorial specializations is for stamina during foraging. Fossorial limbs, designed for digging, are short and stout with heavy muscles and large claws. Fossorial specializations have evolved convergently in several lineages of mammals, including aardvarks, armadillos, moles, and rodents.

Many mammals have returned to the water to become semiaquatic, but only three modern groups of mammals (pinnipeds, sirenians, and cetaceans) are fully aquatic, living in marine environments. Marine mammals share a number of morphological features relating to the demands of underwater locomotion, as well as physiological adaptations for diving, etc. We have an excellent fossil record of whale evolution showing how they evolved from terrestrial species, through amphibious, near-shore forms, to obligatorily aquatic fully marine forms. Whale numbers have decreased drastically in historical times.

Additional Readings

Clauss, M., et al. 2003. The maximum attainable body size of herbivorous mammals: Morphophysiological constraints on foregut, and adaptations of hindgut fermenters. *Oecologica* 136:14–27.

Domning, D. P. 2001. The earliest known fully quadrupedal sirenian. *Nature* 413:625–627.

Fischer, M. S. et al. 2002. Basic limb kinematics of small therian mammals. *Journal of Experimental Biology* 205:1315–1338.

Frank, L. G. 1996. Female masculinization in the spotted hyena: Endocrinology, behavioral ecology, and evolution. In J. L. Gittleman (ed.), *Carnivore Behavior, Ecology, and Evolution*, volume 2. Ithaca, NY: Comstock Publishing Associates, 78–131.

Fordyce, R. E. 2000. The fossil record of whales. *American Paleontologist* 8:2–4.

Gatesy, J., and M. A. O'Leary. 2001. Deciphering whale origins with molecules and fossils. *Trends in Ecology and Evolution* 16:562–570.

Gaucher, E. A. et al. 2004. The planetary biology of cytochrome P450 aromatases. *BMC Biology* 2:19–33.

Gingerich, P. D. et al. 2001. Origin of whales from early artiodactyls: Hands and feet of Eocene Protocetidae from Pakistan. *Science* 293:2239–2242.

Hildebrand, M., and G. E. Goslow. 2001. *Analysis of Vertebrate Structure*, 5th ed., New York: Wiley.

Janis, C. M. 1976. The evolutionary strategy of the Equidae and the origins of rumen and cecal digestion. *Evolution* 30:757–774.

Janis, C. M., and M. Fortelius. 1988. On the means whereby mammals achieve increased functional durability of their dentitions, with special reference to limiting factors. *Biological Reviews* 63:197–230.

Janis, C. M. and P. B. Wilhelm. 1993. Were there mammalian pursuit predators in the Tertiary? Dances with wolf avatars. *Journal of Mammalian Evolution* 1:103–126.

Kirsch, J. A. 1977. The six percent solution: Second thoughts on the adaptedness of the Marsupialia. *American Scientist* 65:276–288.

Ley, R. E., et. al. 2008. Evolution of mammals and their gut microbes. *Science* 320:1647–1651.

Lillegraven, J. A. et al. 1987. The origin of eutherian mammals. *Biological Journal of the Linnean Society* 32:281–336.

Mossman, H. W. 1987. *Vertebrate Fetal Membranes*. New Brunswick, NJ: Rutgers University Press.

Muller, M. N., and R. Wrangham. 2002. Sexual mimicry in hyenas. *Quarterly Review of Biology* 77:3–16.

Renfree, M. B. 1993. Ontogeny, genetic control, and phylogeny of female reproduction in monotreme and therian mammals. In F. S. Szalay, M. J. Novacek, and M. C. McKenna (eds.), *Mammal Phylogeny, Mesozoic Differentiation, Multituberculates, Monotremes, Early Therians, and Marsupials*, New York: Springer Verlag, 4–20.

Renfree, M. B., and G. Shaw. 2000. Diapause. *Annual Review of Physiology* 62:353–375.

Roman, J., and S. R. Palumbi. 2003. Whales before whaling in the North Atlantic. *Science* 301:508–510.

Sears, K. E. 2004. Constraints on the morphological evolution of marsupial shoulder girdles. *Evolution* 58:2352–2370.

Selwood, L., and R. Johnson. 2006. Trophoblast and hypoblast in the monotreme, marsupial, and eutherian mammal: Evolution and origins. *BioEssays* 28:128–145.

Smith, K. K. 2001. The evolution of mammalian development. *Bulletin of the Museum of Comparative Zoology* 156:119–135.

Spoor, F. et al. 2002. Vestibular evidence for the evolution of aquatic behaviour in early cetaceans. *Nature* 417:163–166.

Thewissen, J. G. M. (ed.) 1998. *The Emergence of Whales*. New York: Plenum Press.

Thewissen, J. G. M., and S. Bajpai. 2001. Whale origins as a poster child for macroevolution. *BioScience* 51:1037–1049.

Thewissen, J. G. M., and E. M. Williams. 2002. The early radiations of Cetacea (Mammalia): Evolutionary pattern and developmental constraints. *Annual Reviews of Ecology and Systematics* 33:73–90.

Tyndale-Biscoe, C. H., and M. B. Renfree. 1987. *Reproductive Physiology of Marsupials*. Cambridge, UK: Cambridge University Press.

Werdelin, L., and A. Nilsonne. 1999. The evolution of the scrotum and testicular descent in mammals: A phylogenetic view. *Journal of Theoretical Biology* 196:61–72.

CHAPTER 22

Endothermy: A High-Energy Approach to Life

Endothermy is a derived character of mammals (synapsids) and birds (sauropsids). The two lineages evolved endothermy independently, but the costs and benefits are the same for both. Sustaining activity requires the high levels of aerobic metabolism characteristic of endotherms. Although ectotherms can achieve high levels of activity for short periods, they rely on anaerobic metabolism to do this and become exhausted in a few minutes. Only endotherms can sustain activity at high levels for prolonged periods.

Endothermy is a superb way to become relatively independent of many of the challenges of the physical environment, especially cold. Birds and mammals can live in the coldest habitats on Earth, assuming they can find enough food. That qualification expresses the major problem of endothermy: it is energetically expensive. Endotherms need a reliable supply of food, and the conspicuous interactions of endotherms are often with their biological environment—predators, competitors, and prey—rather than with the physical environment, as is often the case for ectotherms. Because energy intake and expenditure are important factors in the daily lives of endotherms, calculations of energy budgets can help us to understand the consequences of some kinds of behavior.

When all efforts at homeostasis are inadequate, endotherms have two more methods of dealing with harsh conditions: (1) birds and large species of mammals can migrate to areas where conditions are more favorable; and (2) many species of small mammals and some birds can become torpid. This response, a temporary drop in body temperature, conserves energy and prolongs survival—at the cost of abandoning the benefits of homeothermy.

22.1 Endothermal Thermoregulation

Birds and mammals are endotherms. They regulate their high body temperatures by precisely balancing metabolic heat production with heat loss. An endotherm can change the intensity of its heat production by varying metabolic rate; and it can change heat loss by varying insulation. An endotherm maintains a constant high body temperature by adjusting heat production to match heat loss under different environmental conditions.

Endotherms produce metabolic heat in several ways. Besides the obligatory heat production derived from the basal or resting metabolic rate, there is the heat increment of feeding, often called the **specific dynamic action** or specific dynamic effect of the food. This added heat production after eating apparently results from the energy used to assimilate molecules and synthesize protein, and it varies depending on the type of foodstuff being processed. It is highest for a meat diet and lowest for a carbohydrate diet.

Activity of skeletal muscle produces large amounts of heat. This is especially true during locomotion, which can result in a heat production exceeding the basal metabolic rate by 10-fold to 15-fold. This muscular heat can compensate for heat loss in a cold environment, but it can be a problem for animals in warm places. Cheetahs, for example, show a rapid increase in body temperature when they chase prey, and it is usually overheating rather than exhaustion that causes a cheetah to break off a pursuit. **Shivering,** the generation of heat by muscle-fiber contractions in an asynchronous pattern that does not result in gross movement of the whole muscle, is an important mechanism of heat production.

Endotherms usually live under conditions in which air temperatures are lower than the regulated body temperatures of the animals themselves, and in this situation heat is lost to the environment. Hair and feathers reduce the rate of heat loss by trapping air, and a mammal or bird can change its insulation by raising and lowering the hair or feathers to change the thickness of the layer of trapped air. We humans have goose bumps on our arms and legs when we are cold because our few remaining hairs rise to a vertical position in an ancestral mammalian attempt to increase our insulation.

These physiological responses to temperature are controlled from neurons located in the hypothalamus of the brain. The hypothalamic thermostat controls temperature regulation by ectotherms, too; but, in these animals, changes in hypothalamic temperature initiate thermoregulatory behaviors (e.g., orienting to maximize heat loss or gain) rather than physiological processes.

■ Mechanisms of Endothermal Thermoregulation

Body temperature and metabolic rate must be considered simultaneously to understand how endotherms maintain their body temperatures at a stable level in the face of environmental temperatures that may range from −70°C to +40°C. Most birds and mammals conform to the generalized diagram in **Figure 22–1.**

Each species of endotherm has a range of ambient temperatures over which the body temperature can be kept stable by using physiological and postural adjustments of heat loss and heat production. This temperature range is called the **zone of tolerance.** Above this range, the animal's ability to dissipate heat is inadequate; both the body temperature and metabolic rate increase as ambient temperature increases until the animal dies. At ambient temperatures below the zone of tolerance, the animal's ability to generate heat to balance heat loss is exceeded, body temperature falls, metabolic rate declines, and cold death results. The zones of tolerance of large animals usually extend downward to lower temperatures than those of smaller animals because heat is

▶ **FIGURE 22–1** Generalized pattern of changes in body temperature and metabolic heat production of an endothermic homeotherm in relation to environmental temperature. Normal body temperature varies somewhat for different mammals and birds.

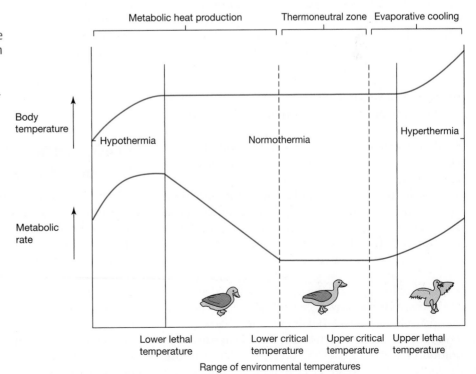

lost from the body surface, and large animals have smaller surface-to-mass ratios than those of small animals. Thus, they retain heat better at lower ambient temperatures. Similarly, well-insulated species have lower zones of tolerance than those of poorly insulated ones.

The **thermoneutral zone (TNZ)** is the range of ambient temperatures within which the metabolic rate of an endotherm is at its standard level, and thermoregulation is accomplished by changing the rate of heat loss. The thermoneutral zone is also called the zone of physical thermoregulation because an animal adjusts its heat loss using processes such as fluffing or sleeking its hair or feathers, postural changes such as huddling or stretching out, and changes in blood flow (vasoconstriction or vasodilation) to exposed parts of the body (feet, legs, face).

The **lower critical temperature** is the point below which an animal must increase metabolic heat production to maintain a stable body temperature. The **zone of chemical thermogenesis** lies below the lower critical temperature. In this zone, the metabolic rate increases as the ambient temperature falls. The quality of the insulation determines how much additional metabolic heat production is required to offset a change in ambient temperature. Well-insulated animals have relatively shallow slopes for the graph of increasing metabolism below the lower critical temperature, and poorly insulated animals have steeper slopes. Ultimately an animal reaches its **lower lethal temperature**. At that point, metabolic heat production is at its maximum rate and is still insufficient to balance the heat lost to the environment. In this situation, the body temperature falls; and, because the rates of the chemical reactions that produce metabolic heat are sensitive to temperature, heat production falls as well. A positive feedback loop is initiated in which falling body temperature reduces heat production, causing a further reduction in body temperature. Death from **hypothermia** (low body temperature) follows.

Endotherms are remarkably good at maintaining stable body temperatures in cool environments, but they have difficulty at high ambient temperatures. The **upper critical temperature** represents the point at which nonevaporative heat loss has been maximized by using all of the physical processes available to an animal—exposing the poorly insulated areas of the body and maximizing cutaneous blood flow. If these mechanisms are insufficient to balance heat gain, the only option vertebrates have is to use evaporation of water to cool the body. The temperature range from the upper critical temperature to the upper lethal temperature is the **zone of evaporative cooling**. Some mammals sweat, a process in which water is released from sweat glands on the surface of the body. Evaporation of the sweat cools the body surface. Other animals pant, breathing rapidly and shallowly so that evaporation of water from the respiratory system provides a cooling effect. Many birds use a rapid fluttering movement of the gular region to evaporate water for thermoregulation. Panting and gular flutter require muscular activity, and some of the evaporative cooling they achieve is used to offset the increased metabolic heat production they require.

At the **upper lethal temperature,** evaporative cooling cannot balance the heat flow from a hot environment. As body temperature rises, the metabolic rate increases, and metabolic heat production raises the body temperature still more, further increasing the metabolic rate. This process can lead to an explosive rise in body temperature and death from **hyperthermia** (overheating).

The difficulty that endotherms experience in regulating body temperature in high environmental temperatures may be one of the reasons that the body temperatures of most endotherms are in the range of 35°C to 40°C. Most habitats seldom have air temperatures that exceed 35°C. Even the tropics have average yearly temperatures below 30°C. Thus the high body temperatures maintained by endotherms ensure that, in most situations, the heat gradient is from animal to environment. (Still higher body temperatures, around 50°C, for example, could ensure that mammals were always warmer than their environment. There are upper limits to the body temperatures that are feasible, however. Many proteins denature near 50°C.)

■ Costs and Benefits of Endothermal Thermoregulation

Endothermy has both benefits and costs compared to ectothermy. On the positive side, endothermy allows birds and mammals to maintain high body temperatures when solar radiation is not available or is insufficient to warm them—at night, for example, or in the winter. The thermoregulatory capacities of birds and mammals are astonishing; they can live in the coldest places on Earth. On the negative side, endothermy is energetically expensive. The metabolic rates of birds and mammals are nearly an order of magnitude greater than those of amphibians and reptiles. The energy to sustain those high metabolic rates comes from food, and endotherms need more food than do ectotherms.

An understanding of the costs of living can be obtained by constructing an energy budget for an animal. An energy budget, like a financial budget, shows income and expenditure but uses units of energy as currency. The energy costs of different activities can be evaluated by converting the energy intake (food) and loss (metabolism, feces, and urine) to a common unit of measurement. Understanding where an animal gets energy and how it uses the energy can help to explain why it does or does not engage in a particular activity, such as migration, or why it does or does not live in a particular area. Although energy budgets are conceptually simple, constructing an energy budget can be difficult. Most animals eat a variety of foods that have different energy contents and engage in activities that have different energy costs. Vampire bats (*Desmodus rotundus*) are an ideal species for an energy budget because they eat only one thing, blood, and they engage in only two activities—resting in a cave and flying between the cave and the feeding site. Studies of vampire bats by Brian McNab (1973) have revealed a clear-cut relationship between energy intake, energy expenditure, and the species' geographic range (**Box 22–1**).

BOX 22–1   Energy Budgets of Vampire Bats

Studies of vampire bats (*Desmodus rotundus*) by Brian McNab (1973) have revealed a clear-cut relationship between energy intake, energy expenditure, and the species' geographic range. Vampire bats inhabit the Neotropics and feed exclusively on blood. Their daily pattern of activity is simple: they spend about 21 hours in their caves, fly out at night to a feeding site, and return after they have fed. Typically, a vampire bat flies about 10 kilometers round-trip at 20 kilometers per hour to find a meal. Thus, a bat spends half an hour per day in flight. The remaining time outside the cave may be spent in feeding (**Figure 22–2**).

The general form of an energy budget equation is

(energy in) = (energy out) ± (biomass change),

and it has five components.

- I = ingested energy (blood). The blood a bat drinks provides the energy needed for all of its life processes: maintenance, activity, growth, and reproduction.
- E = excreted energy. As in any animal, not all of the food ingested is digested and taken up by the bat. The energy contained in the feces and urine is lost.
- M = resting metabolism. This can be subdivided into M_i (metabolism while the bat is inside the cave) and M_o (non-flight metabolism while the bat is outside the cave).
- A = cost of activity (a half hour of flight per day).
- B = biomass increase or decrease. This term is the energy profit or loss a bat shows in its energy budget. It may be stored as fat or used for growth or reproduction (production of gametes, growth of a fetus, or nursing a baby).

The biomass term appears as ± in the equation because an animal metabolizes some body tissues when its energy expenditure exceeds its energy intake. (This is what every dieter hopes to do to lose weight.) Inserting these terms into the general equation gives

$$I - E = M_i + M_o + A \pm B.$$

All of these terms can be measured and expressed as kilojoules per bat per day (kJ/bat • day). For a Brazilian vampire bat weighing 42 g:

- **Ingested energy (I)**—In a single feeding a vampire bat can consume 57 percent of its body mass in blood, which contains 4.6 kJ/g. Thus the ingested energy is

  42 g • 57% • 4.6 kJ/g fluid blood = 110.1 kJ.

- **Excreted energy (E)**—A vampire bat excretes 0.24 g urea in the urine plus 0.95 g of feces daily. Urea contains 10.5 kJ/g, and the feces contain 23.8 kJ/g. Thus the excreted energy is

  0.24 g urea • 10.5 kJ/g + 0.95 g feces • 23.8 kJ/g
  = 2.5 kJ + 22.6 kJ = 25.1 kJ.

- **Resting metabolism**—In a tropical habitat, 20°C is a reasonable approximation of the temperature a bat experiences both inside and outside the cave. While at rest in the laboratory at 20°C, a vampire's metabolic rate is 3.8 cm³ O_2/g • hr. The terms for metabolism can be calculated and converted to joules using the energy equivalent of oxygen (20.1 J/cm³ O_2):

  M_i = 42 g • 3.8 cm³ O_2/g • hr • 20.1 J/cm³ O_2
  • 22 hr/day = 70.6 kJ
  M_o = 42 g • 3.8 cm³ O_2/g • hr • 20.1 J/cm³ O_2
  • 1.5 hr/day = 4.8 kJ.

- **Activity**—The metabolism of a bat flying at 20 km/hr is 11.4 cm³ O_2/g • hr. The cost of the round-trip from the cave to the feeding site is

  A = 42 g • 11.4 cm³ O_2/g • hr • 20.1 J/cm³ O_2
  • 0.5 hr = 4.8 kJ.

- **Biomass change**—The quantities calculated so far are fixed values that the bat cannot avoid except by changing its behavior. The biomass change is a variable value. If the

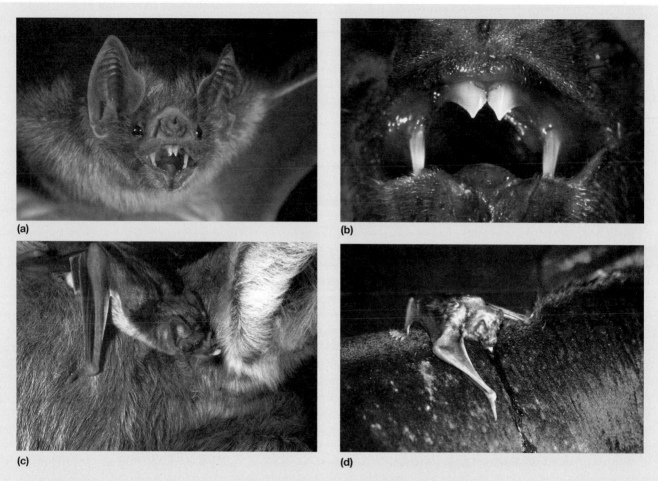

▲ **FIGURE 22–2** Vampire bats can feed on sleeping mammals and birds without awakening them. (a) Vampire bats have small eyes and rely mostly on echolocation to hunt and navigate. (b) A vampire bat uses its sharp incisor teeth to make a shallow incision. (c) The incision is made in an area here where there is a concentration of capillaries, such as behind the ear. (d) Blood flows into the shallow wound, and the bat uses its tongue to lap up the blood. (Contrary to popular belief, vampire bats do not suck blood from their victims.)

assimilated energy is greater than the fixed costs, this energy profit can go to biomass increase. Fixed costs that exceed the assimilated energy are reflected as a loss of biomass. For the situation described, there is an energy profit:

$$I - E = M_i + M_o + A \pm B$$
$$110.1 \text{ kJ} - 25.1 \text{ kJ} = 70.6 \text{ kJ} + 4.8 \text{ kJ} + 4.8 \text{ kJ} \pm B$$
$$B = 4.8 \text{ kJ/bat} \cdot \text{day.}$$

These calculations show that vampires can live and grow under the conditions assumed. What happens if we change some of the assumptions? McNab points out that the northern and southern limits of the geographic range of vampires conform closely to the winter isotherms of 10°C (**Figure 22–3**). That is, bats do not occur in regions where the minimum temperature outside the cave is lower than 10°C. Is this coincidence, or is 10°C the lowest temperature the bats can withstand?

Calculating an energy budget for a vampire bat under these colder conditions provides an answer.

Caves have very stable temperatures that usually do not vary from summer to winter. We will assume that temperature remains constant at 20°C in the cave. Thus, only the conditions a bat encounters outside the cave are altered. Because of limitations of stomach capacity, ingestion cannot increase beyond 57 percent of body mass (the value assumed in the previous calculation). Therefore, we need to recalculate only M_o, A, and B.

- **Metabolism outside**—At 10°C, a bat must increase its metabolic rate to maintain its body temperature, and laboratory measurements indicate the resting metabolic rate increases to 6.3 cm³ O_2/g · hr:

$$= 42 \text{ g} \cdot 6.3 \text{ cm}^3 \text{ O}_2/\text{g} \cdot \text{hr} \cdot 20.1 \text{ J/cm}^3 \text{ O}_2 \cdot 1.5 \text{ hr}$$
$$= 8.0 \text{ kJ.}$$

▶ **FIGURE 22–3** Geographic range of the vampire bat, *Desmodus rotundus.* The area in which vampire bats are found (shaded) closely approximates the 10°C isotherm for the minimum average temperature during the coldest month of the year (dashed line) at the northern limit of its range (in Mexico and the southernmost tip of Texas) and the southern limit of Uruguay, Argentina, and Chile). Positions of the 10°C isotherm and the bats' altitudinal range in the Andes Mountains are not known and are indicated by question marks.

- **Activity**—The cost of activity will not change because the metabolic rate of the bat during flight (11.4 cm^3 O$_2$/g • hr) is higher than the resting metabolic rate needed to keep it warm (6.3 cm^3 O$_2$/g • hr). Only M_o changes, increasing from 4.8 to 8.0 kJ, and the sum of the energy costs becomes 83.4 kJ/bat • day.

Because the assimilated energy remains at 85.0 kJ/bat • day, only 1.6 kJ are available for biomass increase. The assumptions in these calculations introduce a degree of uncertainty,

and probably 1.6 kJ is not statistically different from 0 kJ. Thus, at 10°C, a bat uses all its energy staying alive. A vampire bat could survive under those conditions, but it would have no energy surplus for growth or reproduction. If the temperature outside the cave were lower than 10°C, the bat would have a negative energy balance and would lose body mass with each meal. This match between our calculations and the actual geographic distribution of the bats indicates that energy may be a significant factor in limiting their northward and southward spread.

22.2 Endotherms in the Arctic

Most endotherms (especially small ones) expend most of the energy they consume just keeping themselves warm—even in the moderate conditions of tropical and subtropical climates. Nonetheless, endotherms have proved themselves very adaptable in extending their thermoregulatory responses to allow them to inhabit even Arctic and Antarctic regions. Not even small body size is an insuperable handicap to life in these areas: small birds, such as redpolls and chickadees, weighing only 10 grams, overwinter in central Alaska.

A stable body temperature in extreme cold can be achieved by increasing heat production or by decreasing heat loss. On closer examination, the option of increasing heat production does not seem particularly attractive. Any significant increase in heat production would require an increase in food intake. This scheme poses obvious ecological difficulties in terrestrial Arctic and Antarctic habitats where primary production is extremely low, especially during the coldest parts of the year. For most polar animals the quantities of food necessary would probably not be available.

Because they lack the option of increasing heat production significantly, conserving heat within the body is the primary thermoregulatory mechanism of polar endotherms. Insulative values of pelts from Arctic mammals are two to four times those of tropical mammals. In Arctic species, insulative value is closely related to fur length (**Figure 22–4**). Small species such as the least weasel and the lemming have fur only 1 to 1.5 centimeters long. Presumably, the thickness of their fur is limited because longer hair would interfere with the movement of their legs. Large mammals (caribou, polar and grizzly bears, Dall sheep, and Arctic fox) have hair 3 to 7 centimeters long. There is no obvious reason why their hair

could not be longer; apparently, further insulation is not needed. The insulative values of pelts of short-haired tropical mammals are similar to those measured for the same hair lengths in Arctic species. Long-haired tropical mammals, like the sloths, have fewer hairs per square centimeter of skin, and thus less insulation than on Arctic mammals with hair of similar length.

A comparison of the lower critical temperatures of tropical and Arctic mammals illustrates the effectiveness of the insulation provided (**Figure 22–5**). Tropical mammals have lower critical temperatures, between 20°C and 30°C. As air temperatures fall below their lower critical temperatures, these animals are no longer in their thermoneutral zones and must increase their metabolic rates to maintain normal body temperatures. For example, a tropical raccoon has increased its metabolic rate approximately 50 percent above its standard level at an environmental temperature of 25°C.

Arctic mammals are much better insulated than tropical species, and even small mammals like the least weasel and the lemming have lower critical temperatures that are between 10°C and 20°C. Larger mammals have thermoneutral zones that extend well below freezing. Because of their effective insulation,

▶**FIGURE 22–4** Insulative values of the pelts of Arctic mammals. In air the insulation is proportional to the length of the hair. Pelts from tropical mammals (×) have approximately the same insulative value as those of Arctic mammals (dark circles) at short hair lengths, but long-haired tropical mammals like sloths have less insulation than Arctic mammals with hair of the same length. Immersion in water greatly reduces the insulative value of hair, even for such semiaquatic mammals as the beaver and polar bear (light circles).

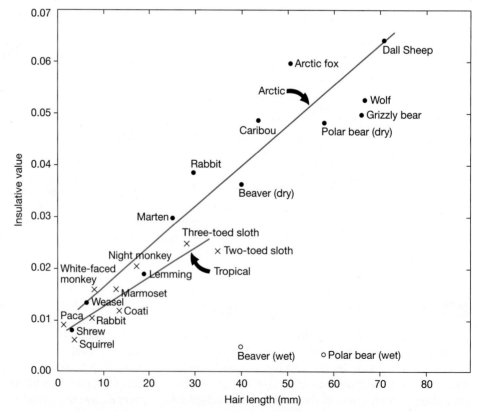

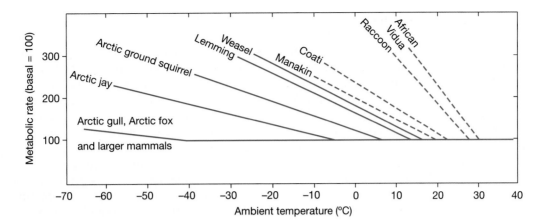

▲ **FIGURE 22–5** Lower critical temperatures for birds and mammals. Solid lines = arctic birds and mammals; dashed lines = tropical birds and mammals. The basal metabolic rate for each species is considered to be 100 units to facilitate comparisons among species. The lower end of the zone of thermoneutrality is at the intersection of the basal metabolic rate (horizontal line) and the increased metabolic rate in the zone of chemical thermogenesis. The steepness of the slope shows how rapidly metabolic heat production increases as ambient temperature falls. The slopes for tropical animals are steeper than those for Arctic animals.

Arctic birds and mammals can maintain resting metabolic rates at lower environmental temperatures than can tropical species, and they show smaller increases in metabolism (i.e., flatter slopes) below the lower critical temperature. The Arctic fox, for example, has a lower critical temperature of −40°C and at −70°C (approximately the lowest air temperature it ever encounters) has elevated its metabolic rate only 50 percent above its standard level. Under those conditions the fox is maintaining a body temperature approximately 110°C above air temperature. Arctic birds are equally impressive, and the Arctic glaucous gull, like the Arctic fox, has a lower critical temperature near −40°C and can withstand −70°C with only a modest increase in metabolism.

Aquatic life in cold regions places still more stress on an endotherm. Because of the high heat capacity and conductivity of water, an aquatic animal may lose heat at 50 to 100 times the rate it would if it were moving at the same speed through air. Even a small body of water is an infinite heat sink for an endotherm; all of the matter in its body could be converted to heat without appreciably raising the temperature of the water. How, then, do endotherms manage to exist in such stressful environments?

Hair and feathers are not good insulation for fully aquatic animals because most of the insulative value is lost when air trapped between hairs or feathers is displaced by water. The insulative value of beaver and polar bear hair falls almost to zero when it is wet through (see Figure 22–4). In water, fat is a far more effective insulator than hair, and fully aquatic mammals have thick layers of blubber on their bodies. This blubber forms the primary layer of insulation; skin temperature is nearly identical to water temperature, and there is a steep temperature gradient through the blubber so that its inner surface is at the animal's core body temperature.

The insulation provided by blubber is so effective that pinnipeds and cetaceans require special heat-dissipating mechanisms to avoid overheating when they engage in strenuous activity or venture into warm water or onto land. This heat dissipation is achieved by shunting blood into capillary beds in the skin outside the blubber layer and into the flippers, which are not covered by blubber. Selective perfusion of these capillary beds enables a seal or porpoise to adjust its heat loss to balance its heat production. When it is necessary to conserve energy, a countercurrent heat-exchange system in the blood vessels supplying the flippers is brought into operation; when excess heat is to be dumped, blood is shunted away from the countercurrent system into superficial veins.

The effectiveness of the insulation of marine mammals is graphically illustrated by the problems experienced by the northern fur seal (*Callorhinus ursinus*) during its breeding season. Northern fur seals are large animals; males attain body masses in excess of 250 kilograms. Unlike most pinnipeds, fur seals have blubber and a dense covering of fur that is probably never wet through to the skin. They are inhabitants of the North Pacific. For most of the year they are pelagic, but during summer they breed on the Pribilof Islands in the Bering Sea north of the Aleutian Islands. Male fur seals gather harems of females on the shore. There they must try to prevent the females from straying, chase away other males, and copulate with willing females.

George Bartholomew and his colleagues have studied both the behavior of the fur seals and their thermoregulation (Bartholomew and Wilke 1956). Summers in the Pribilof Islands (which are near 57°N latitude) are characterized by nearly constantly overcast skies and air temperatures that rise only to 10°C during the day. These conditions are apparently close to the upper limits the seals can tolerate. Almost any activity on land makes them pant and raise their hind flippers (which are abundantly supplied with sweat glands), waving them about. If the sun breaks through the clouds, activity suddenly diminishes—females stop moving about, males reduce harem-guarding activities and copulation, and the adult seals pant and wave their flippers. If the air temperature rises as high as 12°C, females, which never defend territories, begin to move into the water. Forced activity on land can produce lethal overheating.

At the time of the study, seal hunters herded the bachelor males from the area behind the harems inland preparatory to killing and skinning them. Bartholomew recorded one drive that took place in the early morning of a sunny day while the air temperature rose from 8.6°C at the start of the drive to 10.4°C by the end. In 90 minutes, the seals were driven about 1 kilometer, with frequent pauses for rest.

> The seals were panting heavily and frequently paused to wave their hind flippers in the air before they had been driven 150 yards from the rookery. By the time the drive was half finished, most of the seals appeared badly tired and occasional animals were dropping out of the pods (groups of seals). In the last 200 yards of the drive and on the killing grounds there were found 16 "roadskins" (animals that had died of heat prostration) and in addition a number of others prostrated by overheating.

The average body temperature of the roadskins of this drive was 42.2°C, which is 4.5°C above the 37.7°C mean body temperature of adults not under thermal stress.

Fur seals can withstand somewhat higher temperatures in water than they can in air because of the greater heat conduction of water, but they are not able to penetrate warm seas. Adult male fur seals apparently remain in the Bering Sea during their pelagic season. Young males and females migrate into the North Pacific, but they are not found in waters warmer than 14°C to 15°C, and they are most abundant in water of 11°C. Their inability to regulate body temperature during sustained activity and their sensitivity to even low levels of solar radiation and moderate air temperatures probably restrict the location of potential breeding sites and their movements during their pelagic periods. Summers in the Pribilofs are barely cool enough to allow the seals to breed there. An increase in summer temperature associated with global warming might drive the seals from their traditional breeding grounds.

22.3 Migration to Avoid Difficult Conditions

Every environment has unfavorable aspects for some species, and these unfavorable conditions are often seasonal, especially in latitudes far from the equator. The primary cause of migrations is usually related to seasonal changes in climatic factors such as temperature or rainfall. In turn, these conditions influence food supply and the occurrence of suitable breeding conditions.

Long-distance migration is more feasible for birds and marine animals than for terrestrial species, partly because geographic barriers are less of a problem and partly because the energy cost of transport is less for swimming fish and flying birds than for walking mammals. We can consider the costs and benefits of migrating by considering two kinds of animals that represent extremes of body size. The baleen whales are the largest animals that have ever lived, and hummingbirds are among the smallest vertebrate endotherms, yet both whales and hummingbirds migrate.

▮ Whales

The annual cycle of events in the lives of the great baleen whales is particularly instructive in showing how migration relates to the use of energy and how it correlates with reproduction in the largest of all animals. Most baleen whales spend summers in polar or subpolar waters of either the Northern or the Southern Hemisphere, where they feed on krill or other crustaceans that are abundant in those cold, productive waters. For three or four months each year, a whale consumes a vast quantity of food that is converted into stored energy in the form of blubber and other kinds of fat. Pregnant female whales are using energy to support the development of their unborn young, which may grow to one-third the length of their mothers before birth.

Near the end of summer, the whales begin migrating toward tropical or subtropical waters where the females bear their young. The young grow rapidly on the rich milk provided by their mothers, and by spring the calves are mature enough to travel with their mothers back to Arctic or Antarctic waters. The calves are weaned about the time they arrive in their summer quarters. From a bioenergetic and trophic point of view, the remarkable feature of this annual migration is that virtually all of the energy required to fuel it comes from ravenous feeding and fattening during the three or four months spent in polar seas. Little or no feeding occurs during migration or during the winter period of calving and nursing. Energy for all these activities comes from the abundant stores of blubber and fat.

The gray whale (*Eschrichtius robustus*) of the Pacific Ocean has one of the longest and best-known migrations (**Figure 22–6**). The summer feeding waters are in the Bering Sea and the Chukchi Sea north of the Bering Strait in the Arctic Ocean. A small segment of the population moves down the coast of Asia to Korean waters at the end of the Arctic summer, but most gray whales follow the Pacific Coast of North America, moving south to Baja California and adjacent parts of western Mexico. They arrive in December or January, bear their young in shallow, warm waters, and then depart northward again in March. Some gray whales make an annual round-trip of at least 9000 kilometers.

The amount of energy expended by a whale in this annual cycle is phenomenal. The basal metabolic rate of a gray whale with a fat-free body mass of 50,000 kilograms is approximately 979,000 kilojoules per day. If the metabolic rate of a free-ranging whale, including the locomotion involved in feeding and migrating, is about three times the basal rate (a typical level of energy use for mammals), the whale's average daily energy expenditure is around 2,937,000 kilojoules. Body fat contains 38,500 kilojoules per kilogram, so the whale's daily energy expenditure is equivalent to metabolizing over 76 kilograms of blubber or fat per day. Assuming an energy content of 20,000 kilojoules per kilogram for krill and a 50 percent efficiency in converting the gross energy intake of food into biologically usable energy, the energy requirement

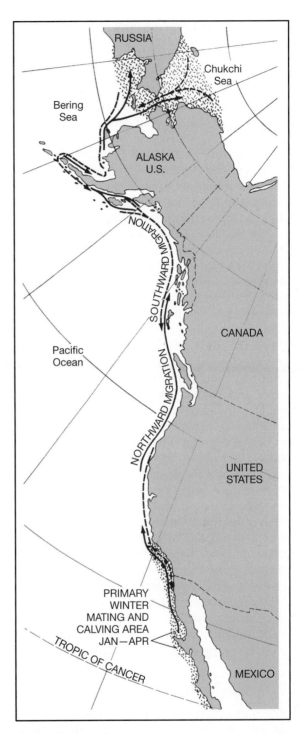

▲ **FIGURE 22–6** Migratory route of the gray whale between the Arctic Circle and Baja California.

for existence is equivalent to a daily intake of 294 kilograms of food.

In addition to satisfying its daily energy needs, a whale on the feeding grounds must accumulate a store of blubber. To live for 245 days without eating, the whale must metabolize a minimum of 18,375 kilograms of fat. Accumulating that amount of fat in 120 days of active feeding in Arctic waters at a conversion efficiency of 50 percent requires the consumption of 70,438 kilograms of krill, or 586 kilograms per day. The total food intake per day on the feeding grounds to accommodate the whale's daily metabolic needs plus energy storage for the migratory period would be not less than 294 + 586 = 880 kilograms of krill per day.

This is a minimum estimate for females because the calculations do not include the energetic costs of the developing fetus or the cost of milk production. Nor do they include the cost of transporting 20,000 kilograms of fat through the water. However, a large whale can do all this work and more without exhausting its insulative blanket of blubber because nearly half the total body mass of a large whale consists of blubber and other fats.

Why does a gray whale expend all this energy to migrate? The adult is too large and too well insulated ever to be stressed by the cold Arctic and sub-Arctic waters, which do not vary much from 0°C between summer and winter. It seems strange for an adult whale to abandon an abundant source of food and go off on a forced starvation trek into warm waters that may cause stressful overheating. The advantage probably accrues to the newborn young, which, though relatively large, lack an insulative layer of blubber. If the young whale were born in cold northern waters, it would probably have to use a large fraction of its energy intake (milk produced from its mother's stored fat) to generate metabolic heat to regulate its body temperature. That energy could otherwise be used for rapid growth. Apparently it is more effective, and perhaps energetically more efficient, for the mother whale to migrate thousands of kilometers into warm waters to give birth and nurse in an environment where the young whale can invest most of its energy intake in rapid growth.

■ Hummingbirds

At the opposite end of the size range, hummingbirds are the smallest endotherms that migrate. Ornithologists have long been intrigued by the ability of the ruby-throated hummingbird (*Archilochus colubris*), which weighs only 3.5 grams to 4.5 grams, to make a nonstop flight of 800 kilometers during migration across the Gulf of Mexico from Florida to the Yucatán Peninsula.

Like most migratory birds, ruby-throated hummingbirds store subcutaneous and body fat by feeding heavily prior to migration. A hummingbird with a lean mass of 2.5 grams can accumulate 2 grams of fat. Measurements of a hummingbird hovering in the air in a respirometer chamber in the lab indicate an energy consumption of 2.89 kilojoules per hour to 3.10 kilojoules per hour. Hovering is energetically more expensive than forward flight, so these values represent the maximum energy used in migratory flight. Even so, 2 grams of fat produce enough energy to last for 24 hours to 26 hours of sustained flight. Hummingbirds fly about 40 kilometers per hour, so crossing the Gulf of Mexico requires about 20 hours. Thus, by starting with a full store of fat, they have enough energy for the crossing with a reserve for unexpected contingencies such as a headwind that slows their progress. In fact, most migratory birds wait for weather conditions that will generate tailwinds before they begin their migratory flights, thereby further reducing the energy cost of migration.

22.4 Torpor as a Response to Low Temperatures and Limited Food

We have stressed the high energy cost of endothermy because the need to collect and process enough food to supply that energy is a central factor in the lives of many endotherms. In extreme situations, environmental conditions may combine to overpower a small endotherm's ability to process and transform enough chemical energy to sustain a high body temperature through certain critical phases of its life. For diurnally active birds, long cold nights during which there is no access to food can be lethal, especially if the bird has not been able to feed fully during the daytime. Cold winter seasons usually present a dual problem for resident endotherms—the need to maintain high body temperature when environmental temperatures are low despite the seasonal scarcity of food energy. In response to such problems, some birds and mammals have mechanisms that permit them to avoid the energetic costs of maintaining a high body temperature under unfavorable circumstances by entering a state of **torpor** (adaptive hypothermia). By entering torpor, an endotherm is giving up many of the advantages of endothermy, but in exchange it realizes an enormous savings of both energy and water. Thus endotherms enter torpor only when they would face critical shortages of energy or water if they remained at normal body temperature.

■ Physiological Adjustments during Torpor

When an endotherm becomes torpid, profound changes occur in a variety of physiological functions. Although body temperatures may fall very low during torpor, temperature regulation does not entirely cease. In deep torpor, an animal's body temperature drops to within 1°C or less of the ambient temperature, and in some cases (bats, for example), extended survival is possible at body temperatures just above the freezing point of the tissues. Arctic ground squirrels actually allow the temperature of parts of their bodies to supercool as low as −2.9°C. Oxidative metabolism and energy use are reduced to as little as one-twentieth the rate at normal body temperatures (**Figure 22–7**). Respiration is slow, and the overall breathing rate can be less than one inspiration per minute. Heart rates are drastically reduced, and blood flow to peripheral tissues is virtually shut down, as is blood flow posterior to the diaphragm. Most of the blood is retained in the body's core.

Deep torpor is a comatose condition, much more profound than the deepest sleep. Voluntary motor responses are reduced to sluggish postural changes, but some sensory perception of powerful auditory and tactile stimuli and ambient temperature changes is retained. Perhaps most dramatically, a torpid animal can arouse spontaneously from this state using heat production by brown fat, a tissue metabolically specialized for heat production. Some endotherms can rewarm under their own power from the lowest levels of torpor; others must warm passively with an increase in ambient temperature until some threshold is reached at which arousal starts.

There are varying degrees of torpor, from the deepest states of hypothermia to the lower range of body temperatures reached by normally active endotherms during their daily cycles of activity and sleep. Nearly all birds and mammals, especially those with body masses under 1 kilogram, undergo circadian temperature cycles. These cycles vary from 1°C to 5°C or more between the average high-temperature characteristic of the active phase of the daily cycle and the average low temperature characteristic of rest or sleep. Small birds (sunbirds, hummingbirds, chickadees) and small mammals (especially bats and rodents) may drop their body temperatures during quiescent periods from 8°C to 15°C below their regulated temperature during activity. Even a bird as large as the turkey vulture (about 2.2 kilograms) regularly drops its body temperature at night. When all these different endothermic patterns are considered together, no really sharp distinction can be drawn between torpor and the basic daily cycle in body temperature that characterizes most small- to medium-sized endotherms.

■ Body Size and the Occurrence of Torpor

Species of endotherms capable of deep torpor are found in several groups of mammals and birds. The echidna, platypus, and several species of small marsupials display patterns of hypothermia, but the phenomenon is most diverse among placentals, particularly among bats and rodents. Certain kinds of hypothermia have also been described for some insectivores, particularly the hedgehog, some primates, and some edentates. Deep torpor, contrary to popular notion, does not occur in bears, even though some of them den in the winter and remain inactive for long periods. Among birds, torpor occurs in some of the goatsuckers or nightjars and in hummingbirds, swifts, mousebirds, and some passerines (sunbirds,

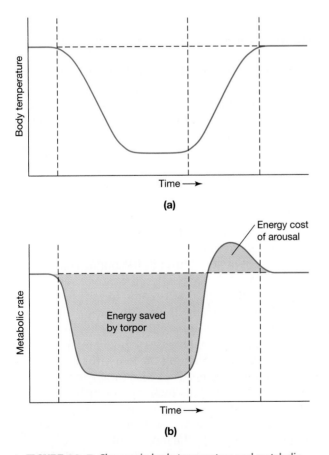

▲ **FIGURE 22–7** Changes in body temperature and metabolic rate during torpor. A decrease in metabolic rate (shown in part b) precedes a fall in body temperature (part a) to a new set point. An increase in metabolism produces the heat needed to return to normal body temperatures; metabolic rate during arousal briefly overshoots resting rate.

swallows, chickadees, and others). Other species show varying depths of hypothermia at rest or in sleep but are not in a semicomatose state of deep torpor.

The largest mammals that undergo deep torpor are marmots, which weigh about 5 kilograms, and torpor and body size are closely related. Torpor is not as advantageous for a large animal as for a small one. In the first place, the energetic cost of maintaining high body temperature is relatively greater for a small animal than for a large one, and as a consequence, a small animal has more to gain from becoming torpid. Second, a large animal cools off more slowly than a small animal, so it does not lower its metabolic rate as rapidly.

Furthermore, large animals have more body tissue to rewarm on arousal, and their costs of arousal are correspondingly larger than those of small animals. An endotherm weighing a few grams, such as a little brown bat or a hummingbird, can warm up from torpor at the rate of about 1°C per minute and be fully active within 30 minutes or less, depending on the depth of hypothermia. A 100-gram hamster requires more than two hours to arouse, and a marmot takes many hours. Entrance into torpor is slower than arousal. Consequently, daily torpor is feasible only for very small endotherms; there would not be enough time for a large animal to enter and arouse from torpor during a 24-hour period. Moreover, the energy required to warm up a large mass is very great. A 4-gram hummingbird needs only 0.48 kilojoule to raise its body temperature from 10°C to 40°C. That is 1/85 of the total daily energy expenditure of an active hummingbird in the wild. By contrast, a 200-kilogram bear would require 18,000 kilojoules to warm from 10°C to 37°C, the equivalent of a full day's energy expenditure. The smaller potential savings and the greater cost of arousal make daily torpor impractical for any but small endotherms.

Medium-sized endotherms are not entirely excluded from the energetic savings of torpor, but the torpor must persist for a longer period to realize a saving. For example, ground squirrels and marmots enter prolonged torpor during the winter **hibernation** when food is scarce. They spend several days at very low body temperatures (in the region of 5°C), then arouse for a period before becoming torpid again. Still larger endotherms would have such large total costs of arousal (and would take so long to warm up) that torpor is not cost effective for them even on a seasonal basis. Bears in winter dormancy, for example, lower their body temperatures only about 5°C from normal levels, and metabolic rate decreases about 50 percent. That small reduction in body temperature, combined with the large fat stores bears accumulate before retreating to their winter dens, is sufficient to carry them through the winter.

Energetic Aspects of Daily Torpor

Studies of daily torpor in birds have emphasized the flexibility of the response in relation to the energetic stress faced by individual birds. Susan Chaplin's work with chickadees provides an example (Chaplin 1974). These small (10 to 12 grams) passerine birds are winter residents in northern latitudes, where they regularly experience ambient temperatures that do not rise above freezing for days or weeks (**Figure 22–8**).

Chaplin found that in winter, chickadees around Ithaca, New York, allow their body temperatures to drop from the normal level of 40°C to 42°C that is maintained during the day down to 29°C to 30°C at night. This reduction in body temperature permits a 30 percent reduction in energy consumption. The chickadees rely primarily on fat stores they accumulate as they feed during the day to supply the energy needed to carry them through the following night. Thus the energy available to them and the energy they use at night can be estimated by measuring the fat content of birds as they go to roost in the evening and as they begin activity in the morning. Chaplin found that in the evening, chickadees had an average of 0.80 gram of fat per bird. By morning the fat store had decreased to 0.24 grams. The fat metabolized during the night (0.56 grams per bird) corresponds to the metabolic rate expected for a bird at 30°C.

Chaplin's calculations show that this torpor is necessary if the birds are to survive the night. It would require 0.92 gram of fat per bird to maintain a body temperature of 40°C through the night. That is more fat than the birds have when they go to roost in the evening. If they did not become torpid, they would starve before morning. Even with torpor, they use 70 percent of their fat reserve in one night. They do not have an energy supply to carry them far past sunrise, and chickadees are among the first birds to begin foraging in the morning. They also forage in weather so foul that other birds, which are not in such precarious energy balance, remain on their roosts. The chickadees must reestablish their fat stores each day if they are to survive the next night.

Hummingbirds, too, may depend on the energy they gather from nectar during the day to carry them through the following night. These very small birds (3 to 10 grams) have extremely high energy expenditures and yet are found during the summer in northern latitudes and at high altitudes. An example of the

▶ **FIGURE 22–8** The black-capped chickadee, *Parus atricapillus*.

lability of torpor in hummingbirds was provided by studies of nesting broad-tailed hummingbirds at an altitude of 2900 meters near Gothic, Colorado (Calder and Booser 1973). Ambient temperatures drop nearly to freezing at night, and hummingbirds become torpid when energy is limiting. Calder and Booser were able to monitor the body temperatures of nesting birds by placing an imitation egg containing a temperature-measuring device in the nest. These temperature records showed that hummingbirds incubating eggs normally did not become torpid at night. The reduction of egg temperature that results from the parent bird's becoming torpid does not damage the eggs, but it slows development and delays hatching. Presumably, there are advantages to hatching the eggs as quickly as possible. As a result, brooding hummingbirds expend energy to keep themselves and their eggs warm through the night, provided they have the energy stores necessary to maintain the high metabolic rates needed.

On some days, bad weather interfered with foraging by the parent birds, so they apparently went into the night with insufficient energy supplies to maintain normal body temperatures. In this situation, the brooding hummingbirds did become torpid for part of the night. One bird that had experienced a 12 percent reduction in foraging time during the day became torpid for two hours, and a second that had lost 21 percent of its foraging time was torpid for 3.5 hours. Torpor can thus be a flexible response that integrates the energy stores of a bird with environmental conditions and such biological requirements as brooding eggs.

■ Energetic Aspects of Prolonged Torpor

Hibernation is an effective method of conserving energy during long winters, but hibernating animals do not remain at low body temperatures for the whole winter. Periodic arousals are normal, and these arousals consume a large portion of the total amount of energy used by hibernating mammals. An example of the magnitude of the energy cost of arousal is provided by Lawrence Wang's study of the Richardson's ground squirrel (*Spermophilus richardsonii*, **Figure 22–9**) in Alberta, Canada (Wang 1978).

▲ **FIGURE 22–9** Richardson's ground squirrel, *Spermophilus richardsonii*.

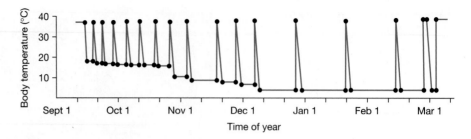

▶ **FIGURE 22–10** Record of body temperature during a complete torpor season for a Richardson's ground squirrel.

The activity season for ground squirrels in Alberta is short: they emerge from hibernation in mid-March, and adult squirrels reenter hibernation four months later, in mid-July. Juvenile squirrels begin hibernation in September. When the squirrels are active, they have body temperatures of 37°C to 38°C, and their temperatures fall as low as 3°C to 4°C when they are torpid. **Figure 22–10** shows the body temperature of a juvenile male ground squirrel from September through March; periods of torpor alternate with arousals all through the winter. Hibernation began in mid-September with short bouts of torpor followed by rewarming. At that time, the temperature in the burrow was about 13°C. As the winter progressed and the temperature in the burrow fell, the intervals between arousals lengthened and the body temperature of the torpid animal declined. By late December, the burrow temperature had dropped to 0°C, and the periods between arousals were 14 to 19 days. In late February the periods of torpor became shorter, and in early March the squirrel emerged from hibernation.

A torpor cycle consists of entry into torpor, a period of torpor, and an arousal (**Figure 22–11**). In this example, entry into torpor began shortly after noon on February 16; 24 hours later, the body temperature had stabilized at 3°C. This period of torpor lasted until late afternoon on March 7, when the squirrel started to arouse. In three hours the squirrel warmed from 3°C to 37°C. It maintained that body temperature for 14 hours and then began entry into torpor again.

These periods of arousal account for most of the energy used during hibernation (**Table 22.1**). The energy costs associated with arousal include the cost of warming from the hibernation temperature to 37°C, the cost of sustaining a body temperature of 37°C for several hours, and the metabolism above torpid levels as the body temperature slowly declines during reentry into torpor. For the entire hibernation season, the combined metabolic expenditures for those three phases of the torpor cycle account for an average of 83 percent of the total energy used by the squirrel.

Surprisingly, we have no clear understanding of why a hibernating ground squirrel undergoes these arousals that increase its total winter energy expenditure nearly fivefold. Ground squirrels do not store food in their burrows, so they are not using the periods of arousal to eat. They do urinate during arousal, so eliminating accumulated nitrogenous wastes may be the reason for arousal, and some time

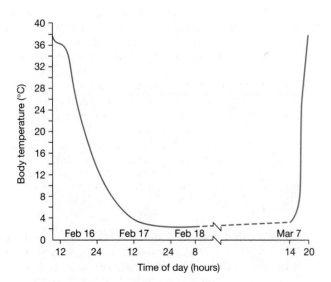

▲ **FIGURE 22–11** Record of body temperature during a single torpor cycle for a Richardson's ground squirrel.

TABLE 22.1 Use of energy during different phases of the torpor cycle by Richardson's ground squirrel

| | Percentage of Total Energy per Month | | | |
Month	Torpor	Warming	Intertorpor Homeothermy	Reentry
July	8.5	17.2	56.5	17.8
September	19.2	15.2	49.9	15.7
November	20.8	23.1	43.1	13.0
January	24.8	24.1	40.0	11.1
March	3.3	14.0	76.4	6.3
Average for season	**16.6**	**19.0**	**51.6**	**12.8**

at a high body temperature may be necessary to carry out other physiological or biochemical activities, such as resynthesizing glycogen, redistributing ions, or synthesizing serotonin. Arousal may also allow a hibernating animal to determine when environmental conditions are suitable for emergence. Whatever their function, the arousals must be important because the squirrel pays a high energy price for them during a period of extreme energy conservation.

22.5 Endotherms in Hot Deserts

Hot, dry areas place more severe physiological demands on endotherms than do the polar conditions we have already discussed. Endotherms encounter two problems regulating body temperature in hot deserts. The first results from a reversal of the normal relationship of an animal to the environment. In most environments, an endotherm's body temperature is warmer than the air temperature. In this situation, heat flow is from the animal to its environment, and thermoregulatory mechanisms achieve a stable body temperature by balancing heat production and heat loss. Very cold environments merely increase the gradient between an animal's body temperature and the environment. The example of Arctic foxes with lower critical temperatures of −40°C illustrates the success that endotherms have had in providing sufficient insulation to cope with enormous gradients between high core body temperatures and low environmental temperatures.

In a hot desert, the gradient is not increased—it is reversed. Desert air temperatures can climb to 40°C or 50°C during summer, and the ground temperature may exceed 60°C or 70°C. Instead of losing heat to the environment, an animal is continually absorbing heat, and that heat plus metabolic heat must somehow be dissipated to maintain the animal's body temperature in its normal range. Maintaining a body temperature 10°C below the ambient temperature can be a greater challenge for an endotherm than maintaining it 100°C above ambient.

To make matters worse, water is scarce in deserts. Evaporative cooling is the major mechanism an endotherm uses to reduce its body temperature. The evaporation of water requires approximately 2400 kilojoules per kilogram. (The exact value varies slightly with temperature.) Thus, evaporation of a liter of water dissipates 2400 kilojoules, and evaporative cooling is an effective mechanism as long as an animal has an unlimited supply of water. In a hot desert,

however, where thermal stress is greatest, water is a scarce commodity; its use must be carefully rationed. Calculations show, for example, that if a kangaroo rat were to venture out into the desert sun, it would have to evaporate 13 percent of its body water per hour to maintain a normal body temperature. Most mammals die when they have lost 10 percent to 20 percent of their body water, so it is obvious that, under desert conditions, evaporative cooling is of limited utility except as a short-term response to a critical situation.

Nonetheless, diverse assemblages of birds and mammals inhabit deserts. The mechanisms they use are complex and involve combinations of ecological, behavioral, morphological, and physiological adjustments that act together to enhance the effectiveness of the entire system. The ancestral structures and physiological characteristics of birds and mammals are remarkably versatile. In many cases, those characters alone are all an animal needs to confront conditions that seem extraordinarily harsh.

The structure and function of the mammalian kidney was described in Section 11.5, and the ability of the kidney to produce urine that has an osmotic concentration several times higher than that of the blood is one key to the success of mammals in deserts. The nephron is the basic functional unit of the kidney in which an ultrafiltrate of the blood is processed to return needed compounds to the blood and secrete waste products into the forming urine. Most mammals have two types of nephrons: those with a cortical glomerulus and abbreviated loops of Henle that do not penetrate far into the medulla, and those with juxtamedullary glomeruli, deep within the cortex, with loops that penetrate as far as the papilla of the renal pyramid. Obviously, the longer, deeper loops of Henle experience large osmotic gradients along their lengths. The flow of blood to these two populations of nephrons seems to be independently controlled. Juxtamedullary glomeruli are more active in regulating water excretion, whereas cortical glomeruli are involved with ion regulation.

Form and function are intimately related in mammalian kidneys. A thick medulla corresponds to long renal pyramids, long loops of Henle, large osmotic gradients, and great concentrating power. Maximum urine osmolalities of mammals are proportional to the relative medullary thickness of their kidneys. Some desert rodents have exceptionally long renal pyramids and urine concentrations that exceed 7000 mmol • kg^{-1}. (The relative medullary thickness of human kidneys is an unimpressive 3.0, and maximum urine concentration is only 1430 mmol • kg^{-1}.)

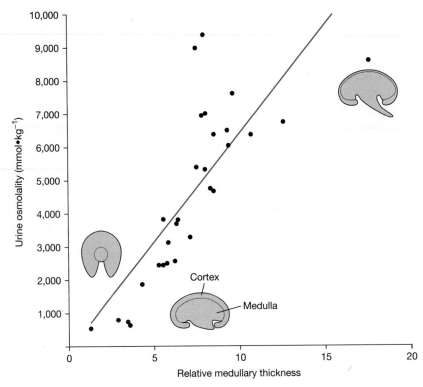

▶ **FIGURE 22–12** Relationship of maximum urine concentration to relative medullary thickness for 29 species of rodents. The sketches illustrate kidneys with different relative medullary thicknesses.

Figure 22–12 shows a strong correlation between relative medullary thickness and maximum urine concentration, but substantial variation is apparent, indicating that other anatomical or physiological factors are involved. For mammals as a group, relative medullary thickness accounts for 59 percent of interspecific variation in maximum urine concentration (Beuchat 1990).

▓ Strategies for Desert Survival

Deserts are harsh environments, but they contain a mosaic of microenvironments that animals can use to find the conditions they need. We can categorize three major classes of responses of endotherms to desert conditions, as follows:

- Relaxation of homeostasis—Some endotherms have relaxed the limits of homeostasis. They survive in deserts by tolerating greater-than-normal ranges of variation in characters such as body temperature or body-water content.
- Avoidance—Some endotherms manage to avoid desert conditions by behavioral means. They live in deserts but are rarely exposed to the full rigors of desert life.
- Specializations—Physiological mechanisms such as torpor in response to shortages of food or water are used by some desert organisms.

These categories are not mutually exclusive; many desert animals combine elements of all three responses.

Relaxation of Homeostasis–Large Mammals in Hot Deserts Large animals, including humans, have specific advantages and disadvantages in desert life that are directly related to body size. A large animal has nowhere to hide from desert conditions. It is too big to burrow underground, and few deserts have vegetation large enough to provide useful shade to an animal much larger than a jackrabbit. On the other hand, large body size offers some options not available to smaller animals. Large animals are mobile and can travel long distances to find food or water, whereas small animals may be limited to home ranges only a few meters or tens of meters in diameter. Large animals have small surface/mass ratios and can be well insulated. Consequently, they absorb heat from the environment slowly. A large body mass gives an animal a large thermal inertia; that is, it can absorb a large amount of heat energy before its body temperature rises to dangerous levels.

The dromedary camel (*Camelus dromedarius*) is the classic large desert animal (**Figure 22–13**). There are authentic records of journeys in excess of 500 kilometers, lasting two or three weeks, during which the

▶ **FIGURE 22–13** Dromedary camels. In the heat of the day, most of these camels have faced into the sun to reduce the amount of direct solar radiation they receive and are pressed against each other to reduce the heat they gain by convection and reradiation.

camels did not have an opportunity to drink. Dromedaries make their longest trips in winter and spring, when air temperatures are relatively low and scattered rainstorms may have produced fresh vegetation that provides them a little food and water.

Camels are large animals—adult body masses of dromedary camels are 400 kilograms to 450 kilograms for females and up to 500 kilograms for males. The camel's adjustments to desert life are revealed by comparing the daily cycle of body temperature in a

camel that receives water daily and one that has been deprived of water (**Figure 22–14**). The watered camel shows a small daily cycle of body temperature with a minimum of 36°C in the early morning and a maximum of 39°C in midafternoon. When a camel is deprived of water, the daily temperature variation triples. Body temperature is allowed to fall to 34.5°C at night and climbs to 40.5°C during the day.

The significance of this increased daily fluctuation in body temperature can be assessed in terms of the

▶ **FIGURE 22–14** Daily cycles of body temperature of camels. A dehydrated camel (*top*) relaxes its control of body temperature compared to a camel with daily access to water (*bottom*).

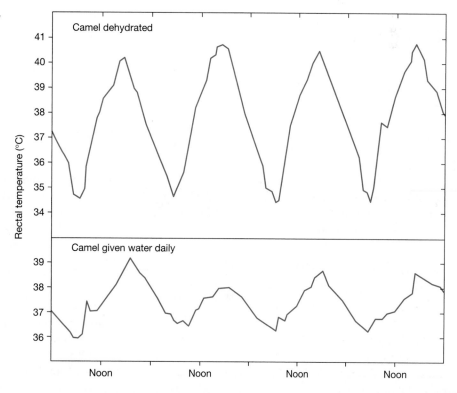

water that the camel would expend in evaporative cooling to prevent the 6°C rise. With a specific heat of 4.2 kJ/(kg • °C), a 6°C increase in body temperature for a 500-kilogram camel represents storage of 12,600 kilojoules of heat. Evaporation of a kilogram of water dissipates approximately 2400 kilojoules. Thus, a camel would have to evaporate slightly more than 5 liters of water to maintain a stable body temperature at the nighttime level, and it can conserve that water by tolerating hyperthermia during the day.

In addition to saving water that is not used for evaporative cooling, the camel receives an indirect benefit from hyperthermia via a reduction of energy flow from the air to the camel's body. As long as the camel's body temperature is below air temperature, a gradient exists that causes the camel to absorb heat from the air. At a body temperature of 40.5°C, the camel's temperature is equal to that of the air for much of the day, and no net heat exchange takes place. Thus the camel saves an additional quantity of water by eliminating the temperature gradient between its body and the air. The combined effect of these measures on water loss is illustrated by data from a young camel (**Table 22.2**). When deprived of water, the camel reduced its evaporative water loss by 64 percent and reduced its total daily water loss by half.

Behavioral mechanisms and the distribution of hair on the body aid dehydrated camels in reducing their heat load. In summer, camels have hair 5 or 6 centimeters long on the back and up to 11 centimeters long over the hump. On the ventral surface and legs the hair is only 1.5 to 2 centimeters long. Early in the morning, camels lie down on surfaces that have cooled overnight by radiation of heat to the night sky. The legs are tucked beneath the body, and the ventral surface, with its short covering of hair, is placed in contact with the cool ground. In this position a camel exposes only its well-protected back and sides to the sun and places its lightly furred legs and ventral surface in contact with cool sand, which may be able to conduct away some body heat. Camels may assemble in small groups and lie pressed closely together through the day. Spending a day in the desert sun squashed between two sweaty camels may not be your idea of fun, but in this posture a camel reduces its heat gain because it keeps its sides in contact with other camels (both at about 40°C) instead of allowing solar radiation to raise the fur surface temperature to 70°C or above.

Despite their ability to reduce water loss and to tolerate dehydration, the time eventually comes when even camels must drink. These large, mobile animals can roam across the desert seeking patches of vegetation produced by local showers and move from one oasis to another; but, when they drink, they face a problem they share with other grazing animals: water holes can be dangerous places. Predators frequently center their activities around water holes, where they are assured of water as well as a continuous supply of prey animals. Reducing the time spent drinking is one method of reducing the risk of predation, and camels can drink remarkable quantities of water in very short periods. A dehydrated camel can drink as much as 30 percent of its body mass in 10 minutes. (A very thirsty human can drink about 3 percent of body mass in the same time.)

The water a camel drinks is rapidly absorbed into its blood. The renal blood flow and glomerular filtration rate increase, and urine flow returns to normal within a half hour of drinking. The urine changes from dark brown and syrupy to colorless and watery. Aldosterone, a hormone produced by the adrenal cortex, helps to counteract the dilution of the blood by the water the camel has drunk by stimulating sodium reabsorption in the kidney. Nonetheless, dilution of the blood causes the red blood cells to swell as they absorb water by osmosis. Camel erythrocytes are resistant to this osmotic stress, but other desert ruminants have erythrocytes that would burst under these conditions. Bedouin goats, for example, have fragile erythrocytes, and the water a goat drinks is absorbed slowly from the rumen. Goats require two days to return to normal kidney function after dehydration.

Large African antelope, such as the 100-kilogram oryx (*Oryx beisa*) and 200-kilogram eland (*Taurotragus oryx*), use heat storage like the dromedary but allow their body temperatures to rise considerably above the 40.5°C level recorded for the camel. Rectal temperatures of 45°C have been recorded for the oryx and 46.5°C for the Grant's gazelle. Body temperatures above 43°C rapidly produce brain damage in most mammals, but Grant's gazelles maintained rectal temperatures of 46.5°C for as long as six hours with no apparent ill effects. These antelope keep brain temperature below body temperature by using a

TABLE 22.2 Daily water loss of a 250-kilogram camel

| Condition | Water Loss (L/day) by Different Routes | | | |
	Feces	Urine	Evaporation	Total
Drinking daily (8 days)	1.0	0.9	10.4	12.3
Not drinking (17 days)	0.8	1.4	3.7	5.9

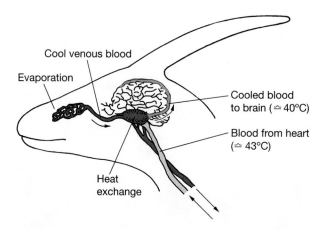

▲ **FIGURE 22–15** The countercurrent heat-exchange mechanism that cools blood going to a gazelle's brain. Blood leaves the heart at about 43°C (light color) and is cooled to about 40°C by cool venous blood (dark) returning from the nasal passages where it was cooled by evaporation of water.

countercurrent heat exchange to cool blood before it reaches the brain. The blood supply to the brain passes via the external carotid arteries; and, at the base of the brain in these antelope, the arteries break into a rete mirabile that lies in a venous sinus (**Figure 22–15**). The blood in the sinus is venous blood, returning from the walls of the nasal passages where it has been cooled by the evaporation of water. This chilled venous blood cools the warmer arterial blood before it reaches the brain. A mechanism of this sort is widespread among mammals. Horses do not have a venous sinus and carotid rete mirabile but use an analogous mechanism that cools the blood in the internal carotid arteries by passing it through the guttural pouches (Baptiste et al. 2000). The guttural pouches are outgrowths from the auditory tubes that envelope the internal carotid arteries and are filled with air that is cooler than the blood.

Large animals illustrate one approach to desert life. Too large to escape the rigors of the environment, they survive by tolerating a temporary relaxation of homeostasis. Their success under the harsh conditions in which they live is the result of complex interactions between diverse aspects of their ecology, behavior, morphology, and physiology. Only when all of these features are viewed together does an accurate picture of an animal emerge.

Avoiding Desert Conditions The mobility of vertebrates allows them to escape some of the stresses of desert conditions, either by finding shelter (such as a burrow) or by moving to shade or water. In general, larger animals can move farther than small

animals, but they also need larger burrows, patches of shade, or quantities of water. Time of activity also provides options for mitigating the harshness of desert conditions—many desert animals are nocturnal, especially during the hottest parts of the year.

Rodents are the preeminent small mammals of arid regions. It is a commonplace observation that population densities of rodents are higher in deserts than in moist situations. Several ancestral features of rodent biology allow them to extend their geographic ranges into hot, arid regions. Among the most important of these characters are the normally nocturnal habits of many rodents and their practice of living in burrows. A burrow provides escape from the heat of a desert, giving an animal access to a sheltered microenvironment while soil temperatures on the surface climb above lethal levels.

Kangaroo rats are among the most specialized desert rodents in North America; Merriam's kangaroo rat (*Dipodomys merriami*) occurs in desert habitats from central Mexico to northern Nevada. A population of this species lives in extremely harsh conditions in the Sonoran Desert of southwestern Arizona (Tracy 2000). During the summer, daytime temperatures at the ground surface approach 70°C, and even a few minutes of exposure would be deadly. Kangaroo rats spend the day in burrows 1 meter to 1.5 meters underground, where air temperatures do not exceed 35°C even during the hottest parts of the year. In the evening, when the kangaroo rats emerge to forage, external air temperatures have fallen to about 35°C.

Not all rodents that live in deserts are nocturnal. Ground squirrels are diurnal and thus conspicuous inhabitants of deserts (**Figure 22–16**). They can be seen running frantically across the desert surface even in the middle of day. The almost frenetic activity of desert ground squirrels on intensely hot days is a result of the thermoregulatory problems that small animals experience under these conditions. Studies of the antelope ground squirrel (*Ammospermophilus leucurus*) at Deep Canyon—near Palm Springs, California—provide information about how the squirrels' behavior is affected by the heat load of the environment (Chappell and Bartholomew 1981a,b).

The heat on summer days at Deep Canyon is intense, and standard operative temperatures in the sun are as high as 70°C to 75°C (**Box 22–2**). Standard operative temperature rises above the thermoneutral zone of ground squirrels within two hours after sunrise, and the squirrels are exposed to high heat loads for most of the day. They have a bimodal pattern of activity that peaks in midmorning and again in the late afternoon. Relatively few squirrels are active in

▶ **FIGURE 22–16** The antelope ground squirrel, *Ammospermophilus leucurus*.

BOX 22–2   How Hot Is It?

Measurements of environmental temperatures figure largely in studies of the energetics of animals. However, the actual process of making the measurements is complicated, and no single measurement is necessarily appropriate for all purposes. The exchange of energy between animals and their environments involves radiation, convection, conduction, and evaporation. In addition, metabolic heat production contributes significantly to the body temperatures of endotherms. The thermal environment of an animal is determined by all of the routes of heat exchange operating simultaneously. The question *How hot is it?* translates to *What is the heat load for an animal in this environment?* Answering that question requires integrating all the routes of energy exchange to give one number that represents the environmental heat load. (It is easier to think of heat load as coming from a hot environment and being represented by a risk of overheating [hyperthermia], but the same reasoning applies to a cold environment. In that situation the problem is loss of heat and the risk of hypothermia.)

Physiological ecologists have developed several measurements of the environmental heat load on an organism, and **Figure 22–17** illustrates four of these. The data come from a study of the thermoregulation of the antelope ground squirrel in a desert canyon in California.

The easiest measurement to make is the temperature of the air (T_a, frequently called ambient temperature). At Deep Canyon, California, in June, air temperature rises from about 25°C at dawn to a peak above 50°C in late afternoon, and then declines. Air temperature is a factor in conductive and convective heat exchange. An animal gains heat by conduction and convection when the air temperature is warmer than the animal's surface temperature and loses heat when the air is

cooler. Conductive heat exchange is usually small, but convection can be an important component of the overall energy

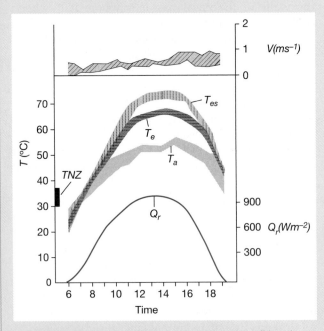

▲ **FIGURE 22–17** Ground-level meteorological conditions in open sunlit areas at Deep Canyon, California, during June. Wind velocity (V) in meters per second, solar insolation (Q_r) in watts per square meter, effective environmental temperature (T_e), and standard operative temperature (T_{es}) in °C. The thermoneutral zone (TNZ) of ground squirrels is indicated.

budget of an organism. However, the magnitude of convective heat exchange depends on wind speed as well as air temperature. Thus, measuring air temperature provides only part of the information needed to assess just one of the three important routes of heat exchange. Consequently, air temperature is not a very useful measure of heat load.

If air temperature is unsatisfactory as a measure of environmental heat load because it makes only a small contribution to the overall energy exchange, perhaps a measurement of the major source of heat is what is needed. Heat from the sun (solar insolation) in this arid habitat is the major source of heat, and the magnitude of the insolation (Q_r) can be measured with a device called a pyranometer. Solar insolation rises from 0 watts per square meter at dawn to about 900 watts per square meter in midday and falls back to zero at sunset. This measurement provides information about how much solar energy is available to heat an animal, but that is still only one component of the energy exchange that determines the heat load.

The effective environmental temperature (T_e) combines the effects of air temperature, ground temperature, solar insolation, and wind velocity. The effective environmental temperature is measured by making an exact copy of the animal (a mannequin), equipping it with a temperature sensor such as a thermocouple, and putting the mannequin in the same place in the habitat that the real animal occupies. Taxidermic mounts are frequently used as mannequins: the pelt of an animal is stretched over a framework of wire or a hollow copper mold of the animal's body. Because the mannequin has the same size, shape, color, and surface texture as the animal, it responds in the same way as the animal to solar insolation, infrared radiation, and convection. The equilibrium temperature of the mannequin is the temperature that a metabolically inert animal would have as a result of the combination of radiative and convective heat exchange. At Deep Canyon the temperatures of mannequins of antelope ground squirrels increased more rapidly than air temperatures and stabilized near 65°C from midmorning through late afternoon. The temperatures of the mannequins were 15°C higher than air temperature, showing that the heat load experienced by the ground squirrels was much greater than that estimated from air temperature alone.

Because the mannequin is a hollow shell—a pelt stretched over a supporting structure—it does not duplicate thermoregulatory processes that have important influences on the body temperature of a real animal. Metabolic heat production increases the body temperature of a ground squirrel, evaporative water loss lowers body temperature, and changes in peripheral circulation and raising or lowering the hair change the insulation. The effects of these factors can be incorporated mathematically if the appropriate values for metabolism, insulation, and whole-body conductance are known. The result of this calculation is the standard operative temperature (T_{es}). An explanation of how to calculate T_{es} can be found in Bakken (1980). For the ground squirrels in our example, the standard operative temperature was nearly 10°C higher than the effective temperature and about 25°C higher than the air temperature. For much of the day, the T_{es} of a ground squirrel in the sun at Deep Canyon was 30°C or more above the squirrel's upper critical temperature of 43°C. Similar calculations can provide values for T_{es} in other microenvironments the squirrels might occupy—in the shade of a bush, for example, or in a burrow. This is the information needed to evaluate the squirrels' behavior to determine if their activities are limited by the need to avoid overheating.

the middle of the day. The body temperatures of antelope ground squirrels are labile, and body temperatures of individual squirrels vary as much as 7.5°C (from 36.1°C to 43.6°C) during a day. The squirrels use this liability of body temperature to store heat during their periods of activity.

High temperatures limit the squirrels' bouts of activity to no more than 9 to 13 minutes. They sprint furiously from one patch of shade to the next, pausing only to seize food or to look for predators. The squirrels minimize exposure to the highest temperatures by running across open areas, and they seek shade or their burrows to cool off. On a hot summer day, a squirrel can maintain a body temperature below 43°C (the maximum temperature it can tolerate) only by retreating every few minutes to a burrow deeper than 60 centimeters, where the soil temperature is 30°C to 32°C. The body temperature of an antelope ground squirrel shows a pattern of rapid oscillations, rising while the squirrel is in the sun and falling when it retreats to its burrow (**Figure 22–18**). Ground squirrels do not sweat or pant; instead, they use this combination of transient heat storage and passive cooling in a burrow to permit diurnal activity. The strategy the antelope ground squirrel uses is basically the same as that employed by a camel—saving water by allowing the body temperature to rise until the heat can be dissipated passively. The difference between the two animals is a consequence of their difference in body size: a camel weighs 500 kilograms and can store heat for an entire day and cool off at night, whereas an antelope ground squirrel weighs about 100 grams and heats and cools many times in the course of a day.

The tails of many desert ground squirrels are wide and flat, and the ventral surfaces of the tails are usually white. The tail is held over the squirrel's back with its white ventral surface facing upward. In this position, it acts as a parasol, shading the squirrel's body and reducing the standard operative temperature. The tail of the antelope ground squirrel

▶ **FIGURE 22–18** Short-term cycles of activity and body temperature of an antelope ground squirrel. The squirrel warms up during periods of activity on the surface (dark circles) and cools off when it retreats to its burrow (light circles).

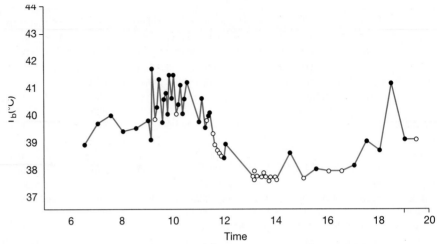

is relatively short, extending only halfway up the back, but the shade it gives can reduce the standard operative temperature by as much as 6°C to 8°C. The Cape ground squirrel (*Xenurus inauris*) of the Kalahari Desert has an especially long tail that can be extended forward nearly to the squirrel's head. Cape ground squirrels use their tails as parasols (**Figure 22–19**), and observations of the squirrels indicated that the shade might significantly extend their activity on hot days.

Use of Torpor by Desert Rodents The significance of daily torpor as an energy conservation mechanism in small birds was illustrated earlier. Many desert rodents also have the ability to become torpid. In

most cases, limiting the food available to an animal can induce the torpor. When the food ration of the California pocket mouse (*Perognathus californicus*) is reduced slightly below its daily requirements, it enters torpor for a part of the day. In this species, even a minimum period of torpor results in an energy saving. If a pocket mouse were to enter torpor and then immediately arouse, the process would take 2.9 hours. Calculations indicate that the overall energy expenditure during that period would be reduced by 45 percent as compared to the cost of maintaining a normal body temperature for the same period. In this animal, the briefest possible period of torpor gives an energetic savings, and the saving increases as the time spent in torpor is lengthened.

(a)

(b)

▲ **FIGURE 22–19** Cape ground squirrel (*Xenurus inauris*) using its tail as a parasol. (a) The erected tail shades the dorsal surface of the animal. (b) The tail is held over the back of a horizontal squirrel, shading its head and body.

The duration of torpor is proportional to the severity of food deprivation for the pocket mouse. As its food ration is reduced, it spends more time each day in torpor and conserves more energy. Adjusting the time spent in torpor to match the availability of food may be a general phenomenon among seed-eating desert rodents. These animals appear to assess the rate at which they accumulate food supplies during foraging rather than their actual energy balance. Species that accumulate caches of food will enter torpor even with large quantities of stored food on hand if they are unable to add to their stores by continuing to forage. When seeds were deeply buried in the sand, and thus hard to find, pocket mice spent more time in torpor than they did when the same quantity of seed was close to the surface (Reichman and Brown 1979). This behavior is probably a response to the chronic food shortage that may face desert rodents because of the low primary productivity of desert communities and the effects of unpredictable variations from normal rainfall patterns, which may almost completely eliminate seed production by desert plants in dry years.

Birds in Deserts Although birds are relatively small vertebrates, the problems they face in deserts are more like those experienced by camels and antelope than like those of small mammals. Birds are predominantly diurnal, and few seek shelter in burrows or crevices. Thus, like large mammals, they meet the rigors of deserts head on and face the antagonistic demands of thermoregulation in a hot environment and the need to conserve water.

Birds are much more mobile than mammals of the same body size. A kangaroo rat or ground squirrel is confined to a home range less than 100 meters in diameter, but it is quite possible for a desert bird with the same body size as those rodents to fly many kilometers to reach water. For example, mourning doves in the deserts of North America congregate at dawn at water holes, some individuals flying 60 kilometers or more to reach them.

The normally high and labile body temperatures of birds give them another advantage that is not shared by mammals. With body temperatures normally around 40°C, birds face the problem of a reversed temperature gradient between their bodies and the environment for a shorter portion of each day than does a mammal. Furthermore, birds' body temperatures are normally variable, and birds tolerate moderate hyperthermia without apparent distress. These are all ancestral characters that are present in virtually all birds. Neither the body temperatures nor the lethal temperatures of desert birds are higher than those of related species from nondesert regions.

The mobility provided by flight does not extend to fledgling birds, and the most conspicuous adaptations of birds to desert conditions are those that ensure a supply of water for the young. Altricial fledglings, those that need to be fed by their parents after hatching, receive the water they need from their food. One pattern of adaptation in desert birds ensures that reproduction will occur at a time when succulent food is available for fledglings. In the arid central region of Australia, bird reproduction is precisely keyed to rainfall. The sight of rain is apparently sufficient to stimulate courtship, and mating and nest building commence within a few hours of the start of rain. This rapid response ensures that the baby birds will hatch in the flush of new vegetation and insect abundance stimulated by the rain.

A different approach, very like that of mammals, has been evolved by columbiform birds (pigeons and doves), which are widespread in arid regions. Fledglings are fed on pigeon's milk, a liquid substance produced by the crop under the stimulus of prolactin, which is a hormone secreted by the anterior pituitary gland. The chemical composition of pigeon's milk is very similar to that of mammalian milk; it is primarily water plus protein and fat, and it simultaneously satisfies both the nutritional requirements and the water needs of the fledgling. This approach places the water stress on the adult, which must find enough water to produce milk as well as meet its own water requirements.

Seed-eating desert birds with precocial young, like the sandgrouse found in the deserts of Africa and the Near East, face particular problems in providing water for their young. Baby sandgrouse begin to find seeds for themselves within hours of hatching. However, they are unable to fly to water holes as their parents do, and seeds do not provide the water they need. Instead, adult male sandgrouse transport water to their broods. The belly feathers of males have a unique structure in which the proximal portions of the barbules are coiled into helices. When the feather is wetted, the barbules uncoil and trap water. The feathers of male sandgrouse hold 15 to 20 times their weight of water, and the feathers of females hold 11 to 13 times their weight.

Male sandgrouse in the Kalahari Desert of southern Africa fly to water holes just after dawn and soak their belly feathers, absorbing 25 to 40 milliliters of water. Some of this water evaporates on the flight back to their nests, but calculations indicate that a male sandgrouse could fly 30 kilometers and arrive with 10 to 28 milliliters of water still adhering to its feathers. As the male sandgrouse lands, the juveniles rush to him, and seizing the wet belly feathers in their beaks, strip the water from them with downward jerks of their heads. In a few minutes, the young birds have satisfied their thirst, and the male rubs himself dry on the sand.

Summary

Endothermy is an energetically expensive way of life. It allows organisms considerable freedom from the physical environment, especially low temperatures, but it requires a large base of food resources to sustain high rates of metabolism. Endothermy is remarkably effective in cold environments; some species of birds and mammals can live in the coldest temperatures on Earth. The insulation provided by hair, feathers, or blubber is so good that little increase in metabolic heat production is needed to maintain body temperatures 100°C above ambient temperatures. In fact, some aquatic mammals, such as northern fur seals, are so well insulated that overheating is a problem when they are on land or in water warmer than 10 or 15°C.

Hot environments are more difficult for endotherms than are cold environments because endothermal thermoregulation balances internal heat production with heat loss to the environment. When the environment is hotter than the animal, movement of heat is reversed. Evaporative cooling is effective as a short-term response to overheating, but it depletes the body's store of water and creates new problems. Small animals—nocturnal rodents, for example—can often avoid much of the daily heat load in hot environments by spending the day underground in burrows and emerging only at night. Larger animals have nowhere to hide and must meet the heat load head-on. Camels and other large mammals of desert regions relax their limits of homeostasis when confronted by the twin problems of high temperatures and water shortage: They allow their body temperatures to rise during the day and fall at night. This physiological tolerance is combined with behavioral and morphological characteristics that reduce the amount of heat that actually reaches their bodies from the environment.

Environments that are both hot and dry—deserts—pose a dual challenge. Animals must have a way of cooling themselves, but water is in short supply. Minimizing the water used to excrete metabolic wastes is an important consideration for desert animals, and birds and mammals achieve water conservation in different ways. Birds have the ancestral sauropsid character of excreting nitrogenous wastes as salts of uric acid; this process releases water as the urate salts precipitate from solution in the urine. Mammals have a unique structure in the kidney, the loop of Henle, that allows them to produce urine with high concentrations of urea and salts.

Mobility is an important part of the response of large endotherms to both hot and cold environments. Seasonal movements away from unfavorable conditions (migration) or regular movements between scattered oases that provide water and shade are options available to medium-sized or large mammals. The great mobility of birds makes migration feasible even for small species.

When environmental rigors overwhelm the regulatory capacities of an endotherm and resources to sustain high rates of metabolism are unavailable, many small mammals (especially rodents) and some birds enter torpor, a state of adaptive hypothermia. During torpor the body temperature is greatly reduced, and the animal becomes inert. Periods of torpor can be as brief as a few hours (nocturnal hypothermia is widespread) or can last for many weeks. Mammals that hibernate (enter torpor during winter) arouse at intervals of days or weeks, warming to their normal temperature for a few hours and then returning to a torpid condition. Torpor conserves energy at the cost of forfeiting the benefits of endothermy.

The most remarkable feature of the ability of birds and mammals to live in diverse climates is not the specializations of Arctic or desert animals, remarkable as they are, but the realization that only minor changes in the basic endothermal pattern are needed to permit existence over nearly the full range of environmental conditions on Earth.

Additional Readings

Aschoff, J. 1982. The circadian rhythm of body temperature as a function of body size. In C. R. Taylor, K. Johansen, and L. Bolis (eds.), *A Companion to Animal Physiology*. Cambridge, UK: Cambridge University Press, 173–188.

Bakken, G. S. 1980. The use of standard operative temperature in the study of the thermal energetics of birds. *Physiological Zoology* 53:108–119.

Baptiste, K. E. et al. 2000. A function for the guttural pouches in the horse. *Nature* 403:382–383.

Barnes, B. M. 1989. Freeze avoidance in a mammal: Body temperatures below 0°C in an Arctic hibernator. *Science* 244:1593–1595.

Bartholomew, G. A., and F. Wilke. 1956. Body temperature in the northern fur seal, *Callorhinus ursinus*. *Journal of Mammalogy* 37:327–337.

Beuchat, C. A. 1990. Body size, medullary thickness, and urine concentrating ability in mammals. *American Journal of Physiology* 258 (*Regulatory, Integrative, Comparative Physiology* 27): R298–R308.

Busch, C. 1988. Consumption of blood, renal function, and utilization of free water by the vampire bat, *Desmodus rotundus*. *Comparative Biochemistry and Physiology* 90A:141–146.

Calder, W. A. 1994. When do hummingbirds use torpor in nature? *Physiological Zoology* 67:1051–1076.

Calder, W. A., and J. Booser. 1973. Hypothermia of broad-tailed hummingbirds during incubation in nature with ecological correlations. *Science* 180:751–753.

Chaplin, S. B. 1974. Daily energetics of the black-capped chickadee, *Parus atricapillus*, in winter. *Journal of Comparative Physiology* 89:321–330.

Chappell, M. A., and G. A. Bartholomew. 1981a. Standard operative temperatures and thermal energetics of the antelope ground squirrel *Ammospermophilus leucurus*. *Physiological Zoology* 54:81–93.

Chappell, M. A., and G. A. Bartholomew. 1981b. Activity and thermoregulation of the antelope ground squirrel *Ammospermophilus leucurus* in winter and summer. *Physiological Zoology* 54:215–223.

Davenport, J. 1992. *Animal Life at Low Temperature*. New York, NY: Chapman & Hall.

French, A. R. 1986. Patterns of thermoregulation during hibernation. In H. C. Heller, et al. (eds.), *Living in the Cold: Physiological and Biochemical Adaptations*, New York, NY: Elsevier, 393–402.

McNab, B. K. 1973. Energetics and distribution of vampires. *Journal of Mammalogy* 54:131–144.

Reichman, O. J., and J. H. Brown. 1979. The use of torpor by *Perognathus amplus* in relation to resource distribution. *Journal of Mammalogy* 60:550–555.

Schleuning, W.-D. 2001. Vampire bat plasminogen activator DSPA-Alpha-1 (Desmoteplase): A thrombolytic drug optimized by natural selection. *Haemostasis* 31:118–122.

Tracy, R. L. 2000. *Adaptive Variation in the Physiology of a Widely Distributed Mammal and Re-examination of the Bases for its Desert Survival*. Unpublished doctoral dissertation, Arizona State University.

Wang, L. C. H. 1978. Energetic and field aspects of mammalian torpor: The Richardson's ground squirrel. In L. C. H. Wang and J. W. Hudson (eds.), *Strategies in Cold: Natural Torpidity and Thermogenesis*. New York, NY: Academic, 109–145.

CHAPTER 23

Body Size, Ecology, and Sociality of Mammals

The origin of some of the derived features of mammalian behavior might lie in the nocturnal habits that are postulated for Mesozoic mammals. If these animals had to rely on scent or hearing instead of vision to interpret their surroundings, ancestral mammals may have benefited from an increased ability to associate information received via the ears and nose and to compare the intensity of stimuli over intervals of time—for example, Are the footsteps of a predator getting louder? Is the scent of the prey getting stronger? In turn, greater associative capacity might contribute to more complex social behavior, and the contact between mother and young during infancy could provide an opportunity to modify behavior by learning.

Social behaviors and interactions between individuals play a large role in the biology of extant mammals. These behaviors are modified by the environment; and relationships between energy requirements, resource distribution, and social systems can often be demonstrated. In this chapter we consider some examples of those interactions that illustrate the complexity of the evolution of mammalian social behavior. In addition, we consider the social behavior of several species of primates. The social behavior of many primates is elaborate but not necessarily more complex than that of some other kinds of mammals, including cetaceans and canids. However, primates have been the subjects of more field studies than have other mammals, and we know a great deal about their social behavior, its consequences for the fitness of individuals, and even a

little about the way some species of primates view their own social systems.

23.1 Social Behavior

Sociality means living in structured groups, and some form of group living is found among nearly all kinds of vertebrates. However, the greatest development of sociality is found among mammals. Much of the biology of mammals can best be understood in the context of what sorts of groups form, the advantages of group living for the individuals involved, and the behaviors that stabilize groups. Mammals may be particularly social animals due to the interaction of several mammalian characteristics, no single one of which is directly related to sociality but which, in combination, create conditions in which sociality is likely to evolve. Thus, the relatively large brains of mammals (which presumably facilitate complex behavior and learning), prolonged association of parents and young, and high metabolic rates and endothermy (with the resulting high resource requirements) may be viewed as conditions that are conducive to the development of interdependent social units.

Of course, not all mammals are social; in fact, there are more species of solitary mammals than social ones. Solitary and social species are known among marsupials and placentals. Monotremes appear to be solitary (as are most placentals of that body size), but the three living monotreme species are too small a sample to form a basis for speculations about the phylogenetic origins of sociality among mammals. Of course, the social behavior of mammals does not operate in a vacuum; it is only one part of the biology of a species. Social behavior interacts with other kinds of behavior (such as food gathering, predator avoidance, and reproduction), with the morphological and physiological characteristics of a species, and with the distribution of resources in the habitat. Our emphasis in this chapter is on those interactions, and we illustrate the interrelationships of behavior and ecology with examples drawn from both predators and their prey.

The social behavior of mammals is an area of active research, and our treatment is necessarily limited. Additional examples can be found in Wilson (1975), Rubenstein and Wrangham (1986), Caro (1994), Connor et al. (1999), Mann et al. (2000), McComb et al. (2001), Packer et al. (2001), and Dunbar (2003). Some examples of the costs and benefits of social behavior were briefly reviewed by Lewin (1987).

23.2 Population Structure and the Distribution of Resources

From an ecological perspective, the distribution of resources needed by a species is usually a major factor in determining its social structure. If resources are too limited to allow more than one individual of a species to inhabit an area, there is little chance of developing social groupings. Thus the distribution of resources in the habitat and the amount of space needed by an individual to fill its resource requirements are important factors influencing the sociality of mammals.

Most animals have a **home range,** an area within which they spend most of their time and find the food and shelter they need. Home ranges are not defended against the incursions of other individuals—an area that is defended is called a **territory**. The value of a home range probably lies in the familiarity of an individual animal with the locations of food and shelter. Many species of vertebrates employ a type of foraging known as traplining, in which they move over a regular route and visit specific places where food may be available. For example, a mountain lion may carefully approach a burrow where a marmot lives, beginning its stalk long before it can actually see whether the marmot is outside its burrow; a hummingbird may return to patches of flowers at intervals that match the rate at which nectar is renewed. This kind of behavior demonstrates a familiarity with the home range and with the resources likely to be available in particular places.

The **resource dispersion hypothesis** predicts that the size of the home range of an individual animal will depend primarily on two factors: the resource needs of the individual, and the distribution of resources in the environment. That is, individuals of species that require large quantities of a resource such as food should have larger home ranges than individuals of species that require less food. Similarly, the home ranges of individuals should be smaller in a rich environment than in one where resources are scarce. The resource dispersion hypothesis is a very general statement of an ecological principle. It applies equally well to any kind of animal and to any kind of resource. The resources usually considered are food, shelter, and access to mates. In Section 17.7 we considered the role of monopolization of resources by individuals in relation to the mating systems of birds, and here we discuss the role of resource dispersion in relation to home range size and sociality of mammals.

■ Body Size and Resource Needs

Studies of mammals have concentrated on food as the resource of paramount importance in determining the sizes of home ranges. The energy consumption of vertebrates increases in proportion to body mass raised to a power that is usually between 0.75 and 1.0. If we assume that energy requirements determine home range size, we can predict that home range size will also increase in proportion to the 0.75 or 1.0 power of body mass. That prediction appears to be correct in general but perhaps wrong in detail (**Figure 23–1**). That is, the sizes of the home ranges of mammals do increase with increasing body size, but the rates of increase (the slopes in Figure 23–1) are somewhat greater than expected. Home ranges appear to be proportional to body mass raised to powers between 1.0 and 2.0. This relationship between energy requirements and home range sizes suggests that energy needs are important in determining the size of the home range but that additional factors are involved. One possibility is that the efficiency with which an animal can find and use resources decreases as the size of a home range or the fragmentation of resources increases (Haskell et al. 2002). If that hypothesis is correct, the sizes of home ranges would be expected to increase with the body sizes of animals more rapidly than energy requirements increase with body size.

The failure of the resource dispersion hypothesis to predict the exact relationship between body size and the size of home ranges indicates that we have more to learn about how animals use the resources of their home ranges. So far we have been assuming

▶ **FIGURE 23–1** Home range size of mammals as a function of body mass on a log-log scale. All groups have slopes greater than 1.00, which indicates a disproportionate (i.e., allometric) increase in home range size as body size increases. Metabolic rates increase with a slope of about 0.75; thus, home range size increases more rapidly than food requirements.

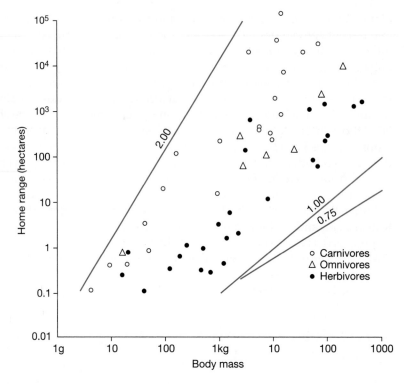

that resources are distributed evenly throughout the home range, but that assumption overlooks the structural complexity of most habitats. What insights can be obtained from a more realistic consideration of how mammals gather food?

■ The Availability of Resources

Three factors seem likely to be important in determining the availability of food to mammals: what they eat, whether their food is evenly dispersed through the habitat or is found in patches, and how they gather their food. We will consider examples of each of these factors.

Dietary Habits Figure 23–1 shows that home range size increases with body size and that dietary habits also affect home range size. Herbivores have smaller home ranges than do omnivores of the same body size, and carnivores have larger home ranges than do herbivores or omnivores. For example, the home range of a red deer (an herbivore) that weighs 100 kilograms is approximately 100 hectares. A bear (an omnivore) of the same body size has a home range larger than 1000 hectares, and a tiger (a carnivore) has a home range of more than 10,000 hectares.

The relationship between home range size and dietary habits of mammals probably reflects the abundance of different kinds of food. The grasses and leaves eaten by some herbivores are nearly ubiquitous, and a small home range provides all the food an individual requires. The plant materials (seeds and fruit) eaten by omnivores are less abundant than leaves and grasses, and their distribution is more fragmented in space and time because different species of plants produce seeds and fruit at different seasons. Thus, a large home range is probably necessary to provide the food resources needed by an individual omnivore. The vertebrates that are eaten by carnivores are still less abundant and more fragmented, and a correspondingly larger home range is apparently needed to ensure an adequate food supply.

Distribution of Resources We have assumed that one part of a home range is equivalent to another part in terms of the availability of food. This assumption may be valid for some grazing and browsing herbivorous mammals, but it is clearly not true for mammals that seek out fruiting trees (which represent patches of food) or for any carnivorous mammal that preys on animals occurring in groups. The sizes of the home ranges of animals that use food occurring in patches should reflect the distribution of patches:

home ranges should be small if patches are abundant and large if patches are widely dispersed.

That relationship is well illustrated by the home ranges of Arctic foxes (*Alopex lagopus*) in Iceland (Hersteinsson and Macdonald 1982). The foxes live in social groups consisting of one male and two females plus the cubs from the current year. The home ranges of the individuals of a group overlap widely with each other, and there is very little overlap between the home ranges of members of different groups. The home ranges of the foxes are located along the coast and do not extend far into the uplands (**Figure 23–2**). Between 60 and 80 percent of the diet is composed of items the foxes find on the shore—carcasses of seabirds, seals, and fishes, and invertebrates from clumps of seaweed washed up on the beach. Little food is available for foxes in the uplands. The foxes concentrate their foraging on the beach during the three hours before low tide, which is the best time for beachcombing. The foxes approach the shore carefully, stalking along gullies. They creep out on the beach carefully, apparently looking for birds that are resting or feeding. If birds

TABLE 23.1 Home ranges of three groups of Arctic foxes in Iceland

	Group 1	Group 2	Group 3	Average
Total area (km²)	10.3	8.6	18.5	12.5
Length of coastline (km)	5.6	5.4	10.5	7.2
Length of productive coastline (km)	5.6	5.4	6.0	5.7
Driftwood productivity (logs/year)	1800	1800	2100	1900

are present, the foxes stalk them. If no birds are present, the foxes search the beach for carrion.

The researchers studied three groups of foxes using radiotelemetry to follow the movements of individuals. The areas of the home ranges varied more than twofold, from 8.6 to 18.5 square kilometers (**Table 23.1**). The sizes of the home ranges were slightly more similar when only the coastline was considered: Each territory included between 5.4 and 10.5 kilometers of coastline.

The coastline was, of course, the main source of the foxes' food, but not all areas of the coastline accumulated floating objects. The distribution of food on the beaches was patchy and depended on the directions of currents. As a result, some parts of the shore were more productive than others. The length of productive coastline occupied by each group of foxes was quite similar—from 5.4 to 6.0 kilometers. Farmers in Iceland use driftwood to make fence posts, and the amount of driftwood that was harvested by the farmers from the coasts in the home ranges of the three groups of foxes varied only from 1800 to 2100 logs per year. Because both driftwood and carrion are moved by currents and deposited on the beaches, the harvest of driftwood by farmers probably reflects the harvest of carrion by the foxes. Thus, the home range sizes of the three groups appear to match the distribution of their most important food resource, and the productive areas of the home ranges of the three groups are very similar despite the more than twofold difference in total areas of their home ranges.

Group Size and Hunting Success It is readily apparent that the average home range size of a species of mammal can influence the social system of that species. Individuals of a species probably encounter each other frequently when home ranges are small,

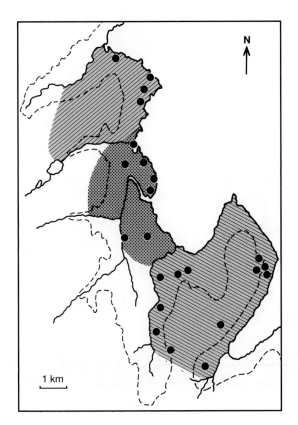

▲ FIGURE 23–2 Map of the territorial boundaries of three groups of Arctic foxes in Iceland. The 200-meter contour line is shown. Black dots mark the sites of dens used by the foxes.

whereas individuals of species that roam over thousands of hectares may rarely meet. Thus the distribution of resources in relation to the resource needs of a species is one factor that can limit the degree to which social groupings can occur. However, sociality may influence resource distribution if groups of animals are able to exploit resources that are not available to single individuals.

The influence of sociality on resource distribution may be seen among predatory animals that can hunt individually or in groups. Some species of prey are too large for an individual predator to attack but are vulnerable to attack by a group of predators. For example, spotted hyenas (*Crocuta crocuta*) weigh about 50 kilograms. When hyenas hunt individually, they feed on Thomson's gazelles (*Gazella thomsoni*, 20 kilograms) and juvenile wildebeest (*Connochaetes taurinus*, about 30 kilograms; **Figure 23–3**). However, when hyenas hunt in packs, they feed on adult

wildebeest (about 200 kilograms) and zebras (*Equus burchelli*, about 220 kilograms). Some species of prey have defenses that are effective against individual predators but less effective with groups of predators. For example, the success rate for solitary lions (*Panthera leo*) hunting zebras and wildebeest is only 15 percent, whereas lions hunting in groups of six to eight individuals are successful in up to 43 percent of their attacks. Groups of lions make multiple kills of wildebeest more than 30 percent of the time, but individual lions kill only a single wildebeest.

The relationship of sociality and body size of a predator to the size of its prey is shown in **Figure 23–4**: social predators (defined as those that hunt in groups of 8 to 10 individuals) attack larger prey than do weakly social predators (average group sizes of 1.6 to 3.1 individuals), and these weakly social predators attack larger prey than do solitary predators (average group sizes of 1.0 to 1.3 individuals).

(a)

(b)

(c)

(d)

▲ **FIGURE 23–3** Spotted hyenas. Spotted hyenas (a) may hunt individually or in packs. They prey on small animals like the Thomson's gazelle (b) when they hunt individually. When they hunt in packs, they attack larger prey such as the wildebeest (c) and zebras (d).

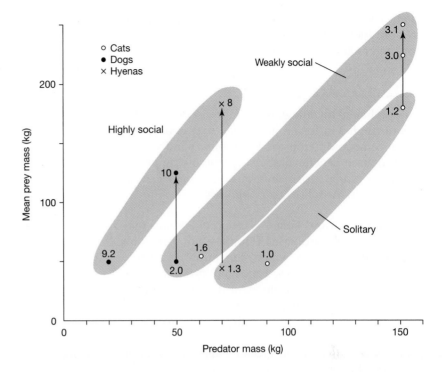

▶ **FIGURE 23–4** Size of prey in relation to predator mass for solitary predators and for predators that hunt in small (weakly social) or large (social) groups. The numbers are average group sizes; vertical lines connect points for species that hunt in groups of variable size.

One consequence of sociality for predatory mammals thus appears to be an increase in the potential food resources of an environment: individual predators may be able to extend the range of prey species they can attack by hunting in groups.

Of course, the major disadvantage to hunting in a group is that there are more mouths to feed when a kill is made. The food requirement of a group of predators is the sum of the individual requirements of the members of the group, and the per capita amount of food obtained by hunting in a group would have to exceed that caught by a solitary hunter to make group hunting advantageous.

Packer and Ruttan (1988) have reviewed factors that could contribute to the evolution of cooperative hunting. The question is, Do predators form groups *because* that allows them to hunt large prey, or *must* they hunt large prey because they live in groups for some other reason? A study of lions on the Serengeti Plains suggests that the second hypothesis is correct (Packer et al. 1990). Female lions are the only female felids to live in social groups. Female lions defend a group territory and protect their cubs from other groups of female lions. The high population densities that are characteristic of lions may have favored group defense of a territory. The presence of large prey makes it possible for lions to hunt in groups, but group hunting does not increase the amount of food available per lion—a female lion hunting by

herself can catch as much prey as her share of a group capture. So groups form because of the advantages they provide in the social structure of the population of lions on the Serengeti, and group hunting is a by-product of that social structure.

A similar interpretation has been suggested for the formation of groups of male cheetahs (Caro 1994). Male cheetahs may live alone or form permanent coalitions of two or three individuals that live and hunt together. In Caro's study, these coalitions were often composed of littermates, and a coalition was more successful in occupying a territory than was a single male. Competition for territories was intense, and territorial disputes were an important source of mortality for male cheetahs. Cheetahs hunting singly concentrated on small prey such as Thomson's gazelles, whereas coalitions attacked larger prey such as wildebeest. Overall foraging success increased with group size for male cheetahs, but Caro concluded that the benefit of a coalition in holding a territory and controlling access to females was probably more important than its effect on food intake.

23.3 Advantages of Sociality

Sociality is not limited to species of mammals that hunt in groups, and the potential advantages of sociality are not confined to predatory behavior.

Mammals may derive benefits from sociality in terms of reproduction and care of young, avoiding predation, and facilitating feeding.

■ Defenses Against Predators

One probable advantage of sociality is the reduced risk of predation for an individual that is part of a group compared to the risk for a solitary individual. The proposed benefits of sociality in avoiding predation take many forms, but most of them can be grouped into the following categories:

- **More Eyes, More Time**—A group of animals may be more likely to detect the approach of a predator than an individual would be, simply because a group has more eyes, ears, and noses to keep watch. As a result of the extra watchers, an individual in a group may be able to devote a larger proportion of its time to feeding and less to watching for predators than a solitary individual can. Mammals that live in groups generally occur in open habitats, whereas solitary species are usually found in forests; that relationship may partially reflect the antipredator aspects of group living.
- **Dilution, Confusion, and the Selfish Herd**— Some of the benefits of sociality in predator avoidance result from a reduced risk of predation for an individual when it is part of a group. The presence of a large number of potential prey animals may exceed the predatory capacity of a limited number of predators if the prey are present only for a limited time. For example, the huge herds of nomadic wildebeest that follow the shifting rains across the African savanna contain far more individuals than a pack of wild dogs can eat during the time it takes the herd to cross the pack's home range—and, when the dogs do attack a herd, the confusion that arises as they try to single out and pursue one animal among a crowd of animals that all look the same may increase a wildebeest's chance of escape. The benefit could be especially strong for individuals that manage to remain in the center of the herd; predators pursue and capture individuals on the periphery of the herd before they reached the ones in the middle. (This is the selfish-herd hypothesis, which proposes that the protection an individual receives from being in a group depends on the behavior of other individuals in the group—not everyone can be in the center.)
- **Group Defenses**—Large social mammals can form a defensive wall when a predator approaches. Typically, the adults confront the predator, with young animals and sometimes females sheltering behind them. Musk oxen (*Ovibos moschatus*) are the one of the most familiar examples of this behavior, sometimes forming a complete circle, with the adults facing outward to confront a pack of wolves, and the young oxen sheltered on the inside of the circle.

■ Sociality and Reproduction

Groupings of animals are important factors in mating systems and in care of young. In Section 17.7, we described mating systems in the context of avian biology, and most aspects of that discussion apply equally well to the mating systems of mammals. In the next section, we extend the analysis by considering the specific relationships among body size, habitat, diet, antipredator behavior, and mating systems of several species of African ungulates.

The extensive period of dependence of many young mammals on their parents provides a setting in which many benefits of sociality can be manifested. Maternal care of the offspring is universal among mammals, and males of many species also play a role in parental care. Group living provides opportunities for complex interactions among adults and juveniles that involve various sorts of **alloparental behavior** (care provided by an individual that is not a parent of the young receiving the care). Collaborative rearing of the young of several mothers is characteristic of lions and of many canids. Frequently, nonbreeding individuals join the mothers in protecting and caring for the young. Among dwarf mongooses (*Helogale undulata*), this kind of behavior extends to the care of sick adults; similar behaviors are reported for mammals as diverse as elephants and cetaceans. Many social groups of mammals consist of related individuals, and these nonbreeding helpers may increase their own fitness by assisting in rearing the offspring of their kin.

23.4 Body Size, Diet, and the Structure of Social Systems

The complex relationships among body size, sociality, and other aspects of the ecology and behavior of herbivorous mammals are illustrated by the variation in social systems of African antelopes (family Bovidae) (Jarman 1974, Leuthold 1977, Estes 1991). The smallest species of antelopes have adult weights of 3 to 4 kilograms (the dik-diks, *Madoqua*, and some

(a)

(b)

(c)

(d)

▲ **FIGURE 23–5** Diet and body size are correlated in African bovids. The smallest species, such as the dik-dik (a) eat leaves and fruit and are highly selective in their choice of the parts of plants they consume (type I diet). Medium-sized antelope, such as the impala (b) are less selective in their browsing (type II diet). Larger species, such as the hartebeest (c), are grazers that select new growth (type III diet). The largest species, such as the Cape buffalo (d), browse and graze unselectively (type IV diet).

duikers, *Cephalophus*); one of the largest (the Cape buffalo, *Syncerus caffer*) weighs 400 kilograms (**Figure 23–5**). The smallest species are forest animals that browse on the most nutritious parts of shrubs, live individually or in pairs, defend a territory, and hide from predators (**Table 23.2**). The largest species (including the 500 kilogram eland, *Taurotragus oryx*, and the Cape buffalo) are grassland animals that feed unselectively, live in large herds, are migratory, and use group defense to deter predators. Species with intermediate body sizes are also intermediate in these ecological and behavioral characteristics. It

seems likely that the correlated variation in body size, ecology, and behavior among these antelopes reveals functional relationships among these aspects of their biology. How might such diverse features of mammalian biology interact?

An antelope's feeding habits appear to provide a key that can be used to understand other aspects of its ecology and behavior. The diets of different species are closely correlated with their body size and habitats. In turn, those relationships are important in setting group size. The size of a group determines the distribution of females in time and space,

TABLE 23.2 Elements of the ecology and social systems of African ungulates

Diet Type	Examples	Body Mass (kg)	Food Habits	Group Size	Mating System	Predator Avoidance
I	Dik-dik, some duikers	3–20	Highly selective browsers	1–2	Stable pair, territorial	Hide
II	Thomson's gazelle, impala	20–100	Moderately selective browsers and grazers	2–100	Male territorial in breeding season, temporary harems of females	Flee
III	Wildebeest, hartebeest	100–200	Grazers, selective for growth stage	Large herds	Nomadic, temporary harems	Flee, hide in herd, threaten predator
IV	Eland, buffalo	300–900	Unselective browsers and grazers	Large herds	Male hierarchy	Group defense

and this is a major factor in establishing the mating system used by males of a species. Group size also plays an important role in determining the appropriate antipredator tactics for a species. Mating systems and antipredator mechanisms are central factors in the social organization of a species.

Body Size and Food Habits

Antelope are ruminants, relying on symbiotic microorganisms in the rumen to convert cellulose from plants into compounds that can be absorbed by the vertebrate digestive system. The effectiveness of ruminant digestion is proportional to body size. Body size is the primary factor determining the anatomical characteristics of the gut of ungulates (Pérez-Barbería et al. 2001). The volume of the rumen is proportional to body mass in species with different body sizes, whereas metabolic rates are proportional to the 0.75 power of body mass. The ecological consequence of this difference in allometric slopes is illustrated in **Figure 23–6**: A large ruminant has proportionately more capacity to process food than does a small ruminant. For animals of very small body size, the metabolic requirements become high in relation to the volume of rumen that is available to ferment plant material.

Because of this relationship, small ruminants must be more selective feeders than large ruminants. That is, a large ruminant has so much volume in its rumen that it can afford to eat large quantities of

food of low nutritional value. It does not extract much energy from a unit volume of this food, but it is able to obtain its daily energy requirements by processing a large volume of food. Small ruminants, in contrast, must eat higher-quality food and rely upon obtaining more energy per unit volume from the smaller volume of food that they can fit into their rumen in a day. In fact, 40 kilograms is the approximate lower limit of body size at which an

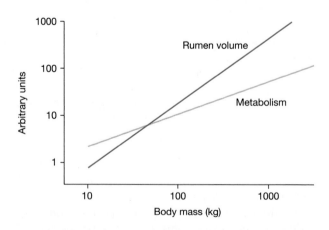

▲ **FIGURE 23–6** Rumen volume and energy requirements in relation to body mass. Rumen volume increases in proportion to body size (a slope of 1), whereas energy requirements are proportional to metabolism (a slope 0.75). Thus, large species are more effective ruminants than are small species. Both axes are drawn with logarithmic scales, and the scale of the vertical axis is in arbitrary units.

unselective ruminant can balance its energy budget; species larger than 40 kilograms can be unselective grazers, whereas smaller species must eat only the most nutritious parts of plants.

The species of antelopes in this example can be divided into four feeding categories:

- **Type I** species are selective browsers. They feed preferentially on certain species of plants, and they choose the parts of those plants that provide the highest-quality diet—new leaves (which have a higher nitrogen and lower fiber content than mature leaves) and fruit. Dik-diks and duikers fall into this category, and they have adult body masses between 3 and 20 kilograms. These animals show little sexual dimorphism in body size and appearance. The males have small horns, while the females are hornless or may also have small horns.

- **Type II** species are moderately selective grazers and browsers. They eat more parts of a plant than do the type I species, and they may have seasonal changes in diet as they exploit the availability of fresh shoots or fruits on particular species of plants. Thomson's gazelle (*Gazella thomsoni*) and the impala (*Aepyceros melampus*) weigh 20 to 100 kilograms and have type II diets. These animals show substantial sexual dimorphism in body size, with the males being about a third again larger than the females in body mass. They are highly dimorphic in appearance: Males have large, elaborate horns and may have a different coat color from females. Females are either hornless or have horns that are much smaller than those of males.

- **Type III** species are primarily grazers that are unselective for species of grass but selective for the parts of the plant. That is, they eat the leaves and avoid the stems. Hence, they are selecting for a growth stage: They avoid grass that is too short because that limits food intake. They also avoid grass that is too long because it has too many stems that are low-quality food. Wildebeest (*Connochaetes taurinus*) and hartebeest (*Alcelaphus buselaphus*), which weigh about 200 kilograms, are type III feeders. Type III species show little sexual dimorphism in size or appearance, and the horns of females are nearly as large as those of males.

- **Type IV** species are very large and are unselective grazers and browsers. They eat all species of plants and all parts of the plant. Eland (*Taurotragus oryx*, 500 kilograms) and buffalo (*Syncerus caffer*,

400 kilograms) are type IV species. Males are substantially larger than females, but there is little dimorphism in horns.

Food Habits and Habitat

The food habits of different antelope species are important in determining what sorts of habitats provide the resources they need. Selective feeding operates at three levels: vegetation type, species and individual groups of plants, and parts of plants eaten. The type of vegetation present largely depends on the habitat—forests contain shrubs and bushes, whereas the plains are covered with grass. The resources needed by species with type I diets are found in forests where the presence of a diversity of species with different growth seasons ensures that new leaves and fruit will be available throughout the year. Species with type II diets are found in habitats that are a mosaic of woodland and grassland, and type III species (which are primarily grazers) are found in savanna and grassland areas. Species with type II and type III diets may move from place to place in response to patterns of rainfall. For example, wildebeest require grass that has put out fresh new growth but that has not had time to mature. To find grass at this growth stage, wildebeest have extensive nomadic movements that follow the seasonal pattern of rain on the African plains.

Type IV feeders eat almost any kind of plant material, and they can find something edible in almost any habitat. They occupy a range of habitats, including grassland and brush, and do not have nomadic movements.

Habitat and Group Size

The habitats in which antelope feed and the types of food they eat set certain constraints on the sorts of social groupings that are possible. For example, species with type I diets live in forests and feed on scattered, distinct items. They eat an entire new leaf or fruit at a bite, and they must move between bites. A type I feeder completely removes the items it eats, so it changes the distribution of resources in its habitat. The upper part of **Figure 23–7** shows the effect of selective browsing on the new leaves on a bush—after the first individual has fed, the bush no longer has new leaves. A second individual cannot feed close behind the first because the food resources of an individual bush are entirely consumed by the first individual to feed there. As a result, the feeding

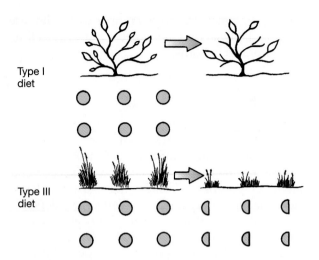

▲ **FIGURE 23–7** The effect of feeding by a selective browser with a type I diet and a grazer with a type III diet.
Top: The browser removes entire food items (new leaves or fruit), thereby changing the distribution of food in the habitat, as well as the abundance of food. *Bottom:* The grazer removes part of a grass clump, changing the abundance of food in the habitat but not its distribution.

behavior of species with type I diets makes it impossible for a group of animals to feed together. If one individual attempts to follow behind another to feed, the second animal must search to find food items overlooked by the individual ahead; consequently, it falls behind. Alternatively, it can move aside from the path of the first animal to find an area that has not already been searched. In either case, small animals in dense vegetation rapidly lose track of each other, and no cohesive group structure is maintained.

Instead of feeding as a group, type I species are solitary or occur in pairs, and the individuals of a pair are only loosely associated as they feed. A type I diet places a premium on familiarity with a home range because a tree or bush is a patch of food that must be visited repeatedly to harvest fruit or new leaves as they appear.

Species of antelope with type II and type III diets are less selective than species with type I diets, and their feeding has less impact on the distribution of resources. These species do not remove all of the food resource in an area, and other individuals can feed nearby. Type III feeders, in particular, graze as they walk—taking a bite of grass, moving on a few steps, and taking another bite. This mode of feeding changes the abundance of food, but not its distribution in space; herds of wildebeest graze together, all moving in the same direction at the same speed and maintaining a cohesive group. Rainfall is a major determinant of the distribution of suitable food in the

habitat of these species. The rainstorms that stimulate the growth of grass are erratic, and the patch sizes in which food resources occur are enormous—hundreds of square kilometers of new grass where rain fell are separated by hundreds of kilometers of old, dry grass that did not receive rain. Instead of having home ranges or territories, species with type II and type III diets move nomadically with the rains. Group sizes change as the distribution of resources changes, from half a dozen to 60 individuals for species with type II diets and from herds of 300 or 400 to superherds of many thousands of individuals during the nomadic movements of wildebeests.

Species of antelope with type IV diets are so unselective in their choice of food that they can readily maintain large groups. Herds of buffalo may number in the hundreds. Because these species can eat almost any kind of vegetation, the distribution of resources does not change seasonally, and the size of the herds is stable.

Group Size and Mating Systems

The mating systems used by African antelope are closely related to the size of their social groups and the distribution of food because those are the major factors determining the distribution of females and the potential for males to obtain opportunities to mate by controlling resources that females need. The females of species with type I diets are dispersed because the distribution of resources in the habitat does not permit groups of individuals to form. A male of a type I species can defend food resources, but individuals must disperse through the territory to feed, and it is not feasible for a male to maintain a territory that attracts a group of females. Males of type I species pair with one female; the male defends its territory year-round, the pair bond with an individual female appears to be stable, and offspring are driven out of the territory as they mature.

A group of individuals of type II species contains several males and females. The evenly distributed nature of food of these species makes it difficult for a male to monopolize resources. Only some of the males are territorial, and even this territoriality is manifested for only part of the year. A territorial male tries to exclude other males from its territory and to gather groups of females. Exclusive mating rights are achieved by holding a patch of ground and containing females within it, driving them back if they try to leave. These species have no long-term association between a male and a particular female.

Type III species are nomadic, and males establish territories only when the herd is stationary. During these periods male wildebeest have access to groups of females within their territories, but the association between a male and females is broken when the herd moves on. However, mothers and their daughters maintain associations for two or three years. Unmated male wildebeest form bachelor herds with hierarchies, and individuals at the top of a bachelor hierarchy try to displace territorial males. If a territorial male is displaced, it joins a bachelor herd at the bottom of the hierarchy and must work its way up to the top before it can challenge another territorial male.

The social structure of buffalo (a type IV species) differs in two respects from that of wildebeests (type III). First, each herd includes many mature males that form a dominance hierarchy. Ability to attain dominance over other males is largely related to body size, and the males of type IV antelope grow throughout life. A mature male may thus be twice as heavy as a female. The individuals near the top of the hierarchy court receptive females, but no territoriality or harem formation is seen. Second, the female membership of the herd is fixed, and this situation results in a degree of genetic relationship among all the members of a herd of buffalo. That genetic relationship among individuals creates situations in which the fitness an individual gains from assisting its relatives may be a factor in the social behavior of the Cape buffalo.

■ Mating Systems and Predator Avoidance

Prey species have various ways to avoid predators, but only some will work in a given situation. In general, a prey species can (1) avoid detection by a predator, (2) flee after it has been detected but before the predator attacks, (3) flee after the predator attacks, or (4) threaten or attack the predator. Body size, habitat, group size, and the mating system all contribute to determining the risk of predation faced by a species, as well as which predator avoidance methods are most effective.

Predators usually attack prey that are the same size as the predator or smaller. Thus, small species of prey animals potentially have more species of predators than do large species. Species of antelope with type I diets are small; consequently, they are at risk from many species of predators. Furthermore, small antelope may not be able to run fast enough to escape a predator after it has attacked. On the other hand, these small antelope live in dense habitats where they are hard to see. They are cryptically colored and secretive, and they rely on being inconspicuous to avoid detection by predators. If they are pursued, they may be able to use their familiarity with the geography of their home range to avoid capture.

Groups of animals are more conspicuous to predators than are individuals, but groups also have more eyes to watch for the approach of a predator. Species of antelope with type II diets live in small groups in open habitats, where they can detect predators at a distance. These antelope avoid predators by fleeing either before or after the predator attacks (**Box 23–1**). Small predators may be attacked by the antelope but usually only when a member of the group has been captured. This sort of defense is normally limited to a mother protecting her young; the rest of the group does not participate.

Species of antelope in the type III diet category are large enough to have relatively few predators, and in a group they may be formidable enough to scare off a predator. Wildebeest sometimes form a solid line and walk toward a predator, a behavior that is effective in deterring even lions from attacking. Many predators of wildebeest focus their attacks on calves, and defense of a calf is usually undertaken only by the mother. Much of the antipredator behavior of wildebeest depends on the similarity of appearance of individuals in the herd to each other. Field observations have shown that the individuals in a group of animals that are distinctive in their markings or behavior are most likely to be singled out and captured by predators.

One of the unavoidable events that makes a female wildebeest distinctive is giving birth to a calf, and the reproductive biology of the species has specialized features that appear to minimize the risks associated with giving birth. For wildebeest, the breeding season and birth are highly synchronized. Mating occurs in a short interval; consequently, 80 percent of the births occur within a period of two or three weeks. Furthermore, nearly all of the births that will take place on a day occur in the morning in large aggregations of females, all giving birth at once. A female wildebeest that is slightly out of synchrony with other members of her group can interrupt delivery at any stage up to emergence of the calf's head in order to join the mass parturition. Presumably, this remarkable synchronization and control of parturition reflects the advantage of presenting predators with a homogeneous group of cows and calves rather than a group with only a few calves that could readily be singled out for attack.

BOX 23–1 Altruism or Taunting?

A distinctive behavior—stotting or pronking, which consists of leaping vertically into the air—is used by some species of antelope with type II diets when they are threatened by a predator (**Figure 23–8a**). The function of stotting is unclear. It may be an alarm signal that alerts other individuals of the species to the presence of a predator, but the advantage to the individual that gives the warning is not clear. Altruistic behavior of this type is usually associated with kin selection, but the individuals in groups of antelope with type II diets are not closely related, and they do not show other types of altruistic behavior, such as group defense of young. It may be that some behaviors of prey species that had been considered to be altruistic alarm signals are really signals directed to the predator by fleet-footed prey.

Alarm signals are given by many other kinds of vertebrates. A familiar example is the white underside of the tail of deer (**Figure 23–8b**). A deer that sees a predator at a distance does not immediately flee but stands watching the predator. It may flick its tail up and down, exposing the white ventral surface in a series of flashes. European hares stand erect on their hind legs when a fox in the open approaches within 30 meters of the hare. In this posture, the hare is readily visible to the fox.

This kind of behavior is not limited to mammals. For example, several related species of fleet-footed lizards that live in open desert habitats have dorsal colors that blend with the substrate on which they live and a pattern of white with black bars on the underside of the tail. These lizards stand poised for flight as a predator approaches, looking back over their shoulder at the predator. The tail is curled upward—exposing the contrasting black-and-white pattern on its ventral surface—and waved from side to side.

Is it possible that behaviors of this sort signal to the predator that it has been detected and that an attack will be unprofitable because the prey is ready to flee? Or is the prey demonstrating its strength and stamina as a way to discourage pursuit? These hypotheses are not mutually exclusive, and both are supported by field observations. For example, cheetahs pursued 50 percent of the gazelles that did not stot, but only 30 percent of those that did stot. Moreover, they captured about 20 percent of the gazelles that did not stot but none of the gazelles that stotted (Caro 1986).

(a)

(b)

▲ **FIGURE 23–8** Enigmatic behaviors in the presence of predators. (a) A springbok stotting. (b) A deer displaying the white ventral surface of its tail.

Type IV species, such as buffalo, are formidable prey even for a pride of lions. They escape much potential predation simply because of their size. When buffalo are attacked, they engage in group defense; if a calf is captured, its distress cries bring many members of the group to its defense. This altruistic behavior probably represents kin selection because the stability of the female membership of buffalo herds results in genetic relationships among the individuals.

23.5 Horns and Antlers

Horns and antlers are conspicuous features of the antelope we have been discussing, and they are characteristic of many large ungulates. Their primary roles appear to be social recognition, sexual display, and jousting between males, although they may also be used for defense. **Figure 23–9** illustrates various types of mammalian cranial appendages, which is the collective term for horns and antlers.

■ Structure and Occurrence of Horns and Antlers

Today all horned ungulates are found among ruminant artiodactyls (deer, giraffes, antelope, and other bovids), but in the past some other types of artiodactyls also had horns. Cervids (deer, caribou, and moose) have antlers. Both horns and antlers are outgrowths from the frontal bone of the skull. A mountain sheep illustrates the typical horn structure of ungulates. The horn consists of a bony core covered by a sheath composed of keratin. The horn grows from its base, and the keratin portion of the horn extends well beyond the bony core. (This is why the tips of cows' horns can be blunted without causing pain to the animal—the keratin is dead material, like fingernails, hooves, and hair.)

Giraffids (giraffes and the okapi) have unusual horns, called ossicones. The bony core is not an outgrowth of the frontal bone; instead, it is a separate bone that fuses with the frontal bone during development. Giraffes' horns are covered with skin rather than being formed by a sheath of keratin.

Caribou (cervids) have antlers rather than horns. Antlers are confined to males of most species of cervids, but female reindeer and caribou have antlers. (Reindeer and caribou are the only species of cervids that form herds containing males and females.) Unlike horns, antlers are branched, consist only of bone, and usually are shed annually. As they grow, antlers are covered by a layer of highly vascularized skin (the velvet). When the antler is mature, blood flow to the velvet is cut off, and the skin sloughs off to reveal the bony antler. (In Asia, velvet is reputed to impart virility. New Zealand has a major deer-farming industry, and dried velvet sells for thousands of dollars per kilo.)

Although the cranial appendages of ruminant artiodactyls appear rather similar, they are not homologous in their mode of growth. They appear to have evolved independently within different ruminant lineages. Modern rhinoceroses are unlike other horned ungulates because their horns are formed entirely of keratin (the epidermal protein that forms hair and fingernails) and are found in both males and females (although some extinct rhinos had bony horns that were present in males only). Rhino horns are single (not paired) structures that form on the midline of the nose region. In contrast, the horns of extant ruminants are paired and form above the eyes, although some fossil artiodactyls had single horns on the nose or on the back of the head in addition to the paired horns over the eyes.

■ The Evolution of Horns and Antlers

The evolution of ruminant horns appears to be tied in with their ecology and social behavior. The evolution of ruminant horns can be understood in the context of changing Cenozoic habitats, which in turn led to changes in diet, body size, behavior, and morphology.

The ancestors of horned ruminants first appeared in the fossil record in the late Oligocene epoch, when they were small, hornless animals with teeth, indicating a diet of fruit and young leaves (i.e., a type I diet). They would have appeared similar to present-day chevrotains (primitive ruminants) and duikers (antelope), which inhabit the tropical forests of Asia and Africa. By the early Miocene epoch, the Eurasian woodlands where these animals lived had become more seasonal and more open in structure. These changes in vegetation changed the availability of food resources. Ruminants responded by becoming somewhat larger (goat-size rather than rabbit-size) and evolving teeth more capable of eating fibrous vegetation such as mature leaves (a type II diet).

This new diet enabled the ruminants to adopt a new type of social behavior. The social behavior of the small, early ruminants was probably like that of the chevrotains: solitary or monogamous, with an individual home range. Mature leaves are much more abundant and concentrated in space than are new leaves and fruits or berries. The new, larger ruminants that ate mature leaves could find their food in a smaller home range. With food concentrated in a smaller area, leaf-eating ungulates could become territorial, defending a territory large enough for several animals. This ecological strategy would not have been practical for smaller ungulates because home ranges large enough to support several animals would be too big to patrol effectively.

Thus territorial ruminants move from a monogamous type of mating system, with only a single female for each male, to a polygynous one with

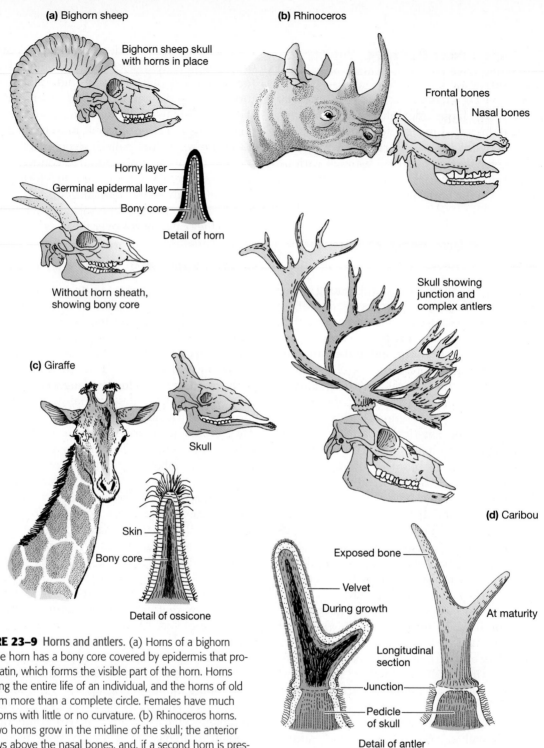

(a) Bighorn sheep

Bighorn sheep skull with horns in place

Horny layer
Germinal epidermal layer
Bony core

Detail of horn

Without horn sheath, showing bony core

(b) Rhinoceros

Frontal bones
Nasal bones

Skull showing junction and complex antlers

(c) Giraffe

Skull

Skin
Bony core

Detail of ossicone

(d) Caribou

Exposed bone

Velvet
During growth

At maturity

Longitudinal section

Junction

Pedicle of skull

Detail of antler

▲ **FIGURE 23–9** Horns and antlers. (a) Horns of a bighorn sheep. The horn has a bony core covered by epidermis that produces keratin, which forms the visible part of the horn. Horns grow during the entire life of an individual, and the horns of old males form more than a complete circle. Females have much smaller horns with little or no curvature. (b) Rhinoceros horns. One or two horns grow in the midline of the skull; the anterior horn grows above the nasal bones, and, if a second horn is present, it is above the frontal bones. When horns are present they are found in both males and females, but not all species of rhinos have horns. (c) Giraffe ossicones. Males and females have horns of approximately the same size. An ossicone has a bony core that is covered by skin and hair. The hair at the tip of the ossicone wears away in old individuals, but the skin covers the bone throughout life. (d) Antlers of a caribou. The antlers of male caribou are much larger than those of females, and females of most species of deer do not have any antlers. Antlers are solid bone and are grown and shed annually. At the end of the breeding season, bone is reabsorbed from the junction region between the base of the antler and the pedicle of the skull, and the antler then falls off.

many potential mates for each male. In this situation, some males could have greater reproductive success (i.e., mate with more females) than other males. Intense male-male competition promoted the evolution of horns or equivalent structures used for social displays. (This is because horns are used for ritual display and stylized combat, and they may actually reduce the incidence of injury during male-male interactions.)

◼ Testing the Hypothesis

That explanation is, of course, historical speculation; we will never know for certain what actually happened. However, three lines of evidence are consistent with this interpretation—the timing of the evolution of horns in African and Eurasian ungulates, sexual dimorphism in the occurrence of horns and antlers, and the failure of North American ungulates to evolve horns.

When Horns Appeared Horns or their equivalents appeared in different families of ruminants in Africa and Eurasia at about the same time in the early Miocene. The evolution of horns was correlated with a change in habitat (seen from the plant fossil record) and an increase in body size (seen in the fossil record of the animals). Among present-day ruminants, the smaller, solitary forms are hornless. Larger forms, in which the females are in groups and the males are territorial, have horns. African and Eurasian ungulates crossed this size threshold as the climate and habitat changed in the Miocene. Thus the evolution of horns correlates with a likely change in behavior, from solitary and monogamous to group-forming and polygamous. What we see today as an ecomorphological shift along a habitat gradient (forest to woodland) may be what happened in evolution over a temporal shift of changing habitats in the Miocene.

Sexual Dimorphism Horns appear to have evolved initially only in males. Fossils show that early members of all horned ruminant lineages included both horned and hornless individuals. Presumably, those with horns were males and those without horns were females. This sexual dimorphism suggests that horns were initially used in male-male interactions. If horns had originally been used for some activity that both sexes engage in, such as defense against predators, both sexes should have evolved them simultaneously.

Large, grazing ruminants (type III and IV species such as wildebeest and buffalo) now have home ranges too large to be defended as territories. These animals are no longer sexually dimorphic, although, instead of the males losing their horns, the females have evolved horns as well. The females use these horns in competition with the males for feeding resources, now that they live with them year-round (Jarman 2000).

North America The failure of horns to evolve in North American ungulates such as camels and horses may be partially explained by the different pattern of vegetational change on that continent. Grasslands rapidly replaced forests in North America, without a persistent stage of open forests. In grassland habitats, camels and horses were unlikely to have passed through an evolutionary stage in which territorial defense would have been a worthwhile ecological strategy. Perhaps this is why they never evolved the type of sexual dimorphism seen in antelopes. Additionally, hindgut fermenters like horses are less efficient feeders than ruminants and so may always have required a home range area that is too large to defend as a territory.

Horses and camelids (camels and llamas) have a different type of social system, called harems. Both these types of ungulates form permanent associations of females and their young, usually accompanied by only a single male. Males that are not part of a harem association form bachelor herds. In this social system, the male defends a group of females from other males rather than defending a piece of real estate. The term *harem* conjures up visions of a male controlling and dominating a group of females, and that was the original interpretation applied by behavioral ecologists. However, more careful observation (and the presence of more female ecologists who have brought new perspectives to the field) has revealed that it is the bond between the females that is the basis of the harem. The females then allow a male to join their social grouping because he keeps other males from constantly pestering them and interfering with the time they can spend feeding.

23.6 Primate Societies

The phylogenetic relationship of humans to other primates has led some biologists to assume that these are the animals that should have the most elaborate social systems and that studying the social systems of primates will provide information about the evolution of human behavior. Both assumptions are controversial. A growing base of information indicates

TABLE 23.3 Social organization of extant primates

Taxon	Social Organization
Prosimii	
Lemuriformes	
Lemuroidea (aye-aye, lemurs, indri, and sifaka)	Largely solitary or monogamous pairs
Lorisoidea (bushbabies, lorises, potto, and the angwantibo)	Largely solitary
Tarsiiformes (tarsiers)	Solitary or monogamous pairs
Anthropoidea	
Platyrrhini (New World monkeys)	Monogamous pairs to large groups
Cebidae (marmosets, tamarins, capuchins, squirrel monkeys, and the owl monkey	
Atelidae (howler monkeys, spider monkeys, sakis, ukaris, and titis	
Catarrhini (Old World monkeys and apes)	Small to large groups
Cercopithecinae (vervet monkey, guenons, mangabeys, macaques, baboons)	
Colobinae (colobus monkeys and langurs)	
Hominidae (apes and humans)	
Gibbons	Monogamous pairs
Orangutan	Solitary
Gorilla	Small groups with a variable number of resident males
Chimpanzee and bonobo	Closed social network containing several breeding males and females
Human	Closed social network containing several breeding males and females

not only that complex social systems exist among many kinds of vertebrates other than primates but also that interpreting primate behavior in the context of human evolution is fraught with difficulty and must be cautiously approached. Nonetheless, some primates do have elaborate and complex social systems, and more long-term research has focused on the social systems of primates than on any other vertebrates. A review of primate behavior emphasizes its variety and complexity and sets the stage for considering the evolution of humans in the next chapter.

The approximately 200 species of primates (**Table 23.3**) are ecologically diverse. They live in habitats ranging from lowland tropical rain forests, to semideserts, to northern areas that have cold, snowy winters. Some species are entirely arboreal, whereas others spend most of their time on the ground. Many are generalist omnivores that eat fruit, flowers, seeds, leaves, bulbs, insects, bird eggs, and small vertebrates. However, many of the colobus (*Colobus*) and howler monkeys (*Alouatta*) are specialized folivores (leaf eaters) with digestive tracts in which bacteria and protozoans ferment cellulose, and some of the small monkeys are insectivores.

▓ Social Systems of Primates

R. W. Wrangham (1982) proposed that the social systems of primates can best be classified based on the amount of movement of females occurring between groups (**Table 23.4**). Four categories can be defined on this basis.

- **Female transfer systems** In species with this type of social organization, most females move away from the group in which they were born to join another social group. Because of this migration of females among groups, the females in a group are not closely related to each other. In contrast, males often remain with their natal groups, and associations of male kin may be important elements of the social behaviors of these species of primates. Male chimpanzees, for example, cooperate in defending their territories from invasion by neighboring males. Most species of primates with female transfer systems live in relatively small social groups.
- **Male transfer systems** Most females of these species spend their entire lives in the group in which they were born. Social relations among the

TABLE 23.4 Characteristics of the social systems of primates

System	Group Size	Number of Males in Group	Male Behavior	Example
Female transfer	Small	One or many	Territorial, harems, sometimes male kin-ship groups	Chimpanzees, gorillas, howler monkeys, hamadryas baboons, colobus monkeys, some langurs
Male transfer	Large	One or several	Male hierarchy, whole group (males and females) may exclude conspecifics from food sources	Most cercopithecines: yellow baboons, mangabeys, macaques, guenon monkeys
Monogamous	Male and female, plus juvenile offspring	One	Both sexes participate in territorial defense and parental care	Gibbons, marmosets, tamarins, indri, titis
Solitary	Individual, or female plus juvenile male off-spring	—	Range of male overlaps ranges of several females	Bushbabies, tarsiers, lorises, orangutans

females in a group are complex and based on kinship. Males of these species emigrate from their natal group as adolescents, and they may continue to move among groups as adults. In some of these species, a single male lives with a group of females until he is displaced by a new male. In other species, several males may be part of the group and maintain an unstable dominance hierarchy among themselves. Cooperation by several adult males may allow them to resist challenges from younger, stronger males that they would not be able to subdue if they acted as individuals. Group size is usually larger for male transfer species than for female transfer species.

- **Monogamous species** A single male and female form a pair, sometimes accompanied by juvenile offspring. These species of primates show little sexual dimorphism, the sexes share parental care and territorial defense, and the offspring are expelled from the parents' territory during adolescence.
- **Solitary species** These species live singly or as females with their infants and juvenile offspring. Male prosimians maintain territories that include the home ranges of several females and exclude other males from their territories, whereas male orangutans do not defend territories. Instead, they repulse other males when a female within the male's home range comes into estrus.

■ Ecology and Primate Social Systems

Three ecological factors appear to be particularly important in shaping the social systems of primates, as they are for other vertebrates:

- **Distribution of food resources** The defensibility of food resources appears to determine whether individuals will benefit from not attempting to defend resources, defending individual territories, or forming long-term relationships with other individuals and jointly defending resources.
- **Group size** The distribution of food in time and in space may determine how large a group can be and whether the group can remain stable or must break into smaller groups when food is scarce.
- **Predation** The risk of predation may determine whether individuals can travel alone or require the protection of a group, whether the benefit of the additional protection provided by a large versus a small group outweighs the added competition among individuals in a large group, and whether the presence of males is needed to protect young.

■ Behavioral Interactions

Life within a group of primates is a balance between competition and cooperation (**Figure 23–10**). Competition is manifested by aggression. Some

(a)

(b)

▲ **FIGURE 23–10** Social behaviors of savanna baboons (*Papio cynocephalus*). (a) A male friend grooming a female baboon in estrus. (b) Aggression among male baboons.

aggression—for example, the defense of food, sleeping sites, or mates—is closely linked to resources. Other types of aggression involve establishing and maintaining dominance hierarchies, which can be an indirect form of resource competition if high-ranking individuals have preferential access to resources.

Cooperation, too, is diverse. Grooming behavior in which one individual picks through the hair of another, removing ectoparasites and cleaning wounds, is the most common form of cooperation. Other types of cooperation include sharing food or feeding sites, collective defense against predators, collective defense of a territory or a resource within a home range, and formation of alliances between individuals. Two-way, three-way, and even more complex alliances that function during competition within a group are common among primates.

Kinship and the concept of inclusive fitness play important roles in the interpretation of primate social behavior. A behavior must not decrease the fitness of the individual exhibiting the behavior if it is to persist in the repertoire of a species. Because fitness is nearly impossible to demonstrate in wild populations, behaviorists normally search for effects that are likely to be correlated with fitness, such as access to females (for males), interbirth interval (for females), or the probability that offspring will survive to reproductive age. Behaviors that increase these measures are assumed to increase fitness. The behaviors may directly benefit the individual displaying the behavior (personal fitness), or they may be costly to the personal fitness of the individual but sufficiently beneficial to its close relatives to offset the cost to the individual (inclusive fitness).

Social Relationships Among Primates

Four general types of relationships among individuals have been described in the social behavior of primates. (For more details, see Watts [1985], Richard [1985], Smuts et al. [1987], Dunbar [1988], and Nishida [1990].)

Adult-Juvenile Associations Primates are born in a relatively helpless state compared to many mammals, and they depend on adults for unusually long periods. The relationship of a mother to her infant is variable within a species—some mothers are protective, whereas others are permissive. Permissive mothers often wean their offspring earlier than protective mothers and may have shorter intervals between the births of successive offspring, although this relationship has not been observed in all species. The offspring of permissive mothers may suffer higher rates of mortality than the offspring of protective mothers, and the incompetence of some inexperienced mothers appears to lead to high mortality among firstborn offspring.

Older siblings often participate in grooming and carrying an infant, but they may also assault, pinch, and bite the infant while it is being fed or groomed by the mother. Allomaternal behavior provided by an adult female who is not the mother includes cuddling, grooming, carrying, and protecting an infant. Several factors seem to influence allomaternal behavior: young infants are preferred to older ones, infants of high-ranking mothers receive more attention and less abuse than infants of low-ranking mothers, and siblings may participate more than unrelated females in allomaternal behavior. Males of the monogamous New World primates participate extensively in caring for infants, carrying them for much of the day and sharing food with them, whereas the relationships of males of Old World primates with infants are more often characterized by proximity and friendly contact than by care.

Female Kinship Bonds The females of some species of semiterrestrial Old World primates live in groups that include several males and females. This social organization is typical of savanna baboons (*Papio cynocephalus*), several species of macaques (*Macaca*), and vervet monkeys (*Cercopithecus aethiops*). Females of these species remain for their entire lives in the troops in which they were born. Female kinship bonds and kin selection play important roles in the behaviors of females in male transfer systems because the females in a group are related to each other (**Box 23–2**).

BOX 23–2   Sociality and Survival

The information gained from field studies often increases enormously when multiple years of data are available. Jeanne Altmann and her colleagues have been studying savanna baboons (*Papio cynocephalus*) in the Amboseli basin at the foot of Mount Kilimanjaro in Kenya for more than 30 years. This is a male transfer species, so the females in a group belong to matrilineages in which they occupy stable positions in dominance hierarchies. The primary social interactions between females are grooming (or being groomed) and resting near each other.

The frequency of these behaviors was calculated for each of 108 females and compared to the survival of her infants between 1984 and 1999 (Altmann et al. 2003). The proportion of infants that survived the first year of life was greater for females that engaged in extensive social behavior than for those that were not as socially integrated in the group (**Figure 23–11**).

The influence of sociality on infant survival can be distinguished statistically from the effect of the dominance status of a female. Although it is true, as would be expected, that high-ranking females have high sociality scores, the correlation between sociality and infant survival remains when the effect of dominance rank is removed in the statistical analysis.

Why do social interactions among female baboons contribute to their infants' survival? Studies of humans may provide insights because studies indicate that social support moderates the effect of stress. Women who have extensive social networks give birth to heavier babies and have lower incidences of disease, accidents, and mental disorders. In contrast, feelings of loneliness are correlated with higher rates of illness and death.

One of the mechanisms by which sociality may enhance fitness is physiological. Supportive social interactions among humans stimulate the release of neurohormones called endorphins, which produce a sense of relaxation, and similar relationships may exist among nonhuman primates. Laboratory studies of other nonhuman vertebrates, including primates, have shown that the presence of familiar individuals of the same species lower heart rate and cortisol levels (indicating reduced stress), delay reproductive senescence, and increase the life span. Thus, social interactions may contribute to a benign environment for infant baboons.

Behavioral mechanisms may also be responsible for some of the effect of sociality on the survival of infant baboons: social interactions among adult baboons increase tolerance from high-ranking individuals and may provide protection from harassment by females from other matrilineages, as well as access to valuable resources such as desirable feeding and drinking sites.

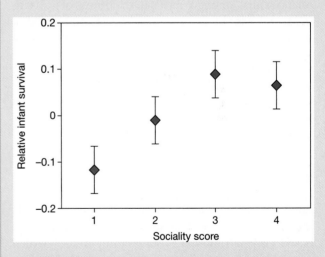

▲ **FIGURE 23–11** Social status and infant survival. Relative infant survival is an index of how well the infants of an individual female baboon survived their first year of life compared to the average survival of all infants in the troop during that year. (Average survival varies substantially from year to year.) The composite sociality index is the frequency of the three primary social interactions for each female and is divided into quartiles. (The first quartile is the 27 females with the lowest sociality index scores, the second quartile the 27 females with the next highest scores, and so on. The total sample is 108 females.) High sociality scores are females that were more socially integrated than the median female, and low scores are females that were less integrated than the median.

The females within a group form a dominance hierarchy and compete for positions in the hierarchy. Related females within a group are referred to as **matrilineages**. Females consistently support their female relatives during encounters with members of other matrilineages. The supportive relationship among females within a matrilineage is an important element of a group's social structure. For example, when their female kin are nearby, young animals can dominate older and larger opponents from subordinate matrilineages. Furthermore, high-ranking females retain their position in the hierarchy even when age or injury reduces their fighting ability. An adolescent female savanna baboon normally attains a rank in the group just below that of her mother, and this inheritance of status provides stability in the dominance relationships among the females of a group. However, the social rank of the matrilineage is not fixed: low-ranking female savanna baboons, with their female kin, may challenge higher-ranking individuals; if they are successful, their entire matrilineage may rise in rank within the group.

Female kinship bonds are clearly important elements of the social structure of male transfer systems, but the exact contribution of the long-term relationships among females to the fitness of individual females is not clear. In some species high-ranking females are young when they first give birth and have short interbirth intervals and high infant survival, but those correlations are not present in all the species that have been studied. Furthermore, female kinship bonds are manifested weakly if at all in female transfer systems, which include most species of apes and many species of monkeys.

Male-Male Alliances Male primates in male transfer species often form dominance hierarchies, but male rank depends mainly on individual attributes and is therefore less stable than female dominance systems based on matrilineage. Young adult males, which are usually recent immigrants from another group, have the greatest fighting ability and usually achieve the highest rank. Some older males achieve stable alliances with each other that enable them to overpower younger and stronger rivals in competition for opportunities to court receptive females. These males probably achieve greater mating success by engaging in these reciprocal alliances than they would achieve on the basis of their individual ranks in the hierarchy.

Cooperative relationships among males are most common in female transfer systems because the males of these species remain in their natal group. As a result, kin relationships exist among the males in a group. In red colobus monkeys (*Colobus badius*), for example, only males born in the group appear to be accepted by the adult male subgroup, and the membership of this subgroup can remain stable for years.

Adult males spend much of their time in close proximity to each other and cooperate in aggression against males of a neighboring group. For example, Goodall (1986) reported the systematic killing of an entire group of male chimpanzees (*Pan troglodytes*) by the males of a neighboring group, which then took over the females in that community. Within a group, male chimpanzees spend more time together than do females, and they engage in a variety of cooperative behaviors, including greeting, grooming, and sharing meat. However, this apparent cooperation is simply a way of cementing relationships that are based on intense and sometimes violent competition over females.

Male-Female Friendships Among Baboons Barbara Smuts's (1985) observations of a group of savanna baboons revealed that interactions between individual male and female baboons were not randomly distributed among members of the group. Instead, each female had one or two particular males called friends. Friends spent much time near each other and groomed each other often. These friendships lasted for months or years, including periods when the female was not sexually receptive because she was pregnant or nursing a baby. Male friends were solicitous of the welfare of their female friends and of their infants. Similar male-female friendships have been described in mountain gorillas (*Gorilla gorilla beringei*), gelada baboons (*Theropithecus gelada*), hamadryas baboons (*Papio hamadryas*), rhesus macaques (*Macaca mulatta*), and Japanese macaques (*Macaca fuscata*).

The advantage of these friendships for a female appears to lie in the protection that males provide to the females and their offspring from predators as well as from other members of the group. The advantage for a male of friendship with a female is less apparent. If the female's offspring had been sired by the male, protecting it would contribute to the male's fitness. However, in Smuts's study of savanna baboons, only half the friendships between males and infants involved relationships in which the male was the likely sire of the infant. The other friendships involved males that had never been seen mating with the mother of the infant. The advantage of friendship for males may depend on long-term associations with females. Smuts noted that males who participated in a friendship with a female had a significantly increased chance of mating with that female many months later, when she was again receptive.

How Do Primates Perceive Their Social Structure?

The preceding summary of primate social structures represents the results of tens of thousands of hours of observations of individual animals over periods of many years. Statistical analyses of interactions between individuals—grooming sessions, aggression, defense—reveal correlations associated with factors such as age, personality, kinship, and rank. Do the animals themselves recognize those relationships?

That is a fascinating but difficult question, especially with studies of free-ranging animals. Observations are accumulating that suggest that primates probably do recognize different kinds of relationships among individuals. For example, when juvenile rhesus macaques are threatened by another monkey, they scream to solicit assistance from other individuals who are out of sight. The kind of scream they give varies depending on the intensity of the interaction (threat or actual attack) and the dominance rank and kinship of their opponent. Furthermore, a mother baboon appears able to interpret the screams of her juvenile and to respond more or less vigorously depending on the nature of the threat her offspring faces. When tape-recorded screams were played back to the mothers, the mothers responded most strongly to screams that were given during an attack by a higher-ranking opponent, less strongly to screams that were given in response to interactions with lower-ranking opponents, and least strongly to screams that were given in interactions with relatives. Baboons were more interested in playbacks that artificially reversed the dominance status of two individuals—that is, when a higher-ranking animal sounded as if it was giving a subordinate response to a lower-ranking individual (Bergman et al. 2003).

In experiments with free-ranging vervet monkeys, the screams of a juvenile were played back to three females, one of them the mother of the juvenile. The mother responded more strongly to the screams than did the other two monkeys, as might be expected if females can recognize the voices of their own offspring. However, the other two monkeys responded to the screams by looking toward the mother, suggesting that they were able not only to associate the screams with a particular juvenile but also to associate that juvenile with its mother.

Observations of redirected aggression also suggest that some primates classify other members of a group by matrilineage and friendships. When a baboon or macaque has been attacked and routed by a higher-ranking opponent, the victim frequently attacks a bystander who took no part in the original interaction. This behavior is known as redirected aggression—and the targets of redirected aggression are relatives or friends of the original opponent more frequently than would be expected by chance. Vervet monkeys show still more complex forms of redirected aggression: They are more likely to behave aggressively toward an individual when they have recently fought with one of that individual's close kin. Furthermore, an adult vervet is more likely to threaten a particular animal if that animal's kin and one of its own kin fought earlier that same day. This sort of feud is seen only among adult vervets, suggesting that it takes time for young animals to learn the complexities of the social relationships of a group.

These sorts of observations suggest that adult primates have a complex and detailed recognition of the genetic and social relationships of other individuals in their group. Furthermore, they may be able to recognize more abstract categories—such as *relative* versus *nonrelative*, *close relative* versus *distant relative*, or *strong friendship bond* versus *weak friendship bond*—that share similar characteristics independent of the particular individuals involved.

Summary

Sociality, the formation of structured groups, is a prominent characteristic of the behavior of many species of mammals. However, social behavior is only one aspect of the biology of a species, and social behaviors coexist with other aspects of behavior and ecology, including finding food and escaping from predators. The size and geography of an animal's home range is related to the distribution and abundance of resources, the body size of the animal, and its feeding habits. Large species have larger home ranges than do small species; and at any body size carnivores have the largest home ranges and herbivores have the smallest.

Social systems are related to the distribution of food resources and to the opportunities for an individual (usually, a male) to increase access to mates by controlling access to resources. Dietary habits, the structural habitat in which a species lives, and its means of avoiding predators are closely linked to body size and mating systems. These aspects of biology form a web of interactions, each influencing the others in complex ways.

The social systems of primates have been the subjects of field studies, and more information about social behavior under field conditions is available for primates than for other mammals. The social systems of primates are complex but not unique among mammals. Some primates are solitary or monogamous; others live in groups and display behaviors that suggest not just recognition of other individuals but also recognition of the genetic and social relationships among other individuals. Studies of other kinds of mammals will probably reveal similar phenomena. Understanding the behavior of mammals requires a broad understanding of their ecology and evolutionary histories.

Additional Readings

Altmann, J. et al. 2003. Social bonds of female baboons enhance infant survival. *Science* 302:1231–1234.

Bergman, T. et al. 2003. Hierarchical classification by rank and kinship in baboons. *Science* 302:1234–1236.

Caro, T. M. 1986. The functions of stotting: A review of the hypotheses. *Animal Behaviour* 34:663–684.

Caro, T. M. 1994. *Cheetahs of the Serengeti Plains.* Chicago, IL: University of Chicago Press.

Cheney, D. and R. Seyforth. 1990. *How Monkeys See the World.* Chicago, IL: University of Chicago Press.

Connor, R. C. et al. 1999. Superalliance of bottlenose dolphins. *Nature* 397:571-572.

Coulson, T. 2007. Group living and hungry lions. *Nature* 449:997.

Dunbar, R. 2003. Evolution of the social brain. *Science* 302:1160–1161.

Dunbar, R. I. M. 1988. *Primate Social Systems.* Ithaca, NY: Cornell University Press.

Estes, R. D. 1991. The Behaviour Guide to African Mammals. Berkeley, CA: University of California Press.

Fryxell, J. M. et al. 2007. Group formation stabilizes predator-prey dynamics. *Nature* 449:1041–1044.

Goodall, J. 1986. *The Chimpanzees of Gombe.* Cambridge, MA: Harvard University Press.

Haskell, J. P. et al. 2002. Fractal geometry predicts varying body size scaling relationships for mammals and bird home ranges. *Nature* 418:527–530.

Hasson, O. 1991. Pursuit-deterrent signals: Communication between prey and predator. *Trends in Ecology and Evolution* 6:325–329.

Hersteinsson, P., and D. W. Macdonald. 1982. Some comparisons between red and Arctic foxes, *Vulpes vulpes* and *Alopex lagopus,* as revealed by radio tracking. In C. L. Cheesman and R. B. Mitson (eds.), *Symposia of the Zoological Society of London,* no. 49. London, UK: Academic, 259–289.

Janis, C. M. 1982. Evolution of horns in ungulates: Ecology and paleoecology. *Biological Reviews* 57:261–318.

Jarman, P. J. 1974. The social organization of antelope in relation to their ecology. *Behaviour* 58:215–267.

Jarman, P. J. 2000. Dimorphism in social Artiodactyla: Selection upon females. In E. S. Vrba and G. B. Schaller (eds.), *Antelope, Deer, and Relatives.* New Haven, CT: Yale University Press, 171–179.

Kelt, D. A., and D. Van Vuren. 1999. Energetic constraints and the relationship between body size and home range area in mammals. *Ecology* 80:337–340.

Leuthold, W. 1977. *African Ungulates: A Comparative Review of Their Ethology and Behavioral Ecology.* New York, NY: Springer.

Lewin, R. 1987. Social life: A question of costs and benefits. *Science* 236:775–777.

Mann, J. et al. 2000. *Cetacean Societies: Field Studies of Dolphins and Whales.* Chicago, IL: The University of Chicago Press,

McComb, K. et al. 2001. Matriarchs as repositories of social knowledge in African elephants. *Science* 292:491–494.

Nishida, T. (ed.) 1990. *The Chimpanzees of the Mahale Mountains. Sexual and Life History Strategies.* Tokyo, Japan: Tokyo University Press.

Packer, C., and L. Ruttan. 1988. The evolution of cooperative hunting. *The American Naturalist* 132:159–198.

Packer, C. et al. 1990. Why lions form groups: Food is not enough. *American Naturalist* 136:1–19.

Packer, C. et al. 2001. Egalitarianism in female African lions. *Science* 293:690–693.

Pérez-Barbería, F. J. et al. 2001. Phylogenetic analysis of stomach adaptation in digestive strategies of African ruminants. *Oecologica* 129:498–508.

Richard, A. F. 1985. *Primates in Nature.* San Francisco, CA: Freeman.

Rubenstein, D. I. and R. W. Wrangham Berkeley, (eds.). 1986. *Ecological Aspects of Social Evolution: Birds and Mammals.* Princeton, NJ: Princeton University Press.

Smuts, B. 1985. *Sex and Friendship in Baboons.* Hawthorne, NY: Aldine.

Smuts, B. et al. (eds.). 1987. *Primate Societies.* Chicago, IL: University of Chicago Press.

Watts, E. S. (ed.). 1985. *Nonhuman Primate Models for Human Growth and Development.* New York, NY: Liss.

Wilson, E. O. 1975. *Sociobiology.* Cambridge, MA: Harvard University Press.

Wrangham, R. W. 1982. Mutualism, kinship, and social evolution. In Kinds College Sociobiology Group (eds.), *Current Problems in Sociobiology.* Cambridge, UK: Cambridge University Press.

CHAPTER 24

Primate Evolution and the Emergence of Humans

Primates have been a moderately successful group for most of the Cenozoic era, although since the end of the Eocene epoch, they have been primarily confined to tropical latitudes (with the obvious exception of humans). Primates include not only the anthropoids, the group of apes and monkeys to which humans belong, but also the prosimians, animals such as bush babies and lemurs. Molecular techniques of studying genetic relationships suggest that chimpanzees are the closest extant relatives of humans, and both genetics and the fossil record suggest that the separation of humans from the African great apes was less than 10 million years ago. Fossils of the genus *Australopithecus*, the sister taxon to our own genus (*Homo*), clearly show that bipedal walking arose before the acquisition of a large brain. A great diversity of new fossils shows that the picture of early human evolution was much more complex and diverse than previously believed. *Homo* and *Australopithecus* lived together in Africa for over a million years, and the extinction of the australopithecines was probably related to climatic changes rather than to competition with our ancestors. Our current situation, in which we—*Homo sapiens*—are the only species of hominin on Earth is new; as recently as 30,000 years ago, we shared the planet with *Homo erectus*, *Homo floresiensis*, and *Homo neanderthalensis*.

24.1 Primate Origins and Diversification

Humans share many biological traits with the animals that are variously called apes, monkeys, and prosimians: we are all members of the order primates (Latin *prima* = first). The first primates were arboreal forms living in early Cenozoic forests. Humans are late-appearing primates, and complex social systems such as those discussed in Chapter 23 are an ancestral feature of the primate lineage.

■ Characteristics of Primates

Features that are typical of primates are listed in **Table 24.1**. Note that many of these characters are not unique to primates: for example, many mammals retain the clavicle, pigs have bunodont molars similar to those of primates, and many ungulates and kangaroos have only a single young per pregnancy.

Most of these traits have been attributed to an arboreal life. All the basic modifications of the limbs can be seen as contributing to arboreal locomotion, as can the stereoscopic depth perception that results from binocular vision, and the enlarged brain that coordinates visual perception and locomotory response. Most primates are arboreal, but some have become secondarily terrestrial (baboons, for example), and humans are the most terrestrial of all. Even so, many of the traits that

TABLE 24.1 Characteristics of primates

Retention of the clavicle (which is reduced or lost in many mammalian lineages) as a prominent element of the pectoral girdle.

A shoulder joint allowing a high degree of limb movement in all directions and an elbow joint permitting rotation of the forearm.

The general retention of five functional digits on the fore- and hindlimbs. Enhanced mobility of the digits, especially the thumb and big toes, which are usually opposable to the other digits.

Claws modified into flattened and compressed nails.

Sensitive tactile pads developed on the distal ends of the digits.

A trend toward a reduced snout and olfactory apparatus, with most of the skull posterior to the orbits.

A reduction in the number of teeth compared to primitive mammals but with the retention of simple bunodont molar cusp patterns.

A complex visual apparatus with high acuity, and a trend toward development of forward-directed binocular eyes and tricolor perception.

A large brain relative to body size, in which the cerebral cortex is particularly enlarged.

A trend toward derived fetal nourishment mechanisms.

Only two mammary glands (some exceptions).

Typically, only one young per pregnancy associated with prolonged infancy and pre-adulthood.

A trend toward holding the trunk of the body upright, leading to facultative bipedalism.

are most distinctively human are derived from earlier arboreal specializations.

However, arboreality cannot be the entire basis for these primate characteristics because they are not shared with many other arboreal mammals. For example, squirrels lack big brains and an opposable thumb, and they retain claws. However, squirrels are rather generalized, all-purpose climbers. Jon Bloch (Bloch and Boyer 2002) has recently discovered nearly complete skeletons of animals called carpolestids, which were probably the sister group to the true primates among the assemblage of protoprimates called plesiadapiformes. The skeletons show that carpolestids were specialized types of climbers with long fingers and an opposable thumb, probably specialized for feeding at the very ends of branches (**Figure 24–1**). True primates may have been derived from a similarly specialized animal.

■ Evolutionary Trends and Diversity in Primates

Figure 24–2 is a simplified representation of the interrelationships of primates, and Table 23–3 presents a traditional classification of extant primates.

Plesiadapiformes The first primatelike mammals (plesiadapiforms) (Greek *plesi* = near; Latin *adapi* = rabbit and *form* = shape) appeared in the earliest Cenozoic. Plesiadapiforms were rather squirrel-like and ranged from chipmunk size to marmot size

▲ **FIGURE 24–1** Reconstruction of the protoprimate *Carpolestes simpsoni*.

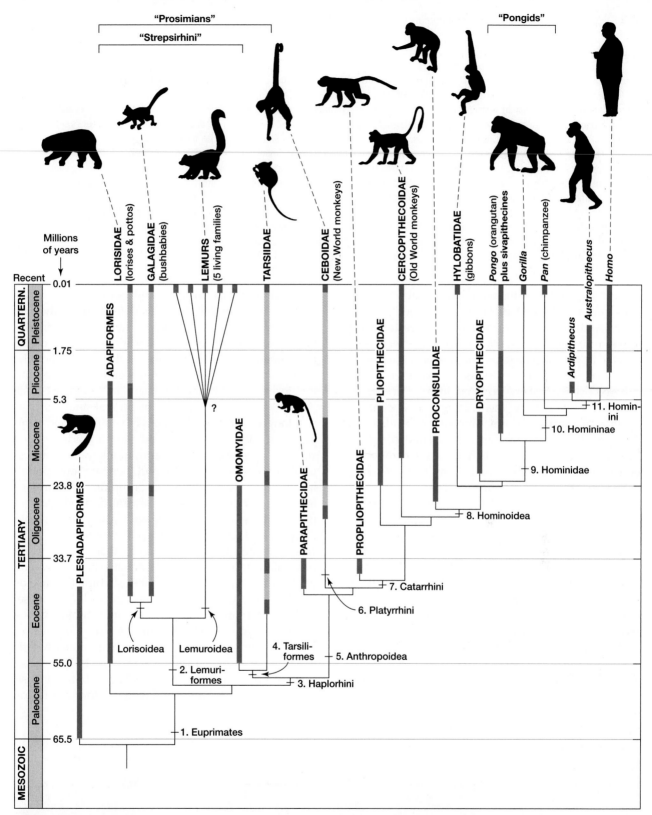

▲ **FIGURE 24–2** Phylogenetic relationships of the primates. This diagram shows the probable relationships among the major groups of primates. Dotted lines show interrelationships only; they do not indicate the times of divergence nor the unrecorded presence of taxa in the fossil record. Light bars indicate ranges of time when the taxon is known to be present but is unrecorded in the fossil record. Numbers indicate derived characters that distinguish the lineages.

(Continued)

◀ **FIGURE 24–2** *Continued*

Legend: 1. Euprimates–cheek teeth bunodont; a nail (instead of a claw) always present in extant forms, at least on the pollex (thumb); postorbital bar present. **2.** Lemuriformes–grooming claw present on second toe; lower front teeth modified into a tooth comb. **3.** Haplorhini–cranium short; orbit and temporal fossa separated ventrally by a postorbital wall; dry nose and free (rather than tethered) upper lip. **4.** Tarsiiformes–eyes greatly enlarged. **5.** Anthropoidea (monkeys and apes)–fused frontal bones; fused mandibular symphysis; lower molars increase in size posteriorly, the third only slightly larger than the second, all with five cusps, the hypoconulid (most posterior cusp) small. **6.** Platyrrhini (New World monkeys)–widely spaced and rounded nostrils; contact between jugal and parietal bones on lateral wall of skull behind orbit; first two lower molars lack hypoconulids. **7.** Catarrhini (Old World monkeys and apes)–narrowly spaced nostrils; number of premolars reduced to two; contact between frontal and sphenoid bones in lateral wall of skull; tympanic bone extents laterally to form a tubular auditory meatus (ear tube). **8.** Hominoidea (apes and humans)–lower molars with expanded talonid basin surrounded by five main cusps; broad palate and nasal regions; enlarged brain; broad thorax with dorsally positioned scapula; reduced lumbar region, with expanded sacrum and the absence of a tail. **9.** Hominidae (new style; great apes and humans)–shortened face with frontal processes of the maxillae, nasals, and the orbits in the same plane; expanded ilium; five or fewer lumbar vertebrae. **10.** Hominini (previously Hominidae; humans and fossil relatives)–skeletal adaptations to bipedality (short, broad ilium, long legs in comparison with arms, big toe not opposable); relatively large molars and relatively small canines.

(**Figure 24–3 a**). They included several different lineages, varying in their diets (as can be judged from their teeth) from generalized omnivores to insectivores and gum eaters. Plesiadapiforms were most diverse in the Paleocene of North America, although they were known from across the Northern Hemisphere. Their numbers declined in the Eocene, and they were extinct by the end of the epoch. The decline in plesiadapiform diversity coincides with the evolution and radiation of rodents in the late Paleocene. There may well have been competition between these two lineages, leading to the eventual extinction of the plesiadapiforms.

Plesiadapiforms shared some derived features of the teeth and skeleton with true primates, but they differed in having smaller brains and longer snouts, in lacking a postorbital bar and an opposable **hallux** (big toe), and in the presence of rodentlike incisors in some forms. Plesiadapiforms also apparently retained claws. In contrast, all true primates have flat nails (like your fingernails and toenails) except for the marmosets of South America, which have secondarily reverted to having claws.

Prosimians The first true primates, or Euprimates (Greek *eu* = good), are known from the early Eocene of North America, Eurasia, and northern Africa and may be represented by isolated teeth from the late Paleocene epoch of Morocco. These early primates belong to a group that has traditionally been called prosimians (Greek *pro* = before; Latin *simi* = ape): the bush babies of Africa; the lemurs of Madagascar; and the lorises, pottos, and tarsiers of Southeast Asia (**Figure 24–4**). Prosimians are in general small, nocturnal, long-snouted, and relatively small-brained compared to the more derived anthropoids (Greek *anthrops* = man), or apes and monkeys. Prosimian diets are more generalized, and there are few specialized herbivores. However, the grouping of prosimians is a paraphyletic one, since several derived features (for example, a short snout with a dry nose rather than a wet, doglike one) indicate that tarsiers are more closely related to the anthropoids than are the other prosimians (see Figure 24–2). An alternative division of the primates is into the Strepsirrhini (lemuroids and lorisoids; Greek *strepsi* = twisted and *rhin* = nose) and the Haplorhini (tarsiers and anthropoids; Greek *haplo* = simple). The Strepsirhini is also paraphyletic, although living strepsirhines form a monophyletic grouping, the lemuriformes (see Figure 24–2).

Most Eocene prosimians were larger than the plesiadapiforms, with larger brains, more forward-facing

▶ **FIGURE 24–3** Reconstruction of fossil skulls of some early primates and primatelike mammals. (The species are not drawn to the same scale.)

(a) *Plesiadapis* (Plesiadapiform)

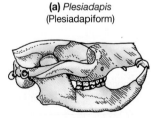

(b) *Notharctus* (Adapid)

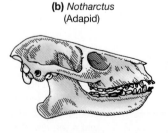

(c) *Tetonius* (Omomyid)

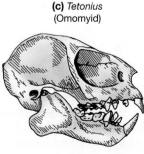

▲ **FIGURE 24–4** Diversity of living prosimians. Lemurs—(a) Ring-tailed lemur, *Lemur catta*. (b) Indri, *Indri indri*. (c) Aye-aye, *Daubentonia madagascarensis*. Lorises—(d) Demidoff's bush baby, *Galagoides demidovii*. Potto—(e) *Perodicticus potto*. Tarsiers—(f) Tarsier, *Tarsius spectrum*.

orbits, and more obviously specialized arboreal features, such as relatively longer, more slender limbs. They can be grouped into two main lineages: the larger adapids, which were long-snouted with teeth specialized for herbivory (see Figure 24–3b), and the smaller omomyids (Greek *omo* = shoulder and *mys* = mouse), which were short-snouted with teeth specialized for insectivory or gum eating (see Figure 24–3c). Judging by the size of the orbits, the adapids were probably diurnal, while the larger-eyed omomyids were probably nocturnal. Adapids are now considered to be more primitive than other euprimates, while omomyids were more derived forms related to tarsiers, and some omomyids show adaptations for tarsier-like leaping.

The diversification of the adapids and omomyids throughout the Northern Hemisphere reflects the tropical-like climates of the higher latitudes during the earlier part of the Eocene. With the late Eocene climatic deterioration in the temperate latitudes, primates declined and eventually disappeared from areas outside of Africa and tropical Asia. The primitive types of prosimians were largely extinct by the end of the Eocene, although some specialized forms (sivaladapids) survived into the late Miocene and early Pliocene epochs in Asia. Even today, almost all nonhuman primates are restricted to the tropics (**Figure 24–5**). This has been true for most of the later Cenozoic, barring some excursions of apes into more northern portions of Eurasia during the warming period in the late Miocene.

Present-day prosimians are a moderately diverse Old World tropical radiation, first known from the late middle Eocene of Africa. The modern lineages are only sparsely known from the fossil record, probably because fossil preservation is rare in tropical forest habitats. The lemurs of the island of Madagascar have undergone an evolutionary diversification into five different families. These include some large (raccoon-size) diurnal specialized herbivores, such as the rather koala-like indri (*Indri indri*) (see Figure 24–4b), and the peculiar aye-aye (*Daubentonia madagascarensis*) (see Figure 24–4c), which uses its specialized long middle finger to probe grubs out of tree bark.

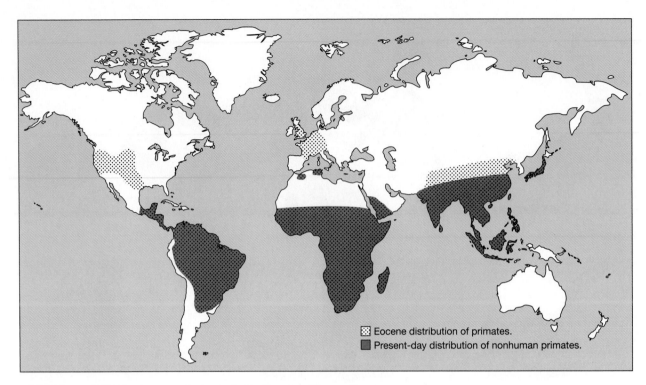

▲ **FIGURE 24–5** Distribution map of primates in the Recent and Eocene times.

Until relatively recent times (only a couple of thousand years ago), there was a much greater variety of lemurs, including much larger forms such as giant arboreal species resembling bear-size koalas and sloths. It seems that the lemurs, in isolation from the rest of the world, evolved their own version of primate diversity including parallels to the anthropoid apes. Unfortunately, much of this diversity is now gone, probably because of the immigration of humans to the island from Asia a few thousand years ago. Eye-witness accounts suggest that some of the giant lemurs may have been alive as recently as 500 years ago. Many present-day lemurs are threatened with extinction as a result of further destruction of forest habitat (Godfrey and Jungers 2003).

Anthropoids Modern anthropoids are in general larger than prosimians, with larger brains housing relatively small olfactory lobes, and have a frugivorous (fruit-eating) or folivorous (leaf-eating) diet rather than an omnivorous or insectivorous one. They also are usually diurnal, with more complex social systems. In arboreal locomotion, they are either quadrupedal above-branch climbers or suspensory under-branch climbers, whereas prosimians are usually clinging and leaping forms. Large anthropoids typically employ a form of suspensory climb-ing in which the animals move relatively slowly by clinging underneath the branches. **Brachiation,** a specialized version of arboreal locomotion in which the animals swing rapidly from the underside of one branch to the underside of the next using their hands to grasp the branches, is seen today in gibbons and spider monkeys.

Modern anthropoids can be distinguished from prosimians by a variety of skull features that reflect a large brain size and by a fibrous diet that requires extensive chewing. Anthropoids have a bony wall behind the orbit (as do tarsiers, a feature linking tarsiers and anthropoids in the Haplorhini), and the bones joining the two halves of the lower jaws are fused, as are the paired frontal bones between the eyes (**Figure 24–6**). They also lack the grooming claw on the second toe that is seen in modern prosimians.

The origin of anthropoids appears to be related to a shift from a nocturnal visual predator to a diurnal one at small body size (less than 100 grams) (Ross 2000). The anthropoids in Asia today are clearly derived from a relatively recent (middle Miocene) African ancestry (note that Africa was an island continent until around 25 million years ago). However, the anthropoid radiation must ultimately have had Asian origins because tree shrews and flying lemurs (mammals related to primates) are both

▶ **FIGURE 24–6** Cranial differences between prosimians and anthropoids.

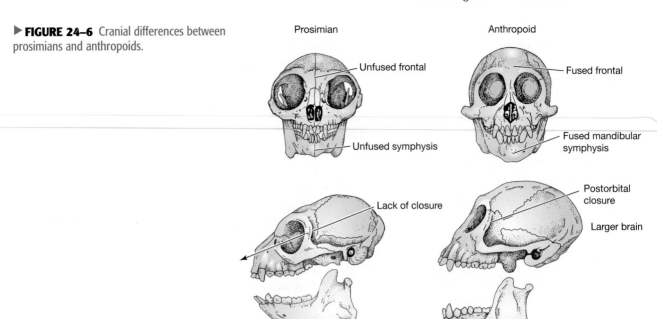

Asian natives. Although today prosimians and anthropoids are both found in Africa and Asia, their present day distributions represent separate evolutionary histories rather than a pattern of ancestor/descendant relationship.

The earliest Asian anthropoid is known from the middle Eocene of China. This is *Eosimias* (Greek *eos* = dawn), a tiny species that was only about 5 centimeters in length and would have weighed about 10 grams. Some larger anthropoid primates (amphipithecines) are known from the late middle Eocene of Myanmar, but these forms left no descendents. There was also a diverse radiation of anthropoids with a variety of body sizes (but none larger than a large cat) known from the Fayum Formation of Egypt, an ancient tropical forest habitat that spans a time range from the late Eocene to the early Oligocene. A newly discovered tiny anthropoid from the earliest Fayum deposits, *Biretia*, may provide a link with earlier African primates reaching back as far as the late Paleocene, suggesting a very early arrival of both anthropoid origins and their arrival on the African continent (Sieffert et al. 2005).

Figure 24–6 shows the difference between modern anthropoids and prosimians. The earliest anthropoids retained small brains and lacked the fused mandibular symphysis. They likely had the highly acute vision with some ability to perceive more than two colors, as these features characterize most living anthropoids, and these fossil forms possessed large, forward-facing orbits.

The early African anthropoids were represented by the oligopithecids and parapithecids, squirrel-like forms that were more primitive than any known anthropoids and retained relatively small brains, and the propliopithecids, which were more apelike forms. Propliopithecids include the cat-size *Aegyptopithecus*, originally thought to be an early true ape. Their anatomy suggests that they were mainly fruit-eaters and somewhat modified for arboreal life. However, while propliopithecids are more closely related to the African anthropoids than are the South American monkeys, they predate the divergence of modern monkeys and apes and thus cannot be classified as true apes (hominoids).

Modern anthropoids can be divided into the Platyrrhini (Greek *platy* = broad)—the New World (broad-nosed) monkeys—and the Catarrhini (Greek *cata* = downward)—the Old World (narrow-nosed) monkeys and apes. All catarrhines have trichromatic color vision, and this is also seen in a few platyrrines, such as the howler monkeys (Surridge et al. 2003).

Platyrrhines, the New World monkeys or ceboids (Greek *cebus* = monkey), first appeared in South America in the Oligocene epoch and were an exclusively New World radiation. They are presumed to have have rafted across the Atlantic Ocean to get to this continent from Africa, and rodents of probable African origin also reached South America at about the same time, probably also by rafting. Platyrrhines are more primitive than catarrhines in retaining three premolars in each jaw half; all catarrhines have only two premolars. Platyrrhines and catarrhines also differ in some details of the skull, especially in the ear region.

Platyrrhines can be divided into the cebids and the atelids (**Figure 24–7**):

- Cebids include the cebines (e.g., the familiar capuchin or organ grinder monkey and the squirrel monkey), the callitrichines (marmosets and tamarins), and the aotines (the owl monkey).
- Atelids include the atelines (wooly, howler, and spider monkeys), the callicebines (titi monkeys), and the pithecines (uakaris and saki monkeys).

Cebids Marmosets and tamarins are small and squirrel-like and have secondarily clawlike nails on all digits except for the big toe. They have simplified molars; and, while a few species are insectivorous, most are frugivores and eat gum exuded from trees. They are also unusual among anthropoid primates in producing twins. The owl monkeys (aotines) are the only nocturnal anthropoids.

Atelids The ateline monkeys are distinguished by having a prehensile tail and a specialized suspensory mode of arboreal locomotion aided by the tail.

The cebid radiation paralleled that of Old World monkeys to a certain extent, although there is no terrestrial radiation equivalent to that of baboons and macaques nor was there ever a cebid radiation equivalent to the anthropoid great apes. The lack of apelike forms among the cebids is surprising, considering that there was an evolutionary radiation of apelike forms (now all extinct) among the Madagascan lemurs. Perhaps the extensive radiation of ground sloths in South America inhibited a terrestrial radiation among the primates. However, a striking parallel does exist between the spider monkey and the gibbon—both are specialized brachiators with exceptionally long arms for swinging through the branches, and they have evolved a remarkable convergence in a wrist joint modification that allows exceptional hand rotation. Spider monkeys can be distinguished from gibbons chiefly by their use of a prehensile tail as a fifth limb during locomotion. (Gibbons, like all apes, lack a tail entirely.)

The catarrhines include the Old World monkeys, apes, and humans. We catarrhines have nostrils that are close together and open forward and downward, and we have a smaller bony nasal opening from the skull than is the case for platyrrhines. There is a trend toward large body size in our lineage; the great apes and humans are the largest living primates, rivaled only by some of the extinct lemurs. The tail is often short or absent, and prehensile tails have never evolved. The group consists of two clades: the Old World monkeys (Cercopithecoidea; Greek *cerco* = tail and *pithecus* = ape) and the apes and humans (Hominoidea; Latin *homini* = man)—the latter including the gibbons (Hylobatidae), the great apes, and humans.

Present-day Old World monkeys include two groups, colobines (Greek *colobo* = shortened) and cercopithecines. Colobines are found in both Africa and Asia, including colobus monkeys, langur monkeys, proboscis monkeys, and the golden monkey. They are more folivorous than are the cercopithecines and have more lophed, higher-cusped molars and a complex forestomach for fermentation of plant fiber. Colobines are primarily arboreal animals, with a long tail and hind legs longer than the fore legs.

Cercopithecines are predominately an African radiation, although the genus *Macaca* (macaques) occurs in Asia (including high latitudes such as Japan and Tibet) and Europe (on Gibraltar, where it is known as the Barbary ape). Cercopithecines include macaques, mangabeys, baboons, guenons, and the patas monkey. They are more omnivorous or folivorous than are the colobines, as reflected in their broader incisors and their flatter, more bunodont molars. Cercopithecines are also more terrestrial, as reflected by their short tail and their equal-lengthed fore and hind limbs. They have cheek pouches for storing food and a hand with a longer thumb and shorter fingers than colobines.

The first Old World monkeys are known from the middle Miocene. These earliest monkeys, victoriapithecines, were more primitive than any known Old World monkey and therefore form the sister group to the modern cercopithecoids. Monkeys are known from a slightly later date than the first true apes, the generalized proconsulids from the early Miocene of Africa. Monkeys are actually more derived than apes in certain respects: they have teeth that are more specialized for herbivory, some have gut specializations for the fermentation of cellulose, and they are more specialized for arboreality than are the generalized Miocene apes.

The radiation of monkeys in the late Miocene and Pliocene coincided with the reduction in diversity of the earlier radiation of generalized apes and apelike forms. Because we ourselves are apes, we often think of the monkeys as being the earlier, more primitive anthropoid radiation. However, among the Old World anthropoids, the converse is actually true: Apes were originally the more primitive, generalized forms, although the extant forms are specialized. The

Platyrrhines—New World

Catarrhines—Old World

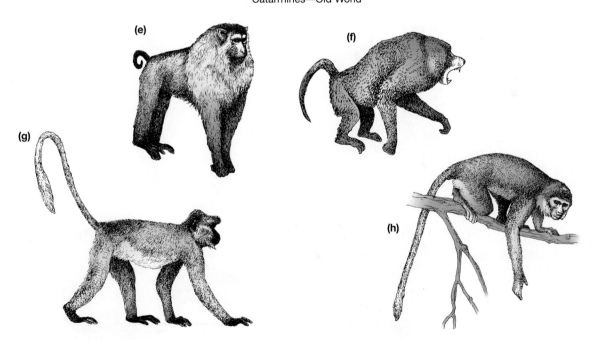

▲ **FIGURE 24–7** Diversity of living monkeys. Platyrrhines—(a) Pygmy marmoset, *Callithrix pygmaea*.
(b) Squirrel monkey, *Saimiri sciureus*. (c) Spider monkey, *Ateles paniscus*. (d) Red howler monkey,
Alouatta seniculus. Catarrhines—(e) Pig-tailed macaque, *Macaca nemistrina*. (f) Savanna baboon,
Papio anubis. (g) Hanuman langur, *Presbytis vetulus*. (h) Red colobus, *Piliocolobus badius*.

radiation of cercopithecoid monkeys is more derived in many respects than that of the apes and ultimately more successful in terms of species diversity.

24.2 Origin and Evolution of the Hominoidea

Apes and humans are placed in the Hominoidea. Hominoids are distinguished morphologically from other recent anthropoids by a pronounced widening and dorsoventral flattening of the trunk relative to body length so that the shoulders, thorax, and hips have become proportionately broader than in monkeys.

The clavicles are elongated, the iliac blades of the pelvis are wide, and the sternum is a broad structure, the bony elements of which fuse soon after birth to form a single flat bone. The shoulder blades of hominoids lie over a broad, flattened back in contrast to their lateral position next to a narrow chest in monkeys (**Figure 24–8**) and most other quadrupeds.

The pelvic and pectoral girdles of hominoids are relatively closer together than in other primates because the lumbar region of the vertebral column is short (**Figure 24–9**). The caudal vertebrae have become reduced to vestiges in all Recent hominoids, and normally no free tail appears postnatally. Balance in a bipedal pose is assisted by the flat thorax, which places the center of gravity near the vertebral column. These and other anatomical specializations of the trunk are common to all hominoids and help to maintain the erect postures that these primates assume during sitting, vertical climbing, and walking bipedally.

The skulls of hominoids also differ from those of other catarrhines in their extensive formation of sinuses—hollow, air-filled spaces lined with mucous membranes that develop between the outer and inner surfaces of skull bones. Chimpanzees, gorillas, and humans share the derived character of true frontal sinuses.

■ Diversity and Evolution of Nonhuman Hominoids

Primates that we can call apes in the broad sense of the word have been around since the late Eocene.

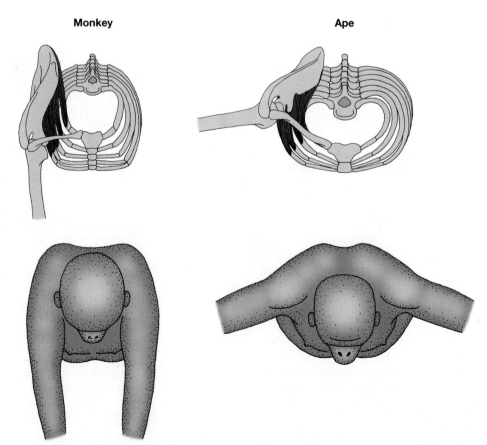

▶ **FIGURE 24–8** Differences in the form of the shoulder girdle in monkeys and hominoids. Upper portion of picture shows the skeleton with the skull and neck vertebrae removed and also the position of the serratus muscle that links the scapula to the ribs. Note the broader chest and the dorsal position of the scapula in the hominoid; curvature of the ribs is also greater, with the vertebral column lying more in the middle of the rib cage, closer to the center of gravity. These features all make it easier for a hominoid to balance in an upright position—this is in contrast to monkeys, which must bend their knees and lean forward to avoid tipping over backward if they are balancing on their hind legs.

Monkey

Ape

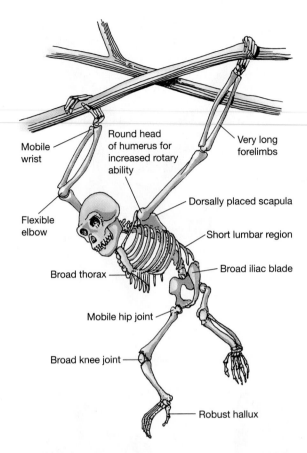

Mobile wrist

Round head of humerus for increased rotary ability

Very long forelimbs

Flexible elbow

Dorsally placed scapula

Short lumbar region

Broad thorax

Broad iliac blade

Mobile hip joint

Broad knee joint

Robust hallux

▲ **FIGURE 24–9** Skeleton of generalized hominoid, showing morphological specializations for suspensory locomotion.

However, primates that can be included in a monophyletic Hominoidea date only from the early Miocene, when the anthropoid lineage diverged into the hominoids and the cercopithecoids (Old World monkeys). Modern apes are a highly specialized radiation of large tropical animals. The Miocene radiation of apes was of more generalized animals, which also radiated into more temperate parts of the Old World. Apes and monkeys can be distinguished by their teeth, which is fortuitous because teeth are frequently the only remains of fossil species. Monkeys have lower molars that have four cusps, whereas those of hominoids have five cusps, which are usually flatter in relief than those of monkeys, with a distinct Y pattern of the grooves between the posterior cusps.

■ Diversity of Present-Day Apes

Present-day apes include the Asian gibbons (including siamangs) and the orangutan as well as the African chimpanzees and the gorilla (**Figure 24–10**).

Evidence of a type of culture, defined as social learning (for example, of tool use) with distinct differences among different biogeographic regions, has been observed in both chimps and orangutans (Whiten and Boesch 2001; van Schaik et al. 2003). All of the great apes are critically endangered. Recent estimates (Walsh et al. 2003) show that the number of chimpanzees and gorillas in western equatorial Africa, a region considered to be the last refuge of their tropical habitat, has declined by more than half in the past 20 years. The reasons relate to commercial hunting, mechanized forest logging, and the spread of the Ebola virus.

Nine species of gibbons (genus *Hylobates*) are found in Southeast Asia, both on the mainland (from India to China) and in the islands (Borneo, Sumatra, Java, and other nearby islands). They are the smallest apes, and they differ from other apes (and, indeed, the great majority of mammals) in their monogamous social system. Gibbons move through the trees most frequently by brachiation. They become entirely bipedal when moving on the ground, holding their arms outstretched for balance like a tightrope walker uses a pole.

There are two subspecies of the orangutan, *Pongo pygmaeus*; one lives on Borneo and one on Sumatra, although their range was greater in prehistoric times. Orangutans are about the same size as humans but are extremely sexually dimorphic, with males being twice the weight of females. Their behavior is fairly solitary, the main groups consisting of females and their offspring. Orangutans are arboreal but rarely swing by their arms, preferring slow climbing among the branches of trees, usually hanging with all four limbs below the branch but sometimes walking on their hind limbs on top of the branch supporting themselves by grasping tree limbs above them.

Gorillas and chimpanzees live in the tropical forests of central Africa. Both are more terrestrial than gibbons and orangutans. On the ground they move quadrupedally by knuckle walking, a derived mode of locomotion in which they support themselves on the dorsal surface of digits three and four rather than placing their weight flat on the palm of the hand, as we do when we walk on all fours (see Figure 24–10c).

Gorillas are the largest and most terrestrial of the apes. Unlike the orangutan, they are highly social and live in groups. Like the orangutan, gorillas are highly sexually dimorphic in body size—the males may weigh up to 200 kilograms, twice the mass of the females—and they are the most folivorous of the apes. Three geographically isolated subspecies of gorilla

▶ **FIGURE 24–10** Diversity of living apes.
(a) Siamang (a type of gibbon), *Hylobates syndactylus*. (b) Orangutan, *Pongo pygmaeus*. (c) Gorilla, *Gorilla gorilla*.
(d) Common chimpanzee, *Pan troglodytes*.

(*Gorilla gorilla*) have been described: the western lowland gorilla, the eastern lowland gorilla, and the mountain gorilla.

There are two (possibly three) species of chimpanzees. The larger and more widely distributed common chimpanzee (*Pan troglodytes*) is known primarily from Central and East Africa. The western subspecies (*Pan troglodytes verus*), from Nigeria and the Cameroon, may be a separate species. There is also a smaller pygmy chimpanzee or bonobo (*Pan paniscus*), known from Central Africa south of the Zaire River, which lives in more forested habitats than other chimpanzees. Chimpanzees are more omnivorous than are the more strictly herbivorous gorillas; they are also more arboreal, exhibiting a greater degree of suspensory locomotion. They are only moderately sexually dimorphic and, like gorillas, live in groups. We now know that chimpanzees use tools for a variety of purposes, including hunting (Pruetz and Bertolani 2007). The bonobo may be more closely related to humans than is the common chimpanzee (Zihlman et al. 1978).

All of the great apes are threatened by habitat destruction and active hunting by humans and are in great danger of extinction in the near future.

■ Diversity of Fossil Apes

Molecular data indicate that the split between apes and Old World monkeys occurred at the start of the Miocene, around 23 million years ago. The first true hominoids were from the early Miocene of East

Africa at around 20 million years ago. These include proconsulids, and the genus *Morotopithecus*, Both types were arboreal forms living in primarily forested habitats, had bunodont molars suggesting a frugivorous diet, and were more derived than the generalized apelike forms of the Fayum Formation (such as *Aegyptopithecus*): for example, like all apes, they lacked a tail. Proconsulids ranged from the size of a small monkey to the size of a female gorilla. While they remained generalized arboreal quadrupeds, not yet showing the specialized suspensory locomotion of many later apes, their hands and feet were more capable of gripping, and the elbow joint was more stable. *Morotopithecus* was the size of a small human and is the first hominoid to have derived features of skeletal anatomy showing a capacity for suspensory locomotion. These features include a highly mobile shoulder joint, a short stiff back resisting lumbar flexion, and a moderately mobile hip joint stable (MacLatchy 2004).

By the middle Miocene, more derived hominoids had diversified into a variety of ecological types. Some of these apes remained in Africa, such as the genera *Afropithecus*, *Kenyapithecus*, and *Equatorius*. Other hominoids spread into Eurasia, following the general middle Miocene warming trend, and both cercopithecine and colobine monkeys are also known from Eurasia in the late Miocene and Pliocene. The later Cenozoic Eurasian hominoids include the dryopithecids (Greek *dryo* = tree) and sivapithecids (after the Hindu god Shiva), which shared anatomical evidence of an upright (orthograde) postures and so can be included within the hominoid crown group (see Figure 24–3). Evidence from paleoclimatic studies and the nature of the other animals in the fauna suggest that these Miocene hominoids primarily occupied woodland or forest habitats.

The middle Miocene *Lufengpithecus* from Thailand appears to be ancestral to the orangutan and related forms, including the late Miocene to Pleistocene *Gigantopithecus* of Asia. The Pleistocene species of *Gigantopithecus* represented the largest primate that has ever lived; at an estimated body mass of 300 kilograms, it would have been twice the size of an average gorilla. Some people have speculated that a surviving lineage of *Gigantopithecus* is behind the legends of the yeti in Tibet and of Bigfoot or Sasquatch of northwestern North America. Bigfoot, if it exists, could have reached North America by migrating across Beringia during the Pleistocene epoch, as did so many other mammals.

A few fossil apes that lived in the mid-Miocene between 13 and 8 million years ago appear to belong to the same clade as the great apes plus humans (now termed the Hominidae). These include European *Graecopithecus* (also known as *Ouranopithecus*; Greek *ourano* = heaven), and the newly discovered *Pierolapithecus* from Spain. *Pierolapithecus* may be the oldest and most primitive member of this clade, and it shares with living members the distinctive shortened face (Moyà-Solà et al. 2004). Fossils of gorillas and chimpanzees have only been found very recently. An animal considered to be a basal gorilla, *Chororapithicus*, has been discovered in Kenyan deposits dated at around 10.5 million years ago (Suwa et al. 2007), and fragments of a chimpanzee are now known from the middle Pleistocene of Kenya (McBrearty and Jablonksi 2005).

It is notable that humans have retained the ancestral characteristics of our clade in numerous dental and associated cranial features, while the lineages of great apes have derived unique specializations. Of course, humans have also evolved their own specializations independently of the various great ape lineages. It remains important to note, however, that all of the living hominoids are derived in comparison with the Mio-Pliocene ape radiation. The fact that gibbons and orangutans are in a less derived position on the cladogram than are humans or African apes (see Figure 24–2) does not imply that earlier apes looked like these modern forms.

24.3 Origin and Evolution of Humans

We now live in an unusual time for hominins because only in the past 30,000 years has just one species of hominin been in existence. As recently as 30,000 years ago our species, *Homo sapiens*, shared the planet with three other species—*Homo neanderthalensis*, *Homo floresiensis*, and *Homo erectus*—and throughout hominin history (with perhaps, the exception of the first half a million years) several species of hominins always coexisted.

Anatomical differences between humans and apes appear in the skull and jaws, the trunk and pelvis, and, to a lesser extent, in the limbs. Apes have long jaws that are rectangular or U-shaped, with the molar rows parallel to each other; the canines are large and pointed; and there is a gap between the canines and the incisors. Hominids have short jaws in association with the shortening of the entire muzzle. The human jaw is V-shaped or bow-shaped, with the teeth running in a curve that is widest at the back

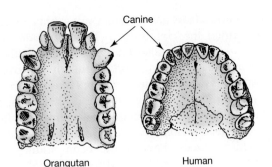

Canine

Orangutan   Human

▲ **FIGURE 24–11** Comparison of ape and human upper jaws.

of the jaw; the canines are small and blunt; and the entire dentition is relatively uniform in size and shape without any gaps between the teeth (**Figure 24–11**). In addition, the hominin palate is prominently arched, whereas the ape palate is flatter between their parallel rows of cheek teeth.

Various evolutionary trends can be identified within the hominins. The points of articulation of the skull with the vertebral column (the occipital condyles) and the foramen magnum (for the passage of the spinal cord through the skull) shifted from the ancestral position at the rear of the braincase to a position under the braincase. This change balanced the skull on top of the vertebral column and signals the appearance of an upright, vertical posture. The braincase itself became greatly enlarged in association with an increase in forebrain size. By the end of the middle Pleistocene, a prominent vertical forehead developed in contrast to the sloping forehead of the apes. The brow ridges and crests for muscular attachments on the skull became reduced in size, in association with the reduction in size of the muscles that once attached to them. The human nose became a more prominent feature of the face, with a distinct bridge and tip.

■ Relationships Within the Hominoidea

Hominoid classification has changed considerably over the past decade or so. Traditionally the Hominoidea was considered to include three families: the Hylobatidae (gibbons), the Pongidae (other apes, or "great apes") and Hominidae (humans). The Hylobatidae, as originally conceived, remains a valid family. However, as shown in Figure 24–2, the notion of the "Pongidae" is a paraphyletic one: molecular data show that chimpanzees are more closely related to humans than are the gorilla and the orangutan, and in turn the gorilla is more closely related to the humans and chimps than is the orangutan.

Despite the fact that humans look rather different from the great apes, and we are certainly different in terms of culture and language, molecular studies show that we are very close to them genetically (Ruvulo 1997). This degree of genetic closeness has led researchers to consider that the great apes and humans belong in a single family, and the term "Hominidae" now includes these apes as well as ourselves. Humans and their immediate ancestors (i.e., extinct taxa more closely related to *Homo sapiens* than to chimpanzees) are now considered to be in the tribe Hominini within the subfamily Homininae (which in turn contains humans, chimpanzees, and gorillas). (The orangutan and its fossil relatives are in the subfamily Ponginae.) Thus, what was called a hominid in fossil studies only a few years ago (and in the previous edition of this book), meaning a fossil human, is now called a **hominin**.

Because until very recently the extant African apes have had no known fossil record, we have depended on molecular studies to understand the history of our closest surviving relatives. They indicate that times of divergence were much more recent than most biologists had previously supposed and indicate a phase of rapid evolutionary change in the lineage leading to the specialized apes, as well as in our own lineage.

Molecular evidence indicates that gorillas separated from the common ancestor of chimpanzees and humans in the late Miocene, between 6 and 8 million years ago. (Note that if the controversy about the status of *Chororapithicus* that was discussed earlier leads to the conclusion that this fossil is truly a gorilla, this date will have to be revised.) Humans separated from their common ancestor with chimpanzees between 5 and 6 million years ago. An intriguing twist to this tale is that genetic data suggest that chimps and humans may have been interbreeding, and thus hybridizing their genomes, before the split between the two lineages became final (Patterson et al. 2006). There is also evidence that chimpanzees have been evolving genotypic changes faster than humans have (Bakewell et al. 2007). The gorillas separated into eastern and western populations about 3 million years ago. The three chimpanzee species are thought to have separated from each other more recently, between about 2.5 and 1.6 million years ago.

■ Primitive Hominids

The earliest well-known hominins were the australopithecines (Latin *australi* = southern), known primarily from East and South Africa in the Pliocene and early Pleistocene. Many new hominins have been described in the past decade, resulting in a confusing plethora of names and arguments about who is related to whom and when bipedality first evolved.

The Generalized Australopithecine Condition In terms of their biology, australopithecines perhaps are best thought of as bipedal apes with a modified dentition that includes the derived hominin features of thick enamel and small canines. Males were usually larger than females, and individuals grew and matured rapidly, unlike the prolonged childhood that characterizes our own species. Microwear analysis of their teeth suggests that early australopithecines, at least, were primarily fruit eaters, perhaps including some meat in their diet as do present-day chimpanzees.

All australopithecines appear to have been capable of bipedal walking. Humanlike footprints were found by Mary Leakey and her associates at Laetoli in Tanzania, in volcanic ash beds radiometrically dated between 3.6 and 3.8 million years in age, and were probably made by *Australopithecus afarensis*, which is known from fossils at the same site. Analysis of these footprints indicates that they do not differ substantially from modern human trails made on a similar substrate, demonstrating the antiquity of bipedalism in hominin ancestry, far earlier than the appearance of an enlarged brain.

The most completely known early australopithecine, and until very recently the earliest known hominin, is *Australopithecus afarensis*. This species is best known from a substantial part of a single young adult female skeleton, popularly known by the nickname Lucy (Johanson and Edgar 1996). Lucy was found in the Afar region in Ethiopia, not far from the Red Sea, in a deposit dated at 3.2 million years. Lucy is an astonishing specimen in several respects. She is the most complete pre-*Homo* hominin fossil ever found, consisting of more than 60 pieces of bone from the skull, lower jaw, arms, legs, pelvis, ribs, and vertebrae. Her overall body size was small. Young but fully grown when she died, Lucy was only about 1 meter tall and weighed perhaps 30 kilograms. Other finds indicate that males of her species were larger, averaging 1.5 meters tall and weighing around 45 kilograms. Lucy's teeth and lower jaw are also clearly humanlike. Her diet may have been rich in hard objects, probably fruits that would have been unevenly distributed in space and time. *A. afarensis* had a brain size of 380 to 450 cubic centimeters, quite close to that of modern chimpanzees and gorillas.

Despite modifications for bipedality, australopithecines retained some apelike features both of limb anatomy and of the semicircular canals in the inner ear (structures responsible for orientation and balance), suggesting the retention of apelike orientation in an arboreal environment rather than a fully terrestrial lifestyle. Arboreality is also reflected in the hands and feet, with the bones of the fingers and toes significantly longer and more curved than in modern humans (**Figure 24–12**). However, the hands of australopithecines were more humanlike than those of fully arboreal apes such as gibbons and orangutans, and they lacked the robust fingers of the knuckle-walking chimpanzees and gorillas. It appears that australopithecines were able to stand and walk bipedally to a certain extent but still spent much of their time in the trees and probably were not capable of sustained running.

Despite the condition of their skeleton, somewhat intermediate between the apelike condition and the human one, it does appear that australopithecines walked more like we do than the caveman image of walking with a stoop and bent hips and knees. That notion of early human walking has more of a basis in Hollywood than in science—it is someone's idea of an intermediate gait between chimpanzees and humans that probably never existed. Computer simulations and anatomical studies indicate that all hominins walked with an upright, straight-hipped, straight-kneed gait. In fact, the virtual Lucy devised by Robin Crompton (Crompton et al. 1998) will inevitably tumble if you try to simulate her walking in caveman style. Note also that, like Lucy, human children have short legs in proportion to their trunk length, but they are as efficient as adults in their walking locomotion (Heglund and Schepens 2003).

Other Early Australopithecines Recently, hominins earlier and even more primitive than Lucy have been discovered. *Ardipithecus ramidus* (formerly *Australopithecus ramidus*) was described in 1994. This genus is based on 17 specimens (all but four of them teeth) from Ethiopian sediments about 4.4 million years old, with associated animals indicating a wooded habitat. *Ardipithecus* (*ardi* = ground floor) has many chimpanzee-like features and lacks some of the derived traits shared by early species of *Australopithecus* and all later hominins. Tentative evaluation of the scant evidence, however, indicates that *Ardipithecus* was bipedal, judging from the position of the foramen magnum. It also had a humanlike rather than apelike arm (especially the shoulder joint

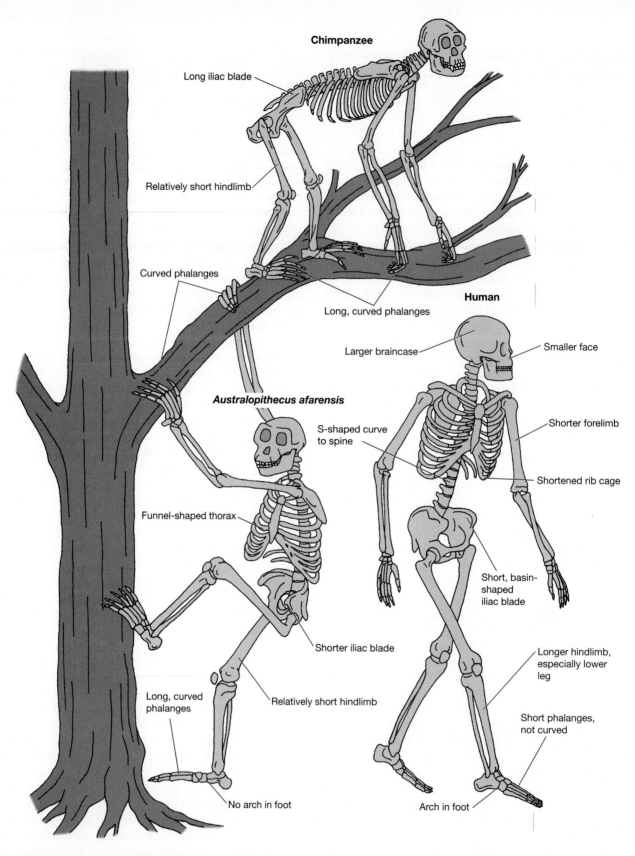

Chimpanzee

Long iliac blade

Relatively short hindlimb

Curved phalanges

Long, curved phalanges

Australopithecus afarensis

S-shaped curve to spine

Funnel-shaped thorax

Shorter iliac blade

Long, curved phalanges

Relatively short hindlimb

No arch in foot

Human

Larger braincase

Smaller face

Shorter forelimb

Shortened rib cage

Short, basin-shaped iliac blade

Longer hindlimb, especially lower leg

Short phalanges, not curved

Arch in foot

▲ **FIGURE 24–12** Comparison of skeletons of a chimpanzee, an australopithecine, and a human. Note the shorter iliac blade, the less funnel-shaped rib cage, the relatively longer legs, and the shorter fingers and toes in *Homo*.

and elbow) and incisiform canines with less sexual dimorphism than modern African apes—traits that are hallmarks of the hominin lineage. A new subspecies, *Ardipithecus ramidus kadabba*, extends the range of this genus back to 5.5 million years ago.

Several new species of *Australopithecus* have been found in the past decade, filling in more gaps in the story of human evolution. **Figure 24–13** shows one current scheme of how the australopithecine species were related to each other and to our own genus *Homo*, although there are nearly as many different phylogenies as there are groups of paleoanthropologists. The earliest known member of the genus *Australopithecus* is now *A. anamensis*, described in 1995, from sites in Kenya and Ethiopia ranging from 4.2 to 3.9 million years old (Leakey et al. 1995). This hominin appears to be intermediate in anatomy between *Ardipithecus* and *Australopithecus afarensis*, with a less apelike dentition than *Ardipithecus* and fragments of limb bones strongly suggesting bipedality. Its estimated body mass, around 50 kilograms, is greater than that of either of these other early hominins. The associated fauna at the Ethiopian site (e.g., woodland antelope rather than savanna ones) reinforce the notion that early hominin evolution took place in the woodlands rather than on the savanna, as once assumed (White et al. 2006).

A possible contemporary of *A. anamensis*, as yet unnamed, was described from South Africa in 2003 (Partridge et al. 2003). There is some debate as to whether the deposits in which this hominin was found were really as old as 4 million years. If this date is correct, however, it would imply that early hominds were more widespread across the African continent than had previously been known.

Australopithecus bahrelghazali is slightly younger than *A. anamensis* and contemporaneous with *A. afarensis*. This hominin, also described in 1995, was found in Chad, in central Africa, in a fossil site with a paleoenvironment interpreted as lakeside woodland. This is the first known australopithecine to be found west of the Rift Valley, suggesting that early hominins were more widespread in Africa—and more diverse in their habitats—than had previously been supposed.

A still younger species from Ethiopia, *Australopithecus garhi*, was described in 1999 (Asfaw et al. 1999). Dated as 2.5 million years old, it is only slightly older than the earliest known specimen of the genus *Homo*. This species, known from a fragmentary skull from Ethiopia, has been hailed as the missing link between the genera *Australopithecus* and *Homo*. The skull was found near some fossilized butchered animal bones, leading to the speculation that *A. garhi* may have been the first species in our lineage to eat meat and use tools.

The latest surviving early australopithecines—which are often called gracile australopithecines because of their relatively small stature and delicate build—was *Australopithecus africanus*, which lived between 2.3 and 3 million years ago. Although *A. africanus* was fully bipedal, it also had very robust arm bones, suggesting that it may have spent even more time in trees than *A. afarensis*. This greater degree of arboreality implies that it may not be in the direct evolutionary line to *Homo*. An alternative viewpoint holds that *A. africanus* is closely related to *Homo* because its skull is more *Homo*-like than that of *A. afarensis*. This hypothesis proposes that bipedalism evolved convergently to some extent among different early hominin lineages (Kingdon 2003). *A. africanus* may also have differed from *A. afarensis* by including more meat in its diet (Sponheimer and Lee-Thorpe 1999).

A final early hominin, known from 3.5 million years ago in Kenya, represents a completely new genus, *Kenyanthropus platyops* (Leakey et al. 2001). This hominin is markedly different from its contemporary, *Australopithecus afarensis*, combining the unique features of a derived face with a primitive cranium (**Figure 24–14**). Although its interrelationships remain uncertain, there is speculation that it may represent an entirely separate lineage of early hominins, perhaps being ancestral to the hominid *Kenyanthropus* [*Homo*] *rudolfensis*.

Extremely Early Possible Hominins We have known for some years about possible late Miocene hominids, such as *Ouranopithecus* (around 8.5 million years old) and *Samburupithecus* (around 9 million years old). The relationship of these taxa to extant hominids has been debated, but no Miocene taxon had been presumed to be a true hominin, a bipedal animal on the direct human lineage. This picture has changed dramatically in just the past few years. In 2000, Martin Pickford and Brigitte Senut announced the discovery in Kenya of *Orrorin tugenesis*, dated at 6 million years old (Senut et al. 2001). In 2002, Michel Brunet and colleagues (Brunet et al. 2002) described an even earlier form, dated at 6.5 million years, *Sahelanthropus tchadensis*, from the central African country of Chad. Both of these possible early hominins are found in areas that were woodland or forest (Leakey and Walker 2003).

Spectacular as these finds are, there is still debate as to their true hominin status. The preserved fragments

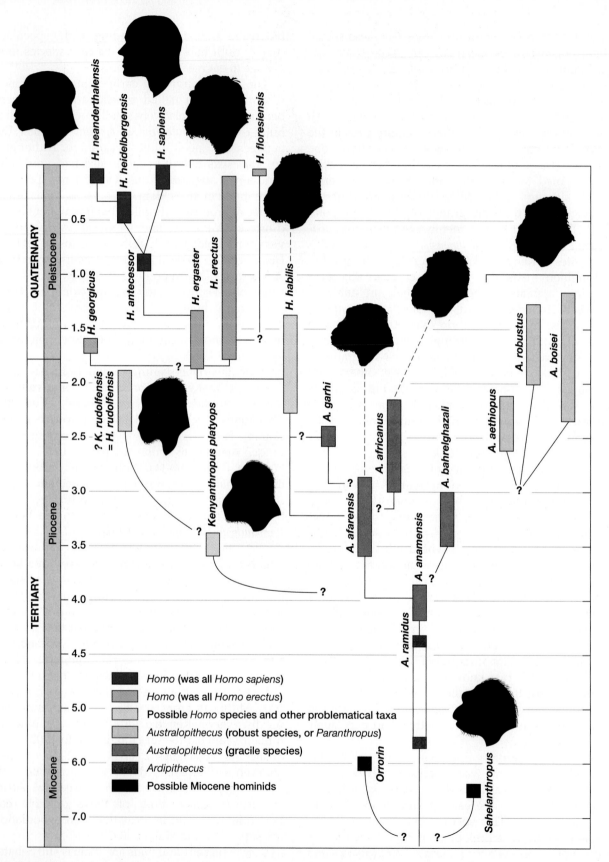

▲ **FIGURE 24–13** A hypothesis of the phylogenetic relationships within the Hominini.

▶ **FIGURE 24–14** Comparison of skulls of hominins (b–e) and a more primitive hominid (a).

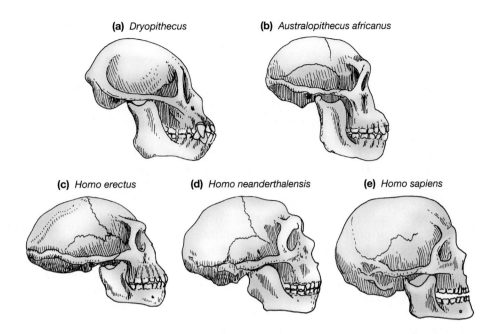

(a) *Dryopithecus*

(b) *Australopithecus africanus*

(c) *Homo erectus*

(d) *Homo neanderthalensis*

(e) *Homo sapiens*

of *Orrorin* combine a femur with features indicating bipedality with large, apelike canines and finger bones indicating climbing abilities. *Sahelanthropus* is known only from a skull, although the ventral position of the foramen magnum suggests a head balanced on top of the vertebral column, as in bipedal hominins. Unlike *Orrorin*, this animal has small, humanlike canines and molars that appear to be hominin in nature (Brunet et al. 2005), but it also has a massive brow ridge and various cranial features that are more apelike than humanlike.

More Derived Australopithecines The later australopithecines were distributed in East and South Africa from 2.5 to around 1.2 million years ago. Called robust australopithecines, they are usually placed in their own genus *Paranthropus*—*Paranthropus robustus* of South Africa and *P. aethiopicus* and *P. boisei* of East Africa. However, it is not entirely clear how these robust australopithecines were interrelated or even whether they represent a single radiation from within the gracile australopithecines. The robust type of australopithecine may have arisen independently on more than one occasion from different gracile australopithecine species.

The robust australopithecines were relatively large, powerfully built forms with pronounced sagittal crests on the skulls and the body proportions of a football player (although they were no more than about 1.5 meters in height). They were terrestrial, savanna-dwelling vegetarians: in comparison with other australopithecines, their huge molars exhibited heavy wear, suggesting a coarse and fibrous diet. These hominins lived sympatrically with early *Homo*, and were a highly successful radiation during the late Pliocene and early Pleistocene. It is likely that their extinction in the mid-Pleistocene was related to climatic changes in Africa rather than to any competition with early humans.

■ Derived Hominids (the Genus *Homo*)

The earliest species of *Homo, H. habilis* (Latin *habilis* = able or "handyman") existed in East Africa from 2.33 to 1.44 million years ago. However, this taxon is rather poorly known and has been the subject of intense debate. *Homo habilis* is best distinguished from australopithecines by its larger cranial capacity (between 500 and 750 cubic centimeters in contrast to 380 to 450 cubic centimeters for *Australopithecus afarensis*). However, this is still small-brained in comparison with later species of *Homo. Homo habilis* also differed from *Australopithecus* in having a smaller face and a smaller jaw and dentition, with smaller cheek teeth and larger front teeth. Like *Australopithecus*, it was relatively small-bodied and retained some specializations for climbing. Some paleoanthropologists consider *H. habilis* to be too primitive to be included in the genus *Homo* and instead place it within the australopithecines (Wood and Collard 1999). Recent finds have extended the range of *H. habilis* (the previous youngest specimen was from 1.65 million years ago) and show that it co-occurred with the more derived *H. erectus* in

East Africa for nearly half a million years (Spoor et al. 2007).

Many specialists have split the original *H. habilis* into two species: *H. rudolfensis*, known from 2.4 to 1.8 million years ago, and *H. habilis* (redefined, known from 1.9 to 1.6 million years ago). Despite being older than *H. habilis*, *H. rudolfensis* was somewhat larger-brained. However, a more recent view, shown in Figure 24–13, places *H. rudolfensis* (now *Kenyanthropus rudolfensis*) as the descendant of *Kenyanthropus platyops*. A South African specimen known from the time of *K. rudolfensis* (around 1.8 million years) has been ascribed to *H. habilis* (Blumenschine et al. 2003), pushing back the age of this species and extending its biogeographic range. This hominin was found in association with stone artifacts and fossil bones of other animals with cut marks, suggesting the use of tools and hunting (or at least scavenging) behavior.

Around 1.9 million years ago a new hominin appeared in the fossil record—*Homo erectus* (Latin *erect* = upright)—originally described in the late nineteenth century as *Pithecanthropus erectus* and known at that time as Java Man or Peking Man. Like modern humans, this hominin was large-bodied, lacked adaptations for climbing, and had relatively small teeth and jaws: There is no doubt that this taxon belongs in the genus *Homo*.

Homo erectus originated in East Africa, where it coexisted for at least several hundred thousand years with two of the robust australopithecines and overlapped with *H. habilis* (see above). *Homo erectus* was the first intercontinentally distributed hominin. It appears to have spread to Asia at least 1.7 million years ago and subsequently perhaps into Europe. This appearance of *Homo* in Asia was once hailed as a landmark in human evolution, with the notion of "out of Africa" being some kind of early human achievement similar to the first man on the moon. In fact, all kinds of other mammals were moving between Africa and Asia at this time, and the discovery of a greater variety of early hominins in Asia (**Box 24–1**) indicates that there was a diversity of hominin migratory events in the early Pleistocene.

Currently the older African version of *H. erectus* is usually called *Homo ergaster*, with the name *erectus* reserved for the Asian hominin (although a true specimen of *H. erectus* is known from the Pleistocene of Ethiopia). *Homo ergaster* is thought to be more closely related to later hominins, and it disappeared around 1.3 million years ago. *Homo erectus* survived for much longer. It was originally thought that the species disappeared between 200,000 and 300,000 years ago, but recent discoveries have shown that it survived for around a quarter of million years longer. Remains from Java have been dated to between 27,000 and 53,000 years old, making *H. erectus* a contemporary of the Neandertals and modern *H. sapiens*. We shall refer to *H. erectus* only, implying inclusion of both *H. erectus* and *H. ergaster* in our discussion of the characteristics of hominins at this evolutionary level.

Four characteristics are especially important features of *Homo erectus* and represent a significant change in the evolutionary history of humans.

- *Homo erectus* was substantially larger than earlier hominins (up to 1.85 meters tall and weighing at least 65 kilograms—the same size as modern humans), with a major increase in female size that reduced sexual dimorphism so that the males were only about 20 percent to 30 percent larger than the females, as in our own species. The reduction in sexual dimorphism in *H. erectus* and later species of *Homo* implies behavioral changes from a polygynous mating system to monogamous pair bonding.

- *Homo erectus* now had body proportions definitively like those of humans. There is debate about the nature of the postcranial skeleton of *H. habilis* and *K. rudolfensis* because of the fragmentary nature of the fossil material, but there is no doubt that *H. erectus* had the short arms, long lower legs, narrow pelvis, and barrel-shaped chest that are characteristic of modern humans.

- *Homo erectus* also had a larger brain than earlier *Homo* species, with cranial capacities ranging from 775 to 1100 cubic centimeters. As the brain is a metabolically demanding organ, this larger brain would imply that *H. erectus* had greater nutritional needs than earlier hominins.

- *Homo erectus* was the first hominin to have a humanlike nose, with downward-facing nostrils. Its other facial features were primitive—a jaw that projected beyond the plane of the upper face (prognathous); teeth that were relatively large; almost no chin; a flat, sloping forehead; prominent, bony eyebrow ridges; and a broad, flat nose. The skeletal proportions were similar to those of modern humans, but the bones were more robust, suggesting a more muscular build.

Homo erectus was also the first hominin to have delayed tooth eruption, with relatively small teeth for its body size. The smaller teeth suggest that *H. erectus* may have cooked its food because cooked food is easier than raw food to chew. The delayed tooth eruption suggests a humanlike extended childhood, which would also imply a humanlike extended life span.

The species *Homo sapiens* (Latin *sapien* = wise), as originally defined, included not only the modern types of humans but also the Neandertals and some earlier forms. More recent evidence has complicated this picture, with the result that the forms originally included in *Homo sapiens* have been split into several different species; the term *H. sapiens* is now reserved for modern humans.

Homo heidelbergensis, known from both Africa and Europe between 500,000 and 300,000 years ago, and *H. antecessor*, known from around 800,000 years ago from Spain, are precursors of modern humans. An even older skull of a similar type of hominin has recently been found from as early as 1 million years ago (Abbate et al. 1998). These hominins had a slightly larger brain, a thicker and more robust skull, larger teeth, and a less prognathous face than *Homo erectus*. Current thinking places *H. heidelbergensis* as the ancestor of the Neandertals (now accorded their own species, *H. neanderthalensis*) and *H. antecessor* as ancestral to both the Neandertal lineage and modern humans (see Figure 24–13).

The Neandertals The first recognized fossils of *Homo neanderthalensis* were found in the Neander Valley in western Germany in 1856, and fossils with Neandertal

---

BOX 24–1   Controversial New Finds of the Genus *Homo*

In the past few years, two new species of our own genus, *Homo*, have been discovered. Both were found outside of Africa, and both have been the subject of intense debate in the scientific community.

The most controversial of the finds is that of a miniaturized humanlike fossil from the island of Flores in eastern Indonesia, named *Homo floresiensis* (Brown et al. 2004; Morwood et al. 2004). *H. floresiensis* was found in deposits ranging from 12,000 to 74,000 years, making it a contemporary of modern humans during the late Pleistocene. It would have stood only about 1 meter tall, much smaller than any other known species of *Homo;* and, along with its small body size, it had a small brain size of only around 400 cubic centimeters, which is no bigger than that of a chimpanzee. Stone tools were also found in association with the bones (Brumm 2006). Because of its small size (and no doubt influenced by the author J. R. R. Tolkien being on people's minds following the success of the *Lord of the Rings* films that were playing in theaters at this time) this creature was promptly nicknamed "the Hobbit" by the popular press (**Figure 24–15**).

The island of Flores has long been a place for unusual animals: today it is home to the largest living lizard, the Komodo monitor lizard (see Section 13.3). Also during the Pleistocene, Flores was home to a dwarfed type of elephant, and dwarfing of large mammals stranded on islands is a common occurrence. Could this be a hominin genus that had followed the similar "Island Rule," whereby small animals get bigger (like the Komodo monitor) and large animals get smaller (see Niven 2006)? One proposal to explain the existence of this small-bodied hominin was that it was an offshoot from earlier Asian *Homo erectus* that had become stranded on Flores long before *Homo sapiens* left Africa. A vigorously debated counterargument was that *H. floresiensis* merely represented some

sort of pygmy, microcephalic version of *H. sapiens* that had reached the island relatively recently. (*Microcephalic* means congenitally small-brained, a pathological condition in modern humans.) The balance of the evidence, however, seems to support the notion that *H. floresiensis* is truly a different species of human, with a relatively derived, though miniature, type of brain and possibly derived from a more primitive hominin than *H. erectus* (see review in Argue et al. 2006). The most recent evidence in this debate comes from the wrist bones (Tocheri et al. 2007), which indicate a more primitive type of wrist than that possessed by either modern humans or Neandertals—further evidence that *H. floresiensis* is not merely some pathological version of *H. sapiens*.

The second controversial new species of *Homo* is a rather different, older hominin, found at the site of Dmanisi, Georgia, dated at around 1.8 million years ago at the beginning of the Pleistocene (Vekua et al. 2002). (Georgia is a country near Turkey, formerly part of the Soviet Union.) This date is similar to that of the first *Homo erectus* found outside of Africa. There has been much debate as to whether this hominin (termed *Homo georgicus* by some people) is some sort of early offshoot from *H. erectus* or from an even earlier hominin species, such as *Homo habilis*. Although it resembles *H. erectus* in a number of ways, a more primitive origin is suggested by its relatively small stature and small brain (which is still larger than that of *H. floresiensis*). Aspects of the elbow and shoulder joints also point to a primitive condition for the genus *Homo;* yet the lower limbs are long, suitable for running and distance travel as seen in early *Homo* (see Box 24–2) (Lordkipandize et al. 2007; see also D. E. Lieberman 2007). This new hominin remains a bit of a mystery, concerning both its taxonomic status (is it really a distinct species?) and whether it represents a separate migration out of Africa to that of *Homo erectus*.

(Continued)

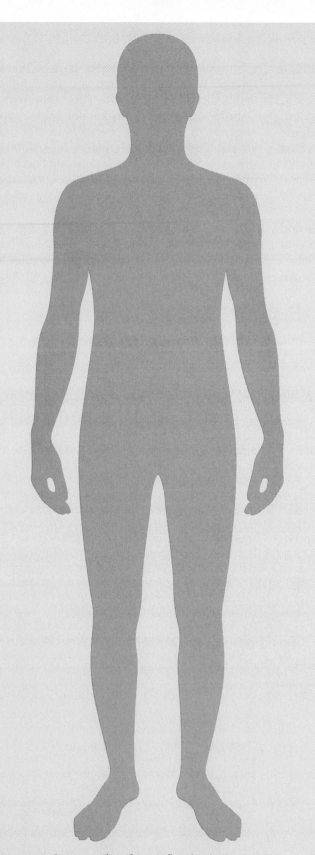
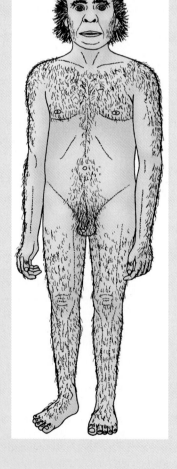

▲ **FIGURE 24–15** Reconstruction of *Homo floresiensis*. The outline of a modern human shows the relative size of *H. sapiens* and *H. floresiensis*.

features first appear about 200,000 years ago. Recent analysis of ancient DNA from the bone of a Neandertal showed considerable genetic difference from modern humans, suggesting that Neandertals were not directly ancestral to modern humans and that the two lineages diverged around 500,000 years ago. Neandertals are often popularly displayed as a primitive caveman, with the unspoken assumption that this is what nonhuman hominins looked like. In fact, the Neandertal features represent a derived condition for hominins. Their stocky build has been interpreted as an adaptation for the cold conditions of Ice Age Europe; however, in fact, they seem to have been more affected by the Pleistocene ice age than modern humans were, and this climate change may have led to their extinction.

Neandertals were short and stocky in comparison with modern humans (**Figure 24–16**). Their body form was robust and muscular, with a barrel chest, large joints, and short limbs. Facially, they had receding foreheads, large protruding noses, prominent brow ridges, and weak chins. Their brains were as large or larger than those of modern-day *Homo sapiens*, but they were enlarged in a slightly different fashion. The Neandertal brain had a larger occipital area (at the back of the head) than ours, while we have a larger, middle temporal region.

Neandertals appear to have been much stronger than extant humans. Their front teeth typically show very heavy wear, sometimes down to the roots. Were Neandertals processing tough food between their front teeth or perhaps chewing hides to soften them, as do some modern aboriginal peoples? The Neandertals were stone toolmakers, producing tools known as the Mousterian tool industry, with a well-organized society and increasingly sophisticated tools. Whether the Neandertals had the capacity for complex speech remains controversial, but they were the first humans known to bury their dead, probably with considerable ritual. Of special importance are burials at Shanidar Cave in Iraq that include a variety of plants recognized in modern times for their medicinal properties.

Neandertals probably hunted the wild horses, mammoths, bison, giant deer, and woolly rhinoceros. Mousterian-style hunting tools were of the punching, stabbing, and hacking type—throwing spears and bows and arrows are unknown. Many skeletal remains of Neandertals show evidence of serious injury during life, and their patterns of injury resembled that of present-day rodeo riders. That similarity suggests that, like rodeo cowboys, Neandertals were in close contact with large, angry animals. Despite the high incidence of injuries, 20 percent of Neandertals

▶ **FIGURE 24–16** Reconstruction of a Neandertal.

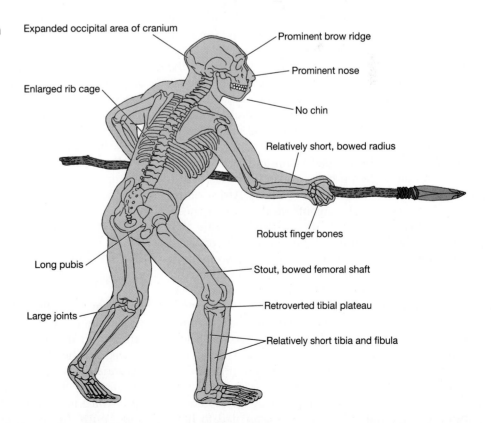

Expanded occipital area of cranium

Prominent brow ridge

Prominent nose

No chin

Enlarged rib cage

Relatively short, bowed radius

Robust finger bones

Long pubis

Stout, bowed femoral shaft

Large joints

Retroverted tibial plateau

Relatively short tibia and fibula

were over 50 years old at the time of death. It was not until after the Middle Ages that human populations again achieved this longevity.

Modern *Homo sapiens* reached Europe and Asia between 40,000 and 50,000 years ago, and European Neandertals and the populations of *Homo erectus* remaining in Asia vanished between 40,000 and 30,000 years ago, with a relict population of Neandertals remaining in Gibraltar (the southernmost point in Europe) until 28,000 years ago (Finlayson et al. 2006). There is much debate about the role of *H. sapiens* in the disappearance of the other species of humans: Modern humans do not have the distinguishing DNA markers of Neandertals, suggesting that interbreeding between these two species did not take place. Did our species gradually outcompete the others in a noncombative fashion, or was there some type of direct conflict? Current evidence indicates that rapidly changing climatic conditions over the interval where the two species intermingled in Europe (from about 45,000 to 30,000 years ago) were behind the Neandertal's demise (Finlayson and Carrión 2007).

Origin of Modern Humans A single African origin of *Homo sapiens* is now supported by both the fossil record and by genetic studies of modern humans, especially the evolution of mitochondrial DNA and the Y chromosome. Few people still adhere to the older multiregional model of *H. sapiens* evolving independently in different areas, each from an already distinctive local population of *H. erectus*. Mitochondrial DNA is inherited only from the mother because it resides in the cytoplasm of the egg, not in the nucleus, and the genome is small—only about 16,000 base pairs. Analysis of mitochondrial DNA allows one to trace the maternal lineage of an individual. A study of mitochondrial DNA from people all over the world (see Cann and Wilson 2003) showed that all living humans can trace their mitochondria to a woman who lived in Africa about 170,000 years ago. This hypothetical common ancestor has been called the African Eve, a phrase that obscures the biological meaning of the discovery. It does not mean that there was only one woman on Earth 170,000 years ago; instead, it means that only one woman has had an unbroken series of daughters in every generation since then.

A similar approach can be used with the Y chromosome, which is passed only from father to son. The Y chromosome has about 60 million base pairs, so it is much more difficult to study than mitochondrial DNA. An analysis of 2600 base pairs from the Y chromosome indicates that all human males are descended from a single individual, who is estimated to have lived 59,000 years ago (Hammer 1995). Naturally, this individual has been called the African Adam.

The difference between the estimates—170,000 years versus 59,000 years—results from uncertainty about the rates of mutation in mitochondrial DNA and the Y chromosome. What is most significant is that both studies indicate that the common ancestor of modern humans lived in Africa, and this conclusion is reinforced by other genetic information. For example, there is more variation in the human genome in Africa than in the rest of the world combined, which is exactly what would be expected if humans originated in Africa. Furthermore, humans have only about one tenth the genetic variation of chimpanzees, and that observation might indicate that human populations were once very small, passing through a genetic bottleneck.

Although they have met with some criticism, these molecular studies generally agree with the fossil record, which shows that modern *Homo sapiens* originated in Africa between 100,000 and 200,000 years ago. Three skulls of hominins have been recently found in Ethiopia, dated at around 160,000 years; they appear to be at the very base of the modern human lineage (White et al. 2003). By 125,000 years ago, anatomically modern humans were widespread across Africa. They crossed over into the Levant region of Asia (the area termed the Middle East today) at around 120,000 years ago, and they first appeared in more northern Eurasia (and also in Australia) between 40,000 and 50,000 years ago.

Somewhere around 40,000 years ago humans acquired modern behavior, with the fossil record showing evidence of art, musical instruments, highly sophisticated tools, record keeping on stone plaques, etc. What happened at this time? Many researchers posit that this represents the time when humans finally acquired the capacity for a modern form of language and symbolic thought.

24.4 Ecological and Biogeographical Aspects of Early Hominid Evolution

For many years the evolution of the hominin lineage was assumed to be related to the appearance of the African savannas, which are grasslands with widely spaced bushes and trees. The spread of African savanna environments was probably related to the formation of the Isthmus of Panama 2.5 million years ago, which blocked the flow of water between North and South America, leading to profound global climatic changes. Paleontological evidence from both the flora and fauna of eastern Africa suggests that the environment changed to a savanna habitat at this time. Grazing antelopes increased in abundance, and new types of carnivores appeared.

It used to be thought that the development of savanna habitats coincided with the split of the human lineage from that of the other apes. The traditional view has long been that the development of the Rift Valley, which extends from north to south in eastern Africa, isolated the human lineage from that of the other apes. Subsequently humans became adapted for these new, open grassland habitats by adopting a bipedal gait, while the apes were relegated to the tropical forests to the west of the Rift Valley and remained primarily arboreal.

We now know this scenario of the evolution of bipedality being an adaptation for savanna environments must be incorrect, although some isolation of humans from other hominoids undoubtedly did occur. The emergence of broad expanses of savanna occurred 2 to 3 million years ago, whereas hominin bipedality certainly extends back at least 4.5 million years, and possibly to 7 million years ago. Thus, it now seems probable that the origins of humans and bipedality took place in more forested environments.

The gracile australopithecines were primarily an early Pliocene radiation, preceding the expansion of savanna habitats. In the late Pliocene, at around 2.5 million years ago and coincident with the major climatic changes, the hominin lineage split into two—one lineage leading to our direct ancestors, the genus *Homo,* and the other leading to the robust australopithecines. Thus during the late Pliocene, from 2.5 million years onward, there were two lineages of hominins—early true humans and robust australopithecines. Further climatic cooling and drying resulted in the reduced abundance of robust australopithecines at the start of the Pleistocene and their extinction later in the epoch. However, had the Pleistocene climatic changes been different—such as reverting to a wetter and warmer regime—it might have been our ancestors who became extinct and the robust australopithecines that survived.

24.5 Evolution of Human Characteristics

Humans are classically distinguished from other primates by three derived features: a bipedal stance and mode of locomotion, an extremely enlarged brain, and the capacity for speech and language. Here we examine possible steps in the evolution of each of these key features and also consider the loss of body hair and the evolution of tool use.

■ Bipedality

Although all modern hominoids can stand erect and walk to some degree on their hind legs, only humans display an erect bipedal mode of striding locomotion involving a specialized structure of the pelvis and hind limbs, thereby freeing the forelimbs from obligatory functions of support, balance, or locomotion. The most radical changes in the hominin postcranial skeleton are associated with the assumption of a fully erect, bipedal stance in the genus *Homo.* Anatomical modifications include the S-shaped curvature of the vertebral column, the modification of the pelvis and position of the acetabulum (hip socket) in connection with upright bipedal locomotion, and the lengthening of the leg bones and their positioning as vertical columns directly under the head and trunk. The secondary curve of the spine in humans is a consequence of bipedal locomotion and forms only when an infant learns to walk. We have by no means perfected our spines for the stresses of bipedal locomotion, which are quite different from those encountered by quadrupeds. One consequence of these stresses is the high incidence of lower back problems in modern humans.

Humans stand in a knock-kneed position, which allows us to walk with our feet placed on the midline and reduces rolling of the hips from side to side. This limb position leaves some telltale signatures on the femur (the thigh bone), both in the articulation with the hip and with the knee joint. This type of bony evidence can aid researchers in deducing whether fossil species were fully bipedal. An unfortunate consequence of this limb position is that humans, especially athletes, are rather prone to knee dislocations and torn knee ligaments. Because women have wider hips than men, their femurs are inclined toward the knees at a more acute angle than men's femurs, and they are especially prone to knee injuries.

The feet of humans show drastic modification for bipedal, striding locomotion. The feet have become flattened except for a tarsometatarsal arch, with corresponding changes in the shapes and positions of the tarsals and with close, parallel alignment of all five metatarsals and digits. In addition, the big toe is no longer opposable, as in apes and monkeys (**Figure 24–17**), although it may still have had some capacity to diverge from the rest of the toes in early hominins. Humans also differ from apes in having a longer trunk region with a more barrel-shaped (versus funnel-shaped) rib cage, resulting in a distinct

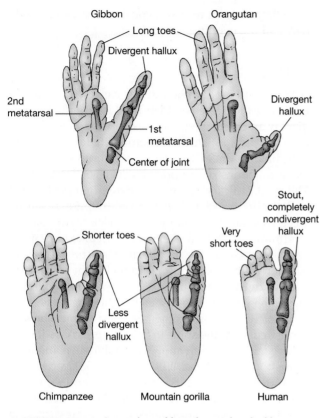

Gibbon

Orangutan

Long toes

Divergent hallux

2nd metatarsal

Divergent hallux

1st metatarsal

Center of joint

Stout, completely nondivergent hallux

Shorter toes

Very short toes

Less divergent hallux

Chimpanzee

Mountain gorilla

Human

▲ **FIGURE 24–17** Comparison of feet of extant hominoids. The positions of the metatarsals are shown for digit II and the positions of both the metatarsals and the phalanges are shown for digit I.

waist (**Figure 24–18**). The humanlike waist may be a specific adaptation for bipedal walking, as it would allow greater rotation of the pelvis in striding without also involving the upper body (Preuschoft 2004).

There are almost as many hypotheses about the reasons for human bipedality as there are anthropologists. Among the suggested reasons are improved predator avoidance (being able to look over tall grass), freeing the hands for carrying objects (either for hunting or collecting foods), thermoregulation (an upright ape presents a smaller surface area to the sun's rays), and energy efficiency of locomotion. An obvious problem with all such hypotheses is that they are difficult to test. Although humans are extremely efficient at bipedal locomotion, especially at walking, it seems unlikely that bipedality evolved specifically for efficient, striding locomotion. Other apes are not nearly as efficient as humans at bipedal walking, but this is the evolutionary condition from which human bipedality must have commenced. That is, a protohominin or early hominin must have walked bipedally in an inefficient way first, before selection could act to increase efficiency, and in fact human locomotion is not particularly efficient when compared to that of quadrupedal mammals (Steudal-Numbers 2003).

In some respects, human bipedal walking may not be such a mystery as it first appears. (Bipedal

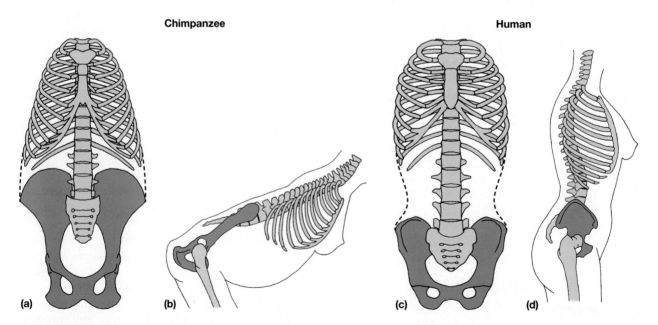

Chimpanzee

Human

(a) (b) (c) (d)

▲ **FIGURE 24–18** Comparison of the trunk regions of a chimpanzee and a human. The oblique, platelike pelvis of a typical ape is changed into an upward-facing basin. This modification increases the mobility of the thorax relative to the pelvis. A shorter rib cage and the larger and more flexible lumbar vertebrae create a waist.

running, however, is a separate issue—see **Box 24–2**) The knuckle-walking of gorillas and chimpanzees is a derived hominoid trait, either evolved independently in the two ape lineages or shared with the common ancestor of great apes and hominins. Other apes, such as gibbons and orangutans, tend to walk in a clumsy bipedal stance on the ground: their upright trunk, adapted for arboreal locomotion, predisposes them to do this—thus, a tendency to walk bipedally appears to be a primitive hominoid feature. Robin Crompton and his colleagues have found that wild orangutans walk bipedally in the trees, supporting themselves by grasping overhead branches above with their hands (Thorpe et al. 2007). This behavior allows them to move out onto narrower supports than they could reach quadrupedally or even by hanging underneath the branch. Although orangutans are not on the direct line to humans, this behavior shows that bipedal walking could have evolved in trees instead of on the ground.

Oreopithecus (Greek *oreo* = mountain), an ape from the late Miocene of Italy, appears to have evolved a type of bipedality convergent with hominins: it had an S-shaped spine, a pelvic girdle rather like *Australopithecus*, and a knock-kneed angle to the femur. However, the foot differed markedly from that of humans, with a widely divergent big toe—apparently providing a broad, tripodal base of support. This foot anatomy, combined with the short legs, indicates a slow, shuffling gait rather than humanlike, efficient striding. In this species, at least, bipedality must have evolved for some other reason besides efficient locomotion, perhaps to increase food-gathering efficiency.

■ Origin of Large Brains

The human brain increased in size threefold over a period of around 2.5 million years. Human brains are not simply larger versions of ape brains but have a number of key differences, such as a relatively much larger prefrontal cortex but relatively smaller olfactory bulbs. We still do not really know what selective pressures led to humans evolving such large brains. Speculations include increasing ability for social interactions, conceptual complexity, tool use, dealing with rapidly changing ecological conditions, and language (and/or a mixture of all of these things) (see Shoenemann 2006).

Brain tissue is a metabolically expensive tissue both to grow and to maintain; on a gram-for-gram basis, brain tissue has a resting metabolic rate 16 times that of muscle (Leonard et al. 2003). Most of the growth of the brain occurs during embryonic

---

BOX 24–2   The Genus *Homo* and the Evolution of Running

We have seen that habitual bipedality evolved with the first hominins and that a variety of reasons have been proposed for this evolutionary event. But in what way was bipedal locomotion different in the genus *Homo* from the australopithecines and their kin?

Denis Bramble and Dan Lieberman (2004) proposed an ingenious hypothesis—that various aspects of the skeleton of *Homo* represent adaptations for endurance running (**Figure 24–19**). One important difference, first seen definitively in *H. erectus* (the condition in *H. habilis* is currently under debate), is a longer lower leg (longer tibia and fibula). This longer leg gives us an increased stride length and a gastrocnemius (calf) muscle with a long tendon, which not only gives us more shapely ankles than apes and earlier hominins but also allows elastic energy storage during running. Various new features of the trunk of *Homo* also aid in stabilizing the trunk to prevent rotation during running, and the shape of pelvic girdle is indicative of a larger gluteus maximus muscle, which in modern humans is highly active during running but not during walking on a level surface.

Perhaps surprisingly, the skulls of *Homo* also reveal running adaptations. The shoulder muscles of apes are connected to the head, but we have largely freed this connection, decoupling movements of the head from those of the pectoral girdle. In addition, we have a nuchal crest down the midline of the back of the skull (not seen in other hominoids) that serves as an anchor for the nuchal ligament that runs from the head to the neck and anterior trunk vertebrae. This ligament stabilizes the head on the neck and may be an adaptation to reduce forward-and-backward pitching of the head during running.

What would have been the evolutionary advantage of long distance running? Bramble and Lieberman note that many of the specific locomotory adaptations of *Homo* are not ones that could have been evolved originally just for long distance walking. Instead they suggest that endurance running may be related either to hunting or for effective scavenging covering long distances.

(Continued)

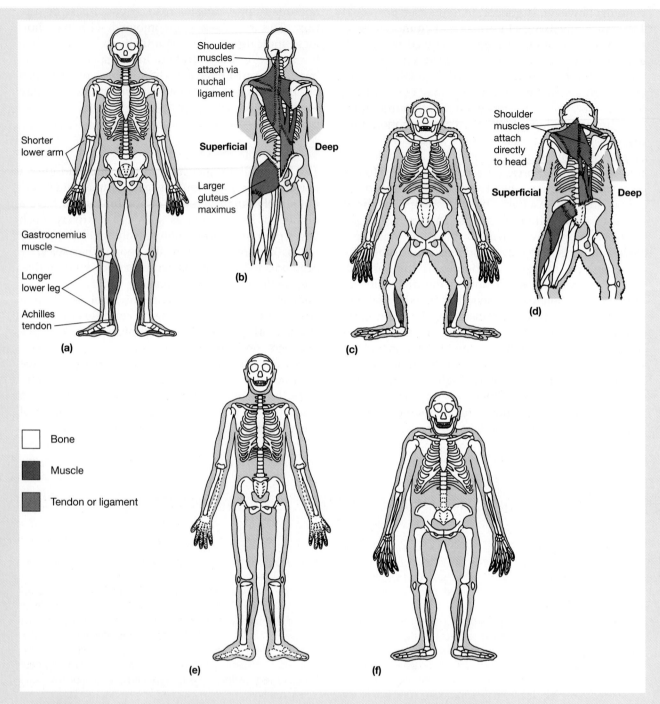

▲ **FIGURE 24–19** Locomotor adaptations for running in the genus *Homo*: (a-b) anterior and posterior views of a modern human, *Homo sapiens*; (c-d) anterior and posterior views of a modern chimpanzee, *Pan troglodytes*; (e) reconstruction of the skeleton of *Homo erectus*; (f) reconstruction of the skeleton of *Australopithecus afarensis*.

development and requires energy input from the mother. Thus, selective pressures for larger brains can be satisfied only in an environment that provides

sufficient energy, especially to the pregnant or lactating female. The evolution of larger brains may have required increased foraging efficiency (partially achieved through larger female size and mobility) and high-quality foods in substantial quantities (partially achieved through the use of tools and fire). The increase in brain size commencing in *Homo erectus* has been ascribed to the development of cooking, as cooked food requires less energy to digest, thus freeing up energy for the brain. However, there is no evidence for cooking with fire until around 5,000 thousand years ago, more than a million years later than the first *H. erectus*.

Larger brains also would have required a change in life-history pattern that probably exaggerated the ancestral primate character of slow rates of pre- and postnatal development, thus lowering daily energy demands and also a female's lifetime reproductive output. Our prolonged period of childhood may allow children to stay with the family long enough to acquire the necessary knowledge for survival from their parents. (Imagine the chaos if human children became sexually mature at around 3 years, as horses do.) There is also speculation that human menopause, a post-reproductive time span that is not seen in other female mammals, may have evolved in order to help a woman assist with the rearing of her long-dependent and thus vulnerable grandchildren. This might be a more effective way to increase her genetic investment than having more children of her own near the end of her life span (see Peccei, 2001, for a review).

The origin of bipedality may be linked to the origin of large brain size and a change in human reproductive biology and life history: only after bipedality had evolved would a female hominin be able to carry the highly altricial type of human newborn. Whereas apes give birth to young that can cling to their mother's fur soon after birth, our large brain necessitates us being born relatively helpless, with much more brain growth after birth than in other mammals (Stanley 1996). (If we were born with brains almost fully formed, birth would be even more of a problem for humans than it is now.)

■ Origin of Speech and Language

Although other animals can produce sounds and many mammals communicate by using a specific vocabulary of sounds (as anyone who has kept domestic pets well knows), the use of a symbolic language is a uniquely human attribute. Although apes and chimpanzees have been taught to use some human words and form simple sentences, this is a long way from the complexity of human language. Yet where in human evolution did language evolve,

and how can we tell this from the fossil record? The first evidence of human writing is only a few thousand years old. Obviously, language evolved before this, but how long before?

Alan Walker argues that controlled speech would not have been possible until a later stage than *Homo erectus* (Walker and Shipman 1996). In *H. erectus*, the spinal cord in the region of the thorax is much smaller than in modern humans. The implication is that *H. erectus* lacked the capacity for the complex neural control of the intercostal muscles that allows modern humans to control breathing in such a way that we can talk coherently. Richard Kay (Kay et al. 1998) pointed out that the exit in the skull (the hypoglossal canal) for the nerve that innervates the tongue muscles (cranial nerve XII) was smaller in other hominins (and is also smaller in chimpanzees and gorillas) than in the modern humans and Neandertals. Additionally, a specific gene involved in the production of language in humans, *FOXP2*, appears to have attained the human condition no later than 200,000 years ago, coinciding with the appearance of anatomically modern humans (Enard et al. 2002).

Even if more derived *Homo* species had evolved the capacity for language, they would not have been able to produce the range of vowel sounds that we can produce until a change in the anatomy of the pharynx and vocal tract had taken place. The primitive position of the mammalian larynx is high in the neck, right behind the base of the tongue. However, in modern humans, the larynx shifts ventrally at the age of one to two years, resulting in the creation of a much larger resonating chamber for vocalization. This change in the vocal tract anatomy is associated with a change in the shape of the base of the skull, so we can infer from fossil skulls when the shift in larynx position occurred. Although there are some differences of opinion, a fully modern condition of the vocal tract was probably not a feature of the genus *Homo* until *H. sapiens* of around 50,000 years ago. Speech also requires a change in neural capacities in order to process the rapid frequency of transmitted sounds, which are decoded at a rate much greater than other auditory signals (P. Lieberman 2007).

Some evolutionary problems with the origin of the vocal tract trait still exist. The descent of the larynx means that the original mammalian seal between the palate and the epiglottis is lost, making humans especially vulnerable to choking on their food. It seems likely that this liability to choke would be a powerful selective force acting in opposition to repositioning the larynx. In addition, the death of infants from SIDS (sudden infant death syndrome) is associated with the developmental period when the larynx is in flux. It is difficult to imagine that the ability to produce a greater range of vowel sounds could counteract these antagonistic selective forces. Perhaps there was another, more immediately powerful reason for repositioning the larynx.

One advantage that repositioning the larynx affords us is the ability to voluntarily breathe through our mouths, an obviously important trait in the production of speech. However, perhaps a more important function of mouth breathing is apparent to anyone with a bad cold. Many of us would suffocate every winter if we were unable to breathe through our mouths. It is tempting to speculate that the human capacity for at least well-enunciated speech owes its existence to a prior encounter of our species with the common cold virus.

Loss of Body Hair and Development of Skin Pigmentation

Humans are unique among primates in their apparent loss, or at least reduction, of body hair and their development of heavily pigmented skin. (Humans have the same number of hair follicles as other apes, but the hairs themselves are minuscule). These features are related, as fur protects animals from deleterious effects of the sun's rays. Chimpanzees have relatively unpigmented skin; but, with the loss of body hair there would be a need to gain skin pigments. It is not clear why humans lost the majority of their body hair. Speculations include increased use of eccrine sweat glands in evaporative cooling (possibly important in hunting) and increased problems with skin parasites such as lice, ticks, and fleas (perhaps in association with a more sedentary way of life with groups of people living together in confined quarters). However, if cooling during hunting was the driving force, then we would expect men to be less hairy than women, rather than the reverse. Rantala (2007) reviews these various hypotheses and concludes that the parasite hypothesis provides the best explanation.

Although we will probably never know precisely why humans lost their covering of hair, genetics enables us to figure out when this might have happened. The genes involved in human skin pigmentation appear to date from at least 1.2 million years ago (Rogers et al. 2004). This implies that hairlessness first came about with the lineage leading to *Homo*

sapiens, perhaps also including the *H. erectus* lineage that branched off around 1.7 million years ago. This corresponds with the time when hominins adopted a home base, which might have made them especially prone to parasitic infections and suggests that we shared hairlessness with other species, such as the Neandertals. This hypothesis would also explain the tendency for females to be less hairy than males, as they would most likely spend more of their time at the home base.

A study of the genetics of skin parasites provides information about when humans started to wear clothes. The human body louse is different from head lice and pubic lice in that it clings to human clothing rather than to hair. This parasite presumably evolved from the head louse after humans started to wear clothes, and this separation occurred between 40,000 and 70,000 years ago, broadly coincident with the emergence of modern *Homo sapiens* in Europe (Kittler et al. 2003).

Humans living close to the equator have dark skin, whereas people farther from the equator have lighter skin. A balance between protection against the damaging effects of ultraviolet light and the need for vitamin D synthesis appears to offer the best general explanation of this phenomenon. Folic acid is essential for normal embryonic development, and ultraviolet light breaks down folic acid in the blood. Melanin in the skin blocks penetration by ultraviolet light, thereby protecting folic acid. Too much melanin in the skin creates a different problem, however, because vitamin D is converted from an inactive precursor to its active form by ultraviolet light. Thus, human skin color probably represents a compromise—enough melanin to protect folic acid while still permitting enough ultraviolet penetration for vitamin D synthesis.

Origins of Human Technology and Culture

Humans are tool users and toolmakers *par excellence*. Some other animals also use tools to a limited extent, usually in rather stereotyped and instinctive ways. Egyptian vultures open ostrich eggs by picking up stones in their beaks and dropping them on the shells, and a Galápagos finch holds twigs and cactus spines in its beak to probe for insects in holes or under bark. Both baboons and chimpanzees use sticks and stones as weapons very much in the way that ancestral humans must have done, and chimpanzees use these same types of materials as tools in obtaining food.

The earliest recognized simple stone tools are found throughout East Africa and date to 2.5 to 2.7 million years ago. They are generally circular and worked on only one side. This suite of stone tools, called the Oldowan culture, remained relatively unchanged for 1 million years, and hunter-gatherer societies such as the Australian Aborigines were still making Oldowan-style tools when they were forcibly thrust into European cultural traditions in the nineteenth and twentieth centuries. Based on the available information about thumb morphology (which shows whether a fossil hominin had the ability for a precision grip), the Oldowan tools could have been made by either *Homo habilis* or *Australopithecus robustus*. However, the earlier gracile australopithecines did not have the hand bone characteristics of precision grip.

Novel tools, especially cleavers and so-called hand axes, appeared in East Africa about 1.4 million years ago. These later tools of *H. erectus*, known as the Acheulean tool industry, differed from Oldowan tools in having a distinct long axis and being chipped on both sides. Acheulean tools changed very little in style for the next 1.2 million years, although the materials chosen changed and later sites have a much higher proportion of small tools than do early sites. The lack of dramatic advance in tool manufacture over this immense period is surprising, especially in light of the spread of these tools to southwest Asia and Western Europe.

Thus, one of the remarkable conclusions from paleoanthropological findings is that the use of tools, chipped stones, and possibly modified bone and antler precedes the origin of the big-brained *Homo sapiens* by at least 1.5 million years. The use of tools by ancestral hominins may have been a major factor in the evolution of the modern *Homo* type of cerebral cortex; in fact, the elaborate brain of *Homo sapiens* may be the consequence of culture as much as its cause.

By at least 750,000 years ago, *Homo erectus* was making advanced types of stone tools and had also apparently learned to control and use fire, although actual hearths in caves are not widely recognized before 500,000 years ago. With fire, humans could cook their food, increasing its digestibility and decreasing the chance of bacterial infection, and preserve meat for longer periods than it would remain usable in a raw state. They could also keep themselves warm in cold weather, ward off predators, and light up the dark to see, work, and socialize. The earliest evidence of wooden throwing spears, found in association with stone tools and the butchered remains of horses, is from 400,000 years ago in Germany.

It is often assumed that much of the evolution of human tool use and culture developed in the context of humans hunting other animals. The term "Man the Hunter" was coined in the 1960s. Anthropologists have pointed out that much of our perception of human evolution as being an upward and onward quest has more to do with western cultural myths of the "Hero's tale" than with anthropological data. Examination of the archeological evidence for hunting by early hominins, such as stone tools and bones with cut marks, has suggested that early humans were scavengers rather than hunters.

A second major shift in our understanding of human evolution results from the increasing number of female anthropologists and paleoanthropologists who have broken from traditional perspectives to emphasize the critical role of women in the development of human tools and human behavior (e.g., Morbeck et al. 1997, Hrdy 1999).

Summary

Evidence for the origin of *Homo sapiens* comes from the Cenozoic fossil record of primates and from comparative study of living monkeys, apes, and humans. The first primatelike mammals are known from the early Paleocene, but the earliest true primates are not known until the early Eocene. These were rather similar to the extant lemurs and were initially present in North America and the Old World. All were arboreal, and some had larger brains in relation to their body size than had other mammals of that time.

After the climatic deterioration of the late Eocene, primates were confined to the tropical regions until the middle Miocene. The higher primates, or anthropoids, evolved in Asia by the middle Eocene and soon spread to Africa. By the Oligocene, the two distinct groups of extant anthropoids had evolved: the platyrrhine monkeys of the New World tropics and the catarrhine monkeys and apes of the Old World. The apes and humanlike species, including *Homo sapiens*, are grouped in the Hominoidea. Many morphological features distinguish the hominoids from other catarrhines, and enlargement of the brain has been a major evolutionary force molding the shape of the hominoid skull, especially in the later part of human evolution.

The first known hominoids occur in the early Miocene, around 25 million years ago. By the late Miocene, hominoids had diversified into a number of environments and had spread widely over Africa, Europe, and Asia. The genetic closeness of humans and great apes (orangutans, gorillas, and chimpanzees) has led to a regrouping of the traditional family Hominidae to include these apes as well as humans. Humans and their fossil relatives have now been assigned to a tribe within the Hominidae, the Hominini. A variety of hominin fossils occurs in late Pliocene and early Pleistocene deposits of Africa. The earliest well-known hominins were the australopithecines, known from 4.4 to 1.2 million years ago. Australopithicines were apparently bipedal but retained considerable arboreal habits and had a relatively small apelike cranial volume.

The earliest member of the genus *Homo* dates from around 2.5 million years ago, concurrent with the earliest stone tools found in East Africa. *Homo* appears to be a primarily terrestrial genus, and its specific postcranial adaptations may relate to the evolution of endurance running. A global climatic change at around this time may have prompted the evolution both of *Homo* and of the robust australopithecines.

Homo erectus ranged across Africa and Eurasia from about 1.8 million to around 30,000 years ago. This hominin had a brain capacity approaching the lower range of *Homo sapiens*, made stone tools, and used fire. *Homo sapiens*, the only surviving species of the tribe Hominini, came into existence around 200,000 years ago. By around 40,000 years ago, some populations of *Homo* had a well-organized society with a rapidly developing culture, especially obvious in their use of stone tools and the development of art. For much of human history, several species of the genus *Homo* have existed in the world together, including some newly discovered forms, such as the dwarfed *Homo floresiensis* that was a contemporary of *Homo sapiens*. The present-day situation, with *Homo sapiens* as the sole existing hominin, is a highly unusual one.

Additional Readings

Abbate, E. et al. 1998. A one-million-year-old *Homo* cranium from the Danakil (Afar) depression of Eritrea. *Nature* 393:458–460.

Argue, D. et al. 2006. *Homo floresiensis*: Microcephalic, pygmoid, *Australopithecus*, or *Homo*? *Journal of Human Evolution* 51:360–374.

Asfaw, B. et al. 1999. *Australopithecus garhi*: A new species of early hominid from Ethiopia. *Science* 284:629–635.

Bakewell, M. A. et al. 2007. More genes underwent positive selection in chimpanzee evolution than in human evolution. *Proceedings of the National Academy of Sciences* 104:7489–7494.

Benefit, B. R., and M. L. McCrossin. 1997. Earliest known Old World monkey skull. *Nature* 388:368–371.

Bloch, J. I., and D. M. Boyer. 2002. Grasping at primate origins. *Science* 298:1606–1610.

Blumenschine, R. J., and J. A. Cavallo. 1992. Scavenging and human evolution. *Scientific American* 267(4):90–96.

Blumenschine, R. J. et al. 2003. Late Pliocene *Homo* and hominid land use from western Oldovai Gorge, Tanzania. *Science* 299:1217–1221.

Bramble, D. M., and D. E. Lieberman. 2004. Endurance running and the evolution of *Homo*. *Nature* 432:345–352.

Brown, P. et al. 2004. A new small-bodied hominin from the Late Pleistocene of Flores, Indonesia. *Nature* 431:1055–1061.

Brumm, A. 2006. Early stone technology on Flores and its implications for *Homo floresiensis*. *Nature* 431:1055–1061.

Brunet, M. et al. 2002. A new hominid from the Upper Miocene of Chad, Central Africa. *Nature* 418:145–155.

Brunet, M. et al. 2005. New material of the earliest hominid from the Upper Miocene of Chad, Central Africa. *Nature* 434:752–755.

Cann, R. L. and A. C. Wilson. 2003. The recent African genesis of humans. *Scientific American* 13(2):54–61.

Cartmill, M. 1998. The gift of the gab. *Discover* 19(11):56–64.

Crompton, R. H. et al. 1998. The mechanical effectiveness of erect and "bent-hip, bent-knee" bipedal walking in *Australopithecus afarensis*. *Journal of Human Evolution* 35:55–74.

Enard, W. et al. 2002. Molecular evolution of *FOXP2*, a gene involved in speech and language. *Nature* 418:869–872.

Finlayson, C. 2005. Biogeography and evolution of *Homo*. *Trends in Ecology and Evolutionary Biology* 20:457–463.

Finlayson, C. et al. 2006. Late survival of Neanderthals at the southernmost extreme of Europe. *Nature* 443:850–853.

Finlayson, C. and J. S. Carrión. 2007. Rapid ecological turnover and its impact on Neanderthal and other human populations. *Trends in Ecology and Evolutionary Biology* 22:213–222.

Fleagle, J. G. 1998. *Primate Adaptation and Evolution*, 2nd ed., San Diego, CA: Academic Press.

Gebo, D. L. et al. 2000. The oldest known anthropoid postcranial fossils and the early evolution of higher primates. *Nature* 404:276–278.

Gibbons, A. 2007. Food for thought. *Science* 316:1558–1560.

Godfrey, L. R., and W. L. Jungers. 2003. The extinct sloth lemurs of Madagascar. *Evolutionary Anthropology* 12:252–263.

Green, R. E. et al. 2006. Analysis of one million base pairs of Neanderthal DNA. *Nature* 444:330–336.

Hammer, M. F. 1995. A recent common ancestor for human Y chromosomes. *Nature* 378:376–378.

Heglund, N. C., and B. Schepens. 2003. Ontogeny recapitulates phylogeny? Locomotion in children and other primitive hominoids. In V. L. Bels et al. (eds.), *Vertebrate Biomechanics and Evolution*. BIOS Scientific Publishers Ltd., Oxford, 283–295.

Hrdy, S. B. 1999. *Mother Nature*. New York, NY: Pantheon.

Jablonski, N. G., and G. Chaplin. 2003. Skin deep. *Scientific American* 13(2):72–79.

Johanson, D. C., and B. Edgar. 1996. *From Lucy to Language*. New York, NY: Simon and Schuster.

Kay, R. F., et al. 1998. The hypoglossal canal and the origin of human vocal behavior. *Proceedings of the National Academy of Sciences, USA* 95:5417–5419.

Kingdon, J. 2003. *Lowly Origin: Where, When, and Why Our Ancestors First Stood Up*. Princeton NJ: Princeton University Press.

Kittler, R. et al. 2003. Molecular evolution of *Pediculus humanus* and the origin of clothing. *Current Biology* 13:1414–1417.

Köhler, M., and S. Moyà-Solà. 1997. Ape-like or hominid-like? The positional behavior of *Oreopithecus bambolii* reconsidered. *Proceedings of the National Academy of Sciences*, USA 94:11747–11750.

Larick, R., and R. L. Ciochon. 1996. The African emergence and early Asian dispersals of the genus *Homo*. *American Scientist* 84:538–551.

Leakey, M., and A. Walker. 2003. Early hominid fossils from Africa. *Scientific American* 13(2):14–19.

Leakey, M. G. et al. 1995. New four-million-year-old hominid species from Kanapoi and Allia Bay, Kenya. *Nature* 376:565–571.

Leakey, M. G. et al. 2001. New hominin genus from eastern Africa shows diverse middle Pliocene lineages. *Nature* 410:433–440.

Leonard, W. R. et al. 2003. Metabolic correlates of hominid brain evolution. *Comparative Biochemistry and Physiology* Part A 136:5–15.

Lewin, R. 1998. *Principles of Human Evolution: A Core Textbook*. Boston, MA: Blackwell Scientific.

Lieberman, D. E. 2007. Honing in on early *Homo*. *Nature* 449:291–292.

Lieberman, P. 2007. The evolution of human speech: Its anatomical and neural bases. *Current Anthropology* 48:39–66.

Lordkipandize, D. et al. 2007. Postcranial evidence from early *Homo* from Dmanisi, Georgia. *Nature* 449:305–310.

MacLatchy, L. 2004. The oldest ape. *Evolutionary Anthropology* 13:90–103.

McBrearty, S., and N. G. Jablonski. 2005. First fossil chimpanzee. *Nature* 427:105–108.

McHenry, H. M., and K. Coffing. 2000. *Australopithecus* to *Homo*: Transformations in body and mind. *Annual Review of Anthropology* 29:125–146.

Morbeck, M. E. et al. (eds.). 1997. *The Evolving Female*. Princeton, NJ: Princeton University Press.

Morwood, M. J. 2004. Further evidence for small-bodied hominins from the Late Pleistocene of Flores, Indonesia. *Nature* 437:1012–1071.

Moyà-Solà, S. et al. 2004. *Pierlapithicus catalunicus*, a new Middle Miocene great ape from Spain. *Science* 306:1339–1344.

Niven, J. E. 2006. Brains, islands and evolution: Breaking all the rules. *Trends in Ecology and Evolutionary Biology* 22:57–59.

Partridge, T. C. et al. 2003. Lower Pliocene hominid remains from Sterkfontein. *Science* 300:607–612.

Patterson, N. et al. 2006. Genetic evidence for complex speciation of humans and chimpanzees. *Nature* 441:1103–1108.

Peccei, J. S. 2001. Menopause: Adaptation or epiphenomenon? *Evolutionary Anthropology* 10:43–57.

Preuschoft, H. 2004. Mechanisms for the acquisition of habitual bipedality: Are there biomechanical reasons for the acquisition of upright bipedal posture? *Journal of Anatomy* 204:363–384.

Pruetz, J. D., and P. Bertolani. 2007. Savanna chimpanzees, *Pan troglodytes verus*, hunt with tools. *Current Biology* 17:412–417.

Rantala, M. J. 2007. Evolution of nakedness in *Homo sapiens*. *Journal of Zoology* 273:1–7.

Rogers, A. R. et al. 2004. Genetic variation at the *MC1R* locus and the time since loss of body hair. *Current Anthropology* 45:105–108.

Ross, C. 2000. Into the light: The origin of Anthropoidea. *Annual Review of Anthropology* 29:147–194.

Ruvulo, M. 1997. Genetic diversity in hominoid primates. *Annual Review of Anthropology* 26:515–540.

Senut, B. et al. 2001. First hominid from the Miocene (Lukeino Formation, Kenya). *Comptes Rendus de l'Academie des Sciences, Paris, Earth and Planetary Sciences* 332:137–144.

Shoenemann, P. T. 2006. Evolution of the size and functional areas of the human brain. *Annual Review of Anthropology* 35:379–406.

Sieffert, E. R. et al. 2005. Basal anthropoids from Egypt and the antiquity of Africa's higher primate radiation. *Science* 310:300–304.

Sponheimer, M., and J. Lee-Thorpe. 1999. Isotopic evidence for the diet of an early hominid, *Australopithecus africanus*. *Science* 283:368–370.

Spoor, F. et al. 1996. Evidence for a link between human semicircular canal size and bipedal behavior. *Evolutionary Anthropology* 30:183–187.

Spoor, F. et al. 2007. Implications of new early *Homo* fossils from Ileret, east of Lake Turkana, Kenya. *Nature* 448:688–691.

Stanley, S. M. 1996. *Children of the Ice Age*. New York, NY: Harmony Books.

Steudal-Numbers, K. 2003. The energetic cost of locomotion: Humans and primates compared to generalized endotherms. *Journal of Human Evolution* 44:255–262.

Stringer, C. B., and R. McKie. 1996. *African Exodus*. New York, NY: Henry Holt.

Surridge, A. K. et al. 2003. Evolution and selection of trichromatic vision in primates. *Trends in Ecology and Evolutionary Biology* 18:198–205.

Suwa, G. et al. 2007. A new species of great ape from the late Miocene Epoch in Ethiopia. *Nature* 448:921–924.

Tattersall, I. 1993. *The Human Odyssey: Four Million Years of Human Evolution*. Upper Saddle River, NJ: Prentice Hall.

Tattersall, I. 2003. Once we were not alone. *Scientific American* 13(2):21–27.

Tattersall, I. 2003. Out of Africa again—and again? *Scientific American* 13(2):38–45.

Thorpe, S. K. S. et al. 2007. Origin of human bipedalism as an adaptation for locomotion on flexible branches. *Science* 316:1328–1331.

Tocheri, M. W. et al. 2007. The primitive wrist of *Homo floresiensis* and its implications for hominin evolution. *Science* 317:1743–1745.

van Schaik, C. P. et al. 2003. Orangutan cultures and the evolution of material culture. *Science* 299:102–105.

Vekua, A. et al. 2002. A new skull of early *Homo* from Dmanisi, Georgia. *Science* 297:85–89.

Walker, A., and P. Shipman. 1996. *The Wisdom of the Bones: In Search of Human Origins*. New York, NY: Alfred A. Knopf.

Walsh, P. D. et al. 2003. Catastrophic ape decline in western equatorial Africa. *Nature* 422:611–614.

White, T. D. et al. 1994. *Australopithecus ramidus*, a new species of early hominid from Aramis, Ethiopia. *Nature* 371:306–312.

White, T. D. et al. 2003. Pleistocene *Homo sapiens* from Middle Awash, Ethiopia. *Nature* 423:742–747.

White, T. D. et. al. 2006. Asa Issie, Aramis, and the origin of *Australopithecus*. *Nature* 440:883–889.

Whiten, A. and C. Boesch. 2001. The cultures of chimpanzees. *Scientific American* 284(1):60–67.

Wong, K. 2003. An ancestor to call our own. *Scientific American* 13(2):4–13.

Wong, K. 2003. Who were the Neandertals? *Scientific American* 13(2):28–27.

Wood, B., and M. Collard. 1999. The changing face of the genus *Homo*. *Evolutionary Anthropology* 8:195–207.

Zihlman, A. L. et al. 1978. Pygmy chimpanzee as possible prototype for the common ancestor of humans, chimpanzees and gorillas. *Nature* 275:744–746.

CHAPTER 25

The Impact of Humans on Other Species of Vertebrates

A review of the biology of vertebrates must include consideration of the effect of the current dominance of one species, *Homo sapiens*, on other members of the vertebrate clade. Never before in the history of the Earth has a single species so profoundly affected the abundance, and even the prospects for survival, of other species.

Some of the influence of humans derives from the size of our population—more than 6 billion now and projected to increase to more than 9 billion in 2050—and our worldwide geographic distribution. However, other species of vertebrates match or surpass humans in numbers, and some at least come close to matching us in geographic distribution. Technology and the consequences of advances in technology are what set humans apart from other vertebrates. Human technology began with stone tools some 2.5 million years ago, and increased consumption of resources has accompanied increased technology. Even existing primitive human societies use more resources than do other species of animals. Extraction and consumption of energy and other resources by human societies has worldwide impacts ranging from oil spills and contamination of soil and water with heavy metals to the loss of wild habitats to agriculture and urbanization. Per capita energy use has tripled since 1850, and most of the increase has occurred in the past few decades. The most highly industrialized societies have the greatest per capita rates of consumption of energy and other resources. The United States, for example, has about 5 percent of the world's population and accounts for about 25 percent of the world's consumption of resources. The developed nations as a group have 20 percent of the world's population and are responsible for 60 percent of resource consumption. Societies with high rates of consumption have correspondingly high rates of production of greenhouse gases and waste products.

Less technologically developed societies are not necessarily ecologically benign, however. They present special problems as growing human populations increasingly impinge on areas that have so far been relatively undisturbed. Wealthy nations at least have the resources to control pollution and create national parks, monuments, and wildlife reserves. Poorer societies struggle daily to meet their basic survival needs and understandably regard conservation as a luxury beyond their reach. Perhaps the most insidious process affecting the relationship between humans and other vertebrates is the spread of Western cultural values that emphasize material possessions, thereby increasing demand for consumer products and expanding the geographic influence of high-impact societies.

This book has emphasized the evolutionary history of vertebrates and their characteristics as organisms, and both kinds of information are essential parts of efforts to conserve natural habitats and to protect endangered species. We tend to assume that the natural state of an environment is the way it was at the time of the first written record, but fossil evidence shows that humans had enormous impacts on vertebrate faunas long before writing was invented some

5000 years ago. Extinctions of vertebrates due to human activities began about 50,000 years ago, and they have accelerated ever since.

The scientific study of vertebrates dates back only a few hundred years, but it, too, has accelerated rapidly. Information about the biology of vertebrates can be used to identify causes of population declines and extinctions and possibly to prevent some species that are currently endangered from becoming extinct.

25.1 Humans and the Pleistocene Extinctions

Starting from the appearance of the earliest vertebrates in the Late Cambrian or Ordovician periods, the diversity of vertebrates increased slowly through the Paleozoic and early Mesozoic eras, and then more rapidly during the past hundred million years. This overall increase has been interrupted by eight periods of extinction for aquatic vertebrates and six for terrestrial forms. Extinction is as normal a part of evolution as species formation, and the duration of most species in the fossil record appears to be from 1 million to 10 million years. Periods of major extinction (a reduction in diversity of 10 percent or more) are associated with changes in climate and the consequent changes in vegetation. But that pattern of reasonably long-lived species and extinctions correlated with shifts in climate and vegetation changed at about the time that humans became a dominant factor in many parts of the world.

For example, the number of genera of Cenozoic mammals reached a peak in the mid-Miocene and a second peak in the early Pleistocene epochs. Only 60 percent of the known Pleistocene genera are living now, and the extinct forms include most species of large terrestrial mammals—those weighing more than 20 kilograms. These large mammals, plus some enormous species of birds and reptiles, are collectively called the Pleistocene megafauna. The term is most commonly applied to North American species such as ground sloths, mammoths, mastodons, and the giant Pleistocene beaver that was the size of a bear, but other continents also had megafaunas that became extinct in the Pleistocene.

The first humans to reach Australia were met by a megafauna that included representatives of four groups: marsupials, flightless birds, tortoises, and echidnas. The largest Australian land animals in the Pleistocene were several species of herbivorous mammals in the genus *Diprotodon*, the largest of

which probably weighed about 2000 kilograms. The largest kangaroos weighed 200 kilograms, and one of them may have been carnivorous. The largest echidnas reached weights of 20 to 30 kilograms and were waist high to an adult human. The flightless bird *Genyornis newtoni* was twice the height of a human, and the horned turtles in the family Meilanidae were nearly as large as Volkswagen Beetle cars. Perhaps the most dramatic species in the Australian megafauna was a monitor lizard at least 6 meters long (as large as a medium-sized *Allosaurus*). The turtles, the monitor lizard, *Genyornis*, and all marsupial species weighing more than 100 kilograms became extinct in the late Pleistocene.

The role that humans played in the extinctions of large animals in the Pleistocene was discussed in Section 19.4. Specialists agree that the arrival of a new, technologically advanced species of predator would inevitably have an impact on other species, but the relative importance of the direct effect of hunting versus indirect effects resulting from changes in the habitat and the introduction of new pathogens remains a subject of debate.

The most striking evidence of the impact of humans on other vertebrates is the apparent correspondence in the times of the arrival of humans on continents and islands and the extinction of the megafauna (**Figure 25–1**). The earliest extinctions appear to have occurred in Australia sometime between 50,000 and 40,000 years ago. In North and South America, at least eight species of large mammals survived until about 10,000 years ago. Humans colonized islands later than continents, and extinctions occurred between 10,000 and 4000 years ago on islands in the Mediterranean Sea, 4000 years ago on islands in the Arctic Ocean north of Russia, 2000 years ago on Madagascar, and only a few hundred years ago on islands in the Pacific Ocean. In each of these cases, the dates of extinctions closely follow the dates when humans are believed to have arrived.

■ The Overkill Hypothesis

Overhunting has been proposed as the primary reason for megafaunal extinctions, and some evidence for overhunting is dramatic. Eleven species of moa (giant flightless birds) existed on New Zealand when the Maori arrived in the late thirteenth century, and all appear to have been extinct within 100 years (Worthy and Holdaway 2002). The role of Maori hunters in the extinction of moa is amply documented. On the North Island of New Zealand a butchering site was discovered on the sand dunes at

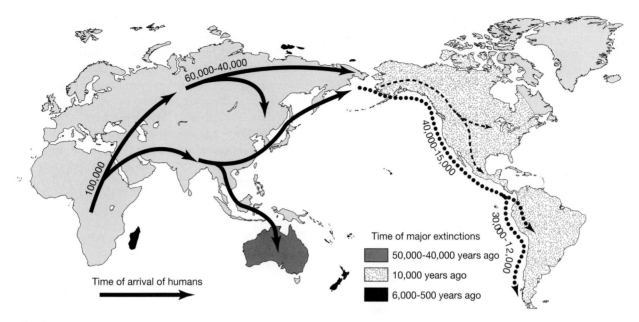

▲ **FIGURE 25–1** The correspondence between the time scale of the spread of human populations and extinction of native species of vertebrates. Modern humans left Africa about 100,000 years ago and spread eastward across Eurasia and into the Americas. The dates of extinctions of Pleistocene megafauna correspond with estimates of when humans arrived in Australia, the Americas, and major islands such as New Zealand, Madagascar, and the islands north of Siberia.

Koupokonui. The remains of hundreds of individuals of three species of moa were found in and around ovens. Uncooked moa heads and necks had been left in piles to rot, while the legs were roasted. At Wairau Bar on the South Island the ground is littered with the bones of moa, an estimated 9000 individuals plus 2400 eggs, and at Waitaki Mouth are the remains of an estimated 30,000 to 90,000 moa (Holdaway and Jacomb 2000).

The best-document examples of the impact of overhunting come from islands and are relatively recent. It is less certain that hunting alone was responsible for continent-wide megafaunal extinctions. Although hunting was probably part of the equation, diseases and habitat changes also played a role.

■ Disease

Although there is no paleontological evidence of disease in the Pleistocene megafauna, modern examples of transmission of disease from domestic animals to wild species abound, and emerging infectious diseases now threaten both wild animals and humans (Daszak et al. 2000). Within the past two decades, lions and wild dogs in Africa have been infected by canine distemper transmitted from domestic dogs. Wild dogs (*Lycaon pictus*) are now nearly extinct in

the Serengeti Plains. Less than a century ago, packs contained 100 or more animals, but today packs consist of only 10 or so adults. The population of wild dogs on the Serengeti is less than 60 animals, and the entire surviving population of the species is no more than 5000 individuals.

Other examples of the transmission of diseases from humans and our pets to wild animals can be cited: rabies has killed a substantial proportion of the approximately 500 remaining Ethiopian wolves, the parasite *Toxoplasma gondii* from domestic cat feces has killed large numbers of sea otters, and Ebola virus has been responsible for the deaths of thousands of gorillas and chimpanzees (Walsh et al. 2003, Whitfield 2003). Wild mountain gorillas in Uganda have contracted mange from parasitic mites in clothing discarded by tourists; and, if measles or tuberculosis gets into the gorilla population, the results will be devastating. A new twist has been added by the discovery that human influenza B virus caused a respiratory infection in harbor seals on the Dutch coast, and the seals are now a reservoir for the virus that could pose a threat to human health.

A newly introduced disease does not kill every individual of a previously healthy population. Some individuals survive, perhaps because they are resistant or maybe just because they are lucky enough to avoid infection. Nonetheless, a disease that drastically reduces the number of individuals of a species may start a process that leads to extinction, and this may be happening to some populations of African wild dogs. The hunting method of wild dogs—prolonged pursuit of antelope until an individual is captured—is energetically expensive. A pack of wild dogs hunts cooperatively, with one individual taking up the chase as another tires. The drastic reduction that has occurred in pack size means that each dog must work harder. To make things worse, spotted hyenas (*Crocuta crocuta*) steal the kills made by wild dogs, and a small pack of dogs probably has more difficulty defending its kills from hyenas than does a larger pack. Wild dogs normally hunt for about 3.5 hours per day, and calculations of the energy cost of hunting and the energy gained from prey show that they just meet their daily energy needs on this schedule. If hyenas steal some of the kills, the wild dogs must increase the time they spend hunting. A 10 percent loss of prey would force the wild dogs to double their hunting time, and a 25 percent loss would force them to spend 12 hours a day hunting. Wild dogs are already working at nearly their physiological limits when they hunt for 3.5 hours per day, and they probably cannot survive if they lose much food to hyenas. Thus, a drastic reduction in pack size sets the stage for a competitive interaction with hyenas, and this interaction could be the factor that drives a pack of wild dogs to extinction (Gorman et al. 1998).

■ Fire

With the arrival of modern humans came the use of fire to manipulate the habitat. *Genyornis newtoni* was a flightless bird that inhabited inland plains and some coastal areas of Australia when the first humans arrived about 50,000 years ago. Although *Genyornis* was a ponderous bird and was probably less fleet-footed than emus (*Dromaius novaehollandiae*), there is only one site known with evidence that humans hunted *Genyornis*. The reason that *Genyornis* became extinct and emus survived may lie in their feeding habits. Emus eat a wide variety of items including grasses, whereas the chemical composition of *Genyornis* eggshells suggests that they were browsers, eating leaves from shrubs. Wildfires were a part of the Australian landscape long before humans arrived, but humans changed the fire regime. Natural fires occur during the dry season and do not recur until the vegetation in a burned area has regenerated. The early human inhabitants of Australia may have set fires at other times of the year and at shorter intervals than the natural fire cycle. A regime of more frequent burning would have converted the shrub lands that *Genyornis* depended on to the grasslands and spinifex that characterize the inland Australian plains today. Thus, habitat change produced by the new fire regime created by humans may have been responsible for the extinction of *Genyornis*. The large herbivorous mammals that became extinct in Australia were also browsers, and the same habitat changes may have been responsible for their disappearance.

25.2 Humans and Recent Extinctions

As we move closer to the present, the role of humans as the major cause of extinction becomes unambiguous. Excavation of fossils preserved in tubelike lava caves formed by volcanoes show that the Hawaiian Islands probably had more than 100 species of native birds when Polynesian colonists arrived about 1700 years ago. By the time European colonists reached Hawaii in the late eighteenth century, that number had been reduced by half.

The Age of Exploration, which began in the fifteenth century, brought sophisticated weapons and commercial trade to areas that had known only stone tools and hunter-gatherer economies. Ships sailed from Europe to all corners of the world, stopping en route to renew their supplies of food and water from oceanic islands. Not surprisingly, extinctions on islands began about two centuries before extinctions on continents (**Figure 25–2**). Notable examples include the dodo (*Raphus cucullatus*), a flightless bird related to pigeons that lived on the island of Mauritius in the Indian Ocean. The dodo was last seen alive in 1662 and was almost certainly extinct by 1690. In the Hawaiian Islands about one-third of the native species of birds that were still surviving when Captain Cook arrived in the late eighteenth century are now extinct.

Animal species continue to become extinct today, and a worldwide survey of extinctions since the start of European colonization reveals two trends: island extinctions began almost two centuries earlier than continental extinctions, and both island and continental extinctions have increased rapidly from the early or mid-nineteenth century through the twentieth century (World Conservation Monitoring Centre 1992). More than 800 species have become extinct in the past 500 years, a rate of extinction that is thought

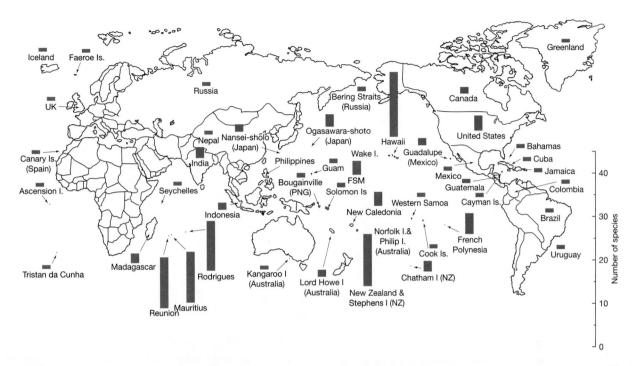

▲ **FIGURE 25–2** The numbers of confirmed extinctions of species of birds since 1600. Islands have suffered more extinctions than continental areas.

to be 1,000 to 10,000 times higher than it would be without the effect of humans. The increasing application of biotechnology and genetic engineering is creating a new category of risks that go beyond the historical concerns with pollution and disease to include direct interference with natural selection and evolution (**Box 25–1**).

The IUCN (World Conservation Union) has summarized the best information available about the conservation status of animals. The 2006 edition of the IUCN Red List of Threatened Species lists a total of 16,118 species of plants and animals currently at risk of extinction. They place species in categories of risk by using criteria that focus on the absolute size of wild populations and changes in the populations in the past 10 years.

Two categories of *Extinct* species are defined:

- **Extinct**—A species is extinct when no living individuals exist. The golden toad (*Bufo periglenes*) from the montane cloud forest of Costa Rica was first described in 1967 and had vanished by 1989. Climate change and infection by chytrid fungi have been proposed as the causes of its extinction.
- **Extinct in the Wild**—A species is Extinct in the Wild when it is known to survive only in cultivation, in captivity, or as a naturalized population

(or populations) well outside the past range. The black-footed ferret (*Mustela nigripes*) was a resident of prairie dog towns that once covered thousands of hectares in North America. Habitat loss as land was converted to agricultural use, prairie dog extermination programs, and bubonic plague has reduced the extent of prairie dog colonies to less that 2 percent of their original area, and black-footed ferrets have suffered from those changes. The last wild ferrets were taken into captivity in the mid-1980s. A captive breeding program was initiated, and more than 200 captive-bred ferrets have been released since 1991. The reintroduced populations of ferrets are growing, and the prospects for successful reestablishment appear to be good (Grenier et al. 2007).

Three categories of *Threatened* species are defined:

- **Critically Endangered**—A species is Critically Endangered when it is facing an extremely high risk of extinction in the wild in the immediate future. The common sturgeon (*Acipenser sturio*) is a large fish (up to 3 meters and 300 kilograms). Its historic range included the entire coastline of Europe from the North Cape to the Baltic, Mediterranean, and Black Seas. The species is

BOX 25–1 Fish Farming

As populations of commercially important fish species have plummeted because of overfishing, the role of fish farming has increased. Fish farms are huge netted enclosures that are anchored in rivers or in coastal waters (**Figure 25–3a**). Young fish from a hatchery are placed in the enclosures and fed until they have grown to harvestable size. Although fish farming is promoted as an ecologically sound alternative to catching wild fish, it creates both short-term and long-term problems.

Pollution and Disease In the short-term, concentrating thousands of fish in a small volume of water and feeding them artificial food creates a tremendous concentration of feces and uneaten food that promotes blooms of bacteria and algae. In addition, the fish crowded into the pens are susceptible to disease and parasites that can be transferred to wild fish. Sea lice are small copepod crustaceans that parasitize fish, and farmed salmon are heavily infected (**Figure 25–3b**). Worse still, when wild salmon swim past the farm pens they become infected with sea lice—a recent study by Martin Krkošek found that a single fish farm in British Columbia increased the infection pressure for juvenile wild salmon by 73 times and this effect extended for 30 km along the migration paths the wild fish followed (Krkošek et al. 2007a).

In a follow-up study, Krkošek and his colleagues analyzed 35 years of records of wild salmon populations, comparing

(a)

(b)

(c)

▲ **FIGURE 25–3** Fish farming. (a) The netted enclosures of an Atlantic salmon farm in Burdwood Islands, British Columbia. (b) Sea lice (small copepod crustacean parasites) on hatchling salmon. The two stringlike structures extending from one of the parasites are paired egg sacks. (c) Comparison of a growth-enhanced, transgenic Atlantic salmon (*Salmo salar*) with a normal fish. At 18 months, the transgenic fish is much larger than the same-age normal fish.

seven rivers that flow into channels with fish farms with 64 rivers in which the wild salmon are not exposed to fish farms (Krkošek et al. 2007b). Until 2001, when sea lice appeared in the fish farms, there were no differences in the year-to-year population changes in salmon from the two groups of rivers. From 2001 onward, however, the wild salmon populations exposed to fish farms shrank every year, whereas the salmon populations that are not exposed to fish farms remained constant. The wild populations exposed to fish farms are shrinking so fast that, if the current trend continues, they will be 99 percent gone in four generations.

Depletion of Commercial Fish Stocks The rationale for fish farming is that the availability of farmed fish will reduce the need to catch wild fish. This reasoning is only partly correct, however, because farmed fish are raised on fish meal that is produced by catching small fish and invertebrates. In 2002, nearly one-third of the total catch of marine life was used to produce fish oil and fish meal, and 70 percent of the oil and 34 percent of the meal were used for aquaculture (Tuominen and Esmark 2003). By 2010, those proportions are expected to rise to more than 80 percent of fish oil and 50 percent of fish meal.

Farmed salmon convert about 25 percent of the biomass of their food into new salmon biomass, so 4 kilograms of fish meal and oil are required to produce 1 kilogram of salmon. In other words, the best one can say of fish farming is that it redirects the target of commercial fishing from the large predatory species of fish that humans eat to the small prey species that the large species feed on. However, that may not be a good exchange because depleting populations of prey species is likely to deprive other species of predatory fish of their food base.

Genetic Change and Evolution The conspicuous problems associated with fish farming, such as pollution, disease, parasites, and depletion of prey species are not the only concerns. Recent work has focused attention on the interaction of evolutionary processes with ecological and commercial elements of fishery biology (Waples and Hendry 2008). For example, the reproductive strategies of farmed and wild salmon are different, and subtle effects on wild fish populations may result when farmed fish escape from the pens and mix with wild fish. One of the general features of the life history of animals is a trade-off between the number of eggs

produced and the size of each egg—large eggs produce large hatchlings that can survive the risks that wild hatchlings face. In contrast, hatchlings of farm-raised salmon are protected from most of the hazards that confront wild hatchlings. As a result, the balance between egg size and egg number in farmed fish is shifted toward many small eggs—in just four generations, the egg size of farmed salmon decreased by 25 percent (Heath et al. 2003).

The change in egg size might not be a problem if farmed fish remained in farms, but not all of them do—escapes are a regular occurrence and sometimes involve large numbers of fish. A hundred thousand farmed salmon escaped from pens in Machias Bay, Maine, during a single storm in December 2000, and the Norwegian Directorate of Fisheries estimated that more than 790,000 fish escaped from Norwegian fish farms in 2006. Fish that escape from farms enter the wild salmon population, bringing with them genotypes that produce small eggs. The influence of this sudden genetic input on wild populations of salmon is unknown, but it is potentially deleterious for the wild genotype that has been shaped by millennia of selection.

Genetic engineering, inserting foreign genes that produce commercially desirable characters, is widely used in crops like corn and soybeans. No genetically engineered fish are farmed commercially yet, but transgenic salmon are being studied in the United States and Europe (Stokstad 2002). Adding a gene for the production of growth hormone from chinook salmon to the genome of Atlantic salmon yields fish that grow as much as six times faster than normal farmed salmon and reach market size a year earlier (**Figure 25–3c**).

What will happen if transgenic fish escape from pens? An experimental study used brown trout with implants that release growth hormone gradually because transgenic fish cannot legally be released into the wild (Johnsson and Björnsson 2001). This study showed that the growth-enhanced fish survived as well as normal fish, increasing concern that escaped transgenics would compete with wild fish. Furthermore, transgenic fish incorporate the foreign gene in their eggs and sperm, so they pass it on to their offspring. A model of the population genetics of a mixed population of wild and transgenic fish shows that a gene that reduces survival of juveniles but increases the reproductive success of the ones that do survive would spread through a population and ultimately cause the extinction of both the wild and transgenic fish (Muir and Howard 1999).

now extinct in some of its former spawning rivers, including the Elbe, Rhine, and Vistula. Breeding populations are restricted to a few European rivers—the Gironde in France, the Guadalquivir in Spain, and the lower Danube in Romania and Bulgaria. Overharvesting is the primary cause of its decline. The flesh is prized and the roe used to

make caviar, so gravid females are especially sought and are killed before they can reproduce.

- **Endangered**—A species is Endangered when it is facing a very high risk of extinction in the wild in the near future. Examples of endangered species include the giant panda (*Ailuropoda melanoleuca*), which was once distributed through Myanmar, northern Vietnam, and a large part of eastern and southern China. Wild pandas now occur only in fragmented populations in mountain ranges in western China; the total population is thought to be about 1200 animals. Attempts to breed giant pandas in captivity outside of China have been generally unsuccessful.

- **Vulnerable**—A species is Vulnerable when it is facing a high risk of extinction in the wild in the medium-term future. The great white shark (*Carcharodon carcharias*) has a worldwide distribution in warm and temperate seas. As a top predator, it has always had a low population density, and overfishing is considered a potential threat.

Three categories of *Lower Risk* are recognized:

- **Conservation Dependent**—These are species that are being sustained by ongoing conservation programs; without those programs the species would likely qualify for one of the threatened categories within five years.
- **Near Threatened**—These are species that are close to Vulnerable status for which no conservation measures are in place.
- **Least Concern**—Species that do not qualify for Conservation Dependent or Near Threatened classification.

We simply lack enough information to reach a conclusion about the status of many species that have been evaluated, and these are placed in a category of their own:

- **Data Deficient**—A species is listed as Data Deficient when information about the species' distribution and abundance is insufficient to assess its risk of extinction. Listing a species in this category emphasizes the need for more information. Only species that have been evaluated can be included in this category, and the selection of a species for evaluation indicates that it may be at risk.

A summary of the status of vertebrates in the *2006 IUCN Red List of Threatened Species* shows that from 12 percent to more than 50 percent of the species of vertebrates evaluated are extinct, endangered, or vulnerable (**Table 25.1**). The proportion of mammals, bird, reptiles, and fishes in these categories has changed little since the 2003 issue of the Red List, but the number of amphibians increased more than 10-fold in three years, from 164 species in 2003 to 1811 species in 2006. This dramatic increase is largely the result of the spread of chytrid infections (see Section 10.7).

It is not easy to calculate the rate at which extinctions are occurring, and different assumptions can

TABLE 25.1 Numbers of species of vertebrates that are listed as extinct, endangered, or vulnerable in the *2006 IUCN Red List of Threatened Species* (http://www.iucnredlist.org/info/stats)

Class	Number of Species Evaluated	Recently Extinct	Extinct in the Wild	Critically Endangered	Endangered	Vulnerable	Percent of Species Evaluated that Are Extinct or Endangered	Data Deficient
Mammals	4856	70	4	162	348	583	22	385
Birds	9934	135	4	181	351	674	12	74
Reptiles	664	22	1	73	101	167	51	62
Amphibians	5918	34	1	442	738	631	31	1426
Sarcopterygians		0	0	1	0	0		0
Actinopterygians	All fishes combined 2914	80	13	232	212	614	40	688
Elasmobranchs (= Neoselachii—sharks, skates, and rays)		0	0	20	25	67		205

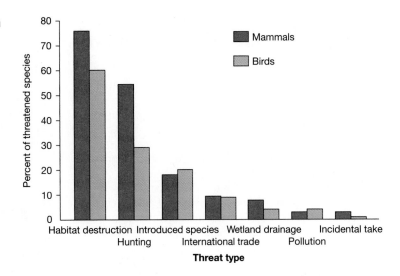

▶ **FIGURE 25-4** The major threats affecting birds (on a worldwide basis) and mammals (Australasia and the Americas). Habitat destruction is the single most important threat for both kinds of animals, affecting 60 percent of birds and 76 percent of mammals. Hunting (for food and sport) is a greater threat to mammals than to birds. Introduced species may be predators or competitors, and *international trade* refers to commercial exploitation for fur, feathers, and the pet trade. *Incidental take* is the term used to designate accidental mortality, such as dolphins that are drowned by boats fishing for tuna.

produce different values. What is clear is that destruction of habitat is the major threat, affecting 60 percent of threatened species of birds and nearly 80 percent of threatened species of mammals (**Figure 25–4**).

25.3 Organismal Biology and Conservation

The Species Survival Commission (SSC) of IUCN organizes much of the scientific information employed in conservation programs. The SSC is a worldwide network of 7000 wildlife biologists, wildlife veterinarians, zoo biologists, marine biologists, and academic scientists with specialized knowledge of groups of animals and plants. The commission publishes SSC Action Plans that describe the current situation for species or groups of species, identifying the major threats, and proposing mechanisms to address those issues. More than 60 Action Plans have been published, and many of them are available online at the IUCN website (http://www.iucn.org).

Developing management plans for endangered species requires enormous amounts of information about the basic biology of the species concerned. Without a thorough understanding of all aspects of a species' biology, well-intentioned management efforts can be ineffective or even have negative effects on the species. In the following sections we describe several examples of situations in which an understanding of ecology, behavior, or physiology of a species is central to effective management.

What Is Critical in a Critical Habitat?

Federal law and some state laws require assessment of the habitat requirements of species that are considered at risk, and wildlife biologists may be charged with the responsibility of determining the critical habitat for a species. In biological terms, the elements of its habitat that are critical for the success of a species are likely to be subtle and complex (**Box 25–2**), and the differences between legislative and biological perspectives can create tension and conflict.

Critical Habitat for Migratory Animals

The problems of identifying and then protecting critical habitat are particularly severe for migratory species, for which even the most superficial definition of critical habitat must include both ends of their migratory paths and the areas in which they stop to rest and feed as they move back and forth. To complicate conservation efforts even more, migratory animals usually move through several national jurisdictions and may pass over or through international waters where no national jurisdiction exists. The legislative and diplomatic effort required to protect these species is enormous and must be based on sound biological information.

Gray Whales The migration of the gray whale (*Eschrichtius robustus*) from its summer feeding grounds north of the Arctic Circle to breeding sites in Ojo de Liebre, Bahía Magdalena, and San Ignacio Lagoons on the Pacific coast of Baja California was

BOX 25–2 Subtle Elements of Critical Habitats

Complex interactions between the elements of a habitat and the needs of a species may determine the survival or extinction of populations of threatened species, but they can be detected only by careful study and application of basic biological information. The desert tortoise (*Gopherus agassizii*) of North America provides an example of this sort of interaction (**Figure 25–5**).

Desert tortoises are herbivores and they lack salt-excreting glands. Plants contain higher concentrations of potassium than do animals, so herbivorous animals must excrete some of the excess potassium they get in their food. Desert tortoises excrete potassium along with nitrogenous wastes in the form of salts of uric acid. Because potassium excretion is chemically linked to nitrogen excretion, an excess of potassium in a tortoise's diet robs the tortoise of nitrogen. That is, a tortoise on a high-potassium diet uses so much nitrogen getting rid of the excess potassium that it does not have enough nitrogen for protein synthesis (Oftedal and Allen 1996).

Feeding-choice trials with captive tortoises showed that tortoises selected the diet with the lowest potassium. When tortoises were offered only high-potassium diets, they reduced the amount of food they ate. Wild tortoises may be able to survive only if plants with favorable ratios of nitrogen to potassium are available.

Plant species vary in potassium content, and two changes that have occurred in the tortoises' habitat in the past 200 years may have reduced the availability of low-potassium plants:

- **Ranching**—The introduction of commercial grazing was the first change. Prior to European settlement, tortoises, rabbits, hares, deer, and antelope shared the open range, and their population densities were low because there was not much food in the habitat. This situation changed in the 1800s when ranchers introduced herds of cattle or sheep at densities far higher than the arid grasslands and deserts could support. The damage that livestock grazing causes in the arid Southwest is obvious to any ecologist, although lobbyists for the ranching industry vehemently deny it at legislative and regulatory hearings. Probably cattle, sheep, and tortoises all prefer plants that have low potassium and

▲ **FIGURE 25–5** A desert tortoise in the Mojave Desert. This tortoise is feeding on yellow fiddlehead (*Amsinckia menziesii*), a native species of plant that is an important component of the tortoises' diet in March and April.

high nitrogen contents; and, if this is the case, high densities of grazing mammals may reduce the availability of high-quality plants for tortoises.

- **Alien Plants**—In many parts of North America, including the southwestern deserts where desert tortoises occur, Eurasian plants that accompanied the spread of European settlers have replaced some native species of plants. These Eurasian plants appear to have higher potassium and lower nitrogen levels than the native plants they have replaced. If this is true, a desert habitat with even a dense growth of Eurasian plants would not provide a diet on which tortoises could survive and grow.

Both of these concerns are in the hypothesis stage now—they are potential problems but we lack data to decide whether they are real problems. Measuring the impact of commercial grazing and alien plants on tortoises and other native desert herbivores will require the combined efforts of ecologists and botanists working in the field and nutritionists and physiologists working in the laboratory.

described in Section 22.3. The adult whales apparently make this 9000-kilometer migration because newborn calves need the warm waters of the lagoons to maintain stable body temperatures (**Figure 25–6**). Thus, the lagoons are an essential part of the gray whale's habitat. More than 300 other species of animals also depend on the lagoons, and a thriving shellfishing industry provides livelihood for local

communities. The United Nations has declared the lagoons a World Heritage Site, and both Ojo de Liebre and San Ignacio have been protected by Mexican legislation since 1972 because of the role they play in the life cycle of whales. In 1988, Mexico included the lagoons in the Vizcaino Biospheric Preserve.

Consequently, in 1994 when Salt Exporters, Inc.— a company jointly owned by Mitsubishi Corporation

▶ **FIGURE 25–6** A mother gray whale and her calf in San Ignacio Lagoon.

and the Mexican government—announced plans to build a plant to extract salt from seawater at San Ignacio Lagoon, the news was greeted with protests and demonstrations locally and by environmentalists around the world. The proposed plant, which would cover 300 square kilometers (116 square miles) in the heart of the lagoon system, would supply all of Japan's industrial salt needs for the manufacture of plastic and chlorine. The proposed plant would draw nearly 23,000 liters (6000 gallons) of water per minute from the lagoon and would have an enormous environmental impact. Salt Exporters already operates a salt plant at nearby Guerro Negro, and environmentalists blame that plant for the deaths of sea turtles by salt poisoning.

Environmental organizations around the world cooperated in opposing plans for the San Ignacio plant, sponsoring a letter-writing campaign that generated nearly a million letters of protest, promoting boycotts of Mitsubishi products by consumers and of Mitsubishi stock by mutual funds, and enlisting the help of governments to apply political pressure. In March 2000, after more than five years of pressure, Mitsubishi and the Mexican government announced that they were abandoning the proposal to build the plant.

Songbirds Birds are the best-known migratory animals, and migration is a central feature of the biology of many species. Loss of habitat in their summer and winter ranges and at stopping places along the migratory routes appears to be contributing to declining populations of shore birds and songbirds that migrate between summer ranges in North America and winter ranges in the tropics.

About 350 species of songbirds occur in North America, and about 250 of them spend their winters in the New World tropics, which extend from southern Mexico through Central and northern South America and into the West Indies. The remaining 100 species of North American songbirds are either year-round residents in northern habitats or migrate only short distances south of their summer ranges. Neotropical migrants compose the majority of species of songbirds in most habitats in North America. In most wooded areas, half of the breeding species are Neotropical migrants and in some northern regions of North America, more than 90 percent of the songbirds are migrants.

Bird watchers have a nationwide system of local bird censuses; and, starting in the 1970s and 1980s, these counts revealed dramatic decreases in the numbers of some Neotropical migrants such as wood thrushes (*Hylocichla mustelina*) and cerulean warblers (*Dendroica cerulea*) in certain areas—as much as 1 percent annually for the past 30 years. In contrast, populations of year-round residents and short-distance migrants, such as chickadees (*Parus atricapillus*) and northern cardinals (*Cardinalis cardinalis*), were stable or even increasing. The pattern of decline is not consistent across species or regions of North America, however. Populations of some species of Neotropical songbirds are declining, but others are increasing. Populations of songbirds in the forested areas of the eastern United States are declining, but those in the West are not.

▲ **FIGURE 25–7** Nest parasitism by brown-headed cowbirds. (a) A female cowbird about to deposit an egg in the nest of a blue grosbeak. (b) Two cowbird eggs in a robin's nest. The parent birds do not remove the cowbird eggs, even though they are speckled brown and white and look nothing like the plain blue robin egg. (c) An adult yellow warbler feeding a cowbird chick that is nearly as large as the foster parent.

This complex picture results from the variety of factors that affect migratory birds in their summer and winter ranges and in stopover points on their routes of migration. Changes in land use are probably the primary reason for declining bird populations. Neotropical migrants breed in their summer ranges, and fragmentation of forests into smaller and smaller patches has reduced the total amount of breeding habitat available and changed the nature of the habitat that remains. Nests near the edges of woodlands generally suffer higher rates of failure than do those nearer the center, and as forest patches grow smaller a greater proportion of the habitat is near an edge.

Part of the increased rate of nest failure near the edges of woodlands may result from nest parasitism by brown-headed cowbirds (*Molothrus ater*), which are obligate nest parasites (**Figure 25–7**). Cowbirds do not build their own nests but lay their eggs in the nests of other species; some 200 species of birds have been reported to be parasitized by cowbirds. A female cowbird lays 20 to 40 eggs, one or two in each nest, and a few pairs of cowbirds can parasitize all the nests in a small woodland. A female cowbird often removes an egg of the host species from a nest when she lays her own, and cowbird eggs develop and hatch more rapidly than the eggs of the host species, giving the cowbird nestling an advantage. Larger and pushier than the nestlings of the host species, cowbird nestlings take so much of the food their unwitting foster parents bring to the nest that the host's nestlings may starve.

Cowbird parasitism is insidious. When a nest with eggs or fledglings is lost to a predator, the parent birds usually build another nest and start a second clutch. They may succeed in reproducing that season despite the loss of their first clutch. In contrast, the parent birds that serve as hosts for the cowbirds do not distinguish the cowbird eggs or nestlings from their own. Thus, when a pair of birds has raised a cowbird to fledging, they behave as if they had successfully fledged their own young and do not nest again that year.

Cowbirds are insectivorous and feed primarily in open fields. Because their movements are not limited by the need to care for their young, they can travel long distances between the open fields where they feed and the woodland edges where which they find the nests of other birds. Populations of cowbirds have increased dramatically as open fields have replaced woodland habitats in North America, and the small remaining woodlands offer songbirds too little interior space to conceal their nests from cowbirds. Thus habitat fragmentation in the summer breeding range of Neotropical migrants appears to be responsible directly and indirectly for some of the decrease in their populations.

Habitat change in the winter range is probably also responsible for some population declines. The huge landmass of North America funnels down to a much smaller area in Central America, where many migrant birds overwinter. Competition for food and space in the small landmass of Central America may be one of the reasons that migrants do not breed in their winter ranges. Because the land area in the winter range is small, habitat changes caused by agricultural practices even on a relatively small scale could affect large numbers of birds. Coffee originated in Ethiopia and was brought to Latin America by Spanish colonists. Traditionally, coffee trees have been grown beneath a canopy of taller trees. This method produces shade-grown coffee. Coffee plantations of this sort are similar in structure to natural forests, although they are less complex and have fewer species of trees, and these traditional coffee plantations provide important habitats for birds. A survey of traditional plantations in Chiapas, Mexico, by the Smithsonian Migratory Bird Center found that more than 150 species of birds live in them, including Neotropical migrants.

In the past 20 years, hybrid coffee trees that grow in full sun have replaced traditional plantations. Sun-grown coffee produces substantially greater yields than shade-grown, and its use has spread rapidly. Currently about 20 percent of cropland planted in coffee in Mexico, 40 percent in Costa Rica, and 70 percent in Colombia has no shade canopy. Sun-grown coffee requires intensive cultivation, with heavy use of chemical fertilizers, insecticides, herbicides, and fungicides. These methods increase soil erosion, acidification, and the amount of toxic runoff. Furthermore, because sun-grown coffee plantations lack the complex structure of shade-grown coffee plantations, they do not provide habitats for forest birds. The diversity of birds plummets when a coffee plantation is converted from traditional shade-grown to sun-grown coffee. Studies in Colombia and Mexico found 94 percent to 97 percent fewer bird species in sun-grown coffee plantations than in shade-grown plantations.

Migratory birds depend on forested habitats in their winter ranges, but the land that can be set aside in parks and reserves in Central and northern South America is insufficient to maintain healthy bird populations. If migratory birds are to survive, human-dominated landscapes in the Neotropics must provide suitable habitats for them, such as traditional shade-grown coffee plantations. The problem is that sun-grown coffee produces a faster and larger return on a financial investment; shade-grown coffee is environmentally friendly but not economically attractive. To promote shade-grown coffee, the Smithsonian Migratory Bird Center has developed the Bird-Friendly® program, which certifies coffee growers that adhere to environmentally sound practices (**Figure 25–8**).

Critical Habitat for Large Mammals

Large animals need a lot of room—they have home ranges that cover hundreds or thousands of hectares, especially in the case of large predatory mammals.

▲ **FIGURE 25–8** The Bird-Friendly™ label for shade-grown coffee.

Furthermore, animals need different habitats for different activities, and the critical habitat must include the resources needed for hunting, mating, and raising young.

For species that have a well-developed social system, the ages and sexes of the individuals in a population may affect the success of a conservation plan. Populations of elephants in African game parks have increased substantially since the parks were established. This population growth testifies to the effectiveness of the parks in protecting the elephants, but it creates serious problems with overcrowding. As their populations increase, foraging elephants tear down mature trees to eat the upper branches and often move outside of the parks, where they raid gardens and orchards. To reduce these problems South Africa has developed a process of selective removal of elephants. Some individuals are trapped and moved to other parks that are below their capacity. It turns out that creating a mixture of juvenile and adult elephants is the key to making this plan work (**Box 25–3**).

Tigers Protecting critical habitats is especially challenging for animals like tigers that occasionally attack humans, yet tigers are among the most highly endangered species of mammals on Earth. A consortium of private and government organizations has developed a plan for Tiger Conservation Landscapes—protected areas that will provide the space and habitats that tigers require while minimizing conflicts with humans (Wikramanayake et al. 2004, Dinerstein et al. 2006, Sanderson et al. 2006, Global Species Program—WWF-International 2008). Tiger Conservation Landscapes are envisioned as networks of core areas surrounded by buffer zones that will allow tigers to raise their young. The core areas are connected by corridors that allow tigers to move among them. Twenty high priority areas that are large enough to accommodate at least 100 tigers have been identified, and the goal of the program is to have at least 100 breeding female tigers in each area by 2010.

25.4 Captive Breeding

In some situations, the threat to a species is so acute that the only alternative to watching the species become extinct is to gather as many individuals as possible and use them to establish a captive breeding program. The Amphibian Ark program that was described in Section 10.7 is an example of this approach: In the face of the rapid southward advance of chytrid fungi a team of scientists collected founder populations of 35 species of amphibians from El Valle de Antón, Panama, and transported them to Zoo Atlanta and the Atlanta Botanical Garden to establish breeding programs.

■ Goals of Captive Breeding

More than 500 species and subspecies of animals are being bred in captivity, and all of the major types of vertebrates are included. About 60 of these breeding programs are guided by Species Survival Plans (SSPs) developed by the Association of Zoos and Aquariums (AZA). A successful captive breeding program is far more than a group of animals that are reproducing in a zoo. The SSP defines a breeding plan that minimizes inbreeding and equalizes the genetic representation of each of the founding members of the captive population.

Genetics of Small Populations If you draw individuals randomly from a population of animals, you will get more individuals with common alleles than with rare ones, and alleles that are very rare may be absent from your sample. This is the familiar genetic bottleneck effect, and it is an inevitable element of captive breeding for most species because only a few wild-caught individuals are available to be the founders of a captive breeding program.

The first requirement of an SSP is that the animals in the captive population must represent a natural genetic component of the population because preserving as much of the genetic diversity of a species as possible is essential for long-term success. Most species show geographic variation in size, color, and behavior that reflects underlying genetic differences. These local varieties are often called subspecies, and the differences among subspecies are the result of local selective pressures. Captive breeding programs can retain this genetic diversity only by breeding pure subspecies lineages.

In addition to geographic variation in genetic characters, there is genetic variation among individuals. Many genetic loci have multiple alleles, and this heterozygosity is the raw material on which natural selection acts. Studbooks record the pedigrees and genetic characteristics of all the individuals of a species in captivity. The AZA currently maintains studbooks for almost 450 species and subspecies; and, when SSPs are developed, this information is used to decide which individuals should be bred to each other.

BOX 25–3 Adolescent Elephants

Between 1981 and 1993, young male and female African elephants (*Loxodonta africana*) that had been orphaned when their mothers were killed were relocated from Kruger National Park to Pilanesburg. There were no elephants already at Pilanesburg, and the orphans matured in the absence of older elephants. When the first elephant calf was born in 1989, the relocation program looked like a success story, except for one problem—the young male elephants were attacking and killing white rhinoceros (*Ceratotherium simum*). By 1997, the elephants had killed more than 40 rhinoceros (Slotow et al. 2000).

The African elephant is an endangered species, and the white rhinoceros is listed as vulnerable. Both species face sufficient risk of extinction without the added complication of having one species attacking and killing the other. The aggressive behavior of the young male elephants at Pilanesburg is not typical of African elephants, but it was not unique to this population. It occurred in some other populations, especially in Hluhluwe-Umfolozi Park, which is in northern KwaZulu Natal province. This was the second population to be established with orphans from Kruger National Park, and the young male elephants at Hluhluwe-Umfolozi began to kill rhinoceros about two years after the behavior appeared in the Pilanesberg population.

The attacks on rhinoceros occurred when the male elephants were in musth. Musth is an annual period during which the level of testosterone circulating in the blood of male elephants rises dramatically. Elephants in musth adopt a distinctive posture that allows them to be recognized from a distance. Their temporal glands swell and secrete an oily material, and sexual and aggressive behavior increases (**Figure 25–9**). In natural populations, males first enter musth when they are 25 to 30 years old. The duration of musth increases as they get older—from a few days to a few weeks for animals between 25 and 30 to two to four months for animals older than 40. Thus, the young males at Pilanesberg had developed musth earlier than would be expected, and their periods of musth lasted far longer than normal for such young males.

The populations of elephants that are killing rhinoceros have one feature in common—all were established by moving only young animals to locations where they matured in the absence of adults. That abnormal situation turned out to be the key to the early and prolonged musth and aggressive behavior of the young males. In natural populations of elephants that include older males, young males are unlikely to be in musth; and, when they do enter musth, the periods are short. Social interactions are responsible for this situation—young males in musth engage in aggressive interactions with the older bulls in musth, and the young males are defeated and driven away by the bulls. Within minutes to hours after being defeated in an aggressive encounter, a young male loses the signs of musth. Repeated interactions with older bulls drive young males out of musth, and the presence of older males in a population may delay its onset in young males.

In 1998, six adult bull elephants were moved from Kruger National Park to Pilanesberg, and their presence had a dramatic effect on the young males. The duration of musth in young males dropped sharply, in most cases falling from weeks to just a few days. Gratifyingly, the young males also stopped killing rhinoceros.

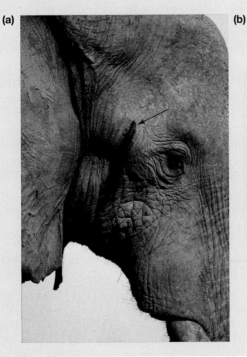

(a)

(b)

▲ **FIGURE 25–9** African elephants in musth. (a) Secretion from the temporal gland of a male in musth (arrow). (b) A large bull sparring with a smaller male.

Several criteria are used to make these pairings: the relatedness of the individuals that are to be mated should be minimized, the bloodlines of different founders should be equally represented, and the rarest bloodlines should have highest priority for breeding.

The relatedness of an individual to the other individuals in the population is expressed as its kinship value. A low kinship value means that an individual's genes are not well represented in the population, and these individuals are good candidates for mating. It often turns out that the best mating involves animals that are in zoos thousands of kilometers apart; when this is the case, one partner must be shipped to the zoo where the other partner lives (**Figure 25–10**).

Inbreeding Depression When the offspring of a small founder population are bred for generation after generation, the average fitness of the population is reduced by two mechanisms. First, every population carries some recessive alleles that have deleterious effects. In a large, genetically diverse population, these genetic loci are usually heterozygous—that is the deleterious recessive allele is usually paired with a normal dominant allele, so the deleterious character is not expressed. As the captive population becomes more inbred, however, more and more of these deleterious recessive alleles appear in the homozygous condition and are expressed.

In addition, individuals that are heterozygous at most of their genetic loci are generally more robust than those that are mostly homozygous. The progressive reduction in heterozygosity that occurs as a

▲ **FIGURE 25–10** Captive breeding. This male tiger is beginning its trip from the South China tiger breeding facility in Suzhou City in Jiangsu Province, China, to the Tiger Valley Reserve in the Free State Province in South Africa, where it will be mated with a female South China tiger.

captive population becomes more inbred lowers the average fitness of the animals in the population. The consequence of these two processes is known as inbreeding depression, and it often appears as reduced rates of reproduction and increased rates of infant mortality in breeding colonies.

Domestication Inadvertent domestication is another genetic pitfall of captive breeding programs. The individuals that adjust well to captivity are more likely to reproduce than the ones that don't adjust, and some of the tendency to adjust is genetically determined. As a result, any captive breeding program selects for genetic characteristics that increase reproductive success in a captive situation, but these characteristics may be quite different from the ones that are beneficial to wild animals. Among antelope, for example, some individuals quickly learn that when a keeper approaches their pen it means they are about to be fed, and they gather eagerly at the feeding area. Other individuals, however, run away when they see the keeper approaching, possibly injuring themselves when they crash into the fence around their enclosure. Extreme wariness is a detrimental character in captivity. However, in the wild, the wary individuals may be less vulnerable to predators than their bolder companions, and loss of wariness in a captive-bred population can spell trouble when captive-bred animals are released into the wild.

■ Designing a Captive Breeding Program

Captive breeding programs are carefully planned to maintain genetic diversity and avoid domestication, but those goals are easier to define than they are to achieve. The first requirement is information about the genetic diversity of the target species, and increasingly this information is coming from genetic studies that use variation in mitochondrial DNA and microsatellites to calculate the relatedness of different wild populations to each other. With this information in hand, breeding programs can determine which lineages must be kept separate. The SSP for tigers provides a good example of how the principles of captive breeding are applied.

Breeding Programs for Tigers The tiger (*Panthera tigris*) may be the most endangered large carnivorous mammal in the world. Tigers once had the broadest geographic distribution of any of the large cats, roaming across all of Asia from south of the equator in Java and Bali north more than 60° to Siberia and from the Pacific Ocean and west more than 100° of

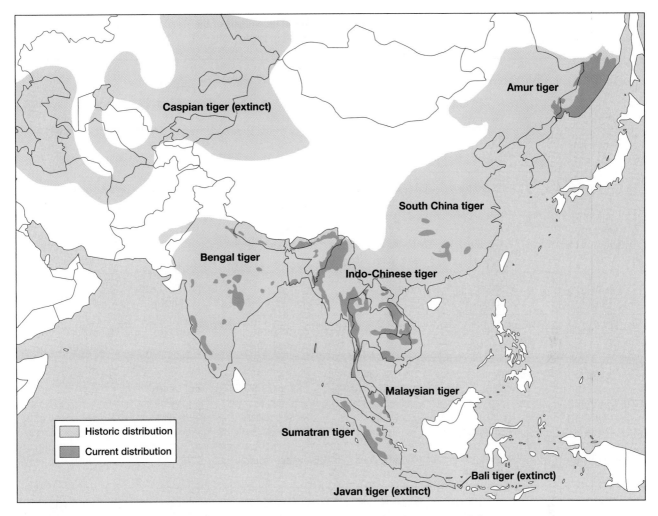

Caspian tiger (extinct)

Amur tiger

South China tiger

Bengal tiger

Indo-Chinese tiger

Malaysian tiger

Sumatran tiger

Historic distribution

Current distribution

Bali tiger (extinct)

Javan tiger (extinct)

▲ **FIGURE 25–11** Tigers once had the largest geographic distribution of any of the large cats, extending from the Caspian Sea to the Pacific Ocean and from Indonesia to Siberia. Three of the original nine subspecies have become extinct since the mid-1900s.

latitude to Iran and the shores of the Caspian Sea (**Figure 25–11**). Nine different subspecies of tigers roamed this vast area from the Late Pleistocene until the mid-1900s (see the color insert), but three of those forms have recently become extinct (**Table 25.2** and **Figure 25–12**).

A genetic analysis of the remaining tigers (Luo et al. 2004) showed that they form six distinct genetic lineages that are identified as subspecies (**Figure 25–13** on page 682). The initial divergence of tigers occurred between 72,000 and 108,000 years ago. Ecological separation and geographic barriers during the Late Pleistocene appear to have been more important than human influences in separating the lineages. This analysis revealed that tigers from Malaysia are genetically distinct from those on the

adjacent Asian continent. In fact, Malaysian tigers are more different from mainland Asian tigers than the Amur (Siberian) tigers are. (This divergence between tigers on the Asian mainland and the Malaysian Peninsula recalls the newly described separation of clouded leopards on Borneo from those on the Asian mainland that was discussed in Section 1.5.)

The Status of Tiger Breeding Programs Captive breeding programs are in progress for Amur and Bengal tigers. The Amur tiger program has established a stable, genetically diverse captive population. Bengal tigers have been bred in India at least since the late 1800s; however, tigers from other genetic lineages have been interbred with Bengal tigers, so many of the Bengal tigers in zoos are

TABLE 25.2 **The nine subspecies of tigers with estimates of the number of individuals of the six surviving subspecies. The number of founding individuals is shown when this information is available.**

Common Name	Subspecies	Geographic Range	Habitat	Estimated Population in the Wild	Number of Individuals in Captivity and Number of Founding Individuals
Bengal tiger	*Panthera tigris tigris*	Indian subcontinent	Dry tropical forest and grassland	3060–4735	201
Caspian tiger	*Panthera tigris virgata*	Turkey through central and west Asia		EXTINCT since the 1970s	0
Amur tiger, Siberian tiger	*Panthera tigris altaica*	Amur River region of Russia and China, and North Korea	Temperate deciduous forest	437–506	About 501, descended from 83 founding animals
Javan tiger	*Panthera tigris sondaica*	Java, Indonesia		EXTINCT since the 1980s	0
South China tiger	*Panthera tigris amoyensis*	South central China	Subtropic to temperate forest	Only one recent sighting	48, descended from 6 founding animals
Bali tiger	*Panthera tigris balica*	Bali, Indonesia		EXTINCT since the 1940s	0
Sumatran tiger	*Panthera tigris sumatrae*	Sumatra, Indonesia	Moist tropical forest	400–500	232, descended from 31 founding animals
Indo-Chinese tiger	*Panthera tigris corbetti*	Continental southeast Asia.	Evergreen tropical wet forest	736–1225	Unknown
Malaysian tiger	*Panthera tigirs jacksoni*	Malay Peninsula	Evergreen tropical wet forest	500	Unknown

Sources: Luo et al. 2004, http://www.wwf.org, http://www.savethetigerfund.org, http://www.aza.org, and http://www.isis.org.

hybrids that are not suitable for conservation programs. A Sumatran Tiger Masterplan has been developed that will draw on more than 200 Sumatran tigers in zoos in Indonesian, Australasia, Europe, and North America.

The captive population of South China tigers is only 48 individuals. To make the situation still worse, these animals are all descended from just six founders, and it has been 20 years since a new wild-caught tiger was added to the South China gene pool. An estimated 20 to 30 wild South China tigers persist, but that number is little more than a guess. No sightings of wild tigers had been reported for decades until late in 2007, when the Shaanxi Forestry Department called a news conference to announce that Zhou Zhenglong, a former hunter, had photographed a wild South China tiger (**Figure 25–14**). The report was initially hailed as evidence that a wild population of the tigers still exists, and Zhou Zhenglong received the 20,000 yuan reward that the Forestry Department had offered for a photograph of a wild South China tiger. Enthusiasm dimmed, however, as experts studied the photographs. Details of the images—the tiger's expression, its coat color, the relative sizes of the tiger and the plants, and the fact that two photographs show the tiger in exactly the same pose in two different settings—raised doubts about the authenticity of the photograph (Holden 2007).

While the captive population of the South China tiger is too small to form an effective captive breeding program, the Indo-Chinese and Malayan tigers present a different problem. In these cases, there are reasonable numbers of tigers in zoos, but until 2004 the two forms were regarded as a single subspecies. As a result, Indo-Chinese and Malaysian tigers were interbred, and extensive genetic analyses of individual animals will be needed before we know whether pure breeding stocks of the two subspecies still exist in the captive programs. If they do not, it will be necessary to capture wild tigers to reestablish the programs.

White Tigers White tigers are a controversial topic in the zoo world. White tigers are Bengal tigers that carry a recessive gene that prevents them from synthesizing orange pigment. (White tigers are not albinos—they can synthesize melanin, and they have black stripes like normal tigers.) The defective allele is recessive, so only individuals that are homozygous for the allele are white. The gene pool of wild Bengal tigers is large enough so that most individuals are heterozygous at that genetic locus, and white tigers occur only rarely in natural populations.

White tigers are much more common in zoos than they are in the wild because it is possible to produce a white bloodline by breeding white cubs to their siblings or back to their parents. Zoo visitors love white tiger cubs, and their popularity places zoos in a difficult situation—on one hand a litter of white cubs is a great box-office attraction that increases gate receipts. But inbreeding reduces genetic diversity—exactly the opposite of the goal of the SSP for Bengal tigers—and most zoo professionals believe that breeding white tigers is unethical.

■ Animal Behavior and Captive Breeding Programs

Animals have complex suites of behaviors that allow them to find appropriate food, evade predators, function in a group, and identify suitable mates. Some of these behaviors are innate—that is, an animal is born with the ability to produce the correct response to certain stimuli—but many behaviors are learned by trial and error. Animals raised in captivity do not experience the same stimuli they would in their natural environment and may not be able to function effectively if they are released. As captive breeding programs are developed to restore species to the wild, increasing effort is being devoted to ensuring that the animals to be released are competent to survive in the wild.

(a)

(b)

(c)

▲ **FIGURE 25–12** The three subspecies of tigers that have become extinct within the past century. (a) Caspian tiger (*Panthera tigris virgata*). This individual was shot in Iran in the 1940s. (b) Javan tiger (*Panthera tigris sondaica*). These two individuals were photographed in the Berlin Zoo in the early 1900s. (c) Bali tiger (*Panthera tigris balica*). This photograph was found among the papers of the hunter who shot it in 1925.

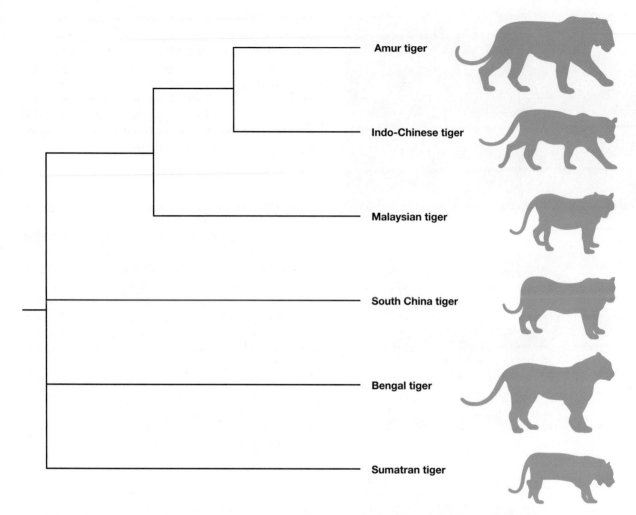

▲ **FIGURE 25–13** A phylogenetic analysis of the surviving subspecies of tigers shows that there are six distinct evolutionary lineages that must be kept separate in breeding programs. Prior to this study the Indo-Chinese and Malaysian tigers were considered to be the same subspecies, but genetic analysis shows that the Indo-Chinese tiger is more closely related to the Amur tiger than it is to the Malaysian tiger. This discovery emphasizes the importance of understanding the genetic relationships among populations of an endangered species. The silhouettes of the surviving subspecies show the relative body lengths of adult males, which range from 2.4 m for Malaysian and Sumatran tigers to 3.3 m for the Amur tiger.

Imprinting, Learning, and the Captive Husbandry of Birds The process known as imprinting is an important feature of the behavior of birds and mammals. Imprinting is a special kind of learning that occurs only during a restricted time in ontogeny called the critical period. Once imprinting is established, it is permanent and cannot be reversed.

Colors, patterns, sound, and movement are the major stimuli for imprinting among birds, whereas scent is the most important stimulus for mammals. Two types of imprinting can be distinguished:

- **Filial imprinting** is the process of learning to recognize the individual characteristics of the parents, and it is responsible for keeping the young with the mother after they move away from the nest or den.
- **Sexual imprinting** refers to learning the characteristics of other members of the species. Sexual imprinting during infancy allows an individual to

▲ **FIGURE 25–14** Paper tiger? This photograph, allegedly of a wild South China tiger, is now thought to be a fake created from a picture of a tiger that appeared on a Chinese calendar.

(a)

(b)

▲ **FIGURE 25–15** Captive husbandry of endangered species of birds. (a) A hatchling California condor in its incubator with the model condor head used to feed it. (b) A northern bald ibis.

recognize a mate of its own species when it matures. Birds and mammals normally imprint on their parents and siblings because those are the only objects in the den or nest that are emitting visual, auditory, and olfactory stimuli. In the absence of their parent, however, infants may imprint on inanimate objects or on members of another species, including humans.

The confusion of species identification by birds that have imprinted on a foster parent or a keeper can be disastrous for programs in which endangered species are reared in captivity and then released. Young birds must recognize appropriate mates if they are to establish a breeding population, and captive-rearing programs go to great lengths to ensure that the young birds are properly imprinted.

For example, hatchling California condors (*Gymnogyps californicus*) in the breeding facility at the San Diego Wild Animal Park are reared in enclosed incubators and fed by a technician who inserts her hand into a rubber glove modeled to look like the head of an adult condor (**Figure 25–15a**).

Still more training may be necessary to produce captive-reared young that can survive after they have been released. A husbandry program for the northern bald ibis (*Geronticus eremita*) is an example of how complicated this process can be (**Figure 25–15b**). Within historic times, the geographic range of the bald ibis extended from the Middle East and North Africa north to Switzerland and Germany, but the wild population has dwindled to fewer than 220 birds in a reserve in Morocco. Bald ibises flourish in

captivity, and there are more than 700 captive individuals. This species seems ideal for reintroduction—it is prolific, and its disappearance from the wild seems to have been caused by human predation rather than by pollution or loss of habitat. Yet two attempts to establish populations by releasing captive-reared birds have failed.

The reason for the failures seems to have been the absence of normal social behaviors in the captive-reared birds. Bald ibises are social birds with extended parental care, and it seems that juveniles learn appropriate behaviors from adults. For some reason, this learning did not occur in captivity. An attempt is now under way to instruct young bald ibises in these social skills—human foster parents are hand-rearing the birds, teaching them to find their way to fields where they can forage, to recognize predators and other dangers such as automobiles, and to engage in mutual preening, which is an important social behavior.

Tameness can be a problem with captive-bred animals, even when extraordinary care is used to avoid imprinting on humans. Asiatic houbara bustards (*Chlamydotis undulate mcqueenii*) being reared in Saudi Arabia succeeded in finding food after they were released, but many of them fell victim to predators, apparently because their sheltered rearing conditions had not taught them to be afraid. That problem may have been solved with the assistance of a young fox named Sophie who is let into a cage with the birds and allowed to chase them. Sophie thinks it's play, but three training sessions apparently convinced the birds to be more wary because survival after release was greater in the trained group than in a control group that had not been exposed to Sophie (van Heezik and Seddon 2001).

25.5 Global Issues in Conservation Biology

We have focused on problems faced by particular species and particular habitats, but global pollution and climate change are a growing area of concern for conservation biology. In general, humans will feel the effects of climate changes working second hand as habitats and plant and animal communities respond to changes in temperature and precipitation, but pollution can affect humans directly (**Box 25–4**).

■ Global Climate Change

Global warming is a reality, and impacts on plants and animals have already been detected (Walther et al. 2002, Root et al. 2003). Growing seasons and flowering times of plants in high latitudes begin earlier in the spring and extend later into the fall than in previous years (Fitter and Fitter 2002). A 90-year record shows that the growth rate of Alaskan white spruce has decreased as temperatures have increased, apparently because of temperature-induced drought stress (Barber et al. 2000). Migratory birds are arriving earlier at their summer breeding ranges, and birds are laying eggs earlier (Jonzén et al. 2006). Complications are occurring for some species because the breeding cycles of birds are matched to the breeding cycles of their prey species, so the birds' eggs hatch at the time of maximum availability of

food for nestlings. Changing temperatures affect the cycles of both the birds and their prey and can throw the cycles out of synchrony. A mismatch of this sort appears to be responsible for a decline in populations of pied flycatchers (*Ficedula hypoleuca*) over the past two decades. The decline in flycatchers corresponds to the degree of mismatch between the time caterpillar densities reach their peak and the time the flycatchers' eggs hatch. In areas where the mismatch is small, the populations of flycatchers have declined only 10 percent, whereas in the areas of greatest mismatch the decline has reached 90 percent.

Why Now? Climate has changed many times in the history of the Earth, and the Pleistocene in particular has been marked by periods of glacial cooling and interglacial warming. What is it about the current episode of global warming that is so alarming? The speed of change is the answer—the mean global temperature is increasing at a rate that exceeds the ability of most plants and animals to adjust.

What Will Happen? Five years ago, discussions of global warming revolved around "Is it real?" and "Are humans responsible for it?" Now we know beyond any reasonable doubt that the answer to these questions is "yes," and discussion focuses on predicting how fast temperatures will increase and what biological, social, and economic changes will accompany increased temperatures. Working Groups I and II of the Intergovernmental Panel on Climate Change established by the World Meteorological Organization and the United Nations Environment Programme, have produced a comprehensive analysis of what we can expect (IPCC 2007a, 2007b).

In general, maximum and minimum temperatures will increase, precipitation will decrease, and extreme weather events will become more frequent. All of the continents will experience drought, and its effects will be especially severe in regions that are already on the edge of aridity, including Australia, sub-Saharan Africa, and the American Midwest. The rate of change in the Arctic has been astonishingly rapid, with the ice cover of the Artic Ocean diminishing by measurable amounts every summer. The Bering Sea is already shifting from an Artic to a sub-Arctic ecosystem (Grebmeier et al. 2006).

Predicting the effects of changes in mean annual air temperature and precipitation on individual species or communities is extraordinarily complicated (Walther 2007). The responses of individual

BOX 25–4   Endocrine Disruptors

Many chemicals mimic the effects of hormones and can disrupt normal physiological functions of vertebrates, and the general name for chemicals with this property is endocrine disruptors. Many plants, for example, contain chemicals that interfere with the reproductive cycles of rodents that feed on the plants; these chemicals are probably part of the plant's defense against herbivores. Synthetic chemicals that are released into the environment by human activities also disrupt the endocrine systems of vertebrates, including humans (Gross 2007).

Endocrine disruptors can alter sexual development when animals are exposed during embryonic development. Waste from a pulp paper mill released into the Fenholloway River in the panhandle of Florida contains androstenedione, an anabolic steroid that has been used by human athletes. In the river, androstenedione has masculinized the mosquitofish (*Gambusia affinis*)—populations downstream from the mill contain only males (McNatt et al. 2000).

Environmental estrogens, chemicals that have a feminizing effect, are frighteningly common. The list of vertebrates that have been feminized by environmental estrogens includes several species of fish (mosquitofish, trout, carp, salmon, and sturgeon), amphibians (frogs and salamanders), reptiles (alligators and turtles), birds (gulls, herons, and birds of prey), and mammals (the Florida panther) (Guillette et al. 2000). In these cases, males have smaller testes than normal and may be infertile. A study of fish populations in a river that receives effluent from a sewage plant near Boulder, Colorado, showed about a 50:50 ratio of males and females upstream of the plant, but 93 percent females downstream.

The concentrations of hormone mimics in water and soil are low compared to the levels normally found in the blood, and representatives of the chemical industry have argued that such low levels could not have physiological effects on humans or other animals. A peer-reviewed evaluation conducted by the National Institute of Environmental Health Sciences discredited the industry position, however, and concluded that many endocrine disruptors do have physiological effects at the concentrations found in the environment (vom Saal et al. 2007).

Humans are not immune from the effects of environmental estrogens. Chemicals called phthalates that leach from plastics (including those used to make infant nursing bottles) have been implicated in the high incidence of thelarche (premature breast development) in girls as young as two years old. Bisphenol A (BPA) is an estrogen mimic that leaches from polycarbonate plastics such as soft-drink bottles, the linings of metal cans, water pipes, and medical equipment. More than 95 percent of Americans have detectable levels of BPA in their blood serum, urine, and reproductive tissues. Between 1938 and 1991, the average volume of seminal fluid of human males in Europe and North America decreased from 3.40 milliliters to 2.75 milliliters, and the average sperm count declined from 113 million sperm per milliliter of seminal fluid to 66 million sperm per milliliter (**Figure 25–16**). Feminization by environmental estrogens such as BPA may be contributing to this phenomenon (Eertmans 2003).

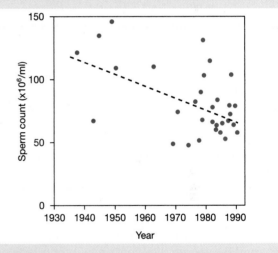

▲ **FIGURE 25–16** The sperm counts (the average number of sperm per milliliter of semen) of human males declined from 1937 to 1992. Investigators believe that at least some of this decline is the result of exposure of humans to feminizing endocrine disruptors, such as bisphenol A, that are widespread in the environment.

species are merely the start of a cascade of interacting events that have positive and negative feedbacks on other elements of the community. In an experimental study, Suttle and his colleagues (2007) found that changes in rainfall that *increased* biodiversity and productivity in the initial two years of the study led to a simplification of the food web, so that the final result was *reduced* biodiversity and productivity.

Despite the difficulties of making predictions about individual species, the magnitude of the changes in habitats that are forecast by the IPCC paint bleak picture. An analysis by an international team predicted that a minimum of 18 percent of the species in the regions sampled—and perhaps as many as 35 percent—will be in irreversible decline toward extinction by 2050 as a result of global warming (Thomas et al. 2004).

25.6 The Paradoxes of Conservation

Although conservation must address biological issues, conservation is not a purely biological issue. Human societies that burn fossil fuels and manufacture products will inevitably produce pollutants that travel far from their sources. It is relatively easy to believe that it is wrong for a multinational corporation to build a salt-extraction plant that will destroy a protected habitat or for a paper mill to release pollutants to save itself the cost of removing them from its waste discharge. Is it equally certain that a subsistence-level farming family in a developing country should not disrupt the habitat of an endangered species to grow the food it needs to survive?

Rich and poor nations respond differently to the often conflicting demands of earning a living versus conserving natural resources. A wealthy nation, especially a large country like the United States, can afford to set aside land to protect habitats and organisms, but that option may not be possible for a poor nation. Effective conservation efforts cannot focus only on biological questions. In the real world, conservation requires intricate balancing of biological, political, economic, and cultural values, and no one response is right for all species and all habitats. Some species need protected areas, but many of the most important areas of relatively undisturbed habitat are in countries where poverty and pressing social problems make complete protection an unrealistic goal. Conservation programs must address social and economic issues as well as biological ones if they are to succeed. The people living near parks and management areas must believe that protecting those habitats and the species they contain will contribute more to their own standard of living than they could gain by clearing the land and killing the animals (Howlett and Dhand 2000, Sinclair et al. 2000, Hulme and Murphree 2001, Balmford et al. 2002).

Efforts of this sort are under way. The National Resources Defense Council is working with the villages around San Ignacio Lagoon to develop sustainable fisheries that are environmentally friendly. These fisheries also will provide an economic basis to make it easier to resist future proposals for industrial development that would damage the lagoon. Project Piaba in Brazil began as a study of fish diversity in the middle Rio Negro basin and has grown into a community-based interdisciplinary project that fosters ecologically sound economic development. The Rio Negro is the primary source of ornamental aquarium fish; more than 20 million fish are exported annually, with a retail value of more than $100 million. If it were not for the fishery, the area would be developed for mining, forestry, and agriculture. Since sustaining the fishery requires sustaining the ecosystem, scientists working with Project Piaba are studying the structure and function of the aquatic systems. They are using that information to develop fishery management procedures to give the local people an incentive to preserve the integrity of the environment that supports the fishery.

The magnitude and complexity of the problems facing conservation biologists are daunting, and the scale on which remedial efforts must be attempted is nearly beyond comprehension. Allison Jolly has calculated that the total world population of all wild primates combined is less than that of any of the Earth's major cities. The entire extant populations of many species of primates are no larger than that of a small town. Conditions are equally critical for many other vertebrates. Nearly four decades ago, Jolly (1980) expressed the problems we face as biologists today:

> This realization has been painful. It began for me in Madagascar, where the tragedy of forest felling, erosion, and desertification is a tragedy without villains. Malagasy peasant farmers are only trying to change wild environments to feed their own families, as mankind has done everywhere since the Neolithic Revolution. The realization grew in Mauritius, where I watched the world's last five echo parakeets land on one tree and knew they will soon be no more. It has come through an equally painful intellectual change. I became a biologist through wonder at the diversity of nature. I became a field biologist because I preferred watching nature go its own way to messing it about with experiments. At last I understood that biology, as the study of nature apart from man, is a historical exercise. From the Neolithic Revolution to its logical sequels of twentieth-century population growth, biochemical engineering of life forms, and nuclear mutual assured destruction, the human mind has become the chief factor in biology . . . . the urgent need in [vertebrate] studies is conservation. It is sheer self-indulgence to write books to increase understanding if there will soon be nothing left to understand.

Reviews of the biodiversity crisis agree that an essential first step in coming to grips with the problem is a clear understanding of what species exist, where they are, and what the critical elements in their survival are. This is an enterprise that must enlist biologists from specialties as diverse as systematics, ecology, behavior, physiology, genetics, nutrition, and animal husbandry. There are so many species about which we know almost nothing, and there is so little time for us to learn.

Summary

The diversity of vertebrates has increased steadily (albeit with several episodes of extinction) for the past 500 million years, peaking in the mid-Miocene, about 15 million years ago. Much of the decline in diversity of vertebrates (and other forms of life) since then can be traced to the direct and indirect effects of humans on the other species with which we share the planet. Major threats to the continued survival of species of vertebrates include habitat destruction, pollution, and over-hunting. At the base of all these phenomena is the enormous increase in human population size. The world's human population is currently about 6 billion people, double what it was only 50 years ago, and is predicted to grow to 9 billion by 2050. A quarter of a million humans are added to the population each day, a population the size of New York City is added each month, and nearly 100 million additional people demand resources each year. Consumption of resources and pollution of the environment increase at rates far greater than the rate of population growth. This differential is increasing as global communication (especially television) exposes the overwhelming majority of humans to a Western lifestyle, raising expectations and aspirations worldwide. Typically the use of resources in a modern technological society increases four to five times faster than population growth, and the release of pollutants rises in proportion to resource use.

Conservation efforts are complicated by political and economic issues, and strategies that are appropriate for developed countries may be impractical in developing nations. Programs that integrate the needs of humans and wildlife have the best chances for long-term success. Information about the basic biology of organisms plays an essential role in conservation by defining the critical elements of a habitat that must be preserved to ensure that a species can survive, by identifying sources of problems, and by guiding management of wild populations and reintroduction programs.

Additional Readings

Balmford, A. et al. 2002. Economic reasons for conserving wild nature. *Science* 297:950–953.

Barber, V. A. et al. 2000. Reduced growth of Alaska white spruce in the twentieth century from temperature-induced drought stress. *Nature* 405:668–672.

Carlsen, E. et al. 1992. Evidence for decreasing quality of semen during the past 50 years. *British Medical Journal* 305:609–619.

Daszak, P., et al. 2000. Emerging infectious diseases of wildlife—Threats to biodiversity and human health. *Science* 287:443–449.

Dinerstein, E. et al. 2006. Setting priorities for the conservation and recovery of wild tigers: 2005–2015. A user's guide. WWF, WCS, Smithsonian, and NFWF-STF, Washington, D.C.

Eertmans, F. et al. 2003. Endocrine disruptors: effects on male fertility and screening tools for their assessment. *Toxicology in Vitro* 17:515–524.

Fitter, A. H., and R. S. R. Fitter. 2002. Rapid changes in flowering time in British plants. *Science* 296:1689–1691.

Global Species Program, WWF-International. 2008. Solving conflicts between Asian big cats and humans: A portfolio of conservation action. Accessed 30 April 2008: <http://assets.panda.org/downloads/abchwcinfdocfromt hewwfglobalspeciesprogramme.pdf>

Gorman, M. L. et al. 1998. High hunting costs make African wild dogs vulnerable to kleptoparasitism by hyaenas. *Nature* 391:479–481.

Grebmeier, J. M. et al. 2006. A major ecosystem shift in the northern Bering Sea, *Science* 311:1461–1464.

Grenier, M. B. et al. 2007. Rapid population growth of a critically endangered carnivore. *Science* 317:779.

Gross, L. 2007. The toxic origins of disease. *PLOS Biology* 5(7):e193, doiL10.371/journal.pbio.0050193

Guillette, L. J. Jr. et al. 2000. Alligators and endocrine disrupting contamintants: A current perspective. *American Zoologist* 40:438–452.

Heath, D. D. et al. 2003. Rapid evolution of egg size in captive salmon. *Science* 299:1739–1740.

Holdaway, R. N. and C. Jacomb. 2000. Rapid extinction of the moas (Aves: Dinornithiformes): Model, test, abd implications. *Science* 287:2250–2254.

Holden, C. 2007. Rare-tiger photo flap makes fur fly in China. *Science* 318:893.

Howlett, R., and R. Dhand. 2000. Nature Insight: Biodiversity. *Nature* 405:205–253.

Hulme, D., and M. Murphree (eds). 2001. *African Wildlife and Livelihoods: The Promise and Performance of Community Conservation*. Portsmouth, NH: Heinemann.

IPCC (Intergovernmental Panel on Climate Change). 2007a. Working Group I: The physical basis of climate change. <http://www. ipcc-wg2.org/index.html>

IPCC (Intergovernmental Panel on Climate Change). 2007b. Working Group II: Climate change impacts, adaptation, and vulnerability. <http://www.ipcc-wg2.org/index.html>

Johnsson, J. I., and B. Th. Björnsson. 2001. Growth-enhanced fish can be competitive in the wild. *Functional Ecology* 15:654–659.

Jolly, A. 1980. *A World Like Our Own: Man and Nature in Madagascar*. New Haven, CT.: Yale University Press.

Jonzén. N. et al. 2006. Rapid advance of spring arrival dates in long-distance migratory birds. *Science* 312:1959–1961.

Krkošek, M. et al. 2007a. Effects of host migration, diversity and aquaculture on sea lice threats to Pacific salmon populations. *Proceedings of the Royal Society* B 274:3141–3149.

Krkošek, M. et al. 2007b. Declining wild salmon populations in relation to parasites from farmed salmon. *Science* 318:1772–1775.

Luo, S.-J. et al. 2004. Phylogeography and genetic ancestry of tigers (*Panthera tigris*). *PLOS Biology* 2:2275–2290. <http://www.plosbiology.org>

McNatt, H. B. et al. 2000. Effects of papermill effluent on a population of eastern mosquitofish (*Gambusia holbrooki*). Center for Bioenvironmental Research at Tulane/Xavier Universities. 2000. October 15–18. New Orleans, LA: Tulane University.

Miller, G. H. et al. 1999. Pleistocene extinction of *Genyornis newtoni*: Human impact on Australian fauna. *Science* 283:205–208.

Muir, W. M., and R. D. Howard. 1999. Possible ecological risks of transgenic organism release when transgenes affect mating success: Sexual selection and the Trojan gene hypothesis. *Proceedings of the National Academy of Sciences* 96:13853–13856.

Oftedal, O. T., and M. E. Allen. 1996. Nutrition as a major facet of reptile conservation. *Zoo Biology* 15:491–497.

Root, T. L. et al. 2003. Fingerprints of global warming on wild animals and plants. *Nature* 421:57–60.

Sanderson, E. et al. 2006. Setting priorities for the conservation and recovery of wild tigers: 2005–2015. The technical assessment. WCS, WWF, Smithsonian, and NFWF-STF,

New York-Washington, D.C. <http://www.worldwildlife .org/tigers/pubs/TCL-technical. pdf>

Silander, J. et al. 2006. Climate change adaptation for hydrology and water resources. FINADAPT Working Paper 6, *Finnish Environment Institute Mimeographs 336*, Helsinki, 52 pp.

Sinclair, A. R. E. et al. 2000. Conservation in the real world. *Science* 289:1875.

Slotow, R. et al. 2000. Older bull elephants control young males. *Nature* 408:425–426.

Smithsonian Migratory Bird Center. <http://nationalzoo.si .edu/ConservationAndScience/MigratoryBirds/Coffee/>

Stokstad, E. 2002. Engineered fish: Friend or foe of the environment. *Science* 297:1797–1799.

Suttle, K. B. et al. 2007. Species interactions reverse grassland response to climate. *Science* 315:640–642.

Thomas, C. D. et al. 2004. Extinction risk from climate change. *Nature* 427:145–148.

Tuominen, T.-R., and M. Esmark. 2003. Food for thought: The use of marine resources in fish feed. Report 02/03 WWF-Norway.

van Heezik, Y. and P. Seddon. 2001. Born to be tame. *Natural History* June 2001, pp. 58–63.

vom Saal, F. S. et al. 2007. Chapel Hill bisphenol A expert panel consensus statement: Integration of mechanisms, effects in animals and potential impact to human health at current levels of exposure. *Reproductive Toxicology* (2007), doi:10.1016/j.reprotix.2007.07.05.

Walsh, P. D. 2003. Catastrophic ape declines in western equatorial Africa. *Nature* 422:611–614.

Walther, G.-R. et al. 2002. Ecological responses to recent climate change. *Nature* 416:389–395.

Walther, G.-R. 2007. Tackling ecological complexity in climate impact research. Science 315:606–607.

Waples, R. S., and A. P. Hendry (eds.). 2008. Special Issue: Evolutionary perspectives on salmonid conservation and management. *Evolutionary Applications* 1:183–423.

Whitfield, J. 2003. Ape populations decimated by hunting and Ebola. *Nature* 422:551.

Wikramanayake, E. et al. 2004. Designing a conservation landscape for tigers in human-dominated environments. *Conservation Biology* 18:839–844.

World Conservation Monitoring Centre. 1992. *Global Biodiversity: Status of the Earth's Living Resources*. London, UK: Chapman & Hall.

Worthy, T. H. and R. N. Holdaway. 2002. *The Lost World of the Moa: Prehistoric Life in New Zealand*. Bloomington, IN: Indiana University Press.

SOUTH DAKOTA

(2 districts)

ALLEGHENY COUNTY

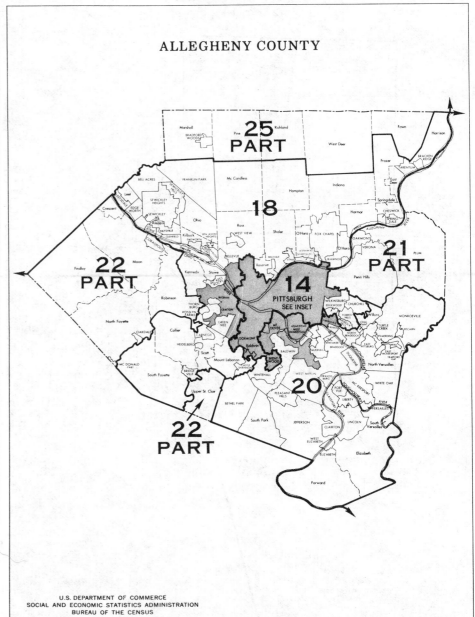

U.S. DEPARTMENT OF COMMERCE
SOCIAL AND ECONOMIC STATISTICS ADMINISTRATION
BUREAU OF THE CENSUS

RHODE ISLAND

(2 districts)

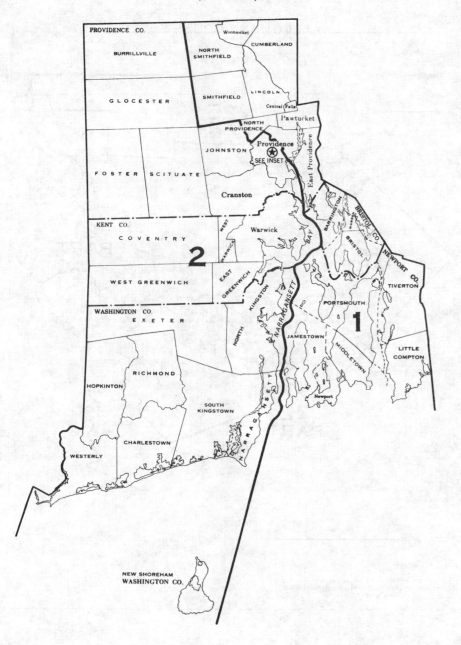

SOUTH CAROLINA

(6 districts)

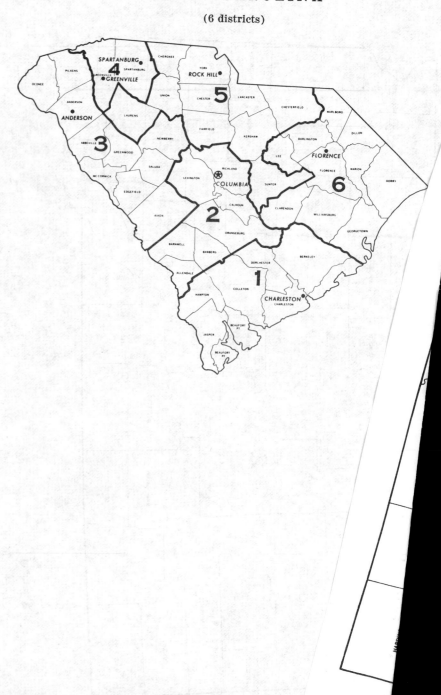

TENNESSEE

(8 districts)

TEXAS

(24 districts)

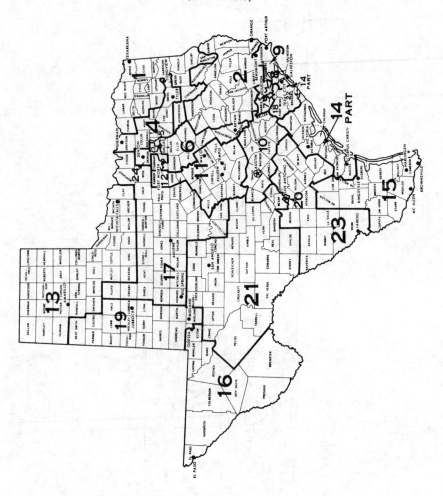

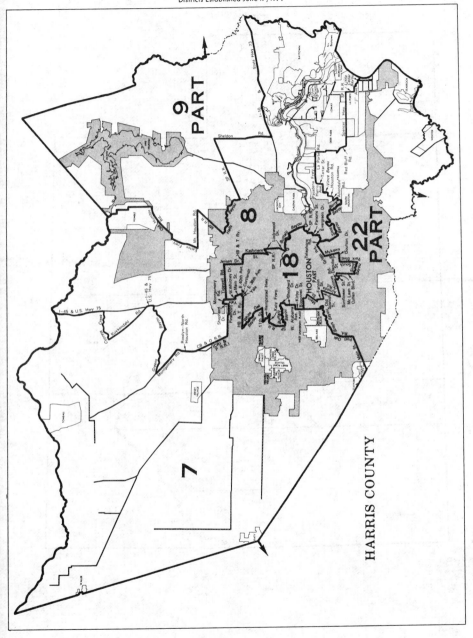

TEXAS
Districts Established June 17, 1971

PART 9

8

22 PART

18

HOUSTON PART

7

HARRIS COUNTY

UTAH

(2 districts)

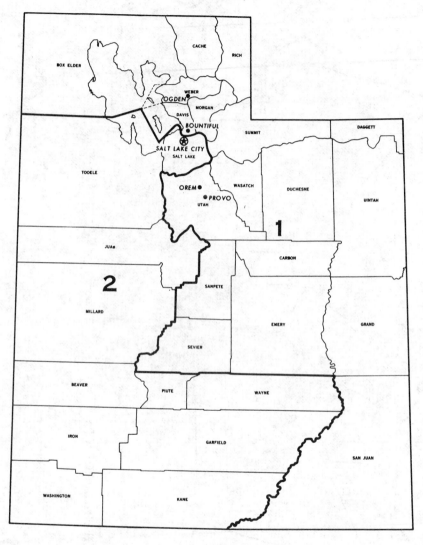

VERMONT

(1 at large)

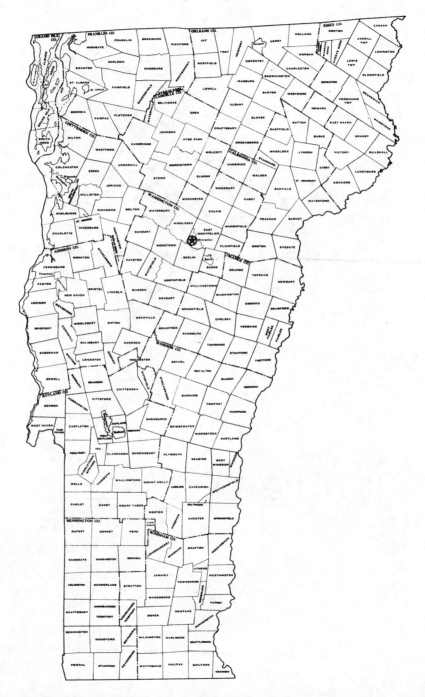

VIRGINIA

(10 districts)

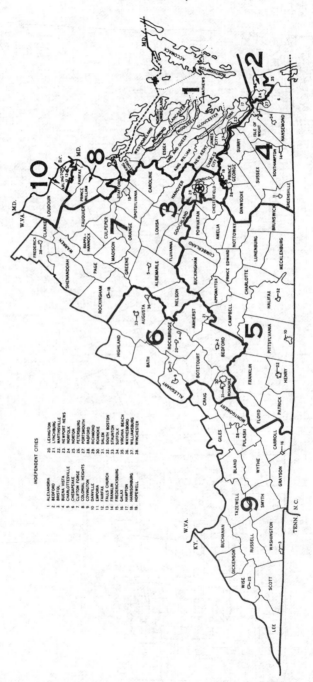

INDEPENDENT CITIES

1 ALEXANDRIA
2 BEDFORD
3 BRISTOL
4 BUENA VISTA
5 CHARLOTTESVILLE
6 CHESAPEAKE
7 CLIFTON FORGE
8 COLONIAL HEIGHTS
9 COVINGTON
10 DANVILLE
11 EMPORIA
12 FAIRFAX
13 FALLS CHURCH
14 FRANKLIN
15 FREDERICKSBURG
16 GALAX
17 HAMPTON
18 HARRISONBURG
19 HOPEWELL

20 LEXINGTON
21 LYNCHBURG
22 MARTINSVILLE
23 NEWPORT NEWS
24 NORFOLK
25 NORTON
26 PETERSBURG
27 PORTSMOUTH
28 RADFORD
29 RICHMOND
30 ROANOKE
31 SALEM
32 SOUTH BOSTON
33 STAUNTON
34 SUFFOLK
35 VIRGINIA BEACH
36 WAYNESBORO
37 WILLIAMSBURG
38 WINCHESTER

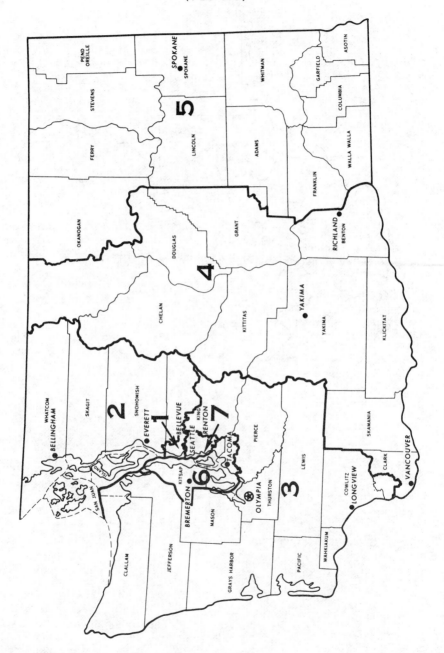

WASHINGTON

(7 districts)

WEST VIRGINIA

(4 districts)

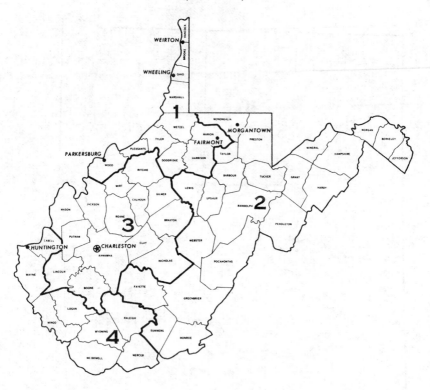

WISCONSIN

(9 districts)

WYOMING

(1 at large)

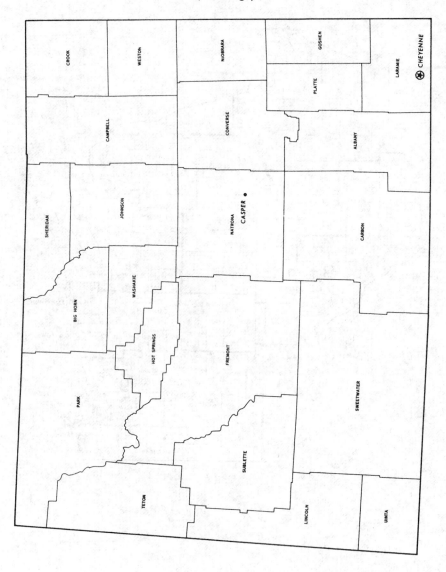

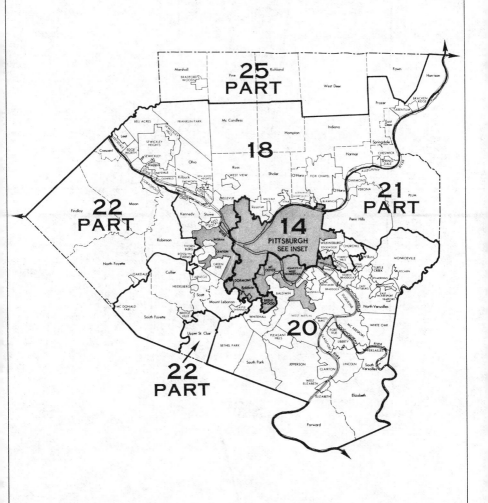

ALLEGHENY COUNTY

RHODE ISLAND

(2 districts)

SOUTH CAROLINA

(6 districts)

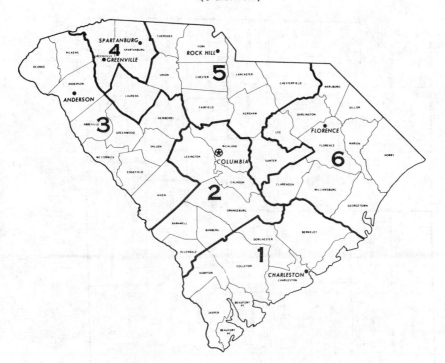

SOUTH DAKOTA

(2 districts)

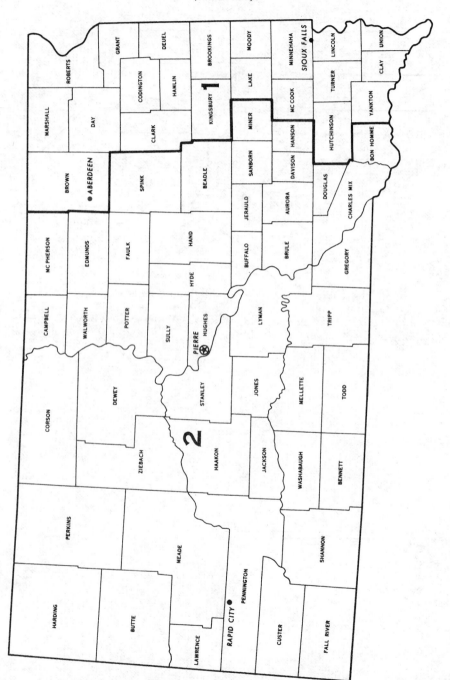

Glossary

abduction Movement away from the midventral axis of the body. *See also* adduction.

acetabulum Socket in the pelvis that receives the head of the femur.

acid precipitation Rain and snow acidified by sulfur- and nitrogen-containing gases produced by burning fossil fuels.

acrodont Teeth loosely attached to crest or inner edge of the jaw. *See also* pleurodont, thecodont.

activity temperature range The body temperatures maintained by an ectothermal animal when it is thermoregulating.

adduction Movement toward the midventral axis of the body. *See also* abduction.

adductor mandibularis The major muscle that closes the jaws.

adipocytes Fat storage cells.

advertisement call The vocalization of a male anuran used in courtship and territorial behavior.

aerobic metabolism Metabolic breakdown of carbohydrates in the presence of oxygen yielding carbon dioxide and water as the end products.

agnathans Jawless vertebrates. *See also* gnathostomes.

allantois The extra-embryonic membrane of amniotes that develops as an outgrowth of the hind gut.

allochthonous Originating somewhere other than the region where found.

allometry Describes a situation in which the proportions of an object change when size changes (contrast with **isometry**).

allopatry Situation in which two or more populations or species occupy mutually exclusive, but often adjacent, geographic ranges.

altricial Helpless at birth or hatching, as in pigeons and cats, for example. *See also* precocial.

alula The tuft of feathers on the first digit of a bird's wing that reduces turbulence in airflow over the wing.

alveoli Small saclike structures that are sites of gas exchange in the lungs.

amble A gait of tetrapods, a speeded-up walk with at least one foot on the ground and two or three feet off the ground at any one time.

ammocetes larva The larval form of lampreys.

ammonotelism Excreting nitrogenous wastes primarily as ammonia.

amnion The innermost extra-embryonic membrane of amniotes that surrounds the embryo.

amniotes Those vertebrates whose embryos have an amnion, chorion, and allantois (i.e., turtles, lepidosaurs, crocodilians, birds, and mammals) in addition to the yolk sac that is present in all vertebrates.

amniotic egg An egg with three extra-embryonic membranes (the **amnion**, **chorion**, and **allantois**).

amphicoelous (amphicelous) The condition in which the vertebral centrum is concave on both the anterior and posterior surfaces.

amphioxus The lancelet, *Branchiostoma lanceolatum*.

amphistylic An upper jaw suspended from the chondrocranium and the hyoid arch.

amplexus Clasping of a female anuran by a male during mating. Axillary amplexus refers to the male clasping the female in the pectoral region; inguinal amplexus is clasping in the pelvic region.

ampullae of Lorenzini Electroreceptors found in the skin of the snout of chondrichthyans.

anadromous Migrating up a stream or river from a lake or ocean to spawn (of fishes). *See also* catadromous.

anaerobic metabolism Metabolic breakdown of carbohydrates in the absence of oxygen, usually yielding lactic acid as an end product.

anapsid A skull lacking temporal fenestrae, or an animal with an anapsid skull. *See also* diapsid, synapsid.

angiosperm Most advanced and recently evolved of the vascular plants, characterized by production of seeds enclosed in tissues derived from the ovary. The ovary and/or seed is eaten by many vertebrates, and the success of the angiosperms has had important consequences for the evolution of terrestrial vertebrates.

anisodactyl The arrangement of toes seen in perching birds, with three in front opposed to one behind.

anisognathy The situation in which the tooth rows in the upper and lower jaws are not the same distance apart. *See also* isognathy.

annulus (plural *annuli*) Rings extending around a structure.

antidiuretic hormone, ADH (also known as vasopressin) A hypothalamic hormone that causes the kidney to conserve water by increasing the permeability of the collecting tubules.

apatite The mineral form of calcium phosphate found in bone.

aphotic "Without light" (e.g., in deep-sea habitats or caves).

apnea "Without breath" (i.e., holding the breath, as during diving).

apocrine gland Type of gland in which the apical part of the cell from which the secretion is released breaks down in the process of secretion. *See also* holocrine gland.

apomorphy (Also known as a derived character) A character that has changed from its ancestral condition. *See also* synapomorphy.

aposematic Device (color, sound, behavior) used to advertise the noxious qualities of an organism.

appendicular Of the limbs (e.g., leg bones or muscles).

apteria Regions of skin without feathers.

arcade Curve or arch in a structure, such as the tooth row of humans.

archaic Of a form typical of an earlier evolutionary time.

archinephric duct The ancestral kidney drainage duct.

archipterygium Fin skeleton, as in a lungfish, consisting of symmetrically arranged rays that extend from a central skeletal axis.

articular Pertaining to a joint, or to a bone at the posterior end of the lower jaw.

artiodactyls Ungulate mammals with an even number of toes, such as cows and sheep (contrast with **perissodactyls**).

atrium (plural *atria*) An entrance. The atria of the heart receive blood from the sinus venosus or the veins.

auditory bulla   An elaboration of bone around the ear region in mammals that may increase auditory acuity.

aural   Refering to the external or internal ear or sense of hearing.

auricle   (also called the pinna) The external flap of the mammalian ear.

australopithecines   Extinct Pliocene/Pleistocene hominids, the sister group of *Homo*.

autapomorphy   An attribute unique to one evolutionary lineage of organisms. *See also* apomorphy.

autochthonous   Originating in the region where found.

autonomic nervous system   The part of the peripheral nervous system that controls glands, smooth muscles, and internal organs and produces largely involuntary responses, including the sympathetic and parasympathetic portions in mammals.

autostylic   A type of jaw suspension in fishes in which the upper jaw is attached to the skull by processes (contrast with **hyostylic**).

autotomy   The voluntary release of a portion of the body to escape a predator, as when a lizard loses its tail. Autotomized structures are subsequently regrown.

axial   Within the trunk region.

bacculum   (also known as *os penis*) A bone in the penis of some eutherian mammals.

baleen   (also known as whalebone) Sheets of fibrous, hornlike epidermal tissue that extend downward from the upper jaw and are used for filter feeding by the baleen whales (mysticetes).

benthic   Living at the soil/water interface at the bottom of a body of water.

bilateral symmetry (bisymmetry)   Characteristic of a body that can be divided into mirror-image halves.

binomial nomenclature   The Linnean system that assigns generic and species names to organisms (e.g., *Homo sapiens* for modern humans).

biomass   Living organic material in a habitat (available as food for other species).

biome   Biogeographic region defined by a series of spatially interrelated and characteristic life forms (e.g., tundra, mesopelagic zone, tropical rain forest, coral reef).

bipedality   Locomotion on two legs, as in humans.

birdsong   The longest and most complex vocalization produced by a bird. The song identifies the species and in many species is produced only by mature males and only during the breeding season.

blubber   An insulating layer of fat beneath the skin, typical of marine mammals.

bone   A mineralized tissue that forms the skeleton of vertebrates. Bone is about 50 percent mineralized. Not to be confused with the anatomical structure called "a bone," which may be composed of bone tissue and/or cartilage tissue.

bound   A gait of tetrapods consisting of jumping off the hind legs and landing on the front legs.

brachial   Pertaining to the forelimb.

brachiation   Locomotion by swinging from the underside of one branch to another.

brachydont   Molar teeth with low crowns. *See also* hypsodont.

branchial   Pertaining to the gills.

branchiomeric   Referring to segmentation of structures associated with, or derived from, the ancestral pharyngeal arches. *See also* metameric.

branchiomeric muscles   Muscles powering the visceral arches.

buccal pumping   Drawing air or water in and out of the mouth region by raising and lowering the floor of the mouth.

bunodont   Molar teeth with low, rounded cusps. *See also* lophodont, selenodont.

calamus   The tubular base of a feather that remains in the follicle.

calcaneal heel   The heel of mammals, formed by the a large tarsal bone, the calcaneus.

calcaneus (or calcaneum)   The large metatarsal bone that forms the heel of mammals.

carapace   Dorsal shell, as of a turtle.

cardiac   Referring to the heart.

cardinal vein   A vein found in embryonic vertebrates and adult non-amniotes.

carnassials   Teeth of eutherian mammals in the order Carnivora that are specialized as shearing blades.

carotid arteries   The large arteries in the neck.

cartilage   A firm and flexible skeletal material.

catadromous   Migrating down a river or stream to a lake or ocean to spawn (of fishes). *See also* anadromous.

catastrophism   Hypothesis of major evolutionary change as a result of unique catastrophic events of broad geographic and thus ecological effect.

cavum arteriosum, cavum pulmonale, cavum venosum   The chambers formed during ventricular contraction in the hearts of turtles and lepidosaurs.

cementum   A bonelike substance that fastens the teeth in their sockets.

centrum   (plural *centra*) Bony portion of a vertebra that surrounds the notochord.

cephalic   Pertaining to the head.

ceratotrichia   Keratin fibers that support the web of the fins of Chondrichthyes.

cerebellum   The dorsal part of the metencephalon of the brain.

cerebrum   The two hemispheres of the brain that form most of the telencephalon.

character/character state   Any identifiable characteristic of an organism. Characters can be anatomical, behavioral, ecological, or physiological.

chloride cells   Cells in the gills of fishes and skin of amphibians specialized to transport sodium and chloride ions.

choana   (plural *choanae*) Internal nares.

chondrification   Formation of cartilage.

chondrocranium   (also known as the neurocranium) A structure that surrounds the brain. Initially formed of cartilage, it is replaced by endochondral bone in most bony fishes and tetrapods.

Chordata   Phylum of animals characterized by having a notochord at some stage of life.

chorioallantoic placenta   A placenta developed from the chorionic and allantoic extraembryonic membranes that replaces the choriovitelline placenta during the embryonic development of all eutherian mammals and some marsupials. *See also* placenta.

chorion   The outermost extra-embryonic membrane of amniotes.

choriovitelline placenta   A placenta developed from the yolk sac that is characteristic of all therian mammals during early development. *See also* placenta.

circadian   "About a day." Circadian rhythms are cycles that have a period of approximately 24 hours.

clade   Phylogenetic lineage originating from a common ancestral taxon and including all descendants. *See also* grade.

cladistic   Pertaining to the branching sequences of phylogeny. *See also* phylogenetic systematics.

cladogram   Branching diagram representing the hypothesized relationships of taxa.

cleidoic egg   One independent of environment except for heat and gas (carbon dioxide, oxygen, water vapor) exchange. Characteristic of amniotes.

cline   Change in a biological character along a geographic gradient.

cloaca   A common opening of the reproductive and excretory tracts.

cochlea   (also known as lagena in nonmammalian tetrapods) The portion of the inner ear that houses the hair cells.

coelom (celom)   Body cavity, lined with tissue of mesodermal origin.

coevolution   Complex biotic interaction through evolutionary time resulting in the adaptation of interacting species to unique features of the life histories of the other species in the system.

collagen   A fibrous protein that contributes to many structures.

columella   The single auditory ossicle of the middle ear of non-mammalian tetrapods, the stapes of mammals. Homologous to the hyomandibula of fishes.

concealed estrus   Estrus that is not revealed by external signals, such as swelling of the genitalia. In the anthropoid lineage concealed estrus is a derived character of humans.

condyle   A rounded articular surface of a bone.

cones   Photoreceptor cells in the vertebrate retina that are differentially sensitive to light of different wavelengths and thus perceive color. *See also* double cone.

conodont   Small spinelike or comblike structures formed of apatite found in marine sediments from the Late Cambrian to the Late Triassic. They are considered to be toothlike elements of an early vertebrate, the conodont animal.

conspecific   Belonging to the same species as that under discussion. *See also* heterospecific.

continental drift   The movement of continental blocks on the mantle of the Earth. *See also* plate tectonics.

conus arteriosus   An elastic chamber in front of the ventricle of some gnathostomes.

convergent evolution   Appearance of similar characters in widely separated evolutionary lineages (e.g., wings in bats and birds). *See also* parallel evolution.

coprolite   Fossilized dung.

coprophagy   Eating the first set of feces that are produced, thereby recycling nutrients that would otherwise be lost.

coronoid process   A vertical flange near the rear of the dentary bone that increases the area for attachment of the temporalis muscle.

corpus luteum   (plural *corpora lutea*) A hormone-secreting structure formed in the ovary from the follicular cells remaining after an egg is released.

cosmine   Form of dentine containing branching canals characteristic of the cosmoid scales of sarcoptoygian fishes.

countercurrent exchange   Fluid streams flowing in opposite directions in adjacent vessels to promote exchange of heat or dissolved substance.

countershaded   Referring to a color pattern in which the aspect of the body that is more brightly lighted (normally, the dorsal surface) is darker colored than the less brightly illuminated surface. The effect of countershading is to make an animal harder to distinguish from its background.

cranial kinesis   Movement within the skull or of the upper jaws independent of the skull.

cranial nerves   The nerves that emerge directly from the brain; 10 pairs in the primitive vertebrate condition, 12 pairs in amniotes.

cranial   Pertaining to the cranium or skull, a unique and unifying characteristic of all vertebrates.

Craniata   Animals with a cranium. Hagfishes have a cranium but lack vertebrae and are sometimes classified in the Craniata but not in the Vertebrata.

cranium   A bony, cartilaginous, or fibrous structure surrounding the brain.

crepuscular   Animals that are active at dawn and dusk.

critical period   The restricted time during the ontogeny of an individual when imprinting occurs.

crop   An enlarged portion of the esophagus that is specialized for temporary storage of food.

cryptodires   Turtles that bend the neck in a vertical plane to retract the head into the shell.

cupula   A cup-shaped, gelatinous secretion of a neuromast organ in which the kinocilium and microvilli are embedded.

cursorial   Specialized for running.

dear enemy recognition   A situation in which a territorial animal responds more strongly to strangers than to its neighbors from adjacent territories.

demersal   More dense than water and therefore sinking, as in the eggs of many fishes and amphibians.

denticles   Small toothlike structures in the skin, as in shark skin.

dentine   A mineralized tissue found in the teeth of extant vertebrates and the dermal armor of some primitive fishes. Dentine is about 90 percent mineralized.

derived character   Also known as an apomorphy. A character that has changed from its ancestral condition. *See also* shared derived characters.

dermal bone   A type of bone that forms within the skin.

dermatocranium   Dermal bones that cover a portion of the skull.

dermis   The deeper cell layer of vertebrate skin of mesodermal and neural crest origin. *See also* epidermis.

detritus   Particulate organic matter that sinks to the bottom of a body of water.

deuterostomy   Condition in which the embryonic blastopore forms the anus of the adult animal; characteristic of chordates. *See also* protostomy.

diapsid   A skull with two temporal fenestrae or an animal with a diapsid skull. *See also* anapsid, synapsid.

diastema   A gap between the cheek teeth and the incisors.

digitigrade   Standing with the heel off the ground and the toes flat on the ground, as in dogs. *See also* plantigrade, unguligrade.

diphyodonty   One replacement of the dentition during an animal's lifetime, as in most mammals.

dispersal   Movements of animal populations via movement of the animals themselves. *See also* vicariance biogeography.

distal   Away from the body. *See also* proximal.

distal convoluted tubule   Portion of a kidney nephron responsible for changing the concentration of the ultrafiltrate by actively transporting salt.

diurnal   Animals that are active during the day.

double cone   Type of retinal photoreceptor in which two cones share a single axon. *See also* cones.

down feathers   Entirely plumulaceous feathers in which the rachis is shorter than the longest barb or entirely absent.

drag   Backward force opposed to forward motion.

dryopithecids   Later Cenozoic Eurasian hominoids, more primitive than any living ape.

durophagous   Feeding on hard material.

eccrine gland   A gland in the mammalian skin that produces a watery secretion with little organic content, forms the sweat glands of humans.

echolocation Determining location in three-dimensional space by sending out a pulse of sound and listening to echoes returning from objects in the environment.

ecosystem Community of organisms and their entire physical environment.

ectoderm One of the embryonic germ layers, the outer layer of the embryo.

ectotherm An organism that relies on external sources of heat to raise its body temperature.

edentulous Lacking teeth.

elastin A fibrous protein that can stretch and recoil.

electrocytes Muscle cells modified to produce an electric discharge.

embryonic diapause Maintaining the embryo in a stage of arrested development before it implants on the wall of the uterus.

enamel A mineralized tissue found in the teeth of extant vertebrates and the dermal armor of some primitive fishes. Enamel is about 99 percent mineralized.

encephalization quotient The ratio of the actual brain size of a species compared to the brain size expected from its body size.

endemism Property of being endemic (i.e., found only in a particular region).

endocasts Fossil impressions of the insides of body cavities.

endochondral bone A type of bone that forms in cartilage.

endocrine disruptor A natural or synthetic chemical that interferes with normal development by duplicating the physiological effect of a hormone.

endocrine glands Glands that discharge hormones into the blood.

endoderm Innermost of the germ-cell layers of late embryos.

endometrium The glandular uterine epithelium of mammals that secretes materials that nourish the embryo.

endotherm An organism that relies on internal (metabolic) heat to raise its body temperature.

eon The largest division of geologic time. The history of Earth has occupied three eons: the Archean, Proterozoic, and Phanerozoic.

epaxial Referring to muscles on the dorsal portion of the trunk. *See also* hypaxial.

epicercal (Also known as heterocercal.) A tail fin with the upper lobe larger than the lower lobe. *See also* hypocercal.

epicontinental sea (epeiric sea) Sea extending within the margin of a continent.

epidermis The superficial cell layer of vertebrate skin of ectodermal origin. *See also* dermis.

epigenetic Pertaining to an interaction of tissues during embryonic development that results in the formation of specific structures.

epiphysis 1. Pineal organ, an outgrowth of the roof of the diencephalon. 2. (plural, *epiphyses*) Accessory center of ossification at the ends of the long bones of mammals, birds, and some squamates. In mammals the epiphyses are the actual articulating ends of the long bones themselves, with the cartilaginous zone of growth between the epiphysis and diaphysis. When the ossifications of the shaft (diaphysis) and epiphysis meet, lengthwise growth of the shaft ceases. This process produces a determinate growth pattern.

epiphyte Plant that grows nonparasitically on another plant.

epipubic bones Bones in noneutherian mammals that project anteriorly from the pubis.

epoch The division of the geological time scale between periods and ages (e.g., Paleocene, Eocene, Oligocene, Miocene, Pliocene, Pleistocene, Holocene).

era The division of the geological time scale between eons and periods (e.g., Paleozoic, Mesozoic, Cenozoic).

estivation (aestivation) Form of torpor, usually a response to high temperatures or scarcity of water.

estrous cycle The normal reproductive cycle of growth, maturation, and release of an egg.

estrus (oestrus) The periodic state of sexual excitement in the females of most mammals (but not humans) that immediately precedes ovulation and during which a female is most receptive to mating. Also known as heat.

estuarine Pertaining to, or formed in, a region where the fresh water of rivers mixes with the seawater of a coast.

euryhaline Capable of living in a wide range of salinities. *See also* stenohaline.

euryphagous Eating a wide range of food items; a food generalist. *See also* stenophagous.

eurythermal Capable of tolerating a wide range of temperatures. *See also* stenothermal.

eurytopic Capable of living in a broad range of habitats.

eusocial Applied to a species or group of animals that display all of the following characters: cooperation in caring for the young, reproductive division of labor, more or less sterile individuals aiding individuals engaged in reproduction, and overlap of two or more generations of life stages capable of contributing to colony labor.

eustachian tube Passage connecting the middle ear to the pharynx.

exocrine glands Glands that discharge through a duct into a cavity or onto the body surface.

explosive breeding A very short breeding season.

extant Currently living; e.g., an extant species has living individuals and an extant lineage has living species.

extra-pair copulation Mating with an individual other than the partner in a monogamous breeding system.

extraperitoneal Positioned in the body wall beneath the lining of the coelom (the peritoneum) in contrast to being suspended in the coelom by mesenteries.

fallopian tube The anterior portion of the oviduct where eggs are fertilized.

fenestra A large opening in a bone (.e.g., the temporal fenestra).

ferment To break down food in the absence of oxygen, as in the stomach of a ruminant.

fever An increase in body temperature in response to infection.

filial imprinting The process by which a young animal learns to recognize its parents.

filoplumes Fine, hairlike feathers with a few short barbs or barbules at the tip.

flow-through ventilation Flow of respiratory fluid (air or water) in one direction, as across the gills of a fish.

foramen (plural *foramina*) An opening in a bone (e.g., for the passage of nerves or blood vessels).

foregut fermenter A mammal in which fermentation of foodstuffs is carried out in a modified stomach (e.g., a cow).

fossa A groove or depression in a bone or organ.

fossorial Specialized for burrowing.

fovea centralis Area of the vertebrate retina containing only cone cells, where the most acute vision is achieved at high light intensities.

free nerve ending A sensory nerve ending in the skin that is believed to sense pain.

furcula Avian wishbone formed by the fusion of the two clavicles at their central ends.

fusiform Torpedo shaped.

gallop A gait of mammals that is a modified bound.

gametes Sex cells—that is, eggs (ova) and sperm.

gastralia Bones in the ventral abdominal wall of some reptiles.

gastrolith A stone swallowed to aid digestion by grinding food in the gizzard.

genetic fitness The contribution of one genotype to the next generation relative to the contributions of other genotypes.

genetic monogamy A social system in which a male and female share parental responsibilities and do not mate with individuals outside the pair.

genus A group of related species.

geosyncline Portion of the Earth's crust that has been subjected to downward warping. Sediments frequently accumulate in geosynclines.

gestation Period during which an embryo is developing in the reproductive tract of the mother.

gill arch Assemblage of tissues associated with a gill; the term may refer to the skeletal structure only or to the entire epithelial muscular and connective tissue complex.

gizzard The muscular stomach of birds and other archosaurs.

glomerulus A capillary tuft associated with a kidney nephron that produces an ultrafiltrate of the blood.

gnathostomes Jawed vertebrates. *See also* agnathans.

gonads The organs that produce gametes—ovaries in females, testes in males.

Gondwana Supercontinent that existed either independently or in close contact with all other major continental landmasses throughout vertebrate evolution until the middle of the Mesozoic and was composed of all the modern Southern Hemisphere continents plus the subcontinent of India.

grade A level of morphological organization achieved independently by different evolutionary lineages. *See also* clade.

Great American Interchange (GAI) Faunal interchange between North and South America when the Central American land bridge (the Isthmus of Panama) was formed about 2.5 million years ago.

gymnosperms Group of plants in which the seed is not contained in an ovary—conifers, cycads, and ginkos.

hallux The big toe.

Harderian gland A gland associated with the eye of modern mammals that secretes an oily substance used to preen the fur.

head-starting Rearing neonatal animals in captivity for a period before they are released in the wild.

hemal arch Structure formed by paired projections ventral to the vertebral centrum and enclosing caudal blood vessels.

hermaphroditic Having both male and female gonads.

heterocercal (also known as epicercal) A tail fin with the upper lobe larger than the lower lobe.

heterocoelus Having the articular surfaces of the vertebral centra saddle-shaped, as in modern birds.

heterodont A dentition with teeth of different sizes and shapes in different regions of the jaw. *See also* homodont.

heterospecific Belonging to a different species from that under discussion. *See also* conspecific.

heterosporous plants Plants with large and small spores; the smaller give rise to male gametophytes and the larger to female gametophytes (equivalent to protogymnosperms).

hibernation A period of torpor in the winter when food is scarce.

hindgut fermenter A mammal in which fermentation of foodstuffs is carried out in the intestine (e.g., a horse).

holocrine gland Type of gland in which the entire cell is destroyed with the discharge of its contents. *See also* apocrine gland.

home range The area within which an animal spends most of its time. A home range is not defended. *See also* territory.

homodont A dentition in which teeth do not vary in size or shape along the jaw. *See also* heterodont.

homologous Inherited via common ancestry.

homology The fundamental similarity of individual structures that belong to different species within a monophyletic group.

homoplasy Similarities that do not indicate common ancestry. Structures resulting from parallel and convergent evolution and evolutionary reversal are examples of homoplasy.

hormone A chemical messenger molecule carried in the blood from its site of release to its site of action.

Hox/**homeobox genes** A sequence of DNA that regulates the expression of genes that control development of body structures.

hydrofoil A water-planing surface, such as the pectoral fins of sharks.

hydrosphere Free liquid water of the Earth—oceans, lakes, rivers, and so on.

hyoid arch The second gill arch.

hyostylic a type of jaw suspension in fishes in which the upper jaw is attached to the skull by the hyomandibula (contrast with **autostylic**)

hyostylic jaw articulation A form of jaw attachment seen in sharks that allows great flexibility.

hypapophyses Sharp processes, on the ventral surface of the neck vertebrae of the egg-eating snake (*Dasypeltis*), that slice through the shell of an egg.

hypaxial Referring to muscles on the ventral portion of the trunk. *See also* epaxial.

hyperdactyly A condition in which the number of digits is increased above the usual tetrapod complement of five.

hyperosmolal Of greater osmotic activity.

hyperphalangy Increase in the number of bones (phalanges) in the digits.

hyperthermia High body temperature.

hypertrophy Increase in the size of a structure.

hypocercal A tail fin with the lower lobe larger than the upper lobe. *See also* epicercal.

hypophysis The pituitary gland.

hyposmolal Of lower osmotic activity.

hypothermia Low body temperature.

hypotreme Having the main gill openings on the ventral surface and beneath the pectoral fins, as in skates and rays. *See also* pleurotreme.

hypselodont Molar teeth with ever-growing crowns. *See also* brachydont.

hypsodont Molar teeth with high crowns.

imprinting A special kind of learning that occurs only during a restricted time (called the critical period) in the ontogeny of an individual.

inclusive fitness The sum of individual fitness plus the effect of kin selection.

index of refraction The amount of deflection of a ray of light as it passes from one medium into another.

individual fitness *See* genetic fitness.

infraorbital foramen A hole beneath the eye through which nerves and blood vessels pass to the muzzle.

infrared The portion of the electromagnetic spectrum with wavelengths from 750 nanometers (just beyond visible light) to 1 millimeter (just before microwave radiation). Often called thermal radiation or heat.

infrasound Sound frequencies below the range of human hearing, approximately 20 hertz.

ingroup The group of organisms being considered. *See also* outgroup.

insolation Solar radiation reaching the Earth's surface.

intercalary cartilage A cartilage lying between the last two bones in the toes of some tree frogs.

intercalary plates Extra elements in the vertebral column of elasmobranchs that protect the spinal cord and major blood vessels.

interspecific Pertaining to phenomena occurring between members of different species.

intraspecific Pertaining to phenomena occurring between members of the same species.

isognathy The situation in which the tooth rows in the upper and lower jaws are the same distance apart. *See also* anisognathy.

isohaline Of the same salt concentration.

isometry Describes a situation in which the proportions of an object remain the same when size changes (contrast with **allometry**).

isosmolal Of the same osmotic activity.

isostasy Condition of gravitational balance between segments of the Earth's crust or of return to balance after a disturbance.

isostatic movement Vertical displacement of the lithosphere due to changes in the mass over a point or region of the Earth.

isotherm Line on a map that connects points of equal temperature.

iteropary Producing several individual babies or litters of young during the lifetime of a female. *See also* semelpary.

Jacobson's organ Also known as vomeronasal organ. An olfactory organ in the roof of the mouth of tetrapods.

keratin A fibrous protein found only in vertebrates that forms epidermal structures such as hair, scales, feathers, and claws.

kin selection Favoring the perpetuation of one's own genes by helping relatives to reproduce.

kinocilium A sensory cell located in neuromast organs.

lactation Producing milk from mammary glands to nourish young; characteristic of mammals.

lagena (also known as cochlea in mammals) The portion of the inner ear that houses the hair cells.

lateral plate mesoderm The ventral part of the mesoderm, surrounding the gut.

lateral line system The sensory system on the body surface of aquatic vertebrates that detects water movement.

Laurentia Paleozoic continent that included most of present-day North America, Greenland, Scotland, and part of northwestern Asia.

lecithotrophy Embryonic development nourished by the yolk when eggs are retained with the reproductive tract until they hatch. *See also* ovoviviparity.

leptocephalus larva Specialized, transparent, ribbon-shaped larva of tarpons, true eels, and their relatives.

lift Vertical force opposed to gravity.

Linnaean system A system of naming living organisms developed by the Swedish naturalist Carl von Linné (Carolus Linnaeus) in the eighteenth century.

lithosphere Crust of the Earth.

littoral Pertaining to the shallow portion of a lake, sea, or ocean where rooted plants are capable of growing.

loop of Henle The portion of the renal tubule of mammals that extends into the medulla. Essential for establishing the concentration gradient that produces a small volume of highly concentrated urine.

lophodont Molar teeth with ridges (lophs) that run in a predominantly internal-external direction across the tooth. *See also* bunodont, selenodont.

lophophorate Pertaining to several kinds of marine animals that possess ciliated tentacles (lophophores) used to collect food (e.g., pterobranchs).

lower critical temperature The point at which an endotherm must increase its metabolic heat production to maintain a stable body temperature.

lower lethal temperature The temperature at which even maximum metabolic heat production is inadequate to maintain a stable body temperature.

mammary gland A gland, found in mammals, that secretes milk. Mammary glands have characters of both apocrine and eccrine glands.

mammilary bodies Organic granules attached to the egg membrane that are the sites of first formation of calcite crystals making up the egg shell.

mandibular arch The most anterior of the gill arches, forming the jaws of gnathostomes.

marginal value theorem The hypothesis that an animal stops foraging in a patch of food when the rate of energy intake falls to the average rate for the habitat as a whole.

marsupium An external pouch in which the young of marsupial mammals develop.

masseter muscle A jaw muscle of mammals originating from the zygomatic arch and inserting on the lower jaw.

masticate To chew thoroughly.

matrilineage Related females within a group that support each other in social interactions.

matrotrophy Embryonic development nourished by materials transferred from the maternal circulation. Placentrophic matrotrophy describes the situation in which a placenta is the site of transfer of nutrients and wastes between the embryo and maternal circulation.

megafauna Species of large terrestrial animals (mammals, birds, and reptiles) that became extinct when human populations expanded, mostly between 50,000 and 10,000 years ago.

Meissner's corpuscle A sensory nerve ending in the skin that is believed to sense touch.

melanocyte A pigment cell containing melanin.

meninges Sheets of tissue enclosing the central nervous system. In mammals these are the dura mater, arachnoid, and pia mater.

menstrual cycle The periodic shedding of the endometrial lining of the uterus; characteristic of humans and some other anthropoid primates.

mesenteries Membranous sheets derived from the mesoderm that envelop and suspend the viscera from the body wall within the coelom.

mesoblast Mesodermal cell.

mesoderm Central of three germ layers of late embryos.

metameric Pertaining to ancestral segmentation, used in reference to serially repeated units along the body axis.

metamorphic climax The period in the life of a tadpole that begins with appearance of the forelimbs and ends with disappearance of the tail.

metamorphosis The developmental transition from larval to adult body form.

metanephric kidney The adult kidney of amniotes (contrast with **opistonephric kidney**).

microvilli Sensory cells located in neuromast organs.

mimicry A tripartite system in which one organism (the mimic) counterfeits the signal of a second organism (the model), thereby deceiving a third organism (the dupe). The signal can be any characteristic of the model that the dupe can perceive—color, pattern, scent, etc.

molting Replacement of old hairs or feathers with new ones.

monogamy A mating system based on a pair bond between a single male and female. *See also* genetic monogamy, social monogamy, and polygamy.

monophyletic Having a single evolutionary origin. *See also* paraphyletic, polyphyletic.

monophyletic lineage A taxon composed of a common ancestor and all its descendants.

monophyly Relationship of two or more taxa having a common ancestor.

monophyodonty No replacement of the dentition during an animal's lifetime (compare to **diphyodonty** and **polyphyodonty**).

morph Genetically determined variant in a population.

morphotypic Referring to a type of classification based entirely on physical form.

musth An annual period of elevated testosterone levels in the blood of male elephants.

myomeres Blocks of striated muscle fiber arranged along both sides of the body, most obvious in fishes.

myrmecophagy Eating ants and termites.

naris (plural *nares*) The external opening of the nostril.

nasolabial groove A channel from the external naris to the lip found in plethodontid salamanders.

Neo-Darwinism *See* New Synthesis.

neonates Newborn individuals.

neopallium (also known as neocortex). The derived expanded portion of the mammalian cerebral cortex.

neoteny Retention of larval or embryonic characteristics past the time of reproductive maturity. *See also* paedomorphosis and progenesis.

nephron The basic functional unit of the kidney.

neural arch Dorsal projection from the vertebral centrum that, at its base, encloses the spinal cord.

neural crest A type of embryonic tissue unique to vertebrates that forms many structures, especially in the head region.

neurocranium (also known as chondrocranium) Portion of the head skeleton encasing the brain.

neuromast organs Clusters of sensory hair cells and associated structures on the surface of the head and body of aquatic vertebrates, usually enclosed within the lateral line system.

neuron The basic functional unit of the nervous system.

New Synthesis (also known as Neo-Darwinism). The combination of genetics and evolutionary biology developed in the early twentieth century.

niche The functional role of a species or other taxon in its environment—the ways in which it interacts with both the living and nonliving elements.

nocturnal Animals that are active at night.

notochord A dorsal stiffening rod that gives the phylum Chordata its name.

occipital Pertaining to the posterior part of the skull.

odontodes Small toothlike elements in the skin. The original toothlike components of primitive vertebrate dermal armor. The denticles of sharkskin are odontodes.

ontogentic Pertaining to the development of an individual.

ontogeny The development of an individual. *See also* phylogeny.

operculum (plural operculae) Flap or plate of tissue covering the gills.

opistoglyphs Venomous snakes with enlarged teeth in the rear of the jaw; rear-fanged snakes.

opistonephric kidney The adult kidney of most non-amniotes.

optimal foraging theory The hypothesis that an animal adjusts its foraging behavior to maximize energy return per unit time.

orobranchial chamber The mouth and gill region of a vertebrate.

orogeny Process of crustal uplift or mountain building.

osmosis Movement of water across a membrane from a region of high water potential (low solute concentration) to a region of low water potential (high solute concentration).

osseous Bony.

osteoderm A bone embedded in the skin; characteristic of crocodilians.

ostracoderm Armored jawless aquatic vertebrates known from the Ordovician to the Devonian.

otolith A mineralized structure in the inner ear of teleost fishes.

outgroup Group of organisms that is related to but removed from the group under study. One or more outgroups are examined to determine which character states are evolutionary novelties (apomorphies).

ovary The female gonad.

oviparity Depositing eggs that develop outside the body.

ovoviviparity Embryonic development nourished by the yolk when eggs are retained with the reproductive tract until they hatch. (Lecithotrophy is the preferred term for this type of embryonic nourishment.)

pachyostosis Increased density of bone; characteristic of diving animals.

Pacinian corpuscle A sensory nerve ending in the skin that is believed to sense pressure.

paedomorphosis Condition in which a larva becomes sexually mature without attaining the adult body form. Paedomorphosis may be achieved by neoteny or by progenesis.

palatoquadrate Upper jaw element of primitive fishes and Chondrichthyes, portions of which contribute to the palate, jaw articulation, and middle ear of other vertebrates.

pancreas A glandular outgrowth of the intestine that secretes digestive enzymes.

pancreatic islets (also known as the islets of Langerhans) Clusters of endocrine cells in the pancreas that secrete insulin.

Pangaea (Pangea) Single supercontinent that existed during the mid-Paleozoic and consisted of all modern continents apparently in direct physical contact with a minimum of isolating physical barriers.

parallel evolution Appearance of similar characters in lineages that have separated recently (e.g., long hind legs in hopping rodents from the North American and African deserts). *See also* convergent evolution.

paraphyletic Referring to a taxon that includes the common ancestor and some but not all of its descendants. *See also* monophyletic, polyphyletic.

parasympathetic nervous system The division of the autonomic nervous system that maintains normal body functions, such as digestion.

parsimonious In evolutionary biology, the hypothesis that requires the fewest changes from ancestral to derived character states.

parthenogenesis Reproduction by females without fertilization by males.

parturition Giving birth.

pectoralis major The large breast muscle that powers the downstroke of the wings in a bird.

pelage The hairy covering of a mammal.

pelagic Living in the open ocean.

pelvic patch   A vascularized area in the pelvic region of anurans that is responsible for uptake of water.

period   The division of the geological time scale between eras and epochs (e.g., Cambrian, Ordovician, Silurian, Devonian, Carboniferous, Permian, Triassic, Jurassic, Cretaceous, Tertiary, Quaternary).

pericardial cavity   The portion of the coelom that surrounds the heart.

pericardium   Thin sheets of lateral-plate mesoderm that line the pericardial cavity.

peritoneum   Thin sheets of lateral-plate mesoderm that line the pleuroperitoneal cavity.

perissodactyls   Ungulate mammals with an odd number of toes, such as horses (contrast with **artiodactyls**).

peritoneal cavity   The portion of the pleuroperitoneal cavity surrounding the viscera.

Phanerozoic   Period of time (Eon) since the start of the Cambrian.

pharyngeal arches   (also known as visceral skeleton) The gill supports between the pharyngeal gill slits.

pharyngeal slits   Openings in the pharynx that were originally used to filter food particles from the water.

pharyngotremy   Condition in which the pharyngeal walls are perforated by slitlike openings; found in chordates and hemichordates.

pharynx   The throat region.

pheromone   A chemical signal released by one individual that affects the behavior of other individuals of the species.

photophore   Light-emitting organ.

phylogenetic   Pertaining to the development of an evolutionary lineage. *See also* ontogenetic.

phylogenetic systematics   (also known as cladistics). A classification system that is based on the branching sequences of evolution.

phylogeny   The evolutionary development of a group. *See also* ontogeny.

physoclistic   Lacking a connection from the gut to the swim bladder in adults (of fishes).

physostomous   Having a connection between the swim bladder and gut in adults (of fishes).

piloerection   Contraction of muscles attached to hair follicles resulting in the erection of the hair shafts.

pinna   The external ear of mammals.

piscivorous   Eating fish.

pitch   Tilt up or down parallel to the long axis of the body. *See also* roll, yaw.

placenta   Extraembryonic tissue that obtains nutrients from the endometrium of the uterus and secretes hormones to signal the state of pregnancy to the mother. *See also* choriovitelline placenta and chorioallantoic placenta.

placentotrophic matrotrophy   Embryonic development nourished by materials transferred from the maternal circulation via a placenta.

placoid scales   Primitive type of scale found in elasmobranchs and homologous with vertebrate teeth.

plantigrade   Standing with the foot flat on the ground, as in humans. *See also* digitigrade, unguligrade.

plastron   Ventral shell, as of a turtle.

plate tectonics   Theory of Earth history in which the lithosphere is continually being generated from the underlying core at specific areas and reabsorbed into the core at others, resulting in a series of conveyor-like plates that carry the continents across the face of the Earth.

plesiomorphic   Pertaining to the ancestral character from which an apomorphy is derived.

plesiomorphy   An ancestral character (i.e., one that has not changed from its ancestral condition). *See also* symplesiomophy, synapomorphy.

pleural cavities   Paired portions of the peritoneal cavity surrounding the lungs.

pleurodires   Turtles that bend the neck in a horizontal plane to retract the head into the shell.

pleurodont   Teeth fused to the inner surface of the jaw bones. *See also* acrodont, thecodont.

pleuroperitoneal cavity   The portion of the coelom that surrounds the viscera.

pleurotremate   Having the main gill openings on sides of the body anterior to the pectoral fins, as in sharks. *See also* hypotremate.

polarity   The direction of evolutionary change in a character.

polyandry   A mating system in which a female mates with more than one male.

polygamy   A mating system in which an individual has more than one mate in a breeding season. *See also* monogamy.

polygyny   A mating system in which a male mates with more than one female.

polymorphism   Simultaneous occurrence of two or more distinct phenotypes in a population.

polyphyletic   Referring to a taxon that does not contain the most recent common ancestor of all the subordinate taxa of the taxon (i.e., not a true taxonomic unit but an assemblage of similar taxa such as "marine mammals"). *See also* monophyletic, paraphyletic.

polyphyodonty   Having more than one replacement set of teeth in a lifetime (see **diphyodonty**).

portal system   Portion of the venous system specialized for the transport of substances from the site of production to the site of action. A portal system begins and ends in capillary beds.

postzygapophysis   Articulating surface on the posterior face of a vertebral neural arch. *See also* prezygapophysis.

powder down feathers   Feathers that break down to produce an extremely fine white powder composed of granules of keratin.

precocial   Well developed and capable of locomotion soon after birth or hatching (e.g., as in chickens and cows). *See also* altricial.

prezygapophysis   Articulating surface on the anterior face of a vertebral neural arch. *See also* postzygapophysis.

progenesis   Accelerated development of reproductive organs relative to somatic tissue, leading to paedomorphosis.

prolonged breeding   A long breeding season.

promiscuity   a breeding system in which both males and females have more than one mate in a breeding season.

proprioception   The neural mechanism that senses the positions of the limbs in space. A derived character of tetrapods.

proteroglyphs   Venomous snakes with permanently erect fangs at the front of the jaw (i.e., cobras and their relatives).

Proterozoic   Later part of the Precambrian, from about 1.5 billion years ago until the beginning of the Cambrian 54 million years ago. *See also* Phanerozoic.

protostomy   Condition in which the embryonic blastopore forms the mouth of the adult animal. *See also* deuterostomy.

prototherians   The monotreme mammals.

protraction   Movement away from the center of the body usually in a forward direction. *See also* retraction.

protrusible   Capable of being moved away (protruded) from the body.

proventriculus The glandular stomach of birds.

proximal Close to the body. *See also* distal.

proximal convoluted tubule Portion of a kidney nephron responsible for changing the concentration of the ultrafiltrate by actively transporting salt.

pseudovaginal canal A midline structure in marsupials through which the young are born.

pterylae Tracts of follicles from which feathers grow.

pygostyle The fused caudal vertebrae of a bird that support the tail feathers.

rachis The central structure of a feather, from which barbs extend.

ram ventilation A respiratory current across the gills; created by swimming with the mouth open.

rectrices (singular *rectrix*) Tail feathers.

refugium Isolated area of habitat fragmented from a formerly more extensive biome.

regional heterothermy Maintaining different temperatures in different parts of the body.

remiges (singular *remex*) Wing feathers.

resource dispersion hypothesis The proposal that the size of an animal's home range will be determined by its needs for resources, such as food, and by the spatial distribution of resources in the environment.

rete mirabile "Marvelous net," a complex mass of intertwined capillaries specialized for exchange of heat and/or dissolved substances between countercurrent flows.

retraction Movement toward the center of the body or in a backward direction. *See also* protraction.

reversal Return to an ancestral feature (e.g., the streamlined body form of whales and porpoises).

ricochet A bipedal hopping gait, as in kangaroos and many rodents.

rod Photoreceptor cell in the vertebrate retina specialized to function effectively under conditions of dim light.

roll Rotate around the long axis of the body. *See also* pitch, yaw.

rostrum Snout; especially an extension anterior to the mouth.

ruminant An herbivorous mammal with a specialized stomach in which microorganisms ferment plant material.

scapulocoracoid cartilage In elasmobranchs and certain primitive gnathostomes, the single solid element of the pectoral girdle.

scutes Scales, especially broad or inflexible ones.

sebaceous gland A gland in mammal skin that secretes oily or waxy materials.

sebum An oily secretion produced by sebaceous glands.

secondary lamellae Projections from the gill filaments where gas exchange occurs.

selenodont Molar teeth with crescentic ridges or lophs rather than cusps that run in a predominantly anterior to posterior direction across the tooth. *See also* bunodont, lophodont.

semelpary Reproducing only once during the lifetime of a female. *See also* iteropary.

semiplumes Feathers intermediate in structure between contour feathers and down feathers.

serial Repeated, as in the body segments of vertebrates.

sexual imprinting The process by which a young animal learns to recognize a mate of its own species.

shared derived characters Derived characters shared by two or more taxa. *See also* synapomorphy.

shivering Generation of heat by asynchronous contraction of muscle fibers.

sinus Open space in a duct or tubular system.

sinus venosus The posteriormost chamber of the heart of non-amniotes, and some reptiles, that receives blood from the systemic veins.

sister group Group of organisms most closely related to the study taxa, excluding their direct descendants.

sivapithecids Later Cenozoic Eurasian hominoids related to the extant orangutan.

sociality Living in structured groups.

social monogamy A mating system in which a male and female share parental responsibility but mate with individuals outside the pair (contrast with **genetic monogamy**).

solenoglyphs Venomous snakes with long fangs in the front of the jaw that are rotated when the mouth is open; vipers.

solute A substance dissolved in a liquid.

somatic nervous system The part of the peripheral nervous system that innervates structures derived from the somatic mesoderm controlling voluntary movements of skeletal muscles and returning sensations from the periphery.

somite Member of a series of paired segments of the embryonic dorsal mesoderm of vertebrates.

spawning The process by which fishes deposit and fertilize eggs.

species In biological time, groups of organisms that are reproductively separated from other groups. In evolutionary time, a lineage that follows its own evolutionary trajectory.

specific dynamic action Increased heat production associated with digesting food.

speciose Referring to a taxon that contains a large number of species.

spermatophore A packet of sperm transferred from male to female during mating by most species of salamanders.

splanchnocranium The visceral or pharyngeal skeleton associated with the gills.

spleen An organ in which blood cells are produced, stored, and broken down.

squamation Scaly covering of the body.

standard metabolic rate The rate of metabolism that sustains vital functions (respiration, blood flow, etc.) in an animal at rest.

stapes Called the *columella* in non-mammalian tetrapods. The single auditory ossicles of the middle ear of tetrapods other than mammals, part of ossicular chain of mammals. Homologous to the hyomandibula of fishes.

stenohaline Capable of living only within a narrow range of salinity of surrounding water; not capable of surviving a great change in salinity. *See also* euryhaline.

stenophagous Eating a narrow range of food items; a food specialist. *See also* euryphagous.

stenothermal Capable of living or being active in only a narrow range of temperatures. *See also* eurythermal.

stratigraphy Classification, correlation, and interpretation of stratified rocks.

stratum (plural *strata*) A layer of material.

suckling Forming fleshy seals against the bony hard palate with the tongue and with the epiglottis, isolating breathing and swallowing during nursing.

supercooling Lowering the temperature of a fluid below its freezing point without initiating crystallization.

surface-to-volume ratio The ratio of body surface area to body volume, often expressed as $cm^2 \cdot cm^3$.

swim bladder (also known as gas bladder) A buoyancy structure of bony fishes. Usually filled with gas, but in coelacanths it is filled with fat.

symbiont   An organism that lives with (usually inside or attached to) another organism, to their mutual benefit.

sympathetic nervous system   The division of the autonomic nervous system that produces largely involuntary responses, which prepare the body for stressful or highly energetic situations.

sympatry   Occurrence of two or more species in the same area.

symphysis   A joint between bones formed by a pad or disk of fibrocartilage that allows a small degree of movement.

symplesiomorphy   Character shared by a group of organisms that is found in their common ancestor (i.e., a primitive character). *See also* plesiomorphy.

synapomorphy   Derived characters (apomorphies) shared by two or more taxa. *See also* plesiomorphy.

synapsid   A skull with a single temporal fenestra or an animal with a synapsid skull. *See also* anapsid, diapsid.

synsacrum   Fused vertebrae and ribs of birds that articulate with the pelvis.

syrinx   The vocal organ of birds, lying at the base of the trachea.

tadpole   The larval form of anurans.

talonid   Basinlike heel on a lower molar tooth, found in therian mammals.

tapetum lucidum   A reflective layer behind the retina that increases sensitivity in low light by directing light back through the retina.

tarsometatarsus   Bone formed by fusion of the distal tarsal elements with the metatarsals in birds and some dinosaurs. *See also* tibiotarsus.

taxon (plural taxa)   Any scientifically recognized group of organisms united by common ancestry.

telencephalon   The front part of the vertebrate brain that contains the cerebral cortex.

temperature-dependent sex determination   Situation in which the sex of an individual is determined by the temperature it experienced during embryonic development. Universal among crocodilians, widespread among turtles, occasional among squamates.

temporal fenestra   An opening in bone of the temporal region of the skull that allows passage of jaw muscles from the skull to the lower jaw.

tentacle   A sensory organ of caecilians that allows chemical substances to be transported from the surroundings to the vomeronasal organ.

territory   An area that is defended against incursion by other individuals of the species. *See also* home range.

testis   (plural *testes*) The male gonad.

tetrapods   Terrestrial vertebrates descended from a four-legged ancestor.

thecodont   Teeth set in sockets in the jaw bones. Also refers to a paraphyletic assemblage of basal, extinct archosaurian reptiles. *See also* acrodont, pleurodont.

therians   Marsupial and eutherian mammals.

thermoneutral zone (TNZ)   The range of ambient temperatures within which an endotherm can maintain a stable body temperature by changing the rate of heat loss to the environment. Also called the zone of physical thermoregulation.

thermophilic   Favoring high temperatures.

thermoregulation   Control of body temperature.

tibiotarsus   Bone formed by fusion of the tibia and proximal tarsal elements in birds and some dinosaurs. *See also* tarsometatarsus.

tidal ventilation   In-and-out flow of respiratory fluid, as in the lungs of a tetrapod.

torpor   A period of inactivity accompanied by a reduction in the regulated body temperature.

tribosphenic molars   Tooth form unique to therian mammals.

troglodyte   Organism that lives in caves.

trophic   Pertaining to feeding and nutrition.

trophoblast   Embryonic tissue of mammals specialized for implanting the embryo on the wall of the uterus, obtaining nutrients from the mother, and secreting hormones to signal the state of pregnancy to the mother.

trot   A gait of tetrapods in which diagonal pairs of limbs are moved together with a period of suspension between each pair of limb movements when all four feet are off the ground.

turbinates   Scroll-like bones in the nasal passages covered by moist tissues that warm and humidify inspired air.

tympanic membrane or tympanum   The eardrum.

ultrafiltrate   A fluid produced in the glomerulus of a nephron; composed of blood with the cells and large molecules removed by filtration.

ultrasound   Sound frequencies above the range of human hearing, approximately 20 kilohertz.

unguligrade   Standing with only the tips of the toes on the ground, as in horses. *See also* digitigrade, plantigrade.

upper critical temperature   The point at which an endotherm must initiate evaporative cooling to maintain a stable body temperature.

upper lethal temperature   The environmental temperature at which cooling mechanisms are insufficient to prevent an explosive rise in body temperature that leads to death.

urea cycle   The enzymatic pathway by which urea is synthesized from ammonia.

ureotelism   Excreting nitrogenous wastes primarily as urea.

ureter   The duct in amniotes that carries urine from the kidney to the urinary bladder or to the cloaca.

urethra   The duct in amniotes that carries urine from the bladder to the outside. In male therian mammals, part of the urethra also carries sperm.

uricotelism   Excreting nitrogenous wastes primarily as uric acid and its salts.

urogenital   Pertaining to the organs, ducts, and structures of the excretory and reproductive systems.

urogenital sinus   Combined opening of the urethra and vagina in most female mammals. (Primates and some rodents have separate openings.)

urostyle   A solid rod formed by fused posterior vertebrae; found in anurans.

vacuoles   Membrane-bound spaces within cells containing secretions, storage products, etc.

vasa deferentia   (singular *vas deferens*) The male reproductive tract of mammals.

vasa recta   The blood vessels surrounding the loop of Henle.

vascular   Relating to blood and blood vessels.

vascular plexus   Intertwined blood vessels in the skin that are the basis for countercurrent blood flow.

vasodilation   Expansion of blood vessels to increase blood flow to a region.

ventricle   A chamber. The ventricle of the heart is the portion that applies force to eject blood from the heart.

Vertebrata   Animals that have vertebrae.

vibrissae   The sensory whiskers of mammals.

vicariance biogeography   Animals and plants being carried passively on moving landmasses. *See also* dispersal.

viscera   Internal organs suspended within the coelom.

visceral arches   Gills and jaws.

visceral nervous system   The part of the peripheral system that innervates portions of the body derived from the lateral plate mesoderm. Includes the autonomic nervous system and sensory nerves that relay information from the viscera and blood vessels. May also include the special branchial motor system of the cranial nerves, although this is now in dispute.

visceral skeleton   Skeleton primitively associated with the pharyngeal arches, uniquely derived from the neural-crest cells and forming in mesoderm immediately adjacent to the endoderm lining the gut.

viviparity   Giving birth to young as opposed to laying eggs.

vomeronasal organ   (also known as Jacobson's organ) An olfactory organ in the roof of the mouth of tetrapods.

vulnerable   A species is considered vulnerable when it is facing a high risk of extinction in the wild in the medium-term future.

Weberian apparatus   A chain of small bones that conducts vibrations from the swim bladder to the inner ear of some bony fishes.

wet adhesion   The process by which arboreal species of frogs stick to smooth surfaces, such as leaves.

yaw   Swing from side to side relative to the long axis of the body. *See also* pitch, roll.

zone of chemical thermogenesis   The range of temperatures within which an endotherm can maintain a stable body temperature via metabolic heat production.

zone of evaporative cooling   Environmental temperatures above the thermoneutral zone; in the zone of evaporative cooling animals pant, sweat, and employ gular fluttering for cooling.

zone of tolerance   The range of ambient temperatures over which an endotherm can maintain a stable body temperature.

Zugdisposition   Preparation for migration by accumulating fat.

Zugstimmung   The condition in which a bird makes migratory flights.

Zugunruhe   Restlessness of caged birds that are prevented from migrating.

zygapophysis   Articular process of the neural arch of a vertebra. *See also* postzygapophysis and prezygapophysis.

zygodactylous   Type of foot in which the toes are arranged in two opposable groups.

zygomatic arch   A temporal bar that is bowed outward to accommodate a large masseter muscle in mammals.

Credits

Photo Credits

Chapter 1 1–2a: Tim Davenport/WCS. **1–2b:** WWF/AP Images. **1–6d:** Matthias Breiter/Minden Pictures/Getty Images. **1–6e:** Lynn Stone/Animals Animals-Earth Scenes.

Chapter 6 6–17: Edward B. Brothers. **6–20:** F. Harvey Pough. **6–23b:** Kathleen Whitlock.

Chapter 10 10–2: Gerhard Roth, University of Bremen, Bremen, Germany. **10–8a,b:** Sharon B. Emerson, University of Utah. **10–14, 10–16a,b:** Theodore L. Taigen, University of Connecticut. **10–20a,e:** Richard J. Wassersug, Dalhousie Medical School. **10–24a,b,c:** F. Harvey Pough.

Chapter 12 12–1: Gene Gaffney. **12–13a,b:** Larry Herbst, University of Florida Health Science Center. **12–14:** Elliott R. Jacobson, University of Florida College of Veterinary Medicine.

Chapter 13 13–1a: Marcus J. Simons, University of Otago, New Zealand. **13–1b:** Alison Cree, University of Otago, New Zealand. **13–10a,b,c:** Benjamin E. Dial, Chapman University. **13–14a,b,c,d:** C.J. Cole, American Museum of Natural History. **13–14e:** Charles M. Bogert, American Museum of Natural History. **13–18a:** Daniel H. Janzen, University of Pennsylvania. **13–18b,c:** F. Harvey Pough.

Chapter 14 14–1a,b: R. Bruce Bury/U.S. Fish and Wildlife Service. **14–4:** F. Harvey Pough. **14–7:** David M. Dennis. **14–10a,b:** J. and K. Storey, Carleton University, Ottawa, Ontario.

Chapter 16 16–7a,b,c: Jeffrey W. Lang, University of North Dakota. **16–14a:** Neg. no. 410765; Photo by Shackleford; From the Department of Library Services, American Museum of Natural History. **16–14b:** Image no. K17088; From the Department of Library Services, American Museum of Natural History. **16–15:** Transparency no. K17360; Illustration by Mick Ellison/DVP; photo by Denis Finnin; From the Department of Library Services, American Museum of Natural History. **16–19:** Barbara Moore, Peabody Museum of Natural History, Yale University. **16–22:** Mick Ellison/American Museum of Natural History. **16–25:** Neg. no. 325097: shot from Sanford Bird Hall; From the Department of Library Services, American Museum of Natural History.

Chapter 17 17–1b: Department of Geology & Palaeontology/Natural History Museum Vienna (NHMW). **17–5b:** Alan S. Pooley. **17–6:** Fred Tilly/Leonard Rue Enterprises/Photo Researchers. **17–8b:** Harry Taylor, ©Dorling Kindersley, from the Natural History Museum, London. **17–27a:** Steven Katovich, USDA Forest Service/Bugwood. org.

17–28, 17–29: Mary Tremaine, Cornell Laboratory of Ornithology. **17–33a:** Allan D. Cruickshank/National Audubon Society/Photo Researchers. **17–33b:** Joan Baron. **17–33c:** Bruce W. Heinemann. **17–33d:** Jen and Des Bartlett/Photo Researchers. **17–34a,b:** David W. Winkler, Cornell University.

Chapter 22 22–2a: Minden Pictures/Getty Images. **22–2b:** Gunter Ziesler/Peter Arnold. **22–2c:** Adrian Warren www.lastrefuge.co.uk. **22–2d:** Jim Clare/NPL/Minden Pictures. **22–8:** Gregory K. Scott/Photo Researchers. **22–9:** Gail R. Michener, University of Lethbridge, Alberta. **22–13:** Peter Ward/Bruce Coleman. **22–16:** George A. Bartholomew and Mark A. Chappell, University of California at Los Angeles. **22–19a,b:** Albert F. Bennett, University of California at Irvine.

Chapter 23 23–3a,b,c,d: Sara Cairns. **23–5a:** Jack A. Cranford, Virginia Polytechnic Institute and State University. **23–5b:** Sara Cairns. **23–5c:** Ric Ergenbright/Corbis. **23–5d:** Sara Cairns. **23–8a:** J & B Photographers/Animals Animals-Earth Scenes. **23–8b:** Stephen J. Krasemann/Photo Researchers. **23–10a,b:** Carol D. Saunders.

Chapter 25 25–3a: Natalie Fobes/Corbis. **25–3b:** Alexandra Morton. **25–3c:** Aqua Bounty Technologies. **25–5:** Kevin Ebi/Living Wilderness. **25–6:** François Gohier/BIOS/Peter Arnold. **25–7a:** Maslowski/Photo Researchers. **25–7b:** Carl R. Sams II/Peter Arnold. **25–7c:** E.R. Degginger/Color-Pic. **25–8:** Smithsonian Institution Photo Services. **25–9a:** M. Harvey/DRK Photo. **25–9b:** M.P. Kahl/DRK Photo. **25–10:** Xinhua/Landov. **25–12b:** Zoological Society of London. **25–15a:** Ron Garrison/San Diego Wild Animal Park. ©1995 Zoological Society of San Diego, San Diego, California, U.S.A. All rights reserved. **25–15b:** David Hosking/Photo Researchers.

Color Insert CI-1LL, CI-1LR: Norbert Wu/DRK Photo. **CI-1UL:** Tom McHugh/Photo Researchers. **CI-1UR:** Bucky Reeves/National Audubon Society/Photo Researchers. **CI-2LL, CI-2ML, CI-2MR:** F. Harvey Pough. **CI-2TL:** Michael Lustbader/Photo Researchers. **CI-2TR:** Cosmos Blank/ National Audubon Society/Photo Researchers. **CI-3BL:** Fred McConnaughey/Photo Researchers. **CI-3BR:** Jany Sauvanet/Photo Researchers. **CI-3TL:** J.H. Robinson/Photo Researchers. **CI-3TR:** Cosmos Blank/National Audubon Society/ Photo Researchers. **CI-16: (Amur CI-4 TL)** Zigmund Leszczynski/Animals Animals-Earth Scenes. **CI-22: (Indochinese CI-4 TR)** Terry Whittaker/FLPA/Minden Pictures. **CI-20: (Malayan CI-4 ML)** Peter Tan Hua Choon (petertan. com). **CI-18: (South China CI-4**

Illustration Credits

Chapter 1 1–6: From A. Wilting et al., 2007, "Clouded leopard phylogeny revisited" *Frontiers in Zoology* 4:15.

Chapter 2 2–3a,b: Figure 2.d from Shu et al, *Nature*, 2001, 414: pp. 419–424. Copyright © 2001 by Macmillan Magazines Ltd. Reprinted with permission. **2–3c:** Fig. 3 from Holland and Chen, 2001, *Bioessays*, 3: 142–151. Copyright © 2001 by John Wiley-Liss. Reprinted with permission of Wiley-Liss, a subsidiary of John Wiley & Sons, Inc. **2–5:** After E. S. Goodrich, 1930, *Studies on the Structure and Development of Vertebrates*. **2–6c:** Based on B. Stahl, 1974, *Vertebrate History*, McGraw-Hill, NY and **2–6d:** After K. V. Kardong, 1998, *Vertebrates*. **2–7:** From J. G. Maisey *Discovering Fossil Fishes*. Reprinted by permission of Nevraumont Publishing Co. **2–9a,b:** From *Functional Anatomy of the Vertebrates*, 2/e by Walker & Liem. Copyright © 1994 Brooks/Cole, a part of Cengage Learning, Inc. Reproduced by permission. www.cengage.com/permissions. **2–12:** Modified from A. G. Kluge et al., 1977, *Chordate Structure and Function*, 2/e, and Bone & Marshall, 1982, *Biology of Fishes*. **2–14:** From *Vertebrates*, 2/e by Kardong et al. Copyright © 1998. Reprinted by permission of McGraw-Hill Companies, Inc. **2–15:** After *Vertebrates*, 2/e by Kardong et al. Copyright © 1998. Reprinted by permission of McGraw-Hill Companies, Inc.. **2–16:** Adapted from *Vertebrates*, 2/e by Kardong, et al. Copyright © 1998. Reprinted by permission of McGraw-Hill Companies, Inc.

Chapter 3 3–1: From *Nature*, Vol. 402, p. 43. **3–2:** Modified from M. A. Purnell, 1994, *Lethaia* 27, pp. 129–138. **3–3:** Based primarily on Donoghue, et al., 2000, with information from J. G. Maisey, 1986 and Mallatt and Sullivan, 1998. **3–5:** Modified from D. Jensen, 1966, *Scientific American* 214(2); 82–90. **3–7:** Modified after J. A. Moy-Thomas and R. S. Miles, 1971, *Paleozoic Fishes*. **3–10a:** From *Analysis of Vertebrate Structures*, 5/e by Hildebrand and Goslow. Copyright © 2001 by John Wiley & Sons, Inc. Used with permission **3–11a:** From *Hyman's Comparative Vertebrate Anatomy* by M. H. Wake. Copyright © 1987 University of Chicago Press. Used with permission.

3–12b: From *The Evolution of Vertebrate Design* by L. B. Radinsky. Copyright © 1987. **3–16:** Modified after J. A. Moy-Thomas and R. S. Miles, 1971, *Paleozoic Fishes*. **3–17:** Modified after J. A. Moy-Thomas and R. S. Miles, 1971, *Paleozoic Fishes*.

Chapter 4 4–1: Modified from *Comparative Physiology of Vertebrate Respiration* by G. M. Hughes. **4–2:** Modified after G. M. Hughes, *Comparative Physiology of Vertebrate Respiration*. **4–4:** Modified after A. Flock, 1967 in *Lateral Line Detectors* and R. F. Hueter et al., 1991, *Underwater Naturalist*,

Special Double Issue, 19(4) and 20(1); 48–55. **4–5:** Modified after E. Schwartzm, 1974, *Handbook of Sensory Physiology*, Vol. 3, Part 3, ed. by A. Fessard. **4–6:** Modified in part from J. Bastian, 1994, *Physics Today*, 47(2); 30–37. **4–8:** Modified from A. J. Kalmijn, 1974, *Handbook of Sensory Physiology*, Vol. 3, Part 3. **4–13:** From *Handbook of Physiology, Adaptation to the Environment* ed. by D. B. Dill, E. F. Adolph and C. G. Wilbur, American Physiological Society, Washington, DC.; **4–13d:** From data in Fry, F. E. J. and P. W. Hochachka, 1970, *Fish*, pp. 79–134 in *Comparative Physiology of Thermoregulation*, Vol. 1, ed. by G. C. Whitlow, Academic Press, NY. **4–14:** Modified from F. G. Carey and J. M. Teal, 1966, *Proceedings of the National Academy of Sciences*, 56:1464–1469.

Chapter 5 5–1a,b: Modified after J. A. Moy-Thomas and R. S. Miles, 1971, *Paleozoic Fishes*. Phil., and R. Lund, 1985, *Journal of Vertebrate Paleontology*, 6, pp. 12–19. **5–3:** Modified after J. A. Moy-Thomas and R. S. Miles, 1971, *Paleozoic Fishes*. **5–5:** Modified after E. S. Goodrich, 1930, *Studies on the Structure and Development of Vertebrates*. **5–7:** Modified from A. P. Kimley, *American Scientist*, 1999, (87), pp. 488–491. **5–8:** Modified in part from J. S. Nelson, 1994; and P. B. Moyle, 1993. **5–9:** Modified after J. A. Moy-Thomas and R. S. Miles, 1971, *Paleozoic Fishes*.

Chapter 6 6–2: Modified after J. A. Moy-Thomas and R. S. Miles, 1971, *Paleozoic Fishes*. **6–5:** From Glenn Northcutt, *Integrative & Comparative Biology*, Vol. 42, 1995. Reprinted by permission. **6–7:** Modified from K. F. Liem in A. G. Kluge, et al, 1977, *Chordate Structure and Function*, 2nd ed., Macmillan, NY. **6–15:** Modified after C. C. Lindsey, 1978, *Fish Physiology*, Vol. 7, p. 1–100. **6–16:** Modified after H. Hertel, 1966, *Structure, Form, Movement*. **6–18:** Modified after Friedrich, 1973, *Marine Biology* and Marshall, 1971, *Explorations of the Life of Fishes*. **6–19:** Modified from Kampa and Boden, 1957, *Deep-Sea Research*, 4: 73–92. **6–23a:** From Paxton & Eschmeyer, *The Encyclopedia of Fishes*, p. 98. Copyright © 1994 Academic Press. Reprinted by permission of Elsevier. **6–24:** Australian data provided by Peter Unmack.

Chapter 7 7–2: Modified from A. Hallam, 1994, *An Outline of Phanerozoic Biography*. **7–5:** Modified from M. J. Benton, 1997, *Vertebrate Paleontology*, 2/e and Benton and Harper, 1997, *Basic Paleontology*.

Chapter 8 8–2: From *The Evolution of Vertebrate Design* by L. B. Radinsky. Copyright © 1987. Reprinted by permission of The University of Chicago Press. **8–3:** From *Vertebrates*, 2/e by Kardong et al. **8–6:** After Clack [2002]. **8–7:** From *Gaining Ground* by J. Clack. Copyright © 2002. Reprinted by permission of Indiana University Press. **8–9:** Fig. 33, p. 51 from *Life's Devices* by A. *Vogel: The Physical World of Plants and Animals* by A. Vogel. Copyright © 1989 by Princeton University Press. Reprinted by permission of the publisher. **8–11:** After Figure 2.14c from *The Vertebrate Body*, 6th edition by Romer and Parsons. Copyright © 1986. Reprinted with permission of Brooks/Cole, a division of Thomson Learning: www.thomsonrights.com. Fax

Table 13.3: Phylogenetic relationships and numbers of species are based on F. H. Pough et al., *Herpetology,* 2nd ed. (Upper Saddle River, NJ: Prentice Hall, 2001). **Table 13.5:** Based on data from L. J. Vitt and J. D. Congdon, *American Naturalist 112* (1978): 595–608; R. B. Huey and E. R. Pianka, *Ecology 62* (1981): 991–999; W. E. Magnusson et al., *Herpetologica 41* (1985): 324–332; and R. B. Huey and A. F. Bennett, in *Predator–Prey Relationships,* M. E. Feder and G. V. Lauder (eds.) (Chicago: University of Chicago Press, 1986), 82–98.

Chapter 14 14–2: Modified from K. A. Nagy, *Stable Isotopes in Ecological Research* ed. by Ehleringres et al., 1998 **14–3:** Based on K. A. Nagy and P. A. Medica, 1986, *Herpetologica,* 42:73-92. **14–5:** From K. A. Nagy, 1973, *Copeia,* 93–102. **14–6:** From K. A. Nagy, 1972, *Journal of Comparative Physiology,* 79:39–62. **14–8:** From B. H. Brattstorm, *Herpetologica,* 18:38–46, 1962. **14–9:** From P. F. Scholander et al., 1957, *Journal of Cellular and Comparative Physiology,* 79:39–62. **14–11:** From F. H. Pough, 1980, *American Naturalist,* 115:92–112. Copyright © 1980 by the University of Chicago Press. Used with permission. **14–12:** Based on data only from F. H. Pough. **Table 14.3:** F. H. Pough, *The American Naturalist 115* (1980): 92–112.

Chapter 16 16–3a,b: From A. S. Romer, 1966, *Vertebrate Paleontology,* 3/e. **16–3c:** From A. S. Romer, 1956, *Osteology of the Reptiles,* University of Chicago Press. **16–5:** Modified from H. Wermuth and R. Mertens, 1961, *Schildkroten, Krocodile, Bruckenechsen.* **16–21:** Modified from K. Padian, 1997, *Nature,* 382:400–401.**16–24:** From J. M. V. Rayner, 1988, *Biological Journal of the Linnean Society,* 34:269–287. **16–25:** From the Sanford Bird Hall, American Museum of Natural History #325097. Courtesy Department of Library Services.

Chapter 17 17–1a: From P. C. Sereneo and R. Chenggang, 1992, *Science,* 255:845–848. Copyright © 1992 American Association for the Advancement of Science. Reprinted by permission of the publisher. **17–3a:** From A Feduccia, 1980, *The Age of Birds,* Harvard University Press. **17–3b:** From L. D. Martin, 1983, *Perspectives in Ornithology* ed. by A. H. Brush and G. A. Clark, Jr., Cambridge University Press, UK. **17–9:** From J. Dorst, 1974, *The Life of Birds.* **17–10:** From A. C. Thompson, 1964, *A New Dictionary of Birds.* **17–17:** From R. T. Peterson, 1978, *The Birds,* 2/e. **17–18:** From R. W. Storer, 1971, *Avian Biology,* Vol. 1. **17–20:** From D. Weihs and G. Katzir, 1994, *Animal Behavior,* 47:649–654. **17–21:** From P. Behler, 1981, *Form and Function in Birds,* Vol. 2. **17–23a-e:** From J. McLelland, 1979, *Form and Function in Birds.* **17–23f:** From A. Grajal et al., 1989, *Science,* 245:1236–1238. **17–26:** Modified from G. R. Martin, 1987, *Nature,* 328:383. **17–32:** From F. B. Gill, *Ornithology.* **17–35:** From A. J. Marshall and D. L. Serventy, 1956, *Proceedings of the Zoological Society of London,* 127:489–510. **17–36a:** Illustration by Adolph E. Brotman, © 1975. Used with permission. **17–37:** From W. T. Keeton, 1969, *Science,* 165:922–928. **Table 17.2:** Modified from E. A. Brenowitz, A. P. Arnold, and R. N. Levin, *Brain Research 343* (1985): 104–112. **Table**

17.3: Modified from M. M. Nice, *Transactions of the Linnaean Society of New York 8* (1962): 1–211.

Chapter 18 18–4: From *Vertebrate Paleontology,* by R. L. Carroll. © 1988 by W. H. Freeman and Company. Used with permission. **18–5:** By Gregory Paul in G. King, 1990, *The Dicynodonts, A Study in Paleobiology.* **18–6a,c:** From P. J. Currie, 1977, *Journal of Paleontology,* 51(5):927–942. **18–6b:** From F. A. Jenkins, Jr., 1970, *Evolution,* 24:230–252, figure 2. **18–7:** From T. S. Kemp, 1982, *Mammal-Like Reptiles and the Origin of Mammals.* **18–8:** From A. W. Crompton and F. A. Jenkins, Jr., 1979, pp. 59–73 in *Mesozoic Mammals.* **18–9:** Figure 28–10 from *The Evolutionary Biology of Hearing* ed. by D. B. Webster et al., 1992. **18–10:** Modified in part from J. A. Hopson, 1994, pp. 190–219 in *Major Features of Vertebrate Evolution.* **18–13:** From K. V. Kardong, *Vertebrates: Comparative Anatomy, Function, Evolution ,* 2/e, 1998.

Chapter 19 19–3: From J. C. Zachos, 2001, *Science,* 292:274–278. Copyright © 2001. Reprinted with permission from AAAS.

Chapter 20 20–1: Modified from drawing by L. L. Sadler in Jenkins & Krause, 1983. **20–3:** Modified from J. Z. Young, 1981, *The Life of Vertebrates,* 3/e. **20–5:** Modified from D. MacDonald, 1984, *The Encyclopedia of Mammals.* Copyright © 1984 by Andromeda Oxford Ltd. Color and line artwork copyright © 1984 by Priscilla Barrett. Reprinted by permission. **20–6a:** After T. E. Lawlor, *Handbook to the Orders and Families of Living Mammals,* 1979. **20–6b:** From *The Vertebrate Body,* 6/e by Romer and Parsons. © 1986. Brooks/Cole, a part of Cengage Learning, Inc. Reproduced by permission. ww.cengage.com/permissions. **20–8:** Adapted from Fig. 9–12, T. A. Vaughan, 1986, *Mammalogy,* 3/e. **20–9:** Modified from various sources, esp. A. W. Ham and D. W. Cormack, 1973, *Histology,* 8th ed., Lippincott and P. J. Harrison and E. W Montagna, 1973, *Man,* 2nd ed., Appleton-Century-Crofts. **20–10:** From *Analysis of Vertebrate Structure,* 5/e by Hildebrand & Goslow. Copyright © 2001 by John Wiley & Sons, Inc. Reprinted by permission. **20–11:** From *Analysis of Vertebrate Structure,* 5/e by Hildebrand & Goslow. Copyright © 2001 by John Wiley & Sons, Inc. Reprinted with permission. **20–12:** From *Analysis of Vertebrate Structure,* 5/e, by Hildebrand & Goslow. Copyright © 2001 by John Wiley & Sons, Inc. Reprinted with permission. **20–13:** *Vertebrates: Comparative Anatomy, Function, Evolution,* 2/e by Kardong. Copyright © 1998 McGraw-Hill Companies, Inc. Used with permission. **20–14:** Modified from *Looking at Vertebrates* by E. Rogers, and *Functional Anatomy of the Vertebrates,* 2/e by Walker and Liem. **20–15:** *Vertebrates: Comparative Anatomy, Function, Evolution,* 2/e by Kardong. Copyright © 1998. McGraw-Hill Companies, Inc. Used with permission. **20–16:** Fig. 13 & Fig. 105 from T. E. Lawlor, *Handbook to the Orders and Families of Living Mammals,* 1979. **20–18:** From G. G. Simpson et al., *Life,* 1957. **20–19:** Modified from L. G. Marshall et al., 1982, *Science 215*:1351–1357 and L. G. Marshall, 1988, *American Scientist,* 76:380–388. **Table 20.1:** L. G. Marshall, J. A. Case, and M. O Woodburne, *Current Mam-*

malogy 2 (1990):433–505; T. A. Vaughan, *Mammalogy,* 3rd ed. (Philadelphia: Saunders, 1986); and R. M. Nowak and J. L. Paradiso, Walker's *Mammals of the World,* 5th ed. (Baltimore: Johns Hopkins University Press, 1991). **Table 20.2:** J. A. Hopson, *Journal of Mammalogy 51* (1970): 1–9; M. C. McKenna, in *Phylogeny of the Primates,* W. P. Luckett and F. S. Szalay (eds.) (New York: Plenum, 1975); T. A. Vaughan et al., *Mammalogy,* 4th ed. (Philadelphia: Saunders, 2000). J. F. Eisenberg, *The Mammalian Radiations* (Chicago: University of Chicago Press, 1981); R. M. Nowak and J. L. Paradiso, Walker's *Mammals of the World,* 5th ed. (Baltimore, MD: John Hopkins University Press, 1991); S. Anderson and J. K. Jones, Jr., *Orders and Families of Recent Mammals of the World* (New York: Wiley–Interscience, 1984); and M. J. Novacek and A. R. Wyss, *Cladistics,* 2 (1987):257–287.

Chapter 21 21–1: Modified from M. Griffiths, 1968, *Echidnas.* **21–2:** Modified from W. W. Ballard, 1964, *Comparative Anatomy and Embryology.* **21–3:** Modified after G. B. Sharman, 1976 in *Reproduction in Mammals.* **21–4:** Modified from M. B. Renfree, 1993 in *Mammal Phylogeny.* **21–6:** Modified after Fig. 2.11 in G. B. Sharman, 1976 in *Reproduction in Mammals,* 6: *The Evolution of Reproduction* ed. by C. R. Austin and R. V. Short, Cambridge University Press, UK. and Fig. 1.6 in R. V. Short, 1972, *Reproduction in Mammals, Reproductive Patterns* ed. by C. R. Austin and R. V. Short, Cambridge University Press, UK. **21–11a,b:** Modified from F. A. Jenkins, Jr. *Primate Locomotion,* 1974. **21–11c:** Modified from Fig. 4 in A. A Biewener, 1989, *Bioscience 39:*776–783. **21–12:** From *Analysis of Vertebrate Structure,* 5/e by Hildebrand & Goslow. Copyright © 2001 by John Wiley & Sons, Inc. Reprinted with permission. **21–13:** From *Analysis of Vertebrate Structure,* 3/e by Hildebrand. Copyright © 1988 by John Wiley & Sons, Inc. Used with permission. **21–14:** From *The New Encyclopedia of Mammals* ed. by D. McDonald. Copyright © 2001. **21–15:** From *The New Encyclopedia of Mammals* ed. by D. McDonald. Copyright © 2001. **21–16:** Reprinted with the permission of The Free Press, a division of Simon & Schuster Adult Publishing Group, from Carl Zimmer, *At the Water's Edge: Macroevolution and the Transformation of Life.* Copyright © 1996 by Carl Zimmer. All rights reserved.

Chapter 22 22–3: Based on B. K. McNab, 1973, *Journal of Mammalogy 54:*131–144. **22–4:** Modified from P. E. Scholander et al., 1950, *Biological Bulletin 99:*237–258. **22–5:** Modified from P. E. Scholander et al., 1950, *Biological Bulletin 99:*237–258. **22–10:** From L. C. H. Wang, 1978, pp. 109–145 in *Strategies in Cold.* **22–11:** From L. C. H. Wang, 1978, pp. 109–145 in *Strategies in Cold.* **22–12:** Based on data only from C. A. Beuchat, 1990, *American Journal of Physiology 258* (Regulatory, Integrative, Comparative Physiology 27):R298–R308. **22–13:** Modified from K. Schmidt-Nielsen, et al., 1957, *American Journal of Physiology 188:*103–112. **22–15:** Modified from C. R. Taylor, 1972, *Comparative Physiology of Desert Animals.* **22–17:** From M. A. Chappell and G. A. Bartholomew, *Physical Zoology 54:*81–93. Reprinted by permission of the University of Chicago Press. **22–18:** From M. A. Chappell and G. A. Bartholomew, 1981, *Physical Zoology 54:*81–93, 1981. Copyright © 1981 University of Chicago Press. Used with permission.

Chapter 23 23–1: From B. K. McNab, *Advances in the Study of Mammalian Behavior* ed. by D. G. Kleiman, Special Publication 7. **23–2:** P. Hersteinsson and D. W. McDonald, 1982, pp. 259–289 in *Telemetric Studies of Vertebrates,* ed. by C. L. Cheeseman and R. B. Mitson, *Symposia of the Zoological Society of London,* No. 49. **23–4:** From B. K. McNab, *Advances in the Study of Mammalian Behavior* ed. by D. G. Kleiman, *Special Publication 7.* **23–6:** From P. J. Van Soest, 1982, *Nutritional Ecology of the Ruminant.* **23–7:** From P. J. Jarman, 1974, *Behavior 58:*215–267. **Table 23.1:** P. Hersteinsson and D. W. Macdonald, in C. L. Cheesman and R. B. Mitson (eds.), *Symposia of the Zoological Society of London,* no. 49 (New York: Academic Press, 1982): 259–289. **Table 23.2:** Modified from P. J. Jarman, *Behaviour* 58 (1974): 215–267. **Table 22.3:** Modified from B. Smuts, *Sex and Friendship in Baboons* (Hawthorn, NY: Aldine, 1985). **Table 23.4:** Based on R. W. Wrangham, in *Current Problems in Sociobiology,* King's College Sociobiology Group (ed.) (Cambridge, UK: Cambridge University Press): 269–289.

Chapter 24 24–1: From J. I. Bloch and D. M. Boyer, 2002, *Science 298:*1606–1610. Copyright © 2002. Reprinted with permission from AAAS. **24–2:** From Szalay et al., *Evolutionary History of the Primates,* p. 76. Copyright © 1980. **24–3:** From *Vertebrate Paleontology,* 3/e by A. S. Romer. Copyright © A. S. Romer. Used by permission of the University of Chicago Press. **24–4:** Modified from J. G. Feagle, *Primate Adaptation and Evolution.* Copyright © 1988. **24–6a:** Fig. 4–9 (b) Fig. 4–12 (c) Fig. 4–15 (d, e) Fig. 4–19 and (f) Fig. 4–25 from J. G. Feagle, *Primate Adaptation and Evolution.* Copyright © 1986. **24–7a:** Fig. 5–25 (b) Fig. 5–12 (c) Fig. 5–16 (d) Fig. 5–14 (e) Fig. 6–5 (f) Fig. 6–7 (g) Fig. 6–16 (h) Fig. 6–15 from J. G. Feagle, *Primate Adaptation and Evolution* by J. G. Fleagle. Copyright © 1986. **24–10a:** Fig. 7–6 (b) Fig. 7–82 from *Primate Adaptation and Evolution* by J. G. Fleagle and Fig. 23–10 in J. Z. Young, 1981, *The Life of Vertebrates,* 3/e. **24–12:** Fig. 17.5 from J. G. Feagle, *Primate Adaptation and Evolution.* Copyright © 1986. **24–16:** Fig. 1 by Stephen Nash in S. E. Churchill, 1998, *Evolutionary Anthropology 7:*46–61. **24–17:** Fig. 4.8 in *Lowly Origin* by J. Kingdon. Copyright © 2003 by Princeton University Press. Reprinted by permission. **24–18:** Modified from S. M. Stanley, 1996, *Children of the Ice Age.* **Table 24.1:** From F. S. Szalay and E. Delson, *Evolutionary History of the Primates* (New York: Academic, 1979).

Chapter 25 25–2: From *Global Biodiversity: Status of the Earth's Living Resources,* 1992. **25–4:** From *Global Biodiversity: Status of the Earth's Living Resources,* 1992. **25-7:** Copyright © Smithsonian Migratory Bird Center. **25-16:** Data from Carlsen et al. 1992.

Name Index

Subject Index

Latin and Greek Lexicon

Many biological names and terms are derived from Latin (L) and Greek (G). Learning even a few dozen of these roots is a great aid to a biologist. The following terms are often encountered in a vertebrate biology. The words are presented in the spelling and form in which they are most often encountered; this is not necessarily the original form of the word in its etymologically pure state.

An example of how a root is used in vertebrate biology can often be found by referring to the subject index. Remember, however, that some of these roots may be used as suffixes or otherwise embedded in technical words and will require further searching to discover an example. Additional information can be found in a reference such as the *Dictionary of Word Roots and Combining Forms*, by Donald J. Borror (Palo Alto, Calif.: Mayfield Publishing Co.)

a, ab (L) away from
a, an (G) not, without
acanth (G) thorn
actin (G) a ray
ad (L) toward, at, near
aeros (G) the air
aga (G) very much, too much
aistos (G) unseen
al, alula (L) a wing
allant (G) a sausage
alveol (L) a pit
ambl (G) blunt
ammos (G) sand
amnion (G) a fetal membrane
amphi, ampho (G) both, double
amplexus (L) an embracing
ampulla (L) a jug or flask
ana (G) up, upon, through
anat (L) a duck
angio (G) a reservoir, vessel
ankylos (G) crooked, bent
anomos (G) lawless
ant, anti (G) against
ante (L) before
anthrac (G) coal
apat (G), illusion, error
aphanes (G) invisible, unknown
apo, ap (G) away from, separate

apsid (G) an arch, loop
aqu (L) water
arachne (G) a spider
arch (G) beginning, first in time
argenteus (L) silvery
arthr (G) a joint
ascidion (G) a little bag or bladder
aspid (G) a shield
asteros (G) a star
atri, atrium (L) an entrance-room
audi (L) to hear
austri, australis (L) southern
avis (L) a bird
baen (G) to walk or step
bas (G) base, bottom
batrachos (G) a frog
benthos (G) the seadepths
bi, bio (G) life
bi, bis (L) two
blast (G) bud, sprout
brachi (G) arm
brachy (G) short
branchi (G) a gill or fin
buce (L) the check
cal (G) beautiful
calie (L) a cup
capit (L) head
carn (L) flesh
caud (L) tail
cene, ceno (G) new, recent
cephal (G) head
cer, cerae (G) a horn

cerc (G) tail
chir, cheir (G) hand
choan (G) funnel, tube
chondr (G) grit, gristle
chord (G) guts, a string
chorio (G) skin, membrane
chrom (G) color
clist (G) closed
cloac (L) a sewer
coel (G) hollow
cornu (L) a horn
cortic, cortex (L) bark, rind
costa (L) a rib
cran (G) the skull
creta (L) chalk
cretio (L) separate
crini (L) the hair
cten (G) a comb
cut, cutis (L) the skin
cyn (G) a dog
cytos (G) a cell
dactyl (G) a finger
de (L) down, away from
dectes (G) a biter
dendro (G) a tree
dent, dont (L) a tooth
derm (G) skin
desmos (G) a chain, tie, or band
deuteros (G) secondary
di, dia (G) through, across
di, diplo (G) two, double
din, dein (G) terrible, powerful

dir (G) the neck
disc (G) a disk
dory (G) a spear
draco (L) a dragon
drepan (G) a sickle
dromo (G) running
duc (L) to lead
dur (L) hard
e, ex (L) out of, from, without
echinos (G) a prickly being
eco, oikos (G) a house
ect (G) outside
edaphos (G) the soil or bottom
eid (G) form, appearance
elasma (G) a thin plate
eleuthero (G) free, not bound
elopos (G) a kind of sea fish
embolo (G) like a peg or stopper
embryon (G) a fetus
emys (G) a freshwater turtle
end (G) within
enter (G) bowel, intestine
eos (G) the dawn or beginning
ep (G) on, upon
equi (L) a horse
ery (G) to drag or draw
erythr (G) red
eu, ev (G) good, true